AF664401

Seabuckthorn (*Hippophae* L.)
A Multipurpose Wonder Plant

The Editor-in-Chief

Prof. Virendra Singh, a world-renowned seabuckthorn scientist, is well known for his valuable contribution in the development of science of seabuckthorn in India. He has been working on seabuckthorn in dry temperate Indian Himalayas, since 1993 and worked on various aspects like genetic diversity and improvement, biochemical characterization, selection and introduction of high yielding local and exotic forms and cultivation technologies of seabuckthorn. He is also instrumental in motivating Indian scientists to start research work on clinical and experimental studies and standardization of various health products of seabuckthorn in his own university and about 35 other research organizations of India. Recently, he worked as Consortium Principal Investigator of Indian Council of Agricultural Research's funded scheme under National Agricultural Innovation Project (NAIP) entitled *"A value chain on seabuckthorn (Hippophae* L.*)"* (0.8 million USD, during 2008-14), a multi-organizational consortium based seabuckthorn R&D programme in dry temperate Himalayas. He is credited with several research papers on seabuckthorn published in reputed international and national journals. He has also edited five books and authored two books on seabuckthorn in collaboration with national and international experts.

Prof. Singh is a Member of Board of Directors of International Seabuckthorn Association (ISA) and also a Member of Scientific Committee of ISA. In order to sensitize the Indian scientists, policy makers, farmers and private companies, Dr. Singh, along with his team, organized the International Workshop on Seabuckthorn, held on February 18-21, 2001 at India International Center, New Delhi. As Organizing Secretary, he has also organized two National Conferences on Seabuckthorn in 2010 and 2011 at Palampur. Dr. Singh is also Secretary of Seabuckthorn Association of India. He has also chaired sessions in ISA's Conferences held at Berlin, (2003), Beijing (2005) and Xining (2011). He is widely traveled and visited several countries of Europe, Asia and North America. For his contribution on seabuckthorn, Dr. Singh was awarded "Outstanding Contribution Award on Seabuckthorn" by "International Seabuckthorn Association" during ISA's Conference on 4th September 2011 at Xining, China. He has also been honoured with "Golden Jubilee Endowment Visiting Fellowship Award" by Institute of Chemical Technology, Mumbai on March, 8, 2013. Recently, as an Organizing Secretary, he successfully organized 7th Conference of International Seabuckthorn Association at New Delhi on November 24-26, 2015.

Seabuckthorn (*Hippophae* L.)
A Multipurpose Wonder Plant

Vol. V: Advanced Technologies for Cultivation, Processing, Health Protection and Environmental Conservation

Editor-in-Chief

VIRENDRA SINGH

Associate Editors

Poonam Thakur, V. K. Sharma, S. P. Taygi, Y. S. Dhaliwal
R. Verma, S. Singh, A. Kumar, A. Singh, Kusum, R. K. Rana
T. Stobdan, S. K. Dwivedi, Madhu Bala & R. C. Sawhney

2018

Daya Publishing House®

A Division of

Astral International Pvt. Ltd.

New Delhi – 110 002

Publisher's Note:

Every possible effort has been made to ensure that the information contained in this book is accurate at the time of going to press, and the publisher and author cannot accept responsibility for any errors or omissions, however caused. No responsibility for loss or damage occasioned to any person acting, or refraining from action, as a result of the material in this publication can be accepted by the editor, the publisher or the author. The Publisher is not associated with any product or vendor mentioned in the book. The contents of this work are intended to further general scientific research, understanding and discussion only. Readers should consult with a specialist where appropriate.

Every effort has been made to trace the owners of copyright material used in this book, if any. The author and the publisher will be grateful for any omission brought to their notice for acknowledgement in the future editions of the book.

Cataloging in Publication Data--DK
Courtesy: D.K. Agencies (P) Ltd. <docinfo@dkagencies.com>

Seabuckthorn (Hippophae L.) : a multipurpose wonder plant / editor-in-chief, Virendra Singh ; associate editors, Poonam Thakur [and twelve others].
volumes cm
Contributed articles.
Includes bibliographical references and indexes.
Contents: vol. V. Advanced technologies for cultivation, processing, health protection and environment conservation.
ISBN 978-93-5124-879-8 (HB)

1. Sea buckthorn. 2. Sea buckthorn--Technological innovations. I. Singh, Virendra, 1961- editor.

LCC QK495.E33S43 2018 | DDC 634.74 23

Published by : **Daya Publishing House®**
A Division of
Astral International Pvt. Ltd.
– ISO 9001:2015 Certified Company –
4736/23, Ansari Road, Darya Ganj
New Delhi-110 002
Ph. 011-43549197, 23278134
E-mail: info@astralint.com
Website: www.astralint.com

Digitally Printed at : **Replika Press Pvt. Ltd.**

त्रिलोचन महापात्र, पीएच.डी.
एफ एन ए, एफ एन ए एस सी, एफ एन ए ए एस
सचिव एवं महानिदेशक

TRILOCHAN MOHAPATRA, Ph.D.
FNA, FNASc, FNAAS
SECRETARY & DIRECTOR GENERAL

भारत सरकार
कृषि अनुसंधान और शिक्षा विभाग एवं
भारतीय कृषि अनुसंधान परिषद
कृषि एवं किसान कल्याण मंत्रालय, कृषि भवन, नई दिल्ली 110 001

GOVERNMENT OF INDIA
DEPARTMENT OF AGRICULTURAL RESEARCH & EDUCATION
AND
INDIAN COUNCIL OF AGRICULTURAL RESEARCH
MINISTRY OF AGRICULTURE AND FARMERS WELFARE
KRISHI BHAVAN, NEW DELHI 110 001
Tel.: 23382629; 23386711 Fax: 91-11-23384773
E-mail: dg.icar@nic.in

Foreword

Seabuckthorn (*Hippophae rhamnoides* L.), a nitrogen fixing multipurpose wonder plant (family *Elaeagnaceae*) of high altitude Himalayas, has been recognized as a rich source of vitamins and antioxidants (vitamin C, E, carotenoids, polyphenols and sterols *etc.*), therefore, it has been utilized in health food, cosmetic and nutraceutical industries in China, Russia, Germany, Finland, India and many other countries. It is also an excellent soil binder and quality fuel wood and fodder plant of cold deserts of Himalayas. While developmental agencies are engaged in cultivation of seabuckthorn for combating climate change and environmental conservation of degraded mountainous lands of Himalayas, CSK Himachal Pradesh Agricultural University, Palampur and many other R&D institutions of India have developed appropriate technologies on seabuckthorn, which can be harnessed for the commercial utilization of this wonder plant for the livelihood improvement of farmers in Himalayan states and health protection. The plant can also be a part of the programmes for providing nutritional security to the people of the country.

I appreciate the efforts of Prof. Virendra Singh, a world renowned Seabuckthorn Scientist of CSK Himachal Pradesh Agricultural University, Palampur, who prepared this very valuable book, which will make it easier for those, who are interested to explore seabuckthorn further. The book entitled *"Seabuckthorn (Hippophae L.)-A Multipurpose Wonder Plant, Vol. V: Advanced Technologies for Cultivation, Processing, Health Protection and Environmental Conservation"* has covered various aspects of seabuckthorn plant, which makes it a useful source on the subject to scientists, students, entrepreneurs and policy makers *etc.* I congratulate Prof. Singh and his co-editors in producing a useful and quality book on seabuckthorn.

Dr. Trilochan Mohapatra
Secretary (DARE) & Director General (ICAR)
Krishi Bhavan, New Delhi – 110 001, India

Preface

Global Seabuckthorn Programme, which was initiated by Russians during 1930s, truly led by Chinese since 1980s, has been joined by many other countries like Germany, Finland, India, Canada, Mongolia and Sweden during the last two decades. Therefore, seabuckthorn program has come of age, a global success story. The potential of seabuckthorn in environmental conservation of mountains and health protection has been realized globally.

Seabuckthorn, one of the greatest gifts of Mother Nature to humanity, was discovered as early as 8th century during Tang dynasty, as mentioned in Tibetan medical classic *"rGyud bzhi"*. Russians discovered its nutritional and medicinal values and used it in its space programme. Development of many high yielding thornless varieties of ssp. *mongolica* by Russian workers made it easier to introduce it in the gardens of Russia. Potential of Russian seabuckthorn varieties so high that even other several countries have introduced Russian seabuckthorn due to their large berries, few thorns, high oil and rich nutrients. Seabuckthorn subspecies like ssp. *sinensis* (China) and ssp. *turkestanica* (Himalayan countries) have great potential in ecological rehabilitation of fragile high altitude mountains, as China has shown.

The discovery of a number of anti-oxidants, like vitamin C, E, carotenoids and polyphenols and sterols led to development of a number of seabuckthorn based health food and cosmetics and some medicines, globally. Preclinical research on seabuckthorn oil and leaf extract has been done on several aspects like anti-oxidant, anti-bacterial, healing of wounds and gastric ulcer, diabetes, cardiovascular and skin diseases *etc.* globally. A number of clinical trials on seabuckthorn have also been carried out in several countries like China, Finland, India, Germany and Russia. Seabuckthorn propagation and cultivation and processing technologies have also been improved. Consequently, seabuckthorn is being cultivated globally by the farmers and a number of health food, skin care and food supplements are in the global market. However, harnessing seabuckthorn as cheap, high quality and effective health product still remains a big challenge before us. Seabuckthorn is an excellent soil binder and it has been planted in over 2.5 million ha marginal mountainous lands, which help in ecological rehabilitation of fragile mountain ecosystems in difficult mountain zones like Loess Plateau, China. Area under seabuckthorn plantations in India has also increased by about 1000 ha through afforestation. Inspired by the success stories of China, Russia and Germany, similarly, other countries like India and Mongolia have also launched national programmes on seabuckthorn. Fruit being multivitamin and rich in anti-oxidants, seabuckthorn is being included in the national programmes of providing nutritional security to people of several countries including India.

Some of us joined hands and decided to compile the global literature on seabuckthorn for use of researchers and students, as there was little information on seabuckthorn in English, most of the information being in Russian and Chinese languages, when we started the work on the book in 1999. Since then, 4 volumes of the book entitled *"Seabuckthorn: A Multipurpose Wonder Plant"* containing about 200 chapters on various subjects of seabuckthorn have been published.

This 5th volume of the book covers the new research findings and emerging advanced technologies on various aspects of seabuckthorn, which will prove useful for seabuckthorn resource development, health application and processing of various seabuckthorn health products.

We recognize and appreciate the cooperation, assistance and guidance of all seabuckthorn experts working globally, who contributed their chapters and made it possible to produce this quality book on seabuckthorn for use of global scientists, students, policy makers, entrepreneurs *etc.*

Prof. Virendra Singh

Editor- in-Chief

(e-mail: virendrasingh1961@yahoo.com)

Contents

PART VI: ENVIRONMENTAL CONSERVATION

PART VII: ANIMAL HUSBANDRY

Part I

CULTIVATION TECHNOLOGIES

Chapter 1

Management of Seabuckthorn (*Hippophae* ssp.) Diseases with Organics in Lahaul Valley: A Dry Temperate Zone of Himachal Pradesh

Pardeep Kumar

CSK Himachal Pradesh Agricultural University,
Krishi Vigyan Kendra Lahaul and Spiti at Kukumseri 175142, Himachal Pradesh, India
(E-mail: pkdogra2007@rediffmail.com)

SUMMARY

Five organics *viz.*, Bio Sanjivani *Trichoderma viride* 1.0 per cent w/w, Bio Sanjivani *Pseudomonas fluorescens* 0.5 per cent w/w, SUDOCEL (*Pseudomonas fluorescens*), NIPROT 0.50 per cent WP (*Trichoderma viride*) and BIOSHIELD (*Pseudomonas fluorescens*) were tested against *Fusarium* wilt and powdery mildew diseases of seabuckthorn. The trail on the management of Fusarium wilt was conducted during 2013 in seabuckthorn nursery. All the organics were used as seedling drenching after the appearance of leaves. For one replication 100 plants were taken and drenched with desired dose of fungicide and per cent incidence was recorded after one month of drenching. It was observed that the combined use of *Trichoderma viride* 1.0 per cent w/w (Bio Sanjivani) and *Pseudomonas fluorescens* 0.5 per cent w/w (Bio Sanjivani) @ 5.0 g/L each gave maximum disease control (52.10 per cent) followed by individual application of *Trichoderma viride* 1.0 per cent w/w (Bio Sanjivani) @ 5.0 g/L (46.96 per cent) and *Trichoderma viride* (NIPROT 0.50 per cent WP) @ 5.0 g/L (44.35 per cent). Similarly another trial for the management of powdery mildew (*Phyllactinia hippophaes*) was conducted in the gene bank of seabuckthorn plantation at HAREC, Kukumseri. Three plants were taken for each treatment. The 2 sprays were given at the intervals of 15 days in the months of August- September, 2013. Minimum disease severity was recorded in *Trichoderma viride* 1.0 per cent w/w (Bio Sanjivani) + *Pseudomonas fluorescens* 0.5 per cent w/w (Bio Sanjivani) treatment (27.33 per cent) followed by *Trichoderma viride* (30.33 per cent) and *Pseudomonas fluorescens* (33.33 per cent).

Keywords: *Seabuckthorn, Hippophae sp., Trichoderma viride, Pseudomonas fluorescens, Diseases management, Lahaul and Spiti and Himachal Pradesh.*

INTRODUCTION

Seabuckthorn (*Hippophae* L.) is known as a golden bush because of its wonderful medicinal, nutraceutical and ecological properties. It has nitrogen-fixing properties and can grow on poor soils with some care in initial couple of years. As such, it is capable of improving the soil health and mitigating the widespread soil erosion. Consequently, it has attracted attention of the planners and the environmentalists alike for its systematic plantations in the valleys. Its berries are commonly harvested for juice making, which is very high in vitamin C and has other healthful properties. In cold desert areas of Lahaul and Spiti, three species of *Hippophae* namely *H. rhamnoides* sub sp. *turkestanica*, *H. salicifolia* and *H. tibetana* are found in various ecological niches. Amongst these species *H. rhamnoides* and *H. salicifolia* are quite common in the Lahaul valley whereas *H. tibetana* is found in some pockets of Spiti valley. Due to long fibrous and tap root system this bush is not only helping in preserving the fragile ecosystem in the cold regions by way of reduced soil erosion but also improving the production of grass growing amidst natural seabuckthorn stands. The grass is a prized

Seabuckthorn (*Hippophae* L.): *A Multipurpose Wonder Plant, Vol. 5: Advanced Technologies for Cultivation, Processing, Health Protection and Environmental Conservation* (Virendra Singh, *Editor-in-Chief*, 2018), p. 3–7, Astral International Pvt. Ltd., New Delhi, India.

commodity throughout Lahaul and Spiti as it is required for the livestock during snow covered winter months when there is no availability of any green grass. As such, planned seabuckthorn orchards with improved temperate grasses planted in between can be very useful to the local populace. Establishing seabuckthorn orchards to exploit its multifaceted usefulness for the benefit of the masses is beset with problem of prevalent diseases in the naturally growing seabuckthorn plants. Although some insect-pests thriving on seabuckthorn plants in the valleys have been reported recently (Anonymous, 2010; Sharma *et al.*, 2010; Sharma and Sharma, 2011). The diseases of seabuckthorn have also been reported by Kumar *et al.* (2011). Keeping in view the economic importance of the diseases, the present study was therefore, carried out against powdery mildew and Fusarium wilt of seabuckthorn during 2013 to generate information on potential of bio-control agents.

MATERIALS AND METHODS

The present study was carried out in Lahaul and Spiti district located within 31°44′34″ N to 32°59′57″ N latitude and 76°46′29″ E to 78°41′34″ E longitude. In order to assess the efficacy of *Trichoderma viride* 1.0 per cent w/w, *Pseudomonas fluorescens* 0.5 per cent w/w, SUDOCEL (*Pseudomonas fluorescens*), NIPROT 0.50 per cent WP (*Trichoderma viride*), BIOSHIELD (*Pseudomonas fluorescens*), experiment was conducted under naturally infested farm of Highland Research and Extension Centre, Kukumseri district Lahaul and Spiti (Himachal Pradesh) during *Kharif* 2013 (Temperature 8.8°C -21°C. relative humidity 30.5 – 32.5 per cent and rain fall 10.7 mm during experimentation). Trail on the management of Fusarium wilt of seabuckthorn was conducted in the month of May in seabuckthorn nursery. All the organics @ 5 g/L were used as seedling drenching after the appearance of leaves. For one replication 100 plants were taken and drenched with desired dose of organics and per cent incidence was recorded after one month of drenching. Similarly for powdery mildew three plants were taken for each treatment. The 2 sprays were given at the intervals of 15 days in the months of August-September, 2013. Then disease percent severity was recorded. In check treatment simple water was applied. Disease incidence and disease severity were calculated with the following formulae.

Disease incidence (PDI) (per cent)

$$= \frac{\text{No. of diseased plants}}{\text{Total no. of plants assessed}} \times 100$$

$$\text{Disease severity} = \frac{\text{Infected tissue area}}{\text{Total tissue area}} \times 100$$

RESULTS AND DISCUSSION

In the present investigation, the efficacy of seven treatments evaluated at HAREC, Kukumseri against seabuckthorn diseases during *Kharif* season 2013. The results of the present study revealed that all the treatments were found effective in managing the seabuckthorn diseases. Among the treatments employed against Fusarium wilt (Table 1.1) it has been observed that when both the bioagents (Bio Sanjivani) *T. viride* and *P. fluorescens* were applied together gave highest disease control (52.10 per cent) as compared to individual application. It is due to the facts that *T. viride* and *P. fluorescens* colonized on emerging roots, provided protection to them and reduced pre-emergence

Table 1.1: Management of Fusarium (*Fusarium* sp.) Wilt of Seabuckthorn with Organic Fungicides

Treatment	*Dose (g/L)*	*Incidence (Per cent)*	*Disease Control (Per cent)*
T1= Bio Sanjivani *Trichoderma viride* 1.0 per cent w/w	5.0	20.33 (26.76)	46.96
T2= Bio Sanjivani *Pseudomonas fluorescens* 0.5 per cent w/w	5.0	22.33 (28.12)	41.74
T3= SUDOCEL (*Pseudomonas fluorescens)*	5.0	24.66 (29.75)	35.66
T4= NIPROT 0.50 per cent WP (*Trichoderma viride)*	5.0	21.33 (27.46)	44.35
T5= BIOSHIELD *(Pseudomonas fluorescens)*	5.0	23.33 (28.84)	39.13
T6= T1+T2	5.0+5.0	18.36 (25.31)	52.10
T7= Control		38.33 (38.22)	-
CD (0.05 per cent)		3.70	-

The figures in parenthesis are arc sine transformed values.

Figure 1.1: Fusarium Wilt (*Fusarium* sp.).

Table 1.2: Management of Powdery mildew (*Phyllactinia hippophaes*) of Seabuckthorn with Organic Fungicides

Treatment	*Dose (g/L)*	*Severity (Per cent)*	*Disease Control (Per cent)*
T1= Bio Sanjivani *Trichoderma viride* 1.0 per cent w/w	5.0	30.33 (33.38)	45.23
T2= Bio Sanjivani *Pseudomonas fluorescens* 0.5 per cent w/w	5.0	33.33 (35.23)	39.81
T3= SUDOCEL (*Pseudomonas fluorescens)*	5.0	35.66 (36.64)	35.60
T4= NIPROT 0.50 per cent WP (*Trichoderma viride)*	5.0	36.33 (37.03)	34.39
T5= BIOSHIELD *(Pseudomonas fluorescens)*	5.0	38.33 (38.22)	30.78
T6= T1+T2	5.0+5.0	27.33 (31.49)	50.65
T7= Control		55.38 (48.07)	-
CD (0.05 per cent)		3.40	-

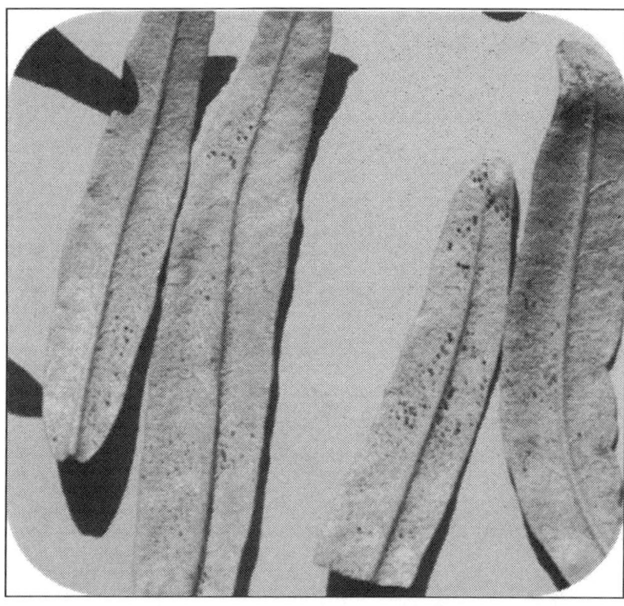

Figure 1.2: Powdery Mildew (*Phyllactinia hippophaes*).

mortality. Soil drenching with Bio Sanjivani *Trichoderma viride* 1.0 per cent w/w and Bio Sanjivani *Pseudomonas fluorescens* 0.5 per cent w/w resulted 20.33 per cent and 22.33 per cent disease incidence, respectively as compared to 38.33 per cent in control treatment. Another treatment NIPROT 0.50 per cent WP (*Trichoderma viride),* BIOSHIELD *(Pseudomonas fluorescens* and SUDOCEL *(Pseudomonas fluorescens)* also gave the significant results 44.35 per cent, 39.13 per cent and 35.33 per cent disease control, respectively.

Similarly results obtained from bioagents against powdery mildew are presented in Table 1.2. The perusal studies revealed that integrated two sprays of *T. viride* and *P. fluorescens* (Bio Sanjivani) at 15 days interval resulted maximum diseases control (50.65 per cent) followed by 45.23 per cent in individual application of Bio Sanjivani *Trichoderma viride* 1.0 per cent w/w @ 5g/L and Bio Sanjivani *Pseudomonas fluorescens* 0.5 per cent w/w (39.81 per cent). Other treatments T3, T4 and T 5 also gave the significant results. Different species of *Trichoderma* are known to produce different kinds of enzymes and active antimicrobial metabolites which have a significant role in biocontrol activity like cell wall degradation, biotic and abiotic stress tolerance, hyphal growth, antagonistic activity and biodegradation of pesticidal residues. *Pseudomonas flourescens* in disease management under field conditions has been greatly attributed to the development of rhizospheric competent. Shamanskaya (2009) had reported *Verticillium* and *Fusarium* fungi from commercial seabuckthorn plantations at Barnaul, Russia. Besides this, the author had also reported endomycosis of fruits caused by saprophytic micro-flora of various species including *Trichoderma viridae, Alternaria alternate, Aspergillus niger, Penicillium* sp. and *Aureobasidium pullulang*. In China, *Fusarium sporotrichiella* is the most common fungus reported to infesting large berry cultivars of seabuckthorn (Luo *et al.*, 2008).

REFERENCES

1. Anonymous. 2010. *Annual Report*. Department of Entomology, CSKHPKV Palampur, p.78.
2. Kumar, P., Sharma, S.K., Sharma, P.C., Rana, R. K., Sharma, S., Manuja, S., Singh, A. and Paul, Y.S. 2011. Diseases of seabuckthorn (*Hippophae* ssp.) in Lahaul and Spiti – A dry temperate zone of Himachal Pradesh. In: *Proceedings of National Conference on "Seabuckthorn (Hippophae L.): Emerging Trends in R&D on Health Protection and Environmental Conservation"* held at CSK Himachal Pradesh Krishi Vishvavidyalaya, Palampur w.e.f. 1-3 December, 2011, pp. 37-40.
3. Luo, Y., Zong, S. and Xu, Z. 2008. Integrated management of the main diseases and pests of seabuckthorn. In: *Seabuckthorn (Hippophae* sp.): *A Multipurpose Wonder* Plant (V. Singh-Editor in Chief). Daya Publishing House, New Delhi, pp.165-177.
4. Shamanskaya, L.D. 2009. Seabuckthorn pests and diseases at Altai. In: *Proceedings of "4th International Seabuckthorn Association Conference (ISA 2009)"* held at Barnaul, Russia, 1-6th September, 2009, pp. 87-88.
5. Sharma, S. K., Sharma, P. C. and Singh, V. 2010-. Efficacy of some insecticides and botanicals against seabuckthorn Psylla in Lahaul valley of Himachal Pradesh. In: *Proceedings of National Conference on "Seabuckthorn: Promoting Health and Protecting Environment"* organized by Defence Institute of Physiology and Allied Sciences (DIPAS), Delhi, 26-27th November, 2010, pp. 35-36.
6. Sharma, S. K. and Sharma, P. C. 2011. An armored scale infesting seabuckthorn (*Hippophae* ssp.) in dry temperate zone of Himachal Pradesh (India)-A new record. In: *Proceedings of "5th International Seabuckthorn Association Conference (ISA 2011)"* held at Xining, Qinghai, China, 3-8th September, 2011, p. 44.

Chapter 2

Pollination Biology in *Hippophae rhamnoides* L. Growing in Cold Desert of Ladakh Himalayas

Amjad Ali[1]*, Veenu Kaul[2] and Sonam Tamchos[2]

[1]Govt. Degree College, Kargil (Ladakh) 194103, J&K, India
[2] Department of Botany, University of Jammu, Jammu 180006, J&K, India
(*E-mail: aaabassi@gmail.com)

SUMMARY

Hippophae rhamnoides L., also known as seabuckthorn, is a reservoir of numerous bioactive compounds of medicinal, nutraceutical and cosmaceutical value. The species has been subjected to large-scale investigation along these lines and huge quantum of literature is available. Extensive studies have also been carried out for its role in ecological management. However, very little is known about its reproductive biology, knowledge of which is an essential tool for species improvement. The present study highlights some interesting features of pollination biology worked out in plants of two populations *viz.*, Khomeni Bagh and Barutsogs from Kargil (Ladakh). Plants of *H. rhamnoides* L. are predominantly dioecious which makes cross-pollination imperative. Floral features of both the sexes at the structural and functional levels point towards a mixture of wind and insect pollination. Flowers, sex not withstanding are small, inconspicuous and marked by the absence of any specialized or pollinator attracting organs like showy sepals or petals and nectary. Despite that, insects belonging to six different orders visit the flowers. Little or no pollen load on their body parts excludes them as pollinators. However, dual modes of anthesis, stigmatic features and abundant pollen speak of the possible biotic element in pollination. While stigma is papillate, moderately wet, yellowish in colour and wavy in outline pollen grains are small, smooth, tricolpate and dry. Stigma attains peak receptivity 72 hour post anthesis. Pollen remains viable till 10 days after anthesis; maximum viability works around 90 per cent at anthesis. Extremely high pollen ovule ratio per plant and ample amount of pollen grains on the hanging slides authenticate its predominantly anemophilous nature. The extent of gene flow through pollen has also been measured effectively reaching upto 30 m.

Keywords: *Hippophae, Protandry, Anthesis and Wind Pollination.*

FLORAL PHENOLOGY AND BIOLOGY

Studies on floral phenology and biology were conducted on the aforesaid two populations of Kargil *viz.*, Barutsogs and Khomenibagh. The two are separated by a distance of 4 km. Inflorescence morphometry was carried out at a stage prior to anthesis when flowers keep compactly grouped together. Data on floral morphometry were generated from 20 freshly opened flowers collected randomly from 10 plants belonging to the two above mentioned populations. Vernier calipers and/or calibrated stage and ocular micrometers were utilized for the purpose.

Phenological events like time of anthesis, anther dehiscence and stigma receptivity were recorded on 40 fully mature yet-to-open floral buds. These were marked one day before anthesis *i.e.* in the evening and monitored the next morning at regular time intervals of 2 hours. Time taken by a flower to open fully and anthers to dehisce and empty their contents were noted down.

For stigma receptivity, about 40 ready to open flowers from each of the 5 plants were selected at random and marked. Pollen grains collected from male plants of the same and/or adjacent populations were applied to the stigma with the help of a clean needle. Five pistils per

Seabuckthorn (Hippophae L.): A Multipurpose Wonder Plant, Vol. 5: Advanced Technologies for Cultivation, Processing, Health Protection and Environmental Conservation (Virendra Singh, *Editor-in-Chief,* 2018), p. 9–18, Astral International Pvt. Ltd., New Delhi, India.

plant were removed at regular intervals of 2 hours after pollination. Of these five, two were kept for fluorescence microscopy and the remaining three were stained in Lewis 2s (1979) stain and examined microscopically to record for pollen germination. Stigmas with germinating pollen grains were treated as receptive.

For fluorescence microscopy, followed the method described in Shivanna and Rangaswamy (1992). These freshly prepared slides were mounted in a 1:1 mixture of aniline blue and glycerine, and examined under fluorescence microscope. Germinating pollen grains with fluorescing pollen tubes on the stigmatic surface were scored.

Pollen Count, Viability and Size

The number of pollen grains in an anther was estimated by squashing mature undehisced anthers in 1 per cent acetocarmine and scoring the same under Olympus OIC microscope as described in Kaul *et al.* (2002). The pollen count per flower was obtained by multiplying the number of pollen grains per anther with total number of anthers per flower. For this, 10 flowers from each of 5 plants belonging to Khomenibagh were utilized.

Pollen viability was determined by squashing freshly dehisced anthers in 1 per cent acetocarmine. Well stained and plump pollen grains were considered viable while small, lightly stained or unstained, shrunken ones as non viable. These slides were also used to measure the size of pollen grains with the help of calibrated ocular and stage micrometers. Pollen viability was also determined by TTC test as recommended by Shivanna and Rangaswamy (1992).

Ovule Count and Size

Ovaries were micro dissected under stereo-microscope (Olympus, OIC) with the help of sharp needles and a pair of fine forceps. Ovules were carefully excised from ovaries and counted. Pistils were also cleared in 5N NaOH and subjected to light and fluorescence microscopies and critically examined for number, shape and size of ovules inside the ovary. Using calibrated ocular and stage micrometers, their size was measured.

Pollen/Ovule ratio

Pollen ovule ratio per flower was calculated by dividing the total number of pollen grains in a male flower to total number of ovules in a female flower. The ratio was also calculated at the levels of inflorescence and plant. For the former, average number of flowers per inflorescence was estimated and multiplied by the average pollen or ovule count per flower. The two counts so obtained were used to estimate P/O per inflorescence. This procedure was repeated at the level of an individual plant. The averages of pollen and ovule counts per male and per female inflorescences were multiplied respectively by the average number of male and female inflorescences per plant to obtain total pollen and ovule production per male and per female plant respectively. Data were generated from five plants of each sex.

Pollination Mechanism

Visitors

The flowers of *H. rhamnoides* L. irrespective of the sex, are minute, light brownish and nectarless. Male flowers produce huge amount of pollen. Once anthesis takes place, the male flowers remain open and finally dry after emptying their contents. Many insects and birds often visit both male and female plants. To know their behavior plants of both the sexes were regularly monitored to categorize the visitors into pollinators, pollen thieves, robbers, *etc.* Visual observations were made regularly at timely intervals of two hours throughout the day on, before and after anthesis. Time of visitation, duration of an individual visit to the flower and foraging behavior of different visitors were noted down. The specimens of each visitor excluding birds were trapped, anaesthetized using chloroform and got identified in the Entomology Laboratory, Department of Zoology, University of Jammu. Some were kept to estimate the pollen load, if any, on their different body parts.

Treatments

Hanging Slide Experiments

In order to ensure its anemophilous nature, hanging slide experiments were conducted. Slides smeared with Mayer 2s albumin (egg albumin and glycerine in 1:1 ratio with a pinch of sodium salicylate) were hanged with the help of thread at different distances from the male plant (5 m, 10 m, 20 m, 30 m), and also at three different levels of the female plants such as upper, middle and lower (Lower- less than 3 feet; Middle- between 3 and 5 feet and Upper- 5 to 7 feet). Twenty four hour later the slides were collected and scanned for pollen load under light binocular microscope (Olympus CH20i) and the quantity recorded.

Bagging

Since the flowers in general and female in particular are minute, it was not possible to bag an individual flower or inflorescence. Therefore approximately 6 inches long twigs bearing pre anthesis floral buds were randomly selected numbers of flowers counted and the whole twig enclosed in butter paper bags. Thorns were clipped to ensure effective bagging. After 90 days or so, the bags were removed and number of fruits formed recorded. The data so obtained were statistically analyzed with one way ANOVA to measure the level of significance between number of flowers bagged and number of fruits obtained.

Manual Pollinations

Dioecism excludes the hassle of emasculation making pollination treatment of any sort easy and manageable. Five female plants each from populations of Khomenibagh and Barutsogs in Kargil were randomly chosen for manual pollination. A total of 40 mature yet to open flowers per plant were marked, branch bearing them tagged and separated into two sets. One set was manually pollinated with pollen grains collected from plants of the same population and the second with that from those of a different population and bagged. Aim was to check difference, if any, in fruit set in flowers pollinated with self (same population) and cross (different population) pollen. At the time of fruit ripening, data could not be properly collected because the populations are unprotected; grazing by cattle and plucking of fruits by villagers, is common. This converted the experiment into an unreliable one. Some plants were even burnt down in Barutsogs.

OBSERVATIONS

Flowers studded with numerous hairs are borne in inflorescences on branches as well as on the main axis; the frequency of inflorescences being higher on the lateral branches. Individual inflorescences are lepidote *i.e.*, densely clothed with scurfy scales and hairs, harbor both vegetative and reproductive structures. The flowers are differentiated in close spirals or in close whorls around the bud axis forming a glomerule and as they age, leaves and branches arise the latter terminate into thorns. Flowers are minute, sessile to sub sessile organized with bract on one side and shoot axis on the other. Number of flowers per inflorescence differs between plants and between sexes. Likewise, the number of inflorescences per branch per unit length varies from 1 to 24 in male and 1 to 17 in female.

Male Inflorescence and Flower

Inflorescence is a raceme, oval to conical in outline varying in color from brown to dark brown in different plants of a population. In some plants, however, it has a silvery surface. Mature inflorescence exhibits plant-to-plant variation in size from 4 to 20 mm. Accordingly number of male flowers also vary (Figure 2.1a). Male flowers usually differentiate on young shoots, exserted away from the main axis. Individual flowers are apetalous, small, each having two brownish tepals and surrounded by a bract. Both the tepals and bract are covered with numerous peltate hairs. The two tepals though similar in length vary slightly in their breadth (Table 2.1). Each flower has four anthers with their individual filaments attached basally. Anther dehiscence is lateral and longitudinal. Inflorescences empty their contents within 10 to 15 days. However, it can take about a month in areas where wind velocity is very low or where trees densely surround the plants. Morphometry of individual flower parts is given in Table 2.1.

Female Inflorescence and Flower

Female inflorescence is a small raceme of 3–12 flowers borne in the axil of a leaf bud. Arranged in relatively loose whorls or spirals they form a diffuse inflorescence that measures 4-12 mm in length, and 2-5 mm in width (Table 2.1). However, either immediately after or prior to anthesis, the floral axis grows or the leaf surrounding the floral bud enlarges widening the distance between flowers.

Female flower is minute varying in size between 4 and 10 mm, non-showy without any attractants and usually surrounded by a single bract from outside. Each comprises of two tepals fused from the base upto more than half of their lengths such that they are free only at the tips. The two enclose the ovary and together with

Table 2.1: Morphometric Details of Male and Female Inflorescences and Flowers of *H. rhamnoides* L.

Sl.No.	Sex	Characters	x ± S.E (Range)	
			Length (mm)	Width (mm)
1	Male (n=50)	Inflorescence	8.2±1.2 (4-20)	5.2±0.6 (3.6-9.0)*
		Bract	3.8±0.4 (2-7)	2.6±0.1 (2.0-3.5)
		Flower	2.1±0.2 (2-4)	2.5±0.1 (2.0.-3.5)
		Tepal	2.8±0.2 (2-4)	2.4±0.1 (2-3.1)/2.6±0.2 (1.5-3.5)
		Anther	1.8±0.04 (1.5-2.0)	0.8±0.2 (0.5-1.0)
2	Female (n=50)	Inflorescence	6.0±0.3 (4-12)	3±0.1 (2-5)
		Bract	5.2±1.0 (3-10)	3.6±0.2 (2-5)
		Flower	6.5±0.5 (4-10)	2.1±0.1 (1.5-3.0)
		Stigma	3.5±0.2 (2.0-6.0)	1.3±0.2 (0.5-2.5)
		Style	0.18±0.02 (0.1-0.3)	0.33±0.06 (0.3-0.5)
		Tepal	4.6±0.3 (2.0-6.0)	1.9±0.1 (1.0-3.0)
		Ovary	0.76±0.02 (0.7-0.9)	0.58±0.02 (0.5-0.7)

*Mean ± S.E, range in parenthesis.

bract have been collectively referred to as perianth. Stellate hairs densely cover all the accessory organs. Style is inconspicuous, whereas stigma is elongate, more or less spathulate broad in the middle and tapering slightly towards the end. Its outline is wavy due to the infoldings at the marginal ends. Morphometric details are enlisted in Table 2.1.

Anthesis

Maturation and anthesis of flowers depends upon the climatic condition of the area and spans second to last week of April. For instance, initiation of anthesis was recorded from April 18th to 22nd in Kargil. Maturation of inflorescence usually proceeds from the base of branch towards the apex in male whereas in females it is the reverse. Flowers marked prior to anthesis were regularly monitored for phenological events. In a population flowers first in male plants; the opening of the flowers of two sexes exhibits a temporal separation of 12 to 48 hours. The differences are elaborated in details below:

Male Flower

The male flowers are particularly peculiar in that two patterns of anthesis have been observed. These are:

I) Opening initiates with the lateral expansion and probable repulsion of the tepals. The apical portions of these tepals, however, are still in overlapping and unexpanded state. Expansion is accompanied by their bending and repelling in a horizontal plane which leads to the formation of openings along the sides. By the time apical portions of the tepals loosen and expand, anthers start dehiscing along longitudinal slits. The apices keep in an overlapped state till the anthers empty their contents. Pollen is shed as and when wind blows. Anther dehiscence is, thus, simultaneous with or occurs immediately before anthesis.

II) In some flowers opening initiates by the expansion of the apices of the tepals, the stimulus of which then spreads laterally. However, what is interesting to note is that the anthers are still in an undehisced state. It takes more than 24 hours for these anthers to dehisce (Ali and Kaul, 2013).

Female Flower

Prior to opening, the floral bud turns dark brown at its apical portion. Opening begins with the formation of a slit at the tip of the flower bud wherefrom the stigma extrusion begins. At this stage the stigma is green, 1mm or less in length. Here, anthesis is marked by mere extension of stigma. At the time of flower opening the stigma is still in the process of development and continues to grow till it attains a length of 7 mm. These changes takes approx. 10 days whereby it undergoes a color change from green to yellow, the latter being the receptive stage. Thereafter, the stigma turns orange followed by brown to dark brown/black. At this stage, stigma shrinks considerably and finally withers out leaving behind the stylar end that remains persistent.

Table 2.2: Manual Pollination Treatments Carried Out in Two Populations of Kargil

Sl.No	Time (in hrs.) after Pollen Deposition	Plant Number/Populations												Overall Stigma Receptivity	Per cent Pollen Germination
		Barutsogs							Khomeni Bagh						
		P1	P2	P3	P4	P5	P6	P7	P1	P2	P3	P4	P5		
1	0-4	–	–	–	–	–	–	–	–	–	–	–	–	–	0
2	5-7	–	–	–	+	–	–	–	–	–	+	–	–	–	2.9 (1.2-4.7)*
3	24	–	–	+	++	++	+	+	+	+	++	+	+	+	20.7 (10.3-34.2)
4	48	+	–	++	+++	+++	+++	–	++	+	+	++	++	++	24.8 (12.7-51.4)
5	72	++	–	+++	+	++	+++	+	+++	++	+	–	+	+++	27.5 (15.6-67.2)
6	96	++	–	++	+	++	++	++	–	++	–	–	+	++	14.7 (5.4-41.2)
7	120	–	+	+	+	+	+	–	–	+	–	–	+	+	11.6 (5.2-29.1)
8	144	–	–	–	–	+	+	–	–	–	–	–	–	–	3.3 (1.3-5.3)
9	168	–	–	–	–	–	–	–	–	–	–	–	–	–	0
10	192	–	–	–	–	–	–	–	–	–	–	–	–	–	0

*Mean; Range in parenthesis.

Female flowers also vary in their behavior vis-à-vis exposing their stigmas. Following are the types:

a) In majority of flowers in a population, female flowers have a persistent perianth. From this tightly closed perianth stigma grows and protrudes out at the time of anthesis (Figure 2.1b).

b) In approximately 10 per cent of the flowers, perianth (bract only) is completely absent or if present withers at an early stage of development leaving (Figures 2.1c, d) behind a naked flower bud and the emergence of stigma marks the beginning of anthesis.

Stigma Receptivity

Stigma is not completely dry. It is moderately wet, with the amount of exudates being less (Figure 2.1e). It is yellowish and wavy in outline; broad in the middle and tapering towards the apex. Its surface is thrown into folds each having numerous sub-spherical unicellular papillae (Figure 2.1f). The stigmatic area in the middle is fragile and splits longitudinally when a little pressure is applied.

Stigma receptivity begins 5 to 7 hour after pollination. However, receptivity increases 48 hour post anthesis and reaches its peak at 72 hour; 96 hour later a steady decline occurs which finally comes to an end after 148 hour. All these observations are based on the extent of pollen germinating on the stigmatic surface. At 72 hour post pollination percentage germinating pollen is mere 27.54 per cent. Plant to plant variation in the receptiveness of stigma was also found (Table 2.2). Natural pollen loads on stigma are very low. In a random sample of 150 (n= 20 plants from each populations) stigmas scanned, average pollen loads range between 5 and 30±7.39. This amount shows no increase even 144 hour after anthesis.

Pollen and Ovule Production

Average flower number per inflorescence works around 9.48 ± 0.63 (4-22) and 5.75 ± 0.24 (3-12) respective for male and female. Each pollen grain is smooth, tricolpate and round to slightly triangular in outline (Figure 2.1g). Pollen count varies from flower to flower. Average pollen produced by an individual flower is 34,821.6 ± 1214.74 and that per unit length works to 37,12,330.77. Pollen grains are small, 25.09 ± 0.09 (20.05-32.12) μm in diameter.

In female, flower is monocarpellary having a single ovule (Figure 2.1h), therefore number of ovules per plant remains the same as that of flower. Each ovule is anatropous measuring 453.07 ± 4.84 (370-580) μm in length and 343.11 ± 7.15 (312-408) μm in width.

Pollen Ovule Ratio

The pollen ovule ratio at various levels is tabulated in Table 2.3. The approx. pollen ovule ratio at the level of plant works at 3.41×10^5.

Pollen Viability

Pollen stain ability at the time of anthesis using 1 per cent aceto-carmine are commensurate with those obtained with TTC assay (Table 2.4).

Pollination Mechanism

Dioecy makes cross pollination imperative in the species under investigation. Flowers of *H. rhamnoides* are small, inconspicuous marked by the absence of any specialized pollinator attracting organs like showy sepals or petals and nectary *etc.* Notwithstanding the lack of

Table 2.3: Estimation of Pollen Ovule Ratio in *Hippophae* L.

Sl.No.	*At the Level of*	*Approximate Number of*		*P/O Ratio*
		Pollen Grains	*Ovules*	
1	Flower	34,821.6	1	3.48×10^4
2	Inflorescence	3,30,108.76	5.75	5.74×10^4
3	Plant	36,31,19,644.8.	1063.63	3.41×10^5

Table 2.4: Results of Pollen Viability in different Populations of *H. rhamnoides* using different Staining Procedures

Area	Populations	Viability at Anthesis using TTC	Stainability at Anthesis using 1 per cent Acetocarmine	Stainability after 10 Day of Anthesis using 1 per cent Acetocarmine
Kargil	Khomeni Bagh	89.94±2.59	88.68±1.63	42.37±1.31
		(75.22-100)*	(80.3-93.8)	(38.5-47.6)
	Barutsogs	93.13±1.46	94.92±1.22	74.15±2.64
		(87.69-100)	(89.7-97.4)	(62.1-82.3)

*Mean±S.E; Range in parenthesis.

Stain ability decreased considerably by about 58 per cent 10 days post anthesis.

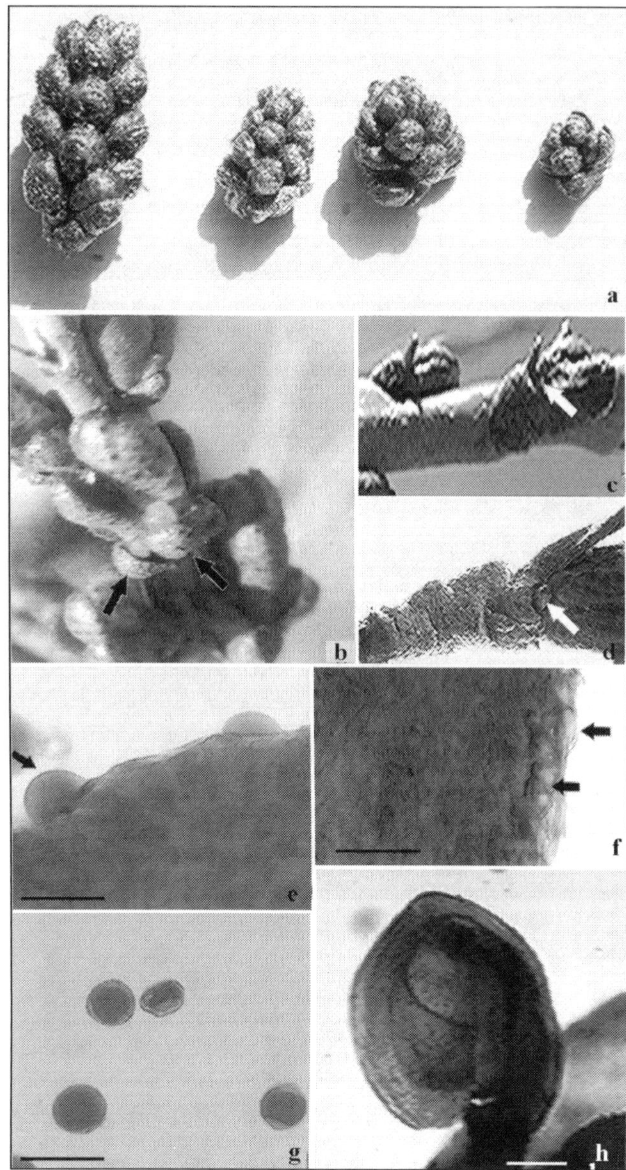

Figure 2.1 a-h: Male Inflorescences at different Developmental Stages.

(a) Female flowers with (b) and without bract (c-d). Receptive stigmatic surface (e-f). Stained pollen grains (g). Monocarpellary single anatropus ovule (h).

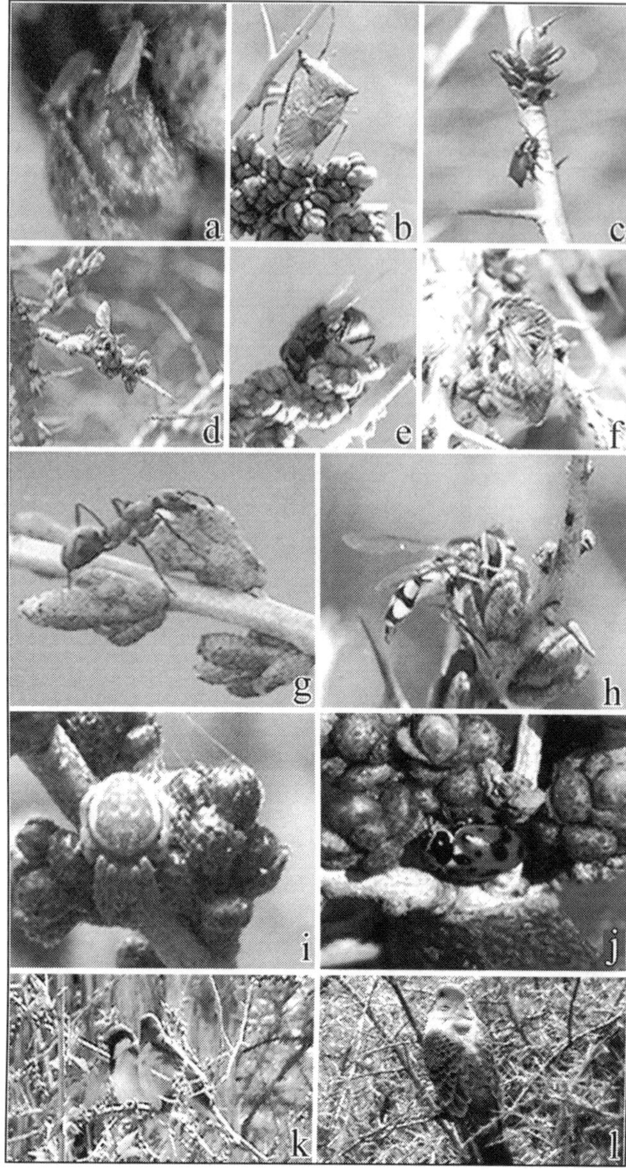

Figure 2.2 a-l: Visitors to Male and Female Plants of *H. rhamnoides* L. Mark the Variety of Biota.

Aphids (a), bug (b), beetle (c), flies (d,e), bee (f), ant (g), syrphid fly (h), spider (i), lady bird beetle (j), sparrow (k) and dove (l).

primary and secondary attractants, flowers are visited by a variety of biota belonging to six different orders/ or class. Nevertheless, as expectedly their number is quite low. House flies (Diptera), spiders (Arachneidae), bees and ants (Hymenoptera), beetles including lady bird (Coleoptera) and bug (Pentatomidae) do visit the flowers (Figures 2.2 a- i). However, no significant pollen load was observed on their body parts which rules out their role as pollinators. Most of the insects are robbers, their sole purpose to visit the flowers is to collect pollen which is in plenty. Birds particularly sparrows visit the flowers for preying on light brown aphids which causes severe damage to the flowers. These aphids belonging to family Aphidae (Homoptera) are in plenty in young developing floral buds.

Hanging Slide Experiments

Above mentioned observations, therefore, points towards prevalence of anemophily in seabuckthorn. This has been confirmed by the presence of ample amount of pollen grains on slide. Pollen count on the slides hanged at 5 m was, on an average, 2242 and 7120 for Khomeni Bagh and Barutsogs respectively. At 10 meter, mean 4982 (Khomeni Bagh) and 12700 (Barutsogs) pollen grains were observed on the slides. Thereafter a gradual decline in the pollen count occurred as the distance from the female plant increased (Table 2.5).

Slides hanged at the upper, middle and lower parts of female plants also trapped pollen evidencing the pollen receipt via wind by female flowers. Maximum

Table 2.5: Results of Hanging Slide Experiments in Two Seabuckthorn Populations

Sl. No.	*Population*	*Distance (m) from Male Plants*					*Position with Respect to Female Plant*		
		5	*10*	*15*	*20*	*30*	*Upper*	*Middle*	*Lower*
1.	Khomeni Bagh	224225± 84.37 (1798-2977)*	4982.46± 1070.39 (2760-6634.3)	3182.4 ±115.30 (2704-4281)	1739 ±72.70 (1369.33-2361.6)	1499.65 ±43.36 (1282.6-1898)	2465.6 ±30.32 (2341-2296)	3226.2 ±54.84 (2746-3501)	2747.18 ±120 (2075-3843.6)
2.	Barutsogs	7120.18 ±204.35 (5546.6-8497.31)	12696.8 ±520.84 (9756-16185)	6135.786 ±742.79 (3965-9516)	2045.13 ±47.08 (1803-2400)	1816.93 ±113.90 (1061-2634)	7636.6 ±388.8 (5750.3-11310)	8676.08 ±628.49 (5312-14196)	2545.33 ±154.43 (1737-3861)

*Mean ±S.E; Range in parenthesis.

pollen was scored from the middle portions of the plants (Table 2.4).

Results of our experiments are likely to be underestimates, since pollen dispersal is related to wind velocity and other environmental factors. During the experimentation poor weather condition and low wind velocity prevailed that are likely to have affected the pollen dispersal rate.

Bagging Experiments

Since the flowers are minute, sessile and could not be bagged individually, therefore butter paper bags were tailored to accommodate inflorescences of individual branches. Care was taken to bag only those flowers, which were un-pollinated and unfertilized by removing those with extruded stigmas. The total number of flowers bagged was estimated by counting the number of flowers per inflorescence and subsequently multiplying it with the number of inflorescences on that branch. The bags were left as such for one and half month until the end of flowering season to make sure that, no pollen ever reaches the stigma.

However, since the populations are not protected, damage to some bags was obvious. A total of 25 bags were used to cover approximately 1500 flowers. Of these, 17 remained intact covering approx. 1066 flowers of which 33 had transformed into fruits *i.e.*, 3.09 per cent of the total flowers bagged.

Fruit Count, Size and Development

Eight to ten days after anthesis, the stigma turns dark brown, becomes dry and finally withers out. However, the portion of stigma proximal to the ovary is persistent and remains attached to the developing fruit. The ovary, green with numerous stellate hairs, begins to swell in the next 6 to 10 days and becomes conspicuous. The surface of the ovary looses stellate hairs significantly and by July, the fruit is left almost glabrous. Colour changes take place in late July or early August from green to orange or red or deep red depending probably upon the genotype of the plant. The approximate fruit set in nature works at 79.46 (51.4-100 per cent) and 116.64 (82-159.4) per unit length for Khomeni Bagh and Barutsogs respectively.

Seed Count and Size

Each fruit harbors a single, shiny, oval to spherical seed. Its colour varies from brown to black, characterized by a single prominent groove that runs longitudinally in the centre of the seed. Overall, seeds have an average length and width of 0.36 and 0.25 cm respectively.

DISCUSSION

Approximately 98 per cent plants under study are strictly dioecious. However, in some, variants exist in the form of individual plants being male only, female only, monoecious, polygamomonoecious and even hermaphrodite (Ali, 2013; Sen *et al.*, 2013). Comprising Ca 2 per cent of the total sample size, phenotypic plasticity of this kind in the sex expression is known to affect the genetic structure by altering the distribution pattern of sexes in space and time (Loveless and Hamrick, 1984). These variants produce good number of seeds (Sen *et al.*, 2013) which germinate normally (Ali, 2013). As per Zapata and Arroyo (1978) they act as additional strategies for enhancing the proficient use of reproductive energy. However, the mode of seed formation in these variants has not been examined critically. There is an equal possibility of these being self pollinated provided they are self-compatible. If so, selfing is likely to promote local sub division in the populations (see Loveless and Hamrick, 1984 and reference therein). Since, tests for self-compatibility have not been carried out; it would be premature to conclude the same for seabuckthorn.

The time of flowering is in April. Male flowers open first while female do so around 24 to 48 hour later. As such, the two are not only separated in space but also in time making the population protandrous. The effect of protandry further gets an impetus by the stigma attaining receptivity 7 to 9 hour later. The pollen grains so released from male flowers get carried by wind to effect pollination in the female plants of the adjoining populations enhancing inter- and at the same time delaying intra-population pollination.

Male flowers of *Hippophae* open by lateral or apical modes; the latter is less common in occurrence. In the former tepals keep folded at their apices but their lateral margins separate out from each other resulting in the formation of a wind tunnel. Impact of wind currents to and fro this tunnel forces escape and dispersal of dry pollen grains. Reverse is true in the apical mode. Flower opening initiates at the tip and spreads till sub apical regions beyond which margins do not unfold. Ali and Kaul (2013) thought the two distinct patterns of anthesis to be adapted to two different modes of pollination; the lateral catering to wind pollination and apical probably for biotic/insect pollination. However, non-significant to negligible pollen load on the body parts of the visitors ruled out this possibility. As of now, the significance of the less frequent apical mode is not clear.

Furthermore, it seems that the buds with apical anthesis are not able to empty their pollen as effectively as their lateral counterparts. A considerable space constraint inherent in the floral architecture itself is the apparent causal factor. The androecial structure does not seem to aid in pollen release as well. Presence of bract on one side and the floral axis on other keep the flowers snuggled/squeezed such that following dehiscence anthers unload their contents inside the tepal cavity. Since the tepals open from top, only strong winds can facilitate pollen release. Normally wind pollinated plants, among other features bear large versatile anthers on slender delicate filaments, which are completely exposed at anthesis. Even low wind currents make the dehisced anthers dangle and release pollen in masses. Such features are altogether lacking in *H. rhamnoides*.

Anthesis in female flowers is indicated by the extension and extrusion of stigma. Majority of the female flowers in a population are characterized by a persistent perianth wherefrom the stigma grows and protrudes out at the time of anthesis. In many flowers, however, the perianth is either absent or if present, withers out at an early stage of development leaving behind a naked flower bud. The likely importance of two types of stigma exposure gains further support from the production of dimorphic pollen. Production of pollen grains in two sizes is particularly advantageous. The small sized pollen grains get dispersed over long distances with ease and require low winds but have difficulty in settling on stigma. On the other hand, large sized pollen gets easily trapped by nearby stigmas. Similar findings have been reported by many workers (Whitehead, 1969; Sharma *et al.*, 1990). Thus, the more exposed stigma can trap small sized pollen, the less exposed ones may settle with large sized pollen.

A noticeable change in stigmatic colour and length occurs with time that coincides with the receptivity. At the time of flower opening, the stigma, still in the process of development is green, ≤ 1mm. It continues to grow till a maximum length of 7 mm is attained and colour changes to yellow, the receptive stage. Thereafter, it turns orange to brown to dark brown, finally shrinks considerably, and withers out except for the persistent short stylar end. Though, this change in intriguing, yellow colour of the stigma can be partly due to the secretions found on its surface at receptivity and partly due to the moderate/considerable load of yellow pollen. Stigma receptivity initiating 24 hr after anthesis and extending for another 120 hours is protracted; with peak recorded at 72 hour. Plant to plant fluctuations within this range has also been noticed. The consequences of prolonged receptivity period are many, both advantageous as well as disadvantageous. There is a possibility of increased pollen loads of heterogeneous composition getting deposited on the stigma and the mode of deposition may become staggered. Both are, however, likely to result in the clogging of stigmatic surfaces with conspecific (intra- and inter-population seabuckthorn pollen) as well as heterospecific (that of other unallied neighboring plants) pollen. This would tend to delay the germination of pollen and hence fertilization. Nevertheless, a direct effect would be intense/fierce competition between conspecific pollen grains of diverse genotypes. Together these tend to increase the fitness and genetic quality of the offsprings (Primark, 1995; Galen *et al.*, 1986; Ashman, 2004; Schemske and Pautler, 1984; Schlichting *et al.*, 1987, 1990; Armbruster, 2002a; Paschke *et al.*, 2002, 2005; Barrett, 2003).

Contrasting yet interesting observation is recorded in nature. Even though pollen produced by a single flower average 34822 that by an individual male plant ($3.41x10^5$) is quite excessive given the number of ovule (s) per flower to be sired is one and that per plant are 1064. Irrespective of such a high P/O ratio the percentage pollen germination in nature is quite low although in reality, a small quantity of pollen may be enough to fertilize a single ovule in the pistil. But since stigmas are small targets, to ensure pollination, density of airborne pollen ought to be quite high (see Proctor and Yeo, 1973). That is why pollen production is enormous.

At 48 hour 12.7-51.4 per cent pollen germinates which increases moderately to 15.6-67.2 per cent at 72 hour. Despite a substantial proportion of pollen germinating, pollen tube of only one was found reaching the ovule. This is made possible by virtue of ca. 50 per cent pollen keeping viable even 10 days after anthesis (Ali, 2013). These observations are further supported by Mangla *et al.* (2013). They reported emergence of 81.88 per cent pollen tubes 24 hour post pollination though it is not clear whether the percentage reported is for naturally or manually pollinated stigmas. Further in 90 per cent instances only one and 10 per cent two pollen tube(s) traversed the compitum but invariably only one entered the ventral pore and caused fertilization (Mangla *et al.*, 2013). Differences in the extent of pollen germination could arise from the local and global effects. Populations worked upon are different (Choglamsar and Sindhu Darshan) and pollinations have been manual also.

Although the flowers are small, inconspicuous, odourless and nectarless with poor reward structure typical of wind pollinated taxa, many features do not conform to this syndrome. Architectural design, structural components of androecium and gynoecium are the ones, which are atypical. Close clustering of both male and female flowers into relatively rigid racemes, lack of versatile adnation, incomplete/non exposure of stamens and lack of feathery or hair like stigmas collectively make it less conducive for wind pollination. Nonetheless, their effect is well balanced by other features. Formation of a wind tunnel in laterally opening male flowers and pollen deposition in tepal cavity of apically opening ones compensate. Unidirectional wind current facilitates pollen movement from the former and extremely strong ones cause the release enmasse from the latter ones. Placement of male flowers away from main body of the shrubs is an incentive to the plant because interference to long distance dispersal of pollen is minimal and that, distance travelled by the released pollen will tend to be greater. It is well known that the density of the surrounding vegetation is directly related to the distance covered by pollen (Willson, 1983; Willson and Burley, 1983; Niklas 1985). Inconspicuous style, elongate, and more or less spathulate stigma characterize the pistil of *H. rhamnoides*. Stigma, three times the size of an ovary, is broad in the middle and tapering bluntly at the apex. It is moderately wet, papillate and wavy due to marginal infoldings. All these co-occurring features increase its surface area and equip the species for effective trapping of wind borne pollen. Together these make up for the plumose stigmas of wind pollinating taxa. The wetness of the stigma is because of oil and phenolics released when the cuticle ruptures at the time of receptivity (Mangla *et al.*, 2013).

In wind-pollinated plants, siring success is likely to be determined by lottery-model competition (Charlesworth and Charlesworth, 1981; Lloyd, 1984) resulting in stigma either receiving many or negligible pollen grains. Willson (1983) advocates that there is "no practical limit to the amount of pollen that wind can carry and wind does not cause pollen to be lost". However, velocity, direction and circulation pattern of wind is important in deciding the distance travelled by and quantum of gene flow via pollen. Pollen flow gets restricted under wet weather conditions, weak directionality and slow speed of wind (Willson and Burley, 1983; Niklas, 1985). Ladakh, during the flowering season, is essentially dry and wind velocity is moderate to very high which produce directional movements. Given that seabuckthorn carpel has predominantly a single ovule and that, a single pollen tube ultimately reaches ovules, afore mentioned conditions are cumulative and provide ambience in optimum pollen load deposition. This is further bolstered by the results of the hanging slide experiments.

Slides hanged at 5, 10 15, 20 and 30 m away from male plants had trapped enormous quantity of pollen confirming the mechanism of wind pollination operative in this species. Maximum pollen load (4982 and 12696 respective for Khomeni Bagh and Barutsogs) on the surface of the slide was observed at 10 m. However, unlike those observed by Mangla and Tandon (2014), the pollen flow in the air crosses beyond 15 m to that of 30 m in present study. At this distance, the pollen count on the slide was 1499 and 1816 at 30 m for Khomeni bagh and Barutsogs respectively.

In another such experiments, when the slides were hanged on the female plant at three levels, maximum deposition was recorded at middle followed by upper level. This observation is coincides with the areas where majority of female flowers differentiate and require pollen for effective fertilization.

Of the two experiments of hanging slide, the former serves as an index of pollen movement only while the latter is equivalent to the realized pollen flow since female plants are the recipients of the carriers of male gametes. While the former is random and likely to lead to over estimation of actual gene flow, the latter is more effective in predicting the amount that reaches the stigma surface.

REFERENCES

1. Ali, A. and Kaul, V. 2013. Dual mode of anthesis in seabuckthorn. *Current Science* 104 (4): 425-426.
2. Ali, A. 2013. Assessment of morphological and cytological variability in seabuckthorn (*Hippophae rhamnoides* L.) from some areas of Ladakh. *Ph.D thesis* submitted to University of Jammu, Jammu (J&K), India.
3. Ashman, T.L. 2004. Floral longevity. In: *Plant Cell Death Processes*. L.D. Nooden (ed), Elsevier Academic Press, San Diego, CA, pp. 349-362.
4. Barrett, S.C.H. 2003. Mating strategies in flowering plants: The outcrossing-selfing paradigm and beyond. *Philosophical Transactions of the Royal Society* B 358: 991-1004.
5. Charlesworth, D. and Charlesworth, B. 1981. Allocation of resources to male and female functions in hermaphrodites. *Biological Journal of Linnean Society* 15: 57-74.
6. Galen, C., Shykoff, J.A. and Plowright, R.C. 1986. Consequences of stigma receptivity schedules for sexual selection in flowering plants. *American Naturalist* 127: 462-476.
7. Kaul, V., Sharma, N. and Koul, A.K. 2002. Reproductive effort and sex allocation strategy in *Commelina benghalensis* L., a common monsoon weed. *Botanical Journal of Linnean Society* 140: 403-413.

8. Lewis, D. 1979. Sexual incompatibility in plants. University Park Press, Baltimore.

9. Lloyd, D.G. 1984. Gender allocation in outcrossing co-sexual plants. In: *Perspectives on Plant Population Ecology*. Dirzo, R., Sarukha´n, J., eds. Sunderland, M.A.: Sinauer Associates, pp. 277-300.

10. Loveless, M.D. and Hamrick, J.L. 1984. Ecological determinants of genetic structure in plant populations. *Annual Review of Ecology and Systematics* 15: 65-95.

11. Mangla, Y., Tandon, R., Goel, S. and Raina, S.N. 2013. Structural organization of the gynoecium and pollen tube path in Himalayan seabuckthorn *Hippophae rhamnoides* (*Elaeagnaceae*). http://aobpla.oxfordjournals.org/plt015.full.pdf.

12. Mangla, Y. and Tandon, R. 2014. Pollination ecology of Himalayan seabuckthorn, *Hippophae rhamnoides* L. (*Elaeagnaceae*). *Current Science* 106 (12): 1731-1735.

13. Niklas, K.J. 1985. The aerodynamics of wind pollination. *Botanical Review* 51: 328-386.

14. Paschke, M., Abs, C. and Schmid, B. 2002. Effect of population size and pollen diversity on reproductive success and offspring size in the narrow endemic *Cochlearia bavarica* (Brassicaceae). *American Journal of Botany* 89 (8): 1250-1259.

15. Paschke, M., Bernasconi, G. and Schmid, B. 2005. Effect of inbreeding and pollen donor provenance and diversity on off spring performance under environmental stress in the rare plant *Cochlearia bavarica. Basic and Applied Ecology* 6: 325-338.

16. Primark, R.B. 1995. *A Primer of Conservation Biology*. Sinauer Ass. Sunderland Mass.

17. Schemske, D.W. and Pautler, L.P. 1984. The effects of pollen composition on progeny fitness components in a neotropical herb. *Oecologia* 62: 31-36.

18. Sen, R., Tamchos, S. and Kaul, V. 2014. Sexual variants in a dioecious taxon, *Hippophae rhamnoides* L. *International Journal of Plant Reproductive Biology* 6: 41-43.

19. Sharma, N., Koul, P. and Koul, A.K. 1990. Reproductive biology of *Plantago* L. III. floral adaptation to pollination in *Plantago lagopus* L. In: *Proceedings of Indian Academy of Sciences (Plant Science)*, 100 (6): 393-398.

20. Shivanna, K.R. and Rangaswamy, N.S. 1992. *Pollen Biology*. A laboratory manual. New Delhi, India.

21. Whitehead, D.R. 1969. Wind pollination in the angiosperms: Evolutionary and environmental considerations. *Evolution* 23 (1): 28-35.

22. Willson, M. F. and Burley, N. 1983. *Mate Choice in Plants*. Princeton University, Press Princeton, New Jersey, USA.

23. Willson, M.F. 1983. *Plant Reproductive Ecology*. John Wiley, New York.

24. Zapata, T.R. and Arroyo, M.T.K. 1978. Plant reproductive ecology of a secondary deciduous tropical forest in Venezuela. *Biotropica* 10 (3): 221-230.

Chapter 3

Frankia Nodulation Study of *Hippophae salicifolia* in Riverine and Non-riverine Areas of Lachen Valley in North Sikkim

Bharat C. Basistha*, Sushen Pradhan and Prena Pradhan

Sikkim State Council of Science and Technology,
Deorali – 737 102, Gangtok, East Sikkim, India
(*E-mail: basisthabharat@gmail.com)

SUMMARY

The study was carried out to compare various factors affecting the nodulation by actinomycetes (*Frankia*) in the roots of *Hippophae salicifolia* growing in riverine and non-riverine areas. The root nodules numbers, weights, soil moisture, plant age, plant height, plant canopy and sizes of the root nodules were determined in a 3-6 years old plant with an altitude ranging from 2540 to 3099 m. Ten trees each from both the areas were selected for the study. Healthy and maximum numbers of nodules were found at a depth range of 4-6 inches. ANOVA of number of root nodules and weight of root nodules found highly significant *P*-0.00096, *P*-0.0066 respectively. Positive correlation was analyzed between number of nodules and the soil moisture content in both the areas. No relationship was found between the weight and the age of the nodules. Soil analysis (organic C, K_2O, P_2O_5, S and Cl) found to be less in riverine areas when compared to non-riverine areas. Maximum number of nodulation was found in riverine area (580 nos.) as compared to non-riverine area (224 nos.). This may be due to deficiency of nutrient in host plant, thus larger nodules are formed to compensate their requirement of nutrients.

***Keywords**: Hippophae salicifolia, Nodulation, Riverine, Non-riverine, Lachen and Sikkim.*

INTRODUCTION

Seabuckthorn (*Hippophae salicifolia* D. Don) belongs to family Eleagnaceae (2n=24) is an actinorhizal angiosperm, which colonizes in a new soil (Basistha *et al.*, 2009). The genus is divided into seven species and nine subspecies based on its morphological variations (Bartish *et al.*, 2002). It is spread almost the entire Eurasian continent from Central China, across the Himalayas, in a more or less continuous belt through the mountain valleys of central, southern and western Asia to the Black Sea. In India, the plant is generally found in the higher Himalayan regions of Himachal Pradesh, Jammu and Kashmir, North East region and Uttarakhand.

H. salicifolia is a thorny, deciduous, dieoecious tree (Singh, 1998; Basistha *et al.*, 2009). It is tallest among all species of seabuckthorn. In addition to its high medicinal and nutraceutical properties, seabuckthorn is also valued for its high ecological roles in nature (Singh, 1998). *H. salicifolia* is an economically important plant that grows in the higher Himalayan region (Bartish *et al.*, 2002). The plant plays important role in preventing soil erosion and soil water loss in regulating microclimate, as well as in retaining stability in the region (Rogen, 1992). *Hippophae* L. has been shown to be superior to any other tree in afforestation, soil conservation, wasteland reclamation, (Wheeler *et al.*, 1986), especially on fragile slopes due to its extensive root system. It is also a

Seabuckthorn (*Hippophae* L.): *A Multipurpose Wonder Plant, Vol. 5: Advanced Technologies for Cultivation, Processing, Health Protection and Environmental Conservation* (Virendra Singh, *Editor-in-Chief*, 2018), p. 19–24, Astral International Pvt. Ltd., New Delhi, India.

multipurpose fast growing species, which is serves as a measure of biodiversity conservation, soil conservation, medicines, food, fodder and fuel wood. *H. salicifolia* is an important genus in the succession of vegetation. It grows in riverine (river side) and non-riverine (non riverside) areas in open natural habitat away from river and stream banks (Basistha *et al.*, 2009).

This actinorhizal species has the capacity of holding soil in fragile slopes, needs no good arable land for its cultivation and has symbiotic relationships with nitrogen fixing actinomycetes known as *Frankia* in its root nodule (Akkermans *et al.*, 1983; Dobrista and Novik, 1992; Rongsen, 1992; Basistha, 2009). Akkermans and Dijk, 1975 reported that the highest nodulation was found in 1 to 2 year old plants, older plants are poorly nodulated, though they were frequently connected by stolons with nodulated young shoots. The nitrogen fixation effect of seabuckthorn root nodules varies with seasons, precipitation, age of plant and environmental factors (Basistha *et al.*, 2009). Oremus, 1979 reported that there was a sharp decrease in the number of nodules in the age classes of four years and older, which indicated that many of them died after three years. Of the total number of living nodule material 89 per cent were younger than four years. Changes occur even in a day, between morning and evening (Zhou *et al.*, 1999). Oremus, 1979 also described variation in the number of nodules in different plots might be partly due to the difference in chemical composition of the soil. However, the results showed that for these factors the differences in the samples were small and they were very unlikely to be responsible for the strong variation in nodule number.

Thus in the present study, Lachen valley was taken as the study area as it has a very interesting admixture of land cover and land use practices. The canopy varies from as low as 10 per cent in shrub areas to almost 80 per cent in dense subtropical forest. The reverine area is mostly dominated by seabuckthorn forest whereas the non-riverine side has mixed forest. As the nodulation of *H. salicifolia* appeared to be highly variable, and also the studies shows that the nodules in homogenous scrub differ widely from place to place, the present study was undertaken to determine its nodulation frequency in the Reverine and Non-Riverine areas and also to check the effect of various factors like plant height, canopy, tillering and soil pH and chemical composition on nodulation.

MATERIAL AND METHODS

Study Area

Seabuckthorn study area consisted of approximately 300 ha, located in Lachen Valley (North District of Sikkim), which lied between 27°47′32.1″ and 27°43′57.2″ North latitude and 88°33′07″ and 88°33′13″ East logititude. The study of nodulation was carried out in the trees aged between 3-6 years in riverine (along side of the river) and non-riverine (away from river side) areas; ten trees each from both the areas were selected for study. GPS locations of all the boundary of seabuckthorn patches were recorded. These locations were sometimes, analyzed to find the distribution area of seabuckthorn using geocoded Indian Remote Sensing 1D LISS III, False Colour Composite (FCC), January 2012 data of 1: 50000 scales.

Mapping of Nodule Position

The area of 1.5 m around the tree and 6-8 inch below (rhizospheric region) the soil surface was dug carefully, the root system was extracted and position of nodules was mapped.

Data Collection

Following a stratified random procedure, ten seabuckthorn trees each from nine diameter classes were selected. The dead dying and malformed trees were avoided. Measurements were taken for plant height using Laser range finder/hypsometer model-400XL/XT/LH (Range-up to 400+yards passive 999 yards reflective), tillering (number of branches from the base) was also recorded through optical counting, girth, plant canopy using standard measuring tape, topography, site cover, vegetation type management, age of plant by Dendrochronology (Oremus, 1979) and were recorded. Simultaneously soil pH (pH meter-Takemura Electric Works LTD –Model No. –DM-15), air temperature (Multi thermometer, CE), aerial humidity (Model LR6 Mignon AA, England) was measured.

Measurement of Nodules

Plants that were 3 to 6 years old and growing in an altitude range of 2540 m to 3099 were selected. In terms of their inter-nodular distance, depth of nodules, nodule location and the growth of nodule were determined. Randomly 30 per cent of the nodules were detached from the main root and their total weight was determined using top pan balance (Docbel, Brawn make) and their size was measured using a measuring tape.

Tagging of the Sample

The tree was then tagged by numbering it using aluminum plant tag and wire for further studies of the same trees in different seasons. All the data were tabulated (Tables 3.1–3.5, Figure 3.1) and evaluated.

RESULTS AND DISCUSSIONS

Results showed that maximum nodulation in *H. salicifolia* were observed in the tap root with few in lateral roots. Healthy and maximum numbers of nodules were found at a depth range of 4-6 inches. The nodules found in the lesser depth region were matured and dried as compared to those found in deeper region. It was observed that the nodules found in sandy soil were larger in size (clumped) and less in number as compared to the nodules found in humus soil, which were scattered and more in number but smaller in size.

Table 3.1: Significance of Root Nodules and Weight in Riverine and Non-riverine Areas

Nodulation	*Source of Variation*	*DF*	*Sum of Squares*	*Mean of Squares*	*F value*	*P value*
Number of nodules	Between Groups	1	6336.8	6336.8	15.51363	0.000963
	Within Groups	18	7552.4	408.4667		
Weight of nodules	Between Groups	1	88112.81	88112.81	9.38889	0.006683
	Within Groups	8	168926.3	9384.796		

Calculations showed that the total number of nodules varied from 4 to 107 in different trees (Tables 3.3 and 3.4). ANOVA of riverine and non-riverine areas of number of root nodules and weight of root nodules found highly significant *P*-0.00096, *P*-0.0066 respectively (Table 3.1).

Figure 3.1 Interprets that in riverine area sandy soil has more moisture content than the humus soil and in non-riverine humus soil has more moisture content. Hence, the graph shows that the soil with more moisture content gives maximum number of nodulation in *Hippophae salicifolia*.

Positive correlation was analyzed between number of nodules and the soil moisture content in both the areas. No relationship was found between the mean nodule weight and the age of the nodules.

Earlier in a study it was showed that the chemical composition of soil is very unlikely to be responsible for the strong variation in nodule number in their respective study as the difference between the samples were small (Oremus, 1979). However in the present investigation the Soil analysis showed that (organic C, K_2O, P_2O_5, S and Cl) was found to be less in riverine areas when compared to non-riverine areas.

The result of the Table 3.2 shows that in the riverine area the mean number of nodules in sandy soil was 56.33 with average weight of 179.16 gm and in the humus soil mean number of nodules was 52.75 with average weight of 13.33 gm. Similarly, in the non-riverine area the mean number of nodules in sandy soil was 18.71 and the mean weight 32.50 gm, and in humus soil it was 23.66 and the mean weight was 42.83 gm.

The data also shows that in the riverine area the moisture content of the sandy soil was 55.83 per cent and of the humus soil it was 14.66 per cent. In the non-riverine area the moisture content of sandy soil was 24.75 per cent and of the humus soil was 28.01 per cent. This may be due to deficiency of nutrient in the growing region of the host plant, thus larger nodules are formed to compensate their requirement of nutrients.

This data interprets that in the riverine area the number of nodules in sandy soil was more compared to the nodules in the humus soil but in non-riverine area the number of nodules in humus soil was more compared

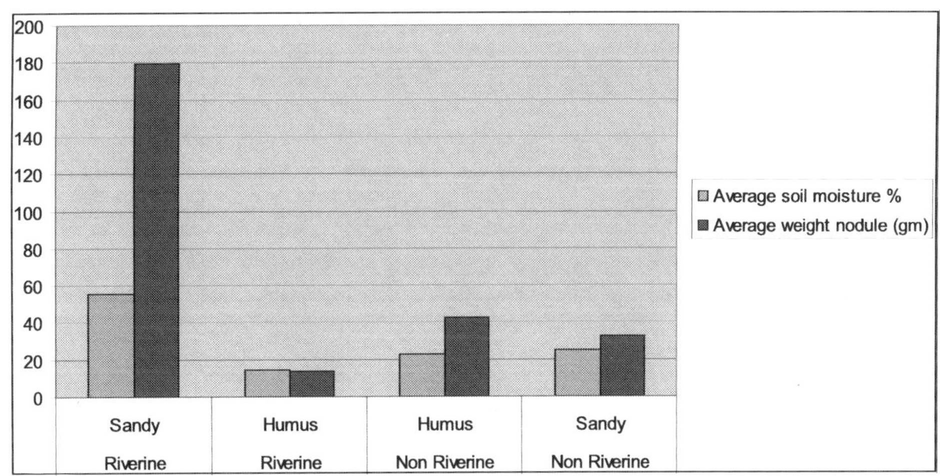

Figure 3.1: Effect of Soil Moisture and Types of Soil in Nodulation in Riverine and Non-riverine Area.

Table 3.2: Nodulation Effect by Soil Moisture and Types of Soil

Area	*Types of Soil*	*Average Soil Moisture Per cent*	*Mean No. of Nodules*	*Average Weight Nodule (gm)*
Riverine	Sandy	55.83	56.33	179.16
Riverine	Humus	14.66	52.75	13.33
Non Riverine	Humus	23.01	23.66	42.83
Non Riverine	Sandy	24.75	18.71	32.50

to sandy soil (Table 3.2). Close correlation was found between number of nodules and moisture content of the soil. From the data it can also be deduced that the moisture content and the type of soil plays an important role in nodulation. Hence the soil that had more moisture content gave high number of nodules compared to soil with low moisture content.

Quantitative study of nodulation was done earlier by some worker like, Stewart and Pearson, 1967 who found variation in nodulation from 98 and 370 nodules per square meter. In a study by Akkermans, 1971, who investigated 32 m^2 one to two year old plant, nodule numbers varied from 9 to 200 per square meter. Same author found low level of nodulation in older plants (4-15 years), which varied from 0 to 25 nodules per square meter. However, these data were based on small study area. Oremus, 1979, found nodulation varied from 30 to 281. All of these workers have done their work on the basis of different age group of plants. Stewart and Pearson, 1967 aswell as Akkermans, 1971, related their nodule number to the age of the vegetation by estimating the age of the shoot. Oremus, 1979 reported a close correlation between the age of the roots and the nodulation, young roots having the highest number. In this respect we based our studies on Quantative analysis of nodulation of different riverine and non riverine areas and the result of Table 3.3 and 3.4 showed nodular quantity widely varied in riverine and non riverine areas. In riverine area nodulation varied from 22 to 107 and in non-riverine area it varied from 4 to 46.

Tables 3.3 and 3.4 when compared, showed that the larger quantity no of nodules were found in the reverine area as compared to the non-riverine area. The data showed that there seemed to be a correlation in height of trees, canopy and soil pH for the nodulation. A negative correlation was observed between the tillering of the plant and nodulation. The sum of total nodules found at the riverine area was 580 and their weight was 1674.50 gm and in non-riverine area number of nodules was 224 and their weight was 347 gm.

Earlier in 1979, Oremus in his studies described that the variation in the number of nodules in different plots might be partly due to the difference in chemical composition of the soil. Thus, in the present study the variations described above between the riverine and non-riverine area with respect to the number and the weight of the nodules can aswell be due to difference in the chemical composition of the soil, therefore from each plot the soil was tested to determine its chemical composition. The result showed (Tables 3.5 and 3.6) that

Table 3.3: Nodulation Data of Riverine Area

Height (feet)	*Canopy (feet)*	*Tillering*	*pH of Soil*	*No. of Nodules*	*Weight of Nodules*	*Inter Nodular Distance*
15	10	1	6.5	61	350	6.7
15	9	1	6.2	29	150	8.2
11	6	1	6.2	22	100	3.9
13	10	1	6.8	41	45	4.5
22	5	1	6.8	107	140	4.5
20	20	2	6.5	76	9.5	3.2
35	20	1	6.8	45	400	6.5
15	10	2	6.8	66	110	13.2
15	12	1	6.6	78	300	4.1
11	5	1	6.8	55	70	2.3

Table 3.4: Nodulation Data of Non-Riverine Area

Height (feet)	*Canopy (feet)*	*Tillering*	*pH of Soil*	*No. of Nodules*	*Weight of Nodules*	*Inter Nodular Distance*
9	4	1	6.8	24	50	14.7
5	3	2	6.5	16	30	5.6
15	9	1	6.8	18	40	5.4
15	6	3	6.9	12	3	3.2
28	15.3	1	6.8	4	7	2.9
14	14.2	1	6.9	12	15	2.8
15	12	5	6.8	46	40	11.3
20	15	1	6.7	36	42	5.6
12	7	6	6.4	36	90	6.0
12	7	1	6.6	20	30	3.6

Table 3.5: Nutrient Analysis of Riverine Soil

E.C m.mho/cm	*Organic Carbon*	*Avg. K_2O ppm*	*Avg. P_2O_5 ppm*	*Avg. Sulfur ppm*	*Chloride mg/gm*	*Clay*	*Slit*	*Sand*
0.11	0.57	38.43	21.76	11.78	0.002	34.00	4.00	85.00
0.14	0.78	46.97	21.68	22.59	0.005	28.00	3.00	88.87
0.14	0.97	42.70	15.23	23.57	0.004	30.00	5.00	90.90
0.16	0.49	64.05	19.56	25.35	0.005	20.00	2.00	93.98
0.06	0.64	67.10	26.88	25.53	0.005	27.00	2.00	75.00
0.23	0.86	71.98	26.87	24.12	0.004	26.00	4.00	78.09
0.23	0.86	76.25	28.29	21.60	0.005	22.00	3.00	88.98
0.19	0.95	80.53	26.66	24.55	0.004	35.00	2.00	86.75
0.23	0.41	78.08	26.67	26.51	0.005	26.00	2.00	83.98
0.23	0.35	78.08	26.11	26.51	0.005	34.00	2.00	96.98

Table 3.6: Nutrient Analysis of Non-Riverine Soil

E.C m.mho/cm	*Organic Carbon*	*Avg. K_2O ppm*	*Avg. P_2O_5 ppm*	*Avg. Sulfur ppm*	*Chloride mg/gm*	*Clay*	*Slit*	*Sand*
0.26	2.19	84.79	29.38	27.50	0.007	34.00	2.00	65.78
0.26	2.08	109.19	43.52	29.46	0.006	28.00	4.00	71.87
0.24	1.03	104.92	65.29	33.38	0.008	30.00	6.00	67.09
0.28	1.08	109.80	27.21	39.28	0.007	30.00	5.00	66.02
0.31	3.24	84.79	26.17	41.24	0.008	30.00	2.00	64.09
0.34	1.24	93.94	76.17	35.35	0.006	26.00	4.00	62.78
0.29	3.56	87.23	29.66	27.50	0.006	31.00	2.00	58.72
0.62	2.70	146.40	36.99	33.38	0.007	34.00	3.00	54.56
0.35	0.99	91.53	38.09	26.51	0.009	31.00	4.00	43.09
0.24	0.97	81.13	30.46	35.36	0.008	33.00	5.00	34.89

the amount of organic carbon, average K_2O, average P_2O_5, average sulfur and chloride mg/gm were found to be less in riverine areas as compared to the non-riverine areas, as these chemicals are leached out in the riverine area due to a continuous flow of water.

CONCLUSION

It can be concluded that the soil of the riverine area which is sandy may be deficient in nutrient content as the nutrient present in the soil may have been leached out by the continuous flowing river water, this may be the cause for less nutrient present in the soil of these area. Hence, it might be due to these reasons that the plants produce large number of nodules to compensate their nutrient requirement from root nodules for their growth. Also the moisture content in the soil might have enhanced the growth of the nodules to fix more N_2 to be supplied to the host plant.

ACKNOWLEDGEMENTS

Authors are grateful to the Department Of Biotechnology, Government of India for finical support and the Department of Science and Technology, Government of Sikkim for providing necessary facilities.

REFERENCES

1. Akkermans, A.D.L. 1971. Nitrogen fixation and nodulation of *Alnus* and *Hippophae* under natural conditions. *Ph.D. Thesis*, Univ. Leiden.
2. Akkermans, A.D.L. and Dijk, C. V. 1975. The formation and nitrogen fixing activity of the root nodules of *Alnus glutinosa* under field conditions. In: *Symbiotic Nitrogen Fixation in Plants* (P.S. Nutman, Ed.). *I.B.P.* 7: 511-520, Cambridge Univ. Press.
3. Akkermans, A.D.L., Roelofsen, W., Blolm, J., Huss-danell, K. and Harkink, R. 1983. Utilization of carbon and nitrogen compounds by *Frankia* in synthetic media and root nodules of *Alnus glutinosa*, *Hippophae rhamnoides*, and *Dastica cannabina*. *Can. J. Bot.* 61: 2793-2800.
4. Bartish, I. V., Jeppson, N., Nybom, H. and Swenson, U. 2002. Phylogeny of *Hippophae* (Elaeagnaceae) inferred from Parsimony analysis of chloroplast DNA and morphology. *Syst. Bot.* 27: 41-54.
5. Basistha, B.C., Sharma, N.P., Lepcha, L., Arrawatia, M.L. and Sen, A. 2009. Ecology of *Hippophae salicifolia* D. Don of temperate and sub-alpine forests

of North Sikkim Himalayas-A case study. *Symbiosis* 9-009-0033-y.

6. Dobrista, S.V. and Novik, S.N. 1992. Feedback regulation of nodule formation in *Hippophae rhamnoides*. *Plant Soil* 144: 45-50.

7. Oremus, P.A.I. 1979. A quantative study of nodulation in *Hippophae rhamnoides* L. ssp. *rhamnoides* in a costal dune area. *Plant and Soil* 52: 59-68.

8. Rongsen, L. 1992. Seabuckthorn a multipurpose plant species for fragile mountain. ICMOD Occasional Paper, No. 20, Katmandu, Nepal, 62p.

9. Singh, V. 1998. Seabuckthorn a wonder plant of dry temperate Himalayas. *Indian Horti*. 43 (2): 6-8.

10. Stewart, W. D. P. and Pearson, M. C. 1967. Nodulation and nitrogen-fixation by *Hippophae rhamnoides* L. in the field. *Plant and Soil* 26: 348-360.

11. Wheeler, C. T., Hooker, J. E., Crose, A. and Berry, A. M. M. 1986. The improvement and utilization in forestry of nitrogen fixation by actinorhizal plants with special reference to *Alnus* in Scotland. *Plant Soil* 90: 393-406.

12. Zhao, J., Zhengtan, M. and Yucia, C. 1999. Research on seabuckthorn health tea and its processing technology. In: *Proceeding of International Symposium on Seabuckthorn* (*Hippophae rhamnoides* L.), Beijing, China.

Chapter 4

Simulation of Vibration Harvesting Mechanism for Seabuckthorn

Longsheng Fu, Jun Peng, Qiang Nan, Dongjian He and Yongjie Cui*

College of Mechanical and Electronic Engineering,
Northwest A and F University, Yangling, 712100 China
(*E-mail: cuiyongjie@nwsuaf.edu.cn)

SUMMARY

Seabuckthorn (SBT) is an ideal plant for ecological management and was thus planted widely in the western of China. Its fruit is of high nutritional and medicinal values. However, its economic value is far from development because SBT fruit is very difficult to be harvested. Mechanical vibration is one of feasible way to make fruit separation. In order to design proper vibratory harvesters for a tree crops, in this paper, vibration harvesting mechanism of SBT was simulated and analyzed by finite element method. Firstly, three-dimensional solid model of SBT tree was built by Pro/E and imported into ANSYS. Next, modal analysis was used to determine natural vibration properties of SBT tree such as natural frequency and vibration mode. Finally, harmonic response analysis was applied to determine steady state response when the SBT tree is added a sine load. The modal analysis results showed that first twenty order natural frequencies of SBT tree varied from 8.8Hz (the first order frequency) to 31.2Hz (the twentieth order frequency). Results of harmonic response analysis showed that vibration force applied to the side branch is effective than to the truck, while with little damage to the tree. In addition, the vibration force applied to the side branch was 180~280Nwith a frequency of 14.0 Hz, it could ensure the majority of SBT fruit to be detached from the tree. The simulation analysis in this paper could provide a basis for designing and development of SBT vibration harvester.

***Keywords**: Seabuckthorn fruit, Vibration harvesting, Finite element model, Modal analysis and Harmonic response analysis.*

INTRODUCTION

Seabuckthorn (SBT) (*Hippophae rhamnoides* L.) is a hardy and deciduous shrub with yellow or orange berries. The wide adaptation, fast growth, strong coppicing, and sucking habits, coupled with efficient nitrogen fixation, make SBT an optimal pioneer plant in soil and water conservation, desertification control, land reclamation and reforestation of eroded areas (Yang and Kallio, 2002). Two million hectares of SBT trees have been planted for the need of ecological management in China, accounting for more than 90 per cent of the world total area of SBT, among which more than 80 per cent are located in the western region (Wu *et al.*, 2000). On the other hand, SBT fruit is rich in vitamins and phenolic compounds and used for medicinal purposes and as food in some parts of the world (Bal *et al.*, 2011; Song *et al.*, 2014).

Unfortunately, it is difficult to harvest the SBT fruit because it does not easily form an abscission layer and the fruit is tightly clustered onthorn-covered branches. In Sakatchewan, Canada, the total labor cost for harvesting an orchard of 4 ha was estimated to be 58 per cent of the total cumulative production cost over 10 years (Li, 2002). In Asia, harvesting is still mainly completed manually or using simple hand-held tools. This difficult and labor intensive process requires about 1500 h/ha (Liang *et al.*, 2008). Therefore, the development of mechanical or other harvesting techniques for SBT have attracted widespread attention.

Seabuckthorn (Hippophae L.): A Multipurpose Wonder Plant, Vol. 5: Advanced Technologies for Cultivation, Processing, Health Protection and Environmental Conservation (Virendra Singh, *Editor-in-Chief*, 2018), p. 25–33, Astral International Pvt. Ltd., New Delhi, India.

The attempts for harvesting include direct juicing harvesters (Stan *et al.*, 1985; Ishii, 2003), tree shakers (Gaetke, 1993), branch shakers (Bantle, 1996; Mann *et al.*, 2001; Olander, 1995) vacuum suctions units (Mu, 2012), hormone treatments (Demenko *et al.*, 1986; Zhu, 1991), and whole branch harvesters (Olander, 2012). Among them, the trunk vibration harvester from Russia was the highest harvesting efficiency reaching 50 kg/h, but its removal rate of 50 per cent is too low to be acceptable. The best harvester is the cutting harvester from Germany, it could remove 80 per cent of the fruit at a harvest rate of 30 kg/h, while only damage 5 per cent of the fruit. Therefore, this method, supplied by the Kranemann Co. Ltd., is the only commercially viable way for mechanical harvesting of SBT fruit. In addition, it was found that some cultivars could be harvested in the field without freezing, such as 'Hergo'. Therefore, it could be possible to breed SBT cultivars suited for harvesting by shaking. For large scale harvesting, the only feasible method is to shake or vibrate the berries off the plant (Fu *et al.*, 2014).

Although the trunk vibration harvester can make the whole bush to be harvested at one time, it is only effective for bushes which have one central trunk with short branches. Bushes which have long and slender branches are more difficult to harvest by shaking the trunk because most of the energy is lost before reaching the berries (Mann *et al.*, 2001; Olander, 1995).

Therefore, there have been some attempts at harvesting SBT berries by vibrating the branch directly (Stan *et al.*, 1985) used a black currant harvester to test seven SBT cultivars. Only one cultivar could be harvested successfully when using a vibration frequency of 18.5 Hz and amplitude of 25 mm. A prototype from Sweden was tested with amplitudes of 40 to 55 mm and frequency of 25 Hz (Olander, 1995). For the 'Indian Summer' cultivar in western Canada, (Mann *et al.*, 2001) found that at frequencies of both 20 and 25 Hz, the percentage of berries removed by shaking increased linearly with the increasing of amplitude. The combination of 25 Hz and 32 mm produced the best effect that 98 per cent of the berries were removed within 15 seconds of shaking during the November harvest period.

Normally, the optimal vibrating speed and amplitude vary from crop to crop, which related to their natural frequency. However, in the current equipment design process, main parameters were commonly acquired by observation and measurement

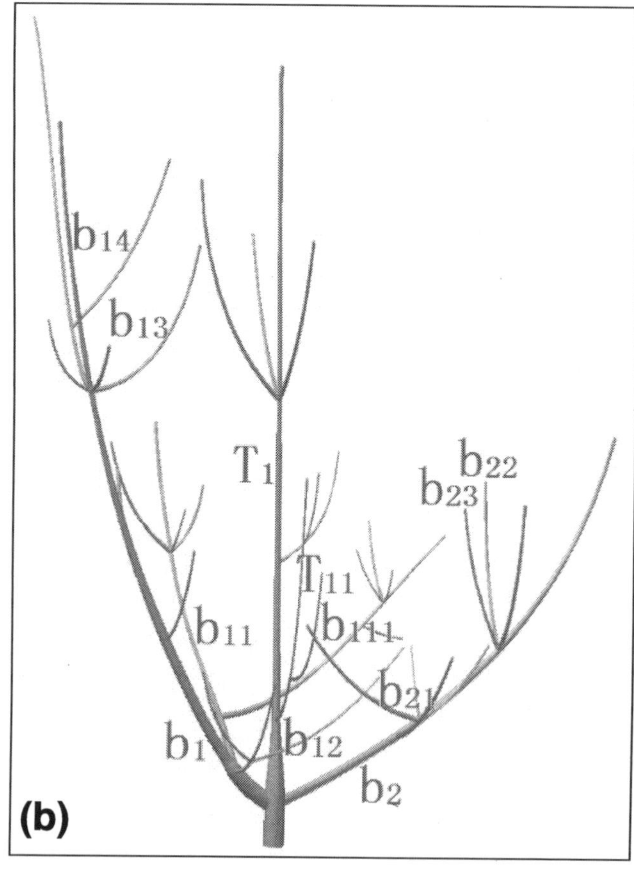

Figure 4.1: A SBT Tree and its 3D Model by Pro/E.

(a) A SBT tree. (b) The 3D model of the SBT tree. The numbering system is that the first letter represents category (b) represents branches, T represents trunk). The number represents grade of branches (branches are classfied three grades, number represents thick branches, thin branches, thiner branches, respectively).

of field experiment which needs to establish high cost testing platformsbut obtains results of randomness. Finite element method had been proved to solve the problem partly (Tang *et al.*, 2006). (Huang *et al.*, 2011) had conducted the finite element simulation of sugarcane cutting process. Quan *et al.* (2011) had carried on the finite element analysis on corn stubble harvesting system.Taking SBT trees of Xinjiang, China as an example, this paper carried out a finite element analysis simulation of vibration harvesting for SBT fruit by combining ANSYS software (ANSYS 15.0, ANSYS, Inc., Canonsburg, USA) with Pro/Engineer (Wild Fire 5.0, PTC Inc., Needham, USA) three-dimensional (3D) modeling software. This work provides a theoretical basis for designing and development of future vibration equipment.

MATERIALS AND METHODS

The Establishment of SBT Model

The SBT tree is usually about 1,500 mm tall, whose trunk is about 1,000 mm tall, crown is around 100 mm in diameter, main branch is about from 20 to 50 mm in diameter, and side branch is about 10 mm or smaller in diameter. Figure 4.1(a) shows a SBT tree in the Qinghe County, Xinjiang in northwest China.

A 3D model of SBT tree was built by Pro/Engineer (Pro/E), as shown in Figure 4.1(b). The shape of SBT tree is very irregular, so its trunk and branches were defined as variable cross-section cylinder. The fruit mainly grows on side branch. There are lots of side branches in a SBT tree, which degrade computational efficiency significantly. Therefore, the structure of SBT tree should be as simple as possible to reduce calculation time of finite element model (FEM) analysis.

Creating Finite Element Model of SBT Tree

The FEM of SBT tree was imported from Pro/E. The unit system of m-t-s was adopted in this research. The solid model needs to be discretized into several sub domains before the finite element analysis. The Solid185 of 8-node 3D entity element in ANSYS was selected for FEM of SBT tree by considering its irregular geometric model.

Table 4.1: Material Property Parameters of SBT Tree (Reference paper)

Density (kg/m^3)	*Young's Modulus (MPa)*	*Poisson Ratio*	*Bending Strength (MPa)*
574	4959	0.348	75

Lignin, as the chief component of SBT tree, is a kind of amorphous structure material but has the isotropic property. Thus,the mechanical properties of the SBT tree wood were determined under the assumption that it is isotropic in nature (Savary *et al.*, 2010). The FEM of SBT tree was meshed by using Smart Size tool of Free Grid method of ANSYS, which was composed of 424,552 elements and 107,843 nodes. The material parameters of SBT tree were shown in Table 4.1 (Shang *et al.*, 2008; Ke, 1989).

Modal Analysis of SBT Tree

Modal analysis is a process for determining dynamic characteristics of a mechanical system (damping, natural frequencies and modes of vibration) with the aim to describe dynamic behaviour of this system. Mode is the natural vibration characteristics of the mechanical system, and each mode has the specific natural frequency, damping ratio and modal shape. This studyis based on the LS-DYNA simulation, to determine the vibration natural frequency and modal shape of SBT tree, which lays the foundation for the further dynamic analysis. This provides a new method for the study on optimal harvest frequency of SBT tree.

The Block Lanczos method was used in this study. All the constraints are concentrated in SBT tree connected to ground roots. In modal analysis, boundary conditions of the roots were zero degrees of freedom constraint, which meant the directions of X, Y and Z were defined as zero displacement constraint. Generally, if the frequency range was unknown, the highest frequency should be defined with a larger value. According to other studies (Tang, 2011), the influential frequency dynamic characteristics mainly concentrated in the low frequency range.Therefore, only the first twenty order modes of SBTtree were extracted and analyzed.

Harmonic Response Analysis of SBT Tree

Harmonic response analysis is used to determine the steady state response when the SBT tree is added a load (sine excitation). Considering the complexity of SBT treestructure, the full method was used in the harmonic response analysis. The full method applies complete system matrix to calculate harmonic response (no matrix reduction).The matrix can be symmetric or asymmetric, so it doesn't involve mass matrix approximation, and using a single process to calculate all the displacement and stress. The essence of vibration harvesting is based on the load applied to trees to produce a resonance effect, which couldseparate the SBT fruit effectively. Figure 4.2 shows a SBT fruit with weight of 0.5 g in average and its binding force to the branches is 0.859 N, where the minimum acceleration needed for separating the fruit is 2*g* (*g* is the acceleration of gravity). According to Newton's second law:

$$F + mg - F_L = ma \tag{1}$$

Where F is the vibration load, m is the mass of fruit, F_L is binding force between the branches and SBT fruits.

A hypothesis that SBT fruit stalk is a cylindrical was made. The fruit stalk diameter is 1.0mm, so the cross-sectional area (A) is 0.785 mm², and the stress (σ) between stalks and branches is 1.101 MPa which is the minimum stress for SBT fruits being separated from branches.

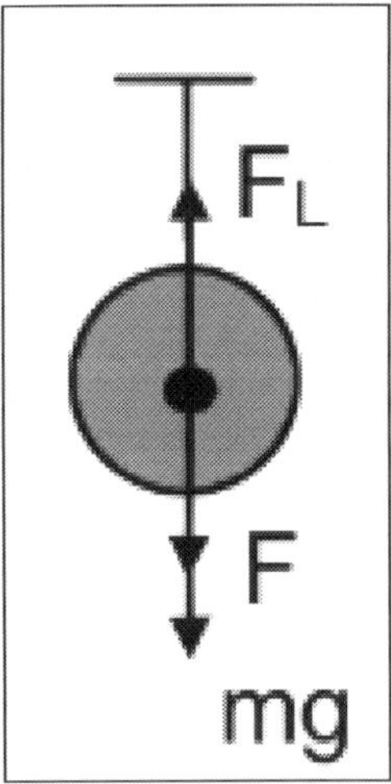

Figure 4.2: Force Model of a SBT Fruit.

RESULTS AND DISCUSSIONS

Modal Analysis of SBT Tree

The FEM of SBT tree includes 107,843 nodes, so it was very difficult to solve all the natural frequencies and vibration modes. In general, the vibration of SBT tree can be expressed as a linear combination of all order natural vibration modes. At the same time, the low order vibration modes have more influence to dynamic response of SBT tree than high order vibration, and vibration modes corresponding with high frequencies will attenuate rapidly because of the structural damping of SBT tree.Therefore, the dynamic characters of SBT tree are mainly decided by the low order vibration modes. The first twenty order natural frequencies of SBT tree were extracted in this study, they are from 8.8 Hz to 31.2 Hz which were increased with increasing of order, as shown in Table 4.2. Among them, eight representative vibration modes were selected and shown in Figure 4.3. The 1storder natural frequency was 8.8 Hz closed with the 2nd order natural frequency of 9.2 Hz, and they mainly reflect vibration of the upper part of left branch b_1, as shown in Figure 4.3(a), which showed a low integrated structure rigidity of SBT tree. The 3rd order 9.7Hz closed to 4th order 10.0 Hz, and they mainly reflect vibration of the upper trunkT_1, as shown in Figure 4.3(b). The 5thorder 10.8Hz neared to 6th order 10.9 Hz, and they mainly describe vibration of branch b_{11}, as shown in Figure 4.3(c). The 7th order 12.6 Hz and 8th order 12.8 Hz mainly reflect vibration of the upper part of right branch b_2, as shown in Figure 4.3(d). The 9th order 15.8 Hz and 10th order 18.1 Hz were mainly described the top part of left branch b_{14} and b_{13}, as shown in Figure 4.3(e). The 11th order 18.8 Hz was almost same to 12th order, and closed to 13th order 19.3 Hz, they mainly reflect vibration of the branch b_{111}, as shown in Figure 4.3(f). The 14th order 19.6 Hz was mainly describe the branch T_{11}, as shown in Figure 4.3(g).The 15thorder 21.2 Hz closed with 16th order 21.7 Hz, and they mainly reflect vibration of the top part of left branch b_{12}, as shown in Figure 4.3(h). The 17th order 25.6 Hz and 18th order 28.1 Hz mainly reflect vibration of the right branches b_{21}, b_{22}, and b_{23}, as shown in Figure 4.3(i).The 19thorder 29.3Hz is closed to 20th order 31.2 Hz and mainly reflects vibration of the top part ofleft branch b_{13}, as shown in Figure 4.3(j). The maximum displacement is occurring in the top branches. When the frequency is 21.2Hz, the left branch b_{12} has the maximum amplitude because of it has the smallest branch of the SBT tree.

Harmonic Response Analysis of SBT Tree

The harmonic response analysis is a technology which can determine the structural response of a structure under known frequency harmonic load. Its purpose is to calculate the response of SBT trees under different frequency and obtain response, and find the frequency where the responses reach peak.Firstly, the primary results of harmonic response analysis were obtained by simulation with a randomforce of 200 N applied on the trunk at position P_2, as shown in Figure 4.4. Secondly, it was the same as the first step, but the location on which force applied on was the side branchesat position P_1 and P_3. Then the harmonic response analysis was conducted to determine the magnitude of force and appropriate vibrating position. Figure 4 showed ten representative points for analyzing and the three force load positions.

Force on Truck

The force of 200N applied on P_2 of trunk, and obtained the Von Mises stress curve of the ten points, as shown in Figure 4.5(a). The frequencies of maximum stress point are 12Hz, 14Hz, 16Hz and 24Hz respectively. The maximum stress of three points (3, 6 and 8) are relatively large (0.137 MPa, 0.250 MPa and 0.126 MPa respectively), while that ofthe remaining seven points are far less than 1.101MPa. It means that the force of 200 N acting on the trunk can't make fruits separated from branches. Modal analysis and harmonic response analysis can be considered as linear approximately. In order to ensure the fruits in the three points (3, 6 and 8)

Table 4.2: First Twenty Order Natural Frequencies

Order	1st	2nd	3rd	4th	5th	6th	7th	8th	9th	10th
Natural frequency (Hz)	8.8	9.2	9.7	10.0	10.8	10.9	12.6	12.8	15.8	18.1
Order	11th	12th	13th	14th	15th	16th	17th	18th	19th	20th
Natural frequency (Hz)	18.8	18.8	19.3	19.6	21.2	21.7	25.6	28.1	29.3	31.2

Figure 4.3 a-j: The Selected Ten Order Natural Frequencies of SBT Tree and their Natural Modes.

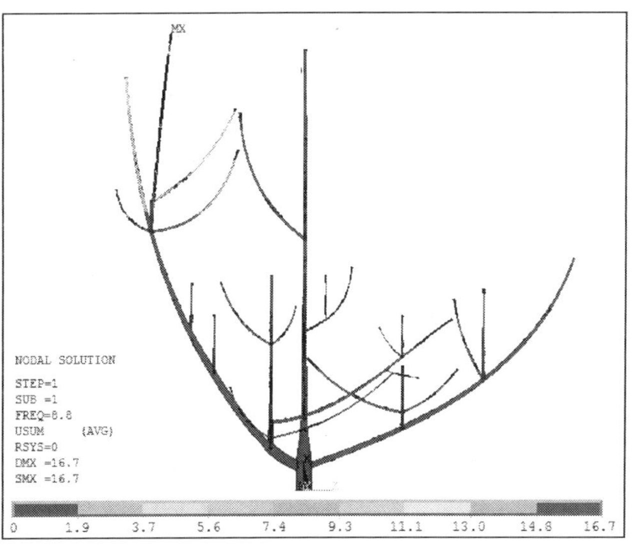

(a) First order (8.8 Hz)

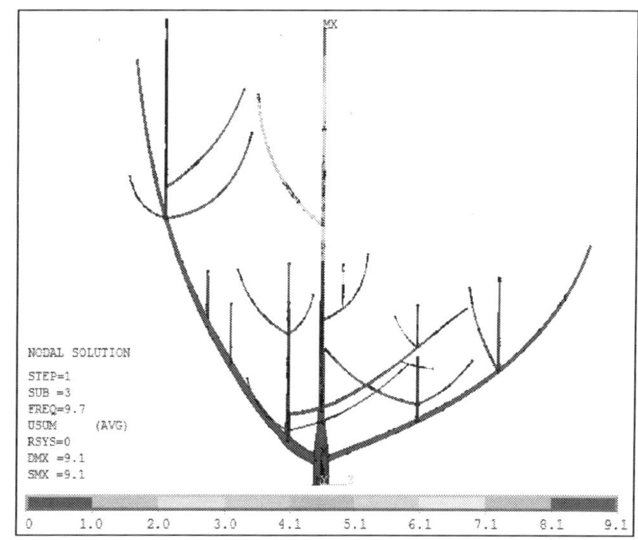

(b) Third order (9.7 Hz)

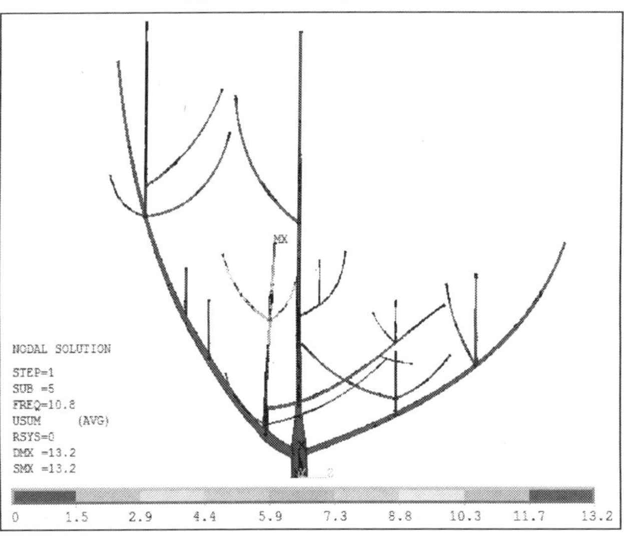

(c) Fifth order (10.8 Hz)

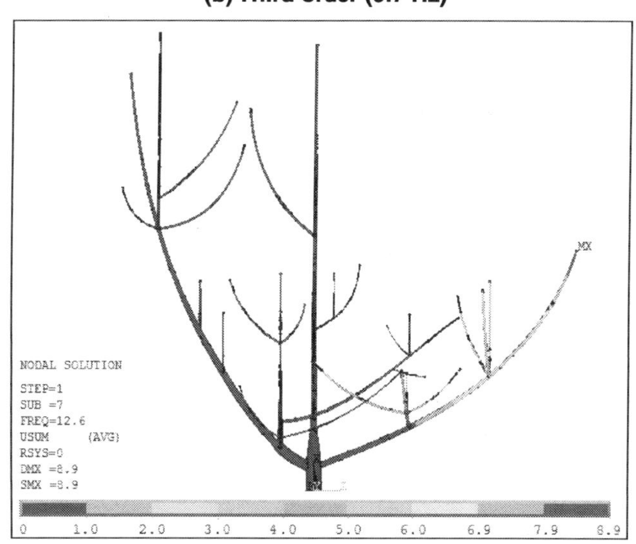

(d) Seventh order (12.6 Hz)

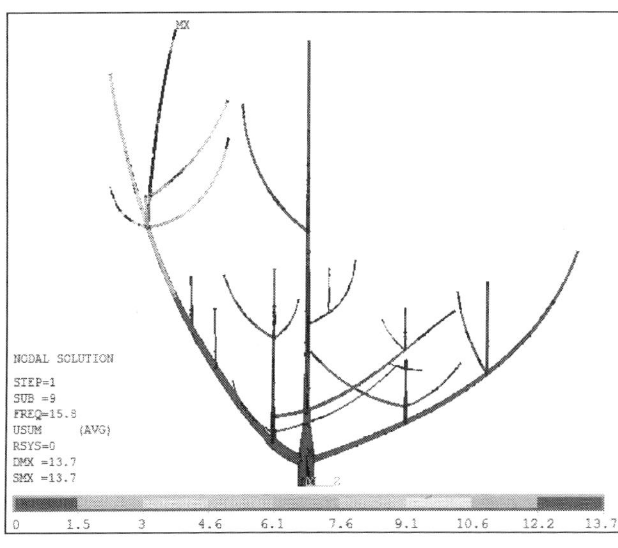

(e) Ninth order (15.8 Hz)

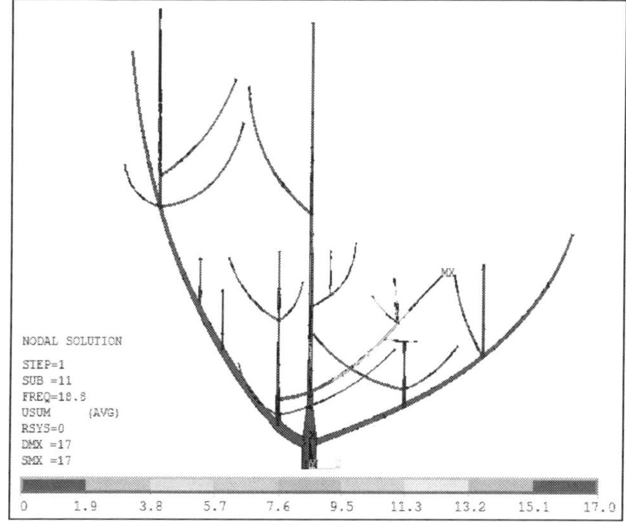

(f) Eleventh order (18.8 Hz)

Contd....

Contd....

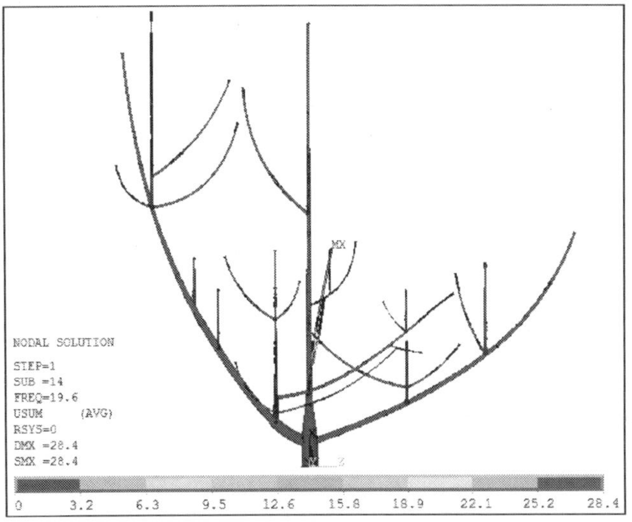

(g) Fourteenth order (19.6 Hz)

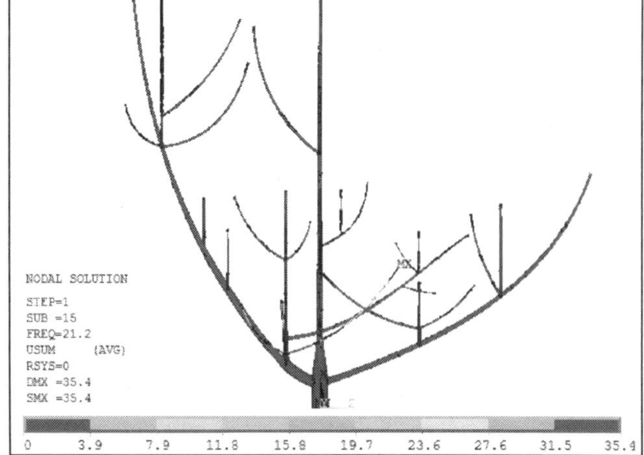

(h) Fifteenth order (21.2 Hz)

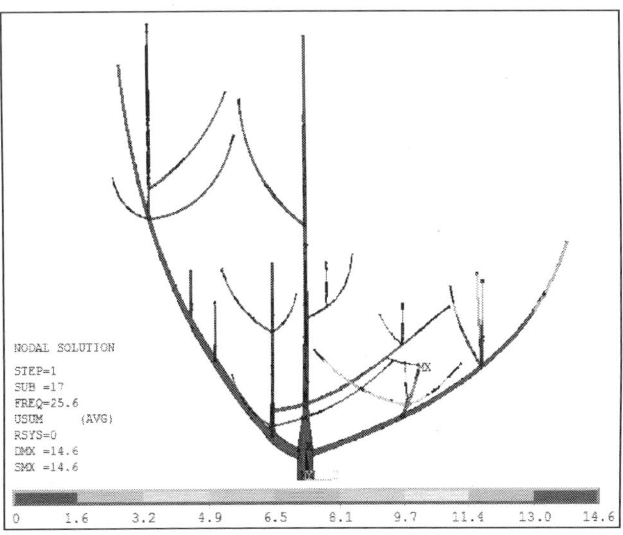

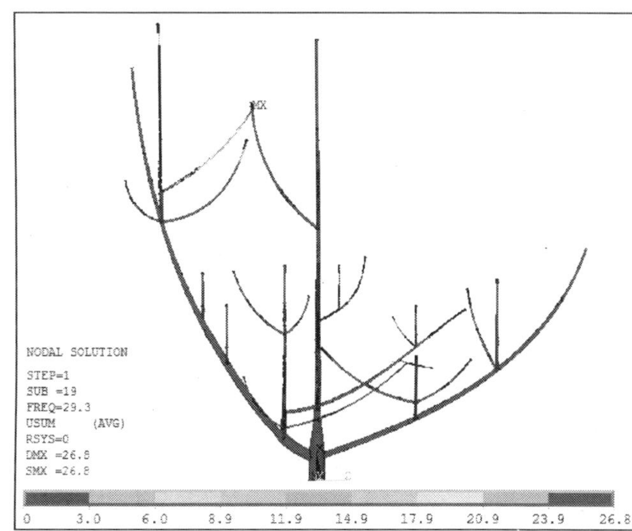

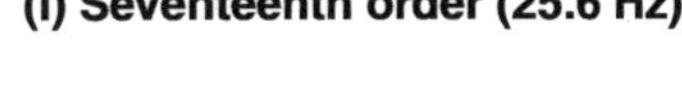

(i) Seventeenth order (25.6 Hz)

(j) Nineteenth order (29.3 Hz)

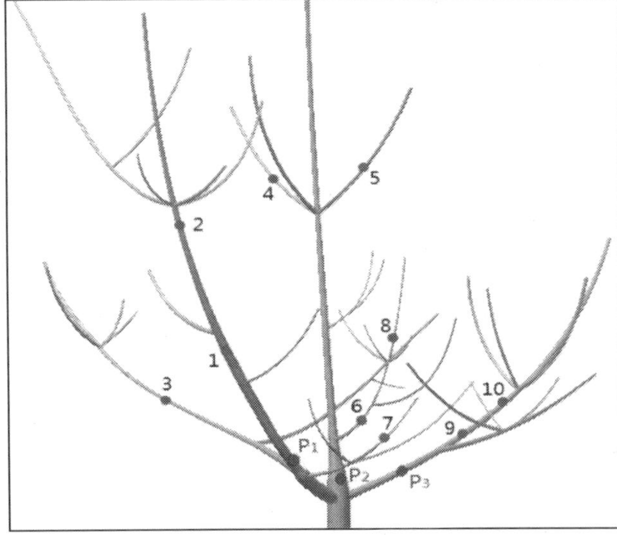

Figure 4.4: Three Force Loading Positions and Ten Representative Points.

being separated from tree, it needs to exert appropriate force.The linear equation can be described as:

$$F / \sigma = F/\sigma_b \quad (2)$$

Where F is the primary vibration force, $F2$ is the appropriate vibration force, σisthemaximum stress of certain point, σ_b is the required stress of fruits detaching from tree.

Therefore, the force needed to make the stress at points 3, 6, and 8 to reach 1.101 MPa is 1748 N. Figure 4.5(b) shows the simulation results when 1748 N applied on SBT tree trunk. By comparing Figures 4.5(a) and 4.5(b), it can be seen the Von Mises stress of each point increase with the increase of applied force, and the variation trend has not changed.The value of stress of point 3, 6, 8 has reached the value of separation stress, but the tree root stress has exceeded the strength limit. It may damage the trunk, so it is not reasonable to apply the force of 1748 N on SBT tree trunk.

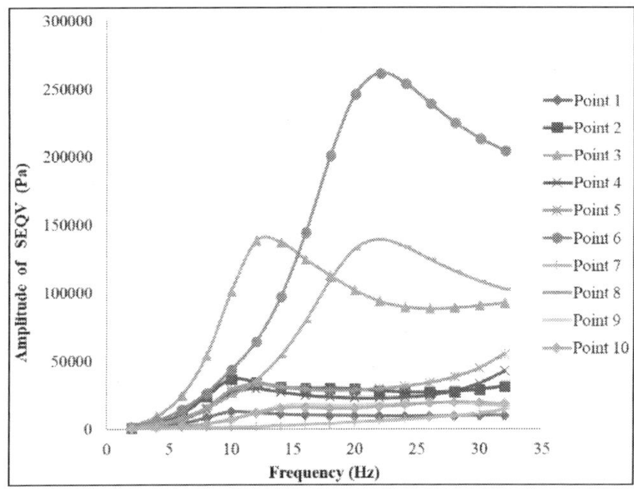

(a) SEQV of the ten representative points when the load force is 200N.

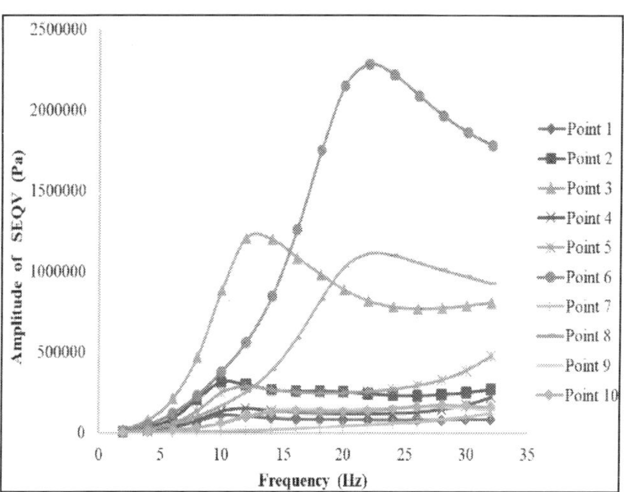

(b) SEQV of the ten representative points when the load force is 1784N.

Figure 4.5 a-b: Harmonic Response Analysis of the Trunk.

Force on Side Branch

When the force was applied on the side branch, only the points on the forced branch can reach a reasonable value for analyzing. Ten points from each side branch was thus selected for a comprehensive analysis, respectively, as shown in Figure 4.6(a).

Firstly, simulation was conducted when the force of 200 N applied on position P_1 of left side branch, as shown in Figure 4.6 (b). It can be seen resonance frequency is about 12 Hz, 14 Hz and 22 Hz. Only the branches exerted on force and the thin branches collateral and close to the acting position had large swing, while the other branches which didn't act on force had very slight swing. Most points achieved the maximum stress value when frequency is 14 Hz; the stress value of points 4, 5, 7, 8 achieved the value of separation stress. At this time, the maximum stress 8.950 MPa is much smaller than the strength limit. In order to make more fruits abscise down and reduce the grasping, so the force can be increased to guarantee most of fruits separated from branches. However, branches where points 1 and 2 locate have very large diameter, and point 10 is far from the force acting point and the force direction is the timber parallel to grain direction, so the stresses of points 1, 2 and 10 are relatively small. According to the growth characteristics of SBT tree, there are no fruits in the branches in large diameter, and the vibration of one position can't make all fruits abscission. So the points 1, 2 and 10 can be ignored, and only need to ensure that the stress of the other seven points are larger than separation stress, and this can make most fruit separated from side branches. In the remaining seven points, the maximum stress of point 9 is smallest, only 0.804 MPa. In order to reach its separation stress, the force can be calculated by the linear approximation equation is 273 N.

Figure 4.6 (c) shows the Von Mises stress curve when the load (273 N) applied on the left branches, and from this figure, the maximum stress of all the points except from points 1, 2 and 10 have achieved minimum separation stress, which is consistent with the expectation. Different branches have different sizes, so the value of vibration force is different. Therefore the diameter is the main influential factor of points stress value.

Therefore, the right branches was selected as the research object, using the same force (273N) to act on the position (P_3) of right branch. In order to eliminate the position of acting point influence the outcome of analysis, P_3 and P_1were selected symmetrical for the trunk. The maximum stresses of points (1, 2 and 10) don't reach the minimumstress, and the stress of other points are larger than 1.101 MPa. Obviously, the force 273 N can make the fruit separation, but the stress of some points is too large, which is not economical and may damage the branches.In the harmonic response analysis, only linear behavior is effective, so it can be assumed that the optimum vibration force is linear relationship with branch's diameter. The average diameter of force point in left branch is 22.9 mm, and that in the right branch is 15.2 mm. The linear equation can be described as:

$$F/D_1 = F2/D_2 \tag{3}$$

Where *F* is the appropriate vibration force applied on left branches, *F2* is the appropriate vibration force applied on right branches; D_1 is the average diameter of force point in left branches, D_2 the average diameter of force point in right branch. By the equation, the vibration load (*F2*) is180 N. Figure 4.6 (d) shows the Von Mises stress curve when 180 N applied on the right branches, only three points 4, 6 and 7 cannot reach the minimum stress for fruit separation. Therefore, the proper force can

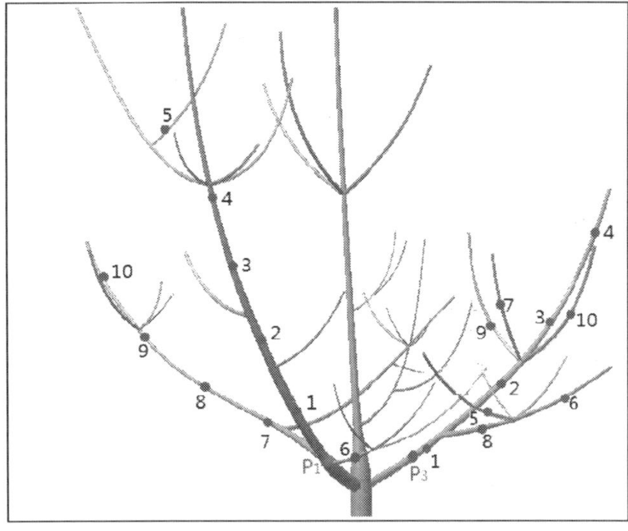

(a) The each ten point selected from left and right branches.

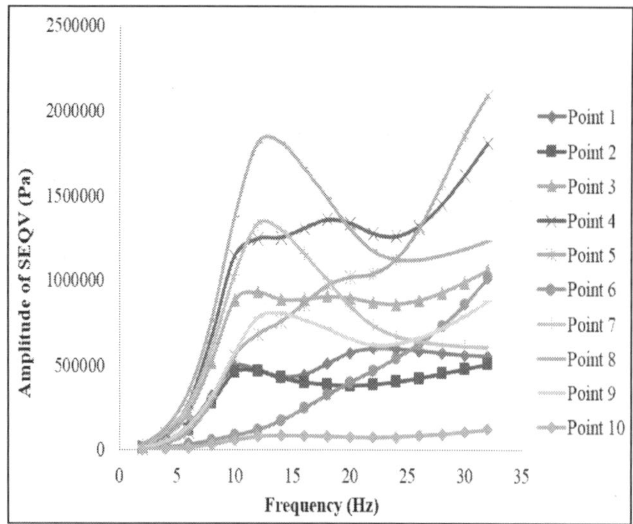

(b) SEQV of the ten points in left branches (200 N).

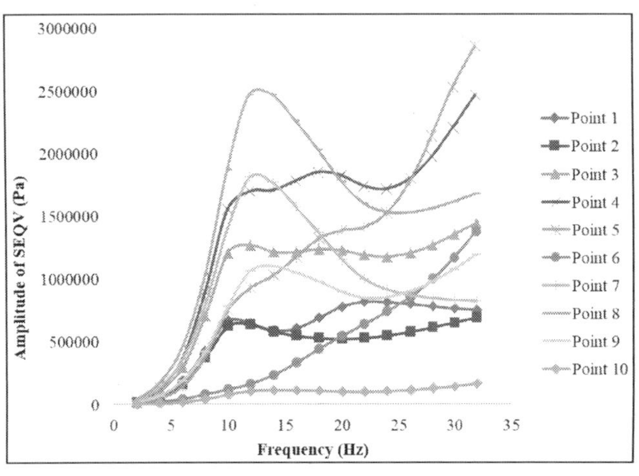

(c) SEQV of the ten points in left branches (273 N).

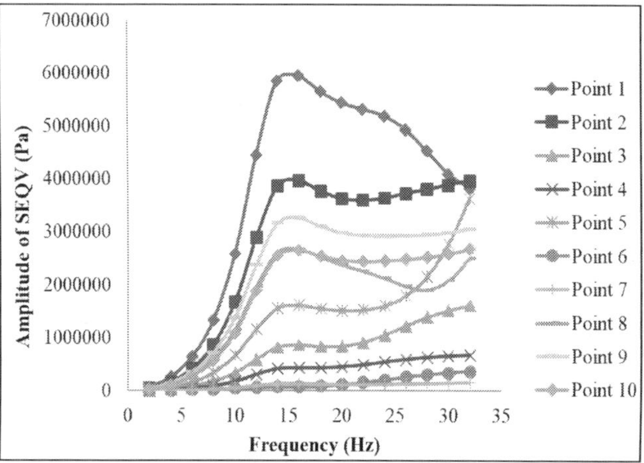

(d) SEQV of the ten points in right branches (180 N).

Figure 4.6: Harmonic Response Analysis of the Branch.

be obtained by the linear relationship, and the vibration force of 180N can make most fruits be separated.

CONCLUSIONS

A vibration harvesting simulation for SBT fruit has been carried out in this paper by combining ANSYS FEM analysis software with Pro/E modeling software. First, a 3D solid model of SBT tree was createdand imported into ANSYS, and then defined finite element type and material type. Second, first twenty order natural frequencies and vibration modes were obtained by modal analysis.A SBT tree has many order natural frequencies in theory, and this is related to the internal degrees of freedom. The number of internal degrees of freedom is equal to the number of natural frequency. Considering the high frequency vibration unsuitable for picking fruits, first twenty order natural frequencies (8.8 Hz to 31.2 Hz) were extracted.Frequency is related to many factors, such as hardness, quality and dimension.The first order modal frequency is 8.8 Hz by harmonic response analysis, which represents the natural frequency of top part in left branch, and each mode represents vibration of different parts.

The harmonic response analyses were carried when the force acted on the trunk and branches. The analysis results show it is inefficient and uneconomical that the vibration force acting on the trunk of SBT tree. It is better that vibration force acting on SBT tree side branches, and its damage is less.From the simulation results, the conclusion that the vibration load ranging from 180 N to 280 N acting on branches can separate most of fruits. The specific value of force acting on branches is related to shape and size of branches. The stress of point in the pedicels and branches connection place is relevant to diameter and distance between the point and point forced.Further experiments, simulation and data analysis are needed to be done for finding relationship between stress and diameter or distance.

ACKNOWLEDGEMENT

The Project was supported by the National Natural Science Foundation of China (Grant No. 31301242), and the Scientific Research Foundation for the Returned Overseas Chinese Scholars, Ministry of Education of China, and the Chinese Universities Scientific Fund, Northwest A and F University (Grant No. QN2013064).

REFERENCES

1. Bal, L. M., Meda, V., Naik, S.N. and Satya, S. 2011. Seabuckthorn berries: A potential source of valuable nutrients for nutraceuticals and cosmoceuticals. *Food Res. Int.* 44: 1718-1727.
2. Bantle, M., Pierre, R. G. S. and Wulfsohn, D. 1996. Mechanical harvesting trials on Western Canadian native fruits. In: *CSAE Paper*, Mansonville, Canada.
3. Demenko, V., Mikityuk, O. and Levinskii, M. 1986. Abscisic acid, ethylene, growth and fruit drop in seabuckthorn. *Fiziologiya Rastenii* 33: 188-194.
4. Fu, L., Su, H., Li, R. and Cui, Y. 2014. Harvesting technologies for seabuckthorn fruit. *Eng. Agric. Environ. Food* 7: 64-69.
5. Gaetke, R. and Triquart, E. 1993. First results with an improved seabuckthorn harvesting technology. In: *Cultivation and Utilization of Wild Fruit Crops*, Braunschweig (Eds.), Bernhard Thalacker Verlag Gmbh and Co., Germany, pp. 37-41.
6. Huang, H. D., Wang, Y. X., Tang, Y. Q., Zhao, F. and Kong, X. F. 2011. Finite element simulation of sugarcane cutting. *Transactions of the CSAE* 27 (2): 161-166.
7. Ishii, G. 2003. Seabuckthorn (*Hippophae rhamnoides* L.) production manual. Miscellaneous publication of the NARO Hokkaido Agricultural Research Center 62: 1-32.
8. Ke, B. and Wang, Y. 1989. A study on the determing methods of bending strength and elasticity. *J. Anhui Agric. Coll.* 1: 1-10.
9. Li, T.S.C. 2002. Product development of seabuckthorn. In: *Trends in New Crops and New Uses*, J. Janick and A. Whipkey (Eds.), ASHS Press, Alexandria, VA, USA, pp. 393-398.
10. Liang, J., Mu, H. and Yang, H. 2008. Harvesting technologies and equipments for seabuckthorn fruit. *Hippophae* 21: 17-20.
11. Mann, D. D., Petkau, D. S., Crowe, T. G. and Schroeder, W. R. 2001. Removal of seabuckthorn (*Hippophaë rhamnoides* L.) berries by shaking. *Canadian Bio-Systems Engineering* 43 (2): 23-28.
12. Mu, H., Yang, H. and Liang, J. 2012. Research and development of dial spring suction type seabuckthorn fruit harvesting machines. *For. Woodwork. Equip.* 40: 30-32.
13. Olander, S. 2012. A review of berry harvest machine development in Sweden. *Acta Horticulturae* 965: 171-177.
14. Olander, S. 1995. Mechanical harvesting of seabuckthorn. In: *International Seabuckthorn Workshop*, Beijing China.
15. Quan, L., *et al.* 2011. Finite element mode analysis and experiment of corn stubble harvester, Trans. Chin. *Soc. Agric. Eng.* 27: 15-20.
16. Savary, S. K. J. U., Ehsani, R., Schueller, J. K. and Rajaraman, B. P. 2010. Simulation study of citrus tree canopy motion. *Transac. ASABE* 53: 1373-1381.
17. Shang, D., *et al.* 2008. Dynamic Performance of Wood before and after CCA Treatment, China. *Wood Ind.* 22: 17-19.
18. Song, Z., Xu, H., Gao, J., Zhang, M., Xiao, R. and Li, W. 2014. Physicochemical properties changes for seabuckthorn cloudy juice during cold crushing, concentrating and storage, Trans. China. *Soc. Agric. Eng.* 30: 264-270.
19. Stan, G., Botez, M. and Stan, S. 1985. Possibilities of vibration to harvest seabuckthorn. Togungsbericht, Akademie der Landwirtschaftswissenschaften der Deutschen Demokranschen Republik 232: 217-223.
20. Tang, X., Ren, J., Liu, C. and Xiao, D. 2011. Simulation of vibration harvesting mechanism for wolfberry. In: *ASABE Annual International Meeting*, v5, Louisville Kentucky USA. pp. 4096-4110.
21. Tang, Y., Wang, Z. and Hu, F. 2006. Analysis of the plated ultrasonic vibrator by finite element method modeling and harmonic response technique. *Piezoelectrics and Acoustooptics* 28: 486-488, 491.
22. Wu, Q. and Zhao, H. 2000. Soil and water conservation functions of seabuckthorn and its role in controlling and exploiting Loess Plateau. *For. Stud. Chin.* 2: 50-56.
23. Yang, B. and Kallio, H. 2002. Composition and physiological effects of seabuckthorn (*Hippophaë*) lipids. *Trends Food Sci.Tech.* 13: 160-167.
24. Zhu, C. 1991. A Study on relations between Abscisic Acid and absciss of fruit of *Hippophae rhamnoides*. *For. Res.* 3: 333-336.

Part II

GENETIC DIVERSITY AND IMPROVEMENT

Chapter 5

Selection of High Yielding Land Races of Seabuckthorn from Wild Seedling Population of Lahaul and Spiti District of Himachal Pradesh, India

R.K. Rana[1]*, Ashok Singh[2], Y.S. Dhaliwal[3] and Virendra Singh[4]

[1]Krishi Vigyan Kendra, Himachal Pradesh Agricultural University, Bajaura, Kullu, HP, India
[2]IHBT Research Station, Ribling, Lahaul HP, India
[3]College of Home Science,
[4]Department of Biology and Environmental Sciences,
CSK Himachal Pradesh Agricultural University, Palampur – 176 062, HP, India
(*E-mail: drrameshrana70@gmail.com)

SUMMARY

A field study was undertaken during 2008-2011 to select high yielding land races of seabuckthorn from existing wild population of seedling origin in Lahaul and Spiti district of Himachal Pradesh. To select the superior land races intensive survey were carried in the district. Initially the selection of promising genotypes was made on the basis of some novel morphological and biochemical characters. The morphological characters like large fruit size, maximum fruit weight, yield per plant, less number of thorns per 10 cm shoot length, higher biomass yield and biochemical characters like higher Vitamin-C (mg/100 g), Oil content (per cent) and β–carotene mg/100 g were taken in consideration. The variation in fruit weight ranged from 10.23 g -36.56 g/100 berries and yield from 0.20 kg–5.5 kg/plant. Among the different species found growing in the area *H. salicifolia* had maximum yield potential followed by *H. rhamnoides* and *H. tibetana*. In the beginning ten land races were selected on the basis of some superior morphological and chemical characteristics out of which three land races are finally identified and selected on the basis of correlation between morphological and biochemical characteristics for specific purpose. Out of these three land races one is *H. salicifolia* selection from Lahaul is a valuable specie which is high yielding (5.5 kg berries/plant), large fruit size (36.56 g/100 berries) and richest in Vitamin C content and having less number of thorns. Another two land Races belongs to *H. rhamnoides* (one from Lahaul and one from Spiti valley) are rich in oil content which may be used for pharmacutical purposes.

***Keywords**: Seabuckthorn, Variation, Land races, Yield, Fruit size and Correlation.*

INTRODUCTION

Seabuckthorn is one of the valuable plant resources with high value of ecology and medicinal properties. Naturally, seabuckthorn is found in Siberia, Central Asia, Caucus, China, Mangolia, Germany, India, Finland and Denmark (Penteleeva and Zubareb, 2001). In India there is no cultivated variety and systematic plantation of seabuckthorn and mostly plants of seedling origin are found growing in Himachal Pradesh, Jammu and Kashmir, Arunachal Pradesh and Uttarakhand. As seabuckthorn is a polymorphic in nature, so in the natural habitat one can find a great diversity of plants which differs in morphological and biochemical features which in turns offers a great scope for selection of promising types.

MATERIALS AND METHODS

The present study was carried out in the Lahaul and Spiti district of Himachal Pradesh in India during the year 2008-2011. The Lahaul and Spiti district lies between

Seabuckthorn (Hippophae L.): A Multipurpose Wonder Plant, Vol. 5: Advanced Technologies for Cultivation, Processing, Health Protection and Environmental Conservation (Virendra Singh, *Editor-in-Chief*, 2018), p. 37–40, Astral International Pvt. Ltd., New Delhi, India.

31° 44 57 to 32° 59 57 N latitudes and 76° 46 29 to 78° 41 34 E longitudes and covers approximately 13,835 km^2 area with an altitudinal range of 2400 to >6600 m above mean sea level. The climate of the area varies from dry temperate, semi-arid and alpine type. The area remains snow covered almost for six months and receives up to 120 to 750 cm average snowfall $year^{-1}$. Rainfall varies from 10- 300 mm $year^{-1}$ and temperature -19° to +32° C. To select the superior land races, intensive survey of the natural seedling population (67 locations) were carried in district during 2008-2012. Initially, the selection of promising genotypes was made on the basis of some novel morphological characters like large fruit size, maximum fruit weight, yield per plant, less number of thorns per 10 cm shoot length and higher yield. These physical parameters were recorded at site. The samples were sent for chemical analysis to the Department of food and Nutrition, College of Home Science CSKHPKV Palampur. Biochemical characteristics like higher Vitamin-C (mg/100 g), oil content (per cent), anthocyanin content (mg/100 g), TSS (°B), acidity (per cent) and β–carotene mg/100 g were analysed. The final selection of the superior genotypes were made on the basis of correlations between morphological and biochemical characteristics.

RESULTS AND DISCUSSION

During survey, three species of seabuckthorn (*H. rhamnoides* ssp. *turkestanica*, *H. salicifolia* and *H. tibetana)* were found in different parts of the district. Distribution pattern showed that in Lahaul and Spiti, these species are found at different locations in isolation as well as in combinations. Among all the species, *H. rhamnoides* ssp. *turkestanica* showed wider range of distribution followed by *H. salicifolia* and *H. tibetana*. In general variation of fruit weight of *H. salicifolia* varied from 13.75 - 36.56 g/100 berries; *H. tibetana* (24.58–29.23 g/100 berries) and *H. rhamnoides* ssp. *turkestanica* (11.57–25.41 g/100 berries) as shown in Figure 5.1. Among all the species, *H. salicifolia* showed maximum fruit weight 36.56 g/100 berries selected from Tinu village followed by *H. tibetana* 29.23 g/100 berries from Lossar and *H. rhamnoides* ssp. *turkestanica* 25.44 g/100 berries from Jispa. The fruit colour and shape of *H. rhamnoides* sub sp. *turkestanica* varied from red, orange and yellow with ovoid, oblong, depressed shape; for *H. salicifolia* yellowish with rounded shape; however for *H. tibetana* orange with rounded shape. *H. rhamnoides* sub sp. *turkestanica* showed highest number of thorns per 10 cm shoot length followed by *H. tibetana* and *H. salicifolia*. Fruit size of

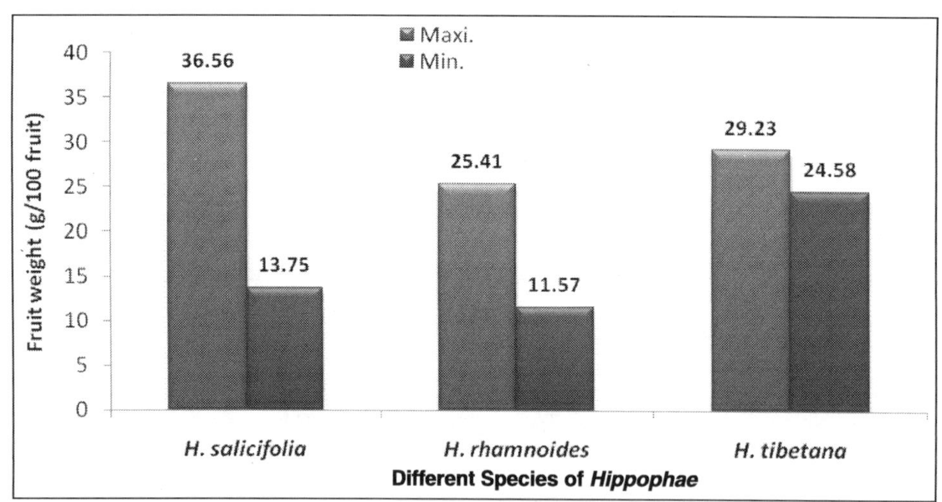

Figure 5.1: Variation in the Fruit Weight for different Species of *Hippophae*.

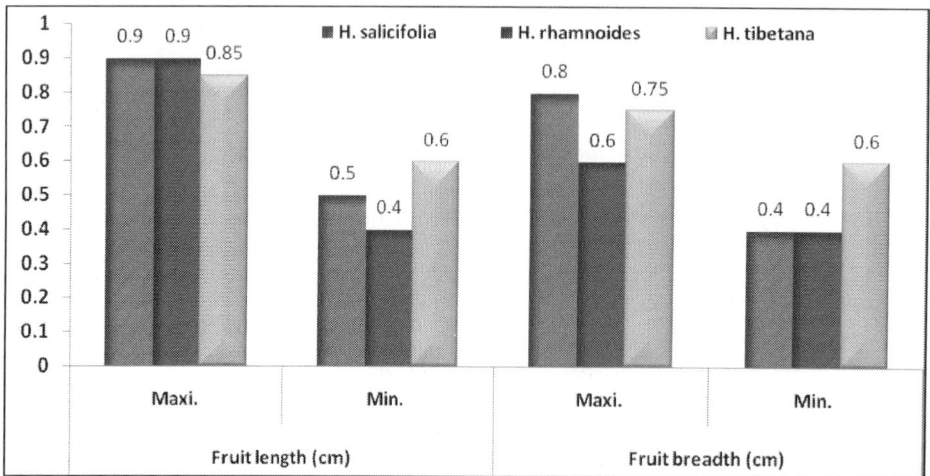

Figure 5.2: Variation in the Fruit Size for different Species of *Hippophae*.

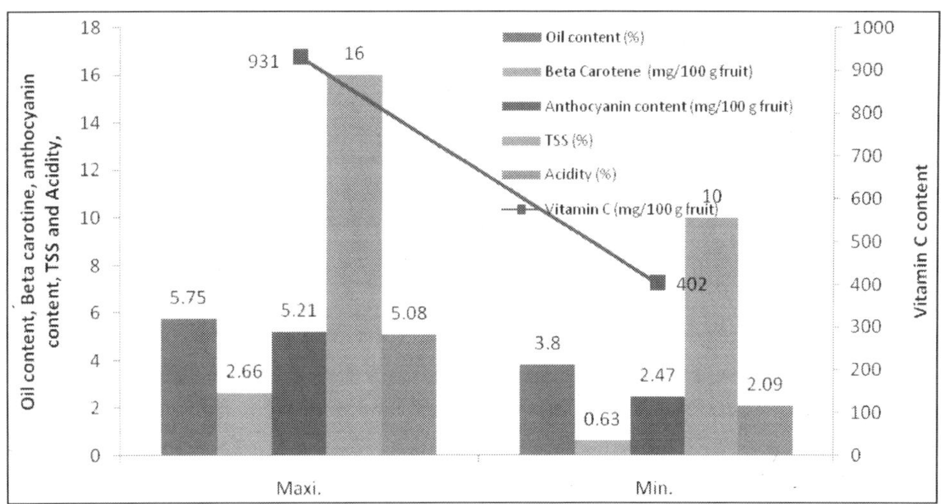

Figure 5.3: Variation in Chemical Characteristics of Seedling Selections of Seabuckthorn from Lahaul and Spiti District of Himachal Pradesh.

H. rhamnoides subsp. *turkestanica*, varied from 0.5 x 0.4 - 0.9 x 0.6 cm; *H. salicifolia* 0.4 x 0.4 - 0.9 x 0.8 cm and for *H. tibetana* from 0.6 x 0.6 - 0.85 x 0.75 cm (Figure 5.2).

The data in Figure 5.3 shows variation among the bio-chemicals characteristics irrespective of the species. The vitamin-C, β–carotene and anthocyanin content varied from 402-930.14, 0.63-2.66 and 2.47-5.21 mg/100 g fruit, respectively. The pulp oil content varied from 3.8 per cent to 5.75 per cent, TSS (10-16 °B) and acidity (2.09-5.08 per cent). It was observed that vitamin-C content was higher in *H. salicifolia* as compared to *H. rhamnoides*. However, percentage of oil content was more in *H. rhamnoides* and compared to selections from *H. salicifolia*. The findings are in line with the findings of Sing and Singh (2001).

Initially five land races from Lahaul valley and 5 from Spiti valley were selected on the basis of superior morphological characteristics like large fruit size, higher yield (> 4 kg/tree) and less number of thorns. Out of these 10 land races, three land races (*H. salicifolia* from

Table 5.1: Correlation between Physical and Biochemical Parameters of *Hippophae*

	Yield (kg/plant)	*Fruit Weight g/100 fruits)*	*No. of Thorns/ 10 cm Shoot*	*Altitude (m)*	*Average Fruit Size LxB (cm²)*	*TSS (°B)*	*Oil Content (Per cent)*	*Vitamin-C (mg/ 100 g)*	*Antho-cyanin (mg/100 g)*	*Flavo-noids (per cent)*	*Juice Yield (per cent)*
Yield (Kg/ plant)	1.000	+0.520	-0.067	-0.083	0.131	-0.312	+0.093	+0.044	-0.037	-0.487	+0.676
Fruit weight g/100 fruit)		1.000	-0.458	-0.731	+0.592	-0.568	+0.662	+0.836	+0.314	-0.365	+0.289
No. of thorns/ 10 cm shoot			1.000	+0.639	-0.377	+0.214	-0.338	-0.613	-0.233	+0.384	+0.218
Altitude (m)				1.000	-0.518	+0.524	-0.609	-0.968	-0.410	+0.484	-0.432
Maximum fruit size Lxb (cm²)					1.000	-0.570	+0.683	+0.618	-0.250	-0.643	-0.204
TSS (°B)						1.000	-0.173	-0.140	+0.257	+0.905	+0.346
Oil content (per cent)							1.000	+0.688	-0.133	-0.846	-0.232
Vitamin-C (mg/100 g)								1.000	+0.455	-0.572	-0.177
Anthocyanin (mg/100 g)									1.000	+0.791	-0.671
Flavonoides (per cent)										1.000	Not cal
Juice yield (per cent)											1.000

Tinu, *H. rhamnoides* from Jispa and *H. rhamnoides* from Spiti) are finally identified and selected on the basis of correlation between morphological and biochemical characteristics.

Correlations between various characteristics of seabuckthorn were worked out (Table 5.1). It was observed that fruit yield is positively related with fruit weight, size, juice yield, oil and vitamin C content of fruit. Further fruit weight has positive correlation with fruit size, oil content, vitamin-C, juice and anthocyanin content but negatively correlated with flavonoids contents. Skundon (2008) has compiled the work of various authors regarding correlations between morphological and biochemical characteristics of seabuckthorn. The various research workers have observed positive correlation between fruit weight and vitamin C. Altitude has also shown positive relation with TSS and flavonoid contents. Further it was observed that oil content in fruit is positively correlated with vitamin-C; vitamin-C with anthocyanin content and anthocyanin content with flavonoids. Present finding are in line with Fefelov (1988) who also observed positive relation of vitamin-C with fruit oil content. This suggests that, the effective selection of high yielding seabuckthorn from seedling population could be achieved by selecting large fruit size. On the basis of these, correlation studies finally three land races were selected. The details characteristics of three best forms for practical use are given below:

Selection-1 (LHST)

The selection belongs to *Hippophae salicifolia*. The plant is of medium size (3.5 m) having less number of thorn (4.2 thorn/10 cm shoot), large leaf size (length 9.3 x width 0.9 cm) and high yielding (5.5 kg/tree). Fruits are round (length 0.9 x width 0.8 cm), yellow with average weight 36.56 g/100 fruits. Chemical composition of fruit: vitamin-C content (930.14 mg/100 g fruit wt.); oil content (4.83 per cent); β-carotene (1.13mg/100g fruit wt.) anthocyanin content (4.75 mg/100 g fruit wt.) and TSS (11.0°B). Fruit matures in last week of October. Suggested use of the selection is for horticulture purpose especially for juice extraction and for breeding for higher yield and large fruit size.

Selection-2 (LHRJ)

The selection belongs to *Hippophae rhamnoides*. The plant is of medium size (2.9 m) having 7 thorn/10 cm shoot, medium leaf size (length 5.8 x width 0.8 cm) and high yielding (5.0 kg/tree). Fruits are small (length 0.8 x width 0.6 cm), red with average weight 25.44 g/100 fruits. Chemical composition of fruit: vitamin-C content (464.80 mg/100g fruit wt.); oil content (5.42 per cent); β- carotene (2.66 mg/100g fruit wt.) anthocyanin content (5.15 mg/100g fruit wt) and TSS (10°B).

Selection-3 (SHRS-II)

The selection belongs to *Hippophae rhamnoides*. The plant is of medium size (2.8 m) having 7 thorn/10 cm shoot, medium leaf size (length 5.1 x width 0.7 cm) and high yielding (5.2 kg/tree). Fruits are (length 0.6 x width 0.5 cm), red with average weight 17.28 g/100 fruits. Chemical composition of fruit: vitamin-C content (475.26 mg/100g fruit wt.); Oil content (5.74 per cent); β- carotene (2.64 mg/100g fruit wt.) anthocyanin content (5.25 mg/100g fruit wt) and TSS (13°B).

CONCLUSION

Selection of *H. salicifolia* is a valuable specie which is high yielding (5.5 kg berries/plant), large fruit size (36.56 g/100 berries) and richest in Vitamin C content and having less number of thorns. Therefore, it is highly suitable for cultivation in Lahaul district and suitable for juice production. Another land races belongs to *H. rhamnoides* are rich in oil content which may be used for pharmacutical purposes. Among the different land-races, *H. salicifolia* selection from Tinu village of Lahaul is the most valuable resources for breeding and cultivation purposes.

ACKNOWLEDGEMENT

The Authors are greatly indebted to ICAR New Delhi for financial support through NAIP project for this research.

REFERENCES

1. Fefelov, V.A. 1998. Interrelation of morphological and biochemical characters of seabuckthorn. *Biologiya, Selektsiya I Agrotekhnika Oblepikhi*, pp. 83-94, Gorky. (In Russian)
2. Penteleeva, E.I. and Zubareb, I.A. 2001. Breeding of high yielding cultivars in Russia. In: *Proceedings of International Workshop on Seabuckthorn*, pp. 83-88, CSK Himachal Pradesh Agricultural University, Palampur, 274p.
3. Singh, V. and Singh R. K. 2001. Evaluation studies on the exotic forms and local genotypes of seabuckthorn (*Hippophae rhamnoides* L). In: *Proceedings of International Workshop on Seabuckthorn*, pp. 107-111, CSK Himachal Pradesh Agricultural University, Palampur, 274p.
4. Skundon, G.M. 2008. Statistical correlation of characteristics in seabuckthorn: Genetic and ecological aspects. In: *Seabuckthorn (Hippophae L.)- A Multipurpose Wonder Plant*. Vol. 3 (V. Singh Edt.), Daya Publishing House, New Delhi, India, pp. 78-99.

Chapter 6

Development of Microsatellite Markers in Seabuckthorn

Ankit Jain, Rajesh Ghangal, Saurabh Chaudhary, Ram Singh Purty and Prakash Chand Sharma*

University School of Biotechnology, Guru Gobind Singh Indraprastha University, Dwarka Sector-16C, Delhi – 110 075
(*E-mail: prof.pcsharma@gmail.com)

SUMMARY

Seabuckthorn is a valued source of many bioactive compounds, having medicinal as well as nutritional significance. Moreover, considering its sustenance in extreme environment, the seabuckthorn genome may provide an important source of elements managing abiotic stress. Considering an increased interest in improvement of seabuckthorn as a potential crop, efforts are being undertaken for genome characterization and assessment of genetic diversity. Despite the advances in genomics, as observed for several crops, the availability of molecular tools such as microsatellite markers is very limited in seabuckthorn. We have developed genic as well as genomic microsatellite (SSR) markers in seabuckthorn and are currently assessing these markers for assessing genetic diversity in seabuckthorn populations. We have identified 56 probable microsatellite marker sites in EST database developed in our laboratory earlier, out of which nine microsatellite markers were found polymorphic in *Hippophae rhamnoides* accessions and also showed cross-amplification in *H. salicifolia* and *H. tibetana*. In an effort to develop genomic microsatellite markers random and microsatellite enriched genomic libraries have been constructed in *H. rhamnoides*. Markers developed by us are being used to assess genetic diversity in geographically diverse seabuckthorn populations collected from Jammu and Kashmir, Himachal Pradesh and Uttarakhand to find association with some morphometric traits. As a byproduct of our current transcriptome study we have identified another more than 6000 unigenes (excluding mononucleotide repeats) carrying microsatellite repeats on screening about 90,000 putative unigenes using MISA script. These new microsatellites sites present in the transcribed regions of seabuckthorn genome will prove to be useful in revealing functional diversity of microsatellite positive unigenes. The identification and further validation of new microsatellite markers will not only help in accessing intra- and inter- specific genetic diversity but also to be enriching the limited markers resources in seabuckthorn.

Keywords: *Seabuckthorn, Microsatellites, Genomic markers, Expressed sequence tags and Cross amplification.*

INTRODUCTION

Seabuckthorn (*Hippophae* sp.) is a hardy, deciduous shrub with orange or yellow berries. All the species of *Hippophae* are diploid with $2n = 24$. It is an important medicinal plant according to traditional Chinese and Tibetan medicine, mainly because of its berries that are rich in vitamins, fatty acids, free amino acids, flavanoids, tocopherol and other secondary metabolites (Tiitinen *et al.*, 2005; Dhyani *et al.*, 2007). Apart from medicinal importance, it is widely used for its ornamental and environmental importance, due to its silvery green foliage and attractive orange/yellow colored berries the shrub has become as a popular garden ornamental plant. Windbreaks made up of Seabuckthorn shrubs are effective in preventing soil erosion in open and flood prone areas due to its extensive root system (Trajkovski and Jeppsson, 1999). Its symbiotic association with actinomycete *Frankia* improves the fertility of the soil due to its nitrogen-fixing capabilities Moreover seabuckthorn has also been used as first inhabitant for the barren lands, which then makes the soil fertile and later on helps other plants and microbial flora to proliferate

Seabuckthorn (Hippophae L.): A Multipurpose Wonder Plant, Vol. 5: Advanced Technologies for Cultivation, Processing, Health Protection and Environmental Conservation (Virendra Singh, *Editor-in-Chief*, 2018), p. 41–45, Astral International Pvt. Ltd., New Delhi, India.

in that area. Hence, seabuckthorn is also known as an ecosystem restorer.

Indian Himalayas are the second richest source of naturally growing seabuckthorn (30,000-40,000 hectares) in the world. In India seabuckthorn mainly grows in high altitude, cold arid conditions of ladakh (Leh and Kargil), Lahaul–Spiti, parts of Chamba and upper Kinnaur district of Himachal Pradesh (Singh *et al.*, 1995), Badrinath and Chamoli in Uttrakhand (Dwivedi *et al.*, 2006). Seabuckthorn has also been reported from Sikkim (Basistha *et al.*, 2001) and Dibang valley in Arunachal Pradesh (Tiwari and Singh, 2001).

The unique features of seabuckthorn, especially its combination of environmental and rural development characteristics (Yang *et al.*, 2005) with nutritional and medicinal application has received a substantial interest and recognition internationally from researchers and development practitioners. Considering its immense importance, improvement of seabuckthorn as a crop plant to harvest the goodness of its nutrients and soil binding abilities, a high-throughput characterization of its genomic features is required. Characterization of agronomically important genes from seabuckthorn as well as development of various marker systems can facilitate effective marker assisted breeding for useful traits. Moreover, it also helps in characterization of total genetic diversity available in germplasm of a species.

With the advent of high through-put sequencing techniques, development of molecular markers has become a thrust area. They have wide range of applications from the perspectives of population biology, behavioral ecology, species evolution, and phylogeny. Molecular markers like RAPD (Williams *et al.*, 1990) and ISSR (Meyer *et al.*, 1993; Gupta *et al.*, 1994) are markers of choice when one is dealing with a totally unexplored experimental system while major disadvantage associated with it is their dominant mode of inheritance as well as poor reproducibility. Other marker system like RFLP (Wyman and White, 1980; Botstein *et al.*, 1980), AFLP (Vos *et al.*, 1995) are expensive, time consuming as well as they generally require radioactive labeling. Among all the molecular marker systems, microsatellites (Litt and Luty, 1989; Tautz, 1989) are the markers of choice which are characterized as robust, reproducible, highly polymorphic and codominant markers.

Three most common approaches used to identify microsatellite loci include 1) Screening of partial genomic libraries (Tautz, 1989; Rassmann *et al.*, 1991) by colony hybridization with repeat containing probes, 2) Enrichment approach based on hybridization capture procedure (Kandpal *et al.*, 1994; Kijas *et al.*, 1994), and 3) *in-silico* mining of microsatellite from the public database. In our laboratory, we have exploited all the three approaches to isolate microsatellites in seabuckthorn.

MATERIALS AND METHODS

Collection of Germplasm

Seabuckthorn samples were collected from the diverse ecological area differing in altitude, soil texture, water availability, sunlight exposure. Various parts of Leh, Lahaul and Chamoili, were the collection sites. Moreover, various phenotypic characters of these individual plants were also recorded. Leaf tissue samples were collected and snap freezed in liquid nitrogen, and transported to laboratory and stored at -80°C.

Genomic DNA Isolation

Genomic DNA was extracted from preserved seabuckthorn tissue using modified CTAB based protocol (Doyle and Doyle, 1990). Addition of Polyvinylpyrrolidone (PVPP) and enhanced concentration of *Beta-mercaptoethanol* (ß-ME) in CTAB–DNA extraction buffer helped in the isolation of high quality DNA from leaf tissues.

Construction of Genomic Library

Genomic DNA (1-2 µg) isolated from the leaves of one plant accession belonging to *H. rhamnoides* was fragmented either by using restriction enzymes or sonication. Size selected DNA was cloned in linearized dephosphorylated pBluescript vector (Stratagene, USA), and further transformed in electro-competent *E. coli* DH10B cells (Invitrogen, USA). A fraction of transformation mixture was used to estimate overall transformation efficiency and remaining transformation mix was stored at -80°C as glycerol stocks until further use.

Construction of Microsatellite Enriched Genomic Library

Simultaneous digestion-linker ligation reaction with seabuckthorn genomic DNA was performed. Product of above reaction was hybridized with biotinylated microsatellite probe and these molecules in turn were captured with streptavidine magnetic beads (Invitrogen). Captured microsatellite positive fragments were amplified using PCR and product of desired size was selected and cloned in TA cloning vector. Ligation mix thus obtained was used to transform Top 10 heat shock competent *E. coli* cells. Plasmids from the recombinant colonies were isolated and sequenced using ABI sequencer. Primer3 software (Rozen and Skaletsky, 2000) was used to design primers from the microsatellite flanking region. Validation of the primers was performed on 14 accessions (5 accessions each of *H. rhamnoides* and *H. salicifolia* and 4 accessions of *H. tibetana*) collected from the premises of Defense Institute of High altitude research (DIHAR), Leh

Development of Genic Markers

A collection of 1584 clustered ESTs of seabuckthorn (*Hippophae rhamnoides*) developed in our laboratory were used to identify EST-SSRs. EST's were screened for microsatellite using MISA (Thiel *et al.*, 2003) at default parameters (mononucleotide repeats were excluded from this search).

As a byproduct of another transcriptome study exploiting next generation sequencing platform we obtained over 6000 microsatellite repeats in around 90,000 putative unigenes by using MISA script using above stated parameters. All microsatellite positive sequences were functionally annotated with Blast2GO (Conesa *et al.*, 2005) software.

RESULTS AND DISCUSSION

Availability of high quality DNA is a prerequisite for all kinds of library construction experiments. Conventional methods generally results in poor quality as well as low yield of genomic DNA in case of seabuckthorn due to presence of high concentration of polyphenols and secondary metabolites. PVPP forms complex with polyphenolics through hydrogen bonding and helps in the removal of polyphenolics and DNA thus obtained (Figure 6.1) with protocol is of high quality and high yield.

While constructing the genomic library overall 70 per cent transformation efficiency was obtained. Transformation efficiency was measured in terms of total number of white colonies, out of total *E. coli* colonies. To confirm the presence of insert in the recombinant clones, colony-PCR was performed taking 5 recombinant and 1 non recombinant clones. Amplicons ranging from 250-850 bp in size were found in the colony PCR (Figure 6.2). We modified the protocols of Nunome *et al.* (2006) and Hamilton *et al.* (1999), for developing enriched genomic library and obtained higher percentage of microsatellite positive clones in comparison to other protocols used earlier in medicinal plants (Tani *et al.*, 2004; Edwards *et al.*, 2007).

By screening of our previous EST database (Ghangal *et al.*, 2011), 56 probable microsatellite loci were identified from 1584 unigenes. 30 primer pairs could be designed for flanking region of these probable microsatellite sites. Primer pairs for remaining 23 loci could not be designed due to presence of microsatellite either too close to 52 or 32 end of the sequence or the sequence was highly AT-rich. Twenty four primers pairs successfully amplified

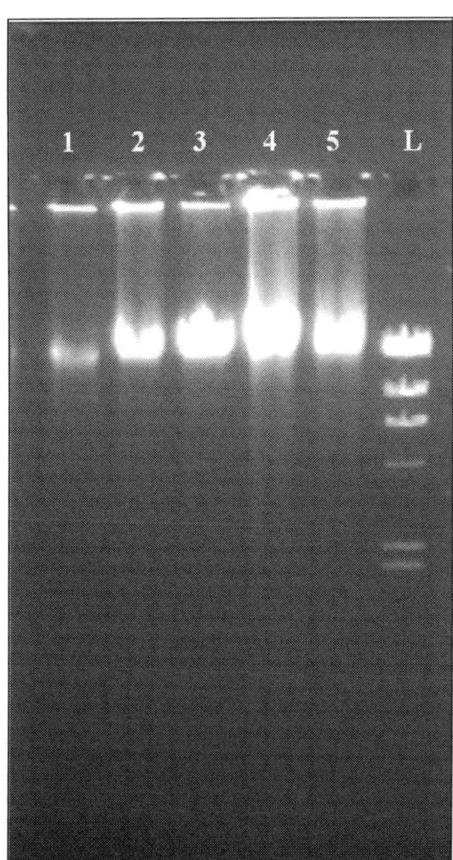

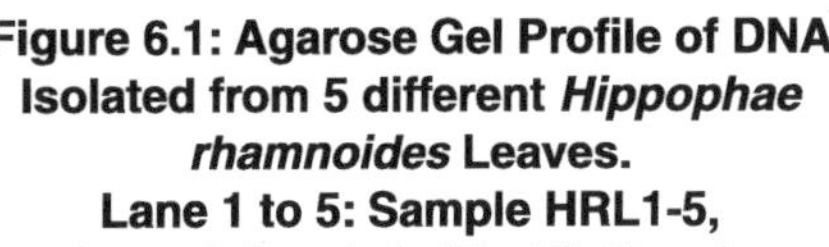
Figure 6.1: Agarose Gel Profile of DNA Isolated from 5 different *Hippophae rhamnoides* Leaves.
Lane 1 to 5: Sample HRL1-5,
Lane 6: Lambda Hin*d*III digest.

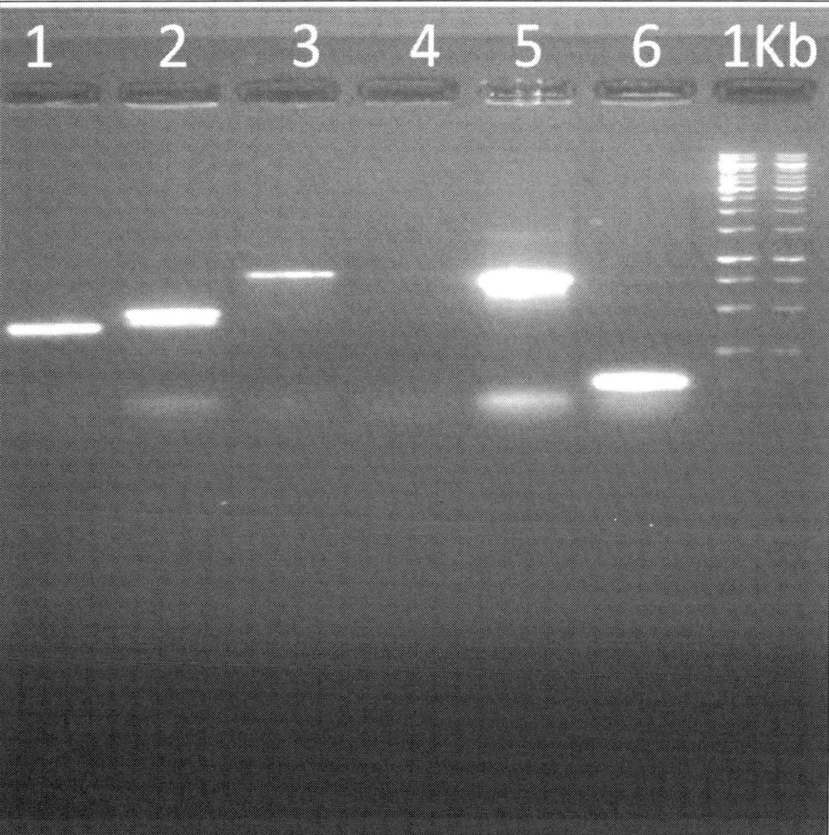

Figure 6.2: Colony PCR with M13 Universal Primer.
Lane 1-5: Random Colonies, Lane 6: Control (Blue Colony),
Lane 7: 1Kb ladder.

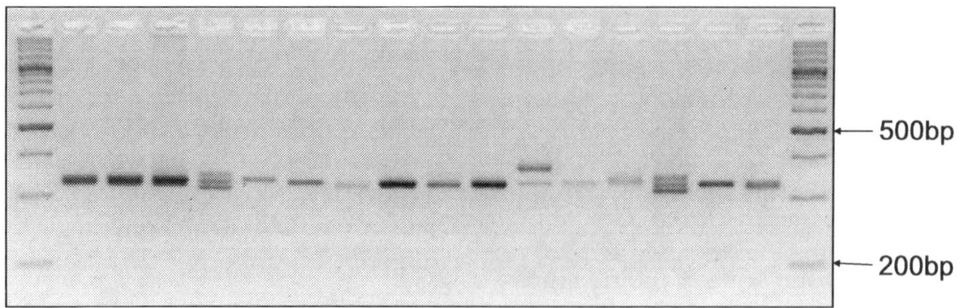

Figure 6.3: PCR Amplicons Obtained Using EST- SSR (HrMS025) Marker in 16 Samples of *H. rhamnoides*.

in *H. rhamnoides*. However, only nine primer pairs were found to be polymorphic in *H. rhamnoides* accessions, and two additional primer pairs were found polymorphic on cross-amplification in *H. salicifolia* and *H. tibetana*. 2–4 alleles were observed within *rhamnoides* genotypes, and 2–5 in all 14 accessions belonging to three *Hippophae* species (Jain *et al.*, 2010). All the loci were found to be polymorphic in *H. salicifolia* but monomorphic in case of *H. tibetana*. We obtained 2-8 alleles with same marker sets when used for large population collected (Figure 6.3) from diverse geographical areas.

By screening the large number of putative unigenes obtained from NGS platform, 6.89 per cent were found to harbor 7270 microsatllites. Out of these, 15.20 per cent unigenes contained more than one microsatellite. It was observed that dinucleotide repeats were most abundant followed by trinucleotide repeats. Detailed functional annotation of microsallite positive unigenes iis under process.

CONCLUSIONS

Modification made in enrichment protocol resulted enhanced recovery of microsatellite positive sequences, thus same protocol can be used in other taxa/species for development of genomic microsatellite markers. Probable microsatellite identification in unigene data has provided insight about abundance and distribution of microsatellite in seabuckthorn transcriptome, it also provide an opportunity to study impact of polymorphism on function of the gene *i.e.* functional diversity. Genic and genomic markers developed by our efforts will be enriching the limited marker resources in seabuckthorn.

ACKNOWLEDGMENT

We are thankful to DIHAR, Leh, India for providing seabuckthorn samples. This work was supported by grants from DRDO (Defence Research and Development Organization), India and DBT (Department of Biotechnology), India.

REFERENCES

1. Basistha, B.C., Adhikari, I.M., Deokata, K.P. and Thapa, K.K. 2001. A case study of *Hippophae* L. with special reference to agro-technique in Sikkim Himalayas. In: *Souvenir of International Workshop on Seabuckthorn*, Feb. 18-21, 2001, New Delhi, India.
2. Botstein, D., White, R.L., Skolnick, M. and Davis, R.W. 1980. Construction of a genetic linkage map in man using restriction fragment length polymorphisms. *American Journal of Human Genetics* 32: 314-331.
3. Conesa, A., Gotz, S., Garcia-Gomez, J.M., Terol, J., Talon, M. and Robles, M. 2005. Blast2GO: A universal tool for annotation, visualization and analysis in functional genomics research. *Bioinformatics* 21: 3674-3676.
4. Dhyani, D., Maikhuri, R.K., Rao, K.S., Kumar, K., Purohit, V.K., Sundriyal, M. and Saxena, K.G. 2007. Basic nutritional attributes of *Hippophae rhamnoides* (seabuckthorn) populations from Central Himalaya, India. *Current Science* 92: 1148-1152.
5. Doyle, J.J. and Doyle, J.L. 1990. Isolation of plant DNA from fresh tissue. *Focus* 12: 13-15.
6. Dwivedi, S.K., Singh, R. and Ahmed, Z. 2006. *The Seabuckthorn*. Field Research Laboratory, DRDO, Leh, Ladakh (J&K), India.
7. Edwards, C. E., Arakaki, M., Quintana-Ascencio, P., Soltis, D. E. and Soltis, P. S. 2007. Isolation and characterization of microsatellite loci from the endangered highlands scrub hypericum (*Hypericum cumulicola*). *Molecular Ecology Notes* 7: 1135-1137.
8. Ghangal, R., Raghuvanshi, S. and Sharma, P.C. 2011. Expressed sequence tag based identification and expression analysis of some cold inducible elements in seabuckthorn (*Hippophae rhamnoides* L.). *Plant Physiology and Biochemistry*, doi: 10.1016/j.plaphy.2011.10.005.
9. Gupta, M., Chyi, Y.S., Romero-Severson, J. and Owen, J.L. 1994. Amplification of DNA markers from evolutionarily diverse genomes using single primers of simple-sequence repeats. *Theoretical Applied Genetics* 89: 998-1006.
10. Hamilton, M.B., Pincus, E.L., Di Fiore, A. and Fleischer, R.C. 1999. Universal linker and ligation procedures for construction of genomic DNA libraries enriched for microsatellites. *Biotechniques* 27: 500-2, 504-7.

11. Jain, A., Ghangal, R., Grover, A., Raghuvanshi, S., Sharma, P.C. 2010. Development of new EST based SSR markers in seabuckthorn. *Physiology and Molecular Biology of Plants* 16: 375-378.

12. Kandpal, R. P., Kandpal, G. and Weissman, S. M. 1994. Construction of libraries enriched for sequence repeats and jumping clones, and hybridization selection for region-specific markers. In: *Proceedings of National Academy of Sciences*, USA, 91: 88-92.

13. Kijas, J. M. H., Fowler, J.C.S., Garbett, C.A. and Thomas, M.R. 1994. Enrichment of microsatellites from the citrus genome using biotinylated oligonucleotide sequences bound to streptavidin-coated magnetic particles. *Bio Techniques* 16: 656-662.

14. Litt, M. and Luty, J.A. 1989. A hypervariable microsatellite revealed by *in vitro* amplification of a dinucleotide repeat within the cardiac muscle actin gene. *American Journal of Human Genetics* 44: 397-401.

15. Meyer, W., Mitchell, T.G., Freedman, E.Z. and Vilgays, R. 1993. Hybridization probes for conventional DNA fingerprinting used as single primers in the polymerase chain reaction to distinguish strains of *Cryptococcus neoformans*. *Journal of Clinical Microbiology* 31: 2274-2280.

16. Nunome, T., Negoro, S., Miyatake, K., Yamaguchi, H. and Fukoka, H. 2006. A protocol for the construction of microsatellite enriched library. *Plant Molecular Biology Reporter* 24: 305-312.

17. Rassmann, K., Schlötterer, C. and Tautz, D. 1991. Isolation of simple sequence loci for use in polymerase chain reaction-based DNA fingerprinting. *Electrophoresis* 12: 113-118.

18. Rozen, S. and Skaletsky, H. 2000. Primer3 on the WWW for general users and for biologist programmers. In: *Bioinformatics Methods and Protocols: Methods in molecular biology*, Krawetz, S., Misener, S. (eds). *Humana*, Totowa, pp. 365-386.

19. Singh, V., Singh, B. and Awasthi, C.P. 1995. Distribution, taxonomy and nutritional values of seabuckthorn growing in dry temperate Himalayas. In: *Proceeding of International Workshop on Seabuckthorn*, pp. 52-59, Dec.12-17, 1995, Beijing, China, 206p.

20. Tani, N., Takahashi, N., Ujino-Ihara, T., Iwata, H., Yoshimura, K. and Tsumura, Y. 2004. Development and characteristics of microsatellite markers for sugi (*Cryptomeria japonica* D. Don) derived from microsatellite-enriched libraries. *Annals of Forest Science 61:* 569-575.

21. Tautz, D. 1989. Hyper variability of simple sequences as a general source for polymorphic DNA markers. *Nucleic Acids Research* 17: 6463-6471.

22. Thiel, T., Michalek, W., Varshney, R.K. and Graner, A. 2003. Exploiting EST databases for the development of cDNA derived microsatellite markers in barley (*Hordeum vulgare* L.). *Theoretical and Applied Genetics* 106: 411-422.

23. Tiitinen, K.M., Hakala, M.A. and Kallio, H.P. 2005. Quality components of seabuckthorn (*Hippophae rhamnoides*) varieties. *Journal of Agriculture and Food Chemistry* 53: 1692-1699.

24. Tiwari, S.C. and Singh, S.S. 2001. A report on seabuckthorn from Arunachal Pradesh, North Eastern India. In: *Seabuckthorn- A Resource of Health and Environment in Twenty First Century, Proceedings of International Workshop on Seabuckthorn*, Feb. 18-21, 2001, New Delhi, India.

25. Trajkovski, V. and Jeppsson, N. 1999. Domestication of seabuckthorn. *Botanica Lithuanica Supplement* 2: 37-46.

26. Vos, P., Hogers, R., Bleeker, M., Reijans, M., Lee, T.V.D., Hornes, M., Frijters, A., Pot, J., Peleman, J., Kuiper, M. and Zabeau, M. 1995. AFLP: A new technique for DNA fingerprinting. *Nucleic Acids Research* 23: 4407-4414.

27. Williams, J.G.K., Kubelik, A.R., Livak, K.J., Rafalski, J.A., and Tingey, S.V. 1990. DNA polymorphisms amplified by arbitrary primers are useful as genetic markers. *Nucleic Acids Research* 18: 6531-6535.

28. Wyman, A.R. and White, R.L. 1980. A Highly polymorphic locus in Human DNA. In: *Proceedings of National Academy of Sciences*, USA, 77: 6754-6758.

29. Yang, Y., Yao, V., Xu, G. and Li, C. 2005. Growth and physiological responses to drought and elevated UV-B in two contrasting populations of *Hippophae rhamnoides*. *Acta Physiologica Plantarum* 124: 431-44.

Chapter 7

Intrataxon Variation in the Internal Transcribed Spacer Sequences of Nuclear Ribosomal DNA in the *Hippophae* L.

Sapna Jain[1], Apekshita Singh[2], Manju Chaudhary[2], Virendra Singh[3] and S.N. Raina[2*]

[1]Laboratory of Cellular and Molecular Cytogenetics, Department of Botany, University of Delhi, Delhi – 110 007,
[2]Amity Institute of Biotechnology, Amity University' Sec 125, Noida – 201 303, U.P.
[3]Department of Biology and Environmental Sciences, CSK Himachal Pradesh Agricultural University Palampur – 176 062, H.P.
(*E-mail: soomr@yahoo.com)

SUMMARY

The genus *Hippophae* L. (seabuckthorn) is a dioecious shrub which is widely distributed throughout the temperate zones of Asia and Europe. It is a storehouse of phytonutrients and bio-active substances. ITS1-5.8S-ITS2 region was amplified in the 16 genotypes of *H. rhamnoides* ssp. *turkestanica, H. salicifolia, H. tibetana, H. rhamnoides* ssp. *mongolica* and *H. rhamnoides* ssp. *carpatica.* The amplification of the ITS 1-5.8 S-ITS 2 region of the 14 *Hippophae* genotypes resulted in a single product of ~750 bp. In the two genotypes of *H. rhamnoides* ssp. *mongolica* two distinct bands of sizes~750 bp and ~550 bp named ITS L and ITS S, respectively were amplified. The length of ITS 1 ranged from 267-290 bp. The length of ITS 2 ranged from 218-220 bp. The ITS 1 and ITS 2 exhibited sequence divergence within the taxon. For example, eleven substitution and two indels were found in ITS sequences of *H. rhamnoides* ssp. *turkestanica* from Uttaranchal. The sequence divergence within *H. rhamnoides* ssp. *turkestanica* in ITS regions (5.2 per cent) was found to be high in comparison with intraspecific sequence divergence in *H. salicifolia* (1.8 per cent) and *H. tibetana* (3.2 per cent).

Keywords*: Seabuckthorn, Intrataxon variation, Amplification and DNA.*

INTRODUCTION

The genus *Hippophae* L. (Elaeagnaceae) includes about seven species (Bartish *et al.*, 2002; Sun *et al.*, 2002) distributed widely from Northern Europe through Central Europe and Central Asia to China, ranging from latitude 27°-69°N and longitude 7°-122°E, with a probable centre of origin in Central Asia (Rousi, 1971; Lu, 1997). It grows in wide range of habitats from seashores to high (~ 5,000 m in altitude) mountains. All the species are diploid (2n=2x = 24), dioecious and wind pollinated, with good seed set (Rousi, 1971; Bartish *et al.*, 2000; Sun *et al.*, 2006).

In India, three species, *H. rhamnoides* ssp. *turkestanica* Rousi (seabuckthorn), *H. salicifolia* D. Don (willow leaved seabuckthorn) and *H. tibetana* Schlecht. (Tibetan seabuckthorn), grow well at higher (~3000-5000 m) altitudes of Indian Himalayas (Singh *et al.*, 1995; Singh, 2003). The taxonomy of *H. rhamnoides* L. is inconsistent. The species has been considered to consist of one to nine subspecies, *H. rhamnoides* ssp. *turkestanica* Rousi, *H. rhamnoides* ssp. *fluviatilis* Rousi, *H. rhamnoides* ssp. *turkestanica* Rousi, *H. rhamnoides* ssp. *sinensis* Rousi, *H. rhamnoides* ssp. *carpatica* Rousi, *H. rhamnoides* ssp. *mongolica* Rousi, *H. rhamnoides* ssp. *caucasica* Rousi,

Seabuckthorn (Hippophae L.): A Multipurpose Wonder Plant, Vol. 5: Advanced Technologies for Cultivation, Processing, Health Protection and Environmental Conservation (Virendra Singh, *Editor-in-Chief,* 2018), p. 47–52, Astral International Pvt. Ltd., New Delhi, India.

H. rhamnoides ssp. *gyantsensis* Rousi and *H. rhamnoides* ssp. *yunnanensis* Rousi, with wide geographical distribution throughout the temperate zones of Asia and Europe, and the subtropical zones of Asia (Paerson and Rogers, 1967; Rousi, 1971; Lian and Chen, 2000; Bartish *et al.*, 2000, 2002). Several authors have designated *H. rhamnoides* in Indian Himalayas as *H. rhamnoides* ssp. *turkestanica* (Singh *et al.*, 1995; Singh, 2003).

The multipurpose seabuckthorn is a winter hardy, deciduous shrub with orange, red and yellow berries. The leaves and berries contain more than 100 different kinds of phytonutrients and bioactive substances including vitamins, fatty acids, free amino acids and minerals (Zeb, 2004). The vitamin C content is 5-100 times higher than any other fruit or vegetable known (Tang, 2002; Ruan *et al.*, 2004). A range of health products are being manufactured from seabuckthorn which finds application in the prevention of cardiovascular diseases and cancer, treatment of skin problems, burns, digestive track disorders, and for anti-inflammation and antiradiation, and for improvement in the immune system (Bernath and Foldesi, 1992; Lian and Chen, 2000; Ruan and Li, 2000, 2005). The plant has a strong ability to fix atmospheric nitrogen.

Studies on the nucleotide sequences of the internal transcribed spacer (ITS-1, ITS-2) regions of 18S-26S nuclear ribosomal DNA (nr DNA) have proven to be extremely useful in taxonomic studies (Baldwin *et al.*, 1995; Ainouche *et al.*, 1997; Lashermes *et al.*, 1997; Kollipara *et al.*, 1997). The ITS regions are flanked by conserved sequences; thus, they can be amplified by universal primers and sequenced. The ITS region have a definite role to play in the processing of nuclear ribosomal RNAs (nrRNAs). Deletion in certain regions of ITS-1 (between the 18S and the 5.8S subunits) has been shown to inhibit maturation of small and large subunit rRNAs (Musters *et al.*, 1990), whereas certain deletions or point mutations in ITS-2 (between the 5.8S and the 26S subunits) prevent processing of large subunit rRNA (Sande *et al.*, 1992). This selection pressure imparts to the ITS regions a characteristic of being relatively conserved compared to the intergenic spacer and external transcribed spacer regions of rDNA. Their sequences, nevertheless, diverge more than nrDNA subunits (26S, 18S and 5.8S) and are sufficiently variable to allow phylogenetic reconstruction (Baldwin *et al.*, 1995). The ITS sequences are in most cases species specific. In other words, the ITS 1 and ITS 2 sequences are similar within a taxon, no matter how many accessions one investigates.

In the present study, we report, first time, explicit intrataxon variation in the internal transcribed spacer (ITS) sequences of nuclear ribosomal DNA in the *Hippophae* taxa.

MATERIALS AND METHODS

Plant Material

The 16 genotypes representing *Hippophae rhamnoides* ssp. *turkestanica* from Uttaranchal (2), *H. rhamnoides* ssp. *turkestanica* from Ladakh (1), *H. rhamnoides* ssp. *turkestanica* from Lahaul (1), *H. salicifolia* from Spiti (4), *H. tibetana* from Spiti (5), *H. rhamnoides* ssp. *mongolica* (2) and *H. rhamnoides* ssp. *carpatica* (1) were investigated. We have maintained the nomenclature informed to us by various taxonomists. The young emerging leaves harvested from the plants in situ were wrapped in aluminum foil before they were immediately immersed in liquid nitrogen. They were stored at -80°C.

DNA Isolation

Total genomic DNA of the 16 genotypes was separately extracted as described by Murray and Thompson (1980).

Amplification of ITS1-5.8S-ITS2 Region

Amplification of ITS1-5.8S-ITS2 region was carried out by using universal primers (White *et al.*, 1990). The primers ITS1 (5′ TCCGTAGGTGAACCTGCGG 3′) and ITS 4 (5′ TCCTCCGCTTATTGATATGC 3′) were used to amplify the entire ITS1-5.8S-ITS2 region. The amplification was carried out following the method used by Goel *et al.* (2002). The 100 µl reaction mixture contained 10µl of 10 X reaction buffer (Biotools), 160 ng template DNA, 10 µl of 2 mM dNTPs (Amersham Pharmacia Biotech, USA), 4µl of 50 mM $MgCl_2$ (Biotools, Spain), 10µl of 10pmols/µl of each primer (Banglore Genei, India), 0.5 µl of Taq DNA Polymerase (5 U/µl) (Biotools). DNA amplification was performed in a My Cycler™ thermal cycler (BIORAD) programmed to 40 cycles each of 1 min at 94°C for template denaturation, 1 min at annealing temperature 50°C, and 2 min at 72°C for primer extension, followed by a final extension cycle of 7 min at 72°C. Following amplification, the products in each case were size separated by gel electrophoresis on 0.85 per cent agarose gels in 0.5 X TBE and stained with ethidium bromide. Lambda DNA digested with HindIII (Bangalore Genei) served as the size marker.

Cloning and Sequencing of Amplified Regions of Nuclear Ribosomal DNA

Amplified ITS fragments were size separated on 1.5 per cent agarose gel containing 0.5 µg/ml of ethidium bromide in 1X TAE buffer. The eluted fragment was purified using MEGA- spin™ Agarose gel extraction Kit (iNTRON BIOTECHNOLOGY INC.) according to the manufacturer's instructions. After this, A-tailing was carried out of the purified product with 6µl of purified DNA, 1µl of 10X PCR buffer, 1 µl of 2 mM dATP, 1µl of 25 mM $MgCl_2$ and 1µl of Taq DNA polymerase

(3U/µl) at 70° C for 30 min in My Cycler™ thermal cycler (BIO RAD). The ligation and cloning was carried out according to the procedure used by Goel *et al.* (2002). The sequencing of the positive clones was done by the Sequencing Centre, University of Delhi, South Campus, New Delhi.

Data Analysis

Sequence data of ITS1-5.8S-ITS2 was aligned with CLUSTAL X programme (Thompson *et al.*, 1997) by use of default setting with a fixed gap penalty of 6.66, and DNA transition weight of 0.5 in the multiple alignment parameter option. Neighbor Joining and maximum parsimony methods were used to analyze the aligned sequence data matrix by PAUP*4.0 b8. Gaps were treated as missing data. Given the large size of the data sets, heuristic searches used the Tree Bisection Reconstruction (TBR) option with MULPARS and ACCTRAN optimization. The amount of support for the branches was assessed using 100 bootstrap replicates with ten random additions per replicate using TBR and MULPARS in PAUP*4.0 b8.

RESULTS AND DISCUSSION

Barring two genotypes of *H. rhamnoides* ssp. *mongolica*, amplification of the ITS1-5.8S-ITS2 region in the remaining 14 genotypes of *Hippophae* taxa resulted in a single product of ~750 bp (Figure 7.1). In *H. rhamnoides* ssp. *mongolica*, two distinct bands of sizes ~750bp and ~550 bp, named ITS L and ITS S respectively, were amplified (Figure 7.1). The presence of two types of ITS sequences, ITS L and ITS S, in *H. rhamnoides* ssp. *mongolica* is an unusual feature. Previous ITS studies on *Hippophae* (Sun *et al.*, 2002) have not reported the presence of more than one ITS type in *H. rhamnoides* ssp. *mongolica*. The average sequence divergence between the two sequences was 39.4 per cent (Figure 7.2). Multiple ITS types is a characteristic feature of polyploid genotypes and (or) hybrid sequences. The species, however, is diploid (2n=2x=24) in constitution (Rousi and Arohonka, 1980; Cao and Lu, 1989).

The boundaries of ITS 1 and ITS 2 sequences for all the *Hippophae* taxa were determined by comparison with published sequences of Asteraceae (Baldwin, 1992, 1993). The ITS sequences of ITS 1 ranged from 267 bp in *H. rhamnoides* ssp. *turkestanica* genotypes from Uttarakhand to 290 bp in *H. tibetana*. The ITS 2 length ranged from 218 bp to 220 bp (Figure 7.2).

The alignment of ITS 1 and ITS 2 showed 49 and 17, and 18 and three substitutions and indels among the taxa, respectively. Interestingly, there was intrataxon sequence divergence in *Hippophae*. For example, *H. tibetana* exhibited 3.2 per cent sequence divergence between five genotypes. Twenty nine substitutions [25 transitions (C/T and A/G) and four transversions (C/G and A/C)] and 14 indels of 1-3 bp were specific to *H. tibetana*. In *H. rhamnoides* ssp. *turkestanica* from Uttaranchal, the genotypes exhibited 5.2 per cent sequence divergence. Eight transitions (A/G and C/T), 3 transversions (C/A) and two indels of one bp were specific to *H. rhamnoides* ssp. *turkestanica* genotypes from Uttaranchal. In the five genotypes of *H. salicifolia*, the sequence divergence was to the extent of 1.8 per cent. In *H. salicifolia* no species specific substitutions were found. The length of ITS L and ITS S was 699 bp and 500 bp in *H. rhamnoides* ssp. *mongolica*. Two hundred and seventy seven (39.4 per cent) substitutions and 12 indels of 1-11 bp indels were detected between ITS L and ITS S. Part of the strict consensus tree is given in Figure 7.3 to highlight the divergence in ITS between *H. tibetana* genotypes.

The noteworthy feature of the present study was that the sequence divergence in ITS regions (5.2 per cent) within *H. rhamnoides* was high when compared with intraspecific sequence divergence in the other two species (1.8 per cent and 3.2 per cent in *H. salicifolia* and the *H. tibetana*, respectively) of the genus and also with intraspecific variation reported in various other plant species (Soltis and Kuzoff, 1995; Lashermes *et al.*, 1997; Baldwin, 1993), and (or) ITS sequence divergence reported at subspecific level in various other genotypes (Sehgal *et al.*, 2009; Goel *et al.*, 2002; Buckler and Holtsford, 1996). Moreover, not even a single synapomorphy was exclusive to *H. rhamnoides*. All the 21 synapomorphies (17 substitutions and 4 indels) of *H. rhamnoides* were also shared with *H. salicifolia*.

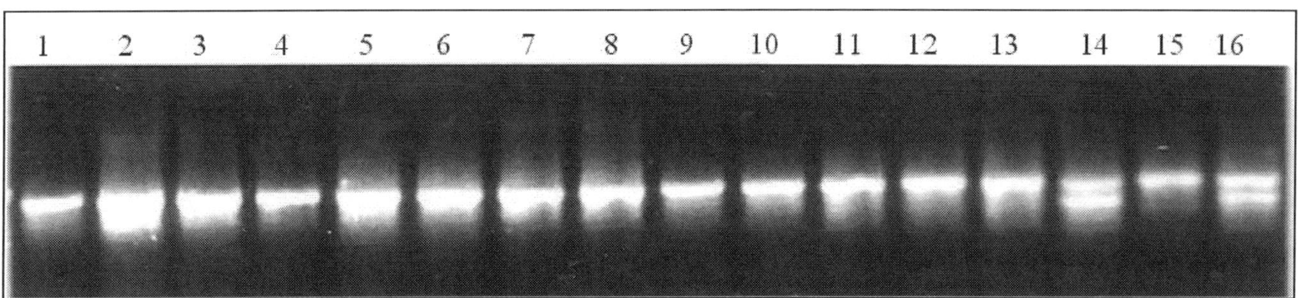

Figure 7.1: Agarose Gel Electrophoresis of the Amplified Nuclear Ribosomal ITS 1 - 5.8 S-ITS2 in *Hippophae* taxa. *H. rhamnoides* ssp. *turkestanica* (Uttaranchal) (lanes 1, 2), *H. rhamnoides* ssp. *turkestanica* Ladakh) (lane 3), *H. rhamnoides* ssp. *turkestanica* (Lahaul) (lane 4), *H. salicifolia* (Spiti) (lanes 5-8), *H. tibetana* (Spiti) (9-13), *H. rhamnoides* ssp. *mongolica* (lanes 14, 16) and *H. rhamnoides* ssp. *carpatica* (lane 15).

```
H.rhamnoides ssp. turkestanica(Uttaranchal)    GACAACCACCATTGTCGCAGCACAGGTTGCCAATGACTAAAATTTTGGGCCAACCACAAG
H.rhamnoides ssp. turkestanica(Uttaranchal)    GACAACCACCATTGTCGCAGCACAGGTTGCCAATGACTAAAATTTTGGGCCAACCACAAG
H.salicifolia (Spiti)                          AACTACCACCATTGTCGCGGCACAGGTCGCCAATGACTAAAAATTTGGGCCAACCGCAAG
H.rhamnoides ssp. mongolic                     AACTACCACCATTGTCGCGGCACAGGTCGCCAATGACTAAAAATTTGGGCCAACCGCAAG
H.rhamnoides ssp. turkestanica (Ladakh)        AACTACCACCATTGTCGCGGCACAGGTCGCCAATGACTAAAAATTTGGGCCAACCGCAAG
H.rhamnoides ssp. carpatica                    AACTACCACCTTTGTCGCGGCACTGGTCGCCAATGACTAAAAATTTGGGCCAACCACAAG
H.salicifolia (Spiti)                          GACAACCACCATTGTCGCAGCACAGGTTGCCAATGACTAAAATTTTGGGCCAACCGCAAG
H.tibetana (Spiti)                             GACAACCACCGTCGTCGCGGCACGGGTCGCCAACGACCTAGATTTTGGGCCAACCACGAG
H.tibetana (Spiti)                             GACAACCACCGTCGTCGCGGCACGGGTCGCCAACGACTAAGATTTTGGGCCAACCACGAG
H.tibetana (Spiti)                             GACAACCACCGTCGTCGCGGCACGGGTCGCCAACGACTAAGATTTTGGGCCAACCACGAG
H.tibetana (Spiti)                             GACAACCACCGTCGTCGCGGCACGGGTCGCCAACGACTAAGATTTTGGGCCAACCACGAG
H.tibetana (Spiti)                             GACAACCACCGTCGTCGCGGCACGGGTCGCCAACGACTAAGATTTTGGGCCAACCACGAG
Elaegnus_Oldhamii                              GACAACCACAATTGTCGTGGCATTGGTCGCCAATGACTAGAA-TTTGGGCCAACCACAAG
                                                ** *****  * ****  ***  *** ***** ***    * ************ * **

H.rhamnoides ssp. turkestanica (Uttaranchal)   ------GCCAGGGGGAACATCATCATGCTTACACCTTTTAACAAAGAAAAGGTGC--AAG
H.rhamnoides ssp. turkestanica (Uttaranchal)   ------GCCAGGGGGAACATCATCATGCTTACACCTTTTAACAAAGAAAAGGTGC--AAG
H.salicifolia (Spiti)                          ------GCCAGGGA-AATGTCATCATGCTTACCACTTTCA-CATAGAAAAGGTGC--AAG
H.rhamnoides ssp. mongolica                    ------GCCAGGGA-AATGTCATCATGCTTACCACTTTCA-CATAGAAAAGGTGC--AAG
H.rhamnoides ssp. turkestanica (Ladakh)        ------GCCAGGGA-AATGTCATCATGCTTACCACTTTCA-CATAGAAAAGGTGC--AAG
H.rhamnoides ssp. carpatica                    ------GCCAGGGA-AATGTCAACATGATTACCACTTTCA-CATAGAAAAGGTGC--AAG
H.salicifolia (Spiti)                          ------GCCAGGGA-AATGTCATCATGCTTACCACTTTCA-CATAGAAAAGGTGC--AAG
H.tibetana (Spiti)                             TTTATTGCCAGGGG-AACGTCGTCGTGCTTACCACTTTCC-TACCGAAAAGGTGCGCAAG
H.tibetana (Spiti)                             TTTATTGCCAGGGA-AACGTCGTCGTGCTTACCACTTTCC-TACCGAAAAGGCGCGCAAG
H.tibetana (Spiti)                             T--ATTGCCAGGGA-AACGTCGTCGTGCTTACCACTTTCC-TACCGAAAAGGTGCGCAAG
H.tibetana (Spiti)                             ---ATTGCCAGGGA-AACGTCGTCGTGCTTACCACTTTCC-TACCGAAAAGGTGCGCAAG
H.tibetana (Spiti)                             ---ATTGCCAGGGA-AACGTCGTCGTGCTTACCACTTTCC-TACCGAAAAGGTGCGCAAG
Elaegnus_Oldhamii                              ------GCAGGAGGGCATGTTACGATGCTTGCCGCGTGCA---TAGAGCAGGGGA--AAG
                                                    **  * *   * *     ** ** *  * *        **  *** *   ***

H.rhamnoides ssp. turkestanica (Uttaranchal)   CACAAA---------AACTAGCAATGG-TAATAATCAT---GGCCAAGAGGCTAGATCA
H.rhamnoides ssp. turkestanica (Uttaranchal)   CACAAA-CAACATTCCAACTAGCAATGGGTAATAATCAT---GGCCAAGAGGCTAGATCA
H.salicifolia (Spiti)                          CACAAA-CAACATTCCAACTAACAATGGGTAATAATCAT---GTCCTTGAGGCTAGATCA
H.rhamnoides ssp. mongolica                    CACAAA-CAACATTCCAACTAACAATGGGTAATAATCAT---GTCCTTGAGGCTAGATCA
H.rhamnoides ssp. turkestanica (Ladakh)        CACAAA-GAACATTCCAACTAACAATGGGTAATAATCAT---GTCCTTGAGGCTAGATCA
H.rhamnoides ssp. carpatica                    CACAAA-CAACATTCCAACTAACAATGGGTAATAATCAT---GTCCTTGAGAACAGATCA
H.salicifolia (Spiti)                          CACAAA-CAACATTCCAACTAACAATGGGTAATAATCAT---GTCCTTGAGGCTAGATCA
H.tibetana (Spiti)                             CGCAAAACGACATTCCAACTAGCAATGGGCGATAATCGTCGTGGCCGAGAGGCCAGATCG
H.tibetana (Spiti)                             CGCAAAACGACATTCCAACTAGCAATGGGCGATAATCGTCGTGGCCGAGAGGCCAGATCG
H.tibetana (Spiti)                             CGCAAAACGACATTCCAACTAGCAATGGGCGATAATCGTCGTGGCCGAGAGGCCAGATCG
H.tibetana (Spiti)                             CGCAAAACGACATTCCAACTAGCAATGGGCGATAATCGTCGTGGCCGAGAGGCCAGATCG
H.tibetana (Spiti)                             CGCAAAACGACATTCCAACTAGCAATGGGCGATAATCGTCGTGGCCGAGAGGCCAGATCG
Elaegnus_Oldhamii                              CACCAAGCAACATTCCAACCAACAACGGG-----GCCATCCAAGCCGAGAGGCTTGAACG
                                               * * **            *** * *** **          * *     **  ***     ** *
```

***Indicates same nucleotides in all sequences and – denotes gap**

Figure 7.2: Part of the Aligned Nucleotide Sequences of ITS I – 5.8S – ITS 2 Region of *Hippophae* Taxa.

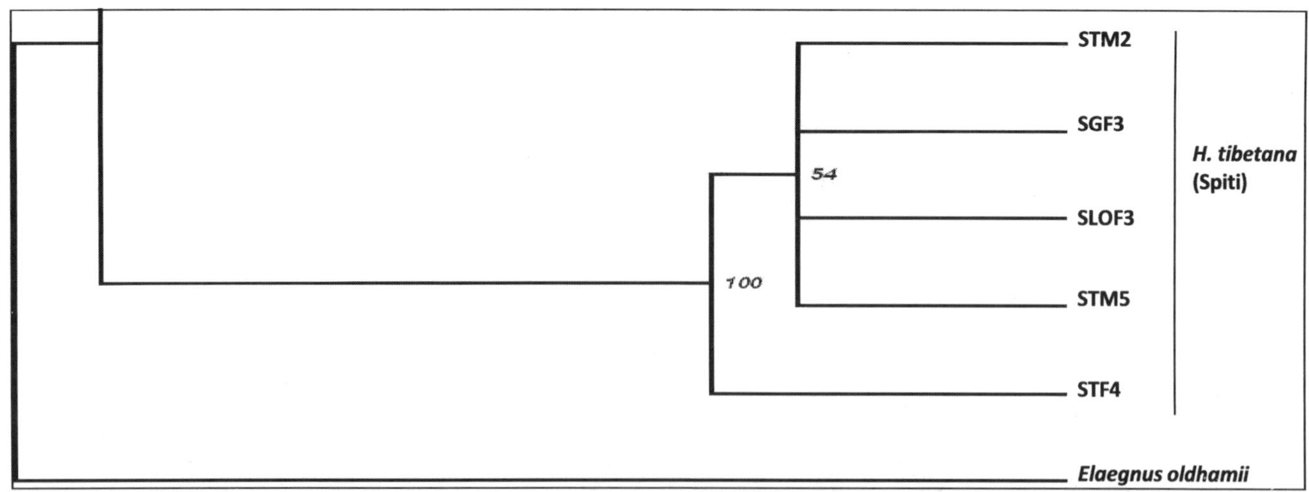

Note: *Elaegnus oldhamii* is an outgroup species. Number at nodes represent bootstrap values.

Figure 7.3: Part of the Strict Consensus Tree Based on ITS 1 and ITS 2 Sequence Data.

ACKNOWLEDGEMENTS

Thanks are due to Defence Research Development Organization (DRDO), Government of India for financial support.

REFERENCES

1. Ainouche, M.L. and Randall, J.B. 1997. On the origins of the tetraploid *Bromus* species (Section Bromus, Poaceae): Insights from internal transcribed spacer

sequences of nuclear ribosomal DNA. *Genome* 40: 730-743.

2. Baldwin, B.G. 1992. Phylogenetic utility of the internal transcribed spacers of nuclear ribosomal DNA in plants: An example from the compositae. *Mol. Phylog. Evol.* 1: 3-16.
3. Baldwin, B.G. 1993. Molecular phylogenetics of *Calycadenia* (Compositae) based on ITS sequences of nuclear ribosomal DNA: chromosomal and morphological evolution reexamined. *Amer. J. Bot.* 80: 222-238.
4. Baldwin, B.G., Sanderson, M.J., Porter, J.M., Wojciechowski, M.F., Campbell, C.S. and Donoghue, M.J. 1995. The ITS region of nuclear ribosomal DNA: A valuable source of evidence on angiosperm phylogeny. *Ann. Missouri Bot. Gard.* 82: 247-277.
5. Bartish, I.V., Jeppsson, N., Bartish, G. L. and Nybom, H. 2000. Inter- and intraspecific genetic variation in *Hippophae* (Elaeagnaceae) investigated by RAPD markers. *Plant Syst. Evol.* 225: 85-101.
6. Bartish, I.V., Jeppsson, N., Nybom, H. and Swenson, U. 2002. Phylogeny of *Hippophae* (Elaeagnaceae) inferred from parsimony analysis of chloroplast DNA and morphology. *Syst Bot* 27: 41-54.
7. Bernath, J. and Foldesi, D. 1992. Seabuckthorn (*Hippophae rhamnoides* L.)-A promising new medicinal and food crop. *J. Herbs Spices Med. Plants* 1: 27-35.
8. Buckler, E.S. 4th, and Holtsford, T.P. 1996. *Zea* Systematics: Ribosomal ITS evidence. *Mol Biol Evol.* 13: 612-622.
9. Cao, Y.L. and Lu, R.S. 1989. Karyotype analysis of *Hippophae* L. in China. *Acta Phytotax Sin* 27: 118-123.
10. Goel, S., Raina, S.N. and Ogihara, Y. 2002. Molecular evolution and phylogenetic implications of internal transcribed spacer sequences of nuclear ribosomal DNA in the *Phaseolus-Vigna* complex. *Mol. Phylo. Evol.* 22: 1-19.
11. Kollipara, K.P., Singh, R.J. and Hymowitz, T. 1997. Phylogenetic and genomic relationships in the genus *Glycine* Wild. based on sequences from the ITS region of nuclear DNA. *Genome* 40: 57-68.
12. Lashermes, P., Combes, M.C., Trouslot, P. and Charrier, A. 1997. Phylogenetic relationships of coffee tree species *Coffea* L. as inferred from ITS sequences of nuclear ribosomal DNA. *Theor. Appl. Genet.* 94: 947-955.
13. Lian, Y.S. and Chen, X.L. 2000. The regular pattern of distribution on the natural components in plants of the genus *Hippophae* L. *J. Northwest Nor. Univ.* 36: 113-128.
14. Lu, R. 1997. Eco-geographical distribution of seabuckthorn and prospects of International co-operation. In: *Worldwide Research and Development of Seabuckthorn*, Lu, S., Li, M., Hu, J., Liu, S. (eds) China Science and Technology Press, Beijing, pp. 11-22.
15. Murray, M.G. and Thompson, W.F. 1980. Rapid isolation of high molecular weight plant DNA. *Nucleic Acid Res.* 8: 4321-4326.
16. Musters, W., Boon, K., Sande, CAFM vander, Heerikhuizen, H. van, and Planta, R.J. 1990. Functional analysis of transcribed spacers of yeast ribosomal DNA. *EMBO J.* 9: 3989-3996.
17. Pearson, M.L. and Rogers, J.A. 1967. *Hippophae rhamnoides* L. *J Ecol.* 50: 501-513.
18. Rousi, A. 1971. The genus *Hippophae* L.-A taxonomic study. *Ann. Bot. Fennici.* 8 (3): 177–227.
19. Rousi, A. and Arohonka, T. 1980. C-bands and ploidy level of *Hippophae rhamnoides. Ann Bot. Fenn.* 2: 1-18.
20. Ruan, C, Qin, P. and Zheng, J., *et al.*, 2004. Genetic relationships among some cultivars of seabuckthorn from China, Russia and Mongolia based on RAPD analysis. *Scien. Hort.* 101: 417-426.
21. Ruan, C.J. and Li, D.Q. 2000. Function and benefits of seabuckthorn improving eco-environment of loess plateau. *Environ Prot* 5: 30-31.
22. Ruan, C.J. and Li, D.Q. 2005. AFLP fingerprinting analysis of some cultivated varieties of seabuckthorn (*Hippophae rhamnoides*). *J. Genet.* 84: 311-316.
23. Sehgal, D., Raina, S.N., Devarumath, R.M., Sasanuma, T. and Sasakuma, T. 2009. Nuclear DNA assay in solving issues related to ancestry of the domesticated diploid safflower (*Carthamus tinctorius* L.) and the polyploid (*Carthamus*) genotypes and phylogenetic and genomic relationships in the genus *Carthamus* L. (Asteraceae). *Mol. Phyl. Evol.* 53: 631-644.
24. Singh, V. 2003. Geographical adaptation and distribution of seabuckthorn (*Hippophae* L.) resources. In: *Seabuckthorn (Hippophae L.). A Multipurpose Wonder Plant, Vol I: Botany, Harvesting and Processing Technologies*, Singh, V. (ed), Indus Publishing Company, New Delhi, pp. 21-34.
25. Singh, V., Singh, B. and Awasthi, C.P. 1995. Distribution, taxonomy and nutritional values of seabuckthorn (*Hippophae* L.) growing in dry temperate Himalayas. In: *Proceedings of International Workshop Seabuckthorn*, ICRTS, Beijing, China, pp. 52-59.
26. Soltis, D.E. and Kuzoff, R.K. 1995. Discordance between nuclear and chloroplast phylogenies in the Heuchera Group (Saxifragaceae). *Evolution* 49: 727-742.
27. Sun, K., Chen, W., Ma, R., Chen, X., Li, A. and Ge, S. 2006. Genetic variation in *Hippophae rhamnoides* ssp.

sinensis (Elaeagnaceae) revealed by RAPD markers. *Biochem. Genet.* 44: 186-197.

28. Sun, K., Chen, X., Ma, R., Li, C., Wang, Q. and Ge, S. 2002. Molecular phylogenetics of *Hippophae* L. (Elaeagnaceae) based on the internal transcribed spacer (ITS) sequences of nrDNA. *Plant Syst Evol* 235: 121-134.

29. Tang, X. 2002. Intrinsic change of physical and chemical properties of seabuckthorn (*Hippophae rhamnoides*) and implications for berry maturity and quality. *J. Hort. Sci. Bio.*, 77: 177-185.

30. Thompson, J.D., Gibson, T.J., Plewniak, F., Mongin, F.J. and Higgins, D.G. 1997. Clustal X Windows interface: Flexible strategies for multiple sequence alignment aided by quality analysis tools. *Nucl. Acids Res.* 25 (24): 4876-4882.

31. van der Sande C.A., Kwa, M., van Nues, R.W., van Heerikhuizen, H., Raue, H. A. and Planta, R.J. 1992. Functional analysis of internal transcribed spacer 2 of *Saccharomyces cerevisiae* ribosomal DNA. *J. Mol. Biol.*, 223: 899-910.

32. White, T.J., Bruns, T., Lee, S. and Taylor, J. 1990. Amplification and direct sequencing of fungal ribosomal RNA genes for phylogenetics. In: *PCR protocols: A Guide to Methods and Applications* Innis, M.A., Gelfand, D.H., Sninsky, J.J., White, T.J. (eds). Academic Press, San Diego, pp. 315-322.

33. Zeb, A. 2004. Important therapeutic uses of seabuckthorn (*Hippophae*): A review. *J. Biol. Sci.* 4: 687-69.

Chapter 8

An Overview of Seabuckthorn Genomics Research at Guru Gobind Singh Indraprastha University, New Delhi, India

P.C. Sharma*, Saurabh Chaudhary and Ankit Jain

University School of Biotechnology, Guru Gobind Singh Indraprastha University, Dwarka, New Delhi, India
(*E-mail: prof.pcsharma@gmail.com)

SUMMARY

Our laboratory has been actively engaged in research on seabuckthorn (SBT) genomics for the last one decade. We have focussed on various lines *viz.* development of microsatellite markers, assessment of genetic diversity in different geographical populations, NGS based transcriptome profiling during cold and freeze stress, identification and expression analysis of abiotic stress responsive genes. In 2005, we started with standardizing a novel RNA isolation method for SBT and successful establishment of first SBT EST database that included 3412 ESTs. Clustering of these ESTs yielded 1665 unigenes comprising of 345 contigs and 1320 singletons. Susequently, we attempted transcriptome profiling in SBT using high throughput next generation sequencing (NGS). Using next generation massive parallel sequencing technology (Illumina platform) and *de novo* assembly, we reported a comprehensive view of the SBT transcriptome with 86,253,874 high quality short reads assembled into 88,297 transcripts (> 100 bp) using different assembly tools representing about 53 Mb of SBT transcriptome. In parallel, DeepSAGE, was employed to study differential gene expression under cold and freeze stress in SBT. In total, 11,922 differentially expressed genes (DEGs) were identified including 6,539 upregulated and 5,383 downregulated genes in response to cold and freeze stress. The Deep SAGE data were further validated using qRT-PCR. Microsatellite markers were also developed by screening genomic libraries and *in silico* analysis of our sequence resources. Our studies on microsatellites provided further insight about the distribution of microsatellites in various genic regions of SBT transcriptome. Markers so developed were evaluated on SBT natural populations from Indian Himalayas to assess genetic diversity. Our findings may provide basic platform for future molecular breeding efforts directed towards SBT improvement.

Keywords: *Seabuckthorn, ESTs, NGS, Transcriptome, DeepSAGE and Microsatellites.*

INTRODUCTION

Seabuckthorn (*Hippophae* sp.), a member of the family Elaegnaceae, is a hardy, deciduous, wind pollinated shrub, having diploid chromosome number of 24 (Elena *et al.*, 2011). Earlier seabuckthorn (SBT) was considered a dioecious plant, however, recent report of Sen *et al.* (2013) suggested the presence of another variants also which may be monoecious or trioecious. Seven *Hippophae* species have been reported, of which two are probably of hybrid origin (Bartish *et al.*, 2002). The natural habitat of SBT is widely spread across Europe, Asia and North America, covering countries like China, Mongolia, India, Russia, Sweden, Norway, Finland and Canada (Lu, 1997; Lian and Chen, 2000). In India, it is generally found growing in hilly areas of Ladakh, Lahaul-Spiti, and some parts of Uttarakhand, Arunachal Pradesh and Sikkim (Stobdan *et al.*, 2008).

In the recent years, medicinal plants have drawn immense attentions of research workers worldwide. The medicinal properties of these plants are mainly due to the occurence of many bioactive compounds including antioxidants, polysaccharides, polyphenols

Seabuckthorn (*Hippophae* L.): *A Multipurpose Wonder Plant, Vol. 5: Advanced Technologies for Cultivation, Processing, Health Protection and Environmental Conservation* (Virendra Singh, *Editor-in-Chief*, 2018), p. 53–61, Astral International Pvt. Ltd., New Delhi, India.

and many secondary metabolites. *Hippophae* sp. is one such plant, having ability to produce a large number of therapeutic molecules. The medicinal value of SBT has been attributed to the presence of high content of these compounds including carotenoides, tocopherols, tannins, ascorbic acid, flavonoides, sterols, polyphenols, triterpenes, organic acids, phytosterols, and many more, in berries, leaves and roots of the plants (Zeb, 2004). SBT preparations have been reported to possess anti-tumor, anti-carcinogenic, anti-atherogenic, and anti-microbial activities, in addition to immuno-modulatory and radio-protective properties (Zeb, 2004; Teng *et al.*, 2006; Suryakumar *et al.*, 2011). In traditional medicinal systems of Tibet, Mongolia, Uzbekistan, China, India, Turkey and Russia, SBT has been used extensively to treat bowel irregularities, skin wounds, gastric ulcers, cough and cold, *etc.* (Sezik *et al.*, 2004; Shinwari *et al.*, 2003; Dhyani *et al.*, 2007; Singh and Lal, 2008). Apart from medicinal and nutritional value, SBT is also known for its ecological importance as the plant helps in preventing soil erosion, conserving water and soil nutrients (Ledwood and Shimwell, 1971; Wei *et al.*, 2007). Moreover, SBT has strong ability to fix atmospheric nitrogen with the help of symbiotic association with an actinomycetes *Frankia*, in root nodules (Trajkovsk and Jeppsson, 1999). SBT has an ability to withstand extreme environmental conditions of temperature and soil pH and grows well in drought conditions (Rousi, 1971; Lu, 1992; Li *et al.*, 2005). Therefore, this plant may be considered as a model plant for the study of tolerance to various abiotic stresses. Although SBT resources have been vastly exploited for medicinal and nutritional purposes, only recently the plant has attracted the attention of molecular biologists. Our laboratory at Guru Gobind Singh Indraprastha University, New Delhi, has been actively engaged in research on various lines of SBT genomics during last one decade. The major outcome of our research includes:

1. Standardization of a protocol for the isolation of good quality RNA (Ghangal *et al.*, 2009).
2. Generation of 3142 expressed sequence tags (ESTs) (Ghangal *et al.*, 2008; Ghangal *et al.*, 2012).
3. Development of EST-based new SSR markers (Jain *et al.*, 2010).
4. Complete transcriptome profiling using Next Generation Sequencing (NGS) technology (Sharma *et al.*, 2012; Ghangal *et al.*, 2013).
5. Mining of microsatellites using NGS of SBT transcriptome (Jain *et al.*, 2013; Jain *et al.*, 2014).
6. DeepSAGE based identification of cold and freeze responsive genes (Chaudhary and Sharma, 2015).

MATERIALS AND METHODS

Isolation of RNA and DNA

SBT leaf tissue was snap freezed in liquid nitrogen and stored in -80pC until use. The leaf tissue, powdered in liquid nitrogen, was transferred to 5 ml of pre-warmed CTAB extraction buffer [2 per cent w/v CTAB, 100 mM Tris–Cl (pH 8.0), 2 M NaCl, 25 mM EDTA (pH 8.0)] containing 100 mg PVPP and 150 µl β-mercaptoethanol. The mixture was vortexed for 1 min, and then incubated at 65pC in water-bath for 20 min. The mixture was brought down to room temperature and equal volume of Chloroform/Isoamyl alcohol (24/1) was added and then mixed well by invert mixing. The contents were centrifuged at 10,000 × g for 25 min at 4pC. Upper aqueous phase was collected carefully and equal volume of phenol (pH 4.5)/Chloroform/Isoamyl alcohol (25/24/1) was added and mixed well, followed by centrifugation at 10,000 × g for 15 min at 4pC. Upper aqueous phase was carefully collected and equal volume of Chloroform/Isoamyl alcohol (24/1) was added, repeating the centrifugation step. To the upper aqueous phase, 0.2 volume of 1 M Glacial acetic acid, 0.1 volume of 3 M sodium acetate and 2 volume of ethanol were added. The pellet was collected by centrifugation after incubating the tubes at –80pC for 2 h. The pellet was washed twice with 5 M NaCl first and once by 70 per cent ethanol by re-suspending the pellet and centrifuging at 12,000 × g for 15 min at 4pC at each of the wash step. After evaporating ethanol, the pellet was resuspended in the minimum amount of DEPC (Diethyl Pyrocarbonate) treated water.

Similarly for DNA isolation, CTAB buffer with same composition as that used in RNA, was used along with equal amount of PVPP and β-mercaptoethanol. Chloroform and isoamyl alcohol (24:1) addition to incubated mixture of extraction buffer followed by centrifugation at 10,000 rpm for 20 mins resulted in the separation of two phases. Aqueous phase was taken and addition of double volume of ethanol along with 3M sodium acetate lead to the precipitation of DNA. Spinning down the precipitate and washing with 5M NaCl yielded better quality DNA. After salt washing, a wash with 70 per cent ethanol and air drying of pellet yielded genomic DNA that was suspended in TE buffer to increase the shelf life of the DNA.

Generation of Expressed Sequence Tags (ESTs)

The cDNA library was constructed using total RNA isolated from SBT leaf tissues as described above. The cDNA was amplified with SMART cDNA Library Construction Kit (Clontech, Japan) and ligated into pDNR-Lib vector (*SfiI* digested). Perfetprep 96 Vac Plasmid Isolation Kit (Eppendrof) was used for plasmid isolation from the recombinant clones. The sequencing was performed on 16 capillary MegaBACE automated DNA sequencer (GE Healthcare) using DYEnamic™ Dye Terminator Kit (GE Healthcare). Phred program (Ewing *et al.*, 1998) was used for base calling of the sequencing chromatograms. A cross-match program and perl script were used for further trimming vector sequences and low quality bases. The CAP3 software (Huang and Madan, 1999) was used to assemble vector trimmed ESTs into unigenes. The assembled ESTs were reported as contigs and singletons. ESTS can (Iseli *et al.*, 1999) and EST Explorer (Nagaraj *et al.*, 2007) were used to determine comprehensive annotation and open reading frames (ORF). NCBI-BLAST tool was used for similarity search. Gene Ontology (GO) terms were also assigned for molecular functions, biological processes and cellular component categories using Blast 2GO suit (Conesa *et al.*, 2005). All the similarity search analyses were performed at e-value ≤ 1E-10, to keep the stringency level high. The data generated during this study were submitted to dbEST of NCBI and can be access with GeneBank Id: EY414394-EY416137, FG35624-FG357288, and GT915047-GT915671.

Transcriptome Analysis using Next Generation Sequencing

SBT seedlings were grown hydroponically in the plant growth chamber as described by Ghangal *et al.* (2012). Total RNA was isolated from leaf and root tissues harvested from 30 days old seedlings. The RNA samples with RIN value (RNA Integrity Number) > 8 were used for further processing. Tag prepations and sequencing was outsourced to a commercial service provider for paired end sequencing using Next Generation Sequencing (NGS) technology (Illumina Platform). The short reads generated were quality checked with NGS QC Tool Lit (Patel and Jain, 2012). Various assembly tools including Velvet (Zerbino and Birney, 2008), ABySS (Simpson *et al.*, 2009), SOAPdenovo (Li *et al.*, 2009), CLC Genomics (Commercially available), Trinity (Grabherr *et al.*, 2011), and Oases (Schulz *et al.*, 2012) were used, employing different parameters and approaches for *de novo* assembly of short reads. Similarity search for annotation of the nonredundant assembled transcripts was done with BLASTx module of BLAST (Altschul *et al.*, 1990). The significant and best hit results were extracted. Gene ontology (GO) terms were assigned using Blast 2GO suite (Conesa *et al.*, 2005).

Identification of Cold and Freeze Responsive Genes

Total RNA was isolated from 30 days old SBT (*H. rhamnoides*) seedlings subjected to cold stress at 4pC, freeze stress at -10pC, and seedlings grown at controlled condition *i.e.* 28pC. RNA samples (RIN value > 8) were outsourced to the commercial service provider for tag preparation and sequencing. Deep SAGE (Nielsen *et al.*, 2006), a combination of classic SAGE and NGS, was employed for the identification of differentially expressed genes under cold and freeze stress. The tags were prepared from all the three mRNA libraries (cold, freeze and control) and sequenced as per manufacturer's protocol. Raw sequence tags generated from Illumina sequencing were filtered into clean tags yielding 21 nt long tags. The clean tags were further processed for alignment and annotation using BLAST and Blast 2 GO suits (Conesa *et al.*, 2005). The tags mapped to reference sequence were filtered and remaining unambiguous tags were determined and normalized to TPM (transcripts per million) (t'Hoen *et al.*, 2008; Morrissy *et al.*, 2009). Further, an algorithm developed by Audic and Claverie (1997), was used for screening differentially expressed genes. The FDR (False Discovery Rate) and log 2 ratio ≥1 were used as threshold to judge the significance of gene expression differences. KEGG (Kyoto Encyclopedia of Genes and Genomes) (Kanehisa *et al.*, 2008), was also done for pathway enrichment analysis of the differentially expressed genes. The 22 randomly selected differentially expressed genes were further validated using qRT-PCR.

Development of Microsatellite Markers

Standardization of DNA extraction methodology resulted in the isolation of high quality DNA used for the construction of microsatellite enriched genomic library. Recombinant clones generated as a result of our modified strategy to reduce DNA loss, were sequenced and screened with perl a script MISA (Thiel *et al.*, 2003). In addition, a new screening methodology was also devised allowing detection of tentative location of microsatellite before sequencing.

In the absence of SBT sequence resources in public domain, in house generated sequences data resulting from ESTs and whole transcriptome sequencing were screened for the presence of microsatellites. In parallel, an attempt was made to study the distribution of microsatellite in various genic regions of SBT. OrfPredictor (Min *et al.*, 2005) was used to predict the open reading frame (ORF) in microsatellite containing unigenes. With the co-ordinates obtained from MISA, genic location of microsatellite could be assigned. Unigenes harbouring microsatellite were subjected to BLAST analysis for similarity search against nr-protein database to identify their putative functions. Gene ontology (GO) terms were also assigned using Blast 2GO suite (Conesa *et al.*, 2005).

Primers for the flanking regions of the selected microsatellites loci were designed using Primer3 (Rozen and Skaletsky, 2000). The amplification of microsatellite loci was performed using 25 µL PCR reaction mixture. Amplifications were carried out in Mastercycler (Ependrof, Germany), with cycling condition as follows: 95pC for 5 min, followed by 35 cycles of 45 s at 95pC, 40 s at primer specific annealing temperature, and 1 min 72pC, the final step being extension at 72pC for 5 min. Amplicons were resolved on 4 per cent metaphor agarose (Lonza, Switzerland) and visualized by ethidium bromide staining on a gel doc system (Alpha Innotech). The data were scored for the presence (1) and absence (0) of alleles. Various tools including DARwin, POPGEN were used to assess various aspects of genetic diversity.

RESULTS

RNA/DNA Isolation

Our modified CTAB method was proved to be efficient for the isolation of total RNA from SBT leaves, having high content of secondary metabolites and phenolic compounds. The RNA isolated was quantified with spectrophotometer. The yield of 600-700 μg/g of leaf tissues was obtained and A_{260}/A_{280} ratio was 1.7. The denatured 1.2 per cent agarose gel electrophoresis was further used to assess the integrity of RNA. The sharpness and brightness of rRNA bands visualized on gel doc system, suggested good quality and quantity of total RNA, isolated from SBT leaf tissues (Figure 8.1).

Similarly for DNA isolation addition of PVPP and β–mercaptoethanol showed marked impact on the quality of DNA. Table 8.1 depicts the improvement of quality by altering the CTAB extraction buffer.

Table 8.1: Absorbance Values Obtained for DNA Sample Isolated using Modified CTAB Method from Seabuckthorn Leaves

Source	A_{230}	A_{260}	A_{280}	A_{260}/A_{280}	A_{260}/A_{230}
CTAB method	1.262	1.804	0.884	2.04	1.43
CTAB Modified method (PVPP+ dippa – mercaptoethanol)	0.72	1.562	0.788	1.98	2.17

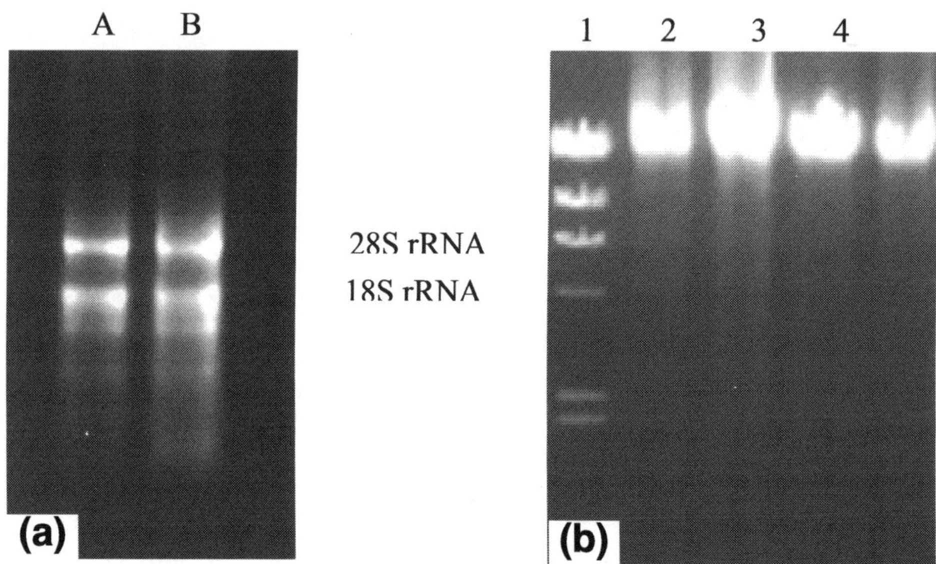

Figure 8.1: (a) Total RNA (A and B) isolated from seabuckthorn leaf tissues on denatured 1.2 per cent agarose gel. (b) Genomic DNA isolation from seabuckthorn leaf tissues samples collection from different sites (1: Lambda HindIII ladder; Lane 2 to 5: Sample HRL1-4).

Expressed Sequence Tags (ESTs) Sequencing

Sequencing of randomly selected clones from 4700 total recombinant clones with desired insert size range yielded 3412 high quality ESTs (≥100 bp). Moreover, 1665 unigenes were assembled from EST sequences with 345 contigs and 1320 singletons (Table 8.2). ESTscan (Iseli *et al.*, 1999) and EST Explorer (Nagaraj *et al.*, 2007) revealed that 1365 unigenes have ORF composed of more than 30 amino acids. The similarity search results revealed maximum sequence homology with *Vitis vinifera* followed by *Populus trichocarpa* and *Arabidopsis thaliana*. In total, 55 per cent unigenes could be assigned GO term. The GO categorization of the unigenes showed 35 per cent, 44 per cent, and 42 per cent in cellular component, molecular function, and biological process category, respectively. KEGG pathway analysis assigned 161 sequences to 46 KEGG pathways, of which majority of the unigenes were related to metabolic pathways. Moreover, 73 putative unigenes having transcription factor (TF) distributed in at least 34 TF families were identified.

Table 8.2: Summary of Expressed Sequence Tags (ESTs) Sequencing

Description	*Data*
Total number of ESTs generated	3412
Average length of ESTs generated	346
Number of contigs	345
Number of singletons	1320
Total number of unigenes	1665

The next generation massive parallel sequencing (Illumina platform), provided a comprehensive view of the SBT transcriptome (Table 8.3). More than 90 millions high quality short reads were generated by sequencing of leaf and root tissues cDNA libraries of SBT. Various assemblers were used to assess the best sequence assembly. ABySS assembler in combination with TGCIL suit was found to be the most appropriate tools for the assembly of short reads. *De novo* assembly of short reads yielded 88,297 transcripts, representing about 53 Mb of the SBT transcriptome. The similarity search revealed 46.8 per cent transcripts showing significant similarity with nr-protein database. Likewise EST database assembled unigenes showed significant homology of SBT with *Vitis vinifera*, followed by *Ricinus communis*, and *Populus trichocarpa*. On the basis of Gene Ontology term assigned to sequences, 43.9 per cent of SBT transcriptome sequences had at least one GO term associated with them. In the different categories *i.e.* biological process, molecular function, and cellular component, the largest group was cellular metabolic process followed by biosynthetic process. The groups such as cellular response to stimulus, external stimulus, biotic stimulus and abiotic stimulus were abundant. In case of transcription factors, around 8.4 per cent of SBT transcriptome sequences were found to have around 7,000 transcription factors representing about 80 different families.

Table 8.3: Summary of Next Generation Sequencing (NGS), Data

Description	*Data*
Total number of reads generated	94,013,936
Read length	91
Total number of HQ reads	86,253,874
Total number of putative unigenes assembled	88,297

Mining of Microsatellites in Seabuckthorn Transcriptome using NGS

The 88297 putative unigenes obtained from *de novo* assembly of SBT transcriptome using NGS were assessed for microsatellite markers. The *in silico* analysis yielded 7.69 per cent unigenes harboured microsatellite repeats in which dinucleotide repeats were found to be most abundant followed by trinucleotide repeats. Most of the microsatellite repeats were found in the coding regions followed by 3′ and 5′ untranslated regions. The most frequently represented repeats were AG and AGG types. Out of the microsatellite positive unigenes, 48.81 per cent were assigned to Gene Ontology (GO) terms to get an overview of association between microsatellite containing unigenes and various biological processes.

Genomic Library Construction and Sequencing

Traditional methods of enriched library construction are time consuming as well as cumbersome and produce a maximum recovery of 60-70 per cent microsatellite positive sequences. In order to reduce the loss of DNA during different steps of library construction like DNA fragmentation followed by gel purification for further ligation to linkers and recovery of single stranded microsatellite positive DNA fragments from magnetic beads. In the current methodology, an attempt was made to reduce the DNA loss. Simultaneous restriction digestion of genomic DNA and ligation with linkers was found more effective in comparison to restriction digestion followed by linker ligation. Alteration of recovery process of microsatellite carrying ssDNA sequences could further reduce the loss of DNA. Further, linker- microsatellite probe was used to assess the possible location of microsatellite in a clone prior to the sequencing and further aided in selection of only those clones wherein primer designing for flanking region of microsatellite was possible. Sequencing of total 64 clones led to the identification of 51 microsatellite containing sequence.

Development of Microsatellite Markers

Forty five primers designed for the flanking regions of microsatellite loci isolated from genomic library construction and sequencing yielded four polymorphic markers for Leh and Lahaul populations. The primers were designed for 30 microsatellites carrying unigenes. Out of 30 primers, 9 pairs displayed 2-5 alleles in *H. rhamnoides* genotypes. Two primers pairs were found polymorphic *H. salicifolia*. On the other hand, no polymorphism was detected in *H. tibetana* genotypes. Similarly, 60 primers were designed from randomly selected microsatellite loci from whole transcriptome sequencing. Of these 19 were found polymorphic on the SBT collection from Leh and Lahaul. All these markers showed cross amplification in limited accession of *H. salicifolia* and *H. tibetana*.

Identification of Cold and Freeze Responsive Genes

DeepSAGE, a tag based high throughput transcriptome profiling technique, was employed to generate differential gene expression profile from the three leaf tissue libraries including control, cold and freeze stressed SBT plants. More than 35 million raw tags were generated which included around 13 million distinct tags representing all the three libraries. After filtering low quality tags and mapping, 11,922 differentially expressed genes in total were identified including 54.9 per cent upregulated and 45.1 per cent down regulted genes. Gene Ontology (GO) terms assigned to differentially expressed genes suggested metabolic, cellular, primary metabolic genes to be most abundant. Moreover, high number of differentially expressed genes were categorised into stress and stimulus responsive genes. Photosynthetic, carotenoide biosynthesis, and plant hormone signal transduction pathways were found to be the most significant enriched pathways under cold stress and freeze stress in SBT, as suggested by KEGG pathway analysis. The differentially expressed genes further mapped back to our existing

SBT transcriptome comprising of 88,297 unigenes revealing 480 SBT specific cold and freeze responsive genes. Out of 480 genes, 22 were randomly selected and their expression was validated using qRT-PCR. The qRT-PCR analysis showed 16 upregulated and 6 down regulated genes under cold stress and 12 were found to be upregulated and 10 downregulated under freeze stress (Chaudhary and Sharma, 2015).

DISCUSSION

The presence of high content of polysaccharides, polyphenols and secondary metabolites in SBT tissues creates problems in isolation of high quality DNA/RNA due to the tendency of these compound to get oxidised and binding with nucleic acids during the extraction procedures (Loomis, 1974). In case of RNA isolation, guanidine based methods were found to be ineffective in the case of SBT. Possible reason behind the failure could be the inability of guanidine salts to act on non-protein complexes such as secondary metabolites (Liu *et al.*, 1998). The modified CTAB protocol (Ghangal *et al.*, 2009) standardized in our laboratory proved to be suitable and efficient for isolation of good quality RNA from SBT leaf tissues. The CTAB used as cell disrupting agent and high content of PVPP helps in removing polyphenols in the very first step of extraction. Same buffer was also found to be effective in DNA isolation. Furthermore the use of 1M glacial acetic acid in combination with ethanol further ensured the recovery of pure RNA in large quantity. However, in case of DNA, only chloroform isolamyl treatment and recovery of aqueous phase followed by precipitation step resulted in better quality which could be further improved by high salt washes.

Our pilot study that aimed generation of expressed sequence tags (ESTs) (Ghangal *et al.*, 2012), provided first ever major valuable resource for SBT transcriptome and an opportunity to understand the physiological and molecular basis of abiotic stress tolerance of the plant. The EST data were further evaluated for the expression analysis of thirteen known and three novel cold and freeze responsive genes in SBT using qRT-PCR. Data thus generated were also used to develop first ever set of thirty EST-SSR markers which were assessed initially over 5 accession each of *H. rhamnoides*, *H. salicifolia*, and *H. tibetana*. Results showed cross species amplification of markers and nine markers were found to be polymorphic among *H. rhamnoides*, whereas two additional markers were found polymorphic in *H. salicifolia* but all markers were monomorphic in *H. tibetana*. These results indicates that *H. rhamnoides* and *H. salicifolia* are more closer species and *H. tibetana* is a distantaly related species. Further advancement in the next generation sequencing technology provided us a time and cost effective technology to generate complete transcriptome profile of SBT. The large amount of data generated from NGS technology using Illumina platform was assembled using two approaches 1). best k-mer strategy, and 2). additive k-mer followed by long read assembler TGICL. ABySS followed by TGICL was found to be more efficient and useful for NGS transcriptome analysis for non model organisms (Ghangal *et al.*, 2013). The similarity search and annotation of unigenes in our study was also supported by our as well as other studies on SBT transcritome profiling (Ghangal *et al.*, 2012; Fatima *et al.*, 2012). The functional annotation of SBT transcripts revealed the conservation of genes involved in various biological processes, molecular functions and cellular component. Whole transcriptome data was proved to be one of the major sequence resources with 88,297 unigenes, of which 6787 were found to represent of 8,037 microsatellite loci. Our study revealed an excess of microsatellites in coding region in comparison to individual untranslated regions. Number of dinucleotide repeats were found to exceed rest of the repeat classes. In total, microsatellite carrying 3313 putative unigenes could be annotated (Jain *et al.*, 2014). Data were further utilized to design primers flanking selected microsatellite loci. Out of selected 60 microsatellite markers, 55 showed amplification at their respective annealing temperatures. Ninteen markers were found polymorphic on SBT accession collected from various geographical locations of Leh and Lahaul.

In a recent study, we employed DeepSAGE, a tag based approach for differential gene expression under cold and freeze stress in SBT (Chaudhary and Sharma, 2015). In total, 11,922 differentially expressed genes (DEGs) were identified that included 6539 upregulated and 5383 downregulated genes in response to cold and freeze stress. The DeepSAGE based technique was proved to be more efficient and powerful for gene expression analysis as large number of differentially expressed genes with rare transcripts have been identified in response to cold and freeze stress in SBT. The functional annotation and pathway analysis of differentially expressed genes suggested their role for sustaining SBT in the harsh environment conditions.

In an attempt to generate more microsatellite markers for SBT, we devised a novel methodology to construct microsatellite enriched genomic library involving simultaneous digestion and ligation of genomic DNA with linker. In addition, following a novel methodology to localize microsatellite in recombinant clones prior to sequencing led to recovery of more than 90 per cent microsatellite positive sequences wherein repeat was present at a reasonable distance from sequence termini such that primers could be designed for desired PCR amplification.

Taken together, our efforts have significantly contributed to the understanding of various aspects of SBT genomics, a neglected area of research till date. We hope that further studies on genetics and genomics of SBT will help in the improvement and better utilization of its medicinal values and understanding of molecular basis of abiotic stress tolerance in this wonder plant, leading to the identification of candidate genes to be explored for the production of abiotic stress tolerant transgenics.

REFERENCES

1. Altschul, S.F., Gish, W., Miller, W., Myers, E.W. and Lipman, D.J. 1990. Basic local alignment search tool. *Journal of Molecular Biology* 215: 403-410.
2. Audic, S. and Claverie, J.M. 1997. The significance of digital gene expression profiles. *Genome Research* 7: 986-995.
3. Bartish, I.V., Jeppsson, N., Nybom, H. and Swenson, U. 2002. Phylogeny of *Hippophae* (Elaeagnaceae) inferred from parsimony analysis of chloroplast DNA and morphology. *Systematic Botany* 27: 41-54.
4. Chaudhary, S. and Sharma, P.C. 2015. DeepSAGE based differential gene expression analysis under cold and freeze stress in seabuckthorn (*Hippophae rhamnoides* L.). *PLoS One* 10: e 0121982.
5. Conesa, A., Gotz, S., Garcia-Gomez, J.M., Terol, J., Talon, M. and Robles, M. 2005. Blast2GO: A universal tool for annotation, visualization and analysis in functional genomics research. *Bioinformatics* 21: 3674-3676.
6. Dhyani, D., Maikhuri, R.K., Rao, K.S., Kumar, K. and Purohit, V.K., Sundriyal, M. and and Saxena, K. G. 2007. Basic nutritional attributes of *Hippophae rhamnoides* (seabuckthorn) populations from Central Himalaya, India. *Current Science* 92: 1148-1152.
7. Elena, T., Capraru, G., Rosu, C.M., Zamfirache, M.M., Olteanu, Z. and Manzu, C. 2011. Morphometric pattern of somatic chromosomes in three Romanian seabuckthorn genotypes. *Caryologia* 64: 189-196.
8. Ewing, B., Hillier, L.D., Wendl, M.C. and Green, P. 1998. Base calling of automated sequencer traces using Pherd. I. Accuracy assessment. *Genome Research* 8: 175-185.
9. Fatima, T., Snyder, C.L., Schroeder, W.R., Cram, D., Datla, R., Wishart, D., Weselake, R. J. and Krishna, P. 2012. Fatty acid composition of developing seabuckthorn (*Hippophae rhamnoides* L.) berry and the transcriptome of the mature seed. *PLoS One* 7: e34099.
10. Ghangal, R., Chaudhary, S., Jain, M., Purty, R. and Sharma, P.C. 2013. Optimization of *de novo* short read assembly of seabuckthorn (*Hippophae rhamnoides* L.) transcriptome. *PLoS One* 8: e72516.
11. Ghangal, R., Raghuvanshi, S. and Sharma, P.C. 2009. Isolation of good quality RNA from a medicinal plant seabuckthorn, rich in secondary metabolites. *Plant Physiology and Biochemistry* 47: 1113-1115.
12. Ghangal, R., Raghuvanshi, S. and Sharma, P.C. 2012. Expressed sequence tag based identification and expression analysis of some cold inducible elements in seabuckthorn (*Hippophae rhamnoides* L.). *Plant Physiology and Biochemistry* 51: 123-128.
13. Ganghal, R., Raghuvanshi, S. and Sharma, P.C. 2008. Exploring seabuckthorn genome using molecular tools. In: *Seabuckthorn-The Wonder Plant*, S. K. Dwivedi, T. Perimalazagan, S. B. Singh and Z. Ahmad (eds.), SSPH Publishers, Delhi, pp. 347-371.
14. Grabherr, M.G., Haas, B.J., Yassour, M., Levin, J.Z. and Thompson, D.A., Amit, I., Adiconis, X., Fan, L., Raychowdhury, R., Zeng, Q., Chen, Z., Mauceli, E., Hacohen, N., Gnirke, A., Rhind, N., Palma, F., Birren, B. W., Nusbaum, C., Toh, K. L., Friedman, N. and Regev, A. 2011. Full-length transcriptome assembly from RNA-Seq. data without a reference genome. *Nature Biotechnology* 26: 644-652.
15. Huang, X. and Madan, A. 1999. CAP3: A DNA sequence assembly program. *Genome Research* 9: 868-877.
16. Iseli, C., Jongeneel, C.V. and Bucher, P. 1999. ESTScan: A program for detecting, evaluating, and reconstructing potential coding regions in EST sequences. In: *Proceedings of International Conference on Intelligent Systems for Molecular Biology*, pp.138-148.
17. Jain, A., Chaudhary, S. and Sharma, P.C. 2013. Application of DNA technologies for improvement of seabuckthorn. In: *Seabuckthorn (Hippophae L.) A Multipurpose Wonder Plant: Emerging Trends in Research and Technologies* (V. Singh, ed.), Daya Publishing House/Astral International, New Delhi, India, pp. 167-178.
18. Jain, A., Chaudhary, S. and Sharma, P.C. 2014. Mining of microsatellites using next generation sequencing of seabuckthorn (*Hippophae rhamnoides* L.) transcriptome. *Physiology and Molecular Biology of Plants* 20: 115-123.
19. Jain, A., Ghangal, R., Grover, A., Raghuvanshi, S. and Sharma, P.C. 2010. Development of EST-based new SSR markers in seabuckthorn. *Physiology and Molecular Biology of Plants* 16: 375-378.
20. Kanehisa, M., Araki, M., Goto, S., Hattori, M., Hirakawa, M., Itoh, M., Katayama, T., Kawashima, S., Okuda, S., Tokimatsu, T. and Yamanishi, Y. 2008. KEGG for linking genomes to life and the environment. *Nucleic Acids Research* 36: D480-D484.
21. Ledwood, J.S. and Shimwell, D.W. 1971. Growth rates of *Hippophae rhamnoides* L. *Annals of Botany* 35: 1053-1058.
22. Lian, Y.S. and Chen, X.L. 2000. The regular patterns of distribution on the natural components in plants of the genus *Hippophae* L. *Journal of Northwest Normal University (Natural Science Edition)* 36: 113-128.
23. Li, C., Yang, Y., Junttila, O. and Palva, E.T. 2005.

Sexual differences in cold acclimation and freezing tolerance development in seabuckthorn (*Hippophae rhamnoides* L.) ecotypes. *Plant Science* 168: 1365-1370.

24. Li, R., Zhu, H., Ruan, J., Qian, W. and Fang, X., Shi, Z., Li, Y., S., Li, Shan, G., Kristiansen, K., Li, S., Yang, H., Wang, J. and Wang, J. 2009. *De novo* assembly of human genomes with massively parallel short read sequencing. *Genome Research* 20: 265-272.

25. Liu, J.J., Goh, C.J., Loh, C.S., Liu, P. and Pua, P.C. 1998. A method for isolation of total RNA from fruit tissues of banana. *Plant Molecular Biology Reports* 16: 1-6.

26. Loomis, W.D. 1974. Overcoming problems of phenolics in the isolation of plant enzymes and organelles. *Methods Enzymology* 31: 528-545.

27. Lu Rongsen, 1992. Seabuckthorn-A multipurpose plant for fragile mountains. International Centre for Integrated Mountain Development (ICIMOD) Occasional Paper No. 20, Kathmandu, Nepal, 62p.

28. Lu Rongsen, 1997. Ecogeographical distribution of seabuckthorn and prospects of international cooperation. In: *Worldwide Research and Development of Seabuckthorn*, Lu, S., Li, M., Hu, J., Liu, S. (eds.) Beijing, China: Science Publishing House and Technology Press, pp. 11-22.

29. Min, X.J., Butler, G., Storms, R. and Tsang, A. 2005. Orf predictor: Predicting protein-coding regions in EST-derived sequences. *Nucleic Acids Research* 33: W677-W680.

30. Morrissy, S., Morin, R.D., Delaney, A., Zeng, T., Mc Donald, H., Jones, S., Zhao, Y., Hirst, M. and Marra, M. A. 2009. Next generation tag sequencing for cancer gene expression profiling. *Genome Research* 19: 1825-1835.

31. Nagaraj, S.H., Deshpande, N., Gasser, R.B. and Ranganathan, S. 2007. EST Explorer: An expressed sequence tag (EST) assembly and annotation platform. *Nucleic Acids Research* 35: W143-W147.

32. Nielsen, K.L., Hogh, A.L. and Emmersen, J. 2006. DeepSAGE digital transcriptomics with high sensitivity, simple experimental protocol and multiplexing of samples. *Nucleic Acids Research* 34: e133.

33. Patel, R.K. and Jain, M. 2012. NGS QC Toolkit: A toolkit for quality control of next generation sequencing data. *PLoS One* 7: e30619.

34. Rousi, A. 1971. The genus *Hippophae* L. a taxonomic study. *Annales Botanici Fennici* 8: 177-227.

35. Rozen, S. and Skaletsky, H. 2000. Primer 3 on the www for general users and for biologist programmers. In: *Bioinformatics Methods and Protocols: Methods in Molecular Biology*, Krawetz, S., Misener, S. (eds). Humana, Totowa, pp. 365-386.

36. Schulz, M.H., Zerbino, D.R., Vingron, M. and Birney, E. 2012. Oases: robust *de novo* RNA-seq. assembly across the dynamic range of expression levels. *Bioinformatics* 28: 1086-1092.

37. Sen, R., Tamchos, S. and Kaul, V. 2014. Sexual variants in a dioecious taxon *Hippophae rhamnoides* L. *International Journal of Plant Reproductive Biology* 6: 41-43.

38. Sezik, E., Yesilada, E., Shadidoyatov, H., Kulivey, Z. and Nigmatullaev, A.M. Aripov, H. N., Takaishi, Y., Takeda, Y. and Honda, G. 2004. Folk medicine in Uzbekistan. I. Toshkent, Djizzax, and Samarqand Provinces. *Journal of Ethnopharmacology* 92: 197-207.

39. Sharma, P.C., Jain, A. and Chaudhary, S. 2012. Transcriptome analysis in seabuckthorn (*Hippophae rhamnoides* L.): A medicinally important plant. In: *Proc. International Conference on Environmental and Biological Sciences*, Dec. 21-22, 2012, Bangkok, pp. 21-24.

40. Simpson, J.T., Wong, K., Jackman, S.D., Schein, J.E. and Jones, S.J. and Birol, I. 2009. ABySS: A parallel assembler for short read sequence data. *Genome Research* 19: 1117-1123.

41. Shinwari, Z.K. and Gilani, S.S. 2003. Sustainable harvest of medicinal plants at Bulashbar Nullah, Astore (Northern Pakistan). *Journal of Ethnopharmacology* 84: 289-298.

42. Singh, K.N. and Lal, B. 2008. Ethnomedicines used against four common ailments by the tribal communities of Lahaul-Spiti in western Himalaya. *Journal of Ethnopharmacology* 115: 147-159.

43. Soares, V.L.F., Rodrigues, S.M., De Oliveira, T.M., Lima, L.S., Hora-Junior, B.T., Gramacho, K.P., Micheli, F., Cascardo, J.C., Otoni, W.C., Gesteira, A.S. and Costa, M.G. 2011. Unraveling new genes associated with seed development and metabolism in *Bixa orellana* L. by expressed sequence tag (EST) analysis. *Molecular Biology Reports* 38: 1329-1340.

44. Stobdan, T., Angchuk, D. and Singh, S.B. 2008. Seabuckthorn: An emerging storehouse for researchers in India. *Current Science* 94: 1236-1237.

45. Suryakumar, G. and Gupta, A. 2011. Medicinal and therapeutic potential of seabuckthorn (*Hippophae rhamnoides* L.). *Journal of Ethnopharmacology* 138: 268-278.

46. Teng, B.S., Lu, Y.H., Wang, Z.T., Tao, X.Y. and Wei, D.Z. 2006. *In vitro* anti-tumor activity of isorhamnetin isolated from *Hippophae rhamnoides* L. against BEL-7402 cells. *Pharmacological Research* 54: 186-194.

47. Thiel, T., Michalek, W., Varshney, R.K. and Graner, A. 2003. Exploiting EST databases for the development and characterization of gene-

derived SSR-markers in barley (*Hordeum vulgare* L.). *Theoretical and Applied Genetics* 106: 411-422.

48. Trajkovski, V. and Jeppsson, N. 1999. Domestication of seabuckthorn. *Botanica Lithuanica* 2: 37-46.

49. t'Hoen, P.A., Ariyurek, Y., Thygesen, H.H., Vreugdenhil, E. and Vossen, R.H., de Menezes RX, Boer JM, van Ommen, G.J. and den Dunnen, J.T. 2008. Deep sequencing based expression analysis shows major advances in robustness, resolution and inter-lab portability over five microarray platforms. *Nucleic Acids Research* 36: e141.

50. Wei, W., Chen, L., Fu, B., Huang, Z. and Wu, D. and Gui, L. 2007. The effect of land uses and rainfall regimes on runoff and soil erosion in the semi-arid loess hilly area, China. *Journal of Hydrology* 335: 247-258.

51. Zeb, A. 2004. Chemical and nutritional constituents of seabuckthorn juice. *Pakistan Journal of Nutrition* 3: 99-106.

52. Zerbino, D.R. and Birney, E. 2008. Velvet: Algorithms for *de novo* short read assembly using de Bruijn graphs. *Genome Research* 18: 821-829.

Chapter 9

Genetic Variation in Growth Characteristics of *Hippophae rhamnoides* L. Grown in Controlled Conditions

Natalia Demidova[1*], Gosta Eriksson[2] and Kenneth Lundkvist

[1]Federal Budget Institution "Northern Research Institute of Forestry", Arkhangelsk, Russian Federation
[2]Department of Plant Biology, Bio Center, Swedish University of Agricultural Sciences, Uppsala, Sweden
(*E-mail: natalia.demidova@bk.ru)

SUMMARY

Seedlings of seabuckthorn (*Hippophae rhamnoides* L.), originating from a factorial cross between three female plants and six male plants, as well as open pollination of the mother plants, were grown in a controlled conditions until bud-set, a period of dormancy and an ensuing bud flush. Variation in growth, duration of growth period, earliness in bud flush and the fraction of frost damaged buds after freezing was studied. Confirming previously reported observations, we found that seabuckthorn (SBT) seedlings do not terminate growth at long nights, but low temperature treatment is needed. This lack of photoperiodic response stands in contrast to the majority of tree species in the temperate region. In this experiment, SBT reveals high estimates of genetic variance for growth and growth rhythm. The female variance component for the characters studied was higher than the male component, probably owing to seed size. The results suggest that estimation of genetic parameters in SBT shall be based upon controlled crosses rather than on open pollinated progenies. The observed maternal effects indicate that seed weight may be a significant covariate in analyses of variance. The freezing test used in this experiment did not disclose any major genetic variance components responsible for the variation in fraction of damaged buds after freezing. The experiments were conducted in the Department of Forest Genetics Swedish University of Agricultural Sciences in 1993-1994.

Keywords*: Seabuckthorn (Hippophae rhamnoides L.), Genetic variation, Growth characters and Controlled conditions.*

INTRODUCTION

The Seabuckthorn (*Hippophae rhamnoides* L.) is known as a useful plant and in many countries of the world it has received increasing attention as a medicine plant with fruits, leaves and bark rich in different vitamins and other biologically active substances (Shjukin, 1850; Darmer, 1952; Gatin, 1963; Trofimov, 1978; Matafonov, 1983; Kallio *et al.*, 2005; Demidova, 2009; Yang *et al.*, 2011; Attrey *et al.*, 2011; Sawhney, 2011; Pandurangan *et al.*, 2011; Maybeck, 2011). It is an important source for essential food ingredients as well (Morsel and Steen, 2005). Seabuckthorn (SBT) also has excellent nitrogen fixing properties with a capacity of fixing 45 kg nitrogen per hectare, which is twice the amount that is fixed by soybean plants (Bond *et al.*, 1954; Lu, 1990). It has, therefore, a wide use for field amelioration, recultivation and for stabilizing sandy areas and slopes (Ruan and Li, 2002; Baoli and Hongjum, 2011).

Hippophae rhamnoides L. occurs over large areas of Europe and Asia, with two discrete distributions in Europe: one European coastal area which stretches along the Baltic and North Sea, along the West coast of England and Ireland, along the East coast of Scandinavia and one Central European area which is directly connected with the mountainous part of Asia (Darmer, 1952; Pearson and Rogers, 1962; Ranwell, 1972). It is a typical pioneer species that can rapidly colonize new areas. SBT is generally a shrub, 5-6 meters in height, but can occasionally reach 11 meters in height (Skogen,

Seabuckthorn (Hippophae L.): A Multipurpose Wonder Plant, Vol. 5: Advanced Technologies for Cultivation, Processing, Health Protection and Environmental Conservation (Virendra Singh, *Editor-in-Chief*, 2018), p. 63–69, Astral International Pvt. Ltd., New Delhi, India.

1972; Sizemskaya and Sapanov, 2005). It forms extensive thickets in habitats where there is no competition from trees or other high vegetation (Rousi, 1976). The plants are sensitive to light conditions, and usually cannot grow under the shadow of other trees. The species is plastic to temperature and soil conditions (Vasilchenko, 1977), level of precipitation, and it can tolerate quite high salt concentrations. Nodule formation is, however, sensitive to soil pH-level (Bond *et al.*, 1954; Skogen, 1972; Lu, 1990).

Knowledge about genetic parameters for important traits is a prerequisite for a successful breeding of any species. There are many reports on among-population variation of markers in *H. rhamnoides* (Yao and Tigerstedt in 1993; Bartish *et al.*, 2000; Ruan *et al.*, 2002; Ercisli *et al.*, 2008; Simon-Gruita *et al.*, 2012). Jeppsson *et al.*, 1999 found more variation within SBT populations and very little variation between populations, what they explained by differences in population sizes and the continuity in spatial distribution. Contrary to this, there are few publications on within-population variation for growth traits and frost hardiness. There are some reports about variation among cultivars (Singh and Zubarev, 2014; Garanovich and Shpitalnaya, 2014). Shan (2014) reported on parental mean heights of eight hybrids and their offspring heights. They estimated the regression coefficient between parent and offspring at 0.17, which in this case is an estimate of the heritability for tree height. The range of heights was 205-233 cm; no tree age was given. It was stated that most of the hybrids were taller than their parental means. A large variation of leaf shape among and within British populations was reported by Pearson and Rogers (1962). Rousi (1965) reported high variation for morphological traits in 46 seedling populations and in adult populations from Asia and Europe. Based on the observed large variation in morphology and growth rhythm he pointed out that the species must have a wide adaptability that can be exploited in breeding. Tang and Tigerstedt (2001) reported low female heritability for winter damage, 0.04 while there were high heritabilities for several flowering and fruit characters, even heritability beyond the permitted 1.00 were reported. Their estimates were based on two sets of disconnected factorial mating between Finnish females and Danish males. One factorial being 3x8 and the other 5x6. Strong non-additive effects were noted for flowering and fruit traits. In a paper addressing the potential for *H. rhamnoides* cultivation in the USA, results as regards growth and frost tolerance of 13 cultivars growing at the Russian Siberia experiment station were presented by Letchamo *et al.* (2007). At age five the tree height varied in the range 1.99 and 3.12 meters. All cultivars were characterized by high general and bud frost tolerance and limited variation among the cultivars.

SBT is a complicated species for breeding since hybridization and accompanying progeny evaluation are required. A major task for SBT breeding is to combine traits of resistance to harsh weather conditions with fruit quality traits. Absence of thorns is one of those quality traits which play an essential role in its cultivars creation. With the course of the development of northern horticulture, there is a need to widen the variety of available berry cultivars, suitable for cultivation under northern conditions (Demidova and Nilov, 1992; Demidova, 1993, 2003). A gene bank of *H. rhamnoides* with large diversity of quality and quantitative characteristics has been collected in the Arkhangelsk Dendrological Garden since 1969. The results from the initial test under Dendrological Garden conditions gave guidelines for transfer of plant material to the North (Demidova, 2003), but it was found that in some years there was a high mortality, which could be attributed to damage of the root systems in mild winters with much snow. Jelisejev (1980) concluded that successful introduction of SBT is not possible without improvement of winter hardiness and frost tolerance.

Frost damage in many tree species is frequently attributed to a growth rhythm unsuitable for the growth conditions such that plants in frost sensitive phases are exposed to late spring frosts or early autumn frosts, alternatively to both of them (Eriksson and Lundkvist, 1986). It is plausible that *H. rhamnoides* plants transferred over wide distances to new and different growth conditions will suffer from maladaptation of the growth rhythm. Nilov (1986) reported that growth cessation in SBT is triggered by low temperatures. This contrasts with the response to long nights of many tree species from the temperature zone (Eriksson *et al.*, 2007). Under temperate climatic conditions there is usually a clinical variation for this response such that northern populations require a shorter night than southern populations for bud set and growth cessation (Eriksson *et al.*, 2007). For future breeding it would be useful to clarify whether *H. rhamnoides* constitutes a deviating species to the general pattern of triggering of growth cessation by long nights. Such a study must be done under strictly controlled conditions.

Published information concerning the genetic parameters for quantitative traits in *H. rhamnoides* is limited. The objective of the present study is to study the genetic variation in a number of characters of adaptive importance of SBT hybrids, by growing seedlings under controlled climatic conditions in a growth chamber. The characters are height growth, number of days to achieve 90 per cent of total growth (D90), earliness in flushing and degree of bud damage caused by frost. In addition, variation in quality characters like stem straightness and number of thorns were assessed.

MATERIALS AND METHODS

The seeds for the study originated from a 3×6 factorial mating carried out in the Arkhangelsk Dendrological Garden in 1989. Seventeen of these 18 families had seeds enough for our experiment. The three females of *H. rhamnoides* spp. *mongolica* were selected in Siberia (Khakasiya) and the six males *H. rhamnoides* spp. *rhamnoides* were selected in the Åland archipelago, Finland. In addition, open-pollinated seeds from the three females were included in the experiment. The Siberian female plants are characterized by large fruit size, absence of thorns but low frost tolerance under Arkhangelsk weather conditions. The Finnish male plants are characterized by good winter hardiness, but high thorniness and small fruits.

Experiments were done under controlled conditions in climate chamber in the Department of Forest Genetics Swedish University of Agricultural Sciences. The cultivation conditions are given in Table 9.1. After germination the seedlings were transplanted to pots and randomized in a block design, 10 blocks × 32 seedlings. A second transplanting to larger pots was done after 10 weeks.

Plant height was measured once a week after the first transplantation of the seedlings and three times a week after the second transplantation. Number of days to reach 90 per cent of final height was interpolated from the individual growth curves.

The stage of bud flushing during the second growth period was assessed every day after plants were transferred to continuous light, using a scale where: 0–dormant bud; 1–buds slightly swollen; 2–buds swollen, scales still closed, brown-green; 3–burst of bud scales, tips of leaf emerging; 4–leaf emerging from the buds and shoot growing; 5–leaf elongation; 6–full leaf, new buds developing.

Freeze testing was carried out during week No 20. One lateral shoot from eight plants of each family were cut and put into separate plastic bags where they remained sealed throughout the testing. This freeze testing of shoots was performed in darkness in freezing chamber under different conditions at -7°C and -17°C, respectively. Control plants were kept at +4°C. The treatment started from a temperature of +10°C and temperature was then slowly reduced by 3°C per hour until final temperature was reached and then maintained for four hours. After the freeze treatment temperature was gradually increased to +10°C. The shoots were stored at +10°C during three days and then visual frost damage was assessed by counting damaged buds. Since no damage was induced by +4°C, only data from -7°C and -17°C were included in the analysis of variance. A separate analysis of bud damage for the -17°C treatment alone was carried out, as well.

The quality characters like thorniness, number of lateral shoots and stem straightness were assessed at the end of experiment. The following scales and methods were used:

1. Stem straightness was estimated after growth cessation, in grades from 1 to 3, where 1– straight, 2–slightly curved and 3–curved.
2. Presence of thorns was recorded according to a scale where: 0 - no thorns, 1 – single thorns, 2 – several thorns.
3. The number of lateral shoots was recorded since it is connected with thorn presence. The more lateral shoots, the higher the thorniness.

Table 9.1: Cultivation Conditions and Subdivision of the First Growth Period into Sub-periods

	Cultivation Conditions			
	Week No	*Night Length (h)*	*Temperature (°C)*	
			Day	*Night*
Germination	1–2	0	20	20
Transplantation 1*				
Sub-period 1 begins	3–5	0	20	20
	6	3	20	20
	7–10	Gradual increase of night by 1h/week until 11 h night	20	20
Transplantation 2**				
Sub-period 2 begins	11–14		20	20
Growth cessation	15–19		20	10
	20–21		10	10
Dormancy period	22	16	4	4
	23–26	16	2	2
Breaking of dormancy and bud flushing	27–30	8	20	10

* A first transplantation was done during the first few days of week 3.

** A second transplantation to larger pots was done during the first days of week 11.

Statistical Methods

Analyses of variance were used to estimate causes of variation in individual characters according to the following model:

$$Y = \mu + b_j + f_k + m_l + (f\times m)_{kl} + å_{jkl}$$

where,

Y = individual plant value

µ = population mean

b_j = effect of box (j=1,.,10) (J=1,.,20 after second transplant)

f_k = effect of female (k=1,.,3)

m_l = effect of male (l=1,.,6)

$(f\times m)_{kl}$ = interaction female × male

$å_{jkl}$ = residual error

When frost damage was analyzed, effect of treatment was included in the model:

$$Y = \mu + t_i + b_j + f_k + m_l + (f\times m)_{kl} + (b\times f\times m)_{jkl} + (t\times f)_{ik} + (t\times m)_{il} + (t\times f\times m)_{ikl} + å_{ijkl}$$

where,

Y = individual plant value

µ = population mean

t_i = treatment (i=1,.,2)

b_j = effect of box (j=1,.,10) (J=1,.,20 after second transplant)

f_k = effect of female (k=1,.,3)

m_l = effect of male (l=1,.,6)

$(f\times m)_{kl}$ = interaction female × male

$(b\times f\times m)_{jkl}$ = interaction box x female × male

$(t\times f)_{ik}$ = interaction treatment × female

$(t\times m)_{il}$ = interaction treatment × male

$(t\times f\times m)_{ikl}$ = interaction treatment × female × male

$å_{ijkl}$ = residual error

Variance components were estimated using the same model, with all effects except treatment considered to be random. SAS statistical package was used for the GLM and VARCOMP procedures (1982).

RESULTS

Figure 9.1 illustrates the large variation among females and males as regards mean percentage of damaged buds after freeze testing at -7°C. The family performance as regards bud damage after freeze testing at -17°C caused significantly higher frost damage than exposure to -7°C.

Regression analysis showed that there was a weak, although non-significant relationship between date of growth cessation and frost damage. Figure 9.2 and Table 9.2 reveal that the additive variance components (CV_A) for females were mostly large for growth and phenology traits. For several of the traits there was a significant female x male interaction. All CV_As for frost damage after exposure to -17°C were close to zero. Bud flushing was recorded at five occasions and showed significant female differences except for one occasion. The largest variance component was obtained at the end of flushing, amounting to 25.1 per cent. There was also a large variation in time of bud flushing between families. Generally the female variance components were a few times larger than both the male and the female×male components.

For the quality characters, thorniness, number of lateral shoots and stem straightness, little or no genetic influence was observed.

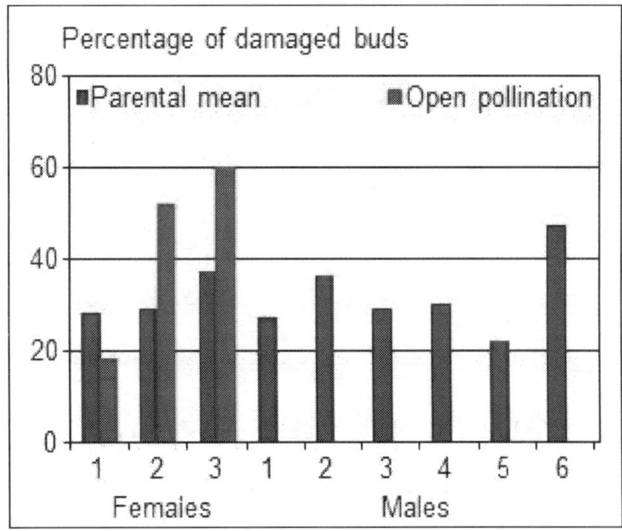

Figure 9.1: Mean Percentage Damaged Buds after Freeze Testing at - 17°C of 18 Full-sib Families from a 3x6 Factorial Mating as well as Mean Percentages for Open Pollination of the Three Female Clones Included in the Mating.

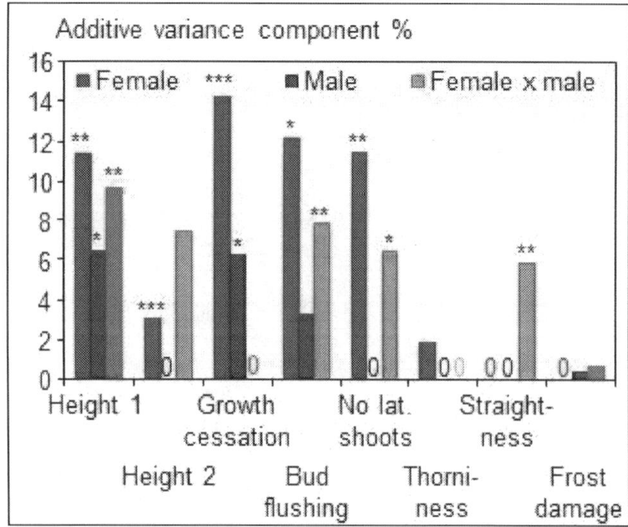

Figure 9.2: Additive Variance Components Estimates for Female, Male, and Female x Male Interaction in Growth Chamber Study of a 3x6 Factorial Mating.

Table 9.2: Mean Values for Bud Damage (-17ºC)

		Male clones						
		1	*2*	*3*	*4*	*5*	*6*	*Open Pollination*
Female clones	1	0.202		0.000	0.376	0.510	0.386	0.176
	2	0.378	0.288	0.312	0.172	0.635	0.439	0.601
	3	0.230	0.293	0.365	0.360	0.274	0.241	0.516

DISCUSSION

Pilot experiment was conducted in present study in controlled conditions, which was not reported earlier. Our results confirm the observations made by Nilov (1986) that long nights do not trigger the growth cessation in *H. rhamnoides*. To obtain growth cessation, the plants must be exposed to low temperature. This is a contrast to the statement by Rousi (1965) that 'the termination of activity in the fall took place in the exact order from north to south', which means that a photoperiodic triggering is important for growth cessation. Our observation also contrasts with the majority of other tree species from the temperate zone. A provenance study under strictly controlled conditions would be useful to get an unequivocal answer to the roles of photoperiod and temperature for growth cessation.

Limited breeding stock samples of SBT were studied in the present study because of practical limitations. Nevertheless, a significant genetic variation in several characters was detected. In spite of the lower number of females than males, the variance component for the female for most traits was few times larger than the male effect (Table 9.2 and Figure 9.2). This suggests that there is a strong maternal effect in these young plants and the true genetic female variance component is presumably far less than shown in Figure 9.2. This is supported by the large drop of the female variance component for plant height from sub-period 1 to sub-period 2. This maternal effect is supposed to originate from the rather large seed size. If our interpretation is correct, it also suggests that open pollinated progenies will not give good estimates of genetic parameters. Based on our data, genetic parameters should be obtained from matings, in which a few females are mated to a large number of males and the genetic parameters should exclusively be derived from the males. Alternatively, individual seed weights must be recorded and used as covariates in the parameter estimations, which though is labor demanding.

The early study of genetic differences in winter hardiness of SBT was done in the Arkhangelsk Dendrological Garden on a limited material (Demidova, 1989). However, there were significant female and male effects for this character but the main factor which had strong influence on frost tolerance was winter weather conditions. In the present study, we were unable to detect any significant genetic effect in frost tolerance after growth cessation. Neither did a regression analysis show any significant relationship between point of time for growth cessation and frost damage.

Besschetnov (1987) reported that bud flushing in SBT was under genetic control. He reported big differences in flushing in plants from different provenances. The significant genetic influence on bud flushing in our study confirms his conclusion.

We did not detect any genetic influences on thorniness. However, Kalinina and Panteleeva (1983) pointed out that usage of highly thorny plants in hybridization gives a thorny progeny, thus suggesting dominance for this trait.

Presence of lateral annual shoots makes berry harvesting difficult. This trait correlates with thorniness (Kalinina and Panteleeva, 1983). In this study we detected significant female and family genetic effect on variation in this quality character.

Our data suggest that it is worthwhile to carry out more comprehensive matings to get more precise estimates of genetic parameters. Such estimates are needed for an efficient breeding of *H. rhamnoides.*

The number of parents is far too low to allow any precision in estimates of genetic correlations between different traits. Such estimates are indispensible for an efficient breeding program for any species.

CONCLUSION

These results of present study indicate that SBT seedlings do not terminate growth at long nights, but low temperature treatment is needed. This lack of photoperiodic response stands in contrast to the majority of tree species in the temperate region. In this experiment, SBT reveals high estimates of genetic variance for growth and growth rhythm. The female variance component for the characters studied was higher than the male component, probably owing to influence from maternal inheritance and seed size. The results suggest that estimation of genetic parameters in SBT shall be based upon controlled crosses rather than on open pollinated progenies. The observed maternal effects indicate that seed weight may be a significant covariate in analyses of variance. The freezing test used in this experiment did not disclose any major genetic variance components responsible for the variation in fraction of damaged buds after freezing. Our data suggest that it is worthwhile to carry out more comprehensive mating to get more precise estimates of genetic parameters. Such estimates are needed for an efficient breeding of *H. rhamnoides.*

REFERENCES

1. Attrey, D. P., Singh, A. K. and Katyal, J. 2011. Effect of solvent strength on total phenol content and DPPH scavenging activity of seabuckthorn extract. In: *Proceedings of the 5th International Seabuckthorn Association Conference*, Xining, China, pp. 67. (Abstract)
2. Baoli, A. and Hongjum, Y. 2011. The analysis on stress resistance of seabuckthorn and cultivation counter measures in the soft sand stone area. In: *Proceedings of the 5th International Seabuckthorn Association Conference*, Xining, China, pp. 54. (Abstract)
3. Bartish, G., Jeppson, N., Bartish V. and Nybom, H. 2000. Assessment of genetic diversity using RAPD analysis in a germplasm collection of seabuckthorn. *Agricultural and Food Science in Finland* 9: 279-289.
4. Besschetnov, V.P. 1987. Specific of seabuckthorn phonology of different geographical provenances in connection with adaptation processes In: *Biology, Selection and Agro Technique of Fruit and Berry Cultures.* Gorkii, Russia, pp. 14-21. (In Russian)
5. Bond, G., Fletcher, W.W. and Ferguson, T.P. 1954. The development and function of the root nodules of *Alnus, Myrica* and *Hippophae. Plant and Soil* 5: 309-323.
6. Darmer, G. 1948. Zur Siedlungsgeschichte des Sandorns *(Hippophae rhamnoides* L.) in Kustengebiet der Ostsee. *Forschungen und Fortschritte* 24: 40-42.
7. Darmer, G. 1952. Den Sandorn als Wild und Kulturpflanze, Leipzig, p. 89.
8. Demidova, N. 1989. Winter hardiness seabuckthorn forms selection for the cultivation on the North. In: *Introduction Questions of Economical Valuable Tree Species to the European North,* Arkhangelsk, pp. 54-62. (In Russian)
9. Demidova, N. 1993. Fruitage of *Hippophae rhamnoides* L. in Arkhangelsk conditions. In: *Proceeding of the International Symposium on Seabuckthorn* (*Hippophae rhamnoides* L.), Novosibirsk, pp. 195-199.
10. Demidova, N. 2003. Research on development of seabuckthorn varieties in the conditions of Russian North. In: *Seabuckthorn* (*Hippophae* L.) - *A Multipurpose Wonder Plant. Vol. I. Botany, Harvesting and Processing Technologies* (V. Singh, Ed. in Chief, 2003), Indus Publishing Company, New Delhi, India, pp. 125-136.
11. Demidova, N. 2009. Review on seabuckthorn research in the Russian Federation and New Independent States (NIS). In: *Proceeding of the 3rd International Seabuckthorn Association Conference,* Laval Universito, Canada, pp. 9-30.
12. Demidova, N.A. and Nilov, V.N. 1992. Introduction of seabuckthorn in the North-West of Russia. In: *Proceeding of the International Symposium*, Bratislava, Russia, pp. 286–293.
13. Ercisli, S., Orhan, E., Yildirim, N. and Guleray, A. 2008. Comparison of seabuckthorn genotypes (*Hippophae rhamnoides* L.) based on RAPD and FAME data. *Turkish Journal of Agriculture and Forestry* 32: 363-368.
14. Eriksson, G., Ekberg, I. and Clapham, D. 2007. An Introduction to Forest Genetics. SLU, Uppsala, 185p.
15. Eriksson, G. and Lundkvist, K. 1986. Adaptation and breeding of forest trees in boreal areas. In: *Proceeding of the Frans Kempe Symposium in Umea,* June 10–11, *Report Dept. of Forest Genetics and Plant Physiology* 6: 67-80.
16. Garanovich, I.M. and Shpitalnaya, T.V. 2014. Role of intraspecific variability in introduction of seabuckthorn (*Hippophae rhamnoides* L.) in Belarus. In: *Seabuckthorn-A multipurpose Wonder Plant Vol. IV. Emerging Trends in Research and Technologies* (V. Singh Editor in Chief 2015), pp. 107-139, Daya Publishing House, 617p.
17. Gatin, G.I. 1963. *Seabuckthorn*. Moscow, 157p. (In Russian)
18. Jeliseev, L.P. 1980. Results and prospects of seabuckthorn introduction in middle zone. In: *Biological Basis of Productivity Increase and Protection of Plant Community*, Gorkii, pp. 121-129. (In Russian)
19. Jeppsson, N., Bartish, I.V. and Persson, H.A. 1999. DNA analysis as a tool in seabuckthorn breeding. In: *Perspectives on New Crops and New Uses.* J. Janick (ed.), ASHS Press, Alexandria, VA, pp. 338-341.
20. Kalinina, L.P. and Panteleeva, E.I. 1983. Seabuckthorn. In: *Progress of Fruit Cultures and Grape Selection*, Moscow, pp. 177-189. (In Russian)
21. Kallio, H., Yang, B. and Halttunen, T. 2005. Flavonol glycosides of berries of three major seabuckthorn subspecies, *Hippophae rhamnoides* ssp. *rhamnoides* ssp. *sinensis* and ssp. *mongolica.* In: *The Proceedings of Invited Speeches of the Second International Seabuckthorn Association Conference,* Beijing, China, pp. 29-35.
22. Letchamo, W., Molnar, T. and Funk, C.R. 2007. Ecogenetic variations in biological activities of seaberry (*Hippophae rhamnoides*). In: *Proc IS on Medicinal and Nutraceutical Plants, Acta Hort.* 756: 229-244.
23. Matafonov, I.I. 1983. Seabuckthorn. Novosibirsk, Nauka, 164p. (In Russian)
24. Meybeck, A. 2011. Cosmetic use of *Hippophae rhamnoides* winter twigs extract to lighten skin pigmentation. In: *Abstract Proceedings of the 5th International Seabuckthorn Association Conference,* Xining, China, pp. 85-86.

25. Morsel, J.T. and Steen, S. 2005. Alternative methods and technologies for seabuckthorn processing. In: *The Proceedings of Invited Speeches of the Second International Seabuckthorn Association Conference*, Beijing, China, pp. 43-49.

26. Nilov, V.N. 1986. Seabuckthorn introduction in the Dendrological Garden. In: *Questions of Artificial Forest Regeneration on the European North*, Arkhangelsk, pp. 91-101. (In Russian)

27. Pandurangan, N., Bose, C. and Banerji, A. 2011. Isolation and synthesis bioactive flavonols from Indian seabuckthorn. In: *Abstract Proceedings of the 5th International Seabuckthorn Association Conference*, Xining, China, pp. 83-84.

28. Pearson, Ì.C. and Rogers, J.A. 1962. Biological Flora of the British Isles *Hippophae rhamnoides* L. *Journal of Ecology* 50: 501-513.

29. Ranwell, D.S. 1972. The management of seabuckthorn (*Hippophae rhamnoides* L.) on selected sites in Great Britain. Report of the *Hippophae* Study Group (The Nature Conservancy). *Mimeographed* 55.

30. Rongsen, L. 1990. Seabuckthorn resources and it's under exploited potential in the Himalayan Region. *Mountain Farming System Series No.* 12.

31. Rousi, A. 1965. Observations on the cytology and variation of European and Asiatic populations of *Hippophae rhamnoides*. ***Annales Botanici Fennici*** 2: 1-18.

32. Rousi, A. 1976. Changes during 35 years in the range of *Hippophae rhamnoides* in the archipelago of Rauma on the Gulf of Bothnia coast. Department of Botany, University of Turku, 232: 1-5.

33. Ruan, C.J. and Li, D. 2002. Role of seabuckthorn (*Hippophae rhamnoides* L.) in improving eco-environment of Loess Plateau of China. In: *Seabuckthorn-A Resource for Health and Environment in Twenty First Century* (V. Singh and P.K. Khosla, Editors), New Delhi, India, pp. 44-53.

34. SAS User's guide: Statistics. 1982. SAS Institute Inc, USA.

35. Sawhney, R.C. 2011. Seabuckthorn for protection against high altitude stress. In: *Abstract Proceedings of the 5th International Seabuckthorn Association Conference*, Xining, China, pp.79-80.

36. Shan, J. 2014. Studies on genetic variation in hybrid offspring's of large fruit seabuckthorn forms. In: *Seabuckthorn-A multipurpose Wonder Plant Vol. IV. Emerging Trends in Research and Technologies* (V. Singh Editor in Chief 2015), Daya Publishing House, pp.140-145.

37. Shchukin, S. 1850. Seabuckthorn berry and the newly discovered oil. In: *Works of the Free Economic Society*. 2: N 4: 41-48. (In Russian)

38. Simon, G. A., Tataru, E., Nicoleta, C., Georgiana, D. C., Ecaterina, P. C., Viorel, R., Luminita, R. and Veronica, S. 2012. The assessment of the genetic diversity of seabuckthorn populations from Romania using RAPD markers. *Romanian Biotechnological Letters*, University of Bucharest, 17 (6): 7749-7756.

39. Singh, V. and Zubarev, Y. 2014. Breading strategies of Russian seabuckthorn (*Hippophae rhamnoides ssp. mongolica*) varieties and their global introduction. In: *Seabuckthorn (Hippophae L.)*: A *Multipurpose Wonder Plant. Vol. IV: Emerging Trends in Research and Technologies*, Daya Publishing House, India, pp. 71-88.

40. Sizemskaya, M. and Sapanov, M. 2005. Ecological conditions of seabuckthorn growing in semidesert area of the Pricaspian Plain in Russia. In: *Abstracts of the Second International Seabuckthorn Association Conference*, Beijing, China, pp. 95-96.

41. Skogen, A. 1972. The *Hippophae rhamnoides* alluvial Forest at Leinora, Central Norway. *Kunglige Norske Videnskabers Selskab Skrifter* 4: 1-115.

42. Tang, X. and Tigerstedt, P.M.A. 2001. Inheritance of flowering, maturity, fruit yield, and winter hardiness of seabuckthorn (*Hippophae rhamnoides*). *J. Amer. Soc. Hort. Sci.* 126: 744-749.

43. Trofimov, T.T. 1978. Seabuckthorn vegetation in natural conditions. In: *Seabuckthorn*. Moscow, Lesnaja promishlennost (Forest Industry), pp. 16-25. (In Russian)

44. Vasilchenko, G.V. 1977. Environmental basis of seabuckthorn culture. In: *Vitamin Plant Resources and there Usage*, Moscow, MGU, pp. 243-245. (In Russian)

45. Yang, B., Zheng, J. and Kallio, H. 2011. Influence of origin, harvesting time, and growth location on contents of inisitols and methylinositols in seabuckthorn berries. In: *Abstract Proceedings of the 5th International Seabuckthorn Association Conference*, Xining, China, pp.57-58.

46. Yao, Y.M. and Tigerstedt, P.M. 1993. Isozyme studies of genetic diversity and evolution in *Hippophae*. *Genetic Resources and Crop Evolution* 40: 153-164.

Chapter 10

A Study on the Geographical Variation of Features of the *Hippophae rhamnoides* L. ssp. *mongolica* Rousi in Mongolia

Ts. Enkhtuul [1]* and Ts. Tsendeekhuu[2]

[1]Mongolian University of Science and Technology, School of Industrial Technology, Ulaanbaatar, Mongolia,
[2]National University of Mongolia, School of Science, Ulaanbaatar, Mongolia
(*E-mail: dug_tuul@yahoo.com, enhtuul_ts@must.edu.mn)

SUMMARY

In this study, the variability of phenotypic features and their geographic variation among Mongolian populations of *Hippophae rhamnoides* L. ssp. *mongolica* Rousi were analyzed and compared. Mongolia is a native range of this subspecies. Also, the Mongolian seabuckthorn has disjunction distribution, occurring in different natural zones of Mongolia and along valleys of eight rivers. *Hippophae rhamnoides* L. ssp. *mongolica* Rousi is a one of the indigenous wild plant of Mongolia. We conducted studies of the morphological features in different natural zones, where desert, semi desert (an annual precipitation 100-150 mm, in regions of Dzungaria Gobi and Creat Lakes depression, Bulgan, Zavhan, Khovd, Tes river) and forest steppe (an annual precipitation 300-400 mm, Selenga, Orkhon rivers) in regions of Mongol daurian and western part Khentei mountain. The vegetative characters of height are for female woody shrubs- 237±0.6sm, CV 21.3 per cent ; age12.1±0.3, CV-29.5 per cent ; length for 2 years shoots 14.2±0.2sm, CV-19.9 per cent ; leaf width 5.97±0.2sm, CV-17.8 per cent. The generative characters of fruits are (weight 100 fruits 30.66±0.08g, length and width 9.25±0.09 sm, 7.58±0.06sm; oval shape frequency 0.17±0.02; spherical-0.56±0.1; cylindrical-0.27±0.04; red color frequency 0.9±0.02; yellow-0.54±0.08; orange- 0.40±0.05) and seeds (weight 1000seeds 13.84±0.08 g; length, width and thick-4.80±0.09mm, 2.64±0.08mm and 1.71±0.04mm respectively, lengthy oval egg shape frequency-0.19±0.05; oval egg-0.54±0.08; egg-0.37±0.06; dark brown color frequency 0.15±0.04; brown egg 0.67±0.09; light brown 0.28±0.08. Coefficient variation (CV per cent) was high for qualitative features (40.65-63.70) and also a portion of broad heritability (H^2 0.51-0.66) than quantitative features. The fruit peduncle varies 2.1±0.01-4.06±0.8 (mean 3.35±0.09) and was more variable in desert and desert steppe zones. The smaller fruits have high oil content (3.7-7.2 per cent) than large fruits (2.4-5.8 per cent) that be confidence report. The morphological characters ssp. *mongolica* in Mongolia distinguished from ssp. *turkestanica* (Singh *et al.*, 2011), and also seed features different from ssp. *turkestanica* ssp. *caucasica* and ssp. *rhamnoides*.

Keywords: *Indigenous population variation, Correlation, Heritability, Phonetic approach and Genetic resource.*

INTRODUCTION

Hippophae rhamnoides ssp. *mongolica* Rousi is one of the Mongolian indigenous rare wild plant because its native locality is Mongolia (Shan *et al.*, 1998), although its distribution range occurs with comparatively limited in adjacent regions: China (Altai mountain), Fast Kazakhstan and Russia (Altai territory, Tuva republic, Irkutsk territory and Buriat Republic) respectively. The subsp. *mongolica* has spread over almost the entire territory of Mongolia where occurs on river lands. In addition, the distribution map ssp. *mongolica* firstly, has been presented in which most of their population range belongs to the western part Mongolia (Figure 10.1).

In Mongolia, the vegetation zones changed, and exactly directed from north to south gradually transferring mountain forest taiga, forest steppe, steppe, desert steppe and desert. According to the zones the Mongolian vegetation divided on 16 botanic-

Seabuckthorn (*Hippophae* L.): *A Multipurpose Wonder Plant, Vol. 5: Advanced Technologies for Cultivation, Processing, Health Protection and Environmental Conservation* (Virendra Singh, *Editor-in-Chief*, 2018), p. 71–78, Astral International Pvt. Ltd., New Delhi, India.

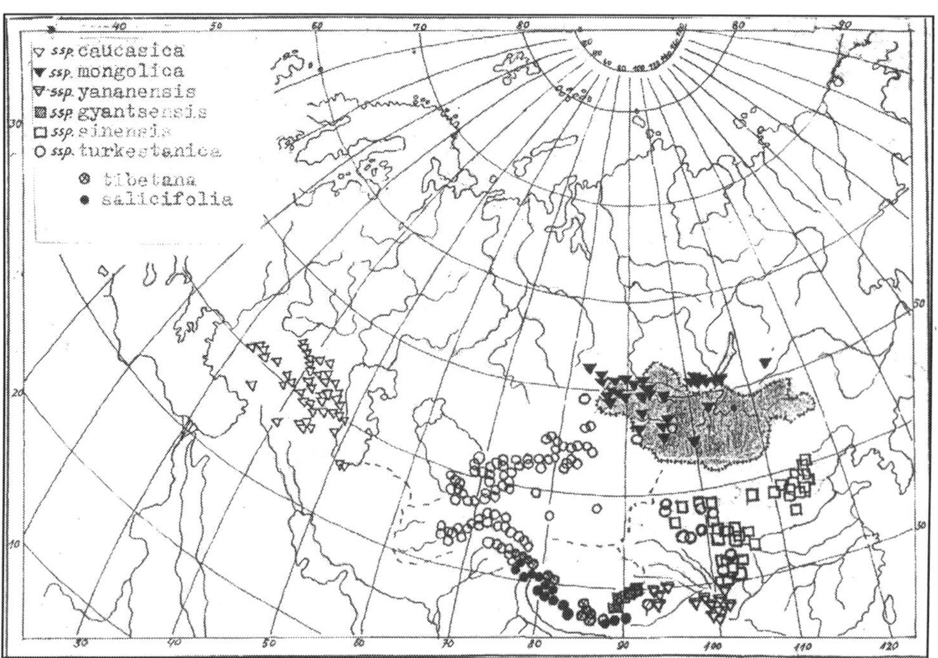

Figure 10.1: The Distribution of the Asiatic including Caucasian Taxa of the Seabuckthorn Specimens Studied (Rousi, 1971).

geographical regions (Grubov, 1982). In this connection, occurrence of the spp. *mongolica* populations within its Mongolian distribution territory located mostly in Western part Mongolia along the valley of rivers follows: Bulgangol, Hovdgol, Bukhmurungol, Zavhangol, (U) Tesgol, Bogd-Borhgol, (Z) Tesgol, and Selengemurungol and others (Figure 10.2).

The botanic-geographical regions to concern ssp. *mongolica* are mostly areas Great Lakes Depression desert steppe, Dzungariangobi desert, Hovd mountain desert steppe, with an annual average precipitation 100-150 mm, and other small area a Mongoldaurian mountain forest steppe (Selenge river Basin) with an annual precipitation 300-400 mm. Those regions isolated by Altai and Khangai mountain ranges; therefore the habitat of wild seabuckthorn is difference (Figure 10.1). The seabuckthorn forest in Mongolia accounted about 23000 ha, and cultivated plantations are about 2000 ha, because some seabuckthorn processing factories mostly used the fruits from natural populations for example

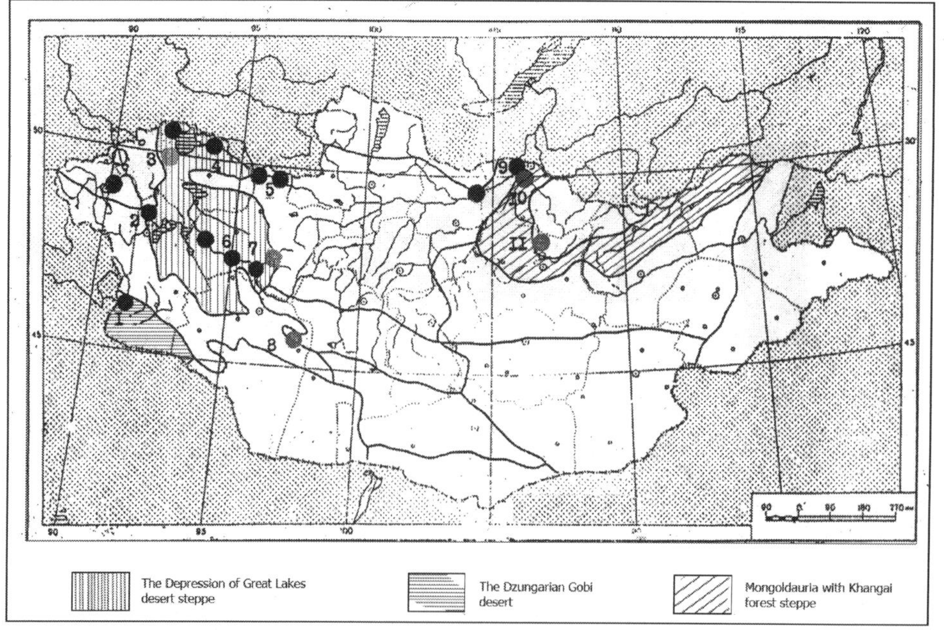

Figure 10.2: Map of the Study Area and the 11 Sites with their Location at the Natural Populations of *ssp. mongolica* within their Mongolian Range: 1. Bulgangol 2. Hovdgol 4. Tesgol (U) 5. Tesgol (Z) 6. Zavhkangol 7. Bogd-Borhgol 9. Selengemurun Cultivated plantations: 3. Ulaangom 7. Bogd-Chigestei 8. Biger 10. Shaamar 11.Sognogor-Bayangol.

Zavhan Province in West Mongolia. The seabuckthorn destructions at natural conditions has caused a series of arid climate conditions such as floods, drought, wind and desertification process, finally, the human's activity use, pasture pressure.

Naturally distributed *Hippophae rhamnoides* is a very considerable genetic recourse in World including ssp. *mongolica.* Its different peculiar range and highly fragmented occurrence and the location of natural populations ssp. *mongolica* between north edge Central Asia and South borderline of Siberian taiga would be great interest in field of the variability inter and middle populations level, of this more polymorphic subspecies.

The polymorphic appearance and morphological variability and correlation analysis of characteristics, seasonal variation, nutritional substances and geographic spatial distribution of genotype. Characters of seabuckthorn populations with their frequency and directional variation of geographical maps within population range Mongolian range has been investigated.

MATERIALS AND METHOD

For description of vegetative and generative features in plants at various natural populations seabuckthorn of Mongolia used the method name. The questionnaire was prepared for 16 questions on seabuckthorn (Kondraschov, 1982).

Variability of characters and their frequency of phenotypical class intervals and their geographic distribution with phenol geographic spatial expansion are based on the latest statistics and phenotypic analysis (Lakin, 1990; Yablokov, 1980). The polymorphism appearance was determined followed Jivotovskii (1982).

The broadsense-heritability in comparison, between populations (Lim_p) and clone (Lim_{cl}) where e^2-elogolical factor and H^2- the share of genetically determination has been defined by Rone (1978). The criteria of share genetical characteristics: if $e^2=0$, $Lim_p=3Lim_{cl}$, $H_2=1$, and $e^{2'''}0$, and limp=$3Lim_{cl}$, $H_2\geq0.4$ that the share of genetical determination is one of pair allele, respectively. If $e^2=0$, $Lim_p<3Lim_{cl}$, $H^2=1$, and $e^{2'''}0$, $lim_p \sim Lim_{cl}$, $H_2\leq0.4$ that the share of genetical determination is allele system of different gene. The oil content defined by Ruschkovskii and starch content-by acid hydrolysis (Pleschkov, 1988).

RESULTS AND DISCUSSION

Variability of the qualitative and quantitative features such as weight, size, shape, color and other characters of seabuckthorn leaves, branch, fruit and seed, comparatively in different populations, and their vegetation zones has been analyzed by biometrical statistic (Figure 10.3).

We have determined the polymorphic degree of weight, in comparison, for fruit and seed, subsp. *mongolica* populations and compared (Table 10.1).

According to the importance of genetically depend characters for selection of seabuckthorn, the broad sense-heritability (H^2) characters of fruit and seed *subsp. mongolica* has been attempt determined.

Oil produced character of seabuckthorn fruit depending from broad sense-heritability (H^2) is different among the populations that it is high (more than of 0.50) in desert steppe, desert regions, than of forest steppe (Table 10.3).

There is an observation that in Mongolia sheep cannot or will not go into areas of seabuckthorn habitat with very high density, and only graze the edge of habitat (Tomas S.C. Li, 2003).

Unfortunately at present the high nutritional and medicine values of the fruit have led to uncontrolled exploration especially in period market economy Mongolia, even destruction of seabuckthorn wild populations in some part its natural distribution. Move

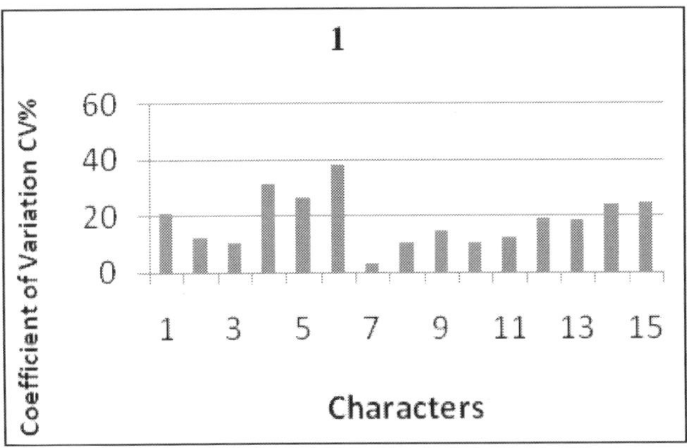

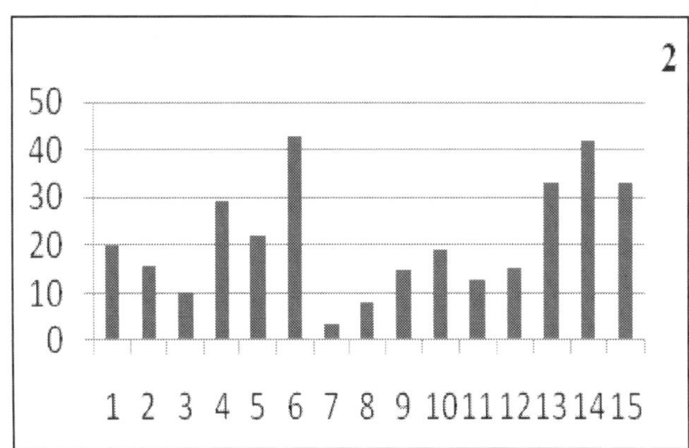

Figure 10.3: The Variation Level of different Characters of Seabuckthorn ssp. *mongolica*.

1. Desert and desert steppe populations: Bulgangol, Hovdgol, Zavkhangol, Bogd-Borhgol, Tesgol (u); 2. Forest steppe populations: Bogd-Chigestei, Selengemurun, Sognogor-Bayangol.

Fruit: 1-weight 2-length 3-diameter 4-length of pedicel 5-pedicel separated force (H) 6- fruit yield, Seed: 7-weight 8-length 9- wide 10-thickness, leaves: 11- length 12- wide, branch: 13-length of two-year-old branch 14-number of two-year-old branch 15-number of fruit on the two-year-old branch.

Table 10.1: Polymorphic Appearance by Weight of Fruit and Seed: µ-number of phenotype h-number of rare phenotype

Populations	Phenotype Class	Fruit		Seed	
		$\mu \pm S_m$	$h \pm S_h$	$\mu \pm S_m$	$h \pm S_h$
Bulgangol	6	4.63±0.8	0.24±0.1	3.85±0.9	0.77±0.1
Hovdgol	6	2.75±1.0	0.54±0.1	1.93±0.9	0.67±0.1
U Tesgol	6	2.95±0.8	0.50±0.14	2.89±0.8	0.50±0.1
Zavhangol	6	4.45±0.4	0.25±0.1	3.68±1.1	0.38±0.2
Bogd-Borhgol	6	3.64±0.2	0.39±0.1	1.55±0.7	0.74±0.4
Selengemurun	6	3.01±0.5	0.59±0.5	2.65±1.0	0.58±0.2
Average	6	3.57	0.41	2.75	0.60

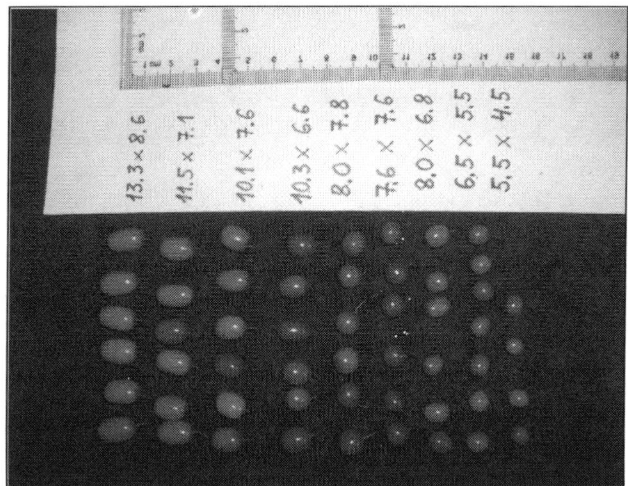

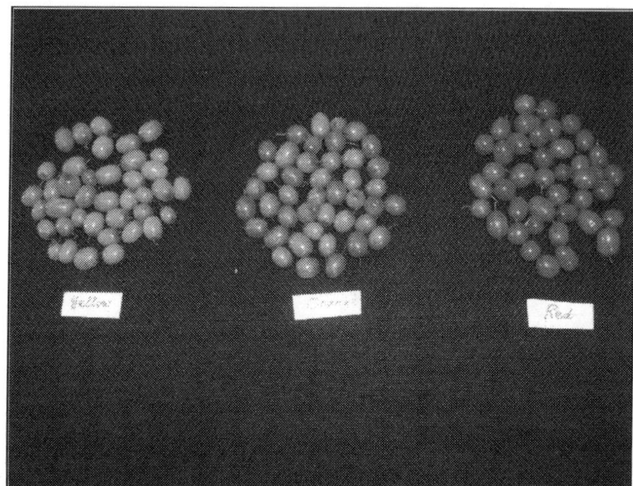

Figure 10.4: Variation in Morphological Characters of different Seabuckthorn Fruits.

Table 10.2: Determination of Broad Sense-Heritability (H^2) some Generative Characters of Seabuckthorn

	Characters	Average of Total Populations				Genetic System of Determination
		E^2	lim_p	H^2	lim_{cl}	
Fruit	Weight	0.75	31.48	0.25	23.66	Allele system of different gene
	Length	0.71	4.95	0.29	3.92	Allele system of different gene
	Diameter	0.66	4.13	0.34	2.76	Allele system of different gene
	Oval shaped	0.49	0.45	0.51	0.18	One of pair allele
	Globe shaped	0.49	0.62	0.51	0.20	One of pair allele
	Cylindrical	0.30	0.31	0.70	0.09	One of pair allele
	Shaped	0.32	0.28	0.68	0.11	One of pair allele
	Red colored	0.46	0.57	0.54	0.24	One of pair allele
	Orange colored	0.40	0.74	0.60	0.09	One of pair allele
	Yellow colored					
Seed	Weight	0.53	7.72.	0.47	4.26	One of pair allele
	Length	0.66	2.52	0.34	2.0	Allele system of different gene
	Wide	0.69	1.58	0.31	1.14	Allele system of different gene
	Thickness	0.79	1.27	0.21	0.86	Allele system of different gene
	Egg shaped	0.63	0.85	0.37	0.59	Allele system of different gene
	Oval-egg shaped	0.54	0.87	0.46	0.52	Allele system of different gene
	Long egg shaped	0.48	0.40	0.52	0.30	Allele system of different gene
	Black brown colored	0.55	0.52	0.45	0.22	One of pair allele
	Brown colored	0.53	0.65	0.47	0.31	One of pair allele
	Light brown colored	0.34	0.31	0.66	0.11	One of pair allele

Table 10.3: Share of Genetical Factor (H^2) for Oil Accumulation Fruit of Seabuckthorn

Natural Zones	*Desert*	*Desert steppe*			*Forest Steppe*		*Total*
Name of Population	*Bulgangol*	*Hovdgol*	*Zavhangol*	*Tesgol (U)*	*Bogd-Chigestei*	*Selengemurun*	
H^2	0.58	0.62	0.72	0.56	0.43	0.38	0.54
Average	0.58	0.64			0.40		

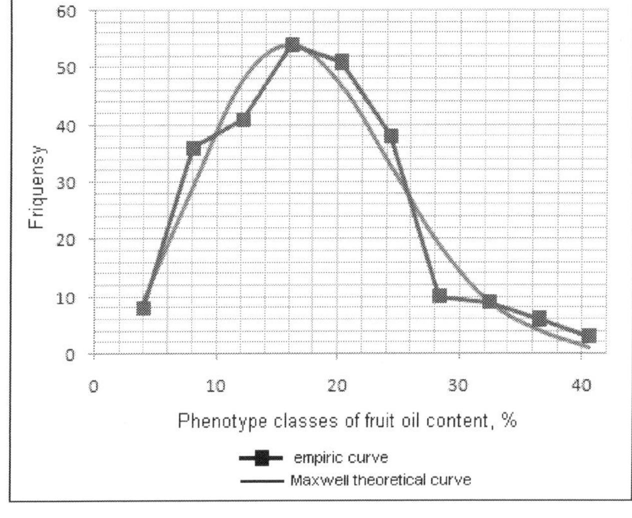

Figure 10.5: Distribution of Variation Intervals of Oil Content in Bulgangol Population.

Figure 10.6: Bimodality Distribution of Oil Content in Fruit Supsp. *mongolica.*

Table 10.4: The Age Variation of Average Oil Content of Fruit (per cent on a dry weight) in Populations of the Seabuckthorn during the 1990-1996

Group of Age	*5-8 age*			*9-12 age*			*Above 13 age*		
Variation	*M±m*	*Variance*	*CV*	*M±m*	*Variance*	*CV*	*M±m*	*Variance*	*CV*
Bulgangol population	14.5	3.1	21.3	16.2	1.9	12.5	23.2	23.2	9.4
Hovdgol population	6.9	1.2	17.3	18.2	2.3	17.4	9.3	3.3	35.5
Biger population	7.8	1.8	24.2	21.8	2.5	21.1	8.8	1.9	21.5
Selengemurun population	17.4	5.7	32.7	22.5	2.7	12.2	14.2	3.9	21.4
Average per cent	11.6±1.5	23.8		19.4±1.0	15.8		13.8±0.9	21.9	

Table 10.5: Seasonal Dynamics of Oil Content Accumulation with Starch Content in Seed of Mongolian Seabuckthorn during 1990-1996

Seasonal date	25.VI	05.VIII	15.VIII	25.VIII	20.IX	20.X	20.XI
Oil content, per cent	4.3	12.0	14.5	14.0	13.5	12.5	10.0
Starch content, per cent	6.2	9.5	2.4	2.5	2.7	3.0	4.8

over need a more effective utilization, development protection and preservation of existing natural seabuckthorn resource and also research on subsp. *mongolica* and their relative selective value forms, ecotypes has become urgent.

Variability of Quantity and Qualitative Characters in Seabuckthorn

Interpopulation analysis showed that the size of leaves of is more variable, and mode on frequency distributions curve for their length displacing right and left side hands for wide to the left. Variation of fruit pedicel which very important for yield harvesting of seabuckthorn is a high changeable in interpopulations (1.90±0.1-6.60±0.1mm) but it was a few distinguish between different populations (Figure 10.3).

The large fruit by weight and size was predominated (71.9 per cent) among of all populations; however, the seed size is a low variable than of fruit. Bimodality distribution of variation curve for weight, length, diameter and thickness both of fruit and seed, always recorded particularly in Bulgangol, Zavhangol, Tesgol (U) populations those located in Western Mongolian desert and desert steppe regions.

The pedicel separated force of fruit is a high variable in each ecological population follows: desert-77.2±2.24H, desert steppe-121.0±1.5 H, forest steppe- 146.7±1.54, respectively. Variability level of our choice all characters of subsp. *mongolica* was almost same, comparatively, in desert, desert steppe and forest steppe zones, but each other is different and quantity vegetative characters are more variation than of fruit and seed, because those seems caused by environmental factors (Figure 10.3).

Correlative relations showed that the number of fruit or yield on the individual plant are depend from amounts of weight, length and diameter of this same fruit with negatively slightly linear correlations (-0.4; -0.25 *etc.*), however it depend from number and length of two-year-old branch with mostly positively linear correlations (0.20; 0.75 *etc.*). Those positive correlated yield of fruit was correspond the Kondraschov's report about the length of branch is a most reliable for yield selection a Altai region Siberia (Li, 2003).

There is a large morphological diversity among seabuckthorn mature natural plants within each subspecies (Rousi, 1971). In this connection the polymorphic degree by number of phenotype variation class for weight of fruit and seed respectively, was different in our studied 6 populations in which the degree is a high for abovementioned Bulgangol Zavhangol populations, than of forest steppe region (Table 10.1).

However, in contrast, the potentiality latent polymorphism of the number of rare phenotype is a high for seed than of fruit (h± S_h: 0.60 versus 0.41) in other hand the polymorphic appearance by weigh character is tendency of reach more in fruit, than of seed (M±S_m 3.57 versus 2.75).

For determination of broad sense-heritability (H^2) used the intervals limit of subsp. *mongolica*, first, homozygous clones growing from vegetative individuals, second, heterozygous individuals growing from seed. Broad sense –heritability adaptive value showing that among of all determined characters the length, diameter, oval and globe shaped and orange and yellow colored of fruit and also weight, oval-shaped egg-shaped, black brown and brown colored seeds are predominated (Table 10.2).

According to the broad sense-heritability value the shape and color of fruit and also weight and color of seed are concern mainly a one of pair allele, in opposite, the other quantities and some qualitative characters are a allele system of different gene. Many characters with more level sense-heritability (H^2) we are considered those a unit marker character to use for our phenol geographical study. The broad sense-heritability as of genetically factor for oil accumulation process in seabuckthorn fruit was a high (more than of 0.50) it was predominated in desert and desert steppe regions than of forest steppe (Table 10.3).

Variability of Oil Accumulation in Seabuckthorn

Oil content of seabuckthorn in both fruit and seed as main valuable product for medicine and those average content (without seed) in all studied 8 populations in Mongolia there is a 5.38-34.15 (on a dry weight) and 3.5-6.5 per cent (on a wet weight) versus 1.6-6.6 per cent in Russia (Schischkina, 1978; Skuridina, 1985).

The smaller fruits contain more average oil content (3.7-7.2 per cent) on a wet weight, than of 2.4-5.8 per cent in subsp. *mongolica* populations that be with confidence the report about fruit size is not a major in breeding program because smaller fruit may contain more value, than large fruits (Tomas S.C.L, 2003). Also, there is a report about the oil content in whole fruit 2-21 per cent (w/w on dry basis) and, the average oil content fruit subsp. *rhamnoides* have higher, than those of subsp. *sinensis* (3.5 V 2.1) (Oomah, 2003).

However, the frequency of phenotypically variation intervals of oil content in all 8 populations of subsp. *mongolica* within their Mongolian range has very widely amplitude that is a 4-40 per cent (on a dry weight) and it was correspond the variation curves of the empiric and theoretical distributions (Figure 10.3).The bimodality variation curve of oil content was defined in Bulgangol population from Dzungarian Gobi desert (Figure 10.5) and this phenomenon seems to may be caused from two subspecies subsp. *mongolica* subsp. *turkestanica* mixed in population range subsp. *mongolica* within western part Mongolia that was already draw up on the map of Asian range subsp. *mongolica* (Rousi, 1971). Definition of oil content, in connection with plant age showed that high oil accumulation period in seabuckthorn fruit was coincided with the 9-12 ages of plant, but only Bulgangol population from Dzungarian Gobi distivinqueshed with above 13 age (Table 10.4).

For the subsp. *mongolica*, oil content in fruit soft tissue is increases from beginning of the ripening, but it declines from middle to the end August, before its maximum content that reaches in September to October. A phenomenon was newly discovered. The maximum of oil content in seabuckthorn seed is another. It is accumulated prior than of fruit and the oil content in seed decreased from middle August maximum period rather small to November, then was constant. The seasonal change of seed oil accumulation was depend as against its starch accumulation, simultaneously at same vegetative period during summer and an autumn, thus in both oil and starch contents was interchangeable (Table 10.5).

Also, correlated variability of oil content changeability in seabuckthorn fruit is depend on precipitation temperature and solar radiation during the current growth year has positively relation (r+0.78), but in interest, from of those climate parameters in previously past growth year has depend slightly negative correlation (r–0.20).

Geographic distribution and the spatial expansion of oil synthetic character of seabuckthorn fruit on the phenol geographical map is a specific boundaries.

Frequency of variation intervals with maximum oil content increased in the stripland around the Great Lakes Depression desert steppe, Dzungarian Gobi desert including Hovsgol mountain forest taiga, with the geographical center point of populations subsp. *mongolica* within is Mongolian range (N48°36′18.0″ E 95°34′13.0″). It should be noted we has been revealed the predicted ancient populations by phenol geographic boundaries in Hovsgol region (Tsendeekhuu, 1996) and, for minimum of oil content frequency intervals was another character of spatial distributions. Thus, minimum frequency intervals of oil content spreading from above mentioned maximum intervals center towards to two opposite direction West and East territorial expansion.

To all appearance of our phenol geographical investigations, if we compared with European and Asian subspecies of genus *Hippophae* L. in my assumption that the subsp. *mongolica* is an initially producer of oil accumulation in its fruit during past and present period within their Mongolian range, concretely, around the Great Lakes Depression desert steppe region in where located the geographical center point concern Mongolian and maybe their adjacent territorial populations in Russia and China (Altai).

The ourpredicted assumption will be required in future a multidisciplinary research on pale geographical and paleoclimatology *etc.* with historical aspect.

Results of our previously phenol geographical studies and variability of natural population subsp. *mongolica* were presented in attempt rather widely consideration (Tsendeekhuu, 1996). The study on geographical distribution and spatial territorial expansion of selective and genetically determined characters in seabuckthorn was correspond with the research of Yao and Tigerstedt (1995) concern to the growth and size of seabuckthorn their geographical diversity that important valuable for selection by geographical area (Thomas S.C.Li, 2003).

ACKNOWLEDGEMENT

We are thankful to the colleagues of Mongolian Association of Seabuckthorn and Ministry of Environment and Green Development of Mongolia for their interest in this work.

REFERENCES

1. Dave, O. B. 2003. Seabuckthorn lipids. In: *Seabuckthorn (Hippophae rhamnoides L.): Production and Utilization*, NRC Research Press, Ottawa, pp. 51-68.
2. Grubov, V.I. 1982. Key to the vascular plants of Mongolia (with an atlas). Leningrad, Nauka, p. 441.
3. Jivotovskii, L.A. 1982. Indications of polymorphic characters in intra populations. *Phenetika Population*, Moscow pp. 38-44. (In Russian)
4. Kondrashov, 1982. Correlation relations of characters in different populations of seabuckthorn. Past, Resource, Leningrad, T. XVIII, 1: 66-73.
5. Lian, Y., *et al.*, 1998. Systematic classification of the genus *Hippophae* L. seabuckthorn research. ICRTS, No.1, Xinxing Printing House, Langfang City, Hebei Province, China, 1: 13-23.
6. Lakin, G.F 1990. *Biometriya*. Moscow, Visshayaschkola, 260p. (In Russian)
7. Pleschkov 1988. *Biochemistry of Plants*. Moscow, Bishchayaschkola, 178p.
8. Rone, V.M. 1978. A genetic analysis of natural populations. A selection of forest wood plants, Riga, Zinatme, pp. 86-91. (In Russian)
9. Rousi, A.1971. The genus *Hippophae rhamnoides* L.: A taxonomic study. *Ann. Bot. Fennici* 177-227.
10. Schischkina, E.E. 1978. Biochemical composition of seabuckthorn fruit. *Oblepiha Moscow Lesnaya Promishlen Nosti*, pp. 173-177. (In Russian and English).
11. Solonenko, L.P. and Skuridin, G.M 1983. Comparatively value of method for chemical composition of seabuckthorn. In: *Biology, Chemistry, and Pharmacology of Seabuckthorn*, Novosibirsk, Science, SO, 57-67.
12. Li, T.S.C. and Thomas, H.J.B. 2003. *Seabuckthorn (Hippophae rhamnoides L.) Production and Utilization.* NRC Research Press, Ottawa, Canada, 133p.
13. Tsendeekhuu, Ts.1996. A phenogeographical study and variability of natural populations of Mongolian seabuckthorn (*Hippophae rhamnoides* L. *ssp. mongolica* Rousi). *The D.Sc. Doctoral Dissertation Thesis*, Ulaanbaatar, 256p. (Text in Mongolian, Autoreferat in Russian)
14. Tsendeekhuu, Ts. 1996. An issue on restoration of historical range *Hippophae rhamnoides* L. ssp. *mongolica* in Hovsgol mountain forest region. The Natural resource of Hovsgol region, pp. 67-72. Press of National University of Mongolia, Ulaanbaata. (In Mongolian)
15. Tsendeekhuu, Ts. 1999. Some results of the study for populations *Hippophae rhamnoides* ssp.*mongolica*

Rousi. Diversity of plant Cover within the Baikal region. In: *Proceeding of the International Conference*, Ulan-Ude, pp. 54-55. (In Russian)

16. Tsendeekhuu, Ts. 2001. The Great Lakes Depression region is one of the hearth territory of origin *Hippophae rhamnoides* ssp. *mongolica* Rousi natural resource and conditions of West Mongolia and adjacent regions. In: *Proceedings of IV International Conference*, Hovd-Tomsk, pp. 33-35.

17. Yablokov, A.B 1980. *Phenetica*. Moskow, Nauka, 135p. (In Russian).

Chapter 11

Improvement of Seabuckthorn at AAFC in Canada

Parul Jain[1], Isobel A.P. Parkin[1], Raju Y. Soolanayakanahally[1,2*] and William R. Schroeder[2]

[1]Saskatoon Research Centre, Agriculture and Agri-Food Canada, Saskatoon, SK, Canada
[2]Agroforestry Development Centre, Agriculture and Agri-Food Canada, Indian Head, SK, Canada
(*E-mail: raju.soolanayakanahally@agr.gc.ca)

SUMMARY

Seabuckthorn (*Hippophae* spp.), native species to the Eurasia, is well adapted to the Canadian Prairies. It has mainly been used to prevent soil erosion in shelterbelts since 1970s. The potential of seabuckthorn to be cultivated commercially has been explored over last few decades. As of 2013, more than three million mature seabuckthorn plants can be found throughout the Canadian prairies. Research efforts at the Agroforestry Development Centre (ADC) of the Agriculture and Agri-Food Canada (AAFC), Indian Head, Saskatchewan, have resulted in release of four elite cultivars: Indian-Summer RC-4, Harvest Moon, AC Autumn Glow and AC Prairie Sunset. In addition, AAFC-ADC has the largest *in-situ* seabuckthorn germplasm collection in North America representing 93 accessions sourced from around the world. Metabolite profile of superior seabuckthorn cultivars developed and released by AAFC-ADC has been found to have higher sugar content and lower titratable acidity. Recently, use of seabuckthorn as value added crop has been pursued under alley cropping systems. New research has been initiated to explore the transcriptome, metabolome and propagation strategies to facilitate future breeding and selection processes to develop elite cultivars.

Keywords: *Hippophae, Seabuckthorn, Vitamin C, Flavonoids and β-carotene.*

INTRODUCTION

Seabuckthorn (*Hippophae* spp.) an indigenous Eurasian species, belongs to the *Elaeagnaceae* family with 2n=24 chromosomes. It is dioecious with berries borne on female trees. It is exceptionally tolerant to extreme temperatures (from -43°C to +40°C), drought, and salinity and requires low fertility inputs. In addition, it has nitrogen fixing ability and develops extensive root systems (Schroeder and Yao, 1995; Li, 2003). Taken together, these attributes make seabuckthorn an excellent plant for soil erosion control, marginal land reclamation, wildlife habitat enhancement and farmyard protection (Schroder *et al.*, 2000). All parts of seabuckthorn plant (bark, leaves, berries, and seeds) have either nutritional or medicinal value and have been used in Eurasia for centuries.

Seabuckthorn was introduced to the Canadian prairies from Siberia originally for ornamental planting, but more importantly since the 1970 sit has been extensively used as shelter belt species to control soil erosion (Schroeder *et al.*, 2000). In mid 1950s, the Shelterbelt Centre of the Prairie Farm Rehabilitation Administration (PFRA) of Agriculture and Agri-Food Canada (AAFC), Indian Head, Saskatchewan, obtained rooted cuttings from AAFC Morden Research Station. Seedlings propagated from these cuttings were planted in two hedge rows at the Centre in 1963. This seed strain, named 'Indian-Summer' has been used for propagation of seabuckthorn seedlings planted in shelter belts throughout Canadian prairies. As of 2013 it was estimated that there were over 3 million mature 'Indian-Summer' seabuckthorn plants in shelterbelts across

Seabuckthorn (*Hippophae* L.): *A Multipurpose Wonder Plant, Vol. 5: Advanced Technologies for Cultivation, Processing, Health Protection and Environmental Conservation* (Virendra Singh, *Editor-in-Chief*, 2018), p. 79–83, Astral International Pvt. Ltd., New Delhi, India.

prairies. Because seabuckthorn was an introduced plant species to North America, it was likely that the plant material had low genetic variability and thus a narrow base for breeding populations. Therefore, a breeding strategy was adopted which included the development of a production population concurrently with a longer term breeding population. We broadened the genetic base of the breeding population by the introduction of seabuckthorn species from multiple sources (Schroeder *et al.*, 2000).

Since 2001, the largest seabuckthorn common garden in North America was located at the AAFC-PFRA Shelterbelt Centre (Franklin *et al.*, 2001). This common garden included seed sources and clonal cultivars obtained from Russia, Finland, China, India and Canada. A total of 93 accessions representing four *Hippophae* species (*H. rhamnoides* L., *Hippophae neurocarpa* S.W. Liu and T.N. He, *H. salicifolia* D. Don and *H. tibetana* Schlechtand). In addition, *H. rhamnoides* subspecies *mongolica*, ssp. *rhamnoides*, ssp. *sinensis*, ssp. *turkestanica* and spp. *yunnanensis* were represented in the garden (Schroeder *et al.*, 2000).

Since 1990s, research at AAFC has provided detailed recommendations for seabuckthorn orchard management. Recommended plant spacing is one meter within row and four meters between rows, with a north-south row orientation to provide maximum sunlight (Li and Schroeder, 2003). Since seabuckthorn is a dioecious plant, distribution of male and female plants influences pollination and berry set. Our recommended male: female ratios are 1:6 to 1:8 depending on plant density and region. Selection of superior male and female cultivars was undertaken as this would have appreciable effect on yield and quality of the berry. For male plants, selection criteria include hardiness, vigour and long flowering period to provide ample pollen. For female plants, selection criteria included berry yield, size and mass, pedicle length and seed mass. Orchard pruning is required to promote growth, train branches and facilitate harvesting. Seed, soft and hardwood cuttings, suckers or tissue culture methods can be used for propagation of seabuckthorn. Generally, mature berries are harvested in late August to September by mechanical harvester or by hand.

In collaboration with the Pacific Agriculture Research Station in Summerland, BC, the biochemical profile of 'Indian-Summer' was quantified (Schroeder and Jana, 1999). Chemical composition of mature 'Indian-Summer' seabuckthorn berries indicates that they are a rich source of vitamins (A, C, E and K), organic acids, amino acids, fatty acids, antioxidants and flavonoids. (Franklin *et al.*, 2001; Beveridge, 2003). In collaboration with the University of Western Ontario, Agroforestry Development Centre (ADC, formerly known as Shelter belt Centre) researchers explored fatty acid composition and transcriptome of mature seed to provide a foundation for research on genes and enzymes involved in biosynthesis of poly unsaturated fatty acids (PFUA). These results demonstrated that Canadian grown seabuckthorn seeds contain essential linoleic acid and α – linolenic acid in 1:1 ratiowhile the pulp is rich in palmitoleic acid making seabuckthorn berries a valuable nutrition source for humans (Fatima *et al.*, 2012).

Seabuckthorn growing in Saskatchewan shelterbelts at six geographically diverse locations along with populations comprised of seedlings originating from seed collected from thornless seabuckthorn cultivars were evaluated. Superior genotypes were selected from a population of 25,000 plants. Selection criteria included: berry yield, berry weight, pedicel length, removal force, absence of thorns, harvesting ease and growth habit. Male cultivars with superior pollen production were also selected. Between 1998 and 2000, the top 47 female and 12 male genotypes from these populations were planted in a replicated cultivar trial.In 2002, five female and one male cultivar were selected from this trial (Schroeder *et al.*, 2002). In 2005, the cultivars Harvest Moon and Orange September were named and released (Schroeder *et al.*, 2014; Tebbitt, 2005). Recently in 2012, the cultivars AC Autumn Glow and AC Prairie Sunset were released. These are the first North American developed seabuckthorn cultivars.

Building on these breeding successes, efforts are being made to better understandthe genetic makeup of the Canadian developed seabuckthorn cultivars, targeting key antioxidant pathways. This would provide an important resource where gene discovery could be exploited for breeding strategies to establish seabuckthorn as a new high value berry plant, in a rapidly emerging functional food and nutraceutical industry.

MATERIALS AND METHODS

Plant Material

A seabuckthorn orchard was planted in a replicated design and maintained since 1999 at Indian Head, Saskatchewan. Weed maintenance was done regularly and trees were pruned when required to remove the dead and damaged branches. Three trees per cultivar were randomly selected and berry samples collected in September, 2013 on same day between 08:00-10:00 am. A sub-sample of 100 berries was removed for measurement of berry weight and pedicle length. The remaining berries, about 300 g, were flash frozen in liquid nitrogen and stored at -80°C for metabolite analysis.

Metabolite Analysis

Berry metabolite analysis was done at the ISO 17025-certified laboratory of Labs-Mart Inc (Edmonton, AB, Canada). Briefly, titratable acidity was quantified using the United States Pharmacopeia (USP) titration method. Samples were titrated with sodium hydroxide following the method of USP monograph for acetic acid. β-Carotene (vitamin A) content was determined by spectrophotometry at 452 nm wavelength. For vitamin C measurement samples were treated with metaphosphoric acid to denature and precipitate protein that would interfere with analysis and stabilize vitamin C for estimation by USP Titration. Vitamin E (α-tocopherol) was estimated by high performance liquid chromatography with ultra-violet detector (HPLC-UV) at 290 nm wavelength following modified USP 35 assay for d-α-tocopherol or d-α-tocopheryl method. Total fat from seed and pulp were measured by gas chromatography with a flame ionisation detector (GC-FID) following AOAC official method 954.02. Total flavonoids were quantified by HPLC-UV at 270 nm wavelength. Total sugars were estimated by HPLC with refractive index detectors (HPLC-RI) at 35ºC, positive polarity. All samples were prepared and analyzed in triplicates.

RESULTS AND DISCUSSION

Research efforts at the AAFC-ADC have focused on improving agronomic traits including yield, berry size, berry sweetness, thornlessness or soft thorns, minimal suckering, pest and disease resistance, pedicel length and retention, skin toughness and ripening period (Figure 11.1). The continuous research effort has resulted in release of four superior seabuckthorn varieties (Figure 11.2) with early maturity and bigger berries (Table 11.1).

AAFC research has demonstrated seabuckthorn is well adapted to Canadian prairies growing conditions. From 1970 to 2013 more than three million seedlings were planted throughout the prairies (Figure 11.3). Planting trends for seabuckthorn, were similar to trends for the native species, buffaloberry (*Shepherdia argentea*), but much higher than pincherry (*Prunus pensylvanica*) and red-elder berry (*Sambucus racemosa*).

Prior research has indicated that Indian-Summer (RC-4), has high oil content, 17.8 per cent w/w basis and high fatty acid content, 17.33 per cent seed weight (Schroeder *et al.*, 2000). Metabolite profile comparison of four AAFC developed cultivars is shown in Table 11.2. Harvest Moonhadtotal sugar content of 38,819 ± 272.7 mg 100 g^{-1} berry which is significantly higher than Indian-Summer (12,405 ± 674.0 mg 100 g^{-1} berry). On the other hand, titratable acidity of Harvest Moon (9,516 ± 136.1 mg 100g^{-1} berry) was significantly lower than Indian-Summer (11,554 ± 262.9). It also showed

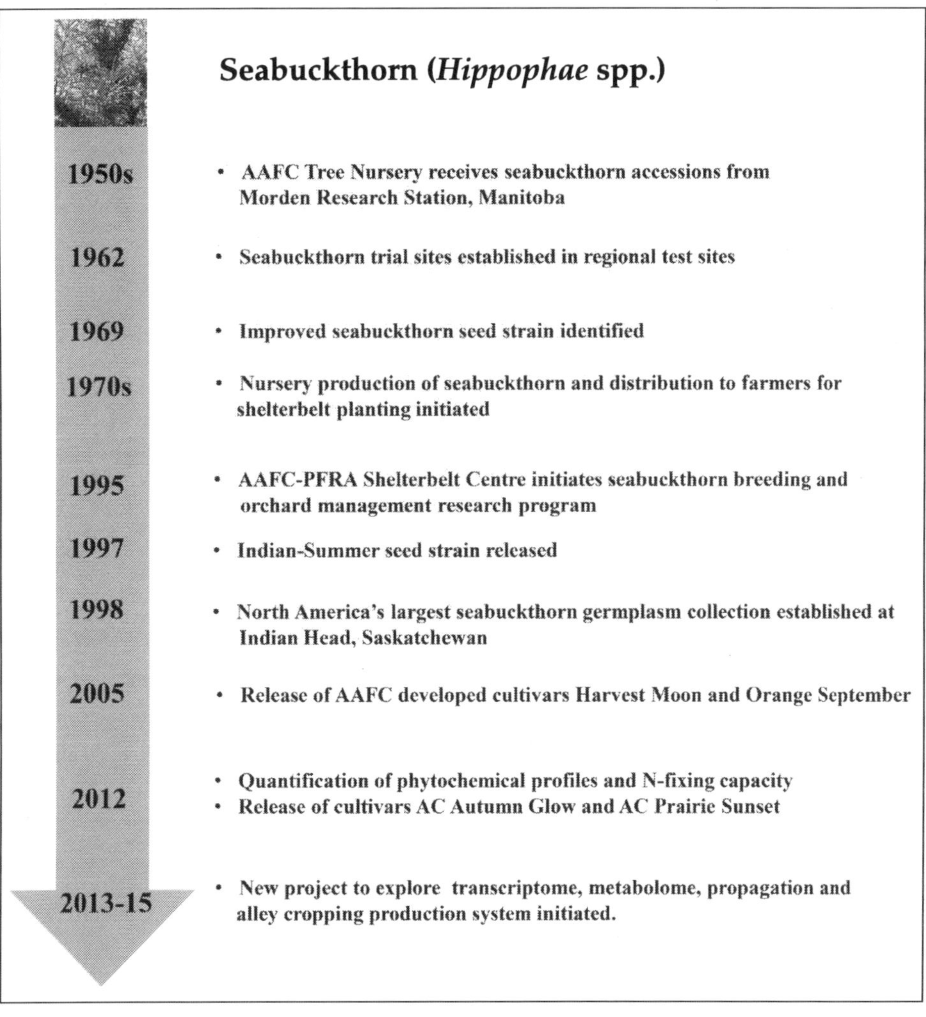

Figure 11.1: Timeline for Seabuckthorn Development at AAFC-ADC, Canada.

Table 11.1: Growth Parameters of AAFC-ADC Released Seabuckthorn Cultivars Growing at Indian Head, Saskatchewan, 2013

Cultivar	*Release Year*	*Flower Date*[1]	*Maturity Date*[2]	*Berry Wt*[3] *(g/100 berries)*	*Pedicel (mm)*
Indian Summer (RC-4)	1997	May, 24	August, 28	32.04 ± 0.3*c*	5.1 ± 0.1*a*
Harvest Moon	2005	May, 26	August, 14	87.01 ± 1.6*a*	3.7 ± 0.4*b*
Prairie Sunset	2012	May, 24	August, 21	84.60 ± 3.1*a*	3.6 ± 0.4*b*
Autumn Glow	2012	May, 24	August, 21	72.43 ± 2.3*b*	4.0 ± 0.2*ab*

1: Date when flowers were receptive; 2: Date when skin colour was fully orange with no green present; 3: Berry weight represents maximum mass during growing season, *a-c* Within a column, values with different letters are significantly different (P<0.05) according to Tukey's test.

Table 11.2: Metabolite Profile of Berries from AAFC-ADC Released Seabuckthorn Cultivars Growing at Indian Head, Saskatchewan in 2013

Cultivar	*Titratable Acidity (mg 100g* $^{-1}$*)*	*β-Carotene (mg 100g* $^{-1}$*)*	*Vitamin C (mg 100g* $^{-1}$*)*	*Vitamin E (mg 100g* $^{-1}$*)*	*Total Fat (Seed) (mg 100g* $^{-1}$*)*	*Total Fat (Pulp) (mg 100g* $^{-1}$*)*	*Total Flavonoids (mg 100g* $^{-1}$*)*	*Total Sugars (mg100g* $^{-1}$*)*
Indian Summer (RC-4)	11.554 ± 262.9*a*	2.6 ± 0.2*b*	546.8 ± 21.2*b*	47.88 ± 5.4*a*	16.816 ± 978.3*a*	20.007 ± 1212.3*c*	127.67 ± 3.5*a*	12.405 ± 674.0*c*
Harvest Moon	9.516 ± 136.1*b*	3.9 ± 0.1*a*	904.3 ± 11.3*a*	22.97 ± 4.5*b*	17.187 ± 845*a*	17.369 ± 813.3*c*	86.81 ± 1.9*b*	38.819 ± 272.7*a*
Prairie Sunset	11.283 ± 387.8*a*	3.8 ± 1.3*a*	578.0 ± 16.4*b*	33.92 ± 1.2*ab*	14.852 ± 234.7*a*	24.030 ± 709.9*b*	132.33 ± 3.6*a*	14.810 ± 381.0*c*
Autumn Glow	8.597 ± 227.7 *b*	1.8 ± 0.2*c*	384.2 ± 30.9*c*	38.08 ± 3.5*ab*	14.971 ± 66.2*a*	30.100 ± 869.9*a*	120.08 ± 6.2*a*	28.387 ± 1840.3*b*

Results presented are on dry weight basis.

a-c Within a column, values with different letters are significantly different (P<0.05) according to Tukey's test.

Figure 11.2: Images of Elite Cultivars Developed and Released by the AAFC-ADC well Adapted to Canadian Prairies Growing Conditions.

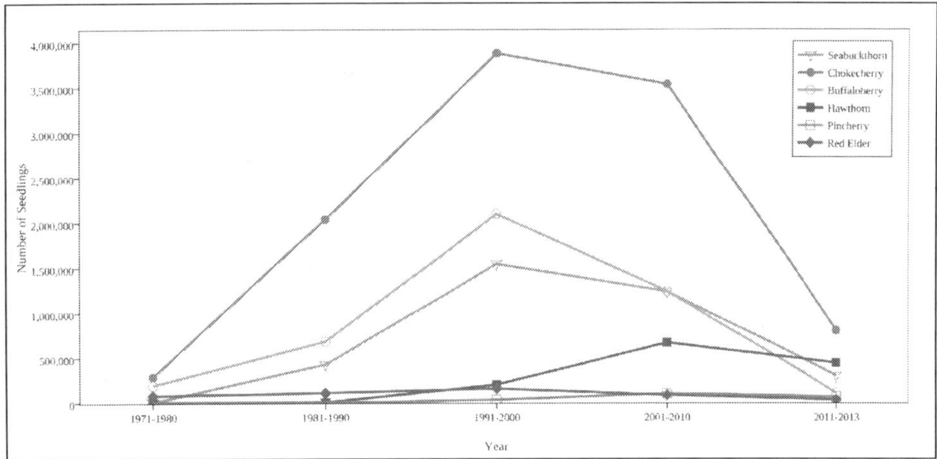

Figure 11.3: Total Number of Berry Seedlings Distributed by AAFC Shelter Belt Centre from 1971-2013.

significantly higher vitamin C and vitamin A content along with high total fat content.

Ongoing research at AAFC-ADC, is studying the transcriptome of developing berries to identify genes related to carotenoids, fatty acids and flavonoids biosynthesis pathway. In addition, metabolite profiles for samples collected over multiple years from multiple locations are being quantified. Together, these results will facilitate future breeding and selection processes to develop elite cultivars suited for value added products acceptable to North American palate. Simultaneously, seabuckthorn berry production integrated with perennial agricultural crops such as forages (Prairie region) and horticultural crops (Atlantic region) is being studied. This diversified production system could be important for marginal lands where conventional agricultural production has a high degree of risk.

REFERENCES

1. Beveridge, T. H. J. 2003. Chemical composition and some physical properties. In: *Seabuckthorn (Hippophae rhamnoides L.): Production and Utilization,* Li, T. S. C. and Beveridge, T. H. J (Eds), NRC Research Press, Ottawa, ON, Canada, pp. 79-89.
2. Fatima, T., Snyder, C. L., Schroeder, W. R., Cram, D., Datla, R., Wishart, D., Weselake, R.J. and Krishna, P. 2012. Fatty acid composition of developing seabuckthorn (*Hippophae rhamnoides* L.) berry and the transcriptome of mature seed. *PLoS ONE* 7 (4): e34009.
3. Franklin, P., Schroeder, W. R. and Jana, S. 2001. In: *Proceedings 7th Biennial Conference on Agroforestry in North America*, Regina, SK, pp. 149-152.
4. Li, T. S. C. 2003. Introduction. In: *Seabuckthorn (Hippophae rhamnoides L.): Production and Utilization,* Li, T. S. C. and Beveridge, T. H. J. (Eds), NRC Research Press, Ottawa, ON, pp. 1-7.
5. Schroeder, W.R., Soolanayakanahally, R. Y. and Naeem, H. 2014. Harvest Moon-Seabuckthorn. *Can. J. Plant Sci.* 94: 1309-1312.
6. Schroeder, W. R., and Li, T. S. C. 2003. Land preparation, orchard design and planting. In: *Seabuckthorn (Hippophae rhamnoides L.): Production and Utilization,* Li, T. S. C. and Beveridge, T. H. J. (Eds.), NRC Research Press, Ottawa, ON, pp. 17-23.
7. Schroeder, W.R., Jana, S. and Barl, B. 2002. Seabuckthorn cultivar and orchard development in Saskatchewan. In: *ADF Final report – 98000273,* Saskatchewan Agriculture and Food, Regina, Saskatchewan.
8. Schroeder, W. R., Jana, S. and Franklin, P. 2000. Development of commercial seabuckthorn cultivars for Canadian orchards. In: *Proceedings of the International Workshop on Seabuckthorn,* Beijing, China.
9. Schroeder, W. R and Jana, S. 1999. Development and demonstration of seabuckthorn for commercial cultivation and processing in Saskatchewan. In: *ADF Final Report-95000329,* Saskatchewan Agriculture and Food, Regina, Saskatchewan.
10. Schroeder, W. R and Yao, Y. 1995. Seabuckthorn: A promising multipurpose crop of Saskatchewan. In: *PFRA-AAFC Annual Report.*
11. Tebbitt, M. C. 2005. *International Registration of Cultivar Names for Unassigned genera,* July 2004-June 2005. *Hort. Science* 40 (6): 1596.

Chapter 12

DNA Quality and Quantity of Seabuckthorn (*Hippophae rhamnoides*) Subspecies in Iran and China

Hamid Ahani[1*], Hamid Jalilvand[2], Jamil Vaezi[3], Seyed Ehsan Sadati[4], Dongrui Jia[5], Xiao Tao Bai[6] and Hadi Bagheri[1]

[1]Natural Resources and Watershed Administration of Khorasan Razavi,
Forest, Ranges and Watershed Management Organization of Iran
[2]Department of Forestry, Faculty of Natural Resources,
Sari University of Agricultural Sciences and Natural Resources, IR-Iran
[3]Department of Biology, Faculty of Sciences, Ferdowsi University of Mashhad, IR-Iran
[4]Research Center of Agricultural and Natural Resources of Mazandaran, IR-Iran
[5]Molecular Lab, Yunnan University, Kunming, Yunnan, China
[6] Molecular Ecology Lab, College of Life Sciences, Lanzhou University, China
(*E-mail: Ahani1977@gmail.com)

SUMMARY

The assessment of the DNA characteristics *of Hippophae rhamnoides* L. species became an important prerequisite for its conservation and marker assisted breeding. The first step in this type of studies is represented by the optimization of genomic DNA extraction in order to get high quality DNA, suitable for generation of molecular markers. We examine quantity and quality differentiation between two ecologically divergent seabuckthorn subspecies, using leaves of Iranian and Chinese forest habitats. We characterized for genetic diversity 100 individuals from 10 populations. The efficiency of a weight, length and ratio of A260/280 by ND software was studied in detail with regard to a variety of factors by GLM, 2t-test independent and cluster analysis employed. The mean values for the mentioned traits observed were 1567 ng/µl, 31 nm, 17.6 nm and 1.79 for Chinese subspecies and 731ng/µl, 14 nm, 7.95 nm and 1.88 for Iranian subspecies, respectively. The results presented here reveal that all of these parameters have a significant effect on yield and quality of DNA extracted from leaves. Precise adjustment of extraction conditions allows for significantly higher yields of high quality DNA from low elevation than high altitude. Ward similarity and Euclidian distance showed nearest population of Iranian Seabuckthorn to Chinese is Qazvin habitat. There are robust different between study traits of two countries. HSD test grouping were similar between all of traits except A260/280 ratio. Two-Sample t-test showed quantifies significant difference between Iran and China. The DNA weight and the mean A260 and A280 values of the samples from China were found to be statistically significantly higher than those of the samples from Iran, whereas the mean A260/280 ratio of the samples from Iran was higher than that of the sample/isolate from China albeit by a non-significant difference.

Keywords: *Hippophae rhamnoides, Genetic, Molecular, China and Iran.*

INTRODUCTION

The Genetic Resources of *Hippophae* are very rich. *Hippophae* includes 6 species and 9 subspecies that are widely distributed in Europe and Asia (Ahani *et al.*, 2014). Phylogeographic analyses for *H. rhamnoides* was followed by allopatric divergence, hybridization and introgression. One study highlighted the complexity of intraspecific evolutions and the importance of the China

Seabuckthorn (Hippophae L.): A Multipurpose Wonder Plant, Vol. 5: Advanced Technologies for Cultivation, Processing, Health Protection and Environmental Conservation (Virendra Singh, *Editor-in-Chief*, 2018), p. 85–94, Astral International Pvt. Ltd., New Delhi, India.

as a center of origin for many temperate plants (Ahani and Sadati, 2014). Study on *Hippophae* is seldom in Iran. Mean germination percentage of this species was 20.5 and 20.1 in nursery and green house reproduction respectively (Ampliquant, 2005; Beer, 1852). As the content available in public databases is increasing rapidly, there has been a great inclination towards development of molecular markers using computational tools (Doyle *et al.*, 1987; Ghangal *et al.*, 2013). The research evolved a significant outcome to start a breeding program for the evolution of seabuckthorn varieties for the mountain areas of Pakistan and Jammu-Kashmir (Grover *et al.*, 2012). Differences in the fingerprints of DNA extracted under different conditions suggest that results could be biased when using gentle extraction procedures. A Spectrophotometer for Nucleic Acid Measurement is used in molecular biology. Analysis of nucleic acids is commonly performed to determine the average concentrations of DNA or RNA present in a mixture, as well as their purity. To date, there are two main approaches used by scientists to quantitate DNA or RNA. These are spectrophotometric analysis and fluorescence tagging. Recent revolutionary advancements in DNA sequencing technology have led to the accumulation of vast amount of sequence data in public domain. One of the more commonly used practices to quantitate DNA or RNA is the use of spectrophotometric analysis using a spectrophotometer like the Nanodrop software (IUPAC, 2010). Spectrophotometric analysis is based on the principles that nucleic acids absorb ultraviolet light in a specific pattern. In the case of DNA and RNA, a sample that is exposed to ultraviolet light at a wavelength of 260 nanometers (nm) will absorb that ultraviolet light. The resulting effect is that less light will strike the photo detector and this will produce a higher optical density (OD). Quantitative DNA analysis is either perfomered by 1) measuring the optical density (OD) at a wave length of 260 nm and 280 nm and/or 2) gel electrophoresis. Though being less accurate in quantitation, gel electrophoresis has the advantage to also allow a qualitative analysis. Reliable measurement of DNA concentration and purity is important for many applications in molecular biology, especially array comparative genomic hybridization (aCGH) where accurate determination of DNA concentration is critical. Impurities in DNA can lead to inaccurate measurement of DNA concentration and could potentially inhibit subsequent labeling reactions. Phylogenetic distance estimated revealed that the ecotypes expressed common heritage for their phylogenetic relationship with a considerable genetic diversity among them as well. Quite a few ecotypes showed close relationship irrespective of their geographic distances and morphological attributes.

For the first time in this study, DNA extracted were employed for characterizing traits variation and region relatedness in seabuckthorn collection located in Iran and China. This article provides recommendations for measuring DNA quantity and quality based on our experience of running hundred of samples in our high-through put genomic services laboratory. In addition, detailed information is provided on the theory and practice behind measuring DNA purity and concentration.

MATERIALS AND METHODS

Case Study

We collected leaves of 100 individuals of *H. rhamnoides* L. from 10 populations of two localities

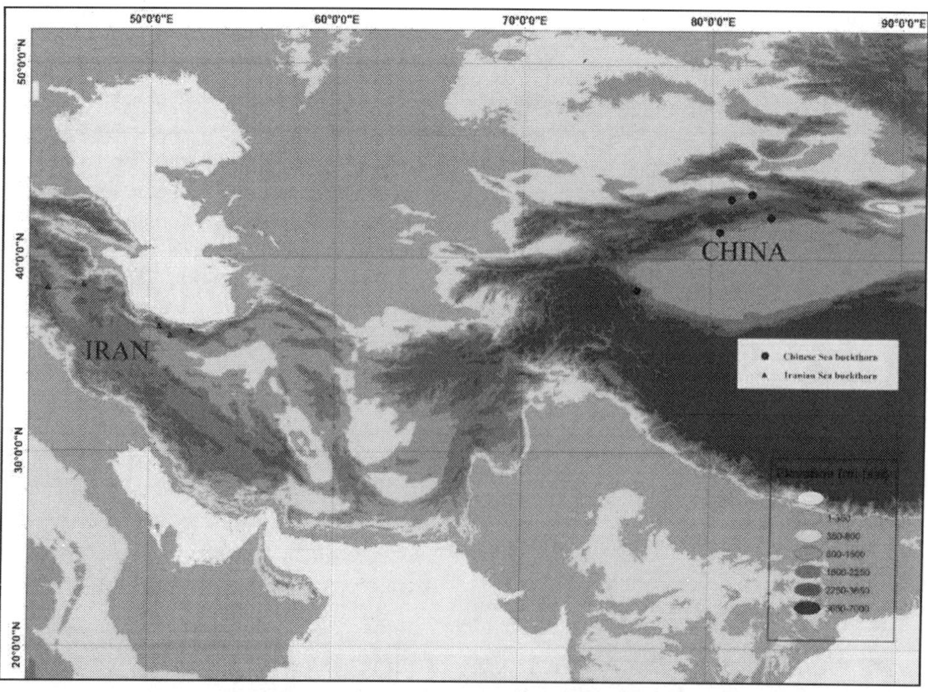

Figure 12.1: Distribution of Seabuckthorn in Iran and China.

Table 12.1: Provenance of Population Samples of Seabuckthorn Species in Iran and China

Popoulation Code	Country	Location	Latitude (N)	Longitude (E)	Altitude (m)
1	Iran	Mazandaran	36°20.253'	52°23.556'	2161
2		Eastern Azerbaijan	38°58.974'	46°48.332'	1776
3		Western Azerbaijan	38°43.669'	44°56.570'	1890
4		Qazvin	36°46.027'	50°52.681'	1701
5		Alborz	35°98.713'	51°10.830'	1859
6	China	Gongliu XJ	43°32.446'	082°12.602	740
7		Zhaosu XJ	43°05.243'	081°05.164'	1730
8		Baicheng XJ	42°13.993'	083°14.045'	1770
9		Akesu XJ	41°37.258'	080°46.639'	1610
10		Yingjisha XJ	38°35.026'	076°07.668'	1945

(Iran and China) (Table 12.1). We sampled 10 trees per population with sampled trees located 100 m apart in a population. Fresh leaves were dried immediately in the field using silica gel and voucher specimens were deposited in Sari Agricultural Sciences and Natural Resources University and Lanzhou University.

DNA Extraction

Genomic DNA was extracted from dried leaf tissue using Plant Mini Kit (Qiagen). Genomic DNA was isolated from approximately 50–100 mg of silica-gel dried, leaf material was sliced in to small pieces and placed in a 30 ml tube containing PVP and enzymatic mix (Jia *et al.*, 2012) and porcelain balls for mechanical grinding; then the balls were removed, tubes centrifuged and the supernatant transferred in to a clean tube; then were incubated for 15 min at 65°C; then the tubes were centrifuged again and over the transferred supernatant incubation and centrifuged; the supernatant was transferred in to a clean tube. DNA was precipitated with cold Isopropanol, washed with 70 per cent ethanol, air dried and rehydrated with 0.1 X TE. Genomic DNA was isolated from young leaves using Cetyltrimethyl ammonium bromide (CTAB) method. (Lu and Ahani, 2013) DNA was quantified on Ethidium Bromide stained agarose gels and spectrophotometer. Formula of DNA quality is (Microarray Shared Resource, 2014) as follow:

Optical Density = Log (Intensity of Incident Light/ Intensity of Transmitted Light)

Weight of DNA

In practical terms, a sample that contains no DNA or RNA should not absorb any of the ultraviolet light. The purity and quantity of extracted DNA were measured with the help of spectrophotometer in 260 and 280 nm wavelengths (Thermo Nicolet 100 UV, London, UK). DNA concentration is estimated by measuring the absorbance at 260 nm, adjusting the A260 measurement for turbidity (measured by absorbance at 320 nm), multiplying by the dilution factor, and using the relationship that an A 260 of 1.0 = 50µg/ml pure dsDNA.

Concentration (µg/ml) = (A260 reading – A320 reading) × dilution factor × 50µg/ml

Total yield is obtained by multiplying the DNA concentration by the final total purified sample volume.

DNA yield (µg) = DNA concentration × total sample volume (ml)

A260 Quality

When using spectrophotometric analysis to determine the concentration of DNA, the Beer-Lambert law (1852) is used to determine unknown concentrations without the need for standard curves. In essence, the Beer Lambert Law makes it possible to relate the amount of light absorbed to the concentration of the absorbing molecule. The following absorbance units to nucleic acid concentration conversion factors are used to convert OD to concentration of unknown nucleic acid samples:

A260 dsDNA = 50 µg/ml

A260 ssDNA = 37 µg/ml

A260 ssRNA = 40 µg/ml

The absorbance at 260 nm is used to calculate the concentration of nucleic acids at a concentration of 1 µg/ml and a 1 cm path length. The absorbance value is also dependent on the amount of secondary structure in the DNA due to hypochromicity.

To ensure the concentration reading is accurate, the absorbance reading should be within the linear range of the spectrophotometers, allow selection of the nucleic acid type and path length such that resultant concentration is normalized to the 10mm path length which is based on the principles of Beer's law. The "A260 unit" is used a quantity measure for nucleic acids. One A260 unit is the amount of nucleic acid contained in 1 ml and producing an OD of 1.

A280 Quality

To evaluate DNA purity, measure absorbance from 230 nm to 320 nm to detect other possible contaminants. The most common purity calculation is the ratio of the absorbance at 260 nm divided by the reading at 280 nm. Good-quality DNA will have an A260/A280 ratio of 1.7–2.0. A reading of 1.6 does not render the DNA unsuitable for any application, but lower ratios indicate more contaminants are present. The ratio can be calculated after correcting for turbidity (absorbance at 320 nm). It is also important to note the phenol and other contaminants can also absorb at 280 nm and can affect the ratio calculation. We derived data at vertical axe at 10 mm absorbance. Horizontal axe is wavelength 280 nm.

A260/280 Ratio

Determination of sample purity is used by this index. The secondary benefit of using spectrophotometric analysis for nucleic acid quantitation is the ability to determine sample purity using the 260 nm: 280 nm calculation. A pure DNA sample will yield an A260/280 of approximately 1.8. For RNA, a pure sample will yield an A260/280 of approximately 2.0. These ratios are commonly used to assess the amount of protein contamination that is left from the nucleic acid isolation process since proteins absorb at 280 nm. In some cases, scientists elect to quantitate DNA and RNA using fluorescence tagging methods with dyes that are highly specific to DNA or RNA. The benefit of fluorescence quantitation of DNA is the improved sensitivity over spectrophotometric analysis. Although, that increase in sensitivity comes at the cost of a higher price per sample and a lengthier sample preparation process. Additionally, there is not currently a fluorescence method to determine protein contamination of a DNA sample that is similar to the 260 nm/280 nm spectrophotometric version.

Purity (A260/A280) = (A260reading – A320reading) ÷ (A280reading –A320reading).

Total yield is obtained by multiplying the DNA concentration by the final total purified sample volume.

DNA Yield (µg) = DNA Concentration × Total Sample Volume (ml)

Strong absorbance around 230 nm can indicate that organic compounds or chaotropic salts are present in the purified DNA. A ratio of length can help evaluate the level of salt carryover in the purified DNA. The lower the ratio, the greater the amount of thiocyanate salt is present. As a guideline, the A260/A230 is best if greater than 1.5. A reading at 320 nm will indicate if there is turbidity in the solution, another indication of possible contamination. Therefore, taking a spectrum of readings from 230 nm to 320 nm is most informative (Jia *et al.*, 2012). DNA yield can be assessed using various methods including absorbance (optical density), agarose gel electrophoresis, or use of fluorescent DNA-binding dyes. All three methods are convenient, but have varying requirements in terms of equipment needed, ease of use, and calculations to consider.

Absorbance Methods

The most common technique to determine DNA yield and purity is measurement of absorbance. Although it could be argued that fluorescence measurement is easier, absorbance measurement is simple, and requires commonly available laboratory equipment. All that is needed for the absorbance method is a spectrophotometer equipped with a UV lamp, UV-transparent cuvettes (depending on the instrument) and a solution of purified DNA. Absorbance readings are performed at 260 nm (A260) where DNA absorbs light most strongly, and the number generated allows one to estimate the concentration of the solution. To ensure the numbers are useful, the A 260 reading should be within the instrument's linear range (generally 0.1–1.0). However, DNA is not the only molecule that can absorb UV light at 260 nm. Since aromatic amino acids present in protein absorb at 280 nm, both contaminants, if present in the DNA solution, will contribute to the total measurement at 260 nm. Additionally, the presence of guanidine will lead to higher 260nm absorbance. This means that if the A260 number is used for calculation of yield, the DNA quantity may be overestimated. Due to their different absorption spectra, the nucleotide composition of the bases present in DNA will have different A260/A280 ratios.

Phenol absorbs with a peak at 270 nm and has an A260/A280 ratio of 2. Nucleic acid preparations uncontaminated by phenol should have an A260/A280 ratio of around 1.8. As well as affecting the ratio calculations, contamination by phenol can significantly contribute to overestimation of DNA concentration and inhibit downstream reactions. The pH of the solution can also affect the A260/A280 ratio, with acidic solutions having a lower ratio of up to 0.2–0.3 and alkaline solutions having an increased ratio by a similar amount (Ogt corporation, 2012).

Table 12.2: Assay Type of DNA Samples (Schuelke, M. 2000; Microarray Shared Resource, 2014)

Assay Type	*Size Range*	*Concentration Range*	*Max # of Samples per chip*	*Recommended DNA Types*
DNA 1000	25-1000 bp	0.5-50 ng/µl	12	PCR products, restriction digest or fragmented DNA
High Sensitivity DNA	50-7000 bp	0.005-0.5 ng/µl	11	Fragmented DNA or DNA libraries in pg/ul range
Qubit dsDNA BR	any	0.1-1000 ng/µl	NA	genomic DNA

Table 12.3: A260/A280 Ratios for Nucleotides

Nucleotide	*A260/A280 ratio*
Adenine	4.50
Cytosine	1.51
Guanine	1.15
Thymine	1.47

Statistical Analysis

General linear model analysis procedure was used with the randomized nested plot design. Two sample t-tests were calculated to determine the relationships between variables for different countries using population plant data. Differences between means were separated using Tukey's Studentized Range (HSD) Test at 0.05 level of probability. We used Bartlett's test for equal variances of data. Euclidian distance with Ward linkage was down by cluster analysis. Statistical analyses were done with the Mini Tab statistical software package. We used ND 2000 software for this study.

RESULTS

Genomic extracted DNA on electrophoresis gel of seabuckthorn ecotypes is shown in Figure 12.2.

The DNA curve of ND 2000 profile among 10 populations of Seabuckthorn obtain of device measurement is compared by curves of 10 individual of each habitat in Figure 12.3 and 12.4.

Ward similarity and Euclidian distance in Figure 5 showed Iran and China are separated in first cluster. Five population of Iran are near together same as China populations except in Yingjisha XJ. Nearest population of Iranian Seabuckthorn to Chinese is Qazvin habitat.

Data were scored result of descriptive statistics is shown in Table 12.4, and phylogenetic relationships by cluster observation were determined and shown in Figure 12.5.

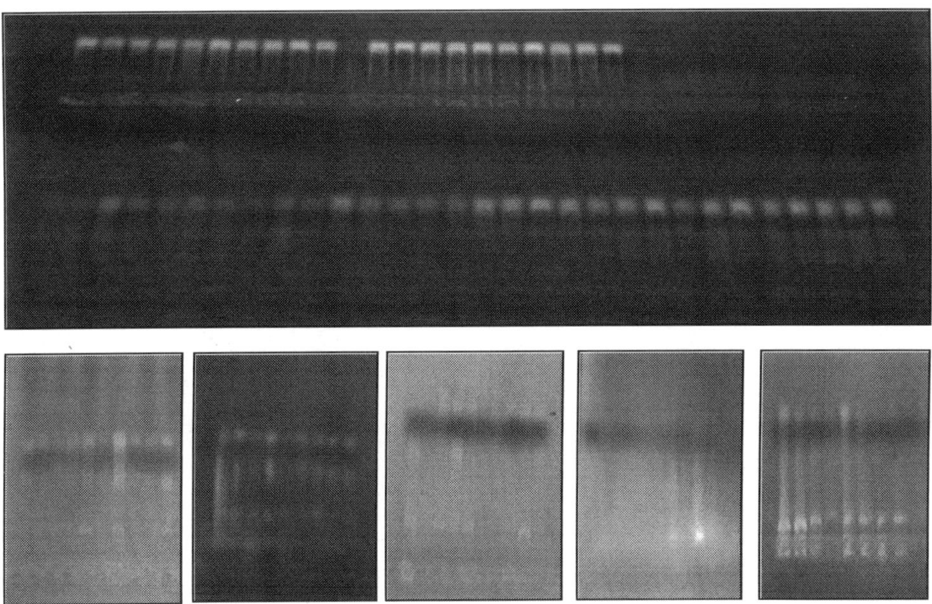

Figure 12.2. Genomic DNA of Seabuckthorn Ecotypes.
(Above photos belonged to Iran, below photo belonged to China individuals).

Table 12.4: Characteristics of DNA Samples in Iran and China

Country	*Location*	*Weight (ng/µl)*		*A260 (mm)*		*A280 (mm)*		*A260/280*	
		Mean	*S.E*	*Mean*	*S.E*	*Mean*	*S.E*	*Mean*	*S.E*
Iran	Mazandaran	509.6	90.1	10.19	1.8	5.13	0.99	2.05	0.07
	Eastern Azerbaijan	1170	413	23.39	8.26	13.52	4.8	1.76	0.12
	Western Azerbaijan	438.4	80	8.76	1.6	4.91	0.99	1.8	0.05
	Qazvin	1007	258	20.13	5.17	10.73	2.78	1.92	0.05
	Alborz	530	129	9.62	2.4	5.44	1.38	1.89	0.12
China	Gongliu XJ	2127	260	42.37	5.22	24.12	2.53	1.75	0.05
	Zhaosu XJ	1514	255	30.1	5.15	18.87	4.16	1.67	0.07
	Baicheng XJ	1506	184	29.79	3.75	17.61	2.02	1.66	0.07
	Akesu XJ	1824	334	35.74	6.61	18.83	3.57	1.93	0.03
	Yingjisha XJ	862	163	17.11	3.28	8.47	1.41	1.95	0.10

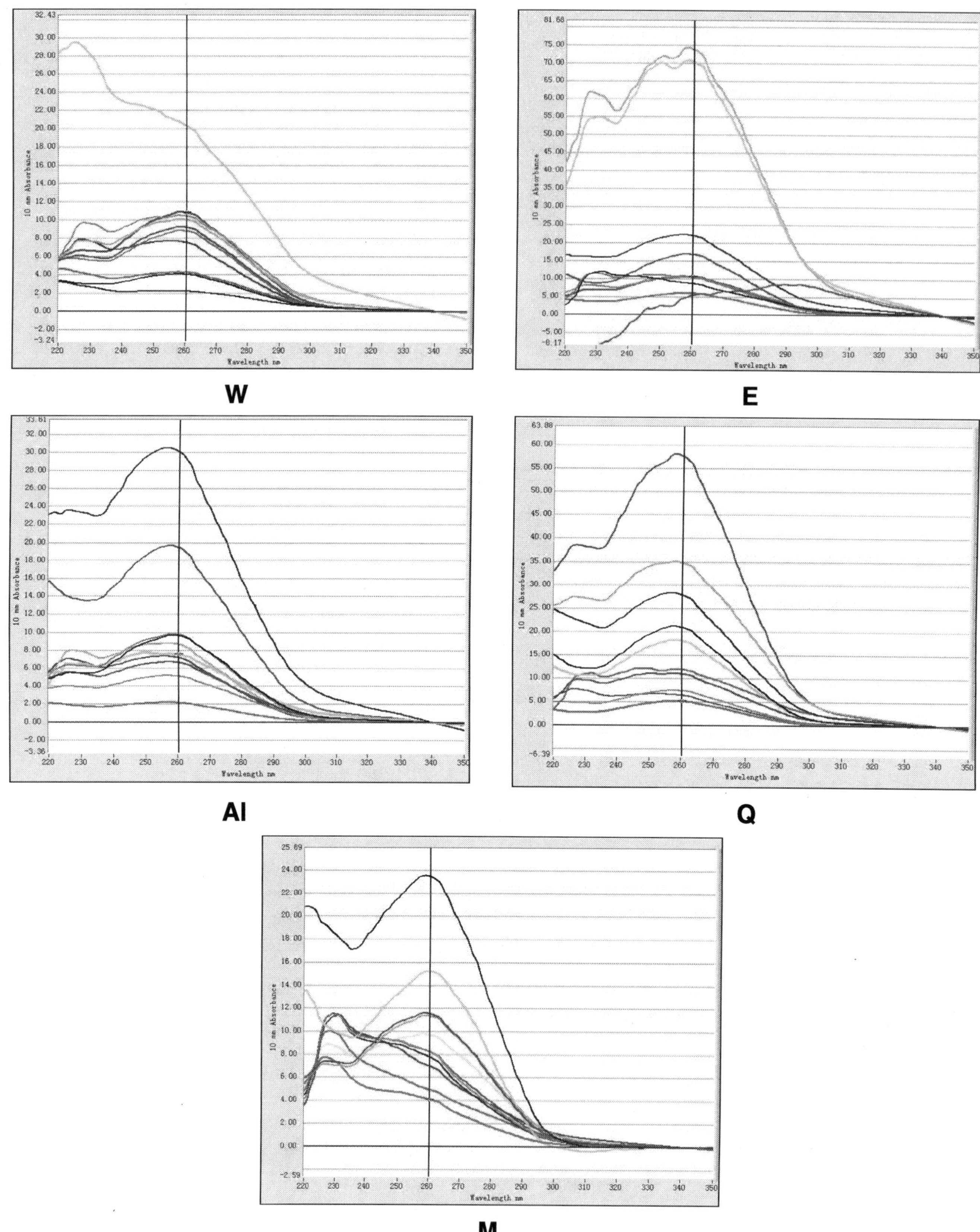

Figure 12.3: Iranian Populations Plots of DNA Quality.

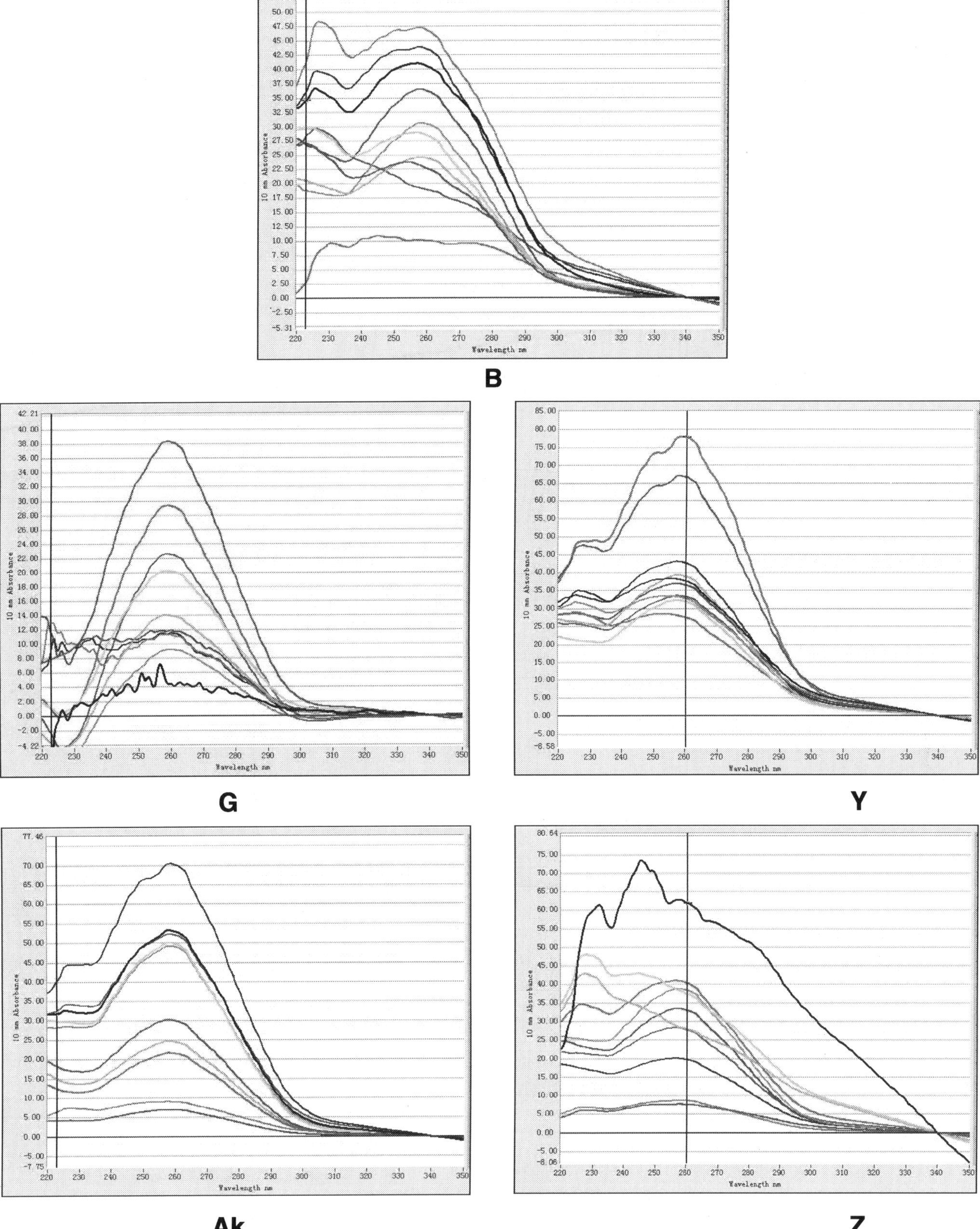

Figure 12.4: Chinese Populations Plots of DNA.

Table 12.5: Grouping Information Using Tukey Method

Country	Location	Weight (ng/µl)	A260 (mm)	A280 (mm)	A260/280
Iran	Mazandaran	c	c	c	a
	Eastern Azerbaijan	abc	abc	abc	ab
	Western Azerbaijan	c	c	c	ab
	Qazvin	bc	bc	bc	ab
	Alborz	c	c	c	ab
China	Gongliu XJ	a	a	a	ab
	Zhaosu XJ	abc	abc	abc	b
	Baicheng XJ	abc	abc	abc	b
	Akesu XJ	ab	ab	ab	ab
	Yingjisha XJ	bc	bc	bc	ab

Bartlett's test statistic was obtained 1.27 with p-value = 0.41. Then variance of data is equal homogeneity. Result of GLM is presented in Table 12.6.

Two-Sample t-test showed (ng; t-value=5.14, p-value=0.000) for quantifies, (A260; t-value=5.10, p-value=0.000), (A280; t-value=5.07, p-value=0.000), for length significant difference between China and Iran, only A260/A280 ratio; t-value=-1.73, p-value=0.08 had not significant difference between China and Iran.

DISCUSSION

We investigate four characteristics of DNA quality and quantity. Optimum values for these traits are:

1. A260 - displays absorbance at 260 nm normalized to a 10 mm path length.
2. A280 - displays absorbance at 280 nm normalized to a 10 mm path length.
3. 260/280 - ratio of absorbance at 260 nm and 280 nm.

The ratio of absorbance at 260 and 280 nm is used to assess the purity of DNA and RNA. A ratio of ~1.8 is generally accepted as "pure" for DNA. If the ratio is appreciably lower in either case, it may indicate the presence of protein, phenol or other contaminants that absorb strongly at or near 280 nm. Inaccurate dilutions can occur if the DNA is not homogeneously re-suspended. For reliable spectrophotometric DNA quantification A260 readings should lie between 0.1 and 1.0. To improve the accuracy of DNA concentration determination allowance should be made for any impurities in the solution. This can be estimated by adjusting the A260 measurement for turbidity which is measured at an absorbance of more. Pure DNA preparations have an A260/A280 ratio of greater than or equal to 1.8. When the A260/A280 ratio is determined for a range of different DNA/protein mixtures it has been shown that the ratio is relatively insensitive to the addition of protein to pure nucleic acid. However the utility of the ratio becomes apparent when DNA is prepared from tissue or blood. These samples have a protein content that greatly exceeds nucleic acid on a weight basis and so in these cases the approximate purity of dsDNA preparations and contamination by protein may be estimated by determination of the ratio of absorbance at 260 nm and 280 nm. De novo short read assembly yielded 88,297 transcripts (> 100 bp), representing about 53 Mb of seabuckthorn transcriptome. The average length of transcripts was 610 bp of the short reads uniquely mapped back to seabuckthorn transcriptome. In total, 7,421 putative seabuckthorn transcription factor genes, distributed in at least 80 different families, were identified representing 8.4 per cent of seabuckthorn transcripts (Ghangal *et al.*, 2013), and then study on new habitats of Iran is suggested. Change in sample acidity: Small changes in the pH of the solution will cause the 260/280 to vary (William, 1997). Acidic solutions will under-represent the 260/280 ratio by 0.2-0.3, while a basic solution will over-represent the ratio by 0.2-0.3. If comparing results obtained using a Nano Drop Spectrophotometer to results obtained using other spectrophotometers, it is important to ensure that the pH of an undiluted sample measured on our instruments is at the same pH and ionic strength as the diluted sample measured on the conventional spectrophotometer. This procedure was first described to measure protein purity in the presence of nucleic acids. However it is now commonly used to assess protein contamination of DNA. It is important to note that the A260/A280 ratio is only an indication of purity rather than a precise answer. We suggest for using florescent markers. This approach not only reduces the cost of marker development but also permits generation of markers at a large scale in a short span of time (*Grover et al.*, 2012). An economic method for the fluorescent labeling (Schuelke, 2000) for this species is proposed for next study. There are two ways to measure repair rates. One is to monitor the removal of the DNA damage; the other is to monitor the restoration of the activity of the damaged DNA.

CONCLUSION

Notwithstanding one method of extraction, on weight, 260 and 280 mean we saw better value for

Table 12.6: Analysis of Variance Using Nested Plot for General Linear Model: Traits versus Country: Population

Traits	Resource	F	p- value	R-Square
Weight ng/µl	Country	10.53	0.01	37.34
	Population (Country)	2.89	0.00	
Length A260	Country	10.34	0.01	37.25
	Population (Country)	2.91	0.00	
Length A280	Country	9.71	0.01	37.88
	Population (Country)	3.1	0.00	
A260/280	Country	1.28	0.29	21.53
	Population (Country)	2.66	0.01	

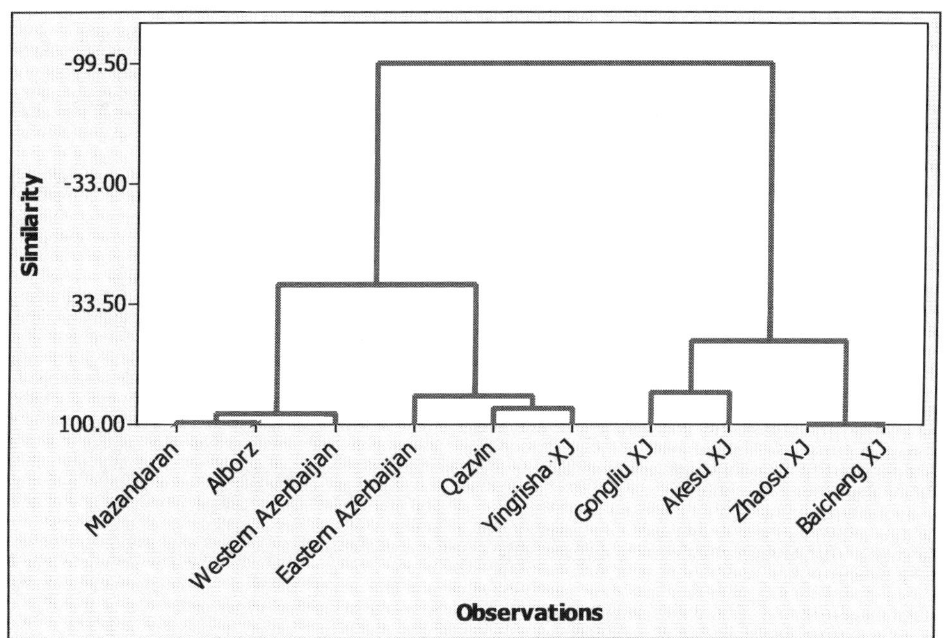

Figure 12.5: Dendrogram of Cluster Analysis (Ward, Euclidian distance).

Chinese samples. Content of DNA was affected for former traits. Some details due to method such as time of shaking, centrifugation, temperature and etc are for different result. The ratio mean of Chinese samples are appreciably lower, this may indicate the presence of copurified contaminants. May be one of reason is for old samples (2 years older than Iranian samples) of Chinese seabuckthorn.

According to the results a high genetic diversity was found in the studied population that mainly related to the seabuckthorn pollination nature that shows sporophytic incompatibility system enhancing out-crossing in natural populations.

Cluster method uses an agglomerative hierarchical that begins with all observations being separate, each forming its own cluster. In the first step, the two observations (Iran and China) closest together are joined. In the next step, either a Yingjisha observation of China joins the first two of Iranian populations; two other observations (China) join together into a different cluster. A function of the linkage criteria Ward: the sum of squared deviations from points to centroids, minimizing the within-cluster sum of squares.

In first three traits by Tukey (HSD), Akesu was unique qroup, for Mazandaran, West Azerbaijan and Alborz were in one group, Qazvin and Yingjisha were together and others were in final group. In ratio of A260/A280 Mazandaran population between 10 habitats are unique. Baicheng and Zhaosu belong to one group and others belong to third group. We conclude most of populations are similar except Mazandaran and two population of China.

Nested plot for General Linear Model showed no significant difference only for country in A260/A280 ratio p- value; other differences were significant.

The DNA weight and the mean A260 and A280 values of the samples from China were found to be statistically significantly higher than those of the samples from Iran, whereas the mean A260/280 ratio of the sample/isolate from Iran was higher than that of the sample/isolate from China albeit by a non-significant difference.

ACKNOWLEDGEMENT

This study was supported by the Lanzhou University of China and Sari University of Agricultural Sciences and Natural Resources of Iran. We are grateful to Askari, Saberi and Mardomi, Haji Yousefi and Najari engineers for supplying some materials. We appreciate for collaborating of Professor Jianquan Liu at sabbatical research in China.

REFERENCES

1. Ahani, H., Jalilvand, H. and Vaezi, V. 2014. Greenhouse treatments on *Elaeagnus rhamnoides* seed germination. *Research Journal of Environmental Sciences* 8 (4): 215-224. DOI: 10.3923/rjes.2014.215.224.
2. Ahani, H., Jalilvand, H., Vaezi, J. and Sadati, S.E. 2014. Investigation of nursery treatments on seabuckthorn (*Elaeagnus rhamnoides* L.) seed germination in the field. *Journal of Biodiversity and Environmental Sciences* 4 (6): 18-8.
3. Ampliquant, 2005. http://ampliquant.com/index.php/aq-07.
4. Beer, L. 1852. Bestimmung der Absorption des rothen Lichts in farbigen Flüssigkeiten" (Determination of the absorption of red light in colored liquids). *Annalen der Physik und Chemie* 86: 78-88.
5. Doyle, J.J. and Doyle, J.L. 1987. A rapid DNA isolation procedure for small quantities of fresh leaf tissue. *Phytochem. Bull.* 19: 1115.
6. Ghangal, R., Chaudhary, S., Jain, M., Singh, P. R. and Sharma, P.C. 2013. Optimization of de novo short read assembly of seabuckthorn (*Hippophae rhamnoides* L.) transcriptome. *Plos One* 8 (6): 1-8.
7. Grover, A., Aishwarya, V. and Sharma, P.C. 2012. Searching microsatellites in DNA sequences: Approaches used and tools developed. *Physiol. Mol. Biol. Plants* 18: 11-19.
8. IUPAC, 2010. Compendium of Chemical Terminology.
9. Jia, D.R., Abbott, R.J., Liu, T.L., Mao, K.S., Bartish, I.V. and Liu, J.Q. 2012. Out of the Qinghai Tibet Plateau: Evidence for the origin and dispersal of Eurasian temperate plants from a phylogeographic study of *Hippophae rhamnoides* (Elaeagnaceae). *New Phytologist* 194: 1123-1133.
10. Lu, R. and Ahani, H. 2013. The Genetic Resources of *Hippophae* and its utilization. *Adv. Agri. Biol.* 1 (2): 27-31, © PSCI Publications.
11. Microarray Shared Resource, 2014. http://www.genome.duke.edu/cores/microarray/, services/sample-qc/dna-qc/index.php.
12. Ogt Corporation, 2012. http://www.ogt.co.uk.
13. Promega, 2011. http://www.promega.com.
14. Schuelke, M. 2000. An economic method for the fluorescent labeling of PCR fragments. *Nature Biotechnology* 18: 233-234.
15. Shah, A. H., Ahmad, S.D., Khaliq, I., Batool, F., Hassan, L. and Pearce, S.R. 2009. Evaluation of phylogenetic relationship among seabuckthorn (spp. *turkestanica*) wild ecotypes from Pakistan using AFLP. *Pak. J. Bot.* 41 (5): 2419-2426.
16. Sharma, P.C., Grover, A. and Kahl, G. 2007. Mining microsatellites in eukaryotic genomes. *Trends Biotechnol.* 25: 490-498.
17. William, W., Wilfinger, K. M. and Piotr, C. 1997. Effect of pH and ionic strength on the spectrophotometric assessment of nucleic acid purity. *Bio. Techniques* 22: 474-481.

Chapter 13

Introduction Research of *Hippophae rhamnoides* L. in Belarus

I. Garanovich and T. Shpitalnaya

The Central Botanical Garden of the NAS of Belarus, Belarus
(E-mail: bel.dendr@gmail.com, t.shpitalnaya@cbg.org.by)

SUMMARY

Complex research of *Hippophae rhamnoides* L. has been conducted in Belarus. First of all primary introductions, testing of seabuckthorn from various introduction centers has been carried out. Alongside species samples, we tested a great number of sorts of Russian selection from the Institute of Horticulture of Siberia (Barnaul city) and the Botanical Garden of Moscow University. Considerable hardiness in winter period has been demonstrated by MSU sorts based on the Baltic climatype genotype. Considerable attention was paid to studying biochemical composition of fruits. An important theoretical part of the work was studying ecological and physiological peculiarities, water regime, soil requirements *etc.*, which was an essential basis for developing recommendations on cultivation. Great attention was paid to working out agricultural techniques for growing seabuckthorn seedlings and especially to mastering vegetative propagation technology. We have also studied seabuckthorn ontogenetic variability and hardiness on the basis of cultivating numerous populations from various parts of the areal different climate types: Syberian, North and South Caucasian, Danube, Baltic and others. A considerable part of the research was devoted to studying diseases and pests. A related institution – Institute of Horticulture – developed recipes of preserves and got several sorts of their own selection. As a result seabuckthorn has been widely spread in amateur horticulture of Belarus. As far as its plantation cultivation is concerned, it faces a number of difficulties connected with fragility of the plant, agrotechnical peculiarities; there are problems with gathering fruit *etc.* At present we are carrying out planned reproduction of seabuckthorn and saturation of consumer market.

Keywords: *Seabuckthorn, Introduction, Ekologo-physiological features and Reproduction.*

INTRODUCTION

The Belarusian territory lies outside the natural area of seabuckthorn (*Hippophae*), although, historically, it used to grow in Belarus, the evidence there of being provided in the palaeobotannical data.

For Belarus seabuckthorn is, above all, an alimentary and medicinal plant which contains oil. Seabuckthorn plays an important role as a soil-stabilising cultivated crop (field-protecting forest belts at reclaimed peat swamps, slopes or fish pond dike dams) and as an ornamental plant, *i.e.* in the sectors of agriculture, healthcare, amenity planting and forest or land reclamation and re-vegetation. Cross-cutting plant studies have been conducted for the introduction conditions, while taking into account of seabuckthorn's specific environmental features and flexibility, dioecism and its vast polymorphism.

MATERIALS AND METHODS

The Central Botanical Garden collection stock's varietal plants and those from their natural populations at F_0, aged 15 to 17 years, characterised by various geographic origins, have served as the research objects, for instance, samples representing the Cis-Baltic climatic type from Russia's Kaliningrad Region (Yantarnyi township); The Danubian climatic type from the Danube river estuary; the North Caucasus one from Kabardino-Balkaria; the South Caucasus one from Seki District (North Asia); and the Siberian one from Altai Territory (Gorny Altai).

Seabuckthorn (*Hippophae* L.): *A Multipurpose Wonder Plant, Vol. 5: Advanced Technologies for Cultivation, Processing, Health Protection and Environmental Conservation* (Virendra Singh, *Editor-in-Chief*, 2018), p. 95–106, Astral International Pvt. Ltd., New Delhi, India.

The F_1 used the seminal progeny of the F_0 representing the above climatypes. All the seeds showcased the same reproduction year and were obtained from the plants grown under the homotypic agroclimatic and technological conditions. The seeds were picked up from the plants (aged 17 years) grown at the Central Botanical Garden (CBG). On aggregate, 1,000 samples representing 27 forms out of 5 climatypes of seabuckthorn (*Hippophae rhamnoides*) have been involved in the experiment.

Seed Sowing Qualities

The seed sowing qualities include germinating capacity, germination energy, good quality and vital capacity which are specified in GOSTs (State Standards) 13056.6-13056.8-97 (GOST, 1995). The seed samples were selected in accordance with GOST 13056.1-67.

Ontogenetic Changes

All ontogenetic changes are age-related and the age classification has been performed under the Rabotnov scale (Rabotnov, 1950).

Phenological Spectrum

The phenological spectrum of the development was studied in accordance with the methods used at the USSR botanical gardens (Method, 1975).

Growth Analysis

The amount of growth of different tree parts, in particular, linear offshoot growth, was recorded under the Molchanov and Smirnov method (1967) alongside recommendations by Maltseva and Fomenko (1990).

Winter Hardiness

The winter hardiness was assessed under the scale used by the CBG Dendrology Department (Lapin, 1967).

Statistical Analysis

All observations were conducted in triplicate and the results of the research were statistically analysed using the mathematical package Match Cad 2000 at $p \leq 0.5$.

RESULTS AND DISCUSSION

Environmental and Climatic Factors of Seabuckthorn's Natural and Cultivation Areas

Belarus is situated in the western part of the East European/Russian plain. The country's territory extends significantly in the latitudinal direction (650 km) and in the longitudinal one (560 km). The Belarusian north-westerly border is close to the Baltic Sea. The geographical location, as well as the former glaciation consequences have caused an apparent zonal nature of the geomorphological, pedological, hydrological and climatic factors (Dziamentsieu *et al.*, 1959).

The national territory is referred to the North Temperate Zone. Its relief is predominantly flat and of glacial origin. The average altitude above the sea level makes 159 metres. Minsk is located in Belarus' central area and is referred to the northern of the western sub-zone on the border between the two agroclimatic regions: the Oshmiany-Minsk-Sventiany and Borisov-Rudensk ones (Shkliar, 1962).

The Botanical Garden territory is situated in the south-eastern part of the Minsk Upland which is featured by a hilly moraine relief. It lies at the watershed of the Svisloch and Volma rivers and has, in general, a western-south-western gradient, the surface flow being provided towards the Svisloch River; the Garden territory altitude averages 211 m above the sea level. The differences among the Garden's individual section elevations reach 15.5 m.

The weather-related factors are known to affect considerably the plants' introduction processes, yields and berry product quality. The Belarusian climate is a temperate continental one which is featured by mild winters, moderate summers and a sufficient amount of precipitation for a normal growth of the tree and shrubbery vegetation. The regional climate is mainly formed under the impact of the air masses coming all way from the Atlantic. The basic thermal parameters change in the north-eastern to south western direction. The meridian vector sees the climate's heat supply changing, while the latitudinal vector sees its continentality change (Agroclimatic, 1969).

In terms of the total temperature changes above 10ºC, when an active plant growth happens, and the precipitation to evaporation ratio, which represents a ratio of the total precipitation amount to the total temperatures during the vegetation period and labels the moisture supply sufficiency, three agroclimatic latitudinal zones have been identified: the Northern one which is moderately warm and humid; the Central one which is warm and moderately humid; and the Southern one which is warm and unsteadily humid (Shkliar, 1962). Every zone, in terms of the climate continentality, is divided into two subzones: the Western and Eastern ones.

The average annual temperature in the research area (Minsk City), according to long-term observations, amounts to 6.5ºC. The temperature regimen in place during the period of interest insignificantly deviated from the multiyear average (6.8ºC).

According to the multiyear data, Minsk's annual precipitation amount totals on average approximately 600 mm. During some moderately cool and rainy years it reaches 685 to 710 mm, while during some dry ones it plummets down to 450 to 470 mm. An analysis conducted into the data related to the average monthly precipitation totals at the Botanical Garden has demonstrated that during the normal watering years the precipitation amount totals within the vegetation period range between 65 and 70 per cent of the annual normal value and their month-by-month distribution patterns are more or less uniform (Climate, 1976).

The snow mantle duration makes for Minsk 113 to 118 days. It is between 28 and 30 cm thick, while the maximum soil freezing depth makes about 1 m, the minimum one is 20 cm and the average multiyear one is approx. 60 cm.

The complete soil defrosting dates are as follows: the average one is 11 April; the early one in 28 March; and the late one is 23 April. The average annual relative air humidity amounts to 81.6 per cent; whereas in May, June and July it falls to 70 to 75 per cent.

The years characterised by an insufficient precipitation amount see the light soil moisture content shrinking rapidly to make a mere 4 per cent by mid-summer in the plough-layer and almost to reach a dead water level in the underlying horizons.

We have to take into account the weather conditions under which the climatypes under scrutiny have been formed. Based on the bibliographical data, we have obtained the following information on the weather conditions of the natural Seabuckthorn growth areas.

The Cis-Baltic climatic type occupies a narrow shoreland belt of the Baltic Sea's south-westerly littoral. The territory's humid and moderately warm climate is greatly influenced by the Atlantic Ocean and the continental cyclonic winds. The average annual temperature in the area makes 7.5ºC, while the vegetation period duration never exceeds 150 days (Agroclimatic, 1972).

The Danubian climatic type of seabuckthorn (Danube river estuary, Vilkovo settlement area) grows in the region of the Atlantic-continental steppe climate, in the micro-climatic sub-region of the Danube lowlands. The summers are hot and the winters are cold and with little snow. Frequent thaws alternate with severe cold snaps. The average annual air temperature is 10.7° C. The vegetation period lasts 245 days (Agroclimatic, 1964; Climate, 1967).

The Caucasian climatic type is referred to the humid climate type. The average annual air temperature ranges between 7.0 and 15.0ºC. The summers are warm and the winters are moderately mild and windy with an unstable snow mantle. The vegetation period totals 156 to 190 days. In the central and eastern parts of the Northern Caucasus region the climate is formed under the influence of the maritime air masses and the continental air streams. In Eastern Georgia and in Azerbaijan the climate is a moderately subtropical one and formed under the impact of the air masses coming from the Asian continent (Agroclimatic, 1978; Climate, 1968; Climate, 1961).

Seabuckthorn of the Siberian climatic type has been formed under the conditions of Eastern Siberia and Altai's continental climate. The average annual air temperature oscillates in the region within the range of –3.5 to 1.2° C and the vegetation period is 100 to 135 days long with an active temperature total of 1495 to 2350ºC (Korovina, 1988).

In summary, the seabuckthorn area is related to the subtropical, temperate, continental and harsh continental climatic zones. A specific trait is represented by low air humidity with considerable diurnal and seasonal temperature differentials (Atlas, 1976). The parent rock materials on the Botanical Garden territory differ as to their mechanical compositions, natures and watering levels (Churilov *et al.*, 1969). The most widespread parent rock materials in the Garden are aggregated fine silt-sand loams. These occupy 84.8 ha, or 86.5 per cent of the entire territory and have the following mechanical compositions: physical clay 15.1 to 19.0 per cent; dust 4.9 to 30.2 per cent; and sand 49.2 to 80.4 per cent. The friable fine silt-sand loams, which in some areas alternate with sands, are less widespread and occupy 7.4 ha, or 7.55 per cent of the territory, they contain 10.0 to 14.3 of physical clay; 10.0 to 28.1 per cent of dust; and 58.2 to 77.5 per cent of sand.

The sod-podzol soils and the sod-podzol marshy ones are featured by increased acidity level and low humus content (1.5 to 2.0 per cent). As the underlying consertal sands, which quite frequently contain gravels and pebbles are close to the surface, the Garden's sod-podzol soils, especially those which develop on loomy sands with underlying consertal sands at the 0.5 m depth, are featured by an unstable and quite often watering level. A more stable, sometimes excessive watering level and a neutral or near-neutral reaction to increased humus content (2.5 to 8.0 per cent) are characteristic of the soddy soils, humic-sod gleys and gley soils. These soils also have a relatively high level of alkali saturation, but are poor in mobile phosphorus and potassium forms (Smeyan *et al.*, 1989; Churilov *et al.*, 1969).

The CBG territory's arable layer (0 to 25 cm) at the experimental sections is dark-grey with a brownish tone, compacted, fine blocky in its structure and densely penetrated with roots; the loamy sands are friable with their transition to the lower horizon being uneven and clearly marked. According to the data provided by laboratory testing and observations, the soil can be referred to the averagely ameliorated type. Analytical testing of the ploughed horizon chemical composition provides evidence of an element composition which is typical of sod-podzol sabulous soils. The soil properties have been considerably influenced by the farming techniques. In a number of cases the upper layers have lost their acidity and built up the humus contents, alkali saturation levels and the mobile phosphorus and potassium form contents.

When collecting and generalising the bibliographical data on the soil mantle of the five *Hippophae rhamnoides* climatic types, the following peculiarities of their natural growth areas have been identified. The soils in the natural seabuckthorn ecotypes are poor in humus, as to their mineral composition they are referred to the hydrocarbonate type, and are sometimes saline. The soil mantle (of the Cis-Baltic climatic type) changes from soddy clay loams with pebble admixture to sandy maritime deposits with an insignificant humus admixture, which soils have been formed on the glacial deposits. The Cis-Baltic soils freeze 20 to 30 cm from the surface down and thaw rapidly (Korovina, 1988).

The soils in the Danube estuary islands are of an alluvial marshy origin, the water table is 1.25 m deep. The wintertime soil freezing is insignificant. The saline soil availability is conditional on the sea proximity (Agroclimatic, 1964). With due regard to the specific peculiarities guiding the formation of the Caucasian climatic type, the local soils are poorly developed and alluvial in their nature, rocky ones with rubble (Climate, 1968). The Caucasian soils freeze 20 to 25 cm deep, but the superficial horizons are exposed to harsh temperature oscillations. At Altai Territory the river valley soils are the alluvial meadow type with rubble. The soil solution pH is close to neutral, saline soils being present occasionally. The soil surface heats on average up to 22 to 24°C, while at the 10 to 20 cm depth up to 17.0 through 22.3°C. In the wintertime the soils freeze down to 151 through 350 cm and in a number of regions the temperature 40 cm deep plummets down to –15.1°C; while locally the seasonal freezing and permafrost joining is observed (Alexeeva *et al.*, 1971; Climate, 1961).

Consequently, Seabuckthorn grows in its natural habitats on humus-poor soils; as to their mechanical composition they are alluvial with rubble and of friable texture. The pH value is near-neutral but more often the reaction is slightly acidic. The Kaliningrad Region and mountainous Caucasus soils are the richest ones in terms of potassium contents and the lowest potassium contents are recorded in the Danube estuary soils. The Caucasus, Kaliningrad Region and Gorny Altai soils are featured by high phosphorus content.

The agrochemical tests conducted at the Botanical Garden indicate that its soils are slightly acidic and well supplied with potassium and phosphorus, although nitrogen-poor; the moisture capacity is low and amounts to 44 to 53 per cent. The soil-related and environmental conditions prevailing on the Garden territory favour seabuckthorn growing and creation of various planting types.

Introduction in Belarus

At the initial stages an active species introduction was performed from the natural populations and breeding centres. Peculiarities of individual development and adaptogenesis, soil and moisture sensitivity, stability under the Belarusian conditions and the biochemical parameters were looked into, agrarian reproduction and planting material growing techniques were developed and plantations set up. A mass-scale sapling production began.

Good seabuckthorn prospects for Belarus were demonstrated. Practical recommendations as to its growing conditions and experimental planting creation were developed.

Seabuckthorn can be aptly described as becoming a popular culture of amateur gardening and gaining an ever growing spread all over the country.

A vital element of the work was strain testing of multiple Russian-bred seabuckthorn varieties. The first varieties included Dar Katuni, Zolotoi Pochatok, Novost Altaya, Maslichnaya and Vitaminnaya which were created at the Siberian Horticulture R&D Institute using the analytical breeding method. In general, they had a good behaviour record in Belarus but suffered from root system asphyxiation, were not sufficiently winter-hardy, were infected by fungoid diseases and the berries were difficult to tear off from their stems. More prospects were associated with the second- and later third-generation varieties which represented hybrids created with the involvement of the natural populations and varieties, as well as such mutants as Chuiskaya, Samorodok, Velikan, Obilnaya, Zolotistaya, Prevoskhodnaya, Obskaya, Sibirskaya and Yantarnaya. These varieties were superior to the old ones in terms of their berry weight and size and stem length. The second-generation seabuckthorn varieties featured by excellent economic and biological characteristics caused a wave of interest at first at the Russian research centres and then globally, including Belarus.

The Altai breeder experience encouraged research to create some zoned seabuckthorn varieties also at other institutions of Russia: Institute of Cytology and Genetics of the Russian Academy of Sciences' Siberian Department (Druzhina, Zarnitsa, Ivushka, Kapriz, Krasny Fakel, Parad, Podruga, Sibirsky Rumianets and Triumf); South Urals R&D Institute of Horticulture (Koster, Lisichka, Ryzhik, Solnyshko and Yantarnoe Ozherelie); Nizhny Novgorod Agricultural Institute (Diuymovochka, Duet and Zolotoi Shar); Michurin All-Russian R&D Institute (Baikal, Karamelka, Zolotaya Kosa, Siurpriz Baltiki and Moriachka); and Buryat Experimental Horticultural Station (Atsula, Ayaganga, Baikalsky Rubin, Bayan-Gol, Sayana and Stepnaya).

The Moscow State University's Botanical Garden bred on the basis of the Cis-Baltic and West-European forms such promising varieties as Aromatnaya, Avgustinka, Botanicheskaya, Botanicheskaya Liubitelskaya, Vorobievskaya, Zheltaya Ranniaya, Krasnoplodnaya, Otradnaya, Podarok Sadu, Perchik, Timofeevskaya, *etc.*

An appreciable number of these varieties have been tested under the Belarusian conditions. The highest adaptivity levels were demonstrated by the Moscow-bred varieties based on the Cis-Baltic climatic type as having large and high-vitamin content fruit; low-growing varieties were selected, these were featured by a desert-grade berry quality and high carotene, oil and ascorbic acid contents. We provide below a comparative tabulated summary of the Russian-bred variety fruits' biochemical compositions based on the bibliographical data and our own studies in the Table 13.1.

Table 13.1: Biochemical Composition of Seabuckthorn Fruit by Varieties

Varieties	*Vitamin, mg per cent*	*Sugar, per cent*	*Carotene, mg per cent*	*Oils, per cent*	*Organic Acids, per cent*
Aromatnaya	125.0	1.7	6.4	6.1	1.3
Botanicheskaya	72.2	1.6	8.6	5.6	3.1
Botanicheskaya Liubitelskaya	152.0	2.8	13.0	4.3	2.65
Vorobievskaya	170.0	3.4	8.6	3.5	3.0
Krasnoplodnaya	70.0-183.0	4.2	7.5	5.7	2.7
Podarok Sadu	101.0	3.4	6.9	5.5	1.7
Otradnaya	176.0	3.2	5.3	5.5	1.9
Perchik	200.0	4.2	7.1	3.5	2.5
Trofimovskaya	183.0	1.8	5.8	3.9	3.5
Avgustina	89.0	6.6	-	-	-
Velikan	157.0	6.6	14.3	6.6	1.5
Vitaminnaya	135.0	4.8	22.0	3.9	2.8
Dar Katuni	66.0-69.0	3.2	3.0	6.3-7.8	1.7
Zolotoi Pochatok	66.0-68.0	5.1	2.8-3.0	3.0	2.1
Maslichnaya	64.0	4.0	26.7	4.7-5.8	1.6
Novost Altaya	50.0	5.5	4.3	5.5-8.2	1.7
Obskaya	135.0-170.0	4.4	2.8-4.1	5.6-8.3	-
Obilnaya	142.0	6.9	2.9	4.9	1.2
Oranzhevaya	240.0-330.0	4.3-5.4	4.3	5.2	1.3-2.3
Panteleyevskaya	86.0	6.9	17.2	5.8	1.2
Prevoskhodnaya	124.0-131.0	6.5	12.6	5.6	2.0
Samorodok	133.0	9.6	14.5	6.9	1.5
Tenga	160.2	9.8	21.5	7.7	1.2
Chuiskaya	134.0	6.4	14.7	6.2	1.3
Yantarnaya	189.4	7.4	10.7	6.6	1.6
Plamennaya	42.0	4.5	12.9	5.1	1.7
Druzhina	167.0	5.6	15.0	3.9-5.6	1.4
Zarnitsa	94.8	3.4	28.8	4.5	2.7
Zolotoi Kaskad	97.2	5.5	9.8	2.6	1.3
Kapriz	89.0	10.2	14.8	4.8	1.2
Ognistaya	73.0	5.0	24.0	5.2	1.3
Parad	142.0	7.9	36.0	3.9	1.7
Zyrianka	110.0	4.9	-	-	-
Podruga	89.0	4.9 -6.5	12.5	2.9	1.4
Ivushka	122.7	3.6-5.4	15.9-18.9	4.8-5.8	1.9-2.7
Krasny Fakel	162.7	5.6	16.2	5.6	3.7
Sibirsky Rumianets	86.3	2.3	27.0	4.1	1.2
Triumf	93.0	3.5	31.4	4.3	1.5
Bayan-Gol	150.0-160.0	7.0	5.5	5.6-6.7	1.2
Baikalsky Rubin	200.0-270.0	7.0	26.0	5.0	1.5
Zaria Dabat	130.0	5.0	4.8	5.7	2.1
Sayana	194.0	8.3	3.7	3.6	-

Individual Development Peculiarities

There is a certain dependence of the plant ontogenesis rhythms on changing rhythms of the ambience and external impacts. The specific rhythms related to the climatic type introduction have not been studied in our country, in practical terms.

An inalienable part of the introduction-related research is represented by studying the phenology of the plants which undergo introduction into a cultivated form under the new growing conditions. The plant life rhythm (phenological phase duration) reflects to a large extent an organism's biological essence and mobile unity with its environment. It goes on and on at a regular phase alteration, while the development phase sequence is shared by all the plants. As the phase-to-phase transition is accompanied by certain conditions, the organism has acquired an ability to transit to a new phase exactly under these conditions. Learning how the plant growth and development rhythms correspond to the climatic factors is, in our opinion, an important and momentous goal within the framework of studying the set of seabuckthorn morpho-biological form peculiarities related to various geographic origins.

Heterogenic phenology manifestation in different climatic types and varieties is indicative of its congenital conditionality and high-level sensitivity to the changing ambience. The modification and genotype variance range ratio is volatile. The latter fact has been proved by some works published after studying the seasonal seabuckthorn development rhythms for the accessions of various geographic provenance and varieties under cultivation. The peculiarities identified (offshoot growth duration, mass leaf fall onset period and vegetation period duration) are determined both by the external factor totality and the plant genotype.

An active seabuckthorn vegetation time begins in the 3rd 10-day period of March through the 1st 10-day period of April (the swelling of the buds phase, or gemmation) (Garanovich *et al.*, 1988). The Siberian climatic type specimens and the Maslichnaya variety begin their vegetation five days earlier than the other experimental objects. Blossoming happens fifteen to twenty-five days after the vegetation starts; this being said, there are also some differences observed in terms of the phase duration in various climatic types and varieties (Figure 13.1). Blossoming begins in the first 10-day period of May, when the daytime temperatures rise up to 16°C and goes on simultaneously with vegetative bud come-out and foliation. The blossoming period depends on the prevailing weather conditions, and in dry and warm weather lasts two to six days.

Under our conditions, the Siberian climatic type and Maslichnaya and Alei variety specimens stand out as to their earlier blossom periods. The Cis-Baltic and Danubian climatic type and Botanicheskaya variety plants occupy an intermediate position. The North-Caucasus climatic type plants lag behind the Siberian ones by five to eight days in terms of their blossoming period. One more prominent fact has to be put to the fore front: leaf individualisation of the male exemplars happens two to three days earlier as compared with the female ones. The blossoms situated along the growing offshoot axis do not open up simultaneously (Figure 13.1).

It is notable that under the Belarusian conditions seabuckthorn has successfully and to the full extent passed through the entire ontogenesis cycle, thus proving its good prospects for the region, ensuring a yield stability and providing evidence, at the same time, of the species' high adaptive potential.

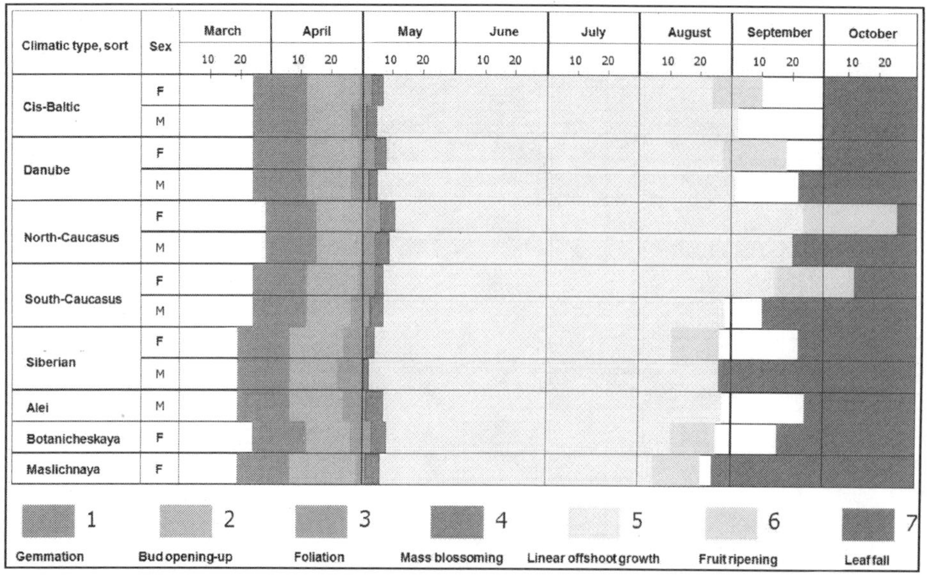

Figure 13.1: Phenological Spectrum of the Seasonal Development in Various Seabuckthorn Climatic Types and Varieties (1989-1990 average).

The species development peculiarities observed under the Belarusian conditions confirm its great flexibility. In our opinion, today's seabuckthorn spreading is limited by the phytocenotic and edaphic factors, whereas the air temperature and precipitation amount play a secondary role.

The Seabuckthorn's biological peculiarity is its initiation of the reproductive period during its third to fifth vegetation period. The preceding vegetation phases, which see cardinal processes going on for the subsequent fruit bearing, *i.e.* latent changes in the internal plant status, determine to a large extent the quantitative and qualitative indicators related to the crop's would-be yields. The start of fructification gives evidence of the plant's good adaptivity to the new conditions. Moreover, the available fruit bearing enables obtaining by way of a further seed reproduction the plants which are more resilient under cultivation. The forms which have a number of specific attributes may be of a breeding interest. Thus, for instance, shrub-like varieties featured by a dense and firm top and good regeneration capabilities, as well as by large and easily separable berries, should be bred.

Intraspecific Variability

An important research into the *Hippophae rhamnoides* intra-species variability, which is deemed crucial from a theoretical viewpoint, has been conducted.

Following the Russian school of thought established by Professor I.P. Yeliseev, the ex-USSR territory has been divided into several major ecological categories, or climatic types, of the species, which predominantly came about under the impact of climatic differences. These do not differ as much in their proliferation, but rather also by a number of external and internal attributes and properties: the Siberian, Central Asian, Caucasian and Cis-Baltic ones. According to its environmental and morphological peculiarities, a climatic type is not homogeneous. For example, within the Siberian climatic type the following ecotypes have been singled out: the Trans-Baikal, Sayan, Altai, East Kazakhstan *et al.* ones (Eliseev, 1982).

The Cis-Baltic specimens are characterised by large and heavy-weight fruit containing a lot of vitamin C, carotene and fatty oils; they are winter-hardy and resilient against asphyxiation. The drawback specific of the climatic type lies in its thorny offshoots. The Caucasian forms are thorny, with small fruit and a low vitamin C and carotene content; however, in terms of their oil content they are close to the Siberian and Cis-Baltic seabuckthorn specimens. The Carpathian form excels in its considerable fruit accumulation of the biologically active substance. The Siberian specimens represent an interest in terms of their large fruit, medium-size height, few thorns, high oil content and winter hardiness; yet, at the same time, they are asphyxiated in the Nonblack Soil Zone soil. Some particularly valuable seabuckthorn forms are to be come across in Altai, Novosibirsk Region, Tuva and Buryatia.

I have studied in greatest detail the environmental, biological and biochemical peculiarities of the Cis-Baltic, Danubian, North-Caucasian, South-Caucasian and Altai climatic types.

The results of the research conducted have been reflected in detail in our publications (Garanovich, 2014; Garanovich *et al.*, 2010; Garanovich *et al.*, 2014). Notably, it has been demonstrated that the Cis-Baltic and Altai (Siberian) climatic types are by far the most promising ones for Belarus. The Cis-Baltic climatic type: the growth processes end simultaneously with those of the Siberian one; it is featured by a superb winter hardiness, large fruits, high vitamin C and seabuckthorn oil contents and resilience against the winter- and spring-time thaws. Its drawbacks lie in its tall height and multiple thorns.

The Altai (Siberian) climatic type: a short vegetation period, high winter hardiness and few thorns. It has the largest fruit with high oil and carotene contents. An important climatic type resilience indicator was represented by their seedling post-wintering preservation rate in the Table 13.2.

As soon as after the three first years of over-wintering natural selection eliminated many introduction population forms of seabuckthorn which had not been adapted to this up growth place. It is worth mentioning that the seabuckthorn seedling winter hardiness is markedly reduced due to the fact that in snowy mild winters the soil does not freeze and a low above-zero temperature is maintained at the root system zone. Plants are eliminated because of the root system asphyxiation, particularly so in respect of pettiest, youngest and most active roots.

Table 13.2: Seabuckthorn (*Hippophae rhamnoides*) Seedling Preservation Rate

Climatic Type	*Preservation Rate, Per cent*				*Over-wintering Grade*
	1	*2*	*3*	*4*	
Cis-Baltic	62.7	65.7	61.3	63.2	I-II
Danubian	50.5	35.9	11.8	32.7	I-III
North- Caucasian	40.9	10.3	41.5	30.9	III
South- Caucasian	52.3	36.1	12.3	33.6	II-III
Siberian	-	38.1	17.7	27.9	I-II

In year 1 the seedling preservation rate ranged between 40.9 per cent and 62.7 per cent. As we can see for ourselves, the amplitude is not too high. Years 2 and 3 saw a higher variation of the winter hardiness indicator. The highest resilience was displayed by the Cis-Baltic climatic type seedlings: their preservation rate made 65.7 per cent and 61.3 per cent, respectively. An expressed resilience reduction was observed among the Danubian, North-Caucasian, South-Caucasian and Siberian climatic type specimens.

In year 2 the North-Caucasian climatic type specimens showed a low resilience (10.3 per cent), while in year 2 so did the Danubian and South-Caucasian climatic type plants: 11.8 per cent and 12.3 per cent, respectively. Changing weather condition both during the vegetation period and the over-wintering time used to lead to changes in the seedling survivability rate. The seedling over-wintering level during the first three years can be assessed in the following way: the Cis-Baltic and Siberian specimens –2 or 3; the Danubian ones –1 through 3; the North-Caucasian one – 3; and the South-Caucasian ones – grades 2 or 3.

The first over-wintering years eliminated the specimens which were unable of surviving even medium parameters, although the meteorological conditions were rather stable and sufficient for the seabuckthorn growth and development. The ambivalent influence of the external impacts on the seedling survival rate allows us to single out among the climatic types under scrutiny the Cis-Baltic accession as by far the most resilient one (63.2 per cent). The remaining specimens continue feeling in ontogenesis a harsh influence of the natural selection.

It is common knowledge that the highest adaptive range is specific of the species featured by vast proliferation areas, including disjunctive ones. Every species' area expands depending on new growth location availability and suitability, as well as on the reproductive potential of its specific populations. The disjunctive areas reflect some century-long changes coming about in the plant existence conditions. According to the concept, the species which are spread in vast spaces occupy a progressive evolutionary position. They are represented by multiple polymorphous populations which, while being an outcome of genetic mutations, have a wide environmental amplitude, a high genetic heterogeneity level and are capable of resisting the influence of varying conditions. The latter is in full true of seabuckthorn, an extremely polymorphous species with a vast disjunctive growth area. Due to the above reasons, the species' adaptive capacity is a considerable one.

Thus wise, based on a population-wide analysis, it has been determined that individual climatic types are capable under the cultivation conditions of forming their artificial introductory populations which differ in their genetic structure from the baseline ones. Almost all the genetic effects are related to maladjustment between the species reaction standard and the indicator levels of some ambient factors. An increased would be generation resilience is possible as a result of the effect produced by the natural and artificial selections and a sufficient quantity of the specimens involved in the reproduction process to support the population genetic pool heterogeneity. The negligible areas occupied by the cultivated introductory populations, as well as the negligible numbers of each generation's surviving specimens might lead to an enhanced inbreeding and a less resilient progeny.

Natural selection fosters preservation of the adaptive attributes through the coming generations, as well as the appearance of the inheritable modifications of a micro-evolutionary nature, thereby strengthening the plant survivability at the population level under the new ambient conditions. Some introduction-stable seabuckthorn climatic types have been identified. It is recommendable to use for cultivation under the Belarusian conditions and as an initial baseline material for breeding to achieve resilience and productivity the first-generation Cis-Baltic climatic type seabuckthorn.

The seabuckthorn (*Hippophae rhamnoides*) which has an extensive proliferation area in different climatic zones forms populations where the plants differ in terms of their bio-morphological attributes. And it is only obvious that the norm of reaction in plants representing different groups will not be identical, when placed in any new conditions. The modification changes affecting the plants will always be limited by the genetically inherent norm of reaction (Goryshina, 1956).

A large number of populations and forms have been singled out within every climatic type in course of expedition-based studies. These materials have provided a basis for a long-term climatic type study under the cultivation conditions in ontogenesis. Some valuable data have been obtained to back up the introduction theory on the peculiarities related to the climatic types, attribute inheritance, adaptive capacity and resilience under the introduction conditions; the latter enabled drawing some conclusions on their breeding prospects in Belarus.

Our domestic breeders can put to their credit the Belarusian Plamennaya variety which has been produced at the Belarusian Horticulture Institute jointly with the Russian researchers as a result of the Otradnaya variety's open pollination. It was passed over to the state-run variety testing system in 1999; it is winter-hardy and drought-resistant. The tree/shrub is of a medium height with a wide top. The variety is an early fruiter, its fruit-bearing period starting at the age of 4 to 5 years. It also produces high yields, approx. 17 tonnes per ha, and is relatively resistant to summertime drying out. The variety is referred to the middle ripening period cohort, the second 10-day period of August. The berries are large (0.8 g) and mainly oval in their shape, orange to red in their colour and moderately pubescent. The variety is universal in its intended purpose. Its advantages are: summertime dry-out resistance, macrocarpous properties, high yields and a high ß-carotene content. Its disadvantages are: a medium thorn supply level, a very compact spadix and a semi-dry fruit disconnection.

Reproduction and Agricultural Engineering

At present, active work is being carried out to reproduce the varietal seabuckthorn and to meet the consumer market demand. When gardening seabuckthorn as a medicinal and horticultural brood, it is vital to apply such reproduction methods which would ensure producing genetically homogeneous seedlings to preserve their varietal allegiance and not to differ from their parental specimens in terms of their commercially valuable attributes. The modern horticulture and ornamental gardening are mainly based on the grafted seeding materials. However, this way is of little effect for seabuckthorn.

The Seabuckthorn is replicated in the seminal and vegetative ways: by way of ratoons and ligneous or herbaceous ("green") grafts. The vegetative reproduction method is used to preserve in the progeny the commercially valuable attributes and properties.

The seminal reproduction method is only acceptable to produce saplings to be used for breeding and introduction purposes, for forest reclamation, field protection and ornamental objectives, as well as to produce parent stock plants. The seabuckthorn, like the majority of the fruit-bearing plants, loses its valuable specific commercial and biological properties, when reproduced by seeds, while due to its dioecious nature approx. 50 per cent of all the seedlings tend to be male ones.

The Seabuckthorn seeds can be planted in late autumn or in early spring. Before planting, they are to be wetted in water and kept for 3 or 4 days under the room temperature. The seeds are then planted into the fertile sandy-loam or slightly clay-loam soil. Sowing is done in 3 to 4 cm deep ribs under the 1 to 4 line pattern, the interlinear distance making 20 cm and that in-between the bands amounting to 70 cm. The little grooves with the seeds sown in them are covered with a friable substrate consisting of organic matter and sand (1:1), the cover layer being 1 to 1.5 cm thick. In case of the springtime sowing, the first sprouts appear 8 to 10 days after.

The Seabuckthorn is characterised by its over-the-ground germination, when the cotyledons are shot up to the soil surface. They represent the first assimilating organs of the germinant. The cotyledons have a wide oval shape, bluntly sharpened at their top and quite often daedalous, somewhat cordate at their base, with small ears, on short stems, fleshy and lustreless on the top and yellowish from the bottom.

The Seabuckthorn sprouts are thinned-out twice: for the first time at the phase when the first two real leaves appear; and for the second time when the fourth pair of leaves appears. The removed saplings are not to be discarded, but rather to be lined out at a new plot of land.

If the care provided proves to be adequate, by autumn time the seedlings grow up to the 30 or 40 cm height and the root crown reach 3 to 4 mm across. In order to achieve a ramified root system, the plant's vertical root is cut off in autumn at the 15 to 18 cm depth.

The most promising method of obtaining genetically homogeneous seabuckthorn planting materials is represented by propagation by herbaceous cuttings. It is based on the herbaceous cutting capability for adventitious root formation and restoration of an integral new organism which does not differ in terms of its biological and genetic attributes from the initial parental plant. The advantages behind propagation by herbaceous cuttings are related to the fact that small-size cuttings are used, whereby a variety of interest can be rapidly propagated given a limited amount of the parental plants. Propagation by herbaceous cuttings is done at the onset of summer after the springtime seeding and planting work is over. The green cuttings strike roots under a special-purpose plastic foil cover with an automatically controlled artificial mist for a subsequent nursery-based completion growing.

The crucial propagation techniques are represented by the following: graft cutting period, selection and cutting techniques, rhizogenous stimulator application, graft cutting condition/environment creation, *etc.* All of these have been fine-tuned in detail and presented in our numerous publications and recommendations.

A critical issue related to introduction of seabuckthorn in cultivation in Belarus was the farming technology aimed at plantation creation and maintenance.

When selecting land plots intended for seabuckthorn plantations, one should take into account that it reacts negatively to the soils of a heavy mechanical composition (it needs light, friable and well aerated fertile soils) and to a close water table. Much of its root system is located in the superficial layer 10 to 40 cm deep with very few roots going as deep as a metre and a half down. The areas of insufficient watering are only suitable for seabuckthorn plantation seeding, if artificial watering is a possibility there. This is very photophilous and does not grow in the shade. An ideal planting option is on an acclivous slope, where the soils are light and where water is not prolific; the trees likewise tend to favour even grounds.

For the majority of the zoned seabuckthorn varieties, when setting up commercial plantations, the most optimal seabuckthorn shrub placement pattern is a 4 × 2 m one. Location of the pollinators (male specimens) among the female plants is alternated with one combined row, where one plant in five is a male one. The pollinator number amounts to 7 to 8 per cent of the total plant quantity.

Organic fertilizers (manure, peat or compost) are introduced in the amount of 10 to 15 kg per sq. m; and the mineral ones –40 g per sq. m for phosphorus and 20 g per sq. m for potassium. Sand is to be added on heavy soils, 10 kg per sq. m. The acidic soils are chalked: for medium-level loamy soils 250 to 400 g per sq. m of agricultural lime is added and for sandy loam soils 250 to 300 g per sq. m. Agricultural lime introduction at the soil preparation stage will suffice for the first 4 or 5 years of the plant life. The saplings are located at the 2 to 2.5 m distance between the plants depending on their growth speed.

All the relevant technological aspects have been fine-tuned in detail and shared with the agricultural experts in the form of special recommendations.

Farm pests and diseases represent a substantial challenge facing the seabuckthorn cultivation success in Belarus. Such seabuckthorn diseases as fruit endomycosis and *Fusarium* or Verticilliaceous wilt have been identified and comparatively well studied. It should be noted that some diseases may be transferred to a new growing area with the planting materials, whereas others can be acquired due to disease transfer from the locally cultivated and local wild flora plants.

Fusarial and Verticilliaceous wilt are seen as the most dangerous fungal diseases which bring death to the plant. Fungi *Verticillium dahliae* Kleb. and *Fusarium sporotrichiella* Bilai have been singled out from the affected branch and root parts. The plant is infected with a fungus through its damaged roots, as well as through its bole, branches and other organs, provided they have any fresh "wounds" to let the infection in.

Verticilliaceous wilt is spread in all the seabuckthorn planting areas. The disease affects the plants once they have entered the fruit-bearing stage, at the age of 5 to 8 years. In early August on some skeletal branches or on whole trees the leaves turn yellow and fall out, just the shoot top leaves remain green. The fruit on such branches prematurely gain the colouring which is characteristic of the variety, but they fail to run in and remain feeble and crinkle. At the end of summer the bark develops reddish swellings and later the bark cracks. Next spring blackened wood can be seen through the cracks. Next year, as a rule, the buds on the affected branch do not open up. Appearing disease becomes apparent also on other branches. An affected tree dies within 1 or 2 vegetation periods. Fungus *Verticillium dahliae* is the pathogenic agent. The fungal mycelium blocks the seabuckthorn conduction system, thus leading to the plants' demise. Fusarial wilt causes the plant drying. Its pathogenic agent is fungus *Fusarium*.

The most affected varieties under the Belarusian conditions are Chuiskaya, Samorodok, Prevoskhodnaya and Kudyrga. The ways to counter the diseases are represented by a timely removal from the land plot of affected shrubs, even if they have individual ill branches. A tree showing any disease signs may not be used for propagation-oriented cutting production.

Endomycosis is a non-infectious disease which appears as a result of the fruit meat being damaged by harsh temperature oscillations between daytime and night or due to mechanical impacts. Novost Altaya variety belongs to one of the most affected ones.

Disease signs are as follows: during the seabuckthorn fruit ripening period some berries having bright spots on their lighted side appear among the brightly coloured berries. The fruit pulp in this spot softens to lose colour subsequently and the entire berry gets soft. The ripe fruit has an unpleasant waft of rank butter. The fruit pulp is penetrated with the *Monilia altaica* fungus mycelia. The ways to counter the disease are represented by picking up ill and partially affected berries; by growing healthy planting material followed by herbaceous cutting disinfection; and by breeding disease-resistant varieties. Green *Capitophoms hippophaes* is the most frequent pest. The aphids and their larvae are light green. The antennae are round shaped and green. The eggs are oval and spend winter on the branches near the buds and in the bark cracks. When the buds open up (in the first half of May) the revived larvae get inside the growing buds, concentrate on the bottom leaf side and suck the young leaf sap. Two weeks after, they develop into wingless females, each of whom giving birth to 40 to 50 larvae. Two or three weeks after the larvae are transformed into winged proliferating females who set up new aphid colonies.

The damaged leaves curl along their central vein, turn yellow and fall down by weakening the shoot and fruit growth. The bark turns black. The *Capitophoms hippophaes* causes a substantial damage to the plants, if its average population exceeds 2 individuals per leaf.

The ways to counter the pest are represented by spraying the plants with a 10 per cent malathion solution (75 g of the agent per 10 l of water), not later than 30 days before the fruit ripens; or with the INTA-VIR agent, 8 g tablets per 10 l of water. It is very effective to spray the plants with an extract of bourtree (*Sambicus nigra*) and tobacco (1 kg and 400 g per 10l of water), as well as with a soapy solution (50 g of laundry soap per 10l of water).

CONCLUSIONS

Thus, when growing such an introduced species, as the seabuckthorn in Belarus is, under the new conditions in place changes of an adaptive nature are coming about both at the specimen level (modification) and at the population-wide one (changing genotype as a result of selection in a number of generations). An introducer's final result or objective is to obtain a group or a whole cultivated population of the plants which are adapted to the new conditions. The intraspecific genetic diversity provides a basis for the species resistance against any unfavourable impacts and is, at the same time, a guarantee of its further evolution.

In the new conditions the introduction work does not only mean growing some acquired materials. Under the impact of isolation and the new climatic environments introductory populations are formed to promote an extended research at the expense of locally reproduced plants. In order to have them preserved, as well as to upgrade the introduced specimens' potential, plantations of subsequent generations have to be set up. The Botanical Garden collection and the entire seminal progeny represent artificially introduced populations. Their structure and characteristic variability have not been sufficiently studied. Natural selection has been in effect during the whole period of the cultivated species studies; due to the latter fact the cultivated population's genetic composition is transformed, certain forms being preserved. When growing the first and subsequent generations, pronounced changes are observed, which changes influence winter hardiness and resilience of seabuckthorn specimens of various climatic types. It is exactly the lack of knowledge in this sphere which provides a constraining factor and is put to the forefront, when introducing subsequent generations of the cultivated species in this specific region.

Thus wise, with due regard to the particularly momentous nature of the issues discussed above, the major vector of our work can be formulated as an overall study of the bio morphological attributes in the generations of artificially introduced populations; their resilience assessment and preservation of the basic seabuckthorn productivity signs.

The research conducted into the entire ontogenetic cycle of *Hippophae rhamnoides* has demonstrated a different-quality nature of its climatic types and populations and a different degree of their resilience and productivity under cultivation.

REFERENCES

1. *Agroclimatic Atlas of the Ukrainian SSR*, 1964. Kiev, 84p. (In Ukrainian)
2. *Agroclimatic Reference Book*, 1969. Minsk, 218p. (In Belarussian)
3. *Agroclimatic Resources of the Georgian SSR*, 1978. Leningrad, 344p. (In Russian)
4. *Agroclimatic Resources of the Lithuanian SSR and Kaliningrad Region of the RSFSR*, 1972. Leningrad, 144p. (In Russian)
5. Alexeeva, S.F. and Shulgin, A.M. 1971. Soil and climatic resources of Altai Territory. In: *Climate and soils*, Leningrad, ðp. 87-94. (In Russian)
6. *Atlas of Medicinal Plant Areas and Resources in the USSR*, 1976. Moscow, 340p. (In Russian)
7. Borodina, I.A. 1970. Joint colchicine and biostimulator action on seabuckthorn (*Hippophae rhamnoides* L.). *Seeds Genetics* 11: 140-142.
8. Churilov, A.K., Bulgakov, N.P., Belokhvost, E.S. and Pashina, I.V. 1969. Essay on the soils of the Botanical Garden of the BSSR Academy of Sciences. In: *Science and Technology*, Minsk, p.36. (In Belarussian)
9. *Climate of Azerbaijan*, 1968. In: *AzSSR Academy of Sciences Publishing House*, Baku. 343p. (In Azerbaijan)
10. *Climate of Minsk*, 1976. Minsk, 288p. (In Belarussian)
11. *Climate of the USSR*, 1961. Leningrad, 290p.; 1962. Leningrad, 360p.; 1961. Leningrad, 300p. (In Russian).
12. *Climate of Ukraine*, 1967. Leningrad, 414p. (In Russian)
13. Dziamentsieu, V.A., Shkliar, A.K. and Yakushka, O.F. 1959. Nature of Belarus. State Educational and Pedagogic Publishing House of the BSSR Ministry of Education, Minsk. 315p. (In Belarussian).
14. Eliseev, I.P. 1982. Some genetic peculiarities of *Hippophae rhamnoides* and their role in selection. In: *Issues of Biochemistry and Physiology of Agricultural Plants Collection of SC*, Works GAI, Gorkiy, pp. 75-77. (In Russian)
15. Garanovich, I.M. 2014. Introduction of seabuckthorn (*Hippophae rhamnoides* L.) in Belarus. In: *Seabuckthorn (Hippophae L.)–A Multipurpose Wonder Plant. Vol. I: Botany, Harvesting and Processing Technologies*, pp. 137-151, Indus Publishing Company, New Delhi, India, 518p.
16. Garanovich, I.M. and Shpitalnaya, T.V. 2010. *Role of Intraspecific Variability and Ontogenetic Peculiarities of the Woody Plants When Introduced in Belarus*, Minsk, 249p. (In Belarussian)
17. Garanovich, I.M. and Shpitalnaya, T.V. 2014. Role of intraspecific variability in introduction of seabuckthorn (*Hippophae rhamnoides* L.) in Belarus. In: *Seabuckthorn (Hippophae L.) –A Multipurpose Wonder Plant. Vol. IV: Emerging Trends in Research and Technologies*, Daya Publishing House, New Delhi, India, pp. 107-139.
18. Garanovich, I. M. and Shpitalnaya, T.V. 1988. Peculiarities of seabuckthorn. BSSR *Academy of Sciences Herald* 5: 10-15.
19. Goryshina, T.K. 1956. Ecological and physiological peculiarities of some acclimatized woody species due to their Geographic Provenance, LSU Herald. *Biological Series* 21: 18-30.
20. GOST (State Standart) 13056.7-93.1995. Seeds of trees and shrubs. *Methods of determining vital capacity*, Minsk, p. 37. (In Belarussian)

21. Korovina, I.A. 1988. An Analysis into the natural seabuckthorn area on the USSR Territory. In: *Biology, Breeding and Agricultural Techniques of Seabuckthorn and Collection of Scientific Works*, Gorky, p. 19-38. (In Russian)

22. Lapin, P.I. 1967. Seasonal rhythm of woody plants development and its role for introduction. *Bull. MBG of the AS of the USSR* 65: 13-18.

23. Maltseva, A.N. 1990. Morphogenesis of *Hippophae rhamnoides* introduced in conditions of the Lower Don. *Abstract of dissertation, Doctor of Biology*: 03.00.05., Moscow, p. 22. (In Russian)

24. *Methods of Phenological Observation in Botanical Gardens of the USSR*, 1975. Moscow, 27p. (In Russian)

25. Molchanov, A.A. and Smirnov, V.V. 1967. *Methods of Studying Growth of Woody Plants*. Moscow, p.9. (In Russian)

26. Rabotnov, T.A. 1950. Aspects of studying population composition for phytocenology purposes. *Issues of Botany* 1: 465-483.

27. Smeyan, N.I., Zinchenko, V.S. and Bogdevich, I.M., *et al.*, 1989. Fertility assessment of Belarusian soils. In: *A Working Edition*, Minsk, p. 7-19. (In Belarussian)

28. Shkliar, A.K. 1962. *Climate of Belarus and Agriculture*, Minsk, 422p. (In Belarussian)

Chapter 14

Distribution and Genetic Resource Study Result of Mongolian Natural Wild Seabuckthorn (*Hippophae rhamnoides* L.)

D. Oyungerel[1*], U. Juuperelmaa[1], Ts. Alimdariya[1], G. Bayaraa[1], S. Battumur[1], G. Altangoo[1], D. Nasanjargal[2], Ts. Tsendeekhuu[2] and K. Otgonbileg[2]

[1]Plant and Agriculture Institute, Darkhan, Uul, Mongolia
[2]Mongolian National Association of Seabuckthorn Growers and Producers, Mongolia
(*E-mail: Oyungerel_2000@yahoo.com)

SUMMARY

Seabuckthorn subspecies (*Hippophae rhamnoides* L. ssp. *mongolica* Rous.) is distributed in confluence of Orkhon-Selenge and basins of Zavkhan Tes, Uvs Tes, Bukhmurun, Tortkhilog Buyant, Khovd, Zavkhan, Borkh and Bulgan rivers. Natural wild seabuckthorn's permanent distribution is in the western part of Mongolia's mountainous region of Altai, Khangai and at the Tes, Zavkhan and Khovd rivers with internal flow to Great Lake concave, which starts from Altai and Khangai ranges' branch mountains and at the Bulgan and other rivers' valley, which have an external flow. By natural and weather condition, Bulgan aimag is located in humid cool, Selenge aimag is in arid cool, Zavkhan aimag is in arid coolish, Govi-Altai and Khovd aimags are in very arid, Uvs aimag is in arid cool region, respectively. Soil humus was in 0-20 sm depth 1.1-2.3 per cent, 20-40 sm depth 0.6-1.7 per cent at the research area. Soil of Zavkhan river basin had lowest humus content (0.76-1.2 per cent), this includes Mongolian sand area and areas with sandy soil. Frame of this study included 18 soums of Selenge, Bulgan, Zavkhan, Govi-Altai, Khovd aimags along six big river basins and revealed resource of 13.5 thousand hectare coverage by Mongolian natural wild seabuckthorn and conducted a mapping. Fruits biochemical analysis was done by oil, vitamin C, sugar content in types of natural population. Analyze result were classified into three groups such as high, medium and low. Oil content was fluctuated between 3.1-4.0 per cent oil content in average 3.5 per cent, sugar 6.0-8.9 per cent, vitamin C content was medium in research sample types. To compare fruit colors with pulp oil and sugar content, fruits with higher content of sugar have from orange to bright yellow color and fruits with higher content of oil have red color. This is explained by synthesis of bio active substance, which belongs to carotenoid group, in the cell of fruit skin. 50-87.6 per cent of total natural wild seabuckthorn has below 10 years age; 13-50 per cent has 10-18 age. Selected 21 forms were taken local MK numbers, registered according to International cultivars, stored in the gene bank of the PAI.

Keywords: *Natural wild seabuckthorn, Distribution, Form, Generation, Resource and Gene bank.*

INTRODUCTION

Seabuckthorn is a plant, which is exclusively adapted in Mongolian climate, soil and geographical condition and has unique resistance to drought freeze; great significance of food, treatment and nature protection. Mongolian seabuckthorn sub species (*Hippophae rhamnoides* L. ssp *mongolica* Rous.) is distributed in confluence of Orkhon-Selenge and basins of Zavkhan Tes, Uvs Tes, Bukhmurun, Tortkhilog Buyant, Khovd, Zavkhan, Borkh and Bulgan rivers.

However, Mongolian seabuckthorn natural wild distribution is in 30,000 ha (Dorjgotov and Zdanevich

Seabuckthorn (Hippophae L.): A Multipurpose Wonder Plant, Vol. 5: Advanced Technologies for Cultivation, Processing, Health Protection and Environmental Conservation (Virendra Singh, *Editor-in-Chief,* 2018), p. 107–113, Astral International Pvt. Ltd., New Delhi, India.

1974). Many years are gone and it's resource has decreased by 50 per cent due to nature and climate change, people's misuse. Lately, substantial works are evolving for seabuckthorn production in many countries of the world. Fruit and berries are started to grow in Mongolia from mid 1950s, and the fruit and berry husbandry is started to develop. Since that time selection of prospective varieties had done and fruit berry orchards were established.

Researcher Ts.Tsendeekhuu (1966) made pheno-geographical map of seabuckthorn population and determined seabuckthorn's distribution origin by pheno-grafical drawing and registered that there are about 30 different forms in Mongolia. He made a hypothesis that Mongolian seabuckthorn was first originated from Asian certain area and it may be migrated to Europe. He determined different natural forms, which differ by maturity period, bush height, weight, color and taste of fruit, pedicle length, oil content (Tsendeekhuu, 1969).

Researcher had stated that the native land of sub species Mongolia (ssp.*mongolica* Ronsi) within the (*Hippophae rhamnoides.* L) biological species is in Mongolia and its western part (Ronsi, 1971; Tsendeekhuu, 1996; Yougchan *et.al*, 1998).

The bushes of natural seabuckthorn are formed in moist meadows and landscapes of river valley in groups with brush (Salix), aspen (Populus), birch (Betula), other bushes and mesophite grassy plants. Subspecies of Mongolian seabuckthorn (*Hippophae rhamnoides* L.ssp *mongolica* Rous.), which are propagated in our country, occur in the basins of rivers Orkhon-Selenge, Tes river of Zavkhan province, Tes, Bukhmurun and Tohilog Buyant rivers of Uvs province, Khovd river, Zavkhan river, Borkh river and Bulgan river. (Tsendeekhuu, 1969) Botanic-Geographical zone includes the Khangai, Mongol Daguur, Khovd, Mongol Altai, depression of Ikh Nuuruud, circles of Gobi-Altai areas.

Natural wild seabuckthorn's permanent distribution is in the western part of Mongolia's mountainous region of Altai, Khangai and at the Tes, Zavkhan and Khovd rivers with internal flow to Great Lake concave, which starts from Altai and Khangai ranges' branch mountains and at the Bulgan and other rivers' valley, which have an external flow.

To see seabuckthorn distribution, natural wild seabuckthorn is distributed in 40-50 per cent of the Orkhon-Selenge rivers' basin, in 30 per cent of the Zavkhan river basin, in 15-20 per cent of the Bulgan river, in 40 per cent of the Khovd river basin, in 40-50 per cent of the Bukhmurun river basin and 20-30 per cent of the Tes river basin. They have been no special succession change revealed (research team 2013).

Research Aim

For determining Mongolian natural seabuckthorn's resource, location, inventory conducting, revealing of its extra ordinary useful form and making out of gene pool following objectives were estimated. Here in:

1. Mapping of resource area, conducting of complete inventory in that area, geo-botanical writings of the population should be done, succession should be determined.
2. Evaluation should be done for the natural forms and extra ordinary useful forms should be revealed.
3. Seabuckthorn aging, regrowth, ratio of male and female bushes, soil structure and water supply should be estimated and guidance for rational utilization, protection and rehabilitation should be processed.

MATERIALS AND METHODS

Research Methodology

1. Use of GIS in Mapping of Natural Seabuckthorn's Resource

High resolution information, which based on the GIS and remote sensing system, was used for the mapping of natural seabuckthorn's resource, determining locations and estimating expansion field amount. Geographical data were processed by two modern vector models (dot, linear, polygon), raster by the GIS, Arc/INFO, Arc/VIEW, ArcGIS programs according to the global standard. Green vegetation spectrum curve was used for checking available vegetation on the remote sensing information and comparative evaluation was done for different plants spectrum curve.

2. Determining of Plants within Seabuckthorn Population, Conducting Geo-botanical Writings and Determining of their Succession

Common method was used for geo-botanical writings of the population forming plants and 13 characteristics are used for evaluation of three, bush and grassy plant species. Succession (autogenic or allogeneic) of the natural bushes was determined. Recreation succession influence was determined in each bush.

3. Estimating of Seabuckthorn Aging, Regrowth and Ratio of Male and Female Bushes

Natural seabuckthorn aging was estimated by following method:

1. Seabuckthorn age from 1 to 10 years was determined by following formula:

$$A = t + 1$$

A: Seabuckthorn age

t: Amount of branch junctions and nodules

2. Seabuckthorn age above 10 years was determined by following formula:

$$A = \frac{t^*H}{H - h}$$

t: Amount of branch junctions and nodules

H: Bush height/m/

h: Main stem height from soil surface to the first branch junctions and nodule/m/

4. Biological and Morphological Evaluation of the Natural Populations

Morphological evaluation was done according to the work program, which elaborated by the I.V. Michurin's Research institute of Fruit and berry, for variety testing and multiplication of fruit and berry cultivars.

Total sugar content was determined by the Bertran's method, oil content by Kokslet method, Vitamin C content by Murr's method, respectively. Samples were taken from each type and forms for storing life genetic resources, registration was done and stored in Agricultural cultivars gene pool.

RESULTS

Natural and Weather Condition

Primary research of the natural wild seabuckthorn conducted in 1974. Therefore we took 1961-1990 periods as basic for our study and compared to last 20 years (1991-2010) and made analysis in climate change in the regions, basing on six weather stations data.

Annual mean temperature is raised by 1.1°C in last 20 years compare to previous 30 years, active temperature sum (above +10°C) raised by 87° in Selenge aimag, 220° in Bulgan aimag, 179° in Zavkhan aimag, 56° in Uvs aimag and by 292° in Khovd aimag, respectively. Maximum raise of active temperature sum was detected in Kovd aimag 2117-2409°C, but this is not an effecting factor to seabuckthorn growth.

Annual mean precipitation was increased by 0.9-29.8 mm in last 20 years. But Bulgan aimag's annual mean precipitation was decreased by 73.9 mm. By natural and weather condition, Bulgan aimag is located in humid cool, Selenge aimag is in arid cool, Zavkhan aimag is in arid coolish, Govi-Altai and Khovd aimags are in very arid, Uvs aimag is in arid cool region, respectively.

Soil Condition of Research Area

Lately soil productive fertility is diminishing so fast, adjacent to it desertification, soil erosion processes are going intensively globally. Among the soil fertility formation factors, organic matter role and significance are very essential and it is stated by the scientists (V.G.Mineev 1978). Seabuckthorn grows in light soil, light soil allows its root distribution and formation of natural plantations as woodland. Therefore seabuckthorn is cultivated for protecting soil erosion.

Table 14.1: Weather Conditions of Studied Sites in Mongolia

Name of Aimag	*Air Mean Temperature, °C*		*Active Temperature Above 10°C*		*Percipitation, mm*		*Number of Frost Free Days*	
	1961-1990	*1991-2010*	*1961-1990*	*1991-2010*	*1961-1990*	*1991-2010*	*1961-1990*	*1991-2010*
Selenge	0.1	1.1	1999	2086	244.5	274.3	190	192
Bulgan	-0.8	0.2	1792	2012	334.7	260.8	194	203
Zavkhan	-2.4	-1.4	1413	1592	218.4	209.4	184	189
Uvs	-3.5	-2.0	2087	2143	137.7	150.7	187	190
Khovd	-0.2	1.5	2117	2409	127.7	128.6	199	209
Govi-Altai	-1.7	-0.6	1186	1344	176.5	180.8	214	222

Table 14.2: Soil Analyze Report of Certified Laboratory, Nr10/01

Name of Aimag	*Depth, sm*	*Humus per cent*	*pH (H_2O)*	*N-NO_3 mg/kg*	*P_2O_5 mg/100g*	*$_2$ mg/100g*
Zavkhan	0-20	1.2	8.1	4.12	2.0	7.3
	20-40	0.76	8.1	2.0	1.5	5.4
Khovd	0-20	1.9	8.2	6.5	2.8	9.2
	20-40	1.3	8.2	5.1	2.1	7.0
Uvs	0-20	2.3	8.8	5.2	2.0	10.0
	20-40	1.7	8.8	4.0	1.4	6.9
Selenge	0-20	2.2	7.7	4.9	2.9	8.3
	20-40	1.2	7.8	4.4	2.3	6.5
Bulgan	0-20	1.9	7.8	6.8	2.6	9.4
	20-40	1.3	7.8	5.7	2.0	8.3
Govi-Altai	0-20	1.8	7.2	2.7	1.7	4.5
	20-40	1.0	7.2	3.2	2.1	5.2

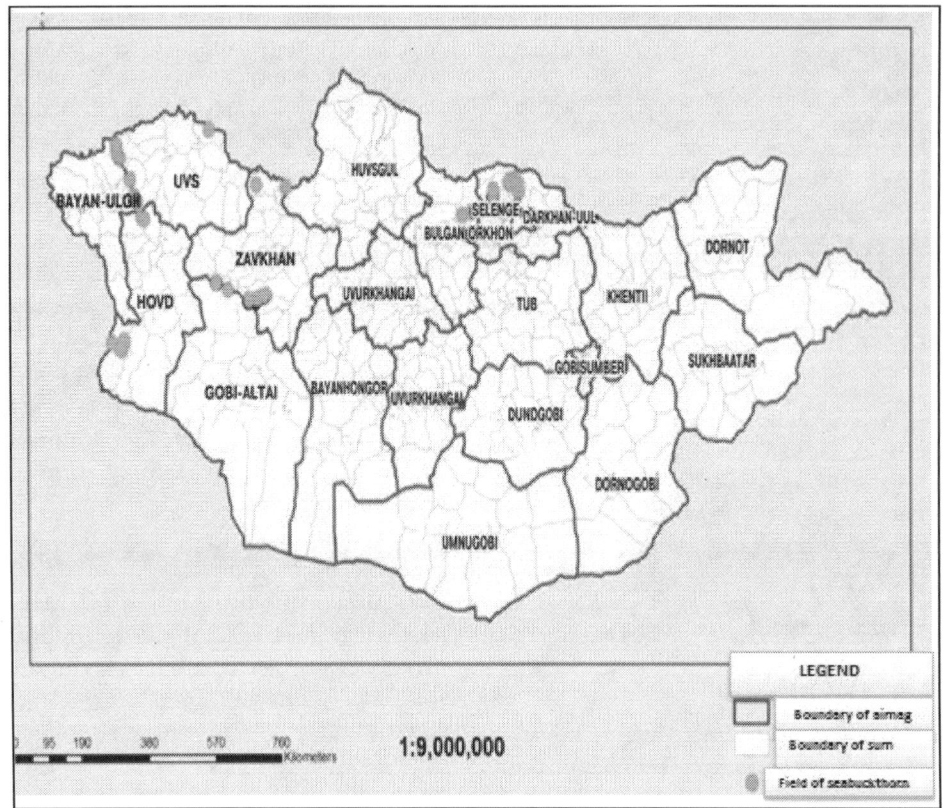

Figure 14.1: Distribution of the Mongolian Natural Wild Seabuckthorn and Field Amount.

As the soil analyzes result, soil humus was 1.1-2.3 per cent at 0-20 sm depth, 0.6-1.7 per cent at 20-40 sm depth.

To compare soil humus content by aimags, soil of Zavkhan river basin has the lowest humus content (0.76-1.2 per cent), that was the Mongolian sand area has sandy soil.

Frame of this study included 18 soums of Selenge, Bulgan, Zavkhan, Govi-Altai, Khovd aimags along six big river basins and revealed resource of 13.5 thousand hectare coverage by Mongolian natural wild seabuckthorn and a mapping was done.

By the natural seabuckthorn woodland location numbers, and fields minimum field was 1.5 hectare the biggest field was 1861.9 hectare. Outlasting big woodlands along big river basins were in Uvs, Zavkhan and Selenge aimag, mainly.

Biological Peculiarity of the Natural Wild Seabuckthorn

Natural seabuckthorn's populations are different from cultivated seabuckthorn. Useful characters are hidden (recessive) in cultivated varieties, but they are dominant in natural seabuckthorn (Tsendeekhuu, 2005). Fruit biochemical characters oil, vitamin C, sugar contents were determined in natural seabuckthorn's populations' forms and each characters were grouped as high, medium and low content.

Oil Content

Oil content was fluctuating between 3.1 and 4.0 per cent, average was 3.5 per cent in selected forms.

Table 14.3: Natural Wild Seabuckthorn Orchards of Mongolia, by Aimags

Sl.No.	*Aimag Name*	*Number of Locations, Piece*	*Field Amount, hec From Low to High*	*Total Field, hec*
1	Govi-Altai	1	53,3	53.3
2	Bulgan	2	77.5-237.0	309.5
3	Selenge	6	17-1162.5	2216.0
4	Uvs	6	161.9-1861.9	5183.7
5	Khovd	6	2,4-1045.6	1171.8
6	Zavkhan	11	1.5-209.9	4579.9
Total		32	-	13513.5

Table 14.4: Fruit Oil Content of Selected Forms

Sl.No.	*Form*	*Oil, Per cent*
1	Uvs-Umnugovi+Buurugiin baruun davkharga 01	4.0
2	Uvs-Bukhmurun-Tsagaanburaa02	3.3
3	Uvs-Bukhmurun-Tsagaanburaa01	3.1
4	Zavkhan-Durvuljin-Khar butnii ekh 01	3.5
5	Khovd-Sumiin ganuu 02	3.2
6	Bulgan-Khangal-Munkhtsagaan 01	3.9
Average		3.5

To compare the river basins, seabuckthorn oil contents of Bukhmurun, Bulgan rivers are high/3.1-4.0/, this is related to the location, weather and elevation (Table 14.4).

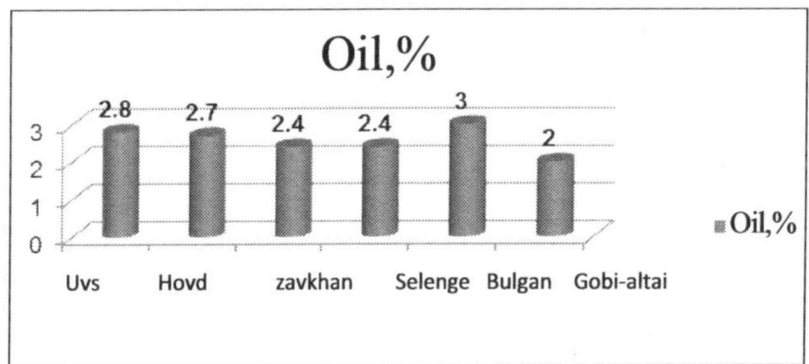

Figure 14.2: Oil Content of Seabuckthorn Fruit.

Sugar Content

Cold resistance doesn't detect in normal condition, but in the case of cold stress the adaptation response arises. Due to cold stress, there formats osmosis regulating substance. Such protecting substance is sugar. General sugar content was fluctuating between 6.0 and 8.9 per cent in selected forms.

Table 14.5: Fruit Sugar Content of Selected Forms

Sl.No.	*Form*	*Sugar, Per cent*
1	Zavkhan-Borkhiingol-Khukhtolgoin bulan 01	8.9
2	Zavkhan-Borkhiingol-Chatsarganatiin tohoi01	8.4
3	Zavkhan-Kharganat-Mongol els 03	8.3
4	Zavkhan-Zuriin bor khoshuu 01	8.3
5	Uvs-Bukhmurun-Arzgariin gol 01	7.2
6	Khovd-Sumiin ganuu 01	6.1
7	Uvs-Dundgol-Dankhariin aral 01	6.0
8	Govi-Altai-Ar uvur chatsarganat	7.8

Sugar content of seabuckthorn forms in Zavkhan and Govi-Altai aimag (along the Zavkhan river basin) was higher (8.3-8.9 per cent) than other seabuckthorn forms, this indicates that sugar content is related to local soil, weather condition, geo-graphical location and elevation (Table 14.5).

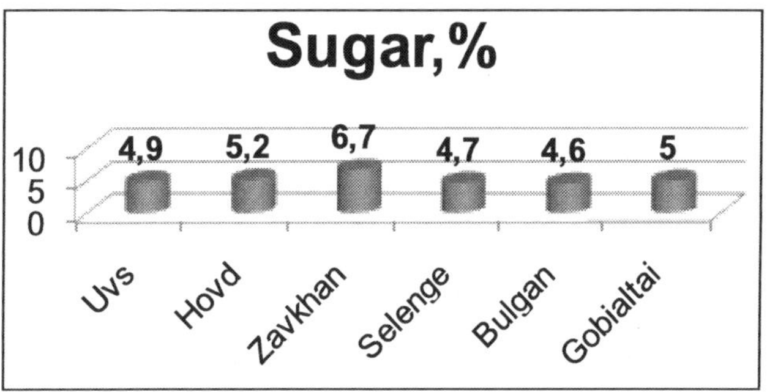

Figure 14.3: Sugar Content of Seabuckthorn Fruit.

Zavkhan and Govi-Altai aimag area is elevated by 1357.5-1537.7 m above sea level, but Orkhon-Selenge basin's elevated by 621.1-771.4 m. Seabuckthorn sugar content, which grown in Orkhon-Selenge area was lower.

Content of Vitamin C

Vitamin C content was between 161.9 and 251.7 mg/ per cent in average 206.8 mg/per cent. To see this result vitamin C content was medium in general.

Table 14.6: Fruit Vitamin C Content of Selected Forms

Sl.No.	*Form*	*Vitamin C mg/per cent*
1	Uvs-Umnugovi-Buurugiin baruun davkharga 01	251,7
2	Selenge-Tsagaannuur-Deed khooloin ekh 01	237,6
3	Selenge-Tsagaannuur-Aral 02	228,8
4	Zavkhan-Borkhiin gol-Khukh tolgoin bulan	211,2
5	Selenge-Zuunburen Narin 02	197,1
6	Selenge-Zuunburen Tami r02	184,8
7	Selenge-Zuunburen Norovbanzadiin tokhoi 03	161,9
8	Zavkhan-Bayan tes-Khosiin gol 01	181,3
Average		206.8

Seabuckthorn of Orkhon-Selenge river basins have higher vitamin C content (Figure 14.4).

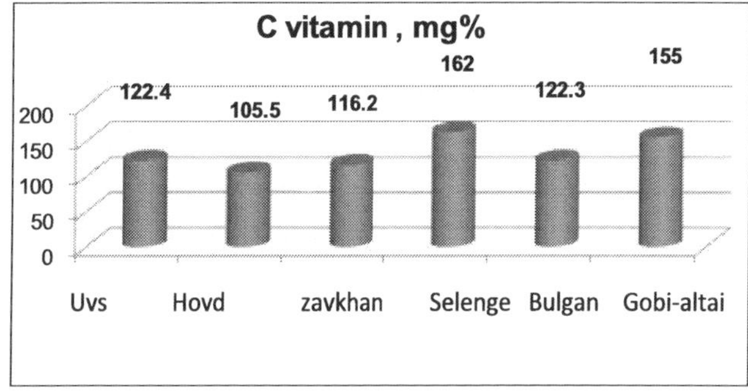

Figure 14.4: Vitamin C Content of Seabuckthorn Fruit.

In all populations of Mongolian areal, yellow color was 58 per cent, orange 35 per cent and red 7 per cent in average respectively. But the red color was different in

Table 14.7: Relation of Fruit some Biochemical Characteristics and Color

Name of Selected from	*Biochemical Characteristic*	*Fruit Color*
Oil Content, per cent		
Uvs-Umnugovi-Buurugiin baruun davkharga 01	4.0	Red
Uvs-Bukhmurun-Tsagaanburaa 02	3.3	Golden yellow
Uvs-Bukhmurun-Tsagaanburaa 01	3.1	Red yellow
Zavkhan-Durvuljin-Khar butnii ekh 01	3.5	Red yellow
Khovd-Sumiin ganuu 02	3.2	Reddish
Bulgan-Khangal-Munkhtsagaan 01	3.9	Reddish
Sugar Content, per cent		
Zavkhan-Borkhiingol-Khukhtolgoin bulan 01	8,9	Yellow
Zavkhan-Borkhiingol-Chatsarganatiin tohoi 01	8,4	Reddish yellow - orange
Zavkhan-Kharganat-Mongol els 03	8,3	
Zavkhan-Zuriin bor khoshuu 01	8,3	Yellowish
Uvs-Bukhmurun-Arzgariin gol 01	7,2	Reddish yellow - orange
Khovd-Sumiin ganuu 01	6,1	Red
Uvs-Dundgol-Dankhariin aral 01	6,0	Yellow

each population. For instance, all seabuckthorn berries in major area of Uvs aimag had red color and higher oil content (Table 14.7).

Sugar content was higher in seabuckthorn of Borkh River and Mongolian sand of Zavkhan aimag, Arzgar river of Uvs aimag, but those seabuckthorn had from orange to bright yellow color (Table 14.7). To see the results, color of forms with higher content of sugar was from orange to bright yellow, but color of higher oil containing forms was red. This is explained by formation of bio-active substances, which synthesis inside the cell of berry's shell.

Seabuckthorn Aging and Statement of Regrowth

Bush aging was determined by the methodology of Forest resource and aging/S.Dashdavaa, 2009. 50-87.6 per cent of total natural wild seabuckthorn has below 10 years age, 13-50 per cent has 10-18 age. To see this result trees and bushes are relatively young. Potential age of fruit productivity for fruit and berry plant is 7-25 years.

Seabuckthorn aging in Selenge aimag varies between 7-20 years and this is related to their good regrowth.

Ex vitro Storing of Selected Forms and Genes

Basic of plant diversity existence is its genetic source. Therefore plant gene pool is called center of plant genetic resource. We evaluated 43 natural wild seabuckthorn forms of 45 soums of six aimags Mongolia and samples were taken from each forms and analysed. 21 forms were selected by their special characters taken local or MK numbers, registered in the PAI's genebank according to International cultivars genebank's rule, and given to the main and work collection.

CONCLUSION

Mongolian natural wild seabuckthorn is covering 13.5 thousand hectare area along six river basins such as Orkhon-Selenge, Bulgan, Khovd, Zavkhan, Tes and Bukhmurun. Within the frame of the study, 43 forms

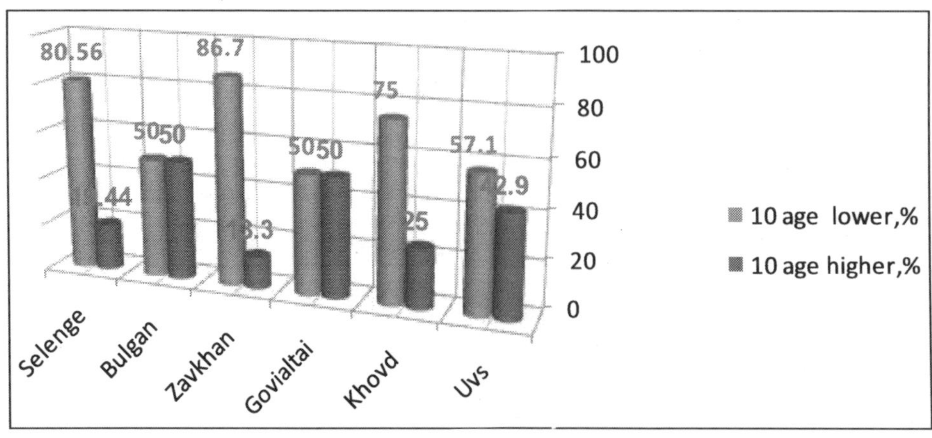

Figure 14.5: Seabuckthorn Aging.

Figure 14.6: Stored Samples of Agricultural Cultivars Gene Bank.

were selected by their morphology biology characters. There determined 21 forms as important selection basic material and genetic resource. By chemical composition, Mongolian natural wild seabuckthorn contains medium sugar, low oil, and medium vitamin C content, respectively. Here high sugar content was revealed in seabuckthorn orchards along Zavkhan rivers, high oil content in orchards in Uvs aimag and high vitamin C content in orhcards along Orkhon-Selenge and Tes rivers but high acidity was revealed in Bayantes river in Zavkhan aimag, respectively. 50-87.6 per cent of total natural wild seabuckthorn has over 10 years age, 13-50 per cent has 10-18 age. To see this result trees and bushes are relatively young at the natural orchards.

REFERENCES

1. Ageeva, L. D. and Cheshuina, V.P. 1983. Oil and carotene content of seabuckthorn varieties. In: *Biology, Chemistry and Pharmacology of Seabuckthorn,* Novosibirsk.
2. Aksenova, N.A. and Dolgacheva, V.S. 1998. New variety samples of seabuckthorn selection at the Botanical Garden Moscow State University, Moscow.
3. Amsheev, Đ.Ì. 1998. Ecosystem ecologation and management of seabuckthorn growers in Buriatii". *Materials of International Symposium on Seabuckthorn,* Novosibirsk.
4. Avdai, C. 2013. Seabuckthorn, UB.
5. Grubov, V.I. 2008. Guide book for distinguishing vascular plants of Mongolia, UB.
6. Goncharaov, P. L. 1998. *Materials of International Symposium on Seabuckthorn,* Novosibirsk.
7. Dashdavaa, 2009. *Handbook for Forest Worker.* UB.
8. Juupelmaa, U. 2011. Seabuckthorn, Darkhan-Uul aimag.
9. Ligaa, U. and Ochirbat, G. 2005. Dictionary for Ethnic Botany, UB.
10. *Proceedings of scientific conference for the "50th Anniversary for Fruit and Berry Branch of Mongolia",* 2005, UB.
11. Nasanjargal, D., Ninj, B. and Ichinkhorloo, D. 2014. *Reference book of the Mongolian fruit and Berry Branch,* UB.
12. Darkhan-Uul Aimag, 2013. *Research Result Report of Distribution and Genetic Resource of Mongolian Natural Wild Seabuckthorn,* Mongolia.
13. Tsendeekhuu, T. 2009. *Plant physiology,* ÓÁ.
14. Purev, G. 2011. *Ecosystem of Mongolia.* Darkhan-Uul aimag.
15. UB. 2014. "Agricultural science". In: *Proceedings of scientific conference for the 60th Anniversary for Fruit and Berry Branch of Mongolia.*
16. Kichina, V.V. 1984. *Genetic and Selection of Fruit Cultivars,* Moscow.

Part III

PHYTOCHEMISTRY AND BIOCHEMISTRY

Chapter 15

Seabuckthorn: A Rich Source of Bioactive Tocopherols and Tocotrienols

Debabrata Acharya*, Pranita Joshi, Chamarthi K. Srikanth, Vishal Menon and Jose Paul

New Product Development and Research, Technology Centre,
Synthite Industries Limited, Kadayiruppu, Kolenchery, Ernakulam – 682 311, Kerala, India
(*E-mail: debabrata@synthite.com)

SUMMARY

Hippophae rhamnoides L., commonly known as seabuckthorn, is a plant of family *Elaeagnaceae*, native to Europe and Asia grows at high altitude under cold climatic condition. The seed and fruit oils of seabuckthorn find use in skin and mucosal disorders in both humans and animals. A large number of bioactive substances including vitamin C, E, carotenoids, omega 3, 6 and 7 fatty acids and organic acids make it a multipurpose plant having high nutraceutical, therapeutic and cosmeceutical potential. Seabuckthorn composed of tocopherol and tocotrienol derivatives render its potent antioxidant activity that can prevent oxidative damage of cells and macromolecules such as proteins, nucleic acids and lipids leading to various clinical disorders. Deficiency of vitamin E in children is known to cause nervous system defects and hemolysis. Vitamin E components of seabuckthorn (α, β, γ, δ tocopherols and α, β, γ, δ tocotrienols) are mentioned to inhibit lipid peroxidation thereby protecting from neuro-degeneration, coronary artery disease and atherosclerosis. Tocotrienols are gaining more focus due to rapid penetration and preferential accumulation in stratum corneum of the skin as well as due to faster free radical neutralization. Current study involves fatty-acid profiling and quantitative analysis of α-tocopherol and total tocotrienol content of supercritical CO_2 extracted seed, pulp and skin oils of *H. rhamnoides* ssp. *turkestanica* obtained from high-altitude of Himalayan region. The analysis method comprises of advanced chromatographic techniques *viz.* GC and HPLC. Comparative study of seed, pulp and skin oil vitamin E components showed presence of maximum content of total tocotrienols 290 mg/100 g in seed oil (with omega 6 fatty acid 41.5 per cent and omega 3 fatty acid 22.42 per cent) followed by skin oil 260 mg/100 g (with 47.17 per cent total omega 7 fatty acids-collective value of palmitoleic acid and cis-vaccenic acid) and pulp oil having 190 mg/100 g of total tocotrienols with 51.78 per cent omega 7 fatty acid. The α- tocopherol content was found to be maximum in skin oil (120 mg/100 g) followed by 100 mg/100 g in seed oil and 80 mg/100 g in pulp oil. These Vitamin E rich extracts can be highly beneficial in health supplements as soft gel and premix forms, functional food and beverages and cosmetic and personal care products.

***Keywords**: Seabuckthorn, Vitamin E, Tocopherols, Tocotrienols, Antioxidant, Omega 6 fatty acid, Omega 3 fatty acid and Omega 7 fatty acid.*

INTRODUCTION

Hippophae rhamnoides L., commonly known as seabuckthorn, is a plant of family *Elaeagnaceae*, native to Europe and Asia (Rousi, 1971; Chen *et al.*, 1990). Seabuckthorn is a nitrogen fixing deciduous shrub that grows at high altitudes in cold climates as reported by Rousi (1971). The high potential of seabuckthorn as phytonutrients has given scientific exploration of the plant for development of various health products. All plant parts of seabuckthorn are considered to be good source of a large number of bioactive substances like vitamins (A, C, E and K), Carotenoids, flavonoids (isorhamnetic, quercetin, myrecitin, kaempferol), organic acids (malic acid, oxalic acid, ascorbic acid *etc.*), steroids (ergosterols, stigma-sterol, lanosterols and amyrins), tannins, fatty acids and some essential amino acids (Kallio *et al.*, 2002a; Yang and Kallio, 2001). These bioactive compounds make it a multipurpose

Seabuckthorn (Hippophae L.): A Multipurpose Wonder Plant, Vol. 5: Advanced Technologies for Cultivation, Processing, Health Protection and Environmental Conservation (Virendra Singh, *Editor-in-Chief*, 2018), p. 117–125, Astral International Pvt. Ltd., New Delhi, India.

plant having high nutraceutical, therapeutic and cosmeceutical potential (Kagliwal *et al.*, 2012).

Seabuckthorn has been used in traditional medicines in China, Tibet, Mongolia for skin diseases, lung disorders (Wu, 1991). Traditionally, the seed and fruit oils of seabuckthorn find use in skin and mucosal disorders in both humans and animals (Pintea *et al.*, 2001). Seabuckthorn is studied for bioactives and their implication for antioxidant activity. Importance of antioxidant compounds such as vitamin E, vitamin C and carotenoids are well mentioned in studies referring to oxidative damage of the cells. Various reasons for oxidative stress are certain drugs used in treatment of diseases and the injury to body tissues. In the latter, release of free radicals lead to activation of phagocytes and hence production of O_2 radicals and H_2O_2, release of catalytic metal ions, damage to mitochondria due to oxidative stress releases O_2 and forms more oxygen radicals (Frei, 1994). Cellular exposure to exogenously and endogenously generated oxidants causes macromolecular damage including protein oxidation, lipid peroxidation and nucleic acid instability and mutation (Halliwell, 1998; Kallio *et al.*, 2002b). Oxidative damage to cells through increasing amounts of superoxide radicals leads to various diseases and disorders (Halliwell, 1997) such as neurodegenerative diseases, metabolic disorders (Halliwell, 1998), cardiovascular diseases *e.g.* atherosclerosis via oxidative modification of low density lipoprotein (Steinberg *et al.*, 1989) and ischemic reperfusion injury to heart tissues, rheumatoid arthritis and stress induced gastric ulcer and inflammatory bowel diseases (Das *et al.*, 1997). Cells however possess different antioxidant defense mechanisms for protection. Antioxidant defense follows several steps *viz.* first line of defense being free radical formation suppression, second line of defence which is suppressing chain initiation and breaking chain propagation reactions, third category is repair and fourth line is adaptation at molecular level for formation and transport of antioxidant to proper site. The limited presence of this oxidative defense yet makes cells go through continuous oxidative damage (Noguchi *et al.*, 2000).

Vitamin E is a non-enzymatic dietary antioxidant compound located in cell membranes and acts through second line of defenseas major chain breaking

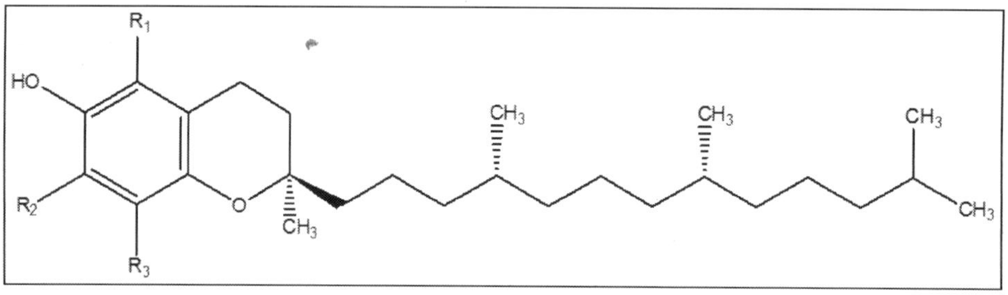

TOCOPHEROLS

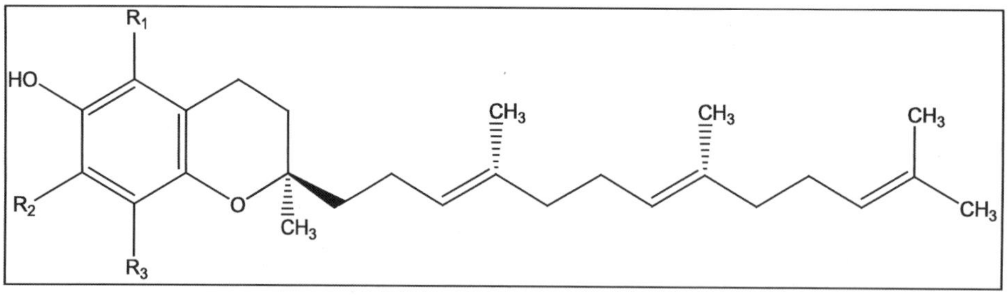

TOCOTRIENOLS

Figure 15.1: A Schematic Presentation of Tocopherols and Tocotrienols.

Compound	R_1	R_2	R_3	*M.W.*	*M.For.*
α-Tocopherol	CH_3	CH_3	CH_3	430	$C_{29}H_{50}O_2$
α-Tocotrienol	CH_3	CH_3	CH_3	424	$C_{29}H_{44}O_2$
β-Tocopherol	CH_3	CH_3	CH_3	416	$C_{29}H_{48}O_2$
β-Tocotrienol	CH_3	CH_3	CH_3	410	$C_{29}H_{42}O_2$
γ-Tocopherol	CH_3	CH_3	CH_3	416	$C_{29}H_{48}O_2$
γ-Tocotrienol	CH_3	CH_3	CH_3	410	$C_{29}H_{42}O_2$
δ-Tocopherol	CH_3	CH_3	CH_3	402	$C_{29}H_{46}O_2$
δ-Tocotrienol	CH_3	CH_3	CH_3	396	$C_{29}H_{40}O_2$

antioxidant. It controls inhibition of lipid peroxidation by converting lipid peroxyl radicals to hydroperoxides and a tocopheroxyl radical which in turn gets oxidized to respective quinine or might be reduced by ascorbic acid and GSH. Tocopherols and tocotrienols, collectively known as Vitamin E are the most effective antioxidants in cardiovascular diseases by inhibition of peroxidation through LDL (Losonczy *et al.*, 1996). Deficiency of the same is known to cause atherosclerosis, cancer (Knekt, 1993) and improper development of nervous system in children (Sokol, 1996). Tocotrienols are gaining more focus due to rapid penetration and preferential accumulation in stratum corneum of the skin as well as due to faster free radical neutralization (tocotrienol.org).

All the four isomers of tocopherols (α, β, γ, δ) and tocotrienols(α, β, γ, δ) are present in seabuckthorn oil obtained from seed and the soft parts of berry (Yoshida *et al.*, 2007). Individual tocopherols and tocotrienols have different level of bioactivity making some of them very important.

Studies report the effectiveness and mechanism of action of seabuckthorn pulp and seed oil. Protection of mitochondria of liver cells from damage caused by cold exposure in rats is reported by Song and Gao (1995) by use of pulp oil. Pulp oil is also mentioned for decline in malondialdehyde (MDA) induced erythrocyte hemolysis, enhanced–SH group levels and Na, K-ATPase and glucose-6 phosphate dehydrase activities of erythrocyte cell membrane in pigs (Rui and Gao, 1989). Seed oil supplementation showed an increase in the glutathione peroxidase activity by 16 per cent, sialic acid content by 46 per cent, sulfhydryl (–SH) group containing compounds by 73 per cent, Na, K-ATPase activity by 44 per cent in rats erythrocyte cell membrane (Ji and Gao, 1991a). Shi *et al.* (1994) reported decline in MDA formation in Cu^{+2} catalyzed *in vitro* oxidation model in presence of seed oil. It is also reported to enhance superoxide dismutase levels, decrease MDA and maintain integrity of mitochondria, endoplasmic reticulum and myofibril in the myocardium levels after exercise. Along with vitamin E, carotenoids and vitamin C of seabuckthorn are mentioned to have antioxidant activity and that the synergistic effect of vitamin E and vitamin C being more than individual compounds.

Various extraction methods have been tried and tested for extraction of oil from seed and pulp of seabuckthorn berry. Supercritical fluid extraction has gained special focus as it is a green process (Kagliwal *et al.*, 2012) and hence gives products without traces of any harmful organic solvents. Carbon dioxide is the most commonly used solvent in supercritical fluid extraction (Risto and Yang, 2003). Supercritical carbon dioxide (SC-CO_2) extracts, oils and oil soluble bioactive compounds without thermal and oxidative changes.

The present work reports the amount of bioactive compounds present in *H. rhamnoides* ssp. *turkestanica* seed, pulp and skin oil obtained by supercritical CO_2 extraction with emphasis on fatty acid profile and vitamin E components *viz.* tocopherol and tocotrienols.

MATERIALS AND METHODS

Raw Material

Seabuckthorn (*Hippophae rhamnoides* spp. *turkestanica*) berries were harvested from high altitude of Himalayan region in September 2014. The seed, pulp and skin were obtained from the berries by mechanical separation. Plant specimen was studied for identification with the help of experts from Defence Institute of High Altitude Research (DIHAR), Leh and a voucher sample is kept in botanical store of quality control department of Synthite Industries Limited, Kerala. The seed is oval in shape and 0.4- 0.5 cm length with brown color. Seabuckthorn pulp was orange in color and pulpy by nature with solids content of 10-11 per cent. Berry skin is the outermost layer of the berry and it appears dark orange when dried.

Chemicals and Reagents

Working standard of vitamin E (73.7 per cent pure) was used. The said standard was calibrated and validated by MPOB standard following AOCS-Ce-8-89 method. n-Hexane, methanol and BF_3 were obtained from Merck Ltd. Other chemicals and reagents were of analytical grade.

SC-CO_2 Extraction of Seabuckthorn Seed, Pulp and Skin

Dried and separated seeds were preprocessed to obtain uniform flakes of raw material. Pulp was processed to separate the juice from the solid content of the pulp. Wet pulp obtained this way was sundried to obtain dried pulp. Dried berry skin was ground and used for supercritical CO_2 extraction. All these materials were subjected to Supercritical CO_2 extraction in a pilot-plant scale supercritical fluid extractor (Nantong Huaxing Petroleum Instrument Co. Ltd., Jiangsu province, China, model SFE 231-50-6) comprising a 5L and a 1L extraction cell and three different separators (S1, S2 and S3), each of 2L, 2L and 1L capacity, with independent control of temperature and pressure. For each extraction, the cell was filled with 1 kg of dried Seabuckthorn seed, pulp and skin. Extractions were performed at 318 K and 38 MPa. Separators were maintained at 318 K and CO_2 flow rate was set to 28 kg/h with total extraction time of 4 hrs.

Determination of Tocopherol and Tocotrienols

Sample preparation of the three different extracts was done by dissolving in *n*-hexane followed by filtration through 0.45 μ syringe filter. Further, these samples were subjected to analysis by HPLC. The amount of α- tocopherol and α, β, γ, δ tocotrienols was analyzed by Shimadzu HPLC using Phenomenex Luna 5μ Silica, 100A° column (250 mm õ 4.6 mm õ 5μ) using Hexane/Isopropyl alcohol/acetic acid (100:0.5:0.1) (v/v/v) mobile phase at 1.0 ml/min. The tocopherol and tocotrienol content were monitored at 295 nm with UV detector. The peak identity was confirmed by the retention time and characteristic UV spectra of respective standard. Peak areas were used to calculate α- tocopherol and α, β, γ, δ tocotrienols.

Fatty Acid Profiling using Gas Chromatography

Seabuckthorn seed, pulp and skin oils extracted using SC-CO_2 were analyzed by GC. The fatty acid methyl esters (FAME) were prepared by alkali (0.5 N methanolic KOH) catalyzed saponification of oils followed by trans-esterification using a methylating reagent (14 per cent BF_3 in methanol). FAME was then extracted with n-hexane/H_2O (1:2 V/V). Hexane layer obtained was concentrated using rotary evaporator (Buchi) and used for GC analysis in an Agilent 6890N GC equipped with HP-FFAP column (25 mm õ 0.20 mm õ 0.33µm) gas chromatograph. The oven temperature was programmed to increase from 50 to 210 pC at a rate of 20 pC/min, with an initial hold of 10 min and a final hold of 7 min. The injector temperature was maintained at 240 p C and the detector at 250 pC. The flow rate of carrier gas (Hydrogen), was adjusted to 1.2 ml/min. The samples were injected using split less mode. The fatty acid composition of oil was determined using GC peak areas. Identification of compounds was carried out according to standardized chromatograms of fatty acids.

RESULTS

Content of Tocopherol and Tocotrienols

Table 15.1: Tocopherol and Tocotrienol Content of Supercritical CO_2 Extracted Seabuckthorn Seed, Pulp and Skin Oil

Component	*Seed Oil (mg/100g)*	*Pulp Oil (mg/100g)*	*Skin Oil (mg/100g)*
α-tocopherol	100	80	120
α-tocotrienol	170	170	137
β-tocotrienol	27	21	27
γ- tocotrienol	90	6	90
δ- tocotrienol	7	n.d.	6
Total tocotrienols	294	197	260

n.d.: No Data.

Skin oil showed highest content of α-tocopherol (120 mg/100 g) followed by 100 mg/100 g in seed oil and 80 mg/100 g in pulp oil. Total tocotrienols were found to be maximum in seed oil (294 mg/100 g). Amount of total tocotrienol was 260 mg/100 g in skin oil, which is lesser in comparison to seed oil but higher than the pulp oil (197 mg/100 g). Skin oil showed highest content of

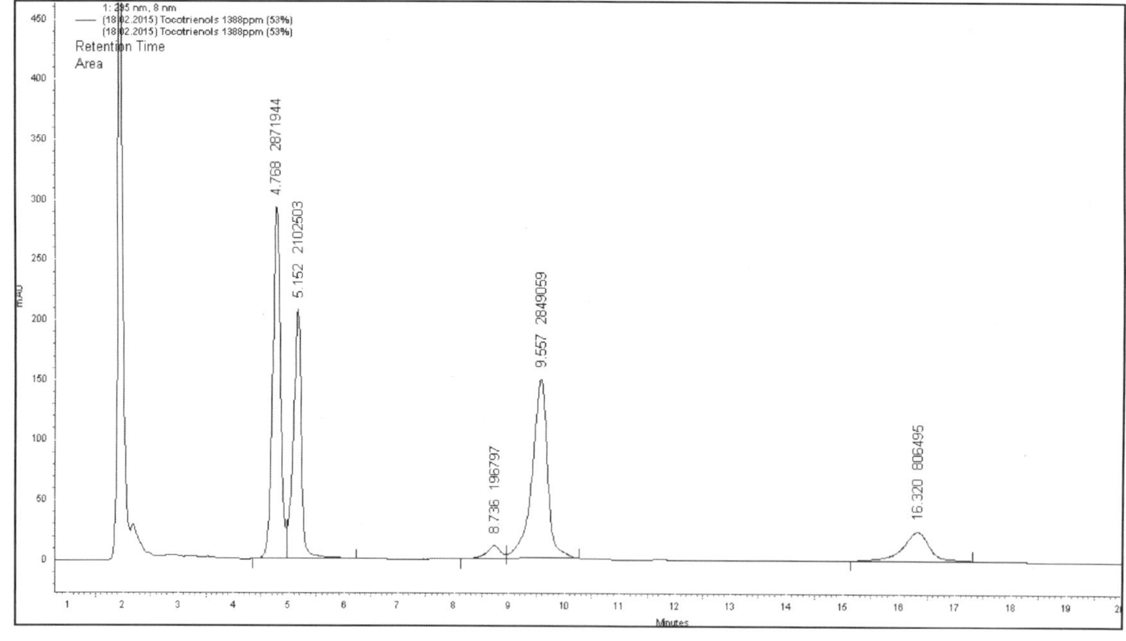

Peak	*Retention Time*	*Compound Name*
1	4.768	α-Tocopherol
2	5.152	α-Tocotrienol
3	8.736	β-Tocotrienol
4	9.557	γ-Tocotrienol
5	16.320	δ-Tocotrienol

Figure 15.2: HPLC Chromatogram of Vitamin E Derivatives.

α-tocopherol (120 mg/100 g) followed by 100 mg/100 g in seed oil and 80 mg/100 g in pulp oil (Table 15.1; Figure 15.2).

Fatty Acid Profiling of Seed, Pulp and Skin Oil

Amount of Linoleic (ω-6) acid is 41.33 per cent and linolenic (ω-3) acid is 22.42 per cent. Thus ratio of ω-6 to ω-3 in tested seed is nearly 2:1. Seed oil contained 88.417 per cent unsaturated fatty acid (USFA) which includes linoleic, linolenic, palmitoleic, oleic and cis-vaccenic acids. USFAs enhance medicinal and nutritional value of seabuckthorn seed oil. Amount of saturated fatty acids (SFA) *viz.* palmitic and stearic was 11.57 per cent. Table 15.2 enlists the composition of fatty acid of seabuckthorn seed oil obtained using SC-CO_2.

Pulp oil showed palmitic, palmitoleic, oleic and cis-vaccenic acid to account for major portion (96.844 per cent) of total fatty acids. Total USFA content (palmitoleic, oleic, cis-vaccenic, Linoleic and linolenic) was noted to be 70.20 per cent in pulp oil. Amount of ω-7 fatty acid *viz.* palmitoleic acid is 41.86 per cent and cis-vaccenic acid is 9.92 per cent ; accounting for total 51.79 per cent ω-7 (Table 15.2). SFA content of pulp oil was 29.79 per cent ; relatively higher than that of seed oil.

Fatty acid profile of SC-CO_2 skin oil resembled to that of pulp oil and contained 68.94 per cent USFA with 47.17 per cent total ω-7; collective value of palmitoleic acid 38.11 per cent and cis-vaccenic acid 7.21 per cent (Table 15.2; Figure 15.3).

DISCUSSION

Seabuckthorn seed and soft parts contain all the four isomers of tocopherol (Kallio *et al.*, 2002a). Kallio *et al.* (2002b) mentioned α-tocopherol as the most abundant of total tocopherols in both soft parts and seeds, accounting 70-80 per cent and 40-50 per cent respectively of the total tocopherols. α-tocopherol to be 96 per cent of total tocopherols in pulp oil and hence is the major isomer. Whereas in seed γ-tocopherol is also present in higher amount (40 per cent of total tocopherols) along with α-tocopherol (Kallio *et al.*, 2002a). Similarly, Barou *et al.* (2001) stated that pulp oil is generally richer in α-tocopherol than the seeds.

Lu Rongsen (2005) reported 93.05 mg/100 g α-tocopherol in pulp oil of *H. rhamnoides* ssp. *turkestanica,* which is closer to the α-tocopherol content (93.6 mg/100 g) reported by Venugopalan *et al.* (2005). α-tocopherol content in both the studies is higher than as determined in present study (80 mg/100 g). *H. rhamnoides* ssp. *turkestanica* seed oil is mentioned to contain 144.1 mg/100 g of α-tocopherol (Lu Rongsen, 2005) whereas total tocopherol 102-143 mg/100 g for *turkestanica* from Xinjiang area, China as reported by Wang and Yang (1990). Value measured for α-tocopherol (100 mg/100 g) is comparable to reported values.

The total vitamin E content for pulp oil as reported by Lu Rongsen (1993) is 87.2 mg/100 g for *H. rhamnoides* ssp. *turkestanica,* 119.2 mg/100 g for ssp. *yunnanensis,* 199.1 mg/100 g for ssp. *sinensis* and 98.2 mg/100 g for ssp. *mongolica.* Lu Rongsen (2005) mentioned 116.16 mg/100 g vitamin E in pulp oil of ssp. *turkestanica,* while Kallio *et al.* (2002b) reported the total tocopherol and tocotrienol content to range from 100-200 in *H. rhamnoides* ssp. *mongolica* and *rhamnoides,* whereas 400-700 mg/100 g in ssp. *sinensis.* The amount of α-tocopherol and total tocotrienols in pulp oil of *H. rhamnoides* ssp. *turkestanica* is determined as 277 mg/100 g in current study.

Studies report total vitamin E content of seed oil as 159.4 mg/100 g for *H. rhamnoides* ssp. *turkestanica,* 103.1 mg/100 g for ssp. *yunnanensis,* 100.2 mg/100 g for ssp. *sinensis,* 120.7 mg/100 g for ssp. *Mongolica* (Lu Rongsen, 1993); total tocopherol and tocotrienol content as 193.7 mg/100 g in ssp. *turkestanica* (Lu Rongsen, 2005), 100-300 mg/100 g in ssp. *mongolica, sinensis* and *rhamnoides* (Kallio *et al.,* 2002 b). The amount of α-tocopherol and total tocotrienols in seed oil of *H. rhamnoides* ssp. *turkestanica* is determined as 390 mg/100 g in current study.

Limited studies are available on vitamin E content of seabuckthorn skin. Jiang (1991) reported total tocopherol content of peel oilof *H. rhamnoides* ssp. *turkestanica* from

Table 15.2: Fatty Acid Composition of SC-CO_2 Extracts of Seed, Pulp and Skin of *H. rhamnoides* ssp. *turkestanica*

Fatty Acid Composition	*Seed Oil*	*Pulp Oil*	*Skin Oil*
Palmitic acid C16:0 (per cent)	9.133	28.843	29.044
Palmitoleic acid C16:1 (per cent)	0.668	41.869	39.073
Stearic acid C18:0 (per cent)	2.448	0.958	1.293
Oleic acid C18:1 (per cent)	22.294	16.208	16.805
Cis-vaccenic (per cent)	1.700	9.924	8.102
Linoleic acid C18:2 (per cent)	41.334	1.622	3.886
Linolenic acid C18:3 (per cent)	22.421	0.579	1.794
USFA (per cent)	88.41	70.20	68.94
SFA (per cent)	11.57	29.79	30.33

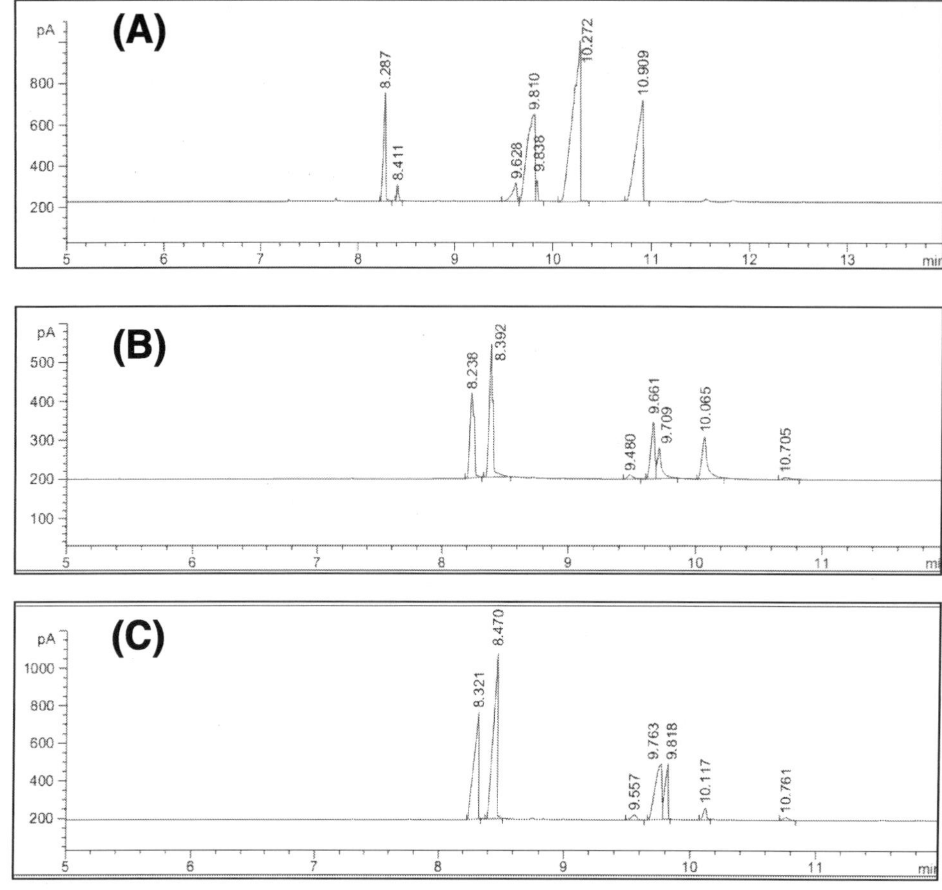

Figure 15.3: Fatty Acid Profiles of (A) SC-CO_2 Extracted Seed Oil, (B) SC-CO_2 Extracted Pulp Oil, (C) SC-CO_2 Extracted Skin Oil.

Xinjiang to be 280 mg/100 g. As the current study does not determine total tocopherol content of oil, comparison of results to reported data is not done.

Variation in content of vitamin E components in seabuckthorn oil is possible based on genetic and environmental factors such as difference of genotype, time of harvesting and altitude of the berries (Antonelli *et al.*, 2005). Kallio *et al.*, 2002b mentioned that amount of total tocopherols and tocotrienols are estimated to be in major proportion in mid-September.

Lu Rongsen (2005) mentioned that seabuckthorn oil shows presence of about 20 fatty acids from C14:1 to C24:1 and that the composition varies from species to species. The study also highlighted that amount of USFA is higher than the SFA in seabuckthorn oil. USFA is mentioned to vary from 82.2 per cent for *H. neurocarpa* ssp. *stallatopilosa* to the highest reported value of 88.3 per cent for *H. neurocarpa* ssp. *neurocarpa* in seed oil. In the present study, USFA content of seed oil is determined to be 88.41 per cent for *H. rhamnoides* ssp. *turkestanica.* This value compared to the reported values as mentioned above. (Lu Rongsen, 2005) also reported that USFA of pulp oil is lesser to that of seed oil and it ranges from 58.5 per cent for *turkestanica* to 78.5 per cent for *H. neurocarpa* ssp. *stallatopilosa.* The results of present study show 70.20 per cent and 68.94 per cent of USFA in pulp and skin oil respectively. These values are higher than the reported values and the reason could be difference informs of ssp. *turkestanica* used. USFA being of biological importance, palmitoleic, Linoleic and linolenic are of more interest.

Fatty acid profiles of *H. rhamnoides* ssp. *turkestanica in* Lahaul form and *H. rhamnoides* ssp. *turkestanica of* Spiti form are studied by (Singh *et al.*, 2008). Amount of USFA is stated to be 65.117 per cent and 86.50 per cent for Lahaul and Spiti form resp. compared to 88.41 per cent obtained in the current study. SFA content is 22.75 per cent and 12.47 per cent for Lahaul and Spiti form resp. compared to 11.57 per cent determined in this study. The profile of USFA for Lahaul (Palmitoleic 16:1- 4.73 per cent, oleic18:1- 23.25 per cent, Linoleic 18:2- 23.79 per cent, α- linolenic18:3- 8.68 per cent), for Spiti (Palmitoleic 16:1- 2.58 per cent, oleic 18:1- 16.33 per cent, Linoleic 18:2- 39.83 per cent, α- linolenic18:3-25.43 per cent) and as determined in current study is palmitoleic 0.668 per cent, oleic 22.294 per cent, Linoleic 41.334 per cent, α- linolenic 22.421 per cent. The result of current study resembles closely to that of Spiti form more than Lahaul. Vereshchagin, A.G. and Tsydendambaev, V.D.1995 reported palmitoleic 2.5 per cent, oleic 24 per cent, Linoleic 36.2 per cent, α- linolenic 25.9 per cent in *H. rhamnoides* ssp. *turkestanica* from central Asia.

Table 15.3: Fatty Acid Composition of Seabuckthorn Seed Oil

Fatty Acid Profile (Per cent)	H. rhamnoides ssp. turkestanica				H. rhamnoides ssp. turkestanica (Ladakh)
	(Present study)	Lahaul form	Spiti form	Natural Distribution: Central Asia	
Palmitic acid C16:0	28.843	27.6	-	34.2	26.95
Palmitoleic acid C16:1	41.869	46.42	38.43	54.8	44.25
Stearic acid C18:0	0.958	0.7	0.75	-	0.692
Oleic acid C18:1	16.208	4.03	6.89	11.00	26.89
Cis-vaccenic	9.924	ND	ND	-	ND
Linoleic acid C18:2	1.622	9.31	6.38	-	1.21
Linolenic acid C18:3	0.579	0.632	0.82	-	n.d.
SFA (per cent)	29.798	28.87	10.59	n.d.	n.d.
USFA (per cent)	70.20	60.68	52.82	n.d.	n.d.

n.d.: No Data.

Fatty acid profile of pulp oil of *H. rhamnoides* from Ladakh obtained by wet process is mentioned by (Venugopalan *et al.*, 2005) which is comparable to the results of current study and has palmitic acid 26.95 per cent, palmitoleic acid 44.25 per cent, stearic acid 0.692, oleic acid 26.89 and Linoleic acid 1.21 per cent. Singh *et al.*, 2008 have studied fatty acid profile of pulp oil from *H. rhamnoides* ssp. *turkestanica* (Lahaul and Spiti forms). It mentions USFA 28.87 per cent and 52.82 per cent for Lahaul and Spiti form respectively compared to USFA content of pulp oil in present study 70.20 per cent. The profile of USFA for Lahaul form (Palmitoleic 46.42 per cent, oleic 4.03 per cent, Linoleic 9.31 per cent, α- linolenic 0.63 per cent), for Spiti form (Palmitoleic 38.43 per cent, oleic 6.89 per cent, Linoleic 6.38 per cent, α- linolenic 0.82 per cent) and as determined in current study is palmitoleic 41.869 per cent, oleic 16.208 per cent, cis-vaccenic 9.924 per cent, linoleic 1.622 per cent and α- linolenic 0.572 per cent. Major SFA in all the three is palmitic acid and ranges from 9.13 per cent in Spiti form to 28.84 per cent as tested in the present study. (Singh *et al.*, 2008) also has mentioned palmitic acid to be the major SFA with amount 26.4-28.2 per cent. The result of fatty acid profile of pulp oil as determined in current study resemble closely to that of Lahaul form. Vereshchagin and Tsydendambaev (1995) reported palmitoleic 54.8 per cent, oleic 11 per cent and 34.2 per cent palmitic acid in *H. rhamnoides* ssp. *turkestanica* from central Asia.

Fatty acid profile of skin oil obtained was closely similar to that of pulp oil which is justified as both the pulp and skin are the components of soft part of the berry.

CONCLUSION

Fatty-acid profiling of *Hippophae rhamnoides* ssp. *turkestanica* by advanced Gas-chromatography technique showed variable quantitative fatty acid composition of unsaturated and saturated fatty acids for seed, pulp and skin oil. With the aid of quantitative analysis, it was found that seed oil is rich in omega-3 and omega-6 fatty-acids whereas, pulp and skin oil are both rich in omega-7 fatty-acids. Quantitative comparative study of tocopherol and tocotrienol content for *H. rhamnoides* ssp. *turkestanica* using HPLC showed highest tocopherol content in skin oil whereas highest tocotrienol content in seed oil.

ACKNOWLEDGEMENTS

Support in collection and taxonomical identification of seabuckthorn plant material by Dr. Tsering Stobdan, Defence Institute of High Altitude Research (DRDO), Leh, is gratefully acknowledged.

REFERENCES

1. Antonelli, M., Raffo, A. and Paoletti, F. 2005. Biochemical changes during ripening of seabuckthorn (*Hippophae rhamnoides*) fruits. In: *Seabuckthorn (Hippophae L.): A Multipurpose Wonder Plant, Vol. 2: Biochemistry and Pharmacology* (V. Singh, Editor-in-Chief, 2005), Daya Publishing House, New Delhi, India, pp. 285-312.

2. Barou, Y., Kallio, H. Peippo, P., Tahvonen, R., Ruilin, P., Xuren, Z., Yuhua, F., and Yicai, H. 2001. Tocopherols in seeds and berried of two subspecies of seabuckthorn (*Hippophae rhamnoides* ssp. *sinensis* and *mongolica*). In: *Proceedings of International Workshop on Seabuckthorn*, Feb.18-21, N. Delhi, pp. 151-153.

3. Chen, J., Qin, N.I. and Li 1990. Research on chemical composition and characteristics of seabuckthorn berry and its oil. *Chem. and Indus of Fores. Pro.* 10: 163-175.

4. Das, D., Bandyopadhyay, D., Bhattacharjee, M. and Banarjee, R.K. 1997. *Free Rad. Biol. Med.* 23: 8-18.

5. Frei, B. 1994. *Natural Antioxidants in Human Health and Disease.* San Diego, Academic Press.

6. Halliwell, B. 1998. Can oxidative DNA damage be used as a biomarker of cancer risk in humans?

Problems, resolutions and preliminary results from nutritional supplementation studies. *Free Radical Res.* 29: 469-486.

7. http://www.tocotrienol.org/.

8. Ji, Y.B. and Gao, Y. 1991a. Effect of feeding seabuckthorn seed oil and seabuckthorn seed oil supplemented with sodium selenite *in vivo* on Na-K-ATPase activity in erythrocyte ghost in rats. *Acta Nutrimenta Sinica* 13 (1): 20-24.

9. Ji, Y.B. and Gao, Y. 1991 b. Effect of feeding seabuckthorn seed oil and seabuckthorn seed oil supplemented with sodium selenite *in vivo* on structural stability of erythrocyte ghosts in rats. *Chinese Biochemical Journal* 7 (4): 441-446.

10. Jiang, Z.R. 1991. Determination of content of fat-soluble vitamin E in seabuckthorn oil and changed during refining process. *Chemistry and Industry of Forest Products* 11 (3): 241-246.

11. Kagliwal, L.D., Pol, A. S., Patil, S.C., Singhal, R. S. and Patravale, V. B. 2012. Antioxidant-rich extract from dehydrated seabuckthorn berries by supercritical carbon dioxide extraction. *Food and Bioprocess Technology* 5 (7): 2768-2776.

12. Kallio, H., Yang, B. and Peippo, P. 2002a. Effects of different origins and harvesting time on Vitamin C, tocopherols and tocotrienols in seabuckthorn (*Hippophae rhamnoides*) berries. *J. Agric. Food. Chem.* 50: 6136-6142.

13. Kallio, H., Yang, B., Peippo, P., Tahvonen, R. and Pan, R. 2002b. Triacylglycerols, glycerophospholipids, tocopherols, and tocotrienols in berries and seeds of two subspecies (ssp. *sinensis* and ssp. *mongolica*) of seabuckthorn (*Hippophae rhamnoides*). *J. Agric Food Chem.* 50: 3004-3309.

14. Knekt, P. 1993. Epidemiology of vitamin E: Evidence for anticancer effects in human. In: *Vitamin E in Health and Disease* (Packer, L. and Fuchs, J.eds), Marcel Dekker, New York, pp. 513-527.

15. Losonczy, K.G., Harris, T.B. and Havlik, R.J. 1996. Vitamin E and vitamin C supplement use and risk of all- cause and coronary heart disease mortality in older persons: The established populations for epidemiologic studies of the elderly. *Am. J. Clin. Nutr.* 64: 190-196.

16. Rongsen, L. 1993. The chemical composition of *Hippophae* fruits in China. In: *Proceedings of International Symposium on Seabuckthorn,* Barnaul, Russia, pp. 398-412.

17. Rongsen, L. 2005. Biochemical characteristics of seabuckthorn (*Hippophae* L.). In: *Seabuckthorn (Hippophae L.): A Multipurpose Wonder Plant,* Vol. II: *Biochemistry and Pharmacology* (V. Singh, Editor-in-Chief, 2005), Daya Publishing House, New Delhi, India, pp. 98-107.

18. Noguchi, N., Watanabe, A. and Shi, H. 2000. *Free Rad. Res.* 33: 809-817.

19. Pintea, A., Marpeau, A., Faye, M., Socacid, C. and Gleizes, M. 2001. Polar lipid and fatty acid distribution in carotenolipoprotein complexes extracted from seabuckthorn fruits. *Phytochem. Anal.* 12: 293-298.

20. Risto, E. and Yang, B. 2003. Seabuckthorn oils: Towards healthy mucous membranes. *Agro Food Industry Hi-Tech.* 53-57.

21. Rousi, A. 1971. The genus *Hippophae* L. a taxonomic study. *Annals Bot.* 8: 177-227.

22. Rui, L.X. and Gao, Y. 1989. Effects of seabuckthorn oil on lipid peroxidation of guinea pigs erythrocyte membrane. In: *Proceedings of International Symposium on Seabuckthorn (H. rhamnoides* L.*)*, Xian, China, pp. 358-364.

23. Shi, H.L., Cai, H.J., Chen, X.Y. and Yang, C.M. 1994. Studies on the antioxidant effects of *Hippophae rhamnoides* L. seed oil. *Acta Nutrimenta Sinica* 16 (3): 292-295.

24. Singh, V., Gupta, R., Tandon, S., and Sawhney, R.C. 2008. Fatty acid composition of fruit pulp and seed oils of some high yielding forms of seabuckthorn in Indian Himalayas. In: *A Multipurpose Wonder Plant,* Vol. III: *Biochemistry and Pharmacology.* (V. Singh, Editor-in- Chief, 2008), Daya Publishing House, New Delhi, India, pp. 215-222.

25. Sokol, R.J. 1996. Vitamin E. In: *Present knowledge in Nutrition* (Ziegler, E.E. and Filer, L.J. Eds), ILSI Press, Washington DC, pp. 130-136.

26. Song, Z.H. and Gao, Y.1995. Effect of seabuckthorn oil and vitamin E on lipid peroxidation of rats after cold exposure. *Acta Mutrimenta Sinica* 17 (1): 27-31.

27. Steinberg, D., Parthasarathy, S., Carew, T.F., Khoo, J.C. and Witztum, J.L.1989. *New Engl. J. Med.* 320: 915-924.

28. Venugopalan, V.V., Rnajith, A., Sarinkumar, K. and Arumughan, C. 2005. A green technology for the integrated processing of fresh seabuckthorn berries. In: *Seabuckthorn (Hippophae L.): A Multipurpose Wonder Plant, Vol 2: Biochemistry and Pharmacology.* (V.Singh, Editor-in- Chief, 2005), Daya Publishing House, New Delhi, India, pp. 522-532.

29. Vereshchagin, A.G. and Tsydendambaev, V.D. 1995. Neutral lipids of mature and developing seabuckthorn (*Hippophae rhamnoides* L.) fruits. In: *Plant Lipid Metabolism* (Kader, J.C. and Mazliak, P., Eds.), Kluwer Academic Publishers, Netherlands, pp. 573-578.

30. Wang, X. S. and Yang, H.R. 1990. Seabuckthorn in Xinjiang are and its biochemical characteristics. *Hippophae* 3 (1): 31-36.

31. Wu, F. 1991. Seabuckthorn medicines in Russia. *Seabuckthorn* 4: 38-41.

32. Yang, B. and Kallio, H.P. 2001. Fatty acid composition of lipids in seabuckthorn (*Hippophae rhamnoides* L.) berries of different origins. *J. Agric. Food Chem.* 49: 1939-1947.

33. Yoshida, Y., Saito, Y., Jones, L.S. and Shigeri, Y. 2007. Chemical reactivities and physical effects in comparison between tocopherols and tocotrienols: Physiological significance and prospects as antioxidants. *J. Biosci. Bioeng.* 104 (6): 439-45.

Chapter 16

Physico-chemical Characterization of Pulp and Seed Oils of Seabuckthorn (*Hippophae salicifolia* D. Don) from Tawang District of Arunachal Pradesh, North-Eastern India

Chitta Ranjan Barik[1, 2], L. Sahoo[1, 3] and Vaibhav V. Goud[1, 2*]

[1]Center for Energy,
[2]Department of Chemical Engineering,
[3]Department of Biosciences and Bioengineering,
IIT Guwahati, Guwahati – 781 039, India
(*E-mail: vvgoud@iitg.ernet.in)

SUMMARY

Seabuckthorn (*Hippophae salicifolia*) is an important medicinal plant of mountainous regions of Asia known for its high valued oils. Seed and pulp oils are considered the most valuable components of the seabuckthorn berries comprising important bioactive phytochemicals. Tawang district of Arunachal Pradesh in Northeast India, with characteristic extremely cold and dry weather, houses seabuckthorn in wild condition. However, no work has been carried out on seabuckthorn populations of this region. This is the first comprehensive report on physico-chemical characterization of seabuckthorn seed and pulp oils of this region. The crude oils yield from pulp, and seeds were found maximum (32.38 ± 0.27 per cent and 12.45 ± 0.14 per cent respectively) when extracted using petroleum ether. Both petroleum ether and n-Hexane extracted pulp oils exhibited high carotenoids content (513.9 ± 10.4 mg/100 g and 579.4 ± 18.7 mg/100 g respectively) than the seed oils in respective solvents (24.12 ± 0.79 mg/100 g and 29.23 ± 1.4 mg/100 g). In contrast, total tocopherols contents of seed oils were higher (186.7 ± 2.5 mg/100 g for petroleum ether and 196 ± 3.9 mg/100 g for n-hexane) than pulp oils (147 ± 1.7. mg/100 g for petroleum ether and 158 ± 2.2 mg/100 g for n-hexane). Both carotenoids as well as tocopherols contents were higher in n-hexane extracted pulp and seed oils than their counterparts. The pulp and seed oils extracted using petroleum ether and n-hexane exhibited identical physical properties of odour, solubility, refractive index, optical rotation, acid value and peroxide value. Refractive indices of all the oils were in the range of 1.5 to 1.73 at 20°C. The oils were found to be dextrorotatory and optically active in the range of 2.11° to 2.36° at 25 °C. Acid value of the petroleum ether extracted seed oil was highest (10.00 ± 0.27 mg KOH/g) whereas it was lowest for n-Hexane extracted pulp oil (8.62 ± 0.2 mg KOH/g). The peroxide values of all the oils were in the range of 1.37 to 1.46 meq O_2/kg. Petroleum ether extracted pulp oil was found to be most viscous (dynamic viscosity of 93 at 25 °C) whereas n-Hexane extracted seed oil was least viscous (dynamic viscosity of 44 at 25 °C) one. The stable physical properties and higher content of carotenoids and tocopherols in solvent extracted seed and pulp oil of Seabuckthorn from Tawang region may find it precious use in nutraceuticals, pharmaceuticals and cosmeceuticals purposes.

***Keywords**: Seabuckthorn, Pulp oil, Seed oil, Physico-chemical, Tawang district and Northeast India.*

INTRODUCTION

Seabuckthorn (*Hippophae* sp.) is a thorny, temperate, deciduous, dioecious, actinorhizal shrub or small tree belonging to the family Elaeagnaceae and native to Eurasean Countries like China, India,Russia, Nepal, Pakistan, *etc.* In India it is mainly confined to high altitude, colder arid zones of Jammu and Kashmir, Himachal Pradesh, Uttarakhand and North Eastern

Seabuckthorn (*Hippophae* L.): *A Multipurpose Wonder Plant, Vol. 5: Advanced Technologies for Cultivation, Processing, Health Protection and Environmental Conservation* (Virendra Singh, *Editor-in-Chief*, 2018), p. 127–133, Astral International Pvt. Ltd., New Delhi, India.

states like Sikkim and Arunachal Pradesh. Although taxonomical classification of genus *Hippophae* is controversial, still there are six species of seabuckthorn like *Hippophae goniocarpa, H. gyantsensis, H. neurocarpa, H. rhamnoides, H. Salicifolia* and *H. tibetana*, and 12 sub-species have been distinctly described (Singh, 2003). Seabuckthorn is a multipurpose plant used for food, fodder, medicines, fuelwood and conservation of soil and biodiversity (Qinxiao and Hongyan, 2003). Berries of seabuckthorn are very rich source of different bioactive phytochemicals like lipids, sugars, amino acids, organic acids, fatty acids, carotenoids (β-carotenes, lycopene, zeaxanthin), tocopherols (α β, γ, δ), sterols (ergosterol, campesterol), Vitamins (A, C, E and K), phenolic compounds and flavonoids (isorhamnetin, kaemferol, quercetin, myricetin) *etc.* (Tiitinen *et al.*, 2005, 2006; Anderson *et al.*, 2009; Lehton *et al.*, 2010; Fatima *et al.*, 2012). The high valued pulp and seed oils of seabuckthorn have very unique combination of phytochemicals like different fatty acids, carotenoids, tocopherols, phytosterols and tocotrienols. Hence these oils have extraordinary nurturing, rejuvenating, anti-aging and therapeutic effects. These can be utilized for curing different skin diseases and skin types like dermatitis, skin burns, skin ulcers, pimples, itching, dry skin, sore skin and in cuts and for cell regeneration (Parimelazhagan *et al.*, 2005). Seabuckthorn oil is also a very good agent against aging symptoms like wrinkles, premature aging *etc.* (Lanev *et al.*, 1995; Zhou, 1998; Xing *et al.*, 2002). Therefore, now a days seabuckthorn oils have a very high demand in different nutraceuticals, pharmaceuticals and cosmeceuticals industries.

Various physico-chemical properties like odour, colour, density, refractive index, viscosity, optical rotation, acid value, peroxide value, *etc.* signify the quality, viability and effectiveness of both the fixed and essential oils. The commercial importance like usability and edibility of oils mainly depends upon these physico-chemical characteristics (Bamgboye and Adejumo, 2010; Parthiban *et al.*, 2011; Barkatullah *et al.*, 2012). Therefore, physico-chemical characterization plays an important role in monitoring the origin, adulteration and the authenticity of particular oil.

Tawang district of Arunachal Pradesh is characteristically represented by its extreme agro-climatic conditions like high altitude, severe cold and arid climates. To combat such harsh weather and to sustain, plant synthesizes various bioactive phytochemicals which may vary region to region and species to species. *H. salicifolia*, which is restricted to the high altitude Himalayan region like Tawang district, is one of the best species of seabuckthorn because of its high quality berries, adoptability to the wide range of soil types, and extreme dry and low temperature tolerance. Flowering of this plant occurs for only one week during June-July and berries mature during the middle of August to October. The yellow or orange-coloured fruit size is usually 1 cm in length (Basistha *et al.*, 2010).

Although almost all the species of seabuckthorn from different regions have been explored, and a lot of work has been carried out on various aspects, still very little efforts have been made to explore Northeast Indian species. Therefore, this study focused on Physico-chemical characterization of pulp and seed oils of seabuckthorn from Tawang district belonging to Arunachal Pradesh of Northeast India.

MATERIALS AND METHODS

Materials

Seabuckthorn (*Hippophae salicifolia* D. Don) berries were collected from Namjang Chu River Valley near Zemithang village located between 27° 712 03.8822N latitude and 91° 722 99.3922 E longitude in Tawang district of Arunachal Pradesh, Northeast India and stored in -20°C. At the beginning seeds were separated by pressing the frozen berries. The collected seeds and pulp residues were dried under shade at room temperaturefollowed by grinding (mechanical grinding) and subsequently used for extraction of oil. All the chemicals used for analysis were obtained from Merck India Pvt. Ltd.

Extraction of Oil

Powdered seed and pulp samples (30 g) were taken in extraction thimbles (43 mm × 123mm, Whatman) for solvent extraction using Soxhlet apparatus. A small amount of cleaned cotton wool is placed on top of the thimble to ensure uniform distribution of the solvent and to prevent any sample loss from top during extraction. Continuous extractions of both the samples were performed using n-hexane and petroleum ether independently at solute (sample) to solvent ratio 1:9 for 8 hrs by heating the solvent just above the boiling point. After completion of extraction, solvent was evaporated employing Vacuum Rotan vapour (R-210 Model, Buchi). Finally different oil samples were collected in separate amber glass vials and stored at 2°C for further analysis.

Physico-Chemical Characterization of Oil

Physico-chemical characterization plays an important role in evaluating the quality of oil. The physico-chemical properties were analysed according to the standard methods (AOCS, 1997). Carotenoids were determined by the method described by Ranjith *et al.* (2006) and tocopherols were analysed according to the procedure reported by Baker and Frank (1968). Statistical analysis of the results were carried out and expressed as mean ± SE.

Determination of Percentage Oil Yield

The percentage oil yield was calculated according to (AOAC, 1997) official method as follows:

Percentage oil yield (W/W) =

$$\frac{\text{Weight of oil}}{\text{Weight of sample used for oil extraction used}} \times 100$$

Carotenoids Content

The determination of carotenoids was carried out following the method described by Ranjith *et al.*, 2006, with slight modifications. In brief seed oil or pulp oil was dissolved in n-hexane (0.1 g/ml) and added to 0.5 per cent NaCl, vortexes for 1min and employed to centrifugationat 6500 rpm for 10 min (Cooling Centrifuge-3K30, Sigma, UK). The supernatants were further diluted with n-hexane and the absorbance at 460 nm was recordedby using a UV/Visible Spectrophotometer (Lambda 45, *Perkin Elmer)*.The carotenoids contentin the oil sample was estimated by plotting β-carotene standard calibration curve (1–10 µg/ml) and expressed as mg/100g β-carotene equivalent.

Tocopherols Content

Tocopherols content were determined according to published method (Baker and Frank, 1968) with some modification. Briefly, 1ml of oil was dissolved in 3 ml of absolute ethanol and vortexed. Further 2 ml of xylene was addedto it and centrifuged at 7000 rpm for 8 min. From the mixture upper xylene portion was pipetted out cautiously. 1ml of xylene extract was mixed thoroughly with 1 ml of 0.12 per cent 2,2-Bipyridyl reagent prepared in absolute ethanol.Absorbance of the mixture was recorded at 460 nm. Further 0.5 ml of 0.2 per cent (w/v) of $FeCl_3$ in absolute ethanol was added to the same mixture and the absorbance was again read out at 520 nm after 15 min. A standard calibration curve was plotted using α-Tocopherol (0 to 100µg). The tocopherols content was calculated and expressed as mg/100 g α-Tocopherol equivalentfrom the following formulae :

Absorbance (α-Tocopherol) = $\text{Absorbance}_{520\ nm} - (0.29\times \text{Absorbance}_{460\ nm})$

α-Tocopherol (mg/100g) = (Absorbanceof unknown sample/Absorbance of Standard) × 10

Odour Determination

Odour is an important character for the acceptance of any kind of oil. Odour of the extracted oil samples were assessed by organoleptic determination according tothe method reported by Evans, (2002).

Colour Determination

In many cases colour plays a vital role in confirming an oil quality and origin. Colour of the extracted oil samples were determined by simple physical observation in normal day light and visualizing under ultraviolet radiation of short wavelength (254 nm) and long wavelength (366 nm) using ultra violet trans-illuminator (Bamgboye and Adejumo, 2010).

Viscosity

Viscosity characterize the rheological properties of an oil. It is the capacity to resists the deformation of a fluid under shear stress. It describe the thickness and internal resistance of oil to flow (Barkatullah *et al.*, 2012). Dynamic viscosity (µ) of the oil was evaluated at25°Cand shear rate of 100 S^{-1}using an Anton Paar Physica MCR 301 Rheometer. Average value of 20 readings measured with an interval of 6 sec and the average shear stress of 0.405 Pa resulted the single viscosity data for an oil sample. The viscosity value of the oil was measured in centipoises (cp).

Refractive Index

Refractive index describe the extent of deflection, during the passage of a beam of light through oil sample. The value increases with an increase in the chain length and carbon atoms present in an oil.Therefore it can also give the evidence of unsaturated long carbon chain (Pearson, 1976). Refractive indices of all the oil samples were measured at 20°C using a Refractometer (Advanced Research Instruments Company, New Delhi). With the help of a micropipette2-3 drops of a particular oil was placed on the prism and the prism was firmly closed. Then lens of the apparatus was suitably adjusted to record the value from the display.

Optical Rotation

Optical activity or optical rotation of an oil sample describes the rotation of the plane of linearly polarized light about the direction of passage as the light passes through it, due to the presence of chiral molecules. Optical activity of the oil samples were measured by taking 10ml of 1 per cent pulp or seed oil in hexane separately, in a 10 ml polarimeter tube, at 25°C, using an automatic polarimeter (Rudolph Research Analytical, Auto pol I). Filling of the tube was carried out very cautiously to prevent formation of any air bubble which could interfere the rotation of light. Nature of the rotation was noted by direction of the rotation of analyzer from the zero position. For clockwise rotation levorotatory (-) and for anti-clockwise rotationdextrorotatory (+) is denoted.

Acid Value

Acid value test determines the free fatty acids contents and also indicates hydrolytic rancidity. The AOCS (1997) standard method was used for the estimation of acid value of all the oil samples. Acid value is the mass in mg of potassium (or sodium) hydroxide require to neutralize 1g of an oil sample. For the estimation, 1g of oil was taken in a conical flask and 20ml of methylated spirit (Methanol: Ethanol, 1:19) was mixed properly. The mixture was then titrated using phenolphthalein indicator against 0.1 N KOH solution. Titration was continued till a pink color was appered.

Then the acid value was calculated employing the following formula and expressed as mg KOH/g:

$$\text{Acid value} = \frac{N \times V \times 56.1}{W \text{ used}}$$

where,

N = Normality of KOH

V = Volume of KOH used

W = Weight in gram of the oil sample

56.1 = Molecular mass of KOH

Free Fatty Acid value

Fatty acids which are not attached to any other molecules like glycerols in an oil or fat sample are called free fatty acids. They may arise due to the degradation or breakdown of triglycerides which is the main reason of hydrolytic rancidity. Half of the acid value was estimated as Free fatty acid value following AOCS (1997) standard method and expressed as mg KOH/g.

Peroxide Value

Peroxide value is a popular test for oxidative rancidity and is used to estimate the concentration of peroxides andhydro peroxidesproduced during the primary lipid oxidation. It is the quantity of peroxide oxygen present in 1 kg of fat or oil. The peroxide value was estimated following AOCS (1997) official method. Concisely, 2 g of oil sample was taken in a conical flask fitted with glass stopper. To this 12 ml mixture of chloroform and glacial acetic acid (1:1.5) was added and shaken till the complete oil sample got dissolved. Further 0.2 ml of saturated potassium iodide solution was mixed and shaken for exactly 1 min. Then 12 ml of deionized water was added into the flask and titrated against 0.01N sodium thiosulphate, adding the titrant gradually till the yellow colour almost disappeared. Again 1ml of starch solution was added as indicator and the titration was continued till the blue gray colour just discharged. One blank test was carried out in the same manner and the Peroxide value was calculated by the following formula and expressed as milli-equivalents per 1000 g of sample or simply meq. O_2/kg.

$$\text{Peroxide value} = \frac{(S \quad B) \times N \times 1000}{W \text{ used}}$$

where,

S = Volume of sodium thiosulphate solution used for the titration of oil sample

B = Volume of sodium thiosulphate solution used for the titration of blank

N = Normality of sodium thiosulphate solution used for the titration

W = Weight in gram of the oil sample

RESULTS AND DISCUSSION

Investigation of physicochemical properties identifies the practical relevance and furnishes reasons for appropriateness and efficacy of numerous oils in day to day life. Physico-chemical characteristics also indirectly signify the quality of oils. Different Physico-chemical characteristics of petroleum ether and n-Hexane extracted seabuckthorn seed and pulp oils like colour, odour, refractive index, optical rotation, acid value, free fatty acid value and peroxide value were investigated. Seabuckthorn oil from the berries is the most important constituent of the plant and has been a center of attraction for its potential for numerous product development. The high oil content in both pulp and seed parts is a very unique feature of seabuckthorn berries. From the results crude oil yield of pulp and seed were higher (32.38 ± 0.27 per cent and 12.45 ± 0.14 per cent respectively) when extracted using petroleum ether as compared to oil yield using n-Hexane as 29.85 ± 0.18 per cent and 10.17 ± 0.22 per cent respectively (Figure 16.1). The results also evidently showed higher yield for both types of oil from this seabuckthorn species and from this region as compared to other species (Kallio *et al.*, 2002; Zeb, 2004; Parimelazhagan *et al.*, 2005). The difference in yield might be due to geo climatic attribution of the region. Also both petroleum ether and n –Hexane were found to be very potent solvents for the extraction of seabuckthorn oil.

Pulp oils extracted using both petroleum ether and n-Hexane exhibited high carotenoids content of 513.9 ± 10.4 mg/100 g and 579.4 ± 18.7 mg/100 g respectively than the seed oils in respective solvents (24.12 ± 0.79 mg/100 g and 29.23 ± 1.4 mg/100 g) (Figure 16.2). Among the solvents studied, n-Hexane extract was found to yield more carotenoids content for both pulp and seed oils. Tocopherols amount also varied depending upon the source of oil and solvent type. Total tocopherols content of petroleum ether and n-Hexane extracted seed oils were higher (186.7 ± 2.5 mg/100 g and 196 ± 3.9 mg/100 g respectively) than respective pulp oils (147 ± 1.7. mg/100 g and 158 ± 2.2 mg/100 g) (Figure 16.2). In this case also n-Hexane contributed higher quantity of tocopherols for both types of oils. In general seabuckthorn pulp oil exhibited high carotenoids content, where as seed oil exhibited high tocopherols (Vitamin E) content. Both the seed and pulp oils are fixed oils. The carotenoids as well as tocopherols are two important natural antioxidants which not only prevent the lipid peroxidation, but also have very high health promoting effect. These two bioactive phytochemicals have very high importance in nutrients, pharmaceuticals and cosmetics industries and their content vary significantly according to genotypes and ecological factors.

All the extracted oils showed almost identical properties of odour, colour, solubility, viscosity,

Table 16.1: Physico-chemical Properties of Seabuckthorn Oils

Parameters	*n-Hexane Extracted Pulp Oil*	*Petroleum Ether Extracted Pulp Oil*	*n-Hexane Extracted Seed Oil*	*Petroleum Ether Extracted Seed Oil*
Odour	Agreeable	Agreeable	Agreeable	Agreeable
Colour	Saddle Brown	Saddle Brown	Chocolate Brown	Chocolate Brown
Viscosity (cp at 25°c)	81	93	44	49
Refractive index	1.67 ± 0.01*	1.73 ± 0.02*	1.5 ± 0.01*	1.53 ± 0.01*
Optical rotation	2.29 ± 0.1*	2.36 ± 0.16*	2.11 ± 0.08*	2.15 ± 0.01*
Acid value (mg KOH/g)	8.62 ± 0.2*	8.7 ± 0.23*	9.75 ± 0.33*	10.00 ± 0.27*
Free fatty acid value (mg KOH/g)	4.31	4.35	4.87	5
Peroxide value (meq. O_2/ kg)	1.37 ± 0.04*	1.39 ± 0.07*	1.46 ± 0.05*	1.42 ± 0.04*

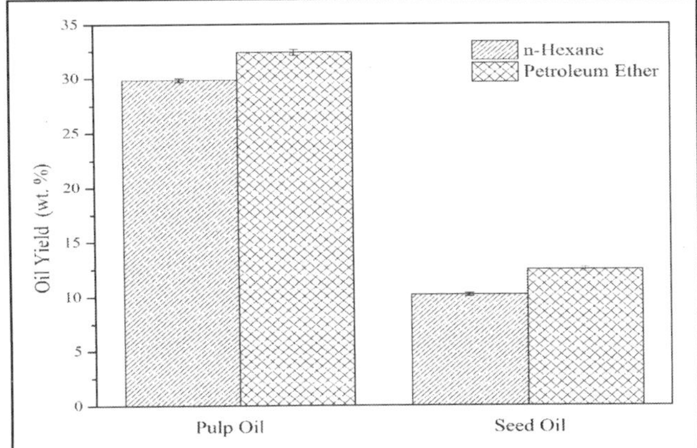

Figure 16.1: Yield Percentage (wt per cent) of n-Hexane and Petroleum Ether Extracted Pulp and Seed Oils.

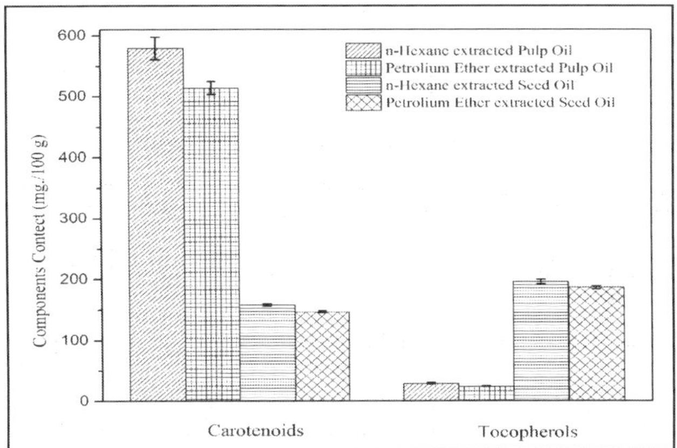

Figure 16.2: Carotenoids and Tocopherols Content of n-Hexane and Petroleum Ether Extracted Pulp and Seed Oils in mg/100 g. The values indicated with '*' are mean ± SE.

refractive index, optical rotation, acid value, free fatty acid value and peroxide value shown in Table 16.1. The pulp oils extracted using petroleum ether and n-Hexane was saddle brown in colour, while the seed oils from both the solvents imparted chocolate brown colour.

Refractive index of petroleum ether extracted pulp oil was highest (1.73 ± 0.02) whereas it was lowest for n-Hexane extracted seed oil (1.5 ± 0.01) at 20°C. Refractive indices in the range of 1.5 to 1.73 at 20°C for all the samples implicated enriched long chain fatty acid contents. All the oils were found to be optically active and dextrorotatory in between 2.11° to 2.36° at 25°C. Petroleum ether extracted pulp oil displayed the maximum optical rotation of 2.36 ± 0.16° and n-Hexane extracted seed oil showed minimum value of 2.11 ± 0.08°. Likewise dynamic viscosity value was maximum (93 cp at 25°C) for petroleum ether extracted pulp oil and was minimum (44 cp at 25°C) for n-Hexane extracted seed oil. Therefore petroleum ether extracted pulp oil was the most viscous contrast to n-Hexane extracted seed oil which was least viscous.

The acid value measures the quantity of acid residues present in an oil sample. It is an indirect approach for the estimation of free fatty acid content in oil samples and their usability and edibility (Akbar *et al.*, 2009). Oil and fats having less free fatty acids have more convenient acceptance (Coenen, 1976). Acid value of the petroleum ether extracted seed oil was highest (10.00 ± 0.27 mg KOH/g) whereas it was lowest for n-Hexane extracted pulp oil (8.62 ± 0.2 mg KOH/g).Free fatty acid value for all the samples were calculated as half of their respective acid values.

Formation of peroxide and hydro peroxides due to primary oxidation processes in oil are generally measured by the peroxide value. Commonly lower peroxide value signifies lower oxidative rancidity and better quality of oil. However peroxide value decreases with the formation of by-products from the secondary oxidation. Peroxide value of very good quality oil is usually less than 2 meq O_2/kg of oil. From the results peroxide values of all the oils were far below the level of 2 meq O_2/kg.Petroleum ether extracted seed oil showed maximum peroxide value of 1.46 ± 0.05 meq O_2/kg, whereas for n-Hexane extracted pulp oil vlaue was found to be 1.37 ± 0.04 meq O_2/kg. The lower peroxide values may be attributed to the higher quantities of antioxidants like carotenoids and tocopherols in the oil samples.

Almost all the values estimated were acceptable according to the standard quality criterions and also agreeswell with the literature data from various sources (Mironov, 1989; Zhang, 1989; Bernath and Foldesi, 1992; Lu, 1993; Parimelazhagan *et al.*, 2005; Ranjithet *al.*, 2006). The desired physico-chemical properties and higher amount of carotenoids and tocopherols in solvent extracted seed and pulp oils of seabuckthorn of Tawang region may be used for commercial applications in various purposes. However more focussed investigation should be carried out on collection of different species as well as sub species and varieties from various agro geo-climatic conditions at different stage of development to optimize bioactive phyto-constituents and oil yield.

CONCLUSION

This work furnishes important informations regarding the physico-chemical properties of the solvent extracted oils of seabuckthorn (*H. salicifolia*). This is the first report regarding the physico-chemical characterization of seed and pulp oils from seabuckthorn species *H. salicifolia* wildly grown in the high altitude river vally of Tawang district. Seed and pulp parts of the species from this region displayed higher oil content in comparison to the seabuckthorn species from other geographical regions when extracted with solvents like petroleum ether and n-Hexane. The extracted seed oils could be regarded as admirable sources of vitamin E due to their higher tocopherols content whereas pulp oils could be considered as marvellous sources of vitamin A due to their high contents of carotenoids (vitamin A precursors). Most of the physico-chemical parameters of the petroleum ether and n-Hexane extracted seed and pulp oils were in the permitted range of AOAC standards. Finally from the results it can be concluded that all the solvent extracted oils imparted very decent and stable physico-chemical properties demanding awide application in various nutraceuticals, pharmaceuticals and cosmeceuticals industries.

ACKNOWLEDGEMENTS

Authors of this study are thankful to Department of Biotechnology (DBT), Ministry of Science and Technology, Government of India for Financial Support to conduct this work and State Institute of Rural Developement (SIRD), Arunachal Pradesh for their help in sample collection.

REFERENCES

1. Akbar, E., Yaakob, Z., Kamarudin, S.K., Ismail, M. and Salimon, J. 2009. Characteristic and composition of *Jatropha curcas* oil seed from Malaysia and its potential as biodiesel feedstock. *Eur. J. Sci. Res.*, 29(3): 396-403.
2. *American Oil Chemists Society* (AOCS) 1997. Official methods and recommended practices. *Commercial Fats and Oils*, 5th Ed, Washington, DC.
3. Andersson, S.C., Olsson, M.E., Johansson, E. and Rumpunen, K. 2009. Carotenoids in seabuckthorn (*Hippophae rhamnoides* L.) berries during ripening and use of pheophytin a as a maturity marker. *J Agric Food Chem* 57: 250-258.
4. Baker, H. and Frank, O. 1968. Determination of serum tocopherol. In: *Practical Clinical Biochemistry* (Ed. A. H. Gowenlock), 6th Edition, London, Heinemann, Medical Books, p. 902.
5. Bamgboye, A.I. and Adejumo, O.I. 2010. Physicochemical properties of Roselle seed oil. *Natural and Food Science* 40 (2): 186-192.
6. Barkatullah, I. M., Rauf, A. and Rahman, I.U. 2012. Physicochemical characterization of essential and fixed oils of *Skimmialaureola* and *Zanthoxylum armatum*. Middle-East. *J. Med. Pl. Res.* 1 (3): 51-58.
7. Basistha, B.C., Sharma, N.P., Lepcha, L., Arrawatia, M.L. and Sen, A. 2010. Ecology of *Hippophae salicifolia* D. Don of temperate and sub-alpine forests of north Sikkim Himalayas-A case study. *Symbiosis* 50: 87-95.
8. Bernath, J. and Foldesi, D.1992. Seabuckthorn (*Hippophae rhamnoides* L.): A promising new medicinal and food crop. *J. Herbs, Spices and Medicinal Plants* 1 (1/2): 27-35.
9. Coenen, J.W.F. 1976. Hydrogenation of edible oils. JAOCS, 53: 338-339.
10. *Codex Alimentarious Commission* 1982. *Recommended Internal Standards, Edible Fats and Oils*, 11, (1st ed.), FAO/WHO: Rome.
11. Fatima, T., Snyder, C. L., Schroeder, W. R., Cram, D., Datla, R., Wishart, D., Weselake, R.J. and Krishna, P. 2012. Fatty acid composition of developing seabuckthorn (*Hippophae rhamnoides* L.) berry and the transcriptome of mature seed. *PLoS ONE* 7 (4): e34009.
12. Kallio, H., Yang, B.R. and Peippo, P. 2002. Effects of different origins and harvesting time on Vitamin C, tocopherols, and tocotrienols in seabuckthorn (*Hippophae rhamnoides*) berries. *J. Agri. Food Chem.* 50: 6136-6142.
13. Lanev, E., Radev, S., Balutsov, M., Klouchek, E. and Popov, A. 1995. The effect of an extract of seabuckthorn on the healing of experimental skin wounds in rats. *Dermatology* 48 (3): 30-33.
14. Lehtonen, H.M., Lehtinen, O., Suomela, J.P., Viitanen, M. and Kallio, H.P. 2010. Flavonol glycosides of seabuckthorn (*Hippophae rhamnoides* sp. *sinensis*) and lingonberry (*Vaccinium vitis- idaea*) are bioavailable in humans and monoglucuronidated for excretion. *J Agric Food Chem.* 58: 620-627.
15. Lu, R. 1993. The chemical composition of *Hippophae* fruits in China. In: *Proceedings of International*

Symposium on Seabuckthorn (Hippophae rhamnoides L.), Barnaul, Russia.

16. Mironov, V.A. 1989. Chemical composition of *Hippophae rhamnoides* of different populations of the USSR. In: *Proc. International Symposium on Seabuckthorn (H. rhamnoides* L.), Xian, China, Oct. 19-23, 1989, pp. 67-70.

17. Parimelazhagan, T., Chaurasia, O.P. and Ahmed, Z. 2005. Seabuckthorn: Oil with promising medicinal value. *Current Science* 88 (1): 8-9.

18. Parthiban, K.T., Selvan, P., Paramathma, M., Kanna, S.U., Kumar, P., Subbulakshmi, V. and Vennila, S. 2011. Physico-chemical characterization of seed oil from *Jatropha curcas* L. genetic resources. *Journal of Economic and Natural Environment* 3 (5): 163-167.

19. Pearson, D. 1976. *The Chemical Analysis of Foods*, 7th Ed. Churchill Livingstone, pp. 6-14.

20. Qinxiao, W. and Hongyan, Z. 2003. Seabuckthorn for the control of soil erosion in mountain lands. In: *Seabuckthorn (Hippophae* L.)*-A Multipurpose Wonder Plant. Vol. I: Botany, Harvesting and Processing Technologies* (V. Singh, Editor-in-Chief, 2003), pp. 401-408, Indus Publishing Company, New Delhi, 518p.

21. Ranjith, A., Kumar, K.S.,Venugopalan, V.V., Arumughan, C., Sawhney, R.C. and Singh, V. 2006. Fatty acids, tocols, and carotenoids in pulp oil of three seabuckthorn species (*H. rhamnoides, H. salicifolia, and H. tibetana*) grown in the Indian Himalayas. *JAOCS* 88: 359-364.

22. Singh, V. 2003. Geographical adaption and distribution of seabuckthorn (*Hippophae* L.) resources. In: *Seabuckthorn (Hippophae* L.)*-A Multipurpose Wonder Plant. Vol. I: Botany, Harvesting and Processing Technologies* (V. Singh, Ed. in Chief, 2003), pp. 21-34, Indus Publishing Company, New Delhi, 518p.

23. Tiitinen, K.M., Hakala, M.A. and Kallio, H.P. 2005. Quality components of seabuckthorn *(Hippophae rhamnoides)* varieties. *J Agric Food Chem.* 53: 1692-1699.

24. Tiitinen, K.M., Yang, B., Haraldsson, G.G., Jonsdottir, S. and Kallio, H.P. 2006. Fast analysis of sugars, fruit acids, and vitamin C in seabuckthorn *(Hippophae rhamnoides* L.) varieties. *J Agric Food Chem.* 54: 2508-13.

25. Xing, J., Yang, B., Dong, Y., Wang, B., Wang, J. and Kallio, H.P. 2002. Effects of seabuckthorn seed and pulp oils on experimental models of gastric ulcer in rats. *Fitoterapia* 73 (7-8): 644-650.

26. Zeb, A. 2004. Chemical and nutritional constituents of seabuckthorn juice. *Pakistan Journal of Nutrition* 3 (2): 99-106.

27. Zhang, W., Yan, J., Duo, J., Ren, B. and Guo. J. 1989. Preliminary study of biochemical constituents of seabuckthorn berries growing in Shanxi Province and their changing trend. In: *Proc. Int. Symp. Seabuckthorn (H. rhamnoides* L.), Oct 19-23, 1989, Xian, China, pp. 96-105.

28. Zhou, Y. 1998. Study on the effect of *Hippophae* seed oil against gastric ulcer. Institute of Medicinal Plants Resource Development, the Chinese Academy of Medical Sciences, Beijing.

Chapter 17

Molecular Characterizations and Bioactivities of Isolated Seabuckthorn Constituents

Chinchu Bose[1*], Virendra Singh[3], Walter Schrenk[2] and Asoke Banerji[1]

[1]Amrita School of Biotechnology, Amrita Vishwa Vidyapeetham, Kollam Distt. – 690 525, India
[2]Amrita-Agilent Analytical Research Centre, Amrita Vishwa Vidyapeetham, Kollam – 690 525 , India
[3]CSK Himachal Pradesh Agricultural University, Palampur – 176 062, India
(*E-mail: banerjiasoke@gmail.com)

SUMMARY

Phytochemical profiles of plants are useful in relating the reported biological activities of the extracts with chemically identified compounds. Standard methods for characterization of biochemicals and nutraceuticals of seabuckthorn (SBT) such as carotenes, unsaturated fatty acids, and ascorbic acid are well established. However the same is not true for the molecular characterization of phytochemicals. We present here our experience in the structural characterization of other phytochemicals of SBT by making use of chemical reactions, UV-Vis spectroscopy in conjunction with shift reagents, HPLC with diode array detector and LC-MS/MS (ESI). Using very small quantities, it was possible to characterize flavonoid and tannin glycosides, their aglycones, sugar residues, point of linkages and sequence of sugar residues. Even isomeric glycosides with differences only in the point of linkage between sugar residues, could be distinguished. Antioxidant activity, DNA protection properties and nitric oxide inhibition activities of SBT constituents have been studied.

Keywords: *Seabuckthorn, Tannins, Flavonoid glycosides, Antioxidant activity, DNA protection, Nitric oxide and Flavonoid glycosides.*

INTRODUCTION

Seabuckthorn (*Hippophae* L.) plays very important role in the preservation of the ecosystems. The extensive network of roots of SBT prevents soil erosion as well as help in the conservation of water. The presence nitrogen fixing bacteria, *Frankia* spp. enriches the soil with nitrogen by fixing atmospheric nitrogen. SBT populations in India are represented by mainly three species *viz. Hippophae rhamnoides* ssp. *turkestanica, H. salicifolia and H. tibetana.* Most of the research work on SBT revolves round the commercially important parts such as seed, fruit *etc.* Till recently, not much attention was given to leaves of SBT. Our work on leaf constituents of SBT was initiated as early as in 2003. There were three main reasons for choosing the leaves for our research work: (i) The leaves are easily available throughout the seasons; (ii) leaves possess good chemo diversity and iii) they are reported have many medicinal activities (Geetha, *et al.*, 2011). Systematic phytochemical investigations on the leaves of *H. rhamnoides* and *H. salicifolia* were undertaken in our laboratory. Our earlier publication (Chinchu *et al.*, 2013) describes the isolation and characterizations of several phytochemicals from the leaves. In continuation, further work on isolation and molecular characterization of intact tannins and flavonoid glycosides are described here. Also biological properties such as radical scavenging activities, nitric oxide inhibition, DNA protection activities of SBT constituents are also described.

MATERIALS AND METHODS

Plant Material

Hippophae salicifolia was obtained from Himachal Pradesh and North Sikkim (Lachen, Lachung).

H.rhamnoides was collected from Ladakh area. The leaves were shade-dried, powdered and extracted sequentially at ambient temperature with appropriate solvents by using a rotary shaker.

General Analytical Methods

TLC was performed using silica gel plates (Kiesel gel, F_{254}, Merck). HPLC Analyses were performed on a Shimadzu-SPD-M20A HPLC, equipped with DAD (Diode Array detector) detector with Phenomenex Luna 5µ C_{18} (2) 100A, size 250 X 4.60 nm. All the compounds were detected at 254 nm at room temperature with an effluent flow rate of 1.2 ml/min and an injection volume of 10µL. The mobile phase consisted of acetonitrile and water containing formic acid (0.03 per cent). ^{1}H NMR spectra were recorded on a Bruker AM 400 spectrometer (400 MHz, for ^{1}H NMR) using TMS as a standard. UV-Vis spectra were recorded on a Shimadzu UV spectrophotometer, Model UV-1800 in methanol. IR spectra were recorded on a Shimadzu IR Affinity-1 in KBr or thin film. LC-MS were recorded on Agilent Technologies 1290 Series HPLC and 6340 Ion Trap LC-MS.

DNA Protection Assay

The ability of plant extracts to protect plasmid DNA from hydroxyl radicals produced by Fenton's reagent was evaluated by DNA nicking assay as described earlier (Lee *et al.*, 2002) with minor modifications. DNA protection assay was performed using super coiled plasmid DNA prepared from pCDH-CMV-FF Luc-EF1a-Puro lenti viral vector. Plasmid DNA (0.5 µg) was added to Fenton's reagents (30 mM H_2O_2, 50 µM ascorbic acid, and 80 µM $FeCl_3$containing different concentrations of the ethanol extract, and the final volume of the mixture was brought up to 20 µL. The reaction mixtures was then incubated for 30 min at 37°C followed by the addition of loading dye bromophenol blue (0.25 per cent in 50 per cent glycerol) and the electrophoresis (90v) was carried out in TAE buffer (40 mM Tris base, 16 mM acetic acid 1mM EDTA, pH 8.0). DNA was analyzed followed by ethidium bromide staining. The gallic acid was used as positive control. The extent of DNA damage protection (per cent) was calculated by comparing the proportion of super coiled DNA with Fenton's reagent to that without Fenton's reagent. Densitometric analysis was carried out using Image software.

Nitric Oxide Scavenging

At physiological pH, nitric oxide generated from aqueous sodium nitroprusside (SNP) solution interacts with oxygen to produce nitrite ions, and is quantified by the Griess Illosvoy reaction (Lo´pez *et al.*, 2004). The reaction mixture contained 10 mM SNP, PBS (pH 7.4) and increasing concentrations of the test solutions in a final volume of 3 ml. After incubation for 150 minutes at 25°C, 1 ml sulfanilamide (0.33 per cent in 20 per cent glacial acetic acid) was added to 0.5 ml of the incubated solution and allowed to stand for 5 minutes. Then 1 ml of napthylethylene diamine dihydrochloride (NED) (0.1 per cent w/v) was added and incubated for 30 minutes at 25°C. The pink chromophore generated during diazotization of nitrite ions with sulphanilamide and subsequent coupling with NED was measured spectrophotometrically at 540 nm against a blank sample. All tests were performed in triplicate. Rutin was used as a standard.

Free Radical Scavenging

For evaluation of free radical scavenging activities of test compounds (2.5-80 µg) in methanol (100 µl) were added to a solution of DPPH (0.025 per cent ; 3.9 ml) in methanol (Chinchu *et al.*, 2013). The mixture was left for 5 min and absorbance of the resulting solution was measured at 540 nm. The inhibition percent was calculated from the following equation: per cent inhibition = (absorbance of control - absorbance of test sample)/absorbance of control) × 100. The results are expressed as IC_{50} (µg).

Extraction and Fractionation

Extraction of SBT leaves was carried out by shaking (3 X 24 hrs) with 70 per cent aq. acetone in a rotary shaker (room temperature). The native tannins could be preferentially precipitated from the 70 per cent aq. acetone extract in methanol by addition of BSA or just sufficient quantity of methanolic lead acetate (10 per cent). The precipitate (the lead complex) was separated by centrifugation. The separated lead complex was suspended in methanol and the tannins were regenerated by the addition of just sufficient quantity of aq. sulfuric acid (8 per cent) to decompose the Pb-complex. HPLC in conjunction with UV spectroscopy revealed the presence of gallic acid ellagic acid derivative(s). The slight excess of lead acetate in the supernatant was removed by precipitating with dil. sulphuric acid till the mixture became acidic (pH 5). The strong colour (green) with methanolic $FeCl_3$, deep brown ring in Molisch test and red colour with Mg-HCl (Shinoda's tests) of the supernatant indicated the presence mainly phenolics and flavonoids (glycosides). On acid hydrolysis, mainly isorhamnetin, quercetin, kaempferol, rhamnose and glucose were identified. Presence of corresponding glycosides is thus inferred.

RESULTS AND DISCUSSION

Literature survey indicates that the polar fractions of SBT contain many bioactive compounds; mainly hydrolysable tannins, flavonoid glycosides, free acids *etc.* Acid hydrolysis of 70 per cent aq. acetone extract of SBT furnished several compounds such as quercetin, isorhamnetin, kaempferol, glucose, rhamnose, quebrachitol, gallic and ellagic acids. These observations suggest that the native extracts contain intricate mixture of glycosides or conjugates with other components. Selective complexation with metals and macromolecules, repeated column chromatography including gel permeation chromatography were used for their fractionation. TLC and HPLC were used for monitoring the progress of isolation procedures.

Isolation and Characterization of Intact Flavonoids and Tannins of SBT Leaves

SBT leaves were extracted exhaustively with chloroform (to remove the non-polar constituents) prior to extraction with 70 per cent aq. acetone. The extraction (3 X 24 hrs) was carried out in a rotary shaker at ambient temperature. The solvents from the extracts were removed using rotary evaporator below 40°C. The individual extracts were subjected to repeated column (SiO_2) chromatography followed by gel-permeation chromatography on Sephadex LH-20. The Sephadex column was eluted with methanol and the fractions were analyzed by TLC, HPLC and chemical reactions. HPLC-DAD of the fractions showed them to be intricate mixtures of many components. UV-Vis spectral data and colour reactions suggested that the extract contain hydrolysable tannins and flavonoid glycoside mixtures. In order to resolve the intricate mixture, the polar extracts were subjected to mild acid hydrolysis. The hydrolysates were extracted with ethyl acetate followed by methyl ethyl ketone. UV -Vis spectra of the hydrolysates, a strong ferric reaction (green) and Shinoda test (Mg-HCl) suggested the presence of flavonoids while gallic (271 nm) and ellagic (254 nm) acids were present in the ethyl methyl ketone extract. Occurrence of substantial quantities of gallic and ellagic acids is very significant as they contribute to the bioactivities of SBT. Our studies indicated that gallic and ellagic acids show radical scavenging activity, anti-diabetic and radioprotective activities of SBT extracts (Vishnuprasad *et al.*, 2010; Poulose *et al.*, 2011; Khan *et al.*, 2014). The fractions obtained from Sephadex LH-20 (gel permeation chromatography) were analyzed by UV-Vis spectra and LC-MS/MS and very valuable structural information could be obtained. Molecular weights of the compounds were obtained from the LC/MS/MS spectra. The identity of sugar residues in the glycosides could be derived from the MS/MS data. The location of position of glycosides was arrived at from the U-Vis spectra of the intact glycosides. It is reported that flavonols (with free hydroxyl at 3-postion) absorb at around 360-370 nm; the blockage of hydroxyls causes marked hypsochromic shifts in the band I. Addition of $AlCl_3$ causes a marked bathochromic shift due formation of Al-complex which is of diagnostic value.

Fraction 1 obtained from the Sephadex column showed a clusters of unresolved signals in the HPLC. However acid hydrolysis of the fraction resulted in the formation of ellagic acid and glucose. Fraction 1, therefore contains ellagi glucotannin. This was supported by UV-Vis spectra. Fraction 2 showed typical flavonoid type absorption. Acid hydrolysis of F2 indicated that it primarily contains isorhamnetin and glucose. Thus it was inferred that fraction 2 is isorhamnetin-3-glucoside. This was supported by LC-MS/MS. UV spectra of Fraction 3 suggest the presence of gallic acid and isorhamnetin with lesser amounts of quercetin and ellagic acid. Fraction 5 mainly contain ellagic acid and closely related compounds.

Determination of Linkage of the Sugars

Both aglycones and carbohydrates are polyhydroxy compounds. Thus there are numerous possibilities for glycosidic links. UV-Vis spectral analyses were very useful for the location of sugars residues in the molecules. Absorption maximum II of flavones (245-270 nm) arise from the benzoyl part of the flavonoids while the maximum I which appears around 330-370 nm represent the cinnamyl moiety. The flavonoids which lack in 3-OH or where the 3-position is blocked, show absorption around 330 nm while those with hydroxyl at 3-positions (flavonols) show absorption at around 350-370 nm. Shift reagents *e.g.* $AlCl_3$, NaOAc, H_3BO_3 are also useful for the determination of the position of sugars in the glycosides (Markham, 1982). Amongst SBT constituents, position 3 seems to be the most preferred for glycosidation. Literature review suggests that there is prominent bathochromic shift in band I when the hydroxyl at 3-position is generated as a result of hydrolysis of 3-glycosides. Thus UV-Vis spectra provide a useful tool for the determination of the position of sugars in glycosides. The UV-Vis spectral studies of the SBT extracts indicate that they are mainly 3-glycosylated. The individual flavonoids were separated by preparative HPLC. Application of the above approach of combining colour reactions, hydrolysis, UV-Vis (with shift reagents) and LC-MS/MS is illustrated by the characterization of some of representative examples as described below:

Quercetin-3-diglucoside (Fraction F4)

Molecular ion of 627 (M+1) was obtained by L C-MS data. MS/MS of the m/e 627 shows the losses of two fragments with m/e 162 successively. Loss of m/e 162 corresponds to glucosyl residue Figure 17.1. Thus fraction F4 was identified as quercetin-3-diglucoside. UV-Vis spectra (in conjunction with shift reagents) and colour reactions (Shinoda test, ferric reaction, Molisch test) were in consistent with the assignment.

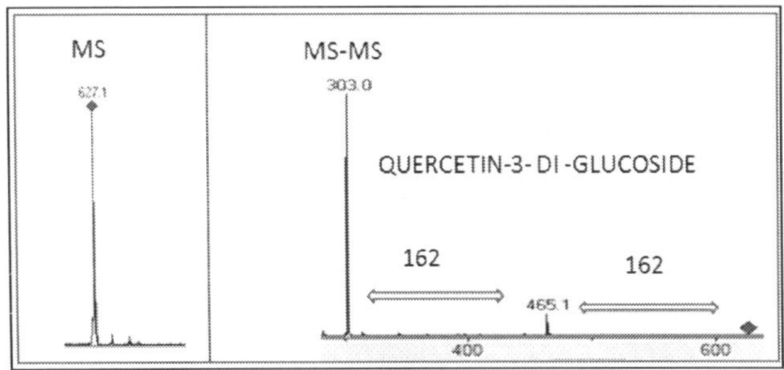

Figure 17.1: LCMS/MS Characterization of Flavonoid Glycoside.

Isorhamnetin-3-glucorhamnoside (F12)

It showed typical UV-Vis spectrum for flavonoid 3- glycoside. The nature of the sugars and sequence of their linkages could be derived from the fragmentation pattern in the mass spectra. The mass peak m/e 625 corresponds to M+1. MS/MS of m/e 625 lost m/e 146 and 162 successively indicting loss of rhamnose followed by glucosyl residues. Thus F12 is thus characterized as Isorhamnetin-3-glucorhamnosides. The assignment was supported by hydrolysis experiment when isorhamnetin, glucose and rhamnose were identified.

Ellagic Glycoside (F17)

Acid hydrolysis of F17 gave ellagic acid. LC-MS showed M+1 at m/e 465 which on fragmentation (MS/MS) lost m/e of 162 giving m/e 303. Fragments obtained from the MS/MS of m/e 303 showed that is due to ellagic acid and not quercetin (which also has m/e 303) Figure 17.2. The UV-Vis (254 nm) was in consistent with assigned structure. The position of the sugar residue could not be deduced on the basis of data obtained.

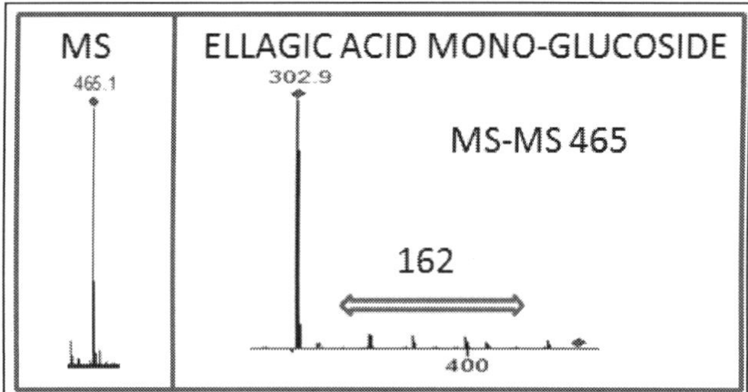

Figure 17.2: LCMS/MS Characterization of Ellagic Glycoside.

Ellagic Rhamnoglucoside (F18)

UV spectrum (254 µm) suggested it to be an ellagic acid derivative. The mass fragmentation of its M+1 m/e 611 show loss of glucose residue followed by loss of rhamnose residue. Therefore it was identified asellagi rhamnoglucoside.

Quercetin-3-rhamnoglucorhamnoside

It gave expected UV-Vis spectra and colour reactions consistent with flavonoid structure. In the LCMS/MS the M+1 at m/e 773 showed the successive losses of rhamnose, glucose and rhamnose. Thus it was assigned the structure: quercetin -3- rhamno glucorhamnoside. UV-Vis spectra in conjunction with the shift reagents supported the structure.

Following the above approach the structures of several flavonoids and tannin glycosides were derived as presented in the Figure 17.3.

Antioxidant Activity

The antioxidant activity was studied as described earlier. (Chinchu *et al.*, 2013). Trolox was used as a reference antioxidant. The antioxidant activities are shown in the following Table 17.1.

Amongst the extracts, 70 per cent aq. acetone extract of SBT leaves from *H. rhamnoides* was the found to be better than those of *H. salicifolia*. Ellagic acid glycosides showed high activities. Quercetin-3-diglucoside and isorhamnetin glucorhamnoside also showed significant activities.

Amongst isolated pure compounds, gallic acid showed the highest antioxygenic activity followed by ellagic acid compared to SBT flavones. Both these may be artifact of hydrolysable tannins present in SBT leaves. Since tannoids including gallic and ellagic acids form major polar constituents of SBT, these should given due importance for their possible role in imparting bioactivities. These constituents have been identified as the active principles of many other medicinal plants, *e.g. Punica granatum, Terminalia chebula.*

DNA Protection Assay

Hydroxyl radicals generated by Fenton's reagent attack supercoiled plasmid DNA and causes single stranded scission (resulting in nicked circular form II) or double stranded breaks (resulting in linear form III). In the present study, the addition of SBT fractions along

QUERCETIN
R=H- GLU-GLU(627[M+]).
R=H- GLU-RHAM-GLU.(773[M+])

KAEMPHEROL
R=H- GLU-GLU (611[M+]).
R=H-GLU-RHAM (595[M+]).

ISORHAMNETIN
R=H- GLU-GLU-RHAM(787[M+]).
R=H- GLU-RHAM-(641[M+]).
R=H-RHAM-GLU -(625[M+]).

ELLAGIC ACID
- GLUC (465)[M+]
-RHAM-GLUC (611[M+]).

Figure 17.3: Some of the Flavonoid Glycosides and Ellagic Acid Derivatives Obtained from SBT. GLU= Glucose, RHAM= Rhamnose.

Table 17.1: Antioxidant Activity of SBT Constituents.

Antioxidant activity studied using DPPH

NO	COMPOUNDS	EC $_{50}$ µg
1	Gallic acid	4
2	Ellagic acid	5
3	Trolox	12.5
4	Ellagicacid-mono glucoside, ellagicacid – rhamno-glucoside	10
5	Ellagic acid mono glucoside	18
6	Quercetin- di -glucoside	21
7	Isorhamnetin-gluco-rhamnoside	25
8	HR leaves 70% acetone ext	40
9	HS leaves 70% acetone ext	50

EC50 value was calculated as concentration(µg) of the compound required for inhibition of 50% of activity. Gallic acid was used as a reference antioxidant

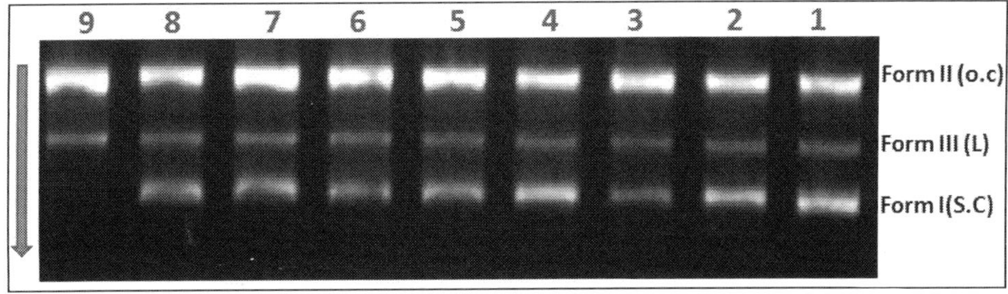

Figure 17.4: Effect of SBT Fractions on the Protection of Super Coiled Plasmid DNA against Hydroxyl Radical Generated by Fenton.

1) DNA (DNA damage control), 2) DNA + Isorhamnetin-glucorhamnoside+Fenton, 3) DNA + HS 70 per cent +Fenton, 4) DNA+ Ellagic acid-mono glucoside, ellagic acid –rhamno-glucoside +Fenton, 5)DNA + Quercetin- di -glucoside+Fenton, 6) DNA+HR70 per cent +Fenton, 7) DNA+ Ellagic acid monoglucoside +Fenton, 8) DNA+ Fenton +Gallic acid, 9) DNA+ Fenton. S.C- Supercoiled, O.C- Open circular, L- Linear.

with Fenton's reagent provided protection to super coiled plasmid DNA. The effect of the fractions was compared with standard antioxidant gallic acid. The concentration of fractions used was 10µg/ml. The extent of DNA damage protection (per cent) was calculated by comparing the proportion of super coiled DNA with Fenton's reagent to that without Fenton's reagent. Tannins and flavonoid glycoside fractions protected the DNA from hydroxyl radicals ellagi tannin. Fraction (F 18) showed high protection (85 per cent) from Fenton. Our results using quercetin did not yield consistent results possibly due limited solubility. However quercetin-3-diglucoside (F4) showed significant (70 per cent) activity. Results obtained are presented in Figure 17.4. Densitometric analysis was carried out using Image J software and graphically represented in Figure 17.5.

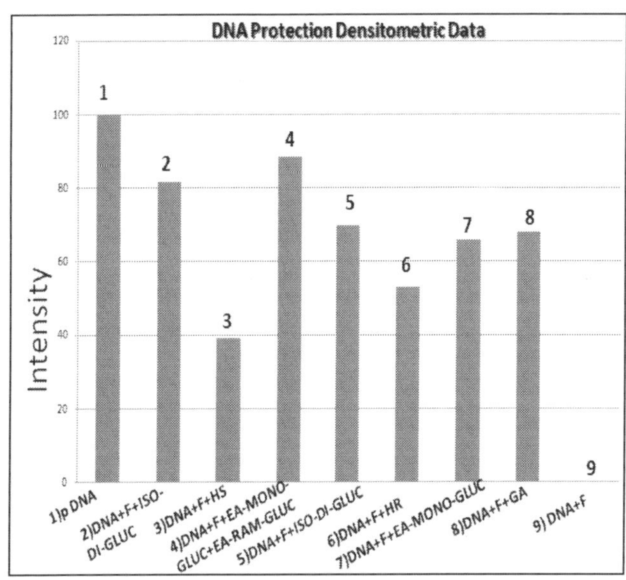

Figure 17.5: Plasmid DNA Protection of SBT Constituents Densitometric Data.

Nitric Oxide Scavenging

Nitric oxide (NO) is an important chemical mediator generated by endothelial cells, macrophages,

Table 17.2: Nitric Oxide Scavenging Properties of SBT Fractions

Nitric oxide scavenging activity

NO	COMPOUNDS	EC_{50} µg
1	Rutin	61
2	HR 70%	182
3	HS 70%	150
4	Isorhamnetin glucorhamnoside	137
5	Ellagic acid mono glucoside	120
6	Ellagic acid mono glucoside, ellagic acid rhamnoglucoside.	115
7	Quercetin di glucoside	98

EC50 value was calculated as concentration (µg) of the compound required for inhibition of 50% of actvity. Rutin was used as a reference.

neurons, *etc.* and is involved in the regulation of various physiological processes. Excess concentration of NO is associated with several diseases. Oxygen reacts with the excess of NO to generate nitrite and peroxynitrite anions, which act as free radicals. Our results suggest that the SBT components such as flavonoid glycosides and tannins show promising NO scavenging effect when compared with standard rutin Table 17.2. EC_{50} value was calculated as concentration (µg) of the compound required for inhibition of 50 per cent of activity.

CONCLUSION

Many interesting biological activities have been attributed to the polar fractions of seabuckthorn (*Hippophae* spp.) leaves. These are intractable complex mixture of many components such as organic acids, glycosides of flavonoids and tannins *etc.* Identification of individual components present difficult proposition. We have developed protocols for separation of individual constituents using selective complexation with metals and macromolecules, column and preparative HPLC chromatography. The molecular characterization of isolated components was carried out using conventional (such as colour tests) and advanced techniques UV-Vis spectroscopy (with diagnostic shift reagents), HPLC-DAD and LCMS/MS. Using combination of these techniques; it was possible to determine the structures of flavonoid glycosides and tannin glycosides. Bioactivities of the isolated molecules were studied.

ACKNOWLEDGEMENTS

We are thankful to the Dean, School of Biotechnology, Amrita Vishwa-Vidyapeetham for his support and keen interest in this work.

REFERENCES

1. Amitava, K., Manna K., Chinchu, B., Das, D.K., Sinha, M., Keshswaraj, B., Das, U., Dey, R., Banerji, A. and Dey, S. 2014. Seabuckthorn (*Hippophae rhamnoides* L.) leaf extract ameliorates the gamma radiation mediated DNA damage and hepatic alterations. *Indian Journal of Experimental Biology* 52: 952-964.
2. Chinchu, B., Pandurangan, N., Shrenk, W. and Banerji, A. 2013. Isolation, characterization of bioactives from the Indian seabuckthorn. Seabuckthorn research for a promising crop: A look at recent developments in cultivation, breeding, technology, health and environment David Eagle, Yury Zubarev, Jörg-Thomas Mörsel (Eds.).
3. Geetha, S. and Asheesh, G. 2011. Medicinal and therapeutic potential of seabuckthorn (*Hippophae rhamnoides* L.). *Journal of Ethnopharmacology* 138: 268-278.
4. Lee, J.C., Kim, H.R. and Kim, J. 2002. Antioxidant property of an ethanol extract of the stem of Opuntiaficus-indicavar, Saboten. *Journal of Agricultural Food Chemistry* 50: 6490-6496.
5. Lopez, L.G., Moreno, L., Cogolludo, L., Galisteo, M., Ibarra, M., Duarte, J., Lodi, F., Tamargo, J. and Vizcaino, F.P. 2004. Nitric Oxide (NO) Scavenging and NO protecting effects of quercetin and their biological significance in vascular smooth muscle. *Molecular Pharmacology* 65: 851-859.
6. Markham, K.R. 1982. *Techniques of Flavonoid Identification*, pp. 36-49, Academic Press, London.

7. Poulose, N., Vishnu, P.C.N., Nidhina, H. and Anilkumar, G. 2011. Ellagic acid stimulates glucose transport in adipocytes and muscles through AMPK Mediated Pathway. *Journal of Diabetes and Metabolic Disorders* 2: 149. doi: 10.4172/2155-6156.1000149.

8. Vishnu, P. C. N., Anjana, T., Banerji, A. and Anilkumar, G. 2010. Gallic acid induces GLUT4 translocation and glucose uptake activity in 3T3-L1 cells. *Febs Letters* 584: 531-536.

Chapter 18

Chemo Profiling of Seabuckthorn (*Hippophae rhamnoides*) Byproducts

C. Varshneya* and Arti Ghabru

Department of Pharmacology and Toxicology, COVAS,
CSK Himachal Pradesh Agricultural University, Palampur – 176 062, H.P., India
(*E-mail: cvarshneya@hotmail.com)

SUMMARY

Seabuckthorn (*Hippophae rhamnoides)* is a wonder plant of temperate Himalayan region rich in various antioxidants, minerals and vitamins. Seabuckthorn byproduct (70 per cent methanolic extracts) *i.e.* leaves, seedcake, pomace with and without seeds were analyzed for their phyto chemical contents. Among all the byproducts, leaves contained highest amount of total phenols (332.49±7.45 mg/g), total flavonoids (271.56±5.41 mg/g), β-carotenoid (262.200±17.48 µg/mg), vitamin C (399.84±4.90 mg/100g), and lycopene (8.50±2.92 mg/100 g) content. The flavonoids found in leaves were rutin and quercetin-3-galactoside, whereas, isorhamnetin was found both in pomace with and without seeds. However, none of these flavonoids were found in seedcake. The leaves were also found to be rich in crude protein (22.09 per cent) and minerals *viz.* Na (160 mg/g), K (7.64 mg/g), Ca (10.32 mg/g), Mg (1.716 mg/g) and Fe (0.720 mg/g). Both pomace with and without seeds were rich in crude fibers. On the basis of the results, it could be concluded that leaves contained highest amount of bioactive phytochemicals.

Keywords: *Seabuckthorn, β-Carotenoid, Flavonoid, Phenols, Lycopene and Lahaul and Spiti.*

INTRODUCTION

Seabuckthorn (*Hippophae rhamnoides)* is a bush species widely distributed throughout the temperate zone of Asia, Europe and all over subtropical zones being found especially at high altitudes. The natural habitat of seabuckthorn extends widely in China, Mongolia, Russia and most parts of Northern Europe. In India the plant inhabits dry temperate region and high altitude regions of Himachal Pradesh, Jammu and Kashmir and Uttrakhand. Seabuckthorn has recently gained in interest for its nutritional and medicinal values (Guliyev *et al.*, 2004; Zeb, 2004). All parts of the plant *i.e.* fruits and leaves are considered to be good source of vitamins, trace elements, amino acids and large number of bioactive substances such as β-carotene, zeaxanthin, lycopene, flavonoid, folic acid, fatty acid, tannic acid *etc.* and these substances are mainly responsible for various pharmacological and biological effects. The chemical and phytochemical composition of seabuckthorn has been reviewed and found to vary with the origin, climate and method of extraction (Beveridge *et al.*, 1999; Zeb and Ahmed, 2004). Its fruit is source of nearly 190 bioactive substances, whereas its oil has nearly 106 such components (Zheng and Song, 1992). There is an ample quantity of quality vitamins in seabuckthorn fruit and leaves. The mineral contents of seabuckthorn make the shrub most important. The presence of polyphenols in seabuckthorn has shown antioxidant properties and they can be used to protect human body against the damaging effect of free radicals (Goa *et al.*, 2000; Yao and Tigerstedt, 1992; Kallio *et al.*, 2002; Yang *et al.*, 2001). Present study was carried out to do the chemo profiling of seabuckthorn byproducts.

*Seabuckthorn (*Hippophae* L.): A Multipurpose Wonder Plant, Vol. 5: Advanced Technologies for Cultivation, Processing, Health Protection and Environmental Conservation* (Virendra Singh, *Editor-in-Chief,* 2018), p. 143–147, Astral International Pvt. Ltd., New Delhi, India.

MATERIALS AND METHODS

Collection of Seabuckthorn Samples

Different byproduct of seabuckthorn *i.e.*, leaves, seed cake, pomace with seeds and pomace without seeds were used in the present study. The leaves and pomace were collected from Keylong region of Himachal Pradesh, India during the month of October, whereas seedcake was collected from the Department of Nutrition, COVAS, CSKHPKV, Palampur, India.

Preparation of Extract

The leaves and pomace were air dried in shade. After shade drying, the pomace was divided into two types *i.e.* pomace with seed and pomace without seed. The dried leaves and pomace were then powdered in the mixer and stored at the room temperature (18-22°C) till further process. The powder of different byproducts were soaked in 70 per cent methanol for 24 hrs and kept at room temperature with intermittent shaking. The mixture was then filtered through filter paper and the extracts were prepared after drying the filtrate in rotator vacuum evaporator at 40°C. The percentage of recovery was calculated after determining the weight of the extracts. Finally, the dried extracts were lyophilized and stored at 4°C till analysis. The chemo profiling of the various seabuckthorn byproducts for the total phenol, total flavonoid, total carotenoids, vitamin E, vitamin C and lycopene content was done by spectrophotometric methods. The proximate analysis and mineral analysis was also done in seabuckthorn byproducts.

Estimation of Total Phenol

Total phenolic content in different extracts of seabuckthorn byproducts was estimated by using Folin-Ciocalteu phenol reagent (FCR) based assay (Gulcin, 2002). The extract (50µl) was added to 950µl of distill water. After proper mixing, FCR (0.5ml) and 20 per cent sodium carbonate solution (2.5ml) were added. The mixture was then kept for 40 minutes at room temperature and absorbance was recorded at 725 nm. Total phenolic content in the extracts of seabuckthorn byproducts were calculated from the standard curve of gallic acid.

Estimation of Total Flavonoid

Total flavanoids were estimated spectrophotometrically according to the method of Zhishen *et al.* (1999) with slight modification. To 1.0 ml of extract (1mg/ml), 4.0 ml of distilled water and 0.3 ml of 5 per cent sodium nitrite was added. After 5 min, 0.3 ml of aluminium chloride and 2.0 ml of sodium hydroxide (1M NaOH) were added to the mixture and total volume was made up to 10.0 ml with distilled water. The solution was mixed thoroughly and absorbance was measured against reagent blank at 510 nm. Rutin was used as a standard for constructing a calibration curve.

Estimation of β-Carotenoids

β-carotenoids were estimated according to the method of Gowenlock *et al.* (1988). To 1.0 ml aliquot of sample, 1.0 ml of ethanol and 3.0 ml of petroleum ether was added. The mixture was shaken vigorously and then centrifuged at 2500 rpm for 15min. The absorbance was read at 450 nm against petroleum ether as blank. The concentration of β-carotenoid in extracts was obtained using standard curve of beta carotene.

Estimation of Vitamin E

Vitamin E was estimated by the method as described by Kayden *et al.* (1973) with slight modification. To extract (1.8 ml), equal amount of absolute alcohol was added and mixed thoroughly. For preparation of standard and blank α-tocopherol and distilled water were used. To standard (1.8 ml), equal amount of distilled water and to blank (1.8 ml) equal amount of absolute alcohol was mixed. To this 1.8 ml of purified xylene was added and vortex for 2 min followed by centrifugation at 800 g for 10 min. 0.4 ml of xylene extract was carefully pipetted out and to this 0.4ml of bathophenanthroline reagent was mixed. Then 0.4 ml of ferric chloride was added followed by 0.4 ml of o-phosphoric acid and 1.0 ml of alcohol. The contents were mixed thoroughly and absorbance was measured at 536 nm.

Estimation of Vitamin C

Vitamin C was estimated by 2, 4 dinitro phenylhydrazine (DNPH) method as described by Baker and Frank (1968).To 2.0 ml of extract 6.0 ml of 5 per cent TCA was added. The solution was mixed thoroughly and centrifuged at 2000 rpm for 10min. 2.0 ml of supernatant was pipette in to each sample tube and 2.0 ml of ascorbate standard was taken in to standard tubes.Then one drop of indophenol reagent was added and mixed well followed by addition of 0.5 ml of DT (2 gm DNPH + 1 gm thiourea in 9N H_2SO_4) mix. All the test tubes were incubated for 1hr in a water bath at 60°C and cooled in ice water. After cooling 2.5 ml of 85 per cent H_2SO_4 was added to tubes and mixed well.The optical density was recorded at 505 nm against distilled water. Ascorbic acid was used as a standard for calculations.

Estimation of Lycopene

Lycopene contents in seabuckthorn byproducts were measured by the method of Ranganna (1976). Samples were extracted with acetone using pestle and mortar until the residue was colorless. The acetone extracts were pooled and transferred to a separating funnel containing about 20 ml petroleum ether and then 20 ml of 5 per cent sodium sulphate solution was added and shaken in the separating funnel. Twenty ml more petroleum ether was added to the separating funnel for clear separation of two layers. Two phases were separated the lower aqueous phase was re extracted with additional 20 ml petroleum ether until the aqueous phase was colorless. The petroleum ether extracts were pooled little more distilled water. The washed petroleum ether extract containing lycopene was poured into a brown bottle containing about 10 g anhydrous sodium sulphate and kept aside for 30 minutes. The petroleum ether extract transferred into 100 ml volumetric flask through funnel containing cotton wool, the sodium sulphate slurry was then washed with petroleum ether until color less. The volume was made up to 100 ml with petroleum ether and absorbance read in a spectrophotometer at 503 nm using petroleum ether as blank.

Absorbance (1 unit) = 3.1206 μg lycopene/ml

mg lycopene in 100 g sample= (31.206×Absorbance)/(Wt. of Sample (g))

Identification and Quantification of Marker Compounds using HPLC

Flavonoid profiling of various Seabuckthorn by products was got done from IHBT, CSIR research center, Palampur, H.P using HPLC system consisted of a Shimadzu HPLC (Model LC-20AT pump, DGU-20A5 degasser) equipped with photo-diode array detector (CBM-20A; Shimadzu, Kypto, Japan) interfaced with an IBM Pentium 4 personal computer. The separation was performed on a Phenomenex Luna C-18 column (250 × 4.6 mm i.d., 5μm). The temperature of the column was set at 25°C. Elution of samples (20 μl) was performed with gradient solvent programme, at a flow rate of 1 ml/min for 30 minutes. The mobile phase consisted of 0.05 per cent triflouroacetic acid in water (A) and acetonitrile (B) with following gradient: 15-60 per cent B in 0-30 min and 15 per cent B in 35 min. The detection was done at 355 nm. The identification of compounds was performed on the basis of retention time, coinjection and spectral matching with standards.

Proximate Analysis

Proximate analysis of seabuckthorn byproduct was done as per method given in AOAC (1995).

Mineral Content

Mineral content was estimated by the method of Shah *et al.* (2007) using atomic absorption spectrophotometer, Analyst 400, Perkin Elmer.

Statistical Analysis

The results of the chemo profiling are present as a mean of 3 determination ± standard error and the significant difference between means was determined using Turkey-Kramer multiple comparison test.

RESULTS AND DISCUSSION

The total phenolic content was expressed as mg equivalent of gallic acid per gram of extract; whereas total flavonoids were expressed as mg equivalent of rutin per gram of extract. Amongst all the seabuckthorn byproducts, leaf extract contained highest amount of total phenols, total flavonoids, vitamin C and lycopene content. However, vitamin E was highest in pomace with seeds and β-carotenoid was highest in leaves which were at par with pomace with seeds (Table 18.1). Upadhyay *et al.* (2010) reported that aqueous and hydroalcoholic extracts of SBT leaves were found to be rich in total phenolics and flavonoids contents.

Flavonoids Profiling of Seabuckthorn Byproducts

The profiling of seabuckthorn byproducts for the different flavonoids was done from IHBT, Palampur, H.P. and the results are present in Table 18.2 and

Table 18.1: Total Phenols, Total Flavonoids, β-crotenoids, Vitamin C, Vitamin E and Lycopene Content of Seabuckthorn Byproducts

Sample	*Total Phenols mg/g*	*Total Flavonoid mg/g*	*β-Carotenoid μg/mg*	*Vitamin C mg/100g*	*Vitamin E mg/100g*	*Lycopene mg/100g*
Leaves	332.49±7.45[a]	271.56±5.41[a]	262.200±17.48[a]	399.84±4.90[a]	216.00±0.02[c]	8.50±2.92[a]
Seedcake	183.75±14.98[c]	126.07±2.51[c]	52.440±15.138[c]	206.27±5.52[d]	76.00±0.02[d]	0.30±0.01[d]
Pomace with seeds	227.04±4.17[b]	202.12±9.31[b]	262.200±15.13[a]	224.40±2.08[c]	234.00±0.02[a]	2.06±0.005[c]
Pomace without seeds	184.28±1.58[c]	128.55±2.19[c]	139.840±23.12[b]	249.33±5.23[b]	224.00±0.03[b]	6.64±0.012[b]

Values are expressed as Mean ± SEM.

The means with same superscripts in between columns do not differ significantly at 5 per cent level.

Table 18.2: Concentration of different Flavonoids in different Extracts of Seabuckthorn Byproducts

Sample	*Different Flavonoid (Yield per cent ± CV per cent)*					
	Rutin	*Quercetin-3-galactoside*	*Myricetin*	*Quercetin*	*Kaempferol*	*Isorhamnetin*
Leaves	0.4678±2.08	0.5712±2.80	–	–	–	–
Seedcake	–	–	–	–	–	–
Pomce with seeds	0.118±3.76	–	–	–	–	0.0172±4.65
Pomace without seeds	0.0644±3.89	0.0346±3.18	–	–	–	0.0123±3.37

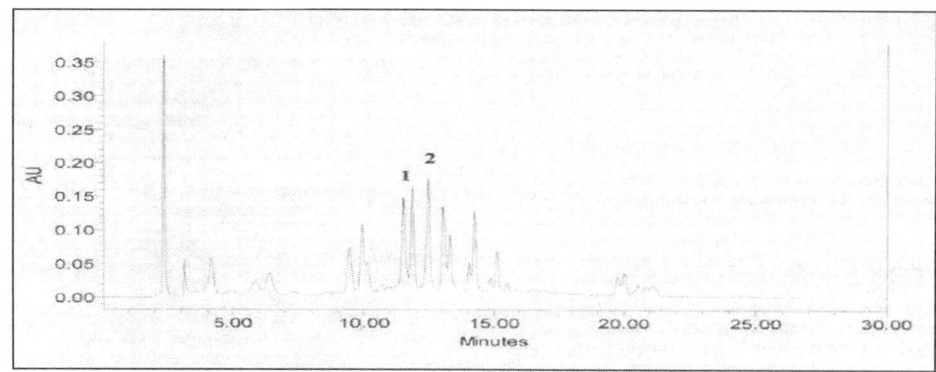

Figure 18.1: High Performance Liquid Chromatogram of Seabuckthorn Leaf Extract (1: Rutin, 2: Quercetin-3-galactoside).

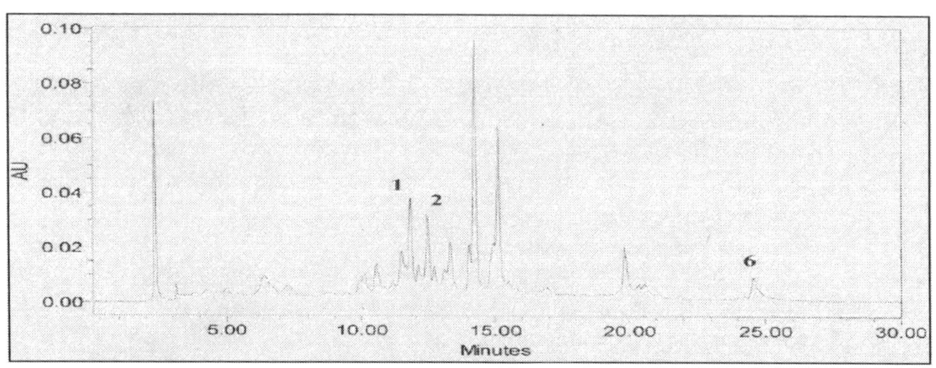

Figure 18.2: High Performance Liquid Chromatogram of Seabuckthorn Pomace Extract (1: Rutin, 2: Quercetin-3-galactoside and 6: Isorhamnetin).

shown graphically in Figures 18.1 and 18.2. It was observed that rutin was present in the leaves, pomace with seeds and pomace without seeds, whereas quercetin-3-galactoside was present in leaves and pomace without seeds. Isorhamnetin was present only in pomace sample. Upadhyay *et al.* (2010) reported that Quercetin–3-O-galactoside, quercetin–3-O-glucoside, kaempferol and isorhamnetin were identified in both aqueous and hydroalcoholic SBT leaf extracts. There was higher quercetin–3-O-galactoside and quercetin–3-O-glucoside content in the aqueous extract, whereas higher kaempferol and isorhamnetin content was observed in hydroalcoholic extract.

Proximate Analysis of Seabuckthorn Byproducts

The proximate analysis of nutritive contents of seabuckthorn byproducts are depicted in Table 18.3. The ash content which is an index of mineral contents, ranged from 1.3 to 4 per cent. The moisture content was significantly highest in seedcake. The crude protein contents ranged from 13.89 to 23 per cent and recorded highest in leaves (22.09 per cent) and seedcake (23 per cent). Ether and crude fibre contents were recorded highest in pomace with seeds (Table 18.3).

Mineral Contents in Seabuckthorn Byproducts

The mineral composition in seabuckthorn byproducts are shown in Table 18.4. High concentrations of sodium (Na) were present, ranging from 40 to 160 mg/g. Among all the byproducts leaves contained high concentration of all the minerals estimated.

CONCLUSION

Total phenols, total flavonoids, vitamin C, vitamin E, lycopene, β-carotenoid, proximate and minerals

Table 18.3: Nutritional Parameters of Seabuckthorn Byproducts

Sample	*Ash (per cent)*	*Moisture Content (per cent)*	*Dry Matter (per cent)*	*Protein (per cent)*	*Ether (per cent)*	*Crude Fiber (per cent)*
Leaves	4^{a}	9^{b}	89^{b}	22.09^{a}	9^{b}	1^{d}
Seedcake	4^{a}	10^{a}	90^{b}	23^{a}	9^{b}	13^{c}
Pomace with seeds	1.3^{c}	6.81^{c}	93.19^{a}	14.01^{b}	20.50^{a}	30.60^{a}
Pomace without seeds	2^{b}	7^{c}	93^{a}	13.89^{b}	19.87^{a}	28.89^{b}

Values are expressed as Mean ± SEM. The means with same superscripts in between columns don't differ significantly at 5 per cent level.

Table 18.4: Mineral Composition of Seabuckthorn Byproducts

Sample	*Na (mg/g)*	*K (mg/g)*	*Ca (mg/g)*	*Mg (mg/g)*	*Zn (mg/g)*	*Fe (mg/g)*
Leaves	160	7.64	10.32	1.716	0.02	0.720
Seedcake	80	4.89	0.155	1.4	0.083	0.398
Pomace with seeds	44	4.43	0.362	0.551	0.029	0.290
Pomace without seeds	40	3.49	0.298	0.421	0.020	0.189

were estimated in seabuckthorn byproducts. The byproducts included leaves, seedcake, pomace with and without seeds. The study showed that the seabuckthorn byproducts are rich in biochemical and nutritional constituents. Leaves were found to be rich in phenolic, vitamin C, protein and mineral contents, whereas pomace was rich in crude fibre. Isorhamnetin was found only in pomace.

REFERENCES

1. AOAC. 1995. Official Methods of Analysis. 16th Edition. Association of official Analytical Chemists, Washington, DC, USA.
2. Baker, H. and Frank, O. 1968. *Clinical Vitaminology*. Interscience, New York.
3. Beveridge, T., Li, T.S., Oomah, B.D. and Smith, A. 1999. Seabuckthorn products: Manufacture and composition. *Journal of Agriculture and Food Chemistry* 47: 3480-3488.
4. Gao, X., Ohlander, M., Jeppsson, N., Bjork, L. and Trajkovski, V. 2000. Changes in antioxidant effects and their relationships to phytonutrients in fruits of seabuckthorn (*Hippophae rhamnoides* L.) during maturation. *J. Agric. Food Chem.* 48: 1485-1490.
5. Gowenlock, A.H., McMurray, J.R. and McLauchlan, D.M. 1988. *Varley's Practical Clinical Biochemistry*, 6th Edition, Heinemann Medical Books, London.
6. Gulcin, I., Oktay, M., Kufrevioglu, O.I. and Aslan, A. 2002. Determinations of antioxidant activity of lichen *Cetraria islndica* (L) Ach. *J. Ethnophrmacol.* 79: 325-329.
7. Guliyev, V.B., Gul, M. and Yildirim, A. 2004. *Hippophae rhamnoides* L.: Chromatographic methods to determine chemical composition, use in traditional medicine and pharmacological effects. *Journal of Chromatography B*, 812 (1-2): 291-307.
8. Kallio, H., Yang, B. and Peippo, P. 2002. Effects of different origins and harvesting time on vitamin C, tocopherol, and tocotrienols in seabuckthorn (*Hippophae rhamnoides*) berries. *J. Agric. Food Chem.* 50: 6136-6142.
9. Kayden, H.J., Chow, C.K. and Bjornson, L.K. 1973. Spectrophotometric method for determination of tocopherol in red blood cells. *J. Lipid Res.* 14 (5): 533-540.
10. Yao, Y. and Tigerstedt, P. 1992. Variation of vitamin C concentration and character correlation between and within natural Seabuckthorn (*Hippophae rhamnoides* L.) populations. *Acta Agric. Scand.* 42: 12-17.
11. Ranganna, S. 1976. In: *Manual of Analysis of Fruits and Vegetable Products*, McGraw Hill, New Delhi. p.77.
12. Shah, A.H., Ahmed, D., Sabir, M., Arif, S., Khaliq, I. and Batool, F. 2007. Biochemical and nutritional evaluations of seabuckthorn (*Hippophae rhamnoides* L. spp. *turkestanica*) from different locations of Pakistan. *Pak. J. Bot.* 39 (6): 2059-2065.
13. Upadhyay, N.K., Yogendra, M.S.K. and Gupta, A. 2010. Antioxidant, cytoprotective and antibacterial effects of seabuckthorn (*Hippophae rhamnoides* L.) leaves. *Food and Chemical Toxicology*. 48: 3443-3448.
14. Yang, B., Karlsson, R.M., Oksman, P.H. and Kallio, H.P. 2001. Phytosterols in seabuckthorn (*Hippophae rhamnoides* L.) berries: Identification and effects of different origins and harvesting times. *J. Agric. Food Chem.* 49: 5620-5629.
15. Zeb, A. and Ahmed, T. 2004. The high dose irradiations affect the quality parameters of dibble oil. *Pak. J. Biol. Sci.*
16. Zheng, X.W. and Song, X.J. 1992. Analysis of the fruit nutrient composition of nine types of seabuckthorn in Liaoning, China. *Northern fruits of China* 3: 22-24.
17. Zhishen, J., Mengcheng, T. and Jianming, W. 1999. The determination of flavonoid content in mulberry and their scavenging effects on superoxide radicals. *Food Chem.* 64: 555-559.

Chapter 19

Comparison of Biochemical Properties of Seabuckthorn Varieties of Different Origins

Jörg-Thomas Mörsel[1], Yury A. Zubarev[2] and Margarita Sordonova[3]

[1]UBF GmbH, An der Mühle 1, D-15345 Altlandsberg, Germany
[2] Institute for Horticulture of Siberia Lisavenko, Smenogorski Trakt 49, 656045 Barnaul, Russia
[3]Institrute of Agriculture at Burjatian Academy of Science, Tretjakov street 25, 3 670045 Ulan-Ude, Russia
(E-mail: thomas.moersel@ubf-research.com, niilisavenko@yandex.ru, sordonova@yahoo.com)

SUMMARY

Seabuckthorn is a highly valuable plant. Products from seabuckthorn have high prices. One of the most important parameter of SBT oil quality is the content of carotenoides, especially in eastern countries. Prices are directly dependent on content of βcaro-tene. In European Community use of some carotenoides as food colorant is allowed. For instance β-carotene and capsanthin are allowed for application in foods generally. These colorants are relatively cheap compared to seabuckthorn pulp oil. We observed that capsanthin has been found in seabuckthorn oil as well as cosmetics produced from seabuckthorn oil. It is well known that different carotenoides occur beside β-carotene as main carotenoid. Neither capsanthin has been reported. But in general the knowledge on SBT composition, especially reporting of geographic differences in seed and pulp oil is low. This research work was primarily initiated by the observance of adulterated SBT pulp oil in European Community market. Because of missing detailed information on composition and comparability of it between European and Asian cultivars it was necessary to have a closer look on this question. Taking into consideration that about 80 per cent of total seabuckthorn market of Russia and quite a few of Chinese represented by varieties bred at the Lisavenko Institute (Barnaul, Russia), it was reasonable to put them in comparative estimation with most promising German varieties. Based on long time cooperation between UBF GmbH Altlandsberg and Lisavenko Institute (Barnaul) the investigations were carried out with authentic cultivars of German and Russian origin. The aim of this work was to collect data on SBT carotenoides, fatty acids and sterols composition. Our co-operation with experimental station of Burjatian Academy of Agro-science in Ulan-Ude gave us the possibility to compare the results with varieties from more eastern sources. In addition to experiments done years before we also investigated 5 German varieties. Seabuckthorn growing and processing is a multinational and diversified business. Raw materials, semi-products and final products are merchandised. Buyer often does not know the origin of raw materials and have to trust their suppliers. Actually there are only a limited number of standards available on national basis, mostly focusing on berries as raw material or seabuckthorn oil. Often quality control on delivery is only coarsely done. Own experimental surveys shows that not really seldom advertised qualities does not really exist. So *e.g.* pesticides are frequently found in berries, juice and oil labelled as to be 'organic'. Price of seabuckthorn oil on world market is high, especially when of organic origin. As in some countries price is calculated based on carotenoids content, adulteration of oils with cheap synthetic carotenoids is a profitable but dishonest procedure. One important question we dealt with was to answer if there is naturally occurring capsanthin in seabuckthorn oil. This was really an important task because there are more than 100 registered varieties worldwide in an area from Far East Asia to Western Europe.

Keywords: *Seabuckthorn, Varieties, Chemical composition, Tocopherols, Carotenoides, Sterols and Vitamins.*

Seabuckthorn (Hippophae L.): A Multipurpose Wonder Plant, Vol. 5: Advanced Technologies for Cultivation, Processing, Health Protection and Environmental Conservation (Virendra Singh, *Editor-in-Chief*, 2018), p. 149–153, Astral International Pvt. Ltd., New Delhi, India.

INTRODUCTION

Griebel and Hess in Albrecht (2003) discovered the high vitamin C content of seabuckthorn fruits. For that reason seabuckthorn became a plant of great importance for the cultivation and processing industries. After the discovery of high vitamin C, the berries were used as a supplementary vitamin supports during the Second World War, as advised by Hoermann (1941). Soon research work began on genetic variability and breeding by Darmer (1947) and Stocker (1948-49) and other breeders. Albrecht and co-workers selected the female cultivars 'Leikora', 'Hergo', 'Frugana', 'Dorana' and 'Askola'. For the development of *seabuckthorn* resources, a working group was formed under leadership of Koch in 1979. The enzyme method, developed by Heilscher and Lerber (1996), which produces a very high quality oil, rich in bioactive substances, has been found useful for the application in cosmetics and medicines. Biochemical characterization of German forms found that German seabuckthorn vareties are not that rich in vitamin C, however they are very rich in oil and other bioactive substances (Singh and Moersel, 2005). Based on long time cooperation between UBF GmbH Altlandsberg and Lisavenko Institute (Barnaul), the studies were carried out with authentic cultivars of German and Russian originS. The aim of this work was to collect data on seabuckthorn carotenoids, fatty acids and sterols composition in seabuckthorn cultivars of different origins.

MATERIALS AND METHODS

Sample material was collected at experimental stations in Barnaul and Ulan-Ude (Russian Federation) directly from plants. The samples are specified by their brands. German samples were collected from plantations in Brandenburg County (Germany) and Mecklenburg-Vorpommern County. Samples are of identified varieties. All samples were cleaned from wood and leaves directly after collecting and frozen at -20 °C. Storage and transport until analysis was under same conditions. All analysis was done within 3 month.

Extraction of Total Lipids

Seed material were finely ground (particle size =2 mm) and Soxhlet extracted with n-hexane for 12 h. Total lipids (TL) recovered were weighed and stored in chloroform at 4°C for further analysis.

Gas Chromatography Analysis of Fatty Acid Methyl Esters

Fatty acids were trans esterified into methyl esters (FAME) using N-trimethylsul-fonium-hydroxide (Macherey-Nagel, Düren, Germany) according to the procedure reported by Arens *et al.* (1994). FAME were identified on a Shimadzu GC-14A equipped with flame ionisation detector (FID) and C-R4AX chromatopac integrator (Kyoto, Japan).The flowrate of the carrier gas helium was 0.6 mL min-1 and the split value with a ratio of 1:40.

A sample of 1μl was injected on a 30 m 90.25 mm 90.2 lm film thickness Supelco SPTM-2380 (Bellefonte, PA, USA) capillary column. The injector and FID temperature was set at 250°C. The initial column temperature was 100°C programmed by 5°C min-1 until 175°C and kept 10 min at 175°C, then 8 until 220°C and kept 10 min at 220°C. A comparison between the retention times of the samples with those of reference compound mixtures, run on the same column under the same conditions, was made to facilitate identification.

Gas Chromatography Analysis of Sterols

Separation of ST was performed after saponification of the oil sample without derivatisation according to Ramadan and Mörsel (2003) and Ramadan *et al.* (2007).

TL (250 mg) were refluxed with 5 mL ethanolic potassium hydroxide solution (6 per cent, w/v) and a few anti-bumping granules for 60 min. The unsaponifiables were firstly extracted 3-times with 10 mL of petroleum ether, the extracts were combined and washed 3-times with 10 mL of neutral ethanol/water (1:1, v/v) and then dried overnight with anhydrous sodium sulfate. The extract was evaporated in a rotary evaporator at 25 °C under reduced pressure, and then ether was completely evaporated under nitrogen. GC analyses of unsaponifiable residues were carried out using a Shimadzu GC-14A equipped with flame ionisation detector (FID) and C-R4AX chromatopac integrator (Kyoto, Japan).The flowrate of the carrier gas helium was 0.6 ml min-1 and the split value with a ratio of 1:40 equipped with FID. The following parameters were performed: DB 5 column (J and W scientific; Falsom, CA, USA) packed with 5 per cent phenylmethyl polysiloxan, 30 m length, 0.25 mm i.d., 1.0 l m film thickness; carrier gas (helium) flow 38 mL min-1. Detector and injector were set at 280 °C. The oven temperature was kept constant at 310 °C and the injected volume was 2 μl. The repeatability of the analytical procedure was tested and the relative standard deviation of three repeated analyses of a single sample was <5 per cent. Quantitative analyses were performed with a Shimadzu (C-R6A Chromatopac; Kyoto, Japan) integrator.

Normal Phase Liquid Chromatography of Tocopherols

Procedure

NP-LC was selected to avoid extra sample treatment (*e.g.*, saponification) according to Ramadan *et al.* (2003) and Ramadan and Mörsel (2007). Analysis was performed with a solvent delivery LC-9A LC (Shimadzu, Kyoto, Japan).

The chromatographic system included a model 87.00 variable wavelength detector and a 250x4 mm i.d. Li-Chrospher-Si 60, 5 μm, column (Knauer, Berlin,

Germany). Separation of tocopherol isomers was based on isocratic elution when the solvent flow rate was maintained at 1 mL min-1 at a column back-pressure of about 65 – 70 bar. The solvent system selected for elution was isooctane/ethyl acetate (96:4, v/v) with detection at 295 nm. 20 µl of the diluted solution of TL in the mobile phase were directly injected into the LC column. Tocopherol isomers were identified by comparing their retention times with those of reference compounds.

Preparation of Standard Curves

Standard solutions were prepared by series of dilution ranged from 25 to 5 mg mL^{-1} with stock solution of each tocopherol isomer. Standard solutions were prepared from a stock solution which was stored in the dark at -20°C. 20 µL were injected and peak areas determined to generate standard curve data.

Quantification

All quantitation was by peak area using Shimadzu C-R6A chromatopac integrator (Kyoto, Japan). Standard curves (concentration vs. peak area) were calculated from six concentration levels by linear regression. Based on the established chromatographic conditions, repeated injections of different concentrations of the standard tocopherols were made 3-times onto the LC system. Injections in triplicate were made at each concentration for both standards and samples. All work was carried out under subdued light conditions. All the experiments were repeated at least thrice when the variation on any one was routinely less than 5 per cent. All experimental procedures were performed in triplicate and their mean values (± standard deviation) were given.

Normal Phase Liquid Chromatography of Carotenoides

Procedure

NP-LC was selected to avoid extra sample treatment (*e.g.*, saponification) according to Ramadan and Mörsel [2003, 2004]. Analysis was performed with a solvent delivery LC-9A LC (Shimadzu, Kyoto, Japan). The chromatographic system included a model 87.00 variable wavelength detector and a 250 x 4 mm i.d. Li-Chrospher-Si 60, 5 µm, column (Macherey-Nagel, Düren, Germany). Separation of Carotenoids was based on isocratic elution when the solvent flow rate was maintained at 1 mL min-1. The solvent system selected for elution was isooctane/ethyl acetate (90:10 v/v) with detection at 295 nm. 20 µl of the diluted solution of TL in the mobile phase were directly injected into the LC column. Carotenoids were identified by comparing their retention times with those of reference compounds.

Preparation of Standard Curves

Standard solutions were prepared by series of dilution ranged from 25 to 5 mg mL^{-1} as described above.

Quantification

All quantitation was by peak area using Shimadzu C-R6A chromatopac integrator (Kyoto, Japan). Standard curves (concentration vs. peak area) were calculated from five concentration levels by linear regression. Based on the established chromatographic conditions, repeated injections of different concentrations of the standard carotenoids were made 3-times onto the LC system. Injections in triplicate were made at each concentration for both standards and samples. All work was carried out under subdued light conditions. All the experiments were repeated at least thrice when the variation on any one was routinely less than 5 per cent.

RESULTS

Figure 19.1 shows the fatty acid composition of the investigated varieties ad cultivars. The important differences are figured out in 19.1a. It can be seen from the figure that important differences are found in 18:1 n-9, 18:1 n-7, 18:2. This may be a result of differences in climate. German cultivars have higher concentration in 18:1 species where Russian have higher in linoleic acid and palmitolenic acid.

As important details are only shown in differences between some fatty acids Figure 19.1b give a better view on this.

Figure 19.2 shows tocopherol pattern observed in the sample materials. Main tocopherol isomer is α-tocopherol followed by γ-tocopherol. This composition

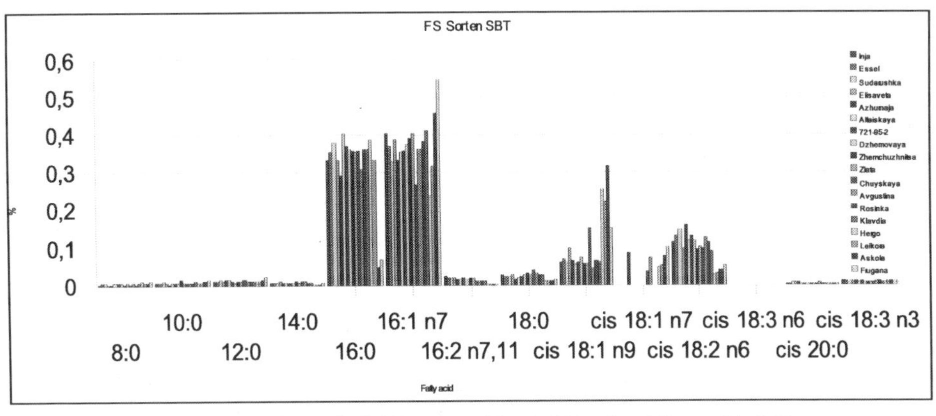

Figure 19.1a: Fatty Acid Composition of Seabuckthorn Varieties.

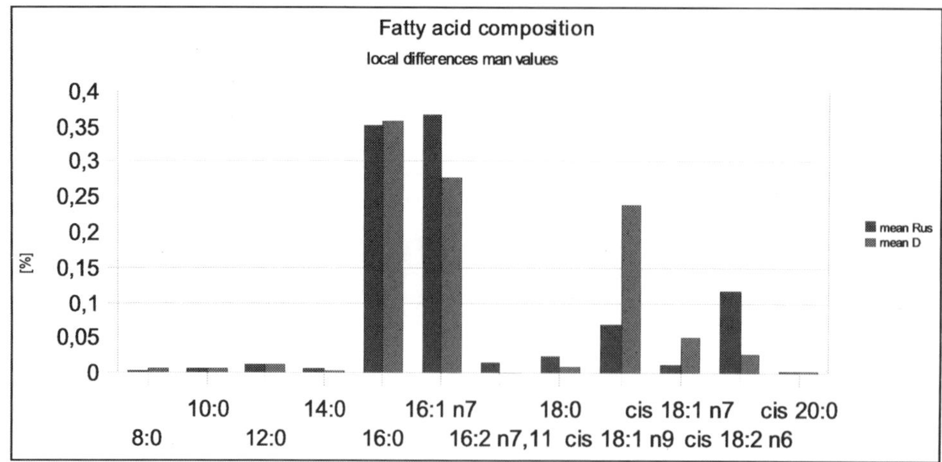

Figure 19.1b: Selected Fatty Acid Composition of Seabuckthorn Varieties.

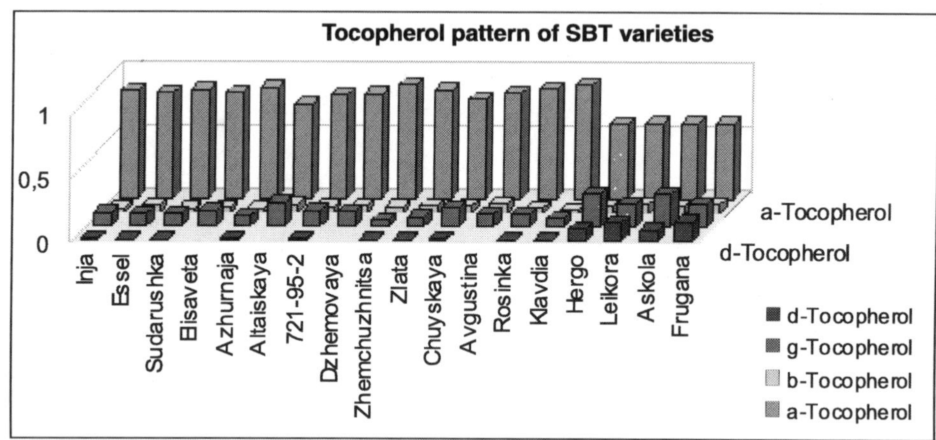

Figure 19.2: Tocopherol Composition of Seabuckthorn Varieties.

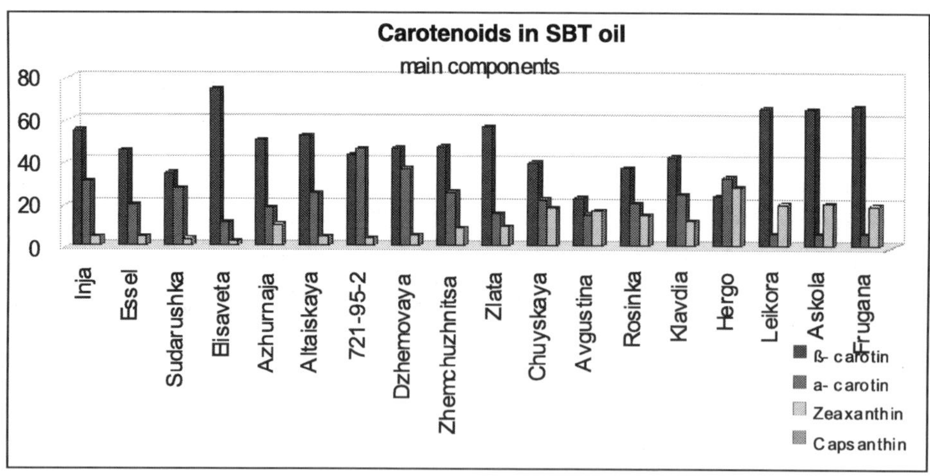

Figure 19.3: Carotenoids in Pulp Oil of Seabuckthorn Varieties.

is also reported in literature and agrees with the common knowledge. A special observation is the generally higher concentration of δ-tocopherol in German varieties. This might be a dependence on the different genetical resources of German and Russian varieties and cultivars.

There are two visible results from this investigation. Neither in Russian nor German varieties capsanthin was found. The second interesting fact is the low concentration of α-carotin in German brands. The β-carotin is also remarkable high.

CONCLUSION

Typical biochemical substances of seabuckthorn oil like fatty acids, tocopherols, sterols can characterize not only nature of Seabuckthorn oil in general, but difference between varieties it was obtain from Russian

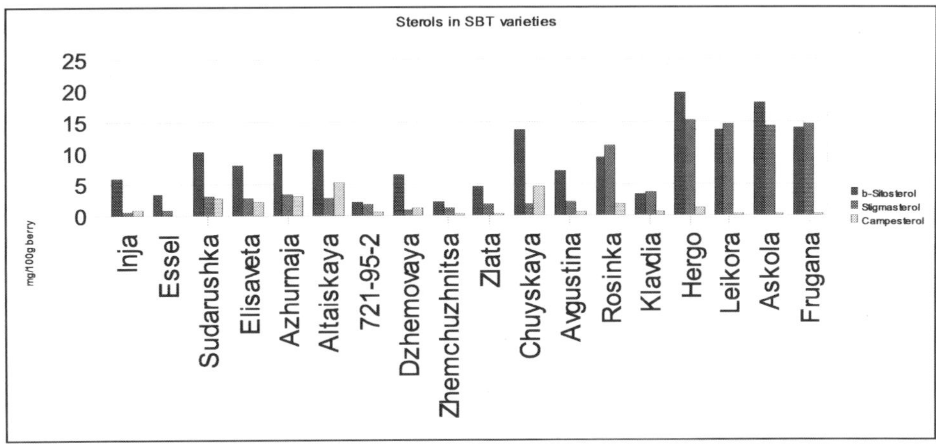

Figure 19.4: Sterol Pattern in Oil of Seabuckthorn Varieties.

and German varieties belong to different subspecies (*H. rhamnoides* ssp. *mongolica*, *H. rhamnoides* ssp. *rhamnoides*). These differences combined with regional and climatic factors can be seen in chemical composition. Tocopherols and sterols are reliable indicators for both subspecies. Fatty acid combination also shows some differences but not so significant.

Oils from authentic berries of both areas didn't contain capsanthin. It means that presence of capsanthin in oil is indicator of adulteration.

ACKNOWLEDGEMENT

This project was partially funded by the Ministry of Economy and European Affairs of Brandenburg and the EU. The responsibility for the content lies with the author.

REFERENCES

1. Albrecht, H.J. 2003. Research on seabuckthorn (*Hippophae rhamnoides*) in Germany. In: *Seabuckthorn (Hippophae L.)-A Multipurpose Wonder Plant.* Vol. I. *Botany, Harvesting and Processing Technologies* (Singh, V., Editor in Chief), pp. 178-186, Indus Publishing Company, 5 FS-Tagore Garden, New Delhi, 519p.
2. AOCS Standard Methods 2009. Official methods and recommended practise of the AOCS, 19th. Ed. Amer Oil Chemists Society, Boulder, Urbana Illinois.
3. Arens, M., Schulte, E. and Weber, K. 1994. *Fat Sci Technol.*, 96: 67-68.
4. Bockisch, M. 1998. Vegetable fats and oils. In: *Fats and oils handbook.* Bockisch M. (ed) AOCS Press, Champaign, 174-344.
5. Darmer, G. 1947. *Hippophae rhamnoides* L.-A new breeding object. Der Zuchter 17/18, 13/15, p. 430-436.
6. DGF Einheitsmethoden 1988. Deutsche Gesellschaft für Fettwissenschaften, Münster.
7. Heilscher, K. and Lerber, S. 1996. Process for cold recovery of clear juice, solids and oil from seabuckthorn berries and use of these products. Patent number DE 44 31 394 C1 (DE 4431394 C1), Germany.
8. Hormann, B. 1941. Cultivation of seabuckthorn in GDR. *Mitt. der AG Sanddorn und Seltene Obstarte,* p. 25.
9. Kallio, H., Yang, B., Peippo, P., Tahvonen, R. and Pan, R. 2002. *J Agric Food Chem* 50: 3004-3009. doi:10.1021/jf011556o.
10. Malter, D., Mörsel, J.T. and Ebert, G. 2002. Zusammensetzung der Kernöle von Wildobstarten, Wildobst.
11. Mörsel, J.T.C. 2003. Mörsel Die Bedeutung der Qualität von Sanddornerzeugnissen, Flüssiges Obst, pp. 272-275.
12. Ramadan, M.F. and Mörsel, J.T. 2002. *Eur Food Res. Technol.* 214: 521-527.
13. Ramadan, M.F. and Mörsel, J.T. 2003. *J. Agric. Food Chem.* 51: 969-974.
14. Ramadan, M.F., Zayed, R. and El-Shamy, H. 2007. *Food Chem.* 103: 885-890.
15. Singh, V. and Moersel, J.T. 2005. Development and commercialization of seabuckthorn: A German experience. In: *Seabuckthorn (Hippophae L.)-A Multipurpose Wonder Plant.* Vol. II. *Biochemistry and Pharmacology* (Singh, V., Editor in Chief), p. 576-584, Daya Publishing House, New Delhi, 600p.
16. Stoker, O. 1948/49. Seabuckthorn in Tirol A plant with very high vitamins C productivity. *Der Zuchter* 19 (1-3): 9-13.

Chapter 20

Essential Quality Components in Seabuckthorn (*Hippophae rhamnoides*) Berries of Different Genetic Backgrounds in China

He Li[1,2], Cheng-jiang Ruan[2*], Rui Chen[2], Li Wang[2], Jian Ding[2] and Xing-jun Tian[1,*]

[1]School of Life Sciences, Nanjing University, Nanjing City, 210023, Jiansu, China
[2]Institute of Plant Resources, Dalian Nationalities University, Dalian City, 116600 Liaoning, China
(*E-mail: ruan@dlnu.edu.cn, tianxj@nju.edu.cn)

SUMMARY

To utilize seabuckthorn (*Hippophae rhamnoides*) berries, biochemical components of 34 cultivars, belonging to ssp. *mongolica*, ssp. *sinensis* and hybrid varieties of two subspecies, growing in northern China were studied with HPLC and GC analysis methods. Results showed the berries of ssp. *sinensis* with small size contained more vitamin C in juice fraction than those of ssp. *mongolica* and hybrid varieties (11.57 *vs* 0.99, and 11.57 *vs* 3.27 g/L, $p < 0.001$). The seeds and lyophilized fruit flesh of ssp. *mongolica* were good source of unsaturated fatty acids (70-85 per cent in seeds, 50-70 per cent in pulps, $p < 0.001$) and total tocopherols and tocotrienols (144.31 µg/g DW in seeds, 508.71 µg/g DW in pulps, $p < 0.05$). Several cultivars (Botanical garden, Xine-2, Wucifeng) of ssp. *mongolica* and hybrid varieties (Zaciyou-10, Chuyi-Shiciyou, Zaciyou-12 and Xinzaci-26) showed good traits in fatty acids, tocopherols and tocotrienols. The cultivars were mainly divided into 2 groups according to the composition and content of carotenoids which has powerful influence on berry colours. The total carotenoids of Siberia rumiane, Botanical garden and Xine-2 (ssp. *mongolica*) exceed 1000 µg/g of DW. These data provide scientific information for food development and breed improvement of seabuckthorn for nutritional purposes.

***Keywords**: Seabuckthorn, Berries, Seeds, Vitamin C, Fatty acid, Tocopherols, Tocotrienols and Carotenoids.*

INTRODUCTION

Seabuckthorn (*Hippophae rhamnoides* L.), a winter hardy and fruit-bearing shrub, naturally distributes over Asia and Europe. *H. rhamnoides* is economic species and is divided into nine subspecies as the most important species, of which ssp. *sinensis* and *mongolica* are abundant and commercially cultivated, mainly distributes in Asia (Rousi, 1971). The berries of seabuckthorn generally contain high concentration of bioactive compounds, for example, ascorbic acid, carotenoids, healthy fatty acids, phenolic compounds and tocopherols (Andersson *et al.*, 2008a; Kallio *et al.*, 2009; Chen *et al.*, 2013), with potentially beneficial effects on human health (Suomela *et al.*, 2006; Suryakumar and Gupta, 2011). Main applications for the berries include food, cosmetics, and pharmaceutical products (Bal *et al.*, 2011). Some products (*e.g.* seed oil) of seabuckthorn berries are used for treatment of the skin, mucosa, and immune systems, especially in cancer and cardiovascular disease therapy, treatment of gastrointestinal ulcers and as a liver protective agent (Ruan *et al.*, 2007; Grey *et al.*, 2010; Xu *et al.*, 2011).

In China, seabuckthorn has been planted as eco-economic plants since 1980s. Its good adaptability, rapid growth, ability to act as protection against wind and sand drift, and its role in soil and water conservation, allow seabuckthorn to be widely used in vegetation construction and restoration of degenerated ecosystems.

Seabuckthorn (Hippophae L.): A Multipurpose Wonder Plant, Vol. 5: Advanced Technologies for Cultivation, Processing, Health Protection and Environmental Conservation (Virendra Singh, *Editor-in-Chief*, 2018), p. 155–166, Astral International Pvt. Ltd., New Delhi, India.

But native cultivars (ssp. *sinensis*) with good adaptability produced low economic value for their small fruits and more thorns, which prevented the large-scale cultivation and market development of bioactive compounds. One breeding approach to overcome these disadvantages is to cross cultivars between two subspecies. The 1st stage of this breeding program of seabuckthorn was implemented to breed new cultivars with promising agronomic traits (big fruit, few thorns, high yield) based on genetic improvement since 1990s. Utilize seabuckthorn cultivars (including introduced and native cultivars) that have been successfully selected and bred in China to implement crossing experiments, *e.g.* hybridizing between cultivars from China and Russia and other countries, etc (Ruan *et al.*, 2007; Li *et al.*, 2009). Some plants with desirable traits of interest were identified from F_1 individuals of several hybrids and introduced a wide scale. Since 2011, the breeding project has entered the 2nd stage, in which the fruit quality (bioactive compounds) and yield combined with ecological adaptability became new targets. The concrete assignment has been built according to various geographical climates of the northern, eastern, western and middle area of China. For example, the breeding task of the northeast of China (semi-humid area) is to improve fruit oil content, bioactive substances and easy picking with machines and promote the economic value of seabuckthorn, whereas the ecological adaptability combined fruit yield is the main target for the northwest of China (semi-arid area). It is necessary and urgent to investigate essential quality components of seabuckthorn cultivars in different areas of China.

In the present study, we collected berries of 17 cultivars of ssp. *mongolica*, 6 cultivars of ssp. *sinensis* and 11 hybrids of ssp. *mongolica* and ssp. *sinensis*, mainly growing in the north of China. The content and composition of bioactive compounds including vitamin C, fatty acids, tocopherols, tocotrienols and carotenoids in berries were investigated. The purpose was to compare these bioactive compounds in seabuckthorn berries of three groups (ssp. *mongolica*, ssp. *sinensis* and inter-sub specific hybrids). A further aim was to identify elite cultivars with more economic value, for the berries with more bioactive compounds were suitable for the manufacturing of some products, for example, nutritional and health products.

MATERIALS AND METHODS

Berries

Berries of 34 seabuckthorn cultivars (Table 20.1) were picked from late-June to early September, 2011. The areas ranged between longitudes 90°222E–127°062E, latitudes 36°932N–47°142N, and altitudes 202–2280 m. The cultivated berries were belonged to ssp. *mongolica* (17 samples) and ssp. *sinensis* (6 samples) and hybrids (11 samples). Of these *rhamnoides* cultivars, 12 were cultivated at the Institute of Selection and Breeding of *Hippophae* in Fuxin, Liaoning Province (121°282E, 42°262N, 380 m). Eight were cultivated at the Research Institute of Berry attached to Heilongjiang Academy of Agricultural Sciences in Suiling, Heilongjiang Province (127°062E, 47°142N, 202 m). Twelve were cultivated at the Jiuchenggong Breeding Base of Seabuckthorn in Dongsheng, Inner Mongolia (110°092E, 39°402N, 1400 m). Wild berries of Zhongguoshaji (ssp. *sinensis*) were collected from Datong, Qinghai Province (101°682E, 36°932N, 2280 m). The berries of Qiuyisike, an introduced cultivar, were collected from Qinghe, Xinjiang Uigur Autonomous Region (90°222E, 46°402N, 1218 m). The berries were hand-picked at the time periods when they were typically harvested for commercial purposes or at the natural maturation. All berries were cut off at the end of fruit stalk from the plants in the eld and taken back to the laboratory within the same day and were stored in the freezer at –50 °C until analyzed.

Measurement of Berry Size, Seed Weight and Berry Shape

Berry size was determined as the weight of 100 fresh berries after they were picked from bushes. Seed weight was measured as the weight of 100 seeds after air drying at room temperature (25°C) for 2 weeks (Tang *et al.*, 2001). Both measurements had three replications. Berry length and width were measured by the micrometer caliper and berry shape was estimated by the ratio of berry length to width averaged over 20 determinations.

Analysis of Vitamin C

Vitamin C was analyzed as ascorbic acid after the conversion of dehydroascorbic acid to ascorbic acid form with dithiothreitol (DTT) (Gökmen *et al.*, 1996). Frozen berries were melted in a microwave oven. The juice was pressed manually diluted 1 to 40 with water (DTT was added at a final concentration of 1 mg/mL). The mixture was allowed to react at room temperature for 2 h. The sample solution was analyzed with an Aligent 1260 Infinity HPLC (High Performance Liquid Chromatography)-system equipped with a G4212B diode array detector (Aligent Technologies Inc., California, USA) and a Shim-pack *VP*-ODS C18 column (250 mm × 4.6 mm, 5 μm) (Shimadzu Corp., Kyoto, Japan). The mobile phase was 0.5 per cent KH_2PO_4 deionized water solution containing 0.1 per cent DTT. The vitamin C concentration was determined according to the absorption peak area at 264 nm, using an external standard method (Ascorbic Acid, 99.9 per cent, Supelco, Sigma-Aldrich Co. LLC., USA).

Lipid Extraction

Frozen berries were lyophilized for 4 days in the freeze dryer (freeze dry system, labconco, USA). The seeds and soft parts (pulps) isolated from freeze-dried berries were crushed separately in a mortar in liquid nitrogen, and the lipids of samples (1 g, dry weight, DW) were isolated using a methanol-chloroform extraction

Table 20.1: Samples of Seabuckthorn Berries of different Genetic Origins

Codes	Cultivars	Pedigree/Background Information	subsp.[a]	Grown Location
M1	Zhuangyuanhuang	seedling of seeds from Russia	mongolica	Fuxin
M2	Wucifeng	seedling of seeds from Russia	mongolica	Fuxin
M3	Siberia rumianes	asexual reproduction materials of cultivar from Russia	mongolica	Fuxin
M4	Fangxiang	seedling of the cultivars from Russia	mongolica	Fuxin
M5	Yalishanda-12	seedling of the cultivars from Russia	mongolica	Fuxin
M6	Jiuyuehuang	seedling of the cultivars from Russia	mongolica	Fuxin
M7	Nanren	germchit of the cultivars from Russia	mongolica	Fuxin
M8	Botanical garden	germchit of the cultivars from Russia	mongolica	Fuxin
M9	Shiyou-1	seedling of varieties from Russia	mongolica	Fuxin
M10	Shiyou-2	seedling of varieties from Russia	mongolica	Fuxin
M11	Shoudu	seedling of varieties from Russia	mongolica	Suiling
M12	Xine-2	seedling of varieties from Russia	mongolica	Suiling
M13	Xine-3	seedling of varieties from Russia	mongolica	Suiling
M14	Fenlan	seedling of varieties from Russia	mongolica	Suiling
M15	Qiuyisike	seedling of varieties (KaTong population) from Russia	mongolica	Qinghe
M16	MZ14	hybrid of 'Chuyi'(ssp. *mongolica* from Russia) x'Mengyou' (ssp. *mongolica* from Mongolia)	mongolica	Suiling
M17	Liusha-1	hybrid of 'Hongsehuoju' x'Zirianka 104' from Russia	mongolica	Fuxin
S1	Zhongguoshaji	seedling of ZGSJ[b] (Fengning population)	sinensis	Suiling
S2	Juda	seedling of ZGSJ[b] (Fengning population)	sinensis	Dongsheng
S3	Jianpingdahuang	seedling of ZGSJ[b] (Jianping population)	sinensis	Dongsheng
S4	Manhanshanci	seedling of ZGSJ[b] (Manhan population)	sinensis	Dongsheng
S5	Zhongxiongyou	seedling of ZGSJ[b] (Yibei population)	sinensis	Dongsheng
S6	Zhongguoshaji[wild]	wild populations of ZGSJ[b]	sinensis	Datong
MS1	Cross-I	progeny of Qiuyisike (ssp. *mongolica*) and ZGSJ[b]	M×S	Fuxin
MS2	Liaofuza	progeny of Qiuyisike (ssp. *mongolica*) and ZGSJ[b]	M×S	Dongsheng
MS3	Zaciyou -1	F1 hybrid of Wulanshalin (seedling of Wulangemu,) and ZGSJ[b] (Fengning population)	M×S	Dongsheng
MS4	Zaciyou-10	F1 hybrid of Wulanshalin (seedling of Wulangemu,) and ZGSJ[b] (Fengning population)	M×S	Dongsheng
MS5	Zaciyou-12	F1 hybrid of Wulanshalin (seedling of Wulangemu) and ZGSJ[b] (Fengning population)	M×S	Dongsheng
MS6	Xinzaci-26	progeny of Qiuyisike and ZGSJ[b] (Manhan population)	M×S	Dongsheng
MS7	Taiyang-Shiciyou	progeny of Taiyang and ZGSJ[b] (Manhan population)	M×S	Dongsheng
MS8	Chuyi-Shiciyou	progeny of Chuyi and ZGSJ[b] (Manhan population)	M×S	Dongsheng
MS9	Yousheng-Shiciyou	progeny of Yousheng and ZGSJ[b] (Manhan population)	M×S	Dongsheng
SM1	Z24	progeny of ZGSJ[b] and Alieyi (from Mongolia)	S×M	Suiling
SM2	HS-56	progeny of ZGSJ[b] and HS-22 (seedling of 'Wulangemu')	S×M	Suiling

a: M×S, hybrid between ssp. *mongolica* (♀) and ssp. *sinensis* (♂); S×M, Cross cultivar between ssp. *sinensis* (♀) and ssp. *mongolica* (♂);
b: ZGSJ, the cultivar of 'Zhongguoshaji'.

procedure (Christie *et al.*, 1989). Lipids were stored in chloroform at –20 °C for a short period until analysis.

Analysis of Fatty Acids

The oil was transesterified by sodium methoxide catalysis (Christie *et al.*, 1982). The fatty acid methyl esters (FAMEs) were dissolved in n-hexane until analysis. The FAMEs were analyzed with a GC-2010 gas chromatograph (GC) equipped with flame ionization detector, controlled by a GC Solution (Shimadzu Corp., Tokyo, Japan). Silica capillary GC column DB-23 (L, 60 m; i.d., 0.25 mm; d_f, 0.25 μm) was used for GC analysis (Agilent technologies Ltd, Palo Alto, U.S.A). The flow rate of the carrier gas nitrogen was 0.8 mL/min, and a split valve with the ratio 1: 30 was opened after one minute. The temperature program was 90 °C, increased at a rate of 7 °C/min to 190 °C, increased at a rate of 3 °C/min to 215 °C and held for 10min, then at a rate of 3 °C/min to 230 °C and held for 5 min. The detector temperature was 270 °C. FAMEs were identified by comparison with analytical standards (Methyl palmitate, Methyl palmitoleate, Methyl stearate, Methyl oleate, Methyl linoleate, Methyl linolenate, Fluka, Sigma-Aldrich Co. LLC., USA) and data from literatures (Yang and Kallo, 2001; George *et al.*, 2007). The fatty acid composition was expressed as a weight percentage of the total fatty acids.

Analysis of Tocopherols and Tocotrienols

The lipids from 1 g lyophilized seeds and the soft parts (dried) were dissolved in n-hexane. Tocopherols and tocotrienols were analyzed with a normal-phase HPLC method, according to methods described by Panli *et al.* (2003) with a slight modication. The instrumentation was a Prominence LC-20A equipped with a CTO-20A oven, and a RF-10A$_{XL}$ fluorescence detector (Shimadzu Corp., Kyoto, Japan). The effluent consisted of a solvent mixture [97.3 per cent (v/v) n-hexane, 1.8 per cent (v/v) ethyl acetate, and 0.9 per cent (v/v) acetic acid]. A Phenomenex Luna 5 µm silica (2) 100A, 250 × 4.60 mm column with a Security Guard silica precolumn was used, with an injection volume of 20 µL. Detection was carried out at 290 nm excitation and 330 nm emission. Evaluation of data was performed by LC solution software (Shimadzu Corp., Kyoto, Japan), using retention times compared against DL-α-, β-, γ-, and δ-tocopherol standards (Supelco, Sigma-Aldrich Co. LLC., USA), and tocotrienol data from literatures (Abidi *et al.*, 2000; Panfili *et al.*, 2003). DL-α-tocopherol 99.5 per cent was used for the quantication of all tocopherols and tocotrienols. Quantication of tocol isomers was correlated to an external calibration. All isomers were expressed as µg/g DW.

Analysis of Carotenoids

The extraction procedure described by Andersson *et al.* (2009) was followed with slight modifications. The lyophilized soft parts were ground in a mortar, and 1 g of each sample was homogenized in 20 mL of ethanol/n-hexane (4:3, v/v) including 0.01 per cent butylated hydroxytoluene (BHT). Samples were sealed and placed in an orbital shaker in darkness for 20 h at 4 °C for further extraction and thereafter centrifuged at 10000 g for 10 min. The extract was concentrated to get rid of solvent under high purity nitrogen, and then dissolved in n-hexane. The samples were analyzed by an Agilent 1100 HPLC system equipped with a diode array detector according to the methods of Andersson *et al.* (2009) with modications of the gradient elution program. The eluent consisted of solvent A [80 per cent acetonitrile, 15 per cent MeOH, and 5 per cent dichlormethane (v/v)] and solvent B [30 per cent acetonitrile, 20 per cent MeOH, and 50 per cent dichlormethane (v/v)]. The binary gradient was as follows: 0-50 per cent B (0–15 min), 50-90 per cent B (15–30 min), 90 per cent B (30–40 min), 90-0 per cent B (40–42 min), and 0 per cent B (42–47 min). A Shim-pack *VP*-ODS RP-18 column (250 mm × 4.6 mm, 5 µm) (Shimadzu Corp., Kyoto, Japan) was used, and detection was carried out at 458 and 665 nm. The identication of carotenoids was done by spectral data of external standards (*all-trans* forms of lutein, zeaxanthin, β-cryptoxanthin, lycopene, Sigma-Aldrich, USA; β-carotene, Dr. Ehrenstorfer, Germany) and from literature data (Andersson *et al.*, 2009). The standard curve of β-carotene was used for quantication of all carotenoids, which was expressed as µg/g DW.

Statistical Analysis

Statistical analysis was conducted using SPSS (SPSS 20.0, SPSS Inc. H, Chicago, IL). Group differences on the traits were established with one-way analysis of variance (ANOVA). LSD test for populations with equal variances and Tamhane's test for those with unequal variances were used for the multiple comparison of the mean intensities of attributes at $p < 0.05$. The heatmap of carotenoids analysis was carried out using HemI (Heatmap Illustrator, version 1.0) (Deng *et al.*, 2014).

RESULTS AND DISCUSSION

Berry Physical Characters

There was significant difference between berry size and seed weight among cultivars of three groups (Table 20.2). The berries of ssp. *mongolica* were at 54.00 ± 12.26 g per cent, much bigger than that of ssp. *sinensis* (10.72 ± 1.54 g per cent, $p < 0.001$) and hybrid varieties (25.41 ± 14.86 g per cent, $p < 0.001$). Seed weights of ssp. *mongolica* (1.69 ± 0.26 g per cent) were higher than that of ssp. *sinensis* (0.79 ± 0.23 g per cent, $p < 0.001$) and hybrid varieties (1.17 ± 0.22 g per cent, $p < 0.001$). The berry size and seed weight of hybrid varieties were intermediate, comparing with cultivars of their parental subspecies ($p < 0.05$). This result implied that berry sizes of native cultivars were improved by crossing with cultivars of ssp. *mongolica*. The berries of Z24 and HS-56 (SM1and SM2) were at 43.67 ± 1.27 g per cent and 59.60 ± 1.19 g per cent, rather similar to some cultivars of ssp. *mongolica*. Variations of physical characters among cultivars may be attributed to differences in genetic backgrounds, geographical location, climate, harvest period and berry maturity (Tang *et al.*, 2001).

The level of vitamin C content varied from the lowest at 0.21 g/L juice of MZ14 (M16) to the highest at 14.31 g/L juice of Zhongguoshaji^{wild} (S6) (Figure 20.1). The berries of ssp. *sinensis* were clearly the best source of vitamin C comparing to those of ssp. *mongolica* and hybrid varieties (11.57 ± 2.02 *vs* 0.99 ± 0.64 *vs* 3.78 ± 1.42 g/L) ($P < 0.001$). Among various cultivars investigated, the berries of ssp. *sinensis* contained the highest vitamin C, but having the smallest size. Zhongguoshaji^{wild} (ssp.

Table 20.2: Berry Size, Seed Weight (g per cent) and Berry Shape of Seabuckthorn Berries[a]

Codes	*Berry Size*	*Weed Weight*	*Berry Shape*	*Codes*	*Berry Size*	*Weed Weight*	*Berry Shape*	*Codes*	*Berry Size*	*Weed Weight*	*Berry Shape*
M1	48.5	1.8	1.0	S1	12.6	0.7	0.9	MS1	32.7	1.5	1.0
M2	59.9	1.9	1.3	S2	8.5	0.6	0.8	MS2	13.0	0.9	1.2
M3	63.5	1.4	1.5	S3	10.0	0.8	0.9	MS3	15.4	1.0	1.1
M4	39.4	2.2	1.4	S4	9.9	0.7	1.0	MS4	28.5	1.3	1.1
M5	62.3	1.8	1.2	S5	11.2	0.6	0.9	MS5	13.3	1.0	1.2
M6	61.5	1.9	1.2	S6	12.2	1.2	0.9	MS6	14.0	1.0	1.0
M7	58.2	1.9	1.7					MS7	21.5	1.3	1.4
M8	48.8	1.7	1.1					MS8	16.3	0.9	1.0
M9	43.9	1.3	1.0					MS9	21.7	1.3	1.0
M10	24.9	1.4	1.1					SM1	43.7	1.3	1.0
M11	35.4	1.8	1.3					SM2	59.6	1.5	1.0
M12	53.6	1.6	1.3								
M13	59.1	1.6	1.4								
M14	67.9	1.7	1.4								
M15	57.9	1.3	1.6								
M16	63.5	1.5	1.2								
M17	69.7	2.1	1.2								
Mean	54.0 a	1.7 a	1.3 a	Mean	10.7 c	0.8 c	0.9 c	Mean	25.4 b	1.2 b	1.1 b
Std.	12.3	0.3	0.2	SD	1.5	0.2	0.0	SD	14.9	0.2	0.1

a: Each value represents the average of analysis results of three subsamples, each analyzed in duplicate.

Values with different lower–case letters (a–c) are significantly different at $p < 0.05$.

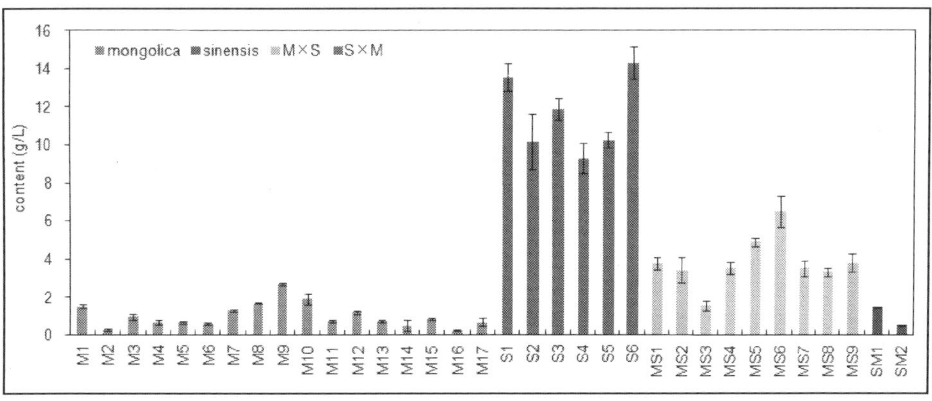

Figure 20.1: Vitamin C in juice (g/L) of Seabuckthorn Berries of different Genetic Backgrounds.

sinensis), as a native and wild cultivar, contained the highest vitamin C at 14.31 ± 0.81. For hybrid varieties, smaller berries of 'M×S' had largely higher vitamin C content than those of ssp. *mongolica* (their female parent), whereas the corresponding values were lower than those of ssp. *sinensis* (their male parent). The vitamin C content of 'S×M' was similar to that of ssp. *mongolica* at 0.96 ± 0.37 g/L juice. These results showed that the genetic background was the most important factor for determining vitamin C contents and the modification of it by inter sub specific hybridization probably influenced vitamin C to certain extent (Kallio *et al.*, 2002a). The breeding program on hybrid varieties 'M×S' in the past years improved vitamin C content of introduced cultivars (ssp. *mongolica*) from other countries. The present studies also showed somewhat a stronger relationship between vitamin C and berry size, which agreed with earlier findings (Tang *et al.*, 2001; Kallio *et al.*, 2002a) and made berry size become a useful indicator of vitamin C.

Fatty Acid Composition in Oil of Seeds and Pulps

Linoleic, α-linolenic, oleic, palmitic, stearic and palmitoleic acids' proportions were found from high to low in seed oils of most cultivars (Figure 20.2). The proportion of linoleic and α-linolenic was typically at 50–70 per cent and it reached 66.35 per cent in seed oil of Botanical garden (M8). In several cultivars [YaLiShanDa-12 (M5), JiuYue Huang (M6), LiuSha-1

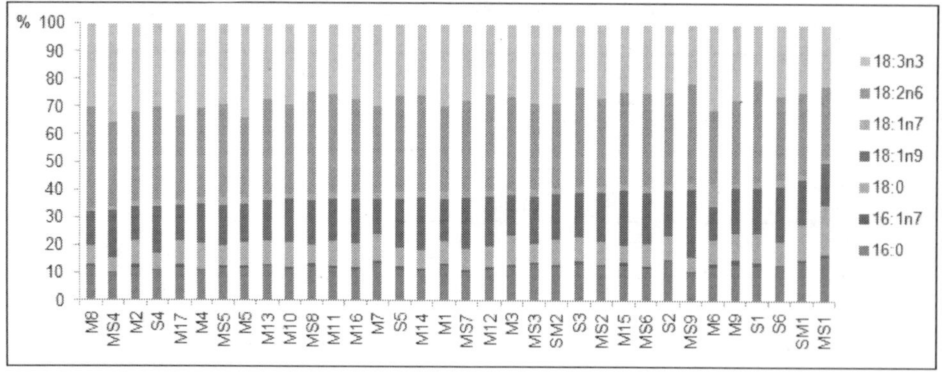

Figure 20.2: Fatty Acid Compositions (Per cent) in Seeds of different Seabuckthorn Cultivars (Sum of Linoleic and α-linolenic acids, maximum to minimum).

(M17), Zaciyou-10 (MS4)], the proportion of α-linolenic acid was even higher than linoleic acid. The proportion of linoleic acid was not significantly different in three groups of seabuckthorn (32.74 per cent in *mongolica* vs 34.22 per cent in *sinensis* vs 32.17 per cent in hybrid varieties, $p > 0.05$) (Table 20.3). α-Linolenic acid of ssp. *mongolica* was higher than that of ssp. *sinensis* (28.67 per cent vs 24.94 per cent, $p < 0.05$). A clear characteristic of the seed oil was the extremely low level of palmitoleic acid (< 2 per cent). The proportions of palmitic (16:0), palmitoleic (16:1), stearic (18:0), and vaccenic (18:1n7) acids showed no significant difference among cultivars of three groups ($p > 0.05$).

Palmitic and palmitoleic acids (45–77 per cent) were dominant fatty acids in oils extracted from pulps (Figure 20.3). The pulp oil of ssp. *mongolica* contained higher proportions of palmitic (35.63 per cent) and palmitoleic acids (33.89 per cent) than ssp. *sinensis* ($p < 0.05$) (Table 20.3). In Fenlan (M14), 76.99 per cent of fatty acids were palmitic and palmitoleic acids. For hybrid varieties, the proportion of palmitic and palmitoleic acids were more approximate to that of ssp. *mongolica*. The highest deviations were recognized in the proportion of palmitoleic acid (13–40 per cent). In several cultivars [Wucifeng (M2), Liusha-1 (M17), Siberia rumianes (M3), Jiuyuehuang (M6), Shiyou-2 (M10) and Chuyi-Shiciyou (MS8)], the proportion of unsaturated palmitoleic acid exceeded that of palmitic acid. A clear characteristic of the pulp oil was the low level of α-linolenic acid (< 8 per cent). The relative levels of olieic, stearic and α-linolenic acids in pulps of ssp. *sinensis* were higher than those of ssp. *mongolica* ($p < 0.01$). Samples of ssp. *sinensis* contained high proportion of oleic acid (19.73 per cent). Stearic acid in Juda (S2) was highest and reached up to 22.25 per cent.

Unsaturated fatty acids are structural components of cell membranes to maintain physiological function of a cell and linoleic and α-linolenic acids play an important role in the physiology of man as essential fatty acids. The oil of seabuckthorn seems to be a good source of unsaturated fatty acids (70–85 per cent in seed oil and 50–70 per cent in pulp oil) and elite cultivars contain a large amount of linoleic and α-linolenic acids, such as Zaciyou-10. The results of the present study and previous investigation also suggested that the berries of ssp. *mongolica* were good source of palmitic acid and

Table 20.3: Comparison between the Fatty Acid Composition in Seed Oil and Pulp Oil of Seabuckthorn Cultivars with different Genetic Backgrounds[a] (Weight percentages)

Fatty Acids	*Seed Oil*						*Pulp/Peel Oil*					
	Mongolica		*Sinensis*		*Hybrid*		*Mongolica*		*Sinensis*		*Hybrid*	
	Mean	*Std.*	*Mean*	*Std*	*Mean*	*Std.*	*Mean*	*Std.*	*Mean*	*Std.*	*Mean*	*Std.*
16:0	11.4	0.8	12.5	1.4	12.2	1.8	35.6 a	3.7	31.0 bc	2.9	34.1 ab	3.3
16:1n7	1.1	0.3	0.9	0.4	0.8	0.4	33.9 a	3.3	20.1 c	4.5	28.4 b	3.9
18:0	8.4	1.0	8.1	1.7	8.6	3.7	6.7 ab	2.9	10.4 a	6.3	8.4 bc	1.6
18:1n9	15.0 c	2.2	17.3 ab	1.4	17.4 a	2.7	7.4 c	1.8	19.7 a	4.0	9.9 b	2.6
18:1n7	2.2	1.5	2.0	0.2	2.0	0.4	5.4 bc	1.0	6.9 a	2.2	6.6 ab	1.6
18:2n6	32.7	2.6	34.2	2.3	32.2	3.0	9.7 ab	1.1	7.7 c	2.3	10.7 a	2.0
18:3n3	28.7 a	2.9	24.9 bc	3.3	26.8 ab	3.9	1.4 bc	0.5	4.3 a	1.1	1.9 b	0.5

a Each value represents the average of analysis results of three subsamples, each analyzed in duplicate. Values with different lower–case letters (a–c) are significantly different at $p < 0.05$.

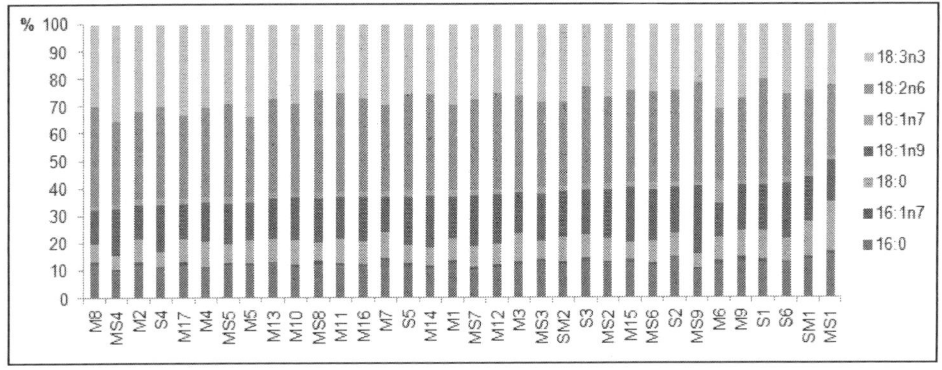

Figure 20.3: Fatty Acid Compositions (per cent) in Pulps of different Seabuckthorn Cultivars (Sum of palmitic and palmitoleic acids, maximum to minimum).

palmitoleic acids and those of ssp. *sinensis* were good source of oleic acid both in seeds and pulps (Yang *et al.*, 2001; Kallio *et al.*, 2002b).

Tocopherols and Tocotrienols in Seeds and Pulps

Four tocopherol isomers and β-, γ- and δ-tocotrienols were found in seeds of all cultivars investigated (Figure 20.4). α- and γ-Tocopherols were the two major isomers, representing typically 60–70 per cent and 10–20 per cent, respectively, of the total tocopherols and tocotrienols. The proportions of β- and δ-tocopherols were typically 5–10 per cent and 3–7 per cent, respectively, while β-, γ-tocotrienol and δ-tocotrienols were 1–3 per cent. In the seeds of different cultivars, the total tocopherols and tocotrienols was typically 100–200 μg/g DW. It varied from 80 to110 μg/g DW in ssp. *sinensis*, from 100 to 180 μg/g DW in ssp. *mongolica*, and from 100 to 220 μg/g DW in 'M×S'. The content of tocopherols and tocotrienols in seeds was low in two cultivars of 'S×M', which was 91.78 ± 12.05 μg/g DW on average. The seeds of *mongolica* and hybrid varieties contained more tocopherols and tocotrienols than those of *sinensis* (144.31 ± 21.04 and 132.06 ± 42.48 vs 103.53 ± 30.98 μg/g DW, $p < 0.01$) (Table 20.4). This result was consistent with previous investigation that the seeds of ssp. *mongolica* were a better source of tocopherols than those of ssp. *sinensis* (Kallio, *et al.*, 2002b). In ssp. *mongolica*, the content of tocopherols and tocotrienols were more than 150 μg/g DW in six cultivars [Zhuangyuanhuang (M1), Shoudu (M11), Xine-2 (M12), Xine-3 (M13), Qiuyisike (M15), Liusha-1 (M17)]. In hybrid varieties the seeds of Zaciyou-12 (MS5) and Xinzaci-26 (MS6) contained high content of tocopherols and tocotrienols, at 182.70 μg/g DW and 222.64 μg/g DW respectively and the content of γ-tocopherol in their seeds exceeded 70 μg/g DW, 3–4 times of mean values of ssp. *mongolica* and ssp. *sinensis*.

In soft parts of berries, α-tocopherol was representing typically 70–80 per cent of total tocopherols and tocotrienols. The proportions of β- and γ-tocopherols were typically 4–9 per cent and 5–10 per cent, respectively. γ- and δ-Tocotrienol were also detected in all samples at the level of 1–5 per cent (Figure 20.5). The berries of ssp. *sinensis* and hybrid varieties also contained α-tocotrienol (2–4 per cent) and

Table 20.4: Major Tocopherol (T) and Tocotrienol (T3) Concentrations in Seed Fraction Oils of Seabuckthorn Cultivars with different Genetic Origins[a] (μg/g DW)

Tocol	*ssp. mongolica*		*ssp. sinensis*		*Hybrid*	
	Mean	*Std.*	*Mean*	*Std.*	*Mean*	*Std.*
α-T	100.8 a	25.7	63.0 bc	22.3	81.8 ab	25.3
β-T	6.5	1.8	6.9	2.0	7.9	2.5
γ-T	24.7	9.0	17.0	4.8	28.2	24.9
δ-T	4.4 bc	1.8	6.1 a	0.9	5.0 ab	1.8
α-T3	nd		nd		nd	
β-T3	3.2	0. 7	2.9	1.6	2.8	1.2
γ-T3	3.2	1.2	2.7	1.7	2.8	1.9
δ-T3	1.5 c	0.6	4.9 a	0.9	3.6 b	1.9
total	144.3 a	21.0	103.5 bc	30.9	132.1 ab	42.5

a Values with different lower-case letters (a–c) are significantly different at $p < 0.05$. nd = not detected.

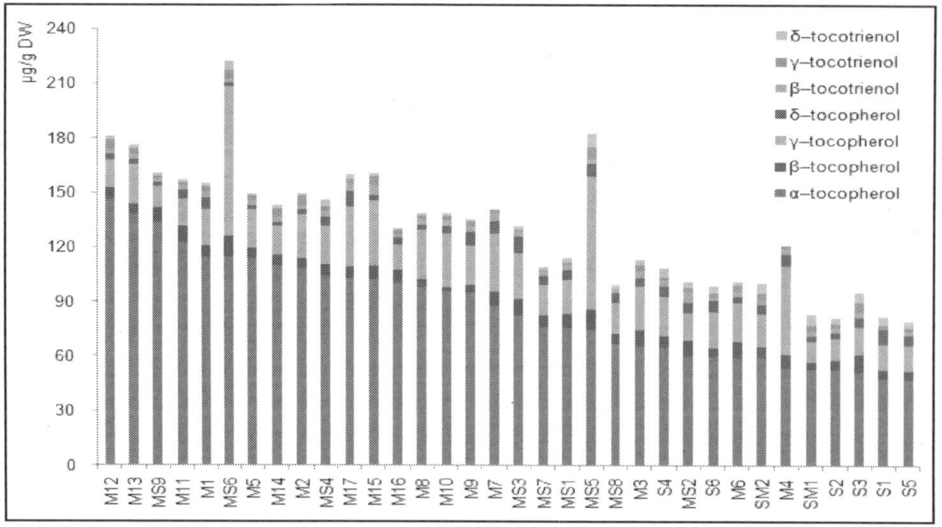

Figure 20.4: Composition of Tocopherols and Tocotrienols in Seeds of different Seabuckthorn Cultivars (α-tocopherol maximum to minimum).

β-tocotrienol (1–3 per cent), respectively, whereas it was almost undetectable in the samples of ssp. *mongolica*. The total tocopherols and tocotrienols varied from 200 to 900 μg based on freeze-dried weight in the samples analyzed. The lyophilized pulps of the berries of ssp. *mongolica* contained higher tocopherols and tocotrienols, comparing with ssp. *sinensis* and hybrid varieties (508.71 ± 208.20 *vs* 317.50 ± 115.63 *vs* 366.77 ± 72.41 μg/g DW, $p < 0.05$) (Table 20.5). The content of α-tocopherol in ssp. *mongolica* was higher than the other groups and reached up to 379.44 ± 190.61 μg/g DW (Table 20.4). In cultivar Wucifeng (M2), the content of α-tocopherol was quite high, up to 828.39 ± 46.49 μg/g DW and the total tocopherols and tocotrienols reached 979.76 ± 76.99 μg/g DW. α-Tocopherol in Fangxiang (M4) was 728.96 ± 76.37 μg/g DW and it contained 884.83 ± 89.43 μg/g DW of total tocopherols and tocotrienols. Andersson *et al.* (2008b) investigated the tocopherols and tocotrienols in berries from four cultivars during ripening and identified the genetic differences were prominent in the content of tocopherols and tocotrienols. In the present study tocopherols and tocotrienols of seabuckthorn are different between subspecies and cultivars.

Tocopherols and tocotrienols are commonly known as vitamin E. The physiological role of it centres on the ability to react with and quench free radicals in cell membranes and other lipid environments, thereby preventing polyunsaturated fatty acids (PUFA) from being damaged by lipid oxidation. α-Tocopherol accounts for most of the lipid-soluble, chain-breaking antioxidant activity in mammalian tissues and plasma (Bramley *et al.*, 2000). Kallo *et al.* (2002a) determined the fresh berries of subsp. *sinensis* were a good source of total tocopherols and tocotrienols. Our results found the lyophilized fruit flesh of ssp. *mongolica* also had good traits of it. These results provided useful information for development of healthy foods using seeds or pulps of seabuckthorn.

Carotenoids in Berries

In the berries investigated, the major identified carotenoids were xanthophylls (lutein; zeaxanthin;

Table 20.5: Major Tocopherol (T) and Tocotrienol (T3) Concentrations in Pulps of Seabuckthorn Berries of different Genetic Origins[a] (μg/g DW)

Tocol	*ssp. mongolica*		*ssp. sinensis*		*Hybrid*	
	Mean	*Std.*	*Mean*	*Std.*	*Mean*	*Std.*
α-T	379.4 a	190.6	224.2 bc	108.8	276.0 ab	64.5
β-T	38.3 a	15.0	16.8 bc	4.5	19.6 b	7.1
γ-T	34.3 a	11.4	29.4 ab	5.7	21.4 bc	4.6
δ-T	13.9 a	9.8	5.8 b	1.7	4.9 bc	1.4
α-T3	nd	—	9.0	2.7	10.8	4.5
β-T3	nd	—	6.1	2.3	4.4	1.8
γ-T3	14.5	11.4	6.8	2.4	9.3	2.9
δ-T3	28.3	18.5	19.3	8.1	20.4	14.3
total	508.7 a	208.2	317.5 bc	115.6	366.8 ab	72.4

a Values with different lower-case letters (a–c) are significantly different at $p < 0.05$. nd = not detected.

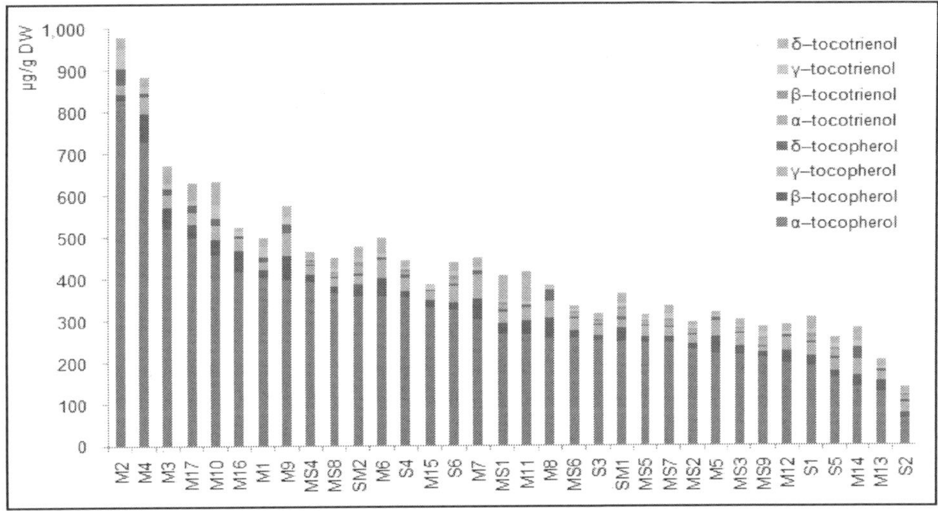

Figure 20.5: Composition of Tocopherols and Tocotrienols in Pulps of different Seabuckthorn Cultivars (α-tocopherol maximum to minimum).

β-cryptoxanthin), carotenes (lycopene; γ-carotene; β-carotene), and esterified carotenoids, whereas β-cryptoxanthin was detectable only in 13 cultivars (Figure 20.6). β-carotene and lycopene were two dominant carotenes in seabuckthorn berries. There was significant difference among cultivars in the composition and content of carotenoids, even in the same genetic background. The total amount of carotenoids varied from 250 to 1270 µg/g of DW in all cultivars. The total carotenoids varied from 270 to 1270 µg/g of DW in ssp. *mongolica*, from 260 to 820 µg/g of DW in ssp. *sinensis*, and from 250 to 960 µg/g of DW in hybrid varieties. Carotenoids in berries of *mongolica* were clearly richer than hybrid varieties and ssp. *sinensis* from an overall perspective. In cultivars of ssp. *mongolica*, Siberia rumianes (M3), Botanical garden (M8) and Xine-2 (M12) had higher contents of total carotenoids, which were 1268.85 ± 206.84, 1149.16 ± 198.46 and 1109.97 ± 178.64 µg/g of DW, respectively. In cultivars of ssp. *sinensis*, Manhanci (S4) had highest carotenoids at 815.63 ± 67.39 µg/g of DW, whereas not more than 450 µg/g of DW for others. In cultivars of hybrid varieties, Zaciyou-10 (MS4), Xinzaci-26 (MS6) and Z24 (SM1) had high contents of total carotenoids exceeding 700 µg/g of DW.

The major carotenoids determined in seabuckthorn berries were zeaxanthin, β-carotene, lutein, β-cryptoxanthin, lycopene, and γ-carotene, of which Zeaxanthin and β-carotene were dominant (Pintea *et al.*, 2005; Adresson *et al.*, 2009). In the present study the content of zeaxanthin and β-cryptoxanthin are relatively low, which probably due to the fact that esterified carotenoids consist of zeaxanthin and β-cryptoxanthin to a high degree (Pintea *et al.*, 2005; Pop *et al.*, 2014). The content of carotenoids may be affected by cultivars, ripening and climate factors (Adresson *et al.*, 2009). The genetic background of cultivars was the direct impact factor for the compositions and contents of carotenoids.

Carotenoids are an important factor to influence colours of fruits and juice (Meléndez-Martínez *et al.*, 2011; Wibowo *et al.*, 2015). The colours of seabuckthorn berries are mainly orange and yellow, also red in a few cultivars. According to carotenoids composition and content, berry samples were mainly classified into 2 big groups (Figure 20.6). One group comprised of 22 cultivars with different extent of orange-red berries. Twelve cultivars with yellow berries were classified into the other group. This result corresponded with colour analysis of berries investigated (data was not shown). Among the carotenoids compounds, the reddish pigment lycopene was the highest in red berries (including M3, M8, M12), about 400-700 µg/g of DW. In yellow berries the content of lycopene was low while more than 70 per cent were yellowish pigments (xanthophylls, γ-, β-carotene). These results confirmed the relationship between the composition of carotenoids and berry colour of seabuckthorn.

CONCLUSION

The results of this study showed that berry size seems to be a useful indicator of vitamin C content in seabuckthorn berries owing to the strong relationship between them, which was mainly accounted for by native cultivars (ssp. *sinensis*) with high vitamin C combined with smallest berries. Berries of ssp. *sinensis* are good materials for fruit juice beverage. The inter-sub specific hybridization improved the composition and content of essential quality components (except vitamin C) in berries of native cultivars (ssp. *sinensis*) and some elite cultivars need to be large-scale cultivated for nutritional purposes. Berry oils of several cross cultivars contain a large amount of unsaturated fatty acids (*e.g.* linoleic, α-linolenic and palmitoleic acids), such as Zaciyou-10, Chuyi-Shiciyou. γ-Tocopherol in Zaciyou-12 and Xinzaci-26 are 3–4 times of mean values of ssp.

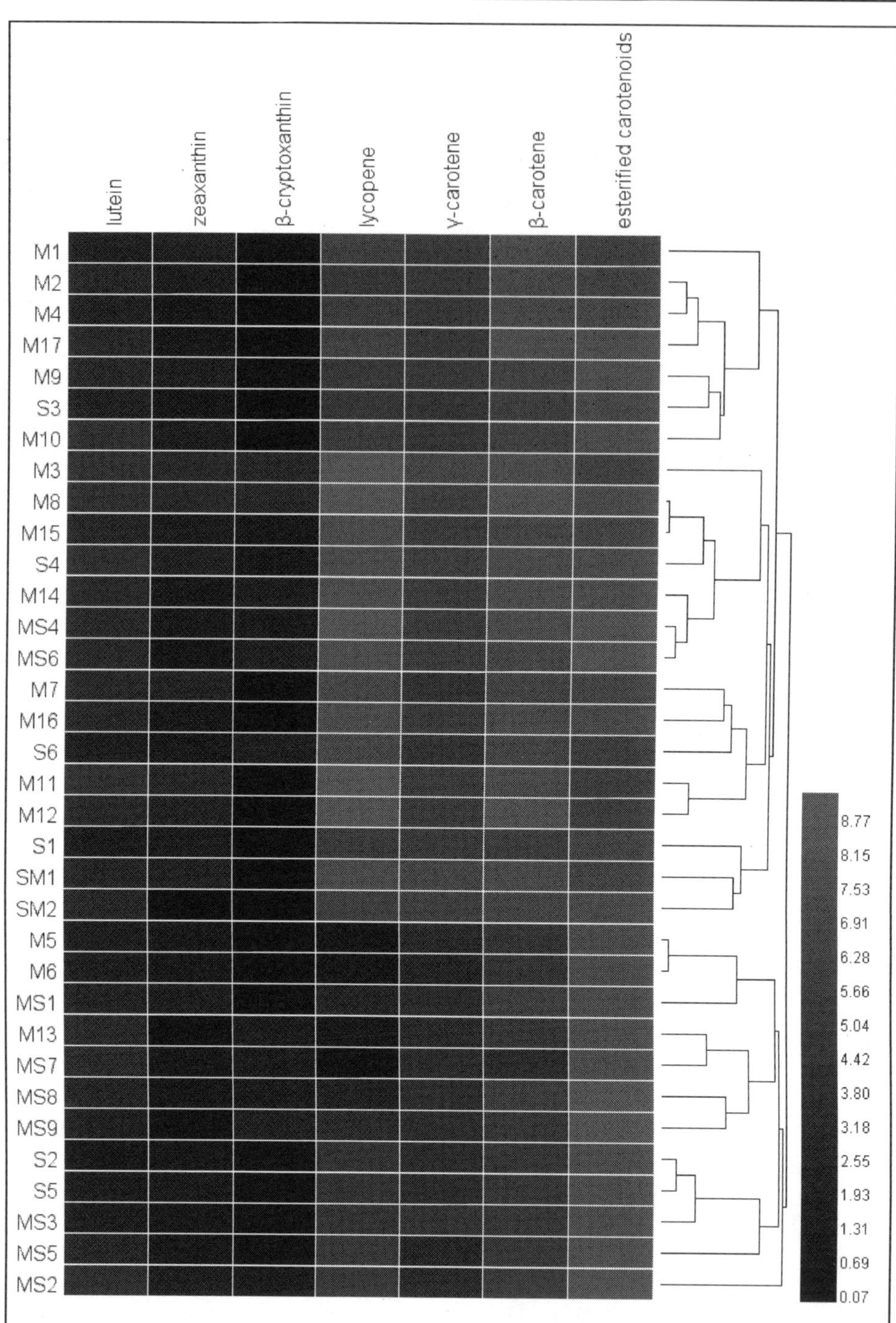

Figure 20.6: Carotenoid Composition and Classification of Berries of different Seabuckthorn Cultivars (Colors indicate content values of components of carotenoids. Red indicates increased content and black decreased).

mongolica and ssp. *sinensis*. Freeze-dried berries of ssp. *mongolica* (introduced cultivars) show good traits of total tocopherols and tocotrienols. The elite cultivars would be utilized to develop healthy foods of seabuckthorn oil.

ACKNOWLEDGEMENT

This research was funded by Natural Science Foundation of China (no. 31100489). The authors would like to express sincere thanks to research associates Guo Hai (Jiuchenggong Breeding Base of Seabuckthorn) and Zhang Jun (Institute of Selection and Breeding of *Hippophae* of Fuxin) for providing the berry samples.

REFERENCES

1. Abidi, S.L. 2000. Chromatographic analysis of tocol-derived lipid antioxidants. *Journal of Chromatography* 881: 197-216.
2. Andersson, S.C., Rumpunen, K., Johansson, E. and Olsson, M.E. 2008a. Changes in tocopherols and carotenoids during ripening in rose hips and seabuckthorn berries. *Acta Horticulturae* 765: 255-261.
3. Andersson, S.C., Rumpunen, K., Johansson, E. and Olsson, M.E. 2008b. Tocopherols and tocotrienols

in seabuckthorn (*Hippophae rhamnoides* L.) berries during ripening. *Journal of Agricultural and Food Chemistry* 56: 6701-6706.

4. Andersson, S., Olsson, M., Johansson, E. and Rumpune, K. 2009. Carotenoids in seabuckthorn (*Hippophae rhamnoides* L.) berries during ripening and use of pheophytin as a maturity marker. *Journal of Agricultural and Food Chemistry* 57: 250-258.

5. Bal, M.L., Meda, V., Naik, N.S. and Satya, S. 2011. Seabuckthorn: A potential source of valuable nutrients for nutraceuticals and cosmeceuticals. *Food Research International* 44: 1718-1727.

6. Bramley, P.M., Elmadfa, I., Kafatos, A., Kelly, F.J., Manios, Y., Roxborough, H.E., Schuch, W., Sheehy, P.J.A. and Wagner, K.H. 2000. Vitamin E. *Journal of the Science and Food Agriculture* 80: 913-938.

7. Christie, W.W. 1989. Fatty acids and lipids: Structures, extraction and fractionation into classes. In: *Gas Chromatography and Lipids* (Christie, W., Ed.). The Oily Press Ltd: Glasgow, Great Britain.

8. Christie, W. W. 1982. A simple procedure for rapid transmethylation of glycerolipids and cholesteryl esters. *Journal of Lipid Research* 23: 1072-1075.

9. Chen, C., Xu, X.M., Chen, Y., Yu, M.Y., Wen, F.Y. and Zhang, H. 2013. Identification, quantification and antioxidant activity of acylated flavonol glycosides from seabuckthorn (*Hippophae rhamnoides* ssp. *sinensis*). *Food Chemistry* 141: 1573-1579.

10. Deng, W.K., Wang, Y.B., Liu, Z.X., Cheng, H. and Xue, Y. 2014. HemI: A Toolkit for Illustrating Heatmaps. *PLoS One* 9: e111988.

11. Gökmen, V. and Acar, J. 1996. A simple HPLC method for the determination of total vitamin C in fruit juices and drinks. *Fruit Process* 5: 198-201.

12. Grey, C., Widén, C., Adlercreutz, P., Rumpunen, K. and Duan, R.D. 2010. Anti-proliferative effects of seabuckthorn (*Hippophae rhamnoides* L.) extracts on human colon and liver cancer cell lines. *Food Chemistry* 120: 1004-1010.

13. Kallio, H., Yang, B.R. and Peippo, P. 2002a. Effects of different origins and harvesting time on vitamin C, tocopherols, and tocotrienols in seabuckthorn (*Hippophae rhamnoides*) berries. *Journal of Agricultural and Food Chemistry* 50: 6136-6142.

14. Kallio, H., Yang, B.R., Peippo, P., Tahvonen, R. and Pan, R. 2002b. Triacylglycerols, glycerophospholipids, tocopherols, and tocotrienols in berries and seeds of two species (ssp. *sinensis* and *mongolica*) of seabuckthorn (*Hippophae rhamnoides*). *Journal of Agricultural and Food Chemistry* 50: 3004-3009.

15. Kallio, H., Lassila, M., Järvenpää E., Haraldsson, G. G., Jonsdottir, S. and Yang, B.R. 2009. Inositols and methylinositols in seabuckthorn (*Hippophaë rhamnoides*) berries. *Journal of Chromatography B* 877: 1426-1432.

16. Li, H., Ruan, C.J., Teixeira, da and Silva, J.A. 2009. Identification and genetic relationship based on ISSR analysis in a germplasm collection of seabuckthorn (*Hippophae* L.) from China and other countries. *Scientia Horticulturae* 123(2): 263-271.

17. Meléndez-Martínez, A. J., Gómez-Robledo, L., Melgosa, M., Vicario, I. M. and Heredia, F. J. 2011. Color of orange juices in relation to their carotenoid contents as assessed from different spectroscopic data. *Journal of Food Composition and Analysis* 24: 837-844.

18. Panfili, G., Fratianni, A. and Irano, M. 2003. Normal phase high performance liquid chromatography method for the determination of tocopherols and tocotrienols in cereals. *Journal of Agricultural and Food Chemistry* 51: 3940-3944.

19. Pintea, A., Varga, A., Stepnowski, P., Socaciu, C., Culea, M. and Diehl, H.A. 2005. Chromatographic analysis of carotenol fatty acid esters in Physalis alkekengi and *Hippophae rhamnoides*. *Phytochemistry Analysis* 16: 188-195.

20. Pop, M.R., Weesepoel, Y., Socaciu, C., Pintea, A., Vincken, J.P. and Gruppen, H. 2014. Carotenoid composition of berries and leaves from six Romanian seabuckthorn (*Hippophae rhamnoides* L.) varieties. *Food Chemistry* 147: 1-9.

21. Rousi, A. 1971. The genus *Hippophaë* L. a taxonomic study. *Annales Botanici Fennici* 8: 177-227.

22. Ruan, C.J., Teixeira da Silva, J.A., Jin, H., Li, H. and Li, D.Q. 2007. Research and biotechnology in seabuckthorn (*Hippophae* sp.). *Medicinal and Aromatic Plant Science and Biotechnology* 1: 47-60.

23. Suomela, J.P., Ahotupa, M., Yang, B., Vasankari, T. and Kallio, H., 2006. Absorption of flavonols derived from seabuckthorn (*Hippophae rhamnoides* L.) and their effect on emerging risk factors for cardiovascular disease in humans. *Journal of Agricultural and Food Chemistry* 54: 7364-7369.

24. Suryakumar, G. and Gupta, A. 2011. Medicinal and therapeutic potential of seabuckthorn (*Hippophae rhamnoides* L.). *Journal of Ethnopharmacology* 138: 268-278.

25. Susan, D., George, S. and Cenkowski, S. 2007. Influence of harvest time on the quality of oil-based compounds in seabuckthorn (*Hippophae rhamnoides* L. ssp. *sinensis*) seed and fruit. *Journal of Agricultural and Food Chemistry* 55: 8054-8061.

26. Tang, X. and Tigerstedt, P.M.A. 2001. Variation of physical and chemical characters within an elite seabuckthorn (*Hippophae rhamnoides* L.) breeding population. *Scientia Horticulturae* 88(3): 203-214.

27. Wibowo, S., Vervoort, L., Tomic, J., Santiago, S.J., Lemmens, L., Panozzo, A., Grauwet, T., Hendrickx, M. and Loey, V.A. 2015. Colour and carotenoid changes of pasteurized orange juice during storage. *Food Chemistry* 171: 330-340.

28. Yang, B. and Kallio, H. 2001. Fatty acid composition of lipids in seabuckthorn (*Hippophaë rhamnoides* L.) berries of different origins. *Journal of Agricultural and Food Chemistry* 49: 1939-1947.

29. Xu Y.J., Kaur, M., Dhillon, S.R., Tappia, S.P. and Dhalla, S.N. 2011. Health benefits of seabuckthorn for the prevention of cardiovascular diseases. *Journal of Functional Foods* 3: 2-12.

Chapter 21

Dynamic Study of some Bioactive Compounds of Seabuckthorn Berries

S. Bayaraa*, B. Battulga, S. Binderya, E. Narmandakh, A. Bayanmunkh, L. Lkhagva and L. Khurelbaatar

Department of Medicinal Chemistry, Drug Research Institute, Mongolia
(*E-mail: bayaraa.s@monos.mn)

SUMMARY

In Mongolia, the seabuckthorn is a nontraditional fruit crop. In 2013-2014, at the Botanical Garden of Drug Research Institute for Monos group, were studied 6 seabuckthorn cultivars: Chuiskaya, Maslichnaya, Vitaminnaya, Oranjivaya, Obilniya and red fruit for their chemical composition depend from harvesting season. The chemical composition of fruit was investigated from 19thAugust to 20th January. We determined total oil, organic acid contents, β-carotenoid, ascorbic acid, polyphenolic content and polysaccharide. The chemical contents were determined by Folin-Ciocalteau assay, colorimetric assay, volume metric analysis and spectrophotometric methods. During the harvesting period, the fruit's chemical composition such as pulp oil content increased and the contents of organic acid, beta carotenoid and ascorbic acid decreased.

Keywords: *Seabuckthorn, Chuiskaya, Maslichnaya, Vitaminnaya, Orangevaya, Obilnaya and Harvesting season.*

INTRODUCTION

In recent years, due to seabuckthorn fruit cultivation in our country to promote the health of national food brands and products based upon market penetration. However, the current condition of the seabuckthorn cultivation increased significantly wild natural resource degradation, cultivation of local varieties derived slowed down markedly in the past, from the outside before there are varieties cultivated image as a mixture of common seabuckthorn. At the international level, the situation is reversed seabuckthorn uniform standardization of raw materials, product ingredients, impacts on human health act pharmacology and clinical research, and the use of proper study of product requirements. Currently, our country several standards related to raw materials/ MNS 664: 88, MNS 783: 88, MNS 916: 85, MNS 2696: 79 is being investigated for the production of biologically active products, direct application of these standards is lacking, so it is necessary to upgrade. Seabuckthorn berries contains a set of biologically active compounds, depending on how the season is changing harvesting products for both raw materials and thus the product can be used as a key measure to standardize. Therefore, we are part of the research, seabuckthorn berry, some types of biologically active compounds dynamics.

MATERIALS AND METHODS

In this study, The botanical garden for "Monos" Group medicinal plants cultivated Red fruit Maslichnaya, Vitaminnaya, Chuiskaya, Obilinaya and Oranjyevaya selected varieties of seabuckthorn berries 19 August, 2013 to 20 January, 2014 conducted study be sampled for up to 10 times. The total phenolic contents of dry medicinal plants were determined with the Folin-Ciocalteau assay by spectrophotometer. We measured at 750 nm UV-Visible spectrophotometer absorbance of samples. Then compare it with standard solution of Gallic acid[3,4].

Seabuckthorn (Hippophae L.): A Multipurpose Wonder Plant, Vol. 5: Advanced Technologies for Cultivation, Processing, Health Protection and Environmental Conservation (Virendra Singh, *Editor-in-Chief*, 2018), p. 167–171, Astral International Pvt. Ltd., New Delhi, India.

The total polysaccharide contents were determined by spectrophotometer. We measured at 488 nm absorbance of samples. Then compare it with standard solution of β-D Glucose[7,8]. Samples ascorbic acid, total organic acids, and total fats meet the general requirements for tablets by Mongolian National First Pharmacopeia, β- carotene, Polyphenol compound, total polysaccharide content determined by spectrophotometric method.

RESULTS

The research work was done at the chemical and pharmaceutical sector in Drug Research Institute, Botanical garden of "Monos" Group. Contents of biologically active compounds in seabuckthorn berries are shown on Table 21.1 depending on how season is changing and shown in Figures 21.1-21.5.

As can be seen from Figure 21.1, Ascorbic acid content was trending to constantly decrease during August to January. The red berry sort's total ascorbic acid content was the lowest concentration in end of the August, Oranjevaya, Chuiskaya sort's total ascorbic acid content was highest concentration on December, January. The berries simple organic acids and water-soluble vitamins/lemon, tartaric, mallic acids, synthesis

Table 21.1: Chemical Constituents of Seabuckthorn Berries

Date	No.	*Ascorbic Acid, mg/100g*	*β-carotene, mg/100g*	*Total Organic Acid, per cent*	*Total Fat, per cent*	*Polyphenols, mg/g*	*Polysaccharide, mg/g*
19 Aug, 2013	1	29.35±0.58	3.83±0.08	1.05±0.03	6.60±0.26	1.94±0.07	34.93±1.40
	2	87.63±1.75	1.87±0.05	1.23±0.03	11.90±0.45	4.95±0.19	23.23±0.93
	3	69.07±1.23	1.64±0.05	1.06±0.03	10.08±0.40	3.57±0.16	36.73±1.45
	4	83.07±1.54	2.49±0.07	1.00±0.02	14.50±0.48	5.38±0.21	52.31±2.01
	5	80.05±1.51	1.78±0.04	1.20±0.03	11.37±0.45	4.22±0.17	31.97±1.40
	6	85.19±1.19	1.73±0.04	1.59±0.03	13.96±0.50	1.82±0.07	25.76±1.01
29 Aug, 2013	1	18.69±0.80	4.05±0.19	1.12±0.03	6.76±0.25	2.35±0.09	36.54±1.44
	2	73.38±3.54	2.19±0.09	1.05±0.03	8.91±0.30	4.92±0.18	25.52±0.95
	3	47.70±2.25	1.51±0.07	1.21±0.03	9.60±0.35	3.82±0.17	40.53±1.62
	4	47.30±2.10	2.66±0.12	0.96±0.02	9.54±0.35	5.32±0.20	62.31±2.40
	5	47.02±1.85	2.05±0.10	0.87±0.02	7.40±0.25	4.02±0.16	36.85±1.50
	6	74.74±3.50	1.97±0.09	1.33±0.03	8.49±0.24	2.08±0.08	29.71±1.18
09 Sep, 2013	1	15.17±0.65	3.34±0.15	1.03±0.02	6.87±0.25	2.16±0.08	39.50±1.62
	2	71.50±3.12	1.57±0.07	0.97±0.02	6.80±0.25	4.90±0.18	28.82±1.12
	3	50.03±2.56	2.12±0.10	1.12±0.03	8.83±0.25	3.41±0.15	42.63±1.62
	4	53.49±5.75	1.57±0.07	0.85±0.02	10.53±0.35	5.08±0.18	66.27±2.50
	5	46.03±2.01	1.89±0.08	1.29±0.03	8.77±0.28	3.97±0.15	37.82±1.58
	6	74.81±5.46	2.13±0.11	1.19±0.03	9.30±0.36	2.35±0.08	31.77±1.20
19 Sep, 2013	1	24.51±0.73	4.24±0.08	0.61±0.02	9.67±0.27	1.94±0.06	40.52±1.50
	2	70.50±2.82	1.94±0.04	0.78±0.02	10.45±0.32	3.58±0.14	29.87±1.21
	3	46.54±0.94	2.17±0.06	0.82±0.02	8.46±0.24	3.16±0.14	45.84±1.43
	4	46.48±0.87	2.53±0.05	0.56±0.01	8.56±0.26	4.15±0.16	69.15±2.82
	5	52.86±0.75	2.14±0.03	0.67±0.02	11.87±0.25	3.05±0.12	40.61±1.65
	6	-	-	-	6.89±0.20	2.16±0.08	35.71±1.35
29 Sep, 2013	1	21.47±0.75	4.16±0.08	0.84±0.02	14.86±0.40	1.24±0.05	45.84±1.74
	2	68.50±2.75	2.34±0.05	0.82±0.02	13.00±0.38	2.83±0.11	35.94±1.36
	3	45.84±1.01	1.82±0.03	0.67±0.02	11.64±0.35	2.64±0.10	53.68±2.14
	4	41.24±1.24	2.37±0.04	0.50±0.02	10.50±0.35	3.48±0.13	69.15±2.81
	5	50.16±1.38	2.08±0.05	0.70±0.02	12.66±0.36	2.57±0.10	45.61±1.82
	6	-	-	-	-	3.17±0.12	36.71±1.52
18 Oct, 2013	1	16.83±1.30	3.07±0.16	0.62±0.02	15.41±0.46	1.48±0.05	48.72±1.95
	2	67.70±3.35	2.29±0.11	0.88±0.02	14.74±0.42	2.65±0.12	38.91±1.48
	3	44.83±2.25	1.51±0.06	0.61±0.02	13.7±0.40	2.27±0.09	61.60±2.40
	4	31.92±1.52	1.24±0.05	0.52±0.01	12.15±0.37	2.86±0.13	62.11±2.45
	5	49.75±2.45	1.83±0.08	0.70±0.02	16.75±0.42	2.73±0.12	48.81±2.08
	6	40.63±2.02	3.03±0.14	0.86±0.02	10.40±0.21	2.28±0.09	36.82±1.44

Contd...

Table 21.1–*Contd...*

Date/No.		*Ascorbic Acid, mg/100g*	*β-carotene, mg/100g*	*Total Organic Acid, per cent*	*Total Fat, per cent*	*Polyphenols, mg/g*	*Polysaccharide, mg/g*
30 Oct, 2013	1	10.21±0.51	3.38±0.11	0.50±0.01	15.61±0.62	1.72±0.06	57.99±2.25
	2	55.35±2.21	2.42±0.07	1.08±0.03	15.12±0.45	2.21±0.08	42.31±1.68
	3	35.80±0.81	2.03±0.04	0.87±0.03	14.25±0.50	4.21±0.17	72.99±2.81
	4	34.87±0.70	1.75±0.06	0.77±0.02	13.05±0.42	2.39±0.10	65.62±2.64
	5	52.60±2.60	1.88±0.07	0.71±0.02	18.08±0.41	2.99±0.12	59.99±2.41
	6	43.64±1.60	2.75±0.07	0.81±0.02	10.58±0.35	3.01±0.12	41.00±1.61
20 Nov, 2013	1	9.24±0.27	3.18±0.08	0.61±0.02	14.87±0.35	1.54±0.06	45.93±1.84
	2	49.22±1.47	2.33±0.07	0.68±0.02	14.82±0.31	2.25±0.09	38.01±1.50
	3	27.58±1.02	2.09±0.04	0.72±0.02	14.18±0.30	2.08±0.08	50.62±1.98
	4	29.77±1.00	1.83±0.06	0.80±0.02	12.86±0.25	2.32±0.08	57.63±2.36
	5	45.12±1.65	1.81±0.07	0.57±0.01	17.76±0.26	2.84±0.12	39.67±1.58
	6	42.82±1.24	2.81±0.07	0.66±0.02	9.75±0.20	1.83±0.07	38.15±1.52
20 Dec, 2013	1	9.08±0.25	3.31±0.08	0.70±0.02	15.28±0.30	1.18±0.05	38.96±1.50
	2	42.54±1.54	2.42±0.07	0.72±0.02	15.64±0.30	2.20±0.09	40.06±1.55
	3	22.43±0.62	2.13±0.04	0.94±0.03	14.96±0.28	1.83±0.06	32.68±1.15
	4	25.13±0.65	1.72±0.06	0.97±0.03	13.17±0.28	2.01±0.07	49.81±1.86
	5	42.52±1.35	1.83±0.07	0.87±0.02	18.53±0.42	2.15±0.07	24.18±0.84
	6	37.21±1.18	2.70±0.07	0.77±0.02	10.92±0.23	1.54±0.06	44.52±1.74
20 Jan, 2014	1	8.52±0.25	3.18±0.08	1.65±0.02	15.19±0.40	1.08±0.05	35.93±1.34
	2	38.21±0.09	2.12±0.07	1.05±0.02	15.35±0.38	1.88±0.07	37.01±1.42
	3	18.09±0.50	2.01±0.04	0.87±0.02	14.87±0.32	1.67±0.07	30.68±1.18
	4	21.77±0.61	1.45±0.06	0.92±0.02	13.52±0.30	1.86±0.08	47.81±1.88
	5	32.54±0.85	1.37±0.07	0.78±0.02	17.95±0.35	1.95±0.08	19.44±0.78
	6	35.14±0.87	2.42±0.07	0.89±0.02	11.05±0.28	1.48±0.05	41.49±1.67

Note: 1-Red fruit, 2-Chuiskaya, 3-Maslichnaya, 4- Vitaminnaya, 5- Obilnaya, 6-Oranjevaya.

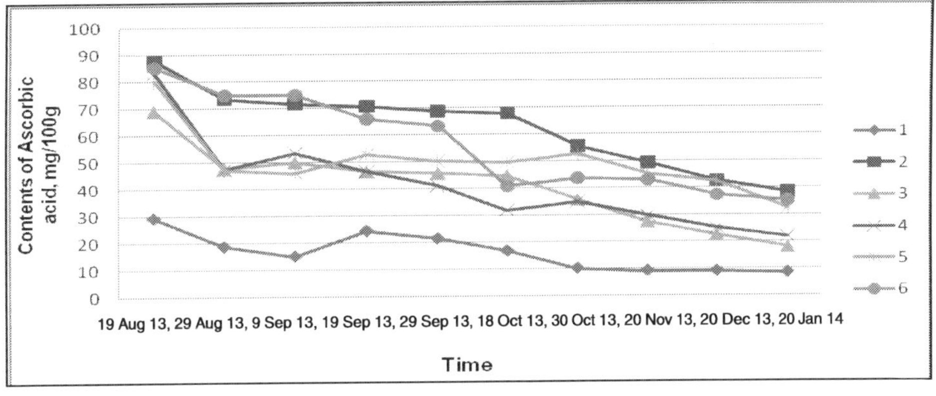

Figure 21.1: Diagram for Contents of Ascorbic Acid Depended by Time.

of vitamin B1, B2, B3, B5, B6 depended on the total sugar and ascorbic acid content, maybe their accumulation and decomposition of changed in temperature. Therefore, in our opinion berries organic acids and vitamin concentration slowly decreased period from August during the total research time. It is necessary to determine sugar content in berries. The concentration of organic acids to that shown in Figure 21.2.

Organic acids known varieties of the highest concentration at the end of August. Drift is expected to decrease the concentration of the cold season. But the 1st sort's berries collected total organic acid concentration was increased of up to 20th November to 20thJanuary. Maybe this sort's contain much amount of sugar other sort's that sugar decomposed to 20th November to 20th January. β-carotene content, depending by the time shown in Figure 21.3.

All of sort's berries β-carotene content was increased in maturation period. However next month β-carotene content was slowly resolution. In our view, the β-carotene is fat-soluble, so maybe this combination changed depending on the content. The amount of at contained in the fruit are shown in Figure 21.4.

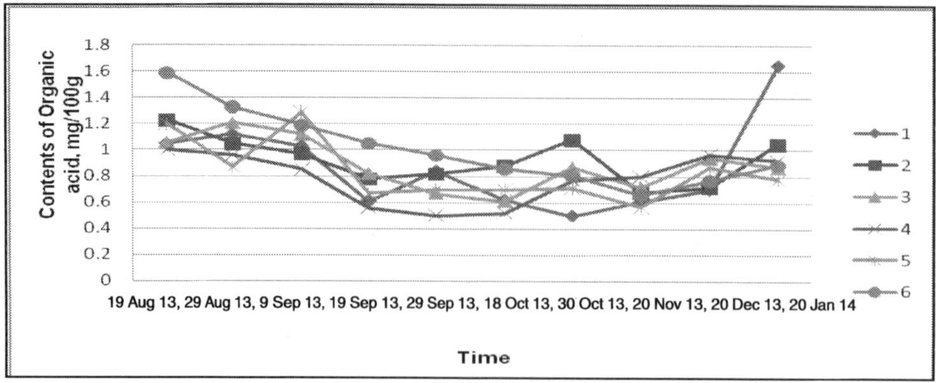

Figure 21.2: Diagram for Contents of Organic Acid Depended by Time.

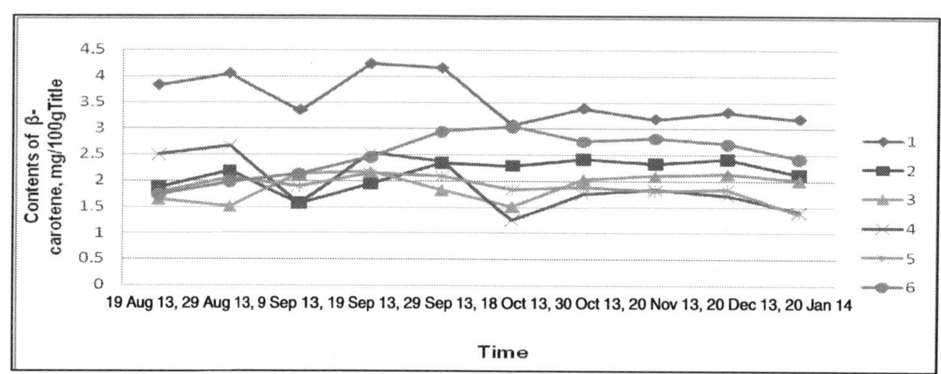

Figure 21.3: Diagram for Contents of β-carotene Depended by Time.

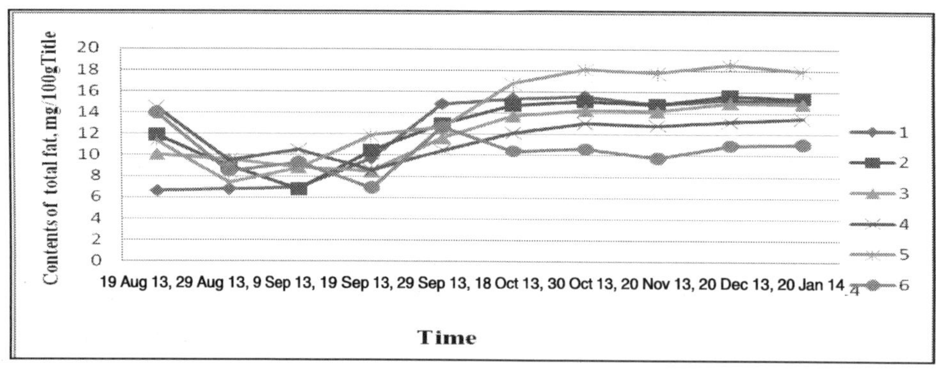

Figure 21.4: Diagram for Contents of Total Fat Depended by Time.

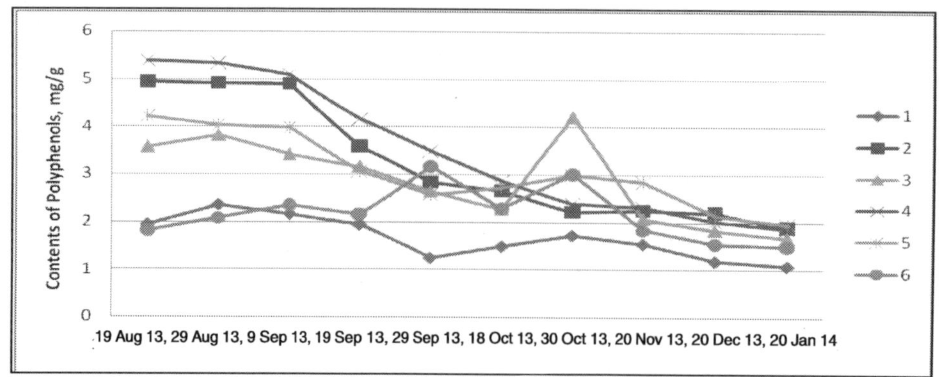

Figure 21.5: Diagram for Contents of Polyphenolic Compounds Depended by Time.

As can be seen from Figure 21.4, the total amount of fat varieties of 6.60–13.96 per cent from 2nd weeks on August in berries maturation period but reduced to the minimum 6, 9-11.87 percent from 19th on September. We determined Red fruit and Oranjevaya sort's total fat constantly increased till 20th on January. This 2 sort's total fat'sconcentration invariable during 20th on October. Accordingly, the berries contain high levels of fat in the cold season. Therefore, if we used all fat's in seabuckthorn when we choose suitable for cold season's/November to January/raw materials.

The results of study seabuckthorn's all sort berries total ascorbic acid 29-85 mg per cent, beta-carotene 1.8-3.8 mg, organic acids 1.0-1.6 per cent of the total, fat 6.6-13.9 per cent concentrations at the end of August. We determination all sort's berries biological active substance changed, ascorbic acid 8.5-35.4 mg per cent, beta-carotene 1.37-2.42 mg, organic acids 0.78-1.05 per cent of the total fat 11.1-17.9 per cent concentrations at the 20th on January. Berries total fat concentrations constantly increased in August to January, but Other 3 kind of biological compounds such as ascorbic acid, beta-carotene, organic acid amounts constantly decreased during in August to January.

REFERENCES

1. *Mongolian National Pharmacopeia*, p. 564, 584.
2. ФСП 42. с.3, ОАО "Нижфарм " 2004,β-каротин.
3. Atanassova, M., Georgieva, S. and Ivancheva, K. 2011. Total phenolic and total flavonoids, antioxidant capacity and biological contaminants in medicinal herbs. *Journal of the University of Chemical Technology and Metallurgy* 46: 81-88.
4. Bhatt, M. and Kampoj, A. and Saluja, A.J. 2013. Spectrophotometric estimation of total polysaccharides in *Kalachoe pinnatum* and *Kalanchoe crenata. International Journal of Pharmacy and Pharmaceutical Sciences* 58: (2).
5. Zubarev, Y.A. Properties and promising ways of seabuckthorn utilization at Altai. The Lisavenko Research Institute of Horticulture for Siberia, Scientific Presentation.
6. Lian, Y.S., Lu, S.G., Su, S.K. and Chen, X.L. 2000. Biology and chemistry of the Genus *Hippophae*. Dansu Scientific and Technological Publishing House, Lanzhou, China, pp. 88-91.
7. Yang, B. and Kallio, H. 2001. Fatty acid composition of lipids in seabuckthorn (*Hippophae rhamnoides* L.) berries of different origins. *J. Agri. Food Chem.* 46: 1939-1947.
8. Alam, Z. 2004. Chemical and nutritional constituents of seabuckthorn juice. *Pakistan Journal of Nutrition* 3 (2): 99-106.

Chapter 22

Glyceride Composition of Seabuckthorn Oil of Altai Territory in Russia

Nathalia V. Goremyikina*, Alexander L. Vereschagin, Yuriy A. Koshelev and Nikolai S. Pershin

Biysk Technological Institute of Polzunov Altai State Technical University
27 Trofimov St., Biysk, Altai Territory, 659305, Russia
(*E-mail: val@bti.secna.ru)

SUMMARY

A procedure to determine seabuckthorn oil glycerides with high-temperature gas chromatography was designed. The seabuckthorn oil glyceride compositions of concentrates from different Altai manufacturers were studied. The effect of the methods used for oil separation and the composition of the used raw materials on seabuckthorn oil triglyceride ratio was studied. A possibility of direct differentiation of monoglycerides, diglycerides and triglycerides was shown. The advantages of the direct method to determine glycerides as compared with the method to determine fatty acid composition with preliminary sample processing were given. It was found the ratio of individual seabuckthorn oil glyceride varieties can be the most important specification that has an effect on oil manufacturing methods and the use of raw materials.

***Keywords**: Seabuckthorn, Oil, Glycerides, Gas-liquid chromatography and Fatty acid composition.*

INTRODUCTION

Seabuckthorn berries are rich in vitamins, carotinoids, flavonoids, proteins, antioxidants, fat acids and phytosteroles (Koshelev and Ageeva, 2004). The most valuable processing product of seabuckthorn berries is seabuckthorn oil having the unique composition of triglycerides so it finds a wide application in medicine. Monoglycerides, diglycerides and triglycerides form the basis of most seed oils. They determine nutritional value of the oil and they are perfect objects to be studied due to various fatty acid compositions. The fat acid composition that vegetal oil glycerides have is interesting with relation to find some differences between the oils of different origination. A lot of authors dealt with the study of seabuckthorn oil composition to standardize it. They found that changes in the composition of seabuckthorn seeds and pulp as compared with a sub species, a breed, soil and climatic conditions, habitats, agricultural methods used, harvesting terms and extraction methods (Vereshchagin *et al.*, 2010; Tsydendambaev *et al.*, 2011). Daineko *et al.* (2003) used the reversed phase HLPC method to study seabuckthorn triglyceride composition and to determine counterfeiting of seabuckthorn oil. There is a theoretical incremental approach for result interpretation but there is no comparison given with the standard substances.

High cost and efficiency made seabuckthorn oil attractive for development of new separating methods, extension of raw material range and also the range of counterfeiting objects. Nowadays there are some different ways to separate seabuckthorn oil. The most common are an extraction method using a volatile solvent (freon 22), centrifugal sedimentation method, and hydrolysis with amylopectin ferments. In production quantities the primary efficacy endpoint of the method is its product yield. The comparison of

Seabuckthorn (Hippophae L.): A Multipurpose Wonder Plant, Vol. 5: Advanced Technologies for Cultivation, Processing, Health Protection and Environmental Conservation (Virendra Singh, *Editor-in-Chief*, 2018), p. 173–180, Astral International Pvt. Ltd., New Delhi, India.

the methods to separate the product with the highest biological potency that is determined by triglyceride composition is not studied enough. The glyceride composition of the seabuckthorn oil that was separated from different parts of the plant was not studied either and is of interest for oil sample identification of different origination.

So far gas chromatography is used to analyze fatty acid composition that disintegrates triglycerides into specific methyl ethers of fat acids. It makes impossible to determine the initial glyceride composition.

The aim of the paper was designing the method to determine seabuckthorn oil glyceride composition and to study the effect of soil and climatic conditions, separation method and starting material composition of different Altai region manufacturers on glyceride composition of seabuckthorn oil.

MATERIALS AND METHODS

Samples Analysis

Effect of Soil and Climatic Conditions

Seabuckthorn oil samples that were grown in different regions of Altai territory (Volchikhinskiy, Krutikhinskiy, Zmeinogorskiy, Mamontovskiy, Pospelikhinskiy and wild bushes in the water-meadow of Katun river of 2013 harvest that are raw material for obtaining oil concentrate were taken for analysis. Seabuckthorn concentrate samples of the following Altai manufacturers were studied: Yagodnoye LLC, Yantarnoye LLC, Altaivitamins ZAO, Alsu LLC, Altaiskiy Sad LLC, Sava LLC.

Separating Method Effect

The seabuckthorn berries of Tchuiskaya breed picked in 2012 on the planting of Lisavenko research institute of Siberia gardening of Russian Agricultural academy were used. Samples of seabuckthorn oil that were obtained from milled press cake (product that is obtained when juicing berries) in different ways: extraction with difluorochloromethane (freon 22) and enzymic hydrolysis. Extraction of seabuckthorn oil was performed in an extractor for 8 hours, then freon 22 was removed. To produce oil by enzymatic hydrolysis were used protosubtilin and celloljuks in the ratio 1:1. Oil produced with centrifugal sedimentor was prepared by rubbing through a sieve to separate seeds and the obtained pulp was centrifuged up to three layer stratification: oil, juice, pulp.

Feed Material Effect

The seabuckthorn berries of Tchuiskaya breed picked in 2013 on the planting of Lisavenko research institute of Siberia gardening of Russian Agricultural academy. Samples of seabuckthorn oil that were produced by extraction with Freon 22 from different parts of seabuckthorn plants: fruit coat (skin), from milled press cake (product that is obtained when juicing berries), kernels (seeds), buds and leaves were taken to be studied.

Method of Analysis

Samples for analysis were prepared in the following way: 0.02 g was dissolved in 5 ml of hexane; 0.2 /l of the obtained solution was put on an evaporator. The analysis was performed by Shimadzu GC-2010 plus gas chromatographer at the following conditions.

Injector

Injector temperature is 370°C, stream in column is 0.75 ml/min, line rate is 33 cm/sec. and division factor is 10. Column is SE-30 «Vitochrome - M» of capillary type, length is 15 m, inner diameter is 0.3 mm, film thickness is 0.25 /m, stationary phase is polymethylsiloxane, programmed column temperature is isothermic mode for 4 min at the initial temperature of 160°C, then it is increased up to 365°C at the rate of 8 °C/min, retention time is 15 min. Total analysis time is 44 min.

Detector

Flame-ionization, detector temperature is 380°C, air stream is 400 ml/min, hydrogen stream is 40 ml/min, carrier gas stream is 30 ml/min, carrier gas is helium.

SUPELCO standards Cat. No. 1787-1AMP; MDT12-1KT were used for determination of glycerides.

The results on glyceride composition were treated as components being synchronous with standard substances in retention time.

Fat acid composition of seabuckthorn oil samples was determined according to the standard technique (FSP [1]002959/01-301210. Seabuckthorn oil concentrates). The objects of research were the same seabuckthorn oil samples of different manufacturers that were produced by different separating methods. The oil samples were obtained from different parts of the plant. Fat acids were transformed into their methyl ethers and analyzed with Kristall Lux 4000 gas chromatograph.

RESULTS

Glyceride Composition

Soil and Climatic Conditions Effect

Characteristic chromatogram of a seabuckthorn oil sample is shown in Figure 22.1.

Glyceride composition data on seabuckthorn oil samples of different manufacturers are shown in Table 22.1.

Separating Method Effect

The analysis results of seabuckthorn oil samples extracted with different methods are shown in Table 22.2.

Table 22.1: Glyceride Composition of Seabuckthorn Oil Samples

Component, Per cent	*Seabuckthorn Oil Composition, per cent*					
	Yagodnoye LLC	*Yantarnoye LLC*	*Altaivitamins ZAO*	*Alsu LLC*	*Altaiskiy sad LLC*	*Sava LLC*
Monolaurin	–	–	–	–	–	–
Monomiristin	–	–	–	0,02±0,01	–	–
Monopalmitin	–	–	–	0,01±0,003	–	–
Monostearin	–	–	–	–	–	0,02±0,01
Dilaurin	0,65±0,13	2,56±0,04	0,69 ±0,14	0,24±0,04	0,16±0,02	0,40±0,06
Dimiristin	0,03±0,01	–	–	–	–	–
Dipalmitin	1,83±0,03	0,81±0,12	–	–	0,37±0,06	–
Distarin	0,07±0,01	0,21±0,03	0,20 ±0,04	–	–	0,06±0,02
Trilaurin	1,57±0,03	0,55±0,08	0,74 ±0,15	0,30±0,05	0,14±0,02	0,18±0,03
Trimiristin	–	–	–	0,02±0,01	–	0,010±0,003
Tripalmitin	38,46±0,40	43,56±0,40	38,16 ±0,40	32,47±0,33	33,39±0,50	44,70±0,67
Triolein	7,48±0,11	7,86±0,12	9,71 ±0,15	13,16±0,20	8,63±0,13	8,20±0,12
Tristearin	1,02±0,15	0,81±0,12	0,53 ±0,10	–	0,16±0,02	–
Trilinolein	–	–	7,67 ±0,15	18,81±0,28	9,20±0,14	0,07±0,02
Tripalmitolein*	32,34±0,48	35,14±0,5	31,28 ±0,5	28,63±0,43	29,06±0,43	36,67±0,55

* Anticipated triglyceride.

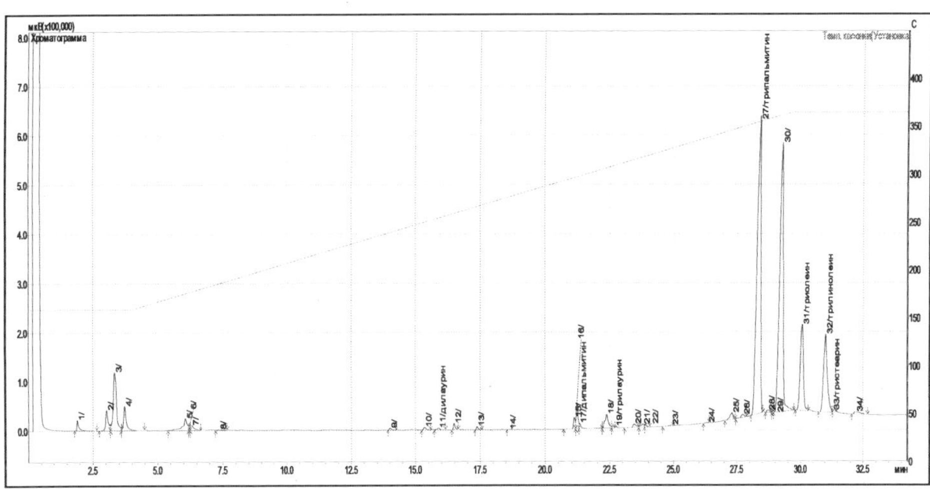

Figure 22.1: Chromatogram of Altaiskiy Sad LLC Seabuckthorn Oil Sample.

Feed Material Effect

Characteristic chromatograms are shown in Figures 22.2-22.3.

The data obtained are shown in Table 22.3.

Fatty Acid Composition

Soil and Climatic Condition Effect

The data obtained are shown in Table 22.4.

Fatty Acid Composition

Separating Method Effect

Fatty acid composition data obtained of the samples are shown in Table 22.5.

Fatty Acid Composition

Feed material effect- The data obtained are shown in Table 22.6.

DISCUSSION

Soil and Climatic Condition Effect

According to the data presented in Table 22.1, seabuckthorn oil glyceride composition of different manufacturers differs. Tripalmitin percentage is typically high from 32 to 44 per cent for all the samples. Negligible amounts of monomiristin (0,02 per cent) and monopalmitin (0,01 per cent) were found only in Alsu LLC sample. Monostearin was found only in Sava LLC

Table 22.2: Glyceride Composition of Seabuckthorn Oil Samples

Glyceride	*Seabuckthorn Oil Composition, Per cent Separated by*		
	Extraction	*Centrifugal Sedimentation*	*Enzymic Hydrolysis*
Monolaurin	-	-	-
Monomiristin	-	-	-
Monopalmitin	-	-	-
Monostearin	-	0,020 ±0,004	-
Dilaurin	0,69 ±0,14	0,40 ±0,8	0,09 ±0,02
Dimiristin	-	-	-
Dipalmitin	-	-	-
Distearin	0,20 ±0,04	0,060 ±0,012	0,020 ±0,004
Trilaurin	0,74 ±0,15	0,18 ±0,03	0,62 ±0,14
Trimiristin	-	0,01 ±0,002	-
Tripalmitin	38,16 ±0,50	44,70 ±0,50	44,37 ±0,50
Triolein	9,71 ±0,15	8,20 ±0,10	9,28 ±0,15
Tristearin	0,53 ±0,10	-	0,82 ±0,16
Trilinolein	7,67 ±0,15	0,07 ±0,01	-
Tripalmitolein	31,28 ±0,50	36,67 ±0,50	41,06 ±0,50

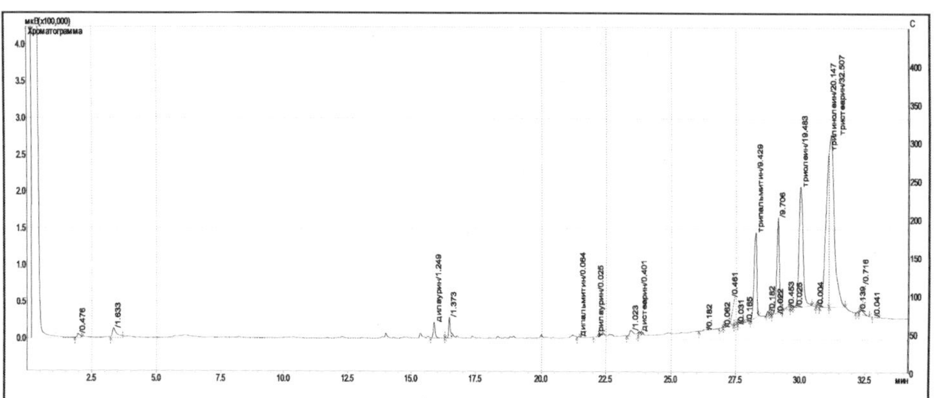

Figure 22.2: Chromatogram of Oil Obtained from Seabuckthorn Berry Seeds.

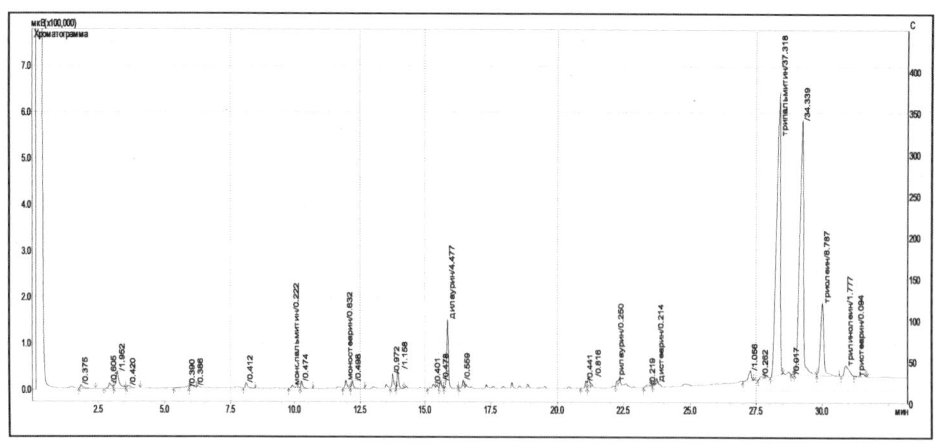

Figure 22.3: Chromatogram of Oil Obtained from Seabuckthorn Berry Coat (Skin).

Table 22.3: Glyceride Composition of Seabuckthorn Oil Samples

Glyceride, Per cent	*Seabuckthorn Oil (100 per cent)*				
	Coat	*Press cake*	*Seeds*	*Buds*	*Leaves*
Monolaurin	-	-	-	-	-
Monomiristin	0,07±0,01	0,020±0,004	0,020±0,004	2,20±0,44	0,44±0,09
Monopalmitin	0,25±0,05	0,040±0,008	0,060±0,012	1,24±0,19	0,620±0,062
Monoolein	0,080±0,016	0,07±0,01	0,020±0,004	3,26±0,05	0,68±0,07
Monolinolein	-	-	-	5,55±0,083	1,35±0,02
Monolinoliat	-	-	-	2,35±0,035	0,17±0,03
Monostearin	0,71±0,14	0,040±0,008	0,060±0,012	-	-
Dilaurin	0,029±0,010	0,01±0,002	-	0,67±0,07	0,56±0,06
Diolein	0,22±0,04	0,21±0,04	0,65±0,13	4,73±0,07	0,74±0,07
Dilinolein	-	-	-	3,37±0,05	1,48±0,02
Dimiristin	0,21±0,4	0,006±0,003	0,020±0,004	0,95±0,09	-
Dipalmitin	0,15±0,03	0,19±0,04	0,07±0,01	1,57±0,02	0,10±0,02
Distearin	0,41±0,08	0,21±0,04	0,15±0,03	8,25±0,12	0,93±0,09
Trilaurin	0,39±0,08	0,24±0,05	0,09±0,02	0,65±0,13	0,31±0,06
Trimiristin	0,07±0,01	0,05±0,01	0,03±0,01	0,99±0,02	-
Tripalmitin	35,75±0,53	34,72±0,52	8,54±0,17	6,90±0,10	30,04±0,30
Triolein	1,08±0,10	-	-	0,40±0,08	8,70±0,13
Tristearin	-	9,80±0,15	-	0,37±0,07	0,92±0,10
Trilinolein	-	-	49,11±0,49	0,19±0,04	14,41±0,22
Tripalmitolein*	32,90±0,49	29,1±0,44	8,58±0,17	1,46±0,22	18,84±0,28

* Anticipated triglyceride.

Table 22.4: Fatty Acid Composition of Seabuckthorn Oil Samples of different Manufacturers

Acid, per cent	*Seabuckthorn oil (100 per cent)*					
	Yagodnoye LLC	*Yantarnoye LLC*	*Altaivitamins ZAO*	*Alsu LLC*	*Altaiskiy Sad LLC*	*Sava LLC*
Unsaturated fattyt acids						
Myristic (C14:0)	0.57±0.11	0.76±0.15	0.87±0.17	0.32±0.05	0.41±0.06	0.38±0.06
Palmitic (C16:0)	36.11±0.54	34.52±0.52	36.65±0.55	24.51±0.37	31.15±0.47	34.83±0.52
Stearic (C18:0)	1.11±0.11	1.24±0.12	1.27±0.13	2.34±0.23	1.46±0.15	1.29±0.13
Arachidic (C20:0)	0.25±0.05	0.42±0.08	0.34±0.07	0.29±0.06	0.83±0.17	0.20±0.04
Behenic (C22:0)	0.23±0.05	0.61±0.12	0.17±0.03	0.33±0.07	0.57±0.11	0.35±0.07
Total	38.27	37.55	39.3	27.79	34.42	37.05
Monounsaturated fatty acids						
Palmitoleic (C16:1)	35.98±0.54	34.57±0.52	34.45±0.52	22.38±0.33	30.83±0.46	35.33±0.53
Oleic (C18:1)	4.34±0.22	3.04±0.15	5.78±0.29	13.21±0.26	12.93±0.26	4.19±0.21
Vaccenic (C18:1)	5.64±0.28	6.81±0.34	5.76±0.29	4.74±0.24	5.19±0.26	7.56±0.38
Total	45.96	44.42	45.99	40.33	48.95	47.08
Polyunsaturated fatty acids						
Linoleic (C18:2)	11.54±0.23	12.65±0.25	12.12±0.24	27.89±0.42	13.80±0.28	11.78±0.24
Linolenic (C18:3)	0.82±0.16	1.09±0.22	0.57±0.11	1.04±0.10	1.59±0.16	0.96±0.19
Total	12.36	13.74	12.69	28.93	15.39	12.74

Table 22.5: Fatty Acid Composition of Seabuckthorn Oil Samples

Acid	*Seabuckthorn Oil Composition Obtained by, Per cent*		
	Extraction	*Centrifugal Sedimentation*	*Enzymic Hydrolysis*
Saturated fatty acids			
Myristic (C14:0)	0.87 ±0.17	0.38 ±0.06	0.43 ±0.06
Palmitic (C16:0)	36.65 ±0.50	34.83 ±0.50	23.34 ±0.40
Stearic (C18:0)	1.27 ±0.12	1.29 ±0.12	2.05 ±0.20
Arachidic (C20:0)	0.34 ±0.06	0.70 ±0.15	0.72 ±0.15
Behenic (C22:0)	0.17 ±0.03	0.35 ±0.04	0.77 ±0.15
Total	39.30	37.55	27.31
Monounsaturated fatty acids			
Palmitoelic (C16:1)	34.45 ±0.50	35.33 ±0.50	19.84 ±0.4
Oleic (C18:1)	5.78 ±0.30	4.19 ±0.30	17.21 ±0.4
Vaccenic (C18:1)	5.76 ±0.30	7.56 ±0.30	7.60 ±0.30
Total	45.99	47.08	44.65
Polyunsaturated fatty acids			
Linoleic (C18:2)	12.12 ±0.20	11.78 ±0.20	24.03 ±0.40
Linoleneic (C18:3)	0.57 ±0.10	0.96 ±0.15	1.76 ±0.18
Total	12.69	12.74	25.79

Table 22.6: Fatty Acid Composition of Seabuckthorn Oil Samples

Acid, Per cent	*Seabuckthorn Oil Composition Separated from, Per cent*				
	Fruit Coat	*Press Cake*	*Seeds*	*Buds*	*Leaves*
Saturated fatty acid					
Myristic (C14:0)	0.65±0.13	0.87±0.17	0.2±0.04	0.53±0.01	0.05±0.01
Palmitic (C16:0)	35.0±0.53	36.65±0.55	10.80±0.22	9.61±0.19	20.83±0.31
Stearic (C18:0)	1.52±0.15	1.27±0.13	2.70±0.14	2.92±0.15	2.18±0.11
Arachidic (C20:0)	0.06±0.012	0.34±0.068	0.23±0.05	0.45±0.09	0.17±0.03
Behenic (C22:0)	0.52±0.10	0.17±0.034	1.10±0.11	13.61±0.27	1.35±0.14
Lauric (C12:0)	0.03±.01	0.03±0.01	0.02±0.01	0.19±0.04	0.05±0.01
Ligniceric (C24:0)	0.31±0.06	1.11±0.11	0.37±0.07	3.89±0.19	0.32±0.06
Total	38.09	40.44	15.42	31.2	24.95
Monounsaturated fatty acids					
Palmitoelic (C16:1)	31.66±0.47	34.45±0.51	2.58±0.13	0.50±0.10	23.42±0.35
Oleic (C18:1)	4.25±0.21	5.78±0.29	12.04±0.24	3.75±0.19	11.33±0.23
Vaccenic (C18:1)	6.15±0.31	5.76±0.29	1.98±0.01	3.95±0.20	4.66±0.23
Total	39.06	45.99	16.6	8.20	39.41
Polyunsaturated fatty acids					
Linoleic (C18:2)	13.50±0.27	12.12±0.24	35.73±0.54	12.17±0.24	25.57±0.38
Linoleneic (C18:3)	1.08±0.11	0.87±0.17	25.71±0.39	12.87±0.26	3.55±0.18
Total	14.58	12.99	61.44	25.04	29.12

sample, but dimiristin in LLC Yagodnoye. Dipalmitin, distearin, trimiristin, trilinolein, tristearin were not found in all the samples. It is assumed the differences are caused by the fact that seabuckthorn oil triglycerides occur both in their pure form, *i.e.*, triglyceride consists of one fat acid as in the standards presented for the identification and in the mixed form *i.e.* triglyceride contains radicals of different fat acids. Nowadays identification of such triglycerides is difficult due to lack of standards.

It can be assumed that at 29th minute tripalmitolein peak appears and fat acid analysis by gas chromatography was performed to confirm it.

According to the data presented fat acid composition of seabuckthorn oil samples of different manufacturers differs. Palmitic acid from 24 per cent to 36 per cent and palmitolic acid from 22 per cent to 36 per cent make the main portion and it confirms our assumption about the significant amount of tripalmitolein presented in the seabuckthorn oil.

Comparative analysis of quantitative content data of triglycerides (tripaltimitin and tripalmitolein) and proper fat acids (Palmitic and Palmitoelic) are shown in Table 22.7.

According to the Table 22.7 the data are in agreement but there are some significant differences. First, tripalmitin percentage is higher than palmitic acid, although palmitic acid percentage in the form of mono- and diglycerides was not taken into account. Second, data on tripalmitolein in Yagodnoye LLC and Altai vitamins ZAO samples are lower than palmitolic acid. It can also be explained due to palmitolic acid presented in the form of other triglycerides. At the same time tripalmitolein percentage in Alsu LLC sample is higher than palmitoelic acid.

Separating Method Effect

According to the data from Table 22.2 glyceride composition of seabuckthorn oil obtained in different ways differs. High percentage of tripalmitin from 38 to 44 per cent and tripalmitolein from 31 to 41 per cent is typical for all the samples. The highest dilaurin and distearin percentage is in the extracted oil, it is lower in the centrifuged oil and negligible amount of them is in fermented oil. Trilaurin percentage in extracted and fermented oils is almost the same (0.7 and 0.6 per cent), whereas it is lower (0,2 per cent) in centrifuged one. All the samples are close to each other in triolein percentage. Trilinolein percentage in extracted and centrifuged oils differs (a hundredfold). Monostearin and trimiristin were found only in centrifuged oil, but tristearin was found in extracted and fermented oils. A lot of peaks of unknown identity were found in the chromatograms, they being more in extracted (11.0 per cent) and centrifuged oils (9,0 per cent) than in fermented one (3,8 per cent).

As it follows from the data presented fatty acid percentage of the seabuckthorn oil of different separating methods differs. Palmitic acid content changed from 23 per cent to 36 per cent and Palmitoelic acid from 19 per cent to 35 per cent make the main percentage that confirms the assumption about significant amount of Tripalmitolein to be found in the seabuckthorn oil. Palmitoelic acid percentage is the lowest in the fermented oil in comparison with extracted and centrifuged one, but oleic and linoleic acid percentage is the highest.

Comparative data analysis of numerical triglyceride composition - tripalmitin and tripalmitolein and fatty acids composition corresponding them is shown in Table 22.8.

Thus, a separating method affects the structure and composition of triglycerides of the seabuckthorn oil.

Feed Material Effect

As it follows from the data presented in Table 22.3 glyceride composition of the seabuckthorn oil produced from different parts of the plant really differs. High tripalmitin percentage from 30 to 35 per cent is typical for the oil obtained from coat (skin), press cake and leaves but in the oil from seeds and buds tripalmitin percentahe is significantly lower from 6.9 to 3.8 per cent. Seabuckthorn seed oil differs from the other samples with high trilinolein percentage 49 per cent. Monolaurin was found in no one of the samples. As for monoglyceride percentage the analyzed samples conform to each other (monopalmitin from 0.02 to 2.2 per cent, monopalmitin from 0.04 to 1.24 per cent), but seabuckthorn seed oil sample contains monoolein up

Table 22.7: Triglyceride and Fatty Acid Percentage in Seabuckthorn Oil Samples of Different Manufacturers

Glyceride or Acid, Per cent	*Seabuckthorn Oil (100 per cent)*					
	Yagodnoye LLC	*Yantarnoye LLC*	*Altaivitamins ZAO*	*Alsu LLC*	*Altaiskiy sad LLC*	*Sava LLC*
Tripalmitin	38.47	43.56	38.16	32.47	33.39	44.70
Palmitic	36.11	34.52	36.65	24.51	31.15	34.83
Tripalmitolein	32.34	35.14	31.28	28.63	29.06	36.67
Palmitoelic	35.98	34.57	34.45	22.38	30.83	35.33

Table 22.8: Triglyceride and Fatty Acid Percentage in Seabuckthorn Oil Samples Separated in Different Methods

Substance	*Seabuckthorn Oil Composition Produced by, Per cent*		
	Extraction	*Centrifugal Separation*	*Enzymatic Hydrolysis*
Tripalmitin	38.16 ±0.50	44.70±0.50	44.37 ±0.50
Palmitic acid	36.65±0.50	34.83±0.50	23.34±0.50
Tripalmitolein	31.28 ±0.50	36.67 ±0.50	41.06 ±0.50
Palmitoelic	34.45±0.50	35.33±0.50	19.84±0.50

Table 22.9: Triglyceride and Fatty Acid Percentage

Glyceride or Acid, Per cent	Seabuckthorn Oil (100 per cent)				
	Coat (Skin)	Press Cake	Seeds	Buds	Leaves
Tripalmitin	35.75±0.53	34.72±0.52	8.54±0.17	6.90±0.104	20.04±0.030
Palmitic	35.00±0.53	36.65±0.55	10.80±0.22	9.61±0.19	20.83±0.31
Tripalmitolein	32.90±0.49	29.10±0.44	8.58±0.17	1.46±0.022	18.84±0.28
Palmitoelic	31.66±0.47	34.45±0.51	2.58±0.13	0.50±0.10	23.42±0.35

to 3 per cent. Monolinolein and monolinoleat are found only in oils from buds and leaves but monostearin only in the oil from coat (skin), press cake and seeds. As for diglyceride percentage the analyzed samples also conform to each other but bud oil differs from the other samples with higher percentage of dimiristin up to 0.95 per cent, dipalmitin 1.57 per cent and distearin 8.25 per cent.

As it follows from the data presented fatty acid composition of the seabuckthorn oil from different parts of the plant differs. Palmic acid (from 21 per cent to 36 per cent) and palmitolic acid (from 23 per cent to 34 per cent) make the main portion in the oil obtained from coat (skin), press cake and leaves that confirms our assumption about significant amount of tripalmitolein present in the seabuckthorn.Bud oil sample distinguishes by higher percentage of unsaturated fat acidin comparison with other samples: behenic acid up to 13 per cent, lauric acid up to 0.2 per cent, lignoceric acid up to 4 per cent, but along with it there is very low percentage of palmitolic acid 0.5 per cent. Seed oil sample distinguishes with its percentage of linoleic acid up to 35 per cent.

Comparative analysis of triglycerides (tripalmitin and tripalmitolein) and their corresponding acids (palmitic and palmitolic) was performed and the data are shown in Table 22.9.

According to the data in Table 22.9 they are in agreement with each other but there are some significant differences.

First, tripalmitin and palmitic acid percentages in the seabuckthorn samples obtained from coat (skin), press cake and leaves are close in values. Palmitic acid percentage in bud and seed oil is higher than tripalmitin. Occurrence of palmitic acid in the form of mono- and di- glyceride can explain the fact.

Second, tripalmitolein and palmitolic acid data in samples of seabuckthorn coat (skin) oil are also close in values. But percentage of palmitolic acid is higher than that of tripalmitolein in the seabuckthorn oil samples obtained from press cake and leaves that is explained with palmic acid that occurs in the form of mono- and di-glycerides. When comparing the data some differences on triglycerides and corresponding fat acids percentage can be explained with some troubles to divide and identify peaks of "pure" and "mixed" triglycerides at small concentrations.

CONCLUSION

A possibility to determine glyceride composition of the seabuckthorn oil by gas chromatography is proved. The data obtained demonstrate different glyceride percentage including the most important bioactive compounds that are polyunsaturated glycerides in the samples that were produced in different soil and climatic conditions of Altai territory. In this regard it is necessary to develop the requirements for seabuckthorn oil glyceride composition.

The data obtained demonstrate different glyceride composition and specifically the most important bioactive compounds that are polyunsaturated glycerides of the samples that were produced by different methods. In this regard the question becomes the grade of biological activity of the seabuckthorn oil depending on separating method and the necessity for standardization of seabuckthorn oil glyceride composition.

It was demonstrated possibility of identification feed material for separating seabuckthorn oil by high-temperature gas chromatography in reliance on glyceride composition.

REFERENCES

1. Dayneka,V.I. and Daineka, L.A. 2003. Study of triglyceride oils *Pinus sibirica* Du Tour. *Chemistry of Natural Compounds* 2: 126-128.
2. FSP [1]002959/01-301210. Seabuckthorn oil concentrates.
3. Koshelev, Y.A. and Ageeva, L.D. 2004. The Seabuckthorn: Monografiya. Bijsk, SIC Pedagogical University, V.M. Shukshin, 320p.
4. Tsydendambaev, V.D., Kuznetsov, E.I., Pchelkin, V.P. and Vereshchagin, A.G. 2011. Distribution of n-octadecenoic acids in triacylglycerols pericarp ripening fruits of seabuckthorn. *Reports of the Academy of Sciences* 436: 121-124.
5. Vereshchagin, A.G. and Tsydendambaev, V.D. 2010. Effect of geographic origin of seabuckthorn on the triacylglycerol composition of its fruits. *Bull. Georg. Natl. Acad. Sci.* 4: 92-100.

Chapter 23

Metabolic Profiles of Berry Quality in Seabuckthorn (*Hippophae* L.) using Untargeted GC-MS Approach

Cheng-Jiang Ruan*, Hai Guo, J Wei Li, ian Ding, He Li, Li Wang, Gui-Jiao Ding and Yue Zhao

Institute of Plant Resources, Dalian Nationalities University, Dalian City, 116600 Liaoning, China
(*E-mail: ruan@dlnu.edu.cn)

SUMMARY

Seabuckthorn (*Hippophae* L.) is a woody perennial shrub or small tree whose berries contain rich bioactive compounds with powerful nutritional and medicinal properties. Untargeted GC-MS metabolic profiling and multivariate analyses were first used to characterize the differences in metabolites between seabuckthorn tissues (pulp *vs.* seeds) as well as between cultivars. Bioactive compounds such as some saturated fatty acids and unsaturated fatty acids could be used to clearly separate pulp and seeds. Cultivars could be clearly separated into two groups using principal component analysis based on untargeted GC-MS metabolic profiling of bioactive compounds in the pulp and seeds. Several metabolic compounds could serve as biomarkers to predict the quality of pulp and seeds and to classify germplasm collections. This dataset provides useful information concerning the metabolic mechanisms underlying quality development of pulp and seeds and contributes towards improving the specialized utilization of these resources and the breeding efficiency of seabuckthorn with an emphasis on quality improvement.

***Keywords**: Hippophae L., Berry quality, GC-MS profiling, Metabolic profiling and Multivariate analysis.*

INTRODUCTION

Seabuckthorn (*Hippophae* L., *Elaeagnaceae* family) is a deciduous woody perennial shrub or small tree. Its berries contain rich bioactive compounds, including unsaturated fatty acids (especially omega-3, omega-6 and omega-7) and saturated fatty acids, vitamin C and E, carotenoids, ursolic acid, polyphenols and phytosterols (Wannes *et al.*, 2010; Bal *et al.*, 2011; Chaman *et al.*, 2011; Zheng *et al.*, 2011). These compounds have powerful nutritional and medicinal properties (Grey *et al.*, 2010; Mathew *et al.*, 2011), and are to an increasing extent being used in food, health products and plant-based medicine in the EU, USA, China, Russia, India and other countries. For example, seabuckthorn oils from seeds and berry pulp have a significant effect on the treatment of dry eyes and mucosa (Larmo *et al.*, 2010), have anti-tumor effects (Yasukawa *et al.*, 2009), reduce hypertensive stress for ventricular microvessels (Koyama *et al.*, 2009) and decrease skin melanin content (Khan *et al.*, 2011).

To utilize and improve seabuckthorn berry quality for human nutrition and health, targeted bioactive compounds and their influencing factors have been studied using different targeted methods, such as high performance liquid chromatography (HPLC) (Andersson *et al.*, 2008; Andersson *et al.*, 2009; Heinaaho *et al.*, 2009), gas chromatography (GC) (Ranjith *et al.*, 2006), liquid chromatography/GC-mass spectrometry (LC-MS, GC-MS) (Yang *et al.*, 2001; Gutzeit *et al.*, 2008; Socaciu, 2011; Giuffrida *et al.*, 2012) and nuclear magnetic resonance (NMR) (Rosch *et al.*, 2004; Deleanu *et al.*, 2009). Pheophytin *a*, which steadily decreased during berry ripening, acted as a marker of the degree of ripening of seabuckthorn berries and was used to convert harvest date into an estimated ripening time (Andersson *et al.*,

Seabuckthorn (Hippophae L.): A Multipurpose Wonder Plant, Vol. 5: Advanced Technologies for Cultivation, Processing, Health Protection and Environmental Conservation (Virendra Singh, *Editor-in-Chief*, 2018), p. 181–192, Astral International Pvt. Ltd., New Delhi, India.

2009). The structures of four flavonol glycoside and oligomeric proanthocyanidin fractions isolated from a seabuckthorn pomace extract were elucidated using NMR (Rosch *et al.*, 2004). In addition, the structure of ethyl glucose in seabuckthorn berries was identified using the combination of HPLC, GC, MS, and NMR (Tiitinen *et al.*, 2006). Recently, untargeted NMR has recently been used in seabuckthorn to study metabolic differences between berry pulp and seeds and to identify potential biomarkers related to bioactive compounds (Li *et al.*, 2013). In addition, the combination of untargeted FTIR (Fourier transform infrared spectrometry) and targeted HPLC-PDA (photodiode array) have also been used to discriminate genotypes (Socaciu, 2011).

The targeted GC-MS has been used to analyze inositols and methylinositols (Yang *et al.*, 2011), phenolic acids (Zadernowski *et al.*, 2005) in the seabuckthorn berries. Dulf (2012) provides useful targeted GC-MS analysis of fatty acid compositions across a number of varietals of seabuckthorn berry subspecies (ssp. *carpatica*) cultivated in Romania, the results showed that the dominating fatty acids in pulp/peel and whole berry oils were palmitic acid (23-40 per cent), oleic (20-53 per cent) and palmitoleic (11-27 per cent); in contrast to the pulp oils, seed oils had higher amount of polyunsaturated fatty acids (65-72 per cent). Fatima *et al.* (2012) used targeted GC-MS analysis to investigate the fatty acid composition of total lipids in four seabuckthorn cultivars and during berry development stages for one seabuckthorn cultivar. The GC-MS profiling of fatty acids in seeds and pulp of berries indicated that the seed oil contained linoleic and alpha linolenic acids at 33–36 per cent and 30–36 per cent, respectively, while the pulp oil contained palmitoleic acid at 32–42 per cent. However, there is no report on metabolic profiling with the purpose of analyzing quality of seabuckthorn berry pulp and seeds using an untargeted GC-MS approach, or even differences between metabolites and correlations within and between pulps and seeds.

Plant metabolomics, based on LC-MS, GC-MS and NMR spectrometry, is an analytical approach that focuses on generating the least biased and most comprehensive qualitative and quantitative overview for the metabolites present in a tissue, organ or whole plant (Deborde *et al.*, 2009; Moco *et al.*, 2009; Hall, 2011; Ruan and Teixeira da Silva, 2011). Multivariate or pattern recognition techniques such as the well-described principal component analysis (PCA) and partial least square (PLS) are important tools for the metabolic analyses (Issaq *et al.*, 2009; Maloney, 2004; Choi *et al.*, 2005; Pereira *et al.*, 2005; Roessner *et al.*, 2001; Wold, 1975; Erikson *et al.*, 2001; Barker and Rayens, 2003; Steinfath *et al.*, 2010). Based on the combination of NMR-based metabolic analyses of the biochemical contents in three contrasting grape cultivars growing in five different locations of the Bordeaux region (France) with PCA and PLS analyses, Pereira *et al.* (2006) found that the compositional differences were more closely associated with vintage than with soil type.

LC-MS- and GC-MS-based metabolomics offers a powerful approach for (i) ascertaining the classification of genotypes (Tikunov *et al.*, 2005), (ii) characterizing metabolic mechanisms of quality development and the importance of environmental conditions (spatial/seasonal/soil type, *etc.*) versus genotype for some desirable quality traits (Carrari *et al.*, 2006; Wolfender *et al.*, 2010; Zhang *et al.*, 2011) and (iii) identifying potential biomarkers for predicting desirable traits (Steinfath *et al.*, 2010; Zhang *et al.*, 2012). Based on GC-MS analyses, Schauer *et al.* (2006) revealed a significant correlation between harvest index and metabolite content in tomato. Based on a combination of untargeted GC-TOF-MS analyses for a large amount of metabolites of potato tubers and PLS analyses with machine learning methods, Steinfath *et al.* (2010) established a method to select a subset of metabolites that correlate with the quality parameters of a given trait.

The purpose of the present study was, therefore, to investigate the difference in metabolic composition between berry tissues (pulp *vs.* seeds) and cultivars. The metabolites from pulp and seeds samples were determined by untargeted GC-MS analyses. All the data underwent multivariate analyses to (i) identify the differences in metabolites between berry tissues (pulp *vs.* seeds) and cultivars and (ii) to identify possible biomarkers for distinguishing the quality of various berry tissues and/or cultivars and for predicting berry quality traits, when combined with the loading plot. We also discuss the potential application of metabolic profiles in special industrial processes, pre-breeding and breeding, and in the identification of seabuckthorn germplasm collections.

MATERIALS AND METHODS

Plant Cultivars and Phenotype Description

Eight seabuckthorn cultivars (Table 23.1, Figure 23.1) growing in the same field were selected from seabuckthorn germplasm collections at the Institute of Fuxing Shaji. This seabuckthorn orchard is located in Fuxing City (E121°012–122°562, N41°412–42°562), Liaoning Province, China, with a mean annual rainfall of 539 mm and an annual evaporation of 1800 mm. Mean annual temperature is 7.5°C and the weather falls into the category of north-temperature continental monsoonal climate (Ruan *et al.*, 2010). These eight cultivars, which are now being planted widely in China, have unique characters such as resistance to dried shrink disease (Ruan *et al.*, 2010), rapid growth, high berry quality and good taste.

Five different individual plants were randomly selected from cutting-derived seedlings of each cultivar and used for sampling. The harvest dates (Table 23.1) for different cultivars were estimated based on our personal

Table 23.1: Seabuckthorn (*Hippophae rhamnoides* L.) Cultivars Used in this Study

Sl.No.	*Cultivars*	*Subspecies*	*Harvest Date*	*Traits*	*Reference*
1	Siberia hongyun	ssp. *mongolica*	July 30, 2010	Bright red berry with a mean weight per berry of 0.37 g, 2.3 per cent sugar, 1.2 per cent acid, 86.3 mg/100 g vitamin C, 27 mg/100 g total carotenoid. Berries mature in middle-July and can be preserved *in loco* for about 20 days on the tree.	In this study
2	Wucifeng	ssp. *mongolica*	July 30, 2010	Length of the berry stalk is about 4.0-5.0 mm, with a mean weight per berry of 0.6 g. Berry pulp: 7.72 per cent sugar, 1.25 per cent acid, 0.61 mg/100 g total vitamin E, 8.42 mg/100 g dippa-carotene. Seeds: 9.48 mg/100 g total vitamin E, 26.39 mg/100 g dippa-carotene, 7.92 mg/100 g total flavonoids. Berries mature in early-August and can be preserved *in loco* for about 30 days on the tree.	Huang and Zhao, 2004; in this study
3	Zhuangyuanhuang	ssp. *mongolica*	July 30, 2010	Few thorns. Berry is an orange ball with a mean weight per berry of 0.48 g. Stronger resistance to dried-shrink disease and drought tolerance. Berries mature in late-August and can be preserved *in loco* for about 32 days on the tree.	Han *et al.*, 2008
4	Fangxiang	ssp. *mongolica*	July 30, 2010	Orange berry with a mean weight per berry of 0.62 g, fleshy, tasty. Berries mature in mid-July and can be preserved *in loco* for about 30 days on the tree.	In this study
5	Liusha-1	ssp. *mongolica*	July 30, 2010	Orange berry with a mean weight per berry of 0.63 g. Berries mature in mid-July and can be preserved *in loco* for about 30 days on the tree.	In this study
6	Alexander-12	ssp. *mongolica*	August 25, 2010	Resistance to dried-shrink disease, height about 2.5–4.0 m, obvious trunk, thornless. Orange columnar berry with a mean fruit weight of 0.462 g, berries mature in mid-August and can be preserved *in loco* for about 23 days on the tree.	Li *et al.*, 2009
7	Cross	Hybrid between Qiuyisike (ssp. *mongolica*) and male Chinese Wucixiong (ssp. *sinensis*)	August 25, 2010	Stronger drought tolerance, berry is an orange ball with a mean weight per berry of 0.33 g. Berries mature in August and can be preserved *in loco* for 90 days on the tree until October.	In this study
8	Jiuyuehuang	ssp. *mongolica*	August 25, 2010	Berry with a mean weight of 0.58 g. Berries mature in September and can be preserved *in loco* for about 20 days on the tree.	In this study

Figure 23.1: Ripe Berries of Eight Seabuckthorn Cultivars Used in this Study. 1: Siberia hongyun; 2: Wucifeng; 3: Zhuangyuanhuang; 4: Fangxiang; 5: Liusha-1; 6: Alexander-12; 7: Cross; 8: Jiuyuehuang.

experience, as well as on ripened color and berry taste. On the same harvest date, the berries (Figure 23.1) were manually collected from inside and outside berry-bearing branches east, west, south and north and top, middle and lower parts of each sampling plant (about 10-20 berries per geographic co-ordinate and for each part of the plant). The harvested berries of each cultivar were mixed and immediately quenched in the liquid nitrogen, and then stored at -80°C until use in sample preparation within two months.

Sample Preparation

Frozen berries of each cultivar were manually divided on ice into pulp and seeds before thawing. After grinding pulp and seeds separately to a fine powder using liquid nitrogen with a mortar and pestle, 100 mg powder of each sample was extracted in 1 mL of chloroform for 1 h on ice to extract non-polar compounds. The extracts were centrifuged for 10 min at 10000 rpm (1°C). After centrifugation, the CH_3Cl extraction was sub-phase of liquid, and the sample's fabrics was collected in surface of liquid (upper-phase). Then, the upper-phase was removed, and the sub-phase (CH_3Cl extraction) was transferred into a new 2-mL Eppendorf tube added 0.2 g NaCl. After mixture 10 min, the CH_3Cl extraction was centrifuged for 5 min at 10000 rpm (1°C). The CH_3Cl extraction was transferred into a new 2-mL Eppendorf tube and mixed with 80 µL methyl nonadecanoate in methanol (non-polar internal standard, 0.2 mg/mL). The solvent was evaporated under nitrogen.

An improved method, previously reported by Broeckling *et al.* (2005), was used for derivatization. The sample was resuspended in 0.8 mL chloroform and hydrolyzed by adding 0.5 mL of 1.25 M HCl in MeOH, and incubated for 4 h at 50°C. Then, the mixed solvent was evaporated under nitrogen. The condensate was re-suspended in 30 µL *n*-hexane and derivatized by adding a standard derivatization solution of 30 µL methyl-*N*-(trimethylsilyl) trifluoroacetamide (MSTFA), and incubated for 1 h at 50°C. After this, the mixed solvent was equilibrated at room temperature, and evaporated under nitrogen. Finally, the precipitate was re-suspended in 1000 µL *n*-hexane and filtered through a 0.22-µm PTEE filter, and transferred into an auto-sampler vial for untargeted GC-MS analyses. In our experiments, methyl nonadecanoate was added as an internal standard, and its retention time was 10.24 min in the GC-MS chromatogram. In addition, we selected a control sample (cv.'Wucifeng') at an interval of two samples to assure the accuracy of the measurements in the GC-MS analyses.

Metabolite Profiling by untargeted GC-MS

GC-MS analyses of the derivatizated sample were performed with a SHIMADZU SSI GCMS-QP2010 Plus (Shimadzu Instruments, Japan). Each sample (1 µL) was injected into the gas chromatograph through a 0.25 µm Rtx-5MS column (30 m × 0.25 mm i.d.). It was operated in a constant pressure mode at 67.2 kPa and the injector temperature was 310°C. The column temperature was held at 85°C for 2 min, increased to 300°C and then held for 12 min. Helium (99.999 per cent) was used as the carrier gas with a flow rate of 1 mL/min. The most important MS operating parameters were: ion source temperature=200°C; interface temperature = 230°C. Solvent cut time was 90 sec, and detector gain was 0.74 kV. Total ion current (TIC) spectra were recorded in the mass range of 40-1000 atomic mass units (amu) with a scan speed of 5000, which started at 1.80 min and ended at 15.00 min. Shimadzu GC-MS solution software (version 2.30) was used for data acquisition.

Data Preprocessing

The resulting data matrix from untargeted GC-MS analysis containing the sum intensity of each mass spectral feature, *i.e.*, the selected cluster of the time groups, and the corresponding sample information such as cultivar, pulp and seeds, were converted into Microsoft Excel format (*.xls) and then exported to SIMCA-P software (version 12.0.1; Umetrics AB, Umeå, Sweden), where all data pre-treatment procedures, including noise reduction, baseline correction, and peak alignment, were carried out. Artifact peaks, such as column bleeding, and solvent peaks were removed. The metabolites were identified by matching peaks against an off-line NIST standard reference database, using retention indices and mass spectral similarities, and based on the published literature (Cakir, 2004; Tian *et al.*, 2004; Singh *et al.*, 2005; Zadernowski *et al.*, 2005; Yang *et al.*, 2011).

Principal Component Analysis and Identification of Metabolic Biomarker

The PCA scatter plot of variables was used to create a hierarchical clustering dendrogram using SIMCA-P software, which showed the separation of samples. In our PCA analyses, the metabolites generated 187 peaks in all samples. In the PCA analyses, the relative content of each component was expressed us the relative area of peak of this component. In our analyses, although there were small differences in retention-time of GC-MS spectrum for some compounds, this did not influence our results. The quality of the models in the PCA analyses was described by R^2_X and Q^2 values. R^2_X was dened as the proportion of variance in the data explained by the models and indicates goodness of t, while Q^2 was dened as the proportion of variance in the data predictable by the model and indicates predictability (Singh *et al.*, 2005). A good prediction model is achieved when $Q^2 > 0.5$, and if $Q^2 > 0.9$, it is regarded as excellent predictive ability (Eriksson *et al.*, 2001).

Then, we generated the loading plot using SIMCA-P software. In the loading plot, some of the peaks are important contributors to variance and the metabolites contained in these peaks are generally identified as biomarkers, each of which is responsible for separating samples (Lee *et al.*, 2009).

RESULTS

Metabolite Profiling of Berry Pulp and Seeds by GC-MS

The untargeted GC-MS analyses of non-polar extracts from berry pulp and seeds of eight seabuckthorn cultivars enabled the detection of a large number of compounds including mainly fatty acids. The metabolites in berry pulp and seeds were significantly different. Cluster analysis allowed the berry tissues and cultivars to be distinguished. In addition, important metabolites in these samples that affected cluster formation could be determined. PCA was applied separately. The distinction between berry tissues and cultivars was obvious in the hierarchical clustering dendrogram (Figure 23.2). The first two highest-ranked vectors from PCA analysis represent the major variance among seed samples.

Pulp

for the berry pulp, PCA analyses of GC-MS spectra showed that the PC1/PC2 plot explained 92.56 per cent of total variance, with high predictability as indicated by a Q^2 value of 0.765, and the two clusters were separated by the second axis. Cluster I consisted of two cultivars ('Siberia hongyun', and 'Liusha-1') (Figure 23.2). The pulp of these two cultivars showed higher intensity GC-MS signals for dodecanoic acid, pentadecylenic acid, palmitic acid, palmitoleate, stearic acid, eicosadienoic acid, docosadienoic acid. Cluster II consisted of the remaining six cultivars ('Wucifeng', 'Zhuangyuanhuang', 'Fangxiang', 'Alexander-12', 'Jiuyuehuang' and 'Cross') (Figure 23.2). Based on the loading plot (Figures 23.3A and 23.4A), the pulp of these six cultivars showed a stronger intensity in the GC-MS signals for methyl-cis-9-hexadecenoate, heptadecenoic acid and erucic acid.

Seeds

For the seeds, the PCA analyses of GC-MS spectra showed that the PC1/PC2 plot explained 97.8 per cent of total variance, with high predictability as indicated by a Q^2 value of 0.946, and the two clusters were also separated, mostly by the second axis. Cluster I comprised three cultivars ('Wucifeng', 'Liusha-1' and 'Cross') on the positive side of axis 2, with stronger intensity GC-MS signals for octanoic acid, nonanoic acid, dodecanoic acid, tetradecanoic acid. Cluster II comprised three cultivars ('Siberia hongyun', 'Zhuangyuanhuang', 'Fangxiang', 'Alexander-12', and 'Jiuyuehuang') on the negative side of axis 2, with stronger intensity GC-MS signals for Methyl-cis-9-hexadecenoate, Oleic acid, eicosenoic acid, Behenic acid, Tricosanoic Acid, erucic acid, Tetracosanoic acid (Figure 23.4B).

Pulp and Seeds

The distinction between berry pulp and seeds was obvious in the PCA analysis. The PC1/PC2 plot described 90.5 per cent of the variance in the 186 buckets of GC-MS spectra, with high predictability as indicated by a Q^2 value of 0.841. Berry pulp and seeds were mostly separated by the second axis. Examination of the PC2 loading plot suggested that this separation was due to berry pulp containing stronger signal intensity in the GC-MS signals than seeds for palmitic acid, Palmitoleate, eicosadienoic acid, Behenic acid, erucic acid. In contrast, seeds contained stronger intense GC-MS signals for

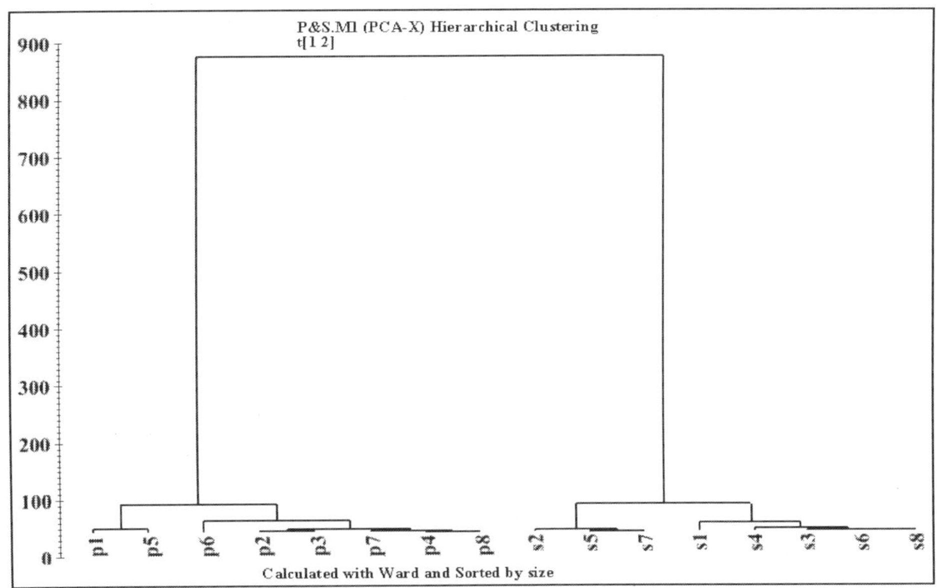

Figure 23.2: Hierarchical Clustering Dendrogram of Berry Pulp Extracts (p1-p8), Seed Extracts (s1-s8) from different Seabuckthorn Cultivars.
1: Siberia hongyun; 2: Wucifeng; 3: Zhuangyuanhuang; 4: Fangxiang; 5: Liusha-1; 6: Alexander-12; 7: Cross; 8: Jiuyuehuang.

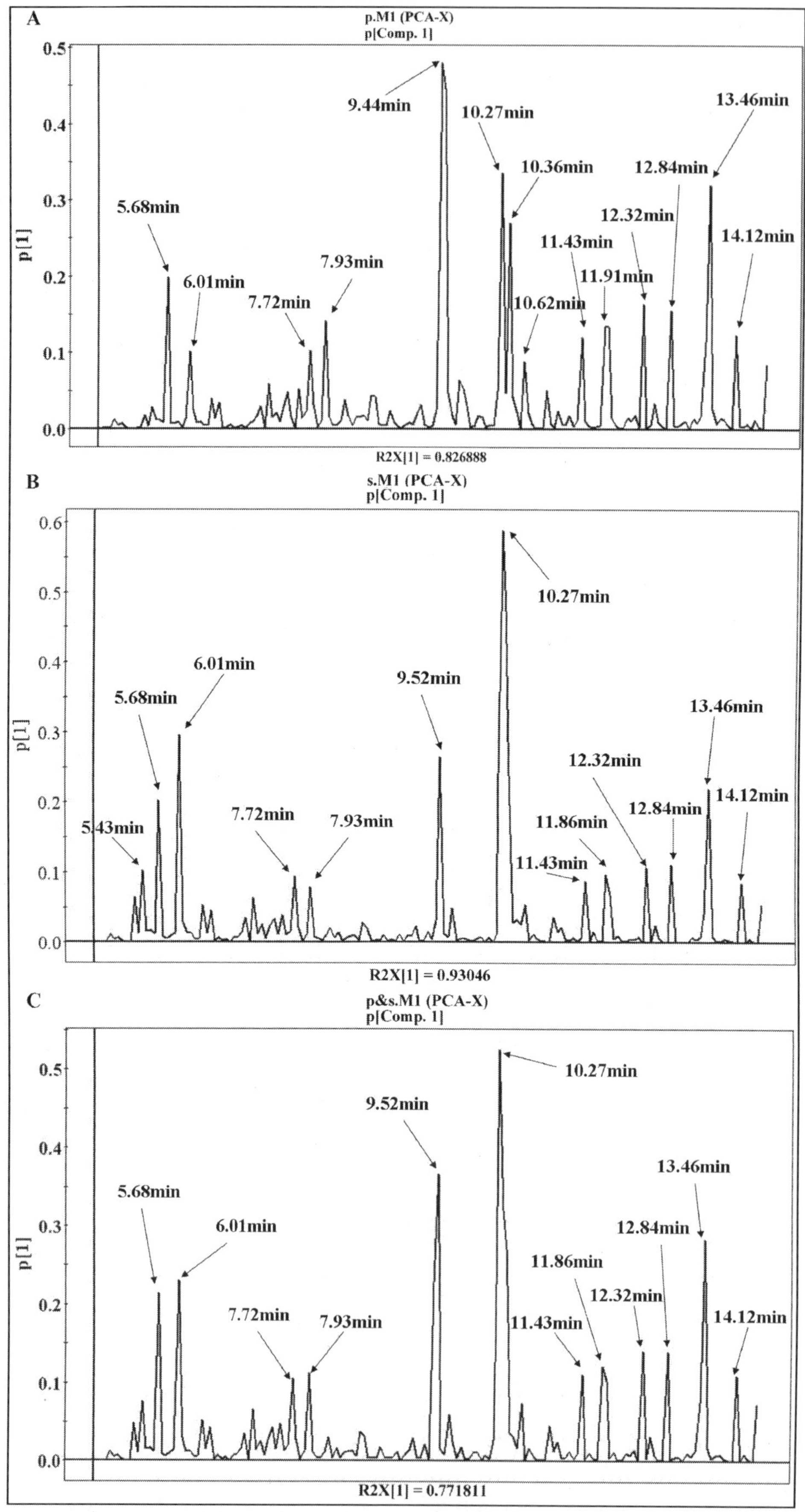

Figure 23.3: The Loading Plots of PC1 in PCA Analyses of the Untargeted GC-MS Analyses for Seabuckthorn Berry Pulp (A), Seed Extracts (B), and their Combination (C).

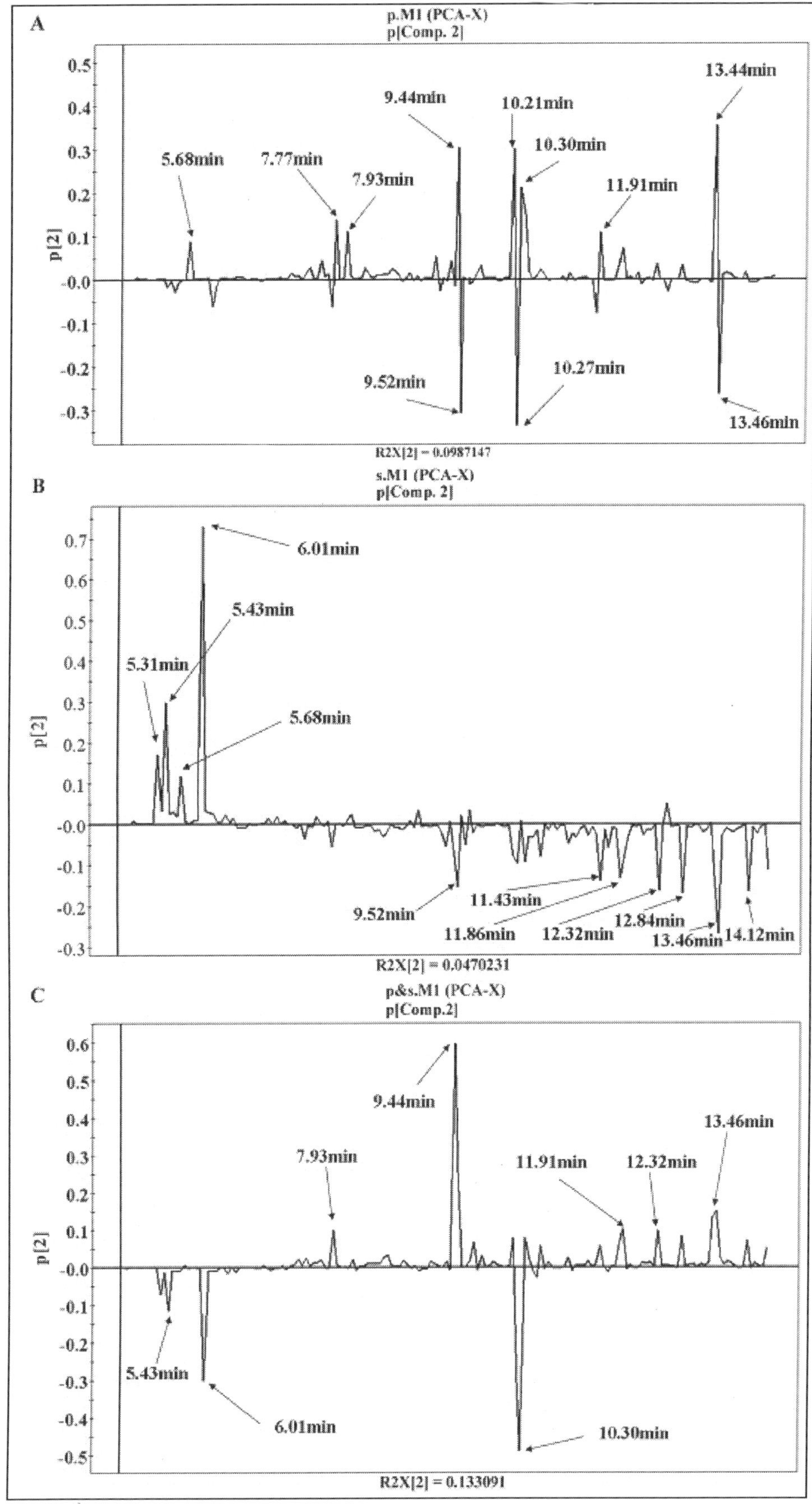

Figure 23.4: The Loading Plots of PC2 in PCA Analyses of the Untargeted GC-MS Analyses for Seabuckthorn Berry Pulp (A), Seed Extracts (B), and their Combination (C).

nonanoic acid, tetradecanoic acid, stearic acid, compared with berry pulp (Figures 23.3C and 23.4C).

Identification of Possible Biomarkers

The loadings plot provides valuable insight into variables that lead to sample clustering and illustrates which metabolites are up- and/or down-regulated. Those variables that were regarded as potential biomarkers for a given trait were represented within a GC-MS metabolite profile by their relative mass fragment intensities at corresponding retention-time windows. These will be referred to as mass features hereafter. The first step of the marker selection procedure was to assign a numerical value which indicates the importance of a mass feature for the prediction of the trait. By the corresponding loadings plot of PCA analyses of the GC-MS spectra for seabuckthorn berry pulp extract (Figure 23.3A), seed extract (Figure 23.3B) and their combination (Figure 23.3C), some metabolites may also be identified as biomarkers for differentiating seabuckthorn cultivars with different quality traits of berry pulp and seeds. These include several special saturated fatty acids and unsaturated fatty acids (Table 23.2), such as tetradecanoic acid, palmitoleate, heptadecenoic acid, docosadienoic acid and erucic acid.

Table 23.2: Compounds Determined to be of Variable Importance in the Multivariate Analysis on the Identified Compounds from the Pulp and Seeds Extract

Sl.No.	*Name*	*Structure*	*Retention Time (min)*
1	Octanoic Acid	C8:0	5.31
2	Nonanoic Acid	C9:0	5.43
3	Dodecanoic Acid	C12:0	5.68
4	Tetradecanoic Acid	C14:0	6.01
5	Pentadecanoic Acid	C15:0	7.72
6	Pentadecylenic Acid	C15:1	7.77
7	Palmitic Acid	C16:0	7.93
8	Palmitoleate	C16:1n7	9.44
9	Palmitoleic Acid	C16:1n-9C	9.52
10	Heptadecanoic Acid	C17:0	10.21
11	Heptadecenoic Acid	C17:1	10.27
12	Stearic Acid	C18:0	10.30
13	Linoleic Acid	C18:2n-6	10.36
14	Methyl DAKUA-Linolenate	C18:3n3	10.62
15	Oleic Acid	C18:1n-9	11.43
16	Eicosenoic Acid	C20:1n-9	11.86
17	Eicosadienoic Acid	C20:2n-9	11.91
18	Behenic Acid	C22:0	12.32
19	Tricosanoic Acid	C23:0	12.84
20	Docosadienoic Acid	C22:2	13.44
21	Erucic Acid	C22:1n-13	13.46
22	Tetracosanoic Acid	C24:0	14.12

DISCUSSION

Several promising functional genomics approaches (Ruan and Teixeira da Silva, 2011), including transcriptomics, proteomics, and metabolomics, are employed in the systematic and comprehensive understanding of complex events of life such as plant quality and resistance formation. Currently, information about the whole genome of seabuckthorn, a woody perennial shrub or small tree, is unavailable, restricting transcriptomic and proteomic studies. Metabolomics is an analytical tool that can function independently of genomic data, identifying sample constituents and the actual biochemical status of the tissues.

Despite the wide applications of untargeted and targeted metabolomics in plant research (*e.g.* targeted analyses for seabuckthorn berry quality and its influencing factors), there are few published reports on the metabolite profiles of seabuckthorn berry pulp and seeds using untargeted approach. Our results show that the combination of untargeted GC-MS spectroscopy with multivariate analysis is a viable tool to: (i) compare the overall metabolite fingerprints of berry pulp and seeds of multiple seabuckthorn cultivars; (ii) rapidly identify differences in bioactive compounds between berry pulp and seeds or cultivars; and (iii) detect possible biomarkers to predict or separate different tissues (pulp *vs.* seeds) and/or cultivars with different quality traits of berry pulp and seeds. Information derived from these analyses would guide industrial utilization of seabuckthorn berry pulp or seeds, and would assist in pre-selection and selection for different aims and traits of cultivars used for their pulp or seeds.

In recent years, seabuckthorn has become an important raw material in health products and cosmetics, especially in China and Russia. This utilization is based on its application in Tibetan, Mongolian, and Chinese traditional medicines for more than one thousand years (Yang *et al.*, 2011). Seabuckthorn oils from the berry pulp and seeds have been shown to lower the risk of cardiovascular and cerebrovascular diseases, regulate immuno-functions, and attenuate inflammation (Ruan *et al.*, 2013). These oils are attracting increasingly more attention because of the increasing interest in the physiological role of monounsaturated fatty acids (Ruan *et al.*, 2013). Of interest, the non-polar extracts of seabuckthorn berries in this study contained most of bioactive components in the seabuckthorn oils (Table 23.1).

As structural components of cell membranes and precursors of other n-6 and n-3 fatty acids and eicosanoids, linoleic acid and α-linolenic acids play an important role in human physiology (Ruan *et al.*, 2013). Even though the contents of these essential fatty acids seem to always be high in berries of all seabuckthorn cultivars (Figure 23.3), the pulp samples seem to have a higher level of palmitoleic acid than the seed samples (Figure 23.4). Palmitoleic acid has fluidity similar to polyunsaturated fatty acids but low susceptibility to lipid peroxidation (Bonanome *et al.*, 1992). An increase in the level of dietary palmitoleic acid has been suggested to improve the metabolism of vascular smooth muscle cells (Fan *et al.*, 2005). Seabuckthorn berry pulp and seeds contain a high level of palmitoleic acid (Figure 23.4), which is not common in the plant kingdom (Yang and Kallio, 2001). By comparing the GC-MS signal intensity of the bioactive compounds in the berry pulp and seeds, the berry pulp of six of eight seabuckthorn cultivars ('Wucifeng', 'Zhuangyuanhuang', 'Fangxiang', 'Alexander-12', 'Cross' and 'Jiuyuehuang') and the seeds of five of eight seabuckthorn cultivars ('Siberia hongyun', 'Zhuangyuanghuang', 'Fangxiang', 'Alexander-12' and 'Jiuyuehuang') had higher contents of palmitoleic acid (Figure 23.4).

The untargeted GC-MS-based metabolomic analyses in this study provided informative details on the quality of seabuckthorn berries (pulp and seed) through simple sample preparation and a short analytical period under identical conditions. This differs from previous studies in which GC-MS was applied to seabuckthorn (Yang *et al.*, 2001; Cakir, 2004; Tian *et al.*, 2004; Singh *et al.*, 2005; Zadernowski *et al.*, 2005; Yang, 2009; Yang *et al.*, 2011), and in which targeted GC-MS spectroscopy was used to identify targeted metabolites and their contents. Based on targeted GC and GC-MS spectroscopy, the composition of the volatile oil isolated from seabuckthorn berries was analyzed, and 30 constituents were identified, representing 94.6 per cent of the oils of berry pulp and seeds (Cakir, 2004). The sugars, acids, sugar derivative and sugar alcohol were identified by comparing the retention times and mass spectra with those of reference compounds, using a targeted GC-MS approach (Yang, 2009). The possible biomarkers identified by a combination of untargeted GC-MS and multivariate analysis (PCA) played important roles for predicting seabuckthorn berry (pulp and seeds) quality, which would allow for the improvement of different industrial process-related aims and breeding.

CONCLUSION

This study provides the use of a GC-MS based metabolomics approach to identify key compounds that are able to discriminate between seed and pulp tissue of seabuckthorn fruit and in combination provide markers for distinguishing between cultivars. The identification of key differential metabolites will be useful for both the identification of key biomarkers for quality attributes of the fruit as well as key markers for genetic improvement initiatives. However, several unidentified metabolites were also detected in untargeted GC-MS spectroscopy. Variable importance plots based on identified metabolites alone could not account for all of the variation among seabuckthorn cultivars, suggesting that these unidentified metabolites may also have been important contributors to variance. The functional validation of possible biomarkers should be further improved through phenotypic data in the field to increase the potential of these metabolic biomarkers to become a readily employed tool to use for seabuckthorn-related industry processes and to advance breeding programs.

ACKNOWLEDGEMENTS

This research was jointly supported by a Marie Curie International Incoming Fellowship within the 7th European Community Framework Programme (PIIF-GA-2010-272048-HIPPOHEALTH, PIIFR-GA-2010-910048-HIPPOHEALTH), and the Technology Foundation for Selected Overseas Chinese Scholar, Ministry of Personnel of China *"Study on breeding seabuckthorn with resistance to dried-shrink disease and selection of new-lines"*.

REFERENCES

1. Andersson, S.C., Olsson, M.E., Johansson, E. and Rumpunen, K. 2009. Carotenoids in seabuckthorn (*Hippophae rhamnoides* L.) berries during ripening and use of pheophytin a as maturity marker. *Journal of Agricultural and Food Chemistry* 57: 250-258.
2. Andersson, S.C., Rumpunen, K., Johansson, E. and Olsson, M.E. 2008. Tocopherols and tocotrienols in seabuckthorn (*Hippophae rhamnoides* L.) berries during ripening. *Journal of Agricultural and Food Chemistry* 56: 6701-6706.
3. Bal, L.M., Meda, V., Naik, S.N. and Satya, S. 2011. Seabuckthorn berries: A potential source of valuable nutrients for nutraceuticals and cosmoceuticals. *Food Research International* 44: 1718-1727.
4. Barker, M. and Rayens, W. 2003. Partial least squares for discrimination. *Journal of Chemometrics* 17: 166-173.
5. Bonanome, A., Pagnan, A., Biffanti, S., Opportuno, A., Sorgato, F., Dorella, M., Maiorino, M. and Ursini, F. 1992. Effect of dietary monounsaturated and polyunsaturated fatty acids on the susceptibility of plasma low density lipoproteins to oxidative modification. *Arterioscler Thromb. Vasc. Biol.*, 12: 529-533.
6. Broeckling, C.D., Huhman, D.V., Farag, M.A., Smith, J.T., May, G.D., Mendes, P., Dixon, R.A. and Sumner, L.W. 2005. Metabolic profiling of *Medicago truncatula* cell cultures reveals the effects of biotic and abiotic elicitors on metabolism. *Journal of Experimental Botany* 56: 323-336.
7. Cakir, A. 2004. Essential oil and fatty acid composition of the fruits of *Hippophae rhamnoides* L. (seabuckthorn) and *Myrtus communis* L. from Turkey. *Biochemical Systematics and Ecology* 32: 809-816.
8. Carrari, F., Baxter, C., Usadel, B., Urbanczyk-Wochniak, E., Zanor, M.I., Nunes-nesi, A.,

Nikiforova, V., Centero, D., Ratzka, A., Pauly, M., Sweetlove, L.J. and Fernie, A.R. 2006. Integrated analysis of metabolite and transcript levels reveals the metabolic shifts that underlie tomato fruit development and highlight regulatory aspects of metabolic network behavior. *Plant Physiology* 142: 1380-1396.

9. Chaman, S., Syed, N., Danish, Z. and Khan, F. 2011. Phytochemical analysis, antioxidant and antibacterial effects of seabuckthorn berries. *Pakistan Journal of Pharmaceutical Sciences* 24: 345-351.

10. Choi, Y.H., Sertic, S., Kim, H.K., Wilson, E.G., Michopoulos, F., Lefeber, A.W.N., Erkelens, C., Kricun, S.D.P. and Verpooret, R. 2005. Classification of *Ilex* species based on metabolomics fingerprinting using nuclear magnetic resonance and multivariate data analysis. *Journal of Agricultural and Food Chemistry* 53: 1237-1245.

11. Deborde, C., Maucourt, M., Baldet, P., Bernillon, S., Biais, B., Talon, G., Ferrand, C., Jacob, D., Helene, F.D., Daruvar, A., Rolin, D. and Moing, A. 2009. Proton NMR quantitative proling for quality assessment of greenhouse-grown tomato fruit. *Metabolomics* 5: 183-198.

12. Deleanu, M., Kiss, C., Nicolescu, A., Socaciu, C. and Deleanu, C. 2009. Characterization of seabuckthorn oil by GC and ^{1}H-NMR. *Bull. UASVM Agric* 66: 559.

13. Dulf, F.V. 2012. Fatty acids in berry lipids of six seabuckthorn (*Hippophae rhamnoides* L. ssp. *carpatica*) cultivars grown in Romania. *Chemistry Central Journal* 6: 106.

14. Eriksson, L., Johansson, E., Kettaneh, N. and Wold, S. 2001. Multi- and megavariate data analysis: Principles and applications. Umetrics AB: Umeå, Sweden, pp. 43-111.

15. Fan, Y.Y., Ramos, K.S. and Chapkin, R.S. 1995. Dietary gamma-linolenic acid modulates macrophage-vascular smooth muscle cell interactions. Evidence for a macrophage derived soluble factor that down regulates DNA synthesis in smooth muscle cells. *Arterioscler Thromb and Vascular Biology* 15: 1397-403.

16. Fatima, T., Snyder, C.L., Schroeder, W.R., Cram, D., Datla, R., Wishart, D., Weelake, R.J. and Krishna, P. 2012. Fatty acid composition of developing seabuckthorn (*Hippophae rhamnoides* L.) berry and the transcriptome of the mature seed. *PloS ONE* 7: e34099.

17. Giuffrida, D., Pintea, A., Dugo, P., Torre, G., Pop, R.M. and Mondello, L. 2012. Determination of carotenoids and their esters in fruits of seabuckthorn (*Hippophae rhamnoides* L.) by HPLC-DAD-APCI-MS. *Phytochemical Analysis* 23: 267-273.

18. Grey, C., Widen, C., Adlercreutz, P., Rumpunen, K. and Duan, R.D. 2010. Antiproliferative effects of seabuckthorn (*Hippophae rhamnoides* L.) extracts on human colon and liver cancer cell lines. *Food Chemistry* 120: 1004-1010.

19. Gutzeit, D., Mönch, S., Jerz, G., Winterhalter, P. and Rychlik, M. 2008. Folate content in seabuckthorn berries and related products (*Hippophae rhamnoides* L. ssp. *rhamnoides*): LC-MS/MS determination of folate vitamin stability influenced by processing and storage assessed by stable isotope dilution assay. *Analytical and Bioanalytical Chemistry* 391: 211-219.

20. Hall, R.D. 2011. Plant metabolomics in a nutshell: Potential and future challenges. Annual Plant Reviews, *Biology of Plant Metabolomics* 43: 1-24.

21. Heinaaho, M., Hegerman, A.E. and Julkunen-tiitto, R. 2009. Effect of different organic farming methods on the phenolic composition of seabuckthorn berries. *Journal of Agricultural and Food Chemistry* 57: 1940-1947.

22. Holmes, E., Loo, R.L., Stamler, J., Bictash, M., Yap, I.K.S., Chan, Q., Ebbels, T., Deiorio, M., Brown, I.J., Veselkov, K.A., Daviglus, M.L., Kesteloot, H., Ueshima, H., Zhao, L., Nicholson, J.K. and Elliott, P. 2008. Human metabolic phenotype diversity and its association with diet and blood pressure. *Nature* 453: 396-400.

23. Issaq, H.J., Van, Q.N., Waybright, T.J., Muschik, G.M. and Veenstra, T.F. 2009. Analytical and statistical approaches to metabolomics research. *Journal of Separation Science* 32: 2183-2199.

24. Khan, B.A., Akhtar, N., Mahmood, T., Shoaib, H.M., Qayum, M. and Saeed, T. 2011. Effects of antioxidants and flavonoids of seabuckthorn on skin whitening and skin erythema. *Asian Journal of Chemistry* 23: 903-906.

25. Koyama, T., Taka, A. and ToGashi, H. 2009. Effects of a herbal medicine *Hippophae rhamnoides* on cardiovascular functions and coronary microvessels in the spontaneously hypertensive stroke-prone rat. *Clinical Hemorheology and Microcirculation* 41: 17-26.

26. Larmo, P.S., Järvinen, R.L., Setälä, N.L., Yang, B., Viitanen, M.H., Engblom, J.R., TahVonen, R.L. and Kallio, H.P. 2010. Oral seabuckthorn oil attenuates tear film osmolarity and symptoms in individuals with dry eye. *Journal of Nutrition* 140: 1462-1468.

27. Lee, E.J., Shaykhutdinov, R., Weljie, A.M., VoGel, H.L. and Facchini, P.J. 2009. Quality assessment of ginseng by ^{1}H NMR metabolite fingerprinting and profiling analysis. *Journal of Agricultural and Food Chemistry* 57: 7513-7522.

28. Li, W., Ruan, C.J., Teixeira da Silva, J.A., Guo, H. and Zhao, C.E. 2013. NMR metabolomics of berry quality in seabuckthorn (*Hippophae* L.). *Molecular Breeding* 31: 57-67.

29. Maloney, V. 2004. Plant metabolomics. *Biology Teacher Journal* 2: 92-99.

30. Mathewa, S., Grey, C., Rumpunen, K. and Adlercreutz, P. 2011. Analysis of carbonyl compounds in seabuckthorn for the evaluation of triglyceride oxidation, by enzymatic hydrolysis and derivatization methodology. *Food Chemistry* 126: 1399-1405.

31. Moco, S., Schneider, B. and Vervoort, J. 2009. Plant micrometabolomics: the analysis of endogenous metabolites present in a plant cell or tissue. *Journal of Combinatorial Chemistry* 8: 1694-1703.

32. Pereira, G.E., Gaudillere, J.P., Van Leeuwen, C., Hilbert, G., Lavialle, O., Maucourt, M., Deborde, C., Moing, A. and Rolin, D. 2005. HNMR and chemometrics to characterize mature grape berries in four wine-growing areas in Bordeaux, France. *Journal of Agricultural and Food Chemistry* 53: 6382-6389.

33. Pereira, G.E., Gaudillere, J.P., Van Leeuwen, C., Hilbert, G., Maucourt, M., Deborde, C., Moing, A. and Rolin, D. 2006. ^{1}H NMR metabolite fingerprints of grape berry: Comparison of vintage and soil effects in Bordeaux grapevine growing areas. *Analytica Chimica Acta* 563: 346-352.

34. Ranjith, A., Kumar, K.S., Venugopalan, V.V., Arumugan, C., Sawhney, R.C. and Singh, V. 2006. Fatty acids, tocols, and carotenoids in pulp oil of three seabuckthorn species (*Hippophae rhamnoides, H. salicifolia,* and *H. tibetana*) grown in the Indian Himalayas. *Journal of the American Oil Chemists Society* 83: 359-364.

35. Roessner, U., Luedemann, A., Brust, D., FiEhN, O., Linke, T., WillMitzer, L. and Fernie, A.R. 2011. Metabolic profiling allows comprehensive phenotyping of genetically or environmentally modified plant systems. *Plant Cell* 13: 11-29.

36. Rosch, D., Krumbein, A., Mugge, C. and Kroh, L.W. 2004. Structural investigations of flavonol glycosides from seabuckthorn (*Hippophae rhamnoides*) pomace by NMR spectroscopy and HPLC-ESI-MS. *Journal of Agricultural and Food Chemistry* 52: 4039-4046.

37. Ruan, C.J., Rumpunen, K. and Nybom, H. 2013. Advances in biotechnological breeding of quality and resistance for multi-purpose seabuckthorn. *Critical Reviews in Biotechnology* 33: 126-144.

38. Ruan, C.J. and Teixeira da Silva, J.A. 2011. Metabolomics: Creating new potentials for unraveling the mechanisms in response to salt and drought stress and for the biotechnological improvement of xero-halophytes. *Critical Reviews in Biotechnology* 31: 153-169.

39. Ruan, C.J., Teixeira da Silva, J.A., Li, Q., Li, H. and Zhang, J. 2010. Pathogenicity of dried shrink disease and evaluation of resistance in a germplasm collection of seabuckthorn (*Hippophae* L.) from China and other countries. *Scientia Horticulturae* 127: 70-78.

40. Schauer, N., Bruedigam, C., Kopka, J., Willmitzer, L., Zamir, D. and Fernie, A.R. 2006. Comprehensive metabolic profiling and phenotyping of interspecific introgression lines for tomato improvement. *Nature Biotechnology* 24: 447-454.

41. Singh, A.K., Nivsarkar, M., Acharya, B.N. and Kaushik, M.P. 2005. Isolation and identification of fatty acids from berries of seabuckthorn (*Hippophae rhamnoides*). *Indian Journal of Chemistry* 44B: 2390-2392.

42. Socaciu, C. 2011. Metabolomics studies applied on different seabuckthorn (*Hippophae rhamnoides* L.) varieties. *Ph.D. Thesis*, University of Agricultural Sciences and Veterinary Medicine, Romania, pp. 20-60.

43. Steinfath, M., Strehmel, N., Peter, R., Schauer, N., Groth, D. and Hummel, J. 2010. Discovering plant metabolic biomarkers for phenotype prediction using an untargeted approach. *Plant Biotechnology Journal* 8: 900-911.

44. Tian, C.J., Nan, P., Chen, J.K. and Zhong, Y. 2004. Volatile composition of Chinese *Hippophae rhamnoides* and its chemotaxonomic implications. *Biochemical Systematics and Ecology* 32: 431-441.

45. Tiitinen, K.M., Yang, B.R., Haraldsson, G. G., Jonsdottir, S. and Kallio, H.P. 2006. Fast analysis of sugars, fruit acids, and vitamin C in seabuckthorn (*Hippophae rhamnoides* L.) varieties. *Journal of Agricultural and Food Chemistry* 54: 2508-2513.

46. Tikunov, Y., Lommen, A., De Vos, C.H.R., Verhoeven, H.A, Bino, R.J., Hall, R.D. and Bovy, A. G. 2005. A novel approach for non targeted data analysis for metabolomics. Large-scale profiling of tomato fruit volatiles. *Plant Physiology* 139: 1125-1137.

47. Wannes, W.A., Mhamdi, B., Sriti J. and Marzouk, B. 2010. Glycerolipid and fatty acid distribution in pericarp, seed and whole fruit oils of *Myrtus communis* var. *italica*. *Industrial Crops and Products* 31: 77-83.

48. Wold, H. 1975. Soft Modelling by Latent Variables. *London, Academic Press.*

49. Wolfender, J.L., Marti, G. and Queiroz, E.F. 2010. Advances in techniques for profiling crude extracts and for the rapid identification of natural products: De-replication, quality control and metabolomics. *Current Organic Chemistry* 14: 1808-1832.

50. Yang, B.R. 2009. Sugars, acids, ethyl β-D-glucopyranose and a methyl inositol in seabuckthorn (*Hippophae rhamnoides*) berries. *Food Chemistry* 112: 89-97.

51. Yang, B.R. and Kallio, H. 2001. Fatty acid composition of lipids in seabuckthorn (*Hippophae rhamnoides* L.) berries of different origins. *Journal of Agricultural and Food Chemistry* 49: 1939-1947.

52. Yang, B.R., Karlsson, R.M., Oksman P.H. and Kallio, H.P. 2001. Phytosterols in seabuckthorn (*Hippophae rhamnoides* L.) berries: Identification and effects of different origins and harvesting times. *Journal of Agricultural and Food Chemistry* 49: 5620-5629.

53. Yang, B.R., Zheng, J. and Kallio, H. 2011. Influence of origin, harvesting time and weather conditions on content of inositols and methylinositols in seabuckthorn (*Hippophae rhamnoides*) berries. *Food Chemistry* 125: 388-396.

54. Yasukawa, K., Kitanaka, S., Kawata, K. and Goto, K. 2009. Anti-tumor promoters phenolics and triterpenoid from *Hippophae rhamnoides*. *Fitoterapia* 80: 164-167.

55. Zadernowski, R., Naczk, M., Czaplicki, S., Rubinskiene, M. and Szlakiewicz, M. 2005. Composition of phenolic acids in seabuckthorn (*Hippophae rhamnoides* L.) berries. *Journal of the American Oil Chemists Society* 82: 175-179.

56. Zhang, H.M., Li, S.L., Zhang, H., Wang, Y., Zhao, Z.L., Chen, S.L. and Xu, H.X. 2012. Holistic quality evaluation of commercial white and red ginseng using a UPLC-QTOF-MS/MS-based metabolomics approach. *Journal of Pharmaceutical and Biomedical analysis* 62: 258-273.

57. Zhang, J.J., Wang, X., Yu, O., Tang, J.J., Gu, X.G., Wan, X.C. and Fang, C.B. 2011. Metabolic profiling of strawberry (*Fragaria xananassa* Duch.) during fruit development and maturation. *Journal of Experimental Botany* 62: 1103-1118.

58. Zheng, J., Kallio, H., Linderborg, K. and Yang, B.R. 2011. Sugars, sugar alcohols, fruit acids, and ascorbic acid in wild Chinese seabuckthorn (*Hippophae rhamnoides* ssp. *sinensis*) with special reference to influence of latitude and altitude. *Food Research International* 44: 2018-2026.

Chapter 24

Production and Standardizing Polyphenols of Seabuckthorn

O.P. Sheichenko, V.I. Sheichenko* and O.N. Tolkachev

Al-Russian Research Institute of Medicinal and Aromatic Plants (VILAR), Grina Str. 7 Moscow, Russia
(*E-mail: vilarnii@mail.ru)

SUMMARY

Seabuckthorn is the popular medicinal plant in Russia; its leaves containing hydrolyzing gallo and ellagi-tannins are used for commercial production of antiviral drug Hiporamin. The drug represent a purified fraction of total monomeric tannins possessing antiviral activity and inactive Guebrachitol. The method of producing standard product was worked out. The technology of enrichment of the tannin fraction of the drug with the active constituents resulted in achieving the tannin contents over 70 per cent (calculated on causarinin) in the drug. The quality control methods included TLC and the UV spectrometry with the shift agents, the ^{1}H, ^{13}C-NMR spectrometry, using saccharose as the comparison specimen, and HPLC. An efficient complex technology of Hiporamin and Quebrachitol production was elaborated at the commercial scale.

Keywords: *Seabuckthorn, Leaves, Hydrolyzing gallo, Ellagi-tannin and Hiporamin.*

INTRODUCTION

Seabuckthorn (*Hippophae rhamnoides* L.) is the popular medicinal plant of the Russian flora, which is used since old times in traditional medicines (Sheichenko and Tolkachev, 2006). Numerous wild populations and breeding varieties cultivated at the Russian plantations are utilized for the purpose currently. All parts of the plants have been extensively studied fortheir chemical composition and biological activities. Seabuckthorn leaves contain hydrolyzing tannins and flavonoid glycosides (up to 10 per cent) and a potent antiviral drug Hiporamin was developed with its purified tannin fraction containing monomeric gallo-ellagi tannins, strictinin, isostrictinin, casuarinin, casuarictin, and hipophenin *B. Hiporaminis* active in respect of wide spectrum of pathogenic viruses (Influenza, Herpes, Cytomegallo-viruses, and HIV *etc.*) (Sheichenko *et al.*, 2006; Sheichenko *et al.*, 1987; Yoshida *et al.*, 1991; Shipulina *et al.*, 2005; Shipulyna *et al.*, 2008; Krepkova *et al.*, 2008).This drug was found to inhibit neuraminidase activity of viral strains and alsopossess interferon-induction effect (Shipulina *et al.*, 2005). In addition, the drug also has a potent antibacterial activity. The maximal total tannin content in seabuckthorn leaves was found at the period of maximal accumulation of biologically active substances; the best time of procurement was determined to be July (over 20 per cent).

According to recent report (Nobre *et al.*, 2006; Deolida *et al.*, 2003), the cyclic polyol Quebrachitol (2-O-methyl-L-inositol) in this fraction possesses antioxidative, cytoprotective and gastro-protective properties bythe mechanisms involving endogenous, nitric oxide release, and the activation of K+ATP channels.

Hiporamin is to date produced at the commercial scale in Russia (Bykov *et al.*, 2008; Bykov *et al.*, 2005). Crude extract from the wild forms of seabuckthorn or their breeding varieties are used for production of the

Seabuckthorn (Hippophae L.): A Multipurpose Wonder Plant, Vol. 5: Advanced Technologies for Cultivation, Processing, Health Protection and Environmental Conservation (Virendra Singh, *Editor-in-Chief*, 2018), p. 193–195, Astral International Pvt. Ltd., New Delhi, India.

drug. Green herbaceous shoots were also shown to be of use as the source for the drug production (Sheichenko *et al.*, 2003).

MATERIALS AND METHODS

Dynamics accumulation study of biological active substances in the wild samples of seabuckthorn and of cultivars hasshown a correlation between the chemical composition and the time of procurement of the plant material. Leaves extracts of the cultivars "Podaroksadu", "Chuiskaya", produced from the plantation of the "All-Russian institute of gardening and plants nursery" were evaluated. It was found that the tannins accumulation in the fraction of polyphenols (yield 27–34 per cent) was significantly greater in the leaves procured in summer (July) when compared to those collected in Autumn (September–October), when Quebrachitol content saccumulated in plants (up to 12–14.4 per cent).

The quality control of the drug and of the plant crude drug wasdonewiththe methods of TLC and UV spectrometry with the shift agents according to normative documentation on the drug substance and on the seabuckthorn leaves (Sheichenko *et al.*, 2001; Sheichenko *et al.*, 2001). Tannins and Quebrachitol contents were also analyzed with the methodsof the ^{1}H-,^{13}C spectrometry and HPLC.

During testing of Quebrachitol content, standard samples of saccharose, $C_{12}H_{22}O_{11}$, possessing related integral intensity of C2 signal of Quebrachitol in the ^{13}C NMR spectrum were used comparing withintegral intensity of the semiacetal carbon of saccharose. A comparison of the integral intensity of the sample mixtures of Quebrachitol-saccharose (1:1) has shown that the saccharose intensity signal was higher than that of saccharose on 7 per cent in the ^{13}C NMR spectrum. This deviation was found as a result of substantial diversity of the corresponding nuclei. Corresponding correction index (1.07) of the systematic error of the Quebrachitol content was included into the calculation formula. Thus, Quebrachitol contents in the drug was found to be 6.95 per cent (Sheichenko *et al.*, 2011).

RESULTS AND DISCUSSIONS

Total tannin contents and their related components composition were estimated on the integral intensity estimation of characteristic aromatic protons of galloyl and hexahydroxydiphenoyl fragments of hydrolyzing tannins. Results of the chemical analysis werecorrelated with the dataof antiviral activity.

According to results of dynamicsdetermination of biologically active accumulation substances in plants there was found dependency of quebrachitol contents increasing on 2.8 per cent in the summer samples (middle July), up to 11-13.4 per cent in the Autumn samples (September-October) (2-3 times more comparing with the summer samples). Thus, relative diminishing tannin contents in the drug in autumn periods of purchase affects its quality. Thus tannin-Quebrachitol contents dependency on the periods of purchase may explain their property and may be used for the pharmaceutical documentations registration on the drug and on related products.

CONCLUSION

The method of quebrachitol and tannins separation in the autumn fractions of the plant source was worked out. Technology of enrichment of the tannin fraction of the drug with the active constituents has allowed to get standard samples of the drug with the tannin contents over 70 per cent (calculated on causarinin). An efficient complex technology of Hiporamin and Quebrachitol production was elaborated at the commercial scale. Computer construction of molecular models has shown that seabuckthorn tannins of casuarinin and causarictine type form stable Quebrachitol supra-molecular associates determined their molecular-biological properties.

REFERENCES

1. Bykov, V. A., Shipulina, L. D., Tolkachev, O. N., Krepkova, L. V. Morozov, V. I. and Val, E. V. 2005. Seabuckthorn leaves as the source for production antiviral drug: Organizing and technological aspects. *Practical Phytotherapy* "INFI", No. 4, pp. 6-7. (In Russian)
2. Bykov, V. A., Tolkachev, O. N., Morozov, V. I., Sheichenko, O. P., Sheichenko, V. I., Shipulina, L. D. and Krepkova, L.V. 2008. Complex study on seabuckthorn as a source for production of the antiviral drug Hiporhamin. In: *Seabuckthorn (Hippophae L.)–A Multipurpose Wonder plant, Vol. III, Advances in Research and Developments,* Daya Publishing House, V. Singh (Ed.-in-Chief), Delhi, pp. 337-356.
3. Deolida, T. M., Lemos, T. M., Machado., T. L. G., Rao, L. L. and Santos, F. A. 2003. Quebrachitol-induced gastroprotection against acute gastric lesions: Role of prostaglandins, nitric oxide and K+ATP channels. *Phytomedicine* 15 (5): 327-333.
4. Krepkova, L.V., Bortnikova, V.V., Shkarenkov, A. A., Shipulina, L. D., Tolkachev, O. N. and Borovkova, M. V. 2008. Experimental toxicological study of Hiporamin: An antiviral drug from seabuckthorn. In: *Seabuckthorn* (*Hippophae* L.)- *A Multipurpose Wonder Plant, Vol. III, Advances in Research and Developments,* Daya Publishing House, V. Singh (Ed.-in-Chief), Delhi, pp. 357-360.
5. Nobre, H. V. Jun., Cunha, G. M. A., Moraes, M. O., Luciana, M. F. D., Oliveira, R. A., Maia, F. D., Nogueira, M. A. S., Lemos, T. L. G. and Rao, V. S. 2006. Quebrachitol (2-O-methyl-L-inositol) attenuates 6-hydroxydopamine-induced

cytotoxicity in rat fetal mesencephalic cell cultures. *Food and Chemical Toxicology* 44 (9): 1544-1551.

6. Sheichenko, O. P., Sheichenko, V. I., Fadeeva, I. I., Zolotarev, B. M. and Tolkachev, O. N.1987. Tannins from the leaves of *Hippophae rhamnoides* L. *Khim. Prirod. Soedin.* No. 6, pp. 902-907. (In Russian)
7. Sheichenko, O. P, Sheichenko, V. I. and Tolkachev O. N. 2001. Dynamics of seasonal accumulations of tannins in leaves of seabuckthorn (*Hippophae rhamnoides* L.) chemical composition. In: *Seabuckthorn-A resource for health and environment in twenty first century.* In: *Proceedings of International Workshop on Seabuckthorn,* New Delhi, February 18-21 2001, V. Singh and P. K. Khosla (Eds.), pp. 162-166.
8. Sheichenko, O. P., Sheichenko, V. I., Tolkachev, O. N., Isaev O. N. and Tsar'kova T. F. 2003. Tannins of green herbaceous shoots of seabuckthorn (*Hippophae rhamnoides* L.). In: *Proceedings of International Scientific Conference "Chemistry, Technology and Medicinal Aspects of Natural Compounds"*, 8-11 October, 2003, Almaty, Kasakhstan, p. 43.
9. Sheichenko, O. P. and Tolkachev, O. N. 2006. Tannins of seabuckthorn (*Hippophae rhamnoides* L.): Chemical properties and biological action. In: *Seabuckthorn (Hippophae L.) – A Multipurpose Wonder Plant, Vol. II, Biochemistry and Pharmacology,* Daya Publ. House, V. Singh (Ed.-in-Chief), Delhi, pp. 197-213.
10. Sheichenko, V. I., Semenova, T. S., Tolkachev, O. N. and Sheichenko, O. P. 2001. Polyphenol fraction from Seabuckthorn leaves: Chemical composition. In: *Seabuckthorn–A resource for health and environment in twenty first century.* In: *Proceedings of International Workshop on Seabuckthorn,* New Delhi, February 18-21 2001, V. Singh and P. K. Khosla (Eds.), pp. 157-161.
11. Sheichenko, V. I., Sheichenko, O. P., Tolkachev, O. N., Shipulina, L. D. and Bykov, V. A. 2011. Quantitative determination of hydrolyzing tannins in seabuckthorn leaves with the NMR method of analysis. *Medicinal and Pharmaceutical Chemistry* No. 11, pp. 15-18.
12. Shipulina, L. D., Fateeva, T. V., Krepkova, L.V., Bortnikova V. V. and Tolkachev, O. N. 2005. Antiviral drug from seabuckthorn: Experimental data of preclinical study. *Practical Phytotherapy "INFI"*, No. 4, pp. 7-16. (In Russian)
13. Shipulina, L. D., Leneva, I. A., Fedyakina, I. I., Krepkova, L.V., Bortnikova, V. V. and Tolkachev, O. N. 2005. Hiporamin–vegetarian antiviral drug: Experimental study and mechanism of its action. In: *Actual Problems of Innovations with Nontraditional Natural Sources and Creation of Functional Products,* Materials of the 3rd Russian Research and Practical Conference (6-7 July 2005), RAEN, Moscow, p. 97. (In Russian)
14. Shipulyna, L. D., Sheichenko, O. P., Sheichenko, V. I., Tolkachev, O. N., Morozov, V. I. and Bykov, V. A. 2008. Seabuckthorn study as the source for production of antiviral drug Hiporamin: Results and prospects. In: *Canada-Russia Agri-Business Knowledge Transfer, Small Fruit Production and Processing Seminar,* Manitoba, July 15, 2008, p. 36.
15. Yoshida, T., Tanaka, K., Chen, X.M. and Okuda, T. 1991. Tannins from *Hippophae rhamnoides. Phytochemistry* 30 (2): 663-666.

Chapter 25

Effect of Dehulling on Chemical Composition of Seabuckthorn Seed and its Main Constituents

I. Von Ulardt[1*], B. Niemeijer[1], R. Valbuena[1] and A. Wähling[2]

[1]Institute of Agricultural and Urban Ecological Projects affiliated to Humboldt-Universität zu Berlin (IASP), Philippstraße 13/Haus 16 10115 Berlin, Germany
[2]NIG Nahrungs-Ingenieurtechnik GmbH, Wasserkunststraße 26 39124 Magdeburg, Germany
(*E-mail: isis.von.ulardt@agrar.hu-berlin.de)

SUMMARY

Commonly seabuckthorn seeds are used as a source of high quality oil. In this study the composition of seeds and their main constituents shell and kernel were analysed in order to assess nutritional properties and utilization as a source of bioactive ingredients additional to oil. The seabuckthorn seed shells were separated from the kernel using definite deformation method and pneumatic separation. The content of fat, soluble and insoluble dietary fibre, protein, dry matter, ashes, carotenoids, total polyphenols and flavonoids were determined on two German cultivars of seabuckthorn seeds (*Hippophae rhamnoides* 'Leikora' and 'Hergo') and their fractions seed shell and seed kernel. Fat, protein and dietary fibre are the major components of seabuckthorn seeds. Significant differences in chemical composition between seed shell and seed kernel were observed. Dehulling of seeds significantly increased the concentration of oil and protein and decreased the fibre content in dehulled seabuckthorn seed, irrespective of cultivars. The hull fraction of both cultivars had the highest fibre, carotenoid and polyphenol contents and the lowest concentration of protein and oil content compared to the whole and dehulled seeds. High protein and oil and low fibre concentration make seabuckthorn seed kernel a potential component for enhanced food and feed products. The shell fraction is high in dietary fibre content as well as polyphenol content and can be considered as a rich source of bioactive ingredient in human nutrition.

Keywords: *Seabuckthorn, Seed, Dehulling and Chemical composition.*

INTRODUCTION

Seabuckthorn (*Hippophaë rhamnoides*) is a berry-bearing, hardy bush plant native to Europe and Asia. A number of cultivars have been developed and cultivated in different parts of the world. The fruits are yellow-orange berries that can be processed into oil, juice and food additives mainly used for production of confectionery and consumer care products (Beveridge *et al.*, 1999; Mörsel and Singh, 2009).

Two different types of oil can be obtained from the fruit: oil from the seed and oil from the pulp and skin of the fruit. The amount of oil mainly depends on the origin of the fruit and other factors (Yang and Kallio, 2002; Tittinen *et al.*, 2005). The quality of edible oils depends not only on the production method, but also from other factors such as the raw material and its characteristics at time of examination (Sharma *et al.*, 2009).

In order to extract the seed oil, the common mechanical compression is mainly done with the whole seabuckthorn seeds. According to other studies, dehulling of seeds from the kernel could improve the quality of the oil and the de-oiled by-product. Evangelista *et al.*, 2010 reported that Hexane-extracted oil from the dehulled cuphea seed contained 70 per cent less chlorophyll content compared with similarly extracted whole seed oil. Chlorophyll leads to a loss of quality in taste and color of oils. Subramanian *et al.* (1990) describe an improved color, a lower wax content in the dehulled extracted sunflower oil as well as lower fibre and higher protein content in the de-oiled press residue. In addition, the amount of damage to the oil expression unit was reduced due to the absence of hulls.

The attention of this study was drawn towards the chemical composition investigation of seabuckthorn seed and its main constituents. Anatomically, seabuckthorn seeds consist of shell and kernel (endosperm layers). The aim of the current program was to determine the effectiveness of the dehulling of seabuckthorn seeds with the objective of raising the nutritional value of its components.

MATERIAL AND METHODS

Materials

Dried seabuckthorn seeds of *Hippophae rhamnoides* 'Leikora' and seeds of *Hippophae rhamnoides* 'Hergo' (standard market quality) were used in the experiments with moisture content of 6.26 g/100 g DM (dry matter) and 5.48 g/100 g DM, respectively. The seed weighed 11.95±0.52 g/1000 seeds for 'Hergo' and 23.45±0.68 g/1000 seeds for 'Leikora'. The outer diameter max and min of 'Hergo' was 4.23±0.32 mm and 2.08±0.17 mm, for 'Leikora' it was 5.96±0.40 mm and 2.89±0.19 mm (Figure 25.1). The samples were kept in light and air tight plastic containers at room temperature until analysis.

Sample Preparation

For measuring hull content, seed kernel (endosperm layers) and seed shell were separated manually after soaking the seeds on moisture filter paper at room temperature for 24 hours. Afterwards, by means of a scalpel, a small longitudinal incision was made. The seed kernel was expressed with a spatula. After air-drying the samples, the dry weight of each fraction was determined (Shireadegan and Robbelen, 1985).

For measuring total ash, dietary fibre, protein, oil, chlorophyll a and b, carotenoid, polyphenol content and flavonoids, the shell of seabuckthorn seeds was separated from the kernel by definite deformation method and pneumatic separation.

After a suitable seed moisture and mill speed combination was determined, seeds (500 g/batch) were broken in a roller mill with an optimum size slit for 'Leikora' seeds of 1.0 mm with corrugating rolls and for 'Hergo' seeds of 0.8 mm with plain roll surfaces. The samples from the mill were categorized into four particle sizes using test sieves (mesh size 2 mm, 1 mm, 0.9 mm and sieve floor) that were stacked on a test sieve shaker, arranged in downwards decreasing mesh sizes. The four granulometric categories were density-graded into two fractions, using an air classifier. The overflow contained the shells and the underflow the seed kernels. For the chemical analysis, the remaining parts of fractions were sorted out manually. The seed shell and seed kernel fractions were kept separately in light and air-tight plastic containers at 4°C until analysis.

Dry Matter and Ash Content

Dry matter (DM) was determined gravimetrically after drying at 105°C for 24 h. Ash content was determined after sample conditioning at 525°C for 4 h. The measurements were done by triple determination.

Fat Content

An aliquot of the sample was boiled with 4 N HCl for 1 h, filtrated, washed with HPLC-grade water and extracted twice with petroleum ether using a Soxtherm (Gerhardt®) by method of LUFA Bd.III 5.1. The measurements were done by triple determination.

Protein Content

For crude protein analysis the Kjeldahl method was applied using standard procedure ISO 1871:2009 "Food and feed products - General guidelines for the determination of nitrogen by the Kjeldahl method".

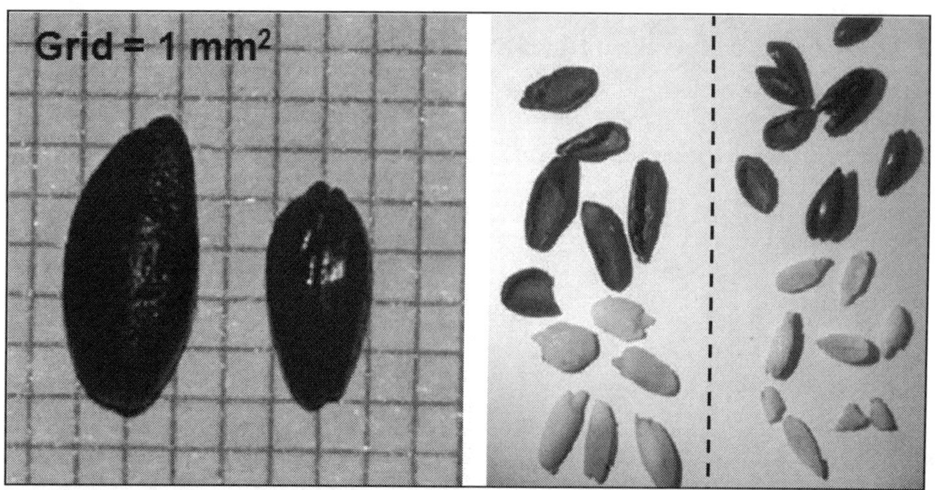

Figure 25.1: Seabuckthorn 'Leikora' (Left) and 'Hergo' (Right) Seed, Seed Kernel and Seed Hull.

Protein was calculated using the general factor (6.25). The measurements were done by triple determination.

Determination of Total, Soluble and Insoluble Dietary Fibre

The samples were analysed to determine the soluble and insoluble fibre content following the Association of Official Analytical Chemists official method "total, soluble and insoluble dietary fibre" (991.43 2007). Both soluble and insoluble fibre residues were corrected for protein, ash and blank, for the final calculation of values. The measurements were done by triple determination.

Total Carotenoids and Chlorophylls a and b

Total amounts of total carotenoids and chlorophylls a and b were determined following the modified method according to the formulas of Lichtenthaler and Wellburn (1985). The pigments have been analysed by spectrophotometer method based on multistep pigments extractions with acetone followed by their determination at wavelengths specific to each category of pigments. (Chlorophyll a at 662 nm, chlorophyll b at 645 nm and total carotenoid at 470 nm). A fourfold determination was performed.

Total Polyphenols

The amount of total phenolic compound was determined by a colorimetric method using the Folin-Ciocalteu assay that was based on a colorimetric oxidation/reduction reaction of phenols (Escarpa and Gonzalez, 2001) (Maksimovic *et al.*, 2005). Measures were carried out at 725 nm using a UV-Vis spectrophotometer and the calibration curve was made with standard solutions of gallic acid in the range 0.04 and 0.2 mg/ml. Results were expressed as gallic acid equivalents. A fourfold determination was performed.

Flavonoids

Selected flavonoids (quercetin, rutin, isorhamnetin-3-rutinosid, isorhamnetin-3-0-glucosid and isorhamnetin) were simultaneously detected from seed shells using a simple High Performance Liquid Chromatographic (HPLC) method after extraction with methanol. The measurements were done by triple determination.

Statistical Analyses

All the given values are means of at least three determinations ± standard deviation. Differences in chemical composition between seabuckthorn seeds of different cultivars, seed kernel and seed shell were estimated applying ANOVA, Tukey´s test and Dunnett-T3 test for independent samples, being statistically significant with P = 0.05 (SPSS 20.0 (2011), SPSS Inc., Chicago, USA). Different letters in rows indicate significant differences between means at P > 0.05.

RESULTS AND DISCUSSION

Ash Content

Tables 25.1 and 25.2 show the content of ash of *Hippophaë rhamnoides* 'Leikora' and 'Hergo' seeds and its main constituents. Ash is the total mineral content of a sample, made up of inorganic components. The results demonstrate that both cultivars studied had a similar ash concentration, with an average of 2.9 g/100 g DM for 'Leikora' and 2.7 g/100 g DM for 'Hergo'. Regarding the ash, the seed kernel ('Leikora' 4.9 g/100 g DM and 'Hergo' 5.7 g/100 g DM) contained significant higher amounts than seed shell ('Leikora' 1.0 g/100 g DM and 'Hergo' 1.1 g/100 g DM) for both cultivars. The mean content of minerals in seabuckthorn seed, reflected by the ash proportion, was found to be between that of seed

Table 25.1: Chemical Composition of Seabuckthorn Seeds 'Leikora' and 'Hergo'

Determination		*Seed*	
		Leikora	*Hergo*
Dry matter (g/100 g)		92.06^{a} ± 0.01	92.20^{b} ± 0.03
Ash [1]		2.90^{a} ± 0.06	2.69^{b} ± 0.02
Fibre [1]	insoluble	46.85^{a} ± 0.75	52.87^{b} ± 0.78
	soluble	1.01^{a} ± 0.39	0.98^{a} ± 0.19
Protein [1]		28.40^{a} ± 0.14	24.70^{b} ± 0.27
Oil [1]		13.98^{a} ± 0.26	11.64^{b} ± 0.22
Chlorophyll [2]	a	0.95^{a} ± 0.06	0.96^{a} ± 0.06
	b	1.59^{a} ± 0.12	1.46^{b} ± 0.02
Carotenoid [2]		1.50^{a} ± 0.18	1.75^{a} ± 0.08
Total polyphenols [1]		6.07^{a} ± 0.98	4.55^{b} ± 0.83

[1] g/100 g dry matter; [2] mg/100 g dry matter. All the given values are means of at least tree determinations ± standard deviation. Means followed by same lower case letter are not significantly different by Tukey Test at the 5% level of probability.

Table 25.2: Chemical Composition of Seabuckthorn 'Leikora' and 'Hergo' Seed Kernel and Seed Shell

Determination		*Constituents*			
		Kernel		*Shell*	
		Leikora	*Hergo*	*Leikora*	*Hergo*
Constituent [1]		54.20^{a} ± 1.48	42.73^{b} ± 2.27	45.80 ± 1.48	57.27 ± 2.27
Dry matter (g/100 g)		93.97^{a} ± 0.18	93.96^{a} ± 0.04	90.63^{b} ± 0.17	91.36^{c} ± 0.02
Ash [1]		4.90^{A} ± 0.22	5.74^{A} ± 0.28	0.98^{B} ± 0.01	1.09^{C} ± 0.01
Fibre [1]	insoluble	18.99^{a} ± 1.70	16.77^{a} ± 1.66	79.33^{b} ± 0.64	78.81^{b} ± 0.24
	soluble	LOD ± 0.00	LOD ± 0.00	0.55^{a} ± 0.42	1.00^{b} ± 0.26
Protein [1]		46.65^{a} ± 0.14	47.63^{b} ± 0.07	8.92^{c} ± 0.14	9.62^{d} ± 0.22
Oil [1]		22.91^{a} ± 0.44	22.91^{a} ± 0.50	1.52^{b} ± 0.09	1.15^{b} ± 0.10
Chlorophyll [2]	a	0.59^{bc} ± 0.03	0.72^{c} ± 0.03	0.47^{ab} ± 0.09	0.39^{a} ± 0.05
	b	0.98^{A} ± 0.01	1.21^{B} ± 0.05	0.97^{ABC} ± 0.24	0.66^{C} ± 0.03
Carotenoid [2]		0.82^{a} ± 0.00	1.76^{b} ± 0.08	1.71^{b} ± 0.04	2.01^{c} ± 0.04
Total polyphenols [1]		6.45^{a} ± 0.07	4.62^{b} ± 0.18	8.17^{c} ± 0.15	5.72^{d} ± 0.19

[1] g/100 g dry matter; [2] mg/100 g dry matter. (LOD) Results below the limit of quantification. All the given values are means of at least tree determinations ± standard deviation. Means followed by same lower case letter are not significantly different by Tukey Test at the 5% level of probability. Means followed by same upper case letter are not significantly different by Dunnett Test at the 5% level of probability.

shell (P > 0.05) and seed kernel (P > 0.05). The fractions could be considered as potential source which could be used to meet part of the nutritional requirement. The ash content in seabuckthorn seed was slightly low compared to the other oil crops (sesame seed 5.3 g/100 g, soya bean 4.7 g/100 g and sunflower seed 3.3 g/100 g) (Souci *et al.*, 1989/90).

Total, Soluble and Insoluble Dietary Fibre

Insoluble dietary fibre was the major fraction of 'Leikora' seeds (46.9 g/100 g DM) and 'Hergo' seeds (52.9 g/100 g DM) (Table 25.1). There were significant differences in the content of insoluble dietary fibre between both cultivars (P > 0.05). The mean insoluble fibre content in the tested constituents (Table 25.2) was being highest in seed shells ('Leikora' 79.3 g/100 g DM and 'Hergo' 78.8 g/100 g DM) and lowest in seed kernels ('Leikora' 19.0 g/100 g DM and 'Hergo' 15.8 g/100 g DM). Dehulling of seeds was found to remove between 60 per cent and 70 per cent of total fibre. The high fibre content of seabuckthorn seeds is largely due to the high proportion of hull. The soluble dietary fibre content in the whole seed was for both cultivars 1.0 g/100 g DM. The content in seed shell was for 'Leikora' 0.6 g/100 g DM and for 'Hergo' 1.0 g/100 g DM.

After dehulling, the oil could be extracted from the kernel fraction. The reduction in fibre content results in a press cake with improved digestibility. Insoluble fibers are characterized by their low density, their porosity, and by their ability to increase faecal bulk and decrease intestinal transit (Elleuch *et al.*, 2011). Fractionated and milled seabuckthorn seed shells contribute beneficial rich sources of dietary fibre to food. Incorporation of rich-fibre by-products has been investigated (Elleuch *et al.*, 2011). Due to removing the shell, further benefits for the handling process of seabuckthorn seed are expected.

Protein Content

The protein is a quantitatively important fraction in the seabuckthorn seed. The protein content of different varieties was the object of early studies, which reported about 30 per cent of proteins in seabuckthorn seeds (Uransanaa, Gerel, Jamyansan and Dash, 2003). The mean protein content of 'Leikora' seeds (28.4 g/100 g DM) was significantly higher (P > 0.05) compared to protein content of 'Hergo' seeds (24.7 g/100 g DM) (Table 25.1).

Due to dehulling, the protein content of both cultivars increased significantly (P > 0.05) by 64 per cent for Leikora and by 93 per cent for Hergo. The seed kernels of both cultivars are a good source of protein, yielding 46.9 g/100 g DM for 'Leikora' and 47.6 g/100 g DM for 'Hergo' (Table 25.2). In comparison to other vegetable protein sources used in human and animal nutrition, seabuckthorn seeds contain high proportions of protein. The protein content in the dehulled seed was two times higher than those reported for sesame (Elleuch *et al.*, 2007), as high as different dehulled lupine cultivars and slightly lower than the values reported for dehulled soy bean (Bähr *et al.*, 2014).

Many oilseed crops contain significant amounts of protein that remain in the meal following oil removal. The dehulling of seabuckthorn seeds, that raises protein levels, will improve the ability to use the resulting kernel for human and animal nutrition.

Oil Content

Tables 25.1 and 25.2 show the amount of total oil found in the samples. The whole 'Leikora' seed contained 14.0 g/100 g DM of oil, while dehulled kernel had 22.9 g/100 g DM and the shell had 1.5 g/100 g DM, only. The whole 'Hergo' seed contained 11.6 g/100 g DM of oil, the kernel 22.9 g/100 g DM and the shell 1.2 g/100 g DM, only. The relative amount of oil differed significantly ($P < 0.05$) among seed cultivars for whole seed. By contrast, no significant differences in oil content among seed cultivars for seed kernel were detected, therefore there are no potential benefits in selection of raw material depending on variety after dehulling. This indicates that the oil was mainly localized in the seed kernel. The oil content in seabuckthorn seeds is commonly about 10 per cent, although higher values (up to 15–16 per cent) have been described in some cultivars and wild berries (Yang and Kallio, 2002).

Dehulling of seeds was found to increase the oil content by 64 per cent for 'Leikora' and by 97 per cent for 'Hergo'. This confirms previous findings by Elleuch *et al.* (2007) and Minkowski (2002) who reported a high oil content of oil for dehulled sesame and rape seeds, respectively.

In order to extract the seed oil, the common mechanical compression is mainly done with the whole seabuckthorn seeds. Seed dehulling performed before oil extraction may reduce the amount of material to be processed, thus increasing the through put of the downstream processing equipment. Dehulling also can reduce maintenance costs associated with the wear of the lining bars and shaft of the screw press. Subramanian *et al.*, 1990 describes an improved color, a lower wax content of the dehulled extracted sunflower oil as well as lower fibre and higher protein content of the de-oiled press residue. This confirms previous findings by Evangelista *et al.* for dehulled cuphea seed.

Total Carotenoids and Chlorophylls a and b

No significant differences in content of total carotenoids of 'Leikora' (1.5 mg/100 g DM) and 'Hergo' (1.8 mg/100 g DM) whole seed were observed, therefore there are no potential benefits in selection of raw material depending on variety (Table 25.1). However, significantly higher amounts were found in shell ('Leikora' 1.7 mg/100 g DM and 'Hergo' 2.0 mg/100 g DM) compared to kernel ('Leikora' 0.8 mg/100 g DM and 'Hergo' 1.8 mg/100 g DM) for both cultivars (Table 25.2). Carotenoids mainly exist in the soft parts of seabuckthorn berry; the concentration in seeds is typically 1/20–1/5 of that in soft parts (Yang and Kallio, 2002).

The average contents of chlorophyll a in whole seeds of both cultivars (1.0 mg/100 g DM) were not statistically different (Table 25.1). The amount of chlorophyll b was significantly greater than chlorophyll a in seeds and constituents of both cultivars. The statistical differences ($P < 0.05$) between the samples are shown in Tables 25.1 and 25.2. Evangelista *et al.* (2010) reported that Hexane-extracted oil from the dehulled cuphea seed contained 70 per cent less chlorophyll content compared to similarly extracted whole seed oil. The chlorophyll leads to a loss of quality in taste and color of oils. Because of small amounts of chlorophyll in seabuckthorn seed, no influence of quality of seed oil will be expected.

Acetone, chloroform, diethyl ether, dimethyl formamid and methanol were used for extraction of pigments in studies with high plant leaves, and it was determined that the extraction rate was various in every solvent (Lichtenthaler and Wellburn, 1985). Other solvents for extraction of pigments in seabuckthorn seeds should be investigated.

Total Polyphenols

Seabuckthorn seeds and its main constituents contained considerably high amounts of polyphenols, shown in Tables 25.1 and 25.2. The content of polyphenols of the 'Leikora' seeds (6.1 g/100 g DM) was significantly higher than for 'Hergo' seeds (4.6 g/100 g DM) ($P > 0.05$). The seeds are generally reported to constitute the major fraction of polyphenols in seabuckthorn berries. These results are in agreement with those reported by Arimboor *et al.* (2008) who found high levels of polyphenols in seed kernel (11.4 g/100g DM) of *H. rhamnoides* ssp. *turkestanica*. The high amount of polyphenols in the seeds can be explained by the fact that the polyphenols are compounds associated with dietary fibre. Studies found that fibre binds polyphenols in fruit and vegetables, which protects the antioxidants from early digestion and effectively transports polyphenols to the colon where these compounds may have a protective effect on colonic health as they are released during plant fibre fermentation by gut bacteria (Padayacheea *et al.*, Part 1, 2012) (Padayacheea *et al.*, Part 2, 2012). On the other hand, Arimboor *et al.* (2008) reported that the amount of polyphenols in seed shell is relatively low. This is in contrast with the present study, where the content of polyphenols in the shell of both cultivars was higher than those of the other constituents. Table 25.2 shows that the high value of polyphenols can be related to the high dietary fibre content of the shells. Other studies correlate with these findings. It was established that coats of seeds, coats of cereal grains and peels of fruits contain higher amounts of polyphenols than the cotyledon, the endosperm and the pulp fractions respectively (Elleuch, Besbes, Roiseux, Blecker and Attia, 2007); (Soong and Barlow, 2004).

The results indicate that seabuckthorn seeds, especially the shells, are very rich sources of antioxidant polyphenols compared with other agri-food secondary products (Arimboor, Kumar and Arumughan, 2008). Therefore, their exploitation as a source of added-value products besides the use as oil source may be highly cost-effective.

Flavonoids

Phenolics are reported as the major contributors to the biological properties such as antioxidant activities of seabuckthorn berries and leaves. In seedless seabuckthorn berry, quercetin has been reported to be the main flavonoid (Häkkinen *et al.*, 1999). According to this study; quercerin was the main flavonoid found in the shell of 'Leikora' and 'Hergo'. However, comparison of the phenolic content of different samples from literature is difficult because of the varying cultivars used. The amount of quercetin in 'Hergo' seed shell (185.7 mg/kg DM) was almost 7 times higher than that of 'Leikora' seed shell (27.0 mg/kg DM). The most frequently studied flavonoid, quercetin, has been shown to may have biological properties consistent with its positive effect on the cardiovascular system. This includes protection against various diseases such as osteoporosis, certain forms of cancer, pulmonary and cardiovascular diseases and also against aging. Especially the ability of quercetin to form bonds with highly reactive substances such as peroxynitrite and the hydroxyl radical is suggested to be involved in these possible positive beneficial health effects (Boots *et al.*, 2008).

There were no detectable levels of rutin, isorhamnetin-3-rutinosid, isorhamnetin-3-0-glucosid and isorhamnetin.

The concentrations of the favonols in whole seeds and seed kernels are under investigation. Other favonoids like kaempferol and myricetin and phenolic acids like p-coumaric, caffeic, ferulic, p-hydroxybenzoic, gallic and ellagic acids should be investigated.

CONCLUSION

The present study provides an insight on the composition of seeds and its main constituents of two selected cultivars of *Hippophaë rhamnoides*, namely seeds, seed kernels and seed hulls of 'Leikora' and 'Hergo'. It quantifies components; either by class (ash, fibre, protein, oil, chlorophyll a and b, carotenoids and total polyphenols) or more detailed for classes of particular interest in terms of uses (flavonoids).

Dehulling has been used as a pre-treatment for a number of oilseeds. This study presents arguments to do this with seabuckthorn seeds as well. The research shows that the process of dehulling seabuckthorn seeds offers promise in modifying the chemical composition to generate highly desirable and specialized products. Dehulling significantly concentrated the major components oil and protein, and lowering fibre content in seed kernel. High protein and oil and low fibre concentration make dehulled seabuckthorn seed a potential ingredient for enhanced food and feed products. The hull fraction high in fibre content can be used as a rich source of dietary fibre in human nutrition. In addition, the results indicate that seabuckthorn seeds, including seed kernel and seed shell are very rich sources of antioxidant polyphenols, flavonoids and carotenoids, and therefore their exploitation as a source of added-value products may be effective and merits a more thorough investigation.

The results suggest that further chemical screening of seabuckthorn seeds and its main constituents could be undertaken in order to assess their interest as a potential source of useful bioactive compounds, and help the valorization of this valuable resource.

ACKNOWLEDGMENTS

The Project is financed by the German Federal Ministry of Economy and Technology, Project executing organisation: EuroNorm GmbH. We acknowledge Sanddorn GbR, Herzberg Germany for providing the seabuckthorn seeds for the study.

REFERENCES

1. Arimboor, R., Kumar, K. S. and Arumughan, C. 2008. Simultaneous estimation of phenolic acids in seabuckthorn (*Hippophae rhamnoides*) using RP-HPLC with DAD. *Journal of Pharmaceutical and Biomedical Analysis* 47: 31-38.
2. Bähr, M., Fechner, A., Hasenkopf, K., Mittermaier, S. and Jahreis, G. 2014. Chemical composition of dehulled seeds of selected lupin cultivars in comparison to pea and soya bean. LWT - *Food Science and Technology* 59: 587-590.
3. Beveridge, T., Li, T. S., Oomah, B. D. and Smith, A. 1999. Seabuckthorn products: Manufacture and composition. *Journal of Agricultural and Food Chemistry* 47: 3480-3488.
4. Boots, A., Haenen, G. and Bast, A. 2008. Health effects of quercetin: From antioxidant to nutraceutical. *European Journal of Pharmacology* 585: 325-337.
5. Elleuch, M., Bedigian, D., Roiseux, O., Besbes, S. and Blecker, C. 2011. Dietary fibre and fibre-rich by-products of food processing: Characterisation, technological functionality and commercial applications: A review. *Food Chemistry* 124: 411-421.
6. Elleuch, M., Besbes, S., Roiseux, O., Blecker, C. and Attia, H. 2007. Quality characteristics of sesame seeds and by-products. *Food Chemistry* 103: 641-650.
7. Escarpa, A. and Gonzalez, M. C. 2001. Approach to the content of total extractable phenolic compounds from different food samples by comparison of chromatographic and spectrophotometric methods. *Analytica Chimica Acta* 427: 119-127.
8. Evangelista, R. L., Cermak, S. C. and Isbell, T. A. 2010. Dehulling of Cuphea PSR23 seeds to reduce color of the extracted oil. *Industrial Crops and Products* 31: 437-443.
9. Häkkinen, S., Heinonen, M., Kärenlampi, S., Mykkanen, H., Ruuskanen, J. and Törrönen, R. 1999. Screening of selected favonoids and phenolic acids in 19 berries. *Food Research International* 32: 345-353.
10. Lichtenthaler, H. K. and Wellburn, A. R. 1985. Determination of total carotenoids and chlorophylls

a and b of leaf in different solvents. *Biochemical Society Transactions* 11: 591-592.

11. Maksimovic, Z., Malencic, D. and Kovacevic, N. 2005. Polyphenol contents and antioxidant activity of Maydis stigma extracts. *Bioresource Technology* 96: 873-877.

12. Minkowski, K. 2002. Influence of dehulling of rape seeds on chemical composition of meal. *Animal Feed Science and Technology* 96: 237-244.

13. Mörsel, J. T. and Singh, V. 2009. Sanddorn: Moderne Anbautechnologien, Norderstedt, Germany, GmbH.

14. Padayacheea, A., Netzelb, G., Netzelb, M., Dayd, L., Zabarase, D., Mikkelsena, D. A. and Gidley, M.J. 2012. Binding of polyphenols to plant cell wall analogues–Part 1: Anthocyanins. *Food Chemistry* 134 (1): 155-161.

15. Padayacheea, A., Netzelb, G., Netzelb, M., Dayd, L., Zabarase, D., Mikkelsena, D. and Gidley, M.J. 2012. Binding of polyphenols to plant cell wall analogues-Part 2: Phenolic acids. *Food Chemistry* 135 (4): 2287–2292.

16. Sharma, R., Sogi, D. S. and Saxena, D. C. 2009. Dehulling performance and textural characteristics of unshelled and shelled sunflower *(Helianthus annuus* L.) seeds. *Journal of Food Engineering* 92: 1-7.

17. Shireadegan, M. and Robbelen, G. 1985. Influence of seed color and hull proportion on quality properties of seeds in *Brassica napus* L. *Fette Seifen Anstrichmittel* 87 (6): 235-237.

18. Soong, Y. and Barlow, P. J. 2004. Antioxidant activity and phenolic content of selected fruit seeds. *Food Chemistry* 88: 411-417.

19. Souci, S. W., Fachmann, W. and Kraut, H. 1989-90. Food composition and nutritional tabels. Stuttgart, 4. Auflage: Wissenschaftliche Verlagsgesellschaft mbH.

20. Tittinen, K. M., Hakala, M. A. and Kallio, H. P. 2005. Quality Components of seabuckthorn (*Hippophae rhamnoides*) variaties. *Journal of Agricultural and Food Chemistry* 53: 1692-1699.

21. Uransanaa, M., Gerel, D., Jamyansan, Y. and Dash, T. 2003. Protein and amino acid composition of seabuckthorn seeds (*Hippophae rhamnoides* ssp. *mongolica* Rouse). *Mongolian Journal of Biological Sciences* 1: 85-88.

22. Yang, B. and Kallio, H. 2002. Composition and physiological effects of seabuckthorn (*Hippophaë*) lipids. *Trends in Food Science and Technology* 13 (5): 160-167.

Chapter 26

Commercially Viable and Biochemically Valuable Seabuckthorn Cultivars in Estonia

Toivo Univer*, Neeme Univer and Krista Tiirmaa

Polli Horticultural Research Centre, Estonian University of Life Sciences, Polli 69108, Karksi-Nuia, Viljandimaa, Estonia
(*E-mail: toivo.univer@emu.ee)

SUMMARY

The common seabuckthorn (*Hippophae rhamnoides* L.) is a relatively recently introduced species in Estonia. It is grown as a decorative plant, a pioneering species in sandy areas and exhausted open pit oil shale mines as well as a promising berry cultivar. There are 856 ha of commercial seabuckthorn orchards in Estonia. Research on seabuckthorn cultivars started in 1973. In the following 40 years, altogether 52 cultivars were introduced from Russia, Byelorussia, Finland and Germany. Cultivars Botanicheskaya, Botanicheskaya Lubitelskaya, Otradnaya, Trophimovskaya, and Vorobevskaya have proved to be better suited to Estonian soil and climatic conditions. Cultivars recommended for commercial growing are Avgustinka, Botanicheskaya, Botanicheskaya Lubitelskaya, Otradnaya and Trophimovskaya. The five-year average yield of these cultivars was 14–16 kg/tree or 17–19 tonnes/hectare. The height of trees in a 7-year orchard was 3.6–5 m. Biochemical analyses showed that seabuckthorn berries are rich in fatty acids. The seeds of the seabuckthorn cultivars analysed contained 8.36 per cent –9.89 per cent and juice 2.43 per cent –3.51 per cent of fatty acids. Fatty acid analysis of the seven seabuckthorn cultivars found that their seeds contained 15 compounds. Statistically important differences ($p<0.05$) in seeds were observed among varieties on C18:0 and in juice on C14:0, C16:0, C16:1, C18:1 and C18:2. The seeds of seabuckthorn berries were rich in linoleic (37.09 per cent –38.52 per cent) and linolenic acid (28.53 per cent –34.12 per cent), while juice contained a high level of palmitic (31.59 per cent –39.02 per cent) and palmitoleic acids (37.74 per cent –43.07 per cent).

Keywords: *Hippophae rhamnoides L., Cultivar, Yield, Biochemical composition, Seed oil, Juice oil and Estonia.*

INTRODUCTION

Many of the substances found in the common seabuckthorn (*Hippophae rhamnoides* L.), a traditional health-promoting plant, are known to have beneficial effects on health. Over 190 compounds have been found in the seeds, pulp, berries and juice of seabuckthorn. These compounds include fat-soluble vitamins (A, K, and E), fatty acids, lipids, organic acids, amino acids, carbohydrates, vitamins C, B1, B2, folic acid, tocopherols, flavonoids, phenols, terpenes and tannins (Li and Wang, 1998; Yang and Kallio, 2002a).

Seabuckthorn seed and pulp oil are considered the most valuable components of the berries, having a unique fatty acid composition as well as fat-soluble vitamins and plant sterols (Yang and Kallio, 2002a). The oil from fruits and seeds is used to treat liver diseases, inflammation, disorders of the gastrointestinal system, eczema, canker sores and other ulcerative disorders of mucosal tissues, wounds, burns, frostbite, psoriasis, rosacea, lupus erythematosus, and chronic dermatosis (Bal *et al.*, 2011).

Seabuckthorn was introduced to Estonia in the 20th century. The first trials with seabuckthorn were carried out in the middle of the 20th century with the goal of revegetation of alvars and sandy areas (Kaar, 1971). Research on seabuckthorn cultivars started in 1973. The

Seabuckthorn (Hippophae L.): A Multipurpose Wonder Plant, Vol. 5: Advanced Technologies for Cultivation, Processing, Health Protection and Environmental Conservation (Virendra Singh, *Editor-in-Chief*, 2018), p. 205–212, Astral International Pvt. Ltd., New Delhi, India.

experience with growing seabuckthorn has revealed that the Estonian climate, where the vegetative period lasts 150 to 180 days and winters are characterized by intermittent frosts and thaws, is conducive to winter damage to seabuckthorn shrubs. Buds are the parts of seabuckthorn plants that are most sensitive to winter damage, followed by bark on the trunk and larger branches (Univer and Univer, 2014). Many seabuckthorn cultivars introduced from Russia (Barnaul, Moscow, Nizhni-Novgorod), Byelorussia, Finland and Germany have not adapted to the climatic conditions in Estonia (Piir, 1996; Siimisker, 1996, 1998; Jalakas *et al.*, 2005).

Nowadays, seabuckthorn is a popular crop in organic farming in Estonia. In 2014, the area under organic orchards was 856 ha in Estonia. Larger orchards in Estonia have adopted a harvesting technology whereby fruit-bearing branches are cut off and frozen at -20 °C to separate the berries. Using this harvesting technology removes a large share of foliage, which interferes with the shrubs' ability to prepare for winter (Univer *et al.*, 2013).

The aim of current Estonian research on seabuckthorn cultivars is to ascertain winter hardy cultivars with stable yield that are adapted to the soil and climatic conditions of Estonia and suited for new methods of harvesting the berries. The biochemical content of the berries of the cultivars grown in Estonia is also an important subject matter. The current paper provides an overview of the experience with growing seabuckthorn as well as of comparison trials and studies of the biochemical content of berries in Estonia.

MATERIALS AND METHODS

Seabuckthorn Cultivars Included in Studies

Seabuckthorn as a crop was first introduced to Estonia as a pioneering species in exhausted oil shale and gravel open pit mines. In spring 1960, 2-year-old shrubs were planted at 3x4 m on the Kundla limestone sod-calcareous soil area on the island of Saaremaa. In spring, 1976 3000 two-year-old seedlings were planted at 3 x 4 m on the banks of oil-shale mining area in northern Estonia.

Seabuckthorn cultivars introduced in Estonia originate from the neighbouring countries: Russia and Finland, but also from Byelorussia and Germany. Comparison trials have been conducted in Southern Estonia, at Polli Horticultural Research Centre of the Estonian University of Life Sciences and at Rõhu Experimental Station.

The seabuckthorn collection in Polli was established in 1973, with the first six cultivars from the Lisavenko Research Institute of Horticulture for Siberia, Barnaul: 'Zolotoy Pochatok', 'Dar Katuni', 'Novost Altaya', 'Maslichnaya', 'Vitaminnaya', 'Cherbinka-1', and a male plant B32. Eight cultivars were added in 1980: 'Yantarnaya', 'Prevoskhodnaya', 'Chuiskaya', 'Obilnaya', 'Orangevaya', 'Samorodok', 'Velikan', and 'Zolotistaya Sibiri'. In 1982–83, 'Bogaterskaya' and 'Aromatnaya' were brought from the Pavlovsk Experimental Station and 'Sibirskaya' and 'Obskaya' from Barnaul. The cultivars introduced later, in 1984–1990 were mainly from the Botanical Garden of the Moscow State University: 'Otradnaya', 'Podarok Sadu', 'Perchik', 'Gibrid Perchika', 'Botanicheskaya', 'Avgustinka', 'Trophimovskaya', 'Vorobevskaya', and one cultivar of German origin, 'Leikora'. The collection was increased in 2005 with Finnish cultivars 'Tarmo', 'Terhi', and 'Tytti', and in 2006 with Byelorussian cultivars 'Grodenskaya', 'Mariya', 'Mendeleevskaya', 'Nivelena', 'Panteleevskaya', 'Plamennaya' ja 'Zolotaya Rannaya'. The seabuckthorn collection was planted in clay loam soil with medium fertility. Two-year-old shrubs, 1 to 4 shrubs per cultivar, were planted at 3x2 m. For the first two years, the ground was clean-cultivated. From the third year onward, a grass crop was planted on the mid-row area and regularly mowed. Comparison trials were established by Robert Piir in 1978. The following data were collected annually: winter hardiness, yield, fruit characteristics and biochemical composition of fruit.

Comparison trials at Rõhu Experimental Station started in 1981 with nine Siberian cultivars: 'Dar Katuni', 'Yantarnaya', 'Obilnaya', Chuiskaya', 'Prevoskhodnaya', 'Samorodok', 'Vitaminnaya', 'Zolotistaya' and 'Velikan'. Two-year-old own-root shrubs were planted with three replications of 21 shrubs each. The shrubs were spaced at 2x4 m. The sod strips between the rows were regularly mown. After the exceptionally rainy summer of 1981, over 18 per cent of shrubs were replaced.

In 1988, a new trial was started with 13 Russian cultivars: nine from the Botanical Garden of the Moscow State University ('Avgustinka', 'Botanicheskaya', 'Botanicheskaya Aromatnaya', 'Botanicheskaya Lubitelskaya', 'Vorobevskaya', 'Gibrid Perchika', 'Otradnaya', 'Podarok Sadu', and 'Trophimovskaya'), one from the State Agricultural Academy in Nizhny-Novgorod ('Priokskaya'), one from the Michurinsk Horticultural Research Institute ('Podarok Chernozemu'), and two from the Lisavenko Research Institute of Horticulture for Siberia, Barnaul. The trial was established in three replications, 21 trees in each replication. The harvest of each tree was weighed with the precision of 0.1 kilograms.

In autumn 2004, a German-Estonian seabuckthorn trial was established with 12 cultivars: 'Askola`, `Dorana`, `Frugana`, `Habega`, `Hergo`, `Leikora`, `Sirola` bred by H.J. Albrecht and `Pollmix` as a pollinator, and Moscowian cultivars `Botanicheskaya Lubitelskaya`, `Gibrid Perchika`, `Otradnaya`, `Podarok Sadu` and `Trophimovskaya`.

The Biochemical Composition of Seabuckthorn Berries Grown in Estonia

Piir (1996) carried out a biochemical analysis of soluble dry-matter, water, sugars, the sum of organic acids, and vitamin C content of seabuckthorn berries at Polli Horticultural Research Centre of the Estonian University of Life Sciences. Soluble dry-matter content was established with a refractometer, organic acids were titrated by the neutralisation method with NaOH (0.1N solution). Vitamin C content was determined by the Tilmans'method, and the content of sugars by the cyanide method. Berry mass was determined as the average of 100 fruits in three portions.

Univer *et al.* (2015) from Polli Horticultural Research Centre of the Estonian University of Life Sciences carried out fatty acid analyses of seabuckthorn berries in 2006 and 2009. The study included seven seabuckthorn varieties bred at the Botanical Garden of the Moscow State University ('Avgustinka', 'Botanicheskaya Lubitelskaya', 'Botanicheskaya', 'Gibrid Perchika', 'Otradnaya', 'Trophimovskaya', 'Vorobevskaya'). The berries were harvested at the commercial maturation stage and frozen at -22 °C. For analyses, 200 g of frozen berries were thawed and juice was manually pressed. The seeds were separated from press cake. Lipids were extracted using chloroform/methanol (2/1 v/v) (Folch *et al.*, 1957). The esterification of seeds was performed as described by Sukhija and Palmquist (1988). The fatty acid composition was analysed using gas chromatography equipped with a flame ionization detector and fused silica capillary CP-7420 column (100 m × 0.25 mm, with film thickness of 25 μm). The content of individual fatty acids was expressed as a percentage of the total fatty acid content. The results were evaluated statistically using the analysis of variance. In the case of significant differences ($p<0.05$), LSD values were calculated and used for comparing treatments' means.

RESULTS

The Characteristics of Seabuckthorn growth and Yield in Estonia

In the earliest trials on limestone sod-calcareous soil, with a planting scheme 3x4 m, the shrubs grew well, reaching an average height and width of 2 m and 1.8 m, respectively, by their eighth year. The majority of shrubs already gave the first crop when the shrubs were six years old. These first experiments showed clearly that as a pioneering species the common seabuckthorn is excellently suited for foresting old mining areas (Kaar, 1971).

In the seabuckthorn collection established at the Polli Horticultural Research Centre in 1973–1980 with cultivars from Siberia, the shrubs started to bear fruit in the third or fourth growing season. All cultivars bore a high yield in 1984, but the following winter (1984/1985) with critically low temperatures caused severe damage to the shrubs. Piir (1996) found that Siberian cultivars were not winter hardy in Estonian climatic conditions. 'Maslichnaya', 'Samorodok', 'Yantarnaya', 'Sibirskaya', and 'Aromatnaya' were the least winter hardy. 'Novost Altaya', 'Dar Katuni' and 'Orangevaya' were relatively hardy. Only 'Orangevaya' may be of interest for its high vitamin C and oil content. Among the Siberian cultivars, the average fruit mass was 0.4–0.7 g, whereas the biggest fruit mass (0.9 g) was noted for 'Velikan'.

The cultivars from Moscow (introduced in 1984–1990) had higher shrubs and smaller fruits with a later maturing time (Piir, 1996). The Moscowian cultivars were also more winter-hardy than the German, Finnish and Byelorussian ones introduced in Estonia in 2004–2006 (Univer and Univer, 2014). The shrubs of the Finnish cultivars that grew in sod cover were severely damaged in the winter of 2012.

The average yields per shrub of seabuckthorn cultivars at the Rõhu Experimental Station were 0.4, 11.1, 6.0, and 7.5 kg in 1983–1986, respectively (Siimisker, 1996). There was no yield in 1987. The average total yield for the whole period (from 1981 to 1987) was 25 kg per shrub. Cold winter temperatures in 1978–1979 caused serious damage to seabuckthorn cultivars. Cultivars that were over-cropped in the year preceding the severe winter suffered greater winter damage than the ones with low yield. For example, 'Vitaminnaya' showed relatively good winter hardiness due to its moderate yield.

In the second comparison trial at Rõhu Station, the shrubs started to bear fruit in the third year after having been planted in the orchard (Siimisker, 1998). The average yield was 7.7 kg per shrub (Table 26.1). In the fourth year, yield per tree fell to 5.2 kg per shrub but increased to 7.6–13.0 kg per shrub in the following years. Based on five year data from the young orchard, the cultivars included in the comparison trial can be divided into two groups: one with low yield and the other with high yield. Cultivars 'Yantarnaya', 'Podarok Chernozemu', 'Chuiskaya', 'Priokskaya' yielded 2.0–4.7 kg per shrub. The average yield of cultivars 'Otradnaya', 'Vorobevskaya', 'Botanicheskaya', 'Trophimovskaya', 'Podarok Sadu', and 'Botanicheskaya Lubitelskaya' was 14–15.6 kg per shrub or 17.5–21.4 t/ha. High-yielding cultivars were characterized by vigorous growth: by their seventh year, the height of the shrubs reached 3.8–4.6 m. Hand-picking berries from such an elevation proved to be very difficult (Jalakas *et al.*, 1998).

The comparison trial of German and Russian seabuckthorn cultivars established in autumn 2004 (Table 26.2) revealed a difference in their adaptive capacity in the Estonian climatic conditions. Moscowian cultivars ('Otradnaya', 'Botanicheskaya Lubitelskaya', and 'Podarok Sadu') proved to have higher yields (30.3–30.5 kg/shrub). The German cultivars suffered considerable winter damage during winters with changeable weather (Hornig *et al.*, 2010).

Table 26.1: The Yield and Height of Seabuckthorn Shrubs at 3–7 Years of Age at Rõhu Experimental Station in 1990–1994 (Siimisker, 1998)

Cultivar/Age	*Yield, kg/shrub*						*Yield t/ha*	*Height (m)*
	1990	*1991*	*1992*	*1993*	*1994*	*Average*		
	3	*4*	*5*	*6*	*7*			
Otradnaya	11.5	21.7	8.0	19.9	16.9	15.6	19.5	4.1
Vorobevskaya	8.3	22.5	12.2	13.4	21.4	15.5	19.4	4.4
Botanicheskaya	13.2	19.6	12.3	13.4	21.4	15.5	19.4	3.6
Trophimovskaya	10.2	19.9	12.1	11.6	20.6	14.9	18.6	4.5
Podarok Sadu	7.2	21.7	10.7	14.3	17.9	14.4	17.9	4.1
Botanicheskaya								
Lubitelskaya	10	15.9	9.9	20.7	13.6	14	17.5	3.8
Avgustinka	9.9	16.2	11	14.8	17.2	13.8	17.3	4.3
Gibrid Perchika	6.9	17	9.9	10.7	20.4	13.4	16.7	4.6
Botanicheskaya								
Aromatnaya	6.2	18.6	6.4	14.8	20.1	13.2	16.5	4.2
Priokskaya	4.5	10.5	3.3	5	0.4	4.7	5.9	2.8
Chuiskaya	5.9	5.2	0.6	0.5	0	2.5	3.1	3.2
Podarok								
Chernozemu	2.7	4	2.4	2.3	0.1	2.3	2.8	3.1
Yantarnaya	3.8	4.9	0.2	1	0	2	2.5	3.1

Table 26.2: The Yield (kg) of Seabuckthorn Shrubs in International Comparison Trial at Rõhu Experimental Station in 2006 – 2014 (Hornig *et al.*, 2010; Toomas Tõrra unpublished data)

Cultivar/Years	*2006*	*2007*	*2008*	*2009*	*2010*	*2011*	*2012*	*2013*	*2014*	*Yield*
	2	*3*	*4*	*5*	*6*	*7*	*8*	*9*	*10*	
Otradnaya	5.7	3.0	7.7	0.5	6.5	4.1	0.3	5.7	0.0	33.5
Botanicheskaya Lubitelskaya	4.4	1.1	4.9	4.2	5.8	3.5	2.8	6.1	0.0	32.8
Podarok Sadu	2.5	1.4	8.7	0.5	7.8	3.1	0.4	5.9	0.0	30.3
Trophimovskaya	1.1	1.2	6.2	0.6	7.2	3.1	1.3	2.5	0.0	23.2
Gibrid Perchika	0.2	1.1	5.6	0.5	5.4	3.9	0.4	4.1	0.0	21.1
Askola	0.6	0.1	2.3	7.2	7.6	0.6	4.2	3.3	0.0	25.9
Sirola	0.6	0.8	4.1	1.9	3.6	1.9	0.6	5.0	0.0	18.5
Hergo	2.9	0.2	1.2	4.3	5.4	0.1	1.1	0.5	0.0	14.5
Habego	0.0	0.0	0.3	4.9	6.7	0.0	0.0	0.0	0.0	11.9
Dorana	2.2	0.0	0.4	4.2	5.4	0.0	0.0	0.0	0.0	12.2
Leikora	0.0	0.0	0.0	3.1	3.3	0.2	1.8	0.9	0.0	9.3
Frugana	0.0	0.0	0.0	2.0	3.7	0.0	0.0	0.0	0.0	5.7
Mean	1.7	0.7	3.5	2.8	5.7	1.7	1.1	2.8	0.0	

The same collection was used to study the effect of rejuvenation pruning on yield. Twelve-year-old seabuckthorn trees selected for this study were four to six meters tall. In the autumn of 1999, branches with berries were cut off. Before pruning, yield ranged to 11.7–21.4 kg/per shrub (Siimisker, 1998). In the year after pruning, there was no yield at all. Yield returned to its pre-pruning level in the third year after pruning. The average yield of the six cultivars was lower in 2001 (5.7 kg per shrub) and 2003 (4.0 kg per shrub) than in 2002 (15.6 kg per shrub) (Table 26.3). 'Trophimovskaya', 'Botanicheskaya', and 'Botanicheskaya Lubitelskaya' had the highest average yields. Cultivars with remarkably big berries were 'Botanicheskaya Lubitelskaya' (0.7–1.1g), 'Trophimovskaya' (0.7–0.9g) and 'Vorobevskaya' (0.7–0.8 g).

The Biochemical Content of Seabuckthorn Berries

Table 26.4 summarizes the results of Piir (1996), who studied the size and biochemical content of the berries of 14 seabuckthorn cultivars bred in Barnaul, Siberia,

Table 26.3: The Yield (kg) of Seabuckthorn Shrubs after Rejuvenation Pruning in 1999 (12th growing season) (Jalakas *et al.*, 2003)

Cultivar	*2001 14th Season*	*2002 15th Season*	*2003 16th Season*	*Total*
Trophimovskaya	6.5 ab	23.9 a	0.9 c	31.3 a
Botanicheskaya Lubitelskaya	4.2 ab	20.2 b	4.1 c	28.5ab
Botanicheskaya	5.6 ab	13.8 cd	8.4 a	27.8 b
Vorobevskaya	9.6 a	14.8 c	1.5 c	25.9 bc
Otradnaya	3.2 ab	11.5 d	6.4 ab	21.1 c
Avgustinka	4.3 ab	9.4 de	2.9 c	16.6 d
Mean	5.6	15.6	4	25.2

Note: Different letters with in a cultivar indicate significant differences at P=0.05 (Duncan's multiple range test).

Table 26.4: The Size of Berries and Biochemical Composition of Selected Seabuckthorn Cultivars from Lisavenko Research Institute of Horticulture for Siberia, Barnaul and in the Botanical Garden of the Moscow State University (Piir, 1996; Jalakas *et al.*, 2005)

Variable	*Unit*	*Natural Subspecies*	*Cultivars*		*Cultivar with Highest Level*
			Average	*Max*	
Average berry mass	g	0.2	0.5 ... 0.6	0.9	Velikan
Dry matter content	per cent	20.1	14.4 ... 15.5	19.1	Obskaya
Solubledry matter	per cent	10.7	9.4 ... 10.1	11.7	Samorodok
Titratedacids	per cent	2.7	2.2 ... 2.9	3.5	Orangevaya
Sugars	per cent	1.9	2.8 ... 4.1	5.6	Velikan
Ascorbicacid	m per cent	33	142 ... 154	365	Orangevaya
Pectines	per cent	1.2	0.9 ... 1.1	1.6	Vitaminnaya
Cellulose	per cent	6.8	4.2 ... 5.0	5.5	Otradnaya
Tannins and bioflavonoids	per cent	0.237	0.126 ... 0.134	0.286	Orangevaya

and 8 cultivars bred in Moscow, all grown at Polli Horticultural Centre. A comparison of the biochemical content of the cultivars berries and natural seedlings of *Hippophae rhamnoides* revealed that breeding had led to higher berry mass, sugar content and vitamin C content. Average berry mass and biochemical content results are similar for cultivars from Barnaul, Siberia, and Moscow; however, the berries of the cultivars from Barnaul had higher content of sugars. Cultivar 'Orangevaya' stood out from among other cultivars included in the study for its high amount of titrated acids, ascorbic acid, tannins and bioflavonoids. Piir and Kelt (1998) also found that cultivars 'Orangevaya' and 'Vitaminnaya' had the highest vitamin C content.

Seabuckthorn berries are rich in oil. Univer *et al.* (2015), who studied seven seabuckthorn cultivars, found that seabuckthorn seeds contained 8.36–9.89 per cent and juice 2.43–3.51 per cent of fatty acids (FA). Fatty acids contained 15 compounds (Tables 26.5 and 26.6). In the oil of seabuckthorn juice, the dominant fatty acids were mono-unsaturated fatty acids (MUPA), which contained mostly (37.16–43.07 per cent) palmitoleic acid (C16:1, ω-7). Palmitoleic acid is a unique fatty acid found in seabuckthorn juice and pulp. The highest levels of oleic acid ω-7 were observed in cultivar 'Gibrid Perchika' and oleic acid ω-9 in cultivars 'Botanicheskaya' and 'GibridPerchika'. The highest amount of palmitic acid (C16:0) was found in cultivar 'Otradnaya' (Table 26.5).

Univer *et al.* (2015) also found that the seeds of seabuckthorn berries were rich in polyunsaturated fatty acids (PUPA). The amounts of linoleic (C18:2, ω-6) and linolenic acids (C18:3, ω-3) were especially high at 37.09–38.52 per cent and 28.53–34.12 per cent, respectively. There was little variation among the seven cultivars in the fatty acid composition of seed oil. Statistically important differences ($p<0.05$) in the seeds of different cultivars were observed only in stearic acid content (C18:0). 'Vorobevskaya' contained the highest amount of stearic acid.

DISCUSSION

Comparison trials of cultivars from Russia, Byelorussia, Finland, and Germany revealed that cultivars from Moscow were best suited to the Estonian climate. Even the shrubs of Finnish cultivars grown in sod-floored orchards were severely damaged in the winter of 2012. However, Heinaaho (2014) has found that Finnish cultivars 'Terhi' and 'Tytti' can be successfully grown in orchards with a plastic mulch floor.

A higher level of different biochemical components indicates that a cultivar is valuable as a component

Table 26.5: Fatty Acid Composition in Seabuckthorn Juice Oil (Univer *et al.*, 2015)

Fatty Acid	*Seabuckthorn Cultivars*						
	Avgustinka	*Bot. Lubitelskaya*	*Botanicheskaya*	*Gibrid Perchika*	*Otradnaya*	*Trophimovskaya*	*Vorobevskaya*
14:0	0.37b	0.24ab	0.26ab	0.20a	0.28ab	0.20a	0.19a
15:0	0.11NS	0.10	0.10	0.09	0.11	0.10	0.11
16:0	36.50ab	36.10ab	33.71a	31.59a	39.02c	32.49a	32.50a
18:0	1.10NS	1.05	0.91	1.35	0.92	0.93	0.94
20:0	0.18 NS	0.23	0.20	0.16	0.20	0.16	0.17
22:0	0.10 NS	0.10	0.07	0.04	0.06	0.07	0.08
24:0	0.12 NS	0.11	0.18	0.09	0.08	0.12	0.09
SFA	38.46 NS	37.91	35.41	33.52	40.65	34.05	34.07
16:1,ω-7	41.82NS	37.74	40.01	37.16	37.97	43.07	42.21
18:1,ω-7	4.68a	5.76b	6.33b	7.58c	4.83a	6.25b	6.28b
18:1,ω-9	3.56a	4.63ab	7.26c	8.23c	4.42ab	5.61bc	6.09bc
20:1,ω-9	0.04 NS	0.03	0.04	0.08	0.03	0.02	0.02
22:1,ω-9	0.19 NS	0.31	0.17	0.15	0.14	0.06	0.06
MUPA	50.29 NS	48.47	53.80	53.20	47.39	55.00	54.65
18:2,ω-6	9.10 NS	11.28	9.49	12.03	10.36	9.46	9.87
18:3,ω-3	0.77 NS	0.92	0.78	0.80	0.84	0.72	0.77
20:2	0.00NS	0.04	0.03	0.00	0.00	0.00	0.00
PUPA	9.87 NS	12.24	10.29	12.83	11.19	10.18	10.64
	98.19	98.49	99.49	99.54	99.55	99.32	99.53

Notes: Data are expressed in percentage of relative fatty acids (per cent).

NS= Not significant, a,b,c,=Different letters with in a cultivar indicate significant differences at P=0.05 (Duncan's multiple range test).

Table 26.6: Fatty Acid Composition in Seabuckthorn Seed Oil (Univer *et al.*, 2015)

Fatty Acid	*Seabuckthorn cultivars*						
	Avgustinka	*Bot. Lubitelskaya*	*Botanic-heskaya*	*Gibrid Perchika*	*Otradnaya*	*Trophi-movskaya*	*Vorobevs-kaya*
14:0	0.12NS	0.12	0.15	0.12	0.12	0.13	0.15
15:0	0.19 NS	0.23	0.23	0.22	0.21	0.21	0.24
16:0	7.70 NS	7.20	7.95	7.69	6.97	7.87	8.71
18:0	2.66a	2.83ab	2.90ab	2.67a	2.78ab	3.10ab	3.48c
20:0	0.47 NS	0.47	0.53	0.49	0.48	0.47	0.57
22:0	0.13 NS	0.12	0.14	0.12	0.11	0.12	0.18
24:0	0.08NS	0.06	0.09	0.06	0.07	0.08	0.11
SFA	11.34 NS	11.02	11.97	11.35	10.73	11.97	13.42
16:1,ω-7	0.55 NS	0.39	0.96	0.65	0.62	1.38	1.58
18:1,ω-7	1.31 NS	1.38	1.73	1.48	1.38	1.6	2.04
18:1,ω-9	16.46 NS	13.13	14.54	14.92	16.19	16.19	15.59
20:1,ω-9	0.22 NS	0.21	0.18	0.20	0.20	0.18	0.18
22:1,ω-9	0.06 NS	0.08	0.08	0.07	0.07	0.10	0.13
MUPA	18.59 NS	15.17	17.48	17.31	18.45	19.44	19.52
18:2,ω-6	37.09 NS	38.52	37.79	37.79	38.01	37.51	37.24
18:3,ω-3	31.79 NS	34.12	31.54	31.76	31.68	29.86	28.53
20:2	0.05 NS	0.07	0.04	0.05	0.04	0.05	0.00
PUPA	68.93 NS	72.71	69.37	69.78	69.73	67.42	65.77
	99.38	99.37	99.19	98.50	99.30	99.21	99.15

Notes: Data are expressed in percentage of relative fatty acids (per cent).

NS=Not significant,a,b,c,=Different letters with in a cultivar indicate significant differences at P=0.05 (Duncan's multiple range test).

in different products as well as a parent in breeding. However, the composition of the seabuckthorn seed and pulp oils varies according to the subspecies, origins, cultivation activities, harvesting time of the berries, and the extraction method (Yang and Kallio, 2002b). In Estonian research, cultivar 'Orangevaya' stood out for its high amount of titrated acids, ascorbic acid, tannins and bioflavonoids; cultivar 'Vitaminnaya' had the highest vitamin C content. Prokkola and Lehto (2001) found a similar level of vitamin C in cultivars 'Terhi' and 'Tytti'.

Currently, some researchers maintain that the best ratio of ω-6/ω-3 fatty acids in food products for human health is 1 (Simopoulos, 2002). The results of Univer *et al.* (2015) were in agreement with earlier reports that the ratio of ω-3 and ω-6 fatty acid in seabuckthorn seed is close to 1:1 (Yang and Kallio, 2001; Wang and Utioh, 2009; Fatima *et al.*, 2012).

There was little variation among the seven cultivars in the fatty acid composition of the seed oil (Univer *et al.*, 2015). Fatima *et al.* (2012) have suggested that the fatty acid composition in seeds does not vary within subspecies and that the growing environment is a major factor controlling fatty acid composition. It has also been found that the oil content of seeds remained relatively constant from green to yellow stage, while the oil content of pulp and whole berries continued to increase (Fatima *et al.*, 2012).

CONCLUSIONS

Winter temperature is changeable in Estonia: freezing weather is followed by longer periods of thaw. Therefore seabuckthorn cultivars from regions with continental climate do not adapt well in Estonia. In addition, propensity to very high yields makes the cultivars less winter hardy in Estonia. The following cultivars are suited for commercial growing in Estonia: 'Avgustinka', 'Botanicheskaya', 'Botanicheskaya Lubitelskaya', 'Otradnaya' and 'Trophimovskaya'. These cultivars had relatively high average yields and are suited to new harvesting methods whereby fruit-bearing branches are cut off. Cultivar 'Orangevaya' is a promising one for breeding due to its outstanding biochemical characteristics.

Oil from the juice and pulp of seabuckthorn berries is rich in palmitic and palmitoleic acids, while oil from the seeds contains unsaturated fatty acids of C18-type oils, linoleic and linolenic acid. Statistically important differences in juice on C14:0, C16:0, C16:1, C18:1 and C18:2 were observed, whereas in seeds of the varieties only on C18:0. These results indicate that although the content of saturated and monounsaturated fatty acids in seabuckthorn juice oil appears to depend on cultivar, there was little variation in the fatty acid composition of the seed oil among the seven cultivars.

ACKNOWLEDGEMENT

We are thankful to Anu Randveer for helpful discussions, translation and editing.

REFERENCES

1. Bal, L.M., Meda, V., Naik, S.N. and Satya, S. 2011. Seabuckthorn berries: A potential source of valuable nutrients for nutraceuticals and cosmoceuticals. *Food Research International* 44: 1718-1727.
2. Fatima, T., Snyder, C.L., Schroeder, W.R., Cram, D., Datla, R., Datla, R., Wishart, D., Weselake, R. J. and Krishna, P. 2012. Fatty acid composition of developing seabuckthorn (*Hippophae rhamnoides* L.) berry and the transcriptome of the mature seed. *PLoS ONE* 7 (4): 1-18.
3. Folch, J., Lees, M. and Sloane Stanles, G.H. 1957. A simple method for the isolation and purification of total lipids from animal tissue. *J. Biol. Chem.* 226: 497-509.
4. Heinaaho, M. 2014. Effect of fertilizers, mulch and land contours on young seabuckthorn cultivars in organic farming. In: *Seabuckthorn* (*Hippophae* L.) *– A Multipurpose Wonder Plant. Vol. IV: Emerging Trends in Research and Technologies,* (Singh, V., Ed. in Chief), Daya Publishing House, New Delhi, pp. 41-58.
5. Hornig, R., Höhne, F. and Jalakas, M. 2010. Results of a German-Estonian seabuckthorn cultivar trial. www.sanddorn.net/euroworks.
6. Jalakas, M., Kelt, K. and Karp, K. 2003. The yield and fruit quality of seabuckthorn (*Hippophae rhamnoides* L.) after rejuvenation cutting. *Agronomy Research* 1: 31-36.
7. Jalakas, M., Piir, R. and Niiberg, T. 2005. *Astelpaju aias ja köögis*, Maalehe Raamat. p. 118. (In Estonian)
8. Kaar, E. 1971. Tasandatud puistangute metsastamine. In: *Põlevkivikarjääride rekultiveerimine*, Tallinn, pp. 77-98. (In Estonian)
9. Li, T.S.C. and Wang, L.C.H. 1998. Physiological components and health effects of ginseng, Echinacea and seabuckthorn. *Functional foods, Biochemical and Processing Aspects* 1: 329-356.
10. Piir, R. 1996. *Hippophae rhamnoides* L. in Estonia for fruit growing. *Agraarteadus* 7 (2): 162-176.
11. Piir, R. and Kelt, K. 1998. Chemical composition on *Hippophae rhamnoides* fruits in Estonia. *Forestry studies*, Estonian Agricultural University Forest Research Institute, 30: 134-139.
12. Siimisker, T. 1996. *Hippophae rhamnoides* cultivar evaluation at the Southern-Estonian cultivar testing farm. *Agraarteadus* 7 (2): 177-180.

13. Siimisker, T. 1998. Seabuckthorn cultivars trials at Rõhu Testing Station. *Forestry studies,* Estonian Agricultural University Forest Research Institute, 30: 162-168.

14. Simopoulos, A.P. 2002. The importance of the ratio of omega-6/omega-3 essential fatty acids. *Biomed Pharmacother*. 56: 253-264.

15. Sukhija, P.S. and Palmquist, D.L. 1988. Rapid method for determination of total fatty acids content and composition of feedstuff and feces. *J. Agric. Food Chem*. 36: 1202-1206.

16. Univer, T., Jalakas, M. and Kelt, K. 2004. Chemical compositions of the fruits seabuckthorn and how it changes during the harvest season in Estonia. *Journal of Fruit and Ornamental Plant Research,* Special Edition, 12: 339-344.

17. Univer, T. and Univer, N. 2014. The hardiness of seabuckthorn cultivars in the Estonian climate. www.sanddorn.net/euroworks.

18. Univer, T., Tiirmaa, K., Univer, N. and Kuusik, S. 2015.The fatty acid composition of the juice and seeds of selected seabuckthorn cultivars (Unpublished Manuscript).

19. Univer, T., Tiirmaa, K., Univer, N. and Kruuv, H. 2013.The results of harvesting the berries of common seabuckthorn in Estonia. *The 6th I.S.A. Congress Potsdam,* October 2013, Germany.

20. Wang, H. and Utioh, A. 2009. Supercritical fluid extraction of seabuckthorn (*Hippophae rhamnoides* L.) seed oil. In: *Proceedings of the 3rd International Seabuckthorn Association Conference,* Quebec, Canada, August 12-16, 2007, pp. 107-113.

21. Yang, B. and Kallio, H. 2001. Fatty acid composition of lipids in seabuckthorn (*Hippophae rhamnoides* L.) berries of different origins. *Journal of Agricultural and Food Chemistry* 49: 1939-1947.

22. Yang, B. and Kallio, H. 2002a. Composition and physiological effects of seabuckthorn (*Hippophae*) lipids. *Trends in Food Science and Technology* 13 (5): 160-167.

23. Yang, B. and Kallio, H. 2002b. Effects of harvesting time on triacylglycerols and glycerol phospholipids of seabuckthorn (*Hippophae rhamnoides* L.) berries of different origins. *Journal of Food Composition and Analysis* 15 (2): 14.

Chapter 27

Dynamics of Oil Content in Dry-Pulp of Seabuckthorn Berries in Half-sib Family and Different Berry Ripe Period

Yue Zhao[1], Cheng-Jiang Ruan[1*], Yang Li[1], Gui-Jiao Ding[1] and Kimmo Rumpunen[2]

[1]Institute of Plant Resources, Dalian Nationalities University, Dalian 116600, China
[2]Swedish University of Agricultural Sciences, Department of Plant Breeding, Balsgård, SE-29194 Kristianstad, Sweden
(*E-mail: ruan@dlnu.edu.cn)

SUMMARY

Seabuckthorn oils contain more than 100 bioactive compounds for human health, but oil contents are lower in seeds and pulp. Genetic improvement for oil content is the main aim of seabuckthorn breeding. In this study, we used micro-extraction to test oil content in dry-pulp of seabuckthorn berries from a half-sib family and different berry ripe time of two cultivars. The results showed there are significant differences in oil contents of dry-pulp among different F_1 individuals of BHi727115 half-sib family, which range from 12.64 to 26.94 per cent, with a mean value of 20.15 per cent. Significant differences in oil contents of dry-pulp did not found in different berry ripe time of two cultivars of BHi72715 and BHi7278, which ranged from 21.08 per cent to 25.09 per cent and from 20.43 per cent to 25.72 per cent, respectively. Our data first presented the change of oil content in berry pulp of seabuckthorn half-sib family, providing scientific basis for seabuckthorn breeding. The data for different ripe time may also provide index for harvest and processing of seabuckthorn berries.

***Keywords**: Seabuckthorn, Oil content in dry-pulp, Half-sib family, Different ripe time and Micro-extraction.*

INTRODUCTION

Seabuckthorn (*Hippophae* L., family *Elaeagnaceae*) is a deciduous woody perennial shrub or small tree, which is an important economic plant, with significant effects of water and soil conservation. This species has been called as "plateau holy fruit", because its berries rich in bioactive compounds (Chen *et al.*, 2009; Gu *et al.*, 2005), such as unsaturated fatty acid (Gu, 2003), flavonoids compounds (Hasiqimige and Gao, 2012), sterols (Chen and Gao, 2004), saccharides (Gu *et al.*, 2005), vitamins and amino acids (Gu *et al.*, 2005; Zhu *et al.*, 1998). Seabuckthorn oils made from its berries, which are also called as "21st century human edible oil", contain rich in bioactive compounds such as carotene, tocopherol, phylloquinone, sterol, polyunsaturated fatty acids (palmitoleic acid, palmitic acid, oleic acid, linoleic acid and linolenic acid *etc.*, Cakir, 2004) as well as Vitamin E and K (Wu and Zhao, 2004). These oils have significant effects in the treatment of cancer (Qiu and Zhang, 1998), tumor, cardiovascular disease (Liu, 2005), digestive system diseases (Zhao, 2000), burns, ulcers and other diseases have a good curative effect.

Oil content in the seeds of seabuckthorn berries is about 7-10 per cent, and only about 1-5 per cent in fresh-pulp. The requirement gap of seabuckthorn oil gap is more than 1200 tons per year in the world, it is only way to solve this huge gap to improve oil contents in the seeds and pulp of seabuckthorn berries. Previous studies have reported the oil contents in dry-pulp of different seabuckthorn cultivars, but there are few researches on oil content in dry-pulp during ripe time, and no report on changes of oil content in dry-pulp of half-sib family.

Seabuckthorn (Hippophae L.): A Multipurpose Wonder Plant, Vol. 5: Advanced Technologies for Cultivation, Processing, Health Protection and Environmental Conservation (Virendra Singh, *Editor-in-Chief*, 2018), p. 213–217, Astral International Pvt. Ltd., New Delhi, India.

Soxhlet extraction method, as first standard of oil extraction method (Dong, 2006), is the most widely used to extract oils from seeds and test oil content. However, this method required more materials and consumed long-time, it doesn't adaptive to carry out samples with less materials. Micro-extraction by the principle of extraction is a method bases on a small amount of samples (Xie and Li, 2012), which has been used in seabuckthorn (Xie and Li, 2012) and other oil plants (Qin, 2005; Dong *et al.*, 2011). In this study, we 1) compared the difference of oil contents in dry-pulp using two extraction methods of micro-extraction and soxhlet extraction, and then determined the oil extraction method adapted to samples with micro-amounts; and 2) analyzed the changes of oil contents in berry pulp among seabuckthorn half-sib family and cultivars during ripe time. These will provide scientific basis for seabuckthorn breeding, harvest and processing.

MATERIALS AND METHODS

Materials

In 2005, the seeds of *H. rhamnoides* subsp. *turkestanica* were collected in natural seabuckthorn forests in Yili of Xinjiang, China. These seeds were used to compare the oil extraction methods.

The samples used to test oil contents in dry-pulp of berries were from Balsgard, Sweden Agriculture University. In 2011, Dr. Cheng-Jiang Ruan harvested the berries of 59 individuals in a half-sib family of BHi727115 (*H. mongolica*). These individuals had the same female parent of BHi727115, but had potential different male parents. During ripe time, Dr. Cheng-Jiang Ruan harvested the berries of two cultivars of BHi72715 (*H. mongolica*) and BHi72782 (*H. mongolica*), harvest data included 3 August, 6 August, 9 August, 12 August, 15 August, 18 August, 21 August, 24 August, 27 August, 30 August, 2 September, 5 September and 8 September. The harvested berries were immediately put in the - 20°C cold room.

Comparison of Oil Extraction Methods

The seeds, which were dried in the 60°C oven for 24 h, were grinded into powder using universal pulverizer machine, which were be continued to dry in the 60°C for 24 h. The powders were used to extract oil using two methods of soxhlet extraction method and micro-extraction method.

The progress of soxhlet extraction method is as follows: 1) each part of the soxhlet extraction instrument were washed and dried in the 80°C oven, until reached to the constant weight; 2) the round-bottom flask's weight were tested and recorded, the seeds powder of about 15 g were put into filtration paper cylinder, and put into the siphon; then, the 15 times petroleum ether solvents (boiling point: 30-60°C)were pour on the top of powder; 3) the oils were extracted at the 56°C for 10 h; the mixed liquids were volatilized at 60°C for 30 min using rotary evaporators, placed in the vacuum desiccators at door temperature for 12 h, and then weighed. The oil content was calculated.

The progress of micro-extraction method is as follows: 1) the seeds powder of about 0.2 g were put into a 1.5 ml centrifuge tube, added 1 ml chloroform: methanol (2:1, v/v) solution and 200 μl glass beads; and then blended on the vortex mixing apparatus for 1 min and placed 1 min, with 5 repeats; 2) the mixed liquids were placed at 4°C refrigerator for 2 h, centrifuged at 7000 rpm for 2 min; then the supernatant were took out and added into to a new centrifugal tube, added 300μl 1 per cent KCl solution and mixed; 3) the mixed liquid were centrifuged at 7000 rpm for 1 min. The lower chloroform phase were transferred into a new centrifuge tube with known weight, and then the solvents were volatilized in the vacuum dryer, the weight were weigh with 24 h interval, until the difference between the two times' weight was less than 0.1 mg. The data were recorded and the oil content was calculated.

There were 15 repeats for each oil extraction method. The differences of seed oil contents for two extraction methods were analyzed using SPSS13.0 version software.

Oil Contents in Dry-Pulp for Half-Sib Family and different Berry Ripe Time

The frozen berries at -20°C cold room were taken out, the berries were cut at ice surface, the seeds were eliminated, and the pulp were grinded using liquid nitrogen grounding method at mortar. Then, the pulp were dried in a -50°C vacuum freezing dryer for 10 days, and the dry powder were kept in a vacuum dryer. Before the oils were extracted using micro-extraction methods, the dry power were dried at 60°C oven until constant weight.

The oil contents in dry-pulp of 59 individuals of BHi727115 half-sib family were tested using the micro-extraction method, with 5 repeats for each sample. For the cultivars of BHi72715 and BHi72782, the oil contents of dry-pulp at 13 ripe time were tested using this method, and also with 5 repeats for each sample. The changes of oil contents in dry-pulp among individuals of half-sib family and different ripe times were analyzed using SPSS13.0 version software.

RESULTS AND DISCUSSION

Comparison of Oil Exaction Methods

The oil content in seeds of *H. rhamnoides* subsp. *turkestanica* was 12.37 per cent ± 0.396 per cent (n = 15) (Figure 27.1) using the micro-extraction method. There was not significant difference among different repetitions, indicating the micro-extraction method was stable for the determination of oil content. By comparison, the oil content was 9.63 per cent ± 1.037

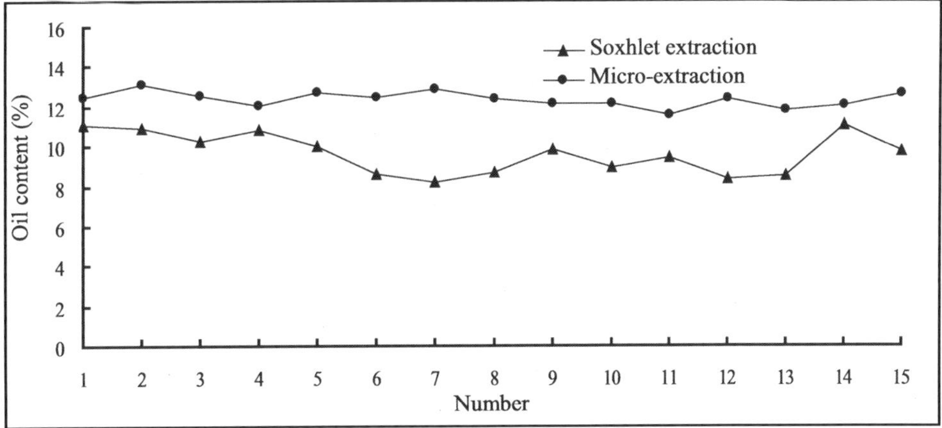

Figure 27.1: Comparison of Oil Contents in Seeds Extracted by Two Methods of Soxhlet Extraction and Micro-extraction.

per cent (n = 15) using the soxhlet extraction method, with a significant difference among different repeats (Figure 27.1).

There was a significant difference in oil contents between two methods of micro-extraction and soxhlet extraction ($F_{1,14}$ = 16.427, $p < 0.001$). For seeds of *H. rhamnoides* subsp. *turkestanica*, the content of oil extracted by micro-extraction was higher than that of oil extracted by the soxhlet extraction. Dong *et al.* (2011) also showed that the soxhlet extraction method couldn't extract the combined state lipids such as phospholipids, glycolipids lipid, lipoprotein and sterol in seeds of seabuckthorn.

The soxhlet extraction is time-consuming and have the requirement of lots of materials; by comparison, the micro-extraction method only need a small amount materials, and its result was stable, which could extract most oil components including grease. Thus, the micro-extraction method is adaptive for the determination of oil content in seabuckthorn berries.

Changes of Oil Contents in Dry Pulp of Berries among Half-sib Family

The oil content in dry pulp of BHi727115 half-sib family ranged from 12.64 per cent to 26.94 per cent (Figure 27.2), with a mean of 20.15 per cent ± 0.399 per cent (n = 59). In 59 individuals of BHi727115 half-sib family, No. 51 individual had the highest oil content in dry pulp (26.94 per cent), the lowest oil content in dry-pulp was found in No. 4 individual (12.64 per cent). The oil contents in dry-pulp for different individuals of this half-sib family showed a normal distribution (Figure 27.3, sig = 0.662 > 0.05, S = 0.228 and K = 0.049 < 1). This indicated that the oil content in dry pulp had the characteristic of quantitative trait, which may be controlled by the multiple genes. The oil content in dry-pulp was lower than that of seabuckthorn berries (34 per cent) reported by Yang and Kallio (2001), which may be from the difference of subspecies.

Changes of Oil Contents in Dry-Pulp of different Berry Ripe Time

The oil contents in dry-pulp of BHi72782 in different ripe time ranged from 20.43 per cent to 25.72 per cent (Figure 27.4), with a mean of 23.12 per cent. Since August 3, the oil contents in dry-pulp increased gradually, and reached the highest value on August 24; after this date, it was stable. This indicated that the berries of cultivar BHi72782 were fully mature at about August 24, which could be harvested in order to obtain berries with high oil content. This change of oil contents in dry-pulp is constant with the results of Kuznetsova *et al.* (2010), in which the oil content was significantly decrease after 80 days seabuckthorn bears berries, in that time the berries were mature.

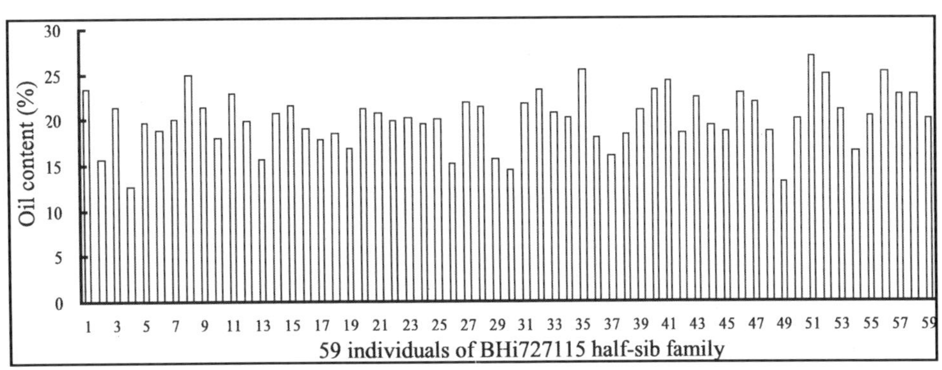

Figure 27.2: Oil Contents in Dry-Pulp of different Individuals of BHi727115 Half-Sib Family.

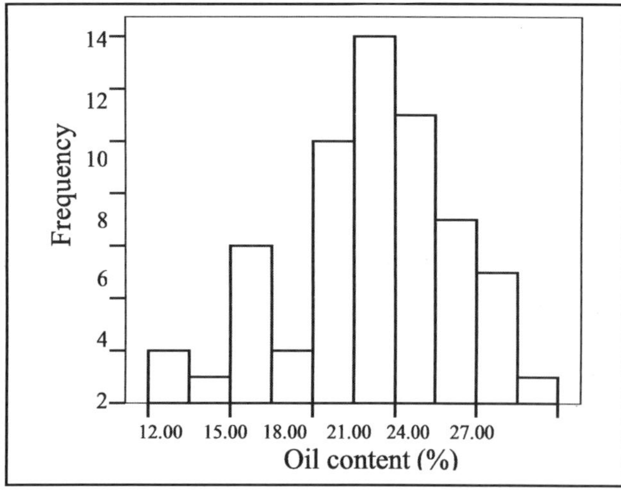

Figure 27.3: Frequency Distribution of Oil Contents in Dry-Pulp of BHi727115 Half-Sib Family.

The oil contents in dry-pulp of BHi72715 ranged from 21.08 per cent to 25.09 per cent in different ripe time (Figure 27.4), with a mean of 22.75 per cent. From August 3 to August 24, the oil contents of dry pulp of BHi72715 decreased gradually; and then it was almost stable from August 27 to September 8.

There are significant differences in the changes of oil contents in dry pulp in different ripe time between two cultivars of BHi72782 and BHi72715, the former had a tendency to increase and the latter had a stable tendency. By combination of berries' color, the BHi72715 cultivar had early mature date; On August 3, its berries were bigger and fully development, the berries' color were yellow; after August 3, the size of berries of BHi72715 was almost stable, only its color changed to orange at the mid-to end of August. In contrast, the BHi72782 cultivar had late mature ate, its berries were small and not fully

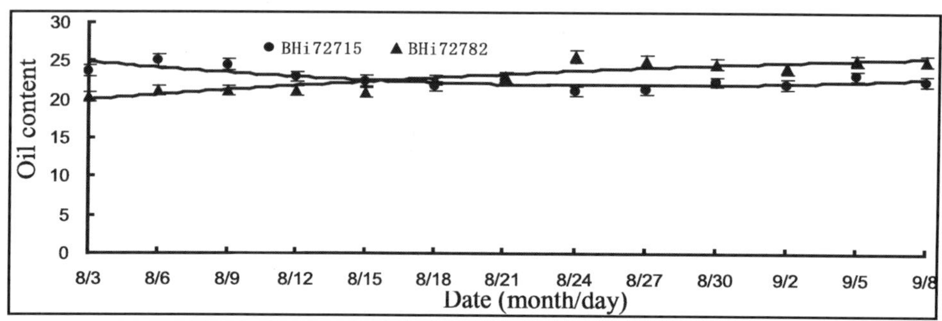

Figure 27.4: Oil Contents in Dry-Pulp of Two Cultivars of BHi72715 and Bhi72782 during Ripe Time.

development on August 3, with color of green-yellow; after then, the berries became bigger and bigger, and the berries' color became yellow at the end of August.

CONCLUSION

There are many methods of oil extraction, mainly including the soxhlet extraction (Singh *et al.*, 2004), ultrasonic extraction and enzyme extraction (Kang *et al.*, 2011), micro-extraction (Xie and Li, 2012) and supercritical CO_2 extraction method (Yin *et al.*, 2002; Xu, 2007). Our study compared soxhlet extraction with micro-extraction methods for extracting oils in the seeds of seabuckthorn seed oils. The oil content measured by micro-extraction was higher than tat of soxhlet extraction. The contents from micro-extraction were stable among different repeats, and its operation was save time and effort, with requirement of a small amount of materials and agents. Thus, the micro-extraction method was adaptive to measure oil content in the berries of seabuckthorn.

The seabuckthorn oil is a special bioactive oil with high nutritional and medical values, but it's demand gap is huge in the world. Thus, it is interesting to find individuals with high oil content in berries and determine the optimal harvest time for gaining more oils. Our data showed the No.54 individuals of F_1 generation half-sib family population of BHi 72715 had highest oil content (26.94 per cent), which could be selected as superior lines for seabuckthorn improvement. There were significant difference in oil content between different cultivars during ripe time, it is necessary to determine optimal harvest dates of different cultivars for gaining the most oil, for example, the optimal harvest date was about August 24 for BHi72785, but about August 3 for BHi72715.

ACKNOWLEDGEMENTS

This research was jointly supported by a Marie Curie International Incoming Fellowship within the 7th European Community Framework Programme (PIIF-GA-2010-272048-HIPPOHEALTH, PIIFR-GA-2010-910048-HIPPOHEALTH), and the Technology Foundation for Selected Overseas Chinese Scholar, Ministry of Personnel of China "Study on breeding seabuckthorn with resistance to dried-shrink disease and selection of new-lines.

REFERENCES

1. Cakir, A. 2004. Essential oil and fatty acid composition of the fruits of *Hippophae rhamnoides* L. (seabuckthorn) and *Myrtus communis* L. from Turkey. *Biochemical Systematics and Ecology* 32: 809-816.

2. Chen, X.Y., Liu, H.Z. and Liu, S.Y. 2009. Extraction and analysis of fatty acid of seabuckthorn. *Journal of Jilin Agricultural University* 31 (5): 628-631,636.
3. Cheng, K.H. and Gao, Y.J. 2004. Comprehensive development and utilization of seabuckthorn oil. *Journal of Chemical Industry of Forest Products* 38 (6): 42-45.
4. Dong, S.Q. 2006. Studies on nutritional assessment of pumpkin seeds: Seed oil extraction and isolation. *Master Thesis,* Hebei Agricultural University, pp. 23-46.
5. Dong, X.L., Bai, P.L., Wang, J.M. and Ruan, C.J. 2011. Comparative study on determination of seed oil content of energy plants by using NMR and Soxhlet extraction. *Renewable Energy Resources* 29 (3): 21-24.
6. Gu, Q.P. 2003. Seabuckthorn development and utilization of medicinal value. *Journal of International Seabuckthorn Research and Development* 1 (2): 28-30.
7. Gu, R., Zhang, Y. and Lai, X.R. 2005. Analysis on composing of fatty acid of seabuckthorn fruit oils by GC-MS and determination of linoleic acid. *Journal of Chengdu University of TCM* 28 (4): 49-51.
8. Hasiqimige, and Gao, B.W. 2012. Advance in the exploitation and utilization of *Hippophae rhamnoides*. *Hubei Forestry Science and Technology* (178): 42-44.
9. Kang, J., Gu, J.J. and Wang, J.G. 2011. Enzymatic extraction of seabuckthorn fruit oil and its fatty acids profile analysis. *Food Science* 32 (2): 260-262.
10. Kuznetsova, E.I., Pchelkin, V.P., Tsydendambaev, V.D. and Vereshchagin, A.G. 2010. Distribution of unusual fatty acids in the mesocarp triacylglycerols of maturing seabuckthorn fruits. *Russian Journal of Plant Physiology* 57 (6): 852-858.
11. Liu, F.Y. 2005. Pharmacological research of seabuckthorn oil. *Bulletin of Biology* 40 (2): 13-15.
12. Qiu, H. and Zhang, Z.M. 1998. The experimental evidence of seabuckthorn preparations resistance of malignant neoformation transfer performance. *Hippophae* 11 (1): 32-33.
13. Qing, R.W. 2005. Identification and characterization of the key enzyme genes in the lipid biosynthesis from *Jatropha curcas* L. *Ph.D Thesis*, Sichuan University, pp. 36-59.
14. Singh, V., Gupta, R.K., Tandon, S. and Sawhney, R.C. 2004. Fatty acid composition of fruit pulp and seed oils of some selected forms of seabuckthorn (*Hippophae rhamnoides* L.). *The Global Seabuckthorn Research and Development* 2 (3): 1-6.
15. Wu, F.H. and Zhao, Y.Z. 2004. The research and development of seabuckthorn oil (*Hippophae rhamnoides* L.) in Russia. *The Global Seabuckthorn Research and Development* 2 (2): 1-5.
16. Xu, X. 2007. Study on the supercritical CO_2 extraction parameters on the contents of functional ingredients. *Post-Graduate Thesis*, China Agricultural University, pp. 13-22.
17. Xie, J.Q. and Li, H. 2012. Micro analysis of fatty acids in seed oil in seabuckthorn (*Hippophae* L.). *Hubei Agricultural Sciences* 51 (15): 3319-3322.
18. Yang, B.R. and Kallio, H.P. 2001. Fatty acid composition of lipids in seabuckthorn (*Hippophae rhamnoides* L.) berries of different origins. *Journal Agricultural and Food Chemistry* 49 (4): 1939-1947.
19. Yin, J.Z., Sun, X.W. and Li, Z.Y. 2002. Experimental research of *Hippophae rhamnoides* L. seed oil with supercritical fluid extraction. *Chemical Engineering* 30 (4): 13-15.
20. Zhao, Y.P. 2000. The refined processes and applications of *Hippophae* oil. *Journal of Yunnan Normal University* 20 (2): 48-50.0
21. Zhu, D., Li, P.B. and Sun, X.C. 1988. Analysis of seabuckthorn seed oils and fruit oils. *Territory and Natural Resources Study* (1): 74-76.

Chapter 28

HPTLC Characterization of Phytoconstituents in Seabuckthorn Leaves: A Comparative Study of Different Species

Jigni Mishra[1], Rakhee[1], Priyanka Sharma[1], Rajkumar Tulswani[1], Virendra Singh[2] and Kshipra Misra[1*]

[1] Defence Institute of Physiology and Allied Sciences Lucknow Road, Timarpur, Delhi – 110 054
[2] College of Basic Science, Himachal Pradesh Agricultural University, Palampur – 176 062, Himachal Pradesh
(*E-mail: kmisra99@yahoo.com)

SUMMARY

For the very first time, this paper reports a comparative study regarding characterization of phytoconstituents from aqueous and alcoholic leaf extracts of three different varieties of seabuckthorn, namely *Hippophae salicifolia, H. rhamnoides mongolica* and *H. rhamnoides turkestanica*. The alcoholic and aqueous extracts from leaves of these varieties of seabuckthorn were prepared by accelerated solvent extraction (ASE). A method for characterizing the extracts in terms of nitrogen bases and flavonoids using high performance thin layer chromatography (HPTLC) was developed. The seabuckthorn extracts were also evaluated for their antioxidant activities with respect to two different free radicals *viz.*, nitric oxide (NO·) and superoxide ($O_2^{-\cdot}$). Results from HPTLC analysis of the extracts indicated the presence of thymine, uracil, adenine, cytosine, guanine and guanosine among nitrogen bases, ranging from 0.23 µg nitrogen base/mg of extract to 7.76 µg nitrogen base/mg of extract. The flavonoids identified in the seabuckthorn extracts were quercetin, gallic acid, ascorbic acid and rutin, ranging from 0.57 µg flavonoid/mg of extract to 15.37 µg flavonoid/mg of extract. Principal component analysis (PCA) technique was used for validation of the results obtained from HPTLC. The antioxidant assays demonstrated substantial free radical scavenging potential of all the three species, where IC_{50} value for $O_2^{-\cdot}$ scavenging activity ranged from 16.18 µg/mL to 84.76 µg/mL and that for NO· scavenging activity ranged from 54.13 µg/mL to 241.07 µg/mL. Though any typical trend could not be observed in the values obtained, still the extracts were found to be significantly rich in nitrogen bases and flavonoids.

***Keywords**: Seabuckthorn, Antioxidant activities and HPTLC fingerprinting.*

ABBREVIATIONS

Hippophae salicifolia aqueous extract- HS1, *Hippophae salicifolia* alcoholic extract- HS2, *Hippophae rhamnoides* ssp. *mongolica* aqueous extract- HRM1, *Hippophae rhamnoides* ssp. *mongolica* alcoholic extract-HRM2, *Hippophae rhamnoides* ssp. *turkestanica* aqueous extract-HRT1, *Hippophae rhamnoides* ssp. *turkestanica* alcoholic extract-HRT2, HPTLC- high performance thin layer chromatography, ROS-Reactive oxygen species, ASE-Accelerated Solvent Extractor, BDL-Below Detection Limit, IC_{50}- amount of extract needed to inhibit free radical concentration by half.

INTRODUCTION

Seabuckthorn (*Hippophae* L., *Elaeagnaceae*), a medicinal deciduous shrub growing plentifully in the cold and arid regions of India, China, Russia and certain North European countries has been extensively studied by researchers in India and abroad alike and has been found to bear a number of beneficial uses,

Seabuckthorn (Hippophae L.): A Multipurpose Wonder Plant, Vol. 5: Advanced Technologies for Cultivation, Processing, Health Protection and Environmental Conservation (Virendra Singh, *Editor-in-Chief*, 2018), p. 219–226, Astral International Pvt. Ltd., New Delhi, India.

right from being an ingredient in traditional medicines and nutraceuticals to carrying out nitrogen fixation and soil conservation. Seabuckthorn is a vast repository of different phytoconstituents that demonstrate bioactivities of varying degrees. Seabuckthorn has been reported to be rich in flavonoids and carotenoids that help in attenuating cardiovascular risks by lowering cholesterol, platelet aggregation and blood pressure (Xu *et al.*, 2011). Berries of seabuckthorn are functional in controlling lipid metabolism and inflammation owing to the presence of polyphenols, carotenoids, fatty acids and vitamins (Sayegh *et al.*, 2014). Michel *et al.* (2012) have reported that the phenolic compounds present in active fractions of seeds, stems and leaves of seabuckthorn impart antimicrobial action against both Gram +ve and Gram –ve microorganisms. Different parts of *Hippophae rhamnoides*, the common seabuckthorn are quite rich in sugars, flavonoids and phytosterols (Tiitinen *et al.*, 2006) that contribute to its appreciable antioxidant activity. Geetha *et al.* (2002) studied extracts obtained from leaves and fruits of seabuckthorn and confirmed that these extracts possess immunomodulating power, hence encouraging proliferation of lymphocytes.

Much of the research pertaining to resourcefulness of seabuckthorn so far has majorly been focused on *Hippophae rhamnoides turkestanica*; whereas two other species of the same genus, *Hippophae salicifolia* and *Hippophae rhamnoides mongolica* are also known to have appreciable bioefficacy (Sharma *et al.*, 2014). Moreover, characterization and biological effects of phytoconstituents like polyphenols, carotenoids, lipids and vitamins in seabuckthorn have been studied vividly but literature pertaining to the identification and characterization of certain other biomolecules like nitrogen bases is quite limited. Nitrogen bases have been documented to have different types of bioactivities, for example, uracil derivatives exert appreciable antiviral and antitumor action (Palasz and Ciez, 2014; Wnuk *et al.*, 2002); cytosine supplementation contributes towards controlling plant diseases like powdery mildew (Huang *et al.*, 2010) and so on. Also flavonoids are known to be quite efficient in treatment of urinary tract infections, cognitive and cardiovascular diseases (Rodriguez-Mateos *et al.*, 2014). Rutin has been reported to possess high antioxidant activity (Chen *et al.*, 2015).

Here in this paper, for the first time, a chemometric analysis of phytoconstituents like flavonoids and nitrogen bases in aqueous and alcoholic leaf extracts of the three aforementioned varieties of seabuckthorn *viz.*, *H. salicifolia*, *H. rhamnoides mongolica* and *H. rhamnoides* ssp. *turkestanica*, on a comparative basis has been attempted using high performance thin layer chromatography (HPTLC). The utilization of HPTLC as a means of identification and characterization has been preferred due to its marked simplicity as an elaborate and automated procedure (El-Gindy *et al.*, 2001). This paper also deals with the antioxidant efficacy of leaf extracts from these three seabuckthorn species. Pomace (Varshneya *et al.*, 2012) and seeds (Negi *et al.*, 2005) from *H. rhamnoides* have already been recently documented to have free radical scavenging effects. Thus here as a novel objective, we examine and compare the antioxidant potentials of extracts derived from the leaves of three different varieties of seabuckthorn.

MATERIALS AND METHODS

Sample Materials

Seabuckthorn plant materials HS, HRM and HRT with specimen numbers CSK HPKV-NAIP-SBT-SP 9, 10 and T1 respectively, were authenticated by Dr. Virendra Singh from CSK, Palampur, Himachal Pradesh. These plants are cultivated in the Seabuckthorn Research Farm of CSK Himachal Pradesh Agricultural University at Kukumseri in Lahaul valley of District Lahaul-Spiti in Himachal Pradesh, India, at an altitude of 2730 m. The age of the plant at the time of leaf collection was 8 years.

Chemicals

The standards and reagents for conducting HPTLC and antioxidant assays were purchased from Sigma-Aldrich and all the chemicals used were of analytical grade. Water used for preparation of reagents and in other purposes was of Millipore grade (Merck, USA).

Preparation of Extracts

Seabuckthorn extracts were prepared by a method as given by Sharma *et al.* (2014). According to this method, a total of six extracts, one aqueous and one alcoholic for each of the three selected seabuckthorn varieties (HS, HRM, and HRT), were prepared using ASE (Dionex 350 ASE model). Extraction was carried out in triplicates where extraction solvent was double distilled water (for aqueous extracts) and 70 per cent ethanol (for alcoholic extracts) at room temperature. Ethanol from the alcoholic extracts was removed by evaporation using Buchi Rotavapor at 40 °C succeeded by lyophilization (Allied Frost FD 5). The extracts collected were preserved at 4 °C, till further use. The aqueous extracts were labeled as HS1, HRM1 and HRT1 while the alcoholic equivalents were named as HS2, HRM2, and HRT2 for *H. salicifolia*, *H. rhamnoides mongolica* and *H. rhamnoides* ssp. *turkestanica* respectively.

HPTLC Analysis of the Extracts

Preparation of Standards and Calibration

Nitrogen Bases

Stock solutions of the nitrogen base standards namely thymine (thy), uracil (ura), adenine (aden), cytosine (cyt), guanine (guan) and guanosine (guanos) were prepared by dissolving 0.5 mg of each standard in 1 mL of methanol and vortexing the same for 10 min. A mixture comprising equal volumes from each of these standards was prepared and was used for further analyses.

Flavonoids

Stock solutions of the flavonoid standards namely quercetin (que), gallic acid (gal), ascorbic acid (asc), hesperidin (hesp) and rutin (rut) were prepared by dissolving 1 mg of each standard in 1 mL of methanol. A mixture comprising equal volumes from each of these standards was prepared, that was used in further analyses.

Preparation of Sample Solutions

20 mg of each seabuckthorn extract (HS1, HS2, HRM1, HRM2, HRT1 and HRT2) under study was weighed and dissolved in 10 mL of methanol, thus resulting in a sample concentration of 2 mg/mL. This sample solution was kept for further use in HPTLC analysis.

Chromatography

Nitrogen Bases

The chromatographic analysis for identifying and quantifying nitrogen bases and nucleosides in the seabuckthorn extracts was brought about on a 20x10 cm glass backed silica gel 60 F_{254} HPTLC plate (Merck) (already activated with methanol and air dried). 60 µL from each sample solution was applied as 6 mm wide bands, using a CAMAG Linomat 5 sample applicator equipped with a 100 µL syringe (Hamilton). Also, three different volumes (5, 6, 7 µL) of the mixture solution containing nitrogen base standards were spotted similarly that are labeled in Figure 28.1(a) as Std(1), Std(2) and Std(3). The plate was then developed at room temperature in a CAMAG twin-trough vertical development chamber, previously saturated for 10 min with the appropriate mobile phase (dichloromethane: methanol: formic acid in a ratio of 8: 2.25: 0.8, by volume). The migration distance for the samples was maintained at 85 mm. Following this, the plate was subjected to densitometric scanning at a wavelength of 254 nm using a CAMAG TLC Scanner 3, utilizing a scanning speed of 20 mm s^{-1} and slit dimension of 6.0 mm x 0.45 mm, with deuterium as the light source. The entire procedure for scanning was realized using Win Cats software (version 1.4.4.6337). Also, quantitation of the standard nitrogen bases that were identified in the extracts was carried out.

Flavonoids

Identification and quantization of flavonoids in the seabuckthorn extracts was performed on a 20x10 cm glass backed silica gel 60 F_{254} HPTLC plate (Merck) following a similar procedure as laid out above for chromatographic analysis of nitrogen bases. Three different volumes (7, 8, 9 µL) of the mixture solution containing flavonoid standards were spotted similarly that are labeled in Figure 28.1(b) as Std(1), Std(2) and Std(3). The mobile phase prepared here was ethyl acetate: dichloromethane: formic acid: glacial acetic acid: methanol in a ratio of 10: 10: 1: 1: 2, by volume. Quantification of the standard nitrogen bases identified in the extracts was carried out.

Chemometric Analysis

PCA (Principal Component Analysis) and Cluster analysis were conducted using the software R: A Language and Environment for Statistical Computing (R Foundation for Statistical Computing, Vienna, Austria. Http://Cran.R-Project.Org/), with Prior Mean Centering.

Evaluation of Antioxidant Activity

Superoxide Anion Radical Scavenging Activity

The superoxide anion radical, ($O_2^{-\cdot}$) scavenging potential for the six seabuckthorn extracts under study was assessed using the method as described by Alkan *et al.* (2014). Superoxide radical was produced by adding sodium hydroxide to dimethyl sulfoxide (DMSO). To 100 µL of alkaline DMSO, 10 µL of nitrobluetetrazolium (NBT) chloride salt and 30 µL of respective seabuckthorn extract in various concentrations (10-100 µg/mL) were added, resulting in a final reaction mixture of 140 µL. The absorbance of this reaction mixture was measured at 560 nm using a microplate reader (BioTek PowerWave XS2, USA). The superoxide anion scavenging activity was expressed as corresponding IC_{50} values in µg/mL. Ascorbic acid in different concentrations (10-100 µg/mL) was taken to generate the standard curve.

Nitric Oxide Scavenging Activity

The nitric oxide radical ($NO^{\cdot}$) scavenging activity of the seabuckthorn extracts was estimated by the method described by Harput *et al.* (2011). To the wells of a 96-well microplate, 60 µL of the respective seabuckthorn extract in various concentrations (10-50 µg/mL) was added, followed by 60 µL of 10 mM sodium nitroprusside dissolved in 1X phosphate-buffered saline (PBS). Then the microplate was incubated at room temperature for 150 min, after which 60 µL of Griess reagent was added, resulting in formation of a chromophore complex whose absorbance was observed at 577 nm. The nitric oxide scavenging activity was expressed as corresponding IC_{50} values in µg/mL. Gallic acid in different concentrations (10-50 µg/mL) was taken to generate the standard curve.

RESULTS

Percentage Yields of The Extracts

The yield of the extracts obtained from ASE was calculated in terms of percentage yield and it was seen that HRT1 gave the highest value (Sharma *et al.*, 2014). The order of percentage yield could be put down as follows: HRT1>HS1>HRM1 and HS2>HRM2>HRT2.

HPTLC Analysis

A method for separation of nitrogen bases as well

Figure 28.1: Structures of the Standard Nitrogen Bases and Flavonoids Identified through HPTLC.

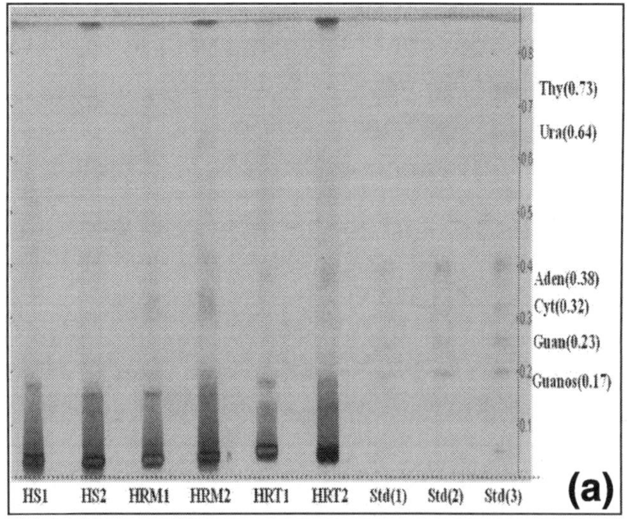

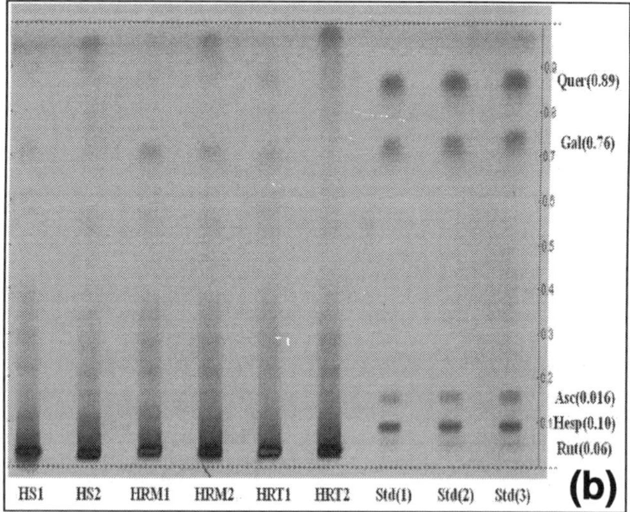

Figure 28.2: HPTLC Profiling Showing R_f Values of Standard N Bases (a) and Flavonoids (b) Found in Seabuckthorn Leaf Extracts at 254 nm.

as for flavonoids using HPTLC was developed after a number of attempts. The structures of the standards selected for identification are given in Figure 28.1. HPTLC fingerprinting (Figure 28.2a) of all the six seabuckthorn leaf extracts confirmed the presence of nitrogen bases namely thymine, uracil, adenine, cytosine, guanine and guanosine in them. The quantities of nitrogen bases identified in the seabuckthorn extracts are represented in Table 28.1 in terms of micrograms (μg) of nitrogen base per milligram (mg) of extract [μg/mg]. Similar analysis for flavonoids in the seabuckthorn leaf extracts demonstrated the presence of quercetin, gallic acid, ascorbic acid and rutin in them (Figure 28.2b) and the corresponding quantities (in μg flavonoid/mg of extract) are given in Table 28.2. Figures 28.3(a) and 28.3(b) depict the corresponding 3D spectra of the nitrogen bases and flavonoids respectively, detected at 254 nm.

Chemometric Analysis

Cluster dendrograms and heat maps for pointing out the occurrence of a particular compound at a given

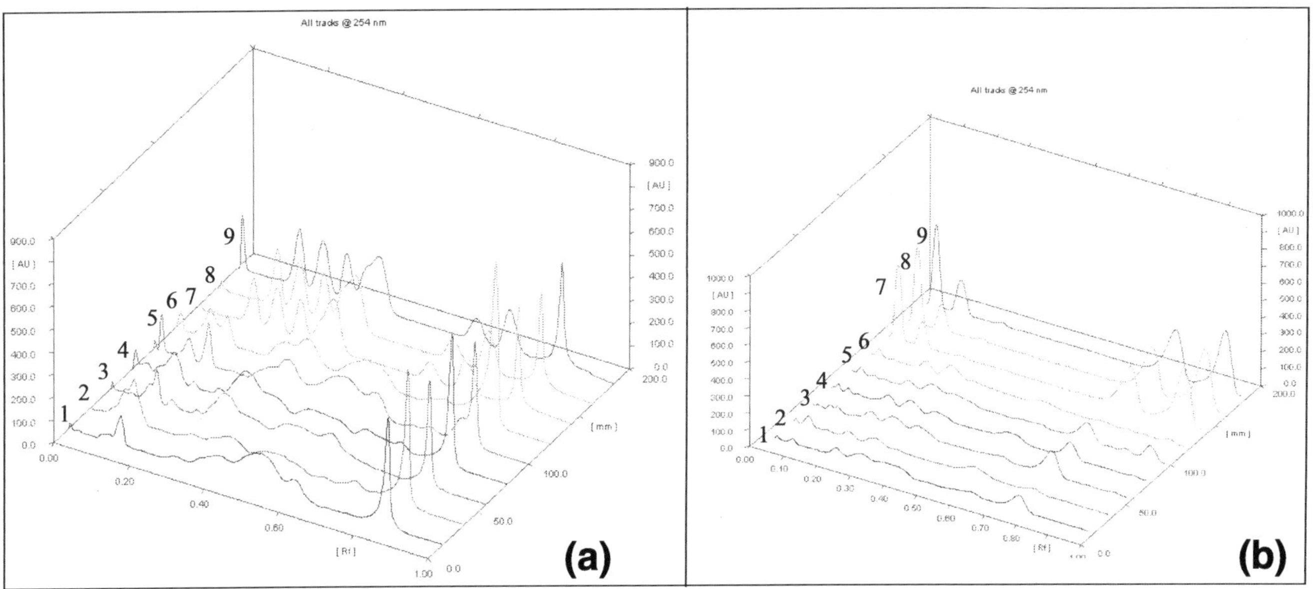

Figure 28.3: 3D Display Spectra of all Tracks at 254 nm of Nitrogen Bases (a) and Flavonoids (b) in Seabuckthorn Leaf Extracts.

28.3(a): Track 1: HS1; 2: HS2; 3: HRM1; 4: HRM2 5: HRT1; 6:HRT2; 7,8,9: Mixture of standard nitrogen bases; 28.3(b): Track 1: HS1; 2: HS2; 3: HRM1; 4: HRM2 5: HRT1; 6:HRT2; 7, 8, 9: Mixture of standard flavonoids.

Table 28.1: Quantities of Nitrogen Bases in the Seabuckthorn Leaf Extracts

Nitrogen base (µg Nitrogen base/ mg extract)	HS1	HS2	HRM1	HRM2	HRT1	HRT2
Thymine	BDL	1.18	BDL	0.80	0.89	0.91
Uracil	2.91	0.76	0.61	3.13	2.83	BDL
Adenine	1.12	1.62	BDL	0.52	1.96	2.15
Cytosine	0.83	BDL	7.3	7.67	0.59	2.60
Guanine	BDL	BDL	0.23	0.32	BDL	BDL
Guanosine	2.56	BDL	0.81	0.48	4.13	2.04

Table 28.2: Quantities of Flavonoids in the Seabuckthorn Leaf Extracts

Flavonoid (µg flavonoid/ mg extract)	HS1	HS2	HRM1	HRM2	HRT1	HRT2
Rutin	0.91	2.84	0.57	0.77	1.33	1.60
Ascorbic acid	9.00	9.38	11.13	14.55	8.55	15.37
Gallic acid	7.14	6.67	6.98	5.70	5.56	BDL
Quercetin	BDL	BDL	BDL	BDL	4.18	0.90

BDL: Below Detection Limit of the instrument.

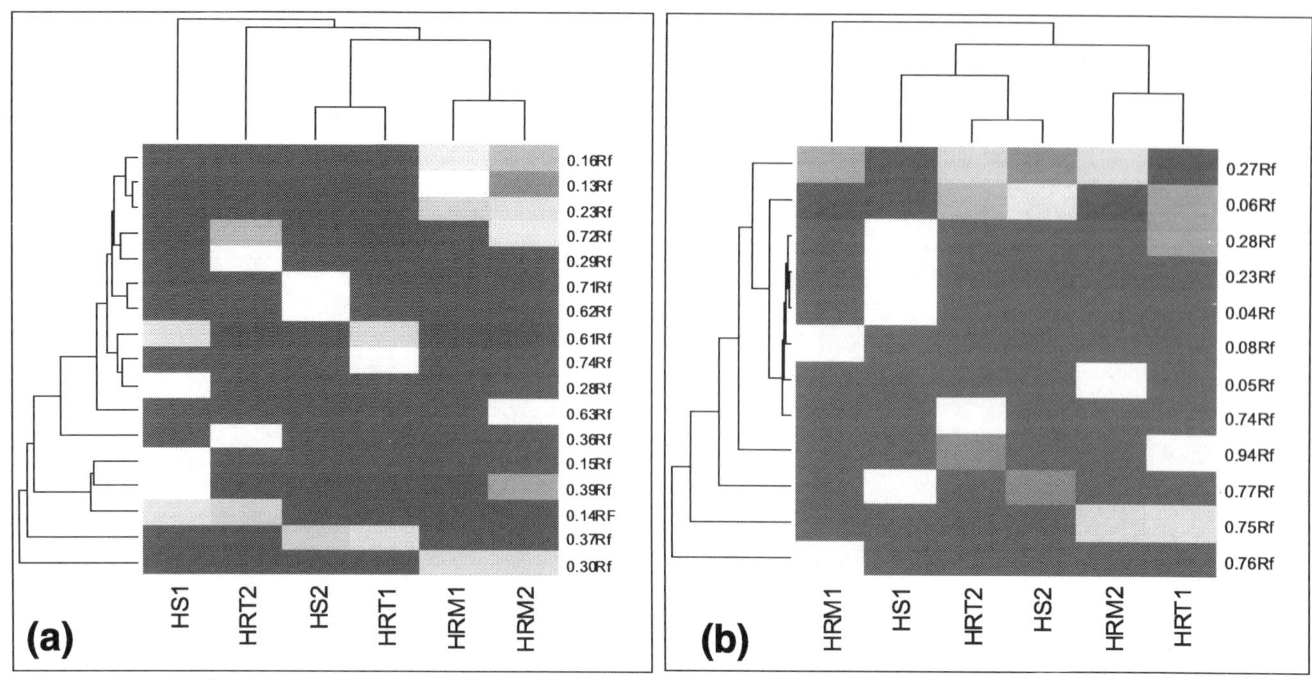

Figure 28.4: Heat Maps Showing R_f Values for Nitrogen Bases (a) and Flavonoids (b) in Seabuckthorn Leaf Extracts.

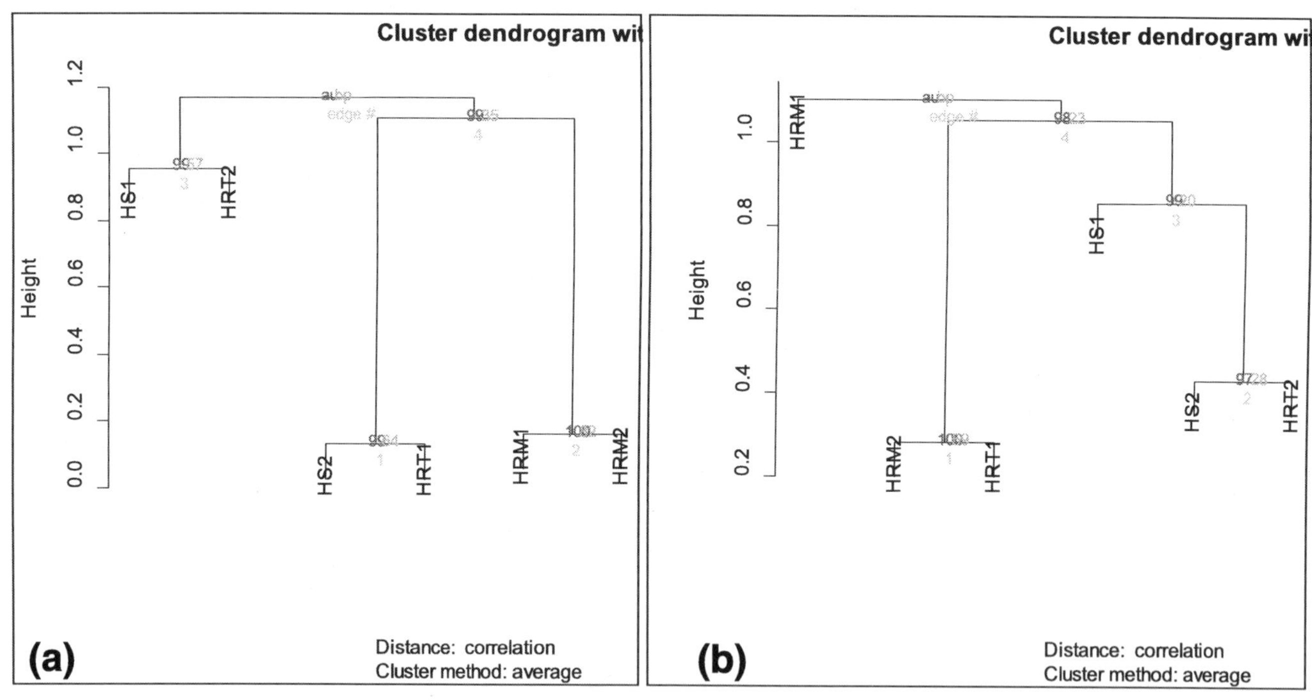

Figure 28.5: Cluster Dendrograms Showing Average Linkage Method of Hierarchical Clustering for Nitrogen Bases (a) and Flavonoids (b) in Seabuckthorn Leaf Extracts.

R_f were generated using R programming. The heat maps showing nitrogen base and flavonoid contents in the seabuckthorn leaf extracts of the three varieties under study are depicted in Figures 28.4(a) and 28.4(b) respectively. The color intensity of a specific grid is representative of the concentration of a certain compound present at a unique R_f. White and red colours indicate high and low concentrations of the compounds respectively. (Bhardwaj *et al.*, 2015). Similarly, the cluster dendrograms constructed for grouping of similar data according to R_f values are shown in Figures 28.5(a) and 28.5(b) for nitrogen bases and flavonoids accordingly.

Antioxidant Activity

The IC_{50} values for both superoxide and nitric oxide radical scavenging activities are mentioned in Table 28.3.

Table 28.3: IC_{50} Values for Free Radical Scavenging Activities of the Seabuckthorn Extracts

IC_{50} value of extract (µg/mL)	HS1	HS2	HRM1	HRM2	HRT1	HRT2
Superoxide radical scavenging activity	16.18	48.49	84.76	42.81	56.09	48.70
Nitric oxide radical scavenging activity	77.39	54.13	75.20	62.19	59.34	241.07

Superoxide Radical Scavenging Activity

All the six seabuckthorn extracts were found to be capable of scavenging superoxide anion radical ($O_2^{\cdot-}$) and thus, minimizing the effect of ROS. Among the aqueous extracts, HS1 showed the best $O_2^{\cdot-}$ scavenging activity (IC_{50}= 16.18 µg/mL) whereas the alcoholic extracts had almost comparable $O_2^{\cdot-}$ scavenging potential.

Nitric Oxide Radical Scavenging Activity

Determination of the nitric oxide radical ($NO^{\cdot}$) scavenging activity for all the seabuckthorn extracts showed that HRT1 had the best $NO^{\cdot}$ scavenging activity (IC_{50}= 59.34 µg/mL) than the other two aqueous extracts, HS1 and HRM1 that had almost similar IC_{50} values. However in case of alcoholic extracts, HS2 had appreciably more $NO^{\cdot}$ scavenging activity (IC_{50}= 54.13 µg/mL) than HRM2 and HRT2.

DISCUSSION

HPTLC fingerprinting confirmed the presence of one or more nitrogen bases and flavonoids in the six seabuckthorn extracts whose quantities are mentioned in the tables 2 and 3 above. But some of these compounds were found to occur below detection limit of the instrument, suggesting that these compounds occur at a very low concentration in the extracts.

Chemometric analysis has recently been adopted as a means for sampling and clustering of data. As an instance, Pierna *et al.* (2012) had used this tool for analysis of components in food materials detected by near infra red spectroscopy. Here as a first attempt, we combine the techniques of chemometrics and HPTLC to identify the flavonoids and nitrogen bases present in seabuckthorn leaf extracts. The heat maps generated from R programming serve the purpose of grouping the extracts according to their functional similarity. From the cluster dendrograms, it was seen that with respect to the nitrogen bases identified in the seabuckthorn leaf extracts, HS1 and HRT2 formed one cluster whereas HS2 and HRT1 formed another cluster, thus indicating that the aqueous extract of *H. salicifolia* shared similar nitrogen base content as alcoholic *H. rhamnoides turkestanica* and vice versa. The aqueous and alcoholic extracts of *H. rhamnoides mongolica* (HRM1 and HRM2) formed a separate cluster. As per the flavonoid content, HRM2 and HRT1 were included in one cluster. HS2 and HRT2 comprised another cluster to which HS1 was fairly similar. HRM1 was an altogether unique cluster in itself.

Superoxide anion radical is one of the reactive oxygen species (ROS) that can get converted into other harmful ROS like hydroxyl free radical ($OH^{\cdot-}$) and peroxide free radical ($O_2^{2-\cdot}$) (Alkan *et al.*, 2014) thus causing damage to biological molecules. Another harmful free radical is nitric oxide radical which gets produced in biological systems by the enzyme nitric oxide synthase. An increased concentration of $NO^{\cdot}$ can enhance nitrosylation reactions that adversely alter structure and functions of protein molecules. (Harput *et al.*, 2011). All the seabuckthorn extracts studied here were found to be capable of scavenging superoxide anion radical and thereby, minimizing the effect of ROS. The superoxide scavenging activity for aqueous seabuckthorn extracts in decreasing order was HS1>HRT1>HRM1 and that for alcoholic extracts was HRM2>HS2>HRT2. Similarly, for nitric oxide radical, the order of scavenging potential in aqueous extracts was seen to be HRT2>HRM1>HS1 and for alcoholic extracts, the order was HS2>HRM2>HRT2. The better antioxidant abilities of some extracts as indicated above could be attributed to the higher content of polyphenolic compounds in the respective extracts (Kumar *et al.*, 2013).

CONCLUSION

HPTLC analysis showed that all the three seabuckthorn varieties *viz.*, *H. salicifolia*, *H. rhamnoides mongolica* and *H. rhamnoides turkestanica* are rich in nitrogen bases and flavonoids. However no defined pattern was observed with regards to quantities of these compounds in the extracts. As far as the antioxidant activity of the seabuckthorn extracts is concerned, all the extracts were found to possess appreciable free radical scavenging potential.

ACKNOWLEDGEMENT

Financial support provided from DRDO, Ministry of Defence, Government of India is gratefully acknowledged. Special gratitude is duly extended to The Director, DIPAS, Delhi for her constant support in the research work. We are also thankful to Miss Anuja Bhardwaj, SRF, DIPAS for her kind help in chemometric analysis represented in this paper.

REFERENCES

1. Alkan, F.U., Anlas, C., Ustuner, O., Bakirel, T. and Sari, A.B. 2014. Antioxidant and proliferative effects of aqueous and ethanolic extracts of *Symphytum officinale* on 3T3 Swiss albino mouse fibroblast cell line. *Asian Journal of Plant Science and Research* 4: 62-68.
2. Bhardwaj, A., Pal, M., Srivastava, M., Tulsawani, R., Sugadev, R. and Misra, K. 2015. HPTLC based chemometrics of medicinal mushrooms. *Journal of Liquid Chromatography and related Technologies.* (In press)
3. Chen, L., Xu, W., Shao, R. and Du, X. 2015. Bioactivities of phytochemicals in *Araiostegia yunnanensis* (Christ). *Food Chemistry* 186: 37-45.
4. El-Gindy, A., Ashour, A., Abdal-Fattah, I. and Shabana, M. M. 2001. Application of LC and HPTLC-densitometry for the simultaneous determination of benazepril hydrochloride and hydrochlorothiazide. *Journal of Pharmaceutical and Biomedical Analysis* 25: 171-179.
5. Geetha, S., Sai Ram, M., Singh, V., Ilavazhagan, G. and Sawhney, R.C. 2002. Antioxidant and immunomodulatory properties of seabuckthorn (*Hippophae rhamnoides*)-An *in vitro* study. *Journal of Ethnopharmacology* 79: 373-378.
6. Harput, U.S., Genc, Y., Khan, N. and Saracoglu, I. 2011. Radical scavenging effects of different *Veronica* species. *Records of Natural Products* 5: 100-107.
7. Huang, L., Wei, P., Fan, L., Ye, D., Zhu, X. and Xu, Z. 2010. The biosynthesis and bioactivity evaluation of the cytosine-substituted mildiomycin analogue (MIL-C) for controlling powder mildew. *World Journal of Microbiology and Biotechnology* 26: 649-655.
8. Michel, T., Destandau, E., Le Floch, G., Lucchesi, M.E. and Elfakir, C. 2012. Antimicrobial, antioxidant and phytochemical investigations of seabuckthorn (*Hippophae rhamnoides* L.) leaf, stem, root and seed. *Food Chemistry* 131: 754-760.
9. Negi, P.S., Chauhan, A.S., Sadia, G.A., Rohinishree, Y.S. and Ramteke, R.S. Antioxidant and antibacterial activities of various seabuckthorn (*Hippophae rhamnoides* L.) seed extracts. *Food Chemistry* 92: 119-124.
10. Palasz, A. and Ciez, D. 2014. In search of uracil derivatives as bioactive agents: Uracils and fused uracils: Synthesis, biological activity and applications. *European Journal of Medicinal Chemistry.* (In press)
11. Pierna, J.A.F., Vermeulen, P., Arnand, O., Tossens, A., Dardenne, P. and Baeten, V. 2012. NIR hyperspectral imaging spectroscopy and chemometrics for the detection of undesirable substances in food and feed. *Chemometrics and intelligent Laboratory Systems.* 117: 233-239.
12. Rodriguez-Mateos, A., Vauzour, D., Krueger, C.G., Shanmuga, D., Reed, J., Calani, L., Mena, P., Del Rio, D. and Crozier, A. 2014. Bioavailability, bioactivity and impact on health of dietary flavonoids and related compounds: An update. *Archives of Toxicology* 88: 1803-1853.
13. Sayegh, M., Miglio, C. and Ray, S. 2014. Potential cardiovascular implications of seabuckthorn berry consumption in humans. *International Journal of Food Sciences and Nutrition* 65: 521-528.
14. Sharma, P., Suryakumar, G., Singh, V., Misra, K. and Singh, S.B. 2014. *In vitro* antioxidant profiling of seabuckthorn varieties and their adaptogenic response to high altitude induced stress. *International Journal of Biometeorology* Doi 10: 1007/s/00484-014-0925-2.
15. Tiitinen, K.M., Yang, B., Haraldsson, G.G., Jonsdottir, S. and Kaliio, H.P. 2006. Fast analysis of sugars, fruit acids, and vitamin C in seabuckthorn (*Hippophae rhamnoides* L.) varieties. *Journal of Agricultural and Food Chemistry* 54: 2508-2513.
16. Varshneya, C., Kant, V. and Mehta, M. 2012. Total phenolic contents and free radical scavenging activities of different extracts of seabuckthorn (*Hippophae rhamnoides*) pomace without seeds. *International Journal of Food Sciences and Nutrition* 63: 153-159.
17. Wnuk, S.F., Ro, B., Valdez, C.A., Lewandowska, E., Valdez, N.X., Sacasa, P.R., Yin, D., Zhang, J., Borchardt, R.T. and Clercq, E.D. 2002. Sugar modified conjugated diene analogues of adenosine and uridine: Synthesis, interaction with S-Adenosyl-l-homocysteine Hydrolase, and antiviral and cytostatic effects. *Journal of Medicinal Chemistry* 45: 2651-2658.
18. Yogendra, M.S., Tirpude, R.J., Maheshwari, D.T., Bansal, A. and Misra, K. 2013. Antioxidant and antimicrobial properties of phenolic rich fraction of seabuckthorn (*Hippophae rhamnoides* L.) leaves *in vitro*. *Food Chemistry* 141: 3443-3550.
19. Xu, Y., Kaur, M., Dhillon, R.S., Tappia, P.S. and Dhalla, N.S. 2011. Health benefits of seabuckthorn for the prevention of cardiovascular diseases. *Journal of Functional Foods* 3: 2-12.

Part IV

MEDICAL RESEARCH

Chapter 29

Synergistic Efficacy of the Combination of Seabuckthorn Seed Oil and Famotidine or Lansoprazole for Treatment of Gastric Ulcerations and Erosions in Dogs

Amit Kumar[1], S.P. Tyagi[1], Piyush Shastri[1], Richa Dogra[1], Adarsh Kumar[1] and Virendra Singh[2]

[1]Department of Surgery and Radiology, College of Veterinary and Animal Sciences
[2]Department of Biology and Environmental Sciences
CSK Himachal Pradesh Agricultural University, Palampur, H.P., India
(*E-mail: drasingla@gmail.com)

SUMMARY

The study was conducted on 16 healthy adult medium-sized mongrel dogs of either sex maintained under uniform environment. Inj. Dexamethasone was administered in all the animals @ 1mg/kg IV *b.i.d.* to create non-fatal gastric ulcerations and erosions (GUE). Dexamethasone was discontinued when there was endoscopic evidence of GUE index reaching 7 on two consecutive observations or 8 in a single observation interval. These dogs were randomly divided into 4 groups of 4 animals each. Animals of group I were treated with famotidine @ 1 mg/kg alone, group II with famotidine @ 1 mg/kg + Seabuckthorn oil @ 1ml/animal, group III with lansoprazole @ 1.5 mg/kg alone and group IV with lansoprazole @ 1.5 mg/kg + Seabuckthorn @ 1ml/animal given orally twice a day till complete healing of GUE. The progression of GUE was evaluated by clinical, haematological, biochemical, faecal occult blood test and gastro-endoscopic observations at different time intervals. The average number of days taken to achieve complete healing of GUE was shortest at 6 days in group 2 (combination of famotidine with seabuckthorn oil) followed by 7.5 days in group 4 (combination of lansoprazole with seabuckthorn oil), 8.25 days in group 1 (famotidine alone) and 9.0 days in group 3 (lansoprazole alone). Subjective assessment of gastric lesions by endoscopy also revealed faster and more uniform healing in group 2. It was concluded that the treatment of GUE in dogs with the combination of famotidine and seabuckthorn oil resulted in faster recovery in comparison to other groups.

***Keywords:** Seabuckthorn, Dog, Gastric erosions, Ulcerations, Famotidine and Lansoprazole.*

INTRODUCTION

The gastric ulcerations and erosions (GUE) are wounds of varying depth in the gastric wall. Defects in gastric mucosa limited to superficial muscularis layers are termed as erosions and those penetrating deeper are termed as ulcers. A combined term 'Gastric ulcerations and erosions (GUE)' is however more appropriate for most clinical cases in animals as erosions may turn into ulcerations anytime and endoscopic visual inspection cannot always differentiate between the two effectively. Gastric ulcers and erosions are formed when gastric mucosal cytoprotective barrier is impaired or overwhelmed by the gastric aggressive factors (Kubiak *et al.*, 2004). Some of these aggressive factors include HCl, proteolytic enzymes (pepsin, trypsin), stimulant of gastric acid secretion, oxygen free radicals, platelet activating factors bile acids and drugs (NSAIDs). Whereas, gastric mucosal protective factors include gastric mucus, tight junctions between epithelial cells, rapid epithelial regeneration, bicarbonate

Seabuckthorn (Hippophae L.): A Multipurpose Wonder Plant, Vol. 5: Advanced Technologies for Cultivation, Processing, Health Protection and Environmental Conservation (Virendra Singh, *Editor-in-Chief*, 2018), p. 229–234, Astral International Pvt. Ltd., New Delhi, India.

secretion, production of prostaglandins, presence of gastrointestinal and pancreatic peptides, the production of epithelial growth factors and gastric mucosal blood flow. Acid and peptic enzymes, secreted in the stomach during initial digestion process, may damage the gastric mucosa if come in direct contact with thelining of the stomach. The maintenance of gastric wall integrity in the face of such gastric acids and enzymes is a specialized function of body accomplished by maintaining a delicate balance in between gastric damaging and gastric protective mechanisms.

The signs of GUE in dogs when present may include vomiting, anorexia, weight-loss, increased salivation, and abdominal pain. But in general the diagnosis of GUE in animals is difficult as not only the clinical signs are mostly non-specific but animals also mostly remain asymptomatic regardless of GUE severity. The treatment of GUE in dogs is thus challenging in the absence of correct and timely diagnosis and the inherent difficulty in such cases to effectively offset the damaging actions of gastric acids and enzymes. Medical management of GUE is directed to promote conductive environment in stomach to facilitate the healing of GUE lesions. This can be achieved by employing one or more methods to reduce gastric acid secretion, neutralize gastric acids, improve gastric mucus secretion, increase mucus bicarbonate secretion and physically protect gastric mucosa. These methods comprise use of antacids, proton pump inhibitors, H_2 antagonists and cyto-protectants.

Studies are being done to investigate different preparations to formulate better therapeutic and preventive agents for management of GUE. Investigation of herbal agents for their anti-ulcerative properties is one of the promising areas in the quest to look for a better therapeutic agent in the management of GUE. A number of herbs and plant products have been found to play a role in protecting or helping to heal gastric ulcers. Common herbs like Turmeric (*Curcuma longa*), *Aloe vera*, ginger (*Zingiber officinales*), Indian gooseberry (*Phyllanthus emblica*) and Neem (*Azadirachta indica*) have shown significant anti-gastroulcerogenic activities in rats in different experimental models of gastric ulcers (Rafatullah *et al.*, 1990; Subramanian*et al.*, 2007; Yamahara *et al.*, 1988; Bandhyopadhyay *et al.*, 2000; Divakar *et al.*, 2001 respectively). Similarly, the seed oil of seabuckthorn (*Hippophae rhamnoides*), an herb growing in cold deserts of Himalayas have shown a potent anti-gastroulcerative activity in rats, dogs and humans (Jiang *et al.*, 1989; Qiu and Qiao, 1997; Tyagi, 2006; Dogra, 2011; Thakur, 2011; Gupta, 2012; Thakur, 2013).

However, in general, though a number of studies have been done world over on various plants to demonstrate their anti-gastroulcerative properties but there is still a scarcity of similar experimental scientific studies on domesticated pets. Nonetheless, in the last few years a number of controlled systematic studies have been conducted in the department of Surgery and Radiology, DGCN COVAS, CSKHPKV, Palampur, India to verify the therapeutic and prophylactic efficacies of a number of drugs and seabuckthorn (SBT) oil in dexamethasone-induced GUE in dogs. In these studies, the SBT seed oil had shown good therapeutic efficacy for GUE in dogs and proved better than omeprazole, famotidine, lansoprazole and sucralfate (Tyagi *et al.*, 2010; Dogra, 2011) in certain cases.

Further, combination therapies consisting of SBT seed oil along with some routine medicine showed better results when compared to each of the agents individually (Gupta, 2015; Thakur, 2013). This necessitated undertaking research studies on other potential combination therapies including synthetic and plant extracts for more effective treatment of GUE.

MATERIALS AND METHODS

The study was done in 16 adult healthy mongrel dogs divided into four equal groups. Intravenous administration of steroidal drug "Dexamethasone" was done in all the animals for creation of non-fatal gastric ulcers and erosions (GUE). Theadministration of the drug was continued till there was endoscopic evidence of GUE index reaching to ulcer index 7 on two consecutive endoscopic observations or 8 in a single observation occasion (Tyagi, 2006).

The different groups of animal were subjected to following treatment protocol:

Table 29.1: The GUE Treatment Protocol

Groups	*Treatment protocols*
I	Famotidine @ 1 mg/kg
II	Famotidine @ 1 mg/kg + 1 ml Seabuckthorn oil
III	Lansoprazole @ 1.5 mg/kg
IV	Lansoprazole @ 1.5 mg/kg + 1 ml Seabuckthorn oil

All the dogs were treated twice a day orally as per above-mentioned protocol. Progress of GUE healing was evaluated and compared on the basis of clinical, haematological, fecal occult blood test and endoscopic observations on day 0, 3, 6, 9, 12... till complete healing. The day '0' was considered the day when endoscopic GUE score reached to desired severity levels of 7/8 and when administration of dexamethasone was stopped as explained above. Clinical parameters were rectal temperature, respiration rate, heart rate, weight of animal, appetite, status of vomiting and melena. Haematological parameters were Haemoglobin (Hb), Packed cell volume (PCV), Total erythrocyte count (TEC), Differential leucocyte count (DLC), Total leucocyte count (TLC). Statistical analysis was done using Analysis of variance (ANOVA) method by Students-Newman (inter-group comparison) and Dunnett's test (intra group comparison) at 5 per cent and 1 per cent level of significance.

RESULTS AND DISCUSSION

On day 0, the gastric lesions showed maximum severity and the endoscopic GUE indices were more or less comparable at 7.50 ± 0.29, 7.75 ± 0.25, 8.00 ± 0.00 and 8.00 ± 0.00 in group's I-IV respectively (Figure 29.1 and Table 29.2). The lesions consisted of multiple linear as well as focal mucosal defects of variable shapes and depths. The lesions were generally large and widely distributed all over the gastrum *i.e.* fundus, body and pylorus. Large adherent multiple blood clots as well as fresh blood streaks were observed inside stomachs of all the dogs giving a typical "waterfall" like character. The mucosa was severely hyperaemic and the mucus layer was appreciably very thin.On day 3, the overall GUE lesions consisted of small multifocal lesions with adherent as well as free floating blood clots primarily over gastric body and pylorus.The mucus layer was best in group II as evidenced by its thickness and luster (Figure 29.1). On day 6, GUE lesions healed completely in group II. In general the mucus layer was much thicker and lustrous in this group as compared to others. The other groups still showed variable number of GUE lesions with or without active hemorrhage and mucosal oedema (Figure 29.1).

GUE lesions healed eventually in all the groups in a gradual manner, the average number of days taken for complete healing of GUE lesions was shortest in group II (combination of famotidine with seabuckthorn oil) at 6.0 days followed by group IV (combination of lansoprazole with seabuckthorn oil) at 7.5 daysfollowed by group I (famotidinealone) and group III (lansoprazole alone) at 9.0 days (Figure 29.1, Tables 98.2 and 29.3). Kumar *et al.* (2013) reported the dose-dependent activity of seabuckthorn seed oil while comparing three different doses of 1ml, 2.5 ml or 5 ml/animal given twice a day orally in similar model of GUE in dogs. He reported the dose of 5 ml/animal to be the most effective one as the GUE lesions healed in an average 7.5 days with such doses. He also reported that the dose of SBT oil 1ml/animal did not show any therapeutic activityand the average healing time of GUE was found to be comparable with control untreated group at 10.5 days. But Thakur (2013) reported that when this ineffective dose of 1ml was combined with sucralfate, it decreased the average healing time in similar models of GUE in dogs from 10.5 to 8.25 days. This was attributed to the therapeutic synergy in between SBT seed oil and sucralfate. In the present study also, same inference of therapeutic synergy in between SBT oil and famotidine or lansoprazole could be drawn as the healing time of GUE in both of the combination therapy groups (II and IV)was shorter than individual therapy groups (I and III).

Use of constituent agents having different gastro-protective actions complementing each other might be the basis of therapeutic synergism in the present study. Borkent and Becker (1998) found the combination of cimetidine, a H_2-blocker and colloidal bismuth subcitrate, a cytoprotective agent much better than either of them alone in the treatment of ulcerative reflux oesophagitis in humans. Similarly, Fandriks *et al.* (2007) reported that combination of famotidine and omeprazole on a once-daily basis was more effective in the treatment of gastritis in comparison to these drugs used alone in human subjects.

The rectal temperature, heart rate and respiration rate did not vary much with the base values and remained within physiological limits throughout the period of study in all the groups. Gradual regaining of body weight and appetite were found in dogs

Table 29.2: GUE Indices of Dogs at various Intervals (Mean± S.E.)

GUE indices of dogs					
Groups / *Days*	*0*	*3*	*6*	*9*	*12*
Group I	7.50±0.29	4.75±0.63	2.25±1.32	1.00±1.0 (n=2)	0.00±0.00 (n=1)
Group II	7.75± 0.25	3.00±0.71	0.0±0.0		
Group III	8.00±0.00	5.00±0.00	2.00±0.71	0.66±0.66 (n=3)	0.00±0.00 (n=1)
Group IV	8.00±0.00	1.75±0.47	0.75±0.47	0.00±0.00 (n=2)	

Table 29.3: Average Number of Days for Complete Healing of GUE Lesions in different Groups

Average Number of Days for Complete Healing of GUE Lesions		
Group I	Famotidinealone	8.25 days
Group II	Famotidine with seabuckthorn oil	6.0 days
Group III	Lansoprazole alone	9.0 days
Group IV	Lansoprazole with seabuckthorn oil	7.5 days

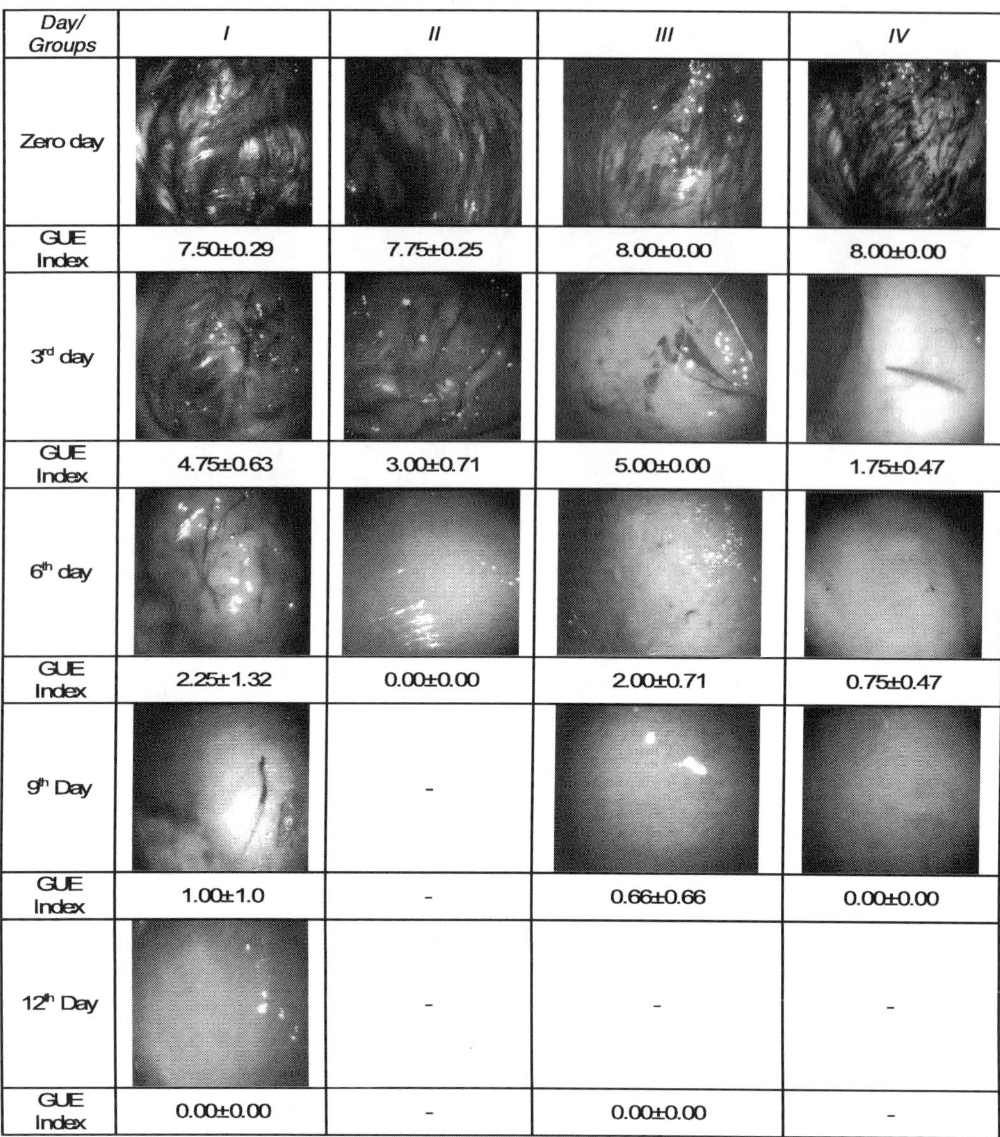

Day/ Groups	I	II	III	IV
Zero day				
GUE Index	7.50±0.29	7.75±0.25	8.00±0.00	8.00±0.00
3rd day				
GUE Index	4.75±0.63	3.00±0.71	5.00±0.00	1.75±0.47
6th day				
GUE Index	2.25±1.32	0.00±0.00	2.00±0.71	0.75±0.47
9th Day		-		
GUE Index	1.00±1.0	-	0.66±0.66	0.00±0.00
12th Day		-	-	-
GUE Index	0.00±0.00	-	0.00±0.00	-

Figure 29.1: Endoscopic View of Gastric Mucosal Surface and GUE Indices of Dogs at Various Time Intervals.

which again correlates with other studies (Tyagi, 2006; Dogra *et al.*, 2013; Kumar *et al.*, 2013) in similar kind of GUE model. Greatest weight-gain was observed in the group II animals because in this group the gastric lesions healing and regaining of appetite was earliest in comparison to other groups (Figure 29.1). Long-term administration of corticosteroids has catabolic effect on the body protein ultimately leading to loss of condition and reduction in body weight (McDonalds, 1982). But once it is discontinued, there was restoration of normal digestive process in the dogs undergoing different treatments which is directly co-related with gain in body weight gain.

The faecal occult blood test was strongly positive in all the groups at day 0. Thereafter, the strength of FOBT reactions gradually decreased but varied within and in between various groups. In general a direct correlation was observed between detection of blood clots or gastric lesions endoscopically and corresponding FOBT reaction. Faecal occult blood tests could detect quantities of blood that were smaller than those required to cause melena (Gilson, 1990). The accuracy of FOBT in cases of dexamethasone-induced GUE in dogs is 100 per cent (Dogra *et al.*, 2013) which was again proved in present study.

In general, a gradual rise in Hb, PCV, and TEC was observed from 0 day till the end of study in all the groups. TLC levels gradually decreased in all the groups to reach towards its normal base values by the end of last observation interval. Dexamethasone induces the characteristic blood leukocyte profile (neutrophilia, lymphopenia, and monocytosis) known as the 'stress leukogram'. However withdrawal of dexamethasone substantially reduces the adverse effect on haematological parameters *i.e.* decreases in leukocyte and granulocyte levels and lymphocytosis in the present study. These finding are in consonance with those of Tyagi 2006; Dogra *et al.* 2013; Kumar, 2013) who also reported a gradual restoration of Hb, PCV and TEC levels towards normal following administration of various gastric ulcer healing medicines and seabuckthorn seed oil in dexamethasone induced GUE in dogs.

CONCLUSIONS

The healing of GUE lesions occur fastest in dogs treated with combination of famotidine and seabuckthorn oil given orally twice daily. The combination of seabuckthorn seed oil with famotidine or lansoprazole showed synergistic therapeutic effect resulting into faster healing of GUE lesions compared with using either of them alone.

REFERENCES

1. Bandyopadhyay, S. K., Satyesh, C. and Pakrashi, A. 2000. The role of antioxidant activity of *Phyllanthus emblica* fruits on prevention from indomethacin induced gastric ulcer. *Journal of Ethnopharmacology* 70 (2): 171-176.
2. Borkent, M.V. and Beker, J.A. 1998. Treatment of ulcerative reflux oesophagitis with colloidal bismuth sub-citrate in combination with cimetidine. *Gut* 29: 385-389.
3. Divakar, M.C., Rao, S.B., Nair, G.R.N. and Hisham, A. 2001.The role of fatty acids on the ulcer healing property of the nimbidin fraction of neem oil. *Journal of Medicinal and Aromatic Plant Sciences* 23 (3): 404-408.
4. Dogra, R. 2011. Therapeutic evaluation of seabuckthorn (*Hippophae sp.*) oil and other medicines in healing of gastric ulcerations and erosions (GUE) in dogs. *MVSc. Thesis*, Department of Veterinary Surgery and Radiology, COVAS, CSKHPKV, Palampur.
5. Dogra, R., Tyagi, S.P. and Kumar, A. 2013. Efficacy of seabuckthorn (*Hippophae rhamnoides*) oil vis-a-vis other standard drugs for management of gastric ulceration and erosions in Dogs. *Veterinary Medicine International*, Article ID 176848, 11 pages, doi:10.1155/2013/176848.
6. Fandriks, L., Lonroth, H., Petterson, A. and Vakil, N. 2007. Can famotidine and omeprazole be combined on a once-daily basis? *Scandinavian Journal of Gastroenterology* 42 (6): 689-694.
7. Gilson, S.D., Parker, B.B. and Twedt, D.C. 1990. Evaluation of two commercial kits for detection of occult blood in faeces of dogs. *American Journal of Veterinary Research* 51 (9): 1385-1387.
8. Gupta, S. 2012. Study on the efficacy of seabuckthorn oil alone and in combination with famotidine in the treatment of gastric ulcerations and erosions (GUE) in dogs. *MVSc. Thesis*, Department of Veterinary Surgery and Radiology, COVAS, CSKHPKV, Palampur.
9. Gupta, S., Tyagi, S.P. and Kumar, A. 2015. Management of Gastric ulcerations and erosions with combination of seabuckthorn (*Hippophae rhamnoides*) seed oil and famotidine in dogs. *Indian Journal of Veterinary Surgery* 36 (1): 29-32.
10. Kubiak, K., Nicpon, J., Jankowski, M. and Spuzak, J. 2004. Gastric ulcers in dogs. *Medycyna-Weterynaryjna* 60 (7): 708-710.
11. Kumar, A., Tyagi, S.P., Dogra, R. and Gupta, S. 2013. Dose-dependent effects of seabuckthorn seed oil on the healing of gastric ulcers and erosions in dogs. In: *Proceedings of 6th Conference of International Seabuckthorn Association*, pp. 9-16, October 14-18, 2013, Potsdam, Germany.
12. Mc Donald, L.E. 1982. Hormones influencing metabolism. In: *Jones Veterinary Pharmacology and Therapeutics* (Booth NH and McDonald LE eds.), pp. 553-592, 5th Edition, Iowa state, Kalyani Publishers, New Delhi.
13. Qiu, G.Q. and Qiao, X. 1997. A preliminary report on the clinical treatment of thirty cases of peptic ulcer with seabuckthorn oil capsules. *Hippophae* 10 (4): 39-41.
14. Rafatullah, S., Tariq, M., Al-Yahya, M.A., Mossa, L.S. and Ageel, A.M. 1990. Evaluation of turmeric (*Curcuma longa*) for gastric and duodenal antiulcer activity in rats. *Journal of Ethnopharmacology* 29: 25-34.
15. Subramanian, S., Sathish, K. D., Arulselvan, P., Senthikumar, G.P. and Mahadeva, R. U.S. 2007. Evaluation of anti-ulcerogenic potential of *Aloe vera* leaf gel extract studies in experimental rats. *Journal of Pharmacology and Toxicology* 2 (1): 85-97.
16. Thakur, A. 2011. Prophylactic efficacies of seabuckthorn (*Hippophae rhamnoides*) oil *vis a vis* other standard drugs for gastric ulceration and erosions in dogs. *M.V.Sc. Thesis*, Department of Veterinary Surgery and Radiology, COVAS, CSKHPKV, Palampur.
17. Thakur, D. 2013. Evaluation of Sucralfate or misoprostol alone and in combination with seabuckthorn (*Hippophae* spp.) oil for the management of gastric ulcerations and erosions in dogs. *M.V.Sc Thesis*, Department of Veterinary Surgery and Radiology, COVAS, CSKHPKV, Palampur.
18. Tyagi, S.P. 2006. Studies on the efficacy of seabuckthorn (Hippophae sp.) oil in the healing of gastric ulcers in dogs. *PhD Thesis*, Department of Veterinary Surgery and Radiology, COVAS, CSKHPKV, Palampur.

19. Tyagi, S.P., Varshney, A.C. and Kumar, A. 2010. Prophylactic efficacy of seabuckthorn oil *vis-a-vis* other gastroprotective agents against gastric ulcerations and erosions. In: "*34th Annual Congress of Indian Society for Veterinary Surgery" and International symposium on "Newer concepts on surgical techniques for farm and companion animal practice"* held at College of Veterinary Sciences, Puducherry, India.

20. Yamahara, J., Mochizuki, M., Rong, H.Q., Masuda, H. and Fujimura, H. 1988. The antiulcer effects in rats of ginger constituents. *Journal of Ethnopharmacology* 23 (2-3): 290-304.

Chapter 30

Flavonoid Constituents, Total Polyphenol and Antioxidant Efficacy of *Hippophae rhamnoides* L. Berries

Nishat Anjum* and Y.C. Tripathi

Chemistry Division, Forest Research Institute, P.O. New Forest, Dehradun – 248 006, Uttarakhand, India
(*E-mail: nishatanjum55@gmail.com, tripathiyc@gmail.com)

SUMMARY

The present study was aimed to isolate and characterize flavonoid compounds from berries of *Hippophae rhamnoides* L. collected from Gangotri valley in Uttarakhand state of India and to examine its total polyphenol and antioxidant efficacy. Systematic phytochemical analysis and chromatographic separation of methanolic extract of *H. rhamnoides* berries led to isolation of altogether three known individual flavonoids and their characterization as quercetin, kaempferol and myricetin by classical chemical degradation as well as spectroscopic methods. The content of total phenolic compounds in the extracts was determined according to the Folin-Ciocalteu method and calculated as gallic acid equivalents (GAE). The antioxidative efficacy of the extract was determined as free radical scavenging activity using DPPH radical scavenging protocol. Overall results showed high concentrations of total phenolics and strong antioxidant activity in the fruits of *H. rhamnoides*. A linear correlation between total phenolic content and antioxidant capacity suggested that berries of the species can be a potent source of natural antioxidant.

Keywords: *Hippophae rhamnoides, Flavonoids, Total phenolics and Antioxidant efficacy.*

INTRODUCTION

Wild edible fruits are consumed frequently by rural and tribal communities across the world especially in the developing countries like India. Often they make an important contribution to the diet particularly among the rural populations. Several of these wild species bear fruits during the time when cultivated fruits are scarce. Wild fruits offer a convenient but cheap means of nutrition providing adequate nutrient supplies to the people living abound. *Hippophae rhamnoides,* a dwarf to tall (2-5 m) plant belonging to the family Elaeagnaceae and commonly known as seabuckthorn is one such plant that grows wild in temperate part of Asia and Europe (Yang and Kallio, 2006). In India, it is found at high altitude of 2500-4000 m in Himachal Pradesh, Uttaranchal, Jammu and Kashmir, Sikkim and Arunachal Pradesh states of the country (Singh, 2003). The plant with yellow or orange berries is attracting considerable attention mainly for its medicinal value and immense economic potential (Li. and Schroeder, 1996). Different parts of seabuckthorn plant have been used for the treatment of diseases in traditional medicine in various countries (Suryakumar and Gupta, 2011). It has been used for the treatment of inflammation, burns and skin ailments such as psoriasis, eczema, lupus erythematosus and dermatosus in Chinese folk medicines (Barkat *et al.*, 2011). All parts of the plant are considered to contain large amounts of compounds that are believed to have beneficial health effects. Berries are rich in carbohydrates, protein, organic acids, amino acids and vitamins (Zeb, 2004). The chemical composition of *H. rhamnoides* berries include high levels of vitamins C, tocopherols and carotenoids especially β-carotene, lycopene, zeaxanthine (Gao *et al.*, 2000; Zeb, 2004). Berries also contain certain other vitamins such as B_1, B_2, folic acid and (Bekker and Glushenkova, 2001), large amounts of sugars, mainly glucose and fructose as well as oxalic citric and tartaric acids (Kumar *et al.*, 2011). In addition, a series of other

Seabuckthorn (Hippophae L.): A Multipurpose Wonder Plant, Vol. 5: Advanced Technologies for Cultivation, Processing, Health Protection and Environmental Conservation (Virendra Singh, *Editor-in-Chief,* 2018), p. 235–244, Astral International Pvt. Ltd., New Delhi, India.

chemical compounds including unsaturated fatty acids (Yang and Kallio, 2001), phytosterols (Yang *et al.*, 2001; Cenkowski *et al.*, 2006; Bal *et al.*, 2011; Kumar *et al.*, 2011), flavonoid (Yuzhen and Fuheng, 1997; Rosch *et al.*, 2003, 2004; Hibasami *et al.*, 2005; Wang *et al.*, 2011), triterpenoids (Novruzov *et al.*, 1979; Salenko *et al.*, 1985; Zheng *et al.*, 2003), etc have also been reported from the plant. Pharmacological studies and clinical trials have confirmed anti-atherogenic, hypocholesteromic, hypotensive, anti-inflammatory properties (Eccleston *et al.*, 2002; Ganju *et al.*, 2005; Kashyap and Sawhney, 2005; Wu *et al.*, 1994; Yang, 1995; Zhang, 1987; Varshney and and Tyagi, 2004), hepatoprotective (Cheng, 1992; Gao *et al.*, 2003), antilipemic, antiarrhythmic (Fengming, 1989), antidiabetic (Zhang *et al.*, 2010; Wang *et al.*, 2011), immunoprotective (Ramasamy *et al.*, 2010), antimicrobial (Kaushal and Sharma, 2011; Michel *et al.*, 2012), cutaneous wound healing (Kiken and Cohen, 2002; Reddy *et al.*, 2012), burn wound healing (Upadhyay *et al.*, 2009), antigastroulcerative (Suleyman *et al.*, 2001; Xing *et al.*, 2002) antitumoral (Zhang *et al.*, 2004), anticancerous (Zeb. 2006) and invigorating blood circulation (Johansson *et al.*, 2000; Cheng *et al.*, 2003) properties of the plant. Several studies have indicated that seabuckthorn products are safe food supplements with no major side effects except for increase in blood glucose with a high dose (Saggu *et al.*, 2007; Upadhyay *et al.*, 2009; Tulsawani, 2010). These therapeutic effects mostly are related to the high amount of compounds like polyphenols possessing antioxidant activity (Kumar *et al.*, 2011).

Like other plant secondary metabolites, content of polyphenol compounds in plants tend to vary with respect to agroclimatic and phytogeographic conditions. A number of studies have shown that the chemical composition of the fruit varies greatly according to the species, and climatic and geological conditions of the areas where the plant is grown (Yao and Tigerstedt, 1994; Kallio *et al.*, 2000; Tang and Tigerstedt, 2001). Also as a matter of fact, different classes of polyphenols have different degree of solubility in different solvents. For instance, more polar phenolic compounds and glycosides are readily soluble in solvents of high polarity and less polar such as isoflavones, flavonones and methoxylated flavones tend to be more soluble in less polar solvents. As such, the successive use of different solvents with increasing order of polarity may extract the high amount of polyphenols. In view of the foregoing facts, the present study was designed to examine extractive values of *H. rhamnoides* berries collected from high altitude regions of Uttarakhand state of India using solvents of varying polarity successively and to evaluate the total polyphenols contents of relevant extract and respective DPPH radical scavenging capabilities. The study also intended to isolate and characterized certain major flavonoid compounds from *H. rhamnoides* berries.

MATERIALS AND METHODS

Chemical and Analytical

The chemicals and solvents used for the extraction, chromatography, phytochemical screening and *in vitro* bioassay of *H. rhamnoides* berries were of analytical grade. Folin-Ciocalteu reagent, gallic acid, 1, 1-Diphenyl-2-picryl hydrazyl (DPPH) were of analytical grade and purchased from Merck. Column chromatography was carried out over SiO_2 gel refers to BDH (60-120 mesh). TLC plates were prepared with SiO_2 gel HF 254 or SiO_2 gel G (BDH) and spots were visualised by iodine vapour/UV 1 amp lamp/Liebermann Burchard reagent/NH_3 vapour and also by subsequent heating at 110°C whenever necessary. Anhydrous Na_2SO_4 was used for drying the solvents. Analytical samples were routinely dried at 800 over P_2O_5 for 24 hours in vacuum.

Melting points of isolated compounds point was taken on Toshniwal apparatus and were uncorrected. UV spectra were recorded on Carry-14 spectrophotometer 1700 using spectral methanol. For other analysis like TPC, TFC and FRS assay, UV/Vis double beam spectrophotometer (Chemito 1700) was used. IR spectra were run on Perkin-Elmer 783 model spectrophotometers in nujol null. ^{1}H NMR spectra were recorded at 90MHz on a Bruker WH 90. Chemical Shifts were reported in 'δ' ppm relative to $SiMe_4$ as internal standard. Mass spectral analysis was performed on Kratos MS-50 mass spectrometers operating at 70eV with the evaporation of the sample in the ion source at about 200°C.

Plant Material

Berries of *H. rhamnoides* were collected from Gangotri valley of Garhwal Hills in the state of Uttarakhand during October 2013 at the stage of maturity, as judged by juiciness and appearance and stored at -20°C. Seeds were separated from frozen berries by pressing and the pulp portion dried at room temperature was used for further extraction.

Extraction

Dried and powdered berries pulp of *H. rhamnoides* (100 g) was successively extracted with n-hexane, chloroform, ethyl acetate, acetone and methanol. The extraction procedure involved adding of solvents to the ground berries pulp, keeping it for 24 hrs with intermittent shaking. Extracts so obtained were separately distilled under reduced pressure to obtained solvent free extracts and the extractive values were determined on dry weight basis with respect to the initial weight of dried and ground berry pulp taken. The resultant crude extracts were transferred into airtight sample bottles and kept at 4°C until they were used.

Phytochemical Screening

Stock solution (1 per cent w/v) of all the extracts including n-hexane, chloroform, ethyl acetate, acetone and methanol of *H. rhamnoides* berries were prepared and subjected to phytochemical screening to detect the presence and or absence of various phytochemicals constituent (Evans, 1989; Harborne, 1998; Kokate, 2001).

Determination of Total Phenolic Content (TPC)

Based on the results of qualitative phytochemical analysis, chloroform, ethyl acetate, acetone and methanol extracts were selected for estimation of total phenolic content. The concentration of phenolics in the extracts was determined using spectrophotometric method (Singleton *et al.*, 1999). Methanolic solution of the extract in the concentration of 1 mg/ml was used in the analysis. The reaction mixture was prepared by mixing 0.5 ml of methanolic solution of extract, 2.5 ml of 10 per cent Folin-Ciocalteu reagent dissolved in water and 2.5 ml 7.5 per cent sodium carbonate. Blank was concomitantly prepared, containing 0.5 ml methanol, 2.5 ml 10 per cent Folin-Ciocalteu reagent dissolved in water and 2.5 ml of 7.5 per cent of $NaHCO_3$. The samples were thereafter incubated at room temperature for 45 min. The absorbance was determined using spectrophotometer at λ_{max} = 765 nm. The samples were prepared in triplicate for each analysis and the mean value of absorbance was obtained. The same procedure was repeated for the standard solution of gallic acid and the calibration line was construed. Based on the measured absorbance, the concentration of phenolics was read (mg/ml) from the calibration line; then the content of phenolics in extracts was expressed in terms of gallic acid equivalent (mg of GA/100 mg of extract). All experiments were replicated thrice and the data were presented as mean values ± standard deviation.

Determination of Total Flavonoid Content

The flavonoid content of chloroform, ethyl acetate, acetone and methanol extracts was measured based on methods described by Ebrahimzadeh *et al.* (2008). Briefly, 0.5 ml of sample was mixed with 1.5 ml of methanol and then, 0.1 ml of 10 per cent aluminium chloride was added, followed by 0.1 ml of potassium acetate and 2.8 ml of distilled water. The mixture was incubated at room temperature for 30 min. The absorbance was measured by a spectrophotometer at λ_{max} = 415 nm. The samples were prepared in triplicate for each analysis and the mean value of absorbance was obtained. The same procedure was repeated for the standard solution of quercetin (QE) and the calibration line was construed. Based on the measured absorbance, the concentration of flavonoids was read (mg/ml) on the calibration line; then, the content of flavonoids in extracts was expressed in terms of quercetin equivalent (mg of QE/100mg of extract). All experiments were replicated thrice and the data were presented as mean values ± standard deviation.

Determination of DPPH Radical Scavenging Activity

The free radical scavenging capacity of the chloroform, ethylacetate, acetone and methanol extracts of *H. rhamnoides* berries was determined following modified 1,1-Diphenyl-2-picrylhydrazyl (DPPH) radical scavenging protocol (Kumarasamy *et al.*, 2007). A stock solution of 1 mg/ml solution of each extract in methanol was prepared. Sixteen serial dilutions of the extract stock solution and one blank sample (control) was made in methanol. 2 ml of freshly prepared 1 mg/ml stock solution of DPPH was added to each test tube. The sample was thoroughly mixed and incubated in the dark at 28-30°C. The absorbance (A) was read at 517 nm after 25 minutes. The percentage DPPH radical scavenging was determined by the formula:-

$$\text{Inhibition (\%)} = [(A_{Control} - A_{Sample})/(A_{Control})] \times 100.$$

The results of each extract was expressed IC_{50} (50 per cent inhibiting concentration value), that is concentration of the extract that causes 50 per cent loss of the DPPH activity (scavenge the initial DPPH concentration by 50 per cent). IC_{50} value was expressed in µg/ml. Lower IC_{50} indicate better DPPH radical scavenging ability. The data were presented as mean values ± standard deviation (n=3).

Isolation and Characterization of Flavonoid Constituents

Dried berries of *H. rhamnoides* were first defatted with hexane and then extracted with methanol in a cold percolator. The methanolic extract was distilled on water bath to remove the solvent. A dark brown semi solid extract thus obtained subjected to chromatographic resolution over silica gel column with different solvents in the order of increasing polarity and the eluents obtained were monitored by TLC. Column chromatographic resolution and crystallization of TLC homogenous single spot eluents $CHCl_3$-MeOH (3:1), $CHCl_3$-MeOH (2:1) and $CHCl_3$-MeOH (1:1) from methanol yielded three pure individual compounds 1, 2 and 3 respectively. Characterization of Compounds 1, 2 and 3 were done by spectroscopic analysis and their identities were finally confirmed by direct comparison (mmp, co-TLC and super imposable IR) with authentic samples.

RESULTS

Extractive Values

The extractive values of *H. rhamnoides* berries in n-hexane, chloroform, ethyl acetate, acetone, and methanol are presented in the Table 30.1.

Table 30.1: Extractive Values of *H. rhamnoides* Berries with different Solvents

Solvents	*Extract Yield (per cent w/w)*
n-Hexane	4.32
Chloroform	3.59
Ethyl acetate	8.65
Acetone	6.20
Methanol	13.27

The percentage yields of n-hexane, chloroform, ethylacetate, acetone and methanol were found to be 4.32, 3.59, 8.65, 6.20 and 13.27 respectively. Extractive values presented in Table 30.1 clearly indicated that the extractability of *H. rhamnoides* berry pulp is affected by different solvents. Methanol as solvents showed high extractability followed by ethylacetate and acetone.

Qualitative Phytochemical Analysis

The results of qualitative phytochemical analysis are summarized in the Table 30.2. It revealed the presence of carbohydrate, protein, amino acids, steroids, terpenoids, phenolics, flavonoids, tannins, saponins, glycosides and fatty oil in the berries of *H. rhamnoides*.

Fatty oil was found only in n-hexane extract. Phenolics and flavanoids were recorded in all extract except n-Hexane. Remarkably, Phenolics and flavonoids were present all the extracts except n-hexane whereas presence of alkaloids was not detected in any of the extracts.

Table 30.2: Phytochemical Screening of Extracts of *H. rhamnoides* Berries

Phytochemicals	*n-Hexane*	*Chloroform*	*EtOAc*	*Acetone*	*Methanol*
Fatty oil	+	–	–	–	–
Steroids	+	–	+		+
Terpenoids	+	+	+	+	+
Alkaloids	–	–	–	–	–
Phenolics	–	+	+	+	+
Flavonoids	–	+	+	+	+
Saponins	–	–	–	–	+
Tannins	–	–	–	+	+
Carbohydrates	–	–	–	–	+
Glycosides	–	–	–	–	+
Protein	–	–	–	–	+
Amino acids	–	–	–	–	+

(+) Present; (-) Absent

Determination of Total Phenolic and Flavonoid Contents

Total polyphenolic content (TPC) and total flavonoid content (TFC) of the chloroform, ethylacetate, acetone and methanol extracts of *H. rhamnoides* berries were determined and the results are given in Table 30.3.

Table 30.3: Total Phenolic and Total Flavonoid Contents in different Extracts

Extracts	*Total Phenolic Content (mg GAE/100 mg of extract)*	*Total Flavonoid Content (mg QE/100 mg of extract)*
Chloroform	8.73±0.12	0.65±0.05
Ethylacetate	14.48±0.16	1.16±0.23
Acetone	20.66±0.27	1.95±0.21
Methanol	44.23±0.23	3.88±0.14

The total phenolic content in chloroform, ethylacetate, acetone and methanol extracts were found to be 8.73±0.12, 14.48±0.16, 20.66±0.27 and 44.23±0.23 mg of GAE/100mg whereas the total flavonoid content was recorded to be 0.65±0.05, 1.16±0.23, 1.95±0.21 and 3.88±0.14 mg QE/100 mg respectively. The values of TPC and TFC considerably differed among different extracts of *H. rhamnoides* berries. Methanol extract exhibited the highest concentration of both TPC and TFC followed by acetone and ethylacetate extracts. Lowest TPC and TFC content however, was recorded in chloroform extract.

Determination of DPPH Radical Scavenging Activity

The free radical scavenging capacity of the chloroform, acetone and methanol extracts of *H. rhamnoides* berries were determine following modified 1,1-Diphenyl-2-picrylhydrazyl (DPPH) radical scavenging protocol and the results are expressed as IC_{50} (µg/ml) are shown in Table 30.4.

The DPPH radical scavenging capacities of the chloroform, ethylacetate, acetone and methanol extracts of *H. rhamnoides* berries, expressed as IC_{50} value were recorded to be 383.56±0.32, 314.31±0.27, 237.24±0.19 and 136.35±0.25 µg/ml respectively. The extracts at the used concentrations exhibited potential effect of DPPH radical scavenging activity as percentage of free radicals inhibition (Table 30.4). A higher DPPH radical scavenging capacity is associated with a lower IC_{50} value. It is evident from the data shown in Table 30.4 that radical scavenging activity is varying according to the polarity vis-à-vis extractability of the extractants. These results suggested that the methanol extract possesses the strongest free radical scavenging capacity.

Table 30.4: DPPH Radical Scavenging Efficacy of different Extracts

Extracts	*DPPH Radical Scavenging Efficacy IC_{50} (μg/ml)*
Chloroform	383.56±0.32
Ethylacetate	314.31±0.27
Acetone	237.24±0.19
Methanol	136.35±0.25

Isolation and Characterization of Compounds

Column chromatography of methanol extract of *H. rhamnoides* berry pulp led to isolation of three pure individual constituents designated as C-1, C-2 and C-3. The phenolic nature of C-1, C-2 and C-3 were detected by positive colour test with phosphomolybdic acid and ammonia vapour (Feigl, 1960). Flavonoid character of all the three compounds was confirmed by their positive response to Shinida Test (Shinoda, 1928) and a characteristic flavonoid UV absorption as well as appearance of conjugated carbonyl bands in the IR spectrum (Mabry *et al.*, 1970).

Compound 1 crystallized from methanol as yellow needles (35 mg) mp 270-72° (dec.), Rf: 0.52 ($CHCl_3$-MeOH, 5:1); UV λ_{max} MeOH : 259, 308 sh, 360 nm, λ_{max} MeOH/NaOMe: 278, 316, 410 (dec.) nm, λ_{max} MeOH/NaOAc: 275, 305, 387 nm, λ_{max} MeOH/NaOAc/H_3BO_3: 267, 297 sh, 320 sh, 366 nm, λ_{max} MeOH/$AlCl_3$: 260 sh, 268, 305 sh, 352, 422 nm, λ_{max} MeOH/$AlCl_3$/HCl: 256 sh, 269, 305 sh, 348422 nm; IR u_{max} nujol : 3320 (-OH), 1672 and 1620 cm^{-1} (conjugated carbonyl); 90 MHz ^1H NMR (DMSO-d_6, "δ"): 6.19 (1H, d, J=2Hz). 6.43 (1H, d, J=2Hz), 6.92 (2H, d, J=9Hz), 8.06(2H, d, J=9Hz), 9.42 (1H, br s, exchangeable with D_2O), 10.12 (1H, br s, exchangeable with D_2O) 10.83 (1H, br s, exchangeable with D_2O), 12.52 (1H, br s, exchangeable with D_2O); MS: m/z 286 (M+, $C_{15}H_{10}O_6$), 284, 258, 257, 229, 213, 153, 152, 143, 134, 129, 124, 121, 93, 78, 59, 65 and 39. It was identified as kaempferol by direct comparison with authentic sample.

Compound 2 crystallized from methanol as dark yellow granules (42 mg) mp 310-12° (dec.), Rf: 0.45 ($CHCl_3$-MeOH, 3:1); UV λ_{max}MeOH: 258, 271 sh, 302 sh, 372 nm, λ_{max} MeOH/NaOMe: 275 sh, 322 (dec.), λ_{max}MeOH/NaOAc: 257 sh, 276, 329, 392 (dec.) nm, λ_{max} MeOH/NaOAc/H_3BO_3: 262, 303 sh, 390 nm, λ_{max}MeOH/$AlCl_3$: 272, 304 sh, 334, 460 nm, λ_{max}MeOH/$AlCl_3$/HCl: 265, 301, 359, 430 nm; IR u_{max}nujol : 3100-3500 (-OH), 1680,and 1630 cm^{-1} (conjugated carbonyl); 90 MHz ^1H NMR (DMSO-d_6, "δ"): 6.21 (1H, d, J=2Hz). 6.37 (1H, d, J=2Hz), 6.85 (1H, d, J=9Hz), 7.23 (1H, d, J=2Hz), 7.26 (1H, dd, J= 2 and 9Hz), 12.48 (1H, br s, exchangeable with D_2O); MS: m/z 302 (M+, $C_{15}H_{10}O_7$), 301, 276, 273, 245, 229, 153, 160, 137, 128, 124, 69. Finally, it was identified as quercetin by direct comparison with authentic sample.

Compound 3 crystallized from methanol as dark yellow granules (18 mg) mp 351-53° (dec.), Rf: 0.39 ($CHCl_3$-MeOH, 3:1); UV λ_{max}MeOH: 256, 274 sh, 302 sh, 375 nm, λmax MeOH/NaOMe: 262 sh, 285 sh, 322, 423 (dec.) nm, λ_{max} MeOH/NaOAc: 271, 338 (dec.) nm, λ_{max} MeOH/NaOAc/H_3BO_3: 258, 305 sh, 393 nm, λmax MeOH/$AlCl_3$: 272, 316 sh, 451 nm, λ_{max} MeOH/$AlCl_3$/HCl: 266, 275 sh, 307 sh, 360 sh, 429 nm; IR u_{max}nujol: 3100-3500 (-OH), 1660 and 1600 cm^{-1} (conjugated carbonyl); 90 MHz ^1H NMR (DMSO-d_6, "δ"): 6.02 (1H, d, J=2Hz). 6.22 (1H, d, J=2Hz), 7.28 (2H, br s), 9.40 (1H, br s, exchangeable with D_2O), 10.62 (1H, br s, exchangeable with D_2O), 12.33 (1H, br s, exchangeable with D_2O); MS: m/z 318 (M+, $C_{15}H_{10}O_8$), 302, 290, 289, 286, 261, 244, 166, 153, 146, 124, 108, 79 and 69. It was identified as myrecetin by direct comparison with authentic sample.

2 Quercetin

1 Kaempferol

3 Myrecetin

DISCUSSION

Extractive Values

Extractive values of *H. rhamnoides* berries determined under sequential extraction (Table 30.1) provided idea about extractability of chemical constituents of berries in different organic solvents. Extractive values showed that the methanol extractive value was greater as compared with other solvents. Plants possess different solubility behaviour in various solvents. The yields of extracts from a plant in different solvents provide information about the solubility of plant chemical constituents in different organic solvents thus suggesting the best solvent for extraction of phytochemicals. It is clear from the Table 30.1 that methanol yielded the highest amount of phytoconstituents in successive extraction.

Qualitative Phytochemical Analysis

The qualitative phytochemical test of the n-hexane, chloroform, ethyl acetate, acetone and methanol extracts of *H. rhamnoides* berries obtained through successive extraction indicated the presence of carbohydrate, protein, amino acids, steroids, alkaloids, terpenoids, phenolics, flavonoids, tannin, saponins, glycosides and fatty oil. Fatty oil was found in n-hexane only. Remarkably, phenolics and flavonoids were present is all the extracts except n-hexane. Carbohydrate, saponins, glycosides, proteins and amino acids were detected in methanol extract only. None of the extract however showed the presence of alkaloids. Different phytochemicals have been found to possess a wide range of activities, which may help in protection against chronic diseases. Phytochemicals such as saponins, terpenoids, flavonoids, tannins, steroids and alkaloids have anti-inflammatory effects (Liu, 2003; Akindele and Adeyemi, 2007; Ilkay Orhan *et al.*, 2007). Glycosides, flavonoids, tannins and alkaloids have hypoglycemic activities (Oliver, 1980; Cherian and Augusti, 1995; Kaur *et al.*, 2012). Polyphenolic compound have been reported to possess antioxidant and antimicrobial efficacies (Rambabu *et al.*, 2011; Tripathi *et al.*, 2015a). Saponins are reported to possess hypocholesterolemic and antidiabetic properties (Acharya *et al.*, 1988; Rupasinghe *et al.* 2003; Tripathi *et al.*, 2015b). Terpenoids have also been shown to decrease blood sugar level in animal studies (Luo *et al.*, 1999). Steroids and triterpenoids showed the analgesic properties (Sayyah *et al.*, 2004; Malairajan *et al.*, 2006). Certain alkaloids, steroids and saponins are responsible for central nervous system activities (Tripathi and Dwivedi, 1990; Argal and Pathak, 2006). The therapeutic actions of phytochemicals validate the traditional uses of various medicinal plants for the treatment of wide range of ailments.

Determination of Total Phenolic and Flavonoid Contents

Phenolic compounds or polyphenols constitute one of the most numerous and ubiquitously distributed group of plant secondary metabolites with more than 8000 phenolic structures currently known. Natural polyphenols can range from simple molecules to highly polymerised compounds. Earlier reports have also mentioned *H. rhamnoides* berries as a good source of polyphenols and larger portion of their biological activities are attributed to their polyphenol content. Further, flavonoids are important class of polyphenols with significant and well documented biological properties. The amount of total polyphenol and total flavonoids in different extracts of berries as quantified and summarized in Table 30.3 clearly suggests chloroform, ethylacetate, acetone and methanol extracts contain varying amount of polyphenolics and flavonoids. In the study, the highest values for total phenol and total flavonoid *i.e.*, 44.23±0.23 mg GAE/100mg and 3.88±0.14 mg QE/100mg respectively were observed in methanol extract. Previous studies have reported varying quantities of TPC and TFC in berries which may be due to seasonal variation, plant age, genetic, agronomic and climatic variations (Kondakova *et al.*, 2009).

Determination of DPPH Radical Scavenging Activity

In the present work, the antioxidant activities of different extracts of *H. rhamnoides* berries were examined by DPPH radical scavenging method. DPPH is a useful reagent for investigating the free radical-scavenging activities of phenolic compounds (Lebeau *et al.*, 2000). The reduction of DPPH absorption is indicative of the capacity of the extracts to scavenge free radicals, independently of any enzymatic activity. The extractracts used for extraction methods resulted in different extracts, with diverse compositions (Table 30.1). All extracts promoted the reduction of DPPH (Table 30.4). The best free radical-scavenger activity was reported for methanol extract with IC_{50} values of 136.35±0.25 µg/ml as IC_{50} value were recorded to be, 314.31±0.27, 237.24±0.19 and respectively The lowest activity *i.e.* was observed for chloroform showing IC_{50} value as 383.56±0.32 µg/ml. The IC_{50} is a parameter, widely used, that describes ability of a compound or extract to scavenge the DPPH radical. It is interesting to note that all the extracts examined possessed moderate to high reactivity and had an important polyphenolic component (Table 30.3). Figure 30.1 illustrates the relationship between DPPH IC_{50} values in relation to total polyphenol.

It can be seen that for all the extracts, the IC50 values decreases with increasing total polyphenol levels. Since lower IC_{50} values indicates better radical scavenging capabilities, this implies that the radical scavenging activity (hence antioxidant activity) increases with increasing total polyphenol level. Also, it can be observed methanol extract had better radical scavenging capability than others. This is probably because methanol extract contains better yield of total polyphenol.

Isolation and Characterization of Compounds

Chromatographic analyses of the methanol extract of berries of *H. rhamnoides* afforded three flavonol compounds 1, 2 and 3. The structures of the isolated compounds (1-3) were elucidated by their respective UV, IR, 1H NMR and Mass spectral data their identity

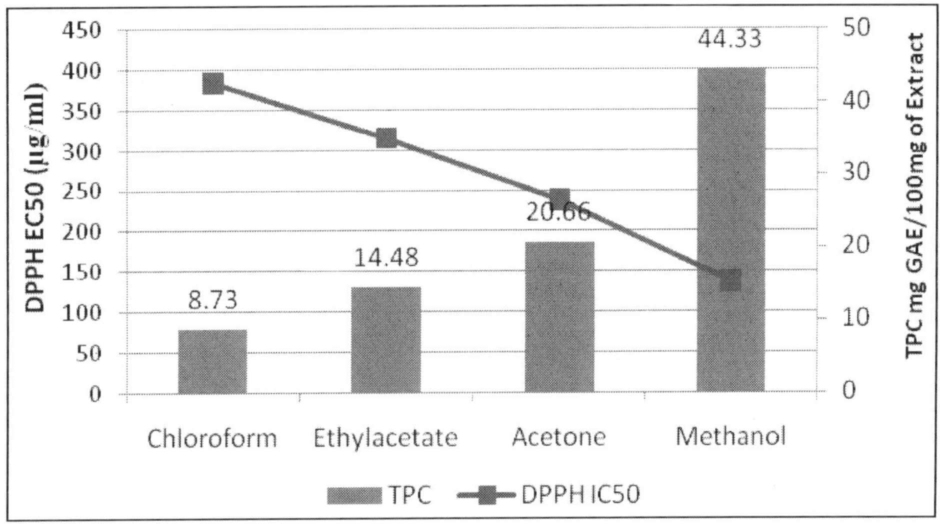

Figure 30.1: Correlation between TPC and DPPH Radical Scavenging Activity.

as kaempferol, quercetin and myrecetin respectively was confirmed by direct comparison with authentic samples. These compounds have already been reported in *H. rhamnoides* berries (Hakkinen *et al.*, 1999; Zhao and Fuheng, 1997) of different geographical origin. Studies have shown that these flavonoids have adaptogenic properties, promote production of red blood cells, reduce allergies, protect against carcinogenesis, decrease cholesterol, improve heart function and are powerful antioxidant and immunomodulator.

CONCLUSION

In the present study, it is noticed that the highest concentration of phenolic compounds in the *H. rhamnoides* berries extracts were obtained using solvents of high polarity; the methanolic extract manifested greater power of extraction for phenolic compounds. The study showed that different extracts of *H. rhamnoides* berries contain moderate to high quantities of phenolic compounds and exhibit antioxidant activity. The phenol content and its linear correlation between the values of the concentration of phenolic compounds and antioxidant activity indicated that these compounds contribute to the strong antioxidant activity. *In vitro* antioxidant efficacy of berries in relation to its chemical composition was investigated in this study. Quest for health friendly phytochemicals as nutraceuticals, food additives, chemopreventive and therapeutic agents is far stronger than ever before. On the commercial point of view not many natural sources have been investigated for their therapeutic and chemopreventive efficacy and chemical composition. Therefore, the plant deserves proper attention towards systematic approach for the collection, storage, processing and value addition that could be helpful in the economic development of tribal areas in the Himalayan foot hills where the species is mostly grown.

ACKNOWLEDGEMENT

Authors are thankful to the Director, Forest Research Institute, Dehradun for providing necessary facilities and the Head, Chemistry Department, Banaras Hindu University Varanasi, India for spectral analysis. Authors are grateful to Indian Council of Forestry Research and Education, Dehradun, Delhi, India for providing funding support for carrying out this study

REFERENCES

1. Akindele, A. J. and Adeyemi, O. O. 2007. Anti-inflammatory activity of the aqueous leaf extracts of *Byrsocarpus coccineus. Fitoterapia* 78: 25-28.
2. Argal, A. and Pathak, A. K. 2006. CNS activity of *Calotropis gigantea* roots. *J. Ethnopharmacoogy* 106: 142-145.
3. Bal, L. M., Meda, V., Naik, S. N. and Santosh, S. 2011. Seabuckthorn berries: A potential source of valuable nutrients for nutraceuticals and cosmoceuticals. *Food. Res. Int.*, 44: 1718-1727
4. Barkat, A. K., Naveed, A., Tariq, M., Haji, M. S. K., Shahiq, U. Z., Akhtar, R., Muhammad, I., Atif, A., Salimullah, K. and Mughal, Q. 2011. In-vivo study of stratum corneum water content and transepideramal water loss using a newly formulated topical cream of *Hippophae rhamnoides* fruit extract. *Afr. J. Pharm. Pharmacol.* 5 (8): 1092-1095.
5. Bekker, N. P. and Glushenkova, A. I. 2001. Components of certain species of the Elaeagnaceae family. *Chem. Nat. Compounds* 37: 97-116.
6. Cenkowski, S., Yakimishen, R., Przybylski, R. and Muir, W.E. 2006. Quality of extracted seabuckthorn seed and pulp oil. *Canadian Biosystems Engineering* 48 (3): 9-16.

7. Cherian, S. and Augusti, K. T. 1995. Insulin sparing action of leucopelargonidin derivative isolated from *Ficus benghalensis* L. *Indian J. Exp. Biol.* 33: 608-611.

8. Cheng, T. J. 1992. Protective action of seed oil of *Hippophae rhamnoides* L. (HR) against experimental liver injury in mice. *Chinese Journal of Preventive Medicine* 26 (4): 227-229.

9. Cheng, J., Kondo, K., Suzuki, Y., Ikeda, Y., Meng, X. and Umemura, K. 2003. Inhibitory effects of total flavones of *Hippophae rhamnoides* L. on thrombosis in mouse femoral artery and *in vitro* platelet aggregation. *Life Sciences* 72 (20): 2263-2271.

10. Ebrahimzadeh, M. A., Pourmorad, F. and Bekhradnia, A. R. 2008. Iron chelating activity, phenol and flavonoid content of some medicinal plants from Iran. *Afr. J. Biotechnol.* 7 (18): 3188-3192.

11. Eccleston, C., Baoru, Y., Tahvonen, R., Kallio, H., Rimbach, G. H. and Minihane, A.M. 2002. Effects of an antioxidant rich juice (seabuckthorn) on risk factors for coronary heart disease in humans. *J. Nutr. Biochem.* 13: 346-354.

12. Evans, W. C. 1989. *Trease and Evan's Pharmacognosy.* 13th edn. Bailliere Tindall, London.

13. Feigl, F. 1960. *Spot Test in Organic Analysis.* Elsevier Publishing Company, Amsterdam, London, New York, Princeton, p. 128.

14. Fengming, Z. 1989. Anti-arrhythmic effects of TFH on the extra corporeal cors. *Journal of Chinese Pharmacology* 5 (1): 44-47.

15. Ganju, L., Padwad, Y., Singh, R., Karan, D., Chanda, S., Chopra, M. K., Bhatnagar, P., Kashyap, R. and Sawhney, R.C. 2005. Anti-inflammatory activity of seabuckthorn (*Hippophae rhamnoides*) leaves. *International Immuno-pharmacology* 5: 1675-1684.

16. Gao, X., Ohlander, M., Jeppsson, N., Bjork, L. and Trajkovski, V. 2000. Changes in antioxidant effects and their relationship to phytonutrients in fruits of seabuckthorn (*Hippophae rhamnoides* L.) during maturation. *J. Agric. Food Chem.* 48: 1485-1490.

17. Gao, Z.L., Gu, X. H., Cheng, F.T. and Jiang, F.H. 2003. Effect of seabuckthorn on liver fibrosis: A clinical study. *World Journal of Gastroenterology* 9 (7): 1615-1617.

18. Hakkinen, S. H., Karenlampi, S. O., Heinonen, I. M., Mykkanen, H. M. and Torronen, A.R. 1999. Content of the flavonols quercetin, myricetin and kaempferol in 25 edible berries. *J. Agric. Food Chem.* 47: 2274-2279.

19. Harborne, J. B. 1998. *Phytochemical Methods - A guide to modern techniques of plant analysis.* 3rd Edn., Chapman and Hall Int. Ed., New York.

20. Hibasami, H., Mitani, A., Katsuzaki, H., Imai, K., Yoshioka, K. and Komiya, T. 2005. Isolation of five types of flavonol from seabuckthorn (*Hippophae rhamnoides*) and induction of apoptosis by some of the flavonols in human promyelotic leukemia HL-60 cells. *International Journal of Molecular Medicine* 5 (5): 805-809.

21. Ilkay, O., Kupeli, E., Sener, B. and Yesilada, E. 2007. Appraisal of anti– inflammatory potential of the clubmoss, *Lycopodium clavatum* L. *J. Ethnopharmacol* 109: 146-150.

22. Johansson, A. K., Korte, H., Yang, B., Stanley, J. C. and Kallio, H. P. 2000. Seabuckthorn berry oil inhibits platelet aggregation. *Journal of Nutritional Biochemistry* 11 (10): 491-495.

23. Kallio, H., Baoru, Y., Tahvonen, R. and Hakala, M. 2000. Composition of seabuckthorn berries of various origins. In: *International Workshop on Seabuckthorn* (IWS-99), Aug. 29-Sept. 2, Beijing, P.R. China, pp. 17-23.

24. Kashyap, R. and Sawhney, R. C. 2005. Anti-inflammatory activity of seabuckthorn (*Hippophae rhamnoides*) leaves. *International Immunopharmacology* 5: 1657-1684.

25. Kaur, N., Tripathi, R. Y. C. and Upadhyay, L. 2012. Phytochemical and antidiabetic evaluation of *Phaseolus trilobus* roots. *Journal of Pharmacy Research* 5 (11): 5202-5205.

26. Kaushal, M. and Sharma, P. C. 2011. Nutritional and antimicrobial property of seabuckthorn (*Hippophae* sp.) seed oil. *J. Sci. Indus. Res.* 70: 1033-1036.

27. Kiken, D.A. and Cohen, D.E. 2002. Contact dermatitis to botanical extracts. *Am. J. Contact. Dermat.* 3: 148-152.

28. Kondakova, V., Tsvetkov, I., Batchvarova, R., Badjakov, I., Dzhambazova, T. and Slavov, S. 2009. Phenol compound qualitative index in small fruits. *Biotechnology Biotechnol. Equip.* 23: 1444-1448.

29. Kokate, C. K. 2001. *Pharmacognosy.* 16th Edn., Nirali Prakasham, Mumbai, India.

30. Kumar, R., Kumar, G. P., Chaurasia, O. P. and Singh S. B. 2011. Phytochemical and pharmacological profile of seabuckthorn oil: A review. *Res. J. Med. Plant* 5: 491-499.

31. Kumarasamy, Y., Byres, M., Cox, P. J., Jaspars, M., Nahar, L. and Sarker, S. D. 2007. Screening seeds of some Scottish plants for free-radical scavenging activity. *Phytother Res.* 21: 615-621.

32. Lebeau, J., Furman, C., Bernier, J.L., Duriez, P., Teissier, E. and Cotelle, N. 2000. Antioxidant properties of di-tert–butylhydroxylated flavonoids. *Free Radical Biology and Medicine* 29: 900-912.

33. Li, T.S.C. and Schroeder, W.R. 1996. Seabuckthorn (*Hippophae rhamnoides* L.): A multipurpose plant. *Hortic. Tech.* 6: 370-380.

34. Liu, R. H. 2003. Health benefits of fruit and vegetables are from additive and synergistic combinations of phytochemicals. *Am. J. Clin. Nutr.* 78: 517S-520S.

35. Luo, J., Cheung, J. and Yevich, E.1999. Novel terpenoid type quinones isolated from *Pycnanthu angolensis* of potential utility in the treatment of type-2 diabetes. *J. Pharmacol. Exptl. Therapy* 288: 529-534.

36. Mabry, T. J. Markham, K. R. and Thomas, M. B. 1970. *The Systematic Identification of Flavonoids*. Springer-Verlag, New York.

37. Malairajan, P., Gopalakrishnan, G., Narasimhan, S. and Veni, K. J. K. 2006. Analgesic activity of some Indian medicinal plants. *J. Ethnopharmacol.* 19: 425-428.

38. Michel, T., Destandau, E., Le Floch, G., Lucchesi, M. E. and Elfakir, C. 2012. Antimicrobial, antioxidant and phytochemical investigations of seabuckthorn (*Hippophae rhamnoides* L.) leaf, stem, root and seed. *Food Chem.* 131: 754-760.

39. Novruzov, E.N, Aslanov, S. M., Imanova, A.A and Gasanova, Z.I. 1979. Ursolic acid from *Hippophae rhamnoides*. *Khimiya Prirodnykh Soedinenii* 6: 868.

40. Ramasamy, T., Varshneya, C. and Katoch, V. C. 2010. Immunoprotective effect of seabuckthorn (*Hippophae rhamnoides*) and glucomannan on T-2 toxin-induced immunodepression in poulry. *Vet Med. Int.* Article ID: 149373.

41. Rambabu, P. Venkata, R. K. and Ganapaty, S. 2011. Anti bacterial and anti fungal potential of roots of *Zizyphus rugosa*. *International Journal of Pharmaceutical Research and Development* 3 (3): 112-114.

42. Reddy, G. A. K., Priyanka, B., Saranya, C. S. and Kumar, A. C.K. 2012. Wound healing potential of Indian medicinal plants. *International Journal of Pharmacy Review and Research* 2 (2): 75-87.

43. Roesch, D., Krumbein, A., Mugge, C. and Kroh, L.W. 2004. Structural Investigations of flavonol glycosides from seabuckthorn (*Hippophae rhamnoides*) Pomace by NMR Spectroscopy and HPLC-ESI-MS. *J. Agric. Food Chem.* 52 (13): 4039-4046.

44. Rosch, D., Bergmann, M., Knorr, D. and Kroh, L. W., 2003. Structure-antioxidant efficiency relationships of phenolic compounds and their contribution to the antioxidant activity of seabuckthorn juice. *J. Agric. Food Chem.* 51(15): 4233-4239

45. Rupasinghe, H. P., Jackson, C. J., Poysa, V., Di Berado, C., Bewley, J. D. and Jenkinson, J. 2003. Soyasapogenol A and B distribution in Soybean (Glycine Max L. Merr) in relation to seed physiology, genetic variability and growing location. *J. Agric. Food Chem.* 51: 5888- 5894.

46. Saggu, S., Divekar, H. M., Gupta, V., Sawhney, R. C., Banerjee, P. K. and Kumar, R. 2007. Adaptogenic and safety evaluation of seabuckthorn (*Hippophae rhamnoides*) leaf extract: A dose dependent study. *Food Chemistry and Toxicology* 45: 609-617.

47. Sayyah, M., Hadidi, N. and Kamalinejad, M. 2004. Analgesic and anti-inflammatory activity of *Lactuca sativa* seed extract in rats. *J. Ethnopharmacol.* 92: 325-9.

48. Salenko, V. L., Kukina, T. P., Karamyshev, V. N., Sidel'nikov, V. N. and Pentegova, V. A. 1986. Chemical investigation of *Hippophae rhamnoides*. II. Main components of the neutral reaction of the saponification products of an extract of the leaves of seabuckthorn. *Chemistry of Natural Compounds* 21 (4): 4810-485.

49. Shinoda, J. 1928. A new biologically active flavones glycoside from the roots of *Cassia fistula* L. *J. Pharm. Soc. Japan* 48: 214-220.

50. Singleton, V. L., Orthofer, R. and Lamuela-Raventos, R. M. 1999. Analysis of total phenols and other oxidation substrates and antioxidants by means of Folin-Ciocalteu reagent. *Methods Enzymol.* 299: 152-178.

51. Singh, V. 2003. Geographical adaptation and distribution of seabuckthorn (*Hippophae* L.) resources. In: *A Multipurpose Wonder Plant-Botany*. Vol. I. *Botany, Harvesting and Processing Technologies* (V. Singh, Editor in Chief), pp. 21-34, Indus Publishing Company, New Delhi, 518p.

52. Suleyman, H., Buyukokuroglu, M. E., Koruk, M., Akcay, F., Kiziltunc, A. and Gepdiremen, A. 2001. The effects of *Hippophae rhamnoides* L. extract on ethanol-induced gastric lesion and gastric tissue glutathione level in rats: a comparative study with melatonin and omeprazole. *Indian Journal of Pharmacology* 33 (2): 77-81.

53. Suryakumar, G. and Gupta, A. 2011. Medicinal and therapeutic potential of seabuckthorn (*Hippophae rhamnoides* L.). *J. Ethnopharmacol.* 138: 268-278.

54. Tang, X. and Tigerstedt, P. M. A. 2001. Variation of physical and chemical parameters within an elite seabuckthorn (*Hippophae rhamnoides* L.) breeding population. *Scientia Horiticulturae* 88: 203-214.

55. Tripathi, Y.C., Tiwari, S., Anjum, N. and Tewari, D. 2015a. Phytochemical, antioxidant and antimicrobial screening of roots of *Asparagus racemosus* Willd. *World Journal of Pharmaceutical Research* 4 (4): 709-722.

56. Tripathi, Y. C., Shulka, P. and Tewari, D. 2015b. Phytochemical evaluation and antihyperglycemic effects of *Elaeocarpus ganitrus* Roxb (Rudraksha) in Streptozotocin Induced Diabetes. *International Journal of Pharmacy and Pharmaceutical Sciences* **7** (1): 280-283.

57. Tripathi, Y. C. and Dwivedi, R. K. 1990. Central nervous system and anti-inflammatory activities of alkaloids of *Fumaria indica. Nat. Acad. Sci. Lett.* 13 (6): 231-233.

58. Tulsawani, R. 2010. Ninety day repeated gavages administration of *Hippophae rhamnoides* extract in rats. *Food Chemistry and Toxicology* 48: 2483-2489.

59. Upadhyay, N. K., Kumar, R., Mandotra, S. K., Meena, R. N., Siddiqui, M. S., Sawhney, R. C. and Gupta, A. 2009. Safety and healing efficacy of seabuckthorn (*Hippophae rhamnoides* L.) seed oil on burn wounds in rats. *Food. Chem. Toxicol.* 47: 1146-1153.

60. Varshney, A. C. and Tyagi, S. P. 2004. Studies on anti-inflammatory activities of seabuckthorn (*Hippophae* sp.) in Canine. In: *Proceedings of the Asian Congress of Zoo and Wildlife Veterinarians*, Lucknow, India, pp. 14-17.

61. Wang, B., Lin, L., Ni, Q. and Su, C. L. 2011. *Hippophae rhamnoides* L. for treatment of diabetes mellitus: A review. *J. Med. Plants Res.* 5: 2599-2607.

62. Wu, J., Yu, X. J., Ma, X., Li, X. G. and Liu, D. 1994. Electrophysiologic effects of total flavones of *Hippophae rhamnoides* L. on guinea pig papillary muscles and cultured rat myocardial cells. *Zhongguo Yao Li Xue Bao* 15: 341-343.

63. Xing, J., Yang, B., Dong, Y., Wang, B., Wang, J. and Kallio, H. P. 2002. Effects of seabuckthorn (*Hippophae rhamnoides* L.) seed and pulp oils on experimental models of gastric ulcer in rats. *Fitoterapia* 73 (7-8): 644-650.

64. Yang, C. 1995. A clinical study of reducing fat and antioxidation of dried *Hippophae* emulsion. *Hippophae* 8: 33-35.

65. Yang, B. and Kallio, H. P. 2001. Fatty acid composition of lipids in seabuckthorn (*Hippophae rhamnoides* L.) berries of different origins. *J. Agric. Food Chem.* 49: 1939-1947.

66. Yang, B., Karlsson, R.M., Oksman, P.H. and Kallio, H. P. 2001. Phytosterols in seabuckthorn (*Hippophae rhamnoides* L.) berries: Identification and effects of different origins and harvesting times. *J. Agric. Food Chem.* 49 (11): 5620-5629

67. Yang, B. and Kallio, H. 2006. Lipophilic components of seabuckthorn (*Hippophae rhamnoides* L.) seeds and berries. In: *Seabuckthorn (Hippophae L.): A multipurpose wonder plant.* (Editor-in-Chief, Singh, V.), Daya Publishing House. Delhi India, 2: 70-97.

68. Yao, Y. and Tigerstedt, P.1994. Genetic diversity in *Hippophae* L. and its use in plant breeding. *Euphytica* 77: 165-169.

69. Zhao, Y. and Fuheng, W. 1997. Seabuckthorn flavonoids and their medicinal value. *Hippophae* 10 (1): 39-41.

70. Zeb, A. 2004. Chemical and nutritional constituents of seabuckthorn juice. *Pakistan J. Nutr.* 3: 99-106.

71. Zeb, A. 2006. Anticarcinogenic potential of lipids from *Hippophaë*–Evidence from the recent literature. *Asian Pac. J. Cancer* 7: 32-34.

72. Zhang, M. 1987. Treatment of ischemic heart diseases with flavonoids of *Hippophae rhamnoides*. *Chinese Journal of Cardiology* 15: 97-99.

73. Zhang, W., Zhao, J., Wang, J., Pang, X., Zhuang, X., Zhu, X. and Qu, W. 2010. Hypoglycemic effect of aqueous extract of seabuckthorn (*Hippophae rhamnoides* L.) seed residues in streptozotocin-induced diabetic rats. *Phytotherapy Research* 24: 228-232.

74. Zhang, P., Sun, B., Yang, H. J., Zhuang, X., Bai, Z., Wang, L. and Qu, W. 2004. Study of flavonoids from seed residues of *Hippophae rhamnoides* L. on apoptosis of human mammary carcinoma Bcap-37 cells. *Huadong Shifan Daxue Xuebao, Ziran Kexueban* 4: 91-96.

75. Zheng, R. X., Xu, X. D., Tian, Z. and Yang, J. S. 2009. Chemical constituents from the fruits of *Hippophae rhamnoides*. *Natural Product Research* 23: 1451-1456.

Chapter 31

Seabuckthorn Leaf Extract counters Jejunum Pathology in Whole Body ^{60}Cobalt-Gamma-Irradiated Mice: A Study on Tissue Histology and Microbiota Towards Development of Safe and Non-Toxic Radiation Countermeasure

Madhu Bala[1*], Manish Gupta[1], C.S. Beniwal, M.Z. Abdin[2] and Jagdish Prasad[1]

[1]Division of Radiation Biology, Institute of Nuclear Medicine and Allied Sciences,
Brig. S.K. Mazumdar Marg, Delhi, India
[2]Centre for Transgenic Plant Development, Department of Biotechnology, Jamia Hamdard,
New Delhi – 110 062, India
(*E-mail: bala44@gmail.com)

SUMMARY

Administration of a single dose (30 mg/kg body weight) of standardized Seabuckthorn leaf extract (SBL-1), 30 min before ^{60}Co-gamma-irradiation (lethal dose, 10Gy), rendered survival to > 90 per cent mice population. Purpose of this study was to investigate the action of SBL-1 on jejunum and the residing microbiota. Thirty day study with 9 weeks old inbred male Swiss albino Strain 'A mice, demonstrated that SBL-1 treatment before ^{60}Co-gamma-irradiation (10 Gy) significantly ($p< 0.05$) countered the radiation induced decreases in crypts; villi number; villus height; villus cellularity, Cryptal Paneth cells. It countered radiation induced increases in cryptal apoptotic cells. These changes were recorded by tissue histology study. Significant disturbances were observed in the native microbiota of jejunum after irradiation. Pre-treatment with SBL-1 countered the radiation induced dysbiosis in mice jejunum as was revealed by metagenomic and bioinformatics approach. This study demonstrated that SBL-1 countered radiation pathology to jejunum by promoting the cryptal stem cells proliferation, by regulating the apoptosis and also maintaining the microbiota homeostasis. Multiple antioxidants were detected in SBL-1 and therefore neutralization of radiation induced oxidative species was expected to be an important mechanism of radioprotection by SBL-1. Seabuckthorn leaves are common constituents of health promoting drinks; and therefore SBL-1 is a promising candidate for developing safe, non-toxic and effective medical radiation countermeasures.

Keywords: *Seabuckthorn, Jejunum pathology, ^{60}cobalt-gamma-irradiated mice, Tissue histology, Non-toxic radiation counter measure and Microbiota.*

INTRODUCTION

Hippophae rhamnoides L. common name seabuckthorn, family Elaeagnaceae is a native plant of high altitude region, commonly found in the north-west Indian Himalayas (8,000-11,000 feet). This plant has wide range of adaptations to extreme temperatures ranging from -43° to + 40°C (Lu, 1992). The high levels of radiations, dry winds and humidity are some of the other features which have been uniquely tolerated by this plant for thousands of years. The phytochemical analysis shows that different parts of seabuckthorn plant are rich in large number of bioactive compounds and anti-oxidants such as vitamin C (Sabir *et al.*, 2005), carotenoids and vitamin E (Pintea *et al.*, 2001), alpha-

Seabuckthorn (Hippophae L.): A Multipurpose Wonder Plant, Vol. 5: Advanced Technologies for Cultivation, Processing, Health Protection and Environmental Conservation (Virendra Singh, *Editor-in-Chief*, 2018), p. 245–256, Astral International Pvt. Ltd., New Delhi, India.

tocopherols, c-tocopherol, beta-tocotrienol (Zeb, 2004) free amino acids (*e.g.*, aspartic acid, threonine, proline, serine, lysine, valine, alanine) (Chen, 1988; Mironov, 1989; Zhang *et al.*, 1989; Kallio *et al.*, 1999); flavonoids (quercetin, myricetin and kaempferol) (Hakkinen *et al.*, 1999); high amount of unsaturated fatty acids (*e.g.* omega-3, omega-6, omega-7), triacylglycerol, glycerophosphopholipids and phytosterols (Yang and Kallio, 2001); phenolic compounds (*e.g.*, isorhamnetin 3-O-glycosides, quercetin 3-O-glycosides, catechins, hydroxybenzoic acids with a catechol structure) (Rosch *et al.*, 2003); large number of mineral elements (*e.g.*, potassium, sodium, calcium, magnesium, iron, phosphorus) (Sabir *et al.*, 2005); monosaccharides sugars (*e.g.*, mannitol, sorbitol, xylitol and xylose) (Makinen and Soderling, 1980); organic acids (*e.g.*, malic acid, oxalic acid, quinic acid) (Kallio *et al.*, 1999) and volatile compounds (*e.g.*, ethyl dodecenoate, ethyl octanoate, decanol, ethyl decanoate, and ethyl dodecanoate) (Kumar *et al.*, 2011; Guliyev *et al.*, 2004). The nutritional and medicinal values of this plant have been known to traditional medicinal systems in India, Tibet, Mongolia, Russia, Germany and many other parts of the world.

Earlier studies demonstrated the unique and effective radioprotective properties of leaves of seabuckthorn. Administration of a single dose of standardized herbal preparation from leaves of seabuckthorn (coded SBL-1, pending patent) intraperitoneally @30 mg/kg body weight, 30 minutes before lethal dose of ^{60}Co-γ-irradiation (10 Gy) rendered >90 per cent survivors in experimental mice population; while all non-SBL-1 treated animals died within 12 days after irradiation (Bala *et al.*, 2009). SBL-1 countered radiation induced mutations, recombinations and high flux of oxidative stress (Tiwari *et al.*, 2009; Saini *et al.*, 2010). The SBL-1 treated irradiated animals had effectively countered the radiation injuries to gastrointestinal (GI) tract (Bala *et al.*, 2015), kidney (Saini *et al.*, 2014) and spleen (Kaur and Bala, 2015). The radiation induced inflammation was countered by modifying the HMGB-1 regulated pathway (Tiwari and Bala, 2011). The radiation induced behavioural changes in rats were also countered by treatment with SBL-1 (Gupta *et al.*, 2011). These findings suggested the promising potential of SBL-1 as an effective radiation countermeasure.

Development of a safe, non-toxic and effective radioprotector has remained a challenge for scientists ever since World War II. Thousands of single molecule drugs screened for radioactive efficacy, so far, have met with little success due to the unacceptable levels of toxicity. The low Linear Energy Transfer (LET) ionizing radiations injure various types of cells present in the body through multiple pathways simultaneously as well as sequentially. These multiple pathways are regulated systemically by thousands of molecules. The SBL-1 is a natural combination of large number of bioactive molecules in a very specific proportion which has been demonstrated to render not only significant (>90 per cent) survival, but also protect the tissue architecture as well as functioning of multiple organs.

The adult mammalian GI tract is colonized by trillions of microorganisms which are collectively 10 times more than the total number of mammalian somatic and germ cells (Eckburg *et al.*, 2005). The microbes present in GI tract help the host in digesting the indigestible components of food (Hooper *et al.*, 2002), in regulating fat storage (Backhed *et al.*, 2004), cholesterol level reduction (Sadzikowski *et al.*, 1977), synthesis of essential vitamins, *e.g.*, B1, B6 (Gill *et al.*, 2006; Kurokawa *et al.*, 2007), regulation of xenobiotic metabolism (Gill *et al.*, 2006; Kurokawa *et al.*, 2007), production of antibodies against the normal flora and related pathogens (Liu and Rhoads, 2013). Several studies have reported that disruption in normal balance of gut microbiota composition (dysbiosis) leads to several metabolic disorders like obesity, inflammatory bowel disease, diabetes, fatty liver, anxiety and colon cancer (Tamboli *et al.*, 2004; Cani *et al.*, 2008; Dumas *et al.*, 2006; Neufeld *et al.*, 2011). Gut microbiota contributes ~36 per cent of small molecules in human blood (Hood, 2012). The gut microbiota, therefore, is now considered as an organ within the living system and it is important to maintain homeostasis of gut microbiota to ensure good health.

The gut microbiota's impact on drug response and metabolism has been explored since the mid 20th century; although, past studies were mostly focused on assessing the metabolic activity of gut microbial communities on antibiotics and botanicals (Lindenbaum *et al.*, 1981; Mathan *et al.*,1989). More recently, the gut microbiota is understood to affect the metabolism, distribution and efficacy of various drugs by multiple mechanisms such as by secretion of enzymes that modify the chemical structure of drug molecules, by secretion of metabolic products that interfere with drug metabolism, by modification of the levels and activities of liver and intestinal enzymes, by modulation of expression of human metabolic genes (Saad *et al.*, 2012; Johnson *et al.*, 2012).

Radiation induced disturbance in population of gut microbiota particularly *Lactobacillus* has been reported in multiple studies performed on radiotherapy patients. The dysbiosis of beneficial microbiota such as *Lactobacillus*, *Bifidobacterium* (Phylum Firmicutes), is considered as one of the important causes of radiation induced GI tract pathologies. The most important niche of the jejunal microbiota is the epithelial layer of the villi. In the healthy animals the epithelial layer is renewed at regular intervals by the newly proliferated cells. The putative multiple intestinal stem cells, located in the crypts of Lieberkühn at the base of the villi, proliferate and give rise to various types of cells present in the epithelial layer (such as enterocytes, goblet cells, entero-endocrine cells and paneth cells). The paneth cells contribute to the innate intestinal defence and regulate the microbial density in the small intestine and protect stem cell from infections. The radiation exposure at lethal doses leads to sterile crypts, denudation and shortening of villi. This leads to increased systemic bacterial infection, and decreased absorption of nutrients.

Present study was undertaken to understand the modifying effects of SBL-1 on radiation induced jejunal architecture as well as on the population of beneficial microbes (*Lactobacillus*) in total body lethally irradiated experimental mice.

MATERIALS AND METHODS

Chemicals and Reagents

The molecular biology grade chemicals were purchased from Merck, India; the DNase and RNase free plastic-ware were from Greiner, Germany. The aluminium chloride, Proteinase K, RNase and lysozyme were purchased from Sigma-Aldrich, USA. 5-bromo-4-chloro-3-indolyl-β-Dgalactopyranoside (X-gal) and Isopropyl β-D-1-thiogalactopyranoside (IPTG) were purchased from Fermentas Germany. High performance thin layer chromatography (HPTLC) plates; RP-18 F254S (20x20 cm), formaldehyde, xylene, ethyl alcohol, paraffin wax were purchased from Merck, India. Reference standards Gallic acid ethyl ester, purity 98 per cent w/w; and Rutin, purity 97 per cent ; were purchased from Acros Organics, Fischer scientific, USA; and Quercetin, purity 98 per cent ; was purchased from Fluka Biochemika, USA.

Plant Extract (SBL-1) and its Characterization

The plant extract was prepared and characterized as per procedure described earlier (Bala *et al.*, 2009). Briefly, the fresh green leaves of *Hippophae rhamnoides* L. (Elaeagnaceae), identified by ethno-botanists and confirmed as *Hippophae rhamnoides* [specimen records preserved at herbarium, Defence Institute for High Altitude Research (DIHAR), Leh, India, voucher specimen No SBTL-2006], were collected from Himalayas. The leaves were shade dried, powdered, soaked in distilled water. The supernatant was collected and lyophilized. The dried powder (yield 0.125 g/g) was coded as SBL-1. To ensure working with standardized extract and also to maintain the quality control, the HPTLC profile of SBL-1 was developed with reference to Quercetin dihydrate, Gallic Acid and Rutin. The separation of each compound was carried out on silica gel 60 F_{254} pre-coated TLC aluminium plates, while allowing linear ascending (9 cm) development at room temperature, in twin trough glass chamber saturated with suitably designed mobile phase. The comparisons with standards were made. Detailed procedure was as described earlier.

Experimental Animals and Treatments

The 8-9 weeks old male, inbred, Swiss albino Strain 'A' mice, weighing 28±2 g, were used after the approval of Animal Experimentation Ethics Committee of the Institute. The animals were maintained under controlled environment at 26±2 °C; 12 h light/dark cycle and offered standard animal food (Golden feed, Delhi) as well as tap water *ad libitum*. All the animal experiments were conducted according to the guidelines of Committee for Protection and Care of Small Experimental Animals (CPCSEA) and as per the approved protocol.

The animals were divided into four groups. Group I was Untreated Control and was administered vehicle (sterile water) only, group II was total body ^{60}Co-gamma-irradiated (10 Gy), group III was Drug alone (administered 30 mg/kg b.w. SBL-1 only) and group IV was Drug+Radiation (administered 30 mg/kg b.w. SBL-1 and after 30 min, animals were total body ^{60}Co-gamma-irradiated, 10 Gy). Each group had three mice and the experiment was repeated two times. The drug was dissolved in sterile water, filtered and administered intraperitoneally. For total (whole) body irradiation, each mouse was placed in a separate wire mesh container and was given one time exposure to deliver 10 Gy radiation dose using ^{60}Co-γ-ray source (GC-220, Atomic Energy of Canada Ltd., Canada, dose rate of 0.31 rad/sec). Fresh air was continuously circulated to avoid hypoxia.

On day 5 after irradiation, the mice were anaesthetized and sacrificed by cervical dislocation. The jejunum portion of the intestine was taken out after leaving approx. 4 cm segment from the pyloric sphincter side of stomach and placed in a cool sterile Petri-dish kept on ice. The jejunum portion was divided into two pieces. One piece was processed for metagenomic studies while another for histological studies.

Metagenomic DNA Isolation

The contents of jejunum portion were flushed out with the help of syringe loaded with 2 ml of TE buffer (10 mM Tris/HCl, 1 mM EDTA, pH 8) and collected in a tube. The jejunum was then cut opened longitudinally and epithelial mucosal layer was superficially scraped using round edged glass rod. Scraped material was pooled with previous collection from jejunum and contents were centrifuged to collect the pellet. The DNA was isolated using DNA isolation buffer (100 Mm Tris/HCl pH 8.0, 50 mM EDTA pH 8.2 and 1.5 M NaCl) and lysozyme (final concentration of 5 mg/ml), as per the procedure described (Hayashi *et al.*, 2002). The extracted DNA was purified using QIAEX II Gel Extraction Kit (Qiagen GmbH) following manufacturer's instructions. Purity and the concentration of the extracted DNA was determined using spectrophotometer absorbance at 260/280 nm. DNA of samples showing absorbance (A260/280 nm) ratios of 1.6-1.7 were accepted as pure. The integrity of DNA was determined after electrophoresis in a 0.8 per cent agarose gel.

PCR Amplification and Amplicon Purification

For the amplification of the 16S rRNA genes, two universal primers, 27f (5′ AGAGTTTGATCAT GGCTCAG 3′, *E. coli* positions 8–27) and 1492r (5′TACGGTTACCTTGTTACGACTT 3′, *E. coli* positions 1492–1513) (Lane, 1991), synthesized from Sigma-Aldrich, India, were used. The PCR reaction mixture contained 0.6 µM of each primer, template DNA, 10x PCR reaction buffer, 1.5 mM of $MgCl_2$ (Fermentas, Germany), 0.1 µg/µL of BSA (Sigma Aldrich, USA), 0.2 mM of dNTPs (Promega Corporation, USA) and 1 Unit of Taq DNA polymerase (Fermentas, Germany) in a final volume of 25 µL. PCR amplification was performed using I-Cycler Thermal Cycler (BioRad Laboratories, USA). The program followed was, 95°C for 10 min, followed by 25 cycles consisting of 94°C for 30s, 50 °C for 1.5 min and 72°C for 1 min, and a final extension period of 72°C for 10 min. To purify the PCR product, the amplified DNA was run on 1.2 per cent agarose gel along with DNA molecular weight marker (O'GeneRuler Express DNA Ladder, Fermentas, Germany). The band of interest was cut and extracted using QIAEX II Gel Extraction kit (Qiagen, Germany). Competent cells were prepared as per procedure described (Seidman *et al.*, 1997). The transformation efficiency of competent cells was checked on Luria agar plates containing Ampicillin, X-gal and IPTG. Competent cells with transformation efficiency of 10^6 transform ants per µg DNA or more were utilized for further experiments. The purified amplicons were ligated and cloned into the pGEM-T Easy Vector System (Promega Corporation, USA) as per manufacturer's instructions. The white colonies growing on Luria agar plate (containing Ampicillin, X-gal and IPTG) were selected for further confirmation. The plasmids were isolated from transformed cells as per the procedure described (Engebrecht *et al.*, 1991). To check the presence of inserts, isolated recombinant plasmids were digested with EcoR1 (Promega Corporation, USA) in the presence of restriction buffer for 1 h at 37°C as per the information supplied with the restriction enzyme. The inserts were then separated by electrophoresis on 1.2 per cent agarose gel containing molecular weight marker.

Sequencing of Inserts and Bioinformatics Analysis

Sequencing and subsequent data analysis was performed on ABI 3130, Genetic Analyzer using software 'Sequence Analysis_v5.2'. All near-full-length sequences were edited to exclude the PCR primer-binding sites using online tool 'Vec Screen v7.1' (https://www.ncbi.nlm.nih.gov/tools/vecscreen/). After removing vector contamination, sequences were tested for possible chimeric artifacts using the DECIPHER's Find Chimeras web tool (http://decipher.cee.wisc.edu/FindChimerasOutputs.html) (Wright *et al.*, 2012). The non-chimeric sequences were analyzed for taxonomical classification using Sequence Match program of the Ribosomal Database Project (RDP), release 11, (http://rdp.cme.msu.edu/index.jsp, 2,809,406 16S rRNA gene sequences). The taxonomic classifications assigned by RDP for sequences having 's_ab' score ≥0.85 were adopted. Since RDP can have error up to 20 per cent for certain regions and read lengths another tool, the Basic Local Alignment Search Tool (BLAST, http://blast.ncbi.nlm.nih.gov/Blast.cgi) as also used, which is reported to achieve accuracy of 100 per cent (Liu *et al.*, 2008). All the sequences were, therefore, also classified using RDP as well as nucleotide BLAST [E-value: 10, database: 16S rRNA gene sequences (Bacteria and Archea)]. All those sequences, which failed to achieve s_ab score of ≥0.85 after sequence match with RDP, but showed more than 85 per cent 'maximum identity' and 'query coverage' after BLAST, were considered for further phylogenetic studies. Sequences which neither had s_ab score of ≥0.85 after RDP sequence match, nor had 85 per cent 'maximum identity' as well as 'query coverage' after BLAST, were not included in the phylogenetic studies. For phylogeny studies the sequences were aligned using the multiple sequence alignment program "Multiple Sequence Comparison by Log- Expectation" (MUSCLE) (Edgar, 2004). The MUSCLE alignment file (in fasta format) was used as input for the MEGA (v5.2) software to determine the phylogeny (Tamura *et al.*, 2011). Phylogenetic trees were generated based on the maximum likelihood algorithm using the MEGA package. Branch stability was assessed by bootstrap analysis (1000 replicates) using the algorithms available in the MEGA package. *Thermus aquaticus* (GenBank Accession: EU682501.1) was used as outgroup to construct rooted tree. The sequences were submitted to GenBank repository (https://www.ncbi.njm.nih.gov/genbank/).

Histology of Jejunum

The jejunum portion was flushed with normal ice cold saline and was placed in 10 per cent neutral buffered formalin for histological studies. The formalin fixed tissue was dehydrated sequentially with different concentration of alcohol and embedded in paraffin wax. Non serial transverse sections (approximately 55m) of jejunum were cut and stained with hematoxylin and eosin for analysis. Each section was separated from the previous one by a minimum of 505m of tissue. Microscopic examinations were done by using Light Microscope Axio Scope Observer D1, from Carl Zeiss, Germany). Only complete sections, which included the opening of crypt and full length of villi from base to the tip, were considered for analyses. Villus height was determined by measuring the distance from the tip of the villus up to the crypt in pixels. A surviving crypt was defined as containing 10 or more adjacent, healthy-looking, non Paneth cells, some Paneth cells, and a lumen (Withers and Elkind, 1970). The number of crypts was counted in each circumference. Only those crypts which were seen directly against the inner muscle layer were counted. Cryptal Paneth cells were identified on the basis of their position within the hemispheric bases of the crypts, their truncated shape, basal nuclei, and deep red staining secretion granules and were counted only if they were contained within longitudinally sectioned crypts resting on the muscle layer. All counts and measurements from each tissue specimen were obtained "blind" from a minimum of 4 coded sections.

RESULTS

Characterization of SBL-1

The concentration of three bioactive and marker constituents of SBL-1 (Gallic acid ethyl ester, Quercetin dihydrate and Rutin trihydrate), as quantified by HPTLC, were found to be 12.09, 4.7 and 8.7 mg/g respectively. The HPTLC chromatogram of the Gallic acid ethyl ester, Quercetin dihydrate and Rutin trihydrate, as separated from SBL-1, is presented in Figure 31.1. For reference pure compounds were used. Gallic acid ethyl ester and Quercetin dihydrate are known for their anti-oxidant properties. Rutin is reported to have tissue regenerative properties.

Isolation of Recombinant Plasmids and *in silico* Analysis

The ethidium bromide stained agarose gel electrophoresis pictures of some of the representative recombinant plasmids, after digestion with EcoR1, are presented in Figure 31.2.

The sequences, which were free from vector contamination, chimeric sequences as well as were of high quality (69 in number), were submitted to GenBank (https://www.ncbi.njm.nih.gov/genbank/) and the GenBank accession numbers were from KF681283 to KF681351. The classification up to genus level, of all 69 sequences was determined by using nucleotide BLAST alignment tool [E-value (expectation value): 10, database: 16S ribosomal RNA sequences (Bacteria and Archea)] of NCBI as well as by using RDP database release 11. The RDP classification up to genus level complemented the Nucleotide BLAST classification in all treatment groups except for the group containing sequences from ^{60}Co-γ-irradiated animals. Genus *Lactobacillus* (Firmicutes) dominated in 3 treatment groups (Untreated Control, Drug alone and Drug+Radiation). In Untreated Control and Drug alone group genus *Lactobacillus* had 100 per cent share of the total population. In Drug+Radiation group the genus *Lactobacillus* had the 89 per cent share of the total population while Clostridium genus had a share of 11 per cent. In ^{60}Co-γ-irradiated group, the genus *Cohaesibacter* and endosymbiont of acanthamoeba (Alphaproteobacteria) dominated and had 55 per cent share of the total population; *Mycoplasma* (Tenericutes) had 27 per cent share and *Lactobacillus* had 18 per cent share (Figure 31.3).

All 69 sequences were used for subsequent phylogenetic analysis. It was observed that 11 recombinant clone sequences [KF681286 (C); KF681322 and KF681330 (D); KF681336 to KF681340 and KF681342 (R); KF681346 and KF681350 (D+R)] showed ≤98 per cent sequence similarity to existing 16S rRNA gene sequences in the GenBank and RDP databases. These clones could be the uncharacterized bacterial species. The phylogenetic tree prepared after analysis of DNA sequence of 16S rRNA isolated from the jejunum of mice

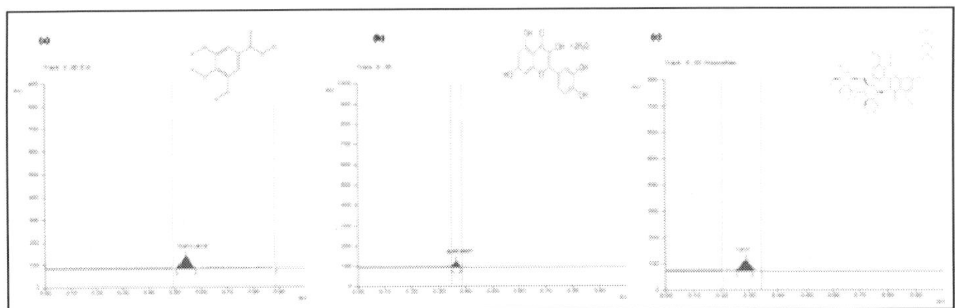

Figure 31.1: (a) HPTLC chromatogram of gallic acid ethyl ester, (b) Chromatogram of quercetin dihydrate, (c) chromatogram of rutin trihydrate as detected in SBL-1.

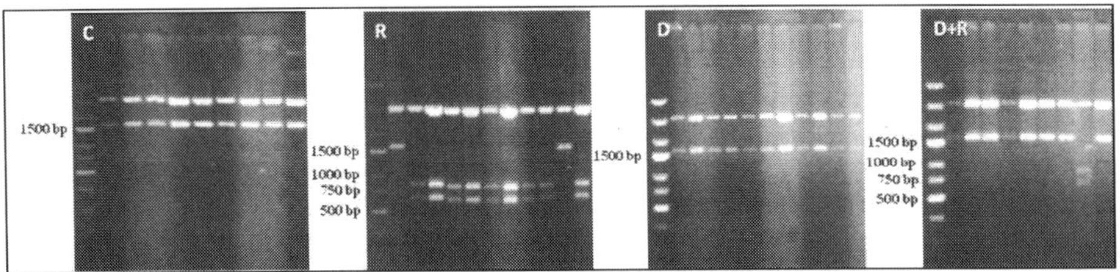

Figure 31.2: Representative Agarose Gel Pictures of Recombinant Clones in pGEM-T Easy Vector as seen after Restriction Digestion with EcoR1 (Stained with ethidium bromide and visualized at 254 nm. Different treatment groups are untreated control (C); ^{60}Co-γ-irradiated (R); Drug alone (D); Drug+Radiation (D+R). Lane 1 in all gels contained the marker (M), molecular weight is mentioned in base pairs.

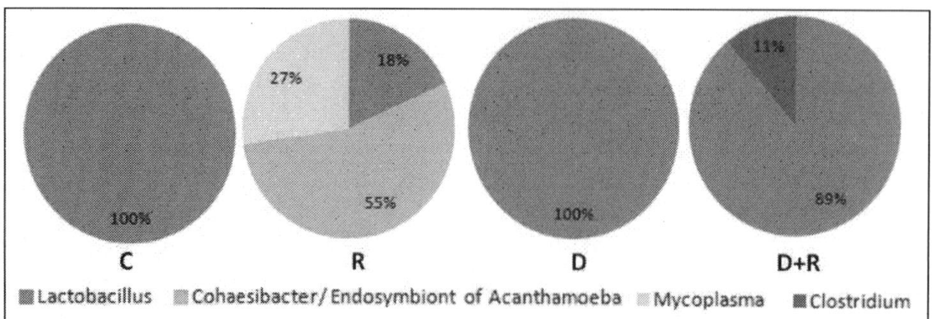

Figure 31.3: Pie Chart Shows Changes in Population of Firmicutes (*Lactobacillus* and *Clostridium*), Alphaproteobacteria [Endosymbiont of Acanthamoeba (as per RDP database)/*Cohaesibacter* (as per NCBI database)] and Tenericutes (*Mycoplasma*) in jejunum of mice belonging to different treatment groups *viz.* untreated control (C); ^{60}Co-γ-irradiated (R); Drug alone (D); Drug + Radiation (D+R).

belonging to different treatment groups is presented in Figure 31.4. Evolutionary history was inferred by using Maximum Likelihood method based on the Kimura 2-parameter model of MEGA package. The analysis involved 78 nucleotide sequences (69 recombinant clones+8 reference sequences+1 outgroup). Condon positions included were 1st+2nd+3rd+noncoding. All positions with less than 95 per cent site coverage were

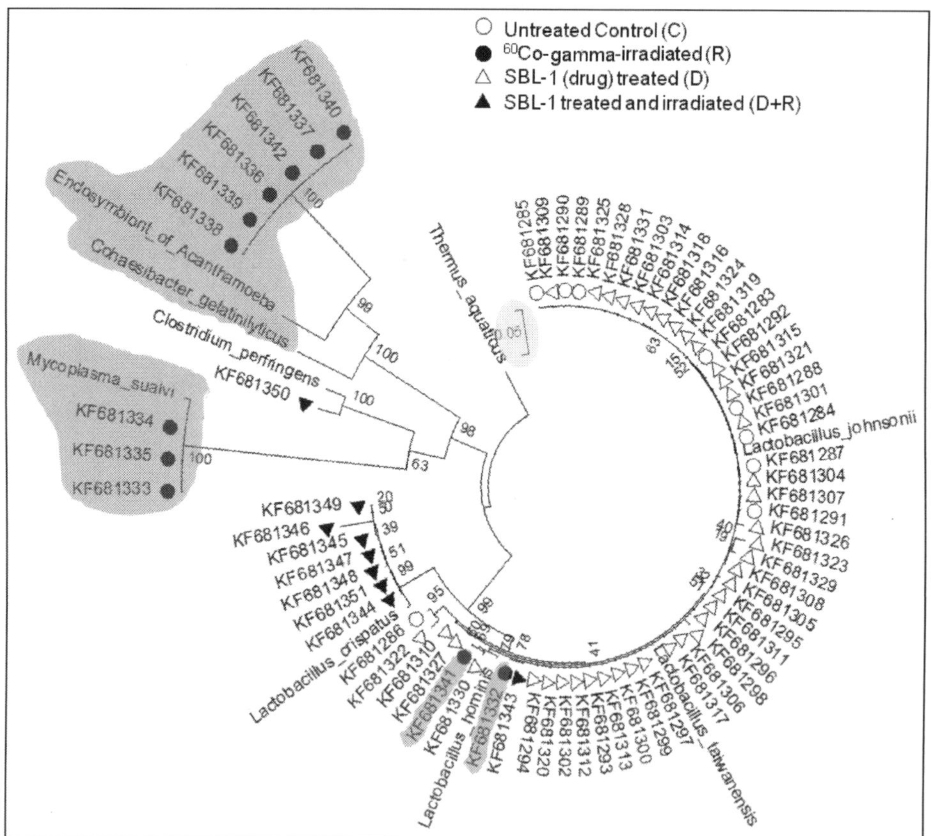

Figure 31.4: The Phylogenetic Tree Prepared after Analysis of DNA Sequence of 16S rRNA Isolated from the Jejunum of Mice Belonging to different Treatment Groups *viz.* untreated control (C); ^{60}Co-gamma-irradiated (R); Drug alone (D); Drug+Rdiation (D+R). The phylogenic tree with highest log likelihood (-9041.7604) is shown. Initial tree(s) for the heuristic search were obtained automatically by applying Neighbor-Join and BioNJ algorithms to a matrix of pair wise distances estimated using the Maximum Composite Likelihood approach, and then selecting the topology with superior log likelihood value. *Thermus aquaticus* (EU682501.1) was used as an out group. Bootstrap values shown at the branches are based on 1000 replicates. Values of 95 per cent or higher were considered significant. A discrete Gamma distribution was used to model evolutionary rate differences among sites [5 categories (+G, parameter = 0.2151)]. The tree is drawn to scale, with branch lengths measured in the number of substitutions per site. The scale bar represents genetic distance (05 substitutions per 100 nucleotides). Reference sequences were from GenBank- *Lactobacillus johnsonii* (EU381129.1), *Lactobacillus taiwanensis* (NR_044507.1), *Lactobacillus hominis* (FR681902.1), *Lactobacillus crispatus* (NR_074986.1), *Mycoplasma sualvi* (NR_041846.1), *Clostridium perfringens* (NR_074482.1), *Cohaesibacter gelatinilyticus* (NR_043777.1), Endosymbiont of Acanthamoeba (AF132139.1).

eliminated. That is, fewer than 5 per cent alignment gaps, missing data, and ambiguous bases were allowed at any position. The jejunal bacterial population of untreated animals (C), SBL-1 treated animals (D) and SBL-1+irradiated (D+R) animals belonged to *Lactobacillus* genus, although there were differences in the species. Bacterial population of animals which were lethally irradiated showed significant divergence in comparison to animals of group C, D and D+R at genus level. Majority of clones of irradiated animals apparently belonged to *Mycoplasma*, *Cohaesibacter* and endosymbionts of Acanthamoeba.

Effects of Radiation and SBL-1 Treatment on Tissue Histology

The villi of jejunum from healthy untreated controls were tall and cylindrical. In irradiated animals there were decreases in the villi height, villi cellularity and cryptal Paneth cells. At Day 5 after irradiation, there was a decrease in villi height to 58 per cent, villi cellularity to 49 per cent and cryptal Paneth cells to 25 per cent in comparison to untreated control. Also there were other gross histological changes such as villi fusion, no recoverable crypt loss and tissue disintegration. Enlargement in lumen and sterile crypts were very evident from histology pictures (Figure 31.5).

Treatment with SBL-1 30 min before irradiation (10 Gy) countered significantly the radiation induced histological changes. The radiation induced decrease in villi height was countered by 1.25-fold, villus cellularity by 2.27-fold, and cryptal Paneth cell number by1.89-fold. The villi number and crypt cellularity were further improved. Animals treated with SBL-1 alone did not show any difference from the untreated animals in the jejunum histology.

DISCUSSION

Increased usage of ionizing radiation in many walks of life has increased the risk of radiation injuries to human population. This has necessitated the need of developing an effective and safe radiation counter-measures for human use. Even after decades of research, this need has not been fulfilled. The synthetic compounds have failed miserably due to severe toxicity. The only drug WR2721, which was approved for human use as an

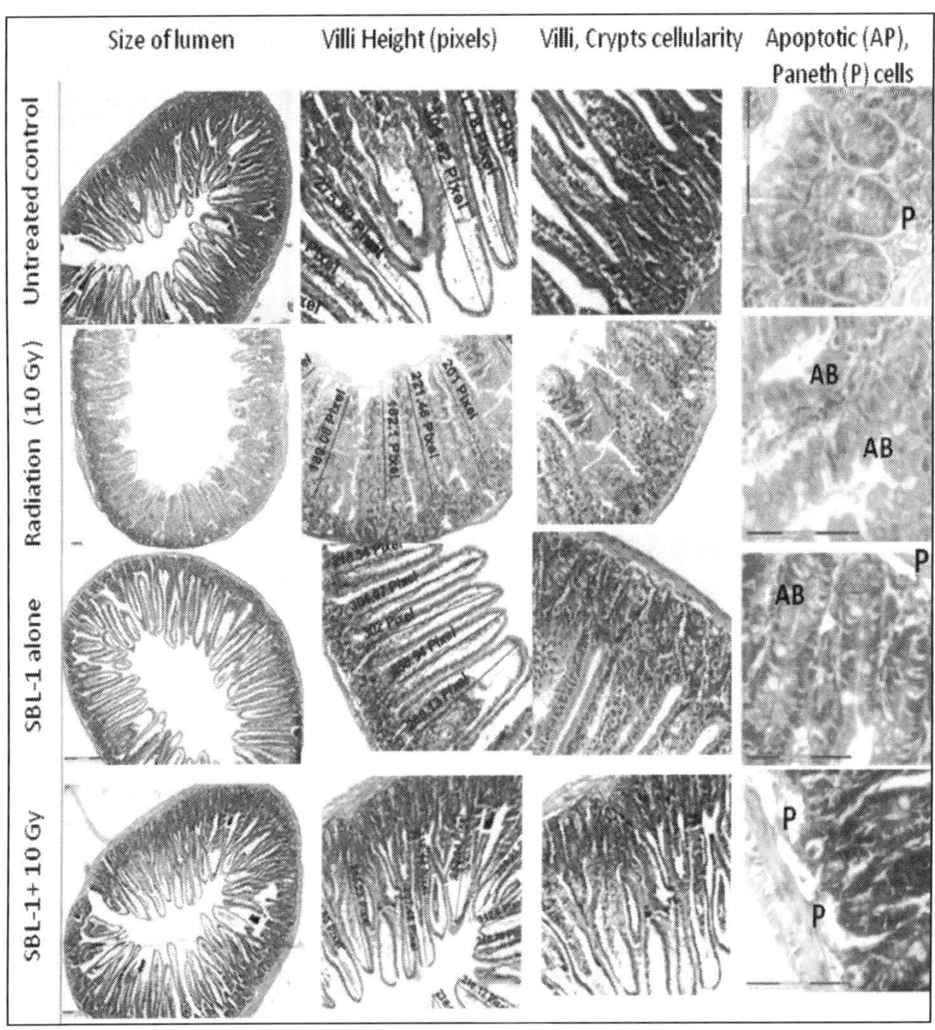

Figure 31.5: The Histological Changes seen in the Total Body Irradiated (10 Gy) Mice and their Modification by Prophylactic Treatment with SBL-1.

adjuvant to radiotherapy, has serious toxic side effects. Herbal drugs are emerging as alternate choice because of their relatively non-toxic nature. The failures of potential radiation countermeasures due to their associated toxicity are compelling enough to consider deploying additional/newer approaches and/or increasing the ambit of pre-clinical studies before zeroing on to the medical preparations for clinical investigations.

The mucosal epithelial layer of jejunum is the niche for resident gut microbial population, particularly Lactobacilli (Hayashi *et al.*, 2005). The gut microbiome project has shown beyond doubt that intestine is not merely an organ for absorption of nutrients and disposal of wastes, but actually performs many more functions owing to the presence of large number of highly specific microbes. The microbiota of intestine plays an important role in determining the pharmacokinetics as well as pharmacodynamics of a drug. Any treatment leading to dysbiosis of intestinal microbiota is therefore, expected to influence the distribution, metabolism, success, failure, efficacy as well as toxicity of the drug.

The mitotically active crypt cells undergo rapid apoptosis or stop dividing temporarily or permanently after irradiation and the damage depends upon the absorbed radiation dose. The loss of crypt cells in irradiated animals impairs the normal regeneration of epithelial layer of intestinal villi, leading to varying degrees of blunting, fusion, attenuation, and denudation of the villi as was seen in Figure 31.5. This causes malabsorption, electrolyte imbalance, diarrhea, inflammation, infections and weight loss. Mucosal atrophy caused by ionizing radiations was predicted to severely alter the mucosal microbiota distribution, primarily due to the loss of their niche. This could also cause translocation of microorganisms or microbial products through the mucosa into blood circulation, leading to sepsis and life threatening complications. It was therefore, considered important to investigate the effects of radio protective drug SBL-1 on jejunal histology as well as the microbiota of jejunum when administered at radioprotective concentration before irradiation with lethal doses of ^{60}Co-γ-radiation (10 Gy).

In this study, per section significant decrease in villi height to 58 per cent, villi cellularity to 49 per cent, cryptal Paneth cells to 25 per cent, together with villi fusion and disintegration of crypt cellularity, indicated the massive cellular damage to the jejunum by lethal dose of ^{60}Co-γ-irradiation (10 Gy) (Figure 31.5). The appearance of increased number of sterile crypts and further decreases in villus and crypts cellularity indicated that most of the stem cells were lost and regeneration of epithelium was not taking place in the irradiated animals. All these histological changes together with significant decrease observed in cryptal Paneth cells and significant increase in lumen enlargement were the typical features of radiation induced GI syndrome and were corroborated with presentations such as diarrhea, weight loss in irradiated animals and was also observed in earlier study (Bala *et al.*, 2009).

The histological protection from radiation toxicity by seabuckthorn leaves (SBL-1) was evident because there was significant countering of radiation induced decreases in crypts cellularity, villi number, villus height, villus cellularity, and cryptal Paneth cell number by day 5 (Figure 31.5). The crypts of Lieberkuhn of GI tract are highly radiosensitive and prone to radiation damage. The protection to crypt cellularity against radiation damage could be attributed to effective countering of radiation induced multiple free radicals and reactive species due to the presence of multiple antioxidants (polyphenols, tannins, flavonoids, and thiols), and more specifically gallic acid, rutin, and quercetin in SBL-1. The flavonoids alone were reported to protect from radiation induced oxidative stress but were effective at low or sublethal doses of radiation only (Uma Devi *et al.*, 2000; Benkovic *et al.*, 2008). Quercetin alone was also reported to protect the human lymphocytes from genetic damage and oxidative stress caused by low doses of radiation only (Devipriya *et al.*, 2008).

The continuous protection of stem cells from pathogens is of paramount importance and the Paneth cells importantly contribute to the innate intestinal defense (Tanabe *et al.*, 2005). The property of SBL-1 to counter radiation induced decrease in Paneth cell number suggested that SBL-1 was helpful in maintaining the innate immune response of jejunum in the irradiated mice. Treatment with SBL-1 alone did not alter the jejunal histology, significantly, in comparison to the untreated control mice. This demonstrated the nontoxic nature of SBL-1 at the doses used in this study. Other studies have also shown the nontoxic nature of *Hippophae* leaf extract (Saggu *et al.*, 2007).

The nutrient availability and absorption is highest in the first two-thirds of the small bowel. But these sites contain relatively small numbers of microbes. Because of acidic pH and relatively higher concentration of oxygen in jejunum, in comparison to distal portions of intestine, presence of facultative gram positive, non-spore forming, aerotolerant, acidophilic anaerobic bacteria belonging to Lactobacillaceae family is frequently reported in jejunum. The abundance of Lactobacilli in unirradiated group of animals is in line with the previous studies. Total body exposure to ^{60}Co-γ-ray (10 Gy) disturbed the microbiota distribution in jejunum thereby creating dysbiosis. Only 18 per cent clones had shown match with *Lactobacillus*. The presence of endosymbionts of Acanthamoeba was predominant (55 per cent). Further 27 per cent match was with *Mycoplasma* (Figure 31.3) the dysbiosis in the jejunum of irradiated mice could be due to the displacement of microbes present in the distal portion of intestine. Presence of Alphaproteobacteria, Tenericutes and Clostridia were reported in large intestine as well as the distal ileum, Mollicutes were reported in the colon as well as faeces of rodents.The radiation injuries such as leakage of vasculature, erosion of epithelial layer and change in pH, may be some of the important reasons facilitating the migration of Alphaproteobacteria, Clostridia and Mollicutes in the jejunum from the lower intestine. The classification based on 16S ribosomal RNA sequence as performed in this study clearly demonstrated that clones detected in the irradiated animals were evolutionarily different than the clones which were detected in the untreated animals. Unlike Lactobacilli detected in unirradiated control, the microbes belonging to Alphaproteobacteria, Clostridia and Mollicutes were not the native inhabitant of jejunum. A recent study, reported increase in population of 12 members of Bacteroidales, Lactobacillaceae, and Streptococcaceae in rat faeces after radiation exposure at lethal doses. The same study also reported decrease in population of Clostridiaceae and Peptostreptococcaceae family members (Lam *et al.*, 2012). In this study, the treatment with SBL-1 alone apparently increased the population of Lactobacilli in jejunum. NCBI classification and bootstrapping values in phylogenetic analysis also showed that the clones isolated from drug alone group were Lactobacilli and were not significantly different in comparison to untreated control; although some newer species of Lactobacilli were detected (Figure 31.4). The microbiota of intestine participates in the metabolism of polyphenols and flavonoids. SBL-1 is rich in polyphenols, flavonoids, tannins and proanthocyanidins. The increase in the population of Lactobacilli in animals treated with SBL-1alone could be to facilitate the metabolism of various constituents of SBL-1 and maintain proper energy balance. The increased amount of fat in the diet of experimental animals was reported to bring out a change in the gut microbiota to ensure the energy balance. In comparison to radiation alone group, the animals treated with SBL-1 at radioprotective concentration before irradiation (10 Gy) preserved the Lactobacilli population to a large extent. As much as 89 per cent clones had shown match with Lactobacillus genus. Complete absence of pathogenic bacteria belonging to *Mycoplasma* or endosymbiont of Acanthamoeba, as was observed in irradiated animals, indicated that pretreatment with SBL-1 protected the native Lactobacilli population of irradiated animals. Appearance of newer species of Lactobacilli such as *L. crispatus, L. gallinarum, L. helveticus*, as predicted by bioinformatic tools, in drug + radiation group of animals needs further verification by wet laboratory techniques.

CONCLUSION

This study shows that the SBL-1 rendered radioprotection by anti-oxidative, tissue regenerative, anti-inflammatory mechanisms, as well as by preserving the symbiosis state of native microbial population in jejunum. The role of intestinal microbiota in multiple functions of body (Kalina and Mohamadzadeh 2005; Corr *et al.*, 2007; Jones *et al.*, 2013) necessitates that drugs meant for complicated pathologies, such as radiation injuries, should also be evaluated for their modifying effect on the intestinal microbiota.

REFERENCES

1. Backhed, F., Ding, H., Wang, T., Hooper, L.V., Koh, G.Y., Nagy, A., Semenkovich, C. F. and Gordon, J. I. 2004. The gut microbiota as an environmental factor that regulates fat storage. In: *Proceedings of the National Academy of Sciences*, USA 101 (44): 15718-15723.
2. Bala, M. and Kaur, P. 2015. Changes in spleen cell populations in total body 60 CO-Gamma irradiated mice and their modification by SBL-1: Implication in radiation protection. *International Journal of Pharmaceutical Sciences Review and Research* (In Press).
3. Bala, M., Gupta, M., Saini, M., Abdin, M. Z. and Prasad, J. 2015. Seabuckthorn leaf extract protects jejunum and bone marrow of ^{60}Cobalt-Gamma-Irradiated mice by regulating apoptosis and tissue regeneration. *Evidence-Based Complementary and Alternative Medicine* 2015: 765705.
4. Bala, M., Prasad, J., Singh, S., Tiwari, S. and Sawhney, R. C. 2009. Whole-body radio protective effects of SBL-1: A preparation from leaves of *Hippophae rhamnoides*. *Journal of Herbs, Spices and Medicinal Plants* 15 (2): 203-215.
5. Benkovic, V., Knezevic, A. H., Dikic, D., Lisicic, D., Orsolic, N., Basic, I., Kosalec, I. and Kopjar, N. 2008. Radioprotective effects of propolis and quercetin in5-irradiated mice evaluated by the alkaline comet assay. *Phytomedicine* 15 (10): 851-858.
6. Cani, P. D., Bibiloni, R., Knauf, C., Waget, A., Neyrinck, A. M., Delzenne, N. M. and Burcelin, R. 2008. Changes in gut microbiota control metabolic endotoxemia-induced inflammation in high-fat diet-induced obesity and diabetes in mice. *Diabetes* 57 (6): 1470-81.
7. Chen, T. 1988. Studies of the biochemical composition of *Hippophae* and its quality assessment in Gansu Province. *Hippophae* 1: 19-26.
8. Corr, S. C., Li, Y., Riedel, C. U., O'Toole, P. W., Hill, C. and Gahan, C. G. M. 2007. Bacteriocin production as a mechanism for the anti-infective activity of *Lactobacillus salivarius* UCC118. In: *Proceedings of the National Academy of Sciences of the USA* 104 (18): 7617-7621.
9. Devipriya, N., Sudheer, A. R., Srinivasan, M. and Menon, V. P. 2008. Quercetin ameliorates gamma radiation-induced DNA damage and biochemical changes in human peripheral blood lymphocytes. *Mutation Research* 654 (1): 1-7.
10. Dumas, M. E., Barton, R. H., Toye, A., Cloarec, O., Blancher, C., Rothwell, A., Fearnside, J., Tatoud, R., Blanc, V., Lindon, J. C., Mitchell, S. C.,

Holmes, E., McCarthy, M. I., Scott, J., Gauguier, D. and Nicholson, J. K. 2006. Metabolic profiling reveals a contribution of gut microbiota to fatty liver phenotype in insulin-resistant mice. In: *Proceedings of the National Academy of Sciences, USA* 103 (33): 12511-12516.

11. Eckburg, P. B., Bik, E. M., Bernstein, C. N., Purdom, E., Dethlefsen, L., Sargent, M., Gill, S. R., Nelson, K. E. and Relman, D. A. 2005. Diversity of the human intestinal microbial flora. *Science* 308 (5728): 1635-1638.

12. Edgar, R. C. 2004. MUSCLE: Multiple sequence alignment with high accuracy and high throughput. *Nucleic Acids Research* 32 (5): 1792-1797.

13. Engebrecht, J., Brent, R. and Kaderbhai, M. A. 2010. Minipreps of plasmid DNA. In: *Current Protocols in Molecular Biology*, Supplement 15 (F.M. Ausubel, R. Brent, R. E. Kingston, D. D. Moore, J. G. Seidman, J. A. Smith and K. Struhl, Ed.), John Wiley and Sons, Chichester, pp. 1.6.1-1.6.10.

14. Gill, S. R., Pop, M., Deboy, R. T., Eckburg, P. B., Turnbaugh, P. J., Samuel, B. S., Gordon, J. I., Relman, D. A., Fraser-Liggett, C. M. and Nelson, K. E. 2006. Metagenomic analysis of the human distal gut microbiome. *Science* 312 (5778): 1355-1359.

15. Guliyev, V. B., Gul, M. and Yildirim, A. 2004. *Hippophae rhamnoides* L. chromatographic methods to determine chemical composition, use in traditional medicine and pharmacological effects. *Journal of chromatography. B, Analytical Technologies in the Biomedical and Life Sciences* 812: 291-307.

16. Gupta, V., Bala, M., Prasad, J., Singh, S. and Gupta, M. 2011. Leaves of *Hippophae rhamnoides* prevent taste aversion in gamma-irradiated rats. *Journal of Dietary Supplements* 8 (4): 355-68.

17. Hakkinen, S. H., Karenlampi, S. O., Heinonen, I. M., Mykkanen, H. M. and Torronen, A.R. 1999. Content of the flavonols quercetin, myricetin and kaempferol in 25 edible berries. *Journal of Agricultural and Food Chemistry* 47(6): 2274-2279.

18. Hayashi, H., Sakamoto, M. and Benno, Y. 2002. Phylogenetic analysis of the human gut microbiota using 16S rDNA clone libraries and strictly anaerobic culture-based methods. *Microbiology and Immunology* 46 (8): 535-548.

19. Hayashi, H., Takahashi, R., Nishi, T., Sakamoto, M. and Benno, Y. 2005. Molecular analysis of jejunal, ileal, caecal and recto-sigmoidal human colonic microbiota using 16S rRNA gene libraries and terminal restriction fragment length polymorphism. *Journal of Medical Microbiology* 54 (Pt 11): 1093-101.

20. Hood, L. 2012. Tackling the microbiome. *Science* 336: 1209.

21. Hooper, L. V., Midtvedt, T. and Gordon, J. I. 2002. How host-microbial interactions shape the nutrient environment of the mammalian intestine. *Annual Review of Nutrition* 22: 283-307.

22. Johnson, C. H., Patterson, A. D., Idle, J. R. and Gonzalez, F. J. 2012. Xenobiotic metabolomics: Major impact on the metabolome. *Annual Review of Pharmacology and Toxicology* 52: 37-56.

23. Jones, R. M., Luo, L., Ardita, C. S., Richardson, A. N., Kwon, Y. M., Mercante, J. W., Alam, A., Gates, C. L., Wu, H., Swanson, P. A., Lambeth, J. D., Denning, P. W. and Neish, A. S. 2013. Symbiotic *lactobacilli* stimulate gut epithelial proliferation via Nox-mediated generation of reactive oxygen species. *The EMBO Journal* 32 (23): 3017-3028.

24. Kalina, W. V. and Mohamadzadeh, M. 2005. Lactobacilli as natural enhancer of cellular immune response, *Discovery Medicine* 5 (26): 199-203.

25. Kallio, K., Yang, B. R., Tahvonen, R. and Hakala, M. 1999. Composition of seabuckthorn berries of various origins. In: *Proceeding of International Workshop on Seabuckthorn*, pp. 17-23, Aug 29-Sep 02, Beijing, China, 97p.

26. Kumar, R., Kumar, G. P., Chaurasia, O. P. and Singh, S.B. 2011. Phytochemical and pharmacological profile of seabuckthorn oil: A review. *Research Journal of Medicinal Plant* 5: 491-499.

27. Kurokawa, K., Itoh, T., Kuwahara, T., Oshima, K., Toh, H., Toyoda, A., Takami, H., Morita, H., Sharma, V. K., Srivastava, T. P., Taylor, T. D., Noguchi, H., Mori, H., Ogura, Y., Ehrlich, D. S., Itoh, K., Takagi, T., Sakaki, Y., Hayashi, T. and Hattori, M. 2007. Comparative metagenomics revealed commonly enriched gene sets in human gut microbiomes. *DNA Research* 14 (4): 169-181.

28. Lam, V., Moulder, J. E., Salzman, N. H., Dubinsky, E. A., Andersen, G. L. and Baker, J. E. 2012. Intestinal microbiota as novel biomarkers of prior radiation exposure. *Radiation Research* 177 (5): 573-583.

29. Lane, D. J. 1991. 16S/23S rRNA sequencing. In: *Nucleic Acid Techniques in Bacterial Systematics* (E. Stackebrandt and M. Goodfellow, Ed.), John Wiley and Sons, Chichester- New York-Brisbane-Toronto-Singapore, pp. 115-175.

30. Lindenbaum, J., Rund, D. G., Butler, V. P.J., Tse-Eng, D. and Saha, J. R. 1981. Inactivation of digoxin by the gut flora: Reversal by antibiotic therapy. *The New England Journal of Medicine* 305: 789-794.

31. Liu, Y. and Rhoads, J. M. 2013. Communication between B-Cells and microbiota for the maintenance of intestinal homeostasis. *Antibodies* 2(4): 535-553.

32. Liu, Z. D., Santis, T. Z. Andersen, G. L. and Knight, R. 2008. Accurate taxonomy assignments from

16S rRNA sequences produced by highly parallel pyrosequencers, *Nucleic Acids Research* 36 (18): e120.

33. Lu, R. 1992. In: *Seabuckthorn: A Multipurpose Plant Species for Fragile Mountains*. Internat. Centre for Integrated Mountain Development, Kathmandu, Nepal, p. 62.
34. Makinen, K. K. and Soderling, E. 1980. A quantitative study of mannitol, sorbitol, xylitol and xylose in wild berries and commercial fruits. *Journal of Food Science* 45 (2): 367-371.
35. Mathan, V. I., Wiederman, J., Dobkin, J. F. and Lindenbaum, J. 1989. Geographic differences in digoxin inactivation, a metabolic activity of the human anaerobic gut flora. *Gut* 30: 971-977.
36. Mironov, V. A. 1989. Chemical composition of *Hippophae rhamnoides* of different populations of the USSR. In: *Proceeding of International Symposium on Seabuckthorn (Hippophae rhamnoides* L.*)*, Oct. 19-23, Xian China, pp. 67-70,.
37. Neufeld, K. M., Kang, N., Bienenstock, J. and Foster, J.A. 2011. Reduced anxiety-like behaviour and central neurochemical change in germ-free mice. *Neurogastroenterology and Motility* 23 (3): 255-64, e119.
38. Pintea, A., Marpeau, A., Faye, M., Socaciu, C. and Gleizes, M. 2001. Polar lipid and fatty acid distribution in caroteno-lipoprotein complexes extracted from seabuckthorn fruits. *Phytochemical Analysis* 12 (5): 293-298.
39. Rosch, D., Bergmann, M., Knorr, D. and Kroh, L. W. 2003. Structure antioxidant efficiency relationships of phenolic compounds and their contribution to the antioxidant activity of seabuckthorn juice. *Journal of Agricultural and Food Chemistry* 51: 4233-4239.
40. Saad, R., Rizkallah, M. R. and Aziz, R. K. 2012. Gut Pharmacomicrobiomics: The tip of an iceberg of complex interactions between drugs and gut-associated microbes. *Gut Pathogens* 4 (1): 16.
41. Sabir, S. M., Maqsood, H., Hayat, I., Khan, M. Q. and Khaliq, A. 2005. Elemental and nutritional analysis of seabuckthorn (*Hippophae rhamnoides* ssp. *turkestanica*) berries of Pakistani origin. *Journal of Medicinal Food* 8 (4): 518-522.
42. Sadzikowski, M. R., Sperry, J. F. and Wilkins, T. D. 1977. Cholesterol-reducing bacterium from human feces. *Applied Environmental Microbiology* 34 (4): 355-362.
43. Saggu, S., Divekar, H. M., Gupta, V., Sawhney, R C., Banerjee, P. K. and Kumar, R. 2007. Adaptogenic and safety evaluation of seabuckthorn (*Hippophae rhamnoides*) leaf extract: A dose dependent study. *Food and Chemical Toxicology* 45 (4): 609-617.
44. Saini, M., Bala, M., Farooqi, H., Abdin, M. Z. and Prasad, J. 2014. Renoprotective activity of *Hippophae* leaf extract in total body ^{60}Co-γ-irradiated mice: An oxidative and histopathology study. *International Journal of Pharmacy and Pharmaceutical Sciences* 6 (3): 161-166.
45. Saini, M., Tiwari, S., Prasad, J., Singh, S., Kumar, M.S. and Bala, M. 2010. *Hippophae* leaf extract concentration regulates antioxidant and prooxidant effects on DNA. *Journal of Dietary Supplements* 7 (1): 60-70.
46. Seidman, C. E., Struhl, K., Sheen, J. and Jessen, T. Introduction of plasmid DNA into cells. In: *Current Protocols in Molecular Biology*, Supplement 37 (F. M. Ausubel, R. Brent, R. E. Kingston, D. D. Moore, J. G. Seidman, J. A. Smith and K. Struhl, Ed.), John Wiley and Sons, Chichester, pp. 1.8.1-1.8.10.
47. Tamboli, C. P., Neut, C., Desreumaux, P. and Colombel, J. F. 2004. Dysbiosis as a prerequisite for IBD. *Gut* 53: 1057.
48. Tamura, K., Peterson, D., Peterson, N., Stecher, G., Nei, M. and Kumar, S. 2011. MEGA5: Molecular evolutionary genetics analysis using maximum likelihood, evolutionary distance, and maximum parsimony methods. *Molecular Biology and Evolution* 28: 2731-2739.
49. Tanabe, H., Ayabe, T., Bainbridge, B., Guina, T., Ernst, R. K., Darveau, R. P., Miller, S.I. and Ouellette, A.J. 2005. Mouse paneth cell secretory responses to cell surface glycolipids of virulent and attenuated pathogenic bacteria. *Infection and Immunity* 73 (4): 2312-2320.
50. Tiwari, S. and Bala, M. 2011. *Hippophae* leaves prevent immune-suppression and inflammation in ^{60}Co-γ-irradiated mice. *Phytopharmacology* 1 (3): 36-48.
51. Tiwari, S., Arya, A., Tyagi, S., Prasad, J., Singh, S., Vats, P., Kumar, D., Jain, S.K. and Bala, M. 2009. Antioxidant, anti-mutagenic and radioprotective properties of seabuckthorn leaf (*Hippophae rhamnoides* L.). *Journal of Medicinal and Spice Plants* 14 (2): 83-89.
52. Uma, D. P., Ganasoundari, A., Vrinda, B., Srinivasan, K. K. and Unnikrishnan, M.K. 2000. Radiation protection by the ocimum flavonoids orientin and vicenin: Mechanisms of action. *Radiation Research* 154 (4): 455-460.
53. Withers, H. R. and Elkind, M. M. 1970. Microcolony survival assay for cells of mouse intestinal mucosa exposed to radiation. *International Journal of Radiation Biology* 17 (3): 261-267.
54. Wright, E. S., Yilmaz, L. S. and Noguera, D.R. 2012. DECIPHER, A search-based approach to chimera

identification for 16S rRNA sequences. *Applied Environmental Microbiology* 78 (3): 717-725.

55. Yang, B. and Kallio, H.P. 2001. Fatty acid composition of lipids in seabuckthorn (*Hippophae rhamnoides* L.) berries of different origins. *Journal of Agricultural and Food Chemistry* 49 (4): 1939-1947.

56. Zeb, A. 2004. Important therapeutic uses of seabuckthorn (*Hippophae*): A review. *Journal of Biological Sciences* 4: 687-693.

57. Zhang, W., Yan, J., Duo, J., Ren, B. and Guo, J. 1989. Preliminary study of biochemical constituents of seabuckthorn berries growing in Shanxi Province and their changing trend. In: *Proceeding of International Symposium on Seabuckthorn (Hippophae rhamnoides* L.*)*, Oct. 19-23, Xian, China, pp. 96-105.

Chapter 32

Oral Administration of Seabuckthorn Fruit Extract Attenuates Repeated Restraint-Induced Behavioural Deficits and Increases Serotonergic Metabolism in Hypothalamus: Evidence for an Antidepressant like Profile

Farhat Batool*[1, 2], Asad Hussain Shah[2,3] and Syed Dilnawaz Ahmad [3]

[1]School of Biomedical Sciences, University of Nottingham, Medical School, Queen's Medical Centre, Nottingham NG7 2UH, UK
[2]Neurochemistry and Biochemical Neuropharmacology Research Unit, Department of Biochemistry, University of Karachi, Karachi – 75270, Pakistan
[2,3]University of Management Sciences and Information Technology, Kotli, Pakistan
[3]Department of Plant Breeding and Molecular Genetics, Faculty of Agriculture, University of Azad Jammu and Kashmir, Rawalakot, Pakistan
(*E-mail: batool@uok.edu.pk)

SUMMARY

The present study was designed to investigate the antidepressant-like effects of aqueous fruit extract of seabuckthorn (*Hippophae rhamnoides* L. spp. *turkestanica*) SBT in animal models of depression. In the first phase of study, test rats were treated with oral administration of SBT-FE (40 mg/kg P.O.; 2 weeks) and controls received an equal volume of fresh water. In the second phase (for one week), two groups of animals were exposed to repeated restraint stress (one group from water treated and other group from SBT-FE treated). All groups of animals were separately submitted to forced swim test (FST), open field test (OFT) and elevated plus maze (EPM) tests for the bio-screening of fruit extract with antidepressant profile. Results revealed that the immobility time in the FST was significantly ($p<0.05$) reduced and prolong struggling (numbers of jumps) was observed particularly in rats orally administered with SBT-FE (40 mg/kg P.O.) following one week stress when compared with their respective controls. Open filed ambulation, numbers of entries in OA and per cent time spent in OA were also significantly ($p<0.01$) increased and were more pronounced in SBT-FE treated rats following exposure to repeated restraint stress when compared with their controls. Significant ($p<0.01$) decreases were observed in weekly food intakes of test rats supplemented with oral SBT-FE following restrained stress when compared with their respective controls. Hypothalamic 5-HT (5-hydroxytryptamine; serotonin) and its metabolite 5-HIAA (5-hydroxyindoleacetic acid) were significantly increased in rats supplemented with SBT-FE following repeated restraint stress. It is therefore, concluded that aqueous fruit extract of seabuckthorn exhibited significant antidepressant-like effects in animal models of depression. This effect is supposed to be mediated via modulation of 5HT-IA receptors in rat brain and possibly explains the antidepressant-like effect of seabuckthorn.

Keywords: *Depression, Seabuckthorn, Behavioral despair syndrome, Plus maze, Serotonin and Hypothalamus.*

INTRODUCTION

Seabuckthorn (*Hippophae rhamnoides* L. spp. *turkestanica*) has been used for centuries in Europe and Asia as a folk medicine for its unbelievable nutritional and medicinal profiles. Natural products exhibiting antidepressant properties and accordingly, natural

Seabuckthorn (*Hippophae* L.)*: A Multipurpose Wonder Plant, Vol. 5: Advanced Technologies for Cultivation, Processing, Health Protection and Environmental Conservation* (Virendra Singh, *Editor-in-Chief*, 2018), p. 257–263, Astral International Pvt. Ltd., New Delhi, India.

medicinal plants may be important sources of novel antidepressant drugs and the usage of plant extracts may be proven better in the management of stress and depression.

Immobilization stress is an uncontrollable stress situation which produces several emotional and neurochemical responses. These responses are accompanied by activation of several brain structures. These stress-induced behavioural changes can be attenuated by systemic or intracerebral administration of anxiolytic and antidepressant drugs (Kennett *et al.*, 1985a; 1987; Guimaraes *et al.*, 1993; Padovan and Guimaraes, 2000; Mechiel Korte and De Boer, 2003). A possible effect of seabuckthorn on these changes, however, has not yet been investigated. Therefore, in the present work we tested the hypothesis that oral administration of seabuckthorn fruit extract would attenuate the neurochemical changes and the behavioural consequences of restraint stress. We also evaluated the involvement of 5-HT1A receptors in the effects of seabuckthorn in this investigation. Hypothalamus is a brain region known to involve in the control of appetite and feeding behaviour. Possible role of 5-HT metabolism in the hypothalamus could be explicable in terms of food intake in stressed situations and this was the key component of this investigation.

Taken together, these data support a possible regulatory role of the hypothalamus on behavioral and neurochemical responses associated with aversive situations such as the restraint stress and feeding behavior.

MATERIALS AND METHODS

Preparation of Seabuckthorn Fruit Extract

The mature and fresh fruits of the *Hippophae rhamnoides* L. were collected from District Rawalakot Kashmir Pakistan in the month of October 2012 when the plant grows wildly under natural conditions and later on they were identified by a Taxonomist.

The whole fruit (50 g) was cleaned and pounded to pieces with a squeezer. The extract was filtered and the filtrate was stored at -20 °C in a refrigerator. The fruit pulp was extracted manually by extracting seeds from 50 berries of seabuckthorn. The calculation was done after weighing 100 berries and 100 sample seeds. The crude extract was then diluted in sterile double-distilled water to make aqueous fruit extract of seabuckthorn (SBT-FE) at a dose of 40 mg/kg for the investigation. The fruit extract was administered orally via mouth (*Par os; P.O.*) to test animals while control animals received an equal amount of fresh tap water under controlled conditions. The extract was kept in dark at 4°C until further analyses.

Animals and Treatment Schedule

Twelve Male Albino Wistar rats with an average weight of 200±10 g on arrival were purchased from International Centre for Chemical and Biological Sciences (ICCBS); Research Institute of Chemistry, University of Karachi, Pakistan. Animals were group-housed (two rats per cage) in an environmentally controlled room temperature (21±1 °C) and relative humidity (55±5 per cent) with a 12:12 h light/dark cycle (light on at 7:00 A.M.) A 5-day familiarization period was allowed before animals were used in the experimental procedures. All experiments were conducted in a balanced design and performed in strict accordance with the Guide for the Care and Use of Laboratory Animals (Institute of Laboratory Animal Resources on Life Sciences, US National Research Council, 1996) and the Institutional Ethical Committee's guidelines for animal research. All solutions and fruit extract dose were freshly prepared on the days of treatment.

Design of Study

Twelve animals were randomly divided into two groups (six animals in each group) and treated as: (1) Water treated and (2) SBT-FE (40 mg/kg P.O.) treated for two weeks. After completion of two weeks, animals were further sub-divided into four groups (three animals in each group) and treated as: (1) Water treated plus Unstressed (2) Water treated plus Stressed (3) SBT-FE plus Unstressed (4) SBT-FE plus Stressed for next one week. All behavioral procedures were carried out in animal models of depression (FST, EPM and OFT) for the evaluation of antidepressant-like effects of SBT-FE in groups of rats orally supplemented with aqueous SBT-FE (40 mg/kg P.O.) following exposure to repeated restraint stress for one week and their comparisons with water treated and unstressed vehicles. Each experiment was conducted in a separate behavioral room in a balanced design with fixed schedule under controlled conditions.

Behavioral Procedures

The rats were assigned as unstressed and stressed for both water and SBT-FE treated groups. The behavioral procedures were carried out after repeated restraint stress in rats prior to oral administration of water and aqueous fruit extract of seabuckthorn at a dose of 40 mg/kg P.O and following 24 h withdrawal from the stress in rats. All animals were submitted to the following behavioral tasks: (a) forced swim test; FST (b) elevated plus maze test; EPM and (c) open field test; OFT.

Assessment of Antidepressant-like Activity

Forced Swim Test (FST)

Behavioural despair was proposed as model to test for antidepressant activity by Porsolt *et al.* (1978). Stressed and unstressed group of animals were individually forced to swim in an open cylindrical container (diameter 10 cm, height 25 cm), containing 19 cm of water (depth) at 25 ± 1 °C; the total amount of time each animal remained immobile during a 5min session was recorded (in seconds) as immobility time, as described previously (Brocardo *et al.*, 2008; Kaster *et al.*, 2005). Each rat was judged to be immobile when it ceased struggling and remained floating motionless in the water, making only those movements necessary to keep its head above water. A decrease in the duration of immobility or struggling time is indicative of an antidepressant-like effect (Porsolt *et al.*, 1977).

Open Field Test (OFT)

To determine the effects of the fruit extract from seabuckthorn on exploratory activity, experimental animals were evaluated in the open-field paradigm as previously described earlier (Batool *et al.*, 2009a). Animals orally administered with SBT-FE (40 mg/kg P.O.) following repeated restraint stress were individually placed in an open field apparatus made up of Perspex plastic with dimensions (40 × 60 × 50 cm) and the floor divided was into 25 equal squares by lines. The latency to move from the centre square (in seconds) and numbers of squares crossed with all paws (crossings) was counted in a 5 min session. All groups of animals were monitored in a balanced design during experiments.

Elevated Plus Maze Test (EPM)

For the assessment of antidepressant-like activity of aqueous fruit extract of seabuckthorn (40 mg/kg P.O.) in stressed and unstressed rats, the elevated plus maze (EPM) apparatus was used. EPM was consisted of two open arms (OA; 16 x 5 cm) and two closed arms (CA; 16 x 5 12 cm) and was elevated (60 cm) above from the floor (Pellow and File, 1986). A 5-min session was used for each rat to determine (1) number of entries in an OA and (2) total time spent in an OA in per cent. Rats orally administered with SBT-FE plus repeated stress and controls were individually placed at the centre of the EPM with heads facing the OA *i.e.*, fear-inducing environment. All behavioural procedures were done in a balanced design.

Figure 32.1: Test Rat Restraint on a Wire Grid for 2h/day in Lab during Experiment.

Restraining Procedure in Animals

The animals selected from water and SBT-FE treated groups were restrained on a wire grid of 10″ x 9″ embedded with a Perspex plate of 9″ x 7″. Restraining procedure was the same as described earlier. Immobilization was affected by pressing the fore limbs of the rats through the gaps in metal grids. Fore limbs and hind limbs both were taped together in a cross manner with Zinc Oxide plaster tape and the head of animals rested on the Perspex plate. After 2h of restraining period the animals were released and housed in their cages.

Statistical Analysis

In this investigation, the results were expressed as mean ± standard deviation (n = 12). The data were analyzed statistically by using two way analysis of variance (2-ANOVA) and different group means were compared by Newman-Keuls test, while $p<0.05$ was considered statistically significant in all cases. The software package *Graphpad Prism 6.0* was used for analysis of data.

RESULTS AND DISCUSSION

Present findings shows that that oral administration of aqueous SBT-FE (40mg/kg P.O.) in rats repeatedly treated with restraint stress for one week exhibited significant (p<0.01) increases in food intake, exploratory activity in an open field arena over a 5-min period when compared with their respective water treated unstressed subjects. It is suggested these increases in locomotors activity in stressed animals in an open field arena indicate the antidepressant-like effects of the extract and can be correlated with possible increases in brain serotonin metabolism. Hence, it is interesting to speculate that stressed-induced or drug-induced decreases in locomotors activity can be reversed or normalized by the oral administration of aqueous SBT-FE as observed from the present results.

Significant increases (p<0.01) were observed in number of jumps in restrained rats follwoed by oral administration of SBT-FE. Less Immobility or increaseed struggling time was observed in the group of animals that are treated with SBT-FE and restraint for 2h/day.

Neurochemical data reveals that administration of SBT-FE is appetite suppressant and decreses in FI can be explainable in terms of enhanced 5-HT metabolism in hypothalamus of rat brain. Possible mechanims underlies to treat depression in the patients by oral administration of SBT-FE as a nutrient supply or food therapy.

Increased availabilty of tryptophan may enhance the serotonin metabolism in brain, therefore, it is suggested that fruit extract may contains the tryptophan as a dietry source to normalise the concentration of 5-HT in the brain of depressive patieents. Neurochemical data

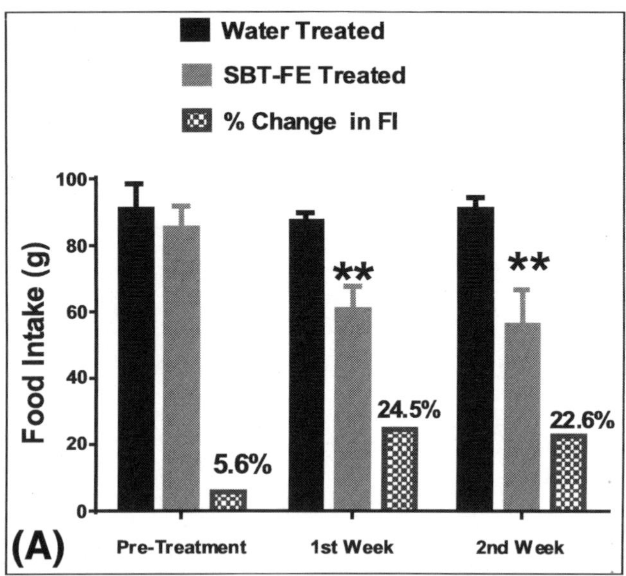

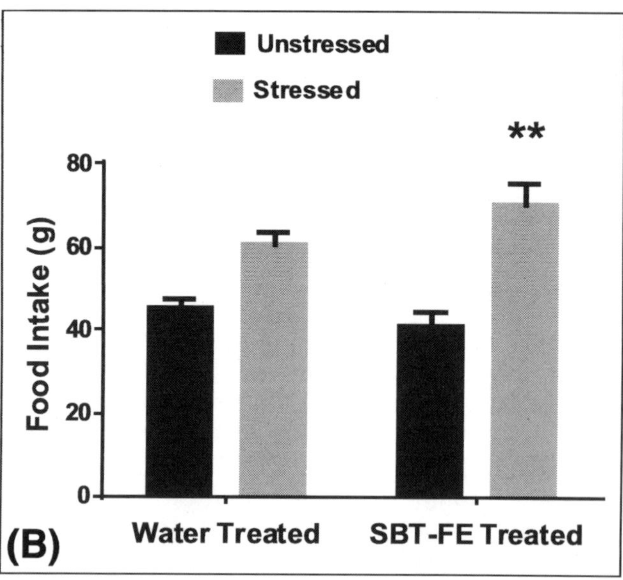

Figure 32.2: Weekly Changes in Food Intake of Rats Treated with Water and Orally Supplemented SBT-FE (40mg/kg) (A) and Group of Rats following Exposure to Repeated Restraint Stress (B).

Values are means ± S.D. (n=12). Significant differences by Newman-Keuls test: *p<0.01 from respective SBT-FE plus stressed animals, +p<0.01 from water plus stressed animals following two-way ANOVA.

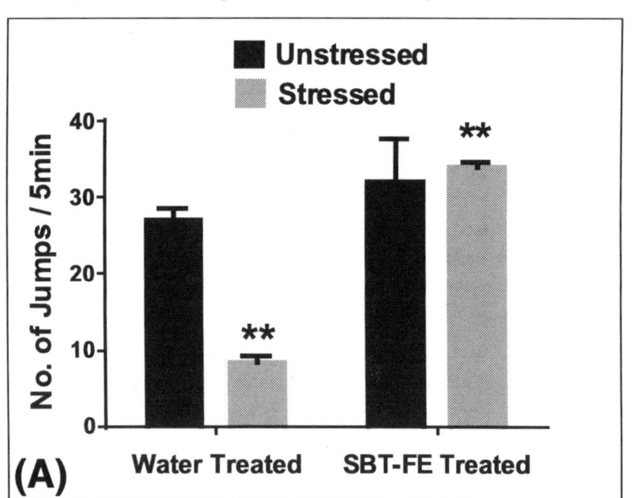

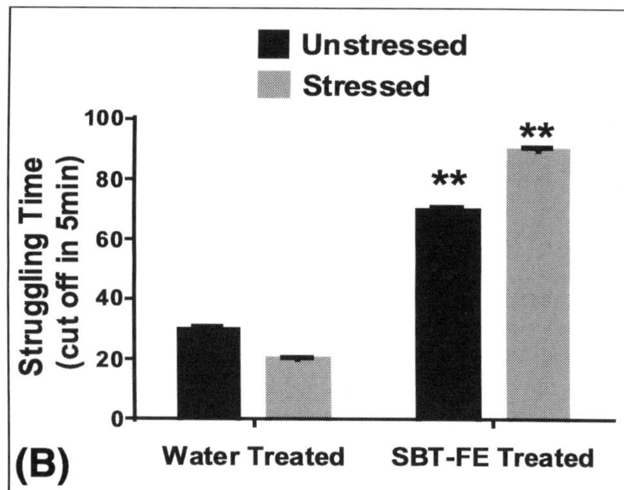

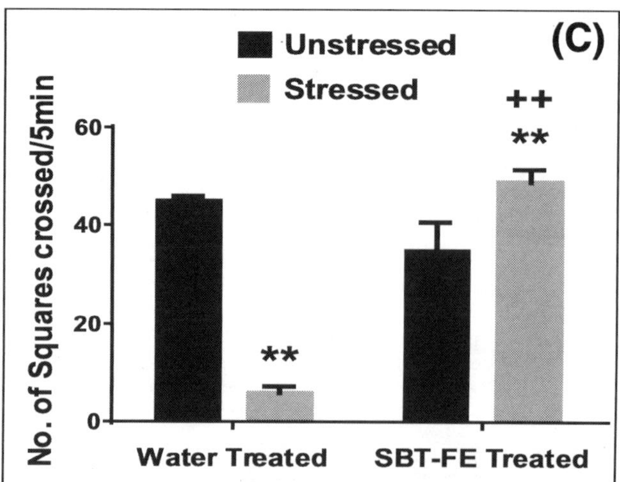

Figure 32.3: Effects of Oral Administration of Seabuckthorn *(Hippophae rhamnoides L.)* Aqueous Fruit Extract (40 mg/kg) on [A] Immobility, [B] Behavioral Despair Time and [C] Exploratory Activity in Rats following Exposure to Repeated Restraint Stress.

Values are means ± S.D. (n=12). Significant differences by Newman-Keuls test: *p<0.01 from respective SBT-FE plus stressed animals, +p<0.01 from water plus stressed animals following two-way ANOVA.

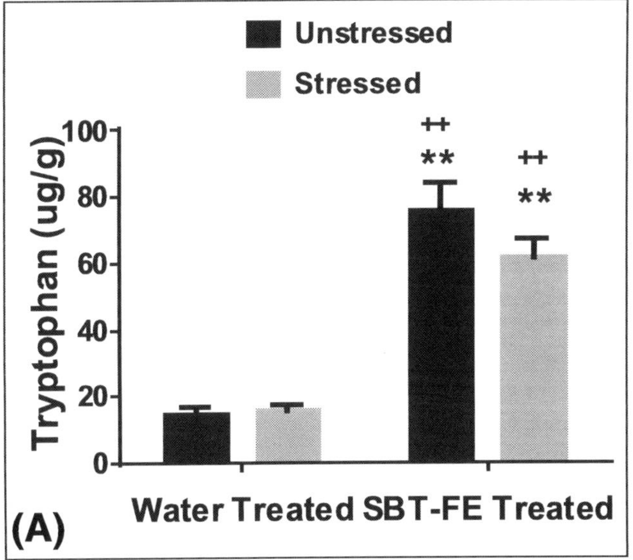

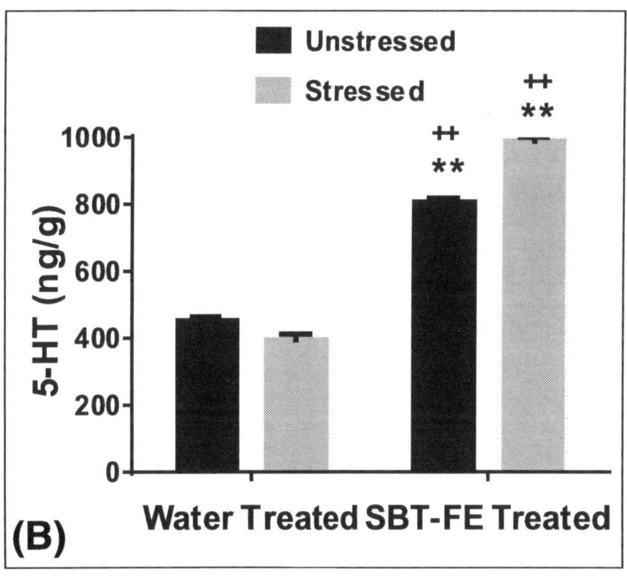

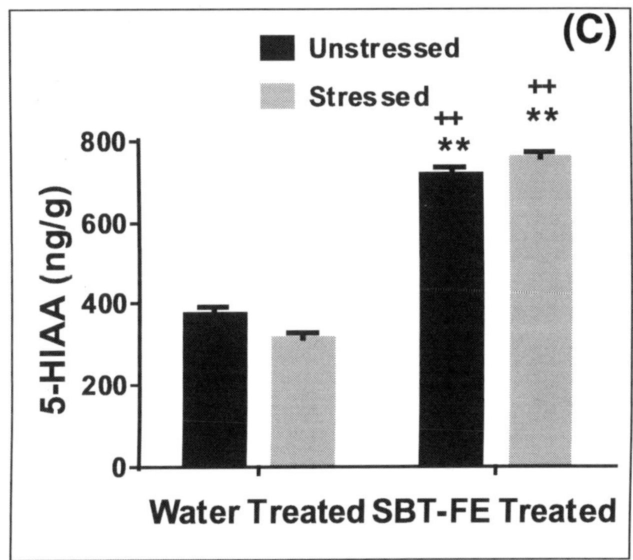

Figure 32.4: Effects of Oral Administration of Seabuckthorn *(Hippophae rhamnoides* L.*)* Aqueous Fruit Extract (40 mg/kg) on Hypothalamic [A] TRP. [B] 5-HT and [C] 5-HIAA. In rats following Exposure to Repeated Restraint Stress.

Values are means ± S.D. (n=12). Significant differences by Newman-Keuls test: *p<0.01 from respective SBT-FE plus stressed animals, +p<0.01 from water plus stressed animals following two-way ANOVA.

reveals significant increases of tryptophan in stressed rats administered with SBT-FE and the consequences of increased 5-HT metabolsim are associated with this effect. Overall findings suggest that seabuckthorn fruit extract has a potentail to treat depression or work a therapeutic agent against this medical condition.

CONCLUSION

The present findings demonstrate that the animals submitted to repeated restraint exhibited behavioral deficits while stressed rats supplemented with seabuckthorn fruit extract exhibited reversal of these behavioral deficits and enhanced serotonergic transmission in rat brain hypothalamus. Increased 5-HT metabolism or attenuation of neurochemical and behavioral deficits in rat model of depression is accountable for the antidepressant like activity of the medicinal plant. Therefore, it is possible to evaluate the consequences of these stress models following repeated restraint sessions.

ACKNOLEDGEMENT

The authors wish to thank University of Karachi, Pakistan for a research grant.

REFERENCES

1. Batool, F., Shah, A. H., Ahmed, S. D., Saify, Z. S. and Haleem, D. J. 2010. Protective effects of aqueous fruit extract from seabuckthorn *(Hippophae rhamnoides* L. spp. *turkestanica)* on haloperidol-induced orofacial dyskinesia and neuronal alterations in the striatum. *Med. Sci. Monit.* 16: BR285-292.

2. Batool, F. and Haleem, D. J. 1997. Neurochemical and behavioral effects of diazepam: Evidences from animal models. *Pakistan J. Pharm. Sci.* 10: 1-8.

3. Batool, F., Shah A. H., Ahmed, S. D. and Haleem, D. J. 2009a. Oral supplementation of seabuckthorn *(Hippophae rhamnoides* L. spp. *turkestanica)* fruit extract modifies haloperidol induced behavioral deficits and increases brain serotonin metabolism. *J. Food and Drug Analysis* 17: 257-263.

4. Batool, F., Shah, A. H., Ahmed, S. D, Saify, Z. S. and Haleem, D. J. 2009b. Possible anxiolytic profile of aqueous fruit extracts of a medicinal plant seabuckthorn *(Hippophae rhamnoides* L. spp. *turkestanica)* in experimental models. *Pakistan J. Bot.* 41: 2791-2800.

5. Batool, F., Siddiqui, S. and Haleem, D. J. 2001. Nurochemical and behavioral profils of 1-(1-Naphthyl) piperazine in rats. *J. College Physicians Surg. Pak.* 11: 50-56.

6. Beveridge, T., Harrison, J. E. and Drover, J. 2002. Processing effects on the composition of seabuckthorn juice from *Hippophae rhamnoides* L. *Cv.* Indian summer. *J. Agric. Food Chem.* 50: 113-116.

7. Brocardo, P. S., Budni, J., Kaster, M. P., Santos, A. R. S. and Rodrigues, A. L. S. 2008. Folic acid administration produces an antidepressant-like effect in mice: Evidence for the involvement of the serotonergic and noradrenergic systems. *Neuropharmacology* 54: 464-473.

8. Detke, M. J., Rickels, M. and Lucki, I. 1995. Active behavior in the rat forced swimming test differentally produced by serotonergic and noradrenergic antidepressants. *Psychopharmacology* 121: 66-72.

9. Dhinghra, D. and Valecha, R. 2007. Evaluation of the antidepressant-like activity of *Convolvulus pluricaulis* choisy in the mouse forced swim and tail suspension tests. *Med. Sci. Monit.* 13: BR155-161.

10. Dhingra, D. and Sharma, A. 2006. Antidepressant-like activity of *Glycyrrhiza glabra* L. in mouse models of immobility tests. *Prog. Neuro-Psychopharmacol Biol. Psychiatry* 30: 449-454.

11. Geetha, S., Sai, R. M., Mongia, S. S. and Singh, V. 2003. Evaluation of antioxidant activity of leaf extracts of seabuckthorn *(Hippophae rhamnoides* L.) on chromium VI induced oxidative stress in albino rats. *J. Ethnopharmacol.* 87: 247-251.

12. Gold, P.W., Goodwin, F. K. and Chrousus, G. P. 1988. Clinical and biochemical manifestations of depression in relation to the neurobiology of stress: Part 1. *N. Eng. J. Med.* 319: 348-353.

13. Goodwin, F.K. and Bunney, W. E. 1971. Depressions following reserpine: A re-evaluation. *Sem. Psychiatry.* 3: 435-484.

14. Kaster, M. P., Santos, A. R. S. and Rodrigues, A. L. S. 2005. Involvement of 5-HT_{1A} receptors in the antidepressant-like effects of adenosine in the mouse forced swimming test. *Brain Res. Bull.* 67: 53-61.

15. Li, T. S. C. and Schroeder, W. R. 1996. Seabuckthorn *(Hippophae rhamnoides)*: A multipurpose plant. *Hor. Technology* 6: 370-380.

16. Millan, M. J. 2004. The role of monoamines in the actions of established and "novel" antidepressant agents: A critical review. *Eur. J. Pharmacol.* 500: 371-384.

17. Murray, C. J. and Lopez, A. D. 1997. Alternative projections of mortality and disability by cause 1990-2020: Global burden of disease study. *Lancet* 24: 1498-1504.

18. Paykel, E. S. 2006. Depression: Major problem for public health. *Epidemiol. Psychiat. Soc.* 15: 4-10.

19. Pellow, S. and File, E. 1986. Anxiolytic and anxiogenic drug effects on exploratory activity in an elevated plus maze: A novel test of anxiety in the rat. *Pharmacol. Biochem. Behav.* 24: 252-259.

20. Posrsolt, R. D., Anton, G., Deniel, M. and Jelfre, M. 1978. Behavioral despair in rats: A new animal model sensitive to antidepressant treatments. *Eur. J. Pharmacol.* 47: 379-391.

21. Posrsolt, R. D., Bertin, A. and Jelfre, M. 1977. Behavioral despair in mice: A primary screening tests for antidepressants. *Arch. Int. Pharmacodyn. Ther.* 229: 327-336.

22. Sanchez-Mateo, C. C., Bonkaka, C. X., Prado, B. and Rabanal, R. M. 2007. Antidepressant activity of some *Hypercarium perforatum* L. fill extracts in the forced swimming test in mice. *J. Ethnopharmacol.* 112: 115-121.

23. Sarko, J. 2000. Antidepressants, old and new: A review of their adverse effects and toxicity in overdose. *Emer. Med. Clin. North A.* 18: 637-654.

24. Schulz, V. 2006. Safety of St. John's Wort extract compound to synthetic antidepressants. *Phytomedicine* 13: 199-204.

25. Shah, A. H., Ahmed, S. D., Khaliq, I., Batool, F., Hassan, L. and Pearce, S. R. 2009. Evaluation of phylogenetic relationship among seabuckthorn *(Hippophae rhamnoides* L. spp. *turkestanica)* wild ecotypes from Pakistan using amplified fragment length polymorphism (AFLP). *Pakistan J. Bot.* 41: 2419-2426.

26. Shah, A. H., Ahmed, S. D., Sabir M., Arif S., Khaliq, I. and Batool, F. 2007. Biochemical and nutritional evaluations of seabuckthorn *(Hippophae rhamnoides* L. spp. *turkestanica)* form different locations of Pakistan. *Pakistan J. Bot.* 39: 2059-2065.

27. Titinen, K. M., Hakala, M. A. and Kallio, H. P. 2005. Quality components of seabuckthorn *(Hippophae rhamnoides)* varieties. *J. Agric. Food Chem.* 53: 1692-1699.

28. Yang, B. and Kallio, H. 2002. Composition and physiological effects of seabuckthorn *(Hippophae)* lipids. *Trends Food Sci. Technol.* 13: 160-167.

29. Zhang, Z. J. 2004. Therapeutic effects of herbal extracts and constituents in animal models of psychiatric disorders. *Life Sci.* 75: 1659-1699.

30. Zeb, A. 2004. Chemical and nutritional constituents of seabuckthorn juice. *Pakistan J. Nutr.* 3: 99-106.

Chapter 33

Composition, Antioxidant and Antiglycemic Activity of Proanthocyanidins from Defatted Seabuckthorn Seed Residues

Hua Chen[1], Cuncun Zhang[1], Zhichao Tan[1], Yanxiang Gao[2], Honggao Xu[1]* and Huibin Wang[1]

[1]Beijing Powdery Food Co. Ltd., Beijing, 101117, P.R. China
[2]Beijing Key Laboratory of Functional Food from Plant Resources,
College of Food Science and Nutritional Engineering, China Agricultural University, Beijing, 100083, P.R. China
(*E-mail: xhg@powdery.com.cn)

SUMMARY

Compositional profile of proanthocyanidins (PAs) in an ethanol extract of seabuckthorn (SB) seed residues were investigated using a combination of UPLC-DAD and UPLC-ESI-MS/MS. Negative ion ESI-MS in full scan and selected ion recording (SIR) mode was adopted for tentatively identify the PAs to consist primarily of dimers and trimers. The established UPLC-DAD fingerprint chromatogram could be employed to reveal the occurrence and composition of PA oligomers for authenticity recognition in a great variety of SB seed extract (SBE)-derived products. Comparative antioxidant assays suggested that SB seed extract was more effective than grape seed extract (GSE) in scavenging hydroxyl radicals and was slightly less potent against DPPH radicals, indicating its potential as a natural antioxidant additive. Moreover, the antiglycemic effect of SBE was evaluated by measuring its inhibitory capacity against carbohydrate-hydrolysing enzymes α-amylase and α-glucosidase. SBE exhibited a stronger inhibitory effect on α-glucosidase compared to the positive controls, acarbose and GSE. Its inhibition on α-amylase initially surpassed that of the positive controls at doses below 50 μg/mL, yet it remained relatively less effective than acarbose at higher doses (50 – 250 μg/mL). Nevertheless, SBE was found to be a more potent inhibitor of α-amylase than GSE at all doses.

***Keywords**: Seabuckthorn seed residues, Flavonoids, Proanthocyanidins, Identification, Antioxidant activity and Antiglycemic property.*

INTRODUCTION

Proanthocyanidins (PAs), also known as condensed tannins, are oligomeric and polymeric conjugates of any combinations of four flavan-3-ol monomeric subunits of (+)-catechin, (–)-epicatechin, (+)-gallocatechin, and (–)-epigallocatechin. PAs are divided into subtypes on the basis of their stereochemistry at C2 and C3 and the hydroxylation pattern of the B-ring (Engstro/m *et al.*, 2014). The most common PAs are: 1) 3′, 4′-dihydroxy-substituted procyanidins, which, with different degrees of galloylation, are predominant in grape seed extract, and 2) 3′, 4′, 5′-trihydroxy-substituted prodelphinidins, which comprise the major PAs in seabuckthorn (SB). While PAs comes with a wide range of degrees of polymerization (DP), the main focus is on PA dimers (DP2) and trimers (DP3), because it has been demonstrated that only PA oligomers with DP < 4 are at least partially absorbed into human intestinal tissue and into blood circulation (Villani *et al.*, 2015). The absorbed PA oligomers can not only exert direct antioxidant and protective effects on the intestinal mucosa, but, once absorbed, can act as anti-oxidative, antiviral, and anti-proliferative agents (Fu *et al.*, 2013;

Nicolosi *et al.*, 2014; Wang *et al.*, 2014), and as potential novel phytotherapeutics to attenuate hyperlipidemia, atherosclerosis, and neurological and cognitive deteriorations (Yamakoshi *et al.*, 1999; Enseleit *et al.*, 2012).

SB seed residue, a by-product from oil extraction of SB seed, has appeared to contain high amounts of PAs, in particular PA oligomers (2 ≤ DP ≤ 6) (Rösch *et al.*, 2004; Fan *et al.*, 2007). However, there is limited body of information on the detailed constitutional profile of the PAs in SB seeds. The most abundant monomeric subunits in SB pomace PA oligomers were found to be (epi) gallocatechins, in particular, (+)-gallocatechin subunits (Rösch *et al.*, 2004). Noticeably, the primary presence of (epi) gallocatechin in prodelphinidins, typical of SB PAs, appeared to greatly enhance their antioxidant capacity compared to grape seed procyanidins (Fan *et al.*, 2007), and may contribute to other distinctive bioactivities such as the anti-cancer effect (Fujii *et al.*, 2013), and taste (Rinaldi *et al.*, 2015). Up till now, more than 60 combinations of (epi) catechins and (epi) gallocatechins of PA dimers and trimers, the two most bioavailable PA oligomers, have been reported recently in SB whole berries. Nevertheless, specific characterization of PA oligomers localized in SB seeds remains scarce (Kallio *et al.*, 2014). In this study, SB seed residue was used for extraction of seed PAs in an effort for economic utilization for value addition.

The use of ultra performance liquid chromatography coupled with electro spray-tandem mass spectrometry (UPLC-ESI-MS/MS) method is becoming preferred in PA studies owning to limited sample preparation, shorter analysis time, low solvent demand, high resolution and sensitivity, and the ability to rapidly resolve a wide array of polar compounds in a complex mixture by polarity disparities (Engstro/m *et al.*, 2014). In this study, possibility of using UPLC coupled with triple-quadrupole mass spectrometer (TQ-MS) for the rapid detection and qualitative analysis of PA was exploited. Selected ion recording (SIR) mode with specific m/z enhances separation of PAs, thus increasing specificity without any clean-up procedures of the extract.

The major objective of the present study was to explore the currently limited knowledge regarding the structural characteristics of the key PA oligomers derivable from SB seed residues and their antioxidant activity (DPPH and HO free radical scavenging assay) as well as their inhibitory effect against α-amylase and α-glucosidase. The tentative identification of oligomeric PAs was carried out using a combination of UPLC-DAD and UPLC-ESI-MS/MS. The UPLC-DAD fingerprinting technique could be employed to reveal the occurrence, content, and composition of PA oligomers for authenticity recognition in a great variety of SB seed-derived products. To our knowledge, this is the first attempt to use UPLC-ESI-MS for the detection and characterization of PAs in SB seeds.

MATERIALS AND METHODS

Materials

LC/MS grade acetonitrile, methanol, and formic acid were purchased from Thermo Scientific (USA). Water was prepared using a Milli-Q water purification system (USA). α-amylase and α-glucosidase were purchased from Sigma-Aldrich Co. LLC (Shanghai, China). All solvents and chemicals used were of analytical grade and purchased from Yuanmu, shanghai, China. GSE was obtained from Tianjin Jianfeng Natural Product R&D Co. Ltd. (Tianjin, China).

Preparation of the Seabuckthorn Seed Residue Extract (SBE)

Extraction of defatted dried seed residues resulted from oil extraction by supercritical CO_2 were as followed: the residue was mixed with 6 times its weight of 60 per cent (v/v) ethanol and settled overnight; the extract was filtered and evaporated; the resulting mixture was settled. The obtained supernatant was subsequently filtered again through a 500 mesh; the filtrate was then purified via a macroporous resin column and rinsed with water until the filtrate was colourless. The resin was then rinsed with 70 per cent (v/v) ethanol until brownish components were eluted. After concentration by a rotary evaporator at 40 °C, the effluent was spray dried and stored in secured aluminium foil at – 20 °C for further usage. The end powder was denoted as seabuckthorn seed residue extract (SBE) in the following experiments.

UPLC-DAD and UPLC-MS/MS Analysis

SBE sample analysis was carried out with an Acquity UPLC system (Waters, USA) for rapid resolution of PAs from other flavonoids such as flavonol glycosides. The UPLC system consisted of a sample manager, a binary solvent manager, a column, and a diode array detector (DAD) for separation. DAD recorded at 200 to 450 nm for PAs and flavonoids. The column used was a 50 x 2.1 mm i.d., 1.7 µm, Acquity UPLC BEH C18 column (Waters, Ireland). The mobile phases consisted of water with 0.1 per cent formic acid (phase A) and acetonitrile (phase B) at a constant flow rate of 0.4 mL/min using the following multi-step gradient: 0–1.02 min, 85 per cent –75 per cent A; 1.02–2.04 min, isocratic of 75 per cent A; 2.04–3.06 min, 75 per cent –60 per cent A; 3.06–4.42 min, isocratic of 60 per cent A; 4.42–6.12 min, 60 per cent –40 per cent A; 6.12–7.48 min, isocratic of 40 per cent A; 7.48–8.84 min, 40 per cent –20 per cent A; 8.84–9.00 min, 20 per cent –85 per cent A; 9.00–10.00 min, column wash and re-equilibration at 85 per cent A before the next injection.

The UPLC system was also coupled with a Xevo TQ triple-quadrupole mass spectrometer (Waters, I class, USA) for rapid qualitative analysis. The Negative ionization mode was used and a UPLC- MS with ESI interface was used for MS analysis by Mass Lynx 4.1. ESI conditions were as follows: capillary voltage, 3.0 kV;

desolvation temperature, 250 °C; source temperature, 150 °C; desolvation and cone gas (N_2), 1000 and 100 L/h, respectively; and collision gas, argon. The scan was set at a mass-to-charge ratio (m/z) of 150 –1700. In the current study, the SIR function of UPLC-ESI-MS was applied for further separation without pre-fractionation steps, which allowed significantly simplified and targeted qualitative analysis.

Determination of Antioxidant Activity

DPPH Radical Scavenging Assay

The potential of the SBE to scavenge 1,1-diphenyl-2-picrylhydrazyl (DPPH) free radicals was performed according to a previous assay with some modifications (Varshneya *et al.*, 2012). A 0.1 mL sample solution (50 to 750 µg/mL) was added to 3.0 mL of DPPH ethanolic solution (6 µM). A control solution was also prepared by adding 0.1 mL of ethanol in water to 3.0 mL DPPH solution. Ascorbic acid and GSE was used as positive controls. The absorbance was determined at 517 nm against water after incubation in the dark for 30 min, and the percentage of DPPH inhibition was calculated as follows:

$$\text{Per cent DPPH inhibition} = \left[1 - \frac{\text{A517 nm sam}}{\text{A517 nm con}}\right] \times 100$$

Hydroxyl Radical Scavenging Assay

Hydroxyl free radial scavenging activity was measured according to the modified procedure (Liu *et al.*, 2015). The reaction mixture contained in a final volume of 100µL of 2-deoxy-D-ribose (28 mM in KH_2PO_4–NaOH buffer, 10 mM, pH 7.4), 100 µL of different concentrations of SBE (25–250 µg/mL) was mixed with 200 µL of 0.5 mM EDTA and of 0.1mM $FeCl_3$ (1:1 v/v), 100 µL of 14 mM of H_2O_2 and 100 µL of 1.0 mM ascorbic acid and incubated at 37 °C for 1 h. After incubation, 1 mL of 8M$Na_2CO_3$was added to stop the reaction. Subsequently, 1 mL of 2.8 per cent thiobarbituric acid (TBA) was added to the reaction mixture and incubated at 100°C for 15 min. After cooling absorbance was measured at 532 nm. Ascorbic acid and GSE was used as positive controls. From the absorbance, the per cent hydroxyl radical scavenging activity was calculated using the following formula:

$$\text{Per cent HO} \bullet \text{ inhibition} = \left[1 - \frac{\text{A532 nm sam}}{\text{A532 nm con}}\right] \times 100$$

α-Amylase Inhibition Assay

The diluted SBE at various concentrations were subjected to α-amylase and α-glucosidase inhibition assays (Arumugam *et al.*, 2014). 8µL of test samples were incubated with 40 µL of the α-amylase for10 min at room temperature. This was followed by the addition of 40 µL of starch solution and incubation at 37 °C for 10 min. Finally, 80 µL of 3,5-dinitrosalicylic acid solution was added and the mixture was incubated for 10 min at 95 °C. Absorbance was read at 540 nm. The reaction mixture without sample was used as a control, and the mixture without substrate was used as a blank.

The per cent inhibition of α-amylase was calculated as:

$$[1 - (A_{sample} - A_{blank})/A_{control}] \times 100$$

where,

A_{sample} represents the absorbance of the experimental sample, A_{blank} represents the absorbance of the blank, and $A_{control}$ represents the absorbance of the control. The experiment was performed in triplicate.

α-Glucosidase Inhibition Assay

α-Glucosidase (50 µL, 0.5 U/mL) and 0.2 M potassium phosphate buffer (pH 6.8, 50 µL) were mixed with test sample (50 µL). After incubation at 37 °C for 15 min, 100 µL3 mM4-nitrophenyl-R-D-glucopyranoside (*p*NPG) was added. The reaction was incubated again at37 °C for 10 min and then stopped by the addition of 0.1 M Na_2CO_3 (750 µL). The absorption of 4-nitrophenol was measured at 405 nm. The experiment was performed intriplicate. The reaction mixture without sample was used as a control, and the mixture without substrate was used as a blank. The per cent inhibition of α-glucosidase was calculated as:

$$[1 - (A_{sample} - A_{blank})/A_{control}] \times 100$$

where,

A_{sample} represents the absorbance of the experimental sample, A_{blank} represents the absorbance of the blank, and $A_{control}$ represents the absorbance of the control. The experiment was performed in triplicate.

Statistical Analysis

All experiments were performed in triplicate. The values were expressed as the mean ± standard deviation.

RESULTS

UPLC-DAD Analysis of Proanthocyanidins

This study proposed the use of a novel method for the detection and identification of possible major PAs in SBE by a combination of UPLC-DAD and UPLC-ESI-MS/MS.

A representative UPLC-DAD chromatogram of SBE recorded at 280 nm was presented in Figure 33.1. The compounds characterized are indicated with numbers according to their elution order. DAD (UV spectra recorded at 200 to 450 nm) allowed satisfactory separation of total flavonoids mainly as PAs (λ_{max} at 280 nm) and flavonol derivatives (λ_{max} at ~ 250 and ~360 nm, data not shown). The UV spectra data revealed the peaks eluted between 0.2 and 1 min to be PA oligomers (λ_{max} at 280 nm), which were later tentatively identified based on their ESI-MS/MS spectra (see further). This chromatographic fingerprint derived from UPLC-DAD could be offered as a source of compositional profile for authentication of the SBE extract.

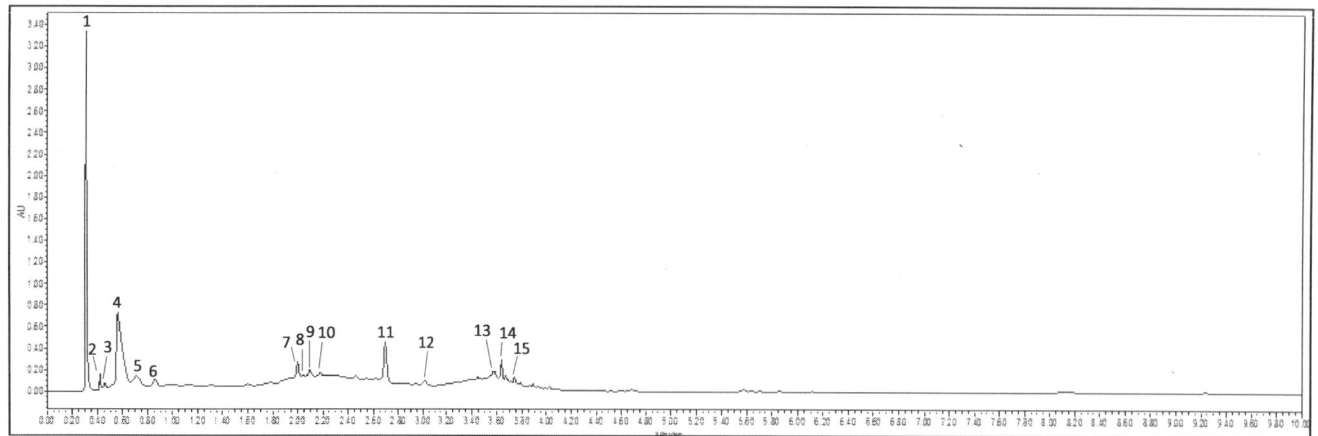

Figure 33.1: Chromatographic Fingerprint of Seabuckthorn Seed Residue Extract (SBE) Obtained by UPLC-DAD Recorded at 280 nm.

Nine peaks (peak 7–15) were eluted later than PAs from 1.8 to 4.3 min. They were tentatively assigned to flavonol derivatives (*i.e.*, flavonol glycosides) based on their characteristic absorption spectra, which exhibited maximum wavelengths around 350 – 370 nm and deep bibliography review (Simirgiotis and Schmeda-Hirschmann, 2010; Fang *et al.*, 2013). Characterization of flavonol derivatives is beyond the scope of this paper.

Identification of Proanthocyanidins by Full Scan MS Spectrum

ESI-MS/MS analysis in the negative mode was used for tentative identification of the previously detected PAs in SBE. UPLC-ESI-MS/MS in full scan mode and data-dependent MS^2 scan mode was used in sequence to provide a qualitative overview of the distribution of the PAs in different oligomeric classes in SBE.

Full scan of SBE from m/z 160 to 1700 was applied to obtain the total ion chromatogram for PA oligomer size screening, based on different molecular weight (Figure 33.2). The base peak was the deprotonated molecule $[M-H]^-$.In this study, full scan MS spectrum confirmed the occurrence of 5 mass clusters with *m/z* values characteristic of the precursor ions $[M-H]^-$of PA DP 2, 3, 4, 5 and 7. To be specific, PA dimers (precursor ion $[M-H]^-$ at *m/z* 577, 593, 609), PA trimers (precursor ion $[M-H]^-$ at *m/z* 865, 881, 897, 913), PA tetramers (precursor ion $[M-H]^-$ at *m/z* 1201, 1217), PA pentamers (precursor ion $[M-H]^-$ at *m/z* 1505, 1521), and PA heptamers ($[M-2H]^{2-}$ at *m/z* 1057, 1065) were detected, respectively. Flavonol-3-ol monomer of (epi) catechin and (epi) gallocatechin exhibited their deprotonated molecular ion $[M-H]^-$ at *m/z* 289 and 305, respectively.

Identification of Proanthocyanidins by MS/MS Fragmentation Patterns

In an effort to corroborate the full scan MS data, the deprotonated MS^1 molecules corresponding to

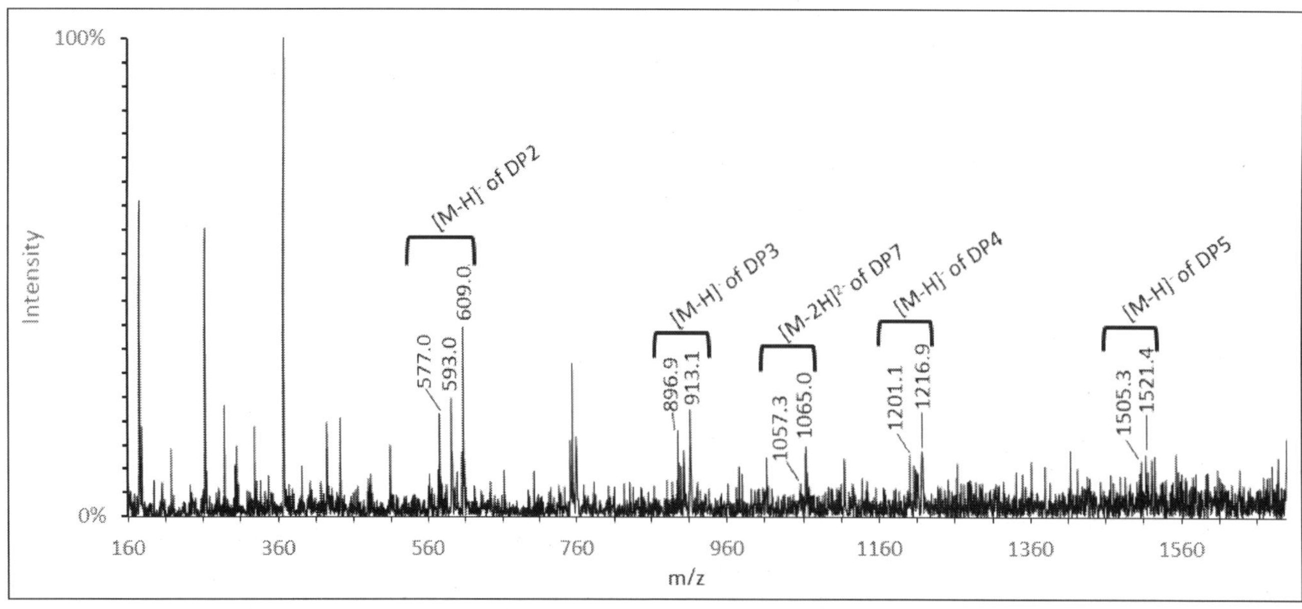

Figure 33.2: Full Scan Mass Spectra of Seabuckthorn Seed Residue Extract (SBE) Analysed by UPLC-ESI-MS in the Negative Mode. The molecular ion peak clusters of PAs with DP 2, 3, 4, 5 and 7 were marked.

Table 33.1: UPLC-ESI-MS/MS Data and Tentative Identification of Major PAs found in Seabuckthorn Seed Residue Extract (SBE). Provided are observed *m/z* of the deprotonated molecular [M-H]⁻ and its MS/MS fragments (product ions), the corresponding elemental composition and calculated mass, and the putative identity.

DP	*Tentative Assignment*	*Molecular Formula*	*No. of Subunits*		*Molecular Weight*	*MS [M-H]⁻ (m/z)**	*MS/MS Fragments (m/z)**
			(E)C	*(E)GC*			
2	PC2 dimers	$C_{30}H_{26}O_{12}$	2	0	578	577	407, 289
	PC1 : PD1 dimers	$C_{30}H_{26}O_{13}$	1	1	594	593	305, 289
	PD1 : PD1 dimers	$C_{30}H_{26}O_{14}$	0	2	610	609	441, 423, 305
3	PC3 trimers	$C_{45}O_{38}O_{18}$	3	0	866	865	577, 425, 407, 289
	PC2 : PD1 trimers	$C_{45}O_{38}O_{19}$	2	1	882	881	697, 593, 577
	PC1 : PD2 trimers	$C_{45}O_{38}O_{20}$	1	2	898	897	609, 593, 305
	PD3 trimers	$C_{45}O_{38}O_{21}$	0	3	914	913	745, 609, 423, 349

DP: degree of polymerization. (E)C: (epi) catechin. (E) GC: (epi) gallocatechin.

[M-H]⁻: negatively charged molecular ion.

* Molecule and Fragment ion detected in negative ion MS/MS.

various PA oligomers with significant intensities were selected as precursor ions in the subsequent tandem MS fragmentation analysis. These ions were fragmented with optimized collision energy and the resulting MS² product ion spectra were monitored.

PA dimers and trimers were tentatively identified based on their respective MS/MS spectra (Table 33.1). Identification of these PAs was based on comparing: 1) their MS data including the experimental *m/z* for the molecular ions and the main characteristic fragments determined by MS/MS, and 2) UV absorption spectra and chromatographic behaviour with those of the authentic standards whenever possible and with data in bibliography (Simirgiotis and Schmeda-Hirschmann, 2010; Liu *et al.*, 2011; Engstro/m *et al.*, 2014; Kallio *et al.*, 2014). Only PA dimers and trimers were analyzed due to relative abundance in the SBE.

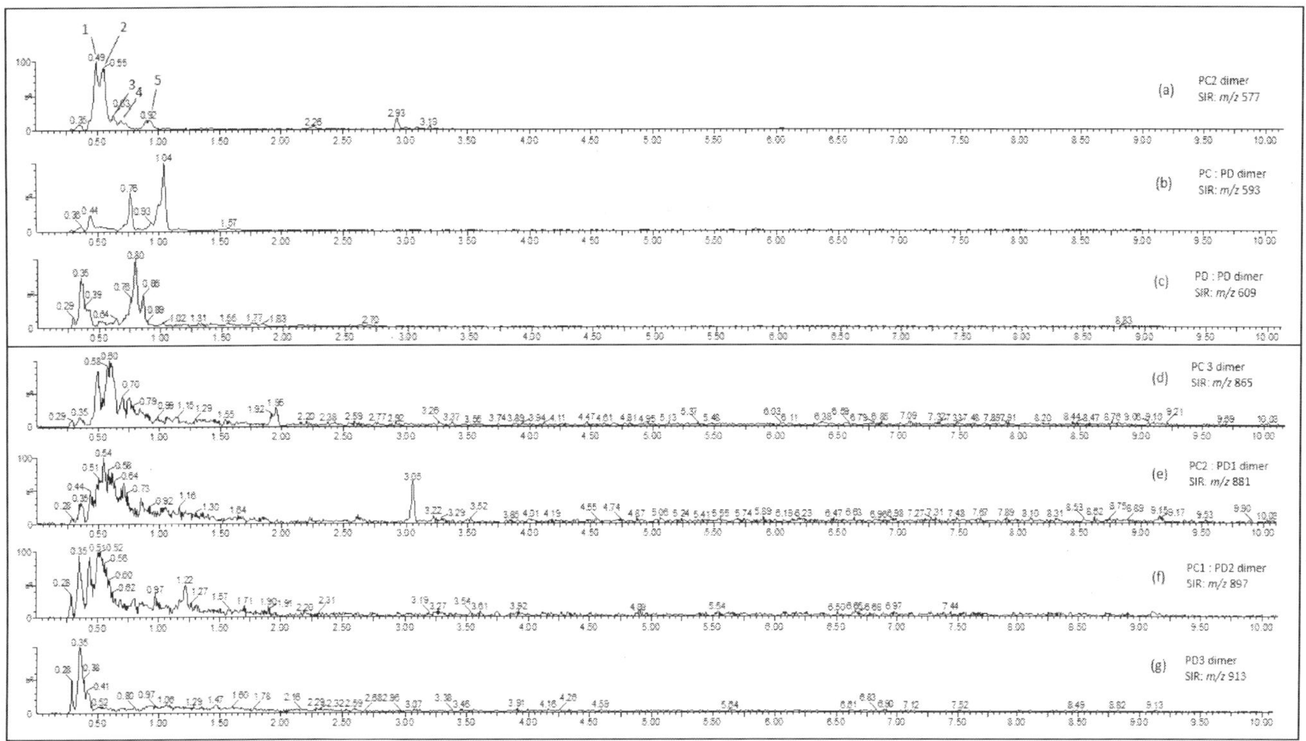

Figure 33.3: UPLC-ESI-MS-SIR Chromatograms of Seabuckthorn Seed Residue Extract (SBE).

Ion traces (a) to (g) corresponding to PA dimers corresponding to m/z 577, 593, and 609, and PA trimers corresponding to m/z 865, 881, 897, and 913 were presented from top to bottom.

PC: procyanidin subunit, the number of constituent PC subunits in the corresponding dimer or trimer are denoted next to PC; PD: prodelphinidin subunit, the number of constituent PD subunits in the corresponding dimer or trimer are denoted next to PD.

It should be noted that definitive confirmation of the exact structure of PAs cannot be achieved only by UPLC-MS/MS, it requires isolation and purification of the unknown compound and NMR spectral data analysis. Thus, only tentative designation of PAs was offered in this work, based on MS spectra and literature data.

Identification of Proanthocyanidins by UPLC-ESI-MS-SIR

The spectra of PAs with specific *m/z* channels for dimers and trimers were recorded by SIR function (Figure 33.3). PA dimers and trimers defined earlier in Table 1 were eluted from 0.2 to 1.9 min as shown. In the case of *m/z* 577, at least 5 characteristic dimeric procyanidin isomers were distinguished (Figures 33.3 a–c), which was in good accordance with a recent study (Kallio *et al.*, 2014). In the cases of *m/z* 593 and 609, a great number of dimeric isomers were also detectable. In comparison, PA trimers (Figures 33.3 d–g) displayed more isomer ion peaks than dimers.

Matching the retention times of peaks in the UPLC-DAD chromatogram recorded at 280 nm and those in the SIR chromatograms of UPLC-ESI-MS revealed that peaks 1 to 6 in Figure 33.1 were very likely to arise from PA dimers and trimers.

Antioxidant Properties of SBE

Scavenging of DPPH Free Radials

After a tentative identification of the composition of SBE, its antioxidant activity were evaluated. Figure 33.4 compared the percentage inhibition between SBE and that of positive controls (ascorbic acid and GSE). The inhibitory effect increased in a dose-dependent manner in all cases, however, SBE exhibited a relatively weak DPPH scavenging capacity than GSE and ascorbic acid.

Scavenging of Hydroxyl Free Radicals

Figure 33.5 compared the hydroxyl free radial scavenging activities between SBE and that of ascorbic acid and GSE. The inhibitory effect of SBE was found remarkable, it increased as the concentration of the extract gradually increased. The inhibition by GSE and ascorbic was slightly less potent compared to SBE.

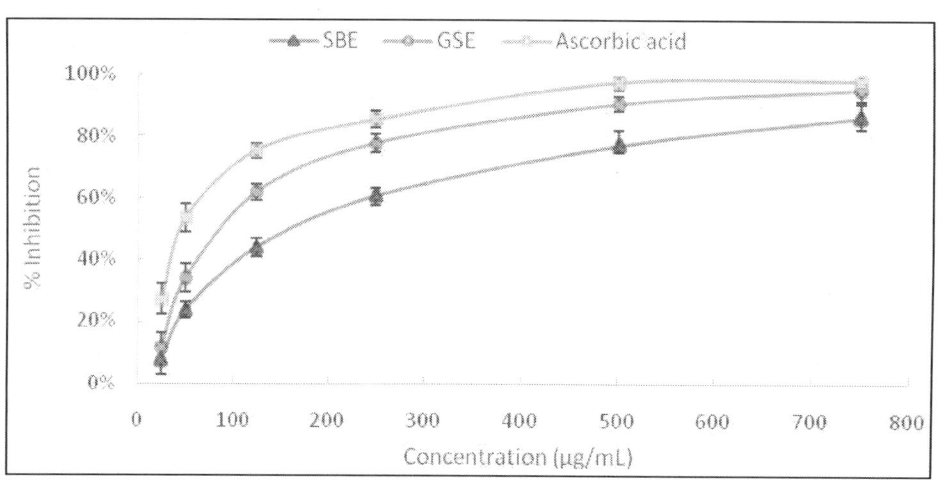

Figure 33.4: DPPH Radical Scavenging Activity of the Seabuckthorn Seed Residue Extract (SBE) in Comparison with that of Positive Controls Grape Seed Extract (GSE) and Ascorbic Acid. Data represented as Mean ± SD of duplicate experiments.

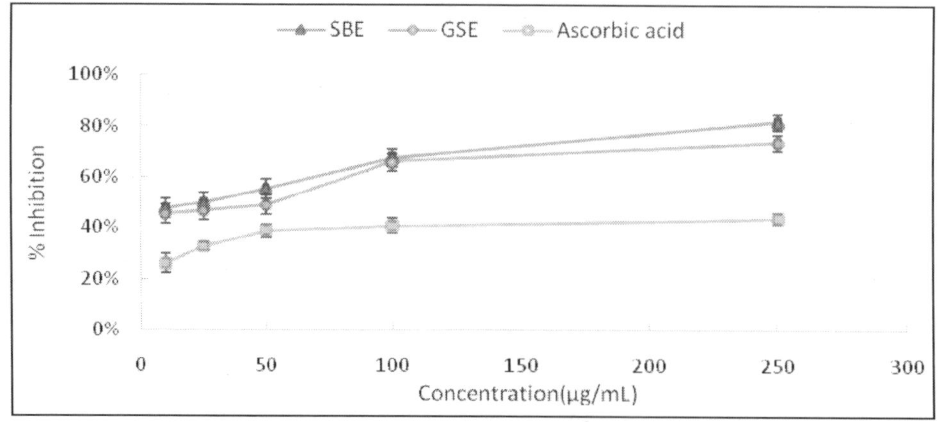

Figure 33.5: Hydroxyl Radical Scavenging Activity of Seabuckthorn Seed Residue (SBE) in Comparison with that of Positive Controls Grape Seed Extract (GSE) and Ascorbic Acid. Data represented as Mean ± SD of duplicate experiments.

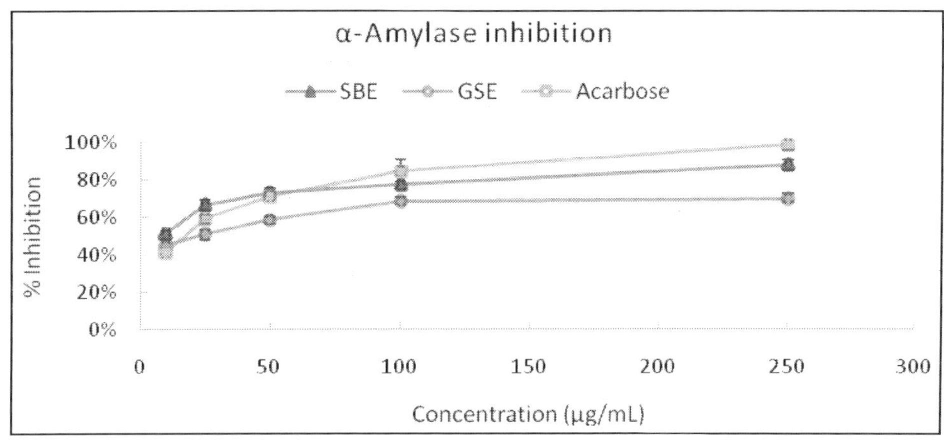

Figure 33.6: α-Amylase Inhibition Activity of Seabuckthorn Seed Residue (SBE) in Comparison with that of Grape Seed Extract (GSE) and Ascorbic Acid. Data represented as Mean ± SD of duplicate experiments.

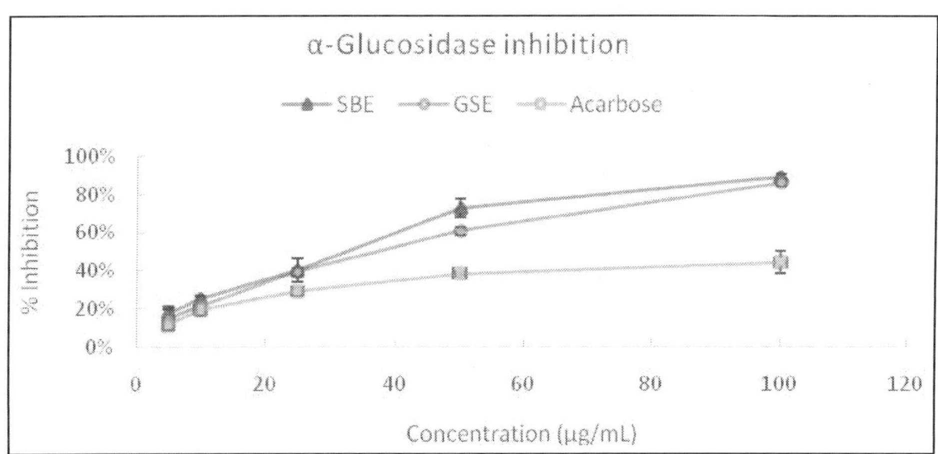

Figure 33.7: α-Glucosidase Inhibition Activity of Seabuckthorn Seed Residue (SBE) in Comparison with that of Positive Controls Grape Seed Extract (GSE) and Ascorbic Acid. Data represented as Mean ± SD of duplicate experiments.

Antiglycemic Potentials of SBE

The carbohydrate digesting enzymes, α-amylase and α-glucosidase have been identified to be the therapeutic targets for management of postprandial hyperglycaemia (Lehtonen *et al.*, 2010). Acarbose is a drug that serves as an inhibitor of the enzymes (Figure 33.6).

In comparison to Acarbose, SBE exerted an effective inhibition on α-amylase at doses below 50 µg/mL, beyond which its inhibitory effect became relatively weaker than that of acarbose. However, it remained a better inhibitor of α-amylase than GSE. A similar effect was observed in its ability to effectively inhibit α-glucosidase, with per cent inhibition higher than that of GSE and acarbose at all concentrations tested, thereby reflecting its antiglycemic potentials (Figure 33.7).

DISCUSSIONS

Recently, a wide range of commercial products have been launched, claiming to consist of PAs with > 95 per cent purity. These products are sold as if they offer the same health benefits regardless of the actual compositional profiles of PAs, especially the content of the most bioactive PA oligomers such as dimers and trimers. The total contents of PAs are commonly determined by colorimetric methods, such as vanillin or butanol-HCl assay and thiolysis and phloroglucinolysis, without detailed information on the profile distribution of each individual PAs, which would mislead and confuse consumers.

Compositional analysis of PAs in plant extract has been confronted with many technical challenges, largely due to the lack of commercial standard compounds and deficient resolution between PAs simply using HPLC (Liu *et al.*, 2011). Advances in LC-ESI-MS/MS analytical techniques have made PA analysis in highly complex mixtures possible. In the present study, we used UPLC for rapid PA separation with excellent resolution. In the subsequent MS/MS analysis, the translational energy in collision-induced dissociation (CID) was optimised to releasethe most informative fragmentation data (Table 33.1). The structural characterization of SBE by UPLC-

DAD and UPLC-ESI-MS indicated that a majority of its PAs were dimers and trimers (Figures 33.1–33.3), which provided an important indicator of the quality and authentication of the prospective SBE products due to their contribution to nutritional properties, bio-activities and taste.

Besides, SBE demonstrated great antioxidant potentials in terms of DPPH and hydroxyl radial scavenging capacities, compared to the conventional PA-rich extract from grape seeds (GSE) and vitamin C, which may be attributed to the significant presence of PAs as the major active compounds. Moreover, SBE displayed remarkable antiglycemic properties, as evidenced by its high α-amylase and α-glucosidase inhibitory activity. Both the antioxidant activity and antiglycemic effect have been widely associated to the occurrence of polyphenolic compounds such as PAs (Fu *et al.*, 2013). This study is the first to describe the dual role of SBE as a potent antioxidant and antiglycemic agent.

CONCLUSIONS

In this study, SB seed residues have been proved as a potential source of PA oligomers, which existed as dimers through pentamers with minor trace of heptamers. Moreover, as can be seen from the UPLC-DAD chromatogram (Figure 33.1), the presence of flavonol glycosides in SB seed was also highly detectable. Additional effort is being taken for in-depth characterization of the flavonol derivatives. The co-existence of these flavonoids may exert synergistic effect with PAs, thus making SBE a more potent antioxidant as revealed by free radial scavenging tests. On other hand, PAs may play an important role in the antiglycemic activity of SBE due to the strong inhibition of α-amylase and α-glucosidase, which was superior to that of GSE. Since the SB seed residues are obtained as a by-product during SB seed oil extraction, this study may contribute to a more economic utilization of SB materials.

It is well known that tandem mass spectrometric method can provide superior sensitivity compared to that of non-specific spectrophotometric methods, making it an essential complimentary tool to elucidate the structural profile of polyphenols such as PAs in relatively complex plant-derived extracts. The present paper demonstrated the high speed of analysis and excellent specificity of this UPLC-EIS-MS/MS method. It ruled out the necessity for expensive, tedious and laborious sample purification and fractionation steps (such as by polyamide Sephadex LH-20 column chromatography, and solid phase extraction columns) before HPLC analysis, by shifting the emphasis to the combination of advanced instrumentation available today. The UPLC-DAD fingerprint indicating the distinguishable PA profiles can be used to evaluate PA-related health claimsof SBE, as well as for the food industry to determine the authenticity and quality of the products.

In the meanwhile, this study broadens our knowledge on the antioxidant bio-compounds in SB seed residues and provides information concerning the ability of SBE to suppress α-amylase and α-glucosidase, therefore implicating its potential hypoglycaemic and anti-diabetic effect.

REFERENCES

1. Arumugam, B., Manaharan, T., Heng, C.K., Kuppusamy, U.R. and Palanisamy, U.D. 2014. Antioxidant and antiglycemic potentials of a standardized extract of *Syzygium malaccense*. *LWT-Food Science and Technology* 59: 707-712.
2. Engstro/m, M.T., Palijarvi, M., Fryganas, C., Grabber, J.H., Mueller-Harvey, I. and Salminen, J.P. 2014. Rapid qualitative and quantitative analyses of proanthocyanidin oligomers and polymers by UPLC-MS/MS. *Journal of Agricultural and Food Chemistry* 62: 3390-3399.
3. Enseleit, F., Sudano, I., Périat, D., Winnik, S., Wolfrum, M., Flammer, A.J., Fröhlich, G.M., Kaiser, P., Hirt, A. and Haile, S.R. 2012. Effects of Pycnogenol on endothelial function in patients with stable coronary artery disease: A double-blind, randomized, placebo-controlled, cross-over study. *European Heart Journal* 33: 1589-1597.
4. Fan, J., Ding, X. and Gu, W. 2007. Radical-scavenging proanthocyanidins from seabuckthorn seed. *Food Chemistry* 102: 168-177.
5. Fang, R., Veitch, N.C., Kite, G.C., Porter, E.A. and Simmonds, M.S. 2013. Enhanced profiling of flavonol glycosides in the fruits of seabuckthorn (*Hippophae rhamnoides*). *Journal of Agricultural and Food Chemistry* 61: 3868-3875.
6. Fu, C., Wang, H., Ng, W.L., Song, L. and Huang, D. 2013. Antioxidant activity and proanthocyanidin profile of Selliguea feei rhizomes. *Molecules* 18: 4282-4292.
7. Fujii, W., Toda, K., Matsumoto, K., Kawaguchi, K., Kawahara, S.I., Hattori, Y., Fujii, H. and Makabe, H. 2013. Syntheses of prodelphinidin B1, B2, and B4 and their antitumor activities against human PC-3 prostate cancer cell lines. *Tetrahedron Letters* 54: 7188-7192.
8. Kallio, H., Yang, W., Liu, P. and Yang, B. 2014. Proanthocyanidins in wild seabuckthorn (*Hippophae rhamnoides*) berries analyzed by reversed-phase, normal-phase, and hydrophilic interaction liquid chromatography with UV and MS Detection. *Journal of Agricultural and Food Chemistry* 62: 7721-7729.
9. Lehtonen, H., Jarvinen, R., Linderborg, K., Viitanen, M., Venojärvi, M., Alanko, H. and Kallio, H. 2010.

Postprandial hyperglycemia and insulin response are affected by seabuckthorn (*Hippophaë rhamnoides* ssp. *turkestanica*) berry and its ethanol-soluble metabolites. *European Journal of Clinical Nutrition* 64: 1465-1471.

10. Liu, P., Kallio, H., Lü, D., Zhou, C. and Yang, B. 2011. Quantitative analysis of phenolic compounds in Chinese hawthorn (*Crataegus* spp.) fruits by high performance liquid chromatography–electrospray ionisation mass spectrometry. *Food Chemistry* 127: 1370-1377.
11. Liu, R., Zheng, W., Li, J., Wang, L., Wu, H., Wang, X. and Shi, L. 2015. Rapid identification of bioactive peptides with antioxidant activity from the enzymatic hydrolysate of Mactra veneriformis by UHPLC–Q-TOF mass spectrometry. *Food chemistry* 167: 484-489.
12. Nicolosi, D., Tempera, G., Genovese, C. and Furneri, P.M. 2014. Anti-adhesion activity of A2-type Proanthocyanidins (A Cranberry major component) on uropathogenic *E. coli* and *P. mirabilis* strains. *Antibiotics* 3: 143-154.
13. Rinaldi, A., Iturmendi, N., Jourdes, M., Teissedre, P.L. and Moio, L. 2015. Transfer of tannin characteristics from grape skins or seeds to wine like solutions and their impact on potential astringency. *LWT-Food Science and Technology*.
14. Rösch, D., Mügge, C., Fogliano, V. and Kroh, L.W. 2004. Antioxidant oligomeric proanthocyanidins from seabuckthorn (*Hippophae rhamnoides*) Pomace. *Journal of Agricultural and Food Chemistry* 52: 6712-6718.
15. Simirgiotis, M. and Schmeda, H. G. 2010. Direct identification of phenolic constituents in Boldo Folium (*Peumus boldus* Mol.) infusions by high-performance liquid chromatography with diode array detection and electrospray ionization tandem mass spectrometry. *Journal of Chromatography A* 1217: 443-449.
16. Varshneya, C., Kant, V. and Mehta, M. 2012. Total phenolic contents and free radical scavenging activities of different extracts of seabuckthorn (*Hippophae rhamnoides*) pomace without seeds. *International Journal of Food Sciences and Nutrition* 63: 153-159.
17. Villani, T.S., Reichert, W., Ferruzzi, M.G., Pasinetti, G.M., Simon, J.E. and Wu, Q. 2015. Chemical investigation of commercial grape seed derived products to assess quality and detect adulteration. *Food Chemistry* 170: 271-280.
18. Wang, Y., Stevens, V.L., Shah, R., Peterson, J.J., Dwyer, J.T., Gapstur, S.M. and McCullough, M.L. 2014. Dietary flavonoid and proanthocyanidin intakes and prostate cancer risk in a prospective cohort of US men. *American Journal of Epidemiology*, kwu006.
19. Yamakoshi, J., Kataoka, S., Koga, T. and Ariga, T. 1999. Proanthocyanidin-rich extract from grape seeds attenuates the development of aortic atherosclerosis in cholesterol-fed rabbits. *Atherosclerosis* 142: 139-149.

Chapter 34

Extracts from Chinese Seabuckthorn Berry have Antimicrobial Activity *in vitro*

Ding Ruojun, Bi Yang*, Li Jixin, Jiang Yumei, Feng Lidan, Zhang Zhong and Wei Juan

College of Food Science and Engineering, Gansu Agricultural University, Lanzhou, China
(*E-mail: biyang@gsau.com.cn)

SUMMARY

Chinese seabuckthorn berry contains functional components. The berry was extracted by ethanol-water solution and MTT assay and growth rate methods were applied to determine antibacterial ability *in vitro* of the extract solution. The antibacterial and antifungal assay indicated that the extract exhibited different degree of inhibitory activity. The result showed that when the concentration of the extract was 1250 mg/L, it had the maximum inhibitory effect against *Escherichia coli, Bacillus subtilis* and *Staphylococcus aureus*. Inhibitory effect of *B. subtilis* was much larger than that of other two bacterium. MIC values (Minimum Inhibitory Concentration) of the extract on three bacteria were 5000, 1250, and 1000 mg/L respectively. IC_{50} (50 per cent inhibiting concentration) evaluation of *Trichoderma viride, Rhizopus stolonifer, Aspergillus niger, Penicillium expansum,* and *Vintage red* indicated that the berry had the strongest inhibitory effect of the *P. expansum* (5520 mg/L), while it had the worst inhibitory effect of *R. stolonifer* (18870 mg/L). The results proved that seabuckthorn berry had high effective and comprehensive antimicrobial function. However, the sort of functional ingredients and studies of its *in vivo* activities need further explored.

Keywords: *Chinese seabuckthorn, Antibacterial activity, Bacteria and fungus.*

INTRODUCTION

Seabuckthorn (*Hippophae rhamnoides* L.), also referred to as sand thorn, sallow thorn, or seaberry (USDA, 2007) was deciduous shrub in the family *Elaeagnaceae*. It was a hardy plant, tolerant against drought and cold, and was useful for land reclamation and farmstead protection (Li and Schroeder, 1996). Seabuckthorn was a native of China and had been domesticated in several areas including Shanxi, Shaanxi, Inner Mongolia, Gansu, and Xinjiang (He *et al.*, 2008) which accounting for about 85 percent of *Hippophae* plants. The Chinese seabuckthorn consisted of three parts, including pulp, seeds and fruit peel. One of the main indicators of berry quality was the content of oil. High amounts of vitamin C, flavonoids, oils and oil-soluble compounds, as well as minerals, were present in the fruit (Kallio *et al.*, 2000). Medicinal value of the berry had been certified in its treatment to radiation damage, burns, oral inflammation and gastric ulcers in China and the former Soviet republic for the past years, and more than 300 medicinal preparations had been reported in literature (Lu, 1992).

The increasing incidence of food-borne diseases meant that there was a constant striving to produce safer food and to develop new natural antimicrobial agents. There was, therefore, still a need for new methods of reducing or eliminating food spoilage and food-borne pathogens, possibly in combination with the existing approaches. For these reasons, microbial contamination of food still posed important public health and economic concerns for the human society (Vivek *et al.*, 2008). Food spoilage was a complex process and it was mainly due to bacterial and fungal infections (Shu, 2006). Chemical compounds were used to combat bacterial and fungus, but the consumer perception of chemical preservatives drove attention towards natural alternatives (Pillai and Ramaswamy, 2011). Interest

Seabuckthorn (Hippophae L.): A Multipurpose Wonder Plant, Vol. 5: Advanced Technologies for Cultivation, Processing, Health Protection and Environmental Conservation (Virendra Singh, *Editor-in-Chief*, 2018), p. 275–280, Astral International Pvt. Ltd., New Delhi, India.

had been focused on the potential application of plant derived extracts. Wilkins and Board (1989) reported that more than 1,340 plants were known to be potential source of antimicrobials; about 60 were mentioned by Nychas (1995) and Beuchat (1994).The use of natural antimicrobial compounds was important not only in the preservation of food but also in the control of human and plant disease of microbial origin (Baratta *et al.*, 1998). Byproducts of plant, such as plant extracts, were studied for their antimicrobial activities and most extracts derived from plants were known to possess antibacterial, insecticidal, antifungal, acaricidal and cytotoxic activities (Faleiro *et al.*, 1999). Therefore, they were intensely screened and applied in the fields of food microbiology, pharmacology, pharmaceutical botany, medical and clinical microbiology, phytopathology and food preservation (Daferera *et al.*, 2000). Antibacterial substances of plants could be divided into inherent and induced-forming antimicrobial substances (Wang, 2014) according to source of compound. They could also be divided into terpenoids, alkaloids, fatty compounds, aromatic compounds according to compound structure (Zhang, 2005). Extraction method of plant antibacterial substances was similar to that of antioxidant substances, including solid-liquid extraction (Pinelo), microwave-assisted extraction (MAE) (Liao, 2007), ultrasound-assisted extraction (Sahika, 1986), and enzymatic extraction (Wang). MAE had been widely used as a mature technology for bioactive ingredient extraction. The MTT (3-(4, 5-dimethylthiazolyl-2)-2, 5-diphenyltetrazolium bromide) assay was widely accepted as a simple and reproducible method for determining cell proliferation or cytotoxicity *in vitro* (Shirahata *et al.*, 2009).

In recent years, a lot of extraction process, antimicrobial properties and antibacterial mechanism of the plant and fruit active ingredient, such as garlic (Booyens *et al.*, 2014), tea (Takahashi *et al.*, 2010), aloe vera (Kumar *et al.*, 2015) and other extracts with different inhibitory effect have been proved. Few studies about antibacterial properties of the berry have been reported, mainly about antibacterial properties of fruit peel and leaves. Jiao *et al.* (2007) expanded bacteriostatic experiment using plate method and tube dilution method of seabuckthorn leaf and proved significant inhibitory effect of leaf extract. Zhang *et al.* (2008) studied antimicrobial activity of leaves flavonoids through filter paper method and the results showed that leaves flavonoids had a good inhibitory effect on *S. aureus, B. subtilis, and E. coli.* Besides, they also indicated that antibacterial activity increasing along with the improvement of total concentration of flavonoids, and inhibition rate increasing as extension of time for the same concentration of total flavonoids. Guan *et al.* (2005) found seabuckthorn polysaccharide had a certain inhibitory effect on *E. coli, S. aureus,* and *B. subtilis.* Negi *et al.* (2005) indicated the possibility of using seeds for medicinal uses and food preservation.

But, berry extracts, the most readily available byproduct of the berry, has not been studied for its antibacterial properties. Therefore, the objective of the present study was to investigate the antibacterial activity of the berry extracts in *vitro* models, and determine their antibacterial activity against different food-borne pathogens for their potential as a natural preservative and for nutraceutical formulations.

MATERIALS AND METHODS

Sample Collection

Seabuckthorn berries were collected from Gansu Conseco Seabuckthorn Co., Ltd (Huachi, Gansu, China); *E. coli, B. subtilis, S. aureus, T. viride, R. stolonifer, A. niger, P. expansum,* and *Vintage red* were all obtained from Microbiological laboratory of Food Science and Technology College, Gansu Agricultural University (Lanzhou, Gansu, China); all solvents/chemicals used were of analytical grade and obtained from TCI Shanghai, China.

Preparation of Seabuckthorn Berry Extracts

Ripe berries were selected and pest and spoiled fruits were excluded. Clean and dried (at 60 °C for 48 h) cherries were stored in -20 °C condition, powdered in mortar quickly and successively extracted in a Soxhlet extractor for 8 with ethanol-water solution according to the procedure of Liang *et al.* (2014). The extraction technology conditions were as following: ethanol concentration 51 per cent, solvent to sample ratio 22:1 (mL/g) extraction time 1h, extraction temperature 86 °C. Kept extracts at 4 °C.

Media and Reagent

The medium were prepared by method of Wang *et al.* (2013). (1) Bacterial media: beef extract and peptone medium(BEP): beef extract 3 g, peptone 10 g, agar 18 g, NaCl 5 g, pH 7.2-7.5, and distilled water 1000 mL were sterilized in 121 °C for 30 min and stored at 4 °C. (2) Fungus media: potato dextrose agar (PDA): potato (peeled and boiled for 30 min) 200 g, sucrose 20 g, and distilled water 1000 mL were sterilized in 121 °C for 30 min and stored at 4 °C.

Solution Preparation

(1) Phosphate buffered saline (PBS) (Dulbecco *et al.*, 1954): Started with 800 mL of distilled water to dissolve NaCl 7.9 g, KCl 0.2 g, NaH_2PO_4 3.12 g, and Na_2HPO_4 7.16 g. Adjusted the pH to 7.4 with HCl. Added distilled water to a total volume of 1000mL and stored at 4 °C.

(2) MTT (3-(4, 5-dimethyl-2-thiazolyl)-2, 5-diphenyl-2-H-tetrazolium bromide): Put 20 mL PBS (0.01mol/L CPH=7.4) to dissolve 25 mg MTT in a flask. Added distilled water to a total volume of 50mL and stored at 4 °C. Expired in 14 days.

Test Bacterial Suspension Preparation (Zhang, 2007)

(1) Activation of Tested Strains

E. coli, *B. subtilis*, and *S. aureus* were grown in BEP agar at 37 °C for 24 h; *T. viride*, *R. stolonifer*, *A. niger*, *P. expansum*, and Vintage Red were grown in PDA at 28 °C for 48 h. Activated strains were stored at 4 °C.

(2) Microbial Suspension Preparation

1×10^8 cfu/mL of each bacterium, $1\sim8\times10^6$ cfu/mL of each mold and $1\sim7\times10^6$ cfu/mL of each yeast which were to be tested were inoculated into the flasks under aseptic conditions by inoculating loop.

Antimicrobial Test of Seabuckthorn Berry Extracts

(1) Antibacterial Activity by MTT Colorimetric Assay

Antibacterial properties of berry extracts were evaluated by MTT colorimetric assay (Huang *et al.*, 2007) in 96- well microlon ELISA plates. Made a twofold dilution of bacterial suspension by sterilized PBS (0.01 mol/L CPH=7.4) into 9 concentrations and took MTT colorimetric assay to all bacterial suspension. A solution of extracts, bacterial suspension, and MTT were added to each well of a 96-well plate by sterilized micropipette. 3 parallel wells at every dosage group and negative control group were set. Table 34.1 showed the quantity of solution added in each well. Suspension medium were mixed by shaking plates to obtain a uniform solution after dosing and cultivated statically at 37°C for 24h. Enzyme-labelling measuring instrument was used to measure the absorbance value at 630nm of different suspension medium for calculating the inhibiting rate of bacterial growth and getting MIC values (Minimum Inhibitory Concentration) of three bacterium. Bacterial inhibitory rate was calculated as following:

Inhibitory Rate (Per cent) =

$$\left(1-\frac{\text{Sample absorption value} \quad \text{Blank control}}{\text{Negative control} \quad \text{Blank control}}\right)\times 100\%$$

(2) Antifungal Activity by Mycelium Growth Rate Method (Xiao *et al.*, 2012)

Mixed berry extracts and PDA (in ratio 1:9), poured the mixture into dishes (5cm in diameter), and cooled for use. A round agar pie (5mm in diameter) of activated fungus/yeast was cut with hole puncher and put on the center of clean dish with mycelium facing down. Put one sample pie on each dish and set three parallels for each sample. Samples were cultured at constant temperature (28°C) and crossing method (Chai *et al.*, 2014) was used to measure diameter of colonies when mycelium was going to overgrow the dish. At the same time, measure extracts antifungal activities to *T. viride*, *R. stolonifer*, *A. niger*, *P. expansum*, and *Vintage Red* in the concentration of 20, 10, 5, 2.5, and 1.25 mg/mL, respectively.

Made the regression line with the concentration of extracts as abscissa and the corresponding inhibition rate as vertical coordinate and obtained regression equation and half maximal inhibitory concentration (IC50) of fungal mycelium growth. Equation to get inhibition rate was as following (D=diameter):

Inhibition Rate (per cent) = ([(Control colony D – Pie D) – (Treatment colony D – Pie D))/((Control colony D – Pie D)) × 100 per cent

RESULTS AND DISCUSSION

Antibacterial Activity of Berry Extracts

Growth inhibition studies were conducted *in vitro* to investigate the antimicrobial action of the berry extracts against the tested microorganisms. The berry extracts were able to reduce the bacterial population completely at or below 1250 mg/L level for all the bacteria tested in

Table 34.1: The Quantity of Solution added in Each Well

	Culture Medium (μL)	*Bacterial Suspension (μL)*	*Extracts Solution (μL)*	*MTT (μL)*	*Sterile Water (μL)*
Experimental Group	100	20	40	20	0
Negative Control	100	20	0	20	40
Blank Control	100	0	40	20	20

Table 34.2: The MIC (Minimal Inhibitory Concentration) of Berry Extracts against different Bacterium

Strains	*Concentrations of Berry Extract mg/L*								
	40000	*20000*	*10000*	*5000*	*2500*	*1250*	*625*	*312.5*	*156.25*
E. coli	–	–	–	+	++	+++	+++	+++	+++
B. subtilis	–	–	–	–	–	+	++	+++	+++
S. cus aureus	–	–	+	++	+++	+++	+++	+++	+++

'–' no bacteria growth; '+' lighter color; '++' darker color; '+++' darkest color.

present investigation (Table 34.2). The extracts inhibited growth to variable extents, depending on the bacterium in question. The growth inhibition pattern showed that berry extract was most effective against *B. cereus* as its complete inhibition growth was at 1250 mg/L, followed by 5000 mg/L against *E. coli* and 10000 mg/L against *S. aureus*. The MIC values were consistent with the study by Guan *et al.* (2005).

In the test of 96- well microlon ELISA plates, added 40μLberry extract into 140μL bacterial suspension, diluted sample 4.5 times with distilled water and cultivated it for 24h. According to absorption values tested by ELISA plates, inhibitory results of berry extract against three bacterium were conducted as shown in Table 34.3. All absorption values of three bacterium were increasing with the improvement of extract concentration. Statistical analysis were performed using Origin 8.6 software to build regression equation between berry extract and three bacterium. Sensitivities differed among different bacterium: the most sensitive bacteria was *B. subtilis* with IC_{50} 590 mg/L, followed by *E. coli* with IC_{50}1290 mg/L and *S. aureus* with IC_{50} 2120 mg/L.

(2) Antifungal Activity of Berry Extracts

Berry extract is rich in nutrients, ingredients which are the most likely environment for the growth of fungi, therefore the selected concentration should not be too high (1250, 2500, 5000, 10000 mg/L, respectively). According to diameters of colony and control colony, inhibitory results of berry extract against fungi and yeast were conducted as shown in Table 34.4.

Antibacterial equation against fungi and yeast was conducted according to different inhibition rates of different extract concentration and IC_{50} was calculated (Table 34.5). The result suggested that with the increase of the concentration, antifungal activity of berry extracts showed a gradual enhancement. The extracts had highest IC_{50} value of *R. stolonifer* (18870 mg/L), followed by *T. viride* (7890 mg/L). The extracts had relatively low and similar IC_{50} value of *A. niger, Vintage red* and *P. expansum* (5990, 5730, and 5520 ml/L successively). The extract had strongest inhibition of *P. expansum* whereas it had weakest suppression of *R. stolonifer*. The results were well fit to the theory conclusion by Zhang (2007).

CONCLUSION

This paper dealt with inhibitive effect of seabuckthorn berry extract by ethanol-water solution on microorganism and certified that antimicrobial activity was selective to different microorganism. Cao *et al.* (2006) reported by acting on the cell wall membrane and cell protoplasm, natural antibacterial substances destroyed the contents of the cell structure or enzyme system, inhibited respiration, and influenced bacteria on mitosis to achieve the inhibitory effect. OH groups, contained in certain active ingredient, were very active, easy to form hydrogen bonds with the active site of enzyme

Table 34.3: Inhibition Rate of Berry Extracts against different Bacterium

Inhibition Rate Per cent	*Concentrations of Berry Extract mg/L*								
	156.25	*312.5*	*625*	*1250*	*2500*	*5000*	*10000*	*20000*	*40000*
E. coli	-3.38	-0.83	-2.01	3.44	20.97	36.53	89.90	105.86	102.33
B.subtilis	0.67	10.58	16.18	34.25	48.64	74.79	98.31	101.10	102.96
S. aureus	2.83	6.80	7.07	11.78	22.42	28.12	56.03	97.07	98.85

Table 34.4: Inhibitory Results of Berry Extract against Fungi and Yeast

Concentration ml/L	*Inhibition Rate per cent*				
	T. viride	*R. stolonifer*	*A. niger*	*P. expansum*	*Vintage Red*
1250	1.02	1.18	7.41	11.54	6.67
2500	22.22	5.56	25.58	15.38	16.07
5000	34.57	16.67	48.84	53.85	35.71
10000	61.22	24.71	77.78	88.46	97.78

Table 34.5: IC50 Values of Berry Extract on Fungi and Yeast

Pathogens	*Antibacterial Regression Equation*	*Correlation Coefficient*	*IC_{50} mg/L*
T. viride	Y = 6.3202x + 0.1315	R^2=0.9466	7890
R. stolonifer	Y = 2.6769x – 0.5208	R^2=0.9471	18870
A. niger	Y = 7.7254x – 3.6896	R^2=0.9660	5990
P. expansum	Y = 9.2040x – 0.8361	R^2=0.9671	5520
Vintage Red	Y = 10.519x – 10.250	R^2=0.9868	5730

and then led to metabolic disorders. Antioxidants in the fruit were mainly some flavonoids and polyphenols, of which structure contained a large number of hydroxyl groups, thus they had some antibacterial effect. Because antioxidants were abundant in the fruit so that the fruits had certain suppression to common food contamination. However, isolation of antibacterial active ingredient and its antibacterial mechanisms are need further study and discussion. As a natural substance, fruit extract is not only nutritional and functional, but also broad-spectrum and highly effective in antimicrobial activity to prevent food poisoning and other diseases caused by microorganisms. At the same time, the fruit extract can prevent food spoilage and extend the shelf life.

ACKNOWLEDGEMENT

This work was funded by the Projects of International Cooperation and Exchanges of Ministry of Science and Technology in China (No. 2014DFR31230). We gratefully acknowledge National Nature Science Foundation of China. At the same time we are sincerely appreciated Mr. Luan from Gansu Conseco Company for providing all the seabuckthorn pulp needed in the experiment. And we are thankful for GE Qian cui-di who introduced the processing of seabuckthorn in the industry which helped us to set the experimental factors. Besides, we wish to express our thanks to Prof. Han Shun-yu for discussions.

REFERENCES

1. Baratta, M.T., Dorman, H.J.D., Deans, S.G., Biondi, D.M. and Ruberto, G. 1998. Chemical composition, antibacterial and antioxidative activity of laurel, sage, rosemary, oregano and coriander essential oils. *Journal of Essential Oil Research* 10: 618-627.
2. Booyens, J., Labuschagne, M. C. and Thantsha, M. S. 2014. *In vitro* antibacterial mechanism of action of crude garlic (*Allium sativum*) clove extract on selected probiotic *Bifidobacterium* species as revealed by SEM, TEM, and SDS-PAGE analysis. *Probiotics and Antimicrobial Proteins* 6: 82-87.
3. Cao, G.Q., Guo, Q. and Huang, M. 2006. Research progress of natural antimicrobial substances. *Storage and Process* 6: 10-12.
4. Chai, D.D., Guo, S. J., Sun, X. and Qin, T.T. 2014. Optimization of culture conditions for 7 *chestnut ectomycorrhizal* fungi in Yanshan Area. *Northwest Agricultural and Forest University (Nat. Sci. Ed.)* 42: 109-115.
5. Daferera, D.J., Ziogas, B.N. and Polissiou, M.G. 2000. GC–MS analysis of essential oils from some Greek aromatic plants and their fungi toxicity on *Penicillium digitatum*. *Journal of Agricultural and Food Chemistry 48: 2576-2581.*
6. Dillon, V. M. and Board, R. G. 1994. Antimicrobial properties of spices and their essential oils. In: *Natural Antimicrobial Systems and Food Preservation*, pp. 167–179, CAB Intl, Wallingford, U.K., 328p.
7. Dulbecco, R. and Vogt, M. 1957. Plaque formation and isolation of pure lines with *Poliomyelitis* Viruses. *Journal of Experimental Medicine* 106: 167-169.
8. Faleiro, L, Miguel, G.M., Guerrero, C.A. and Brito, J.M.C. 1999. Antimicrobial activity of essential oils of *Rosmarinus offýcinalis* L., *Thymus mastichina* L. ssp. *mastichina* and *Thymus albicans*. In: *Proceedings of the Second WOCMAP Congress on Medicinal and Aromatic Plants, Part 2: Pharmacognosy, Pharmacology, Phytomedicine, Toxicology*, Mendoza, Argentina, 501: 45-48.
9. Guan, Q., Yang, W.Z. and Wen, Z.P. 2005. Extraction of polysaccharide from seabuckthorn pericarp and leaves and its bacteriostasis research. *The Global Seabuckthorn Research and Development* 3:17-20.
10. He, S.M., Yuan, X.J. and Wang, J.H. 2008. The plant resource of *Hippophae* and its development and utilization status in China. *Chinese Modern Agricultural Sciences* 15: 87-92.
11. Huang, L.K., Du, J. and Huo, G.C. 2008. Inquiring into the method of counting live germ with MTT. *The Chinese Food Industry* 3: 62-65.
12. Jiao, X., Yin, L.J. and Cheng, Y.Q. 2007. Extraction and antibacterial properties of flavonoids of seabuckthorn leaves. *Chinese Food Science* 28: 124-129.
13. Kallio, H., Yang, B., Tahvonen, R. and Hakala, M. 2000. Composition of seabuckthorn berries of various origins. In: *Proceedings of International Workshop on Seabuckthorn*, Beijing, China, pp. 13-19.
14. Kumar, V. P.P.N., Shameem, U., Pratap, K. P., Kalyani, R. L. and Pammi, S. V. N. 2015. Green synthesis of copper oxide nanoparticles using *Aloe vera* leaf extract and its antibacterial activity against fish bacterial pathogens. *Bio. Nano Science*, DOI 10.1007/s12668-015-0171-z.
15. Liao, Y.T. 2011. The antimicrobial effect of mulberry leaves and the extract and isolate for its antimicrobial activities. *Master Thesis*, Jiangnan University, Wuxi, Jiangsu, China.
16. Ling, K., Jiang, Y.M., Li, J.X., Mi, L. and Zhang, B.J. 2014. Active compound extraction from seabuckthorn fruit and antioxidation activity research. *Science and Technology of Food Industry* 24: 264-269.
17. Li, T.S.C. and Schroeder, W. 1996. Seabuckthorn (*Hippophae rhamnoides* L.): A multipurpose plant. *Hort. Technology* 6: 370-380.
18. Lu, R. 1992. Seabuckthorn: A multipurpose plant species for fragile mountains. Khathmandu, Nepal. ICIMOD Occasional Paper No. 20, pp. 62.

19. Negi, P.S., Chauhan, A.S., Sadia, G.A., Rohinishree, Y.S. and Ramteke, R.S. 2005. Antioxidant and antibacterial activities of various seabuckthorn (*Hippophae rhamnoides* L.) seed extracts. *Food Chemistry* 92: 119-124.

20. Nychas, G.J.E. 1995. Natural antimicrobials from plants. In: *New Methods of Food Preservation (G.W. Gould, Ed in chief)*. Blackie Academics and Professional, Glasgow, pp. 58–89, Springer, Berlin, Germany, 324p.

21. Pillai, P. and Ramaswamy, K. 2011. Effect of naturally occurring antimicrobials and chemical preservatives on the growth of *Aspergillus parasiticus*. *Food Science and Technology* 49: 228-233.

22. Pinelo, M., Ruiz-Rodríguez, A., Sineiro, J., SeCoráns, F. J., Reglero, G. and Nunez, M. J. 2007. Supercritical fluid and solid–liquid extraction of phenolic antioxidants from grape pomace: A comparative study. *European Food Research and Technology 226*:199-205.

23. Sahika, E., Aktug and Mehmet, K. 1986. Sensitivity of some common food poisoning bacteria to thyme, mint and bay leaves. *International Journal of Food Microbiology* 3: 349-354.

24. Shu, X.L., Shi, Q.F. and Ouyang, Y.S. 2006. New developments of the antimicrobial activities of essential oils and its application in food industry. *Chinese Biotechnology* 16: 89-92.

25. Takahashi, T., Aso, Y. and Kondo, W.K.T. 2010. Improving the antibacterial activity against *Staphylococcus aureus* of composite sheets containing wasted tea leaves by roasting. *Journal of Wood Science* 56: 403-410.

26. Terence, N. P. T., Mohamed, B., Sami, S. and Hiroko, I. 2009. MTT reduction by flavonoids in the absence of cells: influence of medium type and serum. In: *Animal Cell Technology: Basic and Applied Aspects*, pp. 317-324, Springer Netherlands, Berlin, Germany, 357 p.

27. USDAR 2007. Plants Profile for *Hippophae rhamnoides* (seaberry). October 8. United States Department of Agriculture. Retrieved from http://plants.usda.gov/core/profile?symbol=HIRH80.

28. Vivek, K. B., Nguyen, T. D. O., Jun, K. and Sun, C. K. 2008. Analysis and the potential applications of essential oil and leaf extracts of *Silene armeria* L. to control food spoilage and food-borne pathogens. *European Food Research and Technology* 227: 1613-1620.

29. Wang, H.B., Wang, S. and Fu, L. 2012. Study on technology of extracting polyphenol from pomegranate peel by enzymatic treatment. *Journal of Chinese Institute of Food Science and Technology* 12: 56-65.

30. Wang, M.L. 2014. Plant antimicrobial peptides and its application in biotechnology. *Chinese Farm Products Processing* 12: 72-74.

31. Wang, S.J., Yang, M. and Yu, X. 2013. Study on antibacterial activity of alcohol-soluble and water-soluble extracts from *Kadsura longepedunculata*. *Journal of Anhui Agricultural Sciences* 41: 1922-1924.

32. Wilkins, K.M. and Board, R.G. 1989. Natural antimicrobial system. In: *Mechanism of Action of Food Preservation Procedures* (G.W. Gould, Ed in chief), pp. 285-362, Elsevier, New York, 448p.

33. Xiao, B.H., Chen, B. and Lei, X.L. 2012. Antibacterial activity of secondary metabolites of marine fungi ZJ27. *Chinese Marine Sciences* 36: 95-99.

34. Zhang, Y.L., Yin, C.P. and Feng, J.T. 2005. Progress in research of botanical fungicide. *Journal of Southwest University for Nationalities (Nat. Sci. Ed.)* 31: 402-409.

35. Zhang, Y.N. 2007. Studies on extraction, purification technology and antimicrobial of Xin Jiang seabuckthorn. *Master Thesis*, Xinjiang Agricultural University, Urumqi, Xinjiang, China.

Chapter 35

Wound Healing and Anti-Stress Prospective of Seabuckthorn (*Hippophae rhamnoides* L.)

Asheesh Gupta* and Gaurav K. Keshri

Pharmacology Division, Defence Institute of Physiology and Allied Sciences (DIPAS), DRDO, Timarpur, Delhi – 110 054, India
(*E-mail: asheeshgupta@dipas.drdo.in)

SUMMARY

Seabuckthorn (*Hippophae rhamnoides* L.), a unique plant has drawn attention in the wide arena of biomedical research owing to its immense nutritional attributes, medicinal and therapeutic potential. The wound repair and anti-stress properties of seabuckthorn (SBT) in the experimental animal models have been discussed. A potent herbal wound healer in form of a topical ointment developed using seabuckthorn aqueous extract, which showed excellent healing properties to augment repair of normal as well as chronic wounds (diabetic and burn). SBT significantly augmented the healing process as evidenced by an enhanced wound contraction, re-epithelialization, cellular proliferation (TGF-β, α SM actin, ECM), collagen type-III deposition, angiogenesis (VEGF) and matrix metallo-proteinases (MMP-2 and 9). Healing activity of SBT was found better than silver sulfadiazine and provide one-iodine based ointments (reference care). Detailed safety and toxicological studies found that SBT ointment is safe for dermal application. Further, a potent anti-stress activity has also been evaluated using SBT extract. SBT decreased the level of stress biomarkers and enhanced hypoxic tolerance in animals indicating its anti-stress adaptogenic activity. Antioxidant potential of the SBT extract was found to be one of the factors observed for its adaptogenic activity. In conclusion, SBT extract possesses potent wound healing and anti-stress properties.

Keywords: *Seabuckthorn antioxidants, Anti-stress, Burns and Wound healing.*

INTRODUCTION

This review explores the medicinal and nutritional applications of seabuckthorn (*Hippophae rhamnoides* L.) family Elaeagnaceae in curtailing various types of maladies. Currently, the wide arena of biomedical research aims to understand and support the traditional uses of seabuckthorn. The known pharmacological effects of seabuckthorn include antioxidant, immunomodulatory, anti-atherogenic, anti-stress, hepatoprotective, radioprotective and tissue repair properties. The recent research outcome suggests that seabuckthorn and its active extract based herbal preparations and nutraceuticals possess immense therapeutic potential. This paper discussed the phytochemical composition, medicinal, therapeutic potential, product development, and give impetus for further research and development to explore the comprehensive benefits of seabuckthorn.

Seabuckthorn: A Unique and Valuable Plant

Seabuckthorn (SBT) is one of the most representative medicinal and economy plant known for its extensive uses of therapeutic, nutritional potential and ecology protection. It is a dwarf to tall, branched and thorny nitrogen-fixing deciduous shrub growing naturally in cold-arid regions of Asia and Europe. Leaves, fruits and seeds from SBT have been found to be a rich source of a large number of bioactive phytoconstituents including flavonoids (isorhamnetin, quercetin, myricetin, kaempferol and their glycoside derivatives), carotenoids

Seabuckthorn (*Hippophae* L.): *A Multipurpose Wonder Plant, Vol. 5: Advanced Technologies for Cultivation, Processing, Health Protection and Environmental Conservation* (Virendra Singh, *Editor-in-Chief*, 2018), p. 281–285, Astral International Pvt. Ltd., New Delhi, India.

(α, β, δ-carotene, lycopene), vitamins (A, C, E and K), tannins, triterpenes, glycerides of palmitic, stearic and oleic acids and some essential amino acids (Beveridge *et al.*, 1999; Zu *et al.*, 2006). The high content of bioactive constituents has been imitated in its extensive exploitation by traditional system of medicine in Tibet, Mongolia, China and Soviet Republics to treat asthma, skin diseases, gastric ulcers and lung disorders. The recent systematic scientific research revealed that SBT has antioxidant (Upadhyay *et al.*, 2010), anti-inflammatory (Ganju *et al.*, 2005) anti-atherogenic (Basu *et al.*, 2007), anti-stress (Saggu *et al.*, 2007) activity, and has been reported to be useful in treating skin wounds (Gupta *et al.*, 2005; 2011; Upadhyay *et al.*, 2009; 2011), cardiovascular diseases (Eccleston *et al.*, 2002), thrombosis and platelet aggregation (Cheng *et al.*, 2003).

Wound Healing Potential of Seabuckthorn

Wound repair is an essential process in maintaining tissue homeostasis in response to injury. Physiological wound repair is characterized by four overlapping phases: hemostatsis, inflammation (infiltration of inflammatory cells), proliferation (granulation tissue formation, neovascularization, collagen accumulation and epithelialization) and tissue remodeling (collagen crosslinking, scar maturation). The normal predictable healing process can be impeded at any step along its path by a variety of factors. Chronic non-healing wound abnormalities cause a great deal of physical and psychological discomfort and morbidity to affected patients (Gupta *et al.*, 2013).

The process of wound healing is promoted by some natural and plant products, which are composed of bioactive phytochemicals *viz.* alkaloids, flavonoids, saponins and phenolic compounds. These agents usually influence one or more phases of the healing process. Phyto constituents represented positive activity at different stages of wound repair process by various mechanisms including anti-microbial, anti-inflammatory, antioxidant, collagen biosynthesis, cell proliferative and angiogenic effects (Kumar *et al.*, 2007). The presence of various life-sustaining constituents in herbal medicaments has urged scientists to examine these plants with a view to explore the potential wound healing properties.

Several studies have demonstrated potential wound healing activity of SBT (Gupta *et al.*, 2005, 2006, 2008; Upadhyay *et al.*, 2009, 2011). The protective and curative effects of SBT against wounds, burns, scalds, ulcers and mucosal injuries have been extensively investigated using different animal models and by clinical trials (Suleyman *et al.*, 2001; Xing *et al.*, 2002; Suryakumar and Gupta, 2011). SBT based preparations have been widely used for treating skin radiation lesions, burn of different etiology, gastric and duodenal ulcers. SBT leaf extract has been scientifically investigated and shown that it enhances both normal and impaired dermal wound healing (burns and diabetic) in experimental rats, and furthers its possible mechanisms of action was also investigated (Gupta and Upadhyay, 2011; Suryakumar and Gupta, 2011). It has been reported that SBT extract possessed strong antioxidant and anti-bacterial properties. *In vitro* studies revealed that SBT extract possessed significant cytoprotective and anti-bacterial activity against tested wound pathogens (*Pseudomonas aeruginosa* and *Staphylococcus aureus*) (Upadhyay *et al.*, 2010), and enhanced angiogenic activity observed in chick chorioallantoic membrane (CAM) model. In-vivo studies demonstrated that SBT extract possessed potent wound healing activity as assessed by various biophysical, biochemical, molecular pro-healing parameters and histopathological observations. The SBT treatment increased cellular proliferation, collagen biosynthesis, neo-vascularization and wound area contraction. The healing efficacy of SBT was found far better than sliver sulfadiazine and povidone-iodine ointments (standard care). The treatment also reduces scar and up-regulates the expression of growth factors (VEFG, TGF-β), ECM protein (collagen type-III) and MMPs. These results were further supported by histopathological examinations. The SBT treatment also increased the endogenous enzymatic (superoxide dismutase, catalase, glutathione peroxidases) and non-enzymatic (reduced glutathione, vitamin C) antioxidants and decrease in lipid peroxide levels and reactive oxygen species in wound granulation tissue (Gupta *et al.*, 2005; Gupta and Upadhyay, 2011; Upadhyay *et al.*, 2011). Further, it has been reported that SBT leaf extract has no cytotoxic, heavy metal contamination and adverse effect after oral and topical administration (Saggu *et al.*, 2007; Upadhyay *et al.*, 2010). On the basis of these findings a potent herbal wound healer in form of a topical ointment developed using seabuckthorn aqueous extract, which showed excellent healing properties. Flavonoids are one of the important biologically active substances present in SBT leaves and fruits. The positive healing effect of topical application of SBT flavone (isolated from fruit pulp) (1.0 per cent, w/v) has been observed on dermal wounds in experimental rats (Gupta *et al.*, 2006).

SBT oil extracted from fruits and seeds combine high levels of beneficial unsaturated fatty acids (omega-3, 6, 7), natural antioxidants and vitamins (E, K), carotenoids, as well as phytosterols (Beveridge *et al.*, 1999). These phytochemicals make it ideal for medicinal and cosmetic industries for giving synergistic power to protect cell membrane and enhance cellular and tissue repair and regeneration. One ingredient, palmitoleic acid, is a component of skin that is considered a valuable topical agent in treating burns and healing wounds. The preventive and curative effects of SBT seed oil have been reported in different types of gastric ulcers, chronic cervicitis and atopic dermatitis (Yang *et al.*, 2000; Xing *et al.*, 2002; Gupta and Upadhyay *et al.*, 2011). A systematic pre-clinical study conducted by Upadhyay *et al.* (2009) demonstrated the safety and

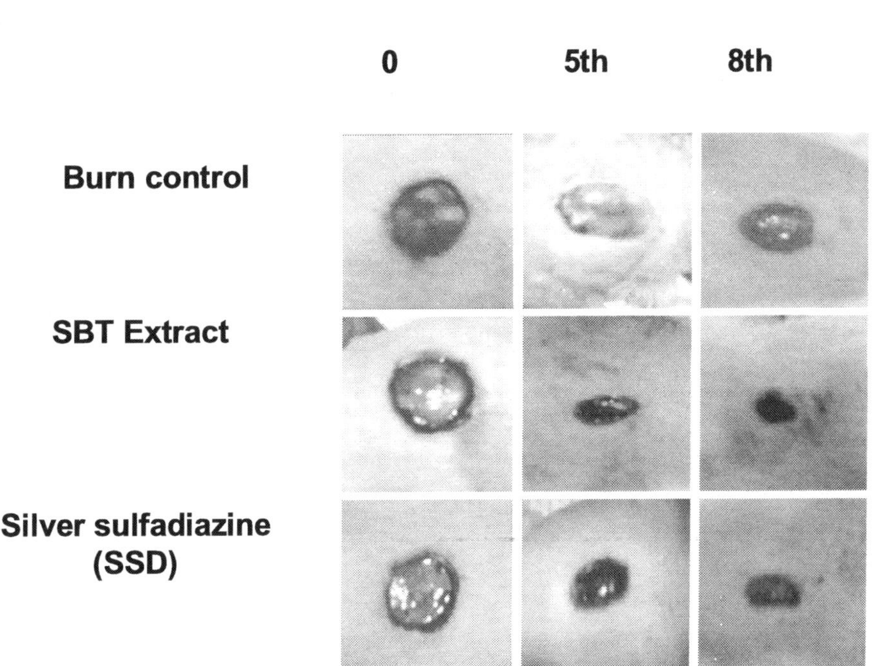

Figure 35.1: Visual Observation of Seabuckthorn (SBT) Extract Treated and Untreated Control and Reference Care (SSD) Treated Burn Wounds in rats. SSD: Silver sulfadiazine.

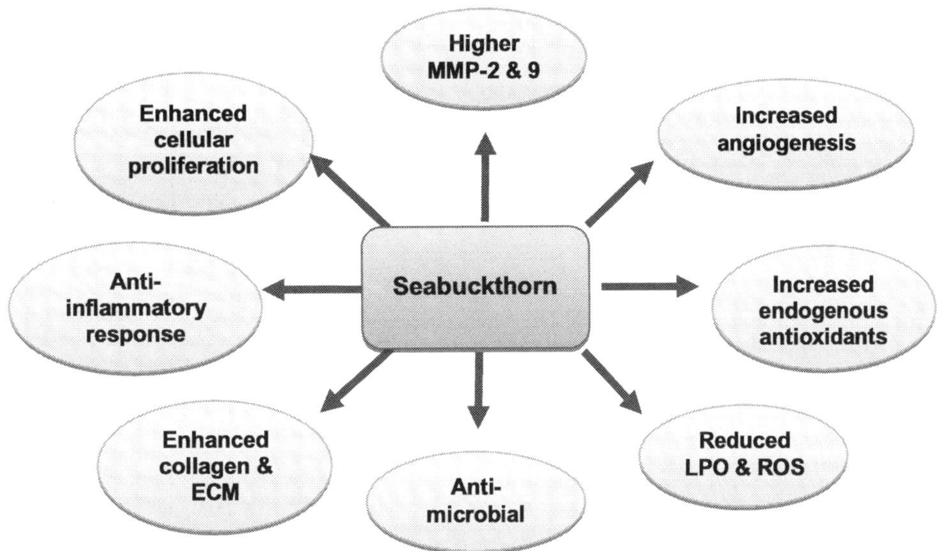

Figure 35.2: The Potential Effects of Seabuckthorn Extract on Wound Healing.
ECM: Extra cellular matrix; MMPs: Matrix metalloproteinases; LPO: Lipid peroxidation; ROS: Reactive oxygen species.

wound healing efficacy of SBT CO_2-supercritical fluid extracted seed oil co-administration by both route *i.e.* topical (200 µl) and oral (2.5 ml/kg BW) on experimental dermal burn wounds in rats. Treatment promotes significant re-epithelialization and wound closure and enhances formation of granulation tissue and collagen biosynthesis in burn wounds. Seed oil treatment showed an increase in endogenous antioxidant and decrease in free radical production in burn wounds. SBT seed oil absorbs strongly in the UV-B range (290–320 nm) and may be used as a natural sunscreen absorber (Beveridge *et al.*, 1999). In conclusion, SBT possesses a significant healing activity for both acute as well as chronic wounds by affecting various phases of tissue repair.

Anti-Stress Adaptogenic Potential of Seabuckthorn

The stress is a part and parcel of modern life. Human mind and body respond to stress, a state of perceived threat to homeostasis, by activating the sympathetic nervous system and secreting the catecholamines *viz.* adrenaline and noradrenaline in the "fight-or-flight" response. The stress response is commonly transient because its accompanying effects like immunosuppression, growth inhibition, and enhanced catabolism, which can be harmful in the long term. The prolonged and persistent stress response can be associated with psycho-physiological, metabolic disorders like hypertension, diabetes, gastric ulcer, cardiovascular ailments and leads to DNA damage. Stress is also broadly defined as a deleterious factor (physical, chemical, or biological), which triggers a series of cellular and systemic events, resulting in restoration of cellular and organismal homeostasis. When the stress is within the threshold limit of the body, then it can leads to optimal stimulation of physiological functions and thereby improved the performance of individual, however, if the stress is beyond the threshold limit of the body, then it can leads to excessive arousal, poor performance which can leads to stress disorders (imbalance of demands/resources).

A non-specific, anti-stress procedure/process (training, cross-adaptation, yoga practice) or herbal, dietary, chemical pharmaco-therapeutic intervention should help in restoration of normal homeostasis of body and boost the endurance and physical performance under stressful conditions. The herbal preparations, which can augment endurance, mental function and non-specific resistance of the body to stress have been termed as adaptogens (Brekhman, 1980). DIPAS, in the last few years, has focused on the development of adaptogens, herbal formulations which can help to enhance non-specific resistance to stress as well as improving physical performance. A potent anti-stress activity has been reported using SBT aqueous extract (Saggu *et al.*, 2007). Efficacy of SBT extract has been evaluated using a passive multiple stress rat experimental model comprising of stressors *viz.* cold (5°c)-hypoxia (428 mmHg)-restraint (C-H-R), and found to possess potent anti-stress adaptogenic activity. In C-H-R model, SBT extract have demonstrated delayed fall in core body temperature and enhanced the recovery process from acute hypothermia. Results suggested that SBT extract supplementation strengthened the oxygen delivery system, better utilization of free fatty acids, improved cell membrane permeability and reversed the hypoxia-induced changes. It has been reported that SBT extract treatment caused a trend for shifting anaerobic metabolism to aerobic during multiple stress C–H–R exposure and post stress recovery (Saggu and Kumar, 2007, 2008). Our recent studies also suggest that SBT supplementation imparts tolerance to hypoxia susceptible animals by facilitating intracellular energy content and augmenting bioenergetics and endogenous antioxidants under acute hypoxia (single stressor) (data not published). Antioxidant potential which, helps to decrease oxidative stress was found to be one of the factors observed for adaptogenic activity of the SBT extract. SBT decreased the level of stress biomarkers and enhanced hypoxic tolerance in animals indicating its anti-stress adaptogenic activity. In conclusion, SBT extract possesses potent wound healing and anti-stress properties.

AKNOWLEGEMENT

The authors are thankful to Director, DIPAS for encouragement and support to the research studies embodied in this paper.

REFERENCES

1. Basu, M., Prasad, R., Jayamurthy, P., Pal, K., Arumughan, C. and Sawhney, R.C. 2007. Antiatherogenic effects of seabuckthorn (*Hippophae rhamnoides* L.) seed oil. *Phytomedicine* 14: 770-777.
2. Beveridge, T., Li, T. S., Oomah, B. D. and Smith, A. 1999. Seabuckthorn products: Manufacture and composition. *Journal Agriculture Food Chemistry* 47: 3480-3488.
3. Brekhman, I.I. 1980. Man and biologically active substances. Pergamon Press, Oxford.
4. Cheng, J., Kondo, K., Suzuki, Y., Ikeda, Y., Meng, X. S. and Umemura, K. 2003. Inhibitory effects of total flavones of *Hippophae rhamnoides* L. on thrombosis in mouse femoral artery and *in vitro* platelet aggregation. *Life Sciences* 72: 2263-2271.
5. Eccleston, C., Baoru, Y., Tahvonen, R., Kallio, H., Rimbach, G.H. and Minihane, A.M. 2002. Effects of an antioxidant rich juice (seabuckthorn) on risk factors for coronary heart disease in humans. *Journal of Nutritional Biochemistry* 13: 346-354.
6. Ganju, L., Padwad, Y., Singh, R., Karan, D., Chanda, S., Chopra, M. K., Bhatnagar, P., Kashyap, R. and Sawhney, R. C. 2005. Anti-inflammatory activity of seabuckthorn (*Hippophae rhamnoides* L.) leaves. *International Immunopharmacology* 12: 1675-1684.
7. Geetha, S., Sai Ram, M., Singh, V., Ilavazhagan, G. and Sawhney, R.C. 2002a. Antioxidant and immune modulatory properties of seabuckthorn (*Hippophae rhamnoides* L.)–An *in vitro* study. *Journal of Ethnophamacology* 79: 373-378.
8. Gupta, A., Avci, P., Sadasivam, M., Chandran, R., Parizotto, N., Vecchio, D., de Melo, W.C., Dai, T., Chiang, L.Y. and Hamblin, M.R. 2013. Shining light on nanotechnology to help repair and regeneration. *Biotechnology Advance* 31: 607-631.
9. Gupta, A., Kumar, R., Pal, K., Banerjee, P.K. and Sawhney, R.C. 2005. A preclinical study of the effects of seabuckthorn (*Hippophae rhamnoides* L.) leaf extract on cutaneous wound healing in albino rats. *International Journal of Lower Extremity Wounds* 4: 88-92.
10. Gupta, A., Kumar, R., Pal, K., Singh, V., Banerjee, P.K. and Sawhney, R.C. 2006. Influence of seabuckthorn

(*Hippophae rhamnoides* L.) flavone on dermal wound healing in rats. *Molecular Cellular Biochemistry* 290: 193-198.

11. Gupta, A. and Upadhyay, N.K. 2011. Seabuckthorn (*Hippophae rhamnoides* L.) seed oil: Usage in burns, ulcers and mucosal injuries. In: *Nuts and Seeds in Health and Disease Prevention* (Preedy, V.R., Watson, R.R., Patel, V.B., Eds.), Academic Press (Elsevier), London, pp. 1011-1018.

12. Gupta, A., Upadhyay, N.K., Sawhney, R.C. and Kumar, R. 2008. A poly-herbal formulation accelerates normal and impaired diabetic wound healing. *Wound Repair and Regeneration* 16: 784-790.

13. Kumar, B., Vijayakumar, M., Govindarajan, R. and Pushpangadan, P. 2007. Ethno-pharmacological approaches to wound healing-exploring medicinal plants of India. *Journal of Ethno-Pharmacology* 114: 103-13.

14. Saggu, S., Divekar, H.M., Gupta, V., Sawhney, R.C., Banerjee, P.K. and Kumar, R. 2007. Adaptogenic and safety evaluation of seabuckthorn (*Hippophae rhamnoides*) leaf extract: A dose dependent study. *Food and Chemical Toxicology* 45: 609-617.

15. Saggu, S. and Kumar, R. 2007. Possible mechanism of adaptogenic activity of seabuckthorn (*Hippophae rhamnoides* L.) during exposure to cold, hypoxia and restraint (C–H–R) stress induced hypothermia and post stress recovery in rats. *Food and Chemical Toxicology* 45: 2426-2433.

16. Saggu, S. and Kumar, R. 2008. Effect of seabuckthorn leaf extracts on circulating energy fuels, lipid peroxidation and antioxidant parameters in rats during exposure to cold, hypoxia and restraint (C–H–R) stress and post stress recovery in rats. *Phytomedicine* 15: 437-446.

17. Suleyman, H., Demirezer, L.O., Buyukokuroglu, M.E., Akcay, M.F., Gepdiremen, A., Banoglu, Z.N. and Gocer, F. 2001. Anti-ulcerogenic effect of *Hippophae rhamnoides* L. *Phytotherapy Research* 33: 77-81.

18. Suryakumar, G. and Gupta, A. 2011. Medicinal and therapeutic potential of seabuckthorn (*Hippophae rhamnoides* L.). *Journal of Ethanopharmacol* 138: 268-278.

19. Upadhyay, N. K., Kumar, R., Mandotra, S. K., Meena, R. N., Siddiqui, M. S., Sawhney, R. C. and Gupta, A. 2009. Safety and healing efficacy of seabuckthorn (*Hippophae rhamnoides* L.) seed oil on burn wounds in rats. *Food and Chemical Toxicology* 47: 1146-1153.

20. Upadhyay, N. K., Kumar, R., Siddiqui, M. S. and Gupta, A. 2011. Mechanism of wound-healing activity of *Hippophae rhamnoides* L. leaf extract in experimental burns. *Evidence Based Complementary Alternative Medicine* 659705-659804.

21. Upadhyay, N. K., Yogendra, K. M. S. and Gupta, A. 2010. Antioxidant, cytoprotective and antibacterial effects of seabuckthorn (*Hippophae rhamnoides* L.) leaves. *Food and Chemical Toxicology* 48: 3443-3448.

22. Xing, J., Yang, B., Dong, Y., Wang, B., Wang, J. and Kallio, P.H. 2002. Effects of seabuckthorn (*Hippophae rhamnoides* L.) seed and pulp oils on experimental models of gastric ulcer in rats. *Fitoterapia* 73: 644-650.

23. Yang, B., Kalimo, K.O., Tahvonen, R.L., Mattila, L.M., Katajisto, J.K. and Kallio, H.P. 2000. Effect of dietary supplementation with seabuckthorn (*Hippophae rhamnoides* L.) seed and pulp oils on the fatty acid composition of skin glycerophospholipids of patients with atopic dermatitis. *Journal of Nutritional Biochemistry* 11: 338-340.

24. Zu, Y., Li, C., Fu, Y. and Zhao, C. 2006. Simultaneous determination of catechin, rutin, quercetin kaempferol and isorhamnetin in the extract of seabuckthorn (*Hippophae rhamnoides* L.) leaves by RP-HPLC with DAD. *Journal of Pharmacy and Biomedical Analysis* 41: 714-719.

Chapter 36

Application of Enriched Fraction of Seabuckthorn (*Hippophae rhamnoides* L.) Leaf Extract as Antimicrobial Finish on Technical Textile

M.S. Yogendra Kumar*, T.S. Raghu, F.V. Varghese, T.M. Kotresh and U.K. Singh

Defence Bioengineering and Electromedical Laboratory,
C V Raman Nagar, Bengaluru – 560 093, Karnataka, India
(*E-mail: msyogendra@debel.drdo.in and msyogendra@gmail.com)

SUMMARY

Flavonoid-rich fraction (FRF) from seabuckthorn leaves extract was prepared by acid hydrolysis process. Total flavonoid content of seabuckthorn leaves extract and FRF estimated as rutin equivalent was found to be 116.98±3.06 and 277.14±6.78 mg/g of extract/FRF respectively. Its major constituents myrcetin, quercetin, kaempferol and isorhamnetin, were determined by reverse phase high performance liquid chromatography (RP-HPLC). Aramid (NomexIIIA) fabric with triethylene tetramine to increase the wicking height of the fabric for better extract uptake of FRF. Surface etching on the amine treated fabric morphology was determined by Scanning electron microscope analysis. Then, FRF was coated using citric acid as cross linking agent on to aramid fabric by pad-dry-cure method for improved wash durability. FRF coated fabric was characterised using Universal attenuated total internal reflection Fourier Transform Infrared spectroscopy. Effect of FRF coating on flammability property of coated fabric was estimated using flammability tester. There was no significant difference in the char produced by FRF coated fabric and control samples. Antimicrobial activity of the FRF coated fabric was assessed by both qualitative (agar diffusion method; AATCC 147-2001) and quantitative (percentage reduction test; (AATCC 100-2001) methods using test organisms for *E. coli* and *S. aureus*. Results suggested that FRF is a valuable ingredient for the development of antimicrobial textiles.

***Keywords**: Seabuckthorn, Nomex IIIA, Antimicrobial activity and Flavonoids.*

INTRODUCTION

Hippophae rhamnoides commonly known as seabuckthorn (SBT), a wild shrub of family Elaeagnaceae, is being used in different parts of the world for its medicinal and nutritional properties (Gupta *et al.*, 2006). It is a dwarf to tall (3-4 m in height), branched and thorny nitrogen fixing deciduous plant, native to Europe and Asia. A wide spectrum of pharmacological effects of SBT have been recently reported including antioxidant, immunomodulatory, antiatherogenic, anti-stress, hepatoprotective and tissue repair (Suleyman *et al.*, 2001; Jayashankar *et al.*, 2012). All parts of the plant are considered to be a rich source of biologically active substances like flavonoids, carotenoids, vitamins (C, E and K), tannins, glycerides of palmitic, stearic and oleic acids and some essential amino acids (Beveridge *et al.*, 1999) which contribute to its wide usage as a natural antioxidant and antimicrobial agent (Nitin *et al.*, 2010; Yogendra *et al.*, 2011; Yogendra *et al.*, 2013).

Functional aspects such as antibacterial and UV protection are playing important role in the development of fabrics (Weijun, Y., 2006; Ramachandran *et al.*, 2004). Because, of their large surface area and ability to retain moisture, textile goods provide an excellent environment for the growth of the microorganisms. Antimicrobial finishing of textile materials aims to

Seabuckthorn (Hippophae L.): A Multipurpose Wonder Plant, Vol. 5: Advanced Technologies for Cultivation, Processing, Health Protection and Environmental Conservation (Virendra Singh, *Editor-in-Chief*, 2018), p. 287–293, Astral International Pvt. Ltd., New Delhi, India.

protect the users from pathogenic microorganisms or from unpleasant odour generators and to ensure a proper functionality of textiles by preventing rotting or aesthetic changes (Simoncic and Tomsic, 2010). An ideal antimicrobial treatment should provide protection against a wide range of bacterial and fungal species, it should be durable to washing, dry cleaning and hot pressing, it should be compatible with textile chemical finishing processes such as dyeing and it should be cost effective (Cristina *et al.*, 2012). A number of chemicals such as, organometallics, iodophors (substances that slowly release iodine), phenols and thiophenols, metallic salts, antibiotics, heterocyclics with anionic groups, nitro compounds, ureas and related compounds, formaldehyde derivatives and amines (Vigo, 1983; Ghoranneviss and Shahidi, 2012) have been employed to impart antimicrobial activity to textile goods. Many of these chemicals, however, are toxic to humans and do not easily degrade in the environment. In recent years, the textile industry looks for eco-friendly natural antimicrobial agents who do not cause toxicity, allergy or irritation to user, does not adversely affect the quality of the textile material and does not produce substances harmful to humans and to the environment (Gao and Cranston, 2008) as substitute for toxic textile chemicals. The herbal antimicrobial finishes overcome the disadvantages of the chemical finishes. Because, they are eco-friendly, non toxic and also non allergic (Wen-Li, 2009; Sun and Worley, 2005). In this sense, some studies have reported the interest of herbs, commonly employed as antimicrobial coatings on the fabrics, since they contain a wide variety of compounds that can have beneficial antimicrobial activity (Rajendran *et al.*, 2012; Rajendran *et al.*, 2013; Yogendra *et al.*, 2014). Although SBT has been known to possess a numerous beneficial properties, its application on textile for antibacterial property enhancement is not reported.

The present investigation primarily aims at developing an eco-friendly natural antimicrobial finish from SBT leaf for textile application. Enriched fraction of SBT leaf extract was applied to Nomex IIIA fabrics and screened for their antimicrobial activity. The coated fabric was characterized by Fourier Transform Infrared Spectroscopy (FTIR). An extensive study was conducted to assess the antimicrobial effectiveness of Enriched fraction of SBT leaf extract by employing standard test methods.

MATERIALS AND METHODS

Fabric Sample

Nomex IIIA (NIIIA) fabric for treatment was purchased from DuPont US. Specifications of fabric are (Blend of 93 per cent Nomex, 5 per cent kevlar, 2 per cent nylon; Count: Warp – 2/40s, weft- 2/40s).

Collection of Plant Material

The SBT leaves were collected from the hilly regions of the North-West Himalayas (the region lies between latitude 32-36° North and longitude 76-79°) in the month of September, where the plant grows widely under natural conditions. Plant material was characterized by an ethno botanist at the Defence Institute of High Altitude Research (DIHAR), Leh, India.

Preparation of Seabuckthorn Leaves Crude Extract

Fresh leaves were cleaned thoroughly with ultrapure water, dried under shade in a clean and dust-free environment. Maceration method was used to prepare the 70 per cent ethanol extract of SBT leaves by soaking the powdered dry leaves with 70 per cent ethanol (1:5 w/v) at room temperature (25 ± 1 °C). After 24 hrs, the supernatant was decanted and the residue was re-soaked in fresh solvent. The process was repeated three times for complete extraction. The supernatant was pooled, filtered through muslin cloth, and centrifuged at 5000*g* (Remi PR 24) for 10 min at 4 °C. Ethanol content of the hydroalcoholic extract was evaporated using rotary evaporator (Buchi R-124 Labortechnik AG, Postfach, CH-9230, Flawil, Switzerland) at 40° C. Finally, both the supernatant solution was lyophilized (Lyophilizer, HITOSICC, Heto-Holten A/S, Denmark) and the dried extract was stored in airtight dark bottle at 4 °C.

Preparation of Flavonoid Rich Fraction (FRF)

0.5 g of obtained crude extract was dissolved in 500 ml of 2N HCl: Methanol (1:1) in a 1000 ml round bottom flask and refluxed at 80°C for 1 hour and 30 minutes. Cooled to the room temperature and removed the methanol content on a rotary evaporator. Then extracted the aglycone part several times using ethyl acetate and finally ethyl acetate content was removed by rotary evaporator to obtain FRF (Markham, 1998).

Determination of Total Flavonoid Content

One ml aliquot of appropriately diluted sample solution was mixed with 2 mL of distilled water and subsequently with 0.15 mL of a 5 per cent $NaNO_2$ solution. After 6 min, 0.15 mL of 10 per cent $AlCl_3$ solution was added and allowed to stand for 6 min, then 2 mL of 4 per cent NaOH solution was added to the mixture. Then the mixture was thoroughly mixed and allowed to stand for another 15 min. Absorbance of the mixture was measured at 510 nm versus prepared blank. Rutin was used as standard compound for the quantification of total flavonoid. All values were expressed as milligram of rutin equivalents per gram of sample (Yanping *et al.*, 2004).

Pre-treatment of Fabrics

Fabrics (size: 30 cm x 30 cm) were pre-treated at 80°C for 60 minutes with 70 per cent triethylene tetramine at the liquid ratio of 15:1. After treatment, the samples were washed in water at 100°C for 30 minutes, dried in the oven at 80°C for 30 minutes and conditioned at 25 at 65°C per cent RH for 60 minutes in the environment chamber (Chyung-Chyung *et al.*, 2003).

Assessment of Hydrophilicity

The hydrophilic behaviour of the treated and untreated fabrics was analysed using wicking test (BS 4554). Fabric strip of 0.5 cm X 8 cm dimension was suspended vertically with its lower edge having contact with distilled water. A spontaneous rise in the height of water in the fabrics due to capillary force was observed. The height in the rise of water for a given time is taken as the direct indication of wicking rate of the fabrics. The rise of water level in the fabric was measured at 20, 40, 60, 120, 180 and 240 seconds. The wicking height measured at 240 seconds was considered for the assessment of hydrophilicity of the fabrics (Ferrero, 2003).

Coating onto Fabrics

SBT extract was applied on the fabric to a wet pick up of 80 per cent using the 'pad-dry-cure' method. The NIIIA fabric (size: 30 cm x 30 cm) was immersed in the solution containing FRF (1 per cent W/V) for 5 min and then it was passed through a laboratory padding mangle (RGE make), which was running at a speed of 20 rpm with a pressure of 1.5 kgf cm^{-2} to remove excess solution. After padding, the fabric was air-dried and then cured for 5 min at 70°C. The treated samples were then evaluated for antimicrobial activity.

Bacterial Strains

Escherichia coli MTCC 25922 as a representative of Gram-negative bacteria and *Staphylococcus aureus* MTCC 25293 as a representative of Gram-positive bacteria were used for this study as both are reference strains used for antimicrobial susceptibility testing according to AATCC standard method. The strains were cultured on nutrient agar (Hi-Media, Mumbai, India) and incubated aerobically at 37°C overnight.

FTIR Characterization

The functional group and absorbency of the SBT-coated fabric was determined using Universal attenuated total internal reflection Fourier Transform Infrared (UATR-FTIR) Spectrophotometer. The spectra of the NIIIA and SBT-coated NIIIA fabric samples was recorded using Perkin Elmer (Spectrum 100) spectrophotometer in the range of 4000-650 cm^{-1}. UTAR consists of Zn-Se single crystal that collects the spectra from the fabrics of dimensions 10mmX10mm placed on to it. Pressure was applied to the samples to ensure the good contact between the sample and crystal to prevent the loss of incident IR radiation. The spectra were obtained for each sample at a resolution of 4cm^{-1} with 32 scans.

45° Flammability Test (ASTM D1230)

Dried NIIIA fabric sample of about 2 x 6 inches of fabric surface is inserted in a Flammability tester at an angle of 45°. A standardized flame (16-18 mm) as per ASTM standard [ASTM – D1230-94] is applied to the surface near the lower end of the fabric for 12 s, and the distance of flame to proceed up the fabric after 12 s is recorded.

Qualitative and Quantitative Antimicrobial Activity Assessment

Antibacterial activity was evaluated by both qualitative and quantitative test methods. The following are the descriptions of test methods employed for this study. Qualitative assessment was made by agar diffusion method (AATCC 147-2001). Treated and untreated control fabric samples placed in intimate contact with AATCC bacteriostasis agar, which has been previously inoculated (Mat culture) with inoculums of test organisms. After incubation, a clear area of uninterrupted growth underneath and along the side of the test material indicates the antibacterial effectiveness of the fabric.

Quantitative assessment was done by percentage reduction test (AATCC 100). Specimens of the test material were shaken in a known concentration of bacterial suspension and the reduction in bacterial activity in standard time was measured. The efficiency of the antimicrobial treatment is determined by comparing the reduction in bacterial concentration of the treated sample with that of control sample expressed as a percentage reduction in standard time.

Per cent Reduction = A-B/A x 100

Where A and B are the surviving cells (CFU/ml) for the flasks containing the control (untreated NIIIA fabric) and test samples (SBT-treated NIIIA fabric), respectively, after 18 hr of contact time.

RESULTS AND DISCUSSION

Extraction of SBT, Determination of Total Flavonoid Content and Application on to NIIIA Fabrics

In our previous study, relatively higher flavonoid compounds were observed from 70 per cent ethanolic extract in comparison with water extract (Upadhyaya *et al.*, 2010). We therefore used 70 per cent ethanol as our extraction solvent and obtained a 23.27 per cent yield of extracted material from SBT leaves. From this crude extract (SLE), flavonoid-rich fraction (FRF) was prepared by acid hydrolysis process and obtained 19.80 per cent of yield. Total flavonoid content estimated as rutin equivalent was found to be 116.98 mg/g of extract in SLE and 277.14 mg/g of fraction in FRF, indicating that total flavonoid content is higher in FRF.

A simple and gradient elution-based reverse phase high performance liquid chromatography (RP-HPLC) method was developed for the analysis and quantification of its major constituent's myrcetin,

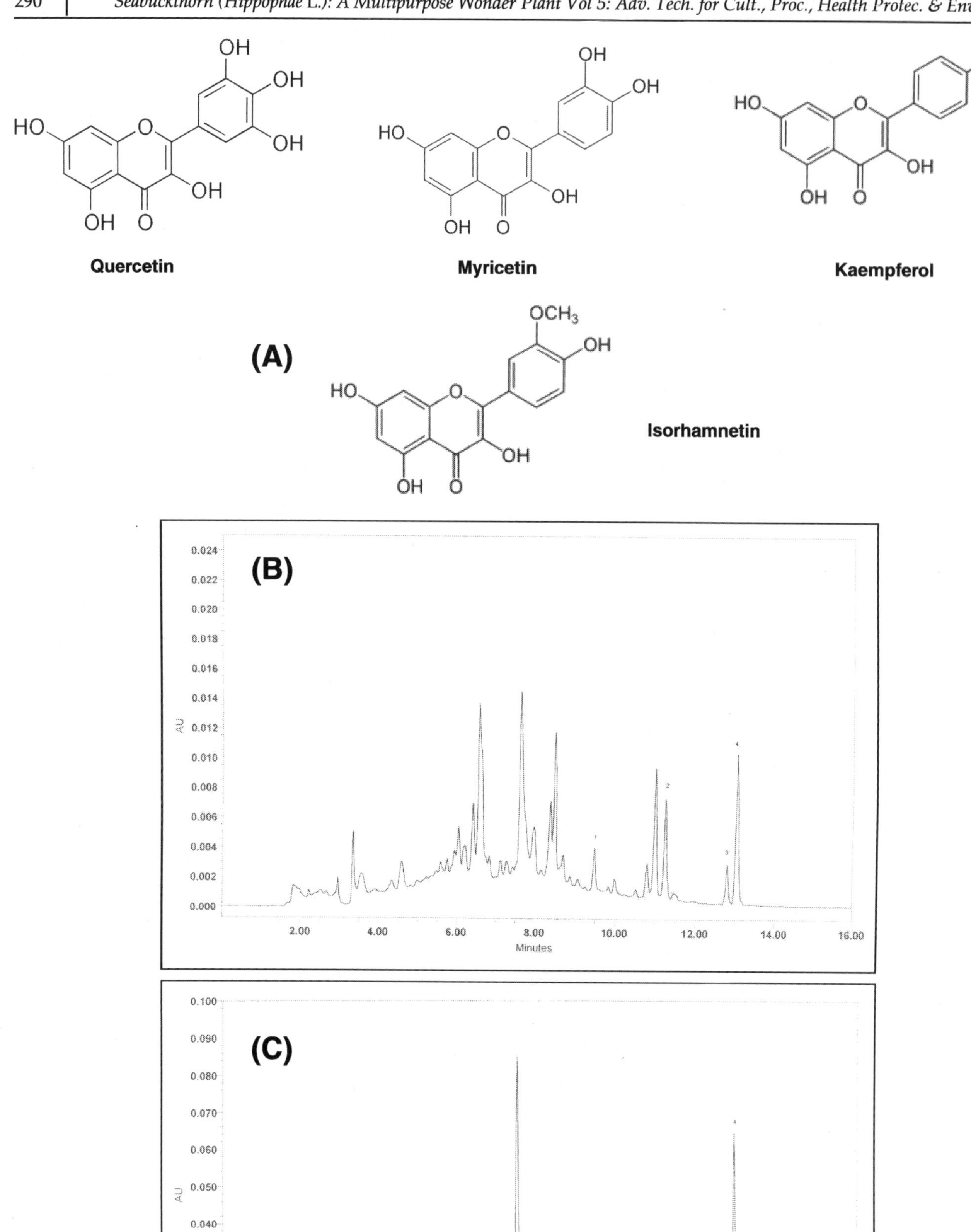

Figure 36.1: Structure of the Quantified Flavonoid Compounds (A), HPLC Chromatogram of SLE (B) and HPLC Chromatogram of FRF of Seabuckthorn Leaves (C).

Table 36.1: Quantitative Determination of Marker Compounds by RP-HPLC Analysis

Sample	*Myricetin*	*Quercetin*	*Kaempferol*	*Isorhamnetin*
	mg (±SD)/g of extract or fraction			
Seabuckthorn leaves extract (SLE)	0.198±0.015	0.257 ±0.005	0.201±0.003	0.857±0.013
Flavonoid rich fraction (FRF)	0.606±0.016	4.632±0.154	12.072±0.271	25.888±8.18

Note: Values expressed as mean ± Standard deviation of three determinations.

quercetin, kaempferol and isorhamnetin. For the development of an effective mobile phase, various solvent systems, including different combinations of acetonitrile, methanol and water with ortho phosphoric acid were tried. Finally, a solvent system consisting of 0.3 per cent ortho phosphoric acid in water and acetonitrile : methanol (75:25) was proved to be successful as it allows the separation of maximum compounds with good resolution. Myricetin, quercetin, kaempferol and isorhamnetin (Figure 36.1A) that might contribute to the antimicrobial behavior were identified in SLE and FRF, as shown in Figures 36.1B and 36.1C. Quantitation was carried out by integration of the peak using an external standard method, and results are presented in mg/g of the FRF or SLE, for three replicate injections. The results indicate that the FRF was found to be rich in myricetin, quercetin, kaempferol and isorhamnetin contents in comparison to SLE (Table 36.1).

Flavonoids form a class of benzo-γ-pyrone derivatives includes flavones, flavanes, flavonols, anthocyanidines, and catechins. They possess a wide spectrum of biological activities such as anticancer, antibacterial, antifungal, antiviral, spasmolytic, hypoglycaemic, antihistaminic and radioprotective potential (Chawla *et al.*, 2007; Markham, 1998). Some of these properties derive from the free radical-scavenging activities of flavonoids. Therefore there are many reports relating to the reactivities of flavonoids with active oxygen species. Recent interest in these substances has been stimulated by the potential health benefits arising from their antioxidant and antimicrobial activities.

Effect of Amine Treatment on the Hydrophilicity of the Fabric

The effect of amine treatment on the hydrophilicity of the fabrics was determined by measuring the wicking height at different interval of time (20 to 240 Sec). Graph is plotted using wicking height and square root of time as shown in the Figure 36.2. It is clear from the figure that, wicking height of the treated fabric considerably improved in comparison with the untreated fabric. This indicates that the amine treatment enhanced the formation of polar species on the surface of the fabric due to etching making it hydrophilic. Similar results

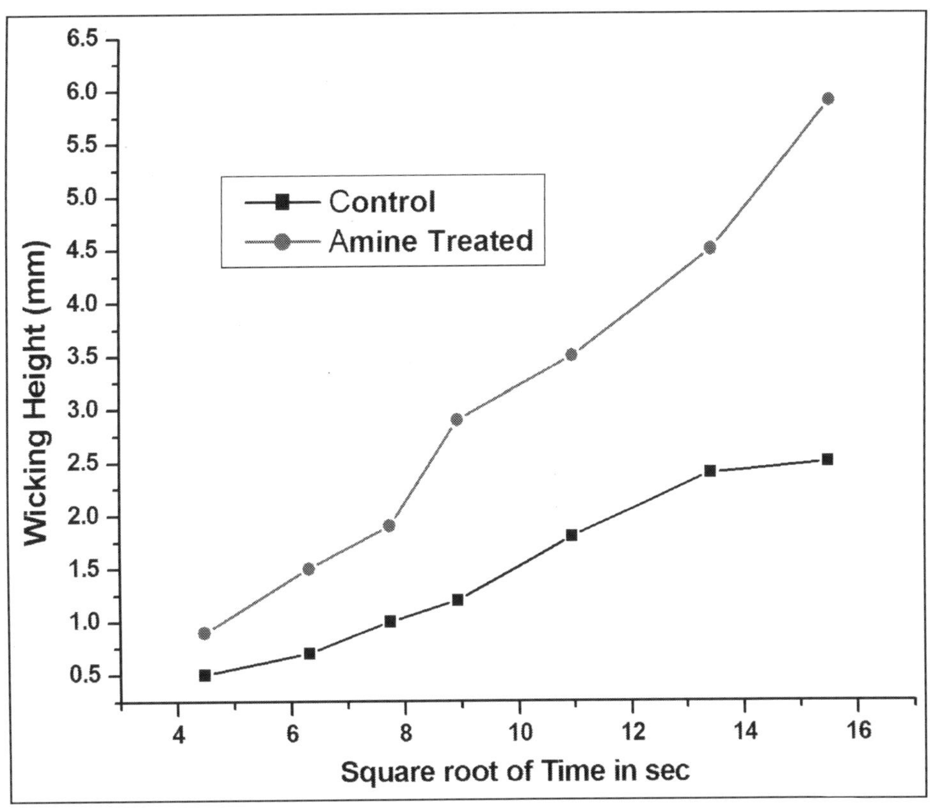

Figure 36.2: Wicking Rate of Treated and Untreated Fabrics.

were reported the effect of amine treatment for Nomex yarns for enhanced the wicking height and dye uptake (Chyung-Chyung and Cheng-Chi, 2003).

FTIR Characterization

UATR FTIR, a surface sensitive technique was used to study the chemical modification on the surface of the treated fabrics. The FRF-treated NIIIA fabric has characteristic peak at wavelength of 3337 cm^{-1} for hydroxyl and at 1728 cm^{-1} for carbonyl group. These characteristic peaks proved the coating of the flavonoids onto the NIIIA fabric as the control NIIIA fabric does not show such characteristic peak. Further, there is much decreased transmittance at 1728 cm^{-1} control NIIIA fabric. The prominence of this feature may be ascribed to the relatively high content of carboxylic acid and hydroxyl containing compounds in FRF.

Similar results were reported on the plant extract containing flavonoids showed the characteristic hydroxyl and carbonyl groups at 3307 cm^{-1} and 1641 cm^{-1} respectively (Jain and Mittal, 2012). In another study also reported the plant extract containing rich flavonoid content shown the bands at 3351 cm^{-1} and 1714 cm^{-1} (Rajeev *et al.*, 2012). Whereas the fabric without FRF coating does not have these characteristic peaks. Further, it was confirmed that peaks at ~3500 cm^{-1}, correspond to –OH stretching and peaks at 1740 cm^{-1}, correspond to C=O groups in another reported study (Chandana and Sanjeeb, 2010).

Flammability Test

This test method used to measure and describe the properties of fabrics in response to heat and flame under controlled laboratory conditions. Char length of treated and control samples was estimated using flammability tester, was found no significant differences in the area of char produced by both the samples. It indicates that the treatment has no effect on the flammability parameters of the fabric.

Qualitative and Quantitative Antimicrobial Activity Assessment

The results of agar diffusion method against the standard test organisms *S. aureus* (Gram positive) and *E. coli* (Gram negative). There was a clear zone of inhibition around the fabric tested with FRF against both the test organisms in contrast with control fabric which allowed the growth of organism. The agar diffusion showed a zone of inhibition which indicated that the antimicrobial activity of FRF-treated sample exhibited a zone of 15.4 ±1.3 mm for *S. aureus* and 18.3 ±1.9 mm inhibition for *E. coli*.

The reduction percentage for E. *coli* and *S. aureus* correspond to the bacterial numbers on the respective control test of 5.52 x 10^6 per millilitre. The reduction percentage was found to be 94.6 per cent for *S. aureus* and 91.2 per cent for E. *coli*.

CONCLUSION

In this work, we extracted bioactive flavonoid rich fraction from SBT leaves by acid hydrolysis process. We focused on studying the antimicrobial property of NIIIA fabric coated with FRF. The FRF-coated fabric showed a high degree of bactericidal activity against test organisms *E. coli* and *S. aereus* used in this study. In the FTIR spectra, it showed distinct hydroxyl and carbonyl peaks. In future, we are interested in increasing the durability of the FRF coated NIIIA fabric using various cross-linking agents, so that the antimicrobial property of SBT is retained in the fabric even after few washes and thus enhancing durability. We are also interested in studying the fungicidal activity of FRF. By this way, we desire to develop an eco-friendly antimicrobial aramid textile with a novel natural antimicrobial agent.

REFERENCES

1. ASTM–D1230-94. Standard test method for Flammability of Apparel Textiles. PA 19428-2959, US.
2. Beveridge, T., Li, T.S., Oomah, B.D. and Smith, A. 1999. Seabuckthorn products: Manufacture and composition. *J. Agric Food Chem.* 47 (9): 3480-8.
3. Chandana, M. and Sanjeeb, K. S. 2010. The *in vitro* stability and *in vivo* pharmacokinetics of curcumin prepared as an aqueous nanoparticulate formulation. *Biomaterials* 31: 6597-6611.
4. Chawla, R., Arora, R., Singh, S., Sagar, R. K., Sharma, R. K., Kumar, R., Sharma, A., Gupta, M. L., Singh, S., Prasad, J., Khan, H. A., Swaroop, A., Sinha, A. K., Gupta, A. K., Tripathi, R. P. and Ahuja, P. S. 2007. Radioprotective and antioxidant activity of fractionated extracts of berries of *Hippophae rhamnoides*. *Journal of Medicinal Food* 10: 101-109.
5. Cristina, R., Cogeanu, A.M., Diaconescu, R.M. and Grigoriu, A. 2012. Antimicrobial treatments of hemp fibers grafted with cyclodextrin derivatives. *Textile Research Journal* 82 (13): 1317-1328.
6. Ferrero, F. 2003. Wettability measurements on plasma treated synthetic fabrics by capillary rise method. *Polymer Testing* 22 (5): 571-578.
7. Gao, Y. and Cranston, R. 2008. Recent advances in antimicrobial treatments of textile. *Textile Research Journal* 78: 60-72.
8. Ghoranneviss, M. and Shahidi, S. 2012. Effect of various metallic salts on antibacterial activity and physical properties of cotton fabrics. *Journal of Industrial Textiles* 42 (3): 193-203.
9. Gupta, A., Kumar, R., Pal, K., Singh, V., Banerjee, P. K. and Sawhney, R.C. 2006. Influence of seabuckthorn (*Hippophae rhamnoides* L.) flavones on dermal wound healing in rats. *Mol Cell Biochem.* 290 (1-2): 193-8.

10. Jain, R. and Mittal, M. 2012. Naringenin: A flavonone from the stem of *Nyctanthes arbortristis* L. *International Journal of Biology, Pharmacy and Allied Sciences* 1 (7): 964-972.

11. Kumar, M. S. Y., Raghu, T. S., Varghese, F. V., Indushekhar, R., Kotresh, T. M., Rajendran, R. K., Babu, M. and Padaki, V. C. 2014. Application of seabuckthorn (*Hippophae rhamnoides* L.) leaf extract as antimicrobial finish on aramid fabric. *Journal of Industrial Textiles* DOI: 10. 1177/15280837Ð4553689.

12. Markham, K.R. 1998. Techniques of flavonoid identification, Academic Press.

13. Nitin, K.U., Kumar, M. S. Y. and Gupta, A. 2010. Antioxidant, cytoprotective and antibacterial effects of seabuckthorn (*Hippophae rhamnoides* L.) leaves. *Food and Chemical Toxicology* 48 (12): 3443-3448.

14. Nema, R., Jain, P., Kare, S., Pradhan, A., Gupta, A. and Singh, D. 2012. Preliminary phytochemical evaluation and flavanoids quantification of *Terminalia arjuna* leaves extract. *International Journal of Pharmaceutical and Phytopharmacological Research* 1 (5): 283-286.

15. Rajendran, R., Radhai, R., Balakumar, C., Hasabo, A., Ahamed, M., Vigneswaran, C. and Vaideki, K. 2012. Synthesis and characterization of Neem chitosan nanocomposites for development of antimicrobial cotton textiles. *Journal of Engineered Fibres and Fabrics* 7 (1): 137-141.

16. Rajendran, R., Radhai, R., Kotresh, T.M. and Emilia, C. 2013. Development of antimicrobial cotton fabrics using herb loaded nanoparticles. *Carbohydrate Polymers* 91: 613-617.

17. Ramachandran, T., Rajendrakumar, K. and Rajendran, R. 2004. Antimicrobial textiles-An overview. *Internet Explorer (India) Journal-Textile* 84 (2): 42-47.

18. Simoncic, B. and Tomsic, B. 2010. Structures of novel antimicrobial agents for textiles-A review. *Textile Research Journal* 80: 1721-1737.

19. Suleyman, H., Demirezer, L.O., Buyukokuroglu, M.E., Akcay, M.F., Gepdiremen, A. and Banoglu, Z.N. 2001. Antiulcerogenic effect of *Hippophae rhamnoides* L. *Phytother Res.* 15 (7): 625-627.

20. Sun, G. and Worley, S. D. 2005. Chemistry of Durable and degenerable biocidal textiles. *Journal of Chemical Education* 82 (1): 60-64.

21. Vigo, T. L. 1983. Protection of textiles from biological attack. In: *Chemical Processing of Fibers and Fabrics, Functional Finishes*, M. Lewin and S. B. Sello (Eds.), Hand book of fiber science and technology: Vol. 2. Part A pp. 367-426. New York, Marcel Dekker.

22. Wang, C.C. and Chen, C. C. 2003. Physical properties of Nomex yarns pretreated with various amines. *Textile Research Journal* 73 (5): 395-400.

23. Weijun, Y., John, H. X., Pei, Li., Kam-Len, Daniel, L. and Tsz-Leung, Kwong. 2006. Durable antibacterial finish on cotton fabric by using chitosan-based polymeric core-shell particles. *Journal of Applied Polymer Science* 102 (2): 1787-1793.

24. Wen-Li, D., Shan-Shan, N., Ying-Lei, X., Zi-Rong, X. and Cheng-Li, F. 2009. Antibacterial activity of chitosan tripolyphosphate nanoparticles loaded with various metal ions. *Carbohydrate Polymers* 75 (3): 385-389.

25. Yanping, Z., Yanhua, L. and Dongzhi, W. 2004. Antioxidant activity of a flavonoid-Rich extract of *Hypericum perforatum* L. *in vitro*. *Journal of Agricultural Food Chemistry* 52: 5032-5039.

26. Yogendra Kumar, M.S., Dutta, R., Prasad, D. and Misra, K. 2011. Subcritical water extraction of antioxidant compounds from seabuckthorn (*Hippophae rhamnoides*) leaves for the comparative evaluation of antioxidant activity. *Food Chemistry* 127: 1309-1316.

27. Yogendra Kumar, M.S., Tirpude, R. J., Maheshwari, D. T., Bansal, A. and Misra, K. 2013. Antioxidant and antimicrobial properties of phenolic rich fraction of seabuckthorn (*Hippophae rhamnoides* L.) leaves *in vitro*, *Food Chemistry* 141 (4): 3443-3450.

Chapter 37

Seabuckthorn Oil has Beneficial Effects on Mucous Membranes of Intimate Area in Oral and Topical Applications

Petra Larmo[1]*, Baoru Yang[2], Heikki Kallio[2] and Risto Erkkola[3]

[1]Aromtech Oy, Veturitallintie 1, 95410 Tornio, Finland
[2]Food Chemistry and Food Development, Department of Biochemistry University of Turku, Finland
[3]Department of Obstetrics and Gynecology, University Central Hospital of Turku, Finland
(*E-mail: petra.larmo@aromtech.com)

SUMMARY

Reduction of estrogen during menopause is associated with urogenital atrophy, the drying and thinning of mucous membranes of intimate area. Atrophic tissue is prone to inflammation and commonly associated with symptoms of dryness and discomfort. Also younger women can be affected during intake of birth control pills, breastfeeding or as a consequence of a medical condition. Here three clinical studies investigating the effects of seabuckthorn oil on urogenital atrophy and dryness are reviewed. Intake of seabuckthorn oil induced beneficial effects on genital dryness and associated symptoms in a double-blind, randomized placebo-controlled study with Sjögren's syndrome patients. A recent double-blind, randomized placebo-controlled study with postmenopausal women indicates that intake of seabuckthorn oil improves the integrity of vaginal mucous membranes. Seabuckthorn oil applied topically in a cream product improved the clinical signs of vulvar mucous membrane and decreased the dryness and other symptoms experienced by healthy women.

Keywords: *Seabuckthorn oil, Vaginal atrophy, Dryness and Mucous membranes.*

INTRODUCTION

During menopause the estrogen secretion from the ovaries decreases and the number of estrogen receptors in the vaginal mucosa falls. These changes affect the amount of acid mucopolysaccharides, hyaluronic acid, collagen and elastic fibres in the urogenital area, and consequently the vulva and vaginal wall thickness, moisture, pH and integrity of the tissue. The collection of signs and symptoms associated with the decrease of estrogen is called vulvovaginal/urogenital atrophy. These typically include genital symptoms of dryness, burning, irritation and susceptibility to infections and inflammation (Palacios, 2009; Castelo-Branco *et al.*, 2005; Neves-E-Castro *et al.*, 2015). Up to more than 40 per cent of postmenopausal women in Western countries are reported to experience symptoms of vaginal atrophy (Nappi and Kokot-Kierepa, 2010). The standard treatment, hormone replacement therapy, is not suitable for all women. Specifically for women with breast cancer or history of venous thromboembolism non-hormonal alternatives are recommended instead (Neves-E-Castro *et al.*, 2015).

Urogenital dryness does not affect women at their menopause only. Among younger women the symptoms can be experienced for example during breastfeeding or use of medication decreasing the estrogen levels (Castelo-Branco *et al.*, 2005). Patients with Sjögren's syndrome, an autoimmune disease associated with damage to exocrine glands, typically experience symptoms of urogenital dryness (Bayetto and Logan, 2010).

Seabuckthorn (Hippophae L.): *A Multipurpose Wonder Plant, Vol. 5: Advanced Technologies for Cultivation, Processing, Health Protection and Environmental Conservation* (Virendra Singh, *Editor-in-Chief*, 2018), p. 295–299, Astral International Pvt. Ltd., New Delhi, India.

Seabuckthorn (*Hippophaë rhamnoides*) oil has traditionally been used for the treatment of inflammations in the genital organs and uterus (Suryakumar and Gupta, 2011). Clinical studies indicate that intake of a standardized combination of seabuckthorn berry and seed oil has beneficial effects on dry eye (Larmo *et al.*, 2010), skin hydration and elasticity (Yang *et al.*, 2009), markers associated with cardiovascular health (Larmo *et al.*, 2013) and inflammation (Lehtonen *et al.*, 2011). In animals, combination of topical application and oral intake of seabuckthorn seed oil has been shown to improve wound healing (Upadhyay *et al.*, 2009).

This article reviews three clinical studies investigating the effects of seabuckthorn oil on genital mucous membranes. Two of the studies concern the effects of oral intake and have been published previously (Larmo *et al.*, 2014; Yang and Erkkola, 2008). One of the studies concerns the effects of topical use in a cream product and has not been reported before.

MATERIALS AND METHODS

Vaginal Atrophy among Postmenopausal Women, Oral Intake

The effects of seabuckthorn oil on vaginal atrophy were investigated in a double-blind, randomized, placebo-controlled, parallel study among postmenopausal women at the Gynecological Center of Turku (Turku, Finland) (Larmo *et al.*, 2014). The 116 participants were 55 to 75 years of age and experienced typical symptoms of vaginal atrophy. During the intervention period of three months they consumed 3 g of seabuckthorn oil or placebo daily as capsules with identical outlook. The seabuckthorn oil used in the study was a standardized combination of seabuckthorn berry and seed oils produced by supercritical carbon dioxide extraction (SBA24 Seabuckthorn Oil, Aromtech Ltd, Tornio, Finland). The placebo oil was fatty acids of medium chain length as triacylglycerols from palm and coconut oils. At the beginning and end of the study the condition of the vaginal mucosa was evaluated and scored by the study gynecologist, and the pH and moisture were measured. Vaginal health index was calculated according to Bachmann *et al.* (1991). Symptoms experienced by the participants were evaluated with symptom questionnaires at study visits and by a phone interview at half-way of the study. During the intervention period the participants also kept a daily logbook concerning their symptoms and the intake of study capsules. From a sample of 30 participants maturation index was analyzed from Pap smear test. Details of the study have been described previously (Larmo *et al.*, 2014).

Women with Sjögren's Syndrome, Oral Intake

The effects of seabuckthorn oil on symptoms of Sjögren's syndrome were studied in double-blind, randomized, placebo-controlled cross-over study of 25 women from 37 to 66 years of age (Yang and Erkkola, 2008). The seabuckthorn oil used in the study was the same standardized SBA24 composition (Aromtech Ltd, Tornio, Finland) that was used in the study with the postmenopausal women (Larmo *et al.*, 2014). Each intervention lasted for three months, during which 3 g of oil was daily consumed. At the beginning and end of interventions the participants reported the severity of their general symptoms of Sjögren's syndrome and the symptoms of genital mucous membranes using a visual analogical scale. Details of the study have been described previously (Yang and Erkkola, 2008).

Dryness of Intimate Area, Topical Application

Effects of seabuckthorn oil on mucous membranes of intimate area in topical use were investigated in an open study of one group at Dermscan Poland during spring 2014. The participants were 22 healthy women from 21 to 65 years of age (mean 49; SD 3 y) experiencing dryness of intimate area. During a study period of one month the participants applied SBA24 Seabuckthorn Oil Cream (Aromtech Ltd, Tornio, Finland) twice per day on their external genital mucous area. The cream contained 3 per cent of seabuckthorn oils produced with supercritical carbon dioxide extraction (A standardized combination of seabuckthorn seed and berry oil SBA24 Seabuckthorn Oil and SHAJIO Seabuckthorn Oil, Aromtech Ltd, Tornio, Finland). At the beginning and end of the study the condition of the mucous membranes of the external genital area was evaluated and scored by a gynecologist using a scale from 0 (no sign/symptom) to 5 (very severe sign/symptom), and the pH was measured using a pH paper strip. The participants attended the study visits without applying any product to the intimate area. At the beginning and end of the study, the participants scored their symptoms of the intimate area using a scale from 0 (no symptom) to 5 (very severe symptom). All included participants finished the study with good compliance. The clinical and subjective scores and the vulvar pH at the beginning and at the end of the study were compared with paired samples T-test/Wilcoxon Signed Rank Test. No corrections for multiple tests were used.

RESULTS

Vaginal Atrophy among Postmenopausal Women, Oral Intake

In the seabuckthorn oil group compared to placebo, there was a significantly ($P<0.05$) better rate of improvement in the integrity of the vaginal epithelium scored by the study gynecologist (Figure 37.1) (Larmo *et al.*, 2014). Seabuckthorn oil induced a beneficial increasing (mean change 0.8, SD 2.8) trend in the Vaginal health index, whereas a small decrease (mean change -0.1, SD 2.0) was observed in the placebo

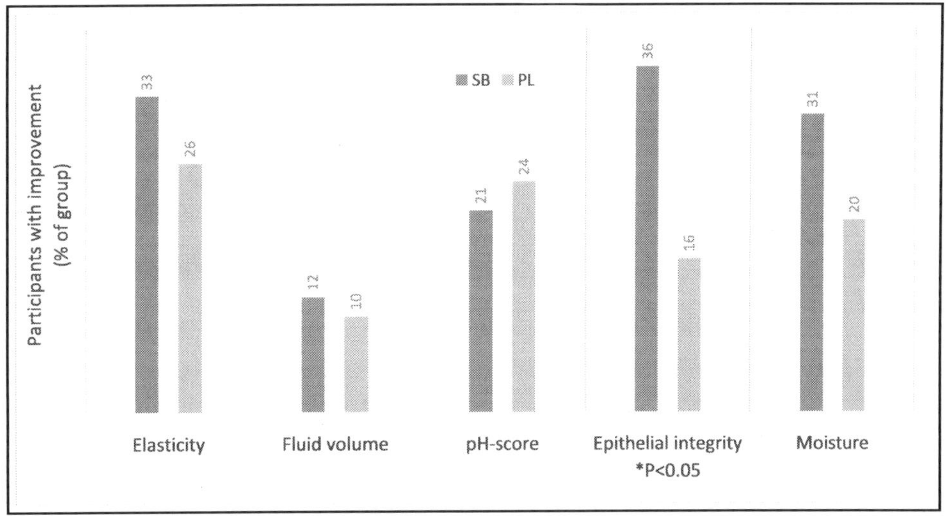

Figure 37.1: Effects of SBA24 Seabuckthorn Oil (SB) and Placebo (PL) Oil Intake for Three Months on the Components of Vaginal Health.

group. The difference between the groups, however, was not statistically significant (P=0.08). According to the questionnaires at study visits, the symptoms of vaginal dryness, burning, and itching improved in more than 50 per cent of participants in both groups and the was no difference between the groups in the experience of symptoms (P>0.05). The other parameters measured at study visits did not differ significantly between the seabuckthorn oil and placebo groups.

Women with Sjögren's Syndrome, Oral Intake

A significantly greater proportion of participants in the seabuckthorn oil group compared to placebo (P<0.05) reported improvement in their symptoms of genital mucous membranes (itching, burning, pain, secretion, dryness) during the intervention (Yang and Erkkola, 2008). The improvement rate in the overall symptoms, including also other typical symptoms of Sjögren's syndrome, was better in the seabuckthorn oil during the first phase of the study (P<0.05), but not when the whole study period was analyzed (P>0.05).

Dryness of Intimate Area, Topical Application

From beginning to the end of the study there was a statistically significant (P<0.05) reduction in the score for dryness of the mucous membranes of the external genital mucosa evaluated by the study gynecologist. Also the scores for integrity of the mucosa, erythema, edema and inflammation improved from baseline to the end of the intervention (P<0.05) (Figure 37.2). pH of the vulvar mucous membrane decreased from mean 5.2 (SD 0.8) to 4.6 (SD 0.6) (P<0.05) indicating a beneficial pH balancing effect by the cream. Symptoms of stinging, itching, burning, pain, discomfort and dryness of vulvar mucous membranes experienced by the participants were less severe at the end of the study compared to the baseline (P<0.05).

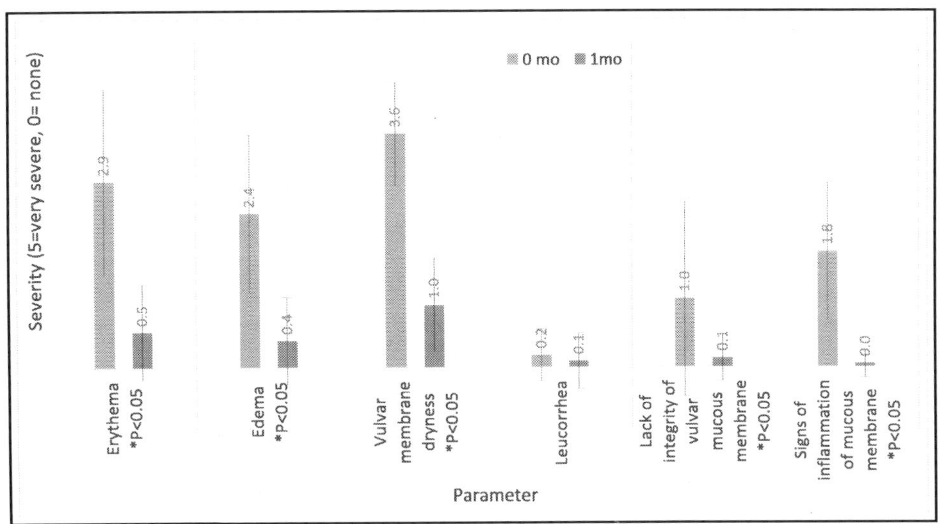

Figure 37.2: Clinical Scores of External Genital Mucous Membranes of Participants at Baseline (0 mo) and after Application of SBA24 Seabuckthorn Oil Cream for One Month (1 mo).

DISCUSSION

In clinical studies intake of combined seabuckthorn seed and berry oil improved the integrity of vaginal mucous membranes among post-menopausal women and relieved the symptoms of intimate area dryness in women suffering from Sjögren's syndrome. Beneficial effects by topical application of seabuckthorn oil cream on dryness and associated signs and symptoms of intimate area were observed in a study with healthy women.

The effects most likely were mediated by several bioactive compounds in the seabuckthorn oils. Seabuckthorn berry oil in particular is rich in β-carotene, a precursor for vitamin A, essential for normal skin and mucosa-associated epithelia, including that of the genital tract (Christian *et al.*, 2011; Ross, 2006; Biesalski and Nohr, 2004). Seabuckthorn seed oil is rich in the essential α-linolenic and linoleic acid, the intake of which is important for the normal structure, hydration and water permeability barrier of skin (Ziboh *et al.*, 2000; McCusker and Grant-Kels, 2010).

Atrophic vaginal epithelium is associated with inflammation (Palacios, 2009) and modulation of inflammation potentially has contributed to the effects seen in the studies. Both seabuckthorn seed and berry oils are rich in tocopherols, tocotrienols, carotenoids and phytosterols, associated with antioxidative and anti-inflammatory effects (Watzl *et al.*, 2005; Reiter *et al.*, 2007; Ku and Lin, 2013). The relatively high content of α-linolenic acid in the oil can support the production of eicosanoids of low pro-inflammatory activity (Ziboh *et al.*, 2000). Even though seabuckthorn oils contain phytosterols, the lack of effect on vaginal pH and maturation index in the study with postmenopausal women (Larmo *et al.*, 2014) suggests the effects on mucosa are not mediated via estrogen receptors.

Topical use of seabuckthorn oil and/or combination of topical use with oral intake have been shown to improve wound healing and skin epithelisation in animal models. Antioxidant and anti-inflammatory mechanisms are suggested to mediate these effects (Upadhyay *et al.*, 2009; Ito *et al.*, 2014). Compared to a base cream without seabuckthorn oil, topical application of seabuckthorn oil containing cream (3 per cent) significantly ($P<0.05$) improved skin hydration and elasticity measured instrumentally in a study of 30 women, indicating beneficial effects in healthy skin (unpublished study by Aromtech Ltd, Tornio, Finland).

In conclusion, clinical studies indicate that seabuckthorn seed and berry oil has beneficial effects on genital mucous membranes in oral intake and in topical applications. The effects most likely are mediated by joint effect of several mechanisms and bioactive compounds in the oils.

REFERENCES

1. Bachmann, G.A., Notelovitz, M., Gonzales, S., Thomson, C. and Morecraft, B. 1991. Vaginal dryness in menopausal women: Clinical characteristics and nonhormonal treatment. *Clinical Practice in Sexuality* 7: 25-32.
2. Bayetto, K. and Logan. R.M. 2010. Sjögren's syndrome: A review of aetiology, pathogenesis, diagnosis and management. *Aust. Dent. J.* 55: 39-47.
3. Biesalski, H.K. and Nohr, D. 2004. New aspects in vitamin A metabolism: The role of retinyl esters as systemic and local sources for retinol in mucous epithelia. *J. Nutr.* 134: 3453S-7S.
4. Castelo-Branco, C., Cancelo, M.J., Villero, J., Nohales, F. and Julia, M.D. 2005. Management of post-menopausal vaginal atrophy and atrophic vaginitis. *Maturitas* 52: S46-52.
5. Christian, P., Labrique, A.B., Ali, H., Richman, M.J., Wu, L., Rashid, M. and West, K.P. 2011 Maternal vitamin A and beta-carotene supplementation and risk of bacterial vaginosis: A randomized controlled trial in rural Bangladesh. *Am. J. Clin. Nutr.* 94: 1643-9.
6. Ito, H., Asmussen, S., Traber, D.L., Cox, R.A., Hawkins, H.K., Connelly, R., Traber, L.D., Walker, T.W., Malgerud, E., Sakurai, H. and Enkhbaatar, P. 2014 Healing efficacy of seabuckthorn (*Hippophae rhamnoides* L.) seed oil in an ovine burn wound model. *Burns* 40: 511-9.
7. Ku, C.M. and Lin, J.Y. 2013. Anti-inflammatory effects of 27 selected terpenoid compounds tested through modulating Th1/Th2 cytokine secretion profiles using murine primary splenocytes. *Food Chem.* 141: 1104-13.
8. Larmo, P., Jarvinen, R., Setala, N., Yang, B., Viitanen, M., Engblom, J., Tahvonen, R. and Kallio, H. 2010. Oral seabuckthorn oil attenuates tear film osmolarity and symptoms in inviduals with dry eye. *J. Nutr.* 140: 1462-8.
9. Larmo, P.S., Yang, B., Hyssälä, J., Kallio, H.P. and Erkkola, R. 2014. Effects of seabuckthorn oil intake on vaginal atrophy in postmenopausal women: a randomized, double-blind, placebo-controlled study. *Maturitas* 79: 316-21.
10. Larmo, P.S., Kangas, A.J., Soininen, P., Lehtonen, H.M., Suomela, J.P., Yang, B., Viikari, J., Ala-Korpela, M. and Kallio, H.P. 2013. Effects of seabuckthorn and bilberry on serum metabolites differ according to baseline metabolic profiles in overweight women: A randomized crossover trial. *Am. J. Clin. Nutr.* 98: 941-51.
11. Lehtonen, H., Suomela, J., Tahvonen, R., Yang, B., Venojärvi, M., Viikari, J. and Kallio, H. 2011.

Different berries and berry fractions have various but slightly positive effects on the associated variables of metabolic diseases on overweight and obese women. *Eur. J. Clin. Nutr.* 65: 394-401.

12. McCusker, M.M. and Grant-Kels, J.M. 2010. Healing fats of the skin: The structural and immunologic roles of the omega-6 and omega-3 fatty acids. *Clin. Dermatol.* 28: 440-51.
13. Nappi, R.E. and Kokot-Kierepa, M. 2010. Women's voices in the menopause: Results from an international survey on vaginal atrophy. *Maturitas* 67: 233-8.
14. Neves-E-Castro, M., Birkhauser, M., Samsioe, G., Lambrinoudaki, I., Palacios, S., Borrego, R.S., Llaneza, P., Ceausu, I., Depypere, H., Erel, C.T., Perez-Lopez, F.R., Schenck-Gustafsson, K., van der Schouw, Y.T., Simoncini, T., Tremollieres, F. and Rees, M. 2015. EMAS position statement: The ten point guide to the integral management of menopausal health. *Maturitas* 81: 88-92.
15. Palacios, S. 2009. Managing urogenital atrophy. *Maturitas* 63: 315-8.
16. Reiter, E., Jiang, Q. and Christen, S. 2007. Anti-inflammatory properties of alpha- and gamma-tocopherol. *Mol. Aspects Med.* 28: 668-91.
17. Ross, C.A. 2006. Vitamin A and carotenoids. In: *Modern Nutrition in Health and Disease*. 10th Edition (Shils ME, Shike M., Ross C.A., Cabarello B., Cousins R., eds.), pp. 351-75Philadelphia (PA): Lippincott Williams and Wilkins.
18. Suryakumar, G. and Gupta, A. 2011. Medicinal and therapeutic potential of seabuckthorn (*Hippophae rhamnoides* L.). *J. Ethnopharmacol* 138: 268-78.
19. Upadhyay, N.K., Kumar, R., Mandotra, S.K., Meena, R.N., Siddiqui, M.S., Sawhney, R.C. and Gupta, A. 2009. Safety and healing efficacy of seabuckthorn (*Hippophae rhamnoides* L.) seed oil on burn wounds in rats. *Food Chem. Toxicol.* 47: 1146-53.
20. Watzl, B., Kulling, S.E., Moseneder, J., Barth, S.W. and Bub, A. 2005. A 4-wk intervention with high intake of carotenoid-rich vegetables and fruit reduces plasma C-reactive protein in healthy, nonsmoking men. *Am. J. Clin. Nutr.* 82: 1052-8.
21. Yang, B., Bonfigli, A., Pagani, V., Isohanni, T., von Knorring. A., Jutila, A-M. and Judin, V-P. 2009. Effects of oral supplementation and topical application of supercritical CO_2 extracted seabuckthorn oil on skin ageing of female subjects. *J. Appl. Cosmetol.* 27: 1-13.
22. Yang, B. and Erkkola, R. 2008. Seabuckthorn oils, mucous membranes and Sjögren's syndrome with special reference to latest studies. In: *Seabuckthorn (Hippophae L.). A Multipurpose Wonder Plant. Vol. III: Advances in Research and Development* (Singh *et al.*, Eds.), Daya Publishing House, New Delhi, India, pp. 254-267.
23. Ziboh, V.A., Miller, C.C. and Cho, Y. 2000. Metabolism of polyunsaturated fatty acids by skin epidermal enzymes: generation of antiinflammatory and antiproliferative metabolites. *Am. J. Clin. Nutr.* 71: 361S-6S.

Chapter 38

Extraction of Antioxidant Compounds from Seabuckthorn Berries and its Antioxidant Activity *in vitro*

Li Jixin, Bi Yang*, Jiang Yumei, Ding Ruojun, Wei Juan, Sheng Wenjun, Liang Kai, Mi Lan and Zhang Bingjin

College of Food Science and Engineering, Gansu Agricultural University, Lanzhou, China
(*E-mail: asl2919_ljx@163.com)

SUMMARY

The berries of seabuckthorn (*Hippophae rhamnoides* subsp. *sinensis*) are rich in polyphenols, flavonoids and other antioxidant substances, which have been used as a traditional Chinese medicine with efficiency of promoting blood circulation to remove blood stasis, resolving phlegm and relieving depression in the chest, invigorating the spleen and strengthening the stomach, producing body fluid to quench thirst, clearing away heat to stop diarrhea. Seabuckthorn berries were used as the material, with the ferric reducing antioxidant power (FRAP) as measure indicator. Based on single factor investigations, using Box-Behnken and response surface methodology, four extraction variables including ethanol concentration, solid-liquid ratio, extraction time, and extraction temperature were studied and optimized for FRAP value. According to the prediction model, the best extraction parameters of seabuckthorn fruit are as followed: ethanol concentration 51 per cent, solid-liquid ratio 1:22, extraction time 1h, extraction temperature 86¡æ. The FRAP value of extract was 87.840 mmol/L under these conditions. The extracts had activity of scavenging several kinds of free radicals. The EC_{50} values was 93.79 μg/mL by DPPH assay, 4.14 mg/mL by ·OH assay, and 0.856 mg/mL by O_2^{-}· assay. And it was estimated at 2.2, 19.1 and 99.5 per cent times of BHT scavenging these free radicals, respectively. The results suggest that extracts of seabuckthorn berries have better antioxidant ability and its ability of the cleaning ·OH is better than that of BHT.

***Keywords**: Hippophae rhamnoides subsp. sinensis, Extraction, Frap and Antioxidant activity.*

INTRODUCTION

Seabuckthorn (*Hippophae rhamnoides* L.), also known as thorn, acid willow, is a deciduous shrub or small tree, as main trees largely planted in China for conservation of water and soil, widely distributed in the North China, Inner Mongolia, northwest and southwest regions with rich development resources and rapid development and good prospects (Tao, 2003; He *et al.*, 2008). Chinese seabuckthorn (*Hippophae rhamnoides* subsp. *sinensis)* is the most appropriated, mainly planted in northwest and southwest of China, about 85 per cent of total amount of *Hippophae* in China (He *et al.*, 2008). Nutritional components of the fruit is very rich, containing a variety of bioactive components like vitamins, organic acids, fatty acids, amino acids, sugars, polyphenols and flavonoid (Zhou *et al.*, 2005), among which flavonoids, polyphenols and other antioxidants can activate blood to remove blood stasis, reduce phlegm and enlarge chest, help produce saliva and slake thrist, clear away heat and toxic materials (Li *et al.*, 2000; Zhou *et al.*, 1991; Xu *et al.*, 1994).

Antioxidants in the fruit can be divided into water soluble antioxidant and fat soluble antioxidant properties with water-soluble antioxidant components including such as flavonoids, polyphenols, procyanidins, catechin, and lipid soluble antioxidants including tocopherols, carotenoids, *etc.* (Hua *et al.*, 2010). At present, the seabuckthorn antioxidant substances extraction methods

Seabuckthorn (*Hippophae* L.): *A Multipurpose Wonder Plant, Vol. 5: Advanced Technologies for Cultivation, Processing, Health Protection and Environmental Conservation* (Virendra Singh, *Editor-in-Chief*, 2018), p. 301–309, Astral International Pvt. Ltd., New Delhi, India.

mainly include the hot water extraction (Wang *et al.*, 2001), organic solvent extraction (Liu *et al.*, 2004), microwave extraction (Gao *et al.*, 2006; Zhao *et al.*, 2004), ultrasonic extraction (Liu, 2007) and enzyme extraction (Choudharis *et al.*, 2007), mainly concentrated in the fruit powder, leaves and *Hippophaë* flavonoids extraction (Chen *et al.*, 2006; Zhang *et al.*, 2011; Jiang, 2006). The research shows that fruit antioxidant extract showed strong scavenging ability to DPPH free radical, hydroxyl radical and superoxide anion, and when the fruit juice reaches a certain concentration, the scavenging effect is in steady state, and especially has a strong inhibitory effect on DPPH and hydroxyl free radicals (Liu *et al.*, 2009). Although there are many reports about seabuckthorn extraction of antioxidant components, no reports are about Chinese seabuckthorn.

The object of this study was to screening the proper solvent to extract the total antioxidant extraction in the Chinese seabuckthorn, than optimized the extracted parameters, and evaluate its antioxidant properties.

MATERIALS AND METHODS

Materials, Reagents and Instruments

Chinese seabuckthorn fruit was harvested in Huachi city of Gansu Province, China, in November 2012. Fruit were cleaned, packed and transported to the lab in the day, and storage at -20°C. Pyridine acridine hydrochloride (tptz) and 1, 1 '-diphenyl-2-Nitro phenyl hydrazine (DPPH) were purchased from Sigma-Aldrich (ST. Louis, MO, USA). Ultraviolet spectrophotometer (Genesys 10S UV-Vis, Thermo Fisher Scientific Co., Ltd., USA).

Determination the Antioxidant Extraction's Total Antioxidant Capacity

FRAP was determined by FRAP with reference to Iris *et al.* (1996). Accurately draw 0.3ml of 0.0, 0.1, 0.2, 0.4, 0.6, 0.8, 1.0 mmol/L $FeSO_4$ standard solution and put it in color tube, then add 2.7 mL TPTZ fluid (by pH3.6 300 mmol/L acetic acid sodium acetate buffer solution, solution of 10 mmol/LTPTZ, 20 mmol/L $FeCl_3$ solution with mixing ratio of 10: 1: 1 ratio). After mixing, 37°C for 10min, determine FRAP, three times, and draw the standard curve.

Samples were determined according to above method, accurate weighs 2.0g grinding crushed fruits and place in 50 mL Erlenmeyer flask, and extract according to above method and take centrifuged supernatant to determine FRAP value.

Screening the Proper Solvent

According to Zhang *et al.* (2013) method with modifications, The fruit 2.0g is grinded and crushed, and add the 40 mL water, ethanol, 75 per cent of ethanol, methanol, acetone, acetic acid, methylene chloride, ethyl acetate as extraction solvent with solid-liquid ratio 1:20, 60°C, refluxing and extracting for 3 h, get centrifugal supernatant and determine FRAP value.

Optimized the Extracted Parameters about Ethanol Concentration, Material-water Ratio, Extracted Temperature and Extracted Time

According to Zhang *et al.* (2013) method with modifications, the extraction solution is 40mL 45 per cent, 55 per cent, 65 per cent, 75 per cent, 85 per cent ethanol solution and determine FRAP value. Add 55 per cent ethanol solution with the ratio of material to liquid 1:10, 1:15, 1:20, 1:25 and 1:30 and determine the FRAP value. The FRAP values were measured at 50, 60, 80, 90 and 70 °C respectively by reflux extraction. FRAP is determined under 60°C for 1h, 2h, 3h, 4h, 5h by reflux extraction.

Interaction of the Extracted Parameters

On the basis of the single factor test, select the ethanol concentration, solid-liquid ratio, extraction time, extraction temperature as independent variables, and FRAP values as response value, and do box Behnken design (Wei *et al.*, 2010), and the coding of factors and levels as shown in the Table 38.1.

Table 38.1: Factors and Levels of Response Surface Experiments

Factors	Levels		
	-1	0	+1
Ethanol concentration A (per cent)	45	55	65
Solid-liquid ratio B (g/mL)	1:15	1:20	1:25
Extraction time C (h)	1	2	3
Extraction temperature D (°C)	70	80	90

Removal of DPPH by Extraction Solution

According to Zheng *et al.* (2005) method, in the dark, at room temperature, add 2 mL anhydrous ethanol and 2 mL DPPH solution, 2 mL extract liquid and 2 mL DPPH solution, 2 mL extract liquid and 2mL anhydrous ethanol, respectively. After reaction for 30 minutes, determined absorption luminosity valuesA_0, A_1, A_2 at 517 nm, with distilled water as reference and BHT as positive control, parallel 3 times.

The calculation formula is as follows:

$$\text{Clearance rate of DPPH} \bullet = \left(1 \quad \frac{A_1 \; A_z}{A_0}\right) \times 100 \text{ per cent}$$

Among which: A_0: 2 mL anhydrous ethanol and 2mL DPPH solution absorption luminosity values, A_1: 2 mL extracting solution and 2mL DPPH solution absorption luminosity values, A_2: 2 mL extracting solution and 2mL anhydrous ethanol absorption luminosity values.

Determination of •OH Clearance by Extracting Liquid

According to Mo *et al.* (2006) method, 8.8 mmol/L H_2O_2 2mL A9 mmol/L $FeSO_4$ 2 mL A9 mmol/L salicylic acid-ethanol 2 mL and 2ml different concentrations of extracted liquid were added in the reaction system, finally add H_2O_2, then reaction is initiated, under 37°C for 0.5 h with distilled water as reference. The absorbance was determined at 510 nm with 9 mmol/L $FeSO_4$2 mL, 2 mL salicylic acid ethanol, 2 mL different concentrations of solution as background absorption, and BHT were the positive control, parallel three times.

Calculation formula

$$\text{Clearance rate of HO}\bullet = \left(1 \quad \frac{A_{A_{x0}}}{A_0}\right) \times 100 \text{ per cent}$$

Among which: A_0: the absorbance of the blank control solution, and Ax-the absorbance of the after adding the extraction solution, A_{x0}: the background absorption without adding the extraction of H_2O_2.

Removal Effect of Extracting Solution on O_2^{-}·

According to Han *et al.* (2009) method, take 0.05 mol/L PH=8.2 tris-HCl buffer 4.5 mL which should be preheated under 25°C in water bath for 20 min. After mix 0.1 mL sample and 0.4 mL 25 mmol/L pyrogallic acid solution, and put it into water bath under 25°C for 5 min. Every 30s absorbance values were measured, and calculate pyrogallol oxidation rate after reaction for 4min, V= (last recorded value - the first recorded values)/4. With Tris-HCl buffer solution as reference, the absorbance value was measured at 299nm, and distilled water was used as blank, and the oxidation rate of V_0 was calculated. BHT as positive control, each concentration of 3 parallel.

Calculation formula:

$$\text{Clearance rate} = \left(\frac{V_0 \quad V}{V_0}\right) \times 100 \text{ per cent}$$

Among which: V_0: pyrogallic acid auto-oxidation rate and V-refers topyrogallic acid auto-oxidation rate after adding extracting liquid.

Statistical Analysis

Statistical analysis was performed by the SPSS 17.0 software (SPSS, Inc., Chicago, IL). All data are expressed as the mean ±SE.

RESULTS AND DISCUSSION

Screening the Proper Solvent

The effect of solvent on the antioxidant capacity of the extracts showed (Figure 38.1) that 75 per cent ethanol extract was the highest, reaching 69.950 mmol/L. Secondly, FRAP value of anhydrous ethanol (57.323 mmol/L), acetone (54.338 mmol/L) and acetic acid (51.596 mmol/L), methanol (46.528 mmol/L) and water (43.421mmol/L) extracts are lower and dichloromethane extraction liquid FRAP values are the lowest. Considering the toxicity, cost and FRAP value and other factors, choose the mixed solution of water and ethanol as the extraction solvent.

Optimized the Extracted Parameters

Data from the influence of the ethanol concentration on the antioxidant capacity of the extracting solution shows (Figure 38.2A) that with the increase of the ethanol concentration, FRAP value of the extracting solution first increases, and then decreases. When the concentration is 55 per cent, the value reaches the top (72.897 mmol/L). When the ethanol concentration is 85 per cent, FRAP value is 57.864 mmol/L, and the total antioxidant capacity decreases by 20.6 per cent, compared with 55 per cent ethanol extract. This may be that the increase of ethanol leads to the increase of the fat-soluble pigments dissolution, to the decrease of the water-soluble antioxidant components and to the decrease of the total antioxidative capability value.

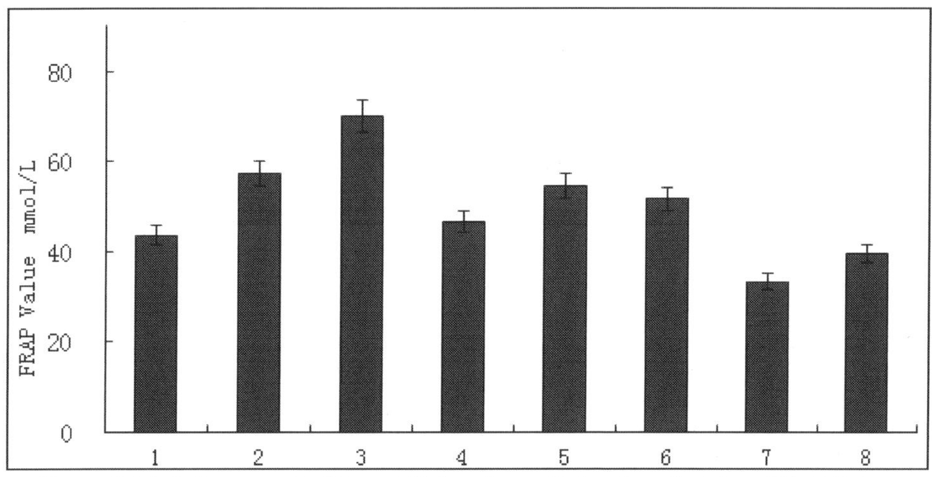

Figure 38.1: Influence of different Solvents on Extracting Solution FRAP Value.
(1: Water, 2: Anhydrous ethanol, 3:75 per cent ethanol, 4: Methanol, 5: Acetone, 6: Acetic acid, 7: Dichloromethane, 8: Ethyl acetate). Bars, ± SE.

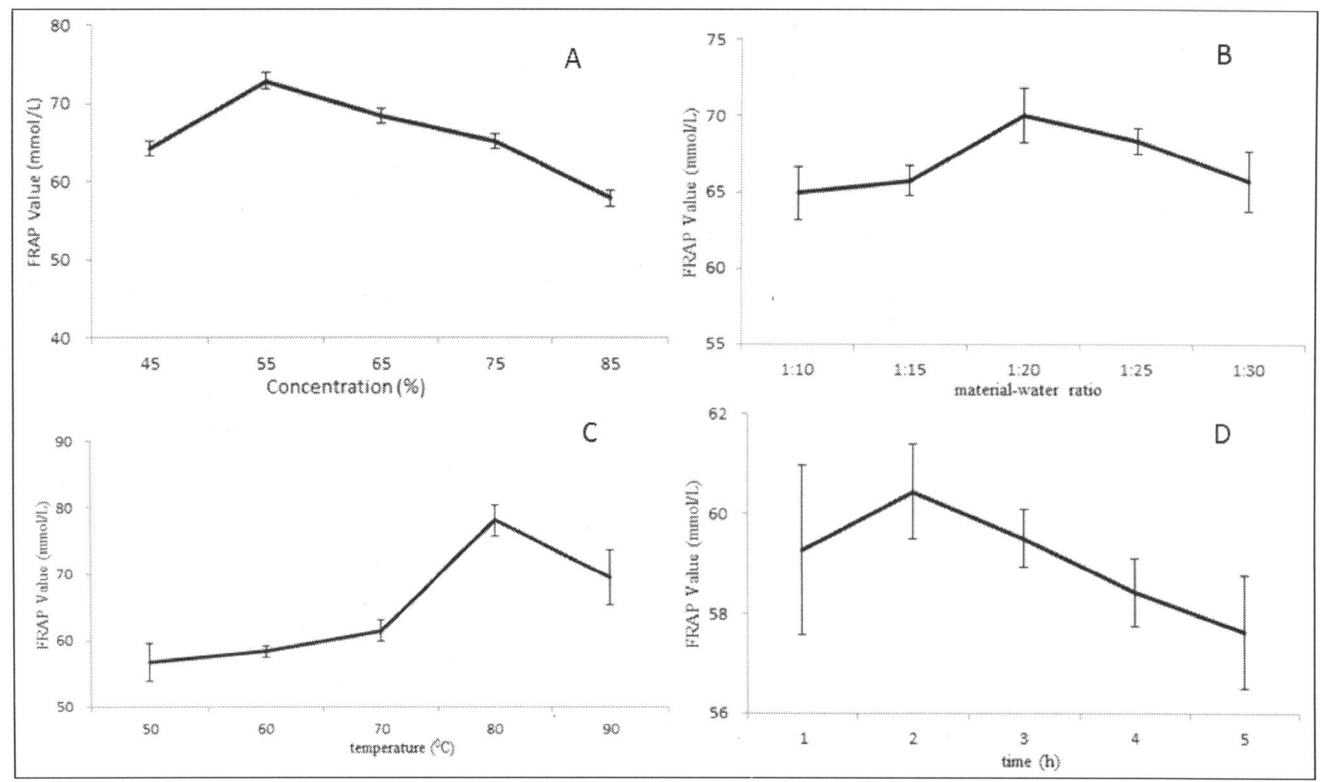

Figure 38.2: Influence of the Parameters on Extracting Solution FRAP Value. (A: Ethanol concentration, B. Material-water ratio, C: Extracting temperature, D: Extracting time). The data represent means ±SE of three replicates.

Data from the influence of material-water ratio on the antioxidant capacity of the extracting solution shows (Figure 38.2B) that with the increase of the material-water ratio, the antioxidant capacity of the extracting solution first increases, and then decreases. When the material-water ratio is 1:20, FRAP value reaches the peak value 70.075 mmol/L. Since then, total antioxidant capacity reduces. The reason may be that since the material-water ratio increases, the concentration difference of the active materials in two-phase increases, the dissolution of the active materials increases, FRAP value increases. but when the dissolution of the active materials achieves balance, the material-water ratio continues to increase, the concentration of the active materials in the extracting solution decreases to lead to the decrease of total anti-oxidant FRAP value of the extracting solution. Therefore, 1:20 feed liquid is more suitable.

Data from the influence of the extracting temperature on the antioxidant capacity of the extracting solution shows (Figure 38.2C) that the temperature has a great influence on antioxidant capacity of the extracting solution. The extracting temperature can change within 50-90°C. With temperature increasing, total antioxidative capability of extracting solution increases rapidly. When the extracting temperature is 80°C, FRAP value reaches the peak value 78.165 mmol/L. When the temperature continues to rise, FRAP value decreases. The reason may be that the role of high temperature for a long time leads to the structure change of the part of oxidation resistance materials due to heating power and leads to the decrease of total antioxidant capacity.

Data from the influence of the extraction time on the antioxidant capacity of the extracting solution shows (Figure 38.2D) that the time influence on antioxidant capacity of the extracting solution changes in unimodal type. When the extraction time is 2h, total antioxidant capacity reaches the maximum 60.436 mmol/L; then it is reduced, with a little change. After extracting 5h (57.637 mmol/L), total antioxidant capacity only decreases by 4.6 per cent.

Interaction of the Extracted Parameters

See the Table 38.2.

The Optimization Results Analysis of Response Surface Method

The data of the response surface analysis test (Table 38.4) matched with a multiple quadratic regression equation of antioxidant capacity of the seabuckthorn fruit is as follows:

$Y = 84.41-1.37A+3.17B-4.91C+8.08D-1.02AB-0.79AC-1.04AD+1.58BC+3.27BD+0.70CD-2.85A^2-4.79B^2-3.34C^2-8.$

According to the variance analysis of the regression equation (Table 38.3), the model is highly significant (P<0.0001), and can better descript the real relationship between various factors and the response value. The

Table 38.2: Results of Response Surface Experiments

Sl.No.	Factor				FRAP Value mmol/L	
	Ethanol Concentration A (per cent)	Solid-Liquid Ratio B (g/mL)	Extraction Time C (h)	Extraction Temperature D (°C)	Actual Value	Predictions
1	-1 (45)	-1 (1:15)	0 (2)	0 (80)	70.26	73.97
2	1 (65)	-1	0	0	70.65	73.26
3	-1	1 (1:25)	0	0	82.68	82.34
4	1	1	0	0	79.00	77.56
5	0 (55)	0 (1:20)	-1 (1)	-1 (70)	69.13	69.94
6	0	0	1 (3)	-1	57.20	58.72
7	0	0	-1	1 (90)	83.96	84.70
8	0	0	1	1	74.83	76.28
9	-1	0	0	-1	68.05	65.15
10	1	0	0	-1	68.79	64.49
11	-1	0	0	1	81.80	83.40
12	1	0	0	1	78.36	78.56
13	0	-1	-1	0	80.52	79.62
14	0	1	-1	0	84.99	82.79
15	0	-1	1	0	67.12	66.63
16	0	1	1	0	77.92	76.13
17	-1	0	-1	0	84.21	83.73
18	1	0	-1	0	80.52	82.55
19	-1	0	1	0	77.09	75.48
20	1	0	1	0	70.26	71.16
21	0	-1	0	-1	63.24	62.98
22	0	1	0	-1	57.69	62.79
23	0	-1	0	1	77.28	72.61
24	0	1	0	1	84.80	85.47
25	0	0	0	0	83.67	84.41
26	0	0	0	0	82.05	84.41
27	0	0	0	0	84.80	84.41
28	0	0	0	0	85.43	84.41
29	0	0	0	0	86.12	84.41

residual equation can be used to determine the best extracting parameter. Among them, B, C, D, A^2, B^2, C^2, D^2 item has a significant effect on FRAP value of extracting solution of the fruit, and other factors are not significant. The absolute value size of monomial regression coefficient of the regression equation is D, C, B and A in turn. This shows that the temperature has the greatest influence on FRAP value of extracting solution, secondly time, material-water ratio and ethanol concentration.

Interaction and Optimization of Response Surface

According to F value in the variance table of the regression equation, the four factors of the most significant interactive effect are AB, AC, AD and CD. The reciprocal effect of the ethanol concentration and material-water ratio shows (Figure 38.3a) that as the increase of ethanol concentration; FRAP value of extracting solution increases gradually. When the ethanol concentration is between 55-65 per cent, FRAP value reaches the maximum. Then it starts to decrease. Under the interaction of ethanol concentration, when the material-water ratio is 1:20, the best FRAP value is 84.795 mmol/L. The reciprocal effect of the extraction time and ethanol concentration shows (Figure 38.3b) that as the increase of extraction time, FRAP value of extracting solution decreases gradually. Under the interaction of extraction time, ethanol concentration reaches the maximum 83.666 mmol/L. The reciprocal effect of the ethanol concentration and extracting temperature shows (Figure 38.3c) that as the increase of ethanol concentration, FRAP value of extracting solution increases gradually. When the ethanol concentration is between 50-60 per cent, Under the interaction of extracting temperature, FRAP value reaches the top

Table 38.3: Analysis Results of Regression and Variance

Source	*Sum of Square*	*df*	*Mean Square*	*F Value*	*P Value*	*Significant*
Model	1846.90	14	131.92	13.85	<0.0001	**
A- ethanol concentration	22.69	1	22.69	2.38	0.1450	
B- solid-liquid ratio	120.42	1	120.42	12.64	0.0032	**
C- extraction time	289.43	1	289.43	30.39	<0.0001	**
D- extraction temperature	783.19	1	783.19	82.24	<0.0001	**
AB	4.15	1	4.15	0.44	0.5197	
AC	2.47	1	2.47	0.26	0.6184	
AD	4.36	1	4.36	0.46	0.5099	
BC	10.04	1	10.04	1.05	0.3220	
BD	42.66	1	42.66	4.48	0.0527	
CD	1.96	1	1.96	0.21	0.6570	
A^2	52.51	1	52.51	5.51	0.0341	*
B^2	148.53	1	148.53	15.60	0.0015	**
C^2	72.19	1	72.19	7.58	0.0155	*
D^2	487.01	1	487.01	51.14	<0.0001	**
Residual	133.33	14	9.52			
Lack of Fit	123.06	10	12.31	7.79	0.0723	not sig
Pure Error	10.27	4	2.57			
Cor Total	1980.23	28				
R-squared=0.9327						
AdjR-squared=0.8653						

Notes: p<0.0100 highly significant (**) Gp<0.0500 notable (*) Gp>0.0500 non-significant.

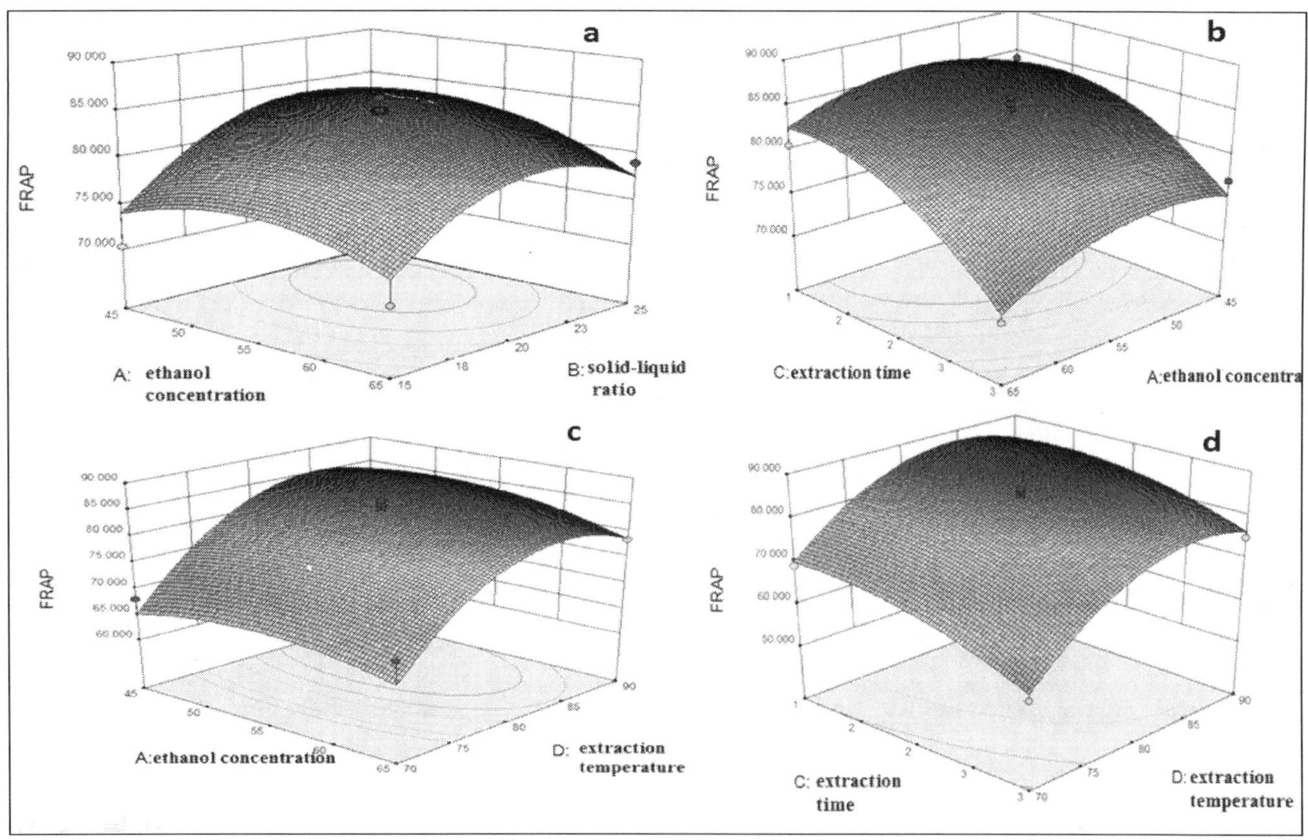

Figure 38.3: Interactive Influence of Response Surface.

(a. Interactive influence of ethanol concentration and solid-liquid ratio, b. Interactive influence of extraction time an ethanol concentration and, c. Interactive influence of ethanol concentration and extraction temperature, d. Interactive influence of extraction time and extraction temperature).

86.458 mmol/L between 80-90°C. The reciprocal effect of extraction time and extracting temperature shows that there is a synergistic effect between temperature and time (Figure 38.3d). Namely, in a certain concentration area, only rise or fall at the same time can improve total antioxidant capacity. When the both are at the temperature 80~90°C and when the time is between 1-2h, total antioxidant capacity of extracting solution reaches the maximum 84.795 mmol/L.

The extraction condition of gaining the maximum FRAP from the regression equation: ethanol concentration 51 per cent, material-water ratio 1:22, extraction time 1h, extracting temperature. At the moment, the predictable total antioxidant capacity is 88.933 mmol/L. On this condition, the result of the test is 87.840 mmol/L, and the relative error between the actual value and the predicted value is 1.12 per cent. the regression equation that is gained after the optimization of the response surface has a certain practical guiding significance.

BHT Scavenging Action on Free Radical

Table 38.4: EC_{50} of Scavenging Effects of BHT on Free Radical

BHT	*DPPH·*	*•OH*	*O_2^-·*
EC_{50} Value	42.70µg/mL	0.217mg/mL	0.86mg/mL

Extracting Solution Scavenging Action on DPPH• Free Radical

The extracting solution of the antioxidant components of the fruit has a certain scavenging activity for DPPH free radical (Figure 38.4), and shows good dose-effect relationship for the eliminating of DPPH free radical in the test range of the concentration 0-0.5 mg/mL. With the increase of concentration, the clearance rate also increases gradually. When the concentration is 0.2 mg/mL, the clearance rate of DPPH reaches 58.41 per cent. Since then, with the increase of concentration, the clearance effect becomes stable. When EC_{50} value is 93.79 µg/mL, it is about 2.20 times of BHT.

Extracting Solution Scavenging Action on Hydroxyl Radical

The extracting solution of antioxidant components of the fruit has the strong scavenging effect on the hydroxyl radical (Figure 38.5). In the range of 0-12 mg/mL concentration, with the increase of extract concentration, the hydroxyl radical scavenging is on the rise. When the concentration is 5 mg/mL, the clearance rate reaches 68.34 per cent. When the concentration is more than 9 mg/mL, the clearance effect becomes mild. EC_{50} value is 4.14 mg/mL, and is about 19.1 times of BHT.

Extracting Solution Scavenging Action on Superoxide Anion

The extracting solution of antioxidant components of the fruit has the strong scavenging effect on the O_2^-· (Figure 38.6). In the range of 0-1.2 mg/mL concentration, there is a good linear relationship. When the concentration is 0.6 mg/mL, the clearance rate reaches 47.41 per cent. EC50 value is 0.856 mg/mL, and is about 99.5 per cent of BHT.

The antioxidant test result shows that the extracting solution has a good ability to scavenging free radical. Different concentration of the extracting solution clearing effect on DPPH· hydroxyl radical, superoxide anion free radical has verified the antioxidant activity of extracting solution. This is related with extracting the high content of flavone and polyphenol. The extracting solution has the best clearing effect on DPPH. Compared with the antioxidant BHT clearing effect on DPPH free radical and hydroxyl radical, the extracting solution clearing effect on the superoxide anion is close to BHT, and the extracting solution clearing effect on DPPH free radical and hydroxyl radical is lower than BHT.

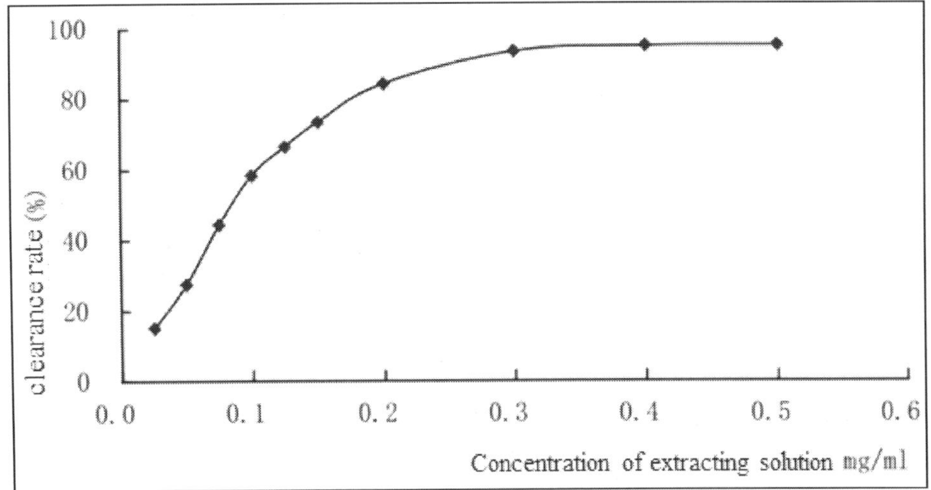

Figure 38.4: DPPH· Scavenging Effects for different Additive Amount of Extracting Solution.

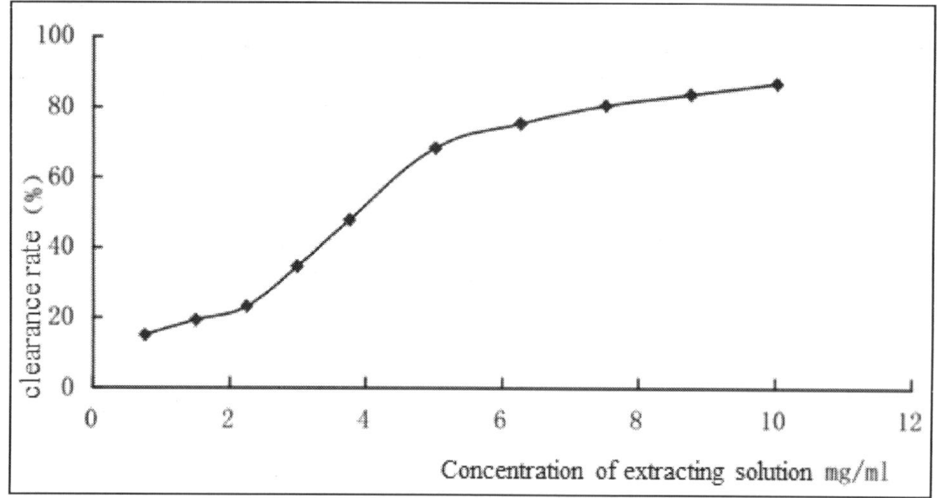

Figure 38.5: ·OH Scavenging Effects for different Additive Amount of Extracting Solution.

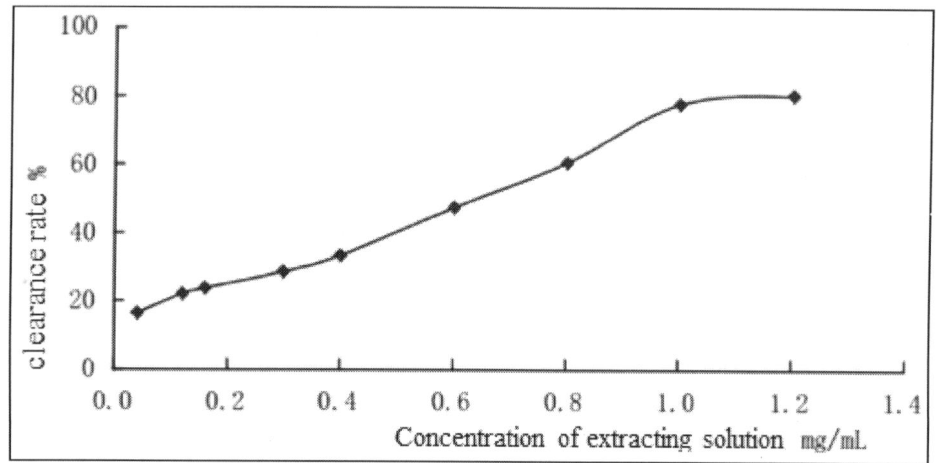

Figure 38.6: O_2^{-}· Scavenging Effects for different Additive Amount of Extracting Solution.

CONCLUSION

The best process parameters of antioxidant components of seabuckthorn fruit from the regression equation: ethanol concentration 51 per cent, material-water ratio 1:22, extraction time 1h, extract temperature 86°C. On this condition, FRAP value of extracting solution is 87.840 mmol/L. The scavenging free radical capacity of antioxidant components extracting solution of seabuckthorn fruit (calculate according to EC_{50}: DPPH· clearance rate (93.79 μg/mL) >O_2^{-}· radical scavenging activity (0.856 mg/mL) >maximum scavenging rate of hydroxyl radical (4.14 mg/mL).

ACKNOWLEDGEMENT

This work was funded by the Projects of International Cooperation and Exchanges of Ministry of Science and Technology in China (No. 2014DFR31230). We gratefully acknowledge National Nature Science Foundation of China. At the same time we are sincerely appreciated Mr. Luan from Gansu Conseco Company for providing all the seabuckthorn pulp needed in the experiment. And we are thankful for GE Qian cui-di who introduced the processing of seabuckthorn in the industry which helped us to set the experimental factors. Besides, we wish to express our thanks to Prof. Han Shun-yu for discussions.

REFERENCES

1. Camelia, P., Cristiana, D. and Nicorescu, V. 2008. Antioxidant activity of seabuckthorn (*Hippophae rhamnoides*) extracts compared with common food additives. *Roumanian Society of Biological Sciences* 13: 4049-4053.
2. Chen, H.F., Zhou, W.M. and Fu, J.X. 2006. Extracting technology of total flavones of leaf of *Hippophae rhamnoides* L. *Acta Agriculturae Boreali-occidentalis Sinica* 15: 148-151.
3. Choudharis, M. and Ananthan, A.L. 2007. Enzyme aided extraction of lycopene from tomato tissues. *Food Chemistry* 102: 77-81.
4. Gao, M., Song, B. and Liu, C. 2006. Dynamic microwave assisted extraction of flavonoids from *Saussurea medusa* Maxim cultured cells. *Biochemical Engineering Journal* 32: 79-83.

5. Han, S.H., Zhu, J.B. and Wang, Y.Y. 2009. Measurement of the antioxidant activity by pyrogallol autoxidation. *China Brewing* 207: 155-157.

6. He, S.M., Yuan, X.J. and Wang, J.H. 2008. The plant resources of *Hippophae* and its development and utilization status in China. *Modern Agricultural Sciences* 11: 87-92.

7. Hua, S.Z., Weng, Z.P. and Teng, X.P. 2010. Study on antioxidant performance of seabuckthorn. *The International Seabuckthorn Research and Development* 8: 34-40.

8. Iris, F.F.B. and Strain, J.J. 1996. The Ferric reducing ability of plasma (FRAP) as a measure of antioxidant power: The FRAP Assay. *Analytical Biochemistry* 239: 70-76.

9. Jiang, S. J. 2006. Extraction and isolation of flavonoids in fruit marc of *Hippophae rhamnoides* L. *Thesis for Masters Degree,* Northwest A and F University, Yangling, 53p.

10. Li, G.Q., Tang, D.R. and Zhao, Y.Q. 2000. The development and utilization of seabuckthorn resources. *Hippophae* 13: 22-26.

11. Liu, C., Wang, C.T. and Li, G. 2009. Study on antioxidant activity of *Hippophae* juice. *Science and Technology of Food Industry* 30: 130-132.

12. Liu, G.B. 2007. Investigation of extraction of seabuckthorn flavonoids and its antioxidant activity. *Thesis for Masters Degree,* Heilongjiang University, Harbin, 53p.

13. Liu, X.J., Wang, Y.H. and Li, P. 2004. Extraction of total flavones from marc of seabuckthorn. *Journal of Beijing University of Chemical Technology* 31: 18-21.

14. Mo, K.J., Liu, S. and Cheng, C. 2006. Study on antioxidant activity of the ginger flavonoid. *Chinese Food Science* 27: 110-115.

15. Tao, X.S. 2003. The development and utilization of seabuckthorn resources in Gansu province. *Journal of Gansu Forestry Science and Technology* 12: 42-44.

16. Wang, S.Y., Zheng, Y.X. and Liu, S.P. 2001. Research of total flavonoids extraction by water from the leaves of seabuckthorn. *Hippophae* 14: 27-29.

17. Wei, L. and Shan, C.H. 2010. Response surface methodology for optimization of alcohol extraction technique of red pigment from sunflower hulls. *Chinese Food Science* 31: 122-126.

18. Xu, M.Y., Sun, X.X. and Tong, W.X. 1994. Medicine research and development of seabuckthorn. *Hippophae* 7: 32-40.

19. Zhang, B., Su, D.Q. and Zeng, D. 2013. Optimization of extraction technology for antioxidant components from seabuckthorn fruits by response surface methodology. *The Food Industry* 34: 75-79.

20. Zhang, Y., Zhang, L.M. and Li, T.D. 2011. Analysis and study on the antioxidant of total flavonoids and quercetin in *Hippophae rhamnoides* L. subsp. *Taishanensis rousi. Chin Hosp Pharm J.* 31: 644-646.

21. Zhao, E.L., Gai, Q.Q. and Zhang, H.R. 2004. Investigation of microwave assisted extraction of seabuckthorn flavonoids. *Food and Fermentation Industries* 30: 148-150.

22. Zhao, WE. and Li, Q.Q. 2011. Determination of total antioxidant capacity of red pigments from Chinese jujube peel by the ferric reducing/antioxidant power assay. *Journal of Zhengzhou University (Engineering Science)* 32: 28-30.

23. Zheng, D.Y. and An, X.N. 2005. Determining method of DPPH free radical scavenging activity of bamboo leaf extractives. *Journal of Fujian Agriculture and Forestry University (Natural Science Edition)* 3: 59-62.

24. Zhou, Y.P. and Jiang, J.L. 1991. The health-care function and application of seabuckthorn. *Hippophae* 4: 35-41.

25. Zhou, Z.Z., Zhou, C.Q. and Kan, J.Q. 2005. Advances in research on chemical components and health protection of seabuckthorn. *Cereal and Food Industry* 15: 15-18.

Chapter 39

Efficacy of Herbal Immunomodulators against Larval Toxocariasis in Swiss Albino Mice

Aman Moudgil[1], S. Mittra[1*], R.K. Agnihotri[1], R.K. Asrani[2], D. Pathak[1], D. Sen[1] and V. Singh[3]

[1]Department of Veterinary Parasitology,
[2]Department of Veterinary Pathology,
DGCN COVAS, CSK HPKV, Palampur , HP, India
[3]Scientist, Seabuckthorn, COBS, CSK HPKV, Palampur – 176 062, HP, India
(*E-mail: susamoymittra@yahoo.co.in)

SUMMARY

A study related to larval toxocariasis in immunomodulated mice was carried out to determine the effect of herbal and chemotherapeutic immunomodulators. There were three herbal immunomodulators (methanolic extracts) *viz., Hippophae rhamnoides* and *Hippophae salicifolia* (seabuckthorn) and *Piper longum* (long pepper) and one chemotherapeutic immunomodulator *i.e.,* levamisole, which had been used for three consecutive days to immunomodulate the experimental mice. Infection with 500 *Toxocara canis* infective ova was given on 14th day of immunomodulatory medication. On the basis of larval recovery, biochemical profile and histopathological observations on day 14 and day 28 post infection, *Hippophae salicifolia* was recorded as superior immunomodulator followed by *Hippophae rhamnoides*, levamisole and *Piper longum* in Swiss albino mice.

Keywords: *Immunomodulator, Seabuckthorn, Toxocara canis and Swiss albino mice.*

INTRODUCTION

In our society, 'dog' is considered as oldest companion and faithful friend of man. Domestic dog (*Canis domestica*) harbours a number of helminthes parasites, of which *Toxocara canis* is one of the most harmful helminthes parasite (Werner, 1782). It has significant zoonotic potential too. Its prevalence in dogs depends on many factors such as age (pup or adult), geographical area and climatic conditions (Pereiraa *et al.,* 2010). Various routes of infection include oral infection, transplacental migration, lactogenic infection and transmission through paratenic host.

Various scientific experimental works had been carried out on *Hippophae rhamnoides* to determine its nutritional and medicinal potential. The ability of fruit and leaf extracts of *Hippophae rhamnoides* to relive the inhibition of lymphocyte proliferation by chromium, indicates that extracts also have immunomodulatory activity (Geeta *et al.,* 2003).

This could be considered as the first study to determine the immunomodulatory efficacy of *Hippophae salicifolia,* in comparison to established herbal immunomodulators *viz., Hippophae rhamnoides* and *Piper longum* and chemotherapeutic immunomodulator levamisole.

MATERIALS AND METHODS

Parasite

The parasite under investigation was *Toxocara canis* (Werner, 1782) of dog. *Toxocara canis* is cosmopolitan in distribution.

Seabuckthorn (Hippophae L.): *A Multipurpose Wonder Plant, Vol. 5: Advanced Technologies for Cultivation, Processing, Health Protection and Environmental Conservation* (Virendra Singh, *Editor-in-Chief,* 2018), p. 311–315, Astral International Pvt. Ltd., New Delhi, India.

Collection of Debris Free Eggs

Toxocara canis eggs from the faeces of donor pups were harvested by floatation concentration technique using Sheather's sugar solution (Soulsby, 1982) and put to embryonation at 27°±1° C. After embryonation completed, these infective eggs were used to infect the experimental animals (Mittra and Sasmal, 1985).

Preparation of Immunomodulators

Collection of Plant Material

Fresh leaves of seabuckthorn (*Hippophae rhamnoides* and *Hippophae salicifolia*) were collected from Kukumseri (Lahaul and Spiti) of Himachal Pradesh (India) in the month of October, 2010. The average altitude of the area of collection was 2730 meters above mean sea level (MSL). The collected leaves of *Hippophae rhamnoides* and *Hippophae salicifolia* were completely shade dried and ground separately with the help of pestle and mortar initially followed by grinder (Cyclotec, 1093 Sample mill, Foss Tecator), to obtain fine powder. Fifty grams of powdered material of either species was weighed and mixed with 400 ml of absolute methanol separately and left for 24 hrs taking into a 500 ml conical flask. The content was then filtered using fine muslin cloth followed by Whatman filter paper no.41.The filterate was concentrated into semisolid form using vacuum evaporation (Rota Vap, Bucchi). The contents were then freeze dried using lyophilizer (Lyoplus) in a round bottom flask for 4-5 hrs until dark brown crystalline form was obtained and then stored at 4°C until used.

Piper longum (Long Pepper)

Good quality whole fruit of *Piper longum* (long pepper) were procured from local market and dried completely in hot air oven at 60°C for half an hour. These dried fruits were then ground with the help of grinder (Cyclotec, 1093 Sample mill, Foss Tecator), to obtain fine powder. Then, methanolic extract of *Piper longum* was prepared in the same manner as that of *Hippophae rhamnoides* and *Hippophae salicifolia.*

Levamisole

Levamisole tablet (150 mg), (Dicaris®, Johnson and Johnson, Jansen Cilag Pharmaceuticals, Mumbai) was used for medication at appropriate dosage.

Dosage of Immunomodulators

Methanolic extracts of *Hippophae rhamnoides, Hippophae salicifolia* and *Piper longum* were administered to the mice of the respective immunomodulated group, orally at a dose rate of 500 mg/kg body weight. Levamisole was orally administered to the mice at a dose rate of 2.5 mg/kg body weight, in aqueous solution. All the immunomodulators were administrered at their respective dose rates for three consecutive days.

Method of Infection

After calculating the dosage (by calculating the number of eggs in each drop), the experimental animals were infected orally with 500 infective ova of *Toxocara canis*. Following each infection, one- two drops of distilled water were sucked into the syringe (without needle), shaken and any remaining egg left, was being flushed into the mouth of experimental animal.

Experimental Design

Sl.No.	Group	Number of Mice	Type of Group
1	I	8	Control with *Toxocara canis* infection *i.e.* infected control
2	II	8	Immunomodulated with levamisole and *Toxocara canis* infected.
3	III	8	Immunomodulated with *Hippophae rhamnoides* and *Toxocara canis* infected.
4	IV	8	Immunomodulated with *Hippophae salicifolia* and *Toxocara canis* infected.
5	V	8	Immunomodulated with *Piper longum* and *Toxocara canis* infected.
6	VI	16	This group contained 16 mice, divided into 4 subgroups (each containing 4 mice and immunomodulated with *Hippophae rhamnoides, Hippophae salicifolia, Piper longum* and Levamisole, but not infected).
7	VII	4	Healthy Control (Non-infected and non-immunomodulated)

Assessment of Immunomodulatory Activity

Recovery of Larvae

Larvae were recovered after 14th and 28th days post infection from various body tissues/viscera of experimental animals.

Method of Recovery of Larvae

Recovery of larvae was done by tissue digestion with artificial gastric juice (pepsin 0.5 g, HCl 0.7 ml and 0.9 per cent Normal saline 99.3 ml) following the method of Soh (1958). The mice were euthanized with excess of chloroform, taking into bell-jar. Various organs/viscera (brain, liver, lungs, kidneys and muscles) were then collected and put in a 50 ml beaker (having 20 ml of artificial gastric juice) after mincing and then left for 24 hours. It was done for each organ/viscera separately. After 24 hours, the whole solution was filtered through a muslin cloth. The filtrate thus obtained was centrifuged at 1000 rpm for 5 minutes. The supernatant was poured off and the sediments were examined under microscope (10X), for the presence of larvae in the respective organ.

Blood Collection

Blood was collected directly from the heart in a sterilized vial without any anti coagulant for serum separation. These vials (containing the blood) were then kept in an inclined position (15°) for 6 hours for proper serum separation. The serum thus separated as a layer, was then aspirated with the help of Pasteur pipette. The collected serum samples were then kept at -20°C in the deep freeze until further investigation.

Total Serum Proteins and Albumin

The values of total serum proteins and albumin were estimated with automatic blood analyzer (Microlab 300 Clinical Chemistry Analyzer by Merck Limited, Mumbai). The total serum protein and albumin values were observed with the help of kit (Agappe Diagnostics Ltd. 'Agappe Hills', Distt. Ernakulam, Kerala) and expressed in g/dl.

Serum Globulin

Serum globulin values were calculated by subtracting the values of serum albumin from corresponding total serum protein values. The serum globulin values were expressed in g/dl.

Albumin-Globulin Ratio

A:G was calculated by dividing albumin values by corresponding globulin values.

Histopathology

The tissue samples (being fixed in 10 per cent neutral buffered formalin) were cut to present the plain surfaces, required to prepare smooth sections. These tissues were then washed overnight in running tap water to remove the formalin absorbed in the tissues. It was then dehydrated by using ascending grades of alcohol starting from 30 per cent to 90 per cent ethanol and finally with dehydrated alcohol. Then cleared with benzene and lastly embedded in paraffin. The sections were cut to 5 μ thickness and stained with routine haematoxylin and eosin stain for the histopathological studies.

Statistical Analysis

The data obtained were subjected to statistical analysis by using computer software Instat from Graphpad software (2008). The mean values of different parameters between healthy (non-infected, non-immunomodulated) control; infected non-immunomodulated control; immunomodulated non-infected and immunomodulated infected group were compared at 5 per cent level of significance using Tukey-Kramer Multiple Comparison test.

RESULTS AND DISCUSSION

Recovery of Larvae

At day 14 post infection, larval recovery from the group immunomodulated with levamisole and infected was significantly ($p<0.05$) less as compared to healthy control group, excluding the larval recovery from kidneys. Whereas, larval recovery from various organs in the group immunomodulated (with *Hippophae rhamnoides*) - infected was also significantly very low as compared to control group except in case of lungs. This observation could be due the ability of *Hippophae rhamnoides* to enhance humoral and cellular immunity of the individual, thereby, shown protective role (Spanu *et al.*, 1994 and Liu *et al.*,1996). Larvae recovered from the brain were also significantly lesser than the infected control group. However, in the group immunomodulated with *Hippophae salicifolia* later infected, the larval recovery was significantly very low as compared to infected control group and also lesser than the group immunomodulated with levamisole and infected, except lungs and kidneys. Although, the group immunomodulated with *Piper longum* later infected, showed almost similar recovery pattern as that of infected control group. Thus, lowest recovery at day 14 post infection was recorded from the *Hippophae salicifolia* immunomodulated infected group.

At day 28 post infection, larvae recovered from brain, lungs, liver and muscles in the groups immunomodulated with levamisole and *Hippophae rhamnoides* later infected were significantly ($p<0.05$) low as compared to infected control group. However, larvae recovered from all the organs of mice of the group immunomodulated with *Hippophae salicifolia* and infected was significantly very low as compared to infected control group and showed significantly very low recovery from the liver as compared to levamisole, *Hippophae rhamnoides* and *Piper longum* immunomodulated-infected groups. The larvae recovered from brain, lungs, liver and muscles in the group, *Piper longum* immunomodulated then infected were significantly lower than the infected control group but significantly higher recovery of larvae was from liver and muscles as compared to the groups, immunomodulated with levamisole, *Hippophae rhamnoides* and *Hippophae salicifolia* and later infected.

Biochemical Profile

Various biochemical parameters *viz.*, total serum proteins, albumin, globulin and albumin globulin ratio were studied following administration of various immunomodulators as well as subsequent larval migration of *Toxocara canis*.

Significantly high values of total serum proteins were recorded at day 14 post infection in the group immunomodulated with levamisole, and then infected. Significantly higher values of globulins were also recorded at day 14 and day 28 post infection in the group immunomodulated with levamisole and infected, as compared to healthy control group. It could be due to the fact that *Toxocara canis* infection leading to the production of gamma-globulins in infected individuals, as significantly ($p<0.05$) high values of globulins were recorded at day 14 and day 28 post infection in both infected control and immunomodulated- infected groups as compared to healthy control group, clearly indicate an augmented immune response.

The values of total serum proteins were higher at day 14 and day 28 post infection in the groups, immunomodulated with *Hippophae rhamnoides* and *Hippophae salicifolia* then infected, as compared to healthy control group. Significantly low globulin values were recorded at day 14 in the group immunomodulated with *Hippophae rhamnoides* then infected and at day 28 post infection in the group immunomodulated with *Hippophae salicifolia* and infected compared to infected control group. These observations were indicative of better immune conference in comparison to *Hippophae rhamnoides*, which establishes *Hippophae salicifolia* a better immunomodulator. Total serum proteins and globulin values were significantly higher in infected control group than *Piper longum* immunomodulated- infected group, whose values were further recorded higher than the values of healthy control group. These results were indicative of mild immune conference.

Histopathology

Histopathological studies revealed inflammatory changes in various organs and these lesions were scored and categorized into four levels: normal (0), mild (+), moderate (++) and severe (+++). However, brain in most of the instances did not show any significant changes.

On day 14 post infection, in infected control group, liver showed granulomatous infiltrations (representing severe inflammatory reaction), moderate degenerative changes and severe necrosis. Also, kidneys, lungs and muscles showed severe inflammatory reaction. These observations were in agreement with the observations of Fan *et al.*(2003), who observed that histopathological changes were moderate to severe in the liver and lungs, mild to moderate in the musculature, and only normal to mild in the brain throughout the trial. However, Agnihotri (1985) reported maximum damage in the liver due to *Toxocara canis* larvae migration at day 2 post infection. In the group, immunomodulated with levamisole followed by infection; liver, kidneys and muscles showed mild inflammatory reaction. Whereas, lungs showed severe inflammatory reaction. There were also mild degenerative changes and necrosis recorded in liver. Various organs of the groups, immunomodulated with *Hippophae salicifolia* and *Hippophae rhamnoides* and then infected; only mild to moderate changes were revealed. However, liver, kidneys and lungs of the group immunomodulated with *Piper longum* later infected, showed severe inflammatory reaction and also moderate degenerative changes and necrosis. There was presence of larvae at certain foci in muscles, in this group.

On day 28 post infection, severe inflammatory reaction was recorded in liver, lungs and muscles of infected control group. There were also moderate degenerative changes and necrosis in liver and muscles only. In the group immunomodulated with levamisole and infected, moderate to severe inflammatory reaction was recorded in every organ, accompanied by mild to moderate degenerative changes and necrosis in liver and muscles. Mild to moderate inflammatory changes were recorded in almost every organ in the groups immunomodulated with *Hippophae rhamnoides* and *Hippophae salicifolia*, followed by infection. Larvae were also recorded at many foci in the group immunomodulated with *Hippophae rhamnoides* and infected. However, mild to severe inflammatory changes along with presence of larvae were recorded in the organs of the mice of *Piper longum* immunomodulated and infected group.

These observations are in partial corroboration with the observations of Agnihotri (1985), who reported lesser degenerative reaction in later phase of the infection in the lungs but more neuronal degeneration, gliosis and satellitosis in later phase of infection in brain tissue in Golden hamster.

CONCLUSION

On the basis of larval recovery, biochemical profile and histopathology, it can be proposed that *Hippophae salicifolia* (methanolic extract) may be a superior immunomodulator followed by *Hippophae rhamnoides*, levamisole and *Piper longum* in larval toxocariasis in Swiss albino mice.

ACKNOWLEDGEMENT

Authors are thankful to the Dean, DGCN COVAS for providing the necessary facilities for undertaking this investigation.

REFERENCES

1. Agnihotri, R.K., Sharma, D. and Sharma, Y. 2008. Incidence of gastrointestinal helminthes in dogs of Himachal Pradesh. *Journal of Veterinary Parasitology* 22 (2): 89-90.
2. Fan, C.K., Lin, Y.H., Du, W.Y. and Su, K.E. 2003. Infectivity and pathogenicity of 14 month cultured embryonated eggs of *Toxocara canis* in mice. *Veterinary Parasitology* 113 (2): 145-155.
3. Geetha, S., Sairam, M., Mongia, S.S., Singh, V., Ilavazhagan, G. and Sawhney, R.C. 2003. Evaluation of antioxidant activity of leaf extract of seabuckthorn (*Hippophae rhamnoides* L.) on chromium (VI) induced oxidative stress in albino rats. *Molecular and Cellular Biochemistry* 87 (2-3): 247-51.
4. Mittra, S. and Sasmal, N.K. 1985. Experimental infection of pups with *Ancylostoma caninum* infected mouse tissues. *Veterinary Parasitology* 18 (3): 51-57.
5. Pereiraa, A.A., Mandarino, F.S., Lopesc, C.W.G. and Pereiraa, M.J.S. 2010. Prevalence of parasites

in soil and dog faeces according to diagnostic tests. *Veterinary Parasitology* 170 (1-2): 176-181.

6. Soh, C.T. 1958. The distribution and persistence of hookworm larvae in the tissues of mice in relation to species and the route of inoculation. *Journal of Parasitology* 44 (2): 515-519.

7. Soulsby, E.J.L. 1982. *Helminths, Arthropods and Protozoa of Domesticated Animals*. Baillere Tindall, London.

8. Werner, P.C.F. 1782. Vermium intestinalium brevis expositionis continunuatio. In: *Systema Helminthum* (Satyu Yamaguti), 1985, 3 (1): 579.

Chapter 40

Protective Effect of Seabuckthorn (*Hippophae rhamnoides*) Leaves Against Ochratoxin A Induced Renal Damage in Japanese Quail

Vikram Patial[1], R.K. Asrani[1*], R.D. Patil[1], Sarvesh Kumar[1], Virendra Singh[2], D.R. Ledoux[3] and G.E. Rottinghaus[3]

[1]Department of Veterinary Pathology, GCN COVAS
[2]Department of Biology and Environmental Sciences, COBS,
CSK HPKV, Palampur – 176 062, H.P., India
[3]Fusarium/Poultry Research Laboratory, University of Missouri, Columbia, 65211, USA
(*E-mail: asranirk@gmail.com)

SUMMARY

The present study was designed to study the protective effect of seabuckthorn (SBT) leaves against ochratoxin A (OTA) induced renal damage in Japanese quail. Day old Japanese quail chicks were divided into 4 groups and fed with basal quail chick mash containing 2 per cent SBT leaf powder (Group SX), OTA at the dietary level of 3 ppm (Group OX), 2 per cent dietary level of SBT leaf powder plus OTA at the dietary level of 3 ppm (Group OS) and standard toxin free feed (Group CX, control) for 21 days. The quail chicks were assessed for various clinico-pathological parameters such as body weights, serum biochemistry and the gross, histopathological and ultrastructural changes in the renal tissues. OTA at 3 ppm level in diet led to anorexia, diarrhoea, marked depression in body weight and some typical nervous signs. Uric acid values were higher in OTA treated birds as compared to control birds, however, partial protection was evident in OS group. The serum total protein and albumin levels in group OX found to be significantly lower throughout the trial; however, in combination group the response was comparatively better. Grossly, mild to moderate renal swelling were recorded in OX group birds and the severity was less in case of OS group birds. Microscopically, degenerative to necrotic changes were consistent in OX group birds, but the changes were less severe in OS group birds. Ultrastructural studies revealed most remarkable and consistent changes in the PCT epithelial cells with severe damage of mitochondria and endoplasmic reticulum in OX group birds, whereas, OS group revealed mild changes in mitochondria with increased number of peroxisomes. In conclusion, the inclusion of 2 per cent SBT leaf powder in feed provided partial protective effects against OTA induced nephropathy in Japanese quail.

***Keywords**: Ochratoxin A, Seabuckthorn, Japanese quail, Kidney, Pathology and Ultrastructure.*

INTRODUCTION

Ochratoxin A (OTA) is a known nephrotoxic and carcinogenic secondary metabolite produced by several species of fungal genera *Aspergillus* and *Penicillium*. It is a wide spread natural contaminant and co-contaminant of various food and feed commodities, resulting in toxicosis in human and animal population. It is more toxic than other ochratoxins (OTB and OTC) and has been reported to be more toxic than Aflatoxin B_1 (Samson, 1992). The agro-climatic conditions in India favour the growth of moulds on feed and grains and production of ochratoxins. *Aspergillus ochraceus* has been found most prevalent in southern, eastern and northern regions whereas *P. cyclopium* has been frequently isolated in the Himalayan regions (Panisup

Seabuckthorn (Hippophae L.): A Multipurpose Wonder Plant, Vol. 5: Advanced Technologies for Cultivation, Processing, Health Protection and Environmental Conservation (Virendra Singh, *Editor-in-Chief*, 2018), p. 317–327, Astral International Pvt. Ltd., New Delhi, India.

et al., 1993). OTA cause significant loses to the poultry industry due to its effects on performance and health. It causes a reduction in growth rate and feed consumption, poorer feed conversion and increased mortality (Pecham *et al.*, 1971). Furthermore, it produces a reduction in total blood proteins (Huff *et al.*, 1988; Stoev *et al.*, 2000), suppression of immune function (Dwivedi and Burns, 1984a and b; Stoev *et al.*, 2000, 2002a; Santin *et al.*, 2002; Alvarez *et al.*, 2004) and impairment of blood coagulation (Raju and Devigoda, 2000). OTA induces degenerative changes and an increase in weight of the kidneys and the liver, as well as decrease in weight of lymphoid organs (Stoev *et al.*, 2000, 2002b).

Seabuckthorn (*Hippophae rhamnoides*) is one of the fascinating plants in the world because of its great medicinal and nutritional value. This plant is native to temperate zones of Europe and Asia including India. This plant, which belongs to the family Elaeagnaceae, is renowned for its versatile pharmacological applications due to the presence of many known and unknown bioactive substances mainly in its fruits and leaves. Fruit and leaves of seabuckthorn are quite rich in vitamins C, B, E, K and other biologically active substance like organic acids, amino acids, carotenoids, tocopherols, flavonoids, sterols *etc.* The fruit and leaf extract of *H. rhamnoides* improved the antioxidant defense system of the cells by increasing the intracellular GSH levels and inhibiting ROS production. The ability of fruit and leaf extracts of *H. rhamnoides* to relive the inhibition of lymphocyte proliferation by chromium, indicates that extracts also have immunomodulating activity (Geetha *et al.*, 2003; Gao *et al.*, 2000).

As OTA is primarily nephrotoxic and one of its toxic mechanisms is due to production of reactive oxygen species and induce lipid peroxidation of cell membranes and on the other hand, *H. rhamnoides* leaves contain various flavonoids with rich antioxidant properties, hence the present investigation has been planned to study the efficacy of SBT leaves in amelioration of ochratoxin induced renal damage in quail.

MATERIALS AND METHODS

Cleaning of Experimental Room and Cages

Prior to arrival of day-old Japanese quail in the department, the experimental room, cages, trays and cage stands were first washed with tap water and then thoroughly cleaned with 2.5 per cent phenol, dried under sunlight and flamed with dry heat using blowlamp. The room was fumigated with formaldehyde gas, just 2 days prior to arrival of quail chicks. The waterers and feeders were daily washed with potassium permanganate solution throughout the experiment.

Collection of Seabuckthorn Leaf and Preparation of Seabuckthorn Leaf Powder

Seabuckthorn (*Hippophae rhamnoides*) leaves were collected from the orchards of Krishi Vigyan Kendra Kukumseri, Lahaul and Spiti, Himachal Pradesh. Leaves were shade dried and then a fine powder of the leaves made with grinder.

Ochratoxin A

Ochratoxin A was produced from the fungus *Aspergillus ochraceous* (NRRL-3174) on maize and the culture material with known level of ochratoxin A was supplied by Dr. G. E. Rottinghaus, University of Missouri, Columbia, USA on gratis.

Experimental Birds

The present studies were conducted on one hundred and eighty, day-old Japanese quail chicks procured from Central Poultry Development Organization, Chandigarh, India. The quail chicks used in the present study were from the same breeding flock. The animal care and experimental protocol were approved and followed by the Committee for the Purpose of Control and Supervision of Experiments on Animals (CPCSEA).

Feeding Schedule

The quail chicks were maintained on chick mash (Quail mash procured from Department of Animal Nutrition, COVAS, CSK HPKV, Palampur) from day one until the end of the experiment. Feed was autoclaved before feeding or mixing with OTA culture material(s). Pre-boiled drinking water was provided during the experiment. Feed and water were given *ad libitum.* Before feeding, the representative samples of chick mash were submitted to Animal Feed Analytical and Quality Control Laboratory, Namakkal, Tamil Nadu, India for analysis of common mycotoxins, if any.

Experimental Design

One hundred and eighty, 1-day-old Japanese quail chicks were randomly divided into 4 groups *i.e.* CX, SX, OX and OS with 45 numbers of birds in each of the groups. The present studies were conducted in three pen replicates of 15 quail per pen assigned to each of three groups. Quail chicks from group CX (control) received standard toxin free feed throughout the experiment. The SBT leaf powder was incorporated in the diet at the rate of 2 per cent (20g per kg feed) in the chick mash in group SX. Chicks from group OX received OTA culture material at the level of 3 ppm OTA in the diet (80 mg OTA per kg was incorporated at the rate of 3.75 per cent in chick mash). Whereas, in combination group, *i.e.* group OS birds received 2 per cent dietary level of SBT leaf powder plus OTA at the level of 3 ppm in diet. Each dietary treatment was started from day one and continued up to 21 days for the study of clinico-pathological and ultrastructural studies.

Clinical Signs, Growth Rate, Mortality and Serum Biochemistry

The birds from all the groups were closely observed, at least thrice daily, for the development of clinical signs and mortality, if any. In order to see the effect of OTA on body weight, 9 randomly selected quail (three quail per replicate) from each treatment group were weighed at 0, 7, 14, and 21 days post- feeding. After weighing, 2 to 3 ml of blood was collected via cardiac puncture from randomly selected 6 birds from each treatment group at 7, 14, and 21 DPF for estimation of total serum proteins (TSP), albumin, and uric acid. All the biochemical estimations were done using diagnostic kits (Reckon Diagnostics Pvt. Ltd., Baroda, India) in a fully automatic Blood Chemistry Analyzer (RA-50 Auto Chemistry System) according to the manufacturer's recommendations.

Gross and Histopathology

Randomly selected six quail chicks (two per replica) were euthanized from each treatment group at 7, 14, and 21 days post-feeding and thoroughly examined for gross lesions in both the kidneys, if any. Representative renal tissue pieces were collected in 10 per cent neutral-buffered formalin for histopathological studies. Fixed tissues were trimmed, embedded in paraffin, sectioned at 3-5 μm, and stained with haematoxylin and eosin stain (Luna, 1968). The gross and histopathological lesions were graded as mild (+), moderate (++) and severe (+++).

Transmission Electron Microscopy

Kidneys pieces of about 1 mm^3 from three birds from each groups (CX, SX, OX, OS) were collected at 21 DPF in chilled Karnovsky fixative (2 per cent paraformaldehyde; 2.5 per cent glutaraldehyde in phosphate buffer, pH 7.2) and fixed for 24 hour at 4°C. The tissues were washed twice (30 min each) with cold 0.2 M phosphate buffer (pH 7.4) and fixed in 1 per cent osmium tetraoxide for 4 h at 4°C. The tissue pieces were then dehydrated in ethyl alcohol, cleared and embedded in epon-araldite resin. Ultra thin sections (600 A°) were cut employing an ultra microtome (Ultracut, Reichert-Jung, Austria), mounted onto copper grids, and stained with uranyl acetate and subsequently with lead citrate. The kidney sections examined with a Phillips CM-10 transmission electron microscope (Phillips, Einhoven, the Netherlands) at the All India Institute of Medical Sciences, New Delhi, India.

Statistical Analysis of Data

Data generated in the present study were subjected to statistical test (Snedecor and Cochran, 1989) for calculation and comparison of mean values among different groups at 95 per cent confidence level ($P \leq 0.05$).

RESULTS

Clinical Signs

Feeding of diets amended with 3 ppm OTA led to anorexia, dullness, diarrhoea, depression in body weight, poor feathering (Figure 40.1) and in coordinated movements. In combination group (Group OS) birds same signs were observed but were comparatively less severe in intensity. In group OX, diarrhoea was more severe from 7 to 21 DPF, when almost (60-70 per cent) of the birds were seen passing out watery faeces with an increase in uric acid and mucus contents. In addition to this, nearly 5-10 per cent of the birds in the group OX showed typical nervous signs included pressing of head on the ground, backward motion to gain balance, ataxia and sudden fall on the ground from day 14 onwards.

Growth Response

Reduction in body weight and stunted growth were observed in almost all the birds fed with OTA culture material amended diets either alone or in combination from one week onward (Figure 40.2). A significant ($P \leq 0.05$) reduction in body weight was observed in OX group birds in comparison to control groups at all interval *i.e.* on 7, 14 and 21 DPF. The body weight of birds from group OS were also comparable with group OX throughout the experiment. In group SX, body weight of birds was more or less similar to plain control group (CX).

Mortality

Mortality in the group OX was observed only up to 14 DPF which was found to be highest during first and second week of the experiment in comparison to

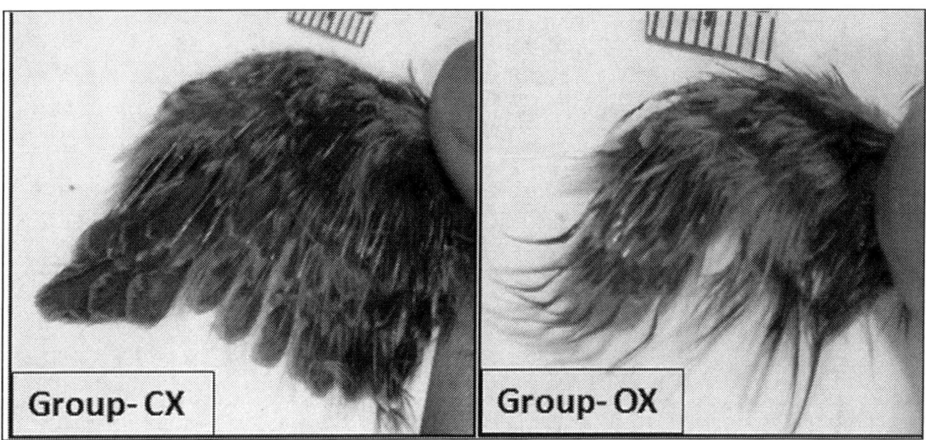

Figure 40.1: Poof Feathering Pattern of a Bird from Group OX at 7 DPF in Comparison to Control Group (CX).

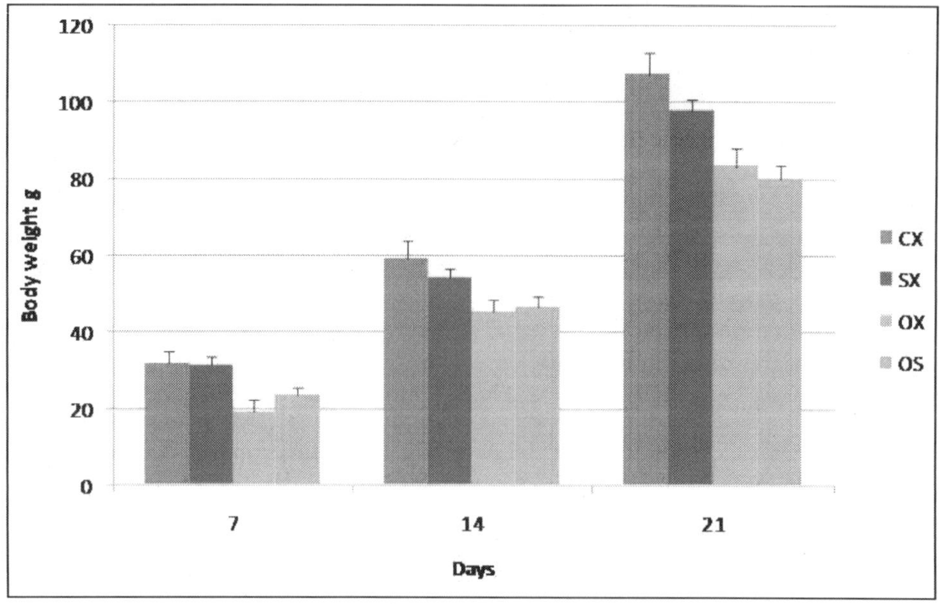

Figure 40.2: Comparative Body Weight of Birds from different Experimental Groups of Japanese Quail at Various Intervals (N=9).

group OS. The overall mortality (51.1 per cent) was found to be highest in the OX group as compared to the OS group where SBT powder given in combination with OTA in diet caused 31.1 per cent mortality during the experiment.

Serum Biochemistry

Amendment of diets with OTA showed a significant effect on total serum protein and the values were, in general, significantly lower in OTA treated groups compared to control groups (CX, SX) (Figure 40.3). At 7 DPF there was a significant decrease ($P \leq 0.05$) in TSP values in groups OX in comparison to control groups (CX, SX) whereas the values in group OS were more or less comparable to controls groups (CX, SX). On 14 DPF, however, no significant difference was found in TSP values between seabuckthorn treated and control groups (CX, SX, OS).There was a significant decrease in albumin concentration in group OX when the corresponding values were compared with groups CX and SX at 7 and 14 DPF (Figure 40.4). There was a drastic reduction in serum albumin concentration at 7 DPF in group OX in comparison to group OS. The birds in group OX showed a significant decrease in serum albumin values on 14 DPF, however, the corresponding values at 14 and 21 DPF in group OS birds were more or less comparable to the control groups (CX, SX). No significant difference in the mean uric acid concentration was observed at 7 DPF among the ochratoxicated and control group birds. At 14 and 21 DPF, a significant ($P \leq 0.05$) increase in

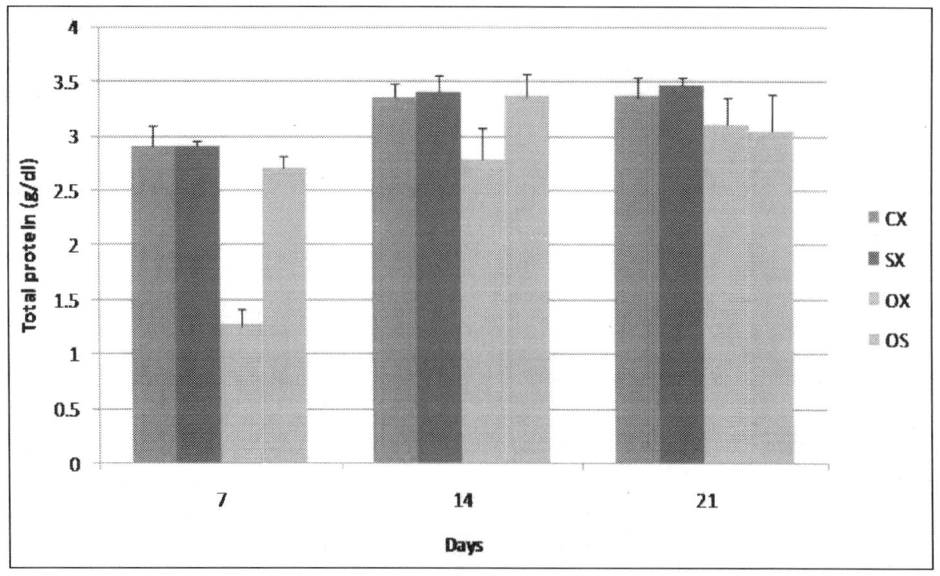

Figure 40.3: Comparative Values of Mean Serum Total Protein in Birds from different Experimental Groups of Japanese Quail at various Intervals (N=6).

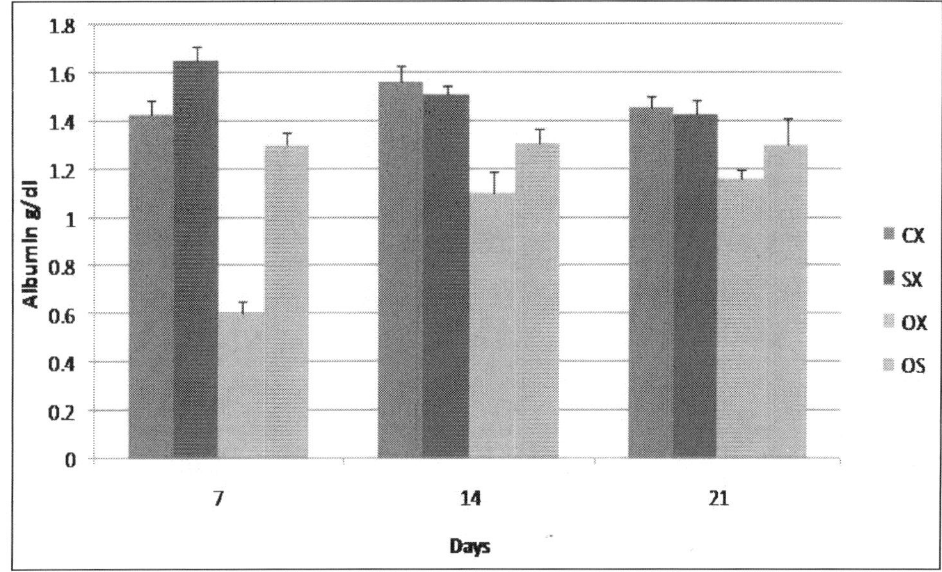

Figure 40.4: Comparative Values of Mean Serum Albumin in Birds from different Experimental Groups of Japanese Quail at Various Intervals (N=6).

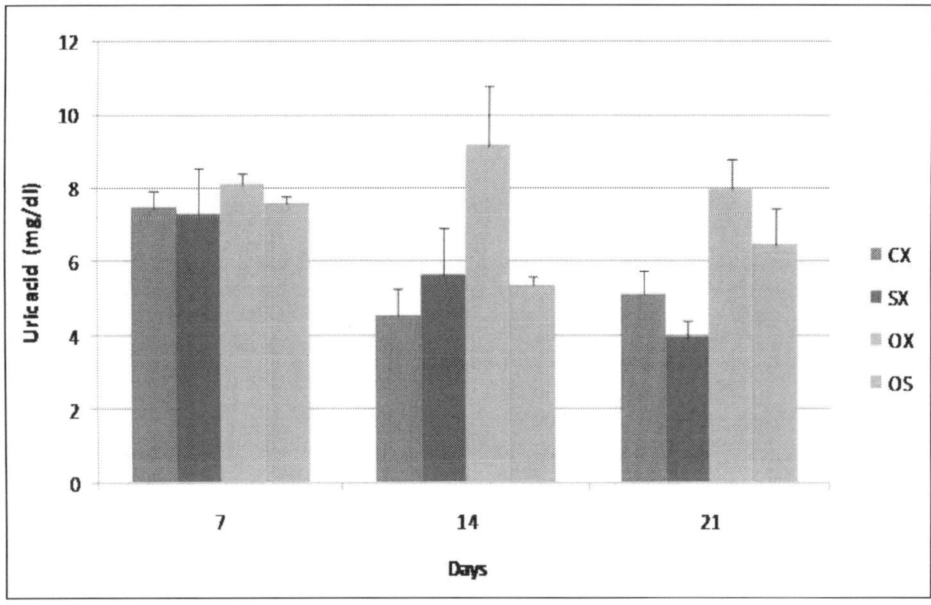

Figure 40.5: Comparative Values of Mean Serum Uric Acid in Birds from different Experimental Groups of Japanese Quail at Various Intervals (N=6).

the uric acid concentration was observed in group OX birds, however, in combination group the values were more or less comparable to the control groups (CX, SX) (Figure 40.5).

Gross Pathology

No gross lesion was observed in the kidneys of quail from control group (CX) when sacrificed at different intervals. In group SX, mild swelling and paleness of kidneys were the only lesions observed in a few birds at various intervals. In group OX, moderate to severely swollen and pale kidneys with prominent renal tubules was evident at almost all the stages of the experiment. Pale discoloration of the kidneys was evident as early as 7 DPF and was a consistent finding with varying severity until the end of the experiment. Occasional mild dilatation of the ureters was also noted. However, mild swelling and paleness of the kidneys was observed at 7 and 21 DPF in group OS birds. Similar changes with moderate intensity were, however, observed at 14 DPF. Mild dilatation of ureters was also observed in isolated birds at 21 DPF. Except for significantly swollen kidneys at 14 DPF, no other significant gross lesions were observed throughout the experiment in group OS.

Histopathology

The microscopic changes in group SX were more or less comparable to control group occasionally showing mild to moderate congestion along with a slight accumulation of hyaline material in the tubular epithelial

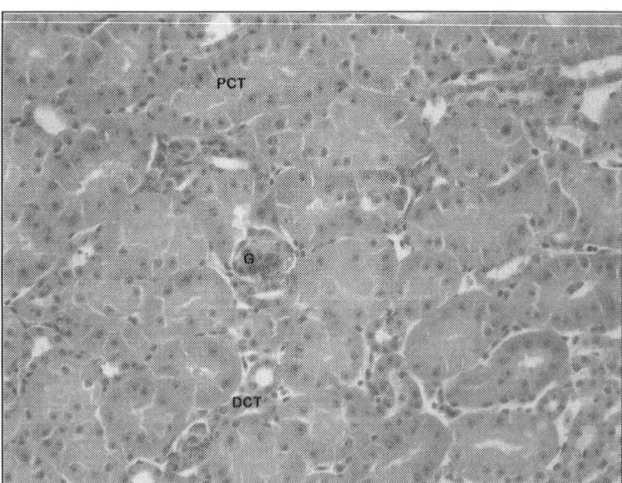

Figure 40.6: Photomicrograph of the Kidney from Control Group (Group CX) Bird at 21 DPF Showing Normal Structure of Glomeruli (G), PCTs and DCTs. H and E x 132.

cells at various intervals. In group OX, the effect of OTA was more pronounced in PCTs. At 7 DPF, mild to moderate swelling of tubular epithelial cells was noticed along with a marked increase in eosinophilic granularity in the cytoplasm. At places, there was marked sloughing of tubular epithelial cells of PCTs. The glomeruli appeared mostly swollen with marked increase in glomerular space. Glomerular capillaries were markedly congested. DCTs were also found swollen but to less extent. In addition, there was a marked increase in the hyaline material both in the epithelial cells and the lumen of PCTs. Marked variation in the size of nucleus *i.e.* karyomegaly was also a distinct finding. Focal mild interstitial infiltration with mononuclear cells was also observed. At 14 DPF, the changes were more pronounced. Some of the nuclei showed karyorrhexis alternating with pyknosis in the tubular epithelial cells. Marked karyomegaly, increased mitotic activity and the proliferative changes in the tubular epithelial cell were other associated lesions. Individual cell necrosis along with scattered heterophilic and mononuclear cell infiltration were consistent findings. In addition, there was also marked dilatation of DCTs in a few birds in the later stages of the experiment. Glomeruli showed moderate to severe proliferation of mesangial cells and swelling of glomerular epithelial cells resulting in complete obliteration of Bowman's space. Mild proliferative changes in tubular epithelial cells were also observed. At 21 DPF, lesions were more or less similar to those observed at 14 DPF but with increased intensity. In group OS, the intensity of lesions was lower in comparison to group OX. At 7 DPF, mild to moderate interstitial congestion, moderate to severe tubular swelling, swollen and disintegrated lining epithelial cells of PCTs along with vacuolation and increased granularity in the cytoplasm were common lesions in group OS. Mild regenerative, necrotic and inflammatory changes were seen in a few birds. In addition, mild to moderate proliferation of mesangial cells led to swollen glomerular tufts and obliteration of Bowman's space. At 14 DPF, the changes were more or less similar to those observed at 7 DPF except the regenerative changes which were more pronounced along with focal tubular swelling. Similarly in addition to these changes, there was distortion of tubular structure and sloughing of lining epithelial cells leading to accumulation of debris the in tubular lumen at 21 DPF.

Ultra Structural Pathology

In Group CX, the ultra thin sections of the kidneys revealed PCT epithelial cells with densely packed clear tall microvilli extending into the lumen. Mitochondria were uniformly distributed throughout the cytoplasm. They were round, oval to elongate in shape with clear outer membrane and inner membrane discernable as folding referred to as cristae. The nuclei of PCT epithelial cells were distally placed near to tubular basement membrane with uniformly distributed euchromatin and heterochromatin material. The rough endoplasmic reticulum was appreciated as thin tube like structures studded with ribosomes in the cytoplasm. The glomeruli showed clear basement membrane and normal podocytes processes.

In group SX, the kidneys revealed PCT epithelial cells with densely packed clear tall microvilli extending into the lumen. Mitochondria were uniformly distributed throughout the cytoplasm, however, mild swelling and reduction in the number of mitochondria was observed. Mild swelling of RER with degranularion and accumulation of ribosomes and polysomes was also evident. One bird also showed moderate increase in the number of peroxisomes in the cytoplasm. The glomeruli showed clear basement membrane and normal podocytes processes.

In group OX, ultra structural changes were characteristic in the epithelial cells of PCTs with mitochondria, the primarily affected organelle. Severely affected epithelial brush border was consistent findings with blunting and variable loss of microvilli. At places, the lining epithelial cells of PCTs showed electron dense cytoplasm with mitochondrial aggregations around indented nuclei. An overall reduction in the number of mitochondria was consistently observed in the epithelial cells of PCTs. Variable shaped mitochondria such round, oval, curved, elongated, dumbbell, banana, pear or tennis racket shapes, 'C' shaped, ring and occasionally forked mitochondria were more or less a consistent feature in the PCT epithelial cells. At many places, intracisternal sequestration of mitochondria undergoing involution and degeneration was also noticed. The mitochondria revealed indistinct outer membrane, swollen cristae substance, loss of cristae appearing as debris and sieve like appearance due to loss of membrane integrity of cristae. There was a slight derangement in the cisternae

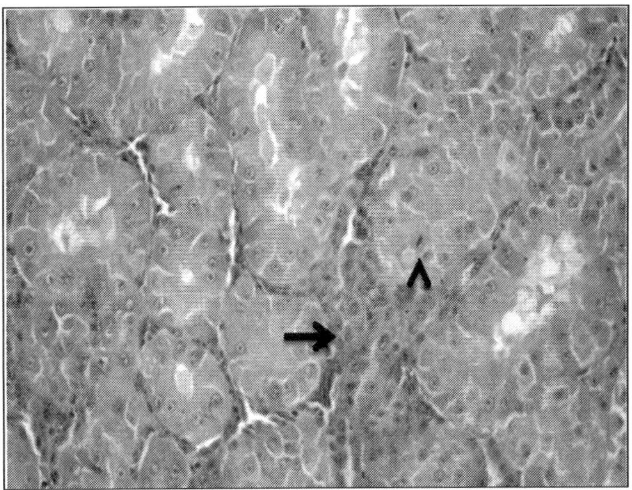

Figure 40.7: Photomicrograph of the Kidney from Group OX Bird at 21 DPF Showing Swollen PCTs, Mitotic Figure (Arrow head) and Interstitial MNCs Infiltration (Arrow). H and E x 132.

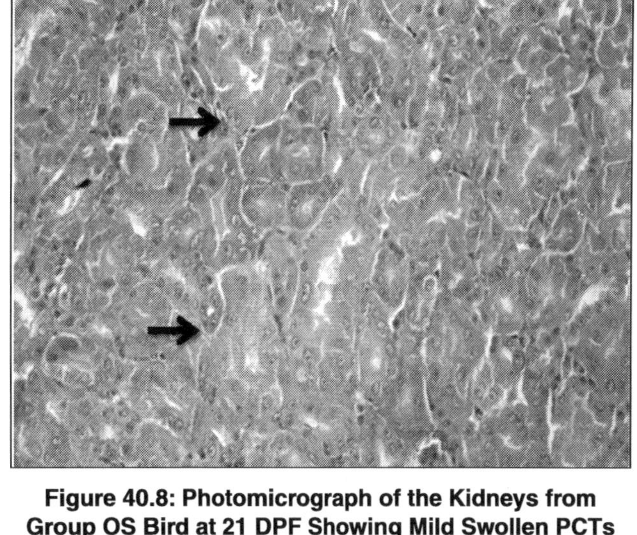

Figure 40.8: Photomicrograph of the Kidneys from Group OS Bird at 21 DPF Showing Mild Swollen PCTs with Presence of Double Nuclei in the PCTs Epithelial Cells (Arrows). H and E x 132.

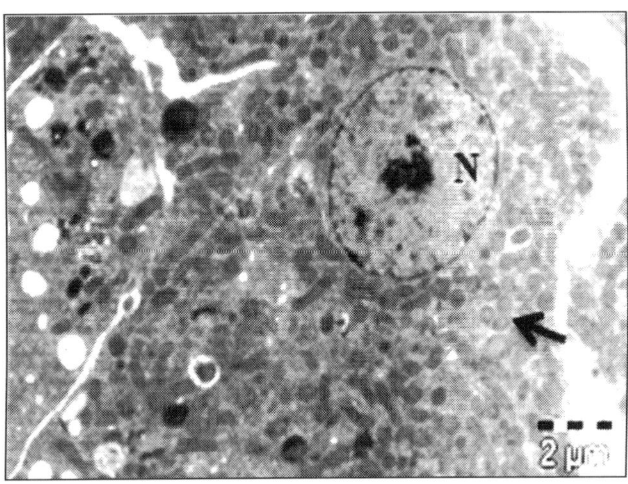

Figure 40.9: TEM Image of the Kidney from Group CX Bird at 21 DPF Showing Even Distribution of Mitochondria (Arrow) Throughout the Cytoplasm.

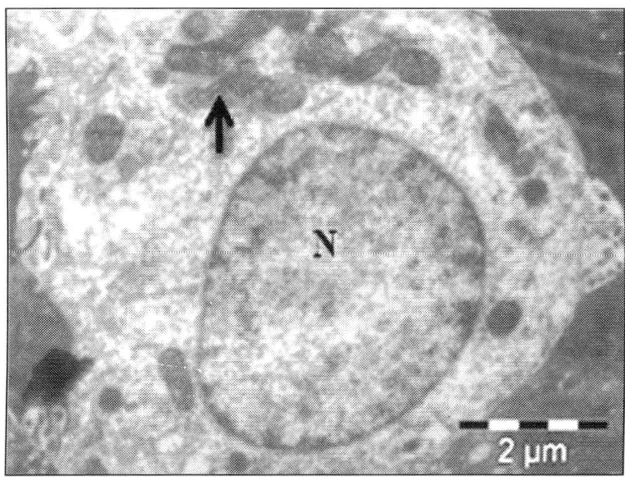

Figure 40.10: TEM Image of the Kidney from Group OX Bird at 21 DPF Showing a Marked Decrease in the Number of Mitochondria (Arrow) along with Disappearance of other Cell Organelles from the Cytoplasm of a PCT Epithelial Cell.

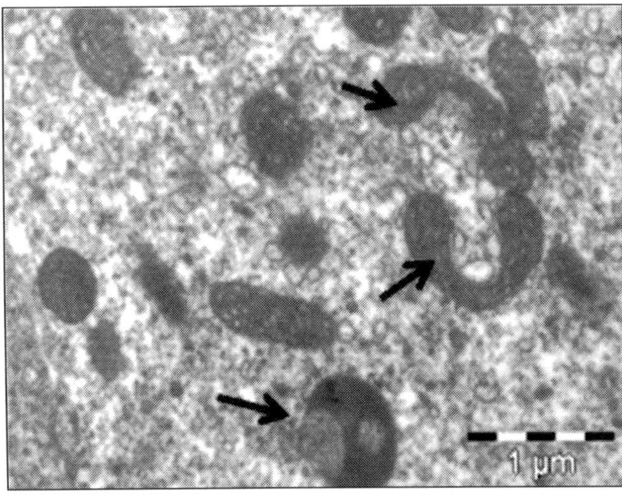

Figure 40.11: TEM Image of the Kidney from Group OX Bird at 21 DPF Showing 'C' Shape and Ring Shape Mitochondria (Arrows).

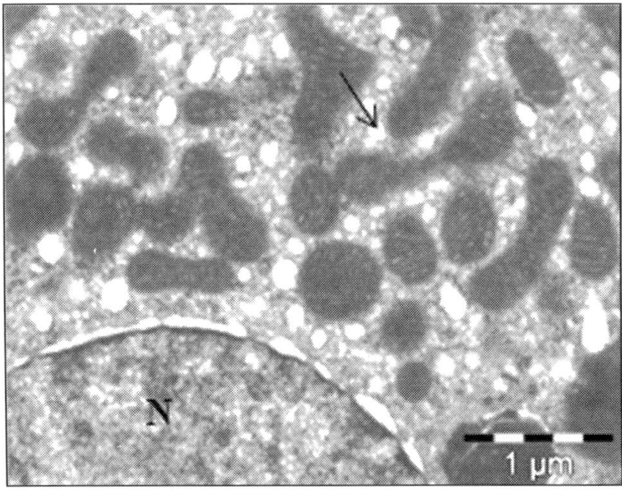

Figure 40.12: TEM Image of the Kidney from Group OS Bird at 21 DPF Showing Degenerating Mitochondria with Indistinct Outer Membrane (Arrow).

structure of ER resulting in the formation of round or even tubular structures in the cytoplasm or occasionally massive uneven dilatation of ER bearing granules over the surface. The moderate degranulation of RER and accumulation of polysomes were also evident. Nucleus invariably showed indentation of nuclear membrane along with uneven distribution of chromatin material with margination, karyomegaly, indistinct nuclear membrane, two nuclei and two nucleoli. A few nuclei of PCT epithelial cells showed clumping of chromatin material with complete loss of distinguishibility of the nuclear membrane. Some of the cells of PCTs were shrunken with almost complete loss of cytoplasmic organelles with indentation of nuclear membrane and formation of cytoplasmic blebs (apoptosis). Glomerular basement membrane showed uneven thickness at various places. Foot processes of podocytes showed thickening and fusion resulting in distorted shapes of podocytes. Urinary spaces were suspected to contain less electron dense proteinaceous material.

In groups OS, the kidneys revealed intact epithelial brush border of PCTs and microvilli were seen extending into the lumen of PCTs. The cytoplasm of PCTs was densely packed with mitochondria in conformity with the control groups (CX,SX). Mitochondria were mild to moderately swollen with presence of elongated mitochondria, however, dumble shape and tennis ball shape mitochondria were also recorded. The overall extent of mitochondrial pleomorphism was less in comparison to group OX. Occasionally, presence of involuting mitochondria was also evident. RER were moderately swollen with moderate to severe degranulation, accumulation of ribosomes and polysomes. Loss of nucleus and nucleolus was present up to milder extent, however, dividing nuclei, double nucleoli and mitotic spindle formation were also not uncommon. Indistinct nuclear membrane with mild to moderate chromatin condensation and margination were also noted. A pronounced increase in the number of peroxisomes was also evident. Glomerular basement membrane was slightly thickened and there was only mild fusion and thickening of podocytes foot processes.

DISCUSSION

Results of the present study showed that feeding of diets amended with 3 ppm of OTA lead to anorexia, diarrhoea, depression in body weight, poor feathering and incoordinated movements. Dwivedi and Burns (1984a) found OTA extremely toxic to young growing broiler chicks (2 ppm and 4 ppm) and the severity of the clinical signs, behavioral changes, growth response and lesion were all dose dependent. Similar clinical signs in chicks fed with 5 ppm OTA levels in feed were observed by Stoev *et al.* (2002a) like weakness and dullness, growth depression and some nervous symptoms such as torticollis, lurch, reeling gait, tremor, side-split, sitting posture or layer fashion with tense or flexion of legs. In combination group (OS) signs were comparatively less severe in intensity further suggesting protective effect of seabuckthorn.

There was a reduction in body weight in OX treated group which was comparatively better in OS group. The reduced feed consumption may well have contributed to the depression of body weight observed during the experiment. According to some workers (Mohiudin *et al.*, 1993), the growth depression in OTA treated chicks may be due to impaired protein synthesis, provoked by OTA. Huff *et al.* (1988) reported lower growth rates in chicks fed OTA at the dose rate of 4 µg per g feed as early as 6 days of age. The growth rates remained lower than control values throughout 21 days of age. The presence of diarrhoea may have been additional factors contributing to reduction in body weight in all the culture material amended groups. In seabuckthorn control group (SX), body weight was more or less similar to plain control group (CX) that showed that addition of seabuckthorn leaf powder in feed had no adverse effect on feed consumption. Moreover, it was clear that control and SBT treated quail chicks utilized better feed than those treated with OTA alone.

Amendment of diets with OTA showed a significant effect on total serum protein and albumin. The values were, in general, significantly lower in OTA treated groups compared to control groups. Manning and Wyatt (1984), Ramadevi *et al.* (2000), Stoev *et al.* (2000) and Kumar *et al.* (2003) reported decreased serum proteins during induced ochratoxicosis in broilers. The hypoproteinemia might be due to the mechanism by which OTA causes disruption in serum protein levels. The inhibition of serum protein synthesis through competitive inhibition of phenylalanyl-t-RNA synthetase with phenylalanine has been reported by Konrad and Roschenthaler (1977) and Bunge *et al.* (1978). The TSP values in group OS were found comparable to control groups at all interval. The results have clearly suggested that SBT leaf powder have some protective effects against the toxicity of OTA as the protein concentration did not change appreciably in relation to both the control groups (CX, SX).

Mean uric acid values were found higher in OTA treated groups but in seabuckthorn combination groups the rise was comparatively lower, indicative of some protective activity. Stoev *et al.* (2002b) suggested that the significant increase in the serum levels of uric acid showed that the kidney function was severely impaired as a result of OTA treatment, which may additionally contribute to the decrease of the serum total protein by increasing of its urinary excretion. Stoev *et al.* (2002a) reported decrease in uric acid concentration in artichoke, sesame seed and OTA (5 ppm) combination groups in comparison to OTA alone group.

The gross lesions recorded in the kidneys were mainly paleness and swelling and were present in all OTA treated groups but the intensity of lesion was less in SBT treated group. Grossly, swollen kidneys noticed in the present study were in conformity with earlier studies conducted in chicken by Elaroussi *et al.* (2006, 2008), Dwivedi *et al.* (1984a), and Kubena *et al.* (1989) and the effect was found to be dose dependent. In a similar study conducted on rats, Prasanna *et al.* (2007) observed gross lesions from 3rd week onwards and the kidneys were markedly enlarged with occasional petechial haemorrhages in the group treated with 0.5 mg/kg b.wt. OTA levels daily. The results of the present study have clearly suggested a partial ameliorating effect of SBT leaf powder in OTA toxicity. Similar to present study, Stoev *et al.* (2002a) reported reduced liver and kidneys weight in antidote treated (artichoke, sesame seed and phenylalanine) groups in comparison to OTA group fed at 5ppm levels to broiler chicks. Srivani *et al.* (2003) reported mild enlargement of the kidneys at 42nd day was observed in the broiler fed with 1 ppm aflatoxin (1 to 14 days) in combination with immunomodulators (tuftsin, stresroak and zinc sulphate) in diet. In other body organs, no visible change was observed grossly.

The intensity of gross lesions in group OS was lower in comparison to group OX and the biochemical parameters also supported this fact. In a similar study, Kurkure *et al.* (2000) observed that liver in turmeric (0.5 g/kg feed) and aflatoxin (1 ppm) combination fed group was less paler in comparison to AFB_1 alone group and also lymphoid organs were less atrophied and paler in turmeric supplemented group. Pathan *et al.* (2006) reported that there was marked improvement in gross pathological changes of various organs after addition of Bio - Bantox (5kg/ton of feed) along with OTA (1.0 mg/kg of feed) in diets of broiler chicks.

The microscopic changes in group SX were more or less comparable to control group and further suggested that supplementation of SBT leaf powder at 2 per cent level has no harmful effect on the kidneys. The enlargement and degenerative changes in the epithelial cells of the kidneys in group OX were probably due to route of elimination of OTA through the kidneys and partly through the liver (owing to enterohepatic recirculation and the hepatobiliary route of excretion of OTA), and to direct toxic action of OTA on these cells (Stoev *et al.*, 2000). Similarly, Elaroussi *et al.* (2008) reported cloudy swelling, granular degenerative changes in the cytoplasm and pyknotic nuclei in the epithelial cells of PCTs when broiler chicks were fed culture material containing 4 and 8 ppm OTA for 1-35 days. The results were also supported by better response in group OS in comparison to group OX on the basis of gross, microscopic and biochemical parameters and also in accordance with Stoev *et al.* (2002a) who reported reduced intensity of microscopic lesions in chicks fed with artichoke extract and sesame seed in addition to OTA (5 ppm) in diets.

In ultra structural studies, the ring form of mitochondria was appeared to be caused by fusion of membranes at the tips of curved and elongated mitochondria (Dwivedi and Burns, 1984). Ring and 'U'-shaped mitochondria have been observed in the liver of rats treated with carbon tetrachloride (Renoylds, 1960) and ammonium carbonate (David and Kettler, 1961). However, the significance of such transformation of mitochondrial morphology is not known (Ghadially, 1982). Urinary spaces were suspected to contain less electron dense proteinaceous material, which indicated the extent of damage to glomeruli and this change was also visible histopathologically as hyaline bodies in tubular lumen as well as in the cytoplasm of PCT epithelial cells. The reason for this is that protein spillover in to the glomerular filtrate is followed by reabsorption in the cells of PCT. Large, round hyaline protein absorption droplets were visible histologically. Ultrastructurally, protein absorption droplet appeared dense, membrane bound bodies. They represent secondary lysosomes that form following endocytosis of protein. Epithelial brush border of PCTs with blunting and loss of microvilli noticed in the present study were consistent with the previous studies made by Solcan *et al.* (2008) in broiler chickens fed 1, 9 and 20 mg/kg OTA for three weeks and by Dwivedi *et al.* (1984 b) after feeding 2 and 4 ppm OTA in broiler chicks for 20 days.

Foot processes of podocytes fuse, slit pores are obliterated, and intercellular junctions are narrowed and lengthened (Sternberg 1970; Venkatachalam and Rennke, 1978). Foot process fusion is a consequence of injury to the podocytes. Studies with the glomerular toxin aminonucleoside have shown that it involves swelling and blunting of foot processes with narrowing of filtration slits; formation of cell junctions between podocytes processes; displacement of slit diaphragm; development of podocytes extensions in to the basement membrane and excessive polymerization of actin microfilaments in foot processes (Caulfield, 1979; Kerjaschki, 1978). The widening of intercellular spaces was also evident in all the birds. Expansion of intercellular spaces indicates the acute cellular swelling and it starts with the separation and shut down of gap junctions, brought about by decreasing pH (cellular acidosis) and increasing calcium concentrations (James *et al.*, 1986). The lost membrane integrity of the cell organelles may be due to a specific inhibition of the protein synthesis by OTA, and subsequent impairment of lipoprotein structures in the cells (Creppy *et al.*, 1983). Moreover, OTA inhibits mitochondrial oxidative phosphorylation by acting as a competitive inhibitor of carrier proteins located in the inner mitochondrial membrane (Meisner and Chan, 1974; Elling *et al.*, 1985).

The earlier studies concluded that nephrotoxic process induced by OTA is driven by two independent pathways: one involving a mainly oxidative pathway and another involving the biotranformation into gentoxic OTA derivatives (Pfohl-Leszkowicz *et al.*, 2002). On the other hand SBT leaf powder, which is rich in antioxidants (Hakkinen *et al.*,1999; Geetha *et al.*, 2003), might have provided the protection against the free radicals generated in the oxidative pathway.

CONCLUSION

From the findings of the present study it is found that the inclusion of SBT powder at 2 per cent level showed no harmful effects on the kidneys as well as on the body weight. Moreover, the inclusion of 2 per cent seabuckthorn leaf powder was found to be beneficial against the harmful effects of ochratoxicosis based on clinico-pathological and ultrastructural studies, although the protection was partial. Further studies are indicated on the use of seabuckthorn leaves preparations or its extracts either alone or in combination with various mycotoxin binders.

REFERENCES

1. Alvarez, L., Gil, A.G., Ezpeleta, O., Garcý-Jalon, J. A. and Lopez de Ceraina, A. 2004. Immunotoxic effects of Ochratoxin A in Wistar rats after oral administration. *Food and Chemical Toxicology* 42: 825-834.
2. Bunge, I.G., Dirheimer and Roschchenthaler, R. 1978. *In vivo* and *in vitro* inhibition of protein synthesis in *Bacillus stearothermophilus* by ochratoxin A. *Biochemistry, Biophysics Research Communications* 83 (2): 398-405.
3. Caulfield, J.P. 1979. Alteration in the distribution of alcian blue staining fibrillar anionic sites in the glomerular basement membrane in aminonucleoside nephrosis. *Laboratory Investigations* 40 (1): 503.
4. Creppy, E.E., Stormer, F.C., Kern, D., Roschenthaler, R. and Dirheimer, G. 1983. Effects of ochratoxin A metabolites on yeast phenylalanyl-tRNA syntheyase and on the growth and "*in vivo*" protein synthesis of hepatoma cells. *Chemical and Biological Interactions* 47 (2): 239-247.
5. David, H. and Kettler, L.H. 1961. Degeneration von Lebermitochondrien nach ammonium-intexikation Zeitschrift flir Zel!forschung und Mikroskopische. *Anatomie* 53: 857-858.
6. Dwivedi, P. and Burns, R.B. 1984a. Pathology of ochratoxin A in young broiler chicks. *Research in Veterinary Science* 36 (1): 92-103.
7. Dwivedi, P., Burns, R.B. and Maxwell, M.H. 1984b. Ultrastructural study of the liver and kidney in ochratoxicosis A in young broiler chicks. *Research in Veterinary Science* 36 (1): 104-116.
8. Elaroussi, M.A., Mohamed, F.R., Barkouky, E.M., Atta, A.M., Abdou, A.M. and Hatab, M.H. 2006. Experimental ochratoxicosis in broiler chickens. *Avian Pathology* 35 (4): 263-269.
9. Elaroussi, M.A., Mohamed, F.R., Elgendy, M.S., Barkouky, E.I., Abdou, E.M. and Hatab, M.H. 2008. Ochratoxicosis in broiler chickens: Functional and histological changes in target organs. *International Journal of Poultry Science* 7 (5): 414-422.
10. Elling, F., Nelsen, J.P., Lillehoj, E.B., Thomassen, M.S. and Stormer, F.C. 1985. Ochratoxin A induced porcine nephropathy: Enzyme and ultrastructural changes after short term exposure. *Toxicon* 23 (2): 247-254.
11. Gao, X., Ohlander, M., Jeppson, N., Bjork, L. and Trajkovsky, V. 2000. Changes in antioxidant effects and their relationship to phytonutrients in fruits of seabuckthorn (*H. rhamnoides* L.) during maturation *Journal of Agricultural Food Chemistry* 48 (5): 1485-90.
12. Geetha, S., SaiRam, M., Mongia, S.S., Singh, V., Ilavazhagan, G. and Sawhney, R.C. 2003. Evaluation of antioxidant activity of leaf extract of seabuckthorn (*Hippophae rhamnoides* L.) on chromium (VI) induced oxidative stress in albino rats. *Molecular and Cellular Biochemistry* 87 (2-3): 247-51.
13. Ghadially, F.N. 1982. *Ultrastructural pathology of the cell and matrix*. 2nd edition London, Butterworths 80: 222-582.
14. Hakkinen, S. H., Karenlampi, S. O., Heinonen, M.I., Mykkanen, H.M. and Torronen, A. R. 1999. Contents of flavanols, quercetin, myriecetin and kaempferol in 25 edible berries. *Journal of Agricultural Food Chemistry* 47 (6): 2274-79.
15. Huff, W.E., Kubena, L.F. and Harvey, R.B. 1988. Progression of ochratoxicosis in broiler chickens. *Poultry Science* 67 (8): 1139-1146.
16. James, J. L., Friend, D.S., MacDonald, J.R. and Smuckler, E.A. 1986. Alteration in hepatocyte plasma membrane in carbon tetrachloride poisoning. *Laboratory Investigations* 54 (3): 268-274.
17. Kerjaschki, D. 1978. Polycation-induced dislocation of slit diaphragm and formation of cell junctions in rat kidney glomeruli. *Laboratory Investigations* 39 (5): 430-440.
18. Konrad, I. and Roschenthaler, R. 1977. Inhibition of phenyalanine to t-RNA synthetase from *Bacillus subtelis* by OTA. *FEBS letter* 83 (2): 341-347.
19. Kubena, L.F., Harvey, R.B., Huff, W.E. and Corrier, D.E. 1989. Influence of ochratoxin A and T-2 toxin singly and in combination on broiler chickens. *Poultry Science* 68 (7): 867-872.
20. Kumar, A., Jindal, N., Shukla, C.L., Pal, Y., Ledoux, D.R. and Rottinghaus, G.E. 2003. Effect of ochratoxin A on *Escherichia coli* challenged broiler chicks. *Avian Disease* 47 (2): 415-424.
21. Kurkure, N.V., Pawar, S.P., Kognole, S.M., Bhandarkar, A.G., Ganorkar, A.G. and Kalorey, D.R. 2000. Ameliorative effect of turmeric (*Curcuma longa*) in induced aflatoxicosis in cockerels. *Indian Journal of Veterinary Pathology* 24 (1): 26-28.

22. Luna, L.G. 1968. *Manual of histological staining methods of the Armed Forces Institute of Pathology*. 3rd edn. McGraw Hill Book Company, New York.

23. Manning, R.O. and Wyatt, R.D. 1984. Toxicity of *Aspergillus ochraceus* contaminated wheat and different chemical forms of ochratoxin A in broiler chicks. *Poultry Science* 63 (3): 458-465.

24. Meisner, H. and Chan, S. 1974. Ochratoxin A inhibitor of mitochondrial transport system. *Biochemistry New York* 13 (14): 2795-2799.

25. Mohiudin, S.M., Warasi, S.M.A. and Reddy, M.V. 1993. Haematological and biochemical change in broiler chickens. *Indian Veterinary Journal* 70 (2): 613-617.

26. Panisup, A.S., Biswas, J.C., Mahmood, S., Singh, Y.P. and Koul, G.L. 1993. Pathoepidemiology of spontaneous ochratoxicosis in Pashmina goats. *Indian Journal of Veterinary Pathology* 17: 26-31.

27. Pathan, A.C., Moregaonkar, Bhikane, A.U. and Gaikwad, N.Z. 2006. Ameliorative efficacy of bio-bantox in experimental ochratoxicosis in broilers. *Indian Journal of Veterinary Pathology* 30 (2): 25-27.

28. Peckham, J.C., Doupnik, B.and Jones, O.H. 1971. Acute toxicity of ochratoxin A and B in chicks. *Applied Microbiology* 21 (3): 492-494.

29. Pfohl-Leszkowicz, A., Bartsch, H., Azémar, B., Mohr, U., Estève, J. and Castegnaro, M. 2002. MESNA protects rats against nephrotoxicity but not carcinogenicity induced by ochratoxin A, implicating two separate pathways. *Facta University Ser. Medical Biology* 9 (1): 57-63.

30. Prasanna, K., Sharma, A.K., Dwivedi, P., Satheesh, C.C. and Telang, A.G. 2007. Pathogenesis of ochratoxin A induced nephropathy in rats with special reference to oxidative damage, cytokine profile and ultrastructural changes. *Indian Journal of Veterinary Pathology* 31 (1): 1-10.

31. Raju, M.V.L.N. and Devegowda, G. 2000. Influence of esterified-glucomannan on performance and organ morphology, serum biochemistry and hematology in broilers exposed to individual and combined mycotoxicosis (aflatoxins, ochratoxin and T-2 toxin). *British Poultry Science* 41 (5): 640-650.

32. Ramadevi, N.R., Gopal, R. and Ravikumar, P. 2000. An assessment of the protective effect of bentonite on ochratoxicosis in broiler with reference to certain haematological profile. *Indian Veterinary Journal* 77 (3): 303-306.

33. Reynolds, E.S. 1960. Cellular localisation of calcium deposition in liver of rat poisoned with carbon tetrachloride. *Journal of Histochemistry and Cytochemistry* 8 (1): 331-332.

34. Samson, R.A. 1992. Mycotoxins: A mycologists perspective. *Journal of Medical and Veterinary Mycology Suppl.* 30 (1): 9-18.

35. Santin, E., Paulillo, A.C., Maiorka, P.C., Alessi, A.C., Krabbe, E. and Maiorka, A. 2002. The effects of ochratoxin/aluminosilicate interaction on the tissues and humoral immune response of broilers. *Avian Pathology* 31 (1): 73-79.

36. Snedecor, G.W. and Cochran, W.G. 1989. *Statistical Methods*. (8th ed), Iowa State University Press, Ames, USA.

37. Solcan, C., Coman, I., Solcan, G., Miron, L. and Oprean, O.Z. 2008. Histological and ultrastructural lesions of the kidneys in experimental ochratoxicosis of broilerchickens. *Bulletin of University of Agricultural Sciences and Veterinary Medicine, Cluj-Napoca-Veterinary Medicine* 65 (1): 91-95.

38. Srivani, M., Anjaneyulu, Y., Rao, A.S., Sarma, B.J.R., and Raju, M.L.V.N. 2003. Efficacy of immunomodulators on induced aflatoxicosis in broiler chicks– A pathological study. *Indian Journal of Veterinary Pathology* 27 (1): 27-29.

39. Sternberg, S.S. 1970. Cross striated fibrils and other ultrastructural alteration in glomeruli of rats with daunomycin nephrosis. *Laboratory Investigation* 23: 39.

40. Stoev, S.D., Anguelov, G., Ivanov, I. and Pavlov, D. 2000. Influence of ochratoxin A and an extract of artichoke on the vaccinal immunity and health in broiler chicks. *Experimental and Toxicological Pathology* 52 (1): 43-55.

41. Stoev, S. D., Djuvinov, D., Mirtcheva, T., Pavlov, D. and Mantle, P. 2002a. Studies on some feed additives giving partial protection against ochratoxin A toxicity in chicks. *Toxicology Letters* 135: 33-50.

42. Stoev, S.D., Koynarsky, V. and Mantle, P.G. 2002b. Clinicopathomorphological studies in chicks fed ochratoxin A while simultaneously developing coccidiosis. *Veterinary Research Communications* 26 (3): 189-204.

43. Venkatachalam, M.A. and Rennke, H.G. 1978. The structure and molecular basis of glomerular filtration. *Circulation Research* 43 (3): 337-347.

Chapter 41

Investigating the Antioxidant Properties and Rutin Content of Seabuckthorn (*Hippophae rhamnoides* L.) Leaves and Branches

Aliye Aras Perk, F. Duygu Ceylan, Onur Yanar and Esra Capanoglu

Istanbul University, Faculty of Science, Department of Botany, Istanbul 34460, Turkey
Istanbul Technical University, Faculty of Chemical and Metallurgical, Food Engineering Department, Istanbul 34469, Turkey

SUMMARY

Present study focused on the antioxidant properties and rutin content of leaves and branches of *Hippophae rhamnoides* (seabuckthorn). Dried leaves, branches and processed branches of seabuckthorn (SBT) were extracted with ethanol and prepared in forms of tea infusions. All samples were analyzed for their contents of rutin, total phenolics (TPC), total flavonoids (TFC) and total antioxidant capacity by using DPPH and CUPRAC methods. TPC of leaves from ethanolic extracts (EE) were significantly higher than branch and processed branch extracts. The DPPH scavenging activity of extracts ranged from 41.93 ± 3.57 and 132.43 ± 6.57 mg TE/g, and the antioxidant capacity measured with CUPRAC method were in between 129.4 ± 18.1 to 538.5 ± 34.8 mg TE/g. Except for the DPPH method, other methoindicated a decrease by the effect of infusion. Both EE and AE of leaf samples had significantly higher rutin content compared to the branches and processed branches.

Keywords: *Seabuckthorn, Hippophae rhamnoides, Leaves, Antioxidant, Phenolic profile and Rutin content.*

INTRODUCTION

Seabuckthorn (SBT) (*Hippophae□ rhamnoides* L.) is a thorny nitrogen-fixing deciduous shrub with good adaptability to various climate conditions and extensive genetic variability with numerous greenish-yellow flowers and bright orange, globular, ellipsoid fruit. It is currently domesticated in several parts of the world (India, China, Nepal, Pakistan, Myanmar, Russia, Britain, Germany, Finland, Romania, France, *etc.*) due to its nutritional and medicinal properties (Khan *et al.*, 2010; Suryakumar and Gupta, 2011; Yogendra Kumar *et al.*, 2011; Pop *et al.*, 2013). In Turkey, SBT is widely distributed throughout North and East Anatolia and known locally as 'Yalancı igde, Cıcırgan, Dijırgan, Cıcılık' or 'Sincan Çalısı' (Aras -Tayhan, 1997). All parts of the plant are considered to be good source of large number of bioactive substances like vitamin, carotenoids, and flavonoids, organic acids, sterols and some essential amino acids. For instance, berries and seeds of SBT are well known for their antioxidative properties, attributed to hydrophilic and lipophilic compounds including ascorbic acid, flavonoids, proanthocyanidins and carotenoids. The leaves of SBT are also considered for their antioxidant potential correlated to flavonoids and phenolic acids derivatives, and they have been used in some countries to make extracts, tea, animal feed, pharmaceuticals and cosmetics (Michel *et al.*, 2012; Wani *et al.*, 2013).

Rutin is one of the bioactive flavonoid compounds, which are present in substantial amounts in various plants (Attanassova and Bagdassarian, 2009). Grapes and buckwheat are the most important rutin containing foods

Seabuckthorn (Hippophae L.): A Multipurpose Wonder Plant, Vol. 5: Advanced Technologies for Cultivation, Processing, Health Protection and Environmental Conservation (Virendra Singh, *Editor-in-Chief*, 2018), p. 329–334, Astral International Pvt. Ltd., New Delhi, India.

between fruits, vegetables and grain crops (Kreft *et al.*, 2006). The natural rutin (32,42,5,7-tetrahydroxy- flavone-3-rutinoside) is one of the attractive phytochemicals because of its pharmacological activities. Rutin has been also reported to have beneficial effects in numerous diseases in humans, including cancer and cardiovascular diseases. In addition, their antioxidant, antiinflammatory, antitumoral and antiviral properties were also noted (Nieoczym *et al.*, 2014).Therefore, it is considered as an important flavonoid in pharmaceutical industry (Chua, 2013).The aim of the present study is to investigate the antioxidant properties and rutin content of seabuckthorn leaves and branches.

MATERIALS AND METHODS

Plant Material

Wild seabuckthorn (*H. rhamnoides* L.) were harvested from Central Anatolia, particularly Sivas region (Figure 41.1 and Table 41.1). The leaves and branches were separated from each other and ventilated. In order to inhibit the activation of oxidation enzyme, leaves and branches were subjected to shocking procedure. For this purpose plant materials were shocked with 90-100 °C vapor in rotary evaporator. Output materials of the shocking unit were cooled with dry air. Cooled materials were subjected to folding procedure for 25-30 min. After folding unit, leaves and branches are transferred into the drying chamber and dried at 110°C. At the end of this treatment leaf tea and processed branch were obtained. These materials were cut into small pieces and stored in the dark at room temperature before use.

Dried leaves, branches and processed branches of *Hippophae rhamnoides* were extracted with ethanol as described below and also prepared in forms of tea infusions and further analyzed. All samples were analyzed for their contents of total phenolics and total flavonoids as well as their total antioxidant capacity by using two different methods including 1.1-diphenyl-2-picrylhydrazyl (DPPH) and CUPRAC (Copper Reducing Antioxidant Capacity). The samples were also analyzed by using HPLC to identify their phenolic profiles.

Extraction Procedure

Two independent extractions were carried out for dried leaves, branches and processed branches samples as described previously by Maheshwari *et al.* (2011) with slight modifications. 2.5±0.01 g of each sample was extracted with 15 ml of 70 per cent aqueous-ethanol in a cooled ultrasonic bath (Azakli, Turkey) for 15 min. After 24 h incubation, the treated samples were centrifuged (Hettich Zentrifugen Universal 32R, UK) for 10 min at 4000 rpm at 4°C and the supernatant were collected.

Table 41.1: Some Characteristics of the Sampling Sites

Sampling Site/ Sampling Period	*Latitude/Longitude*	*Altitude/Longitude*	*Annual Precipitation (mm)*	*Annual Temperature (°C)*	*Climate Type*
Sivas Sincan stream June-August 2013	39054'N–	2000m	794	2.8	Humid

Figure 41.1: Location of *H. rhamnoides* L. Samples from Turkey.

Then 15 ml 70 per cent aqueous-ethanol was added to the pellet and this extraction procedure was repeated two more times. All four supernatants were combined and adjusted to a final volume of 50 mL.

For the preparation of tea infusions, 2±0.01 g of dried leaves were weighed and dissolved in 200 ml of pure water at 90°C. After waiting for 30 minutes, the leaves were removed and the infusion was filtered, and further analyzed.

In the softening process, similar technique to the preparation of wood for microscopic examination was applied. 2±0.01 g of processed branch and branch were weighed and added 75 ml of water. It was heated until boiling. At the boiling point 25 ml of pure water was added and was reboiled. This procedure repeated four more times until final volume of 200 ml. Branches were removed and extract was filtered and further analyzed. Prepared extracts were stored at -20 °C until analysis.

Determination of Total Phenolic Content (TPC) and Total Flavonoid Content (TFC)

The TP of extracts was determined using Folin-Ciocalteu reagent according to the method modified from Velioglu *et al.* (2006) using gallic acid as a standard. A mixture of 100 µl of the extract, 900 µl of distilled water and 1.5 ml of 0.2 N Folin-Ciocalteu reagent was prepared and allowed to react for 5 minutes. Then 1.2 ml per cent 7.5 Na_2CO_3 solution added into the reaction mixture. After incubation for 90 minutes at room temperature, the absorbance was measured at 765 nm using Optima SP-3000 nano spectrophotometer. The TP of extracts was expressed as mg of gallic acid equivalent (GAE) per g sample. The TF was measured colorimetrically as described by Kim *et al.* Absorbance of the mixture was measured at 510 nm versus prepared blank. The TF of extracts was determined by a (+)-rutin standard curve and expressed as mg of (+)-rutin equivalent (RE) per g sample.

Determination of Total Antioxidant Capacity (TAC)

The free radical scavenging activity of seabuckthorn leaves and branches aqueous and ethanolic extracts on DPPH radical were determined according to the method introduced by Kumaran *et al.* (2006) and Rai *et al.* (2006). Decolourisation of purple free radical DPPH solution was measured at 517 nm after 30 min incubation in the dark and at room temperature. A trolox calibration curve was done between 0.01 and 0.2 mg/ml. Results were expressed in mg of trolox equivalents/g of extract (mg TE/g).

As a second antioxidant assay, CUPRAC method was applied according to Apak *et al.* (2005). After 30 minutes, absorbances were measured at 450 nm against reagent blank. Results were expressed in mg of Trolox equivalents/g of sample.

HPLC Analysis of Flavonols

Major flavonols were determined following the method of Capanoglu *et al.* (2008). Extracts were filtered through a 0.45 µm membrane filter and analyzed using a Waters 2695 HPLC system with a PDA (Waters 2996) detector. A Supelcosil LC-18 25 cmx4.60 mm, 5 µm column (Sigma-Aldrich, Steinheim, Germany) was used. Identification was based on the retention times and characteristic UV spectra. Quantification was done using external standards as well as taking the information from the literature into account.

Statistical Analysis

All the experiments were replicated three times and the data were represented as mean ± SD. For multiple comparisons, data were subjected to statistical analysis using SPSS software (version 16.0 for Windows, SPSS Inc.) for the analysis of variance (ANOVA). Duncan's new multiple range test was used to analyze differences between treatments ($p<0.05$).

RESULTS AND DISCUSSION

H. rhamnoides is generally reported to be rich in a wide range of biologically active substances. Due to the antioxidant properties of seabuckthorn leaves (SBTL) the number of studies investigating their potential utilization has grown in recent years, but there are limited number of studies on SBT branches in the literature and therefore, this study was focused mainly on leaves and branches of SBT.

Total Phenolic Content (TPC), Total Flavonoid Content (TFC) and Antioxidant Activity

The TPC of SBT leaves, branches and processed branches were reported in the range of 25.8±2.9 to 75.9±6.5 mg GAE/g (Table 41.2). The TPC of leaves from ethanolic extracts (EE) were significantly higher than branch and processed branch extract ($p<0.05$). It was observed that TPC of processed branch extracts were higher than unprocessed branch extracts. When considering processed and unprocessed branch extracts with respect to the differences between aqueous and ethanolic extraction methods, the TPC of aqueous extracts (AE) were substantially higher than ethanolic extracts (Figure 41.2). This circumstance was interpreted as softening was a more efficient extraction technique vis-à-vis ethanolic extraction. In the present study, relatively higher phenolic compounds were observed from EE in comparison to AE of SBT leaves (Figure 41.2). These results were in accordance with previous studies (Upadhyay *et. al*, 2010; Yogendra Kumar *et al.*, 2013). Leaves were found to contain maximum TPC followed by processed branches and unprocessed branches.

The TFC of SBT leaves, branches and processed branches were indicated in the range of 21.2±3.8 to 74.0±3.0 mg RU/g. In contrast with TPC results, TFC of

Table 41.2: Total Phenolic Content (TPC), Total Flavonoid Content (TFC), Free Radical Scavenging Activity (DPPH) and Copper Reducing Antioxidant Capacity (CUPRAC) of *Hippophae rhamnoides* Ethanolic Extraction, Tea Infusion and Softening (Leaf, branch, processed branch)

	Sample	*Total Phenolics (mg GAE/g)*	*Total Flavonoids (mg Rutin eq./g)*	*DPPH (mg TE/g)*	*CUPRAC (mg TE/g)*
Leaf	Ethanolic extraction	75.9± 6.5^{a}	74.0 ± 3.0^{a}	89.6 ± 2.0^{c}	538.5 ± 34.8^{a}
	Tea infusion	66.9± 5.9^{a}	21.2 ± 3.8^{e}	132.4± 6.6^{a}	293.5 ± 26.7^{b}
Processed	Ethanolic extraction	43.1± 5.7^{c}	59.4 ± 6.8^{b}	83.3 ± 9.8^{c}	291.3 ± 17.2^{b}
Branch	Softening	53.8± 8.9^{b}	45.2 ± 0.6^{c}	113.1± 6.1^{b}	235.5 ± 34.1^{c}
Branch	Ethanolic extraction	25.8± 2.9^{d}	64.6 ± 4.5^{b}	41.9 ± 3.6^{e}	129.4 ± 18.1^{e}
	Softening	43.2± 3.6^{c}	30.5 ± 3.4^{d}	71.1 ± 7.9^{d}	181.7 ± 27.6^{d}

*Data represent average values ± standard deviation of three independent sample.

Different letters in the columns represent statistically significant differences ($p < 0.05$).

all EE was higher than the AE content. The highest level of flavonoids was found in EE of leaves (74.0±3.0 mg/g) followed by branches (64.6±4.5 mg/g) and processed branches (59.4± 6.8 mg/g). But there is no significant difference between EE of branches and processed branches statistically ($p>0.05$). According to the results, except for the DPPH method other methods indicated a decrease in TP and TF by the effect of infusion.

To investigate the antioxidant activities of SBT leaves and branches two different *in vitro* assays including 1.1-diphenyl-2-picrylhydrazyl (DPPH) and Copper Reducing Antioxidant Capacity (CUPRAC) were used. The DPPH scavenging activity of extracts ranged from 41.9 ± 3.6 and 132.4 ± 6.6 mg TE/g. EE of leaf (89.6±2.1 mg TE/g) had significantly higher antioxidant properties than processed branch (83.3±9.8 mg TE/g) and branch (41.9±3.6 mg TE/g). But statistically there is no significant difference between EE of leaves and processed branches ($p>0.05$). This correlation was also valid for AE. When compared to extraction methods, higher scavenging activity was observed in AE of leaves, branches and unprocessed branches (Figure 41.2). The CUPRAC values of extracts ranged from 129.4±18.1 to 538.5±34.8 mg TE/g. Higher antioxidant capacity was

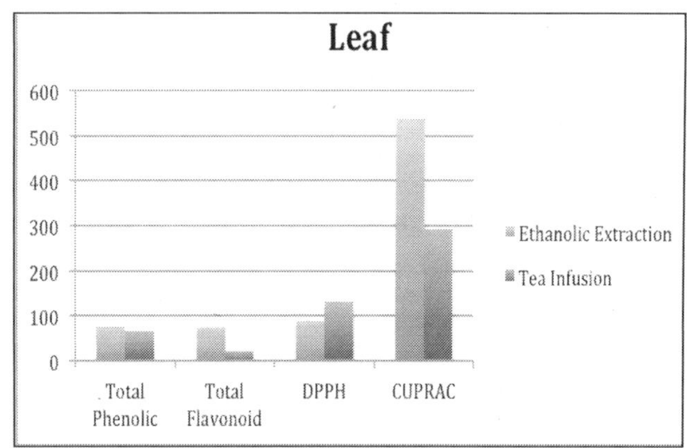

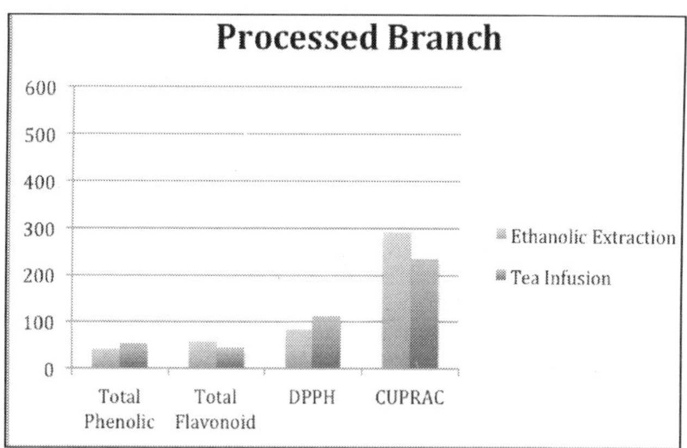

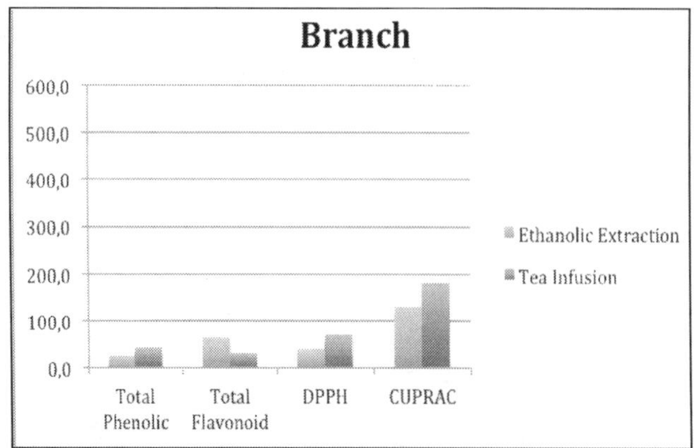

Figure 41.2: Comparing the Total Phenolic Content (TPC), Total Flavonoid Content (TFC), Free Radical Scavenging Activity (DPPH) and Copper Reducing Antioxidant Capacity (CUPRAC) according to Sample Type.

Table 41.3: Rutin Profile of Ethanolic and Aqueous Extract of *Hippophae rhamnoides*

	Leaf		*Processed Branch*		*Branch*	
Compound (μg/g)	Ethanolic	Tea infusion	Ethanolic	Softening	Ethanolic	Softening
Rutin	8377 ± 96[a]	6939± 46[b]	1229 ± 98[d]	1433± 47[c]	757 ± 13[e]	418 ± 108[f]

*Data represent average quantities ± standard deviation (determined by HPLC) of three independent samples.

Different letters in the rows represent statistically significant differences ($p < 0.05$). **nd: not detected.

observed in EE of leaf, branches and AE of unprocessed branches.

As reported from both in our and current study results, antioxidant activity of seabuckthorn organs varies according to different factors such as extraction method, plant sex and harvest time. On the other hand, variations in the obtained values might also be related with the differences in the variety, processing and storage conditions, ripeness and growing conditions of the plants processed (Arts *et al.*, 2000; Goncalves *et al.*, 2004).

According to the results, there are differences between DPPH and CUPRAC assays; higher values were obtained by the CUPRAC method. This might be related with the diversity of the reaction conditions including the required reaction time, and the wavelength at which the measurements are performed (Antolovich *et al.* 2002; Capanoglu *et al.* 2010).

HPLC Analysis

The results as shown in Table 41.3 indicated that the extracts are rich in rutin.

This study also focused on the analysis of rutin from seabuckthorn leaves and branches. In the literature, there is a limited information on the content of rutin in SBT. According to our results, rutin content changed between 418 ± 108 μg/g and 8377 ± 96 μg/g. Rutin content is critical since it was reported to have positive health effects against several types of liver diseases and it has great therapeutic importance. It also causes an increase in pancreatic lipase with consequent reduction in triacylglyceride levels in rats. It has also been described in events associated with the immune system, as seen with other flavonoids (Marcarini *et al.*, 2011). The highest rutin amounts were determined in EE of leaf (8377 ± 96 μg/g). Both EE and AE of leaf had significantly higher rutin content compared to processed branches and branches. On the other hand rutin content of all samples were significantly different ($p<0.005$).

HPLC results indicated that parallel to the spectrophotometric analysis results, infusion resulted with lower rutin content. Ethanolic extraction presented the highest concentration of rutin except processed branches. Martins *et al.* (2014), evaluated and compared the antioxidant and antibacterial activities, and phenolic compounds of the infusion, decoction and hydroalcoholic extract of oregano. According to their results, both preparations, mostly decoction, gave higher antioxidant activity than the hydroalcoholic extract. The antioxidant properties seem to be related to phenolic compounds, mainly flavonoids, since decoction presented the highest concentration of flavonoids and total phenolic compounds, followed by infusion and hydroalcoholic extract, respectively.

CONCLUSION

According to the results of the current study, leaves, processed branches and branches of SBT acquire high amounts of phenolics, flavonoids and also present some degree of antioxidant activities. Taking the high antioxidant and rutin content of leaves into account and with respect to their positive health effects, consumption of SBT in forms of a herbal tea should be investigated. Wide variations in the flavonoid profile and antioxidant activity of samples were observed which is probably related with the effect of type and variety of leaves and branches and other related factors as well as processing conditions including temperature, relative humidity, *etc.* Further studies should be performed in order to establish bioactive properties *in vivo* and *in vitro*.

ACKNOWLEDGEMENT

This study was financially supported by Istanbul University, Scientific Research Project (BAP) Unit, Project no. 35978.45891 and Ministry of Science, Industry and Technology, Techno-initiative Project no. 0239. TGDS.2013.

REFERENCES

1. Antolovich, M., Prenzler, P. D., Patsalides, E., McDonald, S. and Robards, K. 2002. Methods for testing antioxidant activity. *The Analyst* 127: 183-198.
2. Apak, R., Guclu, K., Ozyurek, M., Karademir, S. E. and Altun, M. 2005. Total antioxidant capacity assay of human serum using copper (II)-neocuproine as chromogenic oxidant: The CUPRAC method. *Free Radic. Res.* 39: 949-961.
3. Aras, T. A. 1997. Turkiye'nin *Hippophae rhamnoides* L. Taksonu Uzerinde Ksilolojik ve Palinolojik Arastýrmalar. *PhD. Thesis* (Unpublished).
4. Arts, I. C. W., van de Putte, B. and Hollman, P. C. H. 2000. Catechin contents of foods commonly

consumed in the Netherlands: Fruits, vegetables, staple foods, and processed foods. *Journal of Agricultural and Food Chemistry* 48: 1746-1751.

5. Attanassova, M. and Bagdassarian, V. 2009. Rutin content in plant products. *Journal of the University of Chemical Technology and Metallurgy* 44 (2): 201-203.

6. Capanoglu, E., Beekwilder, J., Boyacioglu, D., Hall, R. and de Vos, R. 2008. Changes in antioxidant and metabolite profiles during production of tomato paste. *Journal of Agricultural and Food Chemistry* 56: 964-973.

7. Capanoglu, E., Beekwilder, J., Boyacioglu, D., De Vos, R. and Hall, R. H. 2010. The effect of industrial food processing on potentially health-beneficial tomato antioxidants. *Critical Reviews in Food Science and Nutrition* 50: 919-930.

8. Chua, L.S. 2013. A review on plant-based rutin extraction methods and its pharmacological activities. *Journal of Ethnopharmacology* 150: 805-817.

9. Goncalves, B., Landbo, A. K., Knudsen, D., Silva, A. P., Moutinho-Pereira, J., Rosa, E. and Mayer, A. S. 2004. Effect of ripeness and postharvest storage on the phenolic profiles of Cherries (*Prunus avium* L.). *Journal of Agricultural and Food Chemistry* 52: 523-530.

10. Khan, B. A., Akhtar, N. and Mahmood, T. 2010. A Comprehensive review of a magic plant, *Hippophae rhamnoides*. *Pharmacology Journal* 2 (16): 65-68.

11. Kumaran, A. and Karunakaran, R. J. 2006. Antioxidant and free radical scavenging activity of an aqueous extract of *Coleus aromaticus*. *Food Chemistry* 97: 109-114.

12. Maheshwari, D. T., Yogendra, M. S., Verma, S. K., Singh, V. K. and Singh, S. N. 2011. Antioxidant and hepatoprotective activities of phenolic rich fraction of seabuckthorn (*Hippophae rhamnoides* L.) leaves. *Food and Chemical Toxicology* 49: 2422-2428.

13. Marcarini, J. C., Tsuboy, M. S. F., Luiz, R. C., Ribeiro, L. R., Hoffmann-Campo, C. B. and Mantovani, M. S. 2011. Investigation of cytotoxic, apoptosis-inducing, genotoxic and protective effects of the flavonoid rutin in HTC hepatic cells. *Experimental and Toxicologic Pathology* 63: 459-465.

14. Martins, N., Barros, L., Santos-Buelga, C., Henriques, M., Silva, S. and Ferreira, I. C. F. R. 2014. Decoction, infusion and hydroalcoholic extract of *Origanum vulgare* L.: Different performances regarding bioactivity and phenolic compounds. *Food Chemistry* 158: 73-80.

15. Michel, T., Destandau, E., le Floch, G., Lucchesi, M. E. and Elfakir, C. 2012. Antimicrobial, antioxidant and phytochemical investigations of seabuckthorn (*Hippophae rhamnoides* L.) leaf, stem, root and seed. *Food Chemistry* 131: 754-760.

16. Nieoczym, D., Socała, K., Raszewski, G. and Wlaz, P. 2014. Effect of quercetin and rutin in some acute seizure models in mice. *Progress in Neuro-Psychopharmacology and Biological Psychiatry* 54: 50-58.

17. Rai, S., Wahile, A., Mukherjee, K., Saha, B. P. and Mukherjee, P. K. 2006. Antioxidant activity of *Nelumbo nucifera* (sacred lotus) seeds. *Journal of Ethnopharmacology* 104: 322-327.

18. Suryakumar, G. and Gupta, A. 2011. Medicinal and therapeutic potential of seabuckthorn (*Hippophae rhamnoides* L.). *Journal of Ethnopharmacology* 138: 268-278.

19. Velioglu, S., Turkmen, N. and Sari, F. 2006. Effects of extraction solvents on concentration and antioxidant activity of black and black mate tea polyphenols determined by ferrous tartrate and Folin–Ciocalteu methods. *Food Chemistry* 99: 835-841.

20. Wani, T. A., Wani, S. M., Shah, A. G. and Masoodi, F. A. 2013. Optimizing conditions for antioxidant extraction from seabuckthorn leaf (*Hippophae rhamnoides* L.) as herbal tea using response surface methodology (RSM). *International Food Research Journal* 20 (4): 1677-1681.

21. Upadhyay, N. K., Yogendra, M. S. and Gupta, A. 2010. Antioxidant, cytoprotective and antibacterial effects of seabuckthorn (*Hippophae rhamnoides* L.) leaves. *Food and Chemical Toxicology* 48 (12): 3443-3448.

22. Yogendra, M. S., Dutta, R., Prasad, D. and Misra, K. 2011. Subcritical water extraction of antioxidant compounds from seabuckthorn (*Hippophae rhamnoides*) leaves for the comparative evaluation of antioxidant activity. *Food Chemistry* 127: 1309-1316.

23. Yogendra, M. S., Tirpude, R. J., Maheshwari, D. T., Bansal, A. and Misra, K. 2013. Antioxidant and antimicrobial properties of phenolic rich fraction of seabuckthorn (*Hippophae rhamnoides* L.) leaves *in vitro*. *Food Chemistry* 141: 3443-3450.

Chapter 42

In vitro and *In vivo* Investigations of Innovative Natural Cosmetic Products, Enriched with Seabuckthorn Leaf Extract

Anna Scheiner[1], Julia Hasdorf[1] and Christoph Schempp[2*]

[1]NIG Nahrungs-Ingenieurtechnik GmbH; Wasserkunststraße 26; 39124 Magdeburg, Germany
[2]University Medical Center Freiburg, Dermatology, Hauptstraße 7; 79104 Freiburg, Germany
(*E-mail: christoph.schempp@uniklinik-freiburg.de, nig.magdeburg@t-online.de)

SUMMARY

The anti-inflammatory, antiviral, antioxidant and antibacterial properties of seabuckthorn leaves respectively seabuckthorn leaf extracts have been verified in experiments. Unlike the oils of seabuckthorn, which are well established in medicine and also in cosmetics, seabuckthorn leaf extracts have not been used in cosmetic products as yet. In cooperation with various partners, the aim was the development of a dermatologically active cosmetic product, based on seabuckthorn leaf extract. Different seabuckthorn leaf extracts with 23, 0 per cent (SDB.0111.01), 18,0 per cent (SDB.0111.02) and 55,0 per cent (SDB.0111.03) polyphenol content (gallic acid equivalent) were developed by NIG GmbH. They were characterized by the determination of *e.g.* total polyphenols, total flavonoids. This was followed by *in vitro* assays of the extracts on ROS (reactive oxygen species) for determining the antioxidant potential, phototoxicity, inflammation inhibition and cytotoxicity. Every extract showed antioxidant and anti-inflammatory potential, no phototoxicity and no cytotoxicity. From the different extracts, SDB.0111.01 was selected for galenic application. It was emulsified in a natural cosmetic cream base. Comparative *in vivo* studies confirmed the compatibility and anti-inflammatory effect of the cosmetic preparation, enriched with the seabuckthorn leaf extract. Furthermore, the stability of the extracts was examined by the determination of RPF-Factor. It was found that SDB.0111.01 presented stability over a period of at least 18 months.

Keywords: *Seabuckthorn leaf extract, Antiinflammatory, Antioxidative, Dermatological application and Cosmetics.*

INTRODUCTION

According to the latest figures from the Federal Statistical Office for the year 2014, seabuckthorn cultures in Germany are cultivated on an area of 436 ha. (Federal Statistical Office, 2014) In 2006 the cultivation area was 250 ha. Harvesting of seabuckthorn berries in Germany is realized by cutting/freezing/shaking system of branches. In addition to approx. 50 per cent seabuckthorn berries and 25 per cent seabuckthorn branches, 25 per cent seabuckthorn leaves are incurred by this technology. Seabuckthorn leaves are deciduous, linear or lanceolate leaves, which have a length of 5 - 8 cm and a width of only 5-7 mm. Depending on variety leaves show a light to dark green color. (von Arnsberg, 2000) The underside of the leaves is covered with star-like silver-grey scales (Figure 42.1). The scales provide protection from evaporation (Albrecht, 2007). The composition and the nutritional values of seabuckthorn leaves were, among others, investigated by Rongsen (1992), our results are shown in Figure 42.2.

As shown in Figure 42.2, seabuckthorn leaves contain 17 per cent crude protein and 5 per cent of crude fat. Because of the nutritional value and availability, the leaves are used as fodder for sheep and cattle in barren regions of China (Lu Rongsen, 1992). In Germany, the not inconsiderable amount of seabuckthorn leaves accruing during harvest, has not been supplied to the value chain till now. The leaves

Seabuckthorn (Hippophae L.): A Multipurpose Wonder Plant, Vol. 5: Advanced Technologies for Cultivation, Processing, Health Protection and Environmental Conservation (Virendra Singh, *Editor-in-Chief*, 2018), p. 335–342, Astral International Pvt. Ltd., New Delhi, India.

Figure 42.1: Leaves from Seabuckthorn.

are mainly waste or by-product. Seabuckthorn leaves contain, in addition to major nutrients, phytochemicals such as polyphenols and flavonoids, whose antioxidant effect has been confirmed in various scientific studies. Thus Zheng *et al.* (2012) showed the influence of an aqueous seabuckthorn leaf extract on the swimming performance in rats. The rats, administered with the extract, showed, compared with the control group, a significant increase in performance capacity as well as reduced oxidative damage. Narayanan *et al.* (2005) induced oxidative stress in glia cells by hypoxia (oxygen deficiency). This caused an increase in cytotoxicity and reduction in the mitochondrial membrane potential and increased DNA damage. The antioxidant activity of used ethanolic seabuckthorn leaf extract was found in significant reduction of DNA damage and cytotoxicity. In 2005 Ganju *et al.* induced arthritis in rats and Gupta *et al.* caused different wounds in rats. Both tests confirmed the antioxidant and antiinflammatory effects of aqueous seabuckthorn leaf extract. Based on these results the seabuckthorn leaf extracts may constitute the base of innovative cosmetic products for dermatological applications. NIG GmbH developed a suitable extraction method and produced 3 seabuckthorn leaf extracts with different contents in total polyphenols. The prepared extracts were analyzed *in vitro* by University Medical Center Freiburg to determine antioxidative potential, phototoxicity and cytotoxicity.

Table 42.1: The Contents of Polyphenols were Determined by the Folin-Ciocalteu Reagent

Seabuckthorn Leaf Extract	*SDB.0111.01*	*SDB.0111.02*	*SDB.0111.03*
Polyphenol content in per cent	23,3	18,1	55,9
Color	yellow	dark-brown	brown
Appearence	powder	powder	powder

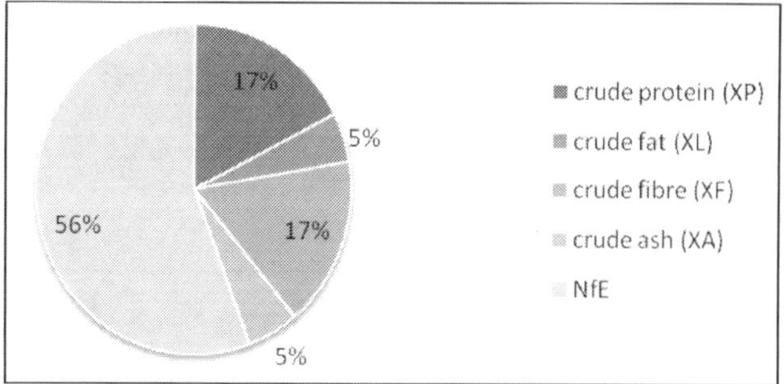

Figure 42.2: Nutritional Information in Seabuckthorn Leaves taken from Chinese Seabuckthorn Varieties (Rongsen, 1992).

Methods Part I - Analyses of Seabuckthorn Leaf Extracts

Determination of the Antioxidant Potential

The assay to determine the antioxidant capacity based on usage of the reagent CM-H2DCFDA (5-(and 6)-chloromethyl-2′, 7′-dichlorodihydrofluorescein diacetate). HaCat keratinocytes were treated with respective seabuckthorn leaf extract in concentration range of 0 - 64µl/ml and incubated for 30min in an incubator. Then the cells were treated with CM-H2DCFDA the reagent, washed after 20 minutes incubation, followed by irradiation with solar simulator. After addition of new cell culture medium the generation of reactive oxygen species (ROS) was measured by fluorescence reader.

Determination of Phototoxicity

The phototoxic effect of seabuckthorn leaf extracts on irradiated HaCaT keratinocytes was investigated. The review of the metabolic activity of the cells was performed by ATP assay (ViaLight).

Determination of Cytotoxicity

The selected seabuckthorn leaf extracts were analyzed to determine the effect on cell viability, and cytotoxicity in primary fibroblasts. The review of the metabolic activity of the cells was performed by ATP assay (ViaLight).

Determination of Anti-inflammatory Effect

In order to determine the antiinflammatory effect of the extracts HaCaT keratinocytes resp. primary fibroblasts were irradiated by solar simulator, and then incubated with the different extracts for 24 hours. After that the concentration of inflammatory mediators in supernatant of cell culture was determined by ELISA.

Results I – Analyses of Seabuckthorn Leaf Extracts

Determination of Antioxidant Potential

By solar simulator massive ROS production could be induced. The results showed that all extracts, in

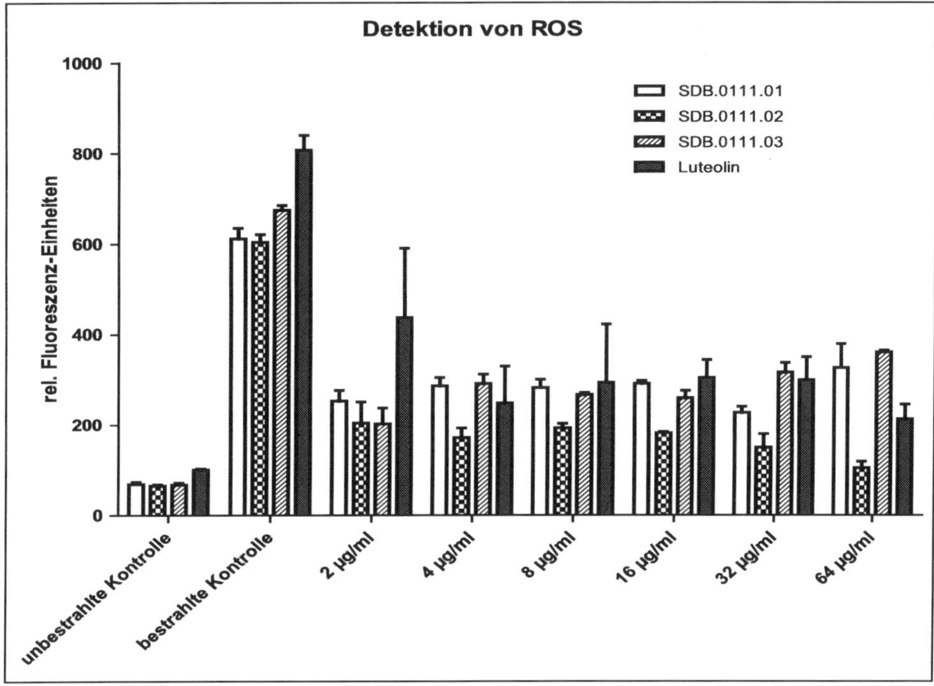

Figure 42.3: Measurement of ROS to Determine the Antioxidant Potential.

depending from concentration, showed antioxidant effects (Figure 42.3).

Determination of Phototoxicity

The extracts showed no phototoxic effect in all tested concentrations. Also the non-irradiated cells showed no phototoxic effects for 4 hours after the addition of extract (Figure 42.4).

Determination of Cytotoxicity

None of the tested extracts showed, up to the tested concentrations, a cytotoxic effect in primary fibroblasts (Figure 42.5).

Determination of Anti-inflammatory Effect

All in the assay examined seabuckthorn leaf extracts showed, regarding to inflammatory mediator IL-6,

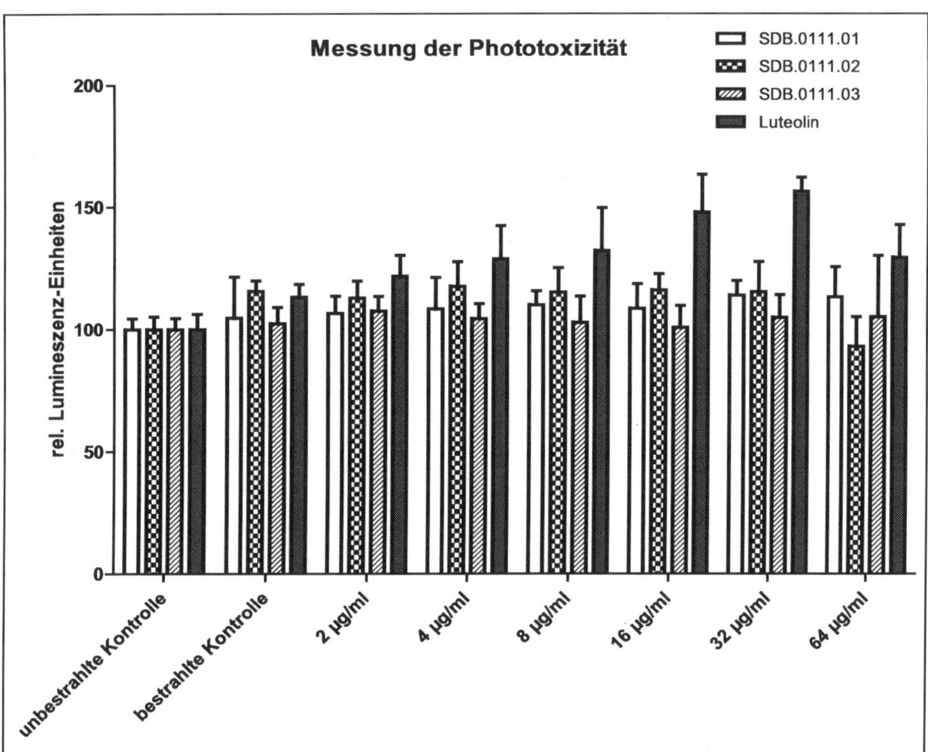

Figure 42.4: Measurement of Phototoxicity of Seabuckthorn Leaf Extracts.

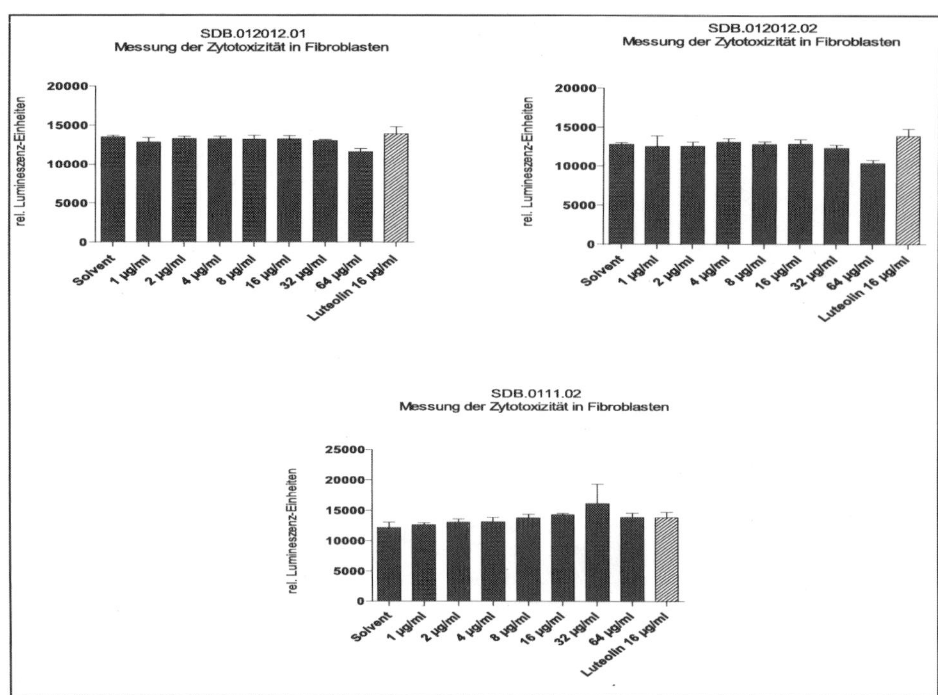

Figure 42.5: Measurement of Cytotoxicity in Primary Fibroblasts.

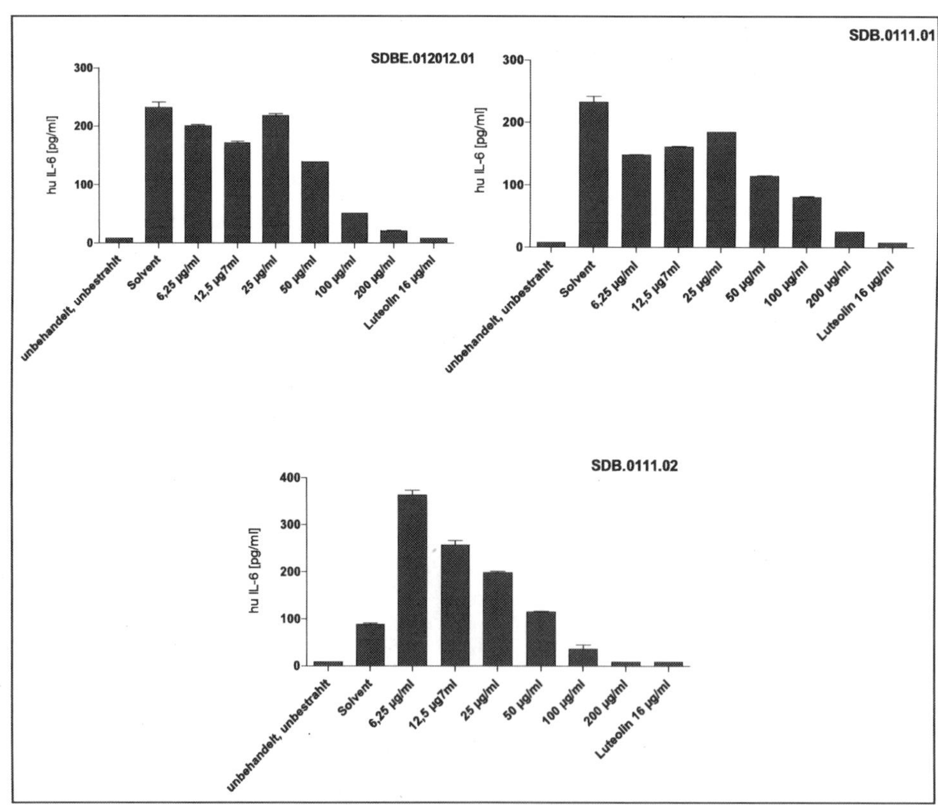

Figure 42.6: Measurement of IL-6 in Keratinocytes.

antiinflammatory effect in HaCaT keratinocytes (Figure 42.6). The University Medical Center Freiburg confirmed the antioxidant, antiinflammatory, non-cytotoxic and non-phototoxic properties of the 3 developed seabuckthorn leaf extracts.

Development of Cosmetic Application Enriched with Seabuckthorn Leaf Extracts

This development was realized in cooperation with a large german company for natural cosmetics. The manufacturer of natural cosmetics complained, the dark color and an unpleasant odor of the seabuckthorn leaf extracts SDB.0111.02 and SDB.0111.03. The company referred to the requirements of the consumer of cosmetic applications where white cream formulations and pleasant scents would be preferred. Then SDB.0111.02 and SDB.0111.03 were excluded for usage in galenic developments. Seabuckthorn leaf extract SDB.0111.01, which contains 23,3 per cent total polyphenols was used for further experiments. Different concentrations of the chosen extract SDB.0111.01 were emulsified into a cream base.

Methods Part II – Analyses of Creams Enriched with Seabuckthorn Leaf Extract

The next steps were *in vivo* analyses of the enriched creams regarding to the antiinflammatory effect and the exclusion of phototoxicity. In parallel, the stability of the extracts had to be guaranteed for dermatological applications over the entire shelf life of 18 months. This study was realized in collaboration with the Institute Galenus Norbert Groth.

The preparations listed below were analyzed *in vivo* by University Medical Center Freiburg:

1. Placebo (pure Cream base)
2. Placebo + 0.1 per cent SDB.0111.01
3. Placebo + 0.2 per cent SDB.0111.01
4. Placebo + 0.4 per cent SDB.0111.01
5. Placebo + 0.2 per cent SDB.0111.01 + Antioxidants
6. Placebo + Antioxidants
7. Untreated skin

Determination of Anti-inflammatory Effect

The anti-inflammatory effect of creams with different concentrations of seabuckthorn leaf extracts were studied in the UVB erythema test compared to placebo. After induction of a defined erythema by UV irradiation, the test preparations were applied on skin and retained occlusive for 48h. The condition of the skin erythema was estimated after 48h with measurements using Mexameter. This method enabled a statement of the anti-inflammatory effect of the preparations.

Determination of Phototoxicity

The phototoxicity of the different semi-solid preparations enriched with different ranges of concentrations extracts of SDB.0111.01 and their skin compatibility were investigated. The exclusion of phototoxicity on irradiated skin, as well as the estimation of tolerance of the test preparations on non-irradiated skin, have been determined by photometric analysis of erythema before and after application of preparations.

Determination of Stability of Seabuckthorn Leaf Extracts

The Radical Protection Factor (RPF) determines the scavenging activity of an antioxidant against a test substance. This test substance consists of a very active semistable radical which reacts with all known antioxidants. The principle of the method is the measurement of the signal amplitude of the test radical by Electron Spin Resonance (ESR). The amplitude of the test radical is measured before and after mixing with the antioxidative product. As a result the scavenging activity of the antioxidant against the test radical with the known concentration of radicals is determined. (Herrling *et al.* 1998)

Results II – Analyses of Cream Base Enriched with Seabuckthorn Leaf Extract

Determination of Anti-inflammatory Effect

The following preparations showed a significant effect *in vivo*: 0,2 per cent SDB.0111.01 and 0,2 per cent SDB.0111.01 plus antioxidants. These two preparations are more effective than placebo and suitable for further development.

None of the preparations showed irritating potential after occlusive application for 48h.

Determination of Phototoxicity

None of the preparations showed phototoxic properties in the long wave UVA range. Rather cream with 0,2 per cent SDB.0111.01 had a soothing effect (significant reduction of UVA erythema).

Occlusive applied none of the preparations had irritating potential.

Determination of Stability of Seabuckthorn Leaf Extracts

Seabuckthorn leaf extract SDB.0111.01 (23 per cent total polyphenols) was stored under room temperature over a period of 18 months. This extract showed a RPF-Factor of 42000 after storage. For purposes of comparison, SDB.0113.01 represents the freshly prepared seabuckthorn leaf extract, which was prepared in analogy to SDB.0111.01. SDB.0113.01 showed a RPF-Factor of 49500. The investigation showed that the antioxidant potential of the selected seabuckthorn leaf extract SDB.0111.01 over a period of 18 months only marginally decreased and thus can be considered as stable. It is further noticed that the content of total polyphenols had direct influence on the level of RPF Factors.

CONCLUSION

During the harvest of seabuckthorn berries by cutting system, an amount of approx. 25 per cent is

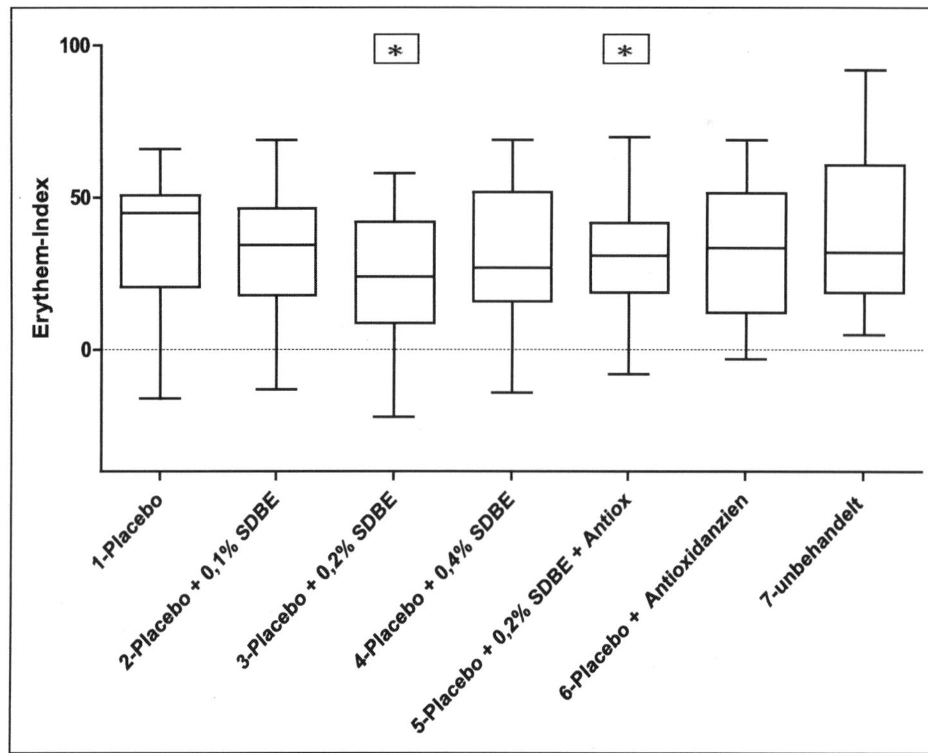

Figure 42.7: Results of UVB Erythema Test on Irradiated Skin.

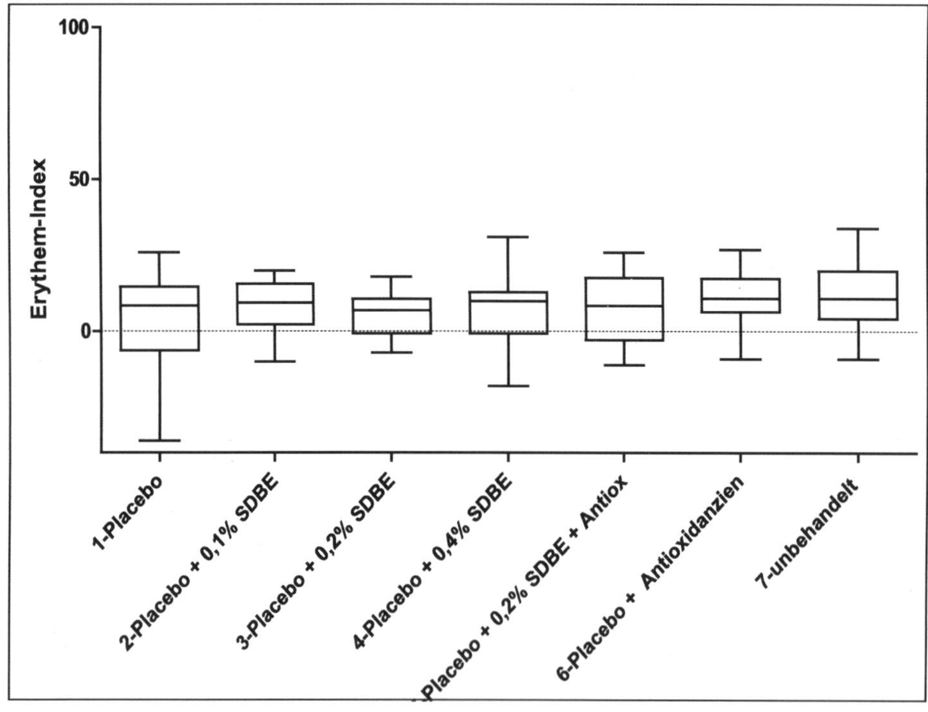

Figure 42.8: Results on Non-irradiated Skin.

formed by seabuckthorn leaves. In the present work 3 aqueous-ethanolic extracts were developed with an optimized extraction process. The extracts were tested for their antioxidant, antiinflammatory effects as well as phototoxic and cytotoxic properties. *In vitro* all 3 extracts showed antioxidant, antiinflammatory effects and did not show photo- and cytotoxic properties in the studied concentrations. Of the 3 extracts extract SDB.0111.01 with 23,3 per cent total polyphenols was selected due to its attractive sensory characteristics for dermatological applications. The stability of SDB.0111.01 over a storage period of 18 months has been demonstrated by the determination of the RPF-Factor. In different

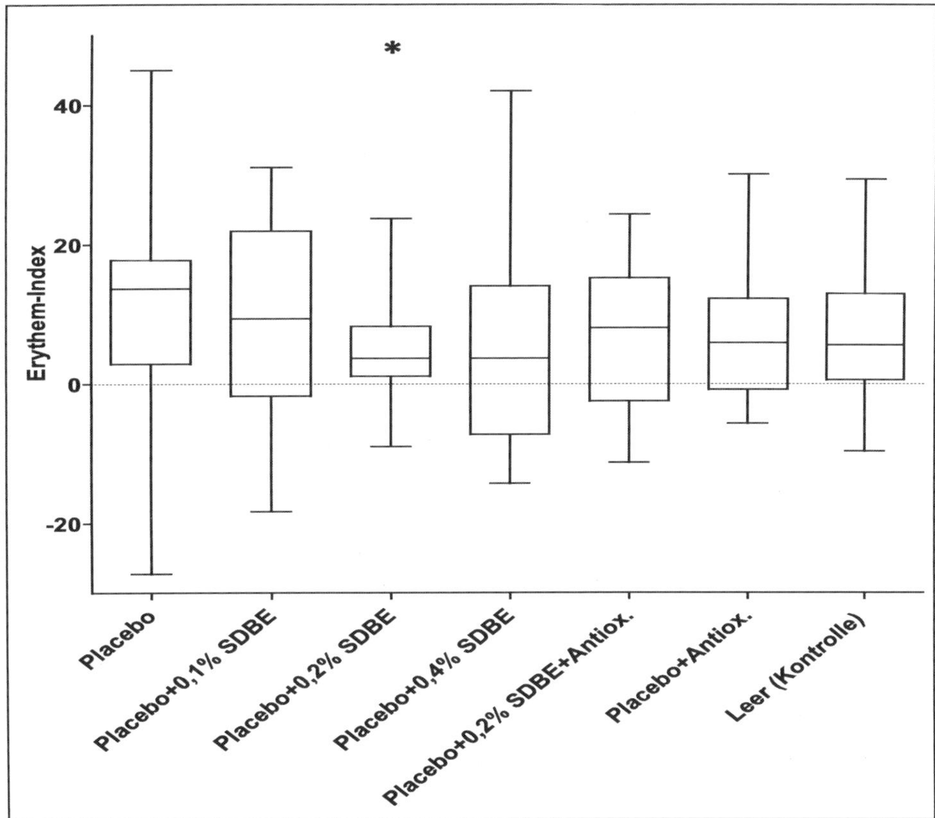

Figure 42.9: Results of UVA Photopatchtest on Irradiated Skin.

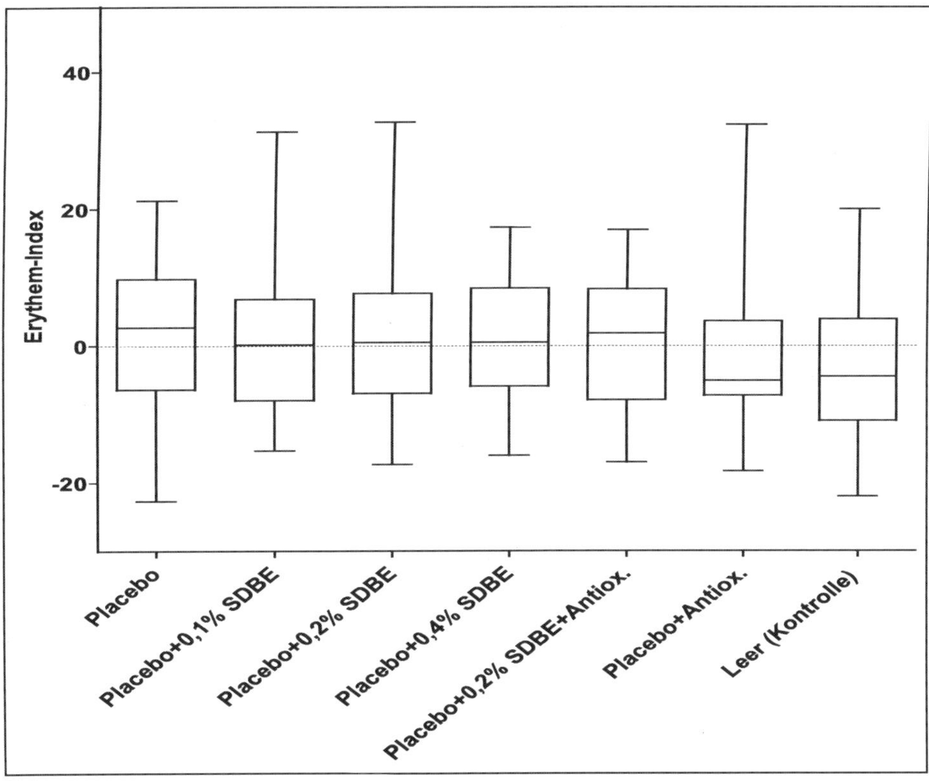

Figure 42.10: Results on Non-irradiated Skin.

concentrations the extract was emulsified in a cream base. The preparations with concentrations of 0,2 per cent SDB.0111.01 and 0,2 per cent SDB.0111.01 plus antioxidants had soothing properties. All creams were well tolerated on skin and phototoxic potential couldn't be determined.

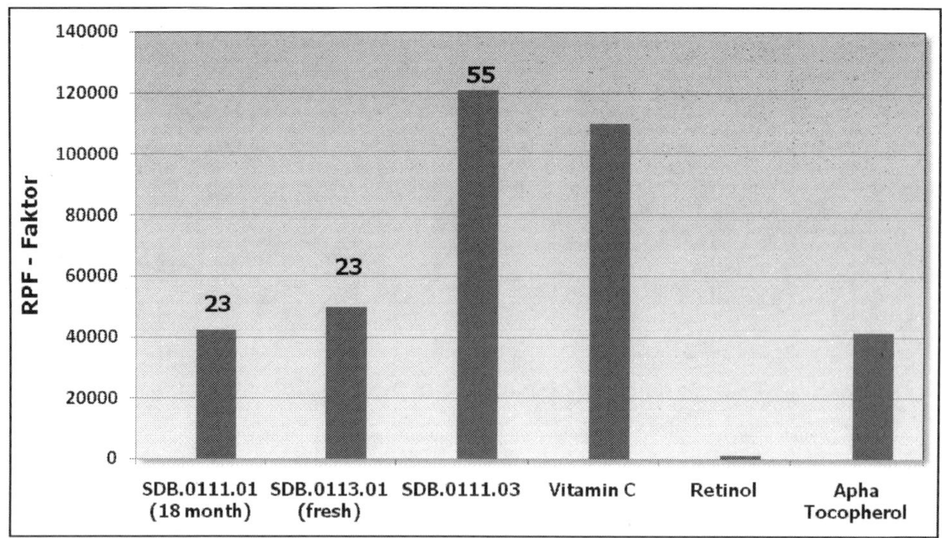

Figure 43.11: RPF-Factor of SDB.0111.01 in Comparison with other Antioxidative Substances.

REFERENCES

1. Albrecht, H.J. 2007. Seabuckthorn- cultivation, harvesting, assortment, Bilsen.
2. Ganju, L., Padwad, Y., Singh, R., Karan, D., Chanda, S., Chopra, M.K., Bhatnagar, P., Kashyap, R. and Sawhney, R.C. 2005. Anti-inflammatory activity of seabuckthorn (*Hippophae rhamnoides*) leaves. *International Immunopharmacology* 5 (12): 1675-1684.
3. Gupta, A., Kumar, R., Pal, K., Banerjee, P.K. and Sawhney, R.C. 2005. A preclinical study of the effects of seabuckthorn (*Hippophae rhamnoides* L.) leaf extract on cutaneous wound healing in albino rats. In: *International Journal of Lower Extremity Wounds*, June 2005, 4: 88-92.
4. Herrling, T., Zastrow, L. and Groth, N. 1998. Classification of cosmetic products-The radical protection factor (RPF). In: *SÖFW International Trade Journal* No.124, 5 (98): 282-284.
5. Narayanan, S., Ruma, D., Gitika, B., Sharma, S.K, Pauline, T., Ram, M.S., Ilavazhagan, G., Sawhney, R.C., Kumar, D. and Banerjee, P.K. 2005. Antioxidant activities of seabuckthorn (*Hippophae rhamnoides*) during hypoxia induced oxidative stress in glial cells. *Molecular and Cellular Biochemistry* 278 (1): 9-14.
6. Rongsen, L. 1992. Seabuckthorn: A multipurpose plant species for fragile mountains. International Center for Integrated Mountain Development (ed.) Nepal, Lithograhing Co. (p) Ld., Kathmandu.
7. Statistisches Bundesamt, 2015. Publikationen im Bereich Obst, Gemüse Gartenbau - Strauchbeerenanbau und -ernte; Fachserie 3 Reihe 3.1.9-2014. https://www.destatis.de/DE/Publikationen/Thematisch/LandForstwirtschaft/ObstGemueseGartenbau/Strauchbeerenanbau.html;08.09.2015.
8. von Arnsberg, H. 2000. Natürliche Heilung mit SFFl-Öl. Altomünster.
9. Zheng, X., Long, W., Liu, G., Zhang, X., Zhang, X. 2012. Effect of seabuckthorn (*Hippophae rhamnoides* ssp. *sinensis*) leaf extract on the swimming endurance and exhaustive exercise-induced oxidative stress of rats. *Journal of the Science of Food and Agriculture* 92 (4): 736-742.

Chapter 43

Effect of Feeding of Seabuckthorn Products on Assessment of Liver and Renal Function in Albino Rat Model

Meenakshisundaram Selvamuthukumaran* and Farhath Khanum

Biochemistry and Nutrition Division, Defence Food Research Laboratory,
Siddartha Nagar, Mysore – 570 011, Karnataka, India
(*E-mail: selkuma@hotmail.com)

SUMMARY

The rats were fed with various seabuckthorn products for a period of 45 days, food intake was recorded daily and body weight gain was recorded on weekly basis. At the end of the experimental period, blood was collected directly from heart, serum separated and analyzed for SGPT, SGOT, urea and creatinine levels in order to assess the various liver and renal functioning in albino rat model. The result shows that there was no significant difference in the food intake and no significant increase in organ weight was found. SGPT and SGOT levels showed no appreciable change upon feeding of seabuckthorn products and the enzyme activities were within the normal range. The urea and creatinine level in blood also exhibited insignificant changes and seabuckthorn products fed rats groups showed a slight reduction in both levels by exhibiting nephroprotective effect to certain extent.

Keywords: *Seabuckthorn, SGPT, Liver function, Urea, SGOT and Creatinine.*

INTRODUCTION

Liver is the major vital organs and one of the largest glands in the human body. They metabolize proteins, stabilize glucose levels within the blood, breaks down fatty acids and stores fat and vitamins. The liver detoxifies the blood by filtering and removing drugs or hormones and destroys old red and white blood cells. Kidney, which is also an important organ in body, which separate urea, mineral salts, toxins and other waste products from the blood. It conserves water, salts and electrolytes.

Liver function test (LFT) and renal function test (RFT) is a biochemical assay designed to give information about the state of a patient's liver and kidney conditions (Lee and Mary, 2009). The parameters measured include Serum glutamate pyruvate transferase (SGPT), Serum glutamate oxaloacetate transferase (SGOT), urea and creatinine and they were the useful biomarkers of liver and kidney injuries in a patient with some degree of intact liver and kidney function (David, 1999; Mcclatchey and Kenneth, 2002; Mengel *et al.*, 2005). The functioning of these organs are mainly based on ascertaining the levels of SGPT, SGOT, urea and creatinine in serum.

The enzymes *viz.*, SGPT and SGOT are released in blood when damage occurs to liver. It is mandatory that the levels of these enzymes should be always within the physiological range of limit. If these enzymes levels increases, diseases like hepatitis and cirrhosis will be noticed in patients (Balldin and Olsson, 2004; Nyblom *et al.*, 2006). Urea is the end product of protein metabolism. It is synthesized in the liver from the ammonia produced by the catabolism of amino acids. It is transported by the blood to the kidneys from where it is excreted. Increased levels of urea are found in renal diseases and urinary obstructions (Morgan *et al.*, 1977). Creatinine is the catabolic product of creatinine phosphate, which is used by the skeletal muscle and their daily production

Seabuckthorn (Hippophae L.*): A Multipurpose Wonder Plant, Vol. 5: Advanced Technologies for Cultivation, Processing, Health Protection and Environmental Conservation* (Virendra Singh, *Editor-in-Chief*, 2018), p. 343–346, Astral International Pvt. Ltd., New Delhi, India.

depends on muscle mass and they are excreted out of the body entirely by the kidneys. Elevated levels were found in the renal dysfunction and reduced renal blood flow in patients (Morgan *et al.*, 1977).

Seabuckthorn (*Hippophae rhamnoides*), is a deciduous shrub with yellow or orange fruits (Li and Schroeder, 1996). It is a unique plant currently being domesticated in several countries like China, Russia, Germany, Finland, Romania, France, Nepal, Pakistan and India. The plant is reported to have considerable medicinal value (Li and Wang, 1998), being useful for the treatment of skin disorders resulting from bed confinement, stomach and duodenal ulcers, cardiovascular diseases and perhaps growth of some tumors. Seabuckthorn berries are among the most nutritious and vitamin-rich fruit found in the plant kingdom because of presence of several natural antioxidants *viz.*, vitamins C, E, carotenoids, anthocyanins and phenols (Chauhan *et al.*, 2001).

Seabuckthorn products were the rich source of antioxidative nutrients, therefore in this study these products were developed in order to assess the effect of feeding of these products on liver and renal functioning using albino rat as a model and moreover before commercialization of these food products it is important to evaluate whether these products are safe for human consumption.

MATERIALS AND METHODS

Raw Materials

Seabuckthorn berries (*Hippophae rhamnoides*) were brought from Field Research Laboratory (Leh, Himachal Pradesh, India) by airlifting and kept frozen at (-20°C) until further studies. All other chemical reagents were procured from E. Merck and Loba Chemicals Limited, Mumbai, India.

Development of Seabuckthorn Products

The seabuckthorn products *viz.* jam and squash were developed as per the procedure of Girdharilal *et al.* 1960, fruit yogurt was developed as per the standard protocol of Sukumar De (1980) and fruit juice powder as per the procedure of Abadio *et al.* (2004).

Animals and Diet

Male albino rats weighing 100-120 g were randomly divided into five groups with six rats each. One group served as control and was fed on balanced casein diet. For the other 4 groups, seabuckthorn products were added at 50 per cent level (50 per cent casein diet + 50 per cent seabuckthorn products). Rats were fed with these products @ 20 g/day *i.e.* 10 g of casein diet and 10 g of seabuckthorn products which is continued up to a period of 45 days. Both food and water were given *adlibitum*. Food intake was recorded daily and body weight gain was recorded on weekly basis. At the end of the experimental period, blood was collected directly from heart, serum separated and analyzed.

Liver and Renal Function Tests

Blood was allowed to stand for 30 min in refrigerator, centrifuged at 500 rpm for 5 min at 5°C and serum was separated. Urea was estimated in serum samples by using kit of Crest Biosystems, Coral Clinical Systems, Goa as per the modified method of Berthelot (1859). The creatinine was determined using creatinine kit (Crest Biosystems) based on the method of Bowers (1980). The SGPT was determined in serum by SGPT (ALAT) kit provided by Crest Biosystems as per the methods of International Federation of Clinical Chemistry (1986). The SGOT was determined in serum by SGOT (ASAT) kit provided by Crest Biosystems as per the modified methods of International Federation of Clinical Chemistry (1986).

Statistical Analysis

All experiments were run in triplicate. Data were subjected to analysis of variance (ANOVA) using Microsoft Excel 2000 (Microsoft Corporation, Washington, USA).

RESULTS

Effect of Feeding of Seabuckthorn Products on Food Intake and Weight Gain of Rats

The Figure 43.1 shows the effect of feeding of seabuckthorn products on food intake (A), body

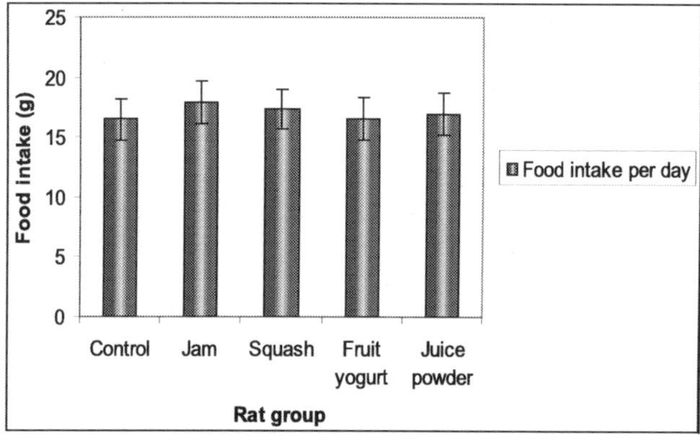

Figure 43.1A: Effect of Feeding of Seabuckthorn Products on Food Intake of Rats.

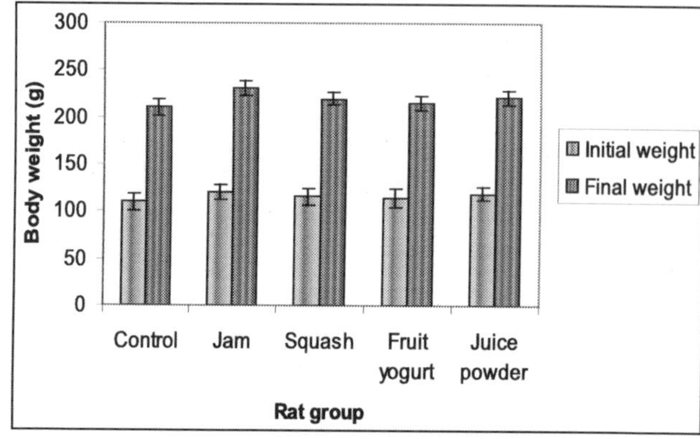

Figure 43.1B: Effect of Feeding of Seabuckthorn Products on Body Weight of Rats.

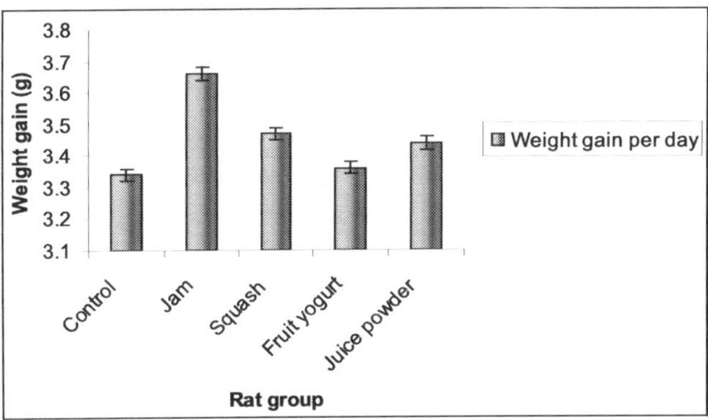

Figure 43.1C: Effect of Feeding of Seabuckthorn Products on Weight Gain of Rats.

weight and weight gain (B and C) of rats. The jam group recorded more food intake and weight gain/ day followed by squash, juice powder, fruit yogurt and control groups at a given quantity of the products fed. There was no significant difference (p>0.05) in the food intake among the groups, which shows that the diet was effectively utilized by rats of all groups.

Effect of Feeding of Seabuckthorn Products on Organ Weight of Rats

The effect of feeding of seabuckthorn products on organ weight/kg body weight of rats were provided in Table 43.1. No significant increase (p>0.05) in organ weight was found among the various rat groups after 45 days of continuous feeding. This study shows that, the effect of feeding of various seabuckthorn products doesn't brought about any remarkable significant changes on the organ weight as a result of feeding, so it can be predicted that the effect of feeding of various seabuckthorn products results in normal functioning of these organs in the body system of albino rats (Greim and Snyder, 2008).

Table 43.1: Effect of Feeding of Seabuckthorn Products on Organ Weight of Rats**

*Group**	*Liver*	*Heart*	*Kidney*
Jam	6.41±0.52 (0.027)	0.94±0.04 (0.004)	1.59±0.08 (0.006)
Squash	6.30±0.35 (0.028)	0.89±0.05 (0.002)	1.51±0.04 (0.006)
Fruit yogurt	6.19±0.89 (0.028)	0.80±0.07 (0.003)	1.41±0.06 (0.006)
Juice powder	6.28±0.74 (0.028)	0.86±0.08 (0.003)	1.48±0.05 (0.006)
Control	6.01±0.58 (0.028)	0.77±0.09 (0.003)	1.35±0.07 (0.006)

* No of rats = 6; ** Mean ± SD of six rats.

Values in parentheses indicates organ/body weight ratio.

Liver and Renal Function

The results of liver and renal function tests in rat fed on seabuckthorn products vis-a-vis control are provided in Table 43.2. The levels of the enzyme markers of liver function *viz.*, SGPT and SGOT showed no appreciable change upon feeding of the seabuckthorn products for a period of 45 days when compared to control rats. The enzyme activities were also within the normal range. This shows that the seabuckthorn products doesn't exhibited hepatic liver injury as a result of continuous feeding and the products were found to be free from causing toxic effects to the body of consumers. (Jayakumar *et al.*, 2006) also reported similar research findings as a result of feeding of *Pleurotus ostreatus* extract.

There was no significant change (p>0.05) in levels of urea and creatinine in blood as well, on contrary the seabuckthorn product fed rat group showed a slight reduction in levels of both urea and creatinine when compared to control group, this may be due to intake of antioxidants continuously for a period of 45 days from seabuckthorn food products which might have slightly reduced their levels in blood by exhibiting a role of nephroprotective effect to certain extent (Ajith *et al.*, 2008). Based on these results, it is precluded that these seabuckthorn based products help in normal functioning of rat liver and kidney. (Ajith *et al.*, 2008) also reported a urea and creatinine level reduction in serum after continuously feeding the extracts of *Zingiber officinale*.

Table 43.2: Effect of Feeding of Seabuckthorn Rroducts on Liver and Renal Function in Albino Rats.**

	LFT		*RFT*	
*Groups**	*SGPT (U/L)*	*SGOT (U/L)*	*Urea (mg/dl)*	*Creatinine (mg/dl)*
Jam	32.5±0.2	30.2±0.4	18.3±0.8	0.7±0.3
Squash	32.9±0.4	30.8±0.5	19.2±0.7	0.8±0.1
Fruit yogurt	33.2±0.3	31.6±0.6	19.8±0.9	0.8±0.1
Juice powder	31.6±0.6	30.0±0.5	17.6±0.4	0.6±0.2
Control	34.5±0.4	32.5±0.3	20.3±0.6	0.9±0.4
Normal range	*** upto 40	*** upto 37	***14-40	*** 0.6-1.2

*No of rats = 6; **Mean ± SD of triplicate analyses; ***Range of value limit adapted from Crest Biosystem Kits.

DISCUSSION

The continuous feeding of seabuckthorn products doesn't bring any remarkable change in weight of organ. The levels of liver enzyme markers *viz.* SGPT and SGOT were found to be within the physiological range after continuous feeding. Urea and creatinine levels in blood were also seems to be normal *i.e.* within the range only. Therefore feeding of seabuckthorn products lead in normal functioning of the liver and kidney in albino rat model system.

CONCLUSIONS

Therefore, based on the above clinical assessment study, the seabuckthorn based products can be safely consumed without posing any threat to the body of consumers and they can also be commercialized to a greater extent.

ACKNOWLEDGEMENT

The authors are thankful to Defence Research and Development Organization for providing grant to carry out this research work.

REFERENCES

1. Abadio, F. D. B., Domingues, A. M., Borges, S.V. and Oliveira, V. M. 2004. Physical properties of powdered pineapple (*Ananas comosus*) juice effect of maltodextrin concentration and atomization speed. *J. Food Eng.* 64: 285-287.
2. Ajith,T. A., Swathy, M. S. and Hema, U. 2008. Protective effect of *Zingiber officinde* against anticancer drug doxorubicin induced acute nephrotoxicity. *Food and Chem Toxicol* 46: 3178-3181.
3. Balldin, J. and Olsson, R. 2004. High AST/ALT ratio may indicate advanced alcoholic liver disease rather than heavy drinking. *Alcohol Alcohol* 39: 336-339.
4. Berthelot, M. P. E. 1859. Urea colorimetric. *Report Chim. Appl.* 2884: 217-284.
5. Bowers, L. D. 1980. Kinetic serum creatinine assays: The role of various factors in determining specificity. *Clinical Chem.* 26: 551-554.
6. Chauhan, A. S., Rekha, M. N., Ramteke, R. S. and Eipeson, W.E. 2001. Preparation and quality evaluation of processed products from seabuckthorn (*Hippophae rhamnoides*) berries. *Beverage and Food World* 27: 31-34.
7. David, J. 1999. Special considerations in interpreting liver function tests. *Am Fam Physician* 59: 2223-2230.
8. Girdharilal, G., Sidappa, S. and Tandon, G. L. 1960. *Preservation of Fruits and Vegetables.* Indian Council of Agricultural Research Publications, New Delhi, India.
9. Greim, H. and Snyder, R. 2008. Toxicology and Risk Assessment: A Comprehensive Introduction. John Wiley and Sons Ltd, West Sussex, England.
10. International Federation of Clinical Chemistry, 1986. IFCC Methods for Measurement of Catalytic Concentrations of Enzymes 24: 481-497.
11. Jayakumar, T., Ramesh, E. and Geraldine, P. 2006. Antioxidant activity of the oyster mushroom, *Pleurotus ostreatus* on CCl_4-induced liver injury in rats. *Food and Chem Toxicol* 44: 1989-1996.
12. Lee and Mary. 2009. Basic skills in interpreting laboratory data. *American Society of Health System Pharmacists*, 259 p.
13. Li, T. S. C. and Schroeder, W. R. 1996. Seabuckthorn (*Hippophae rhamnoides* L.): A multipurpose plant. *Hort Technology* 6: 370-380.
14. Li, T. S. C. and Wang, L. C. H. 1998. Physiological components and health effects of ginseng, echinacea and seabuckthorn. *In: Functional Foods, Biochemical and Processing Aspects* (G. *Mazza*, ed.), Technomic Publishing Company Inc, Lancaster, pp. 329-356.
15. McClatchey and Kenneth, D. 2002. Clinical laboratory medicine, Williams and Wilkins, p. 288.
16. Mengel, M., B., Schwiebert, L. and Peter. 2005. Family Medicine: Ambulatory Care and Prevention. McGraw-Hill Professional, p. 268.
17. Morgan, D. B., Carver, M. E. and Payne, R.B. 1977. Plasma creatinine and urea: Creatinine ratio in patients with raised plasma urea. *Br. Med. J.* 2: 929-932.
18. Nyblom, H., Bjornsson, E., Simren, M., Aldenborg, F., Almer, S. and Olsson, R. 2006. The AST/ALT ratio as an indicator of cirrhosis in patients with PBC. *Liver Int* 26: 840-845.
19. Sukumar, D. 1980. *Outlines of Dairy Technology.* Oxford University Press, Delhi.

Chapter 44

Protective Effect of Seabuckthorn (*Hippophae rhamnoides* L.) in Stz-Induced Lipid Peroxidation and β-Cell Damage in Rat Pancreas

Manju Sharma*, Mohd. Shamim Akhter, Gyanesh Shukla, K.K. Pillai and Zeeba S. Jairajpuri

Department of Pharmacology and Pathology, Hamdard Institute of Medical Science and Research (HIMSR), Hamdard University, New Delhi – 110 062, India
Herbal Drug Research, Ranbaxy Laboratory, Gurgaon, India
(*E-mail: manju_sharma72@yahoo.com)

SUMMARY

Seabuckthorn (*Hippophae rhamnoides* L.) is a branched deciduous shrub native to Europe and Asia. The plant has been used extensively in oriental traditional system of medicine for treatment of different diseases. Seabuckthorn found to have significant antioxidative, antimicrobial, immunomodulatory, cytoprotective, hepatoprotective and tissue regenerative properties. The present study was undertaken to evaluate the antidiabetic and antioxidant effect of seabuckthorn *(Hippophae rhamnoides* L.) in streptozotocin-nicotinamide induced type-2 diabetic rats. Experimental diabetes was induced by a single intraperitonial injection of streptozotocin (60 mg/kg), 15 minutes after the i.p. administration of 120 mg/kg nicotinamide. Seabuckthorn was administered orally to streptozotocin (STZ) diabetic rats. Blood glucose, tissue glutathione (GSH) and thiobarbituric acid reactive substances (TBARS) in pancreas were estimated following the established procedures. Biochemical observations were supplemented with histological examination of pancreatic tissue. The increase of blood glucose, TBARS level with reduction in GSH content were the salient features observed in STZ- diabetic rats. Seabuckthorn produced a significant ($p < 0.05$) reduction in blood glucose levels and TBARS levels in the STZ-diabetic rats. GSH, reduced significantly ($p < 0.05$) in diabetic rats, was brought back to near normal levels by co-administration of seabuckthorn. Degenerative changes of pancreatic beta cells in STZ-diabetic rats were minimized to near normal morphology by administration of seabuckthorn as evident by histopathological examination. The results of the study indicate the role of oxidative stress in the induction of diabetes and suggest a protective effect of seabuckthorn in this animal model.

Keywords: *Antioxidant, Diabetes, Oxidative stress, Seabuckthorn and Streptozotocin.*

INTRODUCTION

India continues to be the 'Diabetic Capital' of the world with 50.8 million diabetics. The pandemic spread of type 2 diabetes, the identification of new therapeutic avenues in the treatment of all pathological aspects of this disorder remains a major challenge for current biomedical research. Plant based herbal drugs are emerging as the primary components of holistic approaches to the diabetes management. Recently, the search for appropriate hypoglycemic agents has been focused on plants used in traditional medicine partly because of leads provided by traditional medicines to natural products that may be better treatment than the currently used drugs (Rates, 2001).

Streptozotocin is often used to induce diabetes mellitus in experimental animals through its toxic effects

Seabuckthorn (*Hippophae* L.): *A Multipurpose Wonder Plant, Vol. 5: Advanced Technologies for Cultivation, Processing, Health Protection and Environmental Conservation* (Virendra Singh, *Editor-in-Chief*, 2018), p. 347–353, Astral International Pvt. Ltd., New Delhi, India.

on pancreatic β-cells. STZ-induced diabetes mellitus is associated with the generation of reactive species causing oxidative damage (Szkudelski, 2001). Diabetics and experimental models exhibit high oxidative stress due to persistent and chronic hyperglycemia, which thereby depletes the activity of antioxidative defense system and thus promotes *de novo* free radicals generation (Baynes, 1997). Over the past few decades there has been increasing scientific and public interest in so called antioxidant hypothesis. Therefore, in addition to control of blood glucose levels, control of oxidative stress offers another avenue for the treatment of the disease. Chemicals with antioxidant properties and free radical scavengers may help in the regeneration of β cells and protect pancreatic islets against the cytotoxic effects of STZ (Coskun *et al.*, 2005)

Seabuckthorn (*Hippophae rhamnoides* L., *Elaegnacea*) is a thorny deciduous shrub native to several countries of Europe and Asia. The distribution ranges from Himalaya regions of India, Nepal, Pakistan, Afghanistan, Britain, Germany, Finland and France. All parts of this plant are considered rich source of a large number of bioactive substances with high medicinal and nutritional properties. Seabuckthorn (SBT) seed oils, berries, leaves and bark are well known for their medicinal properties and have been suggested to be due to high contents of antioxidative substances present in this plant. It has been extensively investigated for its antioxidative, antimicrobial, immunomodulatory, cytoprotective, antiatherogenic, hepatoprotective and tissue regenerative properties (Geetha *et al.*, 2002; Negi *et al.*, 2005; Yu-Wen Hsu *et al.*, 2009; Basu *et al.*, 2007). Furthermore, earlier studies have reported that SBT berries are the rich source of vitamins A, C, E, and K, flavonoids, carotenoids, organic acids and oils (Zeb, 2004). Many medicinal effects of SBT berries, such as, improving functions of digestive system and blood circulation have also been reported (Yang, 2002). Further, it is reported to be well tolerated and without any untoward reaction or side effects in toxicological studies (Upadhyay *et al.*, 2009). It is interesting to note that SBT has recently been reported to possess hypoglycemic activity (Zhang *et al.*, 2010). In view of all the above reports, we chose to evaluate SBT for possible beneficial effect on STZ-induced hyperglycemia and oxidative stress.

MATERIALS AND METHODS

Animals

Male albino Wistar rats (150-200 g) were used for the study. They were housed in polypropylene cages under standard laboratory conditions (12 h light/12 h dark, 21±2° C). The animals were fed on a standard pellet diet (Lipton rat feed Ltd., Pune) and water *ad libitum*. The experimental study was approved by the Institutional Animal Ethical Committee of Jamia Hamdard.

Drugs and Chemicals

Commercially available preparation of Seabuckthorn (*Hippophae rhamnoides* L.), fruit pulp was procured from Panacea International, New Delhi, India. Streptozotocin was procured from Sigma Chemicals Co., St. Louis, MO, USA. All other biochemicals and chemicals used for the experiment were of analytical grade.

Induction of Diabetes

The animal model of type-2 diabetes mellitus (NIDDM) was induced (Pellegrino *et al.*, 1998) in overnight fasted animals by a single intraperitonial injection of 60 mg/Kg STZ, 15 min after the i.p. administration of 120 mg/kg nicotinamide. Hyperglycaemia was confirmed by the elevated blood-glucose levels determined at 72 h.

Experimental Design

The rats were divided into six groups comprising of six animals in each group as follows:

- ✰ Group I: Normal control rats, received normal saline (1 ml/kg)
- ✰ Group II: Diabetic control rats, received STZ in single dose i.p. (60 mg/kg, 15 min after the i.p. administration of 120 mg/kg nicotinamide)
- ✰ Group III: SBT treated rats received (SBT 2 ml/kg. p.o.)
- ✰ Group IV: Diabetic treated rats received (SBT 1 ml/kg. p.o)
- ✰ Group V: Diabetic treated rats received (SBT 2 ml/kg. p.o.)
- ✰ Group VI: Diabetic treated rats received Gliclazide (25 mg/kg p.o.)

Drug treatment was given for duration of 3 weeks.

SBT was given for a period of 3 weeks to group III, IV and V animals. On the last day of the experimental blood samples were collected for biochemical estimations.

Determination of Blood Glucose

Blood glucose level was estimated by glucose oxidase (Braham and Trinder, 1972) method using a commercial diagnostic kit from Span Diagnostic Ltd., Surat, India.

Determination of HbA1c Level

Glycosylated haemoglobin (HbA1c) level was estimated by (Bannon, 1982) method using a commercial diagnostic kit from Monozyme India Limited, Secunderabad, India.

Determination of Oral Glucose Tolerance Test

Oral glucose tolerance test (OGTT) was measured according to the method of Pari and Saravana (Pari and Saravanan, 2002). Glucose solution (2 g/kg) was given to overnight fasted rats. Blood samples were taken at 0,

15, 30, 60 and 120 min. after glucose administration. All the blood samples were collected for glucose estimation.

Determination of Lipid Peroxidation

Lipid peroxidation (LPO) was measured by thiobarbituric acid (TBA) reaction with malondialdehyde (MDA), a product formed due to the peroxidation of membrane lipids (Ohkawa *et al.*, 1979). Tissues were homogenized in chilled phosphate buffer (0.1M, pH 7.4) that contained KCl (1.17 per cent w/v), using motor driven Teflon pestle. Aliquot of 1ml of the suspension medium was taken from the supernatant obtained after the centrifugation of tissue homogenate (10 per cent w/v) at 10,500 × g. About 0.5ml of 30 per cent trichloroacetic acid (TCA) followed by 0.5ml of 0.8 per cent TBA was then added to it. The tubes were kept in shaking water bath for 30 min. at 80 °C. After 30 min. of incubation tubes were taken out and kept in ice cold water for 10 min. These were then centrifuged at 800× g for 15 min. The absorbance of supernatant was read at 540nm at room temperature against appropriate blank. The concentration of MDA was measured from the standard calibration curve prepared by using tetraethoxypropane (TEP). Protein was estimated by the method of Lowry *et al.* (1951). Lipid peroxidation was expressed as n moles of MDA per mg of protein.

Determination of GSH

GSH content was estimated by method of Sedlak and Lindsay (1968).The tissues were homogenized in 0.02M EDTA. Aliquots of 5ml of the homogenates were mixed in the test tube with 4ml of cold distilled water and add 1ml of 50 per cent TCA. The tubes were shaken for 10 min using vortex mixer and then centrifuged at 1200 × g for 15min. Following centrifugation 2ml of supernatant was mixed with 4ml of 0.4M Tris-buffer (pH 8.9). The whole solution was mixed and 0.1ml of 0.01 M DTNB {5, 5′- dithiobis (2-nitrobenzoic acid)} was added to it. The absorbance was read within 5min of addition of DTNB at 412nm using UV-spectrophotometer (Shimadzu, UV-1601, Japan) against a reagent blank with no homogenate.

Histopathological Examination of Pancreas

The pancreas was isolated immediately after sacrificing the animal and washed with ice cold saline. It was then fixed in 10 per cent neutral buffered formalin solution. Sections of 3-5μm thickness were stained with hematoxylin and eosin (H.E.) for histopathological examination.

Statistical Analysis

Data were expressed as the mean ± standard error (S.E.) of the means. For a statistical analysis of the data, group means were compared by one-way ANOVA with *post-hoc* analysis. $P < 0.05$ was considered to be statistically significant.

RESULTS

Effects of SBT on Hyperglyceamia in STZ – Induced Diabetic Rats

Table 44.1 shows the effect of SBT on the blood glucose level. Significant ($P<0.001$) increase in blood glucose level was observed in STZ induced –diabetic rats when compared to normal control rats. Oral administration of SBT at two doses (1 and 2 ml/kg), for 3 weeks reduced the blood glucose level significantly ($P<0.001$) in a dose dependent manner. On its own SBT did not register any significant change blood glucose control values when compared to normal control rats. Gliclazide has its expected glucose lowering effect in diabetic animals.

Effect of SBT on HbA1c in STZ–Induced Diabetic Rats

Table 44.1 shows the effect of SBT on HbA1c level. Significant increase in HbA1c level was observed in STZ diabetic rats when compared to normal control rats. Oral administration of SBT at two doses (1 and 2 ml/kg), for 3 weeks decreased the HbA1c level significantly ($P<0.001$) in a dose dependent manner. There was no significant change in HbA1c level of SBT treated rats when compared to normal control rats.

Table 44.1: Effect of SBT and Gliclazide on Blood Glucose and Glycosylated Haemoglobin Levels in STZ Diabetic Rats

Group	*Treatment*	*Glucose in mg/dl (Day 0)*	*Glucose in mg/dl (Day 21)*	*Glycosylated Haemoglobin (per cent)*
1	Normal saline (1 ml/kg, p.o.)	69.83±3.46	67.83±6.72	5.70±0.6
2	STZ (60 mg/kg, i.p.) + NIC (120 mg/kg, i.p.)	212.5±28.87**	226.08±18.33	12.18±0.32
3	Seabuckthorn fruit pulp (2ml/kg, p.o.)	66.58±7.99ns	71.83±7.42	5.67±0.15
4	STZ ±NIC+SBT (1 ml/kg, p.o.)	226.24±29.57**	103.70±15.64**,ns	9.30±0.12
5	STZ±NIC+SBT (2 ml/kg, p.o.)	230.43±35.27**	93.75±12.07**,ns	6.95±0.7**,ns
6	STZ + NIC + Glic (25mg/kg, p.o.)	198.13±27.8**	79.122±9.57**,ns	5.3±0.7**,ns

Data are expressed in Mean ± SEM, n=6, for Day 0 comparison were made vs Group 1(positive control) and for Day 21 vs Group 2 (Diabetic control) and Group 6 (Gliclazide standard) - ** and ns statistically significance at p<0.01 and non-significance respectively.

Table 44.2: OGTT in Normal and Diabetic Rats Treated with Seabuckthorn Fruit Pulp

GRP	Treatment	0 min	30 min	60min	90 min	120 min
1	Normal saline (1 ml/kg, p.o.)	79.34±3.05	120.94±2.06	107.58±2.70	102.3±2.95	100.6±3.03
2	STZ (60 mg/kg, i.p.) + NIC (120 mg/kg, i.p.)	256.56±5.84	301.51±7.72	328.78±5.69	342.92±9.78	318.18±6.83
3	Seabuckthorn fruit pulp (2ml/kg, p.o.)	85.34±2.69	120.20±3.29	127.77±2.34	124.14±1.69	92.92±3.51
4	SBT (1ml, p.o.)+ glucose load	73.28±1.50	114.35±2.21	103.91±3.01	101.8±3.15	95.83±2.91
5	SBT(2 ml, p.o.) + glucose load	73.43±0.99	92.99±1.62*	89.97±2.37**	82.49±0.93**	78.97±1.59**
6	Gliclazide (25 mg/kg)+ glucose load	74.52±0.84	85.38±1.85**	81.68±2.28**	76.70±1.65**	72.35±1.28**

Data are expressed in Mean ± SEM, n=6, for Day 0 comparison were made vs Group 1(normal control) - ** statistically significance at p<0.01 and * statistical significance at p<0.05.

Effect of SBT on OGTT in STZ-Induced Diabetic Rats

Table 44.2 shows the blood glucose levels of normal control, diabetic control and SBT treated diabetic rats after oral administration of glucose (2 g/kg). In STZ diabetic rats, the peak increase in blood glucose level was observed after 1 hr. The blood glucose level remained high over next 1hr. SBT significantly reduced the blood glucose level in glucose- loaded rats at 30, 60, 90 and 120 min.

Effect of SBT on LPO in STZ-Induced Diabetic Rats

Table 44.3 demonstrates the level of malondialdehyde (MDA), a secondary product of lipid peroxidation in the pancreatic tissue homogenate. SBT treatment on its own did not induce any significant change in the LPO level, whereas STZ treatment resulted in a significant (P< 0.001) increase in MDA level in diabetic rats (group II). SBT treatment significantly (P< 0.001) decreased the level of MDA in diabetic rats when compared with diabetic control rats. Lowering of LPO level was also observed in gliclazide treated diabetic rats.

Table 44.3: Effect of Seabuckthorn Fruit Pulp and Gliclazide on TBARS Thiobarbituric Acid Reactive Substances

Group	Treatment	TBARS Level (nM/mg of protein)
1	Normal saline (1ml/kg, p.o.)	0.197 ± 0.031
2	STZ (60 mg/kg, i.p.) + NIC(120mg/kg,I.P.)	1.17± 0.052**
3	Seabuckthorn fruit pulp (1 ml/ kg, p.o.)	0.24 ± 0.03#
4	STZ±NIC+SBT (1 ml/kg, p.o.)	0.52 ± 0.07#, ns
5	STZ±NIC+SBT (2 ml/kg, p.o.)	0.44 ± 0.03#, ns
6	STZ±NIC+Glic (25mg/kg, p.o.)	0.40 ± 0.04#

Data are expressed in mean ± SEM, n=6, comparison were made vs Group 1 (normal control)-** statistical significance at p<0.01, vs Group 2(diabetic control)-# statistical significance at p<0.01 and vs Group 6(standard) - ns statistical significance at p<0.01.

Effect of SBT on STZ-Induced Changes in GSH Content

Table 44.4 shows the contents of GSH animals treated with STZ, which was significantly (P<0.001) decreased in pancreatic tissue as compared to the normal control rats (group-I). Pancreatic glutathione content increased significantly in SBT treated diabetic rats (group IV and V) when compared with diabetic control rats. Gliclazide has also replenished the GSH contents in STZ diabetic rats.

Table 44.4: Effect of Seabuckthorn Fruit Pulp and Gliclazide on Pancreatic Glutathione

Group	Treatment	Glutathione (μg/mg of protein)
1	Normal saline (1ml/kg, p.o.)	4.33± 0.12
2	STZ (60 mg/kg, i.p.) + NIC (120mg/kg,I.P.)	2.30 ± 0.10**
3	Seabuckthorn fruit pulp (1 ml/ kg, p.o.)	3.93 ± 0.83
4	STZ±NIC+SBT (1 ml/kg, p.o.)	2.36 ± 0.41
5	STZ±NIC+SBT (2 ml/kg, p.o.)	3.22 ± 0.36€
6	STZ±NIC+Glic (25mg/kg, p.o.)	3.67 ± 0.27€€

Data are expressed in mean ± SEM, n=6, comparison were made vs group 1(normal control) ** statistical significance at p<0.01 and vs group 2 (diabetic control) - € statistical significance p<0.05.

Effect of SBT on STZ-Induced Histological Changes in Pancreas

STZ induced diabetes resulted in degenerative and lytic changes in the islets of Langerhans of pancreas (Figure 44.1). The lobules of pancreatic acini were seen with areas of fibrosis. There was depletion of β cells and the dimension of the islet is also considerably reduced and shrunken. The histological picture of pancreas was improved in diabetic rats treated with SBT (Figure 44.1), as evidenced by the regeneration of β cells, which reveals the protective effect of SBT.

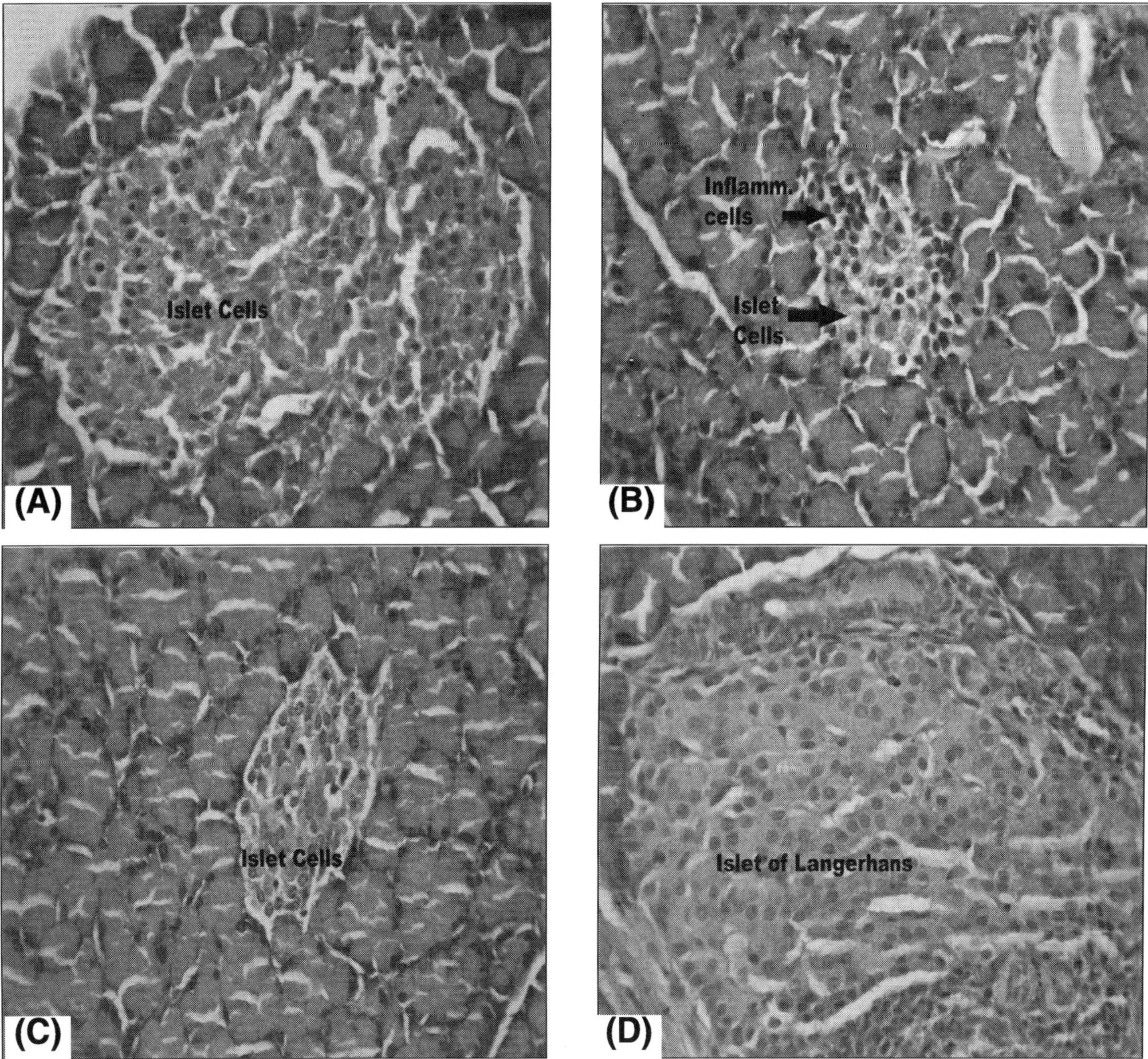

Figure 44.1: Hematoxylin and Eosin-Stained Sections of Rat Pancreas.

(A) Represents normal control group showing numerous beta cells with abundant basophilic cytoplasm. (B): Represents STZ treated rats showing atrophy of the beta cells. The beta cell cytoplasm is scanty and numerous inflammatory cells are seen in the upper part of the islet. (C) Represents pancreatic section of STZ + SBT (2ml) treated rats showing regeneration of beta cells with basophilic cytoplasm. No inflammatory cells are seen. (D) Represents pancreatic section of STZ+ GLC treated rats showing details of islet cells. The islet cells are of normal appearance (HE x 400).

DISCUSSION

Diabetes mellitus is a serious metabolic disorder with micro and macrovascular complications that results in significant morbidity and mortality. Lipid peroxidation has been implicated in the pathogenesis of naturally occurring or induced diabetes (Wolff *et al.*, 1991). There is a wide range of antioxidant defenses which protects against the adverse effects of free radicals production *in vivo* (Halliwell and Gutterridge, 1990).

Disturbances of antioxidant defense systems in Diabetes mellitus have been reported (Strain, 1991). Therefore, treatment with antioxidant may contribute to the prevention and delaying of diabetic complications (Armstrong *et al.*, 1996). The antioxidant activity of the SBT fruit is due to the presence high content of vitamin C and E, carotenoids, phytosterol, tannins as well as antioxidant enzymes (Guliyev *et al.*, 2004). STZ-diabetic rats exhibited persistent hyperglycemia. Seabuckthorn (SBT) treatment to STZ diabetic rats reduced the elevated blood glucose level thereby showing its antihyperglycaemic activity. In diabetes, there is an increased glycosylation of a number of proteins therefore; measurement of HbA1c has proven to be particularly useful in monitoring the effectiveness of therapy (Goldstein, 1995). HbA1c level increased in STZ diabetic rats when compared to normal control rats.

The results showed that SBT decreased blood glucose level, HbA1c level and improved glucose tolerance in diabetic rats. In a recent study antidiabetic effect of an aqueous extract from SBT seed residues has been demonstrated (Zhang *et al.*, 2010).

Some of the recent studies have shown the SBT has antioxidant properties and prevent LPO (Geetha *et al.*, 2002; Negi *et al.*, 2005; Yu-Wen Hsu, *et al.*, 2009; Geetha *et al.*, 2008). Administration of SBT to diabetic rats reduced the glycosylation of Hb by virtue of its free radical scavenging property and thus contribute to decreased level of glycated Hb in SBT treated diabetic rats.

SBT by its ability to scavenge free radicals and to inhibit lipid peroxidation prevents STZ-induced oxidative stress and protects β-cells resulting in decreased blood glucose levels. (Cao *et al.*, 2003) have also reported that the flavonoids from the seed and fruit residue of *Hippophae rhamnoides* L. exhibited hypoglycaemia and hypolipidemic effects. In the oral glucose tolerance test, SBT significantly reduced the blood glucose level in glucose loaded rats at 30, 60, 90 and 120 min. as compared with control rats, loaded only with glucose (Table 44.2). It has been reported in various studies that increased oxidative stress may play a role in pathogenesis and progression of diabetic tissue damage (Baynes, 1991).

In our study, we observed a significant increase in the concentration of MDA, a secondary product of LPO, in pancreas of STZ treated animals. LPO induced by STZ was associated with the decreased enzymatic activities and non-enzymatic (GSH) antioxidant contents. Since an oxidative stress affects the cellular integrity only when antioxidant mechanisms are no longer able to cope with the free radical generation, supplementation of an antioxidant could gear up the detoxification machinery.

SBT berries has been reported to be a rich source of vitamin A,C,E,K, flavonoids, carotenoids, organic acids and oils, was studied for its protective effect against oxidative stress induced by STZ. The cytotoxic effect of STZ in β-cells is mediated by ROS (Szkudelski, 2001). Drugs with antioxidant properties may supply endogenous defense systems and reduce both initiation and propagation of reactive oxygen species (Bergendi *et al.*, 1999). Results of the present study clearly showed that LPO level was decreased in SBT treated diabetic rats. These results suggest that SBT administration in diabetic rats reduce LPO possibly by decreasing free radical formation, increasing antioxidant and associated enzymes content with significant lowering in blood glucose level.

Reduced glutathione (GSH), the cellular antioxidant was lowered in animals treated with STZ. However, in diabetic animals co-treated with SBT, GSH level was increased suggesting protective effect on later by SBT. These findings of elevation of GSH contents by SBT suggest GSH-dependent detoxification of free radicals. In addition, Paolisso *et al.* (1992) have proposed that the ratio of GSH/GSSG plays a critical role in the glucose homeostasis of diabetes. The effects of SBT on plasma glucose and pancreatic LPO produced by STZ may be related to the significant rise in pancreatic glutathione induced by this drug. Numerous studies indicate that dietary supplementation with antioxidant nutrients may be a safe and simple complement to traditional therapies for preventing and treating diabetic complications. Protective effect of SBT is probably due to the counteraction of free radicals by its antioxidant nature and is also evident from the histological observation of pancreas, which shows similar morphology as compared to the normal control rats. The diabetic rats in contrast however showed extensive degeneration and multiple necrotic changes. The present study thus confirms that pancreatic β-cell gets destroyed by STZ and SBT causes its regeneration. In conclusion, our present results demonstrated that SBT, a widely used plant with potent antioxidative activity, can exert anti-diabetic effects by preserving β-cell function in STZ diabetic rats. The protective effects of SBT could be attributed to several constituents with potentially healthy biological properties such as Vitamin C, carotenoids, tocopherols, flavonoids and phytosterols. Nevertheless, detailed mechanistic action on different free radicals by SBT particularly the flavonoids present in SBT requires further study.

REFERENCES

1. Armstrong, A., Chestnutt, J., Gormley, M. and Young, I. 1996. The effect of dietary treatment on lipid peroxidation and antioxidant status in newly diagnosed non-insulin dependent diabetes. *Free Radical Biology and Medicine* 21: 719-726.
2. Basu, M., Prasad, R., Jayamurthy, P., Pal, K., Arumughan, C. and Sawhney, R. C. 2007. Anti-atherogenic effects of seabuckthorn (*Hippophae rhamnoides* L.) seed oil. *Phytomedicine* 14: 770-777.
3. Bannon, P. 1982. Effect of pH on the elimination of labile fraction of glycosylated haemoglobin. *Clinical Chemistry* 28: 2183.
4. Baynes, J. W. and Thorpe, S. R. 1997. The role of oxidative stress in diabetic complications. *Current Opinion in Endrocrinology* 3: 277-284.
5. Baynes, J. W. 1991. Role of oxidative stress in the development of complications of diabetes mellitus. *Diabetes* 40: 405- 412.
6. Braham, D. and Trinder, P.1972. An improved colour reaction for the determination of blood glucose by oxidase system. *Analyst*. 142-144.

7. Bergendi, L., Benes, L., Durackova, Z. and Ferencik, M. 1999. Chemistry, physiology and pathology of free radicals. *Life Sciences* 65: 1865-1874.

8. Cao, Q. H., Qu, W. J., Deng, Y. X., Zhang, Z. C., Niu, W. and Pan, Y.F. 2003. Effects of flavonoids from the seed and fruit residue of *Hippophae rhamnoides* L. on glycol metabolism in mice. *The Journal of Chinese Medicine and Traditional Chinese Medicine* 26: 735-737.

9. Coskun, O., Kanter, M., Korkmaz, A. and Oter, S. 2005. Quercetin, a flavonoid antioxidant, prevents and protects Streptozotocin induced oxidative stress and β- cell damage in rat pancreas. *Pharmacological Research* 51: 117-123.

10. Geetha, S., Jayamurthy. P., Pal, K., Pandey, S., Kumar, R. and Sawhney, R. C. 2008. Hepatoprotective effect of seabuckthorn (*Hippophae rhamnoides* L.) against carbon tetrachloride induced liver injury in rats. *Journal of the science of Food and Agriculture*, 88: 1592-1597.

11. Geetha, S., Sai Ram, M., Singh, V., Ilavazhagan, G. and Sawhney, R. C. 2002. Antioxidant and immunomodulatory properties of seabuckthorn (*Hippophae rhamnoides*)–An *in vitro* study. *Journal of Ethnopharmacology* 79: 373-378.

12. Goldstein, D. E.1995. How much do you know about glycated hemoglobin testing. *Clinical Diabetes* 60-63.

13. Guliyev, V. B., Gul, M. and Yildirim, A. 2004. *Hippophae rhamnoides* L.: Chromatographic methods to determine chemical composition, use in traditional medicine and pharmacological effects. *Journal of Chromatography B* 812: 291-307.

14. Halliwell, B. and Gutterridge, J. M.1990. Role of free radicals and catalytic metal ions in human disease: *An Overview Methods in Enzymology* 186: 1-85.

15. Lowry, O. H., Roenbrough, N. J., Farr, A. L. and Randall, R. J.1951. Protein measurements with the Folin's reagent. *Biological Chemistry* 193: 265-275.

16. Negi, P. S., Chauhan, A. S., Sadia, G. A., Rohinishree, Y. S. and Ramteke, R. S. 2005. Antioxidant and antibacterial activities of various seabuckthorn (*Hippophae rhamnoides* L.) seed extracts. *Food Chemistry* 92: 119-124.

17. Ohkawa, H., Ohishi, N. and Yagi, K.1979. Assay of lipid peroxides in animal tissues by thiobarbituric acid reaction. *Analytical Biochemistry*. 95: 351-358.

18. Paolisso, G., Di Maro, G., Pizza, G. and D'Onofrie, P. 1992. Plasma GSH/GSSG affects glucose homeostasis in healthy subjects and non-insulin-dependent diabetes. *American Journal of Physiology* 263: 435-440.

19. Pari, L. and Saravanan, G. 2002. Antidiabetic effect of Cogent db a herbal drug in alloxan-induced diabetes mellitus. *Comprehensive Biochemistry and Physiology, Part C*. 131: 19-25.

20. Pellegrino, M., Broca, C., Gross, R., Roye, M., Manteghetti, M. and Hillaire-Buys, D. 1998. Development of a new model of type 2 diabetes in adult rats administered with Streptozotocin and nicotinamide. *Diabetes* 47: 224-229.

21. Rates, S. M. 2001. Plants as a source of drugs. *Toxicon* 39: 603-613.

22. Sedlak, J. and Lindsay, R. H. 1969. Estimation of total protein bound and nonprotein sulfhydryl groups in tissue with Ellman's reagent. *Analytical Biochemistry* 25: 192-205.

23. Strain, J. J. 1991. Disturbances of micronutrient and antioxidant status in diabetes. In: *Proceedings of the Nutrition Society* 50: 591-604.

24. Szkudelski, T. 2001. The mechanism of Alloxan and Streptozotocin action in β -cells of the rat pancreas. *Physiological Research* 50: 536-546.

25. Upadhyay, N. K., Kumar, R., Mandotra, S. K., Meena, R. N., Siddiqui, M. S., Sawhney, R. C. and Gupta, A. 2009. Safety and healing efficacy of seabuckthorn (*Hippophae rhamnoides* L.) seed oil on burn wounds in rats. *Food Chemistry and Toxicology* 47: 1146-1153.

26. Wolff, S. P., Jang, Z.Y. and Hunt, J. V. 1991. Protein glycation and oxidative stress in diabetes mellitus and ageing. *Free Radical Biology and Medicine* 10: 339-352.

27. Yang, B. and Kallio, H. 2002. Composition and physiological effects of seabuckthorn (*Hippophae rhamnoides* L.) lipids. *Trends Food Science and Technology* 13: 160-167.

28. Yu-Wen, H., Chia-Fang, T., Wen-Kang, C. and Fung-Jou, L. 2009. Protective effects of seabuckthorn (*Hippophae rhamnoides* L.) seed oil against carbon tetrachloride induced hepatotoxicity in mice. *Food Chemistry and Toxicology* 47: 2281-2288.

29. Zeb, A. 2004. Chemical and Nutritional constituents of seabuckthorn juice. *Pakistan Journal of Nutrition* 3(2): 99-106.

30. Zhang,W., Zhao, J., Wang, J., Pang, X., Zhuang, X., Zhu, X. and Qu, W. 2010. Hypoglycaemic effect of aqueous extract of seabuckthorn (*Hippophae rhamnoides* L.) seed residues in streptozotocin – Induced diabetic rats. *Phytotherapy Research* 24: 228-232.

Chapter 45

Hippophae rhamnoides L. Enhances Hypoxic Tolerance: Role of Hypoxia Responsive Proteins

Geetha Suryakumar*, Kanika Jain, Rajendra Prasad, Lilly Ganju and Shashi Bala Singh

Cellular Biochemistry Division, Defence Institute of Physiology and Allied Sciences, Lucknow Road, Timarpur, Delhi – 110 054, India
(*E-mail: geethasuryakumar@yahoo.com)

SUMMARY

The prevalence of hypoxia-induced maladies in the high altitude milieu potentiates a need to develop supplements and nutraceuticals to enhance physical performance and improve survival. Seabuckthorn (*Hippophae rhamnoides* L.; SBT), a medicinal plant with a wide spectrum of pharmacological effects, has been shown to improve performance in environmental stress models. In the present study, the cardioprotective effect of SBT leaf extract under acute sub lethal hypoxic stress was investigated. SBT considerably increased the hypoxic tolerance, evaluated by exposing the animals to an altitude of 9754 m and observing their gasping time. SBT decreased the lipid peroxidation levels from 384.80 ± 12.26 in the hypoxia alone group to 283.80 ± 14.44 nmol/mg protein in hypoxia + SBT group and reduced tissue damage. SBT led to a 50 per cent reduction in the corticosterone levels. Examining the role of hypoxia responsive proteins in SBT mediated increase in hypoxic tolerance, SBT accentuated the increase in GLUT-1(79.52 ± 3.08) and VEGF (11.10 ± 0.52) levels, as compared to only hypoxia groups (GLUT-1, 73.13 ± 2.43 pg/mg protein and VEGF, 10.42 ± 0.33 pg/mg protein). HO-1 levels increased to 3.55 ± 0.05 in the hypoxia +SBT animals as compared to 2.69 ± 0.06 ng/mg protein in the hypoxia alone animals. SBT elevated the NO levels in hypoxia+SBT groups (3.34 ± 0.12) *vs* hypoxia (2.90 ± 0.03 nmol/ml sodium nitrite) and led to up regulation of eNOS and iNOS. The present study thus provides an insight into the mechanism behind the pro-survival effects of SBT under acute hypoxia.

***Keywords*:** *Hypoxia, Heart, Antioxidants, HIF-1α and Nitric oxide.*

INTRODUCTION

Herbal formulations have been in use for many years globally not only as therapeutic but also as prophylactic and health-promoting agents. Many studies have shown the benefits of plant-derived products in combating the effects of stress and enhancing the endurance to physical stressors in daily life of mankind. Seabuckthorn (*Hippophae rhamnoides* L.), a member of family *Elaeagnaceae*, a unique and valuable plant has gained worldwide attention in the last few decades, mainly for its medicinal and nutritional potential. All parts of this plant are considered to be a good source of large number of bioactive substances such as vitamins, carotenoids, phytosterols, organic acids, polyunsaturated fatty acids and some essential amino acids; all of which possess biological and therapeutic activities against diseases such as obesity, diabetes, cardiovascular disease, cancer, ulcer, inflammation, immune disorders, burn wounds and radiation damage (Suryakumar and Gupta, 2011). While Seabuckthorn berries are rich sources of biochemical compounds with medicinal properties, SBT leaves have also been shown to possess valuable nutrients and bioactive substances including flavonoids. The leaf extract was found to provide protection against chromium-induced oxidative stress (Geetha *et al.*, 2003), besides having

Seabuckthorn (*Hippophae* L.): *A Multipurpose Wonder Plant, Vol. 5: Advanced Technologies for Cultivation, Processing, Health Protection and Environmental Conservation* (Virendra Singh, *Editor-in-Chief*, 2018), p. 355–358, Astral International Pvt. Ltd., New Delhi, India.

potent anti-inflammatory activity (Ganju *et al.*, 2005). The extract showed significant anti-stress and adaptogenic activity when evaluated in cold-hypoxia restrain animal model (Saggu *et al.*, 2008) and also demonstrated anti-hypoxia activity in rats exposed to extreme high altitude conditions (Purushothaman *et al.*, 2008). However investigations into the possible mechanisms by which SBT plays its therapeutic role are not yet complete.

Recent study (Jain *et al.*, 2013) has shown the role of hypoxia inducible factor-1 (HIF-1α) mediated pathways in differential tolerance to acute hypobaric hypoxia. The present study was designed to identify the possible signalling mechanisms by which SBT modulates hypoxic tolerance.

MATERIALS AND METHODS

Ethical Clearance for Animal Study

Male Sprague-Dawley rats (150 ± 15 g) were used for all experiments. Animals were maintained under a 12-h light–dark cycle at temperature 24°C ± 2°C in the Institute's animal house facility. The study was approved by the Animal Ethical Committee of the Institute in accordance with Committee for the Purpose of Control and Supervision of Experiments on Animals (CPCSEA) of the Government of India.

Plant Material and SBT Extract Preparation

The 70 per cent ethanol extract of SBT leaves was prepared and for details please refer to (Ganju *et al.*, 2005). Before administration, the SBT extracts were diluted in sterile saline solution and a dose of 100 mg/kg b.w. was administered orally to the animals using a gastric canula, daily for 7 days.

Hypoxic Exposure and Administration of Seabuckthorn Leaf Extract

Animals were categorized into four groups, with n=10 in each group: Control (unexposed nor moxic, administered normal saline), SBT (Only SBT leaf extract, 100 mg/kg bw), Hypoxia (Only hypoxia, 9754 m at 32pC), Hypoxia + SBT (SBT leaf extract + Hypoxia). Hypoxic tolerance was determined by measuring the time taken for the onset of gasping (GT) as described earlier (Jain *et al.*, 2013). Briefly, animals were exposed one at a time, to simulated hypobaric hypoxia of 9754 m at 32°C in an animal decompression chamber and the time taken for appearance of the first sign of gasping, a characteristic hyperventilatory response, was carefully recorded. Following hypoxia exposure, animals were randomly selected into two groups. While one group was kept at normoxic condition on regular feed, the other group was administered the SBT leaf extract (100 mg/kg bw daily for 7 days). At the end of 7 days, the animals were again exposed to the altitude of 9754 m and the gasping time was noted. Unexposed control group rats were maintained in the normoxic condition within the same laboratory conditions. After the hypoxic exposure, the animals were anaesthetized under sodium pentobarbital and tissue samples were collected. The hearts were perfused in saline and used fresh or snap frozen in liquid nitrogen and stored at -80RC for further use.

Oxidative Stress Parameters

Lipid peroxidation was measured by direct estimation of MDA as described earlier using thiobarbituric acid solution (Doussett *et al.*, 1983). The absorbance was measured at 531 nm using UV-Vis spectrophotometer (BioRad, USA). Results were expressed as nanomoles of MDA formed per mg protein.

Corticosterone ELISA

Corticosterone levels were quantified in the rat plasma, using commercially available kit, rat/mouse Corticosterone ELISA (DRG, Germany) as per manufacturer's instructions. For the assay, the plasma samples were diluted 1:5 using the sample diluent provided in the kit. The concentrations of corticosterone were expressed as ng/ml.

ELISA Based Quantification

Levels of HO-1, GLUT-1, VEGF and Endothelin-1 were quantified using the commercially available rat ELISA kit Cusabio, Wuhan, China as per manufacturer's instructions. The tissues were freshly homogenised in 1X PBS and the supernatants were diluted 1: 5 using the sample diluent provided in the kit.

RESULTS AND DISCUSSION

Effect of SBT on Hypoxic Tolerance and Hypoxia Induced Oxidative Damage in Animals

Exposure to acute sub lethal hypoxic stress, at a simulated altitude of 9754m elicits a distinct response in animals. Administration of Seabuckthorn leaf extract considerably increased the hypoxic tolerance, as evident in the 2-fold higher gasping time (Figure 45.1). Compared to unexposed animals (255.70 ± 4.43 nmol/mg protein), hypoxia increased lipid peroxidation considerably (384.80 ± 12.26 nmol/mg protein, $p<0.001$) (Figure 45.2). SBT administration mitigated this damage and lowered the levels to 283.80 ± 14.44 nmol/mg protein ($p<0.001$

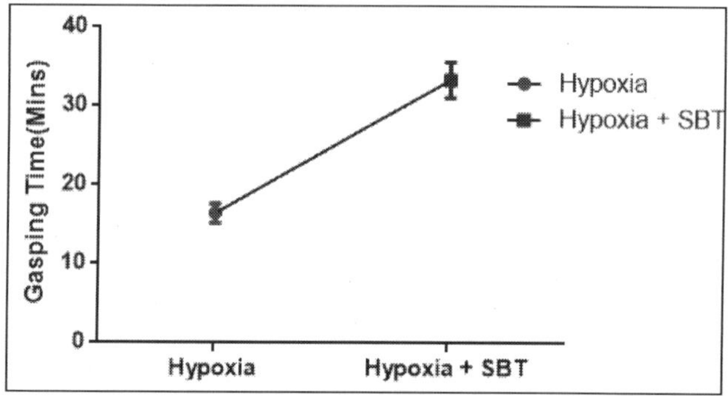

Figure 45.1: Effect of Seabuckthorn on Hypoxic Gasping Time.

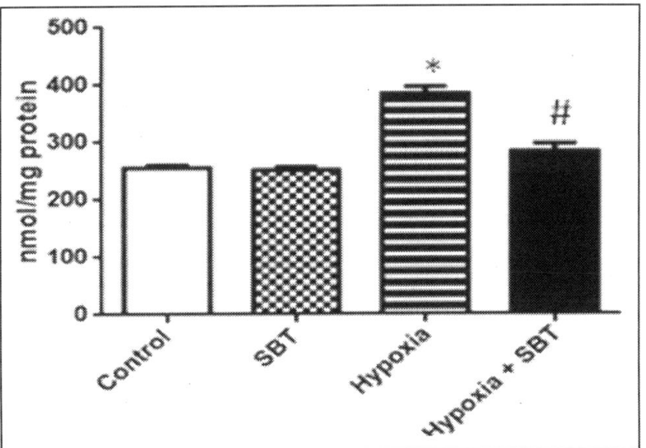

Figure 45.2: Effect of Seabuckthorn on Hypoxia Induced Lipid Peroxidation.

vs hypoxia). While hypoxia led to a manifold increase in the levels of stress hormone, corticosterone, SBT led to a 50 per cent reduction in the corticosterone levels under acute indicating that the administration of the extract substantially decreased the impact of the stress imposed on the animal (Figure 45.3).

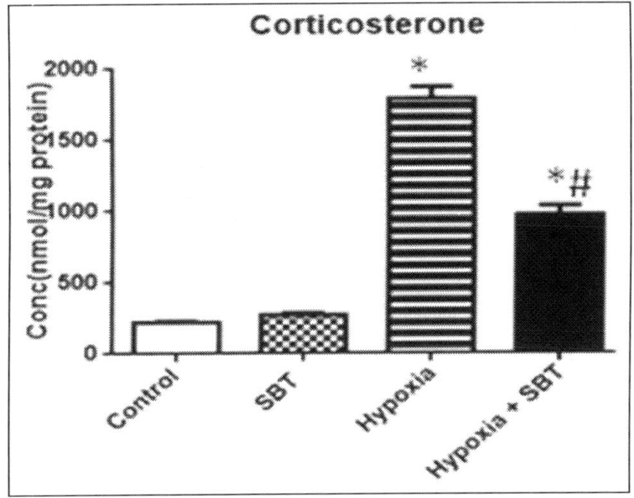

Figure 45.3: Effect of Seabuckthorn on Stress Hormones.

SBT Induces the Expression of Hypoxia Inducible Proteins

Jain *et al.*, 2013 reported the protective role played by HIF-1 α and its regulated proteins in enhancing hypoxic tolerance. To evaluate the role of these hypoxia-responsive proteins in the increased survival under acute hypoxia after SBT treatment, the expression of HIF-1α in the nuclear extracts of the cardiac tissue samples were identified. The results of this study showed that while SBT alone did not mediate a change in the expression of this transcription factor, exposure to hypoxia considerably increased its expression in both the hypoxia ($p<0.001$ vs control) and hypoxia + SBT group ($p<0.05$ *vs* hypoxia). Examining the role of hypoxia responsive proteins in SBT mediated increase in hypoxic tolerance, SBT accentuated the increase in GLUT-1(79.52 ± 3.08) and VEGF (11.10 ± 0.52) levels, as compared to only hypoxia groups (GLUT-1, 73.13 ± 2.43 pg/mg protein and VEGF, 10.42 ± 0.33 pg/mg protein). HO-1 levels increased to 3.55 ± 0.05 in the hypoxia +SBT animals as compared to 2.69 ± 0.06 ng/mg protein in the hypoxia alone animals (Figure 45.4). Thus while earlier studies have emphasized on the benefits of SBT in boosting physical and mental performance, this study drew attention to the modulation of adaptive and maladaptive signalling mechanisms which mediate the defensive action of SBT against hypoxic injury.

CONCLUSION

In summary, this study for the first time sheds light on the mechanistic effect of Seabuckthorn on hypoxia responsive signalling cascades. It was demonstrated that under extreme high altitude conditions, SBT augments hypoxia responsive proteins and hence promotes tolerance to hypoxic stress.

ACKNOWLEDGEMENT

This study was supported and funded by the Defence Research and Development organization, Ministry of Defence, Government of India.

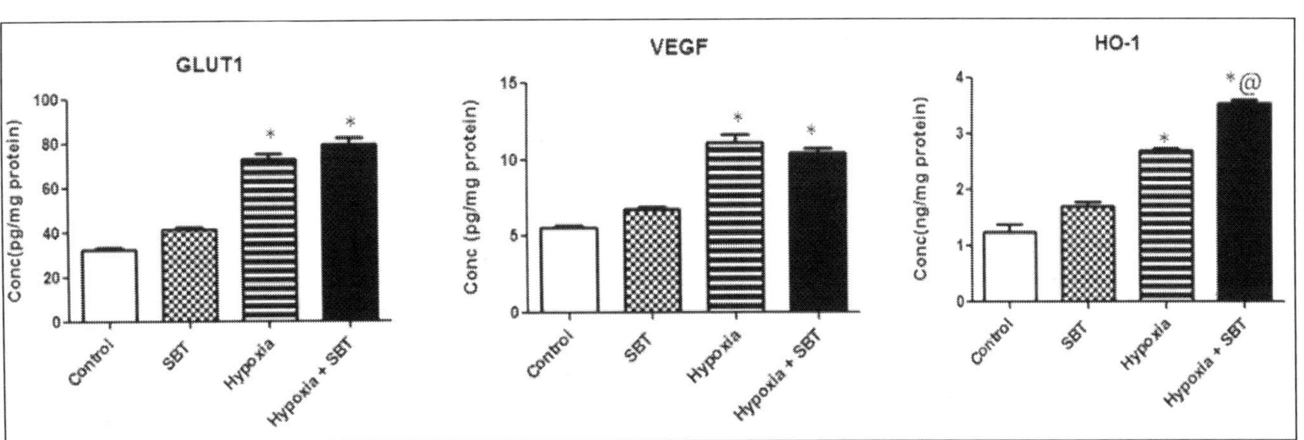

Figure 45.4: Administration of Seabuckthorn Extract Modulates the Expression of Hypoxia Responsive Proteins.

REFERENCES

1. Dousset, J. C., Trouillh, M. and Fogliettis, M. J. 1983. Plasma malondialdehyde levels during myocardial infarction. *Clin. Chim. Acta*. 129: 319-322.
2. Ganju, L., Padwad, Y., Singh, R., Karan, D., Chanda, S., Chopra, M. K., Bhatnagar, P., Kashyap, R. and Sawhney, R. C. 2005. Anti-inflammatory activity of seabuckthorn (*Hippophae rhamnoides*) leaves. *International Immunopharmacology* 5: 1675-1684.
3. Geetha, S., Sai Ram, M., Singh, V., Ilavazhagan, G. and Sawhney, R. C. 2003. Evaluation of antioxidant activity of leaf extract of seabuckthorn (*Hippophae rhamnoides* L.) on chromium (VI) induced oxidative stress in male albino rats. *Journal of Ethnopharmacology* 8: 247-251.
4. Puroshothaman, J., Suryakumar, G., Shukla, D., Malhotra, A. S., Harinath, K., Kumar, R. and Sawhney, R. C. 2008. Modulatory effects of seabuckthorn (*Hippophae rhamnoides* L.) in hypobaric hypoxia induced cerebral vascular injury. *Brain Research Bulletin* 77: 246-252.
5. Saggu, S. and Kumar, R. 2008. Effect of seabuckthorn leaf extracts on circulating energy fuels, lipid peroxidation and antioxidant parameters in rats during exposure to cold, hypoxia and restraint (C–H–R) stress and post stress recovery in rats. *Phytomedicine* 15: 437-446.
6. Suryakumar, G. and Gupta, A. 2011. Medicinal and therapeutic potential of seabuckthorn (*Hippophae rhamnoides* L.). *Journal of Ethnopharmacology* 138: 268-278.

Chapter 46

The Comparative Therapeutic Efficacy of the Combination of Seabuckthorn Seed Oil and Misoprostol or Sucralfate in Gastric Ulcers and Erosions in Dogs

Diviank Thakur, S.P. Tyagi[1]*, Amit Kumar, Shilpa Gupta and Virendra Singh[2]

[1]Department of Surgery and Radiology, College of Veterinary and Animal Sciences
[2]Department of Biology and Environmental Sciences,
CSK Himachal Pradesh Agricultural University, Palampur, H.P., India
(*E-mail: sptyagivet@gmail.com)

SUMMARY

The study was conducted on 16 healthy adult medium-sized mongrel dogs of either sex maintained under uniform environment. Inj. Dexamethasone was administered in all the animals @ 1mg/kg IV *b.i.d.* to induce non-fatal gastric ulcerations and erosions (GUE) of comparable severity. These dogs were then randomly divided into 4 equal groups and treated with Misoprostol @ 10 µg/kg in group 1, Misoprostol @ 10 µg/kg + Seabuckthorn oil @ 1ml/animal in group 2, sucralfate @ 1g/animal in group 3 and Sucralfate @ 1g/animal + Seabuckthorn @ 1ml/animal in group 4 three times a day till complete healing of GUE. The progression of GUE was evaluated by clinical, haematological, biochemical, faecal occult blood test and gastro-endoscopic observations at different time intervals. The average number of days taken to achieve complete healing of GUE was shortest at 8.25 days in group 4 followed by 9.75 days in group 2 and 1 and 10.5 days in group 3. The subjective gross assessment of gastric lesions by endoscopy also revealed faster and more uniform healing in group 4. It was concluded that the treatment of GUE in dogs with the combination of sucralfate and seabuckthorn oil resulted in faster recovery in comparison to other groups.

Keywords: *Seabuckthorn, Dog, Gastric erosions, Ulcerations, Misoprostol and Sucralfate.*

INTRODUCTION

The stomach is an important organ of the digestive tract which acts as reservoir of food, mixes food with gastric juices and begins the process of protein digestion. Secretion of hydrochloric acid and the proteolytic enzyme 'pepsin' is an essential step in this digestive process. However, apart from helping in disintegration of food particles, these substances may damage the gastric wall itself, if come in to its direct contact. Such 'aggressive factors of stomach' are however efficiently kept away from actual gastric wall by a number of gastric processes known as 'gastric protective factors'. Continuous maintenance of a thick mucus layer by gastric mucosa as a physical acid barrier is one of such protective mechanisms. Disturbance of this delicate balance in between gastric aggressive and protective factors, particularly interruption of the gastric mucosal barrier in certain conditions, allows diffusion of gastric acid and pepsins into the deeper layers of stomach leading to localized inflammation, oedema and wounds. Such condition is better known as gastric ulcerations and erosions (GUE). The mucosal defect penetrating through the gastric muscularis mucosa is termed as 'gastric ulcer' whereas, the superficial ulcer that does not extend far into the mucosa is termed as 'gastric erosion'. In dogs, GUE often results from long term use of NSAID'S (non-steroidal anti-inflammatory drugs) and corticosteroids. Gastrointestinal ulceration have also been associated

Seabuckthorn (*Hippophae* L.): *A Multipurpose Wonder Plant, Vol. 5: Advanced Technologies for Cultivation, Processing, Health Protection and Environmental Conservation* (Virendra Singh, *Editor-in-Chief*, 2018), p. 359–365, Astral International Pvt. Ltd., New Delhi, India.

with uraemia (toxins, increased gastrin), liver disease, stress, increased production of HCl (mast cell tumour at any site, gastrin-producing tumour of the pancreas (Zollinger–Ellison syndrome), Hypotension, *e.g.* during surgery, hypo-adrenocorticism, spinal cord disease.

The healing of GUE lesions is slow and complicated due to constant secretion of acid by the stomach for normal digestive process. The back diffusion of gastric acid into the sub-mucosa causes mast cell degranulation resulting in histamine release and subsequent stimulation of acid production by gastric parietal cells, which aggravates the inflammation and oedema at the site of GUE. Hence, the aim of antiulcer therapy is to restore the mucosal barrier rapidly, reduce the amount of gastric acid being produced and/or neutralize its effect readily and thus stopping this vicious cycle of gastric damage. This can be achieved by using a number of drugs like Antacids, H_2 antagonist, Prostaglandin analogues and Cytoprotectants. In small animal practice, all these drugs are used but their doses and frequencies are mostly extrapolated from human medicine. So far only very limited research studies have been done to standardize the doses and frequencies of these drugs in Veterinary medicine. Consequently, at present these medicines have questionable efficacy along with unknown side effects in small animals.

In one of the studies done on the subject, it was reported that Sucralfate @ 1g/animaland misoprostol @ 10 µg/kg twice orally were not as effective for management of GUE as famotidine and lansoprazole in dogs (Dogra*et al.*, 2013). The misoprostol is one of the synthetic prostaglandin analogue which is considered to have a profound stimulatory effect on the gastric mucus layer and sucralfate is considered to act as a coating layer over damaged area of stomach mimicking gastric mucus layer and should have theoretically aided in healing of GUE. This there was a need to investigate and ascertain the effect of higher doses and/or frequency of administration of these drugs in dogs to conclude about their efficacy. In addition to routinely used medicines, large number of herbal preparations are also used for the management of GUE in humans and laboratory animals. These alternative medicines are considered to be effective due to presence of many bioactive materials in them, which directly or indirectly acts on the gastric mucosa. Seabuckthorn (*Hippophae rhamnoides*) is one such plant which has potential to act as gastro-protective in cases of GUE. A number of workers have reported the therapeutic efficacy of seabuckthorn (SBT) oil in the management of GUE in lab animals and dogs (Gupta *et al.*, 2015; Kumar *et al.*, 2013; Dogra *et al.*, 2013; Tyagi, 2006). To increase the effectiveness of treatment, it is presumed that a combination of medicines may show better results. Combination of allopathic and herbal medicines may also work similarly by decreasing the dose of individual medicine which in turn may also reduce their respective side effects, if any. Gupta *et al.* (2015) reported faster healing of GUE lesions in dogs by using a combination of SBT oil and famotidine when compared with either of them alone. So present study was designed to evaluate sucralfate or misoprostol alone, with a dose frequency higher than used previously by other workers and their combination with seabuckthorn oil for the treatment of gastric ulcerations and erosions in dogs.

MATERIALS AND METHODS

The study was conducted in 16 adult healthy mongrel dogs divided into four equal groups. For creation of non-fatal gastric ulcers and erosions (GUE) in dogs, and ascertaining its GUE index, the technique adopted by Tyagi, (2009) was utilized. For this, a steroidal drug "Dexamethasone" was administered intravenously in all the animals @ 1mg/kg b.i.d. for variable periods. The dogs were subjected to gastric endoscopy under anaesthesia every fourth day to observe the development of GUE lesions and to record their GUE index. The drug was discontinued when endoscopic GUE index reached to 7 on two consecutive observation intervals or 8 in a single observation occasion as per the GUE index. Thereafter, the different groups of dogs were subjected to different treatment protocols as given below in Table 46.1.

Table 46.1: The GUE Treatment Protocol Utilized in different Groups of Dogs

Sl.No.	Group	Treatment Protocol
1	I	Misoprostol @ 10 µg/kg orally t.i.d.
2	II	Misoprostol @ 10 µg/kg + Seabuckthorn oil @ 1ml/animal orally t.i.d.
3	III	Sucralfate @ 1gm/animal orally t.i.d.
4	IV	Sucralfate @ 1gm/animal + Seabuckthorn @ 1ml/animal orally t.i.d.

The progress of GUE healing was evaluated and compared on the basis of clinical, haematological, faecal occult blood test and endoscopic observations on day 0, 3, 6, 9, 12, 15.till complete healing. The day '0' was the day on which the endoscopic GUE index of desirable severity was reached and the dexamethasone was discontinued as explained in previous paragraph. Clinical parameters were rectal temperature, respiration and heart rate, weight of animal, status of appetite, vomiting and melena. Haematological parameters were Haemoglobin (Hb), Packed cell volume (PCV), Total erythrocyte count (TEC), Differential leucocyte count (DLC), Total leucocyte count (TLC). Statistical analysis was done using Analysis of variance (ANOVA) method by Students-Newman (inter-group comparison) and Dunnett's test (intra group comparison) at 5 per cent and 1 per cent level of significance.

RESULTS AND DISCUSSION

Gastro-endoscopic study revealed that the severity of GUE lesions reduced gradually leading to their eventual healing in all the groups without any intergroup significance in GUE index. Intra group statistical comparison however, revealed significant reduction of GUE index within as early as day 3 of treatment in groups 2, 3, and 4 and by day 6 in all groups (Table 46.2 and Figures 46.1–46.3). The time taken for complete healing of GUE lesions as evidenced by endoscopic observation ranged from day 6-15 in group 1, day 6-12 in group 2 and 4, and day 9-15 in group 3.Though the healing period range was similar in groups 2 and 4, but, the number of animals showing absence of GUE lesions at particular observation interval differed in these groups leading to different average period of healing. The mean number of days for achieving complete healing of GUE lesions was found to be least in group 4 (combination of sucralfate with seabuckthorn oil) at 8.25 days followed by group 1 and 2 (misoprostol alone or in combination with seabuckthorn oil) at 9.75 days, and group 3 (sucralfate alone) at 10.5 days (Table 46.2).

In a similar kind of study done by Dogra *et al.* (2013), the average healing period of GUE in dogs were calculated as 10.5 days with misoprostol treatment and 13.5 days with sucralfate as against 9.75 days and 10.5 days respectively in the present study. Though, the doses of drugs used in that study of Dogra *et al.* (2013) were same as used in the present study, but the frequency of drug administration was twice a day as against thrice a day in the latter. The reduction in healing time was quite distinct in the group of dogs treated with sucralfate alone in this study, whereas improvement in healing period was only marginal in misoprostol treated group. However, overall this indicates that the misoprostol and sucralfate have better therapeutic efficacy for GUE in dogs with a three times a day dosing regimen.

Table 46.2: GUE Indices of Dogs at Various Observation Intervals during Treatment (Mean± S.E.)

Groups \ Days	0	3	6	9	12	15
Group 1 (Miso)	7.75±0.25 N=4	5.00±1.22 N=4	1.5**±0.86 N=4	1.5**±0.50 N=2	0.5**±0.50 N=2	0.0±0.00 N=1
Group 2 (Miso + SBT)	8.0±0.00 N=4	3.25**±0.85 N=4	2.25**±0.75 N=4	1.33..±0.66 N=3	0.0**±0.00 N=2	
Group 3 (Sucral)	7.25±0.25 N=4	4.5**±0.64 N=4	2.5**±0.64 N=4	0.5**±0.50 N=4	1.0±0.00 N=1	0.0±0.00 N=1
Group 4 (Sucral + SBT)	7.5±0.28 N=4	3.25**±1.03 N=4	1.25**±0.75 N=4	1.0**±1.0 N=2	0.0±0.00 N=1	

p<0.01 (**), p<0.05 (*) intra group comparison.

Table 46.3: Mean Number of Days for Achieving Complete Healing of GUE Lesions in different Groups

Group 1	Misoprostol	9.75 days
Group 2	Misoprostol + Seabuckthorn seed oil	9.75 days
Group 3	Sucralfate	10.5 days
Group 4	Sucralfate + Seabuckthorn seed oil	8.25 days

Kumar *et al.* (2013) compared different doses of seabuckthorn seed oil *i.e.* 1 ml, 2.5 ml and 5 ml per animal orally b.i.d. for the treatment of similarly produced GUE in dogs and reported dose-dependent therapeutic effect of SBT oil on the healing of GUE lesions. However, the 1 ml dose was considered ineffective as the mean healing period of GUE with this treatment was found similar to that of untreated control group. However, the same 1 ml dose of SBT oil when combined with Sucralfate and with an increased frequency of dosing resulted in earlier healing of gastric lesions. This indicates that SBT oil has additive therapeutic effect with Sucralfate even in lower doses.

Qualitatively, the gastric mucus layer was distinctly shinier and thicker in group 4 at day 6 and subsequently when compared with other groups in the present study. Similar findings have also been reported by other workers with the use of seabuckthorn seed oil for the treatment of GUE in dogs (Tyagi, 2006; Dogra *et al.*, 2013; Gupta *et al.*, 2015) and in most cases the early restoration of such healthier mucus layer resulted in faster healing of lesions.

SBT oil has been reported to have anti-gastroulcerative activity due to a variety of biologically active components distributed in different parts of its fruit. Sterols and steryl glycosides in the seed oils are suggested to be the major anti-ulcer compounds (Romero and Lichtenberger, 1990).The concentration

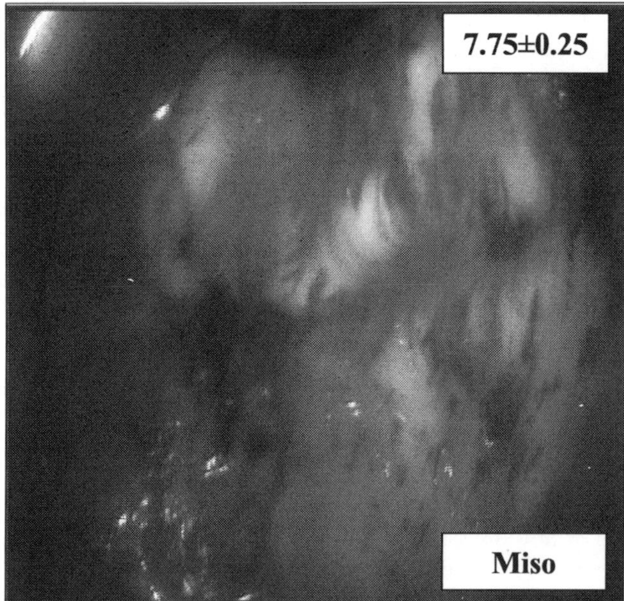

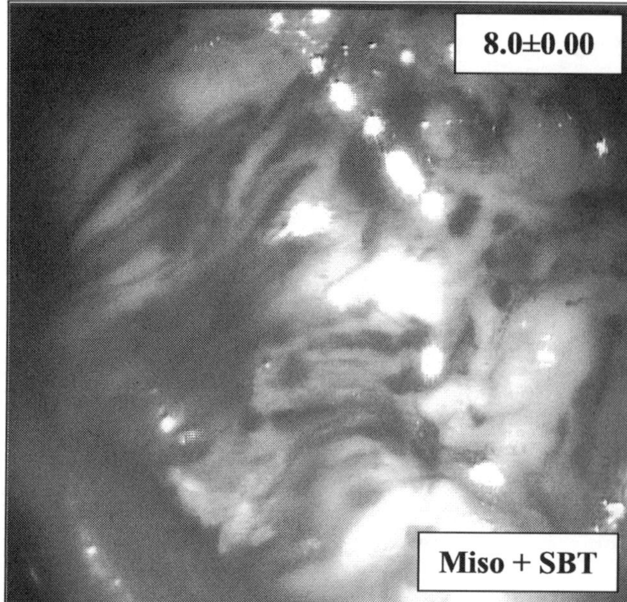

Multiple lesions of variable depths with active haemorrhagic base, thin mucus layer, adherent blood clots, hyperaemic mucosa with typical 'water fall' appearance.

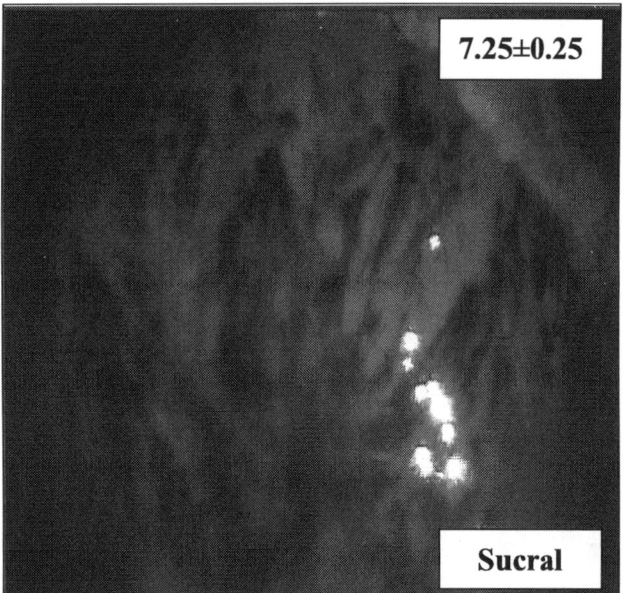

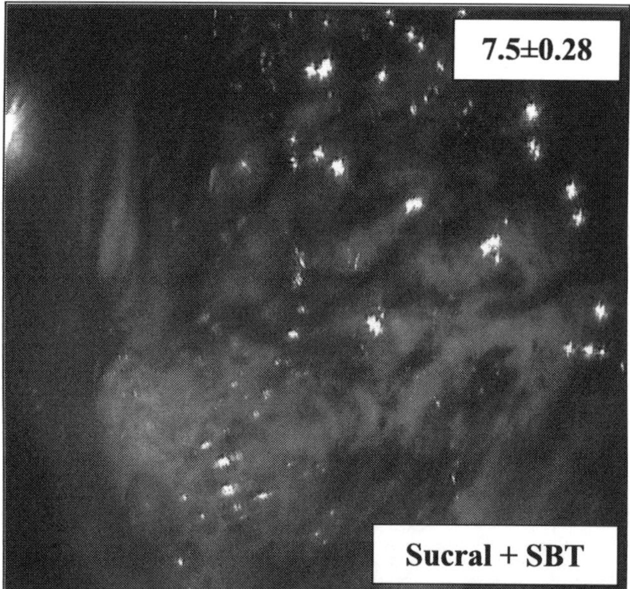

Figure 46.1: Endoscopic View of Gastric Mucosa in different Groups on Day 0.

of sterols in SBT seed oil ranges from 1-2 per cent (Barouand Kallio, 2002). It is also suggested that the anti-ulcer effects of SBT oil may not have been only due to their fatty acids but also because of high content of β -carotene, α-tocopherols and β-sitosterols (Xing *et al.* 2002). β-sitosterol is also reported to increase the gastro protective activity of unsaturated phospholipids (Romero and Lichtenberger, 1990).

The presence of vitamins E, and carotenoides in SBT oil also lend it the anti-oxidant activity which is helpful in any wound-healing (Xing *et al.*, 2002). Evidence exists that administration of SBT oil inhibits lipid-peroxidation in gastric mucosa in experimental models of gastric ulcers in rats (Krichkovskaya *et al.*, 1992). Considering that disruption in the integrity of gastric mucosa is caused due to a large number of inter-dependant causes (Robert and Kauffman 1989), its effective healing can be quicker only when all the causes are addressed simultaneously. SBT seed oilcan effectively manage GUE by a variety of actions *viz.* modification of the prostaglandin synthesis of gastric mucosa, re-epithelialization of damaged mucosa, inhibition of gastric secretion and proteolytic activity within gastric mucosa (Kallio and Baoru 2005; Zhou, 1986; Loginov *et al.*, 1983).

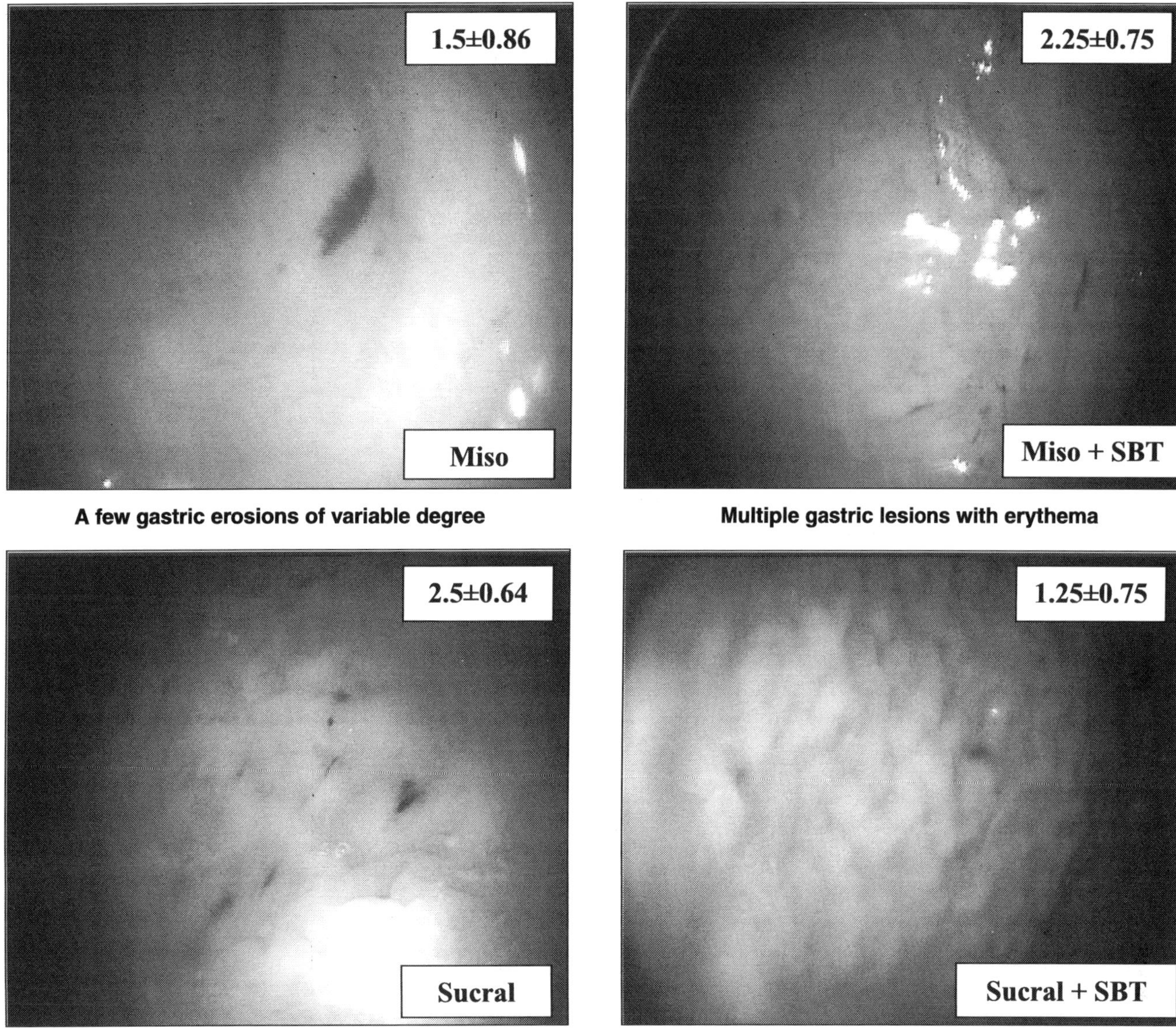

A few gastric erosions of variable degree

Multiple gastric lesions with erythema

Multiple smaller gastric lesions with and without adherent blood clots

A few free floating blood clots

Figure 46.2: Endoscopic View of Gastric Mucosa in different Groups on Day 6.

The rectal temperature, heart rate and respiration rate did not vary much with the base values and remained within normal physiological limits throughout the period of study in all the groups. No statistical difference was observed within or in-between various groups at any observation intervals. A non-significant increase in body weight and appetite *i.e.* restoration towards normal levels was observed during the study. Melena was observed till day 6 in group 1, 3 in group 2, 6 in group 3 and 3 in group 4. The severity of melena gradually decreased towards the end of study in all the groups. In general, gradual changes in all haematological parameters were observed in all the groups and at the end of the study, all parameters returned to their normal baseline levels in all the animals. These findings are similar to those observed by Kumar *et al.* (2013), Tyagi (2006), Dogra *et al.* (2013) and Gupta *et al.* (2015) in similar kinds of studies in dogs and are indicative of normal healing progress of GUE.

CONCLUSION

The healing of dexamethasone-induced gastric ulcerations and erosions (GUE) in dogs occurs faster with a treatment regimen comprising the combination of sucralfate @ 1gm and seabuckthorn oil @ 1ml*per* animal administered three times a day orally when compared with the treatment regimens of sucralfate (@1 gm/animal) alone, misoprostol (@10 μg/kg)alone or in combination with seabuckthorn oil (@1ml/animal) in the same dose frequency.

1-2 sub mucosal localised lesions

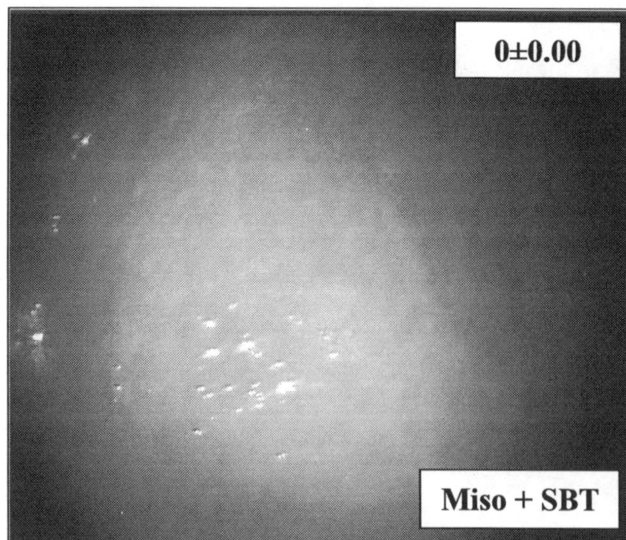

No lesion on gatric mucosa, no blood clots, hick and shiny mucus layer

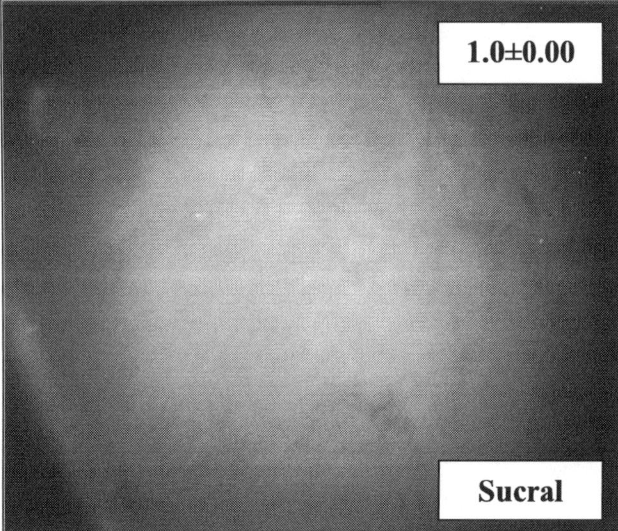

Solitary blood clot with well-formed mucosal layer

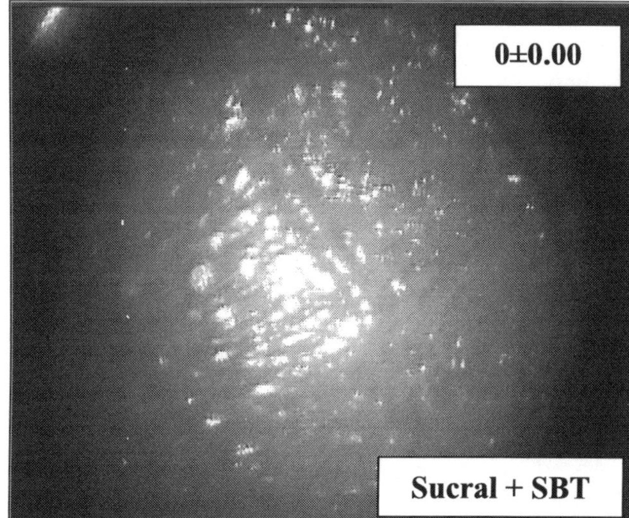

Thick and shiny mucus layer

Figure 46.3: Endoscopic View of Gastric Mucosa in different Groups on Day 12.

REFERENCES

1. Barou, Y. and Kallio, H. 2002. Composition and physiological effects of seabuckthorn (*Hippophae*) lipids. *Trends in Food Science Technology* 13:160-167.
2. Dogra, R., Tyagi, S. P. and Kumar, A. 2013. Efficacy of seabuckthorn (*Hippophae rhamnoides*) oil vis-a-vis other standard drugs for management of gastric ulceration and erosions in dogs. *Veterinary Medicine International*, Article ID 176848, 11 pages. doi:10.1155/2013/176848.
3. Guptam, S., Tyagi, S. P. and Kumar, A. 2015. Management of gastric ulcerations and erosions with combination of seabuckthorn (*Hippophae rhamnoides*) seed oil and famotidine in dogs. *Indian Journal of Veterinary Surgery* 36 (1): 29-32.
4. Kallio, H. and Baoru, Y. 2005. Effects of seabuckthorn (*Hippophae rhamnoides*) oil on the mucosa and skin. In: *Seabuckthorn (Hippophae) A multipurpose wonder plant: Biochemistry and Pharmacology*. (Editor-in-Chief, Singh, V.). Daya Publishing house, New Delhi, 2: 86-90.
5. Krichkovskaya, L.V., Dementij, R.N. and Zyabchenkova, A.K. 1992. Advances in the biochemical and pharmacological research on seabuckthorn. In: *Wugong Agricultural Research Center, Scientific Communication*, Wugong Publishing House, Shannxi Province, China, pp. 141-142.
6. Kumar, A., Tyagi, S. P., Dogra, R. and Gupta, S. 2013. Dose-dependent effects of seabuckthorn seed oil on the healing of gastric ulcers and erosions in dogs.

In: *Proceedings of* 6th A Conference of International seabuckthorn Association, pp. 9-16, 14-18 October, 2013, Germany.

7. Loginov, A. S., Mironov, V. A., Amirov, N. S., Arjun, G. S., Matrosov, V. S., Trubitsyna, I. E. and Chikunova, B. Z. 1983. *Patologicheskaia Fiziologiiaieksperimental' Naiaterapiia* (6): 667-70.
8. Robert, A. and Kauffman, G. L. 1989. Stress ulcers, erosions and gastric mucosal injury. In: *Gastrointestinal Diseases, Pathophysiology, Diagnosis and Management* (Sleisenger M.H. and Fordtran J.S., eds.), 4th edition, Philadelphia WB Saunders, pp. 773-855.
9. Romero, J. J. and Lichtenberger, L. M. 1990. Sterol dependence of gastric protective activity of unsaturated phospholipids. *Digestive Diseases and Sciences* 35 (10): 1231-1238.
10. Tyagi, S. P. 2006. Studies on the efficacy of seabuckthorn (*Hippophae* ssp.) oil in the healing of gastric ulcers in dogs. *PhD Thesis,* Department of Veterinary Surgery and Radiology, COVAS, CSKHPKV, Palampur.
11. Tyagi, S. P. 2009. Utilization of seabuckthorn in healing of gastric ulcer in dogs. In: *Project Report of National Agricultural Innovation Project,* CSKHPKV, Palampur, HP, India.
12. Xing, J. F., Yang, B., Dong, Y. L., Wang, B. W., Wang, J. X. and Kallio, H. 2002. Effects of seabuckthorn (*Hippophae rhamnoides*) seed oil and pulp oil on experimental models of gastric ulcer in rats. *Fitoterapia* 73 (8): 644-650.
13. Zhou, W.1986. The curative effect of seabuckthorn oil on the induced gastric ulcer of rats. *Academic Journal of the Second Military Medical University of China* 7 (6): 468-469.

Chapter 47

The Mycotoxin Protective Action of Seabuckthorn Leaves in Poultry Birds

Ashustosh Thakur, Arti Ghabru and C. Varshneya*

Department of Pharmacology and Toxicology, COVAS,
CSK Himachal Pradesh Agricultural University, Palampur – 176 062, H.P., India
(*E-mail: cvarshneya@hotmail.com)

SUMMARY

The present study was conducted for studying the protective action of seabuckthorn (SBT) leaf extract and glucomannan (GM) in aflatoxin (AF) treated broilers on the basis of growth and haematological parameters. A total of 36 day old chicks were used in this investigation to study the growth performance and haematological parameters for the period of 4 weeks. Two week old chicks were randomly divided into six groups of 6 chicks each. The animals were grouped as: Control (without treatment), Seabuckthorn (SBT), Aflatoxin (AF), Seabuckthorn, Aflatoxin and Glucomannan (SBT+AF+GM), Seabuckthorn and Aflatoxin (SBT+AF) and Aflatoxin and Gucomannan (AF+GM). Seabuckthorn extract, Glucomannan and Mycotoxin were given at the dietary levels of 10000 ppm, 100 ppm and 400 ppb. A significant ($P<0.05$) increase in liver weight, TLC and heterophil count was observed while a significant ($P<0.05$) reduction in weight of bursa of fabricius, TEC, Hb and lymphocyte count was observed in AF treated group as compared to control. In groups where SBT is given with AF and alone showed increase in Hb and lymphocyte count while liver weight and heterophil count got significantly ($P<0.05$) reduced in comparison to AF treated group. In groups where SBT and GM was given alone as well as in combination restored the organ weight, TEC, Hb, lymphocyte and heterophile count thereby, effectively neutralizing aflatoxin effect.

Keywords: *Mycotoxin, Broiler, Seabuckthorn leaves and Haematological parameters.*

INTRODUCTION

The present day intensive animal husbandry practices have made poultry birds susceptible to various environmental contaminants including mycotoxins. The mycotoxins contamination causes 25 per cent of total poultry feed loss. Out of mycotoxins, aflatoxin is responsible for substantial production losses in the poultry industry. The contamination of animal feed is practically unavoidable under moist conditions of midhills of Himachal Pradesh.

Hippophae rhamnoides L (family Elaeagnaceae), commonly known as seabuckthorn (Chharma), is a thorny nitrogen fixing deciduous shrub, native to Europe and Asia (Rousi, 1971). In India, seabuckthorn is widely prevalent in cold desert areas of Himalayan region at altitude of 2500-4000 m above sea level. The plant is extensively used in the traditional system of medicine in Asian Countries. Seabuckthorn leaves are used in the treatment of diarrhea, dermatological disorders and rheumatoid arthritis (Vereshchagin *et al.*, 1959; Khalmatov, 1984). Seabuckthorn is a rich source of carbohydrates, proteins, steroids, flavonoids and vitamins (A and B) (Ambaye and Indap, 1970; Bernath and Foldesi, 1992). The young leaves of seabuckthorn posses high content of carotene and flavonoids (Chen, 1988). The leaves of SBT (*Hippophae rhamnoides* L.) are rich source of proteins and are less fibrous, indicating their potential as a feed resource for the livestock and poultry (Singh *et al.*, 2001). During recent years

Seabuckthorn (Hippophae L.): A Multipurpose Wonder Plant, Vol. 5: Advanced Technologies for Cultivation, Processing, Health Protection and Environmental Conservation (Virendra Singh, *Editor-in-Chief*, 2018), p. 367–373, Astral International Pvt. Ltd., New Delhi, India.

nutraceutical properties of seabuckthorn leaves are being exploited in poultry as feed substitute (Ambatkar, 2009) indicating that 3 per cent substitution of crude protein offers better results in performance of broilers. In the present studies mycotoxin protective action of seabuckthorn leaves is evaluated on the basis of growth and haematological parameters. The investigation was conducted to study the effect of aflatoxin and to study the efficacy of both seabuckthorn leaves extract and glucomannan alone and in combination to ameliorate aflatoxin induced toxic action in broiler chicks.

MATERIALS AND METHODS

Aflatoxin Production

The aflatoxin was produced on rice (Trenk and Hartman, 1970). One hundred grams of rice was taken in 500 ml Erlenmeyer flask and soaked in 75 ml of water overnight. The flasks were autoclaved (15 psi/30 min), cooled and inoculated with *Aspergillus flavus* (MTCC 6672). Thirty ml sterile water was added to each flask and the flasks were incubated at 28°C temperature for 15 days. After 48 hrs of inoculation, whitish moldy growth was seen on the surface of the rice, which later turned yellow followed by green colour. After incubation, the moldy rice was steamed at 100°C for 1 hrs to kill the spores and dried in hot air oven at 60 °C overnight. The dried rice culture was ground to a fine powder and analyzed for its aflatoxin content.

Mycotoxin Estimation

The aflatoxin from the ground rice culture was quantified by using thin layer chromatography (Tapia, 1985) at Animal Feed Analytical and Quality Control Laboratory, Veterinary College and Research Institute, Namakkal- 637001, Tamil Nadu and The levels were further got confirmed from Feed Analysis Lab (Skylark Lab, Pathankot, Punjab).

Collection of Plant Material

The leaves of seabuckthorn were collected from Agricultural Research Extension Centre, Kukumseri, Lahaul-Spiti.

Preparation of Leaves Powder

The leaves were collected, shade dried and ground to obtain a fine powder. The powder was used for the preparation of aqua methanolic extract and lyophilized into fine powder and was mixed in broiler feed to achieve the dietary levels at 10,000 ppm.

Management

The experimental animal house was thoroughly cleaned and disinfected. The chicks were reared in battery cage system throughout the experiment. The chicks were vaccinated with 0.05 ml of New Castle disease vaccine through intra ocular route on days six and fourteen. The chicks were provided standard poultry feed obtained from the Department of Animal Nutrition, COVAS, CSK HPKV, Palampur (H.P.). The feed and water were provided *ad libitum* during the entire period of experimentation. The broiler starter feed was provided up to day 35 of age and thereafter birds were fed finisher feed up to end of the experiment.

Experimental Protocol

The day-old broiler chicks were procured from Uttam Hatchery, Mataur Distt: Kangra (H.P.). On the day of arrival, the chicks were examined for any abnormality and ill health. They were acclimatized in the new environment for two weeks. All the experiments detailed in this study were conducted according to the guidelines of Institutional Animal Ethical Committee. The chronological order of the activities conducted under this study was as follows:

Day	Activity
-10	Preparation of animal house (cleaning, washing and burning of edges)
-3	Fumigation of animal house and Day-old broiler chicks procured
3	Chicks wing banded
7	Vaccinated against Ranikhet (New Castle) Disease with Lasota strain
14	Recording of body weight and booster vaccination with New Castle Disease vaccine (Lasota strain)
15	The birds were divided in to six groups and feeding started as per schedule
21	Body weights recorded
28	Body weights recorded and blood collected for hematological
35	Body weights recorded
42	Body weight recorded and collection of blood for hematological Parameters
42	Collection of organs for determining internal organ weights.

Experimental Design

The birds were fed normal feed and water *ad libitum* till up to two weeks. On the day 15, broiler chicks were divided into six groups (6 in each) as follows:

Evaluation of Effects

Following parameters were recorded to study the effect of aflatoxin and its protection with seabuckthorn leaves extract and glucomannan in chickens.

Growth Response

The body weights of the birds were recorded at weekly interval and expressed in g.

Blood Collection

Blood was collected from the wing vein and was divided into two portions. One portion was collected in vial containing EDTA (20 mg/ml) and the second portion was taken in sterilized vial without EDTA for serum separation. The blood smears were prepared by

Table 47.1: Experimental Design to Evaluate the Effect of Seabuckthorn (*Hippophae rhamnoides*) Leaves Extract and Glucomannan on Toxic Effect of Aflatoxin in Broiler Chickens

Group	Toxin/SBT/GM	No. of Birds	Dose of SBT (ppm)	Dose of GM (ppm)	Mycotoxin (AF) in Feed (ppb)
I	Control	6	-	-	-
II	SBT	6	10000	-	-
III	Aflatoxin	6	-	-	400
IV	AF + SBT+GM	6	10000	100	400
V	AF + SBT	6	10000	-	400
VI	AF +GM	6	-	100	400

AF: Aflatoxin; SBT: Seabuckthorn; GM: Glucomannan.

taking a drop of blood on a clean slide without mixing the anti-coagulant.

Serum Separation

The blood was collected in the disposable syringe and transferred to clean, dry and sterilized glass tube and allowed to stand for clotting at room temperature for 3 to 4 hrs. The serum was separated in sterile tube after centrifugation. The serum samples were stored at -20°c till complete analysis.

Haematological Profile

The blood parameters – Total leucocyte count (TLC), Differential leucocyte count (DLC), Total erythrocyte count (TEC), Hemoglobin (Hb), Packed cell volume (PCV) were estimated according to following methods.

Total Leucocyte Count (TLC)

The total leucocytes count was done by the method of Natt and Herrick (1952) using modified WBC diluting fluid and the values were expressed in thousand per microliter blood.

Differential Leucocyte Count (DLC)

Thin blood smear was prepared using a drop of blood without anticoagulant on day 28 of treatment. Differential Leucocyte Count was done by the method of Lucas and Jamroz (1961) and the results were expressed in per cent.

Total Erythrocyte Count (TEC)

TEC was performed as per the method of Natt and Herrick (1952) using RBC diluting fluid. TEC value was expressed in millions per microliter.

Hemoglobin (Hb)

The hemoglobin values were determined by using acid haematin method (Benjamin, 1985) and the values were expressed in g/dl.

Packed Cell Volume (PCV)

The packed cell volume was determined by using microhaematocrit method (Benjamin, 1985).

Statistical Analysis

The data were analyzed by analysis of variance test using the Graph Pad Instat version 3.00 for windows (GraphPad Software, SanDiego, California, USA, (www. Graphpad.com) and the significant differences between mean values were determined using Turkey-krammer multiple comparison test. The data were presented as Mean ± S E. The inter group comparisons were made at 5 per cent level of significance.

RESULTS AND DISCUSSION

Growth Performance

The effect of seabuckthorn leaves extract on growth performance of broiler chicks was studied in terms of weight gain, feed conversion ratio and internal organ weight of liver, kidney and bursa of Fabricius. The results are presented in Tables 47.2–47.4. The poultry birds fed on aflatoxin showed decrease in weight gain and weight of bursa Fabricius while feed conversion ratio (FCR) and liver weight were increased. The reduced feed intake, lower body weight gain and enlarged liver were observed in chicks fed on feed supplemented with aflatoxin B_1 and these finding of the present investigation were in consistent with previous reports on the performance depressing effects of aflatoxin B_1 (Ledoux *et al.*, 1998; Kalorey *et al.*, 2004). The enlargement of liver could be due to accumulation of lipids in hepatocytes because of general inhibition of lipid transport and interference in lipid synthesis (Tung *et al.*, 1972). The adverse effects of aflatoxin on growth performance were attributable to inhibitory effect of aflatoxin on protein synthesis (Borisova *et al.*, 1987).

The supplementation of feed with seabuckthorn leaves @ 10000 ppm level significantly ($P \leq 0.05$) increased the weight gain in comparison to aflatoxin treated group and control group whereas, liver weight and feed conversion ratio of broilers were decreased. These observations are in consonance with the findings of Geetha *et al.* (2003) who reported a significant increase in body weight following oral feeding of ethanolic leaf extract of seabuckthorn in rats. The body weight

Table 47.2: Effect of Dietary Seabuckthorn Leaves Extract Supplementation and Glucomannan on Body Weight gain (g) in Broilers during Test Period

Groups	*Control*	*SBT*	*AF*	*SBT+AF+GM*	*SBT+AF*	*AF+GM*
Body Weight Gain	736^{ac} ± 47.39	846.66^{a} ±25.25	624.33^{b} ±28.85	799.5^{ac} ± 11.57	688.33^{bc} ± 18.81	697.16^{bc} ± 27.65

Mean values ± S.E. (n=6), The means with same superscripts in between columns do not differ significantly at 5 per cent level.

Table 47.3: Effect of Dietary Seabuckthorn Leaves Extract Supplementation on Feed Conversion Ratio in Broilers

Groups	*Control*	*SBT*	*AF*	*SBT+AF+GM*	*SBT+AF*	*AF+GM*
FCR	1.723^{abc} ± 0.127	1.470^{ac} ±0.449	2.038^{b} ±0.121	1.557^{c} ± 0.046	1.794^{abc} ± 0.047	1.808^{abc} ± 0.081

Mean values ± S.E. (n=6), The means with same superscripts in between columns do not differ significantly at 5 per cent level.

Table 47.4: Effect of Dietary Seabuckthorn Leaves Extract Supplementation in Internal Organ Weight (g) in Broilers

Organs	*Control*	*SBT*	*AF*	*SBT+AF+GM*	*SBT+AF*	*SBT+GM*
Kidney	1.85 ±.076	1.9 ±.057	2.016 ±.030	2.1 ±0.1065	2.05 ±.067	2.116 ±.070
Bursa Fabricius	416^{a} ± 0.09	4.416^{a}± 0.098	2.58^{b} ± 0.151	3.66^{ab} ± 0.22	3.58^{ab} ± 0.38	4.116^{ab} ± 0.037
LIVER	19.16^{a} ± 0.65	22.58^{ace} ± 0.95	35.3^{b} ± 0.67	25.66^{c} ± 1.406	26.66^{e} ± 1.05	22.16^{a} ± 1.08

Mean values ± S.E. (n=6), The means with same superscripts in between columns do not differ significantly at 5 per cent level.

gain was highest in birds fed combination of aflatoxin, seabuckthorn and glucomannan, group IV followed by group VI and group V. The increase in liver weight was highest in group V followed by group IV and group VI. The findings of the present investigation revealed that combination of seabuckthorn with glucomannan has the maximum protective action and the contribution of glucomannan is greater as compared to seabuckthorn alone in the protection.

Haematological Profile

The effect of repeated oral administration of seabuckthorn leaves aqua-methanolic extract and aflatoxin on various haematological parameters like total erythrocyte count (TEC), total leucocyte count (TLC), differential leucocyte count (DLC), hemoglobin (Hb) and packed cell volume (PCV) in broiler chickens are presented in Tables 47.5 to 47.9.

Total Erythrocyte Count

The effect various treatments on Total Erythrocyte Count are shown in Table 47.5. Aflatoxin (group III) significantly (P<0.05) reduced the Total Erythrocyte Count as compared to the control group. Similar results were obtained by Kaloray *et al.* (2004) and Ahamed *et al.* (2006) who reported a decrease in the Total Erythrocyte Count following long term administration of aflatoxin. The aflatoxin induced reduction in TEC value of broilers could be due to reduced protein synthesis resulting from the impaired hepatic metabolism.

There was significant (P<0.05) increase in TEC in seabuckthorn group and Aflatoxin+SBT+GM group in comparison to Aflatoxin group. Similar trend of increased TEC was reported by Saggu *et al.* (2007) by giving aqueous extract of seabuckthorn leaves in rats. There was no significant difference in TEC of group V and group VI. Glucomannan restored the value of Total Erythrocyte Count near to control group showing significant mycotoxins binding activity. The seabuckthorn leaves extract and its combination with glucomannan restored TEC to control values indicating that the combined treatment of seabuckthorn and glucomannan produced better protective action as compared to all other treatments.

Total Leucocyte Count

The results of effect of seabuckthorn and aflatoxin on chicks are depicted in Table 47.6. There was significant (P<0.05) increase in the Total Leukocyte Count in aflatoxin treated group (group III) as compared to the control group (group I). Almost an identical rise in the total leucocyte count has been seen following aflatoxicosis by Wangikar *et al.* (2003) and Kalorey *et al.* (2004).

Table 47.5: Effect of Dietary Seabuckthorn Leaves Extract Supplementation in Total Erythrocyte Count (millions/µl) in Broilers

Groups	*Control*	*SBT*	*AF*	*SBT+AF+GM*	*SBT+AF*	*AF+GM*
TEC	2.57^{ad} ± 0.082	2.78^{a} ±0.079	2.095^{bc} ±0.152	2.55^{a} ± 0.064	2.32^{cd} ± 0.062	2.43^{ac} ± 0.046

Mean values ± S.E. (n=6), The means with same superscripts in between columns do not differ significantly at 5 per cent level.

Table 47.6: Effect of Dietary Seabuckthorn Leaves Extract Supplementation in Total Leucocyte Count in Broilers

Groups	Control	SBT	AF	SBT+AF+GM	SBT+AF	AF+GM
TLC (Thousand/μl)	23.33^{ad} ± 1.022	21.33^{a} ± 0.881	29^{c} ± 1.528	26.33^{cd} ± 1.174	27.33^{cd} ± 0.614	27.16^{c} ± 0.577

Mean values ± S.E. (n=6), The means with same superscripts in between columns do not differ significantly at 5 per cent level.

The Total Leucocyte Count in birds treated with seabuckthorn was similar to control birds. Thus seabuckthorn treatment restored the Leucocyte Count in birds receiving aflatoxin. Saggu *et al.* (2007) reported similar results while studying adaptogenic effects of seabuckthorn on rats. Further the combined treatment with glucomannan and seabuckthorn restored the leucocyte levels near control levels.

Haemoglobin

The effect of various treatments on haemoglobin is presented in Table 47.7. There was significant reduction in the haemoglobin concentration of broilers following the dietary aflatoxin. Similar reduction in haemoglobin was noted by Ahamad *et al.* (2006) and Kalorey *et al.* (2004) following aflatoxin treatment. There was a significant ($P<0.05$) increase in the haemoglobin concentration in the birds treated with seabuckthorm as compared to group III (aflatoxin treatment), whose haemoglobin level was significantly reduced from all other groups. Seabuckthorn increased haemoglobin concentration as compared to aflatoxin treatment group. These finding are in accordance with Saggu *et al.* (2007). However, combination of glucomannan and seabuckthorn restored the Hb concentration to normal level.

Packed Cell Volume

The effect of various treatments on packed cell volume is presented in Table 47.8. There was no change in packed cell volume following aflatoxin treatment as compared to control birds. This finding is in consonance with the findings of Thapa *et al.* (2009). However, at a variance from the present study, an increase in haematocrit values was observed in rats following administration of aqueous leaves extract of seabuckthorn (Saggu *et al.*, 2007).

Differential Leucocyte Count

The effect of various treatments on the differential leucocyte count is presented in Table 47.9. There was a significant ($P<0.05$) lymphocytopenia and heterophilia in the present study. These findings are consistent with the findings of the earlier workers (Tung *et al.*, 1973). The lymphocytopenia is attributed to the immunosuppressive effect caused by aflatoxin. Seabuckthorn in all the treatments restored the lymphocyte counts to control level in the present study.

Table 47.7: Effect of Dietary Seabuckthorn Leaves Extract Supplementation on Haemoglobin Level in Broilers

Groups	Control	SBT	AF	SBT+AF+GM	SBT+AF	AF+GM
Hb (g/dl)	8.76ad ± 0.236	9.33a ±0.399	7.00b ±0.118	8.95acd ± 0.701	7.43d ± 0.224	7.44c ± 0.192

Mean values ± S.E. (n=6), The means with same superscripts in between columns do not differ significantly at 5 per cent level

Table 47.8: Effect of Dietary Seabuckthorn Leaves Extract Supplementation on Packed Cell Volume in Broilers

Groups	Control	SBT	AF	SBT+AF+GM	SBT+AF	AF+GM
PCV (per cent)	24.3 ± 0.7601	24.8 ±0.424	23.3 ±0.022	23.66 ± 0.145	23.5 ± 0.8062	23.2 ± 0.174

Mean values ± S.E. (n=6), The means with same superscripts in between columns do not differ significantly at 5 per cent level

Table 47.9: Effect of Dietary Supplementation of Seabuckthorn Leaves Extract (SBT) and Glucomannan (GM, mycotoxin binder) on differential Leucocytes Count (per cent) in Broiler Birds

Days Post Treatment	Cells per cent	Control	SBT	AF	SBT+AF+GM	AF+SBT	AF+GM
28	H	31^{a}± 1.07	30^{a} ± 0.97	38.83^{b} ±1.43	28.5^{ac} ± 1.82	23.83^{c} ± 1.08	29.33^{ad} ± 1.42
	L	56.5^{af} ± 1.95	60.6^{af} ±1.33	43.91^{b} ± 1.28	58.5^{ade} ± 2.05	45.33^{c} ± 1.87	54.5^{ef} ± 2.36
	M	6.8 ± 1.01	8.5± 0.76	7.3 ± 0.713	8.83 ± 1.01	6.5 ± 1.38	6.33 ±1.12
	E	1.83 ± 0.48	2.43 ± 1.02	2.0 ± 0.365	2.33 ± 0.42	2.16 ± 0.31	2.16 ± 0.60
	B	0.5 ± 0.22	0.3 ± 0.21	0.83 ± 0.307	0.66 ± 0.33	0.33 ± 0.21	0.66 ± 0.33

Mean values ± SE. (n=6), Means bearing same superscripts in between the columns do not differ significantly at 5 per cent level.

Note: L-Lymphocytes, H- Heterophils, M- Monocytes, E- Eosinophils and B-Basophils.

It has been previously reported that seabuckthorn stimulates lymphocyte proliferation due to the presence of some mitogens (Geetha *et al.*, 2003). The combined treatment of glucomannan and seabuckthorn was best in restoring the leucocyte count.

CONCLUSIONS

Aflatoxin fed @ 400 ppm has significantly reduced growth rate, increased FCR of birds and deteriorated the haematological parameters whereas, seabuckthorn fed @ 10000 ppm has improved growth rate and FCR. Seabuckthorn fed @ 10000 ppm in combination with glucomannan @ 1g/kg has shown synergistic effect in amelioration of aflatoxicosis. Also, seabuckthorn has improved growth performance, FCR and haematological status when fed with aflatoxin @ 400 ppb. However, the combination of Seabuckthorn and Glucomannan provided better protection in comparison to singular treatment with either of two.

REFERENCES

1. Ahamad, D.B., Vairamuthu, S., Balasubramaniam, G.A., Muralimanohar, B. and Balachandran 2006. Individual and combined effect of citranin and aflatoxin B_1 in broiler chickens: aclinic-pathological study. *Indian Journal of Veterinary Pathology* 30 (1): 32-35.
2. Ambatkar, S.K. 2009. Evaluation of seabuckthorn leaves (*Hippophae* L.) as protein replacer for efficient broiler production. *M.V.Sc. Thesis*, Department of Animal Nutrition, CSK Himachal Pradesh Krishi Vishvidyalaya, Palampur, India, 80 p.
3. Ambaye, R.Y. and Indap, M.A. 1970. Chemical examination of *Hippophae rhamnoides*. *Indian Journal of Pharmacy* 32 (5): 130-134.
4. Benjamin, M. M. 1985. Outline of Veterinary Clinical Pathology, Kalyani Publishers, Ludhiana.
5. Bernath, J. and Foldesi, D. 1992. Seabuckthorn (*Hippophae rhamnoides* L.): A promising new medicinal and food crop. *Journal Herbs, Spices and Medicinal Plants* 1 (1-2): 27-35.
6. Borisova, L., Duparinova, M., Aleksandrov, M., Tacheva, T., Dzhurov, A., Tikhova, D. and Tsvelkov, A. 1987. Experimentally induced aflatoxicosis in broiler chicken. *Veterinary Nomedit Sinski Nanki* 24 (7): 69-75.
7. Chen, T. 1988. Studies of biochemical composition of *Hippophae* and its quality in Gansu Province. *Hippophae* 1: 19-26.
8. Geetha, S., SaiRam, M., Mongia, S.S., Singh, V., Havazhagan, G. and Sawhney, R.C. 2003. Evaluation of antioxidant activity of leaf extract of seabuckthorn (*Hippophae rhamnoides* L.) on chromium (V) induced oxidative stress in albino rats. *Journal of Ethnopharmacology* 87 (2/3): 247-251.
9. Kalorey, D.R., kurkure, N.V., Ramgoankar, J.S., Sakhare, P.S., Warke, S. and Nighot, N.K. 2004. Effects of toxiroak during induced individual and combined mycotoxicosis in broilers: Growth and haematobiochemical study. *Indian Journal of Veterinary Pathology* 28 (2): 113-117.
10. Khalmatov, K., Kharlamov, I.A., Alimbaeva, P.K., Karraiev, M.O. and Khaitov, I. 1984. Important Medicinal Plants of Central Asia. *Tashkent*. (In Russian)
11. Ledoux, D.R., Rottinghaus, G.E., Bermudez, A.J. and Alonso- Debolt, M. 1998. Efficacy of a hydrated Sodium Calcium alumiosilicate to ameliorate the toxic effects of aflatoxin in broiler chicks. *Poultry Sciences* 78 (2): 204-210.
12. Lucas, A.M. and Jamroz, C. 1961. Atlas of Avian Hematology. Govt. Printer, Washington D.C.
13. Natt, M.P. and Herrick, C.A. 1952. A new blood diluent for counting the erythrocytes and leucocytes to chicken. *Poulty Science* 31: 735-758.
14. Rousi, A. 1971. The genus *Hippophae* L.: A taxonomic study. *Annals Botany Fennici* 8: 177-227.
15. Saggu, S., Divekar, H.M., Gupta, V., Sawhney, R.C., Banerjee, P.K. and Kumar, R. 2007. Adaptogenicand safety evaluation of seabuckthorn (*Hippophae rhamnoides*) leaf extract, a dose dependent study. *Food and chemical Toxicology* 45 (4): 609-617.
16. Singh, B., Bhatt, T.K. and Singh, V. 2001. Nutritional evaluation of seabuckthorn leaves as cattle feed. In: *Proceedings of International Workshop on Seabuckthorn* (Singh V. and Khosla PK, eds), New Delhi, pp. 175-181.
17. Tapia, M.O. 1985. A quantitative thin layer chromatography method for the analyses of aflatoxin, ochratoxin A, zearalenone, T-2 toxin and sterigmatocystin in food stiffs. *Revista Argentina de Microbiologia* 17 (4): 183-186.
18. Thapa, N.K., Muralimonohar, M., Balachandran, C. and Sarathchandran, G. 2009. Effects of aflatoxin mixed feed on haematobiochemical profile in layer chicken. *SAARC Journal of Agriculture* 7 (1): 86-91.
19. Trenk, H.L. and Hartman, P.A. 1970. Effect of moisture content and temperature on aflatoxin production in corn. *Applied Microbiology* 19 (5): 781-784.
20. Tung, H.T., Donaldson, W. and Hamilton, P.B. 1972. Altered lipid transport during aflatoxicosis. *Toxicology and Applied Pharmacology* 22 (1): 97-104.

21. Tung, H.T., Wyatt, R.D., Thaxton, P. and Hamilton, P.B. 1973. Impairment of kidney function during aflatoxicosis. *Poultry Science* 52 (3): 873-878.

22. Vereshchagin, B.A., Sobolevskaya, K.A. and Yakubov, A.N. 1959. Useful Plants of Western Siberia.

23. Wangikar, P.B., Dwivedi, P., Sharma, A.K., Telang, A.G. and Bisht, G.S. 2003. Haematobiochemical alterations in experimental ochratoxicosis and aflatoxicosis alone and in combination in pregnant rats. *Indian Journal of Veterinary Pathology* 27 (1): 30-35.

Chapter 48

Anticancer Activity of Five Chemical Compounds Existed in Seabuckthorn (*Hippophae rhamnoides* L.) against Hela Cells *in vitro*

Wei Juan*, Bi Yang, Su Hailan, Zhang Xitong and Li Jixin

College of Food Science and Engineering, Gansu Agricultural University, Lanzhou 730070, PR China
(*E-mail: weijuan@gsau.edu.cn)

SUMMARY

Seabuckthorn (*Hippophae rhamnoides* L.) has many biologically active compounds that exert antioxidant, antiulcer, anticancer, antibacterial, antihypertensive properties. However, there is no report about the anticancer activities of chemical compounds that existed in Seabuckthorn against Hela Cells. This study investigated the anti-cancer activity of five seabuckthorn components against Hela cells by 3-4,5-dimethyl-2-thiazolyl)-2,5-diphenyl-2-H-tetrazolium bromide (MTT) and Trypan blue exclusion assays. The results showed that isorhamnetin inhibited the proliferation of Hela cells in a dose dependent manner in 24, 48 and 72 h. The IC50 values of isorhamnetin in 24, 48 and 72 h were 81, 251 and 52 umol/L. Quercetin and kaempferide also inhibited the proliferation of Hela cells in 72 h, the IC50 value of quercetin and kaempferide in 72 h were 154 and 117 umol/L. It is suggested that isorhamnetin, quercetin and kaempferide may be potential anti-cancer agents against Hela cells, but the availability and the effectiveness of isorhamnetin and kaempferide *in vivo* is needed to be investigated.

Keywords: *Hippophae rhamnoides L., Anticancer, MTT and Trypan blue exclusion assay.*

INTRODUCTION

Cervical cancer is a cancer arising from the cer-vix. It is due to the uncontrolled growth of cells that have the capability to attack or spread to other parts of the body. Initially there are typi-cally no symptoms. Later symptoms may include: abnormal vaginal bleeding, pelvic pain or pain during sexual intercourse (Parkin *et al.*, 2005). Human papillomavirus (HPV) infection seems to have a key role in the development of more than 90 per cent of cases. However, most people who have had HPV infections do not develop cervical cancer. Other risk factors include: smoking, a weak immune system, birth control pills. Cervical cancer treatment consists of a combination of surgery, chemotherapy and radiotherapy. Because of cancer screening tests and vaccine, cervical cancer has relatively low mortality in Western countries. However, it is still the leading cause of cancer death in women in developing nations (Jemal *et al.*, 2011). Recently Food and Drug Administration (FDA) has approved the use of a combi-nation of two chemotherapy drugs namely hyc-amtin and cisplatin for women with late stage cervical cancer. However, this combination treatment is accompanied with serious side effects including neutropenia, anemia, and thrombocytopenia. Keeping in view this fact, the objective of the current study was to identify some new chemical compounds from plant sources which had anticancer activities against cervical cancer (HeLa) cells.

Seabuckthorn (*Hippophae rhamnoides* L.) is a thorny deciduous bush native to several countries of Eurasia including China, India, Nepal, Russia, Britain, Germany, Finland, and France. *H. rhamnoides* has been used as a traditional medicine for the treatment of cough, indigestion, and blood stasis in China. All parts

Seabuckthorn (*Hippophae* L.): *A Multipurpose Wonder Plant, Vol. 5: Advanced Technologies for Cultivation, Processing, Health Protection and Environmental Conservation* (Virendra Singh, *Editor-in-Chief*, 2018), p. 375–379, Astral International Pvt. Ltd., New Delhi, India.

of the seabuckthorn plant are abundant in various chemical compounds including flavonols, sterols, lipids, triterpenoids, (Rosch *et al.*, 2003; Yang *et al.*, 2001) *etc.* Seeds oil, berries, leaves and bark of seabuckthorn are well known for their medicinal properties. It is demonstrated that seabuckthorn berries and leaves have significant cytoprotective, immunomodulatory and antioxidant activity when studied *in vitro* using immune cells like rat spleenocytes, murine macrophages (J-774), human lymphocytes and C6 glioma cells and *in vivo* in animals fed on heavy metal pro-oxidant like chromium or following exposure to cold-hypoxia-restrain stress (Geetha *et al.*, 2002; Narayanan *et al.*, 2005). Furthermore, earlier studies have reported that seabuckthorn seed oil have significant antioxidant, anti-bacterial, anti-atherogenic, and Cardioprotective activity (Yang *et al.*, 2001; Kallio *et al.*, 2002; Basu *et al.*, 2007). However, there is no report about the anti-cancer activities of SBT in Hela cells.

The objective of this work was therefore to evaluate the anti-cancer activities of several chemical compounds that existed in seabuckthorn in Hela cells. Then to find the new compounds that could be used in the prevention or treatment of cervical cancer.

MATERIALS AND METHODS

Source of Tested Compounds

The tested compounds including kaempferol, kaempferide, isorhamnetin, quercetin and coumarin (Figure 48.1) were purchased from Sigma-Aldrich (ST. Louis, MO, USA). The structures of tested compounds were shown Figure 48.1.

Reagents

Trizol, Dulbecco's modified Eagle's medium (DMEM) and other cell-culture reagents including fetal bovine serum (FBS) were obtained from Invitrogen-Gibco (Grand Island, NY, USA). Dimethyl sulfoxide (DMSO), 3-(4,5-dimethylthiazol-2yl)-2,5-diphenyltetrazolium bromide (MTT) and Trypan Blue were purchased from Sigma-Aldrich (ST. Louis, MO, USA).

Cell Culture

Followed by the method of Kim *et al.* (2013). Cervical cancer (HeLa) cells was obtained from American Type Cell Culture (Bioleaf Company, Shanghai, China) and cultured in DMEM containing 10 per cent FBS and 1 per cent penicillin-streptomycin solution (100U/ml penicillin and 100 ug/ml streptomycin in 0.85 per cent NaCl). The cells were incubated in an atmosphere of 5 per cent CO_2 at 37 °C and were subcultured every 2 or 3 d.

MTT Cytotoxicity Assay

According to the method described by Seo *et al.* (2014). The cytotoxicity of CuEO was assessed by mitochondrial-respiration-dependent 3-(4,5-dimethylthiazol-2-yl)-2,5-diphenyltetrazolium (MTT) reduction method. Hela cells (1×10^4 cells/well) plated on 96 wells plate were treated with different concentrations of chemical compounds for 24, 48 and 72h at 37 °C in 5 per cent CO_2. Then the cells were washed with PBS and incubated with MTT 1 mg/ml in PBS for 4h at 37 °C in 5 per cent CO_2. Afterwards, the plates were centrifuged for 10 min at 2000 rpm, and the supernatants were aspirated. The formazan crystals in each well were dissolved in DMSO. The amount of purple formazan was assessed by measuring the absorbance at 540 nm using a microplate reader (Sunrise Remote, Tecan). The optical density of formazan formed in control cells (without treated with chemical compounds) was taken as 100 per cent viability.

Trypan Blue Exclusion Assay

The cell viability was simultaneously assessed by Trypan blue dye exclusion assay (Strober, 2001). For this, the cells were trypsinized with 0.25 per cent trypsin-EDTA solution, resuspended in phosphate buffer saline (PBS) and stained with 0.4 per cent Trypan blue dye solution (v/v in PBS). Within two minutes, the cells were

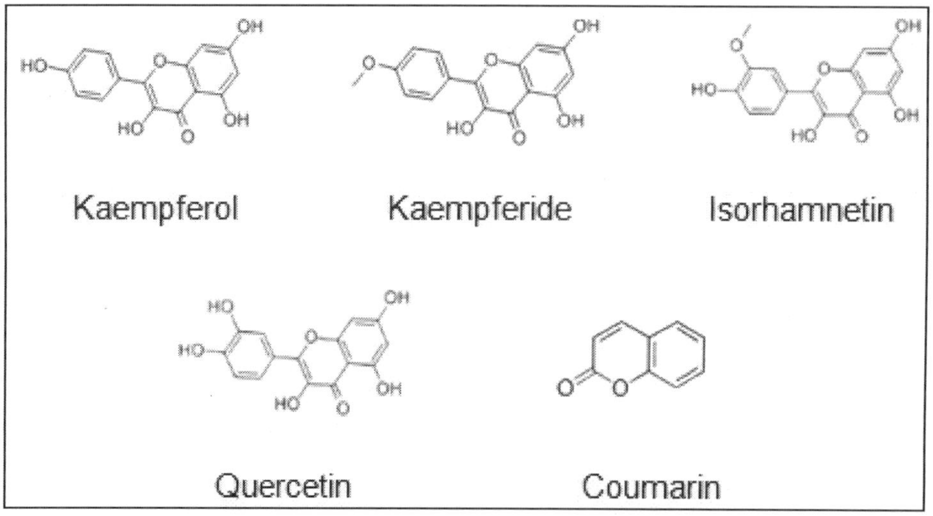

Kaempferol Kaempferide Isorhamnetin Quercetin Coumarin

Figure 48.1: Chemical Structures of the Five Tested Compounds.

loaded in a Neubauer chamber and the number of viable and non-viable cells per 1 x 1 mm square was counted under phase contrast microscope. The dead cells, because of losing the semi permeability of membrane, retained the blue dye and hence are coloured whereas viable or live cells remained unstained. The cells/ml was calculated as average cell count x dilution factor x 104 cells/ml. The per cent cell viability was determined as [(no. of viable cells/total no. of viable + non-viable cells) x 100. The percentage of growth inhibition was represented as {cell viability (control) – cell viability (with extract)}. The concentration at which the growth of cells was inhibited to 50 per cent of the control (IC50) was obtained from this dose-response curve.

Statistical Analysis

Statistical analysis was performed by the SPSS 17.0 software (SPSS, Inc., Chicago, IL). All data are expressed as the mean ± SD. The difference between groups were analyzed by Student's *t* test. Differences at $p < 0.05$ were considered as significant.

RESULTS AND DISCUSSION

The Effect on Cell Viability of Quercetin in Hela Cells

To determine whether quercetin could lead to loss of cell viability, Hela cells were treated with different concentrations of quercetin for 24, 48 and 72 h (Figure 48.2). The cytotoxic effects of quercetin were determined by MTT assay. Quercetin did not inhibit the cell viability of Hela cells in 24 and 48h. But it exhibited a significant cytotoxic effect on human cervical cancer (HeLa) cells in 72 h, the IC50 value of quercetin in 72 h was154 umol/L. Our study revealed that the optimum time for treatment of quercetin was 72h.

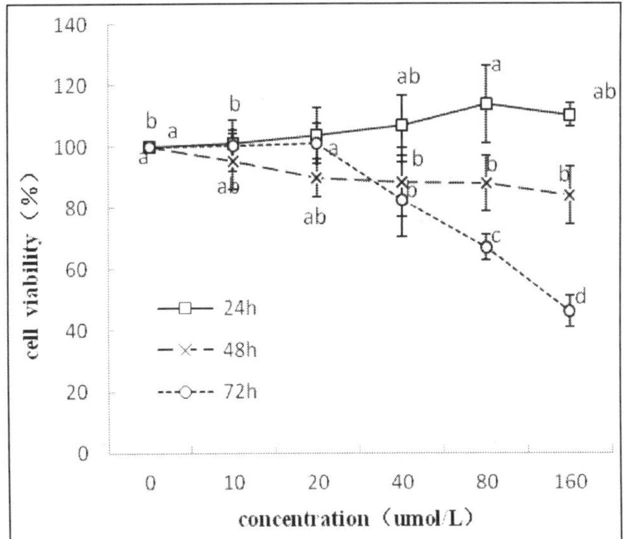

Figure 48.2: The Cell Viability of Hela Cells Treated with different Concentrations of Quercetin for 24, 48 and 72 h. Bars represent SE. Values followed by the same letter are not significantly different.

The Effect on Cell Viability of Kaempferol in Hela Cells

To determine whether kaempferol could lead to loss of cell viability, Hela cells were treated with different concentrations of kaempferol for 24, 48 and 72 h (Figure 48.3). The cytotoxic effects of kaempferol were determined by MTT assay. Kaempferol induced the growth of Hela cells in 24, 48 and 72h. These results suggested that kaempferol was a carcinogen.

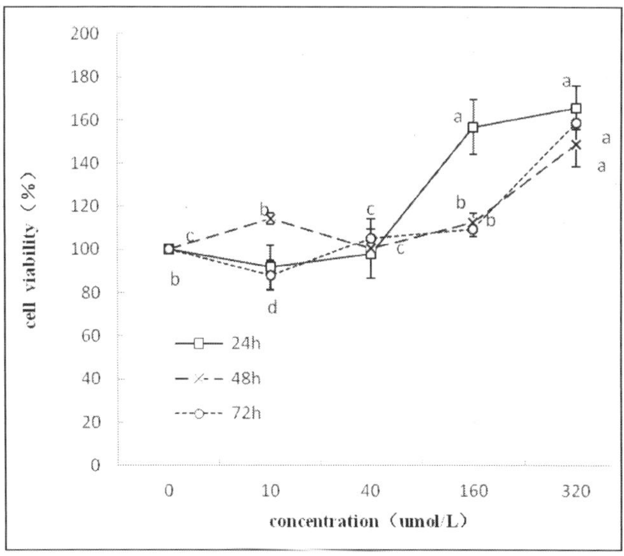

Figure 48.3: The Cell Viability of Hela Cells Treated with different Concentrations of Kaempferol for 24, 48 and 72 h. Bars represent SE. Values followed by the same letter are not significantly different.

The Effect on Cell Viability of Coumarin in Hela Cells

To determine whether coumarin could lead to loss of cell viability, Hela cells were treated with different concentrations of coumarin for 24, 48 and 72 h (Figure 48.4). The cytotoxic effects of coumarin were determined by MTT assay. The results showed that there is no

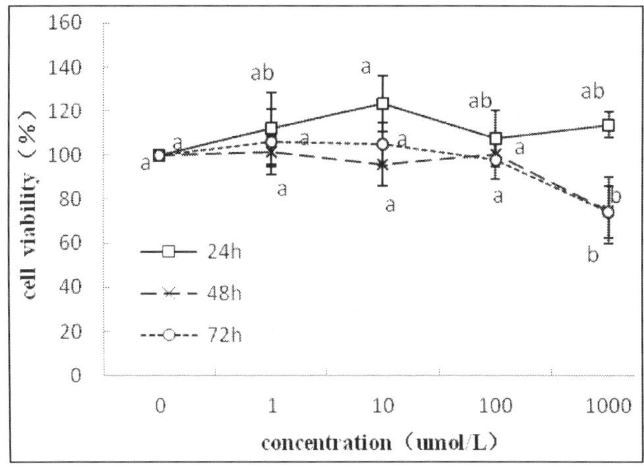

Figure 48.4: The Cell Viability of Hela Cells Treated with different Concentrations of Coumarin for 24, 48 and 72 h. Bars represent SE. Values followed by the same letter are not significantly different.

significant change of the cell viability in Hela cells treated with coumarin in 24, 48, and 72h.

The Effect on Cell Viability of Kaempferide in Hela Cells

To determine whether kaempferide could lead to loss of cell viability, Hela cells were treated with different concentrations of kaempferide for 24, 48 and 72 h (Figure 48.5). The cytotoxic effects of kaempferide were determined by MTT assay. Kaempferide did not inhibit the cell viability of Hela cells in 24 and 48h. But it exhibited a significant cytotoxic effect on human cervical cancer (HeLa) cells in 72 h, the IC50 value of kaempferide in 72 h was: 117 umol/L. Our study revealed that the optimum time for treatment of kaempferide is 72h.

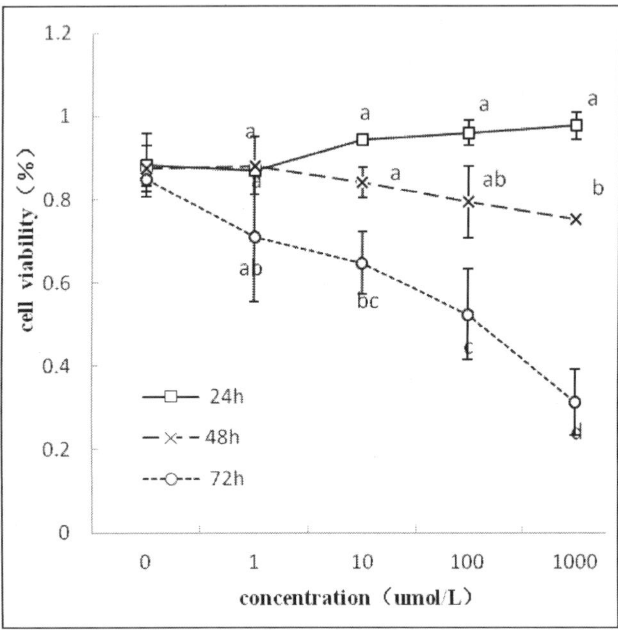

Figure 48.5: The Cell Viability of Hela Cells Treated with different Concentrations of Kaempferide for 24, 48 and 72 h. Bars represent SE. Values followed by the same letter are not significantly different.

The Effect on Cell Viability of Isorhamnetin in Hela Cells

To determine whether isorhamnetin could lead to loss of cell viability, Hela cells were treated with different concentrations of isorhamnetin for 24, 48 and 72 h (Figure 48.6). The cytotoxic effects of isorhamnetin were determined by MTT assay. The results showed that isorhamnetin could inhibit the proliferation of Hela cells in a dose dependent manner in 24, 48 and 72 h. The IC50 values of isorhamnetin in 24, 48 and 72 h were: 81, 251 and 52 umol/L. Thus, isorhamnetin may be potential anti-cancer agents against Hela cells, but the availability and the effectiveness of isorhamnetin *in vivo* is needed to be investigated.

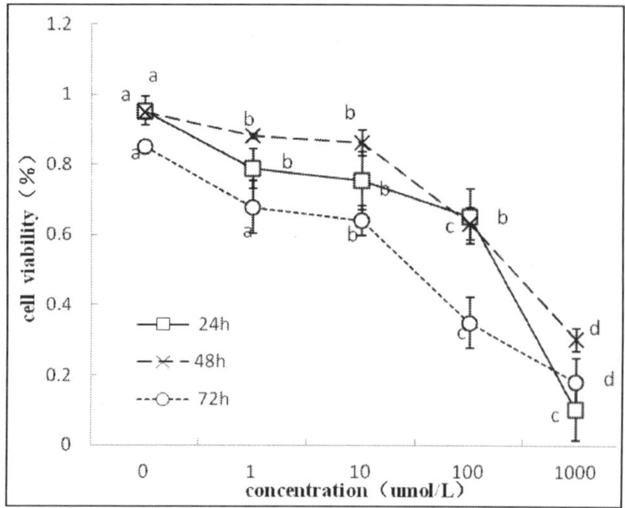

Figure 48.6: The Cell Viability of Hela Cells Treated with different Concentrations of Isorhamnetin for 24, 48 and 72 h. Bars represent SE. Values followed by the same letter are not significantly different.

ACKNOWLEDGEMENT

This work was funded by the Projects of International Cooperation and Exchanges of Ministry of Science and Technology in China (No. 2014DFR31230). We gratefully acknowledge National Nature Science Foundation of China. At the same time we are sincerely appreciated Mr. Luan from Gansu Conseco Company for providing all the seabuckthorn pulp needed in the experiment. And we are thankful for GE Qian cui-di who introduced the processing of seabuckthorn in the industry which helped us to set the experimental factors. Besides, we wish to express our thanks to Prof. Han Shun-yu for discussions.

REFERENCES

1. Basu, M., Prasad, R., Jayamurthy, P. K., Arumughan, C. and Sawhney, R.C. 2007. Anti-atherogenic effects of seabuckthorn (*Hippophae rhamnoides*) seed oil. *Phytomedicine* 14: 770-777.

2. Geetha, S., Sai, R.M., Singh, V., Ilavazhagan, G. and Sawhney, R.C. 2002. Antioxidant and immune modulatory properties of seabuckthorn (*Hippophae rhamnoides*)-An *in vitro* study. *Journal of Ethnopharmacology* 79: 373-8.

3. Jemal, A., Bray, F., Center, M.M., Ferlay, J., Ward, E. and Forman, D. 2011. Global cancer statistics. *Cancer Journal for Clinicians* 61: 69-90.

4. Kallio, H., Yang, B., Peippo, P., Tahvonen, R. and Pan, R. 2002. Triacylglycerols, glycerophospholipids, tocopherols, and tocotrienols in berries and seeds of two subspecies (ssp. *sinensis* and *mongolica*) of seabuckthorn (*Hippophae rhamnoides*). *Journal of Agricultural and Food Chemistry* 50: 3004-9.

5. Kim, K.N., Ko, Y.J., Yang, H.M., Ham, Y.M., Roh, S.W., Jeon, Y.J., Ahn, G., Kang, M.C., Yoon, W.J., Kim, D. and Oda, T. 2013. Anti-inflammatory effect of essential oil and its constituents from fingered citron (*Citrus medica* L. var. *sarcodactylis*) through blocking JNK, ERK and NF-kappa B signaling pathways in LPS-activated RAW 264.7 cells. *Food and Chemical Toxicology* 57: 126-31.

6. Narayanan, S., Ruma, D., Gitika, B., Sharma, S.K., Pauline, T., Ram, M.S., Ilavazhagan, G., Sawhney, R.C., Kumar, D. and Banerjee, P.K. 2005. Antioxidant activities of seabuckthorn (*Hippophae rhamnoides*) during hypoxia induced oxidative stress in glial cells. *Molecular and Cellular Biochemistry* 278: 9-14.

7. Parkin, D.M., Bray, F., Ferlay, J. and Pisani, P. 2002, 2005. Global cancer statistics. *Cancer Journal for Clinicians* 55: 74-108.

8. Rosch, D., Bergmann, M., Knorr, D. and Kroh, L.W. 2003. Structure antioxidant efficiency relationships of phenolic compounds and their contribution to the antioxidant activity of seabuckthorn juice. *Journal of Agricultural and Food Chemistry* 51: 4233-9.

9. Seo, K.H., Park, M.J., Ra, J.E., Han, S.L., Nam, M.H., Kim, J.H., Lee, J.H. and Seo, W.D. 2014. Saponarin from barley sprouts inhibits NF-kappaB and MAPK on LPS-induced RAW 264.7 cells. *Food and Function* 5: 3005-13.

10. Strober, W. 2008.Trypan blue exclusion test of cell viability. *Current Protocols in Immunology* 3: 3 B.

11. Yang, B., Karlsson, R.M., Oksman, P.H. and Kallio, H.P. 2001. Phytosterols in seabuckthorn (*Hippophae rhamnoides* L.) berries: Identification and effects of different origins and harvesting times. *Journal of Agricultural and Food Chemistry* 49: 5620-9.

Chapter 49

Traditional Uses of Seabuckthorn in Trans-Himalayan Regions of India

Padma Gurmet

National Research Institute for Sowa-Rigpa
C.C.R.A.S, Ministry of Health and FW, Govt. of India), Leh-Ladakh – 194 101, India
(E-mail: drgurmet@rediffmail.com)

SUMMARY

"Tsir-mang" Seabuckthorn is one of the widely used wild plant of Trans-Himalayan regions. Due to isolation from major part of the world inhabitants of Ladakh and other Trans-Himalayan regions has been largely dependent on surrounding wild plants for their basic commodities like vegetables, food, fuel, fodder, medicine and ornamental flower *etc.* Seabuckthorn being one of the dominant plant of the region is well-known by the local population for its multifarious uses. *Hippophae* is known with various local names in different areas like *tSer-mang, tSer-Sa-Lu-Lu, Shib-Shu-Lu-Lu* and *sTarbu* followed by half a dozen synonymous in classical *Sowa-Rigpa* (Amchi/Tibetan) medical texts. The earliest reference about medicinal uses of *Hippophae* species are fund in Sowa-Rigpa classical texts, all the three major species of Seabuckthron are mentioned in the pharmacopoeia and formulary of Sowa-Rigpa *i.e. Nam-star* (*Hippophae salicifolia*), *Bar-star* (*Hippophae rhamnoides* ssp. *turkestanica*) and *Sa-star* (*Hippophae tibetana*). There are around two hundred *Hippophae* based formulations in Sowa-Rigpa use for wide range of diseases. *Hippophae* is also one of the most popular fuel and fencing source. It is also occasionally use for building roof and other wooden appliances, children making juice out of the barriers and animals enjoying its leaves are some of the common scenario of this multifarious plant. Since last few years seabuckthorn has been catching attention of Scientist, Environmentalist, Foresters, NGO's and Industry, and it is seen as one of the most potential plants of Ladakh and Trans-Himalayas. A proper scientific research and validation in this field may provide basis for development of many product and effective drugs from seabuckthorn and it can largely contribute for environmental protection too.

Keywords: *Seabuckthorn, Traditional uses, Hippophae salicifolia, Hippophae rhamnoides* ssp. *turkestanica and Sowa-Rigpa.*

INTRODUCTION

Trans-Himalayan regions being one of the most remote and inaccessible region of the world has been a self-sustained region for most part of the history. Due to isolation from major part of the world, inhabitants of these regions has been largely dependent on surrounding wild plants for their basic commodities like vegetables, food, fuel, fodder, medicine and ornamental flower *etc.* '"Seabuckthorn", one of the dominant plant of these region is well-known by the local population for its multifarious uses. Besides its medicinal value, it is used as fuel, fodder, fencing, food and beverage *etc.*

Seabuckthorn is abundantly found in all the eight major valleys of Lahoul, Spiti, Changthang, Indus, Nubra, Daras, Suru and Zanskar of Trans-Himalaya. *tSer-Mang, tSer-Sa-Lu-Lu, Shib-Shu-Lu-Lu* and *sTar-Bu* are the most popular term use for *Hippophae* species followed by half a dozen synonymous in Sowa-Rigpa. The multifarious uses of seabuckthorn can be widely seen in the form of the bio-fences that protect gardens and agriculture lands, fuel stocks for winter, sheep and goat enjoying the leaves and Amchis (traditional doctors) collecting the fruits for medicine, Seabuckthorn grows largely as an unwanted weed and its full potential has

Seabuckthorn (*Hippophae* L.): *A Multipurpose Wonder Plant, Vol. 5: Advanced Technologies for Cultivation, Processing, Health Protection and Environmental Conservation* (Virendra Singh, *Editor-in-Chief*, 2018), p. 381–384, Astral International Pvt. Ltd., New Delhi, India.

been not been acknowledged in an appropriate manner by the natives of this region. Potentials of seabuckthorn for soil conservation, greening and forestation have been used to some extent in few areas but to a minimum level only.

The earliest written information about the medicinal uses of Seabuckthorn is reported in the *rGyud-bZhi*, which is the classical medical text book of Sowa-Rigpa (Amchi/Tibetan medicine). It is one of the most important and widely-used ingredients in various Sowa Rigpa formulations. Three different species of seabuckthorn are mentioned in the classical medical texts, the *dTrimed Shelgong* and the *Shelthring*, namely *Sa-sTar* (*Hippophae tibetana*), *Bar-sTar* (*Hippophae rhamnoides* ssp. *turkestanica*) and *Nam-sTar* (*Hippophae salicifolia*). Of these, *Bar-sTar* (*Hippophae rhamnoides* ssp. *turkestanica*) is the most commonly used species in Tibetan medicines. Recent developments and scientific inputs have added new dimensions and directions for seabuckthorn and the plant has now entered a phase of transition, from traditional uses to modern scientific products and the scope is widening day by day.

Seabuckthorn a Mythological Plant

As per the Tibetan classical texts, the mythology of seabuckthorn goes back to the famous period of the ocean churning, which is considered as the beginning of the universe in Hindu mythology. When nectar came out of ocean, after many days of effort from the gods and *Asuras* (demigods), the clever Rahu (king of *Asura*) ran away with the nectar pot and started drinking it. When Brahma saw this, he threw *Chakara* on Rahu and cut him from the neck into two pieces. Rahu started bleeding and since he had had a sip of nectar, a drop of it fell from his throat onto the earth. It is believed that seabuckthorn and a number of other plants grew out of the nectar that dropped from Rahu's throat and thus that the plant contains some of the properties of nectar (Phuntsog, 1983). Hence seabuckthorn is a very potent plant and has a high medicinal value.

In many villages of Ladakh, seabuckthorn is also regarded as the residence of *Nagas* (Lhu). These are elemental forces or spirits, often associated with water sources and specific components of the environment. For this reason, large and old seabuckthorn trees are regarded as holy, people must be wary of them and will not dare to cut them down or to pollute near to them, as they might well be *Lu-Tser* (residence of *Nagas*).

Traditional Classification of Seabuckthorn and its Distribution in Ladakh and Lahaul-Spiti

The three verities of seabuckthorn mentioned in the *dTrimed Shelgong* and *Shelthring* texts are classified and named according to the variation in size between the different species. *Sa-sTar* is the smallest, with a height of less than 2 ft., so it is given the prefix *Sa* (ground). *Bar-sTar* is given the prefix *Bar*, meaning medium, and has plants growing up to 8 ft. in height. Finally, *Nam-sTar*, (*Nam* meaning 'sky') is the tallest, shaped like a tree. On the basis of this classification, the botanical identification of the different spices can be established as: Sa-sTar -*Hippophae tibetana*; Bar-sTar- *Hippophae rhamnoides* ssp. *turkestanica*; and Nam-sTar - *Hippophae salicifolia*. While all three species are found in Ladakh and Lahaul–Spiti, *Bar-sTar* (*Hippophae rhamnoides*) is the most common species and is found in most parts of Trans-Himalaya and is particularly abundant in the Indus, Nubra, and Suru valleys. *Hippophae salicifolia* is not found in Ladakh region but it is fairly widespread in the Lahoul valley. *Sa-sTar* (*Hippophae tibetana*) is the rarest species and is found in only in a few area of Spiti and in the Rangdum area of the Zanskar valley (Gurmet, Chaurasia and Ballabh, 2000; Chaurasia and Gurmet, 2003).

Traditional Uses of *Hippophae* sp. in Trans-Himalayan Regions

Hippophae is a multi-utility plant for the local population and many of its uses can be easily seen throughout trans-Himalayan regions. Due to its spiky thorns, the most common use is as bio-fencing for plantation areas, orchards and agriculture lands which are effectively protected with seabuckthorn bio-fences. Being one of the coldest regions of the world, fuel is a major problem in Ladakh and Lahaul-Spiti. Seabuckthorn is one of the most useful fuel plants, as it grows without any effort and produces a very warm and long-lasting fire. During community gatherings, social ceremonies and festivals in monasteries, seabuckthorn is widely use as fuel for cooking food, as it is also in most households. *Hippophae* is also a wonderful fodder plant for livestock and is particularly enjoyed by sheep, goat and camels, for whom the leaves are considered very healthy. In the Nubra valley of Leh district, a large number of semi-wild Bactrian Camels survive largely on seabuckthorn and their health is significantly better than the domestic ones.

Due to its hardness, the wood of seabuckthorn is also occasionally used for building roofs, to make handles for agriculture appliances and so on. The use of seabuckthorn as a food substance is not widespread in the region, apart from children making juice out of it and picking the berries to eat. Some sources report that a few people use this plant to make *Chang* (traditional wine), but this has not been verified. Of all of the many uses of seabuckthorn, the most important is as a medicinal plant in the Sowa-Rigpa traditional medical system.

Seabuckthorn an Important Medicinal Plant

The use of seabuckthorn as medicine is described in the text books of Chinese medicine, Ayurveda and Sowa Rigpa. The earliest reference concerning the medicinal uses of *Hippophae* has been described in detail in the "*rGyud bzi*", which is the fundamental text book of Sowa Rigpa. The recently published Chinese Pharmacopoeia of 1977 also describes the medicinal use of seabuckthorn, and it is identified as the Ayurvedic medicinal plant *Amlich* or *Badriphal* (Sharma and Chunekar, 1998). According to these medical systems, seabuckthorn has a wide range of uses including: treatment of respiratory problems (especially asthma); purifying the blood; strengthening digestion; coordinating the balance of functions among various organs (liver, stomach, spleen, kidney and heart); assuaging pain caused by wounds; promoting tissue healing and regeneration; invigorating blood circulation; combating old age; improving strength; lowering blood fat; and preventing cancer. Seabuckthorn is rich in vitamin 'C' and contains a unique combination of other important vitamins such as A, B1, B2, E and K. It is also a rich repository of proteins, organic acids, carotenoids, flavonoids and minerals like iron, cobalt, molybdenum etc (Dwivedi *et al.*, 2006).

"Sowa-Rigpa" Himalayan Art of Healing

It has been already mentioned that the earliest reference to the medicinal uses of seabuckthorn is found in the *rGyud bzi* (Four Medical Tantras) of Sowa Rigpa, therefore it is important to give a brief introduction to this medical system. Sowa-Rigpa is commonly known as Tibetan or Amchi medicine and is among the well-documented, codified classical medical traditions of the world. Originating in India, the system was enriched in Tibet and has a living history of more than 2500 years. Sowa-Rigpa has been popularly practiced throughout the central and south Asian countries of Tibet, Mongolia, Bhutan, China, Nepal, in the Himalayan regions of India and in some parts of the former Soviet Union. In India, it has been popularly practiced in Ladakh, Himachal Pradesh, Arunachal Pradesh, Sikkim, Darjeeling and now in Tibetan settlements all over India (Gurmet, 2005).

Sowa-Rigpa is based on the Ayurvedic theory of the *Panch Mahabhuta* (Jung-wa-lna/five elements) and *Tri-dosh* (Nespa Sum/three humors). According to this theory all the animate and inanimate phenomena of this universe are composed of *Panch Mhabhuta* (Jung-wa-Lna/five elements). It is on the basis of this theory of five basic elements that the sciences of physiology, pathology and pharmacology are established. Being taught by Buddha, philosophically Sowa-Rigpa is based on Buddhism (Anonymous, 2004). The pharmacopoeia of this medicine is based on the skillful use of plants, minerals, animal parts, precious and semiprecious stones and metals. In particular, the competent use of a large number of Himalayan medicinal plants is a unique feature of this medicine.

Seabuckthorn and Sowa-Rigpa

Seabuckthorn is one of the most important and widely used medicinal plants in Sowa Rigpa. Although *sTar-Bu* is the classical name for this medicinal plant there are many synonyms like *gLawa-tsirma, Lai-tsir, Tsar-tsir, Mai-ldan, Stin-ge-sochan, Yalga-gyur etc.* Mostly it is the fruit of seabuckthorn that is used in medicine. The berries are collected during October-November and after cleaning, are left to dry in the sun, before being used to prepare medicines or being made into concentrated decoctions (*Khanda*). *sTar-Bu* is used in compound form and is mixed with other plants and minerals.

Ro (taste), *nus pa* (potency) and *Nowo* (nature) are three important factors that define the efficacy of ingredients in Sowa-Rigpa. Seabuckthorn is sour (*sKur-wa*) in taste (*Ro*), sharp (*rno*) and light (*Yang*) in potency, and neutral (*nyom*) in nature (*Nowo*). Though there are only minor differences in medicinal uses and potencies among the three species of Hippophae, *Hippophae rhamnoides* is by far the most commonly used. According to *Dri med sel gong* and *sel thring* (the main Pharmacognosy and *materia medica* texts in Sowa-Rigpa), *Hippophae salicifolia* (*Nam-sTar*) and *Hippophae tibetana* (*Sa-sTar*) are more effective for treating lung and throat disorders, while *Hippophae rhamnoides* (*Bar-sTar*) is better for dealing with blood disorders.

There are around two hundred popular seabuckthorn-based formulations in the various pharmacopoeias of Sowa Rigpa. These formulations are mixtures of between two and fifty different substances. Out of these, most of the seabuckthorn-based drugs are used in the treatment of lung and phlegm disorders, blood disorders, menstrual problems, throat infections, liver problems, spleen and stomach disorders, peptic ulcer, joint pain, dysentery, appetite stimulant, skin disease, toothache, asthma, against tumors, stone and cancer, diabetes, coughs and colds, as well as being used as a general tonic. Some important formulations are given below, along with the major uses (Anonymous, 2004; Dorjai, 1995; Dawa, Namgail and Phuntsog, 1990; Phuntsog, 1983; Lama, Suresh and Thomas, 2001).

Recent Developments with Seabuckthorn and Public Views in Ladakh

Being a native of a seabuckthorn hotspot area, namely the Indus valley, I have been closely observing the development regarding this plant for the last two decades. Growing wild near riversides, along the borders of agriculture fields, forests and orchards, seabuckthorn has been seen as an unwanted weed, in spite of its multifarious uses. For decades, the great potential held by this plant has been largely ignored by the local population and by policy makers. Fortunately the last 12 years have seen tremendous positive change in this scenario. The first efforts towards the development of seabuckthorn juice was started by Leh Nutrition Project (a Leh-based NGO) with the help of a few foreign volunteers and women's self-help groups around 1993, but was not successful and the production of juice was stopped at the end of the active project period, largely due to the fact that consumers did not enjoy the taste. Some years later, the potential of this plant was identified by some scientists of the Field Research Laboratory (FRL, Defense Research and Development Organisation), Leh, and a wide range of scientific studies were conducted. A multi-vitamin juice was developed by FRL and launched onto the market by a private company. This event created a new direction for the development of seabuckthorn and within a year, this product became popular across India. Seabuckthorn thus showed its potential to become the 'golden bush' for Ladakh.

Herbal tea, made by FRL out of the leaves of seabuckthorn along with a few other plants, has further proved the efficacy of this plant. At the same time people also became aware of the wide range of products developed in China out of this wonder plant. Today, an average village woman can earn twenty rupees per kilogram for the fruits and a good number of women's self-help groups are producing seabuckthorn juice. The local administration has also initiated a few programs for regulation of this plant and the formation of cooperatives is being encouraged for the production of seabuckthorn products. Unfortunately however, the present focus of Seabuckthorn development in Ladakh aims for maximum harvesting volumes, leaving the essential issues of conservation and cultivation to one side. Use of with Seabuckthorn for wasteland management and soil conservation is an area with major scope, which needs to be explored so that maximum use can be made of the large areas of waste and deserted lands in Ladakh. The public perception is changing positively towards Seabuckthorn and this incredible plant's tremendous potentials increasingly realized.

CONCLUSION

Seabuckthorn is one of the most important plant species of the Trans-Himalayan region, with remarkable scope to become a plant of great economic importance. It has tremendous potential to improve the economic and environmental conditions of poor and environmentally harsh desert areas of this region. The vast ethnobotanical knowledge of the local population is an added advantage for the development of seabuckthorn as an economic plant. Particularly Amchis' (Sowa Rigpa practitioners) knowledge about the medicinal uses of seabuckthorn is an important basis for exploring the potential development of new medicines to treat some of the chronic diseases of the modern age. Thus, if the rich ethnobotanical knowledge of the local population could be explored within modern scientific dimensions, seabuckthorn could indeed prove to be the 'golden bush' for the economically backward regions of Ladakh and Lahaul-Spiti. At the same time, it is very important to take concerted action for the conservation and cultivation of seabuckthorn, to ensure its sustainable development and utilization. Keeping in view the vastness and barrenness of the Trans-Himalayan region, *Hippophae* should in particular be adopted for widespread soil conservation and afforestation projects.

REFERENCES

1. Anonymous 2004. *rGyud-bzi*, Tibetan Medical and Astro Institute, Dharamshala, HP, India.
2. Chaurasia and Gurmet 2003. A Checklist on Medicinal and Aromatic Plants of Trans-Himalayan Cold Desert (Ladakh and Lahaul-Spiti). Field Research Laboratory, Leh and Amchi Medicine Research Unit, Leh.
3. Dorjai, G. 1995. *Thungs-Pai-Te Maid-Shel-Gi Melong*, Me-Regs Pai Tun Khang Lhasa, Tibet.
4. Dawa. Pod-Ki-Sowa-Rigpa-Las Sman Zas Jor zoi Lag Lain Sang Sgo Baid Pai Ldai-mig, Rig Drag Publication, New Delhi.
5. Dwivedi, S. K., Stobdan, T. and Singh, S.B. 2009. *Seabuckthorn: The Golden Bush*. Satish Serial Publishing House, Delhi.
6. Dwivedi, S.K., Singh, R. and Ahmad, Z. 2006. *The Seabuckthorn*, Field Research Laboratory, Leh.
7. Gurmet, P., Chaurasia, O.P. and Ballabh, B. 2004. Studies on some important medicinal plants growing along Indus river of Ladakh Himalayas. *Bull. Medico-Ethno-Botanical Research (CCRAS)*, Vol. XXV, No. 1-4, pp.1-11, New Delhi.
8. Gurmet, P. 2005. "Sowa-Rigpa" Himalayan Art of Healing. *Indian Journal of Traditional Knowledge* (CSIR), Vol. 3 (2), April 2004, New Delhi, pp. 212-218.
9. Lama, Y.C., Ghimire, S.K. and Thomas, Y.A. 2001. Medicinal Plants of Dollop, People and Plants Initiative and WWF Nepal Program, Kathmandu, Nepal.
10. Namgail, G. and Phuntsog, S.T. 1990. Amchi Pharmaco-Therapeutics. Central Council for Research in Ayurveda, New Delhi.
11. Phuntsog, D. G. S. 1983. Dri Med Shel Gong/Shel Taing, M/S D. W. Tashigang, BPO Nimo Leh-Ladakh.
12. Sharma, S.K and Chunekar, K.C 1998. Rastriya Ayurveda Vidyapeeth, New-Delhi.

Part V

PROCESSING TECHNOLOGIES

Chapter 50

The Changing of Catechin Quantity in the Products of Seabuckthorn

Yury A. Koshelev[1], Evgeny S. Batashov[1*] and Valery P. Sevodin[2]

[1]Altayvitaminy Pharmaceutical Company, Zavodskaya 69 str., Biysk, Altay Region, Russia,
[2]Biysk techological institute, Trofimova 27, Altay Region, Russia
(*E-mail: evgenczl@mail.ru)

SUMMARY

Seabuckthorn berries are regarded as a very valuable multivitamin raw material containing a lot of catechins (70-230 mg per 100 g) that possess strong anti-oxidant activity. The objects of the research were seabuckthorn juice along with wine and vinegar made of it. It was found that during the production process of wine and vinegar the quantity of catechins was going down. The main component in this group of substances was catechin and this fact was proved by the literature review. The amount of catechins in vinegar was 2 times less in comparison to wine and 5 times less to juice. The products changed their colour during the production process from light-yellow (wine) to brown (wine vinegar). The possible reason of colour changing of seabuckthorn products is the oxidation of biologically active substances. Colour changing and precipitation is common for all the products made of seabuckthorn. However in seabuckthorn vinegar these processes are the most intensive because of strong aeration during the production. Most probably it causes oxidation and polymerization of the individual catechins. The phenomenon of browning and precipitation in wine vinegar is proved by the literary review. In wine vinegar catechin is oxidized due to the presence of iron ions and polymerized with the formation of coloured precipitate. The possible ways to avoid the negative consequences of catechin oxidation are: The reduction of the factors causing oxidation and the inclusion of anti-oxidants into seabuckthorn products.

***Keywords**: Seabuckthorn, Catechins, Juice, Wine, Vinegar and Catechin oxidation.*

INTRODUCTION

Seabuckthorn berries are a very valuable raw product. In Russia it is mostly used to produce seabuckthorn oil for pharmaceutical, cosmetic and food purposes. However seabuckthorn juice being 70 per cent of berries weight hasn't got such a wide application. Its industrial use is rather narrow – the production of blended juices and drinks though the deep juice processing is an insufficiently known as well as a perspective direction (Koshelev *et al.*, 2003).

The possible solving of this problem can be the use of clarified seabuckthorn juice in liquor and spirits industry, the use of seabuckthorn wine material for the composition of fruit and blended wines, the production of vinegar by the fermentation of seabuckthorn wine material. The vinegar can be regarded as a separate product or as a raw material for food industry including canning lines. The significant stage of the development and production startup of any product is the examination of its content changing and stability during the processing.

It's well-known that catechins are physiologically active substances and possess a strong antioxidant activity. Seabuckthorn berries contain a lot of catechins (70-230 mg per cent) (Rosch *et al.*, 2003). However there's a lack of information on the amount of catechins in the final products of seabuckthorn processing. The problem of the catechin content study becomes more

Seabuckthorn (*Hippophae* L.): *A Multipurpose Wonder Plant, Vol. 5: Advanced Technologies for Cultivation, Processing, Health Protection and Environmental Conservation* (Virendra Singh, *Editor-in-Chief*, 2018), p. 387–390, Astral International Pvt. Ltd., New Delhi, India.

significant in view of our theory about the polyphenol essence of the precipitate formed during the storage of the seabuckthorn products.

The purpose of the research is study of the catechin content in the seabuckthorn processing products. The objects of the research are seabuckthorn wine and seabuckthorn vinegar based on the "Chuyskaya" seabuckthorn cultivar (harvested near Biysk, Altay region, Russia on September 2013).

MATERIALS AND METHODS

The determination of the total number of catechins was made by permanganate titration according to the standard method in Russia (State Pharmacopoeia of the USSR. – 11th edition), the determination of polyphenols – according to the Folin-Ciocalteu method (Singleton *et al.*, 1965; ISO 14502-1:2005), when converted into gallic acid by the spectrophotometric method using "Shimadzu" spectrophotometer Model UV-2401 PC.

The determination in individual catechins was made by high-performance liquid chromatography (HPLC) according to the standard method (Handbook on Methods control the quality and safety of biologically active additives to food. 2004). HPLC analyses were performed system: "Waters 2695 Alliance", equipped with UV-Diode array detector using Symmetry 5 µm C 18 column, size 250 × 4,6 mm. All the compounds were detected at 280 nm at room temperature with an eluent flow rate of 1 ml/min and injection volume of 20 µl. The mobile phase consisted of acetonitrile-85 per cent -phosphoric acid mixture (15:85). The test sample was injected without dilution. The estimation of the individual catechin in the test solution was made calculating to the standard sample as given below:

The Time of Retention (Minutes)

Gallic acid	2.9 ± 0.1
Epigallocatechin	4.5 ± 0.1
Catechin	5.8 ± 0.2
Epicatechin	6.8 ± 0.2
Epigallocatechin gallate	9.8 ± 0.2
Gallocatechin gallate	12.0 ± 0.1
Epicatechin gallate	13.5 ± 0.1
Catechin gallate	15.5 ± 0.1

The following standards were used: the standards of catechin, epicatechin, gallocatechin and epigallocatechin by the company "ChromaDex». The standard of the green tea "TeavigoTM" polyphenol by the company "DSM Nutritional Products" was used as the standard of epigallocatechin-3-gallate.

RESULTS

The first step was the analysis of polyphenol and catechin quantity in raw berries. According to the results (Table 50.1) during the processing of seabuckthorn berries the changing of the bioactive substances composition – the quantity of catechins goes 5 times down. We also stated that the products of seabuckthorn processing changed their colour from light-yellow (wine) to brown (wine vinegar). The samples of seabuckthorn processing products are shown on the Figure 50.1.

Table 50.1: The Quantity of Bioactive Substances in Seabuckthorn Berries and Seabuckthorn Vinegar

Name	*Concentration in Berries (mg/100 g)*	*Concentration in Vinegar* (mg/100 ml)*
Polyphenols converted into gallic acid	278.3 g	77.1
Catechins	202.5	42.5

* Produced by *submerged fermentation.*

Figure 50.1: The Samples of Seabuckthorn Processing Products: Vinegar (Left) and Wine (Right).

The possible reason of colour changing of seabuckthorn products is the oxidation of bioactive substances. Further more a more detailed analysis of catechins in seabuckthorn products was carried out by the liquid chromatography method.

As the Table 50.2 demonstrates the main component in this group of substances is catechin. This fact is proved by the literature review, according to foreign researchers the quantity of catechins in seabuckthorn juice is 1.9 mg per 100 ml, epicatechins – 0.28 mg per 100 ml. In comparison to wine the total quantity of catechins in vinegar is 4 times less if produced by surface fermentation and 2 times less if produced by submerged fermentation. This difference can be explained by the peculiarities of technologies. The process of surface fermentation takes place in the devices filled with beechwood shavings. In this case the process vinegar production is going 1.5-2 times slower if compared to submerged fermentation. It causes the oxidation of

Table 50.2: The Quantity of Catechins in Seabuckthorn Products Determined by HPLC

Sample	*Quantity (mg/100 ml)*					
	Epigallocatechin	*Catechin*	*Epicatechin*	*Epigallocatechin Gallate*	*Gallocatechin*	*Total*
Wine	0.44	1.61	1.07	0.07	0.080	3.270
Vinegar produced by surface *fermentation*	0.19	0.43	0.09	0.10	0.014	0.824
Vinegar produced by *submerged fermentation*	0.23	0.98	0.10	0.29	0.005	1.605

bioactive substances due to the increase of the time they contact with oxygen.

DISCUSSION

Colour changing and precipitation is common for all the products made of seabuckthorn. However in seabuckthorn vinegar these processes are the most intensive. The difference between the total catechin quality changing (Table 50.1) and the findings of the high performance liquid chromatography can be explained by the strong aeration during the vinegar production. It possibly can cause oxidation and polymerization of the individual catechins. However the oxidized and polymerized forms can also be determined in the total quantity of catechins by permanganate titration.

This statement is proved by the literature review (Li *et al.*, 2008) in regard of colour changing and precipitation in grape wines. In grape wine catechin is oxidized due to the presence of iron ions (Figure 50.2) whereas polymerization of catechins causes the formation of coloured precipitate (Figure 50.3).

CONCLUSION

As a result we can come to the following conclusion:

- The products of seabuckthorn processing contain catechins and catechin is the main of them.
- During the processing of seabuckthorn berries into wine and further vinegar, the quantity of catechins goes down.
- The small amount of catechins in vinegar can be caused by their intensive oxidation during the production process.
- Aeration causes the formation of oxidized forms and catechin polymerization products what leads to the darkening of seabuckthorn products followed by precipitation.

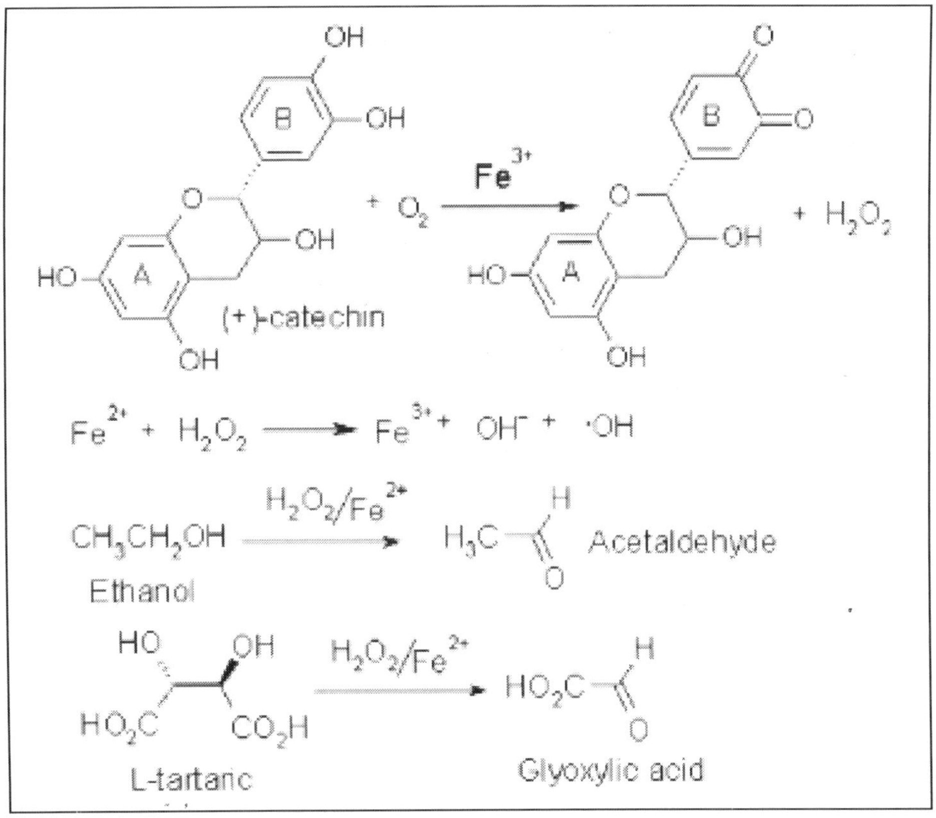

Figure 50.2: The Scheme of the Oxidation Processes taking Place in Wine (Li *et al.*, 2008).

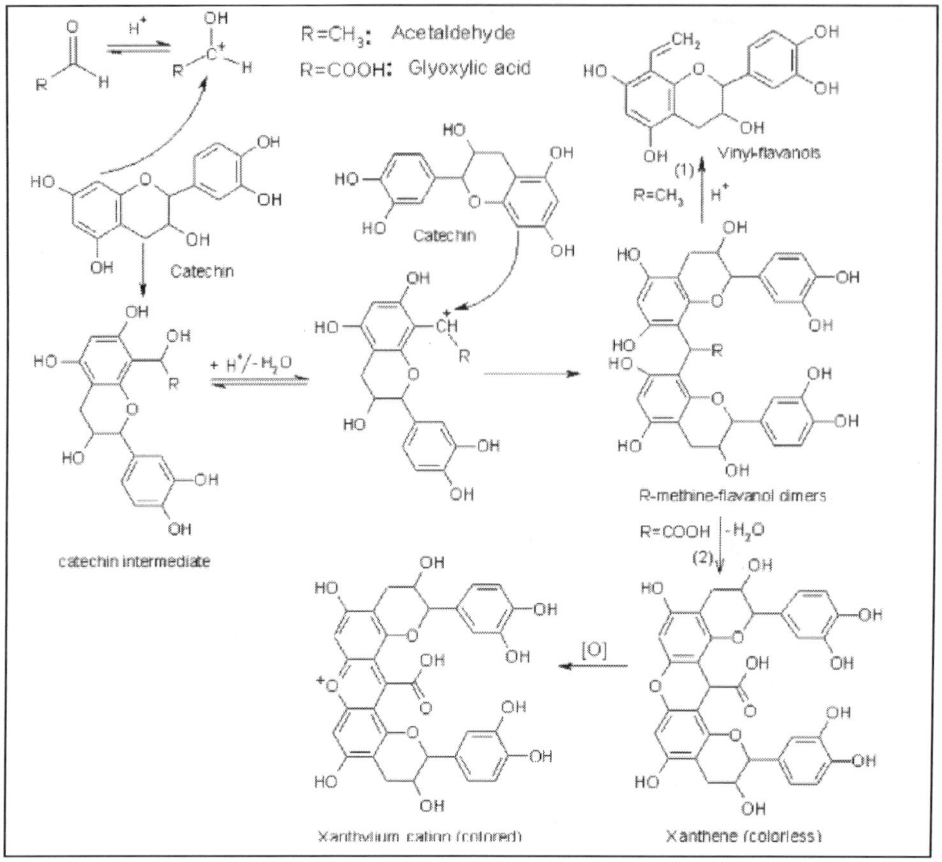

Figure 50.3: The Scheme of Catechin Polarization in Wine (Li *et al.*, 2008).

- The possible ways to avoid the negative consequences of catechin oxidation are:
- The reduction of the factors causing oxidation;
- The inclusion of anti-oxidants into seabuckthorn products.

REFERENCES

1. Handbook on Methods controls the quality and safety of biologically active additives to food 2004. Moscow Cederal center the Gossanepidnadzor of Ministry of Health of the Russian Federation, 204p.
2. ISO 14502-1, 2005. Determination of substances characteristic of green and black tea: Content of total polyphenols in tea. *Colorimetric Method Using Folin-Ciocalteu Reagent*, 18p.
3. Koshelev, Y. and Ageeva, L. 2003. *Seabuckthorn*, Biysk, 423p.
4. Li, H., Guo, A. and Wang, H. 2008. Mechanisms of oxidative browning of wine. *Journal of Food Chemistry* 108: 1-13.
5. Rosch, D., Bergmann, M., Knorr, D. and Kroh, L.W. 2003. Structure antioxidant efficiency relationships of phenolic compounds and their contribution to the antioxidant activity of seabuckthorn juice. *Journal of Agriculture and Food Chemistry* 51: 4233-4239.
6. Singleton, V.L. and Rossi, J.A. 1965. Colorimetry of total phenolics with phosphomolybdic-phosphotungstic acid reagents. *American Journal of Enology and Viticulture* 16: 144-158.
7. *State Pharmacopoeia of the USSR* 1987. 11th Edition, Moscow, Medicine, 328p.

Chapter 51

A Comparison of Separation by Hydrocyclone and Static Methods for Seabuckthorn Processed Residue

Feng Lidan, Sheng Wenjun, Bi Yang*, Ding Ruojun, Li Jie, Hu Jinmin, Li Jixin, Mi Lan, Wei Juan and Zhang Zhong

College of Food Science and Engineering, Gansu Agricultural University, Lanzhou 730070, P.R. China
(*E-mail: biyang@gsau.edu.cn)

SUMMARY

Puree is the main form of processed seabuckthorn in China. The residue of processed berries contains a large number of seeds that can be used to extract seed oil. The seeds were usually separated from the residue by winnowing, a low efficiency approach. In this study two water separation methods (static and hydrocyclone) were compared in the separating effect from wet residue. The results showed that the seeds purity reached 90.80 per cent when it was collected from the underflow mouth by hydrocyclone separation while the seed purity was only 2.02 per cent from overflow mouth. Seeds purity from hydrocyclone separation was similar with that from static separation, while the seed residue rate from hydrocyclone separation was lower than the rate from static separation. The hydrocyclone separation method could start without standing and layering process, and its equipment could be cleaned easily so the working cycle was shortened greatly. Besides, continuous feeding could be applied in hydrocyclone separation method, which could provide a powerful guarantee for the continuous operation of separation. The results suggest that hydrocyclone separation method has obvious advantages.

***Keywords:** Seabuckthorn processed residue, Hydrocyclone separation and Static separation.*

INTRODUCTION

Seabuckthorn (*Hippophae rhamnoides*), also known as *Hippophae* or acid gill, is a shrub of the family *Elaeagnaceae*. China has the largest distribution of seabuckthorn, which accounting for 95 percent of the world total growing area. The seabuckthorn mainly distributes in more than 10 provinces of China, including Northwest, North, Northeast and Southwest (Zang *et al.*, 2015). Seabuckthorn seed oil has physiological functions such as antioxidant (Ting *et al.*, 2011), protecting liver (Hsu *et al.*, 2009), reducing blood lipid (Cheng *et al.*, 2002), and anti depressant (Tian *et al.*, 2013). Pomace produced after fruit juice contains a lot of seeds that can be used to extract seed oil. At present, separation methods of seabuckthorn pomace have no reports available. Traditional winnowing method is always adopted in practical production by which marc needs to go through sunning or drying before winnowing or else mouldy is always likely to be happen and seed oil quality is affected. This method is low in separation efficiency, and likely to cause dust pollution and harm to human health.

As a highly efficient multiphase separation technology, hydrocyclone separation technology use density difference between two-phase and phases under centrifugal force to separate (Jiang, 2010) and has been more widely used in the oil (Fan *et al.*, 2014; Li *et al.*, 2010), chemical, mining, metallurgy and other industries. Applications of hydrocyclone separation technology in food areas include starch production (Cheng *et al.*, 2006), syrup purification (Deng *et al.*, 2001), and rice protein impurity (Sun *et al.*, 2003). Liu Zhe (Liu *et al.*, 2008) separated tomato pomace skin and seeds through using slag generated during the

Seabuckthorn (Hippophae L.): *A Multipurpose Wonder Plant, Vol. 5: Advanced Technologies for Cultivation, Processing, Health Protection and Environmental Conservation* (Virendra Singh, *Editor-in-Chief*, 2018), p. 391–396, Astral International Pvt. Ltd., New Delhi, India.

production of concentrated tomato sauce with water in a 1:35 ratio to remove most of the tomato skin and form uniformly dispersed suspension of tomato skin and seeds. Currently, hydrocyclone separation technology in terms of seabuckthorn marc has not been reported at home and abroad.

In this study, two water selection methods-cyclone and static, are used to separate seabuckthorn pomace. Swirl desander being alternated was conducted whirling tests and static water selection was chosen as a reference. The separation results of two methods were compared in order to provide technical support for the efficient separation of seabuckthorn seed.

MATERIALS AND METHODS

Materials and Equipment

Seabuckthorn marc was collected from Gansu Conseco Seabuckthorn Co., Ltd (Huachi, Gansu, China); fruit stalks which length was greater than 1cm and impurities are removed and the marc were dried. Fruit seeds were evenly distributed in the other impurities prior to use. Swirl desander (Model: LJXL-32, cylinder diameter 219 mm, high 860 mm, the outer diameter of the feed inlet φ = 40 mm, an outer diameter of the overflow opening φ = 40 mm, an outer diameter of the underflow outlet φ = 30 mm. Bucket screw pump (Model LGB-2, flow 1-2t/h, motor power: 380V, 1.5 kw, Yizhoukemei Mechanical Technology Co., Ltd, Yantai, Shandong, China). Separation tank (Model YFJ-200,200 L, stirring speed 19 r/min, motor power 380 v, 1.1 kw Yizhoukemei Mechanical Technology Co., Ltd, Yantai, Shandong, China).

Experimental Design of Hydrocyclone Separation

On the basis of single factor test, the L_9 (3^4) orthogonal test was conducted with solid-liquid ratio, feed temperature, mouth opening degree of underflow as investigation factors, the fruit seeds residual rate of overflow (fruit seeds residual rate), fruit seed purity of underflow mouth (fruit seed purity) as index. The results are shown in Table 51.1.

Table 51.1: Orthogonal Test L_9 (3^4)

Level	*Factor*			
	A *Solid-liquid ratio*	*B* *Mouth Opening Degree of Underflow*	*C* *Feed Temperature/ °C*	*D* *Blank*
1	1:10	30°	25	1
2	1:20	45°	35	2
3	1:30	60°	45	3

Experimental Design of Static Separation

Marc Soaking Stratification

Filled 200 mL water into several 250 mL beaker, respectively, added fruit seeds according to a certain solid-liquid ratio, and stirred 1 min to mix. The test was repeated three times. Clarity, layered effect and the underlying grain cleanliness were used as indexes to record layered effect within 0-23 h. Specific results as listed in Table 51.2

Table 51.2: Evaluation form of Fruit Marc Layered Effect

Program	*Standard*
Clarity 20	Deep color, poor clarity, opaque 0-10
	Light color, clear, transparent 11-20
Layered effect 40	Unstratified 0-10
	Stratified 11-20
	Obviously stratified 21-40
The underlying grain cleanliness 40	More than 50 per cent impurities in underlying seeds 0-20
	More than 25 per cent impurities in underlying seeds 21-30
	Less than 5 per cent impurities in underlying seeds 31-40

Static Water Selection

Solid-liquid ratio and standing time was set and static water selection was conducted with separation tank. Materials were fed according to a certain solid-liquid ratio, stirred sufficiently with separation tank stirring device, still for a certain time, and thus upper suspension and lower sediment were separated by discharge port of the tank. Suspension and sediment were uniformly sampled respectively and were dried to constant weight in 40°C drying oven. The sample was mixed well using quartering sampling and fruit seed purity in the precipitate and fruit seed residual rate in the suspension were counted respectively.

Fruit Seed Purity (Residual Rate) Calculation

Fruit Seed Purity (Residual Rate) =

$$\frac{\text{Amount of dry fruit seed in the material collected}}{\text{Amount of dry materials collected}} \times 100$$

Statistics Analysis

SPSS 19.0 was used to calculate the standard error.

RESULTS AND DISCUSSION

Effect of Solid-Liquid Ratio to Hydrocyclone Separation

The impact of solid-liquid ratio to separation

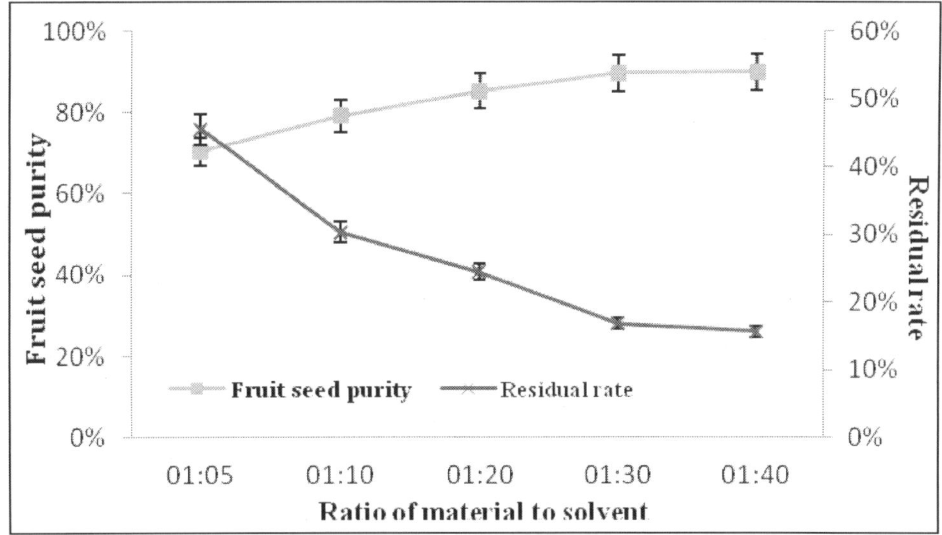

Figure 51.1: Effect of Solid-Liquid Ratio to Separation. Vertical bars represent the standard error.

efficiency was studied in the condition following: a 45 degree as mouth opening of underflow, room temperature as feed temperature. As shown in Figure 51.1, fruit seed purity increases with the solid-liquid ratio increase. The difference was not significant ($p > 0.05$): when solid-liquid ratio was 1: 30, 1: 40, fruit seed purity was 89.6 per cent, 90.0 per cent in consideration of water-saving, environmental protection and other factors, solid-liquid ratio 1:30 was chosen, and the fruit seeds residual rate was 16.8 per cent.

Effect of Mouth Opening Degree of Underflow to Hydrocyclone Separation

Because of stalks with high hardness in the fruit marc, the requirements of pump during hydrocyclone separation was high: (1) capable to convey high hardness marc; (2) certain pressure requirements. Centrifugal pump, concentrated pump, consistency pump and bucket screw pump was test and only bucket screw pump could met requirement (1) but could not meet pressure requirement (all solid and liquid flew from underflow mouth). In order to achieve separation, swirl desander was conducted a partial reconstruction: gauze on the mouth of the desander was removed a ball valve was placed on the underflow port. When the control device separated, internal pressure was controlled by adjusting the valve opening degree.

The impact of underflow mouth openness to separation effect was studied in the condition following: 1: 20 as solid-liquid ratio, room temperature as feed temperature. As shown in Figure 51.2, openness of underflow mouth impacted significantly ($p < 0.05$) on fruit marc separation: when the valve is fully open, the pressure in the swirl desander was zero and all the materials were imported from underflow mouth so that separation was failed. When the valve is fully closed (90 degree), the cyclone pressure was too much and all the materials were derived from the overflow mouth.

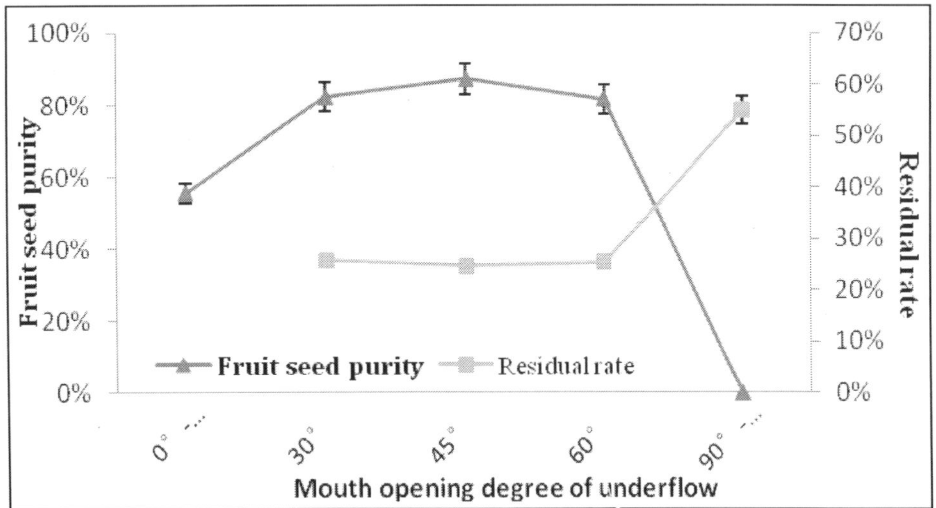

Figure 51.2: Effect of Mouth Opening Level of Underflow to Hydrocyclone Separation. Vertical bars represent the standard error.

When the valve was open to 45 degree, the fruit seeds purity was high to and fruit seeds residual rate was 24.7 per cent.

The Parameter Optimization of Hydrocyclone Separation Technology

On the basis of single factor test, the L_9 (3^4) orthogonal test was conducted with solid-liquid ratio, mouth opening degree of underflow, feed temperature as investigation factors, fruit seed purity of underflow mouth, and the fruit seeds residual rate of overflow as index. Test results and analysis of variance were shown in Tables 51.3–51.5.

Tables 51.3–51.5 shown that the influential factors affecting fruit seed purity and fruit seeds residual rate were both A> B> C (solid-liquid ratio> mouth opening degree of underflow > feed temperature), in which solid-liquid ratio and mouth opening degree of underflow impacted significantly ($p < 0.05$) on the indexes. The optimal combination was $A_3 B_2 C_2$: solid-liquid ratio 1: 30, mouth opening degree of underflow 45, and feed temperature 35°C which led to a result of fruit seed purity 90.02 per cent and fruit seeds residual rate 15.21 per cent. Under the same condition, separating simulation test of wet pomace through rehydration of dried marc was mad and the result was as following: fruit seed purity 90.80 per cent and fruit seeds residual rate 2.02 per cent. The test suggested that moisture content of the material had a significant effect on fruit seed residue ($P < 0.05$). Higher sugar content of fruit marc led to a more easily release of seeds if dried marc was rehydrated because seeds are tightly wrapped in the peel in dried marc.

Experiment of Static Separation

After fully stirring with water, due to differences in specific gravity, heavier components of fruit marc sank to the bottom and lighter components gradually went up to water surface and formed a floating layer. Suspension gradually sank to the bottom layer with increasing immersion time.

As indicated in Figure 51.3, after material mixed with water, different materials finished stratification within 1.5 h and layers became stable state. A steady state maintained until soaked for 6 h while suspension layer of 21 h was reduced compared to that of 1.5 h.

Table 51.3: L_9 (3^4) Orthogonal Test Results Analysis

Test Number		A	B	C	D	Fruit Seed Purity/ Per cent	Residual Rate/ Per cent
1		1	1	1	1	74.24	35.02
2		1	2	2	2	81.23	29.33
3		1	3	3	3	72.31	36.29
4		2	1	2	3	83.55	25.04
5		2	2	3	1	86.23	24.28
6		2	3	1	2	82.21	25.49
7		3	1	3	2	85.26	20.34
8		3	2	1	3	89.80	15.16
9		3	3	2	1	86.58	18.23
Fruit Seed Purity	k_1	75.927	81.017	82.083	82.350	Influential factors: ABCD	
	k_2	83.997	85.753	83.787	82.900		
	k_3	87.213	80.367	81.267	81.887	Optimal combination: $A_3B_2C_2D_2$	
	R	11.286	5.386	2.520	1.013		
Residual Rate	k_1	33.547	26.800	25.223	25.843	Influential factors: ABCD	
	k_2	24.937	22.923	24.200	25.053		
	k_3	17.910	26.670	26.970	25.497	Optimal combination: $A_3B_2C_2D_2$	
	R	15.637	3.877	2.770	0.790		

Table 51.4: Analysis of Variance: Fruit Seed Purity as Index

Factor	DEVSQ	DOF	Fratio	$F_{0.05}$	Significance
A	202.861	2	131.387	19	*
B	51.875	2	33.598	19	*
C	9.919	2	6.424	19	
D	1.544	2	1.000	19	
Deviation	1.54	2			

Table 51.5: Analysis of Variance: Fruit Seeds Residual Rate as Index

Factor	*DEVSQ*	*DOF*	*Fratio*	$F_{0.05}$	*Significance*
A	368.011	2	391.085	19	*
B	29.083	2	30.906	19	*
C	11.771	2	12.509	19	
D	0.941	2	1.000	19	
Deviation	0.94	2			

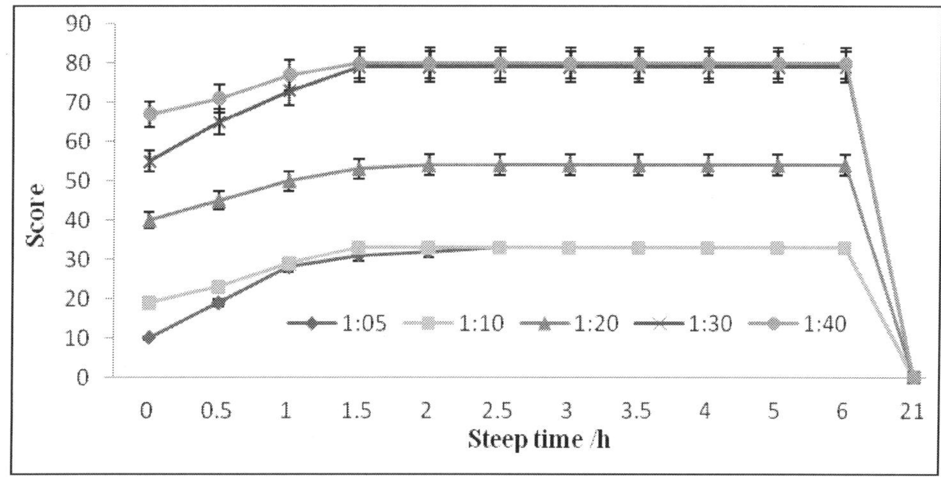

Figure 51.3: Layered Effect of different Solid-liquid Ratio. Vertical bars represent the standard error.

suspension layer sank to the bottom with a slightly agitation.

Different solid-water ratio had significantly effect on stratification ($p < 0.05$). Stratification was worst when solid-water ratio was 1: 5 or 1: 10. A stable stratification was obtained with some suspension and a poor purification of seeds in lower layer. A large number of impurities could not float to the surface because of the viscosity of the solution, especially sugar content of materials. Stratification was best when solid-water ratio was 1: 30 or 1: 40. Seeds in lower layer were finely purified and there were no impurities and little suspension. In practical application, in view of energy saving, economic and social benefits, solid-water ratio of 1:30 were recommended.

Added 90 L water into a 200 L separation tank with stirring device and 30 kg Marc according to solid-liquid ratio of 1: 30, stirred well and let stand for 1.5 h. Phased sampling was conducted on the discharge port of the tank: seabuckthorn seed material came out firstly; the remaining effluent was collected, when clear liquid in the middle poured another container was used to collect remaining effluent.

Comparison of Static Separation and Hydrocyclone Separation

To achieve optimum separation effect, prerequisite of static separation was stratification after material mixing, or fruit seed purity would be affected. Some defects existed in subsequent separation and collection: especially collection pipes were easily to be plug during collection of upper layer suspension so that equipment cleaning difficult seriously impacted on the continuity of materials feeding and working efficiency. Because the marc could be fed directly into swirl desander after uniformly mixed with water and without being soaked, hydrocyclone separation had a short duty cycle. Besides, hydrocyclone separation also processed easier in fruit seeds and pomace collection, and equipment cleaning. But hydrocyclone separation was prone to clogging underflow mouth, so that it remains to be further improved.

Table 51.6: Comparison of Static Separation and Hydrocyclone Separation

	Hydrocyclone Separation	*Static Separation*
Fruit seed purity/ per cent	90.08	92.32
Fruit seeds residual rate/per cent	2.02	5.2
Duty cycle	10 min	1.5-2 h

As indicated in Table 51.6, compared to static separation, the most prominent advantage of hydrocyclone separation was high separation efficiency and possible to achieve continuous production. Through hydrocyclone separation, fruit seed purity was 90.80 per cent, 1.52 per cent lower compared to that of static separation; residual rate of suspended solids in the upper fruit seeds 2.02 per cent, 3.18 per cent lower compared to that of static separation. In the actual production, the first step was to remove long stalks in either separation method in order to improve production efficiency, and maintain normal operation of equipment.

ACKNOWLEDGEMENT

This work was funded by the Projects of International Cooperation and Exchanges of Ministry of Science and Technology in China (No. 2014 DFR31230). We gratefully acknowledge National Nature Science Foundation of China. At the same time we are sincerely appreciated Mr. Luan from Gansu Conseco Company for providing all the seabuckthorn pulp needed in the experiment. And we are thankful for GE Qian cui-di who introduced the processing of seabuckthorn in the industry which helped us to set the experimental factors. Besides, we wish to express our thanks to Prof. Han Shun-yu for discussions.

REFERENCES

1. Cheng, T.J., Gao, L.P., Wang, Y.B., Zhang, J., Sun, Y.F. and Li, D.Y. 2002. The effect of seed oil of *Hippophae rhamnoides* L. (SOHR) on hemorrheology of cerebral infarction in rats. *Pharmacology and Clinics of Chinese Materia Medica* 18: 14-15.

2. Cheng, Y.S., Sun, Y., Lu, Y.W. and Lin, Y.L. 2006. Application of the full cyclone separation equipment in potato starch production. *Grain and Oil Processing and Food Machinery*, pp. 74-77.

3. Deng, L.G. and Hang, Y.G. 2001. Study on the separation of sugarcane molasses by cyclone separation technology. *Light Industry Science and Technology*, pp. 25-29.

4. Fan, X.Y., Wei, X.C., Hu, X.Y., Ou, Y.J.L. and Liu, D.M. 2014. Super heavy oil sewage hydrocyclone separation technology of Fengcheng oilfield. *Xinjiang Petroleum Geology* 35: 713-717.

5. Hsu, Y.W., Tsai, C.F., Chen, W.K. and Lu, F.J. 2009. Protective effects of seabuckthorn (*Hippophae rhamnoides* L.) seed oil against carbon tetrachloride-induced hepatotoxicity in mice. *Food and Chemical Toxicology an International Journal Published for the British Industrial Biological Research Association* 47 (9): 2281-2288.

6. Jiang, M.H. 2010. Research and application of hydrocyclonic separation technology. *Journal of Daqing Petroleum Institute* 34: 101-109,171.

7. Li, D. 2010. Influence factors of oil water separation efficiency of oil and water flow in rod pump. *Oil and Gas Field Surface Engineering* 29: 16-18.

8. Liu, Z. and Gong, J. 2008. Design and experimental of the separation of the hydraulic cyclone with the seed of tomato residue. *Grain and Oil Processing*, pp. 136-138.

9. Sun, Q.J., Yang, M.Y. and Chen, X.Y. 2003. Study on the process conditions of extracting rice protein concentrate by enzymatic method. *Food Science* 24: 83-85.

10. Tian, J.S., Zhen, X.F., Zhang, L.Z. and Qin, X.M. 2013. Antidepressant effect of seabuckthorn seed oil. *Food Science* 34: 279-282.

11. Ting, H.C., Hsu, Y.W., Tsai, C.F., Lu, F.J., Chou, M.C. and Chen, W.K. 2011. The *in vitro* and *in vivo* antioxidant properties of seabuckthorn (*Hippophae rhamnoides* L.) seed oil. *Food Chemistry* 125: 652-659.

12. Zang, Q.Q., Deng, Q.C., Cong, R.H., Ge, Y.Z., Huang, F.H., Wei, X.S., Chen, P. and Zhang, Y. 2015. The seabuckthorn oil efficacy components and pharmacological research progress. *Chinese Grease* 5: 76-81.

Chapter 52

Freezing Characteristics of Seabuckthorn Primary Juice During Freezing Concentration

Mi Lan, Feng Lidan, Zhang Zhong and Bi Yang*

College of Food Science and Engineering,
Gansu Agricultural University, Lanzhou 730070, China
(*E-mail: biyang@gsau.edu.cn)

SUMMARY

Seabuckthorn primary juice contains higher level of physiologically active components, and ordinary thermal concentration would damage the thermo-sensitive components of them. The freeze intensity and ice crystal purity of one liter primary juice after one hour at different temperature freezing was tested with the index of ice crystal mass and its viscosity after thawing. The results showed that with the decreasing of the coolant temperature, the ice crystal quantity increased from 0.08 kg/L at -1°C to 0.2 kg/L of -13°C. The rate of ice crystal formation increased with the decreasing of freezing temperature while the ice crystal purity decreased. The freezing temperature gradually reduced with the increasing of the concentrated juice viscosity. The pH value, total acids, soluble solids and the total flavanoids contents were all increased after freezing concentration of various seabuckthorn primary juice. It is suggested that the freezing can effectively concentrate of seabuckthorn juice and protect active ingredients and mass.

Keywords: *Seabuckthorn primary juice, Freezing concentration, Icing intensity and Ice crystal purity.*

INTRODUCTION

Seabuckthorn (*Hippophae rhamnoides* L.) is a kind of deciduous shrubs. The plant belongs that of *Hippophae* plants of elaedgnus branch. It is mainly distributed in North, Northwest and Southwest of China. Seabuckthorn is a medicine and food homology plant providing very high nutritional value. The fruit contains extreme high vitamin C, while rich bio-active substances in the fruit such as unsaturated fatty acid, functional polysaccharose, amino acid as well as flavonoids even play valuable antioxidant and healthcare functions. Seabuckthorn fruit can be used as medicine with good treatment effect on such diseases as cough with much sputum, indigestion, stomachache and enteritis and so on.

The primary juice of seabuckthorn fruit refers to the original juice from the fruit that is squeezed and extracted; it is commonly used as the raw material of the juice processing. In actual production, the concentrating of seabuckthorn primary juice is conducive to reduce the mass and volume, cost down the expenses on food packaging, storage and transportation, and also inductive to improve the product concentration, increase the osmotic pressure, decrease the water activity, inhibit the growth of microorganism and extend the mass life of seabuckthorn primary juice. At the present stage, there are many concentrating methods for fruit juices in domestic and overseas, evaporation concentration, freezing concentration (Lin, 2001) and reverse osmosis concentration and so on are common methods, among which, heating evaporation concentration has the minimum cost and is widely used for concentrating apple juice, citrus juice, water melon and so on. This method is patented (Xue *et al.*, 2009) on

Seabuckthorn (*Hippophae* L.): *A Multipurpose Wonder Plant, Vol. 5: Advanced Technologies for Cultivation, Processing, Health Protection and Environmental Conservation* (Virendra Singh, *Editor-in-Chief*, 2018), p. 397–402, Astral International Pvt. Ltd., New Delhi, India.

the concentration application for seabuckthorn primary juice, but due to high evaporating temperature, many thermo-sensitive components in the primary juice is easy to coked, decomposed and oxidized and the product mass is reduced (Wang *et al.*, 2007). Reverse osmosis concentration is a new technology developed in recent 20 years, a kind of separation method using semi-permeable membrane film (Zhong *et al.*, 2002). This method has lower evaporating temperature (50-56°C) and good heat transfer effect. Seabuckthorn primary juice is heated in a very short time on the heating face, greatly improving the product mass, reducing Vc loss. However, the production cost is high and the concentration rate is low, and it is used for pre-concentrating fruit juice at the present stage (Ivetta *et al.*, 2007). Freezing concentration is a kind of concentrating method below freezing point, particularly advantageous to thermosensitive food and can avoid the processing defects stated above. The principle is that freeze the seabuckthorn primary juice firstly to make most of water ice up, then remove the ice crystals in a certain way to get the concentrated seabuckthorn primary juice (Sun *et al.*, 2007). Freezing concentration can prevent adverse chemical reaction, providing the advantages of low energy consumption and small losses of flavor, aroma and nutrient composition. It can make the loss of seabuckthorn primary juice to a minimum in concentrating process, and minimize the loss of flavor, aroma and nutrient components of food.

The research studies the laws of effect on seabuckthorn protoplasmic freezing strength and purity of ice crystal from different freezing temps, and change laws of seabuckthorn protoplasmic freezing concentrated solution viscosity and freezing point as extension of freezing time. Finally, it detects nutrition contents in frozen liquid recovery liquid before and after freezing concentration.

MATERIALS AND METHODS

Materials

Seabuckthorn primary juice provided by Gansu Conseco Seabuckthorn Co., Ltd (Huachi, Gansu, China).

Reagent

95 per cent vol ethyl alcohol, absolute ethyl alcohol, aluminum nitrate juice, potassium acetate juice, rutin, primary reagent, solid NaOH, phenolphthanlein indicator, activated carbon.All solvents/chemicals used were of analytical grade and obtained from TCI Shanghai, China.

Instrument

Low temperature refrigerator (BCD-256WDGH, Haier Co., Ltd, Shandong, China), viscometer (DV-II + P,Guangzhou East South Professional Tech Company, Guangzhou, China), low thermometer (WNY02, Changzhou Cheng Heng Instrument Co., Ltd., Changzhou, China), portable digital Refractometer (WYT-4, Quanzhou in Fujian Province Optical Instrument Factory, Quanzhou, China), spectrophometer (WFX-1F2B2, Beijing Reid Analytical Instruments, Beijing, China).

Influence of Temperature and Seabuckthorn Primary Juice on the Freezing Effect

Determination of freezing point seabuckthorn primary juice was accorded to Zhang *et al.* (2010) method. Take a certain amount of juice in the freezing chamber of the refrigerator, and view it every 10 min. When visible small ice crystals in the primary juice appear, measure with low thermometer to determine the temperature is the freezing temperature of seabuckthorn fruit juice. Takethe freezing temperature as the start and conduct post freezing concentration experiment at gradient temperatures.

The influence of the freezing temperature on the freezing effect was accorded to Li *et al.* (2010) method. Take juice of 500 mL and freeze it for 1 hour at -1°C, -4°C, -7°C, -10°C and -13°C respectively and then filter the ice crystals with filter net. Measure the weights of the ice crystals at different temperatures respectively. Melt the ice crystals, and measure with viscometer the viscosity of the melted ice crystals, *e.g.* the purity of the ice crystals, to determine the freezing effect, taking the average value from three parallel experiments.

The influence of seabuckthorn juice concentration on the freezing effect was accorded to Zhang *et al.* (2010) method. Make juice of 500 mL into concentrated juices with different concentration by way of gradient concentrations that is to freeze seabuckthorn primary juice for 1 hour under optimal freezing conditions and take out the ice crystals to get a concentrated seabuckthorn juice as 1. The produced concentrated juice 1 is processed at the same freezing temperature and for the same time to remove the ice and get a concentrated seabuckthorn juice as 2. Then concentrated juices 3, 4 and 5 will be made in the same way. Respectively measure the freezing points and viscosities of juices at different concentration stages (Liu *et al.*, 1994), taking the average value from three parallel experiments.

Component comparisons of seabuckthorn primary juice before-after concentration VS recovered primary juice was accorded to Zhang *et al.* (2010) method. Freezing concentration technology is one of Freezing concentration technology is one of the best methods of keeping the nutrition component of seabuckthorn primary juice, however, nutrition loss still exists in the process of the operation, especially the amount of the primary juice taken away during ice removing can not be ignored. Measurement is conducted on seabuckthorn primary juice, the freeze concentrated juice under -10 °C for 1hour, and the recovered juice of the freeze concentrated juice by adding distilled water to recover to the volume before concentration on pH, total acid, total solid content and total flavanoid content, to study the influence of the freeze concentration treatment on the active components and the quality of seabuckthorn primary juice.

The pH value of solution was determined by pH meter. Total acid content was measured by Zhang *et al.* (2010) method. Adopt titration method, taking the citric acid as the reference acid. Total solid content was measured by a portable digital refractermeter directly (Yang *et al.*, 2004).

Total Flavonoid Content Measurement

Preparation of the standard curve, use rutin calibration to draw the standard curve, prepare the standard juices with concentrations of 0.2 g/L, 0.4 g/L, 0.6 g/L, 0.8 g/L and 1 g/L respectively, measure the absorption values of the juices with a UV spectrophotometer at 415 nm and make the standard curve to get a regression equation.

Take a certain amount of seabuckthorn juices before and after freezing, and the recovery juice that is made by adding the distilled water to the freeze concentrated juice to recover to the volume before concentration, dilute them to certain concentration, then measure the absorption value by a UV spectrophotometer, and introduce the regression equation to calculate the total flavanoids (Chen *et al.*, 2009).

RESULTS AND DISCUSSION

The Freezing Point of Seabuckthorn Primary Juice

Due to more content of the solid content contained in seabuckthorn primary juice, the freezing point is lower than that of the pure water, in this experiment, the temperature measured at the starting point of small visible ice crystals appearing in 500mL seabuckthorn primary juice during freezing is -1 ℃. No ice crystal is produced when it does not reach -1 ℃ during freezing. The possible reason is: The energy provided by the external coolant serving for the main part of the seabuckthorn primary juice is smaller or equal to the energy loss of the phase change occurred by the equipment and seabuckthorn primary juice. From this we learn, ice crystals can be produced at the temperature difference that must be achieved between the coolant temperature and the freezing point of seabuckthorn primary juice, that is, the follow-up freezing concentration experiment is conducted by taking -1 ℃ as the initial freezing temperature.

Effect of the Freezing Temperature on the Ice Crystal Purity and the Icing Intensity

Based on the achieved freezing point of seabuckthorn primary juice, freeze seabuckthorn primary juice of 500mL under the condition of -1 ℃, -4 ℃, -7 ℃, -10 ℃ and -13 ℃ for 1 hour respectively, fish out the ice, and measure the melted liquid viscosity *e.g.* the ice crystal purity, the result is shown as Figure 52.1.

The ice crystals produced during the operation shall not be mixed with any solute as possible, nor mixed with solute in separation so as to prevent much loss of solute. It can be seen from Figure 52.1 that the ice crystal viscosity shows a gradual increasing trend with the decreasing of the freezing temperature. When the external coolant temperature drops from -1 ℃ to -7 ℃, with the coolant temperature decreasing, the soluble solid content mixed in the ice crystals increases gradually, from 149.3 mPa•s to 263 mPa•s. Following, the influence of the temperature decreasing of the coolant on the production amount of the ice crystals becomes flat. With the range of -7 ℃ ~ -13 ℃, the decreasing of the ice crystals is gently, from 263 mPa•s to 290 mPa•s. In the process of freezing concentration, the influence of

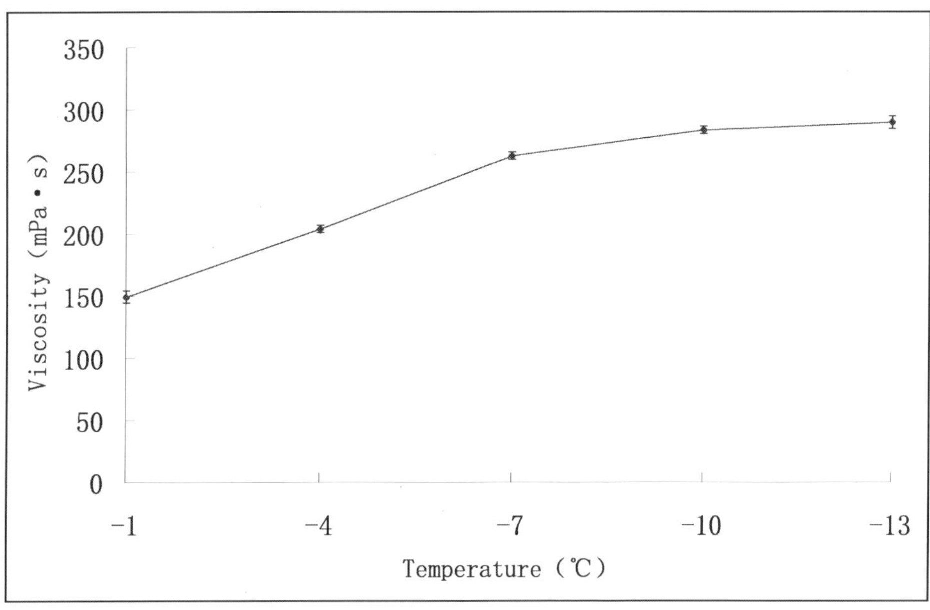

Figure 52.1: Influence of Temperature on Ice Crystal Granularity.

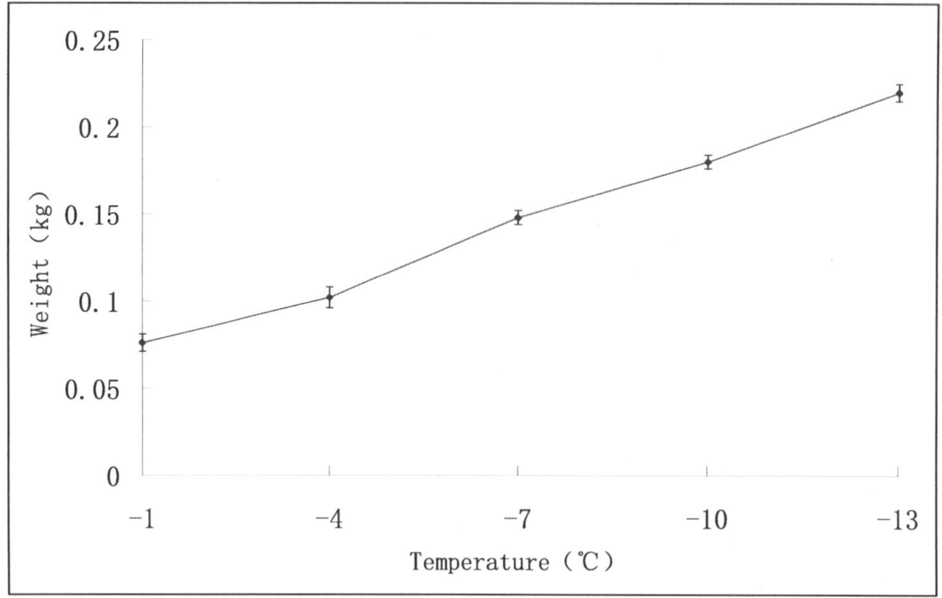

Figure 52.2: Influence of Temperature on Ice Crystal Mass.

the coolant temperature on the crystallizing intensity of freezing concentration is shown as Figure 52.2.

It can be seen from Figure 52.2 that the coolant temperature directly influences the growing rate of ice crystals in the freezing concentration. When the external coolant temperature drops from -1 ℃ to -13 ℃, the ice crystals mass produced is increased at a faster rate, from 0.076 kg to 0.22 kg. It is found that seabuckthorn primary juice is totally frozen after above -13 ℃, and therefore -13 ℃ is selected as the final test temperature.

In comprehensive consideration of the crystallizing intensity of the freezing concentration and the ice crystal purity, so we learn from this: for seabuckthorn primary juice with an initial viscosity of 154 mPa•s, the more relative ideal coolant temperature range shall be: -10 ℃.

The Influence of the Concentration of Seabuckthorn Juice on the Freezing Effect

Conduct the progressive freezing concentration treatment under the same freezing conditions, and achieve different concentrated liquid viscosities, shown as Figure 52.3.

With the increasing of the freezing concentration degree, the viscosity shows a increasing trend, which respectively are 154 mPa•s, 211 mPa•s, 221mPa•s, 246.7 mPa•s, 285.3 mPa•s and 292 mPa•s. This is because the water in the concentrated juice is gradually reduced. However, from the viscosity increasing trend, the increasing at the early stage is relatively large, with the water in seabuckthorn concentrated juice decreasing, the ice crystals amount produced with the same freezing time decreases gradually, therefore the viscosity increasing of seabuckthorn concentrated juice becomes gently gradually.

The relationship between freezing point temperatures of freezing concentrated juice in different degrees and the juice viscosities is shown as Figure 52.4.

It can be seen from Figure 52.4 that with the increasing of the concentrated juice viscosity, within the

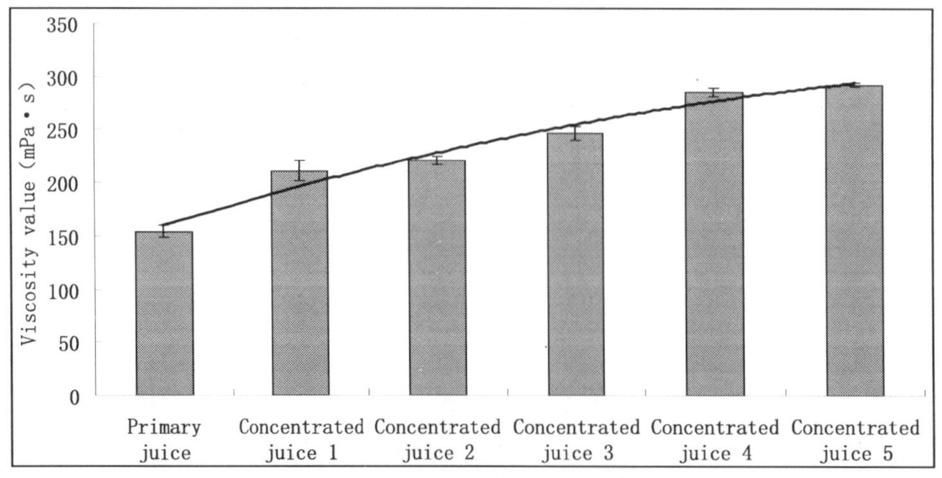

Figure 52.3: The Concentrated Juice Viscosity Changes with the Concentration Degree.

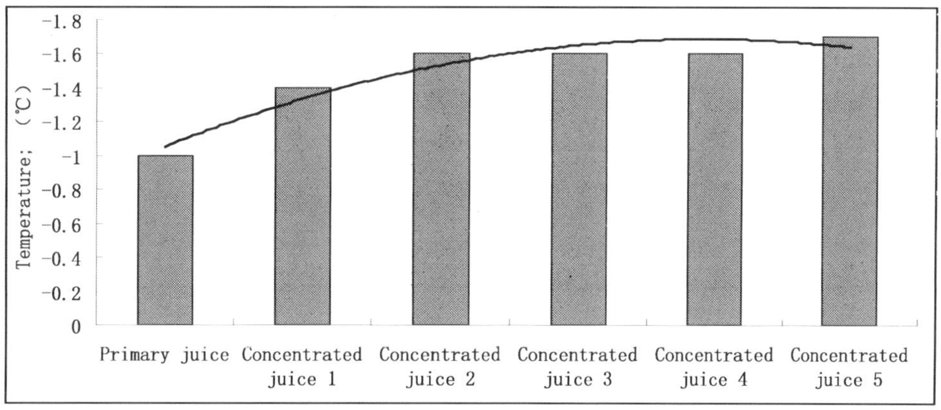

Figure 52.4: Changes of Freezing Points of Concentrated Juices in different Degrees.

same freezing time, the decreasing of the early freezing point of the freezing concentrated juice is distinct. Seabuckthorn primary juice with viscosity of 154 mPa•s has a freezing point of -1 ℃, and the freezing point of the concentrated juice 1 with the viscosity of 211 mPa•s drops to -1.4 ℃. Following that the freezing points of concentrated juices 2, 3 and 4 all are -1.6 ℃ unchanged, while the freezing point of concentrated juice 5 with the viscosity of 292 mPa•s drops to -1.7. This is because that within the viscosity range of the observed seabuckthorn concentrated juice, with the concentration increasing, the substance shifting resistance at the solid(ice crystal)—liquid interface becomes larger, causing bad effect of mass transfer, less ice production of the freezing concentration and the juice freezing point dropping gently.

Measurement Result of Seabuckthorn Primary Juice Components

Measure pH values respectively on seabuckthorn primary juice, the freezing concentrated juice with ice removed under -10 ℃ for 1 hour, and the freezing concentrated juice with adding distilled water to recover to the volume before concentration, the measurement results are shown as Table 52.1, and calculate the nutrition loss degree of the recovered juice relative to seabuckthorn. A loss of 20 per cent is acceptable (Condurso *et al.*, 2009).

Table 52.1. Measurement results of seabuckthorn nutrition components with different processing methods

Juice to be Tested	*Parameter*			
	pH	*Total Acid (g/mol)*	*Soluble Solid Content (Brix)*	*Total Flavoniods (mg/L)*
Primary juice	2.95	0.336	11.4	4307.38
Concen-trated juice	2.75	0.490	12.7	4480.68
Recovered juice	3.03	0.273	9.3	4039.54

It is seen from Table 52.1 that the pH value of the freezing concentrated juice is less than the primary juice in small amount, it meet the characteristic of high acid seabuckthorn primary juice. As for the pH value of the recovered juice increases to 3.03, it is mainly because that the concentration of the juice becomes less due to the loss of the melting mass during freezing concentration.

Titration measurement is conducted on the total acid on three different processing methods. Citric acid is used as the reference acid due to its much content in seabuckthorn. It can be seen from Table 52.1 that the acidity in the freezing concentrated juice is obviously higher than that in seabuckthorn primary juice. While the total acid in the recovered juice is lower, compared with seabuckthorn primary juice, the loss rate of the total acid in the recovered juice reaches 18.75 per cent.

It is seen from Table 52.1 that the soluble solid content in freezing concentrated juice reaches 12.7 Brix, and is more than that of seabuckthorn primary juice, while the content of the soluble solid content in the recovered juice drops to 9.3Brix, the loss rate is 18.42 per cent compared with the primary juice.

A standard curve is made using rutin calibration and achieved the regression equation as y£½3.15x-0.0347, R^2£½0.9991, Have good linear absorption law. Introduce the measured absorption of the juice to be tested into the equation to calculate the total flavanoids, showing that the amount of total flavanoids in the freezing concentrated juice is increased by 4 per cent than the primary juice, while it reduces by 267.84 mg/L, with the loss rate of the soluble solid content is in the recovered juice.

CONCLUSION

The temperature at which seabuckthorn primary juice of 500 mL with viscosity of 154 mPa•s starting to have visible small ice crystals during freezing is -1 ℃. The coolant temperature has a larger influence both on the intensity and the ice crystal purity of the freezing concentration crystallization and the ideal coolant temperature achieved through experiments is

-10 ℃. At this temperature, it is better if the freezing concentration effect can be achieved with relative low energy consumption.

Under the condition of the same freezing concentrating time, with the increase of the freezing concentration, the concentration viscosity increased gradually, showing a change of from fast to slow increasing. This is mainly because that with the icing amount increasing of the seabuckthorn concentrated juice, the water in the soluble melted mass becomes less, the total ice crystals amount produced in the same time reduces, making the viscosity increasing of the concentrated seabuckthorn juice becomes gently gradually. This shows that taking out the ice crystals timely is helpful for improving the freezing concentration efficiency.

Through the components analysis on seabuckthorn primary juices before and after freezing concentration, it is seen that various contents of concentrated seabuckthorn primary juice are higher than that of non-concentrated seabuckthorn primary juice, this is because that the concentration is increased after the water is removed, however, due to the process operation and ice crystals removing, some nutrition contents can be lost. Through test on the recovery juice which is made by adding distilled water to the freezing concentrated juice to recover the volume as before concentration, it is found that, compared with seabuckthorn primary juice, the pH value increase to a certain, the total acid reduces, and various nutrition content show decreasing trend, while the loss rates are all less than 20 per cent. Therefore it is proven that freezing method can make the primary juice concentration effectively and protect the active components and quality of the primary juice.

ACKNOWLEDGEMENT

This work was funded by the Projects of International Cooperation and Exchanges of Ministry of Science and Technology in China (No. 2014 DFR31230). We gratefully acknowledge National Nature Science Foundation of China. At the same time we are sincerely appreciated Mr. Luan from Gansu Conseco Company for providing all the seabuckthorn pulp needed in the experiment. And we are thankful for GE Qian cui-di who introduced the processing of seabuckthorn in the industry which helped us to set the experimental factors. Besides, we wish to express our thanks to Prof. Han Shun-yu for discussions.

REFERENCES

1. Condurso, A., Verzera, V. and Romeo, M. 2008. Solid-phase micro-extraction and gas chromatography mass spectrometry analysis of dairy product volatiles for the determination of shelf-life. *International Dairy Journal* 8: 356-360.
2. Chen, J.M., Li, L.J. and Ma, Q. 2009. Determination and analysis on the total flavonoids of mulberry juice, elaeagnus juice and seabuckthorn juice. *China Brewing* 7: 153-154.
3. Ivetta, V., Eva, B. and Gyula, V. 2007. Concentration of seabuckthorn (*Hippophae rhamnoides* L.) juice with membrane separation. *Separation and Purification Technology* 57: 455-460.
4. Liu, L. 1994. Progressive freeze concentration of liquid food. *Food and Fermentation Industries* 15: 33-36.
5. Lin, X. 2001. Freezing concentration and its application in food industry. *Chinese and Foreign Food Industrial Information* 2: 33-55.
6. Li, Y.M. and Lai, J. 2010. Study on growth speed of ice crystal during freeze concentration of litchi juice. *Cereal and Food Industry* 26: 32-34.
7. Sun, B.B., Ma, H.Q and Chen, S.W. 2007. The influence of the frozen concentration on low sugar juice and wine quality. *Chinese Food Science* 5: 86-89.
8. Wang, H.L., Liu, Z.Y. and Song, J.T. 2007. Nutrient loss dynamics in evaporation concentration process for fruit and vegetable juices (pineapple juice). *Food and Machinery* 5: 31-33.
9. Xue, H.K., Xiao, C.J. and Xie, J.X. 2010. A concentrated seabuckthorn production process. *China Patent*: 201010227303.9.
10. Yang, C.B. and Zhang, X.F. 2004. Dynamic research on the soluble solid content in seabuckthorn in Tibetan Plateau. *Acta Botanica Boreali-Occidentalia Sinica* 2: 698-701.
11. Zhong, H.Y., Yuan, L.J. and Li, Z.H. 2009. Research on reverse osmosis concentration. *Chinese Food Science* 9: 51-54.
12. Zhang, S.M. and Yi, X. 1994. Study on extraction and freeze concentration of watermelon juice. *Refrigeration Technology* 2: 3-9.

Chapter 53

Development of Functional Wines from Seabuckthorn (*Hippophae rhamnoides* L.) Growing in Dry Regions of Indian Himalayas

Rakesh Sharma* and V.K. Joshi

Department of Food Science and Technology, Dr. Y.S. Parmar University of Horticulture and Forestry
Nauni, Solan – 173 230, Himachal Pradesh, India
(*E-mail: drrakes@gmail.com)

SUMMARY

Seabuckthorn (*Hippophae rhamnoides* L.) berries growing wild in the cold and dry regions of Indian Himalayas were explored for the development of wine. The presence of high acid content in seabuckthorn pulp affects the quality of wine adversely. The efforts made to prepare seabuckthorn wine by de-acidifying the pulp with sodium bi-carbonate ($NaHCO_3$ @ 0.6-1.2 per cent), diluting the pulp with water up to 1: 6 dilution and blending the pulp of seabuckthorn with apple pulp in different proportions *i.e.* 25 per cent, 50 per cent and 75 per cent for better fermentation and to improve the nutritive value of wine is reported in this article. The results showed that the seabuckthorn must prepared by dilution had better fermentation behaviour than that prepared by de-acidification with $NaHCO_3$. Addition of DAHP in general, enhanced both the rate of fermentation as well as ethanol content. Highest rate of fermentation (RF=0.80) was recorded in 1: 5 dilution with 0.1 per cent DAHP. In case of de-acidification with blending the '*must*' of SBT + Apple (50 per cent + 50 per cent) had better fermentation behaviour and recorded the highest rate of fermentation (RF= 1.52). Further, with increase in dilution level, the rate of fermentation increased. Further, seabuckthorn pulp blended with apple pulp and diluted with water in the ratio of 1:4 showed better fermentation behaviour. The physico- chemical evaluation of the wines showed that the blending of seabuckthorn pulp with apple pulp has resulted in improvement of ascorbic acid content in the developed wine (122.0 mg/100g) compared to about 75.0 mg/100g in control samples. The ethanol content among different wines ranged from 9.3 to 10.8 per cent (v/v). The total sensory quality of the wine prepared from seabuckthorn pulp by blending with apple pulp in the ratio of 50: 50 and diluting with water (1: 4) was adjudged the best (highest score=17.3) on the basis of characteristics like colour, aroma, bitterness, astringency and overall acceptability. It is concluded that blending of seabuckthorn pulp with apple pulp not only gave an acceptable wine but also increased the nutritional quality of the wine compared to those prepared by diluting seabuckthorn pulp with water as well as de-acidification by $NaHCO_3$. Further, the seabuckthorn-apple blended wine had shown higher ascorbic acid, phenolic contents and antioxidant potential indicating its better nutritional and functional properties compared to apple and seabuckthorn wine.

Keywords: *Seabuckthorn, Hippophae rhamnoides, Wine, De-acidification, Blending, Dilution and Ethanol.*

INTRODUCTION

Seabuckthorn (*Hippophae rhamnoides* L.), belonging to Elaeagnaceae, is a thorny, dioecious bush growing wild in the cold and dry regions of Indian Himalayas mainly in Ladakh (J&K), Lahaul- Spiti and Kinnaur (HP), Kumaon-Garhwal (Uttaranchal) and Sikkim / Arunachal Pradesh in the north-east region of India. The ripe fruits of seabuckthorn are rich source of vitamins (A, B, C, K and E) with antioxidant and anti stress properties (Singh and Singh, 2004). Due to the high perishable nature and acidic taste, fruit cannot be consumed fresh and requires immediate processing. The berries are mainly utilized

Seabuckthorn (*Hippophae* L.)*: A Multipurpose Wonder Plant, Vol. 5: Advanced Technologies for Cultivation, Processing, Health Protection and Environmental Conservation* (Virendra Singh, *Editor-in-Chief*, 2018), p. 403–411, Astral International Pvt. Ltd., New Delhi, India.

for the preparation of beverages by blending with other fruit pulp or by using appropriate addition of sugar; seed oil and many other pharmaceutical products (Sharma *et al.*, 2004; Dwivedi *et al.*, 2004; Kaushal *et al.* 2011; Joshi *et al.* 2011). On the other hand, preparation of wine can also be an outlet for its economic utilization (Negi and Dey, 2009). The basic process of wine making involves the fermentation of grape juice or any other juice or pulp by *Saccharomyces cerevisiae* (Amerine *et al.*, 1980; Sandhu and Joshi 1995; Joshi 1997; Joshi *et al.*, 2005; Joshi *et al.*, 2009; Sharma *et al.* 2009). The quality of wine is known to depend upon a number of factors like cultivars and their characteristics such as adequate sugar level, acid content, colour and aroma (Ethiraj and Suresh 1993). The seabuckthorn wine has been reported to be rich in phenolic compounds (Negi and Dey, 2009). However, the acid content in the seabuckthorn juice is a major factor that affects its fermentation behaviour and the quality of wine (Joshi *et al.*, 2011). Therefore, the de-acidification process is a key step in the wine production procedure. De-acidification by dilution of seabuckthorn pulp with water up to 1: 6 has been reported earlier by Joshi *et al.* (2011). But, such higher level of dilution reduces the fruit content and proportion of other phytochemicals in the wine produced and hence affecting the quality. The chemical de-acidification in grapes to produce wine has also been reported where, the tartaric acid together with the hydrogen tartrate forms a neutral calcium tartrate thereby reducing the acidity level (Joshi and Pandey 1999). Therefore, studies conducted on development of seabuckthorn based functional wines by way of de-acidification with water and chemicals as well as blending with apple pulp in different proportions has been reported in this article.

MATERIALS AND METHODS

Standardization of a Method for De-acidification of Seabuckthorn Pulp

1. **De-acidification by dilution with water and use of sodium bicarbonate:** The fresh seabuckthorn juice is reported to be high in acidity that can be reduced by diluting it with water and by the use of sodium bicarbonate. The musts were prepared by diluting the pulp in 1:5, 1:6, 1:7 and 1:8 ratios with water and with or without DAHP (T_1 to T_8). Similarly, musts were also prepared by treating the pulp with $NaHCO_3$ at the concentrations of 0.6, 0.8, 1.0 and 1.2 per cent (T_9 to T_{16}) to reduce the acid content.
2. **De-acidification by blending of pulp**: In this process, the acidity of the seabuckthorn pulp was reduced by blending the pulp with apple pulp in different proportions *i.e.* 25 per cent, 50 per cent and 75 per cent. The blended pulps were then diluted with water in different ratios. The detail of treatments for preparation of seabuckthorn wine is given below:

T_{17} = SBT 75 per cent +Apple 25 per cent with 1:2 level of dilution, T_{18} = SBT 75 per cent +Apple 25 per cent with 1:3 level of dilution, T_{19} = SBT 75 per cent +Apple 25 per cent with 1:4 level of dilution, T_{20} = SBT 50 per cent +Apple 50 per cent with 1:2 level of dilution, T_{21} = SBT 50 per cent +Apple 50 per cent with 1:3 level of dilution, T_{22} = SBT 50 per cent +Apple 50 per cent with 1:4 level of dilution, T_{23} = SBT 25 per cent +Apple 75 per cent with 1:2 level of dilution, T_{24} = SBT 25 per cent +Apple 75 per cent with 1:3 level of dilution and T_{25} = SBT 25 per cent +Apple 75 per cent with 1:4 level of dilution. All the wines were compared with control (T_1 = 100 per cent SBT, 1:5 dilution).

Fermentation

Fermentation for all the treatment was carried out by the pure wine yeast *Saccharomyces cerevisiae* var. *ellipsoideus* (5 per cent, v/v) at 23±1 °C. *Saccharomyces cerevisiae var. ellipsoideus* strain UCD 595 used in the study was obtained from the department of Oenology and Viticulture (California, Davis, USA). It was maintained on yeast malt extract agar medium and re-cultured after every three months or whenever needed from the stock yeast culture. The procedure for preparation of seabuckthorn wine was the same as reported earlier by Joshi (1997) and illustrated in Figure 1. During fermentation, reduction in total soluble solids (°Brix) was recorded every day. When fermentation was completed, siphoning was done three times during maturation of six months.

Analyses

The wines were analyzed for various physico-chemical characteristics during fermentation. TSS was measured using Erma hand refractometer (0 to 32 °B), and corrected by applying the correction factor for the temperature variation (AOAC, 1980). The titratable acidity (per cent citric acid) was estimated by titrating a known aliquot of the sample against N/10 NaOH solution (AOAC, 1980), while ascorbic acid with 2, 6- dichlorophenol indophenol visual titration method (Ranganna, 1986). Ethanol was estimated by the colorimetric procedure (Caputi *et al.*, 1968). Initial and final TSS were determined and used for calculation of rate of fermentation. The wines were also evaluated for various physico-chemical characteristics before and after maturation in bottles. Sensory evaluation was carried out by a panel of 5 judges on a prescribed proforma given by Joshi (2006). The coded samples were presented to the judges who were asked to evaluate the products. Overall score was determined out of 20 and the wine having overall scores between 17-20 was considered as 'superior', 13-16 as 'standard', 9-12 as 'below standard' and 1-8 as 'unacceptable or spoiled' (Amerine *et al.*, 1980).

Triplicate determinations were made for each attribute. The Complete Randomized Block Design (CRBD) was used to analyze the data of quantitative estimation of physico-chemical characteristics, while the data pertaining to sensory quality was analyzed by Randomized Block Design (RBD) as per the methods given by Panse and Sukhatme (1967).

RESULTS AND DISCUSSION

Changes in Seabuckthorn must as Affected by Method of De-acidification during Fermentation

The musts of all the treatments irrespective of method of de-acidification witnessed a fast reduction in TSS at the initial stages (up to 96 hrs), but after 144 hrs, it changed clearly with the flattening of curve followed by stabilization at 240 hrs (Figures 53.2 and 53.3). Among various treatments pertaining to de-acidification with water, the highest reduction in TSS was observed in treatment T_1 (1:5) with DAHP (Figure 53.2A). Similar results were also seen in sodium bicarbonate treated musts and the highest reduction in TSS was observed in T_7 (1.0 per cent) with DAHP (Figure 53.2B). Amongst the treatments pertaining to dilution, variations in the rate of fermentation (RF) was observed and it ranged from 0.72 to 0.80 °B/24 hours (Figure 53.2A). The highest rate of fermentation was observed in T_1 (1:5) with DAHP (0.80). However, RF of musts prepared by $NaHCO_3$ was lower

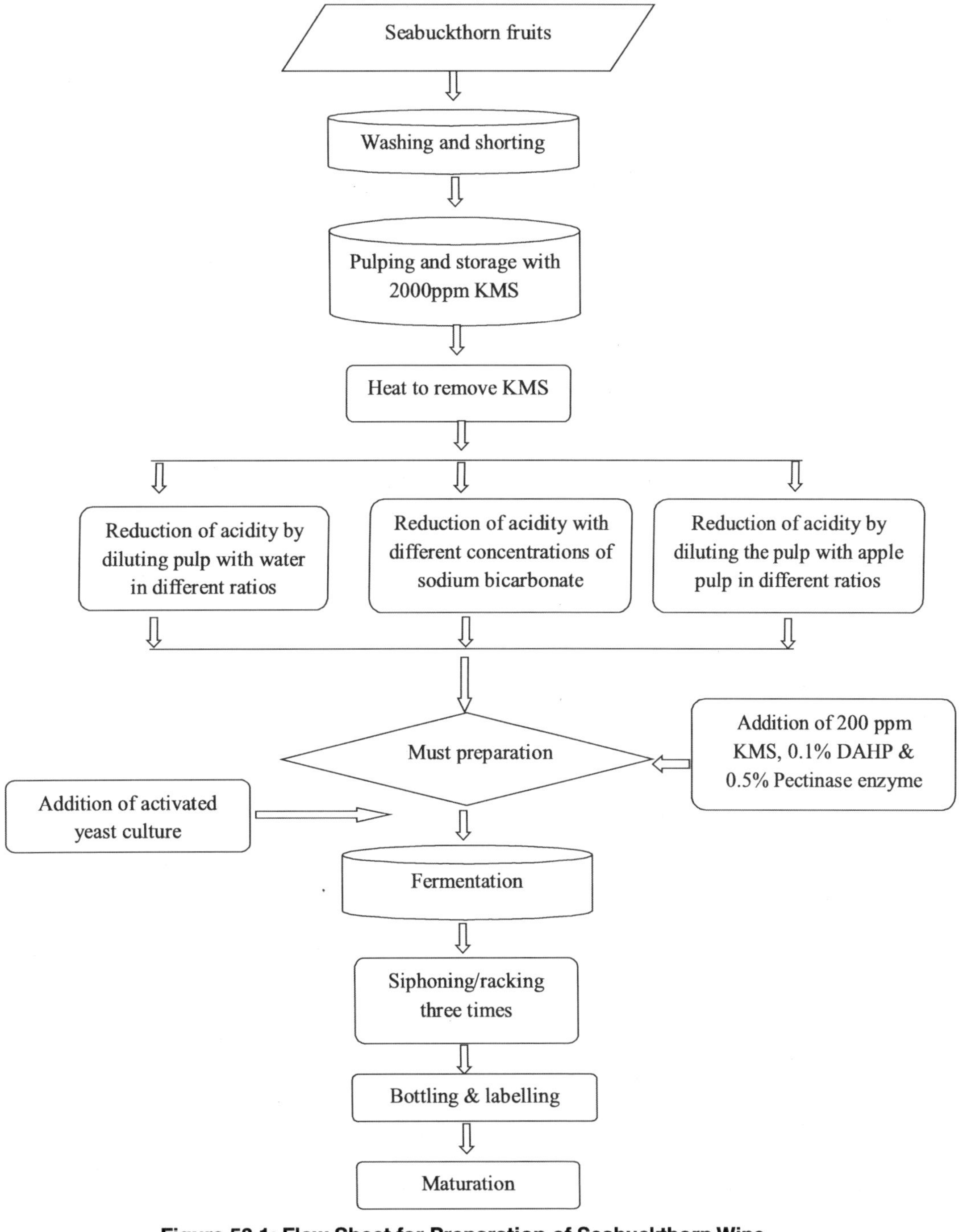

Figure 53.1: Flow Sheet for Preparation of Seabuckthorn Wine.

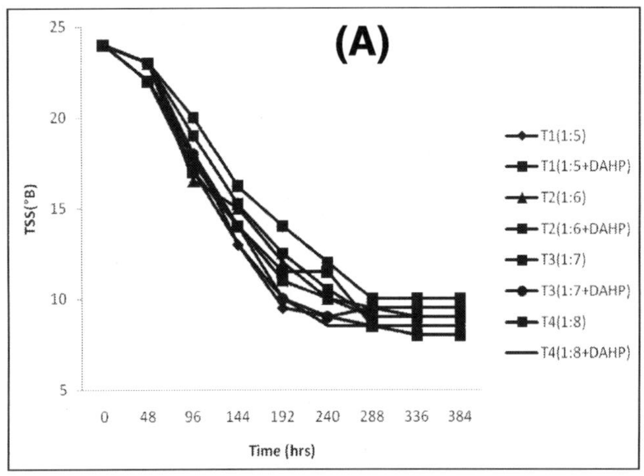

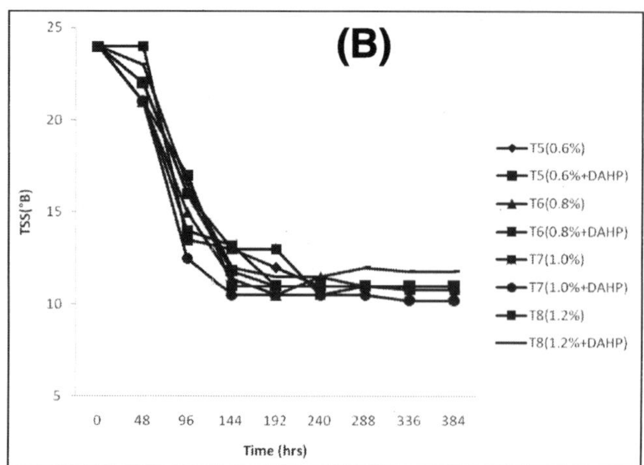

Figures 53.2A and B: Fermentation Behaviour of musts Prepared by different Dilutions of Pulp (A) and De-acidification using different Concentrations Sodium Bicarbonate (B).

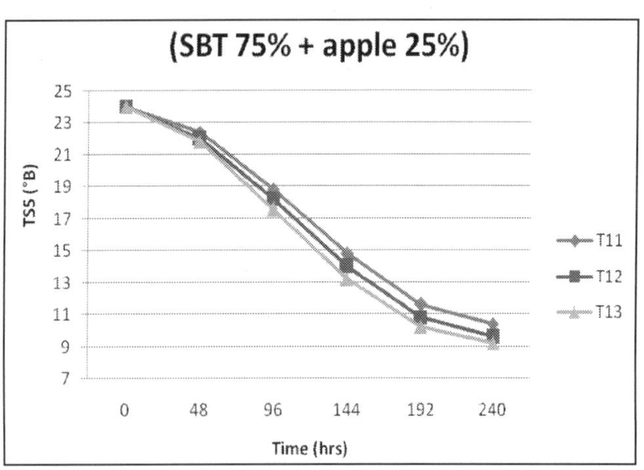

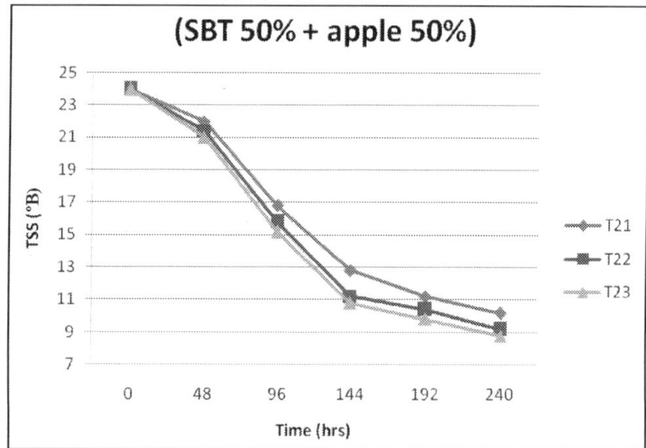

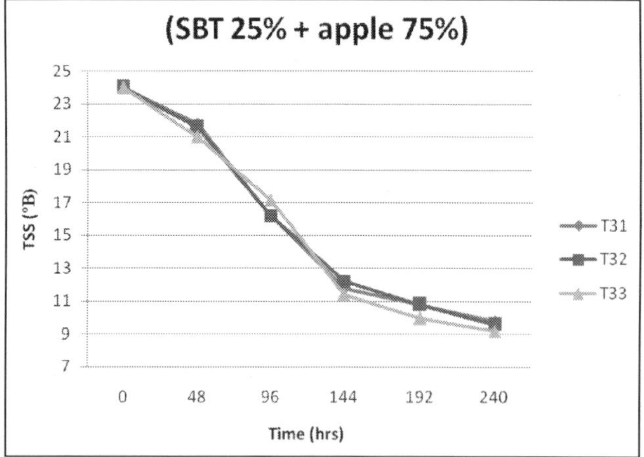

Figure 53.3: Fermentation Behaviour (TSS) of musts Prepared by different Proportions of Blending and Dilutions of Seabuckthorn Pulp with Apple Pulp.

T_{11} (1:2 dilution), T_{12} (1:3 dilution), T_{13} (1:4 dilution); T_{21} (1:2 dilution), T_{22} (1:3d ilution), T_{23} (1:4 dilution) and T_{31} (1:2 dilution), T_{32} (1:3 dilution), T_{33} (1:4 dilution).

than that of diluted musts and varied from 0.65 to 0.69 °B/24 hours. In overall view, the must having DAHP as nutrient source was found to have better RF than that without DAHP. A very distinct difference between the two types of must (one made by dilution and other by neutralizing with sodium bicarbonate) is that those made by dilution, a longer time *i.e.* 288 hrs than those made by sodium bicarbonate *i.e.* 240 hrs was required for fermentation. On the other hand, among various musts prepared by blending with apple pulp, highest reduction in TSS as well as highest rate of fermentation (RF= 1.52) was observed in must of SBT 50 per cent + Apple 50 per cent with 1:4 level of dilution (Figure 53.3). The Further, it was found that with increase in level of dilution the rate of fermentation increased. It can be attributed to the better fermentation conditions such as the initial aerobic conditions needed for the yeast growth and optimum nutrients and the pH provided by the blended pulp. The initial higher loss in TSS is attributed to the higher ferment ability of musts of different treatments because of availability of more sugar and lesser ethyl alcohol in the medium but later decrease is attributed to the increased inhibitory effect of ethyl alcohol concentration (Nishino *et al.* 1985, Sharma and Joshi, 1996). A better fermentation behaviour recorded in SBT-Apple blended musts could be due to better fermentation conditions such as the initial aerobic conditions needed for the yeast growth and optimum nutrients and the pH provided by the blended pulp. Although the TSS (°B) of all the musts was kept at 24°B, the quantity of ethanol differed significantly amongst the wines. The differences in composition of other nutrients in the respective musts (Kulkarni *et al.*, 1980) and the conditions of fermentation *i.e.* anaerobic conditions (Carroll, 1986) might have contributed to these differences; more- over blending and dilution might have exerted its positive effect on ethanol production.

Effect of Method of De-acidification on Physico-chemical Characteristics of Seabuckthorn Wine

Perusal of data pertaining to the effect of de-acidification on various physico-chemical characteristics of seabuckthorn wines, revealed significant differences for TSS among the wines of different treatments (Table 53.1). The TSS ranged from 7.0 to 10.2°B among various treatments. During experimentation, similar trend was also found for the total sugars which varied from 1.50 to 3.5 per cent. It is evident from the data (Table 53.1) that the titratable acidity ranged between 0.76 to 2.60 as per cent citric acid depending on the level of dilution employed or de-acidification by the use of $NaHCO_3$ and extent of blending with apple. De-acidification by blending and dilution was found to be more efficient

Table 53.1: Effect of Method of De-acidification on Physico-chemical Characteristics of Seabuckthorn Wine

Treatments	*TSS (°B)*	*Acidity (Per cent, CA)*	*Ascorbic Acid (mg/100 g)*	*Total Sugars (Per cent)*	*Ethanol Content (Per cent, v/v)*
De-acidification by dilution with water					
T_1 (1:5)- Control	9.0	1.16	75.0	3.35	9.5
T_2 (1:5 +DAHP)	9.8	1.16	80.0	3.30	10.2
T_3(1:6)	7.2	1.08	70.0	3.30	6.91
T_4 (1:6 +DAHP)	7.2	1.04	70.0	3.30	10.1
T_5(1:7)	7.0	1.12	50.0	2.54	10.0
T_6(1:7 +DAHP)	7.0	0.96	53.0	2.50	10.0
T_7(1:8)	7.0	1.12	40.0	1.58	9.92
T_8(1:8 +DAHP)	7.0	1.04	40.0	1.50	10.6
De-acidification by $NaHCO_3$					
T_9 (0.6 per cent $NaHCO_3$)	9.8	2.60	50.0	2.50	9.3
T_{10} (0.6 per cent +DAHP)	9.8	2.48	50.0	2.45	9.3
T_{11} (0.8 per cent)	10.0	2.48	40.0	2.38	8.0
T_{12} (0.8 per cent +DAHP)	10.0	2.48	48.0	2.35	9.3
T_{13} (1.0 per cent)	9.0	1.64	40.0	2.22	8.3
T_{14}(1.0 per cent +DAHP)	9.2	1.72	40.0	2.20	9.6
T_{15} (1.2 per cent)	9.4	1.08	40.0	2.10	9.8
T_{16} (1.2 per cent +DAHP)	10.2	1.80	40.0	1.90	10.2
De-acidification by blending with apple pulp					
T_{17} (S 75 per cent +A 25 per cent, 1:2)	10.0	1.2	122.5	3.5	9.8
T_{18} (S 75 per cent +A 25 per cent, 1:3)	9.6	1.1	110.2	3.3	10.0
T_{19} (S 75 per cent +A 25 per cent, 1:4)	9.2	1.03	105.3	3.0	10.3
T_{20} (S 50 per cent +A 50 per cent, 1:2)	10.2	1.05	112.0	3.6	9.8
T_{21} (S 50 per cent +A 50 per cent, 1:3)	9.2	0.97	108.6	3.4	10.2
T_{22} (S 50 per cent +A 50 per cent, 1:4)	8.8	0.95	102.9	3.2	10.8
T_{23} (S 25 per cent +A 75 per cent, 1:2)	9.8	0.90	93.2	3.6	10.2
T_{24} (S 25 per cent +A 75 per cent, 1:3)	9.6	0.81	89.1	3.5	10.3
T_{25} (S 25 per cent +A 75 per cent, 1:4)	9.2	0.76	83.4	3.2	10.5

in reducing acidity of seabuckthorn wine than by $NaHCO_3$. Further, seabuckthorn wines prepared with 0.1 per cent DAHP found to be lower in acidity than those without DAHP. Ascorbic acid in different wines ranged between 40 to 122.5 mg/100g. Seabuckthorn is rich source of ascorbic acid and has contributed its character in wine also. The wines prepared from apple blended seabuckthorn pulp were still found rich in ascorbic acid content that ranged between 83.4 mg/100g to 122.5 mg/100g, however with the increase in dilution level ascorbic acid content declined, but blending of seabuckthorn pulp with apple pulp increased the nutritional quality and the treatment T_6 (SBT 50 per cent + Apple 50 per cent) recorded 102 mg/100g of ascorbic acid at 1:4 dilution which is much higher than that of control T_1 (100 per cent SBT, 1:5). However non-significant changes in ascorbic acid were observed in treatments with $NaHCO_3$. The ethanol content of different treatments correlated with their respective rates of fermentation and was found in the range of 9.3 to 10.8 (per cent v/v) (Table 53.1). The effect of DAHP can be seen, with the increase in ethanol content within each treatment. Further, blending of seabuckthorn pulp with apple has resulted in better fermentation, hence produced appreciable amount of ethanol.

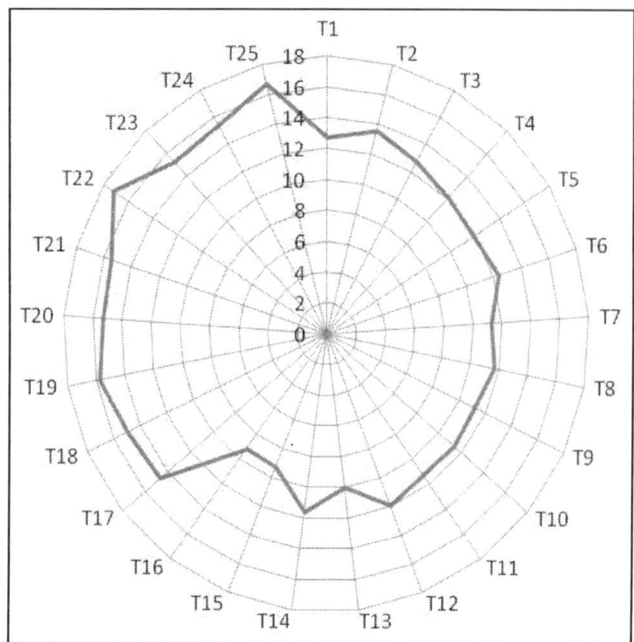

Figure 53.4: Total Sensory Score of different Wines Prepared.

Effect of Method of De-acidification of Sensory Characteristics of Seabuckthorn Wines

The results on sensory analysis of seabuckthorn wine of different treatments (Table 53.2) showed that there were significant differences for different quality attributes. Among different wines, the wine prepared from dilution was found superior to those prepared by $NaHCO_3$ mainly because of suitable reduction in acidity, however the chemical de-acidification process in the production of the wine, might have imparted a salty taste to the wines, thus reducing the acceptability. Further, blending of seabuckthorn pulp with apple had improved the sensory quality of the wine produced as evident from their higher sensory scores compared to other treatments. Based on the total sensory scores (20), wine of treatment T_{22} (SBT 50 per cent + apple 50 per cent with 1:4 dilution) was adjudged as the best product with a score of 17.3 which might be due to better score for colour, aroma, bitterness, astringency and overall acceptability (Figure 53.4). Further, as per the standards prescribed by Amerine *et al.* (1980), it was found to be *'superior'*, whereas rest of wines fall under *'standard'* category. It can be attributed to better taste and flavour of wines produced by blending the seabuckthorn pulp with apple pulp compared those from dilution with water and $NaHCO_3$.

Nutritive and Functional Value of Seabuckthorn-Apple Blended Wine

The effect of blending seabuckthorn pulp with apple pulp for wine making on nutritive and functional values was assessed by analyzing the wine samples for total phenolic contents and antioxidant activity. It is clear from Figure 53.5 that the blending has significantly increased the nutritional value of the blended wine. Seabuckthorn-apple (50:50) wine contained 32.78 mg/100g total phenolic contents and 102.6 mg/100 ascorbic acid compared to 20.36 mg/100g and 15.25 mg/100g (Total phenols) and 70.2 mg/100 g and 84.5 mg/100g (Ascorbic acid) present in seabuckthorn and

Plate 53.1: Seabuckthorn-Apple Blended (50:50) Wine.

Table 53.2: Effect of Methods of De-acidification on Sensory Quality Characteristics of different Seabuckthorn Wines

Treatments	*Color (2)*	*Appearance (2)*	*Aroma (4)*	*Volatile Acidity (2)*	*Titratable Acidity (2)*	*Sweetness (1)*	*Body (1)*	*Flavour (1)*	*Astringency (1)*	*Overall Acceptability (4)*	*Total score (20)*
De-acidification by dilution with water											
T_1 (1:5)	1.5	1.4	3.1	1.4	1.0	0.9	0.3	0.6	0.6	1.9	12.7
T_2 (1:5 +DAHP)	1.4	1.5	3.3	1.5	1.4	0.8	0.2	0.7	0.7	2.1	13.6
T_3(1:6)	1.6	1.7	3.5	0.8	1.0	0.9	0.2	0.6	0.6	1.8	12.7
T_4(1:6 +DAHP)	1.4	1.5	3.4	0.6	1.0	0.8	0.3	0.6	0.6	1.9	12.1
T_5(1:7)	1.2	1.2	3.2	1.0	1.5	0.8	0.5	0.3	0.5	1.7	11.9
T_6(1:7 +DAHP)	1.4	1.4	3.1	1.0	1.1	0.8	0.3	0.5	0.7	2.1	12.4
T_7(1:8)	1.2	1.2	2.9	1.3	0.8	0.6	0.5	0.7	0.6	1.5	11.3
T_8(1:8 +DAHP)	1.4	1.4	3.0	1.1	1.1	0.6	0.3	0.7	0.5	1.6	11.7
De-acidification by $NaHCO_3$											
T_9 (0.6 per cent $NaHCO_3$)	1.6	1.7	2.5	1.3	1.2	0.5	0.5	0.5	0.5	0.9	11.2
T_{10} (0.6 per cent +DAHP)	1.6	1.7	2.6	1.4	1.2	0.5	0.4	0.5	0.5	1	11.4
T_{11} (0.8 per cent)	1.6	1.7	2.5	1.2	1.2	0.5	0.7	0.5	0.5	0.9	11.3
T_{12} (0.8 per cent +DAHP)	1.6	1.7	2.6	1.2	1.2	1.0	0.7	0.5	0.5	0.9	11.9
T_{13} (1.0 per cent)	1.3	1.3	2.5	1.0	1.0	0.4	0.7	0.5	0.5	0.8	10.0
T_{14}(1.0 per cent +DAHP)	1.3	1.7	2.7	1.3	1.2	0.6	0.7	0.6	0.5	1.0	11.6
T_{15} (1.2 per cent)	1.3	1.3	2.3	0.9	0.9	0.3	0.7	0.4	0.4	0.8	9.3
T_{16} (1.2 per cent +DAHP)	1.3	1.3	2.3	0.9	0.9	0.3	0.7	0.4	0.4	0.7	9.2
De-acidification by blending with apple pulp											
T_{17} (S 75 per cent +A 25 per cent, 1:2)	1.4	1.7	3.1	1.4	1.3	0.8	0.8	1.5	0.6	1.4	14.7
T_{18} (S 75 per cent +A 25 per cent, 1:3)	1.4	1.6	3.2	1.4	1.4	0.8	0.8	1.6	0.6	1.5	15.0
T_{19} (S 75 per cent +A 25 per cent, 1:4)	1.8	1.8	3.5	1.3	1.4	0.8	0.7	1.5	0.6	1.7	15.8
T_{20} (S 50 per cent +A 50 per cent, 1:2)	1.5	1.7	3.4	1.4	1.3	0.8	0.8	1.6	0.7	1.5	15.3
T_{21} (S 50 per cent +A 50 per cent, 1:3)	1.6	1.8	3.4	1.3	1.3	0.8	0.7	1.5	0.7	1.7	15.5
T_{22} (S 50 per cent +A 50 per cent, 1:4)	1.9	1.9	3.8	1.4	1.5	0.8	0.8	1.7	0.8	1.9	17.3
T_{23} (S 25 per cent +A 75 per cent, 1:2)	1.5	1.5	3.5	1.4	1.5	0.6	0.9	1.5	0.6	1.5	15.2
T_{24} (S 25 per cent +A 75 per cent, 1:3)	1.6	1.7	3.6	1.3	1.3	0.7	0.7	1.6	0.6	1.6	15.4
T_{25} (S 25 per cent +A 75 per cent, 1:4)	1.8	1.9	3.7	1.4	1.5	0.8	0.7	1.7	0.7	1.8	16.7

apple wine, respectively. Similarly, the apple blended seabuckthorn wine (Plate 53.1) had shown highest antioxidant potential (66.40 per cent) than those of seabuckthorn (56.75 per cent) and apple wine (50.0 per cent).

CONCLUSION

From the present study, it was concluded that among dilution and de-acidification by sodium carbonate treatment, the wine prepared from dilution was observed superior. Addition of DAHP treatment improved both rate of fermentation (RF) as well as ethanol content. But to overcome the drawback of reduction in nutritive value of the seabuckthorn wine with such a high level of dilution, blending seabuckthorn pulp with apple pulp in the ratio of (50: 50) seems to be a best alternative for the development of nutrient rich palatable wine. Even on the basis of physico-chemical and sensory attributes the wines prepared from blended pulp were found better than those of simple dilution with water and de-acidification with $NaHCO_3$. Further, when compared with apple wine and seabuckthorn

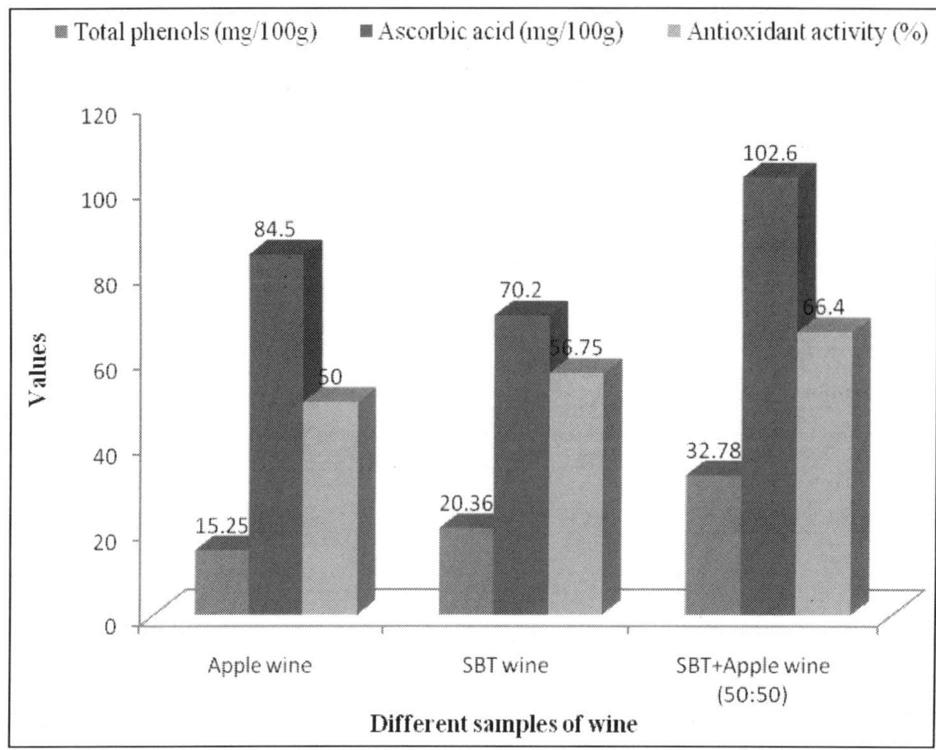

Figure 53.5: Nutritive and Functional Values of Seabuckthorn Based Wine.

wine alone, the seabuckthorn-apple blended wine had shown higher antioxidant potential indicating its better nutritional and functional properties.

REFERENCES

1. Amerine, M.A., Kunkee, K.E., Ough, C.S., Singleton, V.L. and Webb, A.D. 1980. *Technology of Wine Making*, AVI Publ. Co. Inc. Westport, CT, 794p.
2. AOAC 1980. Association of Official Analytical Chemists: Official Methods of Analysis, Hortwitz, W. (ed.), 13th edn, Washington DC, 1015p.
3. Caputi, A., Ueda, M. and Brown, J. 1968. Spectrophtometric determination of ethanol in wine. *American Journal of Enology Viticulture* 19: 160-165.
4. Carroll, D.E.C. 1986. Effects of carbonic maceration on chemical, physical and sensory characteristics of Muscadine wines. *Journal of Food Science* 51: 1195-1196.
5. Dwivedi, S.K., Attrey, D.P. and Chaurasia, O.P. 2004. Studies on preparation and preservation of seabuckthorn squash beverage. *Indian Journal of Horticulture* 61 (1): 78-80.
6. Ethiraj, S. and Suresh, E.R. 1993. Fruit wines. In: *Advances in Horticulture*,Chadha K.L., Pareek O.P. (Eds.), Malhotra Pub. House, New Delhi, India, 4: 1957-1966.
7. Joshi, V.K. and Pandey, S.K. 1999. Biotechnology: Food fermentation (Microbiology, Biochemistry and Technology). Vol. I and II., Educational Publisher and Distributors, Ernakulum, New Delhi.
8. Joshi, V.K. 1997. *Fruit Wines*. 2nd edn, Directorate of Extension Education, Dr YS Parmar UHF, Nauni-Solan, India.
9. Joshi, V.K. 2006. *Sensory science: Principles and Applications in Evaluation of Food*. Agro-Tech Publishers, Udaipur, p. 527.
10. Joshi, V.K., Shah, P.K. and Kumar, K. 2005. Evaluation of different peach cultivars for wine preparation. *Journal of Food Science and Technology* 42 (1): 83-89.
11. Joshi, V.K., Sharma, R., Sharma, S. and Ghanshyam, A. 2011. Effect of dilution and de-acidification on physico-chemical and sensory quality of seabuckthorn wine. *Journal of Hill Agriculture* 2 (1): 47-53.
12. Joshi, V.K., Sharma, S. and Mutum, P.D. 2009. Influence of different yeast strains on fermentation behaviour, physic-chemical and sensory qualities of plum wine. *Natural Product Radiance* 8 (4): 445-451.
13. Kaushal, M., Sharma, P.C., Sharma, R. and Joshi, V.K. 2011. Preparation and evaluation of nutrient rich value added products from seabuckthorn. *Indian Food Packer*, Jan-Feb, pp. 36-41.
14. Kulkarni, J.H., Singh, H. and Chandha, K.L. 1980. Preliminary screening of mango varieties for wine making. *Journal of Food Science and Technology* 17 (4): 218-219.

15. Negi, B. and Dey, G. 2009. Comparative analysis of total phenolic content in seabuckthorn wine and other selected fruit wines. *World Academy of Science, Engineering and Technology* 54: 99-102.

16. Nishino, H., Miyazaki, S. and Tohji, K. 1985. Effect of osmotic pressure on growth rate and fermentation activity of wine yeasts. *American Journal of Enology and Viticulture* 36: 170-174.

17. Panse, V.G. and Sukhatme, P.V. 1967. *Statistical Methods for Agricultural Workers*. ICAR, New Delhi.

18. Ranganna, S. 1986. *Handbook of Analysis and Quality Control for Fruit and Vegetable Products*. Tata McGraw Hill Publ Ltd., New Delhi, 1112p.

19. Sandhu, D.K. and Joshi, V.K. 1995. Technology, quality and scope of fruit wines especially apple beverages. *Indian Food Industry* 11 (1): 24-34.

20. Sharma, P.C., Kaushal, M., Sharma, R. and Kaushal, B.B.L. 2004. Utilization of seabuckthorn (*Hippophae* sp.) for the preparation of some value added products. In: *Souvenir and Book of Abstracts, National Seminar on Cultivation, Harvesting and Scientific Exploitation of Seabuckthorn*, August 26-27, 2004, FRL (DRDO), Leh, India.

21. Sharma, S., Joshi, V.K. and Abrol, G. 2009. An overview on strawberry (*Fragaria* × *Ananassa* (Weston) Duchesne ex Rozier) wine production technology, consumption, maturation and quality evaluation. *Natural Product Radiance* 8 (4): 356-365.

22. Sharma, S.K. and Joshi, V.K. 1996. Optimization of some parameters of secondary fermentation for production of sparkling wine. *Indian Journal of Experimental Biology* 34 (3): 235-236.

23. Singh, V. and Singh, R.K. 2004. Morpho-biochemical variations in seabuckthorn (*Hippophae* L.) populations growing in Lahaul valley, dry temperate Himalayas. *Indian Forester* 130 (6): 663.

Chapter 54

Extraction and Purification of Antifreeze Proteins in *Hippophae rhamnoides* Seedling, Berry and Leaf

Bhavana Sharma and Renu Deswal*

Molecular Plant Physiology and Proteomics Laboratory, Department of Botany,
University of Delhi, Delhi – 110 007, India
(*E-mail: rdeswal@botany.du.ac.in)

SUMMARY

Seabuckthorn (*Hippophae rhamnoides*) belonging to family Elaeagnaceae is a well-known high-altitude shrub (2500-4000 m). It is a cold desert plant having enormous nutraceutical, pharmaceutical and environmental benefits. It is a storehouse of many cold tolerant proteins including antifreeze proteins (AFPs). AFPs are ice binding proteins which prevent the addition of water molecules to growing crystal planes by adsorbing to these in a non-colligative manner. In the present study, our first aim was to optimize the protein extraction procedures as berry and leaf contain high phenolics. A modified phenol extraction procedure allowed retrieval of phenolic less, clear protein extract. The antifreeze activity was measured using Nanolitre osmometer coupled with a phase contrast microscope and Sucrose sandwich splat assay. Many AFPs have been shown to perform dual function by protecting the overwintering plant against low temperature and helping in defence mechanism. One such AFP, Chitinase was purified from the extracellular fluid of *H. rhamnoides* seedlings. A search for other chitinases in leaf and berry resulted in purification of 31 and 34 kDa in leaf and 10, 24 and 46 kDa chitinase in berry. Efforts to validate and analyse the biotechnological potential of these AFPs is underway.

***Keywords**: Seabuckthorn, Antifreeze proteins, Nanolitre osmometer, Ice crystal morphology and Chitinase.*

INTRODUCTION

Hippophae rhamnoides (seabuckthorn) is a deciduous, cold-hardy Himalayan shrub belonging to the family Elaeagnaceae and is native to Eurasia. Seabuckthorn leaves possess high protein (15 per cent), and can be used as source of nutritious food. Berries are rich source of a various antioxidants (vitamins C and E); several carotenoids, including beta-carotene (pro-vitamin A); flavonoids *etc.* The berries can be used to make juice, jams, liquors, jellies and sauces. Various reports have suggested that seabuckthorn oil (from the pulp and seeds) is a good source of essential fatty acids, which is important for maintenance of a healthy skin and can be used clinically for treating a wide variety of skin damage, including burns, eczema, bedsores and radiation injury. Bioactive oil obtained from the leaves and berries can also be used against diseases of the stomach and intestine. The potential of seabuckthorn oils for dermatological applications is extremely promising. Multiple uses in therapeutics, nutraceuticals, cosmetics, and food industries make it a "wonder plant" or "Sanjeevani". Besides possessing medicinal and nutraceutical properties, it can also withstand various abiotic stresses. Seabuckthorn has a capability to withstand temperatures ranging from -43 °C to 40 °C and is considered salinity, UV, frost and drought tolerant. Therefore, a detailed knowledge about its regulatory pathway using genomic and proteomic approaches would help in understanding the tolerance mechanism of this plant and may provide potential for development of cold tolerant crop varieties and its biotechnological applications (Dwivedi *et al.*, 2009; Singh *et al.*, 2003; Singh *et al.*, 2006). It is well known that overwintering plants

Seabuckthorn (Hippophae L.): A Multipurpose Wonder Plant, Vol. 5: Advanced Technologies for Cultivation, Processing, Health Protection and Environmental Conservation (Virendra Singh, *Editor-in-Chief*, 2018), p. 413–422, Astral International Pvt. Ltd., New Delhi, India.

accumulate various proteins involved in signalling and defence in their extracellular spaces like cell wall-modifying proteins which identify the changes at the membrane level and initiate the regulatory pathway to release antifreeze proteins (AFPs) which interact with ice and protect the cells against fatal intracellular and intercellular ice formation and pathogenesis-related (PR) proteins which protect the plants against disease causing organisms (Yu and Griffith, 2001; Griffith *et al.*, 1992). AFPs has gained significant attention due to their potential biotechnological applications in various sectors like food technology by improving the texture and shelf life of frozen food and dairy products and by reducing leakage of nutrients from them during low temperature storage, improving cold tolerance in plants by making transgenics or directly by vacuum infiltration, as cryoprotectant for cryopreservation of cells, tissues, cell lines and organs for transplantation (Arav *et al.*, 1993; Wisniewski *et al.*, 1999), cryosurgery for tumour and cancer cells destruction (Yu and Griffith, 2001), to induce cold tolerance in freeze susceptible organisms, for aquaculture of cold tolerant fishes (Grandum and Nakagomi, 1999; Koushafar and Pham, 1997; Fletcher, 1999; Venketesh and Dayananda, 2008).

Reports have suggested the presence of AFPs in the apoplast of seabuckthorn seedlings (Gupta and Deswal, 2012). Efforts were made to purify these targets in the wild population (Leaf and Berry) of *H. rhamnoides* growing in abundance in the Himalayan regions of India. In the present study, the protein extraction procedure was optimized to obtain good quality protein from leaves and berries of seabuckthorn which contain high phenolics. To analyze the presence of AFPs in leaf and berries their antifreeze activity was determined. Recently, dual functioning AFPs also acting as PR proteins (Class I Chitinases HrCHT 1a of 31 and HrCHT 1b 34 kDa) was purified from the apoplast of *H. rhamnoides* seedling. It shows both antifreeze and hydrolytic activity (Gupta and Deswal, 2014). An attempt was made to purify chitinases in leaf and berries for their potential commercial utilization.

MATERIALS AND METHODS

Biological Material

Leaves and berries of *Hippophae rhamnoides* used in the study were collected from the river side of Rangrik village, Lahaul and Spiti valley, Himachal Pradesh, India. The tissue was collected, washed with double distilled water, frozen in liquid nitrogen and transferred to the laboratory for storage at -80 °C. Seedlings were obtained from seabuckthorn seeds isolated from berries and germinated in laboratory at 24 °C with 270 mmol/ m^2/s, 16 hr light/8 hr dark for 20 days.

Collection of Material at Room Temperature (RT) and in Liquid Nitrogen

To check the stability of antifreeze protein over a range of temperature and to determine whether it is mandatory to carry liquid nitrogen for bringing the research material, berries were collected in two sets. One set was frozen and brought to laboratory in liquid nitrogen and stored at -80 °C while the other set was not frozen in liquid nitrogen and was brought to laboratory at RT, kept at 4 °C for 2 days and then stored at -20 °C. The protein profile in both the samples was checked using 1 DGE and 2DGE.

Heat Treatment to Check Heat Stable Proteins in Berry

To reaffirm the presence of heat stable proteins in berry, proteins were extracted (25 gm tissue) in Tris-glycerol buffer containing 1M Tris pH-7.4, 20 per cent glycerol, 2-5 per cent 1 M Phenylmethylsulfonylfluoride (PMSF) and Polyvinylpolypyrolidine (PVPP). The protein pellet obtained after acetone precipitation was air-dried and solubilised in 50 mM Tris pH 7.4. Protein (3 mg) was loaded on the G-200 gel filtration column and fractions of 5 ml were eluted at the flow rate (0.5 ml/min) in well labelled tubes and their absorbance was taken at 280 nm. Eluted protein fractions were aliquoted into microfuge tubes and boiled for 10 min at 100 °C. Each tube was centrifuged at 10,000 g, 4°C for 10 min to pellet down the heat labile proteins and the supernatant was used to assay the heat stability of AFPs by analysing their antifreeze activity (splat assay) and SDS- PAGE protein profile.

Protein Extraction

For obtaining protein profile without phenolics, phenol extraction followed by methanolic ammonium acetate precipitation was used. The tissue was crushed to powder in Tris pH-8.8 buffered phenol and extraction media (0.1 M Tris-HCl pH 8.8, 10 mM EDTA, 0.4 per cent 2-mercaptoethanol, 0.9 M sucrose). The slurry was transferred to a falcon tube and was agitated for 30 min at 4° C followed by centrifugation for 10 min at 5000 g, 4° C. Phenol phase at the top was removed and transferred into fresh falcon tube. The phenol extracted proteins were precipitated by adding 5 volumes of 0.1 M ammonium acetate in 100 per cent methanol (pre-chilled to -80° C) to the phenol phase and was incubated at -80° C overnight. The precipitated proteins were collected as pellet by centrifugation (30 min, 4000xg, 4° C). The pellet was washed 2x with chilled 0.1 M ammonium acetate in methanol and then 2x with ice-cold 80 per cent acetone containing 10 mM DTT. The final pellet was air dried and dissolved in 50 mM Tris pH 7.4. Total proteins from 20 days old seedlings were isolated in 20 mM ammonium bicarbonate (ABC) buffer pH 8.0.

Resolving Proteins on 1-DGE and 2-DGE

For 1DGE proteins from both the sets of berries (tissue frozen in liquid nitrogen and tissue brought at RT) were dissolved in 1X sample buffer containing β-mercaptoethanol, boiled for 3 minutes and then resolved on 12 per cent SDS PAGE. For 2DGE, 300 μg of proteins from both the sets was loaded on a 13 cm, nonlinear IPG strip, pH 3–10 followed by their overnight rehydration at 20°C. IEF was performed for 8 hours. The strips were equilibrated using equilibration buffer containing 6 M urea, 50 mM Tris pH 8.8, 30 per cent glycerol, 2 per cent SDS, and 1 per cent DTT at the first step for 20 mins and then with 2.5 per cent iodoacetamide at the second step to allow alkylation of proteins. Proteins were then resolved on second dimension 12 per cent SDS-PAGE. Gels were Colloidal CBB stained and their pI and molecular weight were analysed using Image Master 2D Platinum software ver. 6.0 (GE Healthcare).

Detection of Antifreeze Activity using Sucrose Sandwich Splat Assay and Nanolitre Osmometer

Sucrose Sandwich Splat Assay

Splat assay was performed as described in Gupta and Deswal, 2012. In brief, proteins were dissolved in 30 per cent sucrose in 20 mM ABC and 2 μl of this solution was sandwiched between two round cover slips. The sandwich' was snap-frozen in heptane kept at - 80 °C and maintained in a chamber containing heptane at -6 °C for an hour. Ice crystals were allowed to anneal, the growth of a single ice crystal was observed and images were photographed using a phase-contrast microscope fitted with a camera. For the quantification of the IRI activity, Image J software was used to measure the diameters of ice crystals (Smallwood *et al.*, 1999).

Detection Antifreeze Activity

Nanoliter osmometer is used to measure TH activity and to observe ice crystal morphology. In brief, nanoliter volume of protein sample was loaded in the well of the disc after applying mineral oil in it. The sample was frozen rapidly at -20 °C to form multiple ice crystals and the temperature was then increased to allow melting until a single ice crystal is left in the well. The ttemperature was further decreased and the change in the morphology of the ice crystal was observed. In the presence of AFPs the ice crystal are hexagonal whereas these are disc shaped in their absence.

Chitin Affinity Chromatography

Colloidal chitin was prepared using bleached chitin and the chitin beads dissolved in 20 mM ABC pH 8.0 were added to the protein solution until the optical density of the mixture reached 0.8 - 0.9 at 590 nm. The mixture was left for overnight binding at 4 °C with constant slow stirring. It was then centrifuged at 12000 rpm for 20 minutes to settle the chitin beads and the supernatant was collected as flow through. The beads were washed using 20 mM ABC pH 8.0 and 10 mM acetic acid to remove the non specific binding. Chitinase elution was done using 20 mM and 250 mM acetic acid pH 3.0 for 30 min at ice. The elutes were neutralized immediately using Tris pH-9.0 and the purified chitinase was resolved on a 12 per cent SDS-PAGE. (Hon *et al.*, 1995; Kang *et al.*, 1999).

Chitinase Hydrolytic Activity Assay

Chitinase activity was determined colorimetrically by measuring the release of N-acetylglucosamine (GlcNAc) from the substrate colloidal chitin. The reaction mixture containing 100 μL enzyme, 200 μL of 2 per cent colloidal chitin in (50 mM sodium acetate buffer pH 5.0) and 200 μl buffer was incubated at 37°C for 30 min with constant agitation. The reaction was then terminated by adding 1ml Potassium ferricyanide solution (0.5 g potassium ferricyanide in 1 liter of 0.5 M sodium carbonate) and boiling the mixture for 15 minutes. The reducing end group produced was determined colorimetrically by measuring the decrease in absorbance at 410 nm (Imoto and Yagishita, 1971).

RESULTS

The wild parts of seabuckthorn are rich in phenolics which hinder in extracting protein, various extraction procedures (Tris-Glycerol, TCA with and without extraction buffer and Phenol) were compared for extracting the proteins. Phenol extraction followed by ammonium acetate precipitation was found to be the best method to obtain high protein yield and better protein profile without phenolic contamination. Colloidal CBB stained SDS-PAGE gel of total protein extract showed total of in 27 polypeptides in leaf and 22 polypeptides in berry (Figure 54.1). Rubisco being a photosynthetic protein was present only in leaf and was absent in berry. Polypeptides of 66, 10 kDa in berry while Rubisco large and small subunit in leaf (marked by) were found to be unique in each sample. Polypeptides of 46, 39, 32, 28 and 24 in berry while 20 and 14 kDa in leaf show higher intensity (marked by +) in respective samples.

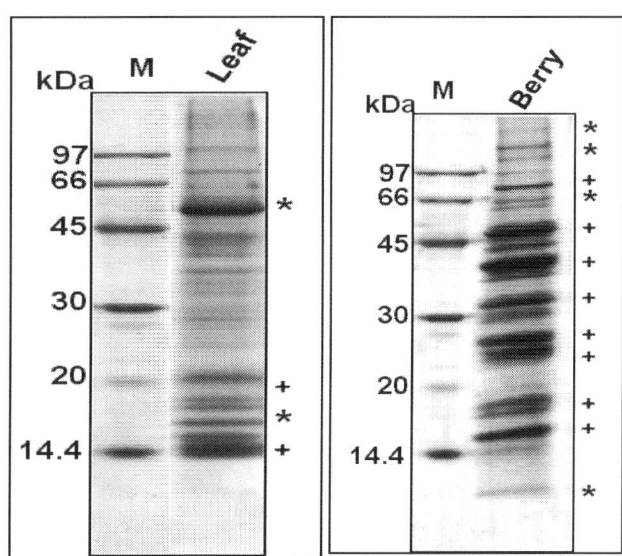

Figure 54.1: Colloidal CBB Stained 1 DGE (12 per cent) Showing Protein Profile of *H. rhamnoides* Leaf and Berry.

As presence of antifreeze proteins have been shown in apoplast of *H. rhamnoides* seedlings, their presence was also examined in the leaf and berry. Antifreeze proteins (AFPs) are a class of polypeptides which bind to interact with ice crystals and modify their growth and morphology owing to the presence of two unique properties, thermal hysteresis (TH) and ice recrystallisation inhibition (IRI) (Antikainen and Griffith, 1997). By virtue of its TH activity AFPs protect the plants by depressing the freezing point but once ice is formed another property, IRI comes into action (Griffith and Ewart 1995; Zachariassen and Kristiansen, 2000; Raymond *et al.*, 1989). For analyzing the antifreeze activity Sucrose sandwich splat assay and nanoliter osmometer coupled with phase contrast microscopy were used. Ice crystal morphology provides a qualitative analysis of AFP activity (Atici and Nalbantoglu, 2003). In the presence of AFPs, ice crystals are hexagonal or needle shaped while they are circular disc shaped in the absence of AFPs. Therefore, by monitoring the shapes of ice crystals, presence or absence of AFPs can be detected.

In leaf and berry ice crystals grow as flower shaped indicating the presence of antifreeze activity in both (Figure 54.2B). To further confirm the presence of antifreeze activity splat assay which observe the ice recrystallisation inhibition activity of the AFPs was used. Quantification of IRI activity showed 63.45 per cent inhibition in leaf while 59 per cent in berry as compared to control (Figure 54.2A) thereby confirming the presence of AFPs in both leaf and berry. AFPs further control the size, shape and aggregation of smaller ice crystals which could cause physical damage to the tissue.

Usually the material collected from wild is brought to laboratory by freezing the tissue in liquid nitrogen, but carrying liquid nitrogen to the hilly areas is very cumbersome and inconvenient therefore an experiment was planned to check whether it is mandatory to bring the tissue frozen in liquid nitrogen and how it affects the stability of antifreeze proteins. The material was collected in two sets and was brought to laboratory either by freezing it in liquid nitrogen or directly at RT. The proteins were extracted from both the sets and were resolved on 1-DGE and 2-DGE. It was observed that the proteins in both the conditions were stable. Although differences in the intensities of few polypeptides of 60, 59, 44, 43, 32 and 26, 22, 16 and 14 kDa (marked as and + in respective samples) was observed but these were not very significant on 1DGE (Figure 54.3a). Noteworthy differences in spot intensities were observed when the proteins from different sets were resolved on 2-DGE (Figures 54.3c1; 54.3c2). The proteins from the tissue frozen in liquid nitrogen were resolved on 2DGE and were analysed using Image Master 2-D Platinum software, a total of 125 spots were observed while only 95 spots were detected in the proteins extracted from tissue without liquid nitrogen. Decrease in the intensity of various spots was observed in the 2-DGE profiles of tissue frozen without liquid nitrogen showing down regulation of these proteins and suggesting their labile nature. However, analyses of the antifreeze activity showed the presence of hexagon shaped ice crystals in both the sets suggesting the stability of antifreeze proteins in both the samples (Figures 54.3b1; 54.3b2). Heat stable antifreeze proteins had been identified from a perennial rye grass *Lolium perenne* (Sidebottom *et al.*, 2000; Pudney *et al*, 2003) and have been purified from *Ammopiptanthus mongolicus* leaves (Wei *et al.*, 1999), which could survive -20 ° C in winter to 40 °C in summer and bark of *Prunus persica* (Arora and Wisniewski,

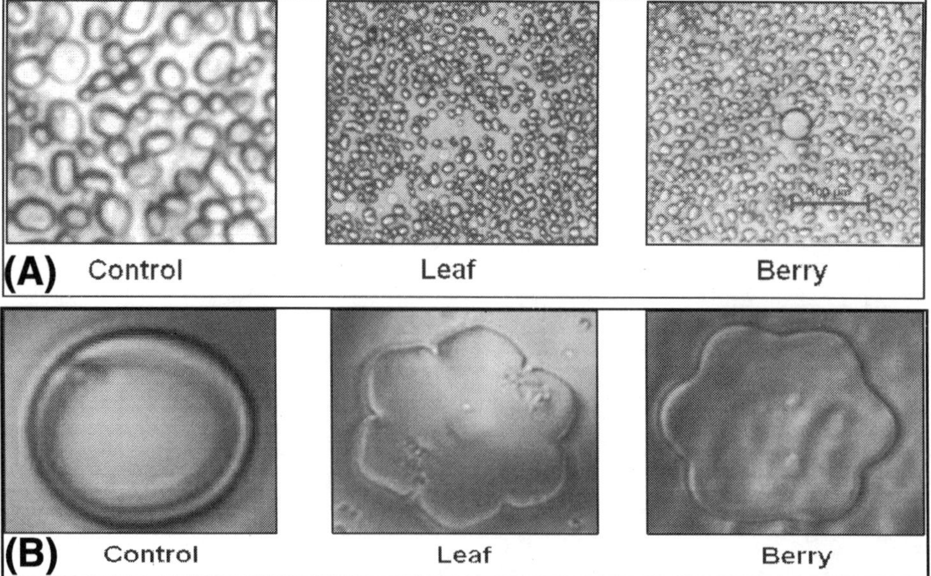

Figure 54.2: (A) Ice recrystallization inhibition assay observed with sucrose sandwich splat assay for analysis of antifreeze activity in seabuckthorn leaf and berry crude protein sample. Magnification scale bar in splat assay represents 100 µm. (B) Ice crystal morphologies of control, leaf and berry as observed by phase contrast microscopy and nanoliter osmometer.

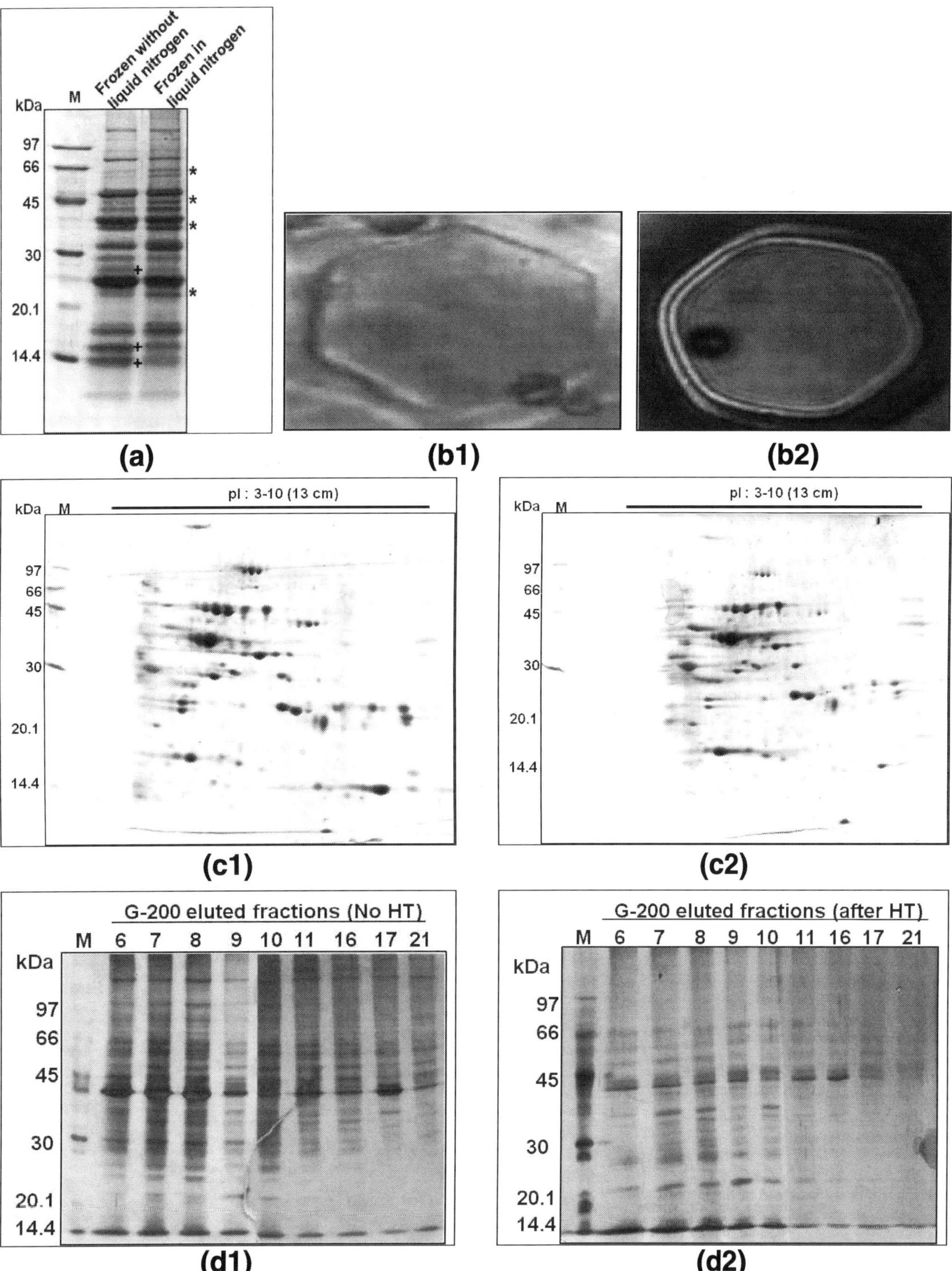

Figure 54.3: Colloidal CBB Stained 1DGE and 2 DGE (12 per cent) of *H. rhamnoides* Berry Brought to the Laboratory c1) with and c2) without freezing in liquid nitrogen. Ice crystal morphologies of berry samples b1) frozen with and b2) without liquid nitrogen using nanoliter osmometer. Silver stained 1DGE (12 per cent) of G-200 eluted Berry protein fractions d1) before and d2) after heat treatment.

1994). Seabuckthorn being capable of withstanding temperature ranging from -43 °C to 40 °C shows frost and drought tolerance. To confirm the presence of heat stable AFPs in berry, the proteins were separated on G-200 chromatography column, eluted fractions showing high IRI activity were further resolved on 1DGE to obtain the protein profiles and IRI activity was calculated before and after heat treatment (Figures 54.3d1; 54.3d2). Many heat stable antifreeze proteins were observed in the range of 66-14 kDa with polypeptides of size 66, 54, 50, 45, 41, 36, 30, 26, 20 and 16 kDa retaining upto 35 per cent IRI activity as compared to 67 per cent IRI activity without heat treatment. Similar proteins were also observed in the same range on 2-DGE and spots of size 45, 41, 36, 30, 26, 20 and 16 kDa were present, suggesting that the data is in accordance with the 2DGE profiles of the material collected at RT and confirmed the presence of various heat stable AFPs in *H. rhamnoides* berry. Identification and further characterisation of these targets could provide a better picture about the heat labile and heat stable AFPs. However, the experiment needs to be repeated again for further conclusion of the results.

Chitinases are glycosyl hydrolases which catalyze the hydrolysis of chitin, β-1,4-linked polymer of *N*-acetyl-D-glucosamine (GlcNAc), present in a wide range of organisms like bacteria, fungi, viruses, yeasts, plants, arthropods and humans. The size ranges from 20 to 90 kDa. These are pathogenesis-related (PR) proteins which help the plants in defense response against various phytopathogens and also in response to environmental stresses like salinity, low temperature and drought. Chitinases are tissue specific and have been found to be constitutively present in various parts of the plants like stems, seeds, flowers, fruit and tubers. It has been found that AFPs are homologous to PR proteins including β-1,3-glucanases, chitinases, thaumatin-like proteins (Hon *et al.*, 1995; Huang *et al.*, 2001) and polygalacturonase inhibitor proteins (Meyer *et al.*, 1999; Worrall *et al.*, 1998). In winter rye, AFPs with antifungal, hydrolytic activities and antifreeze activity have been found suggesting their dual role in providing

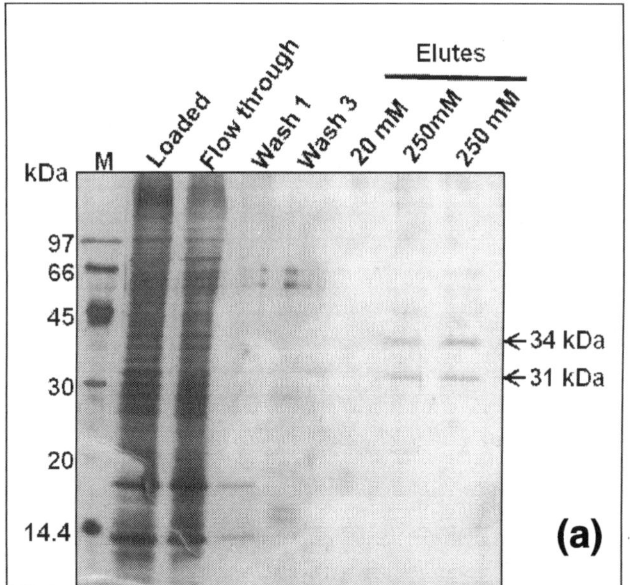

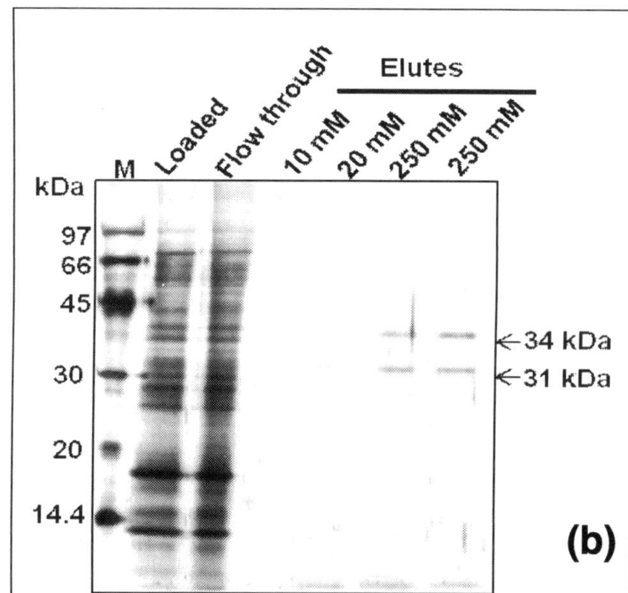

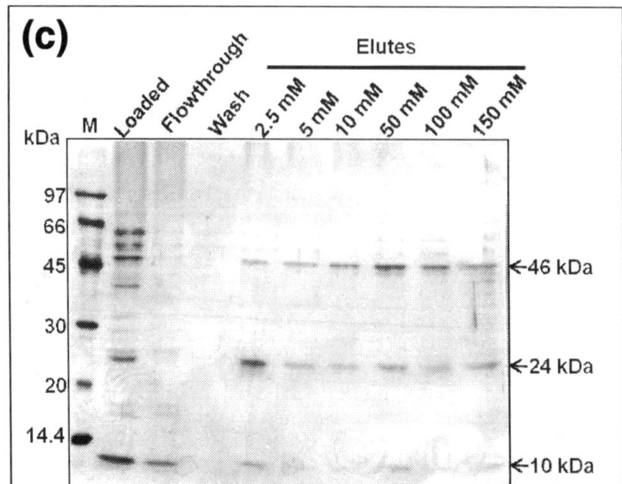

Figure 54.4: Silver Stained 1DGE (12 per cent) Showing Partially Purified Chitinases in *H. rhamnoides* a) Seedling (31, 34 kDa), b) Leaf (31, 34 kDa) and c) Berry (10, 24, 46 kDa).

freeze tolerance and disease resistance in overwintering plants. Chitinases possess wide range of applications (Nagpure *et al.*, 2014) in addition to providing defence against fungus (Dahiya *et al.*, 2006). Chitinases could be utilized commercially as a potential biocontrol agent or biopesticide/fungicide/insecticide against various plant-pathogens, in raising fungal resistant transgenics (Baranski *et al.*, 2008; Distefano *et al.*, 2008; Shah *et al.*, 2009; Xiao-Jing *et al.*, 2005) and for the production of various biologically active substances (Rattanakit *et al.*, 2007) Reports have also suggested their role in human diseases including Gaucher disease (GD), asthma and cancer by acting as a diagnostic marker for monitoring disease progression and a target for various therapeutic treatments of human healthcare. Acid mammalian chitinase (AMCase): a 50 kDa protein has been identified as a novel therapeutic target for asthma (Boot *et al.*, 2001; Elias *et al.*, 2005; Zhu *et al.*, 2004). Similarly, YKL-40 (chitinase III like protein) is a biomarker linked with type 2 diabetes, cancer and cardio-metabolic disorders (Ling and Recklies, 2004; Johansen, 2006). Reports have indicated that biological hydrolysis of chitin to N-acetyl-glucosamine (GlcNAc) using purified chitinase could be (Rattanakit *et al.*, 2003) utilized for producing bioactive compounds, biofuel production, in cosmetics, as nutritional and food supplement (Chen *et al.*, 2010).

Chitinases of 31 and 34 kDa possessing both antifreeze and hydrolytic activity has been purified from the extracellular fluid of *H. rhamnoides* seedlings. Therefore an effort was made to scrutinize the presence of chitinases in various parts of seabuckthorn which could provide a clear picture of various chitinases and their tissue specific role. Proteins isolated from leaf, berry and 20 days old seedlings of *H. rhamnoides* were used to purify the chitinase using chitin affinity chromatography. The proteins were kept overnight to allow binding with chitin in 20 mM ABC pH 8.0. The non chitin binding proteins were collected as flow through and 3-4 washes with 20 mM ABC were given to remove binding of any non-specific proteins with chitin. To further elute the chitin binding proteins from the column various ranges of acetic acid pH 3.0 was used which lead to the partial purification of various chitinases in seedlings, leaf and berry (Figures 54.4a–c). In leaf and seedlings no protein was eluted at lower concentration (20 mM) of acetic acid while higher concentration (250 mM) lead to the elution of 34 and 31 kDa polypeptides suggesting strong binding affinity of these proteins with chitin. In berry, it was observed that using even very lower concentration of acetic acid (2.5 mM) lead to the purification of three polypeptides of 10, 24 and 41 kDa suggesting the weak affinity of these polypeptides. The purity of the proteins eluted was checked by resolving these on 12 per cent SDS PAGE.

Chitinase hydrolytic assay and antifreeze activity analysis were done to further confirm that the proteins eluted from chitin affinity column are chitinases and whether they posses any hydrolytic or antifreeze activity. The proteins purified from all the samples possess both hydrolytic and antifreeze activity. The highest antifreeze and hydrolytic activity (58 Units/mg) was observed in the leaf while lowest hydrolytic activity (36 Units/mg) was observed in the berry. It was observed that although the antifreeze activity was lowest in the seedlings but it posses higher hydrolytic activity (51 Units/mg) than berry (Table 54.1).

Table 54.1: Chitinases Present in Seabuckthorn Seedling, Leaf and Berry and a Comparative Account of their Antifreeze and Hydrolytic Activity

Material	*Chitinases Mol wt. (kDa)*	*Ice Crystal Morphology*	*Hydrolytic Activity (Units/mg)*
Seedling	31, 34	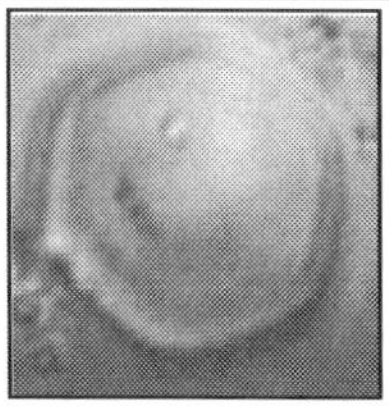	58
Leaf	31, 34	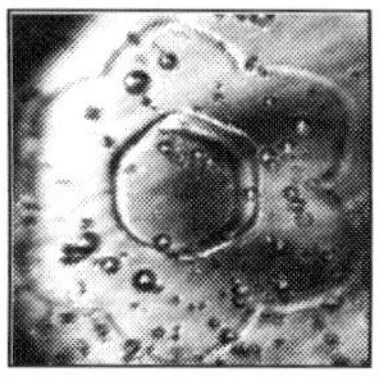	51
Berry	10, 24, 41	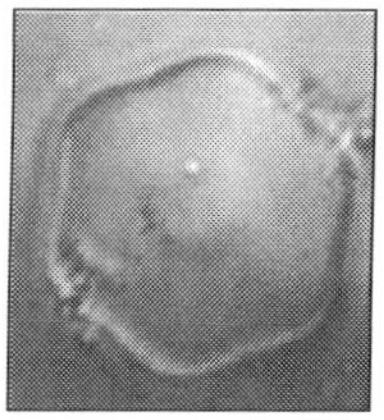	36

DISCUSSION

Overwintering and many freeze tolerant plants secrete antifreeze proteins to survive the freezing conditions. AFPs functions to protect the plant by binding to ice crystals and preventing their recrystallisation which could damage the tissue. AFPs are expressed in almost every part like seeds, barks, leaf, flowers, berries, roots and tubers of many cold tolerant plants (Sharma and Deswal, 2014). Seabuckthorn being a cold hardy plant can survive temperatures upto - 40 °C and investigation for the presence of antifreeze activity has shown accumulation of antifreeze proteins in the apoplast of *H. rhamnoides* seedling. To further examine the presence of antifreeze activity in the wild parts of seabuckthorn, proteins were isolated from leaf and berry of *H. rhamnoides* and were resolved on 1DGE to obtain their protein profiles. Their antifreeze activity analysis showed the presence of antifreeze activity in both leaf and berry. Moreover, the experiments suggested the stability of AFPs over a wide range of temperature. Dual functioning AFPs also acting as PR proteins providing defence against psychrophilic organisms have been isolated from various plants. In winter rye 3 such proteins have been isolated including Class I and Class II endochitinase AFPs, thaumatin-like proteins and β-1,3-endoglucanase AFPs (Griffith and Yaish, 2004). Moreover, Chitinase genes encoding AFPs of 31.7 kDa and 24.8 kDa were cloned from winter rye leaves (Yeh *et al.*, 2000). PR proteins with antifreeze activity have also been reported in *Bromus inermis* (Nakamura *et al.*, 2008) and *Daucus carota,* (Meyer *et al.*, 1999; Worrall *et al.*, 1998). Dual functioning AFPs, Class I Chitinases have been purified from apoplast of seabuckthorn seedlings. Similar efforts to purify these in leaf, berry and seedlings lead to the purification of various tissue specific chitinases, possessing both antifreeze and hydrolytic activity suggesting their role in providing cold tolerance and defence against psychrophilic organisms. Chinolytic Class I chitinases of 33 and 39 kDa and Class III chitinase of 25 kDa have been purified from pineapple leaf (Taira *et al.*, 2004) and of 39.6 kDa and 55.6 kDa in *Ipomoea batatas* leaf possessing only antifungal activity (Liau *et al.*, 2008). Antifungal and Cryoprotective Chitinase of 43 kDa have been purified from *Vitis vinifera* Cv. Cardinal (Caballero *et al.*, 2009) and of 24, 48 kDa from *Annona cherimola* (Goni *et al.*, 2013) berries.

CONCLUSION

The results observed in the study confirmed the presence of antifreeze proteins in *H. rhamnoides* leaf and berry as suggested by observing their ice recrystallisation inhibition activity and ice crystal morphology. Moreover, heat stability of AFPs was observed as shown by the protein profiles of berry samples brought at RT without freezing them in liquid nitrogen and providing heat treatment. Identification and purification of these targets could lead to commercial utilization of heat stable AFPs in food technology where pasteurization is required before cold storage. Chitinases partially purified in seedlings, leaves and berries of seabuckthorn confirmed the importance of these dual functioning AFPs in providing cold tolerance and defense resistance against psychrophilic organisms. Further purification and characterization of these proteins would help in better understanding of their differential role in different parts of seabuckthorn. An exhaustive investigation of other AFPs from the repertoire of seabuckthorn would help in obtaining other potential candidates for their commercial utilization in various biotechnological sectors.

ACKNOWLEDGEMENT

Financial assistance to RD from DBT and Delhi University (R&D grant) are duly acknowledged. BS acknowledges UGC, Government of India for research fellowship.

REFERENCES

1. Antikainen, M. and Griffith, M. 1997. Antifreeze protein accumulation in freezing-tolerant cereals. *Physiologia Plantarum* 99: 423-432.
2. Arav, A., Rubinsky, B., Fletcher, G. and Seren, E. 1993. Cryogenic protection of oocytes with antifreeze proteins. *Molecular Reproduction and Development* 36: 488-493.
3. Arora, R. and Wisniewski, M.E. 1994. Cold acclimation in genetically related (sibling) deciduous and evergreen peach (*Prunus persica* Batsch). *Plant Physiology* 105: 95-101.
4. Atici, O. and Nalbantoðlu, B. 2003. Antifreeze proteins in higher plants. *Phytochemistry* 64: 1187-1196.
5. Baranski, R., Klocke, E. and Nothnagel, T. 2008. Chitinase CHIT36 from *Trichoderma harzianum* enhances resistance of transgenic carrot to fungal pathogens. *Journal of Phytopathology* 156: 513-21.
6. Boot, R.G., Blommaart, E.F.C., Swart, E., Vlugt, K.J.V.D., Bijl, N., Moe, C., Place, A. and Aerts, J. M. F. G. 2001. Identification of a novel acidic mammalian chitinase distinct from chitotriosidase. *Journal of Biological Chemistry* 276: 6770-8.
7. Caballero, C.F., Romero, I., Goni, O., Escribano, M.I., Merodio, C. and Sanchez-Ballesta, M.T. 2009. Characterization of an antifungal and cryoprotective class I chitinase from table grape berries (*Vitis vinifera Cv. Cardinal*). *Journal of Agriculture and Food Chemistry* 57: 8893-8900.
8. Chen, J.K., Shen, C.R. and Liu, C.L. 2010. N-Acetylglucosamine: Production and applications. *Marine Drugs* 8: 2493-516.
9. Dahiya, N., Tewari, R. and Hoondal, G.S. 2006. Biotechnological aspects of chitinolytic enzymes: A review. *Applied Microbiology and Biotechnology* 71: 773-82.
10. Distefano, G., Malfa, S.L., Vitale, A., Lorito, M., Deng, Z. and Gentile, A. 2008. Defence-related gene expression in transgenic lemon plants producing an antimicrobial *Trichoderma harzianum* endochitinase during fungal infection. *Transgenic Research* 17: 873-79.
11. Dwivedi, S.K., Parimelazhagan, T. and Singh, S.B. 2009. *Seabuckthorn (Hippophae* spp.*) The Golden Bush,* Satish Serial Publishing House, New Delhi, India.
12. Elias, J., Homer, R., Hamid, Q. and Lee, C. 2005. Chitinases and chitinase-like proteins in T (H) 2 inflammation and asthma. *Journal of Allergy and Clinical Immunology* 116: 497-500.
13. Fletcher, G.L., Goddard, S.V. and Wu, Y.L. 1999. Antifreeze proteins and their genes: From basic

research to business opportunity. *Chemtech* 29: 17-28.

14. Goni, O., Sanchez-Ballesta, M.T., Merodio, C. and Escribano, M.I. 2013. Two cold induced family 19 glycosyl hydrolases from cherimoya (*Annona cherimola*) fruit: An antifungal chitinase and a cold-adapted Chitinase. *Phytochemistry* 95: 94-104

15. Grandum, S., Yabe, A., Tanaka, M., Takemura, F. and Nakagomi, K. 1997. Characteristics of ice slurry containing antifreeze protein for ice storage applications. *Journal of Thermophysics and Heat Transfer* 3: 461-466.

16. Griffith, M., Ala, P.I., Yang, D.S.C., Hon, W.C., and Moffat, B.A. 1992. Antifreeze protein produced endogenously in winter rye leaves. *Plant Physiology* 10: 593-596.

17. Griffith, M. and Ewart, K.V. 1995. Antifreeze proteins and their potential uses in frozen foods. *Biotechnology Advances* 13: 375-402.

18. Griffith, M. and Yaish, M.W.F. 2004. Antifreeze proteins in overwintering plants: A tale of two activities. *Trends in Plant Science* 9: 399-405.

19. Gupta, R. and Deswal, R. 2012. Low temperature stress modulated secretome analysis and purification of antifreeze protein from *Hippophae rhamnoides*, a Himalayan wonder plant. *Journal of Proteome Research* 11: 2684-2696.

20. Hon, W.C., Griffith, M., Mlynarz, A, Kwok, Y.C., and Yang, D.S.C. 1995. Antifreeze proteins in winter rye are similar to pathogenesis-related proteins. *Plant Physiology* 109: 879-889

21. Huang, T. and Duman, J.G. 2001. Cloning and characterization of a thermal hysteresis (antifreeze) protein with DNA-binding activity from winter bittersweet nightshade, *Solanum dulcamara*. *Plant Molecular Biology* 48: 339-350

22. Imoto, T. and Yagishita, K. 1971. A simple activity measurement of lysozyme. *Agricultural and Biological Chemistry* 35: 1154-1156.

23. Johansen, J.S. 2006. Studies on serum YKL-40 as a biomarker in diseases with inflammation, tissue remodelling, fibroses and cancer. *Danish Medical Bulletin* 53: 172-209.

24. Kang, S.C., Park, S. and Lee, D.G. 1999. Purification and characterization of a novel Chitinase from the entomopathogenic fungus, *Metarhizium anisopliae*. *Journal of Invertebrate Pathology* 73: 276-281.

25. Koushafar, H., Pham, L., Lee, C. and Rubinsky, B. 1997. Chemical adjuvant cryosurgery with antifreeze proteins. *Journal of Surgical Oncology* 66: 114-121.

26. Liau, C.Y. and Chung-Sheng, L. 2008. Detection of chitinolytic enzymes in *Ipomoea batatas* leaf extract by activity staining after gel electrophoresis. *Journal of the Chinese Chemical Society* 55: 678-681.

27. Ling, H., and Recklies, A.D. 2004. The chitinase 3-like protein human cartilage glycoprotein 39 inhibits cellular responses to the inflammatory cytokines interleukin-1 and tumour necrosis factor-A. *Biochemical Journal* 380: 651-9.

28. Meyer, K., Keil, M. and Naldrett, M.J. 1999. A leucine rich repeat protein of carrot that exhibits antifreeze activity. *FEBS Letters* 447: 171-178

29. Nagpure, A., Choudhary, B., Rajinder K. and Gupta, R.K. 2014. Chitinases in agriculture and human healthcare. *Critical Review in Biotechnology* 34 (3): 215-232.

30. Nakamura, T., Ishikawa, M., Nakatani, H. and Oda, A. 2008. Characterization of cold responsive extracellular chitinase in Bromegrass cell cultures and its relationship to antifreeze activity. *Plant Physiology* 147: 391-401.

31. Pudney, P.D.A., Buckley, S.L., Sidebottom, C.M., Twigg, S.N., Sevilla, M.P., Holt, C.B., Roper, D., Telford, J.H., McArthur, A.J. and Lillford, P.J. 2003. The physico-chemical characterization of a boiling stable antifreeze protein from a perennial grass (*Lolium perenne*). *Archives of Biochemistry and Biophysics* 410: 238-245.

32. Rattanakit, N., Yano, S., Wakayama, M., Plikomol, A. and Tachiki, T. 2003. Saccharification of chitin using solid-state culture of *Aspergillus* sp. S1–13 with shellfish waste as a substrate. *Journal of Bioscience and Bioengineering* 95: 391-6.

33. Rattanakit, N., Yano, S., Plikomol, A., Wakayama, M. and Tachiki, T. 2007. Purification of *Aspergillus* sp. S1–13 chitinases and their role in saccharification of chitin in mash of solid state culture with shellfish waste. *Journal of Bioscience and Bioengineering* 103: 535-41.

34. Raymond, J.A., Wilson, P.W. and De Vries, A.L. 1989. Inhibition of growth on non basal planes in ice by fish antifreeze. In: *Proceedings of the National Academy of Sciences, U.S.A.* 86: 881-885.

35. Shah, J.M., Raghupathy, V. and Veluthambi, K. 2009. Enhanced sheath blight resistance in transgenic rice expressing an endochitinase gene from *Trichoderma virens*. *Biotechnology Letters* 31: 239-44.

36. Sharma, B. and Deswal, R. 2014. Antifreeze proteins in plants: An overview with an insight into the detection techniques including nano biotechnology. *Journal of Proteins and Proteomics* 5 (2): 89-107.

37. Sidebottom, C., Buckley, S., Pudney, P., Twigg, S., Jarman, C., Holt, C., Telford, J., McArthur, A., Worrall, D., Hubbard, R. and Lillford, P. 2000. Heat stable antifreeze protein from grass. *Nature* 406: 256.

38. Singh, V., Kallio, H., Sawhney, R.C., Gupta, R.K., Lu, R., Eliseev, I.P., Khabarov, S.N., Korovina, M.A., Skuridin, G.M. and Shchapov, N.S. 2003. *Seabuckthorn (Hippophaë L.)-A Multipurpose Wonder Plant. Vol. I: Botany, Harvesting and Processing Technologies* (V. Singh, Ed. in Chief, 2003), Indus Publishing Company, New Delhi, 518 p.

39. Singh, V., Yang, B., Kallio, H., Bala, M., Sawhney, R.C., Gupta, R.K., Morsel, J.T., Lu, R. and Tolkachev, O.N. 2006. *Seabuckthorn* (*Hippophae* L.)-*A Multipurpose Wonder Plant. Vol. II: Biochemistry and Pharmacology* (V. Singh, Ed. in Chief, 2006), Daya Publishing House, New Delhi, 600p.

40. Smallwood, M., Warrall, D., Byass, L., Elias, L., Ashford, D., Doucet, C.J., Holt, C., Telford, J., Lillford, P. and Bowles, D.J. 1999. Isolation and characterization of a novel antifreeze protein from carrot (*Daucus carota*). *Biochemical Journal* 340: 385-391.

41. Taira, T., Toma, N. and Ishihara, M. 2005. Purification, characterization and antifungal activity of chitinases from Pine apple (*Ananas comosus*) leaf. *Bioscience Biotechnology and Biochemistry* 69 (1): 189-196.

42. Venketesh, S. and Dayananda, C. 2008. Properties, potentials, and prospects of antifreeze proteins. *Critical Review in Biotechnology* 281: 57-82.

43. Wei, L.B., Yong, J., Nian-Hong, S., Su-Qin, G. and Yun-Biao, F. 1999. Biological Characterization of Heat-stable antifreeze proteins from leaves of *Ammopiptanthus mongolicus. Acta Botanica Sinica* 41 (8).

44. Wisniewski, M., Webb, R., Balsamo, R., Close, T. J., Yu, X.M. and Griffith, M. 1999. Purification, immunolocalization, cryoprotective, and antifreeze activity of PCA60: A dehydrin from peach (*Prunus persica*). *Physiologia Plantarum* 105: 600-608.

45. Worrall, D., Elias, L., Ashford, D., Smallwood, M., Sidebottom, C., Lilliford, P., Telford, J., Holt, C. and Bowles, D. 1998. A carrot leucine rich repeat protein that inhibits ice recrystallisation. *Science* 282: 115-117.

46. Xiao-Jing, X., Li-Qun, Z., You-Yong, Z. and Wen-Hua, T. 2005. Improving biocontrol effect of *Pseudomonas fluorescence* P5 on plant diseases by genetic modification with chitinase gene. *Chinese Journal of Agricultural Biotechnology* 2: 23-7.

47. Yang, B. and Kallio, H. 2006. Physiological effects of seabuckthorn (*Hippophaë rhamnoides*) fruit pulp and seed oils. In: *Seabuckthorn-A Multipurpose Wonder Plant Vol. II. Biochemistry and Pharamacology*, Singh, V. (Ed. in Chief), pp.363-189, Daya Publishing House, New Delhi, 600p.

48. Yeh, S., Moffatt, B.A., Griffith, M., Xiong, F., Yang, D.S.C., Wiseman, S.B., Sarhan, F., Danyluk, J., Xue, Y.Q., Hew, C.L., Kirby, A.D. and Lajoie, G. 2000. Chitinase genes responsive to cold encode antifreeze proteins in winter cereals. *Plant Physiology* 124: 1251-1263.

49. Yu, X.M. and Griffith, M. 2001. Winter rye antifreeze activity increases in response to cold and drought, but not abscisic acid. *Physiol. Plant* 112: 78-86.

50. Zachariassen, K.E. and Kristiansen, E. 2000. Ice nucleation and antinucleation in nature. *Cryobiology* 41: 257-279.

51. Zhu, Z., Zheng, T., Homer, R.J., Kim, Y.K., Chen, N.Y., Cohn, L., Hamid, Q. and Elias, J.A. 2004. Acidic mammalian chitinase in asthmatic Th2 inflammation and IL-13 pathway activation. *Science* 304: 1678-82.

Chapter 55

Parameters Optimization of High Hydrostatic Pressure Processing on Microbial Inactivation in Seabuckthorn Pulp and Comparison of Polyphenols Preservation after Optimized Processing

Sheng Wenjun, Bi Yang, Li Jixin, Zhang Zhong, Feng Lidan, Mi Lan, Wei Juan and Hu Jinmin

College of Food Science and Engineering, Gansu Agricultural University, Lanzhou, China
(E-mail: shengwj@126.com)

SUMMARY

The process of seabuckthorn either in China or other countries is still limited in several products in which the original pulp occupies the top of the range. However, the thermal sterilization which was unavoidable in the traditional producing of the seabuckthorn pulp always causes the most decomposition of polyphenols. To preserve the antioxidants like flavonoids, the trial was exerted to test the feasibility of using ultra high pressure processing instead of thermal sterilization. For this, the high hydrostatic pressure (HHP) process with the pressure from 100MPa to 600MPa was employed in the experiment. Firstly, the influential parameters of HHP on the sterilization of the seabuckthorn pulp were optimized. In the experiment, three chosen parameters of HHP listed as the hydrostatic pressure, the temperature of specimen and the processing time. Then a preliminary comparison of polyphenols preservation between two samples successively after the HHP and the thermal sterilization was analyzed. The results proved a maximum decrease of the aerobic bacterial count from 1700 CFU/mL before processing to 40 CFU/mL afterwards when the processing conditions were controlled as follows: the hydrostatic pressure 500Mpa, the processing time 15 min and the initial temperature of the specimen 30 °C. Furthermore, ANOVA showed the hydrostatic p7ressure and the treating time exerted highly significant influences on the sterilization of seabuckthorn pulp ($P<0.05$). The comparison of polyphenols preservation between two mentioned samples showed the index of the sample from the HHP was as two times as that from the thermal sterilization. It is suggested that the HHP could be potentially more effective in the preservation of polyphenols compared with the thermal treatment, in the premise of satisfying the sterilization.

Keywords: *Seabuckthorn pulp, High hydrostatic pressure, Sterilization and Polyphenol.*

INTRODUCTION

Seabuckthorn, with the alternative name as the acid willow, is categorized to *Elaeagnaceae Hippophae* family and known being rich in trace elements, vitamins and bioactivities (Wang *et al.*, 2010). It is a unique plant currently being domesticated in several countries like China, Russia, Germany, Finland, Romania, France, Nepal, Pakistan and India. (Zhang *et al.*, 2001). Given its strong antioxidant capacity, the berry could be used as medical or edible material to inhibit the chromium-induced free radicals, apoptosis and DNA damage (Beveridge *et al.*, 1999). Meanwhile, the seabuckthorn has a significant assistance in curing of burns and ulcers besides its potential function to prevent the heart disease and the cancer (Geetha *et al.*, 2002). The plant is also reported to have considerable medicinal value to be useful for the treatment of skin disorders

Seabuckthorn (Hippophae L.)*: A Multipurpose Wonder Plant, Vol. 5: Advanced Technologies for Cultivation, Processing, Health Protection and Environmental Conservation* (Virendra Singh, *Editor-in-Chief*, 2018), p. 423–430, Astral International Pvt. Ltd., New Delhi, India.

(Li *et al.*, 2003). All of the functions above are believed relating to the contents of antioxidants. For this reason, the seabuckthorn are among the most nutritious and vitamin-rich fruit found in the plant kingdom because of presence of several natural antioxidants *viz.*, vitamin C, vitamin E, carotenoids, anthocyanins and phenols (Chauhan *et al.*, 2001).

HHP processing is defined as the using of the static liquid pressure above 100 MPa (100 MPa to 1000 MPa) to sterilize the packaged food at a comparatively low temperature. For many kinds of foods, the HHP were applied successfully. Recent twenty years, it has been noticed in the beverage producing (Vanniivi *et al.*, 2004). The main effects of it on the foods include the enzyme inactivation, the textures change and the chemical reaction rate change would be expected besides the sterilization. Furthermore, the HHP is a physical process, with the instantaneous compression, the low energy consumption and the non-chemical additive, is considered to be the environmentally friendly treatment (Yang *et al.*, 2011). The HHP generally brings the destruction of the three-dimensional structures of microbes. At the same time, the influence on the vitamins, the pigments and the flavor compounds are small. So the HHP technology is especially ideal for the foods which content the thermal-sensitive components (Lin *et al.*, 2005). In this field, it was reported that the presence of the colony count of the kiwi juice after the processing of HHP was dropped evidently. The results showed that the pressure had a linear relationship with the decreased colonies. When the pressure reached 400MPa and lasted for 15 min, *Escherichia coli* could be completely inactivated (Zhao *et al.*, 2006). In another research, the total plate count of bacteria in the litchi juice was significantly decreased with the rising of pressure. When the pressure rose up to 200MPa, a sharp decline of the total plate count was observed. The effect of HHP treatment on the activity of the enzyme was mainly manifested on the change of protein three-tier structure. The study showed that the HHP treatment could break the secondary bonds which maintain the protein tertiary structure and this destruction depended not only on the change amplitude of the pressure also on the temperature, the treatment time, the pH value and the chemical compositions. Now in China, the seabuckthorn original pulp is more and more used as the important ingredient of the beverage which aimed the improvement of human health. But in the traditional process of seabuckthorn pulp, the thermal treatment is unavailable. So the reservation rate of antioxidants in the seabuckthorn pulp is considerable low. And by far, the application of the HHP in the seabuckthorn pulp producing has not been reported yet then the technicians and the managers from the manufacture of the seabuckthorn products really hope relative research would be done.

For this reason, the experiment was exerted to test the feasibility of using the HHP instead of thermal sterilization in the seabuckthorn pulp producing. The triple objectives of the experiment were set up as following: firstly, the influences of three chosen factors listed as the hydrostatic pressure, the temperature of specimen and the processing time on the aerobic bacterial count of the seabuckthorn pulp would be analyzed and the optimal level of the individual factor would be selected; secondly, the parameters optimization of the HHP on the sterilization of the seabuckthorn pulp would be carried out; finally, the content of polyphenols from the optimized HHP sterilized seabuckthorn pulp and the thermal sterilized one would be preliminary compared.

MATERIALS AND METHODS

Instruments

FR-900 vacuum package machine: Shanghai Maya Multiple Functional Food Machine Co ltd; PYX-DHS thermo water bath: Shanghai Yuejin Medical Instrument factory; SW-CJ-1F clean bench: Su jing group of Ante Corp manufacturing; 600MPa-2L ultra-high pressure food processing equipment: Hua Tai Sen Miao Co ltd.

Inoculated Seabuckthorn Pulp

The sterilized seabuckthorn pulp which was produced by Gansu Conseco Company with soluble solids content as 20 per cent was rewarmed to 20°C in the water bath. After the rewarming, the seabuckthorn pulp was quantified inoculated with the fixed concentration of the bacterium suspension. In order to ensure the microbial species and the quantity are same, all the samples were inoculated simultaneously. The samples were packaged in the disinfected polyethylene bags (20ml/bag) and stored at 4°C to be prepared for the further experiments in 24 hours.

Afterwards, the aerobic bacterial count of the sterilized seabuckthorn pulp was immediately detected as blank control data. Before each ultra-high pressure treatment, the same index was detected from randomly selected three test samples as the negative control data.

Strain and Medium

Penicillium expansum (CICC: 40356): China Center of Industrial Culture Collection (CICC); *Staphylococcus auras* (ATCC: 25923): Beijing Branch of Global Bioresource Center (ATCC); *P. expansum* inoculated in Czapek's solution as 10^{-9} concentration; *S. auras* inoculated in Luria-Bertani solution as 110^{-9}concentration.

Regularities of the Influences from Three Factors on the Aerobic Bacteria Count of the Sample

Pressure was respectively arranged as 100MPa, 200 MPa, 300 MPa, 400 MPa, 500 MPa and 600 MPa and the treatment time was fixed as 20 min after the sampled was rewarmed to 25°C. Through the Chinese national standard method, the aerobic bacterial counts were detected as the indices to explore the effect of the processing pressure on the sterilization of seabuckthorn pulp. Similar to the above process, the pressure was kept as 400MPa and the sample temperature was rewarmed to 25°C then the treatment time were respectively arranged as 10min, 15min, 20min, 25min, 30min, 35min to find the law of index change. In the temperature single factor test, the pressure was kept as 400MPa and the treatment time as 25min then the temperatures were set as 25°C, 30°C, 35°C, 40°C, 45°C, 50°C. All the treatments took triple replicas and reported the average data.

Optimization of the HHP Parameters to Enhance the Effect of Sterilization on Seabuckthorn Pulp

After the single factor tests, the Box-Behnken Design was employed to find the relations between the index and the three chosen factors. Taken the initial sample temperature, the pressure, and the treating time as the independent variable respectively, the best level of them in the single factor test were put as the "zero" code. Then, with the same "step" the "-1" code level and "+1" code level were designed and the real parameters referred to the codes were presented as follows.

Table 55.1: The Codes and the Values of Factors in BBD Experiment

Factor	Code			Value		
Pressure (MPa)	-1	0	1	400	500	600
Temperature (°C)	-1	0	1	20	25	30
Treating time (min)	-1	0	1	10	15	20

Detection of Aerobic Bacterial Count

The detection of aerobic bacterial count was followed the reference of the Chinese national standard method. For each treatment in the experiment, the seabuckthorn of 20ml was weighed in 225ml sterilized buffer liquid. Then the aliquots of 1ml of the dilutions to be analyzed were placed into Petri dishes and added with 15ml of liquid NA (cooled to 46±1 °C in water bath) within 15 minutes after preparation of dilutions. When the inoculum and the medium were mixed carefully, the plates were placed stable allowing the agar to cool and solidify. Test two parallel plates for each dilution. The plates were inverted and incubated for 48±2h at 36±1 °C. After incubation, the colonies were counted. The amount as the weighted mean from the successive dilutions were calculated, which contain between 30 and 300 colonies. The calculated result is the weighted means of the successive dilution multiply by dilution factor.

Determination of Polyphenol

The determination of polyphenol was assisted with the Folin-Ciocalteu method. The 10mL of seabuckthorn pulp after the HHP treatment was added with 40ml anhydrous ethanol. Then the ultrasonic extraction sustained 30min was used to extract the polyphenol from the samples. The mixture was centrifuged under 4°C with the rotary speed as 1000r/min for 20 min. The supernatant was separated for the determination. To detect the polyphenol, the prepared supernatant was taken as much as 0.4ml by adding 2ml of Folin ciocalteu reagent (l mol/L, 10 times diluted by distilled water) and reacted for 5min then the 1.8ml sodium carbonate concentrated as 7.5 per cent was added too. Then another reaction in the darkness for one hour was ensured. Using the wavelength of 765nm, the absorbance of samples was determined. The standard curve was made by using the pyrogallic acid- ethanol solution (10-150 mg/L) and the corresponding absorbance at 765 nm were determined. The polyphenolic content was reported as the milligram of pyrogallic acid per 100ml seabuckthorn pulp. The distilled water was used as the blank control.

Statistics Analysis

The correlation and the significance were analyzed aided with the software Design expert 7.1.3. The replicated data was presented as the mean.

RESULTS AND DISCUSSION

Analysis of Influences from Three Factors upon the Aerobic Bacteria Count of Seabuckthorn Pulp

When the treatment time was kept as 20 min and the initial sample temperature was kept as 25°C, the aerobic bacterial counts of the samples after the HHP treatment were shown in Figure 55.1. When the pressure was below 300 MPa, the aerobic bacterial counts around 1.5×10^3 CFU/ml were almost same to the negative control group. Then the pressure went up to 400 MPa, the aerobic bacterial counts dropped to 1.4×10^2 CFU/ml. compared with the negative control group, the mortality reached 91 per cent. Afterwards, the pressure increased further to 500 MPa and the aerobic bacterial counts dropped to 40 CFU/ml with the mortality as 99.9 per cent. Upper limit in the experiment, the pressures continuously went up to 600 MPa but the aerobic bacterial counts stayed as 30 CFU/mL which meant the mortality was still reported as 99.9 per cent.

According to results of the single factor test, the optimal pressure "zero code" level in the RSM was chosen as 500 MPa.

In the time factor experiment, the aerobic bacterial count never dropped sharply like which happened in the pressure experiment when the variable went up to 400MPa. However, the level as 15min still could be chosen as the optimal level in the RSM. The results showed in the Figure 55.2.

In the temperature single factor experiment, the aerobic bacterial count's dropping was almost linear

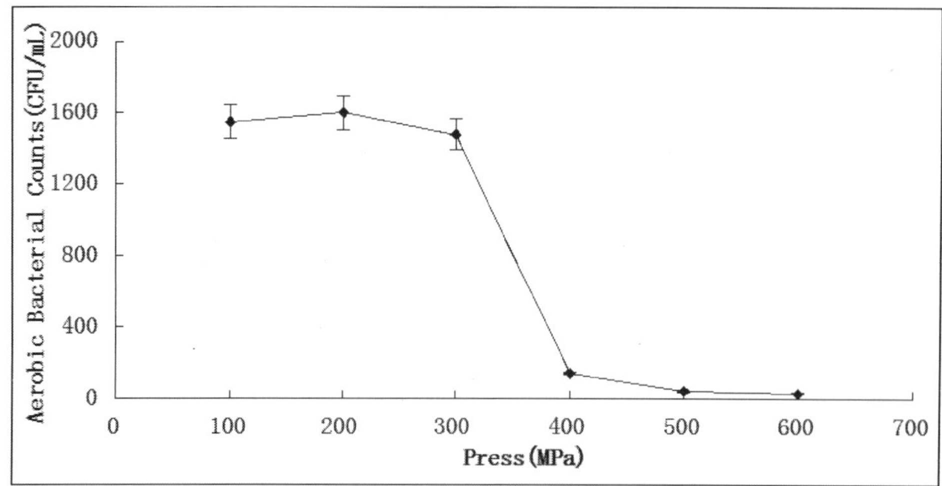

Figure 55.1: Effect of the Pressure on the Aerobic Bacterial Counts. Bars represent SE.

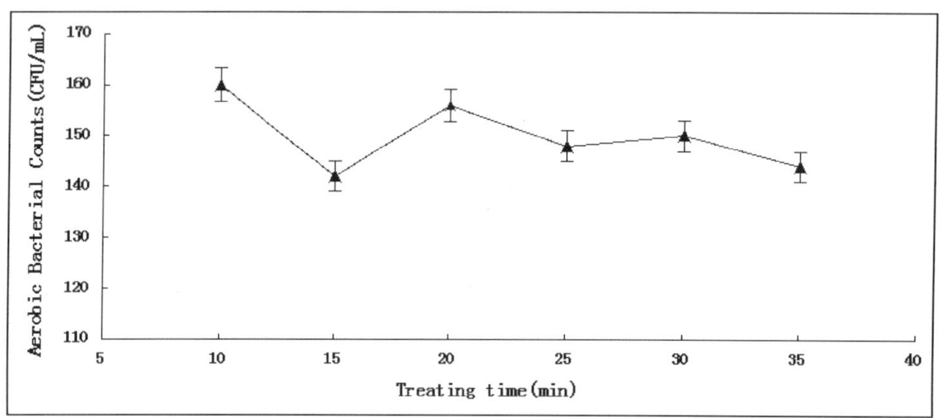

Figure 55.2: Effect of the Treating Time on the Aerobic Bacterial Counts. Bars represent SE.

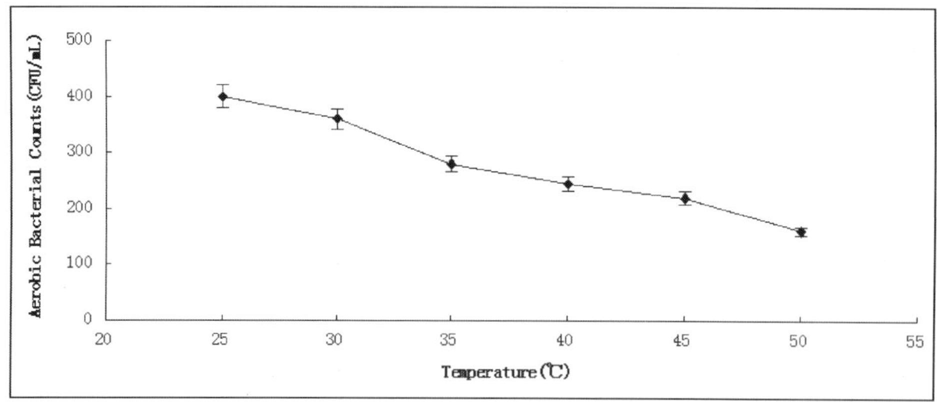

Figure 55.3: Effect of the Initial Sample Temperature on the Aerobic Bacterial Counts. Bars represent SE.

to the rising of the temperature. In this way, it may be believed that the comparatively higher temperature would inspire the bacteria to be sensitive. However, in this experiment, the lower temperature was originally expected to protect the thermal sensitive components. So the temperature as 25 °C was entitled as the optimal level in the RSM as showed in Figure 55.3.

Optimization of the HHP parameters to enhance the effect of sterilization on seabuckthorn pulp

Based on the single factor tests, the influential factors were optimized through the BBD experiment. The experiment run sheet and the results were showed in the Table 55.2.

Table 55.2: Running of Box-Behnken Design and Results

Sl.No.	*Factors*			*Aerobic bacterial counts (CFU/mL)*
	Pressure (MPa)	*Temperature (°C)*	*Time (min)*	
	X_1	X_2	X_3	*Results*
1	400	20	15	140
2	600	25	20	30
3	500	20	10	45
4	500	25	15	30
5	400	25	20	120
6	500	25	15	40
7	500	20	20	30
8	600	25	10	35
9	600	20	15	30
10	500	30	20	35
11	500	25	15	80
12	400	30	15	160
13	500	25	15	80
14	600	30	15	30
15	500	30	10	80
16	400	25	10	200
17	500	25	15	50

The regression equation was obtained from the results of running sheet.

$$Y=56.00-61.88 X_1+7.50X_2-18.13X_3-5.00X_1X_2 +18.75X_1X_3-7.50X_2X_3+41.38X_1 2-7.38X_2 2-1.12 X_3 2$$

Derived from multiple regression equations fitting analysis, the p-value of the model was far lower than 0.05 and manifested the remarkable significance. Meanwhile, the p-value of the 'lack of fit' was 0.9569 which was larger than the criterion 0.10 and the adjusted R-squared was 0.9495 which also satisfied the least requirement as 0.80. (H. N. Sin, 2006). The Pred R-Squared of 0.8709 was in reasonable agreement with the Adj R-Squared of 0.8845.

When the regression equation has been verified acceptable, the significance of other variation sources demonstrated that X_1, X_3 and X_1^2 had remarkable significance. The interaction terms X_1X_3 showed significant variance. The deceasing sequence of factor's influences on the index was counted to be the pressure, the processing time then the temperature. The analysis showed in the Table 55.3.

The influences of the interactions between three factors were showed in the Figure 55.4. When the temperature varied from 20^{oC} to 30^{oC} and the pressure went up from 400MPa to 500 MPa, the aerobic bacterial counts dropped sharply and reached to an optimal value as 57.5 CFU/mL and became stable. In the same way, the synergy effect existed between the pressure and the time because the aerobic bacterial count dropped steeply when the pressure ranged from 400 MPa to550 MPa meantime the time ranged from 10min to 20min. The interaction between the time and the temperature showed that the time was comparatively influential since the prolonged time still proved the drop of the aerobic bacterial counts even the temperature was in the higher level.

Verification of Model Parameters

The optimal parameters obtained from predication through the BBD experiment showed that when the hydrostatic pressure was set as 500Mpa, the processing time was 15.5 min and the initial temperature of the sample was rewarmed to 30.2°C, the aerobic bacterial count could theoretically dropped to 30 CFU/mL. In the actual verification, when the operating parameters were controlled as the hydrostatic pressure 500Mpa, the processing time 15min and the initial temperature of the specimen 30°C, the aerobic bacterial count was calculated as 40 CFU/mL.

Comparison between Preservation of Polyphenol from different Sterilized Samples

The effects of sterilization methods on the content of polyphenol of the seabuckthorn pulp were showed in the Figure 55.5. There was a significant difference between the HHP sterilization and the thermal sterilization. Contrasted to the control group, the polyphenol content of the samples after the HHP still went down and dropped to the 113mg/100mL which was 86.08 per cent of the control group. Compared with the thermal sterilization, the dropping amplitude was small. The content of polyphenols from the sample after the HHP

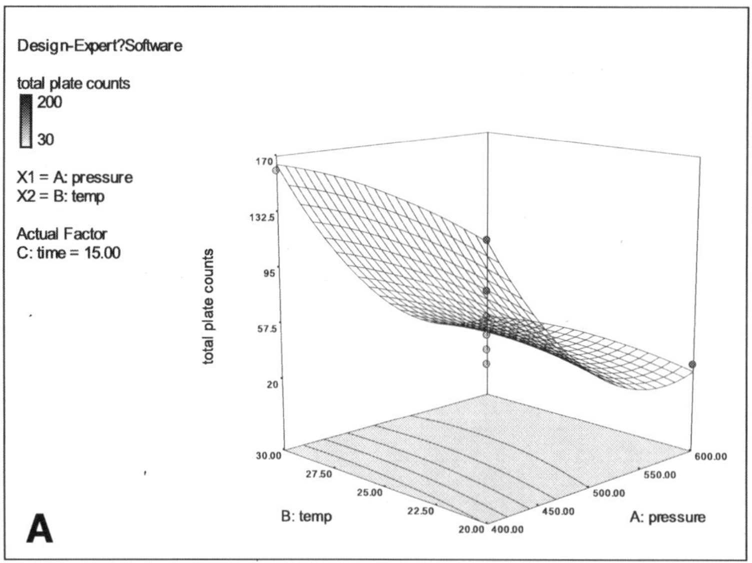

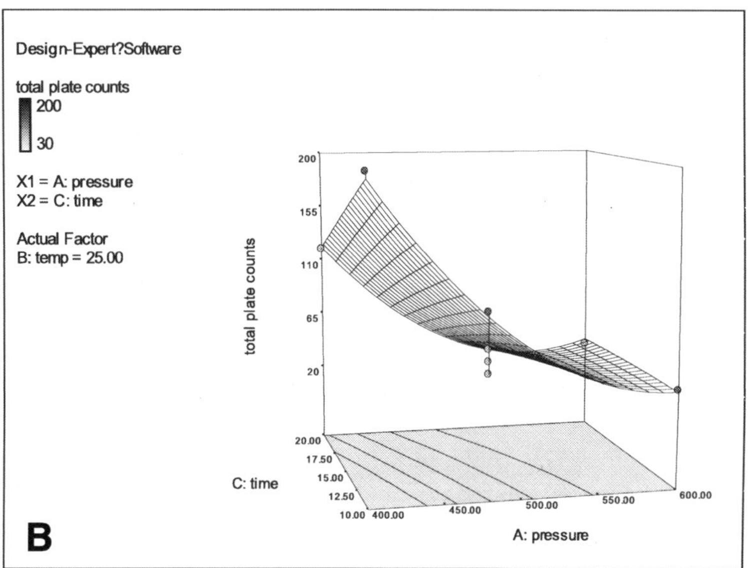

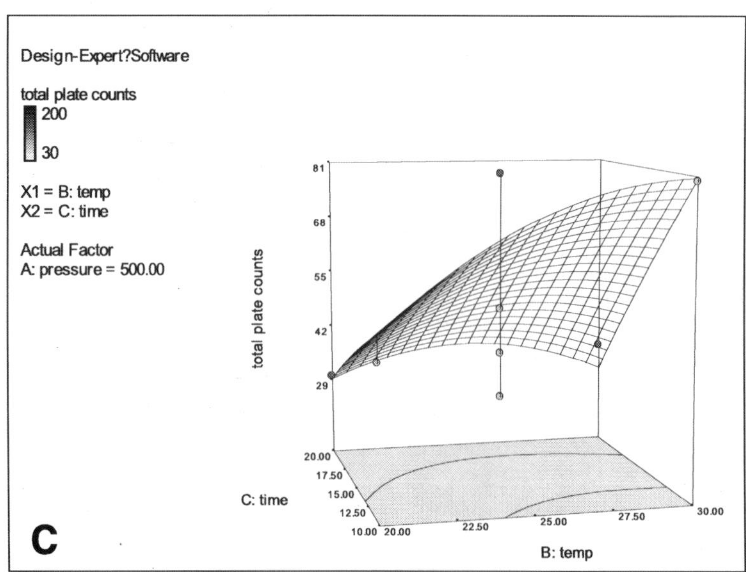

Figure 55.4: Surface Response Plots of the Aerobic Bacterial Counts to Interaction Terms. The colored dot pointed the optimal index value and the lines of contour (ground area) described the trend of index's varying with the fixed interaction terms changed along levels. A: the interaction effect between the pressure and the temperature. B: the same index between the pressure and the treating time. C: the index between the temperature and the time.

Table 55.3: ANOVA for Response Surface Quadratic Model

Source	*Sum of Source*	*df*	*Mean Square*	*F Value*	*P Value*	*Significance*
Model	42761.99	9	4751.33	14.61	0.0009	Significant
X_1	30628.13	1	30628.13	94.19	< 0.0001	**
$_{x}2$	450.00	1	450.00	1.38	0.2779	
$_{x}3$	2628.13	1	2628.13	8.08	0.0249	**
X_1X_2	100.00	1	100.00	0.31	0.5965	
X_1X_3	1406.25	1	1406.25	4.32	0.0761	*
X_2X_3	225.00	1	225.00	0.69	0.4330	
X_1^2	7207.96	1	7207.96	22.17	0.0022	**
X_2^2	229.01	1	229.01	0.70	0.4291	
X_3^2	5.33	1	5.33	0.016	0.9017	
Residual	2276.25	7	325.18			
Lack of Fit	156.25	3	52.08	0.098	0.9569	Not significant
Pure Error	2120.00	4	530.00			
Cor Total	45038.24	16				
R-squared=0.9495						
Adj *R*-squared=0.8845						
Pred R-Squared=0.8709						

* $P<0.05$ noted as **$P<0.10$ noted as *.

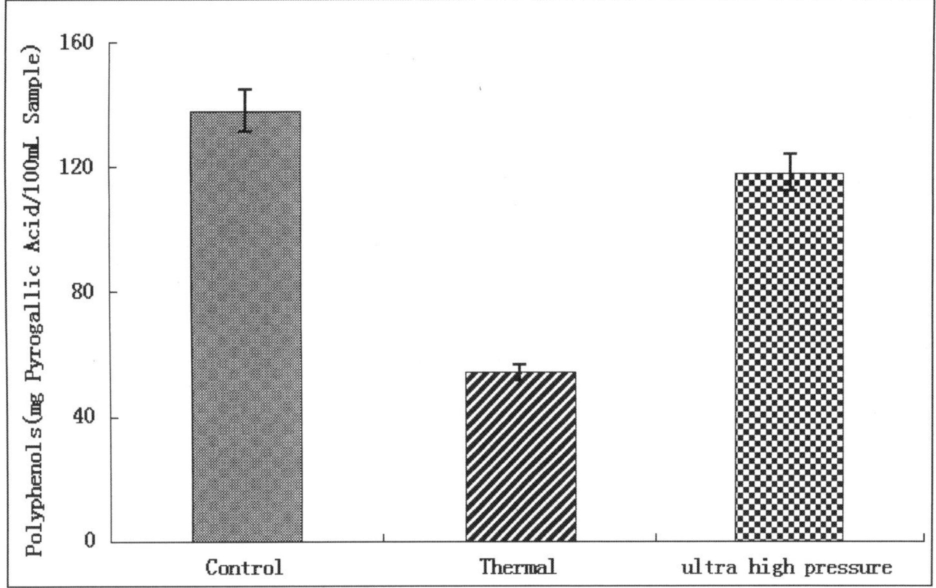

Figure 55.5: Comparison between Polyphenol from different Sterilized Samples. Bars represent SE.

sterilization was close to two times of the index from the thermal sterilized sample. The difference between the two sample groups was significantly. The retention rate of polyphenol proved the HHP sterilization was more available for the preservation of the polyphenol.

CONCLUSION

The optimal parameters of HHP sterilization were approved. When the pressure reached to 500Mpa accompanied with the processing time as 15min and the sample temperature as 30°C, the aerobic bacterial count could drop from the initial 1700 CFU/mL in the control group to the experimentally lowest 40 CFU/mL. Meanwhile, the preservation of polyphenols from HHP sample was two times as the index from the thermal sample. The outcomes of the experiment indicated that HHP could be potentially available in the preservation of polyphonels in the premise of satisfying the microbial inactivation.

ACKNOWLEDGEMENTS

This work was funded by the Projects of International Cooperation and Exchanges of Ministry of Science and Technology in China (No. 2014DFR31230). We gratefully acknowledge National Nature Science Foundation of China. At the same time we are sincerely appreciated Mr. Luan from Gansu Conseco Company for providing all the seabuckthorn pulp needed in the experiment. And we are thankful for GE Qian cui-di who introduced the processing of seabuckthorn in the industry which helped us to set the experimental factors. Besides, we wish to express our thanks to Prof. Han Shun-yu for discussions.

REFERENCES

1. Bao, Z.Y. 2004. Study on the lethal effect of the ultra high pressure technology on the pathogenic microorganisms in cow milk. *Thesis for master degree*, Inner Mongolia Agricultural University, Hohhot, Inner Mongolia, China.
2. Ding, X.R. and Wang, J.M. 2012. Effects of ultra high pressure on the microbial indicators in sauce food. *Chinese Food Additives*, p. 143-147.
3. Gao, Y.X. and Cui, Y.J. 2006. Research progress of flavonoids in seabuckthorn. *Chinese Beverage Industry* 1: 17-23.
4. Jiang, H.T. and Zhong, L. 2009. Effect of ultra high pressure treatment on the quality of orange juice. *Chinese Food Science* 17: 24-29.
5. Lin, X.Y, Chen, J.H. and Zheng, D.D. 2005. Application of ultra high pressure sterilization in food process. *Agricultural Products Process* 1: 9-12.
6. Li, X.H. 2003. Study on the dynamic changes of active components of seabuckthorn. *Thesis for Master Degree*, Ji Ling Agricultural University, Changchu, China.
7. Lu, H.X., Hang, Y.Y. and Li, J.R. 2012. Optimization of ultra high pressure sterilization technology in sausage. *Chinese Food Science* 12: 89-92.
8. Lu, R.R., Yu, J.Y., Qi, W.M. *et al.*, 2010. Dynamic model of pressure resistant strains in high pressure inactivation of low acid food. *Agricultural Engineering Journal* 9: 350-356.
9. Mei, Z. 2010. Using response surface method to optimize seabuckthorn pigments extraction and antioxidant analysis. *Chemical Industry of Forest Products* 4: 78-84.
10. Shen, H.P. 2012.The development and application of food safety by the ultra high pressure sterilization technology. *Chinese Food Industry* 7: 44-46.
11. Sun, T., Fu, X.Y. and Zhang, B. 2009. Aability of scavenging DPPH used in determination of pro-anthocyanidins from seabuckthorn seed. *Journal of Ningxia Medical College* 1: 26-28.
12. Wang, D.P., Si, Q.J. and Chen, Q. 2010. Study on the seabuckthorn beverage process. *Chinese Food Engineering* 1: 27-30.
13. Yang, J., Cheng, L.H. and Du, Y.L. 2011. Research on the ultra high pressure processing of soft packaged food. *Modern Agricultural Science and Technology* 3: 351-353, 356.
14. Zhang, F.S., Xu, X.Y. and Chen, F. 2012. Key process of the ultra high pressure sterilization on white radish juice. *Chinese Food Science* 4: 8-12.
15. Zhao, Y.S. and Zhao, J.F. 2006. Effect of ultra high pressure on the content of Vc in freshly squeezed Kiwi Juice. *Chinese Food Science* 11: 24-25.
16. Zu, Y.G., Chun, J.Z. and Fu, Y.J. 2005. Orthogonal test was adopted to optimize ultrasonic extraction of total flavones from *Hippophae rhamnoides*. *Chemical Industry of Forest Products* 3: 85-88.

Chapter 56

Drying Behavior and Radical Scavenging Activity of Sucrose and Sugar Alcohols on Seabuckthorn Pulp

Sonika Singh* and Hardeep Singh Gujral

Department of Food Technology, Guru Nanak Dev University, Amritsar – 143 005, Punjab, India
(*E-mail: sonika24900@gmail.com)

SUMMARY

Seabuckthorn (SBT) is the perennial, deciduous plant known as the miracle berry owing to its healing and medicinal properties. It finds a wide mentioning in ancient Chinese medicine and is popular for the high vitamin C content, omega fatty acids and antioxidant properties. The aim of the present study is to prepare a low calorie base food product which is rich in antioxidant even after drying. For the purpose of the study two varieties of pulps of seabuckthorn *i.e. Hippophae salicifolia* (S) and *Hippophae rhamnoides* (R) were used. The pulps were mixed with different ratios of sugar and different sugar alcohols and their drying kinetics were observed along with the free radical scavenging effect on the pulps. The comparison between the effects of drying on the aesthetics was determined using sensory analysis to conclude the best combination of pulp and sugar alcohol. The comparisons were conducted among all the sugar/sugar alcohol and pulp combinations to find out the end product with highest antioxidant activity to settle on the best base product.

Keywords: *Seabuckthorn, Hippophae salicifolia, Hippophae rhamnoides, Drying rate, Free radical scavenging activity, Antioxidant activity and Sugar alcohols.*

INTRODUCTION

Seabuckthorn is a perennial, woody, nitrogen fixing, thorny, deciduous shrub of 1-5 m height. It grows generally in nutritionally poor sites of river side, slopes, hill tops and farmer's fields in cold regions of high altitude Indian Himalayas and China and low altitude areas of Europe and North America. The dry temperate and cold desserts, spread in over 75,000 sq. km area in north-west Himalayas, comprising Lahaul-Spiti, parts of Chamba and Kinnaur districts of Himachal Pradesh and Ladakh district in Jammu and Kashmir, are characterized by extreme climatic conditions, high rates of soil erosion, shortage of fuel wood, timber and fodder. The area has a very low productivity of agricultural lands and sparse vegetation (Singh *et al.*, 1995). Seabuckthorn has emerged as a plant of national interest during the last two decade due to awareness about the high potential of its health protection and environmental conservation potential. The food products and nutraceuticals processed from seabuckthorn are widely consumed in China and European countries like Finland, Germany and Russia. Seabuckthorn is also a popular fodder, fuel wood and soil binder of cold deserts in Himalayas (Gupta *et al.*, 2005). Fruit of seabuckthorn (the berry) is very rich in vitamin C (300-2750 mg/100 g), vitamin E, vitamin K, carotenoids, sterols, omega fatty acids, organic acids *etc.* and therefore has high potential in health food, cosmetic and drug industries. The berry is of pea size, often less than 1 cm in diameter, varying in weight of 100 fruits from 10 to 25 g. Seabuckthorn products made from the fruit of seabuckthorn in global market are present in the form of table purpose fruit, juices, concentrate, wines, beer, jams, jellies, hard and soft candies, tea (from leaves and extract), flavoured cakes and confectionaries, nutraceuticals capsules *etc.*

Seabuckthorn (Hippophae L.): A Multipurpose Wonder Plant, Vol. 5: Advanced Technologies for Cultivation, Processing, Health Protection and Environmental Conservation (Virendra Singh, *Editor-in-Chief*, 2018), p. 431–438, Astral International Pvt. Ltd., New Delhi, India.

In India a considerable amount of study has been done on the value chain of seabuckthorn and product development. D.R.D.O in year 2002 developed a health juice by the name of "Leh Berry". Seabuckthorn RTS, squash and juices are made by cottage industries available locally in Ladhak and Manali region of Himachal Pradesh. To improve the acceptability of the prepared product blending of seabuckthorn juice/ pulp with different fruits like mango, papaya, guava was done. Acceptable blends of mango pulp had been identified for the preparation of RTS, Squash, Syrup, Fruit toffee and Jam (Dhaliwal *et al.*, 2010).

Although a number of food product development studies have been carried out in this field, there is a considerable amount of gap to be narrowed down to bring the products to market shelves and for consumer to enjoy health benefits. The product development in India is in an experimental stage and fruit leather can be a potential health food product because of its high antioxidant content and high Vitamin C content (Chaurasia *et al.*, 2010). Seabuckthorn is available in raw form as berries or pulp, which has a lot of moisture and is bulkier to transport and work upon. The dried form of the product is malleable, is easy to manufacture into different shapes and doesn't require large storage space.

The objective of present investigation was to study the effect of Sucrose and Sugar Alcohols on the antioxidant activity and Vitamin C retention in seabuckthorn leather prepared from two varieties of seabuckthorn *i.e. Hippophae rhamnoides* ssp. *turkestanica* and *Hippophae salicifolia* and to prepare a low calorie base material for further development of a food product which is rich in terms of antioxidant and is workable to be further processed and modified into a desired product. As seabuckthorn is high in vitamin C content which is heat and light sensitive, a very high treatment of temperature will render the pulp to damage. On the other hand, a very low temperature will not be able to dry the product; hence 60°C was considered the optimum temperature to incorporate drying process.

MATERIALS AND METHODS

Seabuckthorn Samples

The Two varieties of seabuckthorn pulp *viz. Hippophae salicifolia* and *Hippophae rhamnoides* were collected from the Lahaul Spiti Seabuckthorn Co-operative society, Keylong (Himachal Pradesh). The seabuckthorn variety *Hippophae salicifolia* has its origin from Tinu village where as *Hippophae rhamnoides* originates from Jispa village; both in Keylong district. The pulp was extracted and packed in the month of October 2012. Potassium metabisulphite (KMS) was added as the preservative in the pulp at rate of 0.05 per cent. KMS prevents growth of bacteria in the pulp and hence protects the pulp from fermentation. The pulp was kept at refrigeration temperature (5°C) at all times *i.e.* during transportation and storage so that pulp doesn't lose vitamin C content due to heat. The pulps *H. salicifolia* and *H. rhamnoides* were respectively contained in a plastic food grade 5 L container at all times.

Sucrose and Sugar Alcohols

Sucrose (white granulated Sugar) was bought from the local market. The sugar purchased is free from dust particles and is hermetically sealed in food grade High Density Poly Ethylene package. Sugar Alcohols Sorbitol, Mannitol, Maltitol and Xylitol were purchased from local lab chemical supplier.

Drying of Seabuckthorn Pulp

The drying was carried out for duplicates of both the varieties of seabuckthorn pulps *i.e. H. rhamnoides* (R) and *H. salicifolia* (S). Sucrose and the sugar alcohols *i.e.* Mannitol, sorbitol, Xylitol and Maltitol were added to the pulps at 10 per cent and 20 per cent (w/w) and were mixed thoroughly to ensure proper distribution. The mixture was taken in aluminum round trays for drying at the tray load of 7kg/m^2. The drying was carried out at the temperature of 60±°C in a laboratory cabinet dryer. Trays were weighted at 30 min intervals for the first 2 hours of drying and then after every hour till the moisture content of the pulp reached at 25-30 per cent wet basis. The pulp that changed to leather after drying was cooled at room temperature and packed in polythene pouch and stored in airtight jar at -18°C. The drying rate was determined in kg moisture removed/ hour of drying/kg dry matter. The moisture rate was determined from the ratio of moisture content at any level and at anytime in (per cent dry basis) to initial moisture content. The page model was adequately used to describe the moisture ratio and drying rate constants. The page model is a semi-theoretical model among the thin-layer drying models. It is generally used to study the drying behavior of thin layer of food products. The semi-theoretical thin layer models deals with the external resistance offered to dry *i.e.* between product and the drying air unlike the internal resistance where orientation of the food particles plays major role. Following equation has been used to study the drying behavior in this experiment-

$$MR = m - m_e / m_o - m_e = \exp(-kt^n)$$

Where MR= Moisture ratio; m= moisture content at any given time t (per cent dray basis); m_e= equilibrium moisture content (per cent dry basis); m_o= initial moisture content; t= time (hours); k= Drying rate constant (h^{-1}); n= dimensionless coefficient.

'k' is used to measure the slope in the drying model which is related to the effective diffusivity when drying happens in the falling rate period and liquid diffusion pedals the process. R^2 was taken as the coefficient of co-relation to measure the strength of liner relationship between Drying rate and Moisture ratio. The moisture content was measured by gravimetric method using digital balance.

Antioxidant activity

Free radical scavenging activity was measures using DPPH (1.1-diphenyl-2- picrylhydrazyl). 100mg of sample (on dry basis) was taken in eppendoff tubes and was extracted with 1 ml of methanol for 2 hours in shaker. The mixture was centrifuged at 10,000 rpm for 8 minutes and the supernatant was collected. 100µL of supernatant was taken and 3.9 ml of DPPH solution was added in a glass tube. The glass tube was covered properly and was kept for incubation at room temperature for 30 minutes. Absorbance was taken at 515 nm using UV-vis spectrophotometer (Shimadzu 1800, Japan). Methanol was taken as a blank for the transmittance and absorbance values. The radical scavenging activity of DPPH was calculated using following formula-

Per cent scavenging activity = [(Absorbance of control – Absorbance of sample)/Absorbance of control] × 100

RESULTS AND DISCUSSIONS

Two cultivars of seabuckthorn pulp were used in drying *i.e. Hippophae rhamnoides* and *Hippophae salicifolia*. The cultivars differ on the basis of moisture content, total solids, vitamin C content and colour. *H. rhamnoides* is cultivar of deep orange colour whereas *H. salicifolia* is light yellow to light orange in colour. *H. rhamnoides* has more total solids and hence the drying time was generally lower for the *H. rhamnoides* pulp. The physical characteristics of the end dried product were studied by visually studying the appearance (Tables 56.1 and 56.2).

Drying Behaviour

The drying trend of different samples highlights the reasons for the anomalous and normal behavior of pulp-sucrose/sugar alcohol mixtures. The drying was carried out for pulps from both the cultivars. A standard control was taken as a reference for comparison of solutions from both the cultivars. The samples were taken in duplicates and the result was inferred in the form of average of the duplicates. The drying of samples was carried out with ratio of 1:10 and 2:10 for sucrose/sugar alcohol and pulp (w/w) (Figure 56.1). The ratio of the sucrose and sugar alcohols were selected on the basis of previous similar studies and care were taken for the fact that amount of sugar alcohols present in the food shall be in a particular range as guided by FDA.

Table 56.1: Remarks on the Sensory Analysis of End Dried Product from *Hippophae rhamnoides*

Solution	*10 per cent Solvent*	*20 per cent Solvent*
Sucrose	Sticky consistency paste	Sticky consistency paste
Sorbitol	Formation of crystals on the surface after cooling, sticky paste	Formation of crystals on the surface after cooling, sticky paste
Mannitol	Dry consistence with crystals	Dry consistency with crystals, little sticky
Maltitol	Formation of ripples on surface, brittle appearance	Formation of ripples on surface, brittle
Xylitol	Smooth cake, uniform appearance	Smooth cake, uniform appearance

Table 56.2: Remarks on the Sensory Analysis of End Dried Product from *Hippophae salicifolia*

Solution	*10 per cent Solvent*	*20 per cent Solvent*
Sucrose	Sticky paste, non workable	Sticky paste, non workable
Sorbitol	Smooth paste	Smoother paste, small crystals on surface
Mannitol	Dry consistency	Dry consistency with ripples on the edges
Maltitol	Brittle and dry	Brittle and very dry
Xylitol	Smooth cake with adequate moisture	Smooth cake with adequate moisture

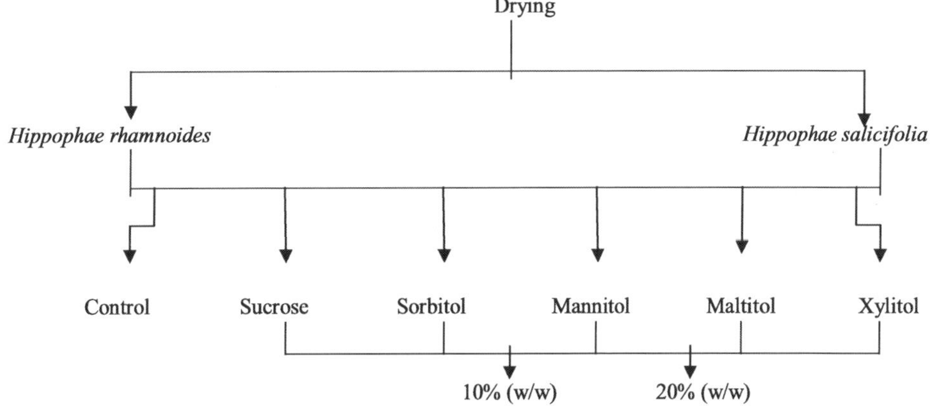

Figure 56.1: The Chart Describing the Sampling Procedure for Both the Cultivars of Seabuckthorn.

A separate control was taken for each pulp along with 10 and 20 per cent quantity of sucrose and sugar alcohol in the pulps. The drying time for the control pulp of *H. rhamnoides* to reach a moisture content of 30 per cent from 86 per cent was 6 hours whereas *H. salicifolia* took about 8 hours to reach from moisture content of 90 per cent to 30 per cent (wb). Clearly the cultivar *H. rhamnoides* has more total solids than *H. salicifolia.* The drying trend after the addition of sucrose for both the cultivars was almost similar (Figures 56.2 and 56.3). After addition of sucrose at 10 per cent and 20 per cent in *H. rhamnoides* and *H. salicifolia* resepectively the moisture content came down but drying time almost took the same time as control as sucrose binds water and hence the drying is slower. But as total solids increased, it compensates for the water binding capacity of sucrose and sample gets dried in almost the same drying time as control.

Moisture ratio is an important parameter that informs about the ratios of moisture removed in a particular time period. The value of moisture ratio lies between 0 and 1. It is generally observed in thin layer drying process that the moisture ratio decreases significantly initially during falling rate period. After the falling rate period the moisture ratio will mostly be constant. For control samples R and S, the moisture ratio followed the normal trend as observed in thin layer drying. However the moisture ratio deviated from normal (a dip was observed for 10 per cent and 20 per cent sucrose solutions) midway (Figure 56.4). This could be described by association with the drying curve. Sugar being very hygroscopic releases water with a tug after being dried for a certain amount of time and gets crystallized.

Similar graph for moisture ratios was observed for S cultivar samples for control, 10 per cent sucrose and

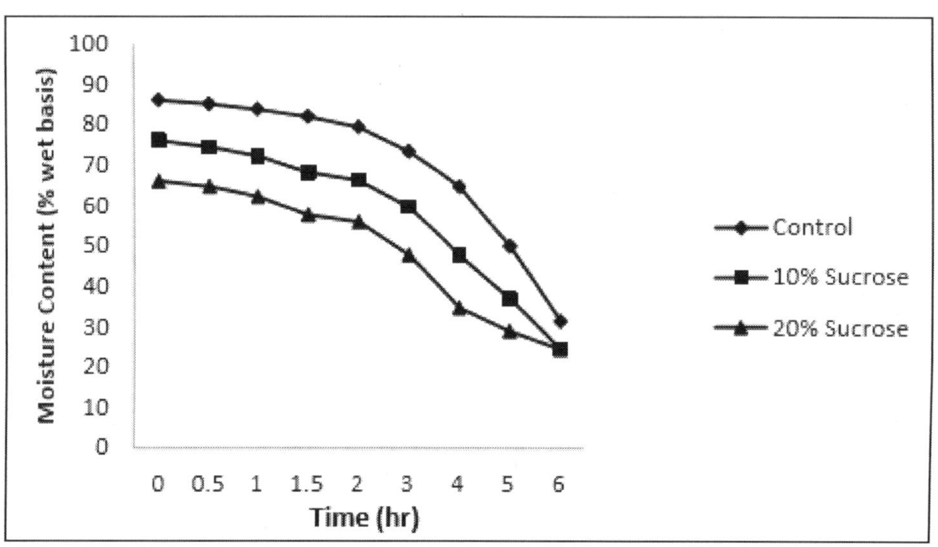

Figure 56.2: Moisture Content (per cent wb) of *H. rhamnoides* Pulp (control, 10 per cent sucrose, 20 per cent sucrose).

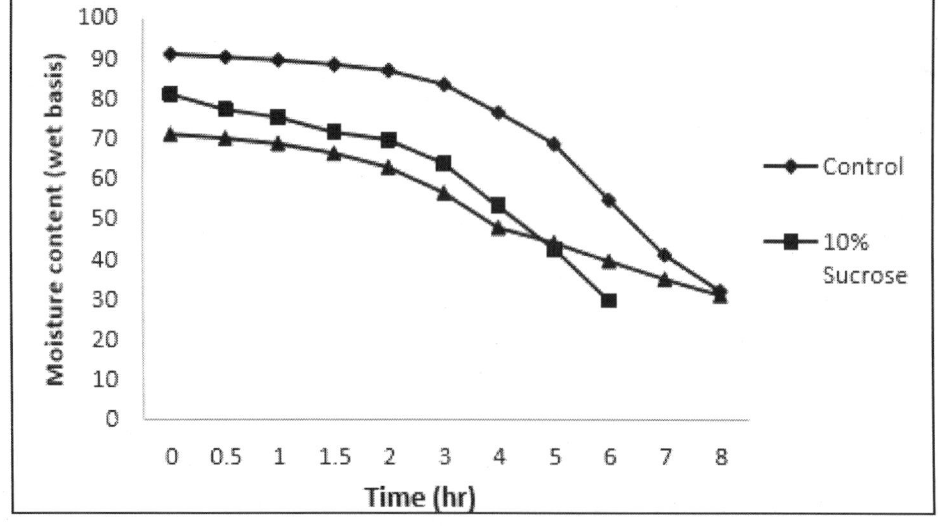

Figure 56.3: Moisture Content (per cent wb) of *H. salicifolia* Pulp (control, 10 per cent sucrose, 20 per cent sucrose).

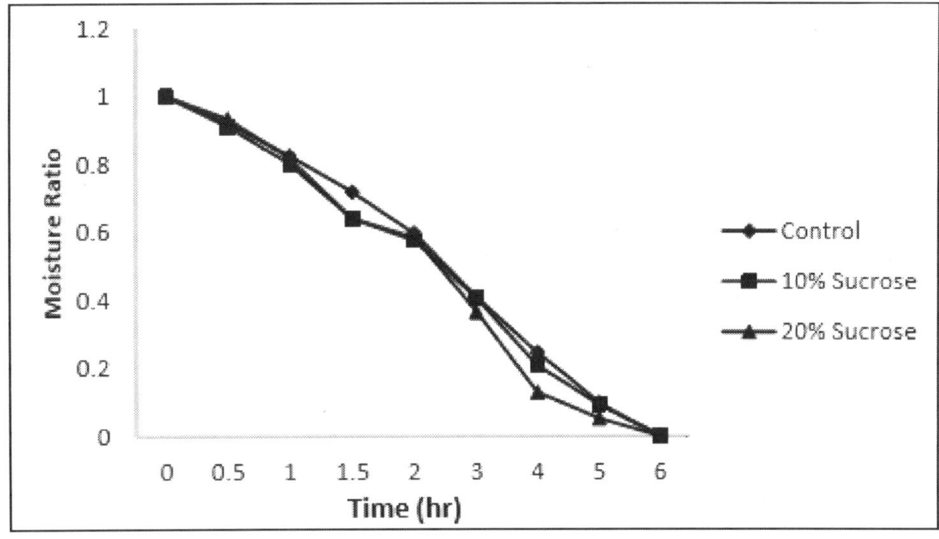

Figure 56.4: Moisture Ratio of *H. rhamnoides* Pulp (control, 10 per cent sucrose, 20 per cent sucrose).

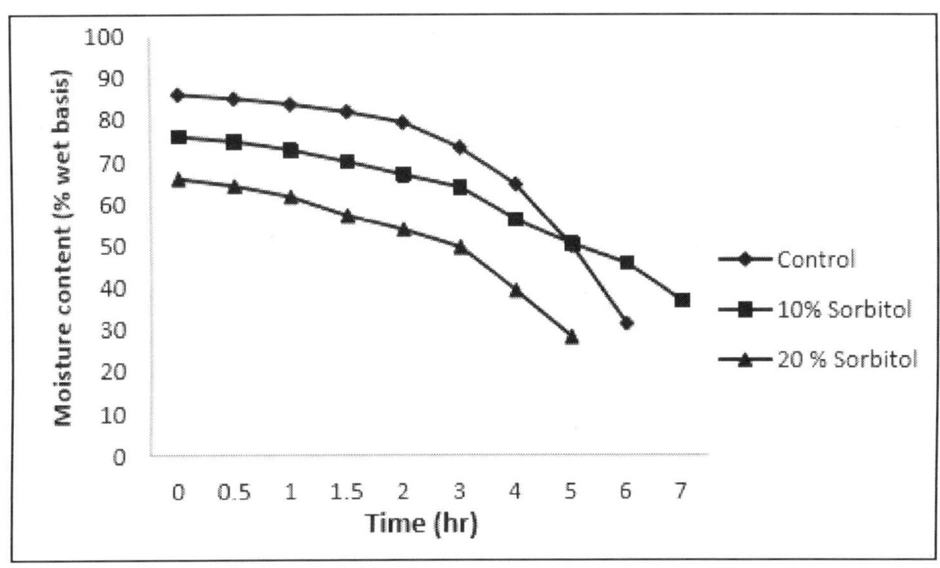

Figure 56.5: Moisture Content Values for *H. salicifolia* Cultivar Solution (10 per cent and 20 per cent sorbitol).

20 per cent sucrose. However due to larger moisture content a sharper dip was observed in ratio just after the falling rate period (Figure 56.5). Sorbitol samples (S and R) showed similar attributes as compared to sucrose samples in terms of moisture content and drying time however addition of sorbitol lead to a greater decrease in drying rate of seabuckthorn leathers as compared to sucrose. Sorbitol is more soluble (69 per cent at 20°C) than sucrose however it is less sweet (Pepper, 1990). Moisture ratio decreases with increasing drying time. The drying rate constant was increased on addition of sorbitol as compared to sucrose; the reason could be attributed to increase in humidity of the air during drying.

Mannitol is manufactured by hydrogenation of fructose using either invert sugar or high fructose corn syrup as raw material (Pepper, 1990). The drying rate for Mannitol samples for both the cultivars followed a normal drying curve trend (Figure 56.6). As compared to control the sample (R) containing 10 per cent Mannitol overlapped the curve in moisture content graph. This deviation could be explained by the hygroscopic nature of sugar alcohols. The moisture loss seizes during drying after equilibrium Moisture Content (EMC) is achieved. The moisture ration increased between 2nd and 4th hour of drying for samples containing 10 per cent and 20 per cent Mannitol for both S and R cultivars solutions.

Mannitol is a disaccharide sugar alcohol comprising glucose and sorbitol (Pepper, 1990). The Maltitol syrups as a result are thicker in consistency. The appearance of the dried product obtained was not appealing. Being thick provides more water holding capacity and

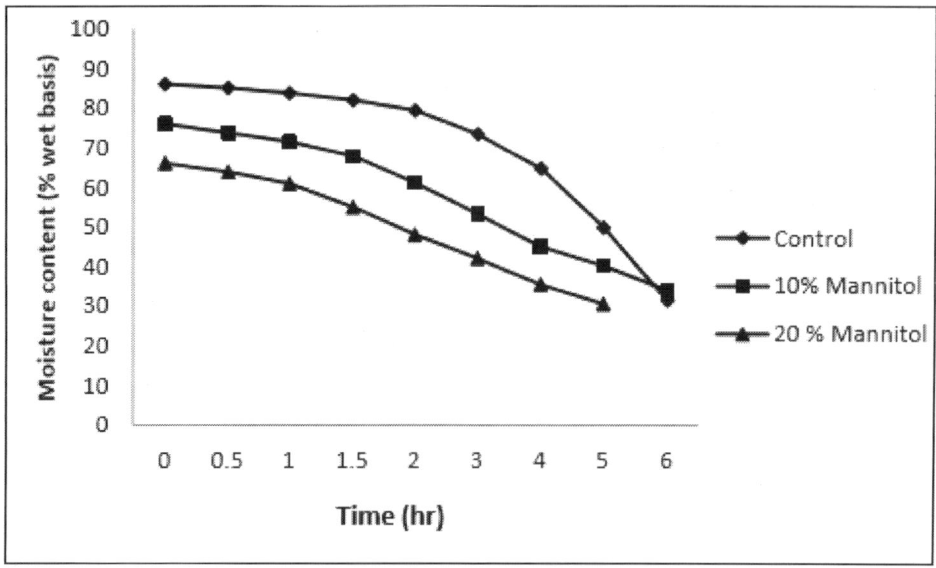

Figure 56.6: Moisture Content of Sample Containing 10 per cent and 20 per cent Mannitol for *H. rhamnoides* Solution.

lesser drying in certain drying time for the solutions. A constant fall in moisture content was observed up till last hours in moisture content for R sample as opposed to lesser decrease in control sample. In S Maltitol sample a sharp decrease for 10 per cent Maltitol sample was observed between 4th and 5th hour. This could be attributed to the capacity for Maltitol to segregate and settle down (owing to heavier molecular weight) in the solution. Normal trends for moisture ratio were observed as compared to control sample.

Xylitol is a five carbon sugar alcohol and is commercially manufactured from the hemicelluloses xylan. Xylitol is sweetest of all the sugar alcohols and has the lowest calorific value too (Pepper, 1990). The visual appearance of the dried product contacting 10 per cent and 20 per cent Xylitol was appealing. The taste of the product was also sweeter as compared to control, sucrose and other sugar alcohol containing samples. The drying rate of the sample decreased with increasing the amount of Xylitol in the sample as the total solid content of the sample increases. The only aberration in the Xylitol drying kinetics was for a significant change in moisture ratio between 2.5th to 3rd hours of drying. This behaviour of the sample was consistent in duplicates and could be explained from bound water theory of foods.

Antioxidant Activities

The percentage antioxidant activity of seabuckthorn pulps (*H. rhamnoides* and *H. salicifolia*) was determined using DPPH and it was found that the radical scavenging effect in seabuckthorn pulps A and B were was 95.79 per cent and 94 per cent. When the pulp was dried at 60°C over a period of 6 h into leather, decrease in antioxidant activity was observed (Figures 56.7 and 56.8). Lohachoompol *et al.* (2004) reported that there was no significant difference in the antioxidant activity of blueberries after drying from the fresh berries. Diaconeasa *et al.* (2015) studied the antioxidant activity of blueberries and reported that antioxidant activity decreased when blueberries were dried at 70°C. Sogi *et*

Table 56.3: Drying Rate Constant (k), Dimensionless Coefficient (n) and R^2 Values for Seabuckthorn (*Hippophae rhamnoides*) Solution in Presence of Sucrose and Sugar Alcohols

Hippophae rhamnoides	*Drying Rate Constant (k)*	*n (Dimensionless coefficient)*	*R^2*
Control	0.202	1.419	0.9935
10 per cent Sucrose solution	0.233	1.368	0.9923
20 per cent Sucrose solution	0.205	1.602	0.9920
10 per cent Sorbitol Solution	0.208	1.331	0.9953
20 per cent Sorbitol Solution	0.252	1.373	0.9898
10 per cent Mannitol solution	0.335	1.328	0.9924
20 per cent Mannitol Solution	0.341	1.487	0.9918
10 per cent Maltitol solution	0.234	1.281	0.9954
20 per cent Maltitol solution	0.319	1.402	0.9895
10 per cent Xylitol Solution	0.300	1.408	0.9873
20 per cent Xylitol solution	0.322	1.570	0.9938

Table 56.4: Drying Rate Constant (k), Dimensionless Coefficient (n) and R^2 Values for Seabuckthorn (*Hippophae salicifolia*) Leather Samples in Presence of Sucrose and Sugar Alcohols

Hippophae salicifolia	*Drying Rate Constant (k)*	*n (Dimensionless coefficient)*	R^2
Control	0.191	1.396	0.9834
10 per cent Sucrose solution	0.175	1.597	0.9872
20 per cent Sucrose solution	0.154	1.538	0.9970
10 per cent Sorbitol solution	0.281	1.209	0.9501
20 per cent Sorbitol solution	0.262	1.542	0.9738
10 per cent Mannitol solution	0.165	1.480	0.9982
20 per cent Mannitol solution	0.231	1.339	0.9826
10 per cent Maltitol solution	0.160	1.876	0.9856
20 per cent Maltitol solution	0.317	1.131	0.9845
10 per cent Xylitol solution	0.256	1.431	0.9885
20 per cent Xylitol solution	0.317	1.259	0.9952

al. (2013) carried out the study on antioxidant activity of mango peel and kernel as affected by drying methods and reported a decrease in antioxidant activity of mango kernel with the application of heat during drying.

The addition of sucrose and sugar alcohols prevented any decrease in the antioxidant activity with xylitol having the most significant effect for *Hippophae*

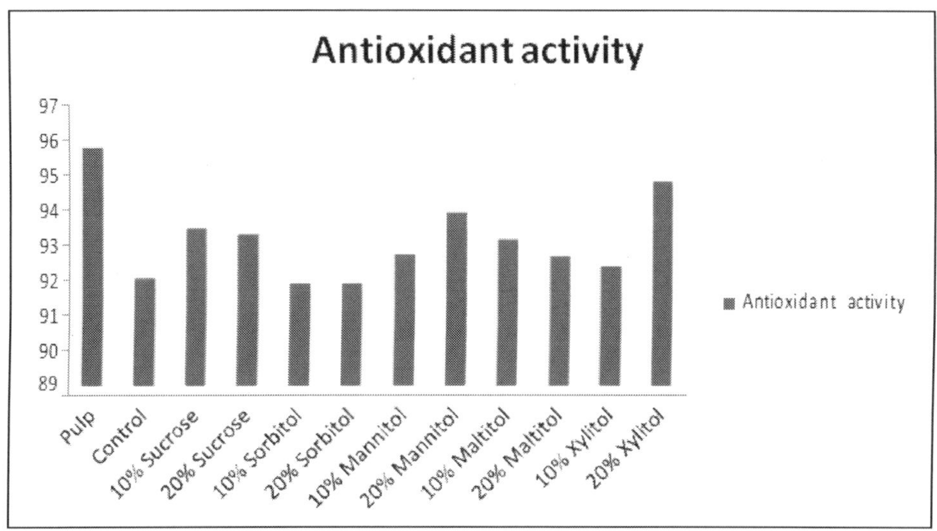

Figure 56.7: Antioxidant Activity of Control Seabuckthorn *H. rhamnoides* Leather Containing Sucrose and Sugar Alcohols.

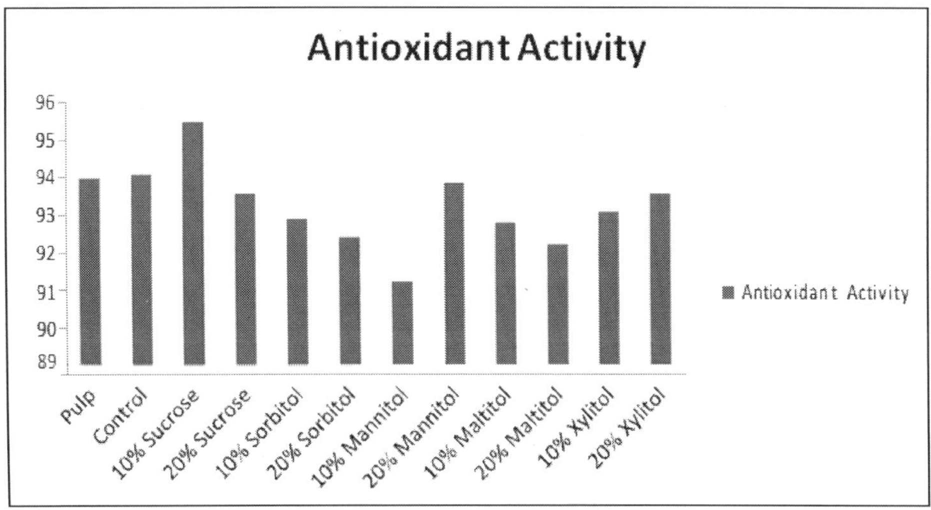

Figure 56.8: Antioxidant Activity of Control Seabuckthorn *H. salicifolia* Leather Containing Sucrose and Sugar Alcohols.

rhamnoides variety of seabuckthorn leather followed by mannitol, sucrose, maltitol and sorbitol. In case of *Hippophae salicifolia* variety of seabuckthorn leather, the most significant of all sugars was sucrose followed by mannitol, xylitol, sorbitol and maltitol. The better retention of antioxidant activity in the presence of sugar and sugar alcohols can again be attributed to the decrease in the availability of free water which leads to greater retention of the compounds that contribute to the antioxidant activity of the pulp. Benjakul *et al.* (2005) reported that when sugars are subjected to heat, the caramelisation products that are formed have antioxidant activity and this may have attributed to the increase in antioxidant activity of seabuckthorn leathers. They also reported that during heating of sugar solutions, reducing compounds might be formed and these could exhibit the antioxidant activity of maillard reaction products associated with reducing power (Yen and Hsieh, 1995). The reducing power of caramelisation products might be due to hydrogen donating ability (Shimada, Fijikawa, Yahaa and Nakamura, 1992). Tsai (2005) studied the effect of sugar on anthocyanin during high temperature heating and they reported that at low pH, the hydrolysis of sucrose increase the antioxidant capacity.

CONCLUSION

Upon the addition of sucrose and sugar alcohols at the concentration of 10 per cent and 20 per cent, there is some reduction in drying time to make seabuckthorn (*H. rhamnoides* and *H. salicifolia*) leather which is attributed to the higher total solids and less moisture present in the pulp containing sucrose and sugar alcohols. However, it was observed that drying rate of the pulp containing sucrose and sugar alcohols was less as compared to control pulp. Moisture ratio of seabuckthorn (*H. rhamnoides* and *H. salicifolia*) leather decreased with increase in drying time. Antioxidant activity of seabuckthorn (*H. rhamnoides* and *H. salicifolia*) leather was increased in the presence of sucrose and sugar alcohols due to their protective effect on the bioactive compounds responsible for antioxidant activity and breakdown of sugars into products that contribute to the antioxidant activity.

REFERENCES

1. Benjakul, S., Visessanguan, W., Phongkanpai, V. and Tanaka, M. 2005. Antioxidative activity of caramelisation products and their preventive effect on lipid oxidation in fish mince. *Food Chemistry* 90: 231-239.

2. Dhaliwal, Y.S., Verma, R., Anand, S. and Singh, V. 2010. Formulation and acceptability of seabuckthorn based value added products. In: *Souvenir of National Conference on "Seabuckthorn, Emerging Trends in Production to Consumption"*, C.S.K.H.P.K.V., Palampur Feb.16-18, 2010.

3. Diaconeasa, Z., Cuibus, L. and Socaciu, C. 2015. Phenolic content and their antioxidant activity in various berries cultivated in Romania. *Bulletin UASVM Food Science and Technology* 72 (1).

4. Gupta, R.K. and Singh, V. 2005. Minerals dynamics during fruit ripening in seabuckthorn (*Hippophae* L.) growing in dry temperate Himalayas. In: *The 2nd International Seabuckthorn Association Conference on "Seabuckthorn: A Multipurpose Plant, Solution to Nutrition, Health and Environment"*, Aug. 26-29, Beijing, China.

5. Lohachoompol, V., Srzednicki, G. and Craske, J. 2004. The change of total anthocyanins in blueberries and their antioxidant effect after drying and freezing. *Journal of Biomedicine and Biotechnology* 5: 248-252.

6. Pepper, T. 1990. *Alternative Bulk Sweeteners in Sugar Confectionary Manufacture First Edition*, Blackie and Son Ltd, London, pp. 17-26.

7. Chaurasia, O.P., Stobdan, T., Kumar, R., Kumar, P., Sharma, D., Tayade, A. and Singh, S.B. 2010. Seabuckthorn of Trance-Himalaya and its nutritive health products, In: *National Conference on Seabuckthorn, Emerging Trends in Production to Consumption*, C.S.K.H.P.K.V., Palampur Feb 16-18, 2010, p.59.

8. Shimada, K., Fujikawa, K., Yahaa, K. and Nakamura T. 1992. Antioxidative properties of xanthan on autoxidation of soybean oil cyclodextrin emulsion. *Journal of Agricultural and Food Chemistry* 40: 945-948.

9. Singh, V. 1995. Rooting rates of hardwood cuttings of seabuckthorn (*Hippophae rhamnoides*). *Journal of Tree Sciences* 14: 87-88.

10. Sogi, D.S., Siddiq, M., Greiby, I. and Dolan, K. 2013. Total phenolics, antioxidant activity, and functional properties of 'Tommy Atkins' mango peel and kernel as affected by drying methods. *Food Chemistry* 141: 2649-2655.

11. Tsai, P.J., Delva, L., Yu, T.Y., Huang, Y.T. and Dufosse, L. 2005. Effect of sucrose on the anthocyanin and antioxidant capacity of mulberry extract during high temperature heating. *Food Research International* 38: 1059-1065.

12. Yen, G.C. and Hsieh, P.P. 1995. Antioxidative activity and scavenging effects on active oxygen of xylose-lysin Maillard reaction products. *Journal of Food and Agriculture* 67: 415- 420.

Chapter 57

Formulation and Quality Evaluation of Seabuckthorn Blended Sand Pear Squash

R. Verma*, S. Sharma, Y.S. Dhaliwal and Virendra Singh

Department of Food Science, Nutrition and Technology
CSK Himachal Pradesh Agricultural University, Palampur – 176 062, H.P., India
(*E-mail: ranjana3in@yahoo.com)

SUMMARY

Sand pear (*Pyrus pyrifolia*) also known as Asian pear or Apple pear is among the underutilized fruits of Himachal Pradesh. Besides having good nutritional and medicinal profile, sand pear fruit has hard grit cells which impart coarse and grainy texture to it. Presence of these hard cells reduces the palatability of fruit juice/pulp and its value added products. The taste of the juice is bland with low acid content. On the other hand, seabuckthorn juice is acidic in taste and cannot be consumed as such. So, blending of these two juices can improve the taste and texture of the value added products. Hence, attempts were made to standardize various blending proportions of seabuckthorn juice to improve the quality attributes of sandpear based squash. Squash as per FPO specifications was prepared by blending sand pear juice with seabuckthorn pulp in varying proportions *viz.* 75:25, 50:50 and 25:75 and stored up to six months of storage. Chemical evaluation of the squash revealed that TSS, total sugars, reducing sugars and acid brix ratio of pure sand pear squash decreased significantly ($P \leq 0.05$) as the blending proportion of seabuckthorn pulp increased in the squash. Whereas, the ascorbic acid content of fresh sand pear squash increased significantly ($P \leq 0.05$) from 4.56 to 14.64, 24.68 and 34.75 mg/100ml in seabuckthorn blended squash with 25, 50 and 75 per cent blending proportions respectively. The sensory scores for various attributes increased significantly ($P \leq 0.05$) with increase in the blending proportions of the seabuckthorn pulp in the squash. It was observed that irrespective of the blending proportions, the sensory scores of all the treatments decreased with increase in storage of squash but remained within the acceptable range of liked moderately to liked very much on 9 point hedonic scale. Results of the study revealed that nutritional and sensory quality of sand pear squash can further be improved by blending it with seabuckthorn- a miraculous berry which is blessed by nature with the goodness of antioxidants and medicinal properties.

INTRODUCTION

Sand pear commonly known as *Patharnakh* belongs to family *Rosaceae*. The fruit is round and green. When fully ripened, it is firm and juicy and has high water content and a crisp, grainy texture. It is highly nutritious and contains protein, fat, carbohydrates, thiamine, riboflavin, niacin, malic acid, citric acid, fructose, sucrose, ascorbic acid and other organic elements. Sand pear in addition to fiber also contains potassium, calcium, magnesium, selenium, iron, manganese, and other compounds (USDA, 2008). In India, *Patharnakh* pear is largely grown for consumption as fresh fruit, while in other countries; it is processed to a considerable extent. The commercial processing in our country has been limited to canning of pear slices in sugar syrup, however it can also be utilized for preparation of various value added products such as jams, pickles, candies and beverages.

Pear juice is low in acid, bland in flavor and pulpy juice and is so thick that it does not appeal to most people. So, blending of the sand pear juice with seabuckthorn which has bright orange colour with acidic taste can influence the flavor, taste and consistency of the product. Utilization of sand pear by blending with other suitable fruits for the preparation of value added products will not only provide a wide spectrum of

Seabuckthorn (*Hippophae* L.): *A Multipurpose Wonder Plant, Vol. 5: Advanced Technologies for Cultivation, Processing, Health Protection and Environmental Conservation* (Virendra Singh, *Editor-in-Chief*, 2018), p. 439–445, Astral International Pvt. Ltd., New Delhi, India.

processed products at household as well as commercial level but will also add to the economy of the farmers who has to sell their produce at throw away prices.

MATERIALS AND METHODS

The sand pear (*Pyrus pyrifolia*) fruits were procured from Regional Research Station, Bajaura (Kullu), CSK HPKV, Palampur. The procured fruits were sorted, washed, graded according to the size and thereafter pulping was done. The seabuckthorn pulp was procured from the Mantra Ayurveda, Faridabad, U.P and chemicals and other ingredients were procured from the local market.

Squash is a type of fruit beverage containing at least 25 per cent fruit juice or pulp and 40 to 50 per cent total soluble solids, commercially. It also contains about 1.0 per cent acid and 350 ppm sulphur dioxide or 600 pm sodium benzoate as per FPO specifications. It is diluted before serving. In the present study, squash was prepared by blending sand pear juice with seabuckthorn pulp in varying proportions *viz.*100:00, 75:25, 50:50 and 25:75. To the requisite amount of sugar, acid and water was added for preparation of pure sand pear squash with 45°B TSS, 1.0 per cent acid and 25 per cent fruit juice and with the same proportion of ingredients squash blended with seabuckthorn pulp was prepared (Figure 57.1, Plate 57.1).

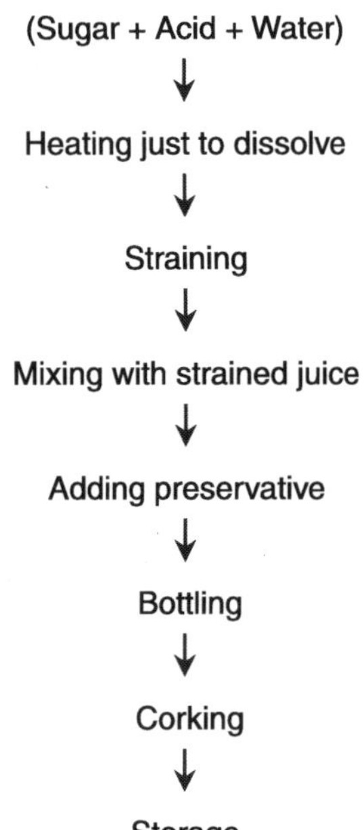

Figure 57.1: Steps Involved in Preparation of Squash.

The chemical and nutritional characteristics of the formulated squash were ascertained by using standard methods (Ranganna, 2007) and the sensory evaluation was conducted by using a panel of ten judges for evaluating the samples on a 9 point hedonic scale (Larmond, 1977). The recorded data were subjected to statistical analysis with the help of computer software. Data were compared at 5 per cent level of significant.

Plate 57.1: Physical Appearance of Squash Prepared by using Various Proportions of Sand Pear and Seabuckthorn.

RESULTS AND DISCUSSIONS

The initial TSS content of sand pear squash was 45.30°B which decreased significantly ($P \leq 0.05$) to 44.95, 44.00 and 43.00°B with addition of 25, 50 and 75 parts of seabuckthorn pulp in sand pear squash respectively (Figure 57.2). With storage, the mean TSS content of squash decreased significantly from 44.31 to 44.07 and 43.48°B after storage of 3 and 6 months respectively. Conversion of sugars into acid could be the reason for decrease in TSS (Sahu *et al.*, 2005; Sirohi *et al.*, 2005; Kumari, 2007).

The data pertaining to the total, reducing and non reducing sugars of sand pear based squash blended with seabuckthorn pulp are presented in Table 57.1.

As is evident from the data that with the addition of seabuckthorn pulp in sand pear juice for preparation of squash, there was a declining effect on the total and reducing sugars content of squash. The values for total and reducing sugar of pure sand pear squash were 19.50 and 13.48 per cent which significantly decreased to 19.20 and 13.25 in 75:25, 18.70 and 13.00 in 50:50, 18.50 and 12.93 per cent in 25:75 Sand pears: seabuckthorn pulp respectively. On the contrary, with increase in storage period the mean total and reducing sugar values increased significantly ($P \leq 0.05$) from 18.98 and 13.17 to 20.31 and 14.10 per cent after 6 months of storage respectively. The increase in total sugars content with storage duration may be due to the breakdown of polysaccharides like pectin and starch to the simple sugars Krishnaveni *et al.* (2001) and Singh and Nath (2004) and an increase in reducing sugars may be due to conversion of non-reducing sugars owing to the process of hydrolysis and lactose hydrolysis to glucose and galactose (Sethi, 1993; Kannan and Thirumaran, 2004; Kumari, 2007). A perusal of data revealed that as the proportion of seabuckthorn pulp increased in squash, the non-reducing sugars decreased from 6.02 to 5.95, 5.70

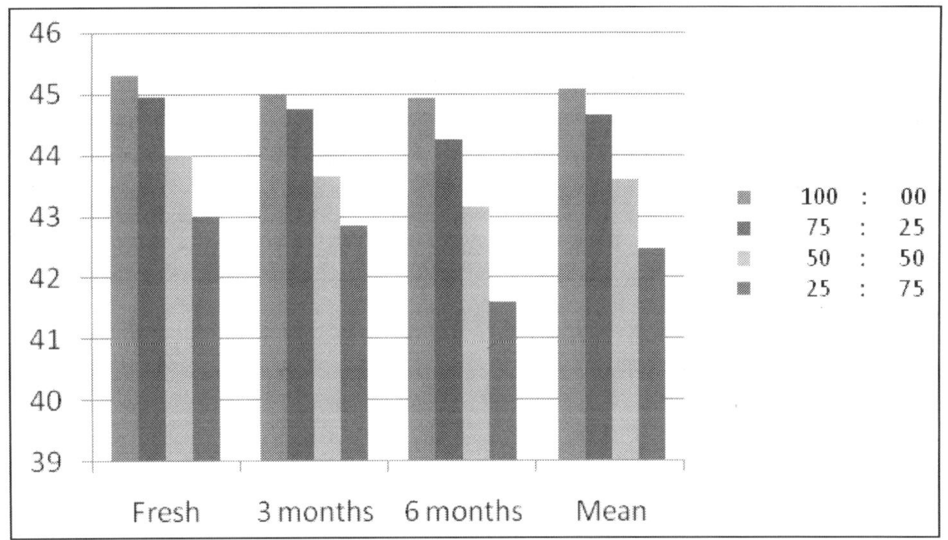

Figure 57.2: Effect of Blending and Storage on TSS Content of Squash.

Table 57.1: Effect of Blending and Storage on the Sugars of Sand Pear Based Squash Blended with Seabuckthorn Pulp

Parameter/Proportions	*Storage Intervals*			
Sand pear: Seabuckthorn	*Fresh*	*3 Months*	*6 Months*	*Mean*
Total Sugars				
100 : 00	19.50	20.00	20.71	20.07
75 : 25	19.20	19.87	20.58	19.88
50 : 50	18.70	19.45	20.00	19.38
25 : 75	18.50	19.25	19.97	19.24
Mean	18.98	19.64	20.31	19.64
Reducing Sugars				
100 : 00	13.48	13.98	14.25	13.90
75 : 25	13.25	13.78	14.13	13.72
50 : 50	13.00	13.50	14.00	13.50
25 : 75	12.93	13.49	14.01	13.48
Mean	13.17	13.69	14.10	13.65
Non Reducing Sugars				
100 : 00	6.02	6.02	6.46	6.17
75 : 25	5.95	6.09	6.45	6.16
50 : 50	5.70	5.95	6.00	5.88
25 : 75	5.57	5.76	5.96	5.76
Mean	5.81	5.96	6.22	6.00
CD ($P \leq 0.05$)	*Total Sugars*	*Reducing Sugars*	*Non Reducing Sugars*	
Storage intervals (A)	0.17	0.11	0.14	
Blends (B)	0.19	0.12	0.16	

and 5.57 per cent in 100:00, 75:25, 50:50 and 25:75 sand pear: seabuckthorn proportions respectively. Whereas, with storage, the mean non-reducing sugars content increased significantly from 5.81 to 5.96 per cent and 6.22 per cent after 3 and 6 months respectively.

Table 57.2 indicates the effect of blending seabuckthorn pulp and storage on the chemical characteristics of sand pear based squash. The mean value of pH in sand pear squash was 3.33 which decreased significantly ($P \leq 0.05$) to 3.28, 3.16 and 3.23 in 75:25, 50:50 and 25:75 proportion of sand pear and seabuckthorn pulp in squash respectively. Similarly, with storage the pH values decreased significantly in all the blending proportions. The pH value of fresh squash was 3.35 which decreased to 3.14 after 6 months of storage. The decrease in pH might be due to certain

Table 57.2: Effect of Blending and Storage on the pH, Acidity, Brix/Acid Ratio and Ascorbic Acid of Sand Pear Based Squash Blended with Seabuckthorn Pulp

Parameter/Proportions	*Storage Intervals*			
Sand pear: Seabuckthorn	*Fresh*	*3 Month*	*6 Months*	*Mean*
pH				
100 : 00	3.40	3.34	3.25	3.33
75 : 25	3.36	3.29	3.20	3.28
50 : 50	3.34	3.21	2.94	3.16
25 : 75	3.30	3.23	3.15	3.23
Mean	3.35	3.27	3.14	3.25
Acidity (per cent)				
100 : 00	1.05	1.09	1.12	1.09
75 : 25	1.23	1.26	1.27	1.25
50 : 50	1.32	1.33	1.35	1.33
25 : 75	1.41	1.42	1.44	1.42
Mean	1.25	1.28	1.30	1.27
Ascorbic Acid (mg/100g)				
100 : 00	4.56	2.88	2.25	3.23
75 : 25	14.64	9.52	6.88	10.34
50 : 50	24.68	16.50	11.40	17.55
25 : 75	34.75	22.58	16.71	24.68
Mean	19.66	12.88	9.32	13.95
CD ($P \leq 0.05$)	*pH*	*Acidity*	*Ascorbic Acid*	
Storage intervals (A)	0.15	0.01	0.04	
Blends (B)	0.18	0.01	0.05	

chemical reactions taking place during storage, thus leading to the production of organic acid (Sahu *et al.*, 2005; Saravana and Manimegalai, 2005; Kumari, 2007).

The acidity content of squash increased significantly as the proportion of seabuckthorn increased in sand pear squash. The acidity content of 1.05 per cent in pure sand pear squash increased to 1.23, 1.32 and 1.41 per cent as the proportion of seabuckthorn increased from 25 to 50 and 75 in sand pear squash respectively. With storage, the acidity content increased in sand pear based squash blended with seabuckthorn pulp. The mean acidity content of squash (1.25 per cent) increased significantly (1.28 and 1.30 per cent) after 3 and 6 months of storage. Organic acid formed by ascorbic acid degradation might have increased the acidity on storage. Similar results have been reported by Krishnaveni *et al.* (2001), Sirohi *et al.* (2005) and Kumari (2007).

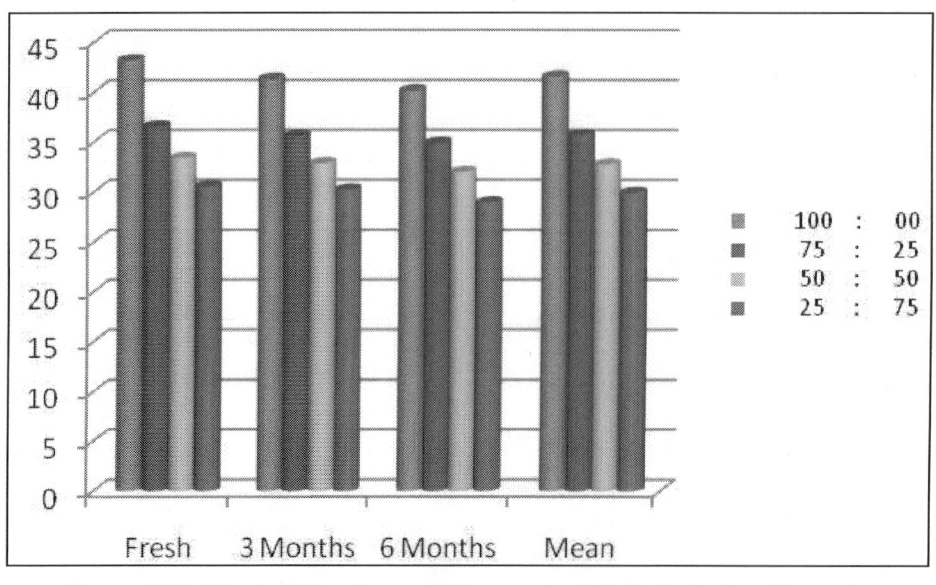

Figure 57.3: Effect of Blending and Storage on Brix/Acid Ratio of Squash.

As the proportion of seabuckthorn increased from 25 to 50 and 75 per cent in squash, the ascorbic acid content of sand pear squash increased significantly from 4.56 to 14.64, 24.68 and 34.75 mg/100ml respectively. On the contrary, the mean ascorbic acid content decreased from 19.66 to 9.32 mg/100 ml after 6 months of storage. The results are in conformity with those reported by Krishnaveni *et al.* (2001), Saravana and Manimegalai (2005) and Kumari (2007).

As is evident from the Figure 57.3 the brix/acid ratio content of pure sand pear squash was 43.14 which decreased significantly ($P \leq 0.05$) to 36.54 in 75:25 proportion and 33.33 to 30.50 in 50:50 and 25:75 proportion respectively.

Sensory Evaluation

The sensory scores of sand pear squash blended with seabuckthorn as affected by storage are presented in Table 57.3. The colour score of sand pear squash was recorded as 6.67 which increased significantly ($P \leq 0.05$) from 7.00 to 8.00 and 8.33 as the proportion of seabuckthorn pulp increased from 25 to 50 and 75 per cent respectively. With storage, the mean colour scores of squash prepared with all the blends decreased significantly ($P \leq 0.05$) from 7.50 to 7.42 and 7.17 with increase in storage duration of 3 and 6 months. Scrutiny the data revealed that as the proportion of seabuckthorn pulp increased in squash the colour scores increased considerably. The bright orange colour of the seabuckthorn imparted bright colour to the squash prepared by using different proportions of seabuckthorn and squash with 75 parts of seabuckthorn pulp was rated as best in terms of colour. Irrespective of the storage intervals, the colour scores in all the treatments increased and irrespective of the treatments the colour scores of all the treatments decreased with increase in storage of squash. The reduction in color scores reduced during storage might be due to effect of storage on color pigments. The results are in concurrence with those reported by Saravana and Manimegalai (2005) and Kumari (2007).

Similarly, the flavor and taste scores of sand pear squash blended with seabuckthorn increased with increase in blending proportion of seabuckthorn

Table 57.3: Effect of Blending and Storage on the Sensory Scores of Sand Pear Based Squash Blended with Seabuckthorn Pulp

Parameter/Proportions	*Storage Intervals*			
Sand pear: Seabuckthorn	*Fresh*	*3 month*	*6 months*	*Mean*
Colour				
100 : 00	6.67	6.67	6.33	6.56
75 : 25	7.00	7.08	7.00	7.03
50 : 50	8.00	7.92	7.67	7.86
25 : 75	8.33	8.00	7.67	8.00
Mean	7.50	7.42	7.17	7.36
Flavor				
100 : 00	6.58	6.42	6.33	6.44
75 : 25	7.00	6.92	6.67	6.86
50 : 50	7.67	7.25	6.67	7.19
25 : 75	7.00	6.67	6.33	6.67
Mean	7.06	6.81	6.50	6.79
Taste				
100 : 00	6.67	6.67	6.33	6.56
75 : 25	7.00	6.92	6.67	6.86
50 : 50	7.67	7.67	7.00	7.44
25 : 75	7.00	6.92	6.67	6.86
Mean	7.08	7.04	6.67	6.93
Consistency				
100 : 00	6.67	6.67	6.67	6.67
75 : 25	7.00	7.00	6.58	6.86
50 : 50	8.00	7.92	7.75	7.89
25 : 75	8.33	8.00	7.75	8.03
Mean	7.50	7.40	7.19	7.36
CD ($P \leq 0.05$)	*Colour*	*Flavor*	*Taste*	*Consistency*
Storage intervals (A)	0.31	0.36	0.36	0.30
Blends (B)	0.36	0.42	0.42	0.35

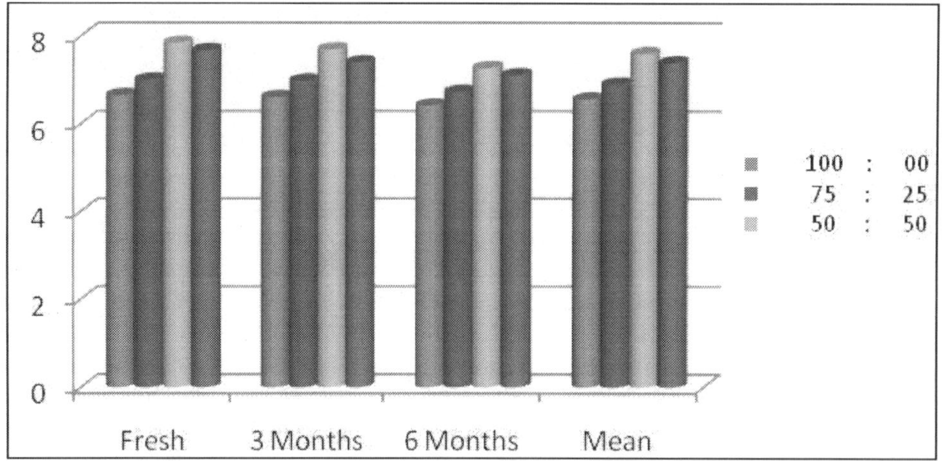

Figure 57.4: Overall Acceptability of Squash on 9 Point Hedonic Scale.

pulp. The mean flavor and taste scores of sand pear squash were 6.44 and 6.56 which increased to 6.86, 6.86 and 7.19, 7.44 in squash prepared with sand pear: seabuckthorn blends of 75:25 and 50:50 respectively. The mean flavor and taste scores in squash prepared with 25:75 parts of sand pear and seabuckthorn were 6.67 and 6.86 respectively and were less than those recorded for squash prepared by using 25:75 and 50:50 parts of sand pear and seabuckthorn pulp. The squash prepared by blending equal proportion of sand pear and seabuckthorn scored maximum scores (7.19 and 7.44) in respect of flavor and taste. In all the treatments, the flavor scores decreased with storage and the mean values of 7.06 of freshly prepared squash decreased to 6.81 and 6.50 after 3 and 6 months of storage respectively. The reduction in flavor scores might be due to degradation of volatile substances of flavor constituents upon storage time. Similar observations have been reported by Nath *et al.* (2005) and Kumari (2007). Similarly, the taste scores also decreased from 7.08 to 7.04 and 6.67 with storage intervals of 3 and 6 months respectively. The decrease in taste scores during storage may be due to reduction of volatile flavoring components during storage which are ultimately responsible for decreased taste. Sirohi *et al.* (2005) and Kumari (2007) have also reported similar observations.

Similar observations were recorded for the consistency scores of seabuckthorn blended sand pear squash where the scores increased significantly as the proportion of seabuckthorn increased in blends for squash. The mean scores for consistency irrespective of storage in pure sand pear squash were 6.67 which increased to 6.86, 7.89 and 8.03 in squash prepared by using 25, 50 and 75 parts of seabuckthorn pulp. The mean scores for consistency were 7.50 which decreased to 7.40 and 7.19 after 3 and 6 months of storage thereby signifying that in all the treatments the consistency scores declined significantly with increase in storage interval. This declining trend might be due to decrease in TSS with the increasing storage period. Similar observations have been reported by Kumari (2007).

The overall acceptability scores also followed similar trends where the squash prepared by using equal proportions of sand pear juice and seabuckthorn pulp was rated best with maximum scores of 7.84 and after this the scores declined (7.67) as the proportion of seabuckthorn increased (Figure 57.4). In general it was observed that blending of seabuckthorn pulp improved the acceptability scores from 6.56 to 6.90 and 7.60. A decrease in scores with storage was observed where the mean overall acceptability scores of 7.29 declined to 7.17 and 6.88 after 3 and 6 months of storage respectively but the values ranged between liked moderately to liked very much.

CONCLUSIONS

Sand pear is good source of nutrients. In general, a considerable proportion of the fruit goes waste between harvest and consumption due to the lack of proper and adequate infrastructure and processing facilities. Extraction of juice and its conversion into different value added products like beverages by blending with other suitable fruits can be one of the alternatives for checking postharvest losses of fruit crops. Such value-added products can significantly enhance farmers' businesses, affording them a steady income throughout the year when their sales of fresh produce tend to dip.

REFERENCES

1. Kannan, S. and Thirumaran, A. S. 2004. Studies on the storage life of jamun fruit products. *Journal of Food Science and Technology* 41 (2): 186-188.
2. Krishnaveni, A., Manimegalai, G. and Saravana, K.R. 2001. Storage stability of jack fruit RTS beverage. *Journal of Food Science and Technology* 38 (6): 601-602.
3. Kumari, A. 2007. Nutritional quality evaluation of the products prepared from locally available mango

(*Mangifera indica*) supplemented with whey. *M Sc. Thesis.*, Department of Food Science and Nutrition, CSK Himachal Pradesh Krishi Vishvavidyalaya Palampur, India.

4. Larmond, E. 1977. Laboratory methods for sensory evaluation of foods. Canada Department of Agriculture Publications 1637. Food Research Institute, Ottawa: Ontario 73 p.

5. Nath, A., Yadav, D. S., Sharma, P. and Dey, B. 2005. Standardization of ginger-kinnow squash and its storage. *Journal of Food Science and Technology* 42 (6): 520-522.

6. Ranganna, S. 2007. *Hand Book of Analysis and Quality Control for Fruits and Vegetables Products*, 3rd Edition: Tata Mc. Graw, Hill Publishing Company, New Delhi.

7. Sahu, C., Patel, S. and Choudhary, P.L. 2005. Technology for manufacture of whey based mango herbal (lemon grass) beverage. *Journal of Food Science and Technology* 42 (5): 421-424.

8. Saravana, K.R. and Manimegalai, G. 2005. Studies on the stability of the whey based papaya juice blended RTS beverage. *Journal of Food Science and Technology* 42 (2): 185-189.

9. Singh, K.A. and Nath, N. 2004. Development and evaluation of whey protein enriched bael fruit beverages. *Journal of Food Science and Technology* 41 (4): 432-436.

10. Sirohi, D., Patel, S., Choudhary, P. L. and Sahu, C. 2005. Studies on preparation and storage of whey based mango herbal pudina beverage. *Journal of Food Science and Technology* 42 (2): 157-161.

11. USDA 2008. *National Nutrient Database for Standard Reference*, Release 21.

Chapter 58

Survival and Fermentation Ability of Indigenous Yeast Isolates in Seabuckthorn (*Hippophae rhamnoides* L.) Juice

Sohini Walia, M.K. Gupta*, S.S. Kanwar and Aditi Sourabh

Department of Microbiology, College of Basic Sciences,
CSK HPKV, Palampur – 176 062, H.P., India
(*E-mail: mkgupta@hillagric.ernet.in)

SUMMARY

Four indigenous yeast isolates *viz.*, Sc01, Sc02, Sc03 and Sc04 belonging to *Saccharomyces cerevisiae* were used in the present study in immobilized form by using calcium alginate as a carrier. The alcoholic fermentative ability of these four yeast isolates in their free and immobilized state was tested to identify most efficient isolate of yeast for fermentation of seabuckthorn (*Hippophae rhamnoides* L.) juice in its natural and amended form. The chemical evaluation of fermented product was done on the basis of total titrable acidity, pH, residual total soluble solids and amount of ethanol produced. On the basis of ethanol production, Sc02 was found to be the most effective isolate in fermentation of seabuckthorn juice in its free as well as immobilized form among all the yeast isolates. The highest ethanol content of 10.46 per cent (v/v) was produced by Sc02 immobilized yeast cells by fermentation of amended seabuckthorn juice diluted with water in a ratio of 1:1 at pH 3.5 and total soluble solids of 20°B. Sc04 produced minimum amount of ethanol with free or immobilized cells.

***Keywords**: Fermentation, Immobilization, Survival and Yeast.*

INTRODUCTION

Seabuckthorn (*Hippophae rhamnoides* L.) has acquired more and more national economic significance in recent years. A rich biochemical composition, variety of species, fast fruiting, high productivity, ecological adaptation are main properties which ensure wide utilization of this plant in industry, pharmacology, forest and land rehabilitation. Seabuckthorn is a nitrogen fixing plant of the family Elaeagnaceae which occurs as a native plant in many temperate zones of the world (27°– 69° N latitude, 7° W and 22° E longitude) including Asia and Europe (Lu, 1992; Alam, 2004). The distribution ranges from India, Nepal, Bhutan, Pakistan to China, Mongolia, Russia, Hungry, Romania, Switzerland, Germany, France and Britain, and northwards to Finland, Sweden (Jeppsson, 1999) and Norway (Yao, 1993). In India, it grows very widely in Himalayan states of Himachal Pradesh, Ladakh region in Jammu Kashmir, Uttaranchal, Sikkim and Arunachal Pradesh (Dwivedi and Amed, 2008).

Although whole seabuckthorn plant is important, berries are much popular for extracting juice. The nutritional value of the berries is very high. Seabuckthorn berries are an excellent source of bioactive substances such as carotenoids, tocopherols, vitamin C, organic acids and polyphenols (Gupta and Singh, 2006; Li *et al.*, 2007). Antioxidants (including superoxide dismutase, vitamin C, carotenoids, vitamin E, flavonoids and phenolics) in seabuckthorn berries have free radical scavenging activity (Singh, 2006) and also help in preventing cardiovascular diseases, mucosa injuries, and skin problems (Kallio *et al.*, 1999).

Seabuckthorn juice has extremely low pH, low sugar level and high content of organic acids which

Seabuckthorn (*Hippophae* L.): *A Multipurpose Wonder Plant, Vol. 5: Advanced Technologies for Cultivation, Processing, Health Protection and Environmental Conservation* (Virendra Singh, *Editor-in-Chief*, 2018), p. 447–452, Astral International Pvt. Ltd., New Delhi, India.

may be problematic in industrial processes especially for fermentation. Therefore, the fermentation of this fruit juice needs to be evaluated to enhance the quality of the product especially in view of its low pH. Immobilization is one of the methods to utilize enzymes or microorganism more efficiently for fermentation processes and to augment the fermentation ability of cells (Kaur and Kocher, 2002). These systems not only increase the productivity of the process but also reduce the overall cost of the process. Synthetic, natural, organic and inorganic materials have been used for immobilization of cells in wine industry (Bakoyianis *et al.*, 1997; Bardi *et al.*, 1997; Mallios *et al.*, 2004). The immobilized cells can be used for several times without the loss of activity (Cysewki and Wilke, 1978) and their use facilitates the easy conversion of batch process into a continuous mode (Pandey *et al.*, 2000). Above all, immobilization processes facilitate the easy separation of product and the catalytic cell biomass which otherwise is often a tedious, time consuming and expensive step (D' Souza, 1989). Since, immobilized cells are relatively resistant to harsh environment; therefore, the present study was designed to study the impact of immobilized cells on fermentation of seabuckthorn juice in its natural as well as amended form.

MATERIALS AND METHODS

Procurement of Seabuckthorn Juice

The Seabuckthorn juice without preservatives was procured from Lahaul and Spiti area through Post Harvest Technology Unit of College of Home Science, CSK HPKV, Palampur. The juice was stored at 4±2°C.

Chemical Analysis of Seabuckthorn Juice

Juice samples were analyzed chemically by the following standard techniques:

Determination of Total Soluble Solids (TSS)

Total Soluble Solids of the juice samples was determined with portable hand refractometer (0 to 32°B) by putting a drop of juice on the prism and taking the readings. Temperature corrections were applied when the readings were made at temperature other than 20°C (A.O.A.C., 1980). Results were expressed in degree Brix (°B).

Determination of pH

pH of the juice was determined by digital pH meter (Cyberscan, EUTECH Instruments, Germany). Before taking the pH of juice, pH meter was calibrated using standard buffer solution at pH 4, 7 and 10.

Determination of Total Titrable Acidity

The total titrable acidity of the juice was determined using the method described by Ranganna, (1995). Briefly, ten ml of juice sample was diluted with distilled water to make volume up to 100 ml and filtered through whatman No. 1. Ten ml from this diluted solution was titrated against N/10 NaOH solution using phenolphthalein as an indicator.

Yeast Strains

The four yeast strains *viz.* Sc01, Sc02, Sc03 and Sc04 available in the Department of Microbiology belonging to *Saccharomyces cerevisiae* were used in the present study.

Yeast Cells Immobilization using Calcium Alginate as a Carrier

Yeast cells were immobilized by entrapping in calcium alginate by the process described by Kierstan and Bucke (1977). Yeast cells from 48 hr old culture in Potato Dextrose Broth were harvested by centrifugation at 10,000 rpm (REMI compufuge) for 15 min at 4°C. The harvested cells were mixed with 2 percent aqueous solution of sodium alginate. The resulting slurry was extruded dropwise into 0.1 M cold and stirred aqueous solution of calcium chloride. Beads thus formed were allowed to stand for 30 min at 4°C to allow complete gelation and were collected through filtration using a sterile muslin cloth.

Survival Studies of Free and Immobilized Yeast Cells in Seabuckthorn Juice

The survival studies were carried out with free yeast cells and the cells immobilized in calcium alginate. Seabuckthorn juice was inoculated with the free and calcium alginate immobilized yeast cells and incubated at 28°C for 48 hours. Cell growth was observed by making serial dilutions (10^{-1} to 10^{-8}) in 0.8 M sodium phosphate buffer (pH 5.5) at different intervals of time (0, 4, 24 and 48 hours). In case of immobilized cells, beads were first dissolved in 1 per cent solution of sodium citrate. Spread plate technique was performed to know viable count on Potato Dextrose Agar. The plates were incubated at 28°C and cell count was taken at 0, 4, 24 and 48 hours of incubation.

Fermentation of Natural Seabuckthorn Juice

Seabuckthorn juice was sterilized at 115°C for 20 minutes. The four yeast strains were grown in Potato Dextrose Broth for 48 hrs and were used as inocula to ferment seabuckthorn juice. In case of immobilized cells beads obtained after immobilization were used to ferment the juice for five days.

Fermentation of Amended Seabuckthorn Juice

The natural Seabuckthorn juice was amended to raise its total soluble solids to 20°B and pH to 3.5. The juice was diluted with water in 1:1 ratio to reduce its thickness. The free and immobilized yeast strains were used to ferment the amended juice for five days.

Determination of Ethanol Concentration

Per cent of ethanol in alcoholic samples was estimated by the method of Caputi *et al.* (1986). One ml of alcoholic sample was added directly to the distillation flask, diluted to 30 ml with water and then distilled. Distillation was started at 70° ± 2°C and 20 ml of distillate was collected in a 50 ml volumetric flask containing 25 ml of potassium dichromate solution. The contents in the volumetric flask were heated at 60°C in a water bath for 20 minutes and the final volume was made to 50 ml with distilled water. After mixing and cooling the contents of the flask, the absorbance was recorded at 600 nm wavelength (GEAESYS 10 uv scanning, Thermo electron corporation). The amount of ethanol in each sample was determined by using the standard curve for ethanol.

Statistical Analysis

The results obtained from survival, and fermentation studies of the yeast strains under their free and immobilized conditions were subjected to the appropriate statistical analysis (Two–way ANOVA test).

RESULTS

Chemical Analysis of Seabuckthorn Juice

The total soluble solids content of the seabuckthorn juice was in the range of 11.08 to 12.0°B and pH of the juice samples was in the range of 2.70 to 2.78. Total titrable acidity in seabuckthorn juice samples varied between 3.01 to 3.10 per cent (malic acid).

Survival of Free Yeast Cells in Seabuckthorn Juice at different Intervals of Time

The tolerance of cells to various parameters influences their ability to carry out efficient fermentation with high productivity. Sugar concentration, pH, and ethanol concentration are found to have a profound effect on the growth, metabolism and survival of microorganisms. The cell number and growth are largely influenced by changing conditions in the growth medium. All types of cells exhibit different tolerance to various conditions prevailing in the environment which play an important role in controlling the fermentation process. Strains tolerant to these extreme conditions are beneficial and desired in the fermentation technology. In the present study, the survival of free as well as immobilized yeast cells was studied in the seabuckthorn juice which had extremely low pH.

The data on survival of free and immobilized yeast cells in seabuckthorn juice is depicted in Tables 58.1 and 58.2 respectively. All yeast strains exhibited a good growth after 24 hrs and 48 hrs. Sc02 showed significantly higher growth than other yeast cells after 48 hrs followed by Sc01 strains while strains Sc03 and Sc04 exhibiting minimum growth which were statistically at par with each other.

Fermentation of Natural Seabuckthorn Juices with Free and Immobilized Yeast Cells

After fermentation of seabuckthorn juice with free and immobilized yeast cells for five days under natural conditions {27°C, pH 3.07 and 11.8°B TSS},Yeast strains Sc01, Sc02, Sc03 and Sc04 also the cell types *i.e.* free and immobilized cells were adjudged better in comparison to other strains on the bassis of ethanol percentage. The ethanol content varied between 0.81 to 1.13 per cent (v/v) among all yeast isolates (Table 58.3). The immobilized cells produced more ethanol as compared to free cells. Sc02 produced highest ethanol content of 1.13 per cent (v/v) followed by Sc01 strain, Sc03 and Sc04 being statistically lower and at par with each other.

Table 58.1: Survival of Free Yeast Cells (log cfu/ml) in Seabuckthorn Juice at different Intervals of Time

Yeast Strains	*Time of Incubation*				*Mean*
	0 hr	*4 hrs*	*24 hrs*	*48 hrs*	
Sc01	4.07	4.18	6.33	6.69	5.32
Sc02l	4.16	4.20	6.42	6.83	5.37
Sc03	4.06	4.14	6.29	6.41	5.23
Sc04	4.08	4.10	6.30	6.40	5.23
Mean	4.097	4.15	6.34	6.55	

C.D. (5 per cent); Incubation time 0.020; Yeast strains 0.020 and Incubation time x yeast strains 0.036

Table 58.2: Survival of Immobilized Yeast Cells (log cfu/ml) in Seabuckthorn Juice at different Intervals of Time

Yeast Strains	*Time of Incubation*				*Mean*
	0 hr	*4 hrs*	*24 hrs*	*48 hrs*	
Sc01	5.40	5.40	6.75	7.71	6.31
Sc02	5.59	5.62	6.92	7.93	6.52
Sc03	5.31	5.56	6.46	7.72	6.26
Sc04	5.01	5.48	6.49	7.02	6.00
Mean	5.33	5.51	6.65	7.59	

C.D. (5 per cent); Incubation time 0.014; Yeast strains 0.014 and Incubation time x yeast strains 0.028.

Table 58.3: Percent Ethanol Content in Natural Seabuckthorn Juice Fermented with Free and Immobilized Yeast Cells

Yeast Strains	*Percent Ethanol (v/v)*		*Mean*
	Free Cells	*Immobilized Cells*	
Sc01	0.82	0.92	0.87
Sc02	1.10	1.13	1.12
Sc03	0.72	0.81	0.77
Sc04	0.74	0.74	0.74
Mean	0.84	0.90	

C.D. (5 per cent); Type of cells 0.49; Yeast strains 0.75 and Type of cells x yeast strains NS.

Fermentation of Amended (TSS 20°B, pH 3.50, seabuckthorn juice and water in 1:1 ratio) Seabuckthorn Juice with Free and Immobilized Yeast Strain

Ethanol concentration for different yeast isolates was in the range of 6.07 to 10.46 per cent (v/v). Immobilized cells of Sc02 strain produced highest ethanol content of 10.46 per cent (v/v) followed by Sc01 and Sc03 strain, with free cells of Sc04 strain producing minimum amount of ethanol 6.07 per cent (v/v) (Table 58.4).

Table 58.4: Percent Ethanol Content in Amended Seabuckthorn Juice Fermented with Free and Immobilized Yeast Cells

Yeast Strains	*Percent Ethanol (v/v)*		*Mean*
	Free Cells	*Immobilized Cells*	
Sc01	7.17	8.80	7.98
Sc02	8.09	10.46	9.28
Sc03	6.13	8.87	7.50
Sc04	6.07	8.50	7.28
Mean	6.87	9.16	

C.D. (5 per cent); Type of cells 0.125; Yeast strain 0.177 and Type of cells x yeast strains 0.249.

DISCUSSION

Chemical Analysis of Seabuckthorn Juice

The total soluble solids content of the seabuckthorn juice was in the range of 11.08 to 12.0°B. The results obtained are in agreement with other studies (Chauhan *et al.*, 2001; Dwivedi *et al.*, 2002), where TSS was reported in the range of 10 to 12°B. The pH of juice samples was in the range of 2.70 to 2.78. This is very much close to an average pH of 3.13 reported by Beveridge *et al.* (2002) for Indian seabuckthorn juice. However, whatever little difference is observed, that might be due to growth habits, cultivar and time of harvesting. Similar results were reported by Tiitinen *et al.* (2005).

The titrable acidity in the juice was expressed in terms of per cent malic acid. Total titrable acidity in seabuckthorn juice samples varied between 3.01 to 3.10 per cent (malic acid). The level of acidity (per cent malic acid) in seabuckthorn berries grown in different areas has been reported as 2-4 per cent in Himachal Pradesh (Singh, 1998). Thus, the acidity of juice used in present study falls within the range reported for the plants growing in Himachal Pradesh.

Survival of Free Yeast Cells in Seabuckthorn Juice at different Intervals of Time

The tolerance of cells to various parameters influences their ability to carry out efficient fermentation with high productivity. Sugar concentration, pH, and ethanol concentration are found to have a profound effect on the growth, metabolism and survival of microorganisms. The cell number and growth are largely influenced by changing conditions in the growth medium. All types of cells exhibit different tolerance to various conditions prevailing in the environment which play an important role in controlling the fermentation process. Strains tolerant to these extreme conditions are beneficial and desired in the fermentation technology. In the present study, the survival of free as well as immobilized yeast cells was studied in the seabuckthorn juice which had extremely low pH.

All type of yeast cells, free as well as immobilized were able to grow in seabuckthorn juice. However, calcium alginate immobilized yeast cells exhibited better tolerance than free cells. The tolerance of yeast cells to low pH is beneficial as it prevents the bacterial contamination and is an important attribute in fermentation process (Basappa, 2002). These results are fully supported by the findings of Bajaj *et al.* (2001) and Prema *et al.* (2005) who found that immobilized cells of *Saccharomyces cerevisiae* exhibit better tolerance to different pH as compared to free yeast cells. Williams and Munnecke (1981) reported that immobilized cells survived at low pH due to the pH gradient formed with in the cells. They noticed that *Saccharomyces cerevisiae* immobilized in calcium alginate showed a broad range of pH tolerance from pH 3 to 7.5. Similarly, Bajpai and Margritis (1986) and Yadav *et al.* (1997) reported that optimum pH of cells is broadened after their immobilization.

Fermentation of Natural Seabuckthorn Juices with Free and Immobilized Yeast Cells

After fermentation of seabuckthorn juice with free and immobilized yeast cells for five days under natural conditions {27°C, pH 3.07 and 11.8°B TSS},Yeast strains Sc01, Sc02, Sc03 and Sc04 also the cell types *i.e.* free and immobilized cells were adjudged better in comparison to other strains on the basis of ethanol percentage. The ethanol content varied between 0.81 to 1.13 per cent (v/v) among all yeast isolates. The immobilized cells produced more ethanol as compared to free cells. Sc02 produced highest ethanol content of 1.13 per cent (v/v) followed by Sc01 strain, Sc03 and Sc04 being statistically lower and at par with each other.

Fermentation of natural seabuckthorn juice resulted in very low ethanol production (Table 58.3). This may be due to low pH, low sugar levels and high acidity of seabuckthorn juice. Only immobilized cells of Sc02 yeast isolate were able to ferment natural seabuckthorn juice with ethanol production of 1.13 per cent (v/v). This may be due to the high cell density (Kiyohara *et al.*, 2003), high enzymatic activity, high pH and osmotic tolerance associated with the immobilized cells (Norton *et al.*, 1995).

Fermentation of Amended (TSS 20°B, pH 3.50, seabuckthorn juice and water in 1:1 ratio) Seabuckthorn Juice with Free and Immobilized Yeast Strain

Ethanol concentration for different yeast isolates was in the range of 6.07 to 10.46 per cent (v/v). Immobilized cells of Sc02 strain produced highest ethanol content of 10.46 per cent (v/v) followed by Sc01 and Sc03 strain, with free cells of Sc04 strain producing minimum amount of ethanol 6.07 per cent (v/v). Lebeda (2001) obtained 13 per cent ethanol concentration on dilution of seabuckthorn juice with water and its amendments with addition of sugar. The ethanol concentration obtained in the present study was lower than reported by Lebeda (2001). This may be due to the fact that yeast species and the strains of particular species may differ considerably with regard to the ethanol production (Bajaj *et al.*, 2001).

The ethanol content of 10.46 per cent (v/v) by fermentation of amended seabuckthorn juice with immobilized yeast cells is comparable to ethanol concentration {11 per cent (v/v)} obtained by Tsakiris *et* al. (2004) using *Saccharomyces cerevisiae* immobilized in calcium alginate.

CONCLUSION

Natural seabuckthorn juice was fermented by only immobilized cells of Sc02 strain with small amount of alcohol production {1.13 per cent (v/v)}. However, when seabuckthorn juice was diluted with water and adjusted to sugar level of 20°B at 3.5 pH, it produced maximum amount of alcohol [10.46 per cent (v/v)] at dilution ratio of 1:1. On the basis of ethanol production, Sc02 was found to be the most efficient strain, both in free as well as immobilized state. The highest ethanol content of 10.46 per cent (v/v) was produced by Sc02 immobilized yeast cells by fermentation of diluted seabuckthorn juice (1:1 ratio) at pH 3.50 and TSS 20°B. Sc04 produced minimum amount of ethanol in all the cases, whether free or immobilized. In brief, immobilized Sc02 cells were found to be suitable for fermentation of seabuckthorn juice. However, for better results, amendments of natural juice are required.

REFERENCES

1. A.O.A.C. 1980. *Official Methods of Analysis of Association of Analytical Chemists*. Benjamin Franklin Station, Washington. D. C.
2. Alam, Z. 2004. Chemical and nutritional constituents of seabuckthorn juice. *Pakistan Journal of Nutrition* 3 (2): 99-106.
3. Bajaj, B. K., Sheema, Y., Thakur, R. L. and Yousuf, S. 2001. Selection and characterization of yeast for desirable fermentation characteristics. *Indian Journal of Microbiology* 41 (2): 107-110.
4. Bajpai, P. K. and Margaritis, A. 1986. Effect of temperature and pH on immobilized *Zymomonas mobilis* for continuous production of ethanol. *Biotechnology and Bioengineering* 24: 824-828.
5. Bakoyianis, V., Kanellaki, M., Koutinas, A. A., Agelopoulos, K., Kanellaki, M. 1997. Comparative study of kissiris, Y–alumina and calcium alginate as support of cells for batch and continuous wine making at low temperature. *Journal of Agriculture and Food Chemistry* 45: 4884-4888.
6. Bardi, E., Koutinas, A. A., Psarianos, C. and Kanellaki, M. 1997. Volatile byproducts formed in low temperature wine making using immobilized yeast cells. *Process Biochemistry* 31: 579-584.
7. Basappa, S. C. 2002. Investigation on Chhang from finger mellet (*Eleucine coracena* Gairtn.) and its commercial prospectus. *Indian Food Industry* 21 (1): 46-53.
8. Beveridge, T., Harrison, J. E. and Drover, J. 2002. Processing effects on the composition of seabuckthorn juice from *H. rhamnoides* L. Cv. Indian Summer. *Journal of Agriculture and Food Chemistry* 50: 113-116.
9. Caputi, A., Ueda, M. and Brown, T. 1968. Spectrophotometric determination of ethanol in wine. *American Journal of Enology and Viticulture* 19: 160-165.
10. Chauhan, A. S., Rekha, M. N., Ramteke, R. S. and Eipeson, W. E. 2001. Preparation and quality evaluation of processed products from (*Hippophae rhamnoides* L.) berries. *Beverages and Food World* 27 (1): 31-34.
11. Cysewki, G. R. and Wilke, C. R. 1978. Process design and economic studies of alternative fermentation methods for the production of alcohol. *Biotechnology and Bioengineering* 20: 1421-1427.
12. D'Souza, S. F. 1989. Immobilized cells: Techniques and applications. *Indian Journal of Microbiology* 29 (2): 83-87.
13. Dwivedi, S. K. and Amed, Z. 2008. Seabuckthorn (*Hippophae* sp.): A potential underutilized fruit plant for cold Arid India. *Acta Horticulture* 769: 213-218.

14. Dwivedi, S. K., Attrey, D. P. and Kareem, A. 2002. Postharvest management of seabuckthorn (*Hippophae* sp.) in cold arid regions of Ladakh. In: *National Workshop on Postharvest Management of Horticultural Produce*, Feb. 7-8, 2002, Chandigarh.

15. Gupta, R. K. and Singh, V. 2006. Mineral composition of seabuckthorn (*Hippophae* L.). *Seabuckthorn (Hippophae* L.) *A multipurpose wonder Biochemistry and Pharmacology* 2: 272-284.

16. Jeppsson, N. 1999. Progress in research and development of seabuckthorn in Sweden. In: *Proceedings of International Symposium on Seabuckthorn (Hippophae rhamnoides* L.), Beijing, China.

17. Kallio, K., Yang, B. R., Tahvonen, R. and Hakala, M. 1999. Composition of seabuckthorn berries of various origins. In: *Proceedings of International Symposium on Seabuckthorn* (*Hippophae rhamnoides* L.), Beijing, China.

18. Kaur, M. and Kocher, G. S. 2002. Ethanol production from molasses and sugarcane juices by an adapted strain of *Saccharomyces cerevisiae*. *Indian Journal of Microbiology* 42: 255-257.

19. Kierstan, M. and Bucke, C. 1977. The immobilization of microbial cells, sub cellular organelles and enzymes in Calcium Alginate Gel. *Biotechnology and Bioengineering* 19 (3): 87-89.

20. Kiyohara, P. K., Lima, U. A., Santos, H. S. and Santos, P. S. 2003. Comparative studies between yeast immobilized on alumina leads and on membranes prepared by two routes. *Brazilian Journal of Microbiology* 34: 129-137.

21. Lebeda, A. 2001. Danube seabuckthorn population: source of forms valuable for selection. In: *International Workshop on Seabuckthorn*, International Centre, Feb. 18–21, 2001, New Delhi, India, pp. 112-118

22. Li, X., Kong, L. and Li, H. 2007. Advances on effective compositions of seabuckthorn. *Journal of Jilin Agricultural University* 29 (2): 162-167.

23. Lu, R. 1992. Seabuckthorn: A multipurpose plant species for fragile mountains. ICIMOD occasional Paper No. 20. Kathmandu, Nepal, 62p.

24. Maiorella, B. L., Wilke, C. R. and Blanch, H.W. 1981. Alcohol production and recovery. *Advances in Biochemistry Engineering* 20: 43.

25. Mallios, P., Kourkoutas, Y., Iconomopoulou, M., Koutinas, A. A., Psarianos, C., Marchant, R. and Banat, I. M. 2004. Low temperature wine making using yeast immobilized on pear pieces. *Journal of Science and Food Agriculture* 84: 1615-1623.

26. Norton, S., Watson, K. and D'Amore, T. 1995. Ethanol tolerance in immobilized brewer's yeast cells. *Applied Microbiology and Biochemistry* 43 (1): 18-24.

27. Ocloo, F.C.K. and Ayernor, G. S. 2008. Physical, chemical and microbiological changes in alcoholic fermentation of sugar syrup from cassava flour. *African Journal of Biotechnology* 7 (2): 164-168.

28. Pandey, A., Soccol, C. R. and Soccol, V. T. 2000. Biopotential of immobilized amylases. *Indian Journal of Microbiology* 40: 1-14.

29. Prema, P., Palavesm, A., Kumar, R. R. and Hameedu, A. S. 2005. Fermentation of pretreated substrates for ethanol production with immobilized yeast cells, *Saccharomyces cerevisiae*. *Asian Journal of Microbiology, Biotechnology and Environmental Sciences* 7 (3): 371-376.

30. Ranganna, S. 1995. *Handbook of Analysis of Quality Control for Fruit and Vegetable Products*, 2nd Ed, Tata McGraw-Hill Publishing Company Ltd., New Delhi.

31. Singh, V. 1998. Seabuckthorn: A wonder plant of dry temperate Himalayas. *Indian Horticulture* 43 (2): 6-9.

32. Singh, V. 2006. Free radicals, diseases, antioxidants and antioxidant properties of seabuckthorn (*Hippophae rhamnoides* L.). Seabuckthorn (*Hippophae* L.) a multipurpose wonder plant. *Biochemistry and Pharmacology* 2: 3-69.

33. Thoukis, G., Ueda, M. and Wright, D. 1965. The formation of Succinic acid during alcoholic fermentation. *American Journal of Enology and Viticulture* 16 (1): 1-8.

34. Tiitinen, K. M., Hakala, M. A. and Kallio, H. P. 2005. Quality components of seabuckthorn (*Hippophae rhamnoides*) varieties. *Journal of Agriculture and Food Chemistry* 53: 1692-1699.

35. Tsakiris, A., Sipas, V., Bekatorou, A., Mallouchos, A. and Koutinas, A. A. 2004. Red wine making by immobilized cells and influence on volatile composition. *Journal of Agriculture and Food Chemistry* 52 (5): 1357-1363.

36. William, D. and Munnecke, D. M. 1981. The production of ethanol by Immobilized Yeast cells. *Biotechnology and Bioengineering* 23: 1813-1825.

37. Yadav, B. S., Sheoran, A., Rani, U. and Singh, D. 1997. High ethanol productivity in an immobilized cell reactor. *Indian Journal of Microbiology* 37 (2): 65-67.

38. Yao, Y. 1993. Effects of temperature sum on vitamin C concentration and of seabuckthorn (*H. rhamnoides*) fruit: Optimal time of harvest. *Agricultural Science in Finland* 2: 497-505.

Chapter 59

Optimization of Extraction Technique of Ursolic Acid from Seabuckthorn by Response Surface Methodology

Lu Chang-Zheng, Shan Yong-Kai, Li Shu-Zhi and Ren Bei-Lei

Qinghai Tsinghua Biology Bio-Tech Co., Ltd.,
Xi'ning 810016, China

SUMMARY

RSM was used to optimize ursolic acid extraction process from seabuckthorn. Used marc of seabuckthorn fruit and which had extracted juice and seeds as the raw material, on the basis of single factor experiments, selected levels and factors which had greater impact on ursolic acid extraction, according to the central composite design, three factors and three levels of response surface method (RSM) were used, leaching rate as the reference indicators, and ursolic acid optimum extraction parameters: extraction pressure 22.5 Mpa, extraction temperature 40.4!, CO_2 flow rate 23.3 L•h^{-1}. Under these conditions, leaching rate of ursolic acid was 374 mg/100g.

Keywords: *Seabuckthorn, Ursolic acid, Extraction and RSM.*

INTRODUCTION

Hippophae, the seabuckthorn's, are deciduous shrubs in the family Elaeagnaceae, which is also called sandthorn, sallowthorn or seaberry. Seabuckthorn is widely distributed in Eurasia's temperate zone. China is one of the places of origin of seabuckhorn where seabuckthorn is naturally grown in Northwest and Southwest China. Research has shown that seabuckthorn berries are high in vitamins C, B1, B2, E, F, K, P, pro-vitamin A, sugars, flavonoids and organic acids. Ursolic acid is one of the bio-active compounds that is identified in seabuckthorn berries. Ursolic acid is believed to have the potential in anti-inflammatory, anti-hepatitis, anti-oxidation and reduction of blood sugar, thus it attracts a lot of attention regarding seabuckthorn ursolic acid. However, study on extraction of ursolic acid from seabuckthorn marc is not seen. This research utilised seabuckthorn pumace including seabuckthorn skin and seed residues after production as raw material. The extraction method parameters are discussed and the optimum extraction condition is also confirmed in this research.

MATERIALS AND METHODS

Raw Material

Seabuckthorn berry marc. Seabuckthorn was picked up in Qinghai Province (in Northwest China). The marc was placed in the oven with a temperature of 105! after being dried, grounded, and sieved.

Reagents

95 per cent ethanol, acetone, ethyl acetate, distilled water, ursolic acid (control sample), methanol, ethanol. All reagents are analytic standard.

Equipment

FW-400A grinder, BS323S weighing scale, HS.2 electronic water-bath, PB-10 pH meter RE-52Arotary evaporator, SHB-3 water-circulated vacuum pump, DG/20-002 oven.

Seabuckthorn (Hippophae L.): A Multipurpose Wonder Plant, Vol. 5: Advanced Technologies for Cultivation, Processing, Health Protection and Environmental Conservation (Virendra Singh, *Editor-in-Chief*, 2018), p. 453–458, Astral International Pvt. Ltd., New Delhi, India.

Diagram

Co-solvent, temperature, pressure

CO_2 flux, particle size ↓ Vacuum concentration ↓

Raw material→sieve→defat→CO_2 extraction→centrifugation→supernatant→drying→extract

↑

Determine ursolic acid content

METHODOLOGY

Weighed 5g raw material (seabuckthorn marc), added chloroform by 1:40 based on ratio, then the sample was de-fatted by refluxing re-greasing; reagent was collected, the residue was remained after volatilising. The collected residue and co-solvent were added at a ratio of 1:20 for critical super CO_2 extraction. Supernatant was collected after critical super CO_2 extraction. The supernatant was the raw ursolic acid.

Determination of Ursolic Acid

Chromatograph Condition

Chromatographic column: Luna 5u C_{18} (2), 250 mm×4.6 mm; Shimpack VP-ODS, 150 mm×4.6 mm, 5 µm; wavelength 215nm; mobile phase 0.1mol/L, methanol: acetic acid (volume 85:15); mobile phase velocity 0.8ml/min, temperature 25 °C; sample input: 20µl; data analysis was carried out by peak area external standard quantitative method.

Preparation of Standard Solutions

Weighed 24.0 mg ursolic acid control sample, added it into a 10 ml volumetric flask, methanol diluted to required scale, hence the standard solution of ursolic acid 2.40 mg/ml. Absorbed 0.10, 0.50, 1.00, 2.00, 3.00, 4.00, 5.00 ml standard solution respectively and accurately, placed into 5ml graduated flask using methanol to dilute until it reached the graduation, injected 20µl sample and measured the peak area. The mean of peak area is X while the injection volume of sample is Y, a linear regression is listed as below:

$Y=201.55X-2070.9, r=0.9997$

Ursolic acid content is between 35 and 1120µg/ml and the linear relationship is obtained.

Preparation for Sample Solution

Weighed 1ml ursolic acid extract and added into volumetric flask, then added methanol to set up the volume to 25 ml, then the solution was filtered using Millipore filter (0.45µm), then absorbed 10µl of the solution. Chromatograph measurement was carried out based on the conditions as mentioned above. All samples are triplicated, mean was calculated, the content of ursolic acid was calculated according to the linear regression.

Experiment Design

Single Factor Experiment

Single factor experiment was focusing on how cosolvent, extraction pressure, extraction temperature, CO_2 flux and raw material granularity affect the yield of ursolic acid. Details of Single factor experiment please see Table 59.1.

Response Surface Analysis Experiment

According to Box-Benhnken central group and experiment design rationale [4, 5], utilised 3 to 3 levels of the factor-analytical laboratory response, then selected the factor level from single factor experiment. Please see Table 59.2 for details.

RESULTS AND DISCUSSIONS

One Factor Experiment Outcome

Effect of Cosolvent on Ursolic Acid Yield

In experimental trial, co-solvent was not initially added, as a result, it showed a poor performance on ursolic acid extraction. Testing thereafter confirmed the yield of ursolic acid was very low without co-solvent.

Table 59.1: Single Factor Experiment

Factors	*Levels*				
	1	*2*	*3*	*4*	*5*
Cosolvent	95 per cent Ethanol	Acetone	Acetic ether	—	—
Extraction Pressure (Mpa)	5	10	15	20	25
Extraction temperature (°C)	25	30	35	40	45
CO_2 flux ($L \cdot h^{-1}$)	5	10	15	20	25
Raw material granularity (per cent)	20	40	60	80	100

Table 59.2: Factor-Analytical Experiment Table

Factors	*Codes*		
	-1	*0*	*1*
X1 extraction pressure (MPa)	15	20	25
X2 extraction temperature (°C)	35	40	45
X3 CO_2flux(L·h^{-1})	15	20	25

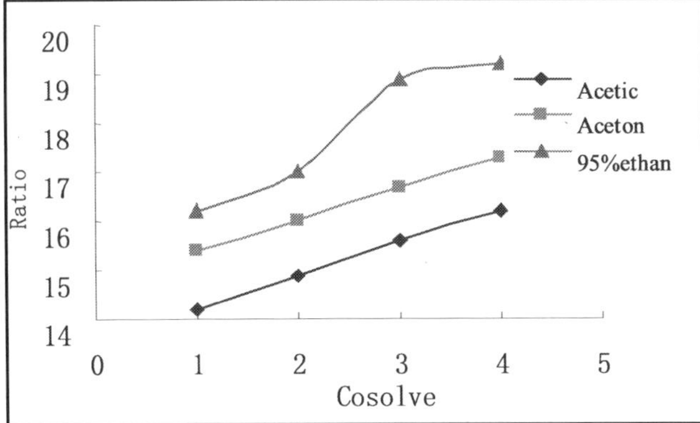

Figure 59.1: Effect of Cosolvent on Ursolic Acid Extraction Ratio.

Therefore, co-solvent must be added in the experiment. The amount of cosolvent should be adequate, if it is too less, it will not have a significant effect on ursolic acid extraction; while it is too much, the extraction will be conducted in a not critical liquid condition where is yet the best condition. From Figure 59.1, it can be seen that 95 per cent ethanol extracted ursolic acid has a higher extraction ratio. On the other hand, 95 per cent ethanol is a cheap, widely-used and reliable reagent, thus 95 per cent ethanol can be used as a co-solvent with a 30 per cent of the raw material by dry basis.

Effect of Extraction Pressure on Ursolic Acid Extraction Ratio

From Figure 59.2, if all other parameters are fixed, extraction ratio will be increased while the extraction pressure is higher. The extraction ratio has a linear relationship with the extraction pressure. When the pressure is between 5 and 20 MPa, it shows a significant effect. However, when the pressure is over 20 MPa, the extraction ratio does not have a significant effect. Taking into account of the requirement in equipment and operation while high pressure is in use, particularly the cost will also increase, the best extraction pressure is chosen to be 20 MPa.

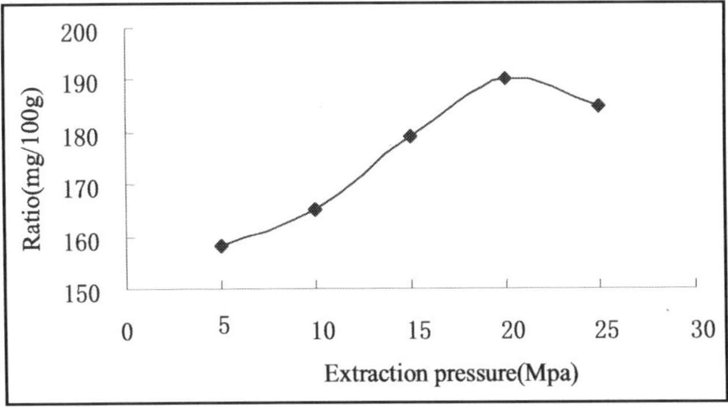

Figure 59.2: Effect of Extraction Pressure on Ursolic Acid Extraction Ratio.

Effect of extraction temperature on ursolic acid extraction ratio

From Figure 59.3, it can be seen that, when the extraction temperature is between 25 °C and 40 °C, when the temperature is higher, the more ursolic acid is extracted. However, when the extraction temperature is greater than 45 °C, the extraction ratio is decreasing. This is possibly because when the temperature is continuously going up, the density of CO_2 is decreasing and its dissolving capacity is lowing, as a result, the extraction ratio is going down. High temperature also plays a negative effect on ursolic acid extraction, so the best extraction temperature is 40 °C.

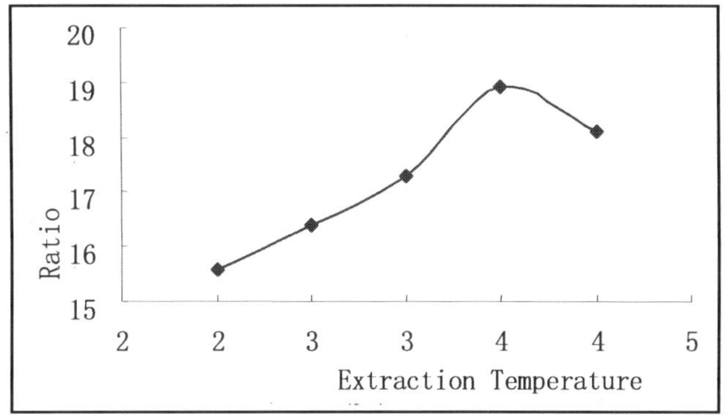

Figure 59.3: Extraction Temperature on Ursolic Acid Extraction Ratio.

Effect of CO_2 Flow Rate on Ursolic Acid Extraction

CO_2 flow rate is another parameter on ursolic extraction ratio. When the flow rate of CO_2 increases, the time of CO_2 remained in the extraction equipment will reduce. As a result, the time of effective reaction will decrease, so the extraction balance will be affected. On the other hand, as the flow rate hikes, it will stir the fluid and speed up mass transfer. In other words, it will assist extraction. As the flow rate of CO_2 increases,

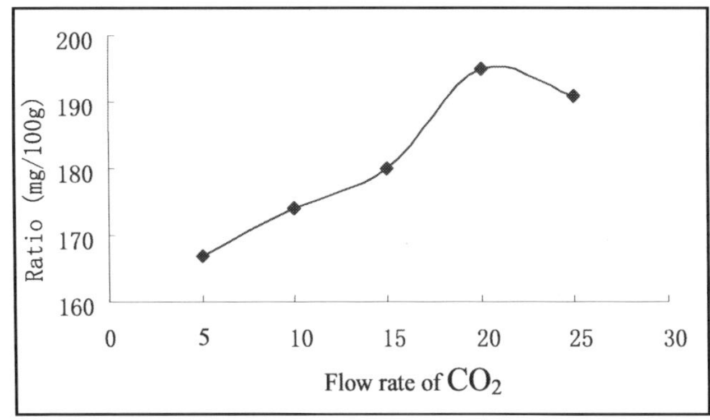

Figure 59.4: Effect of CO_2 Flow Rate on Ursolic Acid Extraction.

extraction rate will also go up. However, if the flow rate is too high, and does not have an effective reaction, the extraction ratio will gradually go down and the cost will go up if the flow rate is continuously jump. So taking into account of all these factors, the premium flow rate is 20 $L{\cdot}h^{-1}$ (Figure 59.4).

Effect of Raw Material Particle Size on Ursolic Acid Yield

The particle size of raw material will directly impact on ursolic acid's extraction time and rate. The particle size is associated with its total surface and solid phase mass transfer coefficient. The smaller particle size has a wider surface area, so the contact area will relatively expand which will increase its extraction rate. From Figure 59.5, it can be seen that as the particle size of raw material decreases, the extraction rate will increase. However, when the particle size reaches 20-100 mesh, the increase of extraction rate is not significant. So taking into account of the cost in particle size reduction, the best particle size is 80 mesh.

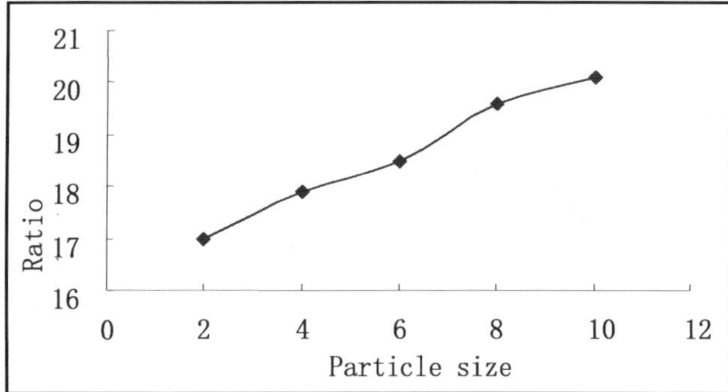

Figure 59.5: Effect of Particle Size on Ursolic Acid Extraction.

Results and Discussion on Response Surface Testing

Testing Results

Extraction pressure, temperature and CO_2 flow rate are variables, while ursolic acid extraction rate is response variable Y, the trial is designed as Table 59.3 below. From 1 to 12, it is factorial test while 13-15 is core test to assess the test error[5] (Table 59.3).

Table 59.3: Experiment Design and Results

Sl.No.	X1	X2	X3	Extraction Rate (mg/100g)
1	-1	-1	0	126
2	-1	1	0	245
3	-1	0	1	312
4	-1	0	-1	156
5	0	1	1	305
6	0	-1	-1	156
7	0	1	-1	125
8	0	-1	1	303
9	1	1	0	324
10	1	0	1	338
11	1	0	-1	196
12	1	-1	0	290
13	0	0	0	316
14	0	0	0	358
15	0	0	0	354

Note: 2.2.2 Analysis of variance and regression equation.

The Surface fitting equation of ursolic acid extraction rate (Y) and extraction pressure (X1), extraction temperature (X2) flow rate of CO_2 (X3):

$$Y = 342.6667 + 38.625{*}X1 + 15.5{*}X2 + 78.125{*}X3 - 34.08333{*}X1{*}X1 - 62.33333{*}X2{*}X2 - 58.08333{*}X3{*}X3$$

From Table 59.4, it can be seen that it has a significant impact on formula's first time and second time, but no significant effect on interaction items, so it can be neglected. So there is no simple linear relationship between factors and its relative response values. The formula R^2=90.89 per cent indicates that 90.89 per cent of the value changes depend on its variables change, so this model can be used to analyse ursolic acid extraction rate.

Table 59.4: The Variance Analysis of Regression Equation

Sources of Variance	DF	SS	MS	F	Pr>F
X1	1	11935.13	11935.13	6.460907	0.051778
X2	1	1922	1922	1.040447	0.354508
X3	1	48828.13	48828.13	26.4324	0.003642
X1*X1	1	4289.256	4289.256	2.321927	0.188067
X1*X2	1	1806.25	1806.25	0.977787	0.368152
X1*X3	1	49	49	0.026525	0.877002
X2*X2	1	14346.26	14346.26	7.766138	0.038591
X2*X3	1	272.25	272.25	0.147379	0.716836
X3*X3	1	12456.64	12456.64	6.743222	0.048438
Regression	9	92130.52	10236.72	5.541502	0.036954
First time	3	62685.25	20895.08	11.31125	0.011474
Second time	3	27317.77	9105.922	4.929359	0.059156
Interaction item	3	2127.5	709.1667	0.383897	0.769716
Residual	5	9236.417	1847.283		
Lack of fit	3	8161.75	2720.583	5.06312	0.169347
Error	2	1074.667	537.3333		

Value F is utilized, when the value P is smaller, its relative variable will increase.

In all factors, the effect on extraction rate from greater to less is X3>X1>X2, that means flow rate of CO_2>extraction pressure > extraction temperature.

Isoheight and Response Surface Graph Analysis

Compare 3 different figures, it can be confirmed that flow rate of CO_2 has the greatest impact on ursolic acid extraction rate and the second is extraction pressure as

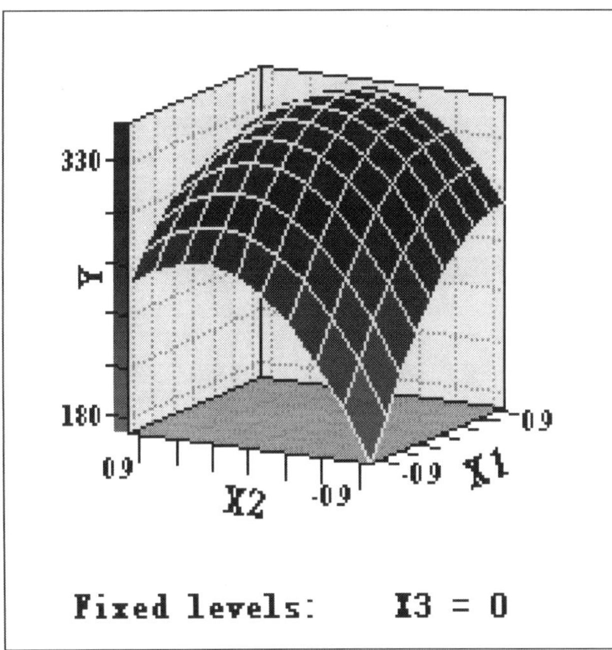

Figure 59.6: Response Surface Y=f(X1, X2).

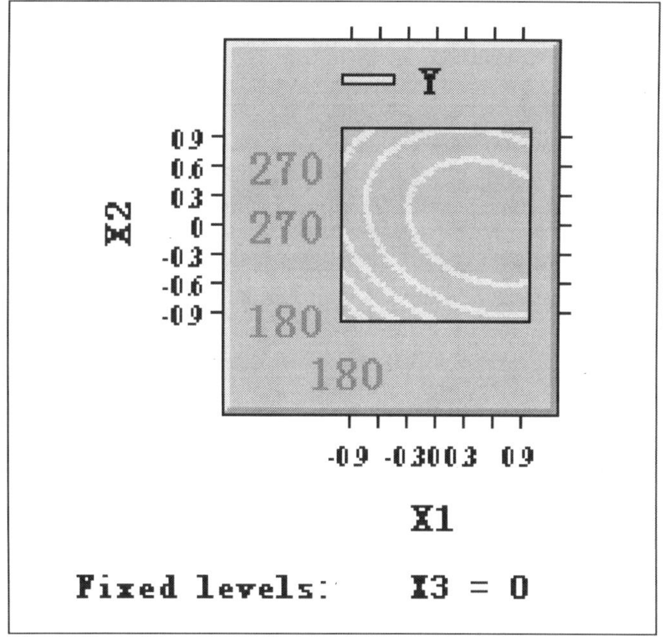

Figure 59.7: Isoheight Y=f(X1, X2).

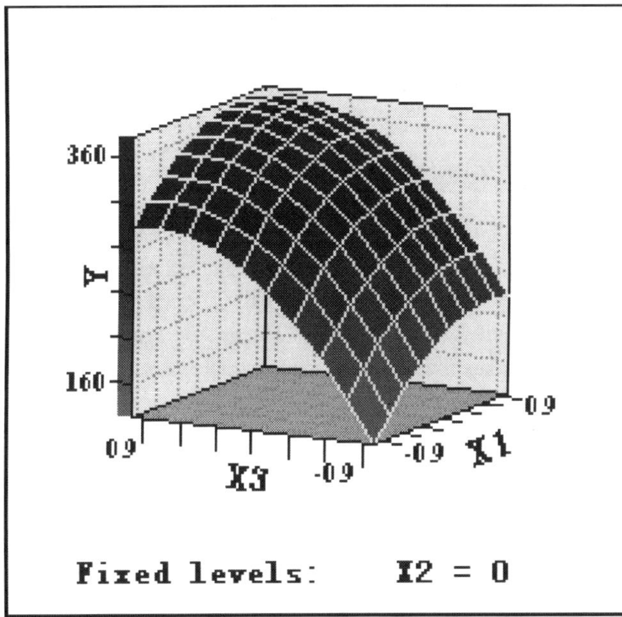

Figure 59.8: Response Surface 8 Y=f(X1, X3).

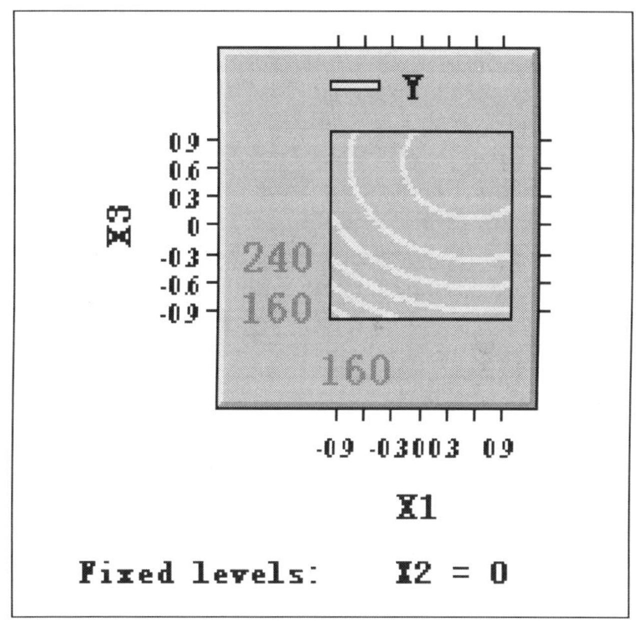

Figure 59.9: Isoheight Y=f(X1, X3).

their curve is comparably steep. Extraction temperature has a small impact on ursolic acid rate as the curve is flat.

Discussion on Response Based on the Experiment

Table 59.5: Analysis Sheet on Response

Factors	*Code*	*Value*
X1	0.50709	22.5
X2	0.08177	40.4
X3	0.66305	23.3

Estimated value on stable point is 378.9941.

Table 59.6: Characteristic Values

Characteristic Value	*X1*	*X2*	*X3*
-30.2138	0.93977	-0.32462	-0.10706
-57.1268	0.21112	0.30493	0.92868
-67.1594	0.26882	0.89534	-0.35509

From Table 59.6, it can be seen that all characteristic values are negative and the stable point falls at (0.50709, 0.08177, 0.66305) where the response value is the greatest. So the optimum condition of use super critical CO_2 to extract ursolic acid is extraction pressure

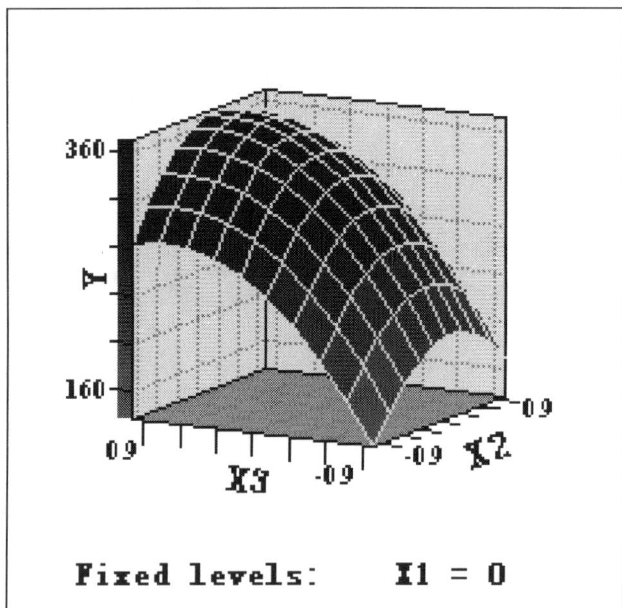

Figure 59.10: Y=f(X2, X3).

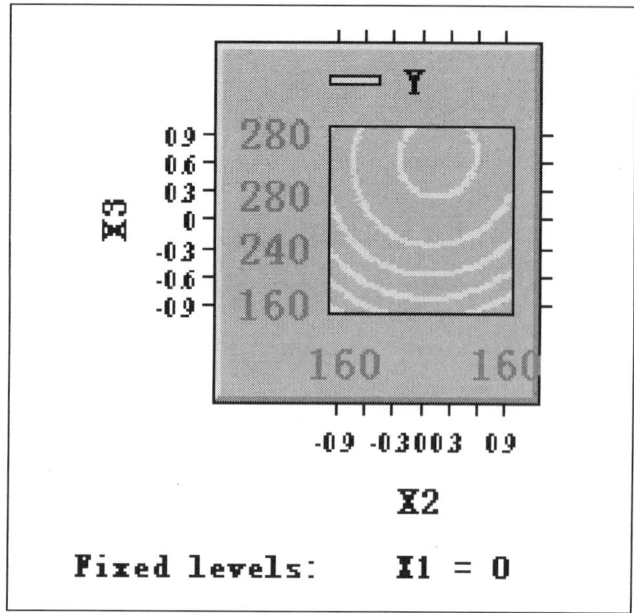

Figure 59.11: Y=f(X2, X3).

22.5Mpa, temperature 40.4°C, flow rate of CO_2 23.3L·h-1. Under this condition, the extraction rate is 374mg/100g which is close to estimated value 378.9941.

CONCLUSIONS

The experiment confirmed that utilizing RSM method is effective on extraction of ursolic acid from seabuckthorn. The method of response surface analysis can establish the statistic model between factors and response value, and its effect between each factor can be identified. Thus it is helpful to confirm the optimum conditions of extraction and improve efficiency.

The optimum condition of ursolic acid extraction is extraction pressure 22.5Mpa, temperature 40.4°C, flow rate of CO_2 23.3L·h-1. Under this condition, the extraction rate is 374 mg/100 g which is close to estimated value 378.9941.

REFERENCES

1. Fuliang, Y., Peilei, R. and Ying, X. 2007. Study on physiological active components in North Shaanxi buckwheat hull. *Food Science* 12 (28): 156-161.
2. Jingjing, G., Qingai, W. and Jian, K. 2008. Xinjiang wine brewing process of *Hippophae rhamnoides* L. by response surface method. *Food Industry Technology* 8: 224-226.
3. Lingyun, Y., Yuqiao, S. and Jiaoshe, Li, 2002. Study on the chemical constituents of the leaves of the China Seabuckthorn. *Seabuckthorn Hippophae* 2 (16): 33-34.
4. Mingdi, Z. and Zhangcun, W. 2008. Study on Optimization of extraction technology of corn germ protein by response surface method. *Agrotechny*. 7 (142): 174-178.
5. Tong, L., Yanan, W. and Zilong, W. 2007. Study on extraction and purification of ursolic acid and oleanolic acid in Hawthorn. *Food Science* 7 (28): 141-144.
6. Yan, Z. and Dan, Z. 2008. Study on the extraction technology of ursolic acid in Fructus Ligustri Lucidi. *Journal of Shaanxi University of Science and Technology* 8: (4): 85-88.

Chapter 60

Effect of Processing Techniques on Seabuckthorn Juice and Seabuckthorn Beverages

Galina Zvaigzne[1*], J. Thomas Moersel[1], Sasha Kuehn[1], Daina Karklina[2] Dalija Seglina[2] and Inta Krasnova[2]

[1]Untersuchungs-, Beratungs-, Forschungslaboratorium GmbH, Altlandsberg 15345
[2]Latvia State Institute of Fruit Growing, Graudu Str.1, Dobele, LV-3701, Latvia
(*E-mail: galina.zvaigzne@ubf-reseach.com)

SUMMARY

Seabuckthorn (*Hippophae rhamnoides* L.) juice, one of the important products obtained from fresh seabuckthorn berries, is commercially very much in demand. A good understanding of the properties of seabuckthorn juice is required to design and operate processing plants that maintain high juice quality. The properties of fresh berries and the initial processing conditions influence the quality of the final end-use product. Heat treatment of juice is essential for obtaining the desired product shelf life, but the process should be designed to minimize unwanted quality degradation. The impact of different processing technologies on bioactive compounds of three varieties of seabuckthorn juices was investigated. Seabuckthorn juices were treated by high-temperature short-time (HTST) pasteurization (94 °C 30 s) and ultra-high temperature (UHT) sterilization (130 °C 2 s). Sensory attributes (*i.e.* colour, odour, sweetness, acidity, flavour and overall acceptability) of different variety seabuckthorn mix juices processed with UHT pasteurisation was evaluated by 12 panellists using a hedonic scale (1-9 points). All sensory attributes were rated equivalent for UHT- and HTST-treated juice ($P > 0.05$). No significant changes in the biochemical properties were detected after HTST or UHT treatment. Sensory results showed differences between HTST pasteurised juice. Although promising results have been obtained further optimisation is required in order to maximise the quality of the resulting product. The study was done at UBF GmbH Institute, Altlandsberg, Germany and Latvia State Institute of Fruit Growing, Latvia.

Keywords: *Seabuckthorn, Biochemical compounds and Pasteurisation.*

INTRODUCTION

Seabuckthorn (*Hippophae rhamnoides* L.) is a multipurpose plant, due to the typical biochemical composition of its fruits; it's mainly used in medicine, cosmetic and food industry. Seabuckthorn beverages can be used as functional foods in our daily diet as their consumption is known to be beneficial for human body. Seabuckthorn from a little known wild shrub has evolved into a well-established fruit crop (Singh, 2003). Fresh seabuckthorn fruits contain significant amount of vitamins C, E and B, carotenoids, organic acids, polyunsaturated fatty acids, phenolic compounds and essential elements. It has been used in traditional medicines in Tibet and Indian Himalayas (Singh, 2006; Abid *et al.*, 2007). During the second half of the 20th century, numerous animal experiments and clinical studies have been carried out to investigate the physiological effects of different fractions and products such as juice and oil from the berry, especially in China and Russia. The health effects of seabuckthorn covers antioxidative, anti-inflammatory effect, promotion of cardiovascular, risk reduction of type 2 diabetes have been reported (Geetha *et al.*, 2002; Gupta *et al.*, 2006; Singh *et al.*, 2006; Basu *et al.*, 2007; Chuhan *et al.*, 2007; Geetha *et al.*, 2009; Suryakumar, 2011). The seabuckthorn juice and beverages are assumed as the optimal food

Seabuckthorn (Hippophae L.): A Multipurpose Wonder Plant, Vol. 5: Advanced Technologies for Cultivation, Processing, Health Protection and Environmental Conservation (Virendra Singh, *Editor-in-Chief*, 2018), p. 459–464, Astral International Pvt. Ltd., New Delhi, India.

for use in any human diet (Zeb, 2004). One of the most popular products of seabuckthorn it is juice and nectars. The juice and beverages of seabuckthorn are nutritious increasingly gaining popularity among people.

Seabuckthorn juice is a product of high commercial and nutritional value due to its rich vitamin C content, vitamin E and B, carotenoid, organic acids, polyunsaturated fatty acids, phenolic compounds and essential elements, and its desired sensory characteristics. Heat treatment of juice is essential for obtaining the desired product shelf life, but the process should be designed to minimize unwanted quality degradation. Commercially, the juice is rapidly heated at 95 – 98 °C: the exact temperature depends on the type of equipment used and on velocity of juice flow. Heating usually takes 10 to 30 seconds or less and the juice is heated rapidly without local overheating. Ultra-high temperature processing (UHT), or ultra-heat treatment, sterilizes food by heating it above 135 °C - the temperature required to kill spores in juices - for 1 to 2 seconds. UHT is most commonly used in milk production, but the process is also used for fruit juices, cream, soy milk, yogurt, wine, soups, honey, and stews. Tetra Pak recommended for juice, nectar and still drinks process 138 °C for 4 seconds (http://www.tetrapak.com).

Seabuckthorn juices are sensitive to heat. Their vitamin content and fresh aroma and flavour may be lost or damaged by undue exposure to heat. Currently, the food industry is looking at replacing the traditional well established preservation technologies with novel thermal and no thermal technologies that may produce high-quality food products with improved energy efficiency and to be more environments friendly. The objective of this work was to comparatively evaluate the effect of high-temperature short-time (HTST) pasteurization and alternative ultra-high temperature (UHT) processing chemical and biochemical quality parameters directly after treatment of three seabuckthorn varieties juices and its beverages.

MATERIALS AND METHODS

Juice Samples

Seabuckthorn *(Hippophae rhamnoides* L.*)* juices of German cultivars 'Leikora' and 'Hergo', from Sanddorn GbR, Herzberg/Elster production line of the commercial juice plant in Germany; Seabuckthorn (*Hippophae rhamnoides* L.) juice of Latvian cultivar 'Botanicheskaya-Lubitelskaya', from Latvian State Institute of Fruit – Growing experimental processing Laboratory; orange juice of Greek 'Navel' variety *(Citrus sinensis)* from S.A. Biofresh production line of a commercial juice plant in Laconia Southern Greece were obtained. No thermal treatment was applied. The juices were immediately frozen in a forced circulation freezer and kept at -20 °C until use. All juices were compared in terms of biologically active compounds and consumer acceptance.

UHT Sterilization

The UHT sterilization was carried out using the laboratory pilot scale pasteuriser with a tubular heat exchanger (Armfield FT74, HTST/UHT Processing Unit, Hampshire, England) at 130 °C for 2 s and 2.26 bar back pressure. The schematic diagram of the UHT process is illustrated in Figure 60.1.

The frozen seabuckthorn and orange juices were defrosted and mix beverages (90 percent orange juice and 10 percent seabuckthorn juice, without sugar and without water) were prepared. The UHT sterilization was conducted on the same day for each experiment. All prepared juices (Orange juice, Seabuckthorn juices and Orange-Seabuckthorn juices) were UHT processed at 130 °C for 2 seconds.

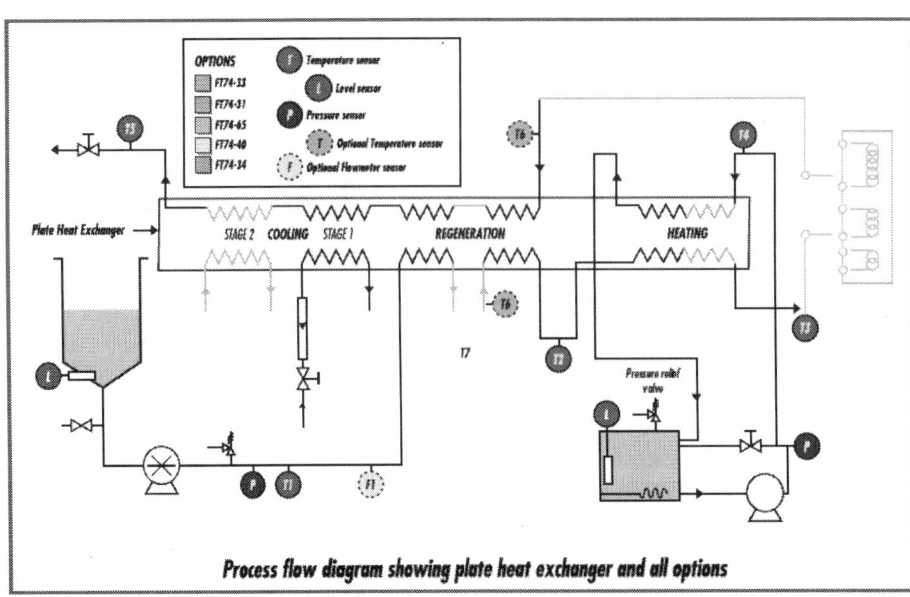

Figure 60.1: Process Flow Diagram UHT Juice Sterilisation (Source Armfield).

The UHT processed juices were filled in glass bottles of 150 ml capacity with screw -cup closures. The six bottles were kept for sensory evaluation and six samples were immediately sent to laboratory for the determination of total solid, pH, titration acidity, vitamin C, vitamin E, total phenols, total carotenoids and antioxidant activity. The sensory evaluation was conducted within one week after the production of UHT samples. The sensory was evaluated using hedonic scale test by 14 panellists who were the analysts and students in the UBF GmbH Company.

HTST Pasteurisation

High-temperature short-time (HTST) pasteurised samples of Seabuckthorn and orange juices were pasteurised with a tubular heat exchanger (Armfield FT74, HTST/UHT Processing Unit, Hampshire, England) at 94 °C for 30 s and 2.26 bar back pressure.

Analytical Methods

The total soluble content (°Brix) and pH were measured by 'ATAGO N20' hand refractometer (the mistake ± 0.1 per cent) the temperature 20 °C (ISO 2173:2003); 'JENCO' pH meter respectively. The total titration acidity content was determined titrated with 0.1N NaOH (ISO 750:1998); L-ascorbic acid (vitamin C) (mg/100 ml) content was determined with the enzymatic calorimetric method by R-Biopharm; The total content of phenols (mg/100 ml) was determined by using a spectrometer UV-1650-PC at wave length 760 nm (ISO14502-1:2005); The total carotenoid and β-carotene content was determined by using a spectrometer UV-1650-PC at wave length 450 nm (DGF standard methods, C-III). Antioxidant activity was analysed by HPLC – UV/Vis-online TEAC detection by ABTS radical-scavenging method.

Sensory Evaluation

Seabuckthorn juices and Orange - Seabuckthorn juices were evaluated using the hedonic rating and line scale according to ISO 4121:1987 by 14 trained panellists aged 25 to 60 years. Hedonic rating test was done using scale, in which 9 for like extremely; 5 for neither like or dislike and 1 is for dislike extremely. The taste, aroma, sourness, sweetness, and colour of juices were recorded using a line scale.

Statistical Analysis

The data processing was carried out by MS Excel variance analysis.

RESULTS AND DISCUSSION

To compare the impact of different processing technologies, the main biochemical compounds present in seabuckthorn juices were quantified.

Effect of Processing Technologies on Biochemical Properties of Seabuckthorn Juices

The soluble solids (°Brix), sugar composition, pH, titration acidity values of fresh seabuckthorn juices, HTST pasteurisation and UHT heat treated seabuckthorn juices are provided in Table 60.1. There were no significant differences in these values after HTST and UHT treatments compared to the fresh juices ($P < 0.05$). The content of °Brix of seabuckthorn juices was not significant increased by both heat treatment compared to fresh juices ($P < 0.05$).

Seabuckthorn juice is a rich source of bioactive compounds such as flavanones, vitamins C and E, carotenoids. Antioxidant capacity of fruit juice mainly results from its ascorbic acid content while flavanone

Table 60.1: The Effect of HTST and UHT Processing on Physico-chemical Characteristic in Seabuckthorn Juices

Treatment	*Soluble solids*	*Glucose*	*Fructose*	*Titration acidity*	*pH*
	(°Brix)		*(percent)*		
			Leikora		
Fresh (untreated)	7.34 ± 0.15	1.89 ± 0.05	0.82 ± 0.02	3.64 ± 0.04	2.79 ± 0.02
HTST (94 °C 30 s)	7.36 ± 0.15	1.88 ± 0.05	0.84 ± 0.02	3.61 ± 0.04	2.79 ± 0.02
UHT (130 °C 2 s)	7.34 ± 0.15	1.89 ± 0.05	0.86 ± 0.02	3.66 ± 0.04	2.8 ± 0.02
			Hergo		
Fresh (untreated)	5.98 ± 0.15	0.07 ± 0.00	0.12 ± 0.01	2.77 ± 0.04	2.84 ± 0.02
HTST (94 °C 30 s)	6.12 ± 0.15	0.07 ± 0.00	0.12 ± 0.01	2.77 ± 0.04	2.83 ± 0.02
UHT (130 °C 2 s)	6.11 ± 0.15	0.07 ± 0.00	0.12 ± 0.01	2.77 ± 0.04	2.83 ± 0.02
			Botanicheskaya-Lubitelskaya		
Fresh (untreated)	8.84 ± 0.15	1.96 ± 0.05	0.91 ± 0.02	3.18 ± 0.04	2.85 ± 0.02
HTST (94 °C 30 s)	9.01 ± 0.15	1.96 ± 0.05	0.91 ± 0.02	3.2 ± 0.04	2.85 ± 0.02
UHT (130 °C 2 s)	8.98 ± 0.15	1.94 ± 0.05	0.91 ± 0.02	3.18 ± 0.04	2.85 ± 0.02

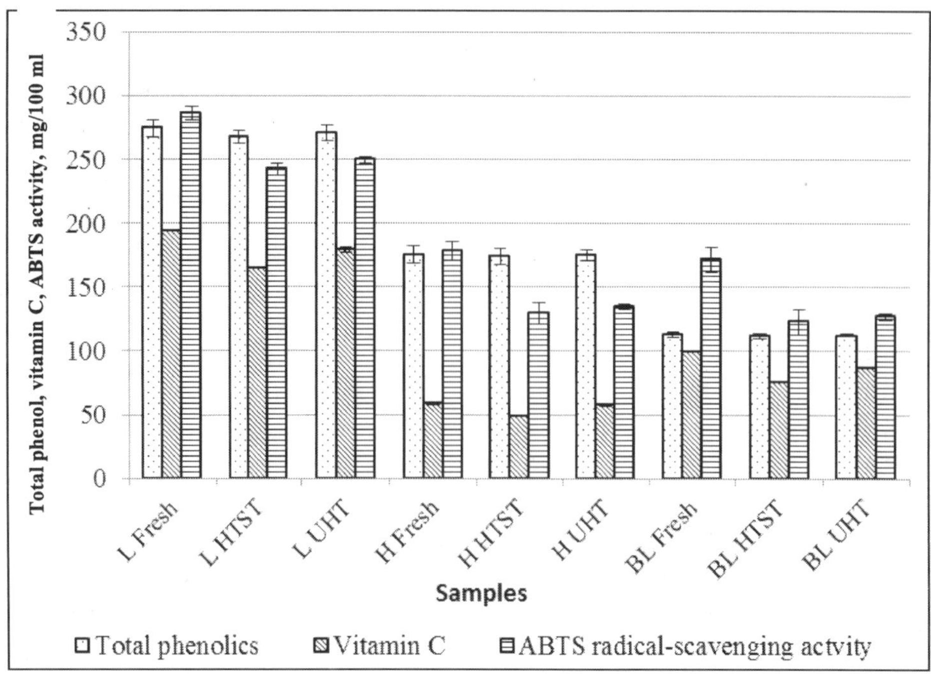

Figure 60.2: The Content of Total Phenolic, Vitamin C and ABTS Radical-Scavenging Activity of Fresh, HTST and UHT Seabuckthorn juices; L: Leikora; H: Hergo; BL: Botanicheskaya-Lubitelskaya.

glycosites have a limited contribution to its antioxidant capacity (Gorinstein, *et al.*, 2006; Rongsen, 2014). The effect of processing on total phenolic compounds, vitamin C and antiradical activity are presented in in Figure 60.2.

Significantly higher total phenolic content, vitamin C and ABTS radical-scavenging activity of fresh 'Leikora' seabuckthorn juice were 274.76 mg/100 ml, 194.36 mg/100 ml and ABTS radical-scavenging activity 286.86 mg troloks ekv/100 ml, respectively. Total phenolic content, vitamin C and ABTS radical-scavenging activity of fresh 'Hergo' and 'Botanicheskaya-Lubitelskaya' Seabuckthorn juices were 175.8 mg/100 ml, 113.3 mg/100 ml, 58.47 mg/100 ml, 99.25 mg/100 ml, and ABTS radical-scavenging activity 178.2, 172.02 mg troloks ekv/100 ml, respectively. The total phenolic content did not change significantly after the HTST and UHT treatments ($P > 0.05$). Changes in contents of ascorbic of UHT and thermal HTST are seen in Figure 60.2. There were signicant differences between UHT and HTST treated seabuckthorn juices in terms of vitamin. Vitamin C reduced in 'Leikora', 'Hergo' and 'Botanicheskaya-Lubitelskaya' varieties (7.58 per cent, 9.45 per cent, 12 per cent) respectively after UHT treated and (14.85 per cent, 16.75 per cent, 23 per cent) respectively after HTST treated compared with the fresh juices ($P < 0.05$). There is no study in the literature reporting effects of UHT treatment on total phenol content, on vitamin C and antioxidant capacity of seabuckthorn juice. However, there are several studies on the effects of UHT treatment on biochemical contents of apple, pomegranate and sugarcane juices (Lewis, *et al.*, 2000; Qudsieh, *et al.*, 2002; Margherita and Giussani, 2003). Plant pigments (carotenoids) ensure

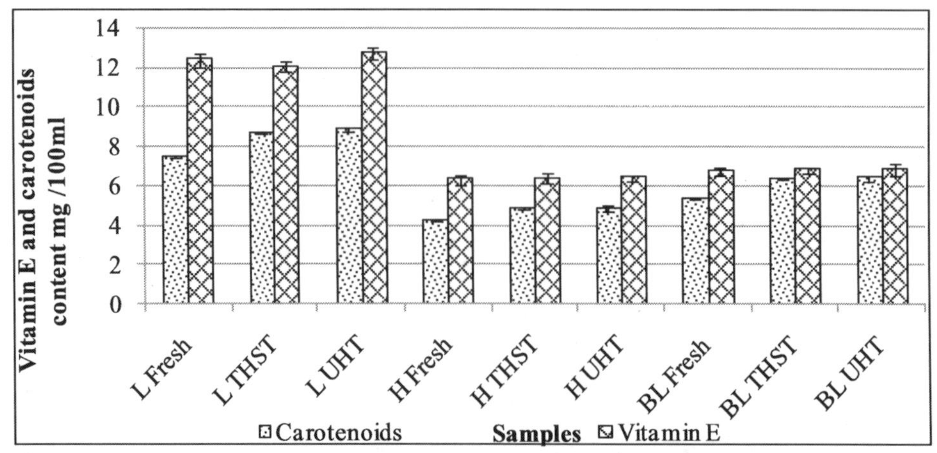

Figure 60.3: The Content of Carotenoids and Vitamin E of Fresh, HTST and UHT Seabuckthorn Juices; L: Leikora; H: Hergo; BL: Botanicheskaya-Lubitelskaya.

Table 60.2: Biochemical Characteristic of Orange-Seabuckthorn Juices UHT Treated

Samples	*Total phenolics*	*Carotenoids*	*Vitamin E*	*Vitamin C*	*ABTS radical-scavenging activity mg Trolox eqv/100 ml*
	(mg/100 ml)				
	Average				
Control	99.46 ± 3.77	1.53 ± 0.01	0.09 ± 0.36	57.02 ± 0.00	88.84 ± 2.79
Orange-Leikora SB juice	117.66 ± 3.89	2.30 ± 0.00	1.23 ± 0.33	90.95 ± 0.00	117.14 ± 3.72
Orange-Hergo SB juice	114.20 ± 2.51	2.01 ± 0.00	0.64 ± 0.35	109.67 ± 0.01	101.35 ± 3.72
Orange-Botan-Lub. SB juice	99.55 ± 4.65	1.57 ± 0.05	078 ± 0.023	50.98 ± 0.05	98.38 ± 3.28

colour intensity from yellow to red in Seabuckthorn fruits. To compare the impact of different processing technologies, the carotenoids and vitamin E properties in seabuckthorn juice where quantied (Seglina *et al.*, 2014). Figure 60.3 shows the carotenoid content and vitamin E content in 'Leikora', 'Hergo' and 'Botanicheskaya-Lubitelskaya' varieties in fresh and treated (HTST and UHT) seabuckthorn juices.

Just after treatment, HTST and UHT seabuckthorn juices showed a signicant increase in the carotenoids content. Regard fresh juices 'Leikora', 'Hergo' and 'Botanicheskaya-Lubitelskaya' on carotenoid content the carotenoid content was higher in 'Leikora' variety juice (7.47 mg/100 ml). After HTST and UHT heat treated the carotenoids content were 8.67 and 8.89 mg/100 ml in 'Leikora' 4.83 and 4.85 mg/ml in 'Hergo' and 6.8 and 6.68 mg/100 ml 'Botanicheskaya-Lubitelskaya' seabuckthorn juice respectively. The carotenoid content average (13 per cent) increase on all seabuckthorn juices compared fresh juices. Moreover, UHT juices showed the highest carotenoid content among all tested juices.

The vitamin E content of fresh, HTST, UHT heat treated seabuckthorn juices samples shown in Figure 60.3. There were no significant differences in vitamin E content after HTST and UHT treatment compared to fresh juices.

Sensory Properties

Fruit juices are an important source of bioactive compounds. Seabuckthorn juice in trading is presented as a drink or nectar with added sugar or other sweeteners, so they do not always provide the benefits expected by the consumer. As an alternative for the production we have used the mix of Orange juice with seabuckthorn juices. The Ratio sugar-acid of orange juice was 15. The changes of biologically active compounds in mix of Orange juice with seabuckthorn juices are represented in Table 60.2.

Acceptability of mix fresh orange juice and seabuckthorn juices ('Leikora', 'Hergo' and 'Botanicheskaya-Lubitelskaya') of fresh, UHT and HTST heat treated samples were investigated using hedonic scale. The results of the consumer acceptability study are shown in Table 60.3.

Table 60.3: Sensory Evaluation of Orange – Seabuckthorn Juices

	Samples		
	Orange - Leikora Seabuckthorn Juices Mix	*Orange - Hergo Seabuckthorn Juices Mix*	*Orange - Botanicheskaya - Lubitelskaya Seabuckthorn Juices Mix*
	**Hedonic scale*		
Fresh (control)	7.8	7.6	7.5
UHT treated	7.4	7.1	7
HTST treated	6.1	6.2	6.2

* Consumer Acceptance Testing (n=14); 9 point hedonic scale-9=like extremely; 8=like very much; 7=like moderately; 6=like slightly; 5=neither like nor dislike; 4=dislike slightly; 3=dislike moderately; 2=dislike very much; and 1=dislike extremely.

As seen in Table 60.3, fresh Orange-Leikora, Orange-Hergo, Orange-Botanicheskaya-Lubitelskaya seabuckthorn juice samples in terms of overall characteristics were located in the " very much liking" part of the hedonic scale with a 7.8, 7.6, 7.5 score, while the UHT and HTST treated Orange-Leikora, Orange-Hergo and Botanicheskaya-Lubitelskaya seabuckthorn samples were evaluated in the " like moderately" and " like slightly" part of the hedonic scale with 7.4, 7.1, 7. 0 and 6.1, 6.2, 6.2 scores, respectively. These results indicated that although sensorial characteristics of Orange-Seabuckthorn juices changed after the UHT treatment samples were more liked than the HTST heat treated samples. One of the problems faced by the consumer was delamination UHT juices in the short time. More stable to delamination was Orange-Leikora juice.

CONCLUSION

The chemical composition of seabuckthorn juice and Orange-Seabuckthorn beverage varies with the fruit cultivar, as well as the method of pasteurization. Significantly higher total phenolic content, vitamin C and ABTS radical-scavenging activity of fresh seabuckthorn cultivar 'Leikora' juice was found.

There were significant differences between UHT and HTST treated seabuckthorn juices in terms of vitamin C. Vitamin reduced less (7.6 per cent - 12 per cent) after UHT treatment but more (14.9 per cent - 23 per cent) after HTST treatment compared with the fresh juices ($p < 0.05$), while carotenoid content increase (on average 13 per cent) on all seabuckthorn juices compared fresh juices. Moreover, UHT juices showed the highest carotenoid content among all tested juices. The total phenolic and vitamin E content did not change significantly after the HTST and UHT treatments ($P > 0.05$). Fresh seabuckthorn juice samples in terms of overall characteristics were located in the "very much liking" part of the hedonic scale (7.5-7.8 score). Obtained results indicated that although sensorial characteristics of Orange-Seabuckthorn juices changed after the UHT treatment samples were more liked (7-7.4 score) than the HTST heat treated samples (6.1-6.2) score. As shows results from sensory investigation Orange-Seabuckthorn mix juices can be recommended as alternative for the production of the functional beverages.

REFERENCES

1. Abid, H., Hussain, A. and Ali, S. 2007. Physicochemical characteristics and fatty acid composition of seabuckthorn (*Hippophae rhamnoides* L.) oil. *Journal of Chemical Society of Pakistan* 29: 256-259.
2. Geetha, S., Ram, M. S., Singh, V., Ilavazhagan, G. and Sawhney, R.C. 2002. Antioxidant and immunomodulatory properties of seabuckthorn (*Hippophae rhamnoides*): An *in vitro* study. *Journal of Ethnopharmacology* 79: 373-378.
3. Gupta, A., Kumar, R., Pal, K., Singh, V., Banerjee, P.K. and Sawhney. R.C. 2006. Influence of seabuckthorn (*Hippophae rhamnoides* L.) flavone on dermal wound healing in rats. *Molecular and Cellular Biochemistry* 290: 193-198.
4. ISO 4121. 1987. *Sensory Analysis Methodology Evaluation of Food Products by Methods Using Scales.*
5. Lewis, M.J. 2003. Improvements in the pasteurisation and sterilisation of milk. In: *Dairy Processing Improving Quality*, Smit, G. (ed.), Woodhead Publishing Ltd., Cambridge, p. 93.
6. Qudsieh, H.Y.M., Yusof, S., Osman, A. and Rahman, R.A. 2002. Effect of maturity on chlorophyll, tannin, color, and polyphenol oxidase (PPO) activity of sugarcane juice (*Saccharum officinarum* var., *yellow cane). Journal of Agriculture and Food Chemistry* 50: 1615-1618.
7. Rongsen, L. 2014. Studies on correlation of berry quality parameters of seabuckthorn in relation to increasing altitude in China. In: *Seabuckthorn-A Multipurpose Wonder Plant Chemistry and Biochemistry* (V. Singh, Ed. in Chief, 2014), pp: 281-285, Daya Publishing House, New Delhi, 610p.
8. Seglina, D., Olsteine, A., Krasnova, I., Juhnevicca, K. and Sne, E. 2014. Effect of Spices on the quality of seabuckthorn nectar. In: *Seabuckthorn-A Multipurpose Wonder Plant* Vol. VII: *Processing Technology* (V. Singh, Ed. in Chief, 2014), pp. 507-517, Daya Publishing House, New Delhi, 610p.
9. Singh, V. 2003. Geographical adaption and distribution of seabuckthorn (*Hippophae* L.) resources. In: *Seabuckthorn-A Multipurpose Wonder Plant* Vol. I: *Botany, Harvesting and Processing Technologies* (V. Singh, Ed. in Chief, 2003), pp. 21-34, Indus Publishing Company, New Delhi, 518p.
10. Singh, V. 2006. Free radicals, disease, antioxidants and antioxidant properties of seabuckthorn. In: *Seabuckthorn-A Multipurpose Wonder Plant* Vol. II: *Biochemistry and Pharmacology* (V. Singh, Ed. in Chief, 2006), pp. 3-69, Daya Publishing House, New Delhi, 600p.
11. Suryakumar, S. and Gupta, A. 2011. Medicinal and therapeutic potential of seabuckthorn (*Hippophae rhamnoides* L.). *Journal of Ethnopharmacology* 138: 268-278.
12. Zeb, A. 2004. Chemical and nutritional constituents of seabuckthorn juice. *Pakistan Journal of Nutrition* 3: 99-106.
13. http://www.tetrapak.com.

Chapter 61

Nutritional Quality and Acceptability of Food Products Prepared from Seabuckthorn Berries (*Hippophae salicifolia*) grown in Lahaul Valley of Himachal Pradesh

Y.S. Dhaliwal*, R. Verma, S. Anand and V. Singh[1]

Department of Food Science, Nutrition and Technology
[1]Department of Biology and Environmental Sciences
CSK Himachal Pradesh Agricultural University, Palampur – 176 062, H.P., India
(*E-mail: ysdhaliwal44@yahoo.co.in)

SUMMARY

Seabuckthorn berries (*H. salicifolia)* procured from the different locations of Lahaul district of Himachal Pradesh were analyzed for physico-chemical and nutritional characteristics. The average weight of berries ranged between 122.00- 260.00 g/1000 berries. The average length and width ranged between 0.63 to 0.75 cm and 0.66 to 0.86 cm, respectively. The berries were round in shape with yellowish orange colour. The moisture content of berries ranged between 63.38 to 72.20 per cent and the total soluble solid content ranged between 9.00 to 10.00°B. The values for acidity and ascorbic acid content ranged between 2.03 to 2.11 per cent and 1200 to 1209 mg/100gm respectively. The pulp was extracted from the seabuckthorn berries as per the method standardized in the laboratory. Attempts were made to utilize the pulp of seabuckthorn berries for preparation of value added products *viz.* Ready-to-Serve (RTS) beverage, Squash, Syrup and Toffee. The different formulations for the preparation of different seabuckthorn based products have been standardized/developed and prepared products were evaluated for chemical characteristic and sensory attributes. The efforts have been made to retain the original anti-oxidant properties of the seabuckthorn berries in the specialty products. The prepared products were evaluated for sensory acceptability on 9 point hedonic scale. Results of the study revealed that acceptable quality RTS, squash, syrup and toffees can be prepared using pure seabuckthorn pulp with 18.00, 41.00, 71.00 and 85°B total soluble solids and can be popularized for commercially for economic upliftment and nutritional improvement of the target areas.

Keywords: *Seabuckthorn berries, Formulations, Nutritional profile and Sensory acceptability.*

INTRODUCTION

Seabuckthorn (*Hippophae* spp.), family-*Elaeagnaceae,* is one of important temperate plant in Northern India. Seabuckthorn are reported from Jammu and Kashmir, Himachal Pradesh, Uttarakhand, Sikkim and Arunanchal Pradesh (Singh, 2003). Seabuckthorn is one of the nutritious fruits in comparison to other underutilized temperate fruits and has worldwide distribution. Fruits of seabuckthorn contain high quantity of vitamin C, A, B, E and K and has been used for the production of various food products like juices, Squash, Syrup, Beer, Jam and Jelly besides its use in pharmaceutical industry as reported by Chauhan *et al.* (2003). This future food source, has been gaining attention because of its nutritional benefits as it has been reported to contain more than 190 compounds in the seed, pulp, fruit and juice, Bal *et al.* (2011). The main aim of the study was to analyze the physic-chemical characteristics of the berries

Seabuckthorn (*Hippophae* L.): *A Multipurpose Wonder Plant, Vol. 5: Advanced Technologies for Cultivation, Processing, Health Protection and Environmental Conservation* (Virendra Singh, *Editor-in-Chief,* 2018), p. 465–472, Astral International Pvt. Ltd., New Delhi, India.

so that the value added products can be developed and can be popularized commercially for economic upliftment and nutritional improvement of the target areas. The fruit of seabuckthorn represent one of the important vitamin sources for people living in cold and long winter regions of the world. Because of its multiple uses, seabuckthorn is known as wonder plant as reported by Bawa *et al.* (2002).

MATERIALS AND METHODS

Procurement of Sample

The ripe fruit of seabuckthorn berries (*H. salicifolia*) were procured from the different locations of Lahaul district of Himachal Pradesh and were brought to the Laboratory of Department of Food Science and Nutrition, College of Home Science, CSK Himachal Pradesh Agricultural University, Palampur for further analysis. The berries were cleaned under running water and whole berries were sorted out.

Physico-chemical Characteristics

Seabuckthorn berries were analyzed for different physical parameters *viz.*, 1000 fruit weight, length, width, colour and shape as per standard methods. The fruit length and width were measured using vernier calipers while fruit weight was recorded by digital balance. The moisture content was determined by standard method (AOAC, 1990). TSS was analyzed with the help of hand refractometer. The pH, acidity, Vitamin C was determined by standard methods of Rangana (2007).

Standardization of Products

The pulp was utilized for the preparation of value added products *viz.*, Ready to serve beverage, Squash, Syrup and Toffee. The different formulations were standardized and the prepared products were evaluated for chemical characteristics and sensory attributes as per standard methods.

RESULTS AND DISCUSSION

Physico-chemical and Nutritional Quality of Seabuckthorn Samples

The physical as well as chemical characteristics of the samples procured from different locations of Lahaul valley of Himachal Pradesh were assessed. The colour of samples varied from yellow to yellowish orange and the shape was oblong to round. The average weight of berries ranged between 122.00- 260.00 g/1000 berries. The average length and width ranged between 0.63 to 0.75 cm and 0.66 to 0.86 cm, respectively. The moisture content of berries ranged between 63.38 to 72.20 per cent and the total soluble solid content ranged between 9.00 to 10.00°B. Chauhan *et al.* (2003) and Dwivedi *et al.* (2006) reported 58.76 and 74.58 per cent moisture content in Indian seabuckthorn berries. Katoch *et al.* (2001) also reported a moisture level of 75.55 per cent in samples of berries collected from Lahaul valley of Himachal Pradesh. The values for acidity and ascorbic acid content ranged between 2.03 to 2.11 per cent and 1200 to 1209 mg/100gm respectively. Vitamin C being highly instable gets deteriorated and the areas where seabuckthorn berries grow is far away from the university and it take at least two days for samples to reach the lab. Also due to tough terrain of the area it is a cumbersome job to pluck and collect the berries and the process of collection might take a few days to weeks. Juice percentage recovered ranged from 50 to 71 per cent where Chauhan *et al.* (2001) reported the juice content varied from 65- 68 per cent. According to Katoch *et al.* (2006) the length and width of seabuckthorn berries varied from 6.9-7.2 mm and 5.9- 6.2 mm, respectively and the average weight of 100 berries was approximately 12.1g. Figure 61.1 shows the mean values of anthocyanin and β-carotene of seabuckthorn berries procured from different locations of Lahaul region.

Plate 61.1: Sample of *H. salicifolia*.

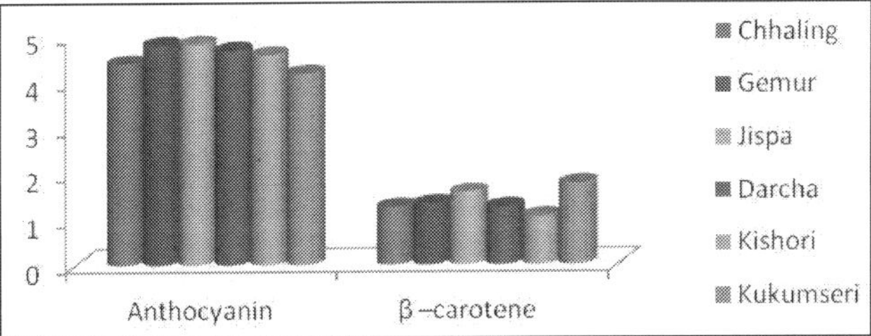

Figure 61.1: Mean Values of Anthocyanin and β-carotene (mg/100g) Content of Seabuckthorn Berries.

Fine Tuning of the Processing Technologies for Food Products

Fine tuning of methodologies for preparation of various value added products *viz.* ready to serve (RTS), Squash, Syrup, Fruit Toffees, Jam, Biscuits and Fruit leather was done using the pulp procured in bulk. In order to improve the sensory attributes of the seabuckthorn based products, peach fruit was used in different blending proportions (100:0, 75:25, 50:50, 25:75 and 0: 100). The prepared products were evaluated for consumer's acceptability on 9 point hedonic scale by the panel of ten judges. On the basis of sensory scores the best prepared products were stored in different packaging materials and storage study was carried out. RTS was stored in glass bottles whereas squash and syrup were stored in glass as well as plastic bottles. Fruit leather was stored in aluminium foil, cling film and butter paper. Jam was stored in glass jars, plastic jars, poly packs and laminated pouches.

Nutritional Quality and Stability of the Developed/Standardized Foods

1. Ready to Serve Beverage

On the basis of sensory scores pure seabuckthorn, pure peach and 50:50 seabuckthorn: peach RTS was kept for storage study in glass bottles for a period of six months at two different temperatures *viz.* room and refrigerator and was evaluated for different chemical parameters at an interval of 1.5 months. The parameters *viz.*, TSS, pH, Acidity, Ascorbic acid and sugars were recorded during storage. The ascorbic acid content of the RTS beverages slightly decreased with the increase in the storage period. The retention of ascorbic acid content was slightly higher in all beverages which were stored at refrigerated conditions. The sensory scores decreased with the increase in storage period and were rated between liked slightly and liked very much (Figure 61.2).

2. Squash

The squash was prepared in different proportions of seabuckthorn and peach pulp (100:0, 75:25, 50:50, 25:75 and 0:100) and after sensory evaluation three different squashes *viz*, pure seabuckthorn, 50:50 (Seabuckthorn: Peach) and pure peach squash were kept for storage study in glass and plastic bottles at room and refrigerator conditions. Effect of packaging material, storage intervals and storage conditions on the nutritional parameters of the squash prepared by blending seabuckthorn pulp with peach pulp was assessed. The overall acceptability scores of the peach blended seabuckthorn squash are shown in Figure 61.3.

3. Syrup

The syrup was stored in glass and plastic bottles and kept at two different temperatures (room and refrigerator) for the storage study. The products were evaluated fresh and after every 3 months till six months of storage for different chemical parameters. After the sensory evaluation the best combination was selected

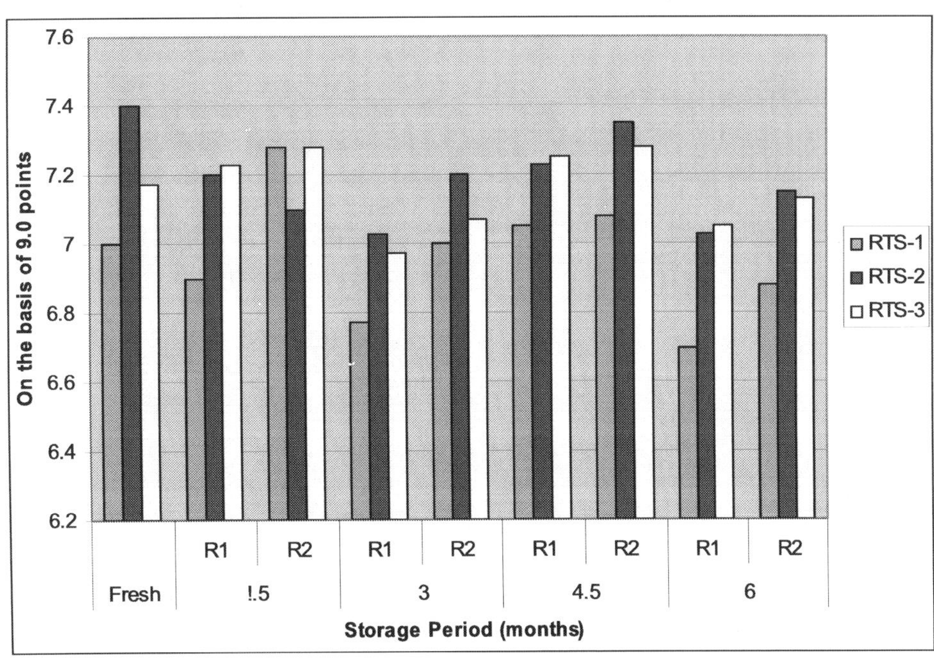

Figure 61.2: Overall Acceptability Scores of Peach Blended Seabuckthorn RTS. (SBT: Peach, 1-100:0, 2-50:50, 3-0:100 and R1- Room, R2- Refrigerator).

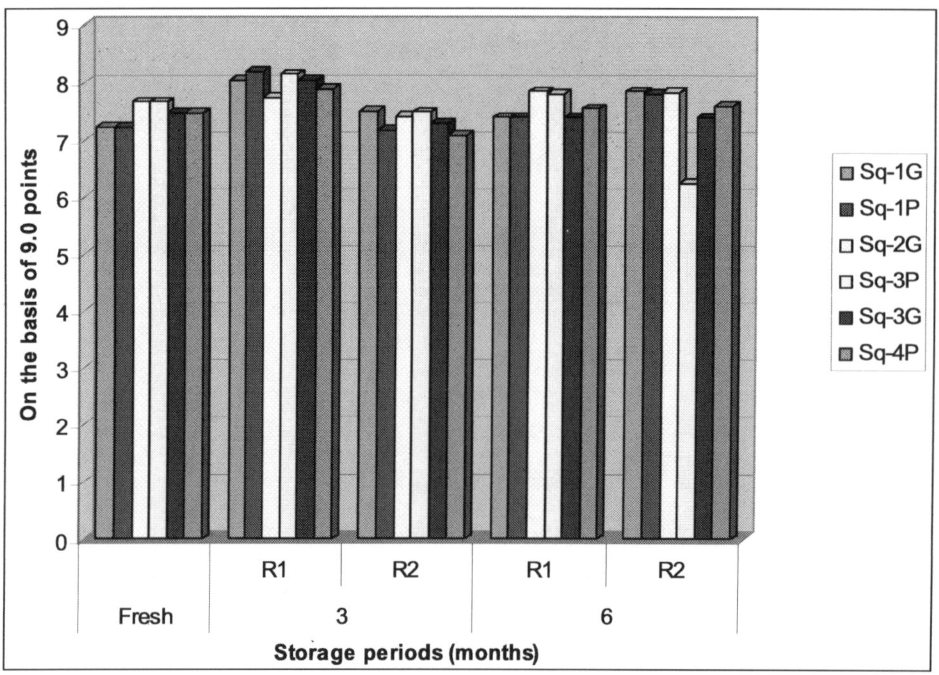

Figure 61.3: Overall Acceptability Scores of Peach Blended Seabuckthorn Squash. (SBT: Peach, 1-100:0, 2-50:50, 3-0:100 and R1- Room, R2- Refrigerator and G-Glass, P-Plastic bottles).

for further storage along with pure seabuckthorn and pure peach syrup.

4. Jam

Jam was prepared by blending seabuckthorn pulp with the pulp of different fruits *viz.* apple, mango, papaya and peach. Fruits were blended in equal proportions and then the pulp prepared was mixed in different proportions with seabuckthorn pulp. The jam for storage study was in the proportions of Seabuckthorn: Mixed fruit pulp (100:0, 50:50 and 0:100). The packaging materials used had significant effect on the TSS. The maximum values of TSS were observed in the jam packed in glass jars followed by plastic jars. The highest value of TSS was observed in pure seabuckthorn jam (69.7°B). The storage of jam had decreasing effect on the TSS. The values decreased from 70°B to 68.2°B after 6.0 months of storage. Figure 61.4 illustrated the values of TSS of seabuckthorn jam contained different proportions of mixed fruit pulp and stored in different packaging materials. The overall acceptability scores of jam are presented in Figure 61.5.

5. Fruit Leather

Six different formulations of fruit leather were standardized for the preparation of leather using pulp, sugar and pectin in different proportions in order to get non sticky and smooth leather. The prepared leather was packed in different packaging material like aluminium foil, cling film and butter paper for storage study at room temperature as well as in refrigerator.

6. Fruit Toffee

Methodology for preparation of fruit toffee has been standardized. The seabuckthorn and peach pulp in different proportions was blended (100:0, 75:25, 50:50, 25:75 and 0:100) and toffees were prepared. The prepared toffees were wrapped in butter paper and packed in cellophane paper. The prepared products were evaluated for different sensory parameters on 9 point hedonic scale (Figure 61.6).

7. Seabuckthorn Powder

The seabuckthorn berries were dried at 55°C and then ground to a fine powder and was utilized in the formation of biscuits. The method used for preparation of biscuits was standardized in the laboratory. Biscuits were prepared by using different proportions of refined wheat flour and seabuckthorn powder. The results of overall acceptability scores of biscuits contained different proportions of seabuckthorn powder can be utilized for preparation of different types of biscuits as well as other battery items. The addition of the seabuckthorn powder improved the nutritional profile of the bakery products. The colour of the seabuckthorn supplemented bakery products was also acceptable and improved the consumers acceptability.

Mineral Estimation of Samples from different Locations of Lahaul and Spiti Regions

The different samples of seabuckthorn berries collected from different locations of Lahaul and Spiti

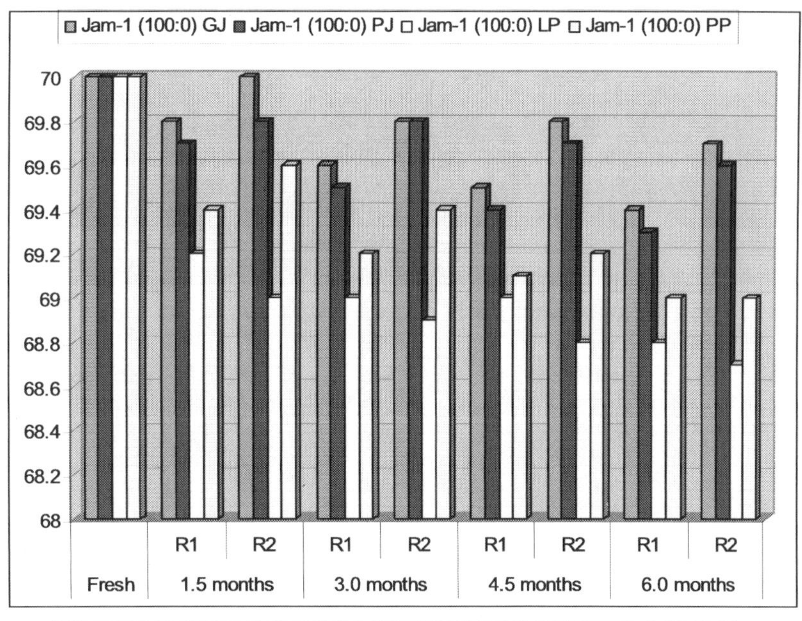

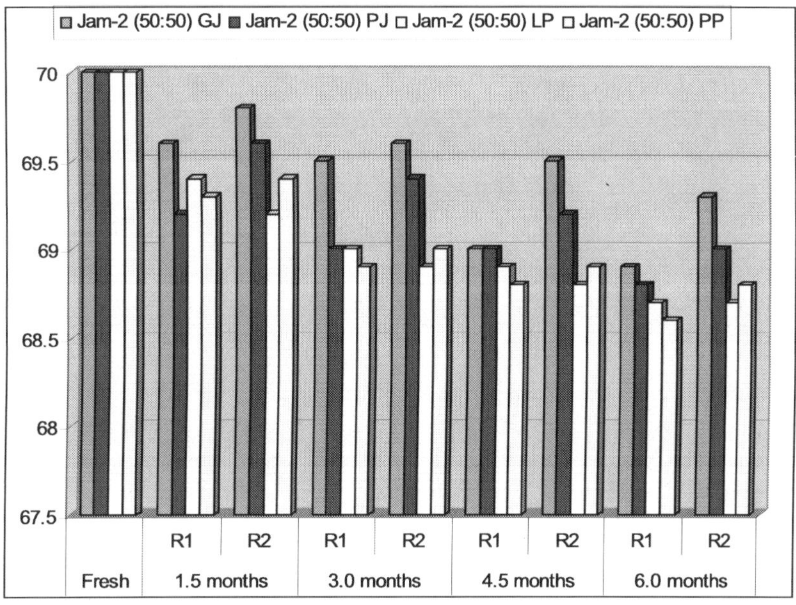

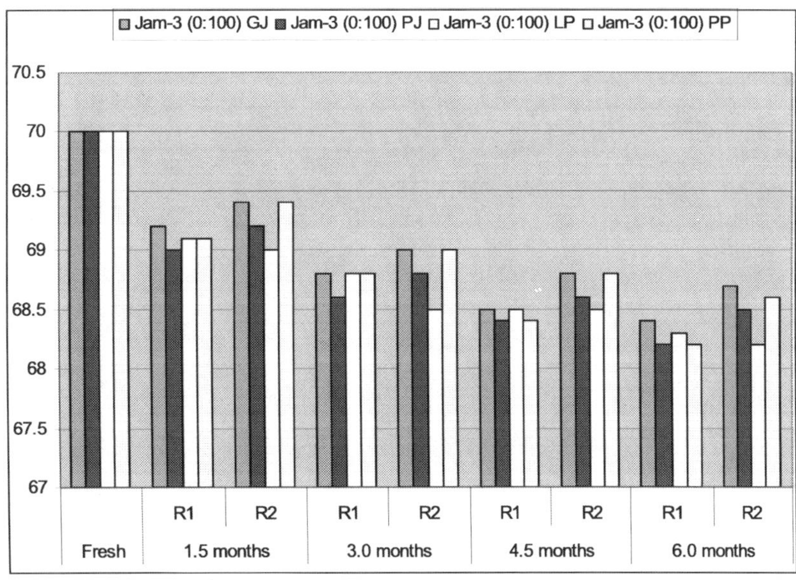

Figure 61.4: TSS Content of Seabuckthorn Jam Prepared from Blends Contained different Proportions of Mixed Fruit Pulp and Stored in Different Packaging Materials.

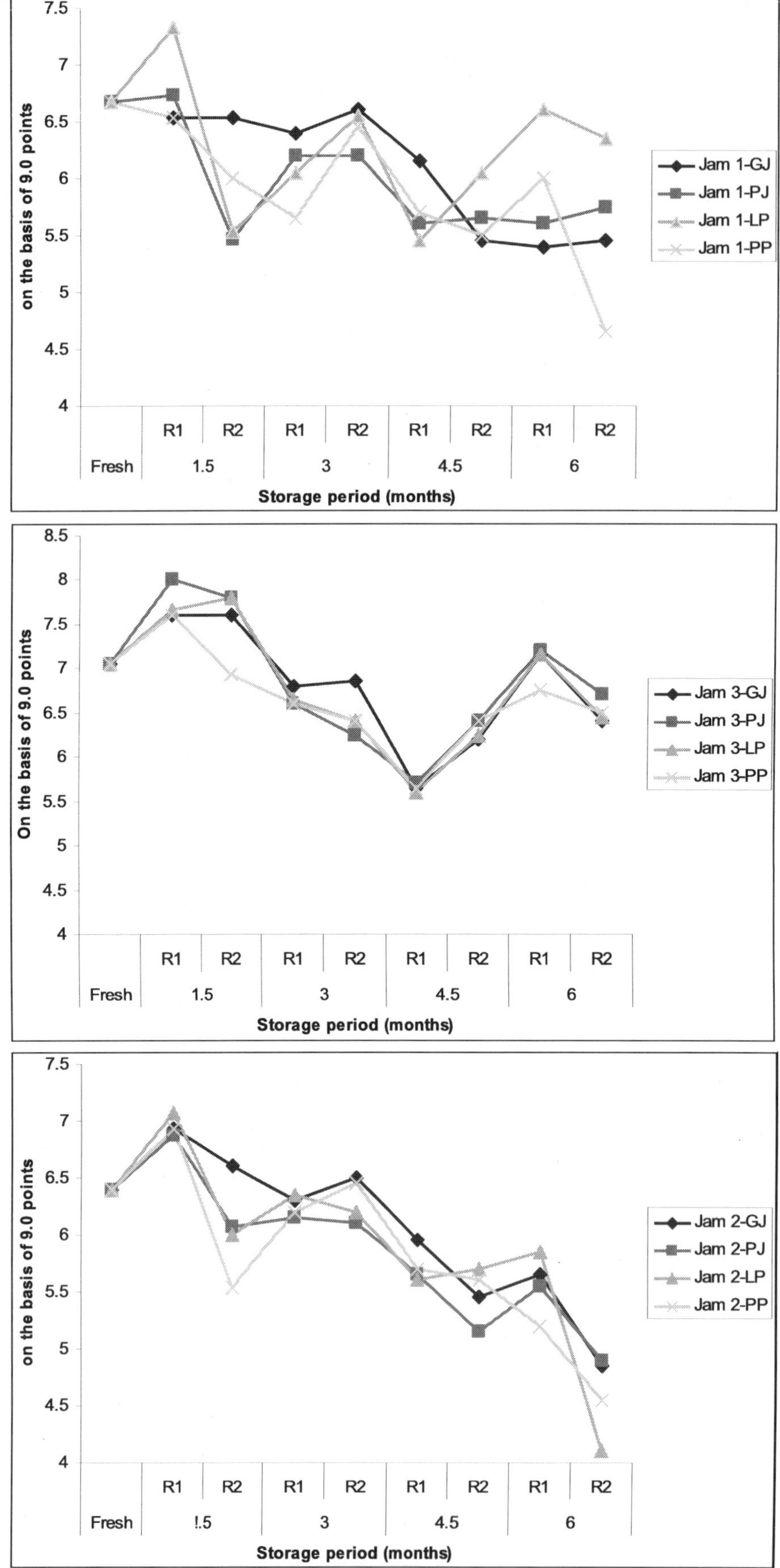

Figure 61.5: Overall Acceptability Scores of Seabuckthorn Jam Blended with Mixed Fruit. (100:0, 50:50 and 0:100, R1- Room, R2- Refrigerator and GJ-Glass Jar, PJ-Plastic Jar, LP-Laminated Pouches and PP-Poly Packs).

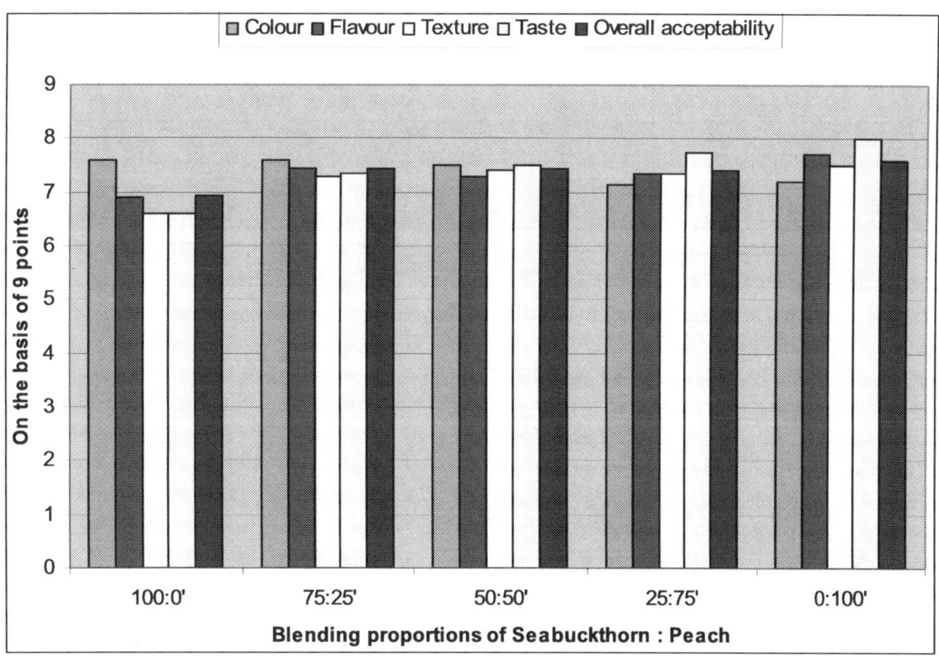

Figure 61.6: Sensory Scores of Fruit Toffee Prepared from Blends of Seabuckthorn: Peach.

areas of Himachal Pradesh were analyzed for different minerals using and results are presented in Figures 61.7.

CONCLUSION

The average weight of berries of specified locations ranged between 122.00- 260.00 g/1000 berries. The average length and width ranged between 0.63 to 0.75 cm and 0.66 to 0.86 cm, respectively. The berries were round in shape with yellowish orange colour. The moisture content of berries ranged between 63.38 to 72.20 per cent and the total soluble solid content ranged between 9.00 to 10.00°B. The ascorbic acid content ranged

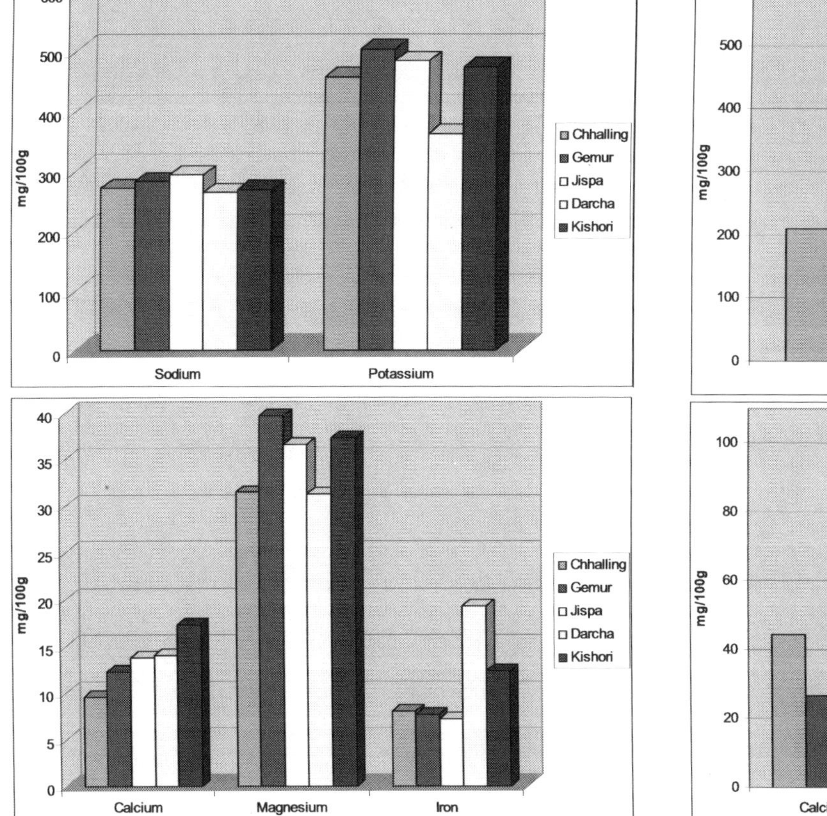

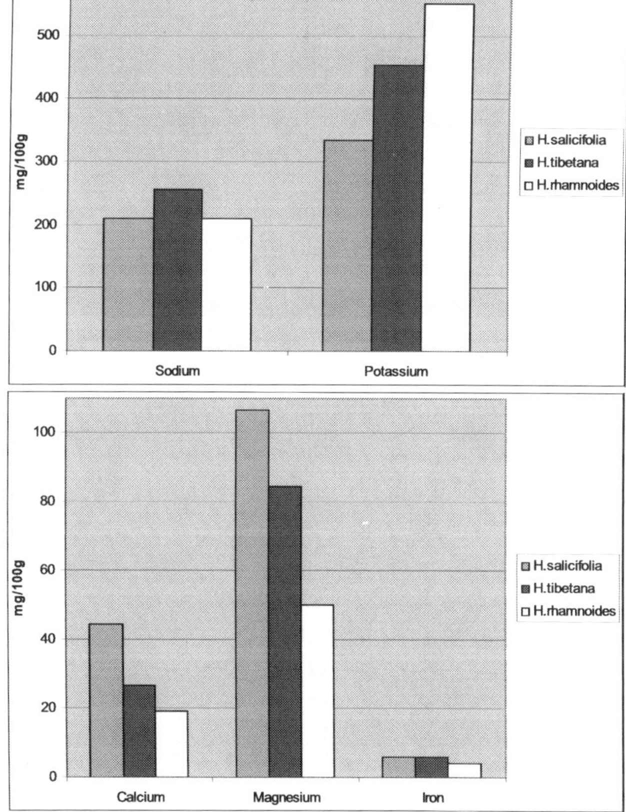

Figure 61.7: Mineral Content of Seabuckthorn Samples Procured from Different Locations of Lahaul and Spiti Regions.

between and 1200 to 1209 mg/100gm. Owing to high acidity (2.03 to 2.11 per cent) and astringency in case of seabuckthorn, the utilization of this fruit for preparation of various processed products becomes limited, despite having high nutritional qualities. Therefore, blending of two fruit juices is thought to be a convenient alternative for its utilization in order to have some value added fruit drinks which are of high quality in respect of both sensory and nutritional aspects. Blending of seabuckthorn pulp with locally available fruits like apricot, apple, mango, litchi, peach and papaya etc in varying proportions can improve the sensory quality of the prepared value added products. Blending with other fruits can further improve the sensory attributes. Fruit blends present a series of advantages, such as the possibility of combining different aromas and flavours, plus the sum of nutritionally different components. Commercial potential of the fruit is needed to be exploited in the form of its use in the preparation of value added products. This can boost the economical status of the tribal people of Himachal Pradesh and the country as a whole.

ACKNOWLEDGEMENT

The facilities generated under the NAIP sub-project on "Nutritional quality and value addition of seabuckthorn" with the assistance of grant from ICAR is fully acknowledged for research and development component of the project in the Department of Food Science, Nutrition and Technology, CSKHPKV, Palampur, India.

REFERENCES

1. AOAC. 1990. Approved Methods of Association Of Official Analytical Chemists. AOAC, Washington.
2. Bal, L. M., Meda, V., Naik, S. N. and Satya, S. 2011. Seabuckthron berries: A potential source of valuable nutrients for nutraceuticals and cosmeceuticals. *Food Research International* 44: 1718- 1727.
3. Bawa, A.S., Khanum, F. and Singh, B. 2002. Seabuckthorn A Wonder Plant. *Natural Product Radiance*, July-August, p. 8-14.
4. Chauhan, A.S., Rekha, M.N., Ramteke, R.S. and Eipeson, W.E. 2003. Seabuckthorn (*Hippophae rhamnoides* Lin.) berries: Harnessing its potential for processing. *J Food Sci. Technol*. 40 (4): 349-356.
5. Chen, G., Liu, B. and Yu, Y. 1995. Studies on pigments of seabuckthorn. *Hippophae* 7: 34-40.
6. Dwivedi, S.K., Singh, R. and Ahmed, Z. 2006. The Seabuckthorn. Field Research Laboratory (DRDO) p. 8-24.
7. Ranganana, S. 2007. Handbook of Analysis and Quality Control for Fruits and Vegetables products. 4th Edn., Tata McGraw Hill Pub. Co. Ltd., New Delhi, India.
8. Singh, V. 2003. Performance of some exotic forms of seabuckthorn. In: *International Congress on Seabuckthorn*, p. 47-53, 14-18th September, 2003, Berlin.

Part VI
ENVIRONMENTAL CONSERVATION

Chapter 62

Role of Seabuckthorn in Improvement of Soil Fertility in Cold Desert of Himalayas

Abhay Sharma[1]*, Virendra Singh[2], S.P. Dixit [2] and Kanika Baghla[3]

[1]Department of Biology and Environmental Sciences, College of Basic Sciences
[2]Department of Soil Sciences, College of Agriculture
[3]Department of Agricultural Engineering, College of Agriculture
CSKHPKV Palampur – 176 062, Himachal Pradesh, India
(*E-mail: sharma.abhay5@gmail.com)

SUMMARY

Seabuckthorn (*Hippophae* L.) is a winter hardy, deciduous shrub belonging to the family Elaeagnaceae. It is an actinorhizal plant characterized by its ability to form a symbiosis with the nitrogen fixing actinobacteria *Frankia* in the root nodules. It is found in the cold desert of north-western Himalayas at elevations ranging between 3200-4500 m above sea level. Extreme climate, scant rainfall, massive snowfall, high wind velocity, extreme temperature condition, sparse vegetation, high UV radiation, and extremely xeric conditions are the common features of this region. As it develops extensive root system making it ideal for soil binding and its strong root system and nitrogen fixing ability improves soil quality or fertility. An attempt has been made to study the effect of seabuckthorn on soil fertility in comparison to willow and wasteland in relation to increasing altitudes in cold desert of western Himalayas in Himachal Pradesh (India). In order to achieve the objective of the present study, soil samples were collected at six different altitudes, where *Hippophae rhamnoides* was found naturally and were chemically analyzed. The result showed that soil pH ranged from 8.2 to 7.8 in the soil top layer (0-10 cm) and decreased with an increase in altitude and increased with an increase in soil depth. Soil organic carbon ranged from 1.05 to 2.35 per cent in the soil top layer and increased with an increase in altitude and decreased with an increase in soil depth. The values of soil available macronutrients also varied, *i.e.* N (125 to 205 kg ha^{-1}), P (15 to 31 kg ha^{-1}), K (94 to 284 kg ha^{-1}), S (28 to 53 kg ha^{-1}), exchangeable Ca (8.56 to 10.20 c mol (p+) kg^{-1}) and Mg (2.3 to 3.6 cmol (p+) kg^{-1}), respectively, in the soil top layer and increased with an increase altitude and decreased with an increase in soil depth. Also, status of available micronutrients varied *i.e.* Fe (3.9 to 11.4 mg kg^{-1}), Cu (1.2 to 3.8 mg kg^{-1}), Zn (1.8 to 1.1 mg kg^{-1}) and Mn (4.5 to 2.3 mg kg^{-1}), in the soil top layer and decreased with an increase in soil depth. Available Fe and Cu increased and available Zn and Mn decreased with an increase in altitude. Values of organic carbon, available N, S, exchangeable Ca, Mg and available micronutrients Fe, Cu, Zn and Mn were in higher contents under seabuckthorn vegetation than willow and wasteland. Hence, seabuckthorn with its rich nodules and its increasing nutrient quality especially organic carbon, available N, P and K contents proved to be a wonder plant, which plays a significant role in increasing soil nutrients and improving the fertility in the soils of high altitude cold desert of Himalayas.

***Keywords**: Seabuckthorn, Nitrogen fixing, Soil fertility and Cold desert of Himalayas.*

INTRODUCTION

In India, cold desert comes under the trans Himalayan zone, which is approximately 1,03,113 sq. km. These are arid areas not affected by the Indian monsoons because they lie in the rain shadow of the Himalayan mountain systems. The bulk of the Indian cold desert lies in Ladakh region of Jammu and Kashmir, Lahaul and Spiti, Pooh sub division of Kinnaur, Pangi area of Himachal Pradesh and some areas of northern Uttarakhand and Sikkim. Extreme climatic conditions,

Seabuckthorn (Hippophae L.): A Multipurpose Wonder Plant, Vol. 5: Advanced Technologies for Cultivation, Processing, Health Protection and Environmental Conservation (Virendra Singh, *Editor-in-Chief*, 2018), p. 475–484, Astral International Pvt. Ltd., New Delhi, India.

scant rainfall, massive snowfall, high wind velocity, extreme temperature conditions from low to high, sparse vegetation, high UV radiation, intense solar radiation and extremely xeric conditions are the common features of the this region (Devi and Thakur, 2011).

The diverse climate and the varied environmental conditions prevailing in cold desert of Himalayas support diverse habitat and ecosystems with equally diverse life forms. But this unique and the most beautiful corner of the Himalayas is under threat due to various environmental problems prevailing in the region, which include the utilization of the slope, land resources for agriculture, poor water management, soil erosion and degradation of soil fertility. Erosion results in the degradation of productivity of soil in a number of ways.

It was found that seabuckthorn (*Hippophae rhamnoides* L.) plantation played a beneficial role in controlling soil erosion and improving soil fertility in Loess Plateau area of China (Qian, 1986). Zhao *et al.* (1990) proposed that utilization of seabuckthorn is superior to any other tree in afforestation, soil conservation and wasteland reclamation.

Seabuckthorn (*Hippophae* L.) a beautiful plant with multipurpose properties is a perennial, woody, nitrogen fixing and thorny, deciduous shrub or a form of tree covered with silvery scales belongs to family *Elaeagnaceae* (Rousi, 1971). It naturally grows in Europe and cold regions of Asia (Rongsen, 1992). Therefore, plant is naturalized in snowfall or low temperature conditions, which meet the chilling requirement of the plant. It tolerates temperature as low as -40°C during winter. There is natural distribution of seabuckthorn in the regions with the temperature isolines over 25°C on the average in the summer (Gupta *et al.*, 2001). It has been reported growing in about 40 countries like Russia, China, Indian Himalayas and Central Asia, Germany, Finland, Sweden, *etc.* It has also been introduced in Canada, USA, Bolvia, Korea and Japan (Singh, 2003).

Seabuckthorn is one of the most magical plant resources with higher value of economy and ecology. Its berry is a very rich source of vitamins and is called treasure of bio-activity substance because of its over 190 bioactivity substances possessing unique medicinal properties (Maertz, 2006). With help of scientific studies and researches, seabuckthorn has become the source of important medicinal and nutritional products in China, India, Russia, Germany, Finland, Sweden, *etc.* Yet seabuckthorn is much more than just a pretty plant with nutritious berries. It also has significant characteristics such as wide adaptation, fast growing rate, strong coppicing, suckering habits and efficient nitrogen fixation, this species is often the first woody species colonizing open areas (Rousi, 1965). Seabuckthorn is ecologically sound for soil and water conservation, land reclamation and rehabilitation, reforestation, establishment of wildlife habitats, fuelwood production, soil improvement, fixing of atmospheric nitrogen, and farmland protection (Schroeder, 1990). This plant is also equally important for firewood, fodder and serves as a soil binder species. Therefore, seabuckthorn should serve as a measure to safeguard medicinal and nutritional plants, to conserve biodiversity and environment and to generate sustainable income source for local people (Rousi, 1971).

Seabuckthorn has an outstanding ability to take roots even in poor soils, because of its ability to fix atmospheric nitrogen through the presence of symbiotic bacteria *Frankia* (*Actinomycetes*) in the root nodules of the plant. Root nodules of seabuckthorn have been found to fix 180 kg of nitrogen/ha/year, which generally improves soil fertility. The results improved root growth enhances the entire soil ecosystem in the form of rich organic matter, more oxygen and more soil organism, which means more soil biodiversity (Jike and Xiaoming, 1992).

Seabuckthorn (*Hippophae rhamnoides* L.) because of its strong canopy has the ecological function of protecting against wind and preventing sand from drifting, improving the biological diversity and also recovering and rebuilding the ecological system. Branching in seabuckthorn is intense and upward oriented. So its dense canopy intercepts rainfall and lessens the speed of rain drops. This releases the rainwater slowly on the ground canopy and thick layer thus, protect the soil from erosion (Chengjian and Daiqiong, 2002). For the farmers living in the mountains, seabuckthorn offers the opportunity to maintain a sustainable livelihood, providing healthy foods, variety of medicines and protecting their land from soil erosion (Lu Rongsen, 1992; Ansari, 2003).

It is well understood that the key to long-term success in increasing soil fertility in cold deserts is to encourage the seabuckthorn plantations. The present study summarizes the effect of seabuckthorn plantation on soil fertility. The results will help to provide a platform for future research into the role of seabuckthorn in improvement of soil fertility in cold desert of Himalayas. The main aim of the study was to examine the effect of seabuckthorn on soil fertility in the cold desert regions of Himalayas.

MATERIALS AND METHODS

Study Area

An investigation was conducted in Spiti cold desert of western Himalayas, which has a geographical area of 5,582 sq.km., situated between 32°05′55.1″-78°21′99.8″ N latitude and 32°26′72.0″-77°53′81.1″ E longitude in the extreme North-East region of Himachal Pradesh, India.

Field Study

Experimental Details

The field experiment was carried out in 6 places at different altitudes in an elevation range of 3390-4040 m

Figure 62.1: Seabuckthorn (*Hippophae rhamnoides*) Forest, Spiti, HP.

above sea level in Spiti valley. Soil samples at the depths of 0-10 cm, 10-20 cm and 20-30 cm were collected under seabuckthorn (*Hippophae rhamnoides*), willow (*Salix daphnoides*) vegetation and control land. Soil samples were collected from the different locations such as Tabo (3390 m asl), Shichling (3520 m asl), Lingti (3560 m asl), Shego (3615 m asl), Rangrik (3790 m asl) and Hansa (4040 m asl) with increasing altitude.

Soil Sampling

Soil samples were collected from various sites at soil depths of 0-10 cm, 10-20 cm and 20-30 cm from 5 different locations within 3 plots, each of 10 m × 10 m size, chosen randomly in every site with the help of a soil auger and were thoroughly mixed together making 5 replicates of each samples from each soil depth weighing about 200 g and subsequently stored in polythene bags for the determination of chemical properties.

Laboratory Analysis

The processed soil samples were analyzed chemical properties (soil organic matter, available N, P, K, S, exchangeable Ca and Mg available micro-nutrients Fe, Cu, Zn and Mn) from the following standard methods.

a) Soil pH

The pH of soil water suspension was determined using electrode pH meter. 10 g soil sample was mixed with 25 ml distilled water in 1: 2.5 ratios. The suspension was stirred intermittently with glass rod and wait for 30 minutes. Then electrode was inserted into supernatant and pH was recorded (Jackson, 1967).

b) Organic Carbon

A 1 g finely ground soil sample was taken into 500 ml conical flask, to which 10 ml of 1 N $K_2Cr_2O_7$ and 20 ml conc. H_2SO_4 were added. Then content was shaken for a minute and allowed to set aside for exactly 30 minutes. Then about 1.0 g sodium fluoride or 10 ml orthophosphoric acid, 200 ml distilled water and 10 drops of diphenylamine indicator were added and shaken vigorously to mix. The solution was titrated against N/2 ferrous ammonium sulphate from a burette, till the colour changes from violet to bright green and quantity of organic carbon of the soil was estimated (Walkey and Black, 1934).

c) Available N

A 2 g of soil sample was taken in Kjeldahl's distillation flask, to which 20 ml of $KMnO_4$ and 20 ml of NaOH was added and closed with the cork. Approximately 10 ml of Boric acid solution was taken in the 125 ml conical flask with help of pipette. Distill the ammonia gas at a steady rate and collect about 50 to 60 ml of distillate. Titrate the boric acid solution containing dissolved ammonia against 0.01 N H_2SO_4. The end point is reached when the colour changes from bluish green to wine red. Available N was calculated by alkaline potassium permanganate method (Subhaiah and Asija, 1956).

d) Available P

Available P was estimated in the soil by Alkali extraction. Alkali solution of 0.5 M $NaHCO_3$ (pH 8.5) is used, known as Olsen's reagent prepared by dissolving 42 gm $NaHCO_3$ in distilled water and making volume to 1 litre and adjusting pH of the solution to 8.5 by adding either NaOH or HCl solution (Olsen *et al.*, 1954).

e) Available K

The most commonly used method for K determination is by flame photometry. A 5 g soil sample was taken in a pH bottle, to which 25 ml 1N NH_4OAc (pH 7.0) with a pipette (1: 10 ratio) was added. The contents were shaken for 5 minutes on a mechanical shaker and filtered immediately. Prepare a standard curve by adjusting 40 ppm K solution to 100 and 0, 5, 10, 15, 20 ppm K solution accordingly. Take the reading and compare this reading with standard curve and note down the concentration of K from it. Available K was calculated by neutral normal ammonium acetate extraction method (Jackson, 1967).

f) Available S

A 10 g soil sample was taken into a 150 ml Erlenmeyer flask and 50 ml of 0.15 per cent $CaCl_2$ extraction solution was added in the flask. The contents were shaken for 30 minutes on a mechanical shaker and filtered immediately. Then, 10 ml of aliquot from filtered solution and 10 ml of Morgan's reagent were transferred in 25 ml flask. Addition of 2 ml gum acacia and 1 g $BaCl_2$ was done in the flask and again shaken for 1 minute. After 5 minutes sulphur was calculated with reference to standard curve (Jackson, 1973).

g) Exchangeable Ca and Mg

A 5 g of air dried soil sample was taken in 500 ml conical flask, to which 50 ml of neutral normal ammonium acetate was added and was allowed to stand for 30 minutes. The contents were shaken on a mechanical shaker and filtered immediately and determine the Ca and Mg by using an Atomic Absorption Spectrophotometer after the removal of ammonium acetate (Jackson, 1967).

h) Available Micronutrients

A 10 g soil sample was weighed in clean and dry pH bottle and 20 ml DTPA extracting solution was added in the same bottle. The contents were shaken on a mechanical shaker with a speed of approximately 120 cycles per minute continuously for 2 hours. Then soil suspension was filtered through and the estimation of micronutrients (Fe, Cu, Zn and Mn) in the soil was done by the DTPA method in the Atomic Absorption Spectrophotometer (Lindsay and Norvell, 1978).

i) Statistical Analysis

Soil chemical properties were subjected to (ANOVA) analysis of variance. Simple correlation was worked out to establish the relationships between selected soil fertility parameters and along and within increasing altitudes and soil depths as per the procedure outlined by Gomez and Gomez (1984). Relationship was tested at 5 per cent level of significance.

RESULTS AND DISCUSSION

Variation of different Soil Nutrients with Altitude under Seabuckthorn in Spiti

A perusal of Table 62.1 and Figures 62.2 and 62.3 showed that soil pH under seabuckthorn in Spiti significantly decreased with an increase in altitude and increased with an increase in soil depth, whereas, soil organic carbon significantly increased with an increase in altitude and decreased with an increase in depth of soil. Similarly, soil available N significantly increased with an increase in altitude and decreased with increasing depth.

Soil available P, available K and available S under seabuckthorn vegetation at different places in Spiti significantly increased with an increase in altitude and decreased with an increase in soil depth. Similarly, soil exchangeable Ca and Mg significantly increased with an increase in altitude and decreased with an increase in soil depth.

Soil available Fe and available Cu under seabuckthorn vegetation in Spiti significantly increased with an increase in altitude and decreased with an increase in soil depth. Whereas, soil available Zn and available Mn significantly decreased with increasing altitude and soil depth.

Variation of different soil nutrients with altitude under seabuckthorn in Spiti

From Table 62.2 it was found that different altitudes and the species interactions were significant ($p \leq 0.05$) for the soil pH organic carbon, available N, P, K, S and exchangeable Ca, Mg and available Fe, Cu, Zn, Mn. Soil characteristics varied greatly under seabuckthorn than willow and wasteland soils. Compared with willow and wasteland, soil under seabuckthorn vegetation had higher values of pH, organic carbon, available N, S, exchangeable Ca, Mg and available micro-nutrients Fe, Cu, Zn and Mn. Also, the soil nutrients increased with an increase in altitude, except soil pH and available Zn and Mn.

As we know seabuckthorn (*Hippophae rhamnoides* L.) has an extraordinary capacity to grow and survive under adverse conditions (-40° to 40° C) and has extensive subterranean rooting system with strong soil binding ability useful for soil stabilization, river bank control and water retention and also fixes nitrogen which improves soil quality or fertility. Above results indicated that much more soil nutrients enriched in the seabuckthorn soils than willow and wasteland soils. Soil pH was alkaline in reaction and the similar findings have been reported by Sharma and Kanwar (2010) and Sharma (2011). The soil organic carbon decreased with an increase in soil depth and increased with an increase in altitude and the results were corroborated with findings of Singh and Datta (1988), Lu (2009) and Gong *et al.* (2007).

Available macronutrients (N, P, K and S) increased with an increase in altitude and decreased with an increase in soil depth and similar results were shown by Guo (2003), Sharma (2011) and Zhao *et al.* (2013). Exchangeable ions (Ca and Mg) increased with an increase in altitude and decreased with an increase in soil depth and the similar findings have been reported by Singh and Raman (1982) and Xiaoning *et al.* (2002). Available micronutrients (Fe, Cu, Zn and Mn) decreased with an increase in soil depth, among which available Fe and Cu increased and available Zn and Mn decreased with an increase in altitude and the similar findings

Table 62.1: Integrated Effects of the Seabuckthorn (*Hippophae rhamnoides*) on the Soil pH, Organic Carbon and other available Nutrients

Locations		Depth (cm)									Depth (cm)		
	Altitude m asl	(0-10)	(10-20)	(20-30)	(0-10)	(10-20)	(20-30)	(0-10)	(10-20)	(20-30)	(0-10)	(10-20)	(20-30)
		pH			OC (per cent)			Ca (cmol(p+)kg^{-1})			Mg (cmol(p+)kg^{-1})		
Tabo	3390	8.2±0.15	8.2±0.10	8.3±0.09	1.05±0.03	1.03±0.02	1.00±0.03	8.56±0.03	8.53±0.03	8.51±0.02	2.3±0.24	2.2±0.29	1.9±0.41
Shichling	3520	8.1±0.12	8.2±0.08	8.3±0.10	1.18±0.04	1.15±0.02	1.13±0.03	9.15±0.03	9.12±0.01	9.09±0.01	2.6±0.17	2.4±0.38	2.3±0.19
Lingti	3560	8.0±0.21	8.0±0.26	8.1±0.18	2.21±0.05	2.19±0.04	2.17±0.02	9.87±0.02	9.83±0.02	9.81±0.01	2.8±0.38	2.5±0.07	2.3±0.20
Shego	3615	7.7±0.18	7.8±0.18	7.9±0.19	2.24±0.03	2.22±0.02	2.2±0.06	9.98±0.01	9.95±0.03	9.93±0.02	2.9±0.30	2.7±0.11	2.4±0.44
Rangreek	3790	7.8±0.13	7.9±0.15	8.0±0.20	2.29±0.02	2.27±0.02	2.24±0.04	9.99±0.02	9.97±0.02	9.95±0.03	3.3±0.28	3.1±0.08	2.8±0.24
Hansa	4040	7.8±0.29	7.9±0.12	8.0±0.19	2.35±0.02	2.30±0.05	2.28±0.02	10.20±0.02	10.18±0.02	10.16±0.01	3.6±0.33	3.3±0.22	3.2±0.29
CD (P ≤ 0.05)			0.105			0.020			0.013			0.135	
		N (kg ha^{-1})			P (kg ha^{-1})			Fe (mg kg^{-1})			Cu (mg kg^{-1})		
Tabo	3390	125±4.03	124±1.98	119±3.87	15±3.02	13±1.97	10±3.00	3.9±0.23	3.5±0.49	3.2±0.30	1.2±0.23	0.9±0.30	0.7±0.23
Shichling	3520	142±2.13	138±4.98	132±3.94	18±4.03	15±2.04	13±2.07	4.1±0.16	3.6±0.58	3.3±0.23	1.7±0.43	1.5±0.49	1.3±0.30
Lingti	3560	153±3.03	149±1.99	144±4.02	20±4.01	18±2.10	15±2.01	4.7±0.22	4.3±0.30	4.0±0.10	2.0±0.21	1.7±0.30	1.4±0.18
Shego	3615	156±2.24	153±3.02	144±3.97	23±3.12	20±2.98	16±2.11	5.7±0.24	5.3±0.32	5.0±0.50	2.5±0.48	2.1±0.12	1.9±0.40
Rangreek	3790	168±2.28	165±1.96	162±2.03	26±3.00	22±3.96	19±1.99	9.1±0.10	8.8±0.21	8.5±0.48	3.0±0.51	2.4±0.22	1.9±0.30
Hansa	4040	205±4.95	198±5.12	195±1.97	31±1.01	26±5.04	20±2.03	11.4±0.39	11.2±0.40	10.9±0.31	3.8±0.40	3.4±0.38	2.8±0.22
CD (P ≤ 0.05)			1.630			1.520			0.175			0.149	
		K (kg ha^{-1})			S (kg ha^{-1})			Zn (mg kg^{-1})			Mn (mg kg^{-1})		
Tabo	3390	94±3.98	91±3.02	88±2.03	28±3.02	25±2.98	21±1.00	1.8±0.19	1.5±0.20	1.3±0.32	4.5±0.52	4.1±0.11	3.8±0.20
Shichling	3520	136±3.03	131±0.99	126±2.10	34±4.10	31±1.97	28±2.03	1.6±0.23	1.3±0.33	1.1±0.30	4.0±0.29	3.7±0.38	3.3±0.28
Lingti	3560	145±4.98	139±3.00	133±3.04	39±1.00	36±3.02	32±1.99	1.5±0.30	1.3±0.28	1.0±0.48	3.6±0.40	3.2±0.23	2.8±0.32
Shego	3615	225±2.98	219±2.01	214±2.03	45±2.05	40±4.96	36±4.01	1.3±0.32	1.0±0.22	0.7±0.20	3.2±0.22	2.9±0.10	2.3±0.31
Rangreek	3790	264±3.96	259±1.02	254±4.10	50±5.02	45±3.05	43±3.17	1.2±0.21	0.8±0.19	0.5±0.18	2.7±0.37	2.2±0.21	1.7±0.29
Hansa	4040	284±4.04	279±2.98	273±2.98	53±2.78	49±1.23	45±4.79	1.1±0.10	0.7±0.13	0.6±0.21	2.3±0.21	2.0±0.19	1.6±0.52
CD (P ≤ 0.05)			4.720			1.590			0.122			0.165	

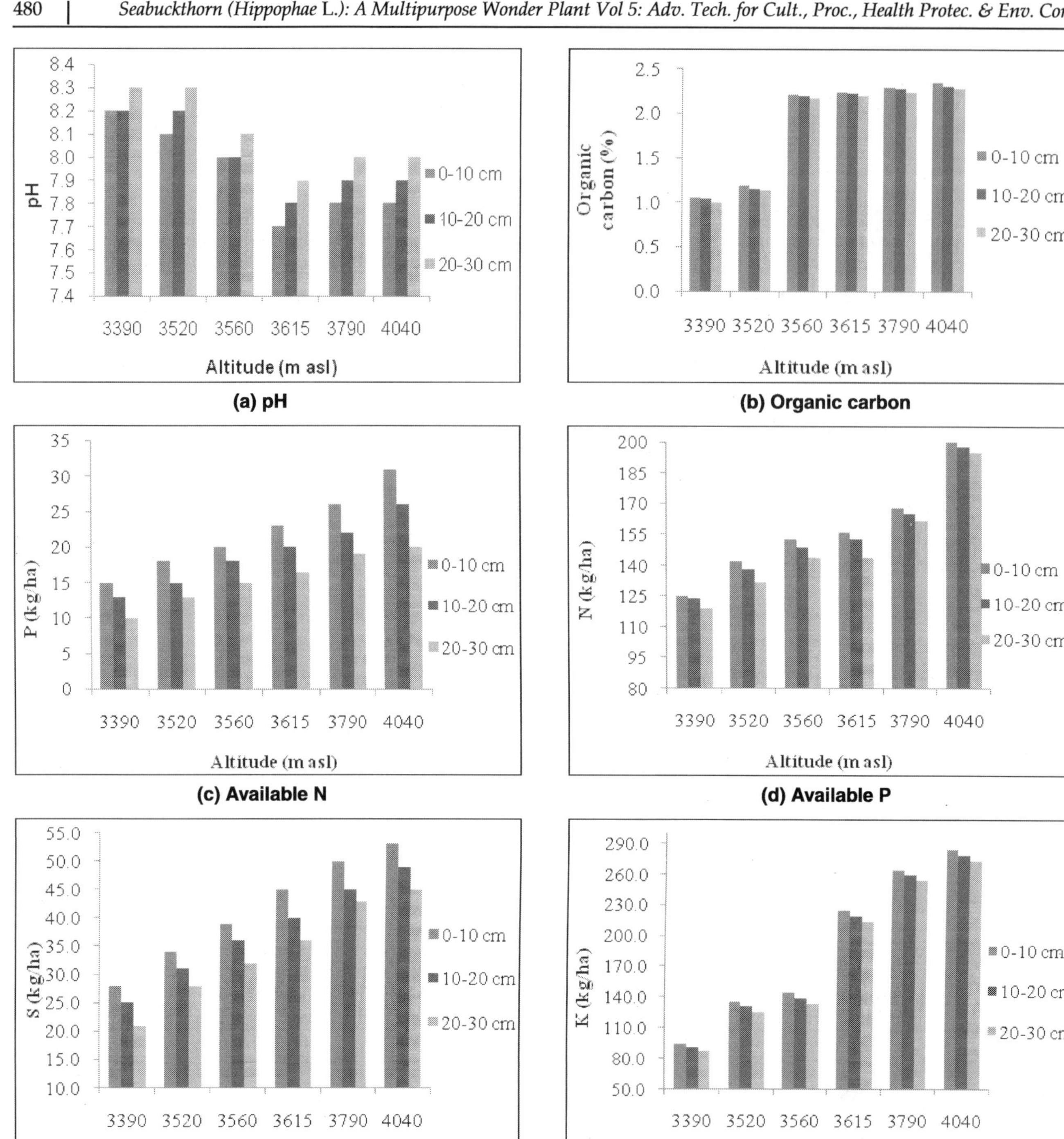

(a) pH **(b) Organic carbon**

(c) Available N **(d) Available P**

(e) Available P **(f) Available P**

Figure 62.2: Variations of different Soil Nutrients in Relation to Altitude under Seabuckthorn in Spiti. a) pH, b) Organic carbon, c) Available N, d) Available P, e) Available K, f) Available S.

have been given by Jalali *et al.* (1989), Parmar (1999) and Sharma (2011).

A comparison of willow and seabuckthorn shows that seabuckthorn had better results than willow and wasteland. This implies that seabuckthorn is more suitable to increase soil fertility status in the cold desert area of Spiti, while willow may be less suitable.

As indicated in Table 62.3, soil pH was found negatively and significantly correlated with organic carbon ($r = -0.653$, $p \leq 0.05$), available N ($r = -0.504$, $p \leq 0.05$), available P ($r = -0.689$, $p \leq 0.05$), available K ($r = -0.608$, $p \leq 0.05$), available S ($r = -0.598$, $p \leq 0.05$), exchangeable Ca ($r = -0.648$, $p \leq 0.05$), exchangeable Mg ($r = -0.506$, $p \leq 0.05$), available Fe ($r = -0.455$, $p \leq 0.05$), available Cu ($r = -0.558$, $p \leq 0.05$) and was found

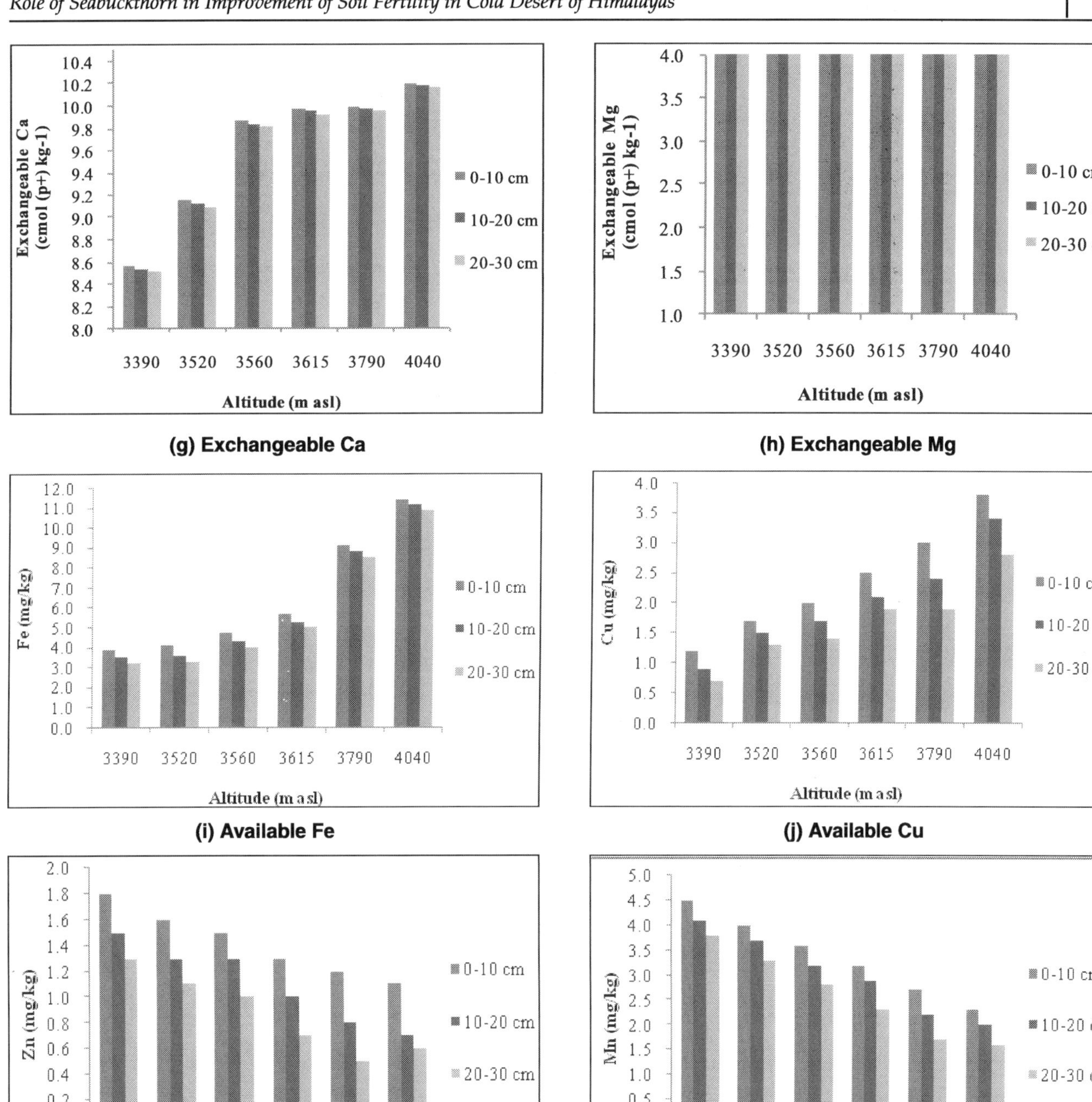

(g) Exchangeable Ca

(h) Exchangeable Mg

(i) Available Fe

(j) Available Cu

(k) Available Zn

(l) Available Mn

Figure 62.3: Variations of different Soil Nutrients in Relation to Altitude under Seabuckthorn in Spiti. g) Exchangeable Ca, h) Exchangeable Mg, i) Available Fe, j) Available Cu, k) Available Zn, l) Available Mn.

positively and significantly correlated with available Zn ($r = -0.322, p \leq 0.05$) and available Mn ($r = -0.356, p \leq 0.05$).

CONCLUSION

Planting seabuckthorn in the cold desert has played a very significant role as the seabuckthorn soils accumulated higher values of soil nutrients than the willow and wasteland. Hence, has great potential for sustainable environmental protection in cold desert areas of Himalayas and because of its strong root system and nitrogen fixing ability improves soil quality or fertility. Overall results revealed that seabuckthorn is a wonder plant and suggest that planting seabuckthorn vegetation is more advantages for soil improvement and increasing soil fertility in the cold desert regions of Himalayas.

Table 62.2: Integrated Effect of Seabuckthorn, Willow and Wasteland on the Soil pH, Organic Carbon and other Soil Nutrients

Locations	Altitude m asl	Seabuck-thorn	Willow	Wasteland	Seabuck-thorn	Willow	Wasteland	Seabuck-thorn	Willow	Wasteland	Seabuck-thorn	Willow	Wasteland
			pH			OC (per cent)			N (kg ha^{-1})			P (kg ha^{-1})	
Tabo	3390	8.2±0.10	8.1±0.23	8.0±0.18	1.02±0.032	0.96±0.041	0.86±0.034	122.7±4.09	104.7±4.06	91.0±3.96	12.7±3.20	16.0±3.20	8.7±2.00
Shichling	3520	8.2±0.10	8.0±0.18	8.0±0.22	1.15±0.035	1.11±0.036	0.92±0.057	137.3±5.50	120.7±4.41	111.0±5.20	15.3±3.27	17.7±3.16	11.7±2.12
Lingti	3560	8.0±0.23	8.1±0.24	8.0±0.22	2.19±0.038	2.08±0.047	1.20±0.03	148.6±4.74	137.7±3.67	134.0±4.61	17.7±3.27	19.3±3.77	12.3±3.00
Shego	3615	7.7±0.15	7.9±0.15	7.9±0.19	2.22±0.039	2.17±0.04	1.26±0.037	151.0±6.04	149.3±3.53	141.0±4.61	19.7±3.84	22.0±3.57	16.0±2.87
Rangrik	3790	7.9±0.25	7.8±0.25	7.8±0.27	2.26±0.033	2.17±0.042	1.54±0.049	165.0±3.31	157.3±3.96	151.3±4.06	22.3±4.33	25.0±4.03	16.0±3.64
Hansa	4040	7.9±0.20	7.7±0.15	7.8±0.18	2.31±0.042	2.29±0.037	2.01±0.045	199.3±5.76	167.3±4.41	157.3±3.84	25.7±5.50	32.3±3.42	17.7±3.27
CD (P ≤ 0.05)			0.180			0.030			2.820			2.630	
			K (kg ha^{-1})			S (kg ha^{-1})			Ca (cmol(p+)kg^{-1})			Mg (cmol(p+)kg^{-1})	
Tabo	3390	91.0±3.74	101.3±6.38	85.1±4.93	24.7±3.74	20.7±2.87	18.0±3.31	8.53±0.031	7.35±0.023	6.50±0.021	2.1±0.30	1.9±0.33	1.7±0.27
Shichling	3520	131.0±4.71	151.3±3.27	147.3±4.87	31.0±3.57	25.3±4.45	22.6±4.00	9.12±0.016	9.03±0.026	7.92±0.018	2.4±0.27	2.1±0.20	1.8±0.13
Lingti	3560	139.0±6.14	161.2±3.56	150.0±4.03	35.7±3.57	27.7±4.21	25.3±4.09	9.84±0.020	9.05±0.024	8.28±0.021	2.5±0.23	2.3±0.15	2.0±0.27
Shego	3615	207.6±3.12	252.0±3.84	163.3±4.66	40.3±5.14	34.3±4.61	27.3±4.21	9.95±0.021	9.67±0.021	8.51±0.025	2.7±0.27	2.7±0.27	2.1±0.20
Rangrik	3790	259.0±5.19	280.3±4.27	180.0±3.90	46.0±4.52	39.7±4.61	36.0±3.77	9.96±0.023	9.75±0.025	8.80±0.026	3.1±0.20	2.8±0.17	2.6±0.17
Hansa	4040	278.7±5.59	300.3±5.14	203.7±5.47	49.0±4.55	43.3±4.69	42.0±3.00	10.18 ± 0.016	10.09 ± 0.018	9.62±0.021	3.4±0.28	3.0±0.20	3.1±0.23
CD (P ≤ 0.05)			8.190			2.760			0.022			0.140	
			Fe (mg kg^{-1})			Cu (mg kg^{-1})			Zn (mg kg^{-1})			Mn (mg kg^{-1})	
Tabo	3390	3.53±0.43	3.30±0.33	2.6.0±0.36	0.93±0.30	0.83±0.26	0.70±0.23	1.53±0.30	1.20±0.32	0.93±0.24	4.13±0.40	3.20±0.33	2.80±0.30
Shichling	3520	3.70±0.47	3.50±0.35	2.97±0.40	1.50±0.39	1.07±0.25	0.80±0.25	1.33±0.32	1.13±0.33	0.80±0.28	3.67±0.42	2.98±0.32	2.37±0.36
Lingti	3560	4.33±0.35	3.93±0.40	3.47±0.36	1.70±0.33	1.33±0.38	0.80±0.20	1.27±0.39	0.97±0.30	0.73±0.20	3.20±0.43	2.53±0.28	2.00±0.30
Shego	3615	5.33±0.43	4.50±0.37	3.47±0.36	2.17±0.41	1.90±0.33	1.47±0.40	1.00±0.33	0.73±0.26	0.67±0.18	2.80±0.43	2.10±0.26	1.64±0.39
Rangrik	3790	8.80±0.37	5.97±0.36	5.37±0.32	2.43±0.56	2.13±0.39	1.70±0.32	0.83±0.35	0.70±0.22	0.57±0.17	2.20±0.51	1.60±0.37	1.37±0.32
Hansa	4040	11.17±0.38	8.23±0.46	6.47±0.33	3.33±0.52	3.10±0.30	1.80±0.36	0.80±0.26	0.70±0.25	0.53±0.15	1.97±0.41	1.17±0.32	1.30±0.26
CD (P ≤ 0.05)			0.300			0.260			0.210			0.290	

Table 62.3: Correlation between pH, OC and other available Soil Nutrients with each other Under Seabuckthorn Vegetation

Correlation	*OC*	*N*	*P*	*K*	*S*	*Ca*	*Mg*	*Fe*	*Cu*	*Zn*	*Mn*
pH	-0.653**	-0.504*	-0.689**	-0.608**	-0.598**	-0.648**	-0.506**	-0.455**	-0.558**	0.322	0.356**
OC		0.749**	0.653**	0.787**	0.815**	0.965**	0.659**	0.675**	0.715**	-0.527**	-0.758**
N			0.798**	0.884**	0.882**	0.819**	0.868**	0.938**	0.912**	-0.473**	-0.746**
P				0.756**	0.791**	0.700**	0.839**	0.748**	0.858**	-0.159**	-0.441**
K					0.883**	0.836**	0.807**	0.916**	0.856**	-0.601**	-0.831**
S						0.851**	0.883**	0.847**	0.922**	-0.398**	-0.674**
Ca							0.709**	0.715**	0.780**	-0.563**	-0.786**
Mg								0.834**	0.900**	-0.202	-0.518**
Fe									0.866**	-0.501**	-0.758**
Cu										-0.284	-0.590**
Zn											0.805**

*Significant level: 0.01, r = > 0.589; **Significant level: 0.05, r = > 0.468.

REFERENCES

1. Ansari, A.H. 2003. Seabuckthorn (*Hippophae* L.)-A Potential resource for biodiversity conservation in Nepal Himalayas. *International Workshop on Underutilized Plant Species*, 6-8th May 2003, Leipzig, Germany.
2. Chengjiang, R. and Daiqiong, L. 2002. Function and benefit of *Hippophae rhamnoides* L. improving eco-environment of Loess Plateau of China. In: *12th Conference of Institute of Soil and Water Conservation*, Shannxi, China.
3. Devi, U. and Thakur, M. 2011. Exploration of ethnobotanical uses of some wild plants from cold desert of Himachal Pradesh. *Asian Journal of Experimental Biological Sciences* 2: 362-366.
4. Dwivedi, S.K., Singh, R. and Ashmed, Z. 2006. Morpho-biochemical characteristics of seabuckthorn (*Hippophae rhamnoides*) growing in cold arid Ladakh Himalayas. In: *Seabuckthorn Hippophae L.: A Multipurpose Plant. Vol. II. Biochemistry and Pharmacology*, pp. 151-158.
5. Dwivedi, S.K., Singh, R. and Raut, B. 2003. Studies on distribution of seabuckthorn through remote sensing and its genetic diversity in Ladakh. In: *International Congress on Seabuckthorn*, Berlin, Germany, 14-18 September, p. 29.
6. Dwivedi, S.K., Stobdan, T. and Singh, S.B. 2009. Seabuckthorn in Ladakh. In: *Seabuckthorn (Hippophae sp.): The Golden Bush*. Satish Serial Publishing Houses, Delhi, pp. 35-51.
7. Gomez, K.A. and Gomez, A.A. 1984. Statistical procedures for agricultural research. Second Ed. John Wiley and Sons, New York, USA, 680p.
8. Gong, J., Chen, L.D., Fu, B.J. and Wei, W. 2007. Integrated effects of slope aspect and land use on soil nutrients in a small catchment in a hilly loess area China. *International Journal of Sustainable Development and World Ecology* 14: 307-316.
9. Gou, Z.S. 2003. Seabuckthorn for the improvement of the microclimate and soil properties of mountainous wastelands in arid and semi- arid China. In: *Seabuckthorn (Hippophae L.): A Multipurpose Wonder Plant*. Vol. I. (V. Singh, Editor-in-Chief, 2003). Indus Pub. Co, New Delhi, pp. 417-432.
10. Gupta, V.N., Poundyal, V.P. and Subedi, C.K. 2001. Ecology and distribution of seabuckthorn (*Hippophae* sp.) resources in north-west Nepal. A report submitted to TISC/NARMSAP, Hattisar and Kathmandu, Nepal.
11. Jackson, M.L. 1967. Soil Chemical Analysis. Prentice-Hall of India, LTD., New Delhi, pp. 219-221.
12. Jackson, M.L. 1973. Soil Chemical Analysis. Prentice Hall, India Pvt. Ltd., New Delhi.
13. Jike, Z. and Xiaoming, Z. 1992. Progress of study on *Frankia* in nodules of seabuckthorn. *Hippophae* 2: 4-10.
14. Jalali, V.K., Talib, A.R. and Takkar, P.N. 1989. Distribution of micronutrients in some benchmark soils of Kashmir at different altitudes. *Journal of Indian Society of Soil Science* 37: 465-469.
15. Lindsay, W.L. and Norwell, W.A. 1978. Development of DTPA soil test for Zinc, Iron, Manganese and Copper. *Soil Science Society of American Journal* 42: 421-428.
16. Lu, J.L. 2009. Study on Ecological effect of artificial seabuckthorn forest with different ages in Loess Hilly region. *M.Sc. Thesis*, Agricultural Sciences, China, 180p.
17. Maertz, J. 2006. Seabuckthorn nutritional properties: Meet the little orange berry from the Himalayas that's interesting in your well-being. Sibu: The

Seabuckthorn Company; Accessed on http://www.mysibu.com (25th August, 2006).

18. Olsen, S.R., Cole, C.V., Watanave, F.S. and Dean, L.A. 1954. Estimation of available phosphorus in soils by extraction with sodium carbonate. United States Department of Agriculture, *Circular No.* 939.

19. Parmar, D.K., Sharma, V., Sharma, K.D. and Sharma, T.R. 1999. Micronutrients status of vegetable growing pockets in cold desert area of Himachal Pradesh. *Journal of Indian Society of Soil Science* 47: 280-283.

20. Pradeep, S., Kumar, K. and Singh, V.1999. A vegetational study in the different regions of the cold temperate regions of H.P. *Journal of Indian Society of Soil Science* 36: 334-336.

21. Qian, Z. 1986. By taking seabuckthorn as a breakout for accelerating treatment on Loess Plateau. *Soil and Water Conservation* 4: 3-6.

22. Rongsen, L. 1992. Seabuckthorn- A multipurpose plant for fragile mountains. International Centre for Integrated Mountain Development ICIMOD. *Occasional Paper No. 20*, Kathmandu, Nepal, 62p.

23. Rousi, A. 1965. Observation on the cytology and variation of European and Asiatic populations of *Hippophae rhamnoides*. *Annales Botanici Fennici* 2: 1-18.

24. Rousi, A. 1971. The genus *Hippophae* L.: A taxonomic study. *Annales Botanici Fennici* 8: 177-227.

25. Schroeder, W.R. 1990. Shelter belt planting in the Canadian prairies. In: *Protective Plantation Technology*. Publishing House of Northeast Forestry University, Harbin, China, pp. 35-43.

26. Schroeder, W.R. 1995. Improvement of conservation trees and shrubs. *PFRA Shelterbelt Centre Supp. Rpt.* 95-1, 42p.

27. Sharma, A. 2011. Impact assessment of seabuckthorn (*Hippophae* L.) on soil fertility status. *M.Sc. Thesis*, Department of Biology and Environmental Sciences, COBS, CSK Himachal Pradesh Krishi Vishvavidyalaya, Palampur, India, 86p.

28. Sharma, V. and Kanwar, B.B. 2010. Copper status and its relation with soil properties in pea growing soils of high hills dry temperate zone of the Himachal Pradesh. *Journal of Indian Society of Soil Science* 44: 32-37.

29. Singh, B. and Raman, S.S. 1982. Altitudinal pattern of the soil characteristics under forest vegetation in eastern Himalayan region. *Journal of Indian Society of Soil Science* 38: 93-99.

30. Singh, O.P. and Datta, B. 1988. Organic carbon and nitrogen status of some soils of Mizoram occurring at different altitudes. *Journal of Indian Society of Soil Science* 36: 414-420.

31. Singh, V. 2003. Geographical adaptation and distribution of seabuckthorn (*Hippophae* L.) resources. In: *Seabuckthorn (Hippophae L.): A Multipurpose Wonder Plant. Vol. I. Botany, Harvesting and Processing Technologies* (V. Singh, Ed. in Chief, 2003), pp. 21-34, Indus Publishing Company, New Delhi, 518p.

32. Subbiah, B.V. and Asija, G.L. 1956. A rapid procedure for estimation of available nitrogen in soils. *Current Science* 25: 259-260.

33. TISC. 2001. Proceeding of workshop on Ecology and Distribution of Seabuckthorn (*Hippophae* spp.) Resource in Northwest Mountains of Nepal. HMG/ DANIDA NARMSAP, Tree Improvement and Silviculture Component, Hattisar, Kathmandu, Nepal.

34. Walkley, A. and Black, C.A. 1934. An estimation of the method for determination of soil organic matter and a proposed modification of the chromic acid titration method. *Soil Science* 37: 29-39.

35. Xiaoning, T., Baoli, A., Xiufend, W. and Jing, N. 2002. Soil Improvement of seabuckthorn (*Hippophae rhamnoides*) plantations and its characteristics of the roots in Loess Plateau. 12th ISCO Conference, China, pp. 403-409.

36. Zhao, H. 1990. Seabuckthorn root nodule quantity and origin. *Hippophae* No. 2.

37. Zhao, Z., Shahrour, I., Bai, Z., Fan, W., Feng, L. and Li, H. 2013. Soils development in open cast coal mine spoils reclaimed for 1-13 years in the West-Northern Loess Plateau of China. *European Journal of Soil Biology* 55: 40-46.

38. Zhiben, M., Huanjin, L., Hualong, S., Jingyi, F. and Yansheng, C. 1989. International between Chinese seabuckthorn and some birds and wildlife in the Loess Plateau region. In: *Proceedings of the International Symposium on Seabuckthorn*, pp. 298-299, Xian, China, 414p.

Chapter 63

Ecological Restoration and Soil Improvement Performance of the Seabuckthorn Flexible Dam in the Pisha Sandstone Area of Northwestern China

Yang Fang-She[1*], Cao Ming-Ming [1], Li Huai-En[2], Wang Xin-Hong[2] and Bi Ci-Fen[3]

[1] School of Urban and Environment Sciences, Northwest University, Xi'an, 710069, China
[2] Key Laboratory of Northwest Water Resources and Environment Ecology of Ministry of Education, Xi'an University of Technology, Xi'an, 710048, China
[3] Institution of Soil and Water conservation, Bureau of Upper and Middle Reaches of Yellow River, Xi'an, 710043, China
(*E-mail: yangfangshe978@126.com)

SUMMARY

A field experiment was conducted to assess the ecological restoration and soil improvement performance of the seabuckthorn (*Hippophae rhamnoides*) flexible dam (SFD) by examining its sediment trapping capacity and ecological effects within the Pisha Sandstone area of Loess Plateau of China. The Pisha Sandstone is very hard when it is dry, while it is very frail when wet. The SFD was proposed to control soil erosion and water loss and meliorate soil in the Pisha sandstone area. To evaluate the effectiveness of the SFD, we compared the ecological status and soil quality between a gully with the SFD and a gully without the SFD. The results show that the SFD exhibits excellent performance for ecological restoration and soil melioration by the strong sediment trapping capacity. The results are as follows: (i) Deposition commonly occurred in the SFD gully; (ii) The deposition patterns are obviously different from upper to lower gully; (iii) More surprisingly, unlike trees and other shrubs, the seabuckthorn has good cloning capacity by its root nodule; (iv) Soil moisture, soil organic matter (SOM), soil nutrients, vegetation coverage and biodiversity in the gully vegetated with the SFD are all strikingly increased. The SFD is a both effective and novel biological measure for ecology restoration and soil melioration within the Pisha Sandstone area.

Keywords: *Seabuckthorn flexible dam (SFD), Pisha sandstone, Soil erosion and Ecological restoration.*

INTRODUCTION

The Pisha Sandstone area is located in the Loess Plateau of China, which is a major source of coarse sediment entering the upper and middle reaches of Yellow River. The Pisha Sandstone covers an area of 17, 500 km^2, with a considerable sediment yield of 0.214 billion ton per year (Jin, 2003; Ran, 2006). Coarse sediment > 0.05 mm (Qian *et al.*, 1980) derived from the Pisha Sandstone area accounts for 71.1 per cent of the total coarse sediment yield (0.301 billion ton per year) in the middle reaches of Yellow River (Zheng, 2005; Xu and Gao, 2007). The Pisha Sandstone is extremely hard like stone when it is dry, but very soft and friable when wet. Therefore, erosion commonly occurred in many gullies during and after rainfall events (Ziadat and Taimeh, 2013). The large amount of soil and water loss leads to soil nutrients loss (Zhao *et al.*, 2013). Undoubtedly, this further worsened environment of this area, which makes many types of vegetation hardly live in this area. The threat of the Pisha Sandstone to vegetation is often compared with the toxin of arsenic as a metaphor and local people conventionally call it as Pisha (Chinese Pinyin of Arsenic) Sandstone.

Seabuckthorn (Hippophae L.): A Multipurpose Wonder Plant, Vol. 5: Advanced Technologies for Cultivation, Processing, Health Protection and Environmental Conservation (Virendra Singh, *Editor-in-Chief*, 2018), p. 485–500, Astral International Pvt. Ltd., New Delhi, India.

The Pisha Sandstone belongs to the continent fragment bedrock, which is composed of thick layer Sandstone, erinaceous shale and mudstone (Bazhenov *et al.*, 1993; Mark *et al.*, 1999), there by easily being eroded by wind and water due to the diagenesis. The maximal erosion module of this area is up to 3×10^4 $t{\cdot}km^{-2}{\cdot}year^{-1}$, by which this area was often compared to "the cancer of the earth" or "the most severe soil and water loss in the world" (Bi, 1998, 2003). At present, the engineering measure, such as the check dam and some small-scale sediment retention reservoirs, was mainly applied to this area to control soil and water loss. However, rare vegetation measures are there used to restore the ecosystem of the area. Due to the special topography and extremely poor soil, some measures used successively to the other regions of Loess Plateau, including terrace or bench terrace, hedge fence, forest shelter belt, grass protective strip, *etc.* (Cao *et al.*, 2007a, 2007b; Zhao *et al.*, 2013), are not suitable to this area. Current studies showed that soil and water conservation measures are different in various regions, including some man-made ones, which is mainly associated with local climate condition, soil physical property, autochthonous types of vegetation and their coverage, agricultural farming system structure, land use change, hydro-mulching, and etc (Fernández *et al.*, 2012; Sérgio *et al.*, 2013; Prokop and Poreba, 2012; Cerdà, 1999). In fact, with the lapse of time, the negative effects of the engineering measure are gradually exposed due to the emergence of some new environmental problems. Therefore, people have recognized that vegetation is still the fundamental measure for soil erosion control and ecosystem rehabilitation in most regions of the world. Thus, seeking a type of vegetation, which can adapt the Pisha Sandstone area well, becomes the key to control soil erosion and restore the ecosystem of the area.

Hippophae is commonly known as seabuckthorn, a multi-use plant which can be used to control soil erosion, biological nitrogen fixation and medicinal purposes (Lu Rongsen, 1992). There are two common species of seabuckthorn in China, namely, *Hippophae salicifolia* D. Don and *Hippophae rhamnoides* L. *H. salicifolia* has a shrub-to-tree habit and is restricted to the Himalaya region, whereas *H. rhamnoides* is bushy, mainly growing at high altitude in China. Additionally, it is also widely distributed in Asia and Europe (Rousi, 1965, 1971). The seabuckthorn is a native plant of Loess Plateau in northwestern China, and has excellent biological features such as drought tolerance, barren soil tolerance and rapidly cloning capacity by its root nodule. To control soil erosion and restore the ecosystem of the Pisha Sandstone area, based on the ancient Chinese Lao Zi's general philosophical view points of "using soft to overcome hard" and "using flexibility to dissipate energy", the concept of the seabuckthorn flexible dam (SFD) was proposed by Bi and Li (1998; 2003) and Li *et al.* (2009). Actually, this is an active response to the call of "using the seabuckthorn to control Loess Plateau of China" put forward earlier by Qian (1986). The SFD is composed of the preferentially selected 2-3a seabuckthorn seedlings, and the seabuckthorn seedlings are grown, in an interlaced way, in the transects of certain large or small gully within the Pisha Sandstone area, according to specific row spacing and plant spacing within a row. The SFD can effectively slow flow velocity and dissipate flow energy by large numbers of branches and leaves of the seabuckthorn thus being able to efficiently trap coarse sediment transported by flood. Studies using different kinds of vegetation to control soil erosion or to prevent soil and water loss have been made for a long time in arid and semi-arid or tropical areas (Fang *et al.*, 2012; Hussein *et al.*, 2007; Mishra *et al.*, 2006; Zhang *et al.*, 2004; Yu, 2002; Yu *et al.*, 2003; Gomi *et al.*, 2006; Sidle, 2006). Great attention has been paid to vegetation reducing soil loss by decreasing runoff volume or by changing runoff-sediment yield relationship (Slattery and Phillips, 2011; Kumar *et al.*, 2008; Pan *et al*, 2011; Li *et al.*, 2008; Ide *et al.*, 2009; Hartanto *et al.*, 2003). The measures include vegetation barriers, vegetation filter strips, and riparian forest buffer system, tree/grass hedge system and *etc.* Kiepe (1995) studied effects of the hedge barrier system on trapping soil and retarding slope runoff and found that little runoff and soil loss occurred. Tree/grass hedges had also obvious effect on soil erosion control in the Central Kenyan highland and other regions (Angima *et al.*, 2000). Rao *et al.* (1991) managed *Leucaena leucocephala* as hedgerows for trapping soil and sediment in semiarid area of India. Detailed studies of vegetation and erosion process interactions were undertaken within an ephemeral channel in SE Spain at three scales (channel network, reach and patch) by repeat surveys and mapping after floods by Sandercock and Hooke (Sandercock and Hooke, 2011).

Moskalski and Sommerfield (2012) examined sediment deposition and retention in a section of salt marsh in the St. Jones River estuary in Delaware. Spaan *et al.* (2005) studied vegetation barrier by using the *Andropogon gayanus* (a species of dense grass) with remarkable effectiveness in diminishing soil loss in the Africa sub-Saharan semi-arid area and found that hedgerow intercropping was beneficial in terms of soil richness and water conservation. These vegetation barriers offer a direction for researchers in the management of erosion control (Oteroa *et al.*, 2011; Reeder *et al.*, 2005; Dorioz *et al.*, 2006; Rasmussen *et al.*, 2011a, b). Current studies showed that vegetation barriers have good soil and water conservation, ecological restoration and pollution control effects (Udawatta *et al.*, 2010; Ghebremichael *et al.*, 2008; Lee *et al.*, 2003; Arorak *et al.*, 1996; Lowrance *et al.*, 1985). Chen (2010) pointed out that landscape restoration has many positive effects on soil water storage and effective water use. Further, forest ecosystem can saliently increase soil nutrients, thus being beneficial to the growth of forest and vegetation community (Amazonas, *et al.*,

2011; Mohammad and Alseekh, 2012). Another study also showed forest restoration has strong improvement effect on soil physical properties and the growth of seedlings on steep slopes after coal surface mining (Fields-Johnson *et al.*, 2012). Related studies indicated vegetation structure and density, species diversity, and ecosystem processes were the main impacting factors for ecological restoration (Ruiz-Jaen and Aide, 2005 b; Cao *et al.*, 2008). All the vegetation measures or treatments and good forest management mentioned above have better soil and water conservation and ecological effects in the entire watershed (Lu *et al.*, 2001).

Compared to the check dam or concrete dam, the SFD has many advantages. On the one hand, storage capacity of the check dam or concrete dam is gradually reducing with sediment deposition. Eventually, they become death or no use. On the other hand, the habitat and the migration pathway of flora and fauna in the gully were completely blocked off by the check or concrete dam, which is rather adverse to the restoration of regional ecological environment. However, the SFD is composed of the seabuckthorn, which can create a migration passage for flora and fauna thus being able to improve local ecological environment.

As mentioned above, the ecological environment of the Pisha Sandstone area is extremely poor. To restore ecology and improve soil of this area, the SFD-a novel vegetation measure and its concept was proposed. Subsequently, in order to examine the effectiveness and performance of the SFD on the ecological restoration and soil improvement within the Pisha Sandstone area, a field prototype experiment has been systematically conducted since 1996. In this paper, we analyzed the sediment retention capacity, the ecological restoration and soil melioration effects of the SFD within the Pisha Sandstone area in many aspects. This study has an important value for understanding the SFD efficacy and can provide the theoretical evidence for widespread planting of the SFD within the Pisha Sandstone area or even in other similar regions of the world.

MATERIALS AND METHODS

Outline of Study Area

The Pisha Sandstone area is close to Shaanxi Province, Inner Mongolia Autonomous Region and Shanxi Province (39°37223-40°412483N, 109°42303-110°152403E) (Figure 63.1). Due to the special hilly topography and poor vegetation cover, the Pisha Sandstone was eroded into many deep gullies and hummocks, resulting in the present barely hilly landscape (Figure 63.2). The study area is located 5 km west of Jungar county, Ordos city, Inner Mongolia, China, at 39°372N, 110°92E (Figure 63.1).

An investigation was mainly conducted at east gully one (EG1), a tributary of the xi-zhao gully in Jungar County. The xi-zhao gully covers an area of 15 km^2, and it is a tributary of ku-ye River, a second-order tributary of Yellow River. Investigation was also conducted in the other several gullies such as EG2, EG3, EG4, WG6 (west gully 6) and *etc.* Jungar County situated southeast of Erdos plateau, typical of Pisha Sandstone area of Loess Plateau of China, is one of the most serious sand-dust storm sources. Also, it is one of the most dominant coarse sediment sources into the upper and middle reaches of Yellow River in China. Jungar County belongs to the typical low hilly zone of Loess Plateau. The study area has a continental and monsoon climate. Mean annual rainfall is 390 mm and mean annual potential evaporation is 2265 mm. Rainfall in July and August accounts for approximately 70-80 per cent of the whole year. Average annual temperature is 7.4 The mean elevation is around 1400 m above sea level. The EG1 where the experiment was mainly conducted, with a 1628 m length and average 6.8 m width, covers an area of 1.67 km^2. Mean slope of main gully channel is 0.04 (4 m vertical: 100 m horizontal) and the average

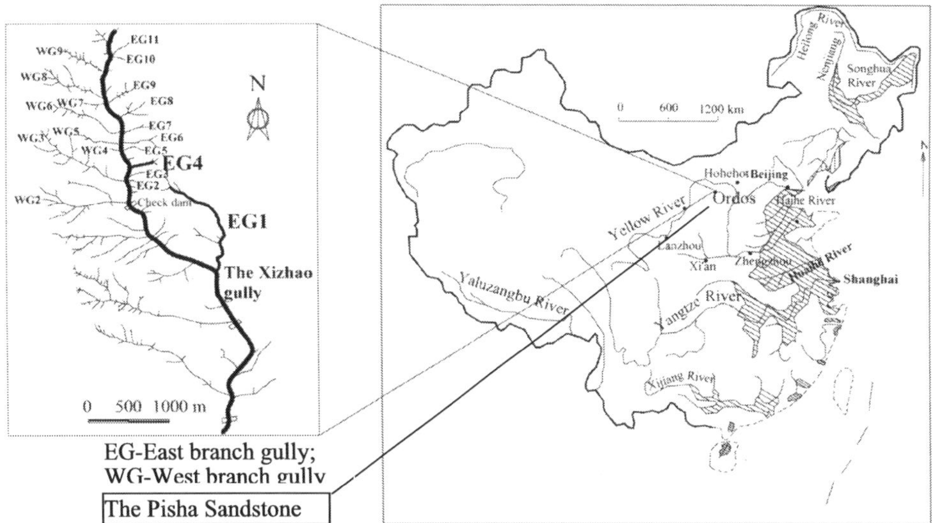

Figure 63.1: Location Map of the Pisha Sandstone Area, Xi-Zhao Gully and EG1 (*i.e.*, East branch Gully one).

Figure 63.2: Easily Eroded Bare Pisha Sandstone.

gully side slope is 0.74. The EG1 has two tributary gullies. The left tributary gully is approximately 294 m long and averagely 2.66 m wide, with a mean channel slope of 0.066 and an average side slope of 0.83. The right tributary gully is approximately 246 m long and averagely 2.45 m wide, with a mean channel slope of 0.058 and an average side slope of 1.16. The bare gully similar to the EG1, approximately 100 m below the EG1, was selected as the contrast gully (non-vegetated), which is located downstream of the Xi-zhao gully. The contrast gully has a length of 400 m, an average width of 4.5 m, a mean channel slope of 0.039, and a mean side slope of 0.72. Nine monitoring sections were designed in the contrast gully from down- to up-stream, with a 50 m interval between each other. It should be declared that during the course of comparison, the few SFDs within the WG6 are also selected. However, the topography of the WG6 is nearly similar to the EG1, here no special description was done on it. On average, soils in the investigated EG1 and WG6 had 8.3 g kg^{-1} gravel (2.0>size >1.0 mm), 800 g kg^{-1} coarse sand (1.0>size>0.25 mm), 178 g kg^{-1} fine sand (0.25>size>0.05 mm), 13.7 g kg^{-1} including coarse silt, middle silt, fine silt, coarse clay and fine clay (size<0.05 mm), 0.5 g kg^{-1} organic C, and 8.8 pH_W.

The Seabuckthorn Flexible Dam System

The diagram of single seabuckthorn flexible dam is presented in Figure 63.3. The SFD is substantially a small-scale man-made seabuckthorn forest (Figure

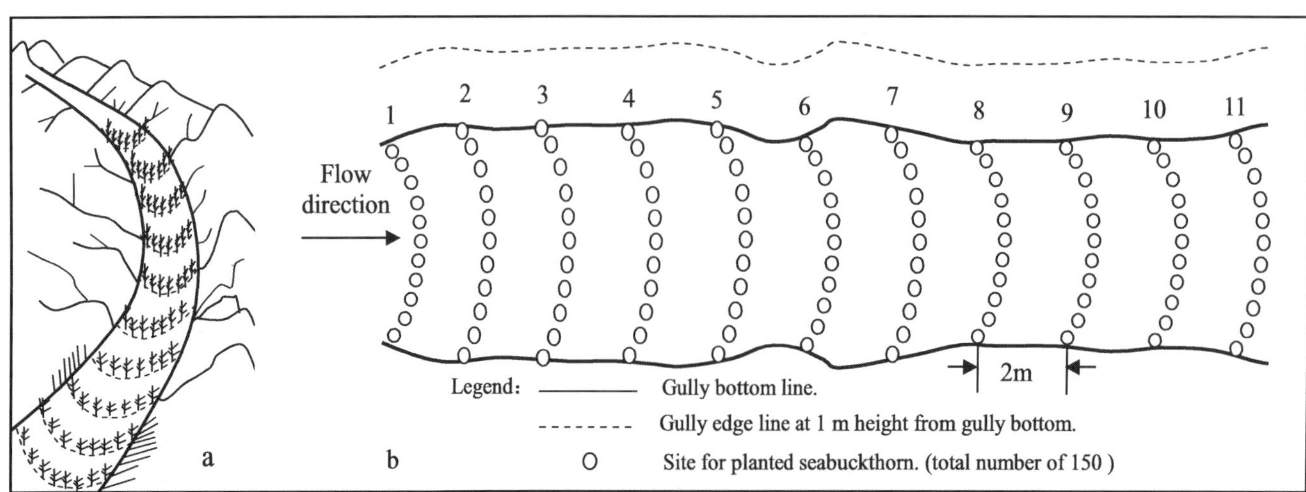

Figure 63.3: Schematic Diagram of the SFD.
(a) Sketch map of the SFD and (b) Layout diagram of the SFD.

Figure 63.4: Growth State of Several SFDs in 2005 (Planted in 1996).

63.4). The seabuckthorn flexible dam system (SFD system) (Figure 63.5) composed of a series of seven dams was vegetated in the upper EG1 in spring 1996. Some SFD systems were also sparsely vegetated in the other branch gullies of the Xi-zhao gully. The seven seabuckthorn flexible dams (SFDs) were named as No.0-6. Four SFDs were planted in the major gully while three were vegetated in the left and right tributary gullies of the EG1. Each of the SFDs was planted in the gully at 0.4 m depth, with 2.0 m spacing between rows and 0.3 m spacing within a row. There are some shrub and herbaceous plants including *Festuca arundinacea Schreb, Clinelymus dahurcus Turcz, Youngia japonica, Artemisia arenaria* DC, *Salsola collina, Heteropappus altaicus,* and *etc.* in the gully.

Sampling Locations and Measurement Methods

Growth Measurements

The recorded growth parameters on the SFD include basic diameter, height and canopy. The basic diameter was measured at 10 cm height above ground; the height was measured using normal method by leveling rod. As for canopy, the mean in north-south direction and that in west-east direction were respectively recorded. These were measured in specific sampling plots designed in the upper, middle and lower sections of each SFD within the EG1, respectively. Each sampling plot is approximately

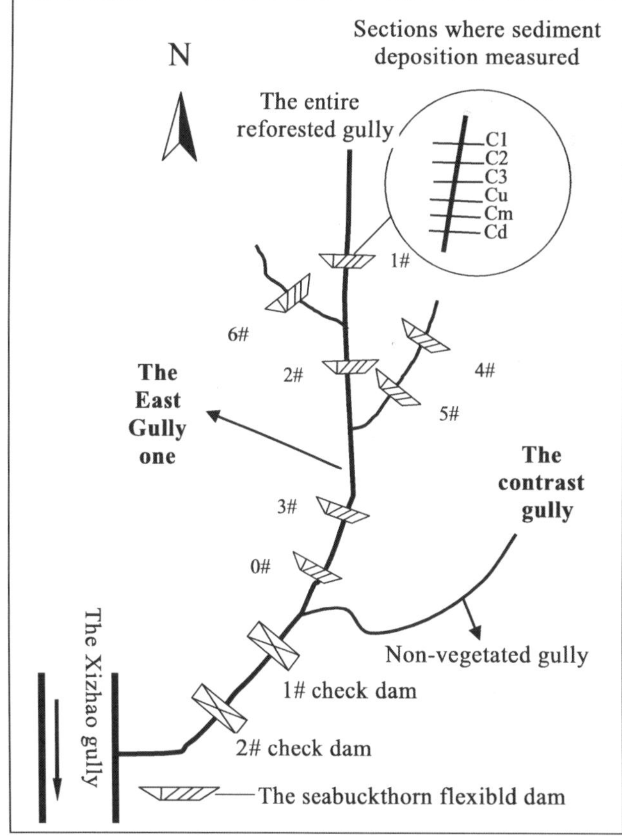

Figure 63.5: Sketch Map of the SFD System in the EG1.

5.0 m by 3.0 m. In addition, the other nine sampling plots are located in the three SFDs at the mouth of the EG4. The total number of the sampling plots is 30.

Sediment Deposition Measurements and Grain Size Distribution

The deposition elevation in different sections was measured within the SFD in EG1 in 2005. Six cross-sections (C_d, C_m, C_u, C_3, C_2, C_1, respectively) (Figure 63.5) were marked by concrete stakes, and the relative channel elevation was measured to compare the changes of sediment deposition from 1997 to 2005. The grain size distribution of deposited sediment within the SFD was measured by sieving method (Lao, 1988), because the deposited sediment is coarser and the grain size is commonly greater than 0.05 mm.

Measurements of Soil Moisture Contents

Soil moisture data are collected from the several major flexible dams located in the EG1. Sampling sections were generally distributed in the front, upper, middle and lower location of the SFD during the period 1996 to 2010. The soil moisture measurements of the 2#, 4# and 5# SFDs as well as the contrast gully were conducted on 8th August 1998, 10th August 1999 and 24th August 2000, five days later after rainfall event; those of the 6# SFD and the contrast gully are only conducted on 16th August 2005, five days later after rainfall event; and those of the 0#, 1#, 2#, 5# and 6# SFDs together with the contrast gully were performed on 9th October 2006, five days later after rainfall event. The first section is located at the first row of the SFD, the second section at the middle, and the third section at the last row. For each section within the SFD, we firstly dig a hole with a diameter of 20 cm and a total depth of 100 cm at the centre of each section within the entire deposition body caused by the SFD, then samples were collected every 10 or 20 cm interval from the surface downwards along the inner profile of the hole (10 cm interval at the 0-40 cm depth; 20 cm interval at the 40-100 cm depth), nonetheless, the surface sample was taken at 5 cm depth. The weight of each soil sample is approximately 0.5 kg. All soil samples (the total number of 240) were preserved in a sealed plastic bag and rapidly taken back to the laboratory. All data are of volumetric soil moisture (here always expressed as per cent m^3 water/m^3 soil) in the top 100 cm of soil profile, which represents the majority of the active root zone of these sites. The soil moisture contents were measured after each sample was oven dried at 105° C for 8 h.

Measurements of Soil Organic Matter and Nutrients

The collected samples were gently broken by hand, passed through a 2-mm sieve and air dried. Then, the large pieces of stubble and root that passed through the sieve were removed by hand. In the laboratory, a 0.5-g sub-sample was removed to determine organic matter using the standard oil bath heating and potassium dichromate method. Soil nutrients here were only included ammonium nitrogen, Available phosphorus, Available potassium (K_{avail}), which were measured in 2010. Ammonium nitrogen (N_{amm}) was measured according to indophenol blue colorimetric method. Available phosphorus (P_{avail}) was extracted with sodium bicarbonate and measured with colorimetry. Available potassium (K_{avail}) in each soil sample was extracted with ammonium acetate (Lao, 1988).

Determining of Species Diversity and Vegetation Coverage

A vegetation investigation was conducted at five different sites in the EG1 and contrast gully (without the SFD) on 10th August 2005, respectively, to assess the species diversity and vegetation coverage of the SFD. We selected a sampling grid of 4 m × 2 m at certain location near each chosen section. These locations are as follows: (1) the monitoring sections within the SFD in the EG1 (at the upper and lower sections, two quadrats); (2) the depositional section at the tail of the SFD at the gully outlet (one quadrat); (3) within the No. 2 SFD in the EG1 (at the upper and lower sections, two quadrats); (4) the corresponding sections of the contrast gully (five quadrats), named as S1-S5, which are spaced 100 m apart from each other from downstream towards upstream.

RESULTS AND DISCUSSION

General Growth Characteristics of the SFD

Growth differences were evident in all these SFDs due to some undetermined factors such as topography, soil property, soil moisture, soil organic matter, soil nutrients, and *etc.* Growth parameters of the SFD were summarized in Table 63.1.

As it can be seen from Table 63.1, a marked change in growth is found in the 2#, 3# and 5# SFDs from 1998 through 1999 to 2005, that is, all the SFDs grow rapidly

Table 63.1: Growth Parameters of the Selected Seabuckthorn Flexible Dams

Parameters	*No. 5# SFD*			*No. 2# SFD*			*No. 3# SFD*		
Age (a)	12	5*	6**	14	7*	8**	12	5*	6**
Height (m)	2.75	1.58*	2.06**	2.98	1.53*	2.05**	2.04	1.48*	1.69**
Canopy (m)	1.86	1.40*	1.72**	2.41	1.36*	1.69**	1.61	1.33*	1.69**
Basic diameter (cm)	--	5.00*	--	--	5.39*	--	3.72	--	--

Notes: Survey time of the No. 5#, No. 2# and No. 3# SFD is on 6th, August, 2005. The sign '*' denotes the survey time is on 30th, October, 1998. The sign '**' denotes the survey time is on 20th, November, 1999. The SFD denotes the seabuckthorn flexible dam.

in different ages. Among them, the change in growth of 2# SFD is greater than those of 3# and 5# SFDs, whatever the period from 1998 to 1999 or from 1999 to 2005. This is related to the radiations exposure and topography factors such as slope, indigenous soil property, the vegetation photosynthesis and transpiration, and etc, which is a complicated mechanism. Based on field observation, the topography of the gully reach with the 2# SFD is beneficial to the seabuckthorn growth, where the slope is gentle, sunshine radiation exposure is much, the understory coverage is better and thus the soil moisture is sufficient. In sum, these parameters describe the good growth of the SFDs. As shown in Table 63.2, the extending capacity of the typical SFDs was summarized. We clearly found the SFD has a very good horizontal extending capacity due mainly to its rapidly self-reproduction ability by developed nodules. From Table 63.2, we found that the mean length extension rate of the 1# and 3# SFDs within WG6 is about 1.24 m yr^{-1}, indicating that they have more rapid extension rate as compared to the 0# and 2# SFDs in the EG1. However, the area extension rate of the SFDs in the EG1 is greater than that of those in the WG6. Overall, the mean length extension rate of the surveyed SFDs is 1.09 m yr^{-1} and the average area extension rate is 5.61 m^2 yr^{-1}. From the case of the 1# and 3# SFDs in the WG6, we found that although the length extension rate is rapid, the area extension rate is not necessarily rapid, which is mainly subject to the complicated vegetation growth mechanism. However, the developed root system and its strong extending capacity are also helpful for stabilizing the slope toe to prevent collapse.

The Sediment Trapping Effect of the SFD

Based on the field observed data in 1997-2005, we found that deposition commonly occurred in all the SFDs in the EG1, and the deposited sediment showed different patterns characterized by different slopes and forms. The field investigation also demonstrated that the coarse and fine sediment from upper gully and side slopes were effectively separated through the joint use of the SFD system and lower rigid check dam. We called the phenomena as "trapping coarse sediment and silting fine sediment", that is, the coarse sediment can be just trapped in the gully but the finer sediment silted in the lower check dam thus greatly trapping sediment from the gully.

1). The naturally sorting effects of the SFD on sediment and forming of the small reservoir in gully: Figure 63.6 presented the grain size distributions of deposition sediment of the typical SFDs (0#, 1#, 2# and 5#) and the 1# check dam. As can be seen from Figure 6, the mass percents of coarse sediment greater than 0.05 mm of 2# and 5# SFDs are smaller than the 0# and 1# SFDs due mainly to sediment retention effect of the SFD system. The mass percent of the coarse sediment decreased from 99.7 per cent of 1# SFD to 86.6 per cent of 0# SFD, and the sediment reduction rate is approximately 1/7, which showed the reduction in sediment is quite remarkable because of the trapping effect of the whole SFD system. Furthermore, the finer sediment was almost deposited into the lower 1# check dam.

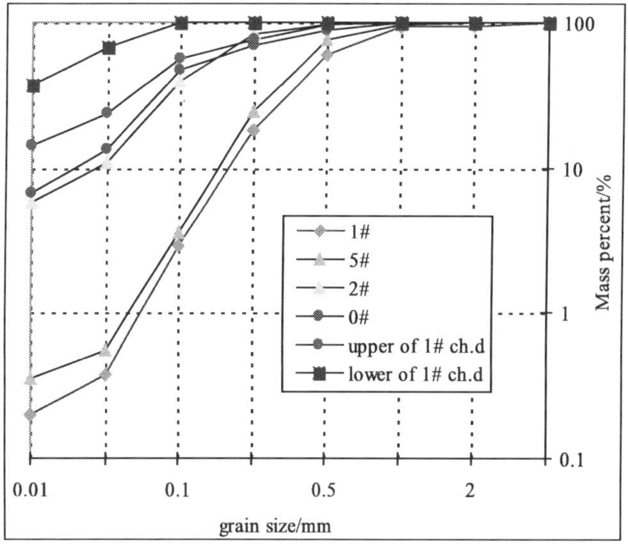

Figure 63.6: Distributions of Sediment Grain Sizes in the Typical SFDs. (Note: The ch. d. -the check dam.)

The field data showed that the mass percent of coarse sediment is about 37 per cent in the upper section of 1# check dam and that of the lower section is approximately 11 per cent, which demonstrated that

Table 63.2: Outline of Extension of the Selected SFDs via Self-Reproduction along Gully Length

Survey Sites		*Plant Time*	*Seedling Age at Planting*	*Survey Width of Dam (m)*	*Extension (m)*	*Extension per Year (m·yr⁻¹)*	*Extension Area (m²)*	*Extension Area per Year (m²·yr⁻¹)*
No.2# dam at EG1	upper	1997.5	2 year	5.85	9.10	1.14	53.24	6.65
	lower			7.55	9.50	1.19	71.73	8.97
No.0# dam at EG1	lower	1996.5	3 year	9.90	5.70	0.63	56.43	6.27
No.1# dam at WG6	lower	2002.5	2 year	1.89	3.75	1.25	7.07	2.36
No.3# dam at WG6	upper	2002.5	2 year	3.64	3.65	1.22	13.27	4.42
	lower			4.41	3.40	1.13	14.99	5.00
Average				5.54	5.85	1.09	36.12	5.61

Note: Survey time is on 8th, May, 2005.

the coarse and fine sediment were well separated. In addition, the flood filtered by the SFD system can be stored in the lower rigid check dam. Thus, a small reservoir was formed providing irrigation water for local agriculture. The field investigation fully showed that the SFD system can not only trap sediment but also store a large proportion of surface runoff by the joint use of the SFD and lower check dam. Due to the strong sediment retention capacity by the SFD, the soil and sediment loss in the gully would be greatly controlled. First of all, the sediment was trapped and would thus form the deposition body, which is able to lift the erosion benchmark level of the gully. This can allow the gully bed slope to become gentle or flat so as to lessen the gully erosion and protect gully bed, thus being helpful for the restoration of vegetation community as well as soil improvement. The most main reason for the SFD trapping sediment is the SFD can greatly increased the roughness of the gully together with understory and other subordinate or attendant vegetation community.

2). The various deposition slopes and forms by the SFD: Based on our field observation, the coarser sediment deposited commonly occurred within the SFDs in the gully during flood season of every year. According to the reservoir and general riverbed deposition theory, the deposition patterns consist of the delta and parallel deposition forms. In general, the delta deposition has three segments of front slope, top slope and tail slope. The deposition forms of 1#, 2#, 5# and 0# SFD were showed in Figure 63.7. The deposition forms of 1#, 2# and 0# SFD are roughly the delta deposition, but that of the 5# SFD belongs to the parallel deposition. Here, as a case of 0# SFD, the front segment is steep with slope 4.1 per cent, the top segment is more flat with slope 0.6 per cent, and the tail segment is slightly steep with slope

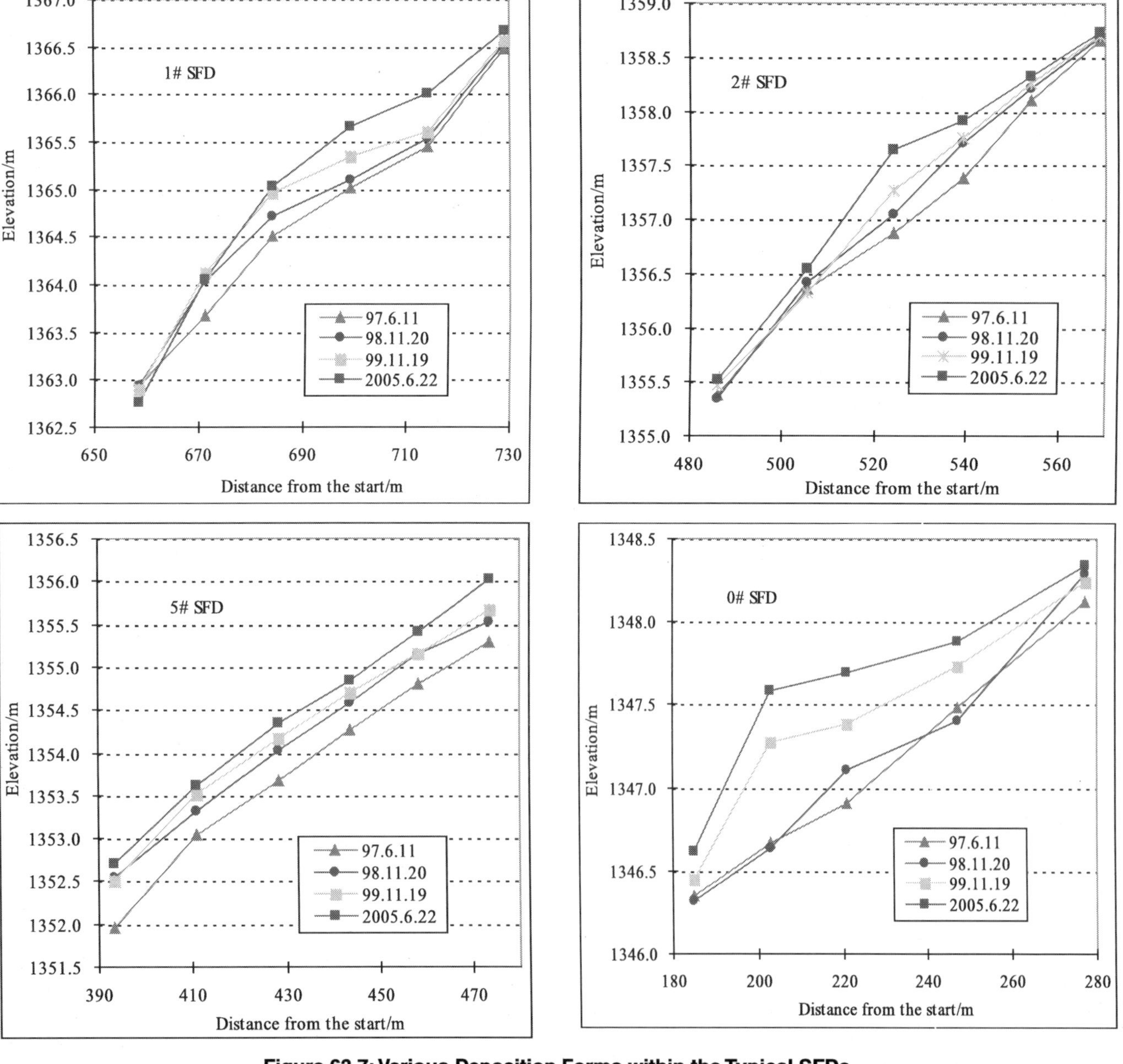

Figure 63.7: Various Deposition Forms within the Typical SFDs.

1.5 per cent, so the two slopes of top and tail segments of the deposition were smaller than the initial slope of 2.8 per cent of the gully except for the front slope. The deposition form of 5# SFD presents a little parallel deposition. This is because the 5# SFD is just located in the confluence of the left branch gully into the EG1. Thus, the flood flowing into the EG1 is easily resisted by that occurred in the EG1 (*i.e.* the main gully), that is, the resistance action can simultaneously slow velocity of flood flowing into the EG1 and even result in local reverse flow, which made the back water rapidly occur in the 5# SFD. Thus, the parallel deposition form is eventually produced.

As afore mentioned, the SFD can greatly increase the roughness of the gully bed, and thus slow down the velocity of the flood to compel the sediment to deposit within the gully. Gradually, with the rising of the deposition level, the gully channel slope is becoming more and more gentle, which keeps the gully bed not be easily eroded and therefore protect the surface soil nutrients in the gully. In addition, the litters under the SFD also contribute to the conservation of the nutrients. The humus, produced by the senescence, death, secretion of the roots of the understory vegetation, can improve the soil quality as well. Moreover, the decomposition of the litters can accrue the humus too. Hence, the soil was improved greatly in the gully, which is definitely beneficial to the growth and restoration of the understory vegetation. Like that, a benignant circle of the interaction of between the SFD and the soil quality improvement was naturally developed within the SFD gully. These discussed above powerfully indicated that the strong sediment retention capacity of the SFD is the basis of using it to restore the ecosystem and improve soil quality.

The Characteristics and Variations of Soil Moisture

As shown in Figures 63.8a–c, the comparison of soil moisture of 2#, 4#, 5# SFDs with the contrast gully is evident in August, 1998, 1999 and 2000. As can be seen from Figure 63.8d, the variations of soil moisture

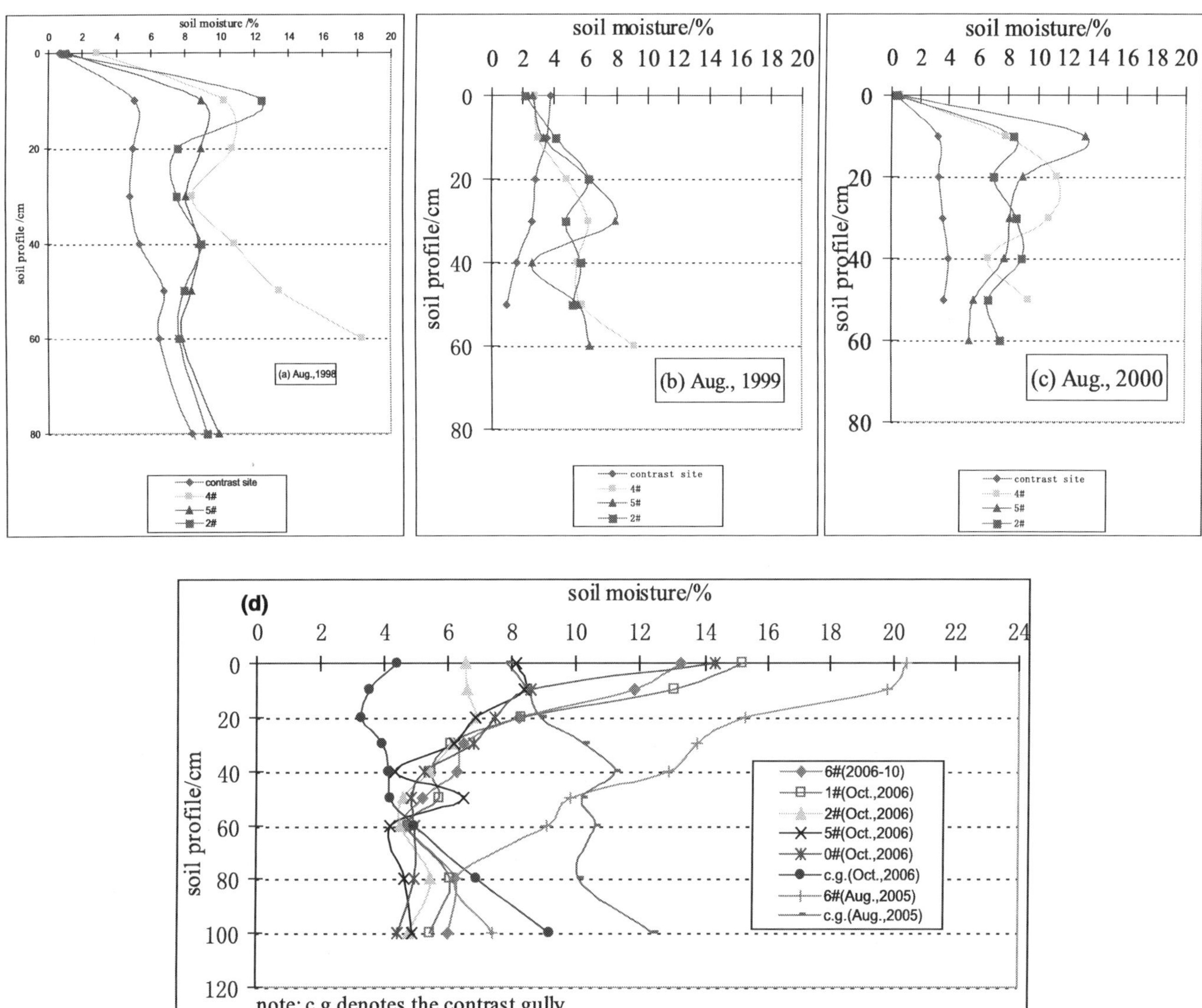

Figure 63.8: Comparison of Soil Moisture Profile Distribution between Typical SFDs and the Contrast Gully.

between the 0#, 1#, 2#, 5# and 6# SFDs (respectively) and the contrast gully were also obvious in October 2006. Compared to the contrast gully, the change in soil moisture of the 6# SFD is also marked in August 2005. Soil moisture of each typical SFD is obviously greater than the contrast gully (Figure 63.8). Moreover, soil moisture variations in profile are also quite striking. Basically, soil moisture in top 0-40 cm depth is rapidly decreased with an increase of depth, and the mean soil moisture in this depth is approximately 9 per cent, while the mean soil moisture of the contrast gully is only 4 per cent. However, the soil moisture in top 40-100 cm depth of each SFD has roughly the same variation as the contrast gully. Based on our field investigation, this is attributed to the process of the seabuckthorn roots distributed in 0-60 cm soil depth using soil moisture. According to our field observation, the depth of major rhizosphere of the seabuckthorn is not over 60 cm because of the strong horizontal extending and developing capacity of its root system. The interaction of the horizontally developed roots and soil moisture of 0-60 cm depth leads to the variation. However, no variation is evidently found in the contrast gully due to no the SFD in the gully bed. Another reason is the much litter covered in the SFD gully can greatly reduce evaporation of soil moisture.

As presented in Figures 63.8a–c, soil moisture of the three SFDs of 2#, 4# and 5# was respectively greater than the contrast gully, especially in 1998 and 2000. The soil moisture in profile is sharply increased in top 0-10 cm depth, however, soil moisture variation in 10-80cm depth is slight. It is no doubt that the mean soil moisture in 0-100 cm depth in the SFD gully was also greater than the contrast gully. Only the average soil moisture in profile in the SFD gully in 1999 is slightly smaller than that in 1998 and 2000. This is because of different rainfall and climate condition during the same period.

The soil moisture comparison in profile between the SFD gully and the contrast gully showed that the seabuckthorn plant can strikingly increase the mean soil moisture of the Pisha Sandstone gully, particularly in 0-40cm soil layer. Thus this can relatively increase the water storage of soil reservoir, adjust soil water resource in the gully, and be beneficial to the growth and restoration of vegetation. Increase of soil moisture in the gully has a significant effect on the restoration of the ecological environment within the Pisha Sandstone area.

The Characteristics and Variations of Soil Organic Matter

Soil organic matter (SOM) is the most important component of soil fertility and SOM is often used to assess the soil melioration. The seabuckthorn growth has an important effect on SOM which can improve the gully soil quality. The comparison of SOM in the upper and middle sections of the typical SFD and that of the contrast gully was shown in Figure 63.9a. It was clear that SOM of the gully with the SFD is greater than that of the corresponding sections in the contrast gully. For example, the average SOM of the 2# SFD increased by 96 per cent compared to the contrast gully, the average SOM of the 6# SFD increased by 145 per cent, and the average SOM of the 0# SFD increased by 148 per cent. Obviously, the SFD can improve SOM of the Pisha Sandstone gully, which can promote the vegetation growth and increase the roughness of the gully bed. In terms of the investigation in August 2005, many kinds of plants grow quite well in the gully (See Table 63.4 hereinafter).

As shown in Figure 63.9b, the SOM distributions in profile at various sections of the typical SFDs were quite different from those of the contrast gully. The contrast section the 0# dam is located at the inlet of it. We found that SOM of all typical SFDs (1#, 2#, 5#, 6# and 0#) is greater than the contrast gully. And the surface SOM of the contrast gully is minimal while the SOM of bottom soil layer is rather greater, but that of the middle soil layer is relatively smaller, which is in accordance

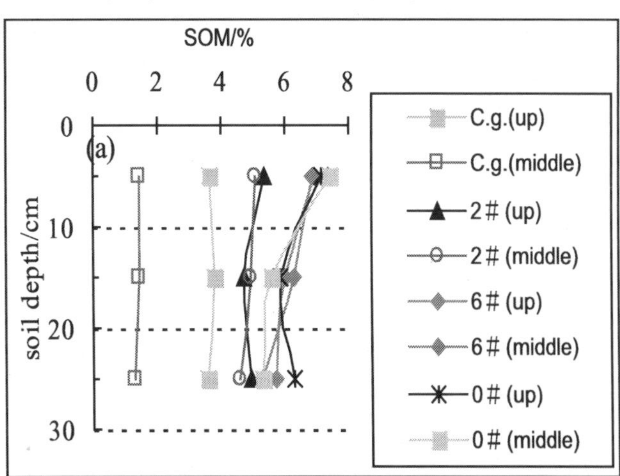

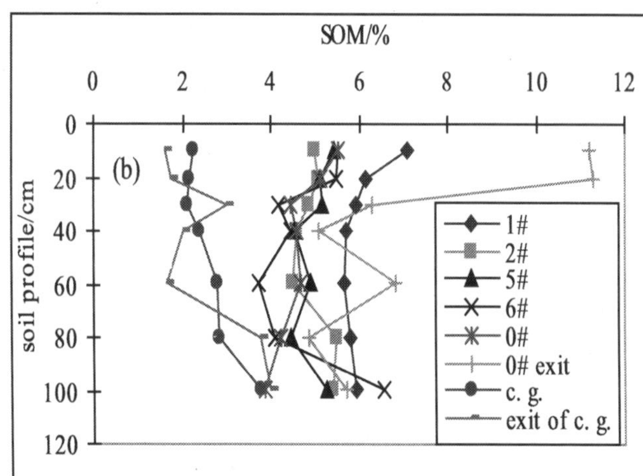

Figure 63.9: Comparison of Soil Organic Matter of Typical SFDs.
(a) SOM comparison between typical SFDs and the contrast gully (Aug., 2005) and
(b) Layer-averaged organic matter distribution in profile of each typical SFD and the contrast gully (Oct., 2006).
(Notes: up-upper section; middle-middle section; c. g. – the contrast gully.)

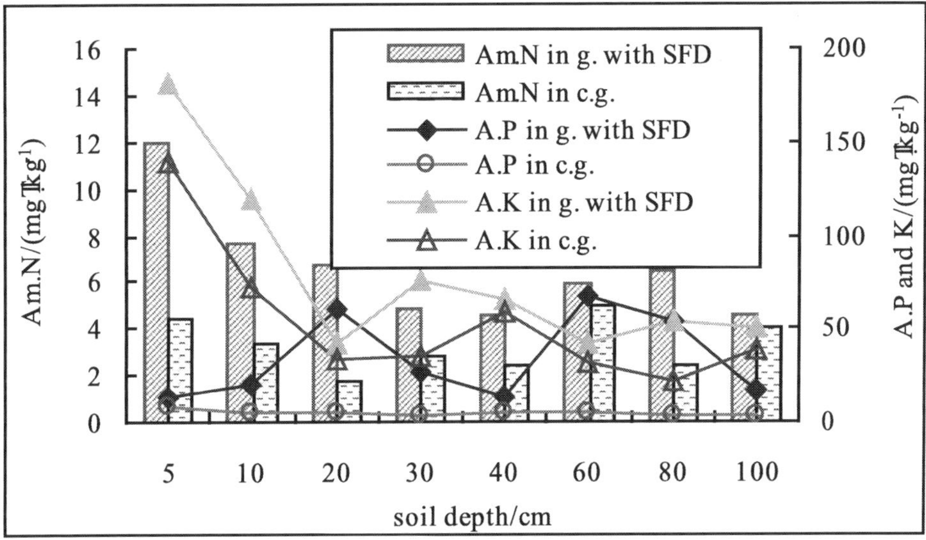

Figure 63.10: Comparison of Ammonia Nitrogen, available Phosphorus and Potassium between the SFD Gully and the Contrast Gully (Aug., 2010).
(Notes: Am. N-Ammonia Nitrogen; A. P-Available Phosphorus; A. K-Available Potassium; c. g. -the contrast gully.)

with the evolution of SOM in the course of vegetation community restoration. Based on the statistical results, the mean SOM of each SFD is greatly higher than the contrast gully. Furthermore, the SOM variations in profile of the typical SFDs are relatively smaller than the contrast gully. These results showed that the SFD indeed helps increase SOM and meliorate the Pisha Sandstone gully soil.

The Characteristics and Variations of Soil Nutrients

Soil nutrients have a close relationship with the soil organic matter. The SFD can also increase the content of soil nutrients and further improve soil fertility by improving soil organic matter. The distributions and variations of ammonium nitrogen, available phosphorus and potassium in soil profile in the EG1 with the SFD and the contrast gully are shown in Figure 63.10. It is evident that soil nutrients in the EG1 with the SFD were greater than the contrast gully. In addition, in top soil layer of 0-30 cm, ammonium nitrogen and available potassium rapidly became small with an increase of soil depth, but the change in soil layer of 30-100 cm is relatively slight, which was consistent with the SOM variations mentioned above. We also found available phosphorus in the SFD gully varied little in the whole profile of 0-100 cm, not only the ammonium nitrogen in the contrast gully is smaller, but also the available phosphorus is even smaller, and the variation of the available phosphorus is much slighter in profile.

In total, the soil nutrients of the gully with the SFD are much higher than the contrast gully. The soil nutrients comparison of the SFD gully and the contrast gully is shown in Table 63.3. In the soil layer of 0 -30 cm, ammonium nitrogen content of the SFD gully is 2.5 times of the contrast gully, available phosphorus content of the SFD gully is 5.73 times of the contrast gully, and the available potassium of the SFD gully is 1.5 times of the contrast gully. Moreover, in the soil layer of 30-100 cm, soil nutrient content of the SFD gully is still higher than the contrast gully. The above comparison indicated that the SFD can effectively improve soil fertility. It is well known that the litter and some dead roots can provide or supply nutrients for the soil, which is beneficial to the vegetation growth. Certainly, the interaction of the seabuckthorn plant and soil nutrients may bring about those variations in profile, which is mainly related to soil physical and chemical properties.

Above these variations and comparisons of soil nutrients between the SFD gully and the contrast gully showed that the SFD can strikingly improve soil fertility of the gully, which is helpful for the growth

Table 63.3: Comparison of Nutrients at 0-30 cm and 30-100 cm Soil Layers between the SFD Gully and the Contrast Gully

Locations	*Soil Layer (cm)*	*Ammonia Nitrogen ($mg{\cdot}kg^{-1}$)*	*Available Phosphorus ($mg{\cdot}kg^{-1}$)*	*Available Potassium ($mg{\cdot}kg^{-1}$)*
The gully with SFD	0-30	7.78	30.13	104.63
	30-100	5.13	37.43	52.88
The contrast gully	0-30	3.13	5.26	70.13
	30-100	3.45	4.16	37.85

and restoration of vegetation community as well as the ecosystem rehabilitation within the Pisha Sandstone area.

Changes in Species Diversity and Vegetation Coverage

1) Species Diversity

Species diversity is an index describing richness extent of species in some ecological community. There are 18 species of vegetation within the SFD in the EG1 gully (Table 63.4), and the cover of 88 per cent is Clinelymus dahurcus Turcz. However, there are only 5 plant species in the corresponding study locations. Plant species richness meant the number of total species per area, and the difference of the understory should be taken into consideration. The species richness of the understory within the SFD is obviously higher than the contrast gully. It can be inferred that the SFD can greatly improve the growth of other species.

2) Vegetation Coverage

Development of the forest ecology of different areas at certain period is widely evaluated by the forest thickness (FT) (Fan *et al.*, 2008). The index is usually expressed as FT (mm) = total forest storage volume (m^3)/total land area (m^2)×1000 ($mm{\cdot}m^{-1}$)= Forest storage volume per unit area ($m^3{\cdot}ha^{-1}$)×forest coverage×1000 ($mm{\cdot}m^{-1}$). According to the calculation formula of FT, the mean thickness coverage is often used to evaluate the vegetation cover and health state of the ecological environment of some area. The mean thickness coverage is usually expressed as:

$$VC = \frac{A \times C_U \times \sum_{j=1}^{n} P_j \times H_j}{A} \quad (1)$$

where VC denotes the mean thickness coverage, n is the total species of understory vegetation, A is the square of study area, Cu is the understory coverage of study area, P_j is the proportion of the i-th species, and H_j is the mean height of the i-th species. The higher the value of the mean thickness coverage is, the better the growth of vegetation. As shown in Table 4, the mean thickness coverage of the SFD in the EG1 is 0.376, and the corresponding value of the contrast gully is only 0.056. It is very clear that the mean thickness coverage of understory vegetation community within the SFD is 6.76 times of the contrast gully (non-vegetated gully). This indicated that the other kinds of vegetation within the SFD also grow better.

3) Shannon-Weaver diversity index

The Shannon-Weaver diversity index is commonly used to evaluate the diversity of plant species (Begon *et al.*, 1986; Colin, 1986). The greater the species diversity is, the better the stability of the plant community is. The Shannon-Weaver diversity index (Eq. 2) is expressed as:

Table 63.4: Vegetation Investigation in the EG1

Sites	*Area (m^2)*	*Understory Coverage*	*Species Richness and Proportion*		*Mean Height (m)*	*Mean Species Richness (m^2)*	*Mean Thickness Coverage (m)*	*Shannon-Weaver Index*
			Understory Community	*Proportion*				
Under the seabuckthorn shrub	8	96 per cent	*Clinelymus dahurcus* Turcz	88 per cent	0.4	2.3	0.376	0.817
			Artemisia arenaria DC	2 per cent	0.2	2.3	0.376	0.817
			Milulotus fficinalis-(Madnle) Len	4 per cent	0.3			
			Youngia japonica	3 per cent	0.4			
			Artemisia gmelinii	2 per cent	0.5			
			Salsola collina	1 per cent	0.2			
			Echinops latifolius-Tausch					
			Heteropappus altaicus					
			Stipa capillata					
			Paris tetraphylla A. Gray Setaria					
			Artemisina arayi					
			Others[a]					
			Total species:18	100 per cent	0.33			
The contrast gully	8	35 per cent	*Heteropappus altaicus*	90 per cent	0.15	0.63	0.056	0.267
			Stipa capillata	3 per cent	0.26			
			Medicago sativa	2 per cent	0.12			
			Setaria	3 per cent	0.18			
			Paris tetraphylla A. Gray	2 per cent	0.42			
			Total species: 5	100 per cent	0.23	0.63	0.056	0.267

Note: [a] Contains 5 species.

$$H = -\sum_{i=1}^{n} P_i(\log_2 P_i) \tag{2}$$

where n is the number of individuals of the i-th species, P_i is the proportion of the i-th species number to the total species number of plant community.

H is usually less than 1, and the H is proportional to the biodiversity. The H within the SFD is 0.817, however, H value of the contrast gully is only 0.267. The comparison between the two H values showed that the seabuckthorn plant can indeed increase the number of plant species, thus being beneficial to the vegetation community restoration.

CONCLUSIONS

The results showed that the SFD plays a predominant role in the ecological restoration and soil improvement within the Pisha Sandstone area, and exhibits good performance. As discussed above, on the one hand, the SFD has a strong sediment retention capacity, particularly for the coarse sediment greater than 0.05 mm from the gully head and side slope, which provides a good basis for using it to restore the ecosystem and improve soil quality within the Pisha Sandstone area. On the other hand, the SFD is able to stabilize the slope toe to prevent collapse. Water stored in the deposition is helpful for the vegetation growth in the gully and usable for local agricultural irrigation. Moreover, the soil moisture distribution in profile within the SFD gully was quite different from the contrast gully (i. e. non-vegetated gully). This may be attributed to the interaction of the seabuckthorn and the other species of vegetation. The SFD can greatly improve SOM and the other nutrients, which is clearly demonstrated by the comparison of the SFD gully and the contrast gully. In addition, based on the field investigation and our analysis, the SFD is obviously capable of increasing the species diversity and taxa richness of the Pisha Sandstone gully, while that of the contrast gully was quite low. These findings provide the important theoretical proof for widespread planting of the SFD. Up to now, the SFD has been planted about 1500,000 ha within the Pisha Sandstone area by Chinese government. This action has made a great success in improving the ecological environment of this area.

As a whole, the SFD plays a significant role in local ecological restoration and soil melioration by soil erosion control, soil moisture storage, soil nutrients increase, and improvement of species diversity as well as vegetation coverage, which indicates it is very effective. By virtue of the SFD and its planting convenience, the SFD is progressively spread to many other similar severe soil erosion areas in northwest of China.

Although we have achieved some research results on the ecological restoration and local soil improvement by examining the ecological effects of the SFD, more complicated issues still need to be further studied, such as the intrinsic interaction of certain ecological processes, impacts on the local climate, and *etc.* Moreover, the studies on these issues will have a significant value for development of the ecology discipline and flexible vegetation engineering. Currently, based on our previously accumulated data, the authors set out to study the intrinsic interaction mechanism between the SFD and the soil in the gully from the biological viewpoint. Thus, the study on the SFD is very interesting, and the authors hope more researchers to focus on the SFD.

ACKNOWLEDGEMENTS

This study is supported by both National Natural Science Foundation (Grant No: 51279163) and Shaanxi Provincial Innovation Team Project (Grant No: 2014KCT-27), and is also financed by Natural Science Foundation for young scholars of Shaan'Xi Province (Grant No: 2010JQ5003). In the course of writing this manuscript, Professor Liu Kang, Bai Hong-ying and Ma Jun-jie give us many good advices and suggestions for improving this paper. And Doctor Ross from Australia, being as a visiting Professor of college of foreign language of Northwest University, revised initial manuscript. And Doctor Smith from America, being as an exchange scholar of Northwest University, also revised this manuscript. We really thank all of them for their cordial help very much.

REFERENCES

1. Amazonas, N.T., Martinelli, L.A., Piccolo, M.D.C. and Rodrigues, R.R. 2011. Nitrogen dynamics during ecosystem development in tropical forest restoration. *Forest Ecol. Manage.* 262: 1551-1557.

2. Angima, S.D., O'Neill, M.K., Omwega, A.K. and Stott, D.E. 2000. Use of tree/grass hedges for soil erosion control in the Central Kenyan highland. *J. Soil Water Conserv.* 55 (4): 478-482.

3. Arorak, K., Mickelson, S.K., Baker, J.L., Tieeney, D.P. and Peters, C.J. 1996. Herbicide retention by vegetative buffer strips from runoff under natural rainfall. *Trans. ASAE* 39: 2155-2162.

4. Bazhenov, M.L., Chauvin, A., Audibert, M. and Levashova, N.M. 1993. Permian and Triassic paleomagnetism of the southwestern Tien Shan: Timing and mode of tectonic rotations. *Earth Planet. Sci. Lett.* 118: 195-212.

5. Begon, H. and Townsend, 1986. *Ecology: Individuals, population and communities.* Oxford: Blackwell Scientific Publications.

6. Bi, C.F. and Li, G.F. 1998. Special function of seabuckthorn on controlling soil erosion in the soft rock region. Express Water Resource. *Hydropower Information* 19 (18): 1-3. (In Chinese)

7. Bi, C.F. and Li, G.F. 2003. Field model research on sediment retention by seabuckthorn plant "flexible

dam" in gully channel in the soft rock region. *The Global Seabuckthorn Res. Develop.* 1 (1): 6-12. (In Chinese with English summary)

8. Bi, C.F., Wang, F.G. and Li, G.F. 2003. Experiment on sediment retention by plant "flexible dam" in gully in soft rock region. *J. Sediment. Res.* 23 (2): 14-25. (In Chinese with English summary)
9. Cao, S., Chen, L., Feng, Q. and Liu, Z. 2007a. Soft-riser bench terrace design for the hilly loess region of Shaanxi Province, China. *Landscape and Urban Planning* 80: 184-191.
10. Cao, S., Chen, L. and Liu, Z. 2007b. Impact of three soil types on afforestation in China's Loess Plateau. *Landscape and Urban Planning* 83: 208-217.
11. Cao, S., Chen, L., Liu, Z. and Wang, G. 2008. A new tree-planting technique to improve tree survival and growth on steep and arid land in the Loess Plateau of China. *Journal of Arid Environments* 72: 1374-1382.
12. Cerda, A. 1999. Parent material and vegetation affect soil erosion in eastern Spain. *Soil Science Society of America Journal* 63: 362-368.
13. Chen, L.D., Wang, J.P., Wei, W., Fu, B.J. and Wu, D.P. 2010. Effects of landscape restoration on soil water storage and water use in the Loess Plateau Region, China. *Forest Ecol. Manage.* 259: 1291-1298.
14. Colin, V.P. 1986. *Ecology*, New York: John Wiley and Sons.
15. Dorioz, J.M., Wang, D., Poulenard, J. and Trevisan, D. 2006. The effect of grass buffer strips on phosphorus dynamics-A critical review and synthesis as a basis for application in agricultural landscapes in France. *Agric. Ecosyst. Environ.* 117: 4-21.
16. Fan, B.M., Li, Z.Y. and Chen, J. 2008. A new indicator for modern forestry development and forest thickness. *Chinese Forestry Sci. Tech.* 7 (4): 53-60. (In Chinese with English Summary)
17. Fang, N.F., Shi, Z.H., Li, L., Guo, Z.L., Liu, Q.J. and Ai, L. 2012. The effects of rainfall regimes and land use changes on runoff and soil loss in a small mountainous watershed. *Catena* 99: 1-8.
18. Fernández, C., Vega, J. A., Jimenez, E., Vieira, D. C. S., Merino, A., Ferreiro, A. and Fonturbel, T. 2012. Seeding and mulchingandseeding effects on post-fire runoff, soil erosion and species diversity in Galicia (NW Spain). *Land Degradation and Development* 23: 150-156. DOI 10.1002/ldr.1064.
19. Fields-Johnson, C.W., Zipper, C.E., Burger, J.A. and Evans, D.M. 2012. Forest restoration on steep slopes after coal surface mining in Appalachian USA: Soil grading and seeding effects. *Forest Ecol. Manage.* 270: 126-134.
20. Ghebremichael, L., Veith, T.L., Hamlett, J.M. and Gburek, W.J. 2008. Precision feeding and forage management effects on phosphorus loss modeled at a watershed scale. *J. Soil Water Conserv.* 63: 280-291.
21. Gomi, T., Sidle, R.C., Noguchi, S., Negishi, J.N. and Rahim, A. N. 2006. Sediment, wood, and detritus accumulations in tropical headwater streams: Effects of logging and riparian management. *Forest Ecol. Manage.* 224: 166-175.
22. Hartanto, H., Prabhu, R., Widayat, A.S.E. and Asdak, C. 2003. Factors affecting runoff and soil erosion: Plot-level soil loss monitoring for assessing sustainability of forest management. *Forest Ecol. Manage.* 180: 361-374.
23. Hussein, J., Yu, B., Ghadiri, H. and Rose, C. 2007. Prediction of surface flow hydrology and sediment retention upslope of a vetiver buffer strip. *J. Hydrol.* 338: 261-272.
24. Ide, J.I., Kume, T., Wakiyama, Y., Higashi, N., Chiwa, M. and Otsuki, K. 2009. Estimation of annual suspended sediment yield from a Japanese cypress (*Chamaecyparis obtusa*) plantation considering antecedent rainfalls. *Forest Ecol. Manage.* 257: 1955-1965.
25. Jin, Z.P. 2003. Soil and water conservation and the agricultural and animal husbandry development within the Pisha Sandstone area. Zhengzhou: Water resources of Yellow River Press. (In Chinese)
26. Kiepe, P. 1995. No runoff, no soil loss: soil and water conservation in hedgerow barrier systems. *Tropical Resource Management Papers* No. 10, p.158.
27. Kumar, S., Anderson, S.H., Bricknell, L.G. and Udawatta, R.P. 2008. Soil hydraulic properties influenced by agroforestry and grass buffers for grazed pasture systems. *J. Soil Water Conserv.* 63: 224-232.
28. Lao, J.C. 1988. *Manual of Agricultural Chemistry Analysis on Soil.* China Agriculture Press, Beijing. (In Chinese)
29. Lee, K.H., Isenhart, T.M. and Schultz, R.C. 2003. Sediment and nutrient removal in an established multispecies riparian buffer. *J. Soil Water Conserv.* 58 (1): 1-8.
30. Li, H.E., Yang, F.S., Zhang, R.J., Bi, C.F. and Tang, Y.J. 2009. Field experiment on the effects of the seabuckthorn flexible dam on water flow. *J. Hydroelec. Engi.* 28 (1): 124-129.
31. Li, Z.B., Zhu, B.B. and Li, P. 2008. Advancement in study on soil erosion and soil and water conservation. *Acta Pedologica Sinica* 45 (5): 802-809.
32. Lowrance, R., Leonard and Sheridan, R. J. 1985. Managing riparian ecosystems to control nonpoint pollution. *J. Soil Water Conserv.* 40: 87-91.

33. Lu, S.Y., Cheng, J.D. and Brooks, K.N. 2001. Managing forests for watershed protection in Taiwan. *Forest Ecol. Manage.* 143: 77-85.

34. Mark, W.M., Jorge, C.R. and Constantino, M.M. 1999. Late Paleozoic to Early Jurassic tectonic development of the high Andean Principal Cordillera, E1 Indio Region, Chile. *J. S. Am. Earth Sci.* 12: 33-49.

35. Mishra, S.K., Tyagi, J.V., Singh, V.P. and Singh, R. 2006. SCS-CN-based modeling of sediment yield. *J. Hydrol.* 324: 301-322.

36. Mohammad, A.G. and Alseekh, S.H. 2012. The effect of *Sarcopoterium spinosum* on soil and vegetation characteristics. *Catena* 100: 10-14.

37. Moskalski, S.M. and Sommerfield, C.K. 2012. Suspended sediment deposition and trapping efficiency in a Delaware salt marsh. *Geomorphology* 139-140: 195-204.

38. Oteroa, J.D., Figueroa, A., Munoz, F.A. and Pena, M.R. 2011. Loss of soil and nutrients by surface runoff in two agro-ecosystems within an Andean paramo area. *Eco. Eng.* 37: 2035-2043.

39. Pan, C.Z., Ma, L., Shangguan, Z.P. and Ding, A.Z. 2011. Determining the sediment trapping capacity of grass filter strips. *J. Hydrol.* 405: 209-216.

40. Prokop, P. and Porêba, G. J. 2012. Soil erosion associated with an upland farming system under population pressure in Northeast India. *Land Degradation and Development* 23: 310-321.

41. Qian, N., Wang, K.Q., Yan, L.D. and Fu, R.S. 1980. Effects of coarse sediment source region in the middle Yellow River on scour and siltation in the lower Yellow River. In: *First International Academic Conference Proceedings on River and Sediment*. China Waterpower Press, Beijing, pp. 33-62. (In Chinese with English summary)

42. Qian, Z.Y. 1986. By taking seabuckthorn resource as a breakout for accelerating treatment on Loess Plateau. *J. Soil Water Conserv. Technical Infor.* 4: 3-6. (In Chinese)

43. Ran, D.C. 2006. Effect of check dam on retaining and reducing coarse grain sediment in middle reaches of Yellow River. *J. Hydraul. Eng.* 37 (4): 443-450. (In Chinese with English summary)

44. Rao, M.R., Ong, C.K. and Pathak, P. 1991. Productivity of annual cropping and agroforestry systems on a shallow Alfisoi in semi-arid India. *Agroforest. Syst.* 15: 51-63.

45. Rasmussen, A.B.P., Peter, W.L., Ursula, S. and Mc Knight, B.K. 2011a. Buffer strip width and agricultural pesticide contamination in Danish lowland streams: Implications for stream and riparian management. *Ecol. Eng.* 37: 1990-1997.

46. Rasmussen, J.J., Annette, B.P., Peter, W.L., Ursula, S.M. and Brian, K. 2011b. Buffer strip width and agricultural pesticide contamination in Danish lowland streams: Implications for stream and riparian management. *Ecol. Eng.* 37 (12): 1990-1997.

47. Reeder, K.F., Debinski, D.M. and Danielson, B.J. 2005. Factors affecting butterfly use of filter strips in Midwestern USA. *Agric. Ecosyst. Environ.* 109: 40-47.

48. Rongsen, A. 1992. Seabuckthorn a multi-purpose plant species for fragile mountains. *ICIMOD Occasional Paper No.20,* Kathmandu, Nepal, 62 p.

49. Rousi, A. 1965. Observation on the cytology and variation of European and Asiatic populations of *Hippophae rhamnoides*. *Annales Botanici Fennici* 2: 1-18.

50. Rousi, A. 1971. The genus *Hippophae* L. a taxonomic study. *Annales Botanici Fennici* 8: 177-227.

51. Ruiz Jaen, M.C. and Aide, T.M. 2005b. Vegetation structure, species diversity, and ecosystem processes as measures of restoration success. *Forest Ecol. Manage.* 218: 159-173.

52. Sandercock, P.J. and Hooke, J.M. 2011. Vegetation effects on sediment connectivity and processes in an ephemeral channel in SE Spain. *J. Arid Environ.* 75: 239-254.

53. Sergio, A.P., Maruxa, C.M., Diana, C.S.V., Donald, L.M. and Jan, J.K. 2013. Effectiveness of hydromulching to reduce runoff and erosion in a recently burnt pine plantation in central Portugal. *Land Degradation and Development*. Article first published online: 14 Aug. 2013, DOI: 10.1002/ldr.2236.

54. Sidle, R.C., Ziegler, A.D., Negishi, J.N., Nik, A.R., Siewc, R. and Turkelboom, F. 2006. Erosion processes in steep terrain-Truths, myths, and uncertainties related to forest management in Southeast Asia. *Forest Ecol. Manage.* 224: 199-225.

55. Slattery, M.C. and Phillips, J.D. 2011. Controls on sediment delivery in coastal plain rivers. *J. Environ. Manage.* 92: 284-289.

56. Spaan, W.P., Sikking, A.F.S. and Hoogmoed, W.B. 2005. Vegetation barrier and till age effects on runoff and sediment in alley crop system on a Luvisol in Burkina Faso. *Soil Till. Res.* 83: 194-203.

57. Udawatta, R.P., Garrett, H.E. and Kallenbach, R.L. 2010. Agroforestry and grass buffer effects on water quality in grazed pastures. *Agroforest. Syst.* 79: 81-87.

58. Xu, J.H. and Gao, Y.J. 2007. Impact of controlling to sediment-laden and coarse sand area in middle Yellow River on water resource of lower Yellow River. *J. Soil Water Conserv.* 21 (6): 47-50. (In Chinese with English summary)

59. Yu, X.X. 2002. Distribution patterns and hydrological effects of moss and litters in *Abies fabri* forests on eastern slope of Gongga mountain. *J. Beijing Forestry University* 24 (5): 14-18.
60. Yu, X.X., Niu, J.Z. and Niu, J.L., 2003. Study on ecology rehabilitation on small watershed in mountainous area. In: *Proceedings of Symposium on Soil and Water Conservation Between China and USA*, pp. 48-55.
61. Zhang, B., Yang, Y.S. and Zepp, H. 2004. Effect of vegetation restoration on soil and water erosion and nutrient losses of a severely eroded clayey Plinthudult in southeastern China. *Catena* 57: 77-90.
62. Zhao, G., Mu, X., Wen, Z., Wang, F., and Gao, P. 2013. Soil erosion, conservation, and eco-environment changes in the Loess Plateau of China. *Land Degradation and Development* 24: 499-510, DOI 10.1002/ldr. 2246.
63. Zheng, X.M. 2005. Research on management of concentrated coarse sediment source area of the Loess Plateau. *J. Soil Water Conserv. China* (12): 5-6. (In Chinese with English summary)
64. Ziadat, F.M. and Taimeh, A.Y. 2013. Effect of rainfall intensity, slope and land use and antecedent soil moisture on soil erosion in an arid environment. *Land Degradation and Development* 24: 582- 590.

Chapter 64

Seabuckthorn: An Ecosystem Based Adaptation Strategy for Arid Mountain Areas of Pakistan

Babar Khan[1*], Abdul Wahid Jasra[2] and Hassan Ali[3]

[1]World Wide Fund for Nature Pakistan (Gilgit-Baltistan), GCIC complex, NLI colony Jutial, Gilgit 15100, Pakistan,
[2]International Centre for Integrated Mountain Development (ICIMOD), Country Office Pakistan, Islamabad,
[3]Khyber-Pakhtunkhwa Agricultural University, Peshawar, Pakistan
(*E-mail: bkhan@wwf.panda.org)

SUMMARY

Seabuckthorn (*Hippophae rhamnoides*) has been used for the prevention of soil erosion and the production of food and medicine products in arid mountain ecosystems of the world. This study was conducted to evaluate seabuckthorn for strengthening the marginalized mountain farming systems incold arid Karakoram-Hindu Kush- Pamir region, in the face of changing climate, rising populations and food insecurity. Results revealed that seabuckthorn fixed atmospheric nitrogen (NO_3) was high in older plantations near root nodules while its concentration was low in deeper soils of younger plantations. NO3 concentration was maximum (7.96 mg/L) at furrow slice depth (12 cm) and minimum (2.41 mg/L) in 25 cm deep soils. At 50 cm, it was slightly higher (2.43 mg/L) than the intermediate depth. Likewise, total combustible organic matter increased with age of plantations but decreased with the depth of soils. Its value was maximum (1.65 per cent) at 12 cm in 10 years old seabuckthorn plantation and minimum (1.06 per cent) in 50 cm deep soil of the 8 years old plantation. pH value increased with increasing depth but decreased with the age of plantation whereas, electrical Conductivity (EC) value of sampled soil decreased with increasing depth and the age of plantation. Changes in sandy loam texture of sampled soils were observed neither with changing depth nor with the age of seabuckthorn plantations. Results concluded that seabuckthorn, a non-leguminous nitrogen fixer, with vitamin rich fruits and extensive root system could increase soil fertility, combat soil erosion and stabilize active slopes in arid and semi-arid mountain areas. Therefore, as part of an ecosystem based adaptation strategy, seabuckthorn based agroforestry models need to be promoted at large scale enabling mountain people adapt to the adverse effects of climate change, while securing healthy food and adequate water for their families.

Keywords*: Seabuckthorn, Nitrogen fixation, Soil properties, Climate adaptation and Gilgit-Baltistan.*

INTRODUCTION

Over the next decades, it is predicted that billions of people, particularly those in developing countries, face shortages of water and food and greater risks to health and life as a result of climate change. The urgency for adaptation is highlighted by projections from the three reports produced by the IPCC in 2007 (UNFCCC, 2013). Under a business as usual scenario, greenhouse gas emissions could rise by 25-90 per cent by 2030 relative to 2000 and the Earth could warm by 3°C this century. Even with a temperature rise of 1-2.5°C the IPCC predict serious effects including reduced crop yields in tropical areas leading to increased risk of hunger, spread of climate sensitive diseases such as malaria, and an increased risk of extinction of 20-30 per cent of all plant and animal species. By 2020, up to 250 million people in Africa could be exposed to greater risk of waterstress (UNFCCC, 2013).

Mountains are among the regions most affected by climate change. Mountain areas are typically exposed to multiple hazards. Climate change is likely to increase this exposure, as extreme events such as storms,

Seabuckthorn (*Hippophae* L.): *A Multipurpose Wonder Plant, Vol. 5: Advanced Technologies for Cultivation, Processing, Health Protection and Environmental Conservation* (Virendra Singh, *Editor-in-Chief*, 2018), p. 501–506, Astral International Pvt. Ltd., New Delhi, India.

landslides, avalanches, and rockfalls are expected to become more common and more intense in mountain areas, threatening both livelihoods and infrastructure (Kohler *et al.*, 2010). Millions of people living in the catchment areas of the Himalayas are predicted to face increased risk of floods as glaciers retreat followed by drought and water scarcity and all associated risks to life and livelihoods (UNFCCC, 2013). Adapting to adverse climatic changes will thus entail adjustments and changes at every level. Communities must build their resilience, including adopting appropriate technologies while making the most of traditional knowledge, and diversifying their livelihoods to cope with current and future climate stress (IPCC, 2014). Local coping strategies and traditional knowledge need to be used in synergy with government and local interventions where choice of adaptation interventions depends on local and national circumstances (UNFCC, 2013) as Indigenous and local knowledge and practices have great potential to address climate change vulnerabilities and should be viewed as an invaluable knowledge resource for climate change adaptation and resilience building (MoSTE, 2015).

Seabuckthorn, a woody, fruit-bearing, deciduous, dioecious, spiny shrub or tree with silvery scales, widespread across Himalayas (Lu, 1992) can be one such highly effective local adaptation strategies to cope with emerging stressors of ever changing climatic patterns for the mountain areas, as it is highly effective and useful in retraining, reclaiming and conserving soil, especially on fragile slopes, due to its extensive root system. It tolerates low temperatures to -43°C, high soil pH up to 8.0 and salt from seawater intrusion in coastal areas (Lu, 1992; Bond, 1983). Introduction of seabuckthorn in Nepal for firewood has also produced environmental benefits. The Nepalese used to cut tree branches from forest trees for firewood, demolishing the forest and causing severe soil erosion. The quickly re-growing seabuckthorn has reduced harvesting pressure on native woody plants (Li and Schroeder, 1996).

Seabuckthorn can be planted on marginal lands due to its symbiotic association with nitrogen fixing actinomycetes on its root system, which is characterized by nitrogen fixing nodules (Akkermans *et al.*, 1983; Dobritsa and Novik, 1992). It is a pioneer species, and often is the first woody species colonizing open areas, such as abandoned farmland, wastelands, slopes and rocky islands. Riverbanks, lakeshores, steep slopes, and other susceptible terrain can benefit from the establishment of seabuckthorn. Wind breaks made up of seabuckthorn are effective at preventing wind erosion in open areas (Li and Schroeder, 1996).

Mountain areas of Pakistan in general and that of Gilgit-Baltistan and Chitral show distinct signs of un-sustainability, decreasing soil fertility and a high degree of instability, particularly in the face of changing climatic patterns and socio-economic and ecological vulnerability. This study was therefore designed to determine role of seabuckthorn in improving nitrogen, organic matter, pH and electrical conductivity of marginal soils to ascertain if seabuckthorn plantations can be raised as a part of an overall ecosystem based adaptation strategy to help people get adapt to adverse effects of climate change in the arid mountainous areas of Pakistan.

MATERIALS AND METHODS

Study Area

Gilgit-Baltistan (GB), formerly named as the Northern Areas of Pakistan, encompassing 72,496 km^2 land area is home to over 1.6 million people. Located in the extreme north of Pakistan (77° 41′ 20.403½ E and 35° 27′ 24.81½ N to 72° 30′ 26.932½E and 35° 54′ 58.338½N), GB borders internationally with the Xinjiang Uygur Autonomous Region of China, Wakhan corridor of Afghanistan and India (Khan *at el.*, 2004). Gilgit, the administrative capital of Gilgit-Baltistan, is situated above 3,000 m ASL, with landscape dominated by mountains of the Himalaya, Karakoram and Hindu Kush ranges. The climate of the region varies widely from monsoon influenced moist temperate to arid and semi-arid cold desert in the north. Below 3,000 m precipitation is less than 200 mm per annum. However, there is a sharp precipitation gradient along the altitude and over 2,000 mm annual snowfall is observed above 6,000 m ASL. Temperatures in the lower parts of valleys vary from extreme hot (40°C) in summer to many degrees below freezing point (-10°C) during winter (Khan *et al.*, 2010). Only 2 per cent of the total area is arable and > 5 per cent is under forest cover. Vegetation is classified into four distinctive zones *viz.*, Mountain Sub-Tropical Scrub Forests, Mountain Dry Temperate Coniferous Forests, Mountain Dry Temperate Broad leaved Forests and Northern Dry Scrub Forests, each having peculiar biota (Khan *et al.*, 2014).

Sampling

Twelve composite samples were collected randomly from four different strata, including three different sites and one control (barren land) from Jalalabad area in Gilgit district, using stratified random sampling technique following Soil Testing Manual (RoP, 2008). Samples were collected from various depths *i.e.*, 12 cm (furrow slice), 25 cm and 50 cm from three different age forests *viz.*, barren (T0), 06 years (T1), 08 years (T2), and 10 years (T3) seabuckthorn plantations, using soil Augar. Collected samples were air dried and mixed thoroughly to prepare composite samples, and stored in polythene bags, tagged with key information for lab analysis.

Analysis

Sampled soil were analyzed for organic matter, Nitrate-Nitrogen (NO_3-N), pH, Electrical conductivity (EC), and soil texture at Mountain Agriculture Research Centre (MARC) Gilgit, using standard soil analytical procedures (Page *et al.*, 1982). Statistical analysis was performed in MS Excel for simple calculations, monographs and tables (Mc-Cullough and Heiser, 2008).

RESULTS

Samples from four different aged, seabuckthorn plantations were collected to analyses the pH value, EC m/L, contents of nitrogen nitrate and organic matter contents. The samples were collected from 12 cm, 25 cm and 50 cm deep seabuckthorn forest soils for investigation.

Area and Distribution

Seabuckthorn is common in trans-Himalayas. The plant is mostly found in the Northern parts of Pakistan including Gilgit-Baltistan, Kurram Agency, Chitral and upper Swat. In Gilgit-Baltistan alone, seabuckthorn forests and plantations cover more than 200 km^2 area in Gilgit, Ghizer, Hunza, Skardu and Ganche districts.

Climate Trends

A careful analysis of climate data for the past 1-3 decades shows an overall increase in temperatures and rainfall but a decrease in snow fall as shown in the Figures 64.1a–d.

N-NO$_3$

Average mean score of 10.32 shows the higher contents of Nitrogen Nitrate in the data sample in the ten-year plantation as compared to others, which is much lower (Table 64.1) while the standard deviation results show better reliability for T2 (0.267).

Organic Matter

The sample data was also tested for organic matter content was high in samples taken from 12 cm depth of 10-year-old plantation as compared to deeper soils in younger plantations (Table 64.2). However if we see the standard deviation scores of our data set, it shows more reliability for T2 sample, which is an 8-year-old plantation.

Electrical Conductivity

The sample data was analysed for Electrical conductivity, which shows higher mean value for T0, which means the electrical conductivity in the soil sample was higher before the plantation while it is more consistent in a ten year old plantation where the standard deviation results shows more consistency than rest of the sample data (Table 64.3).

Soil pH

For pH value of the sample data, the mean score showed that T0 has the highest average mean with 7.58 of pH value. It means that pH value is higher at different depths for T0 than the other mean values for T1, T2 and T3, which are 7.51, 7.41 and 7.55 respectively (Table 64.4).

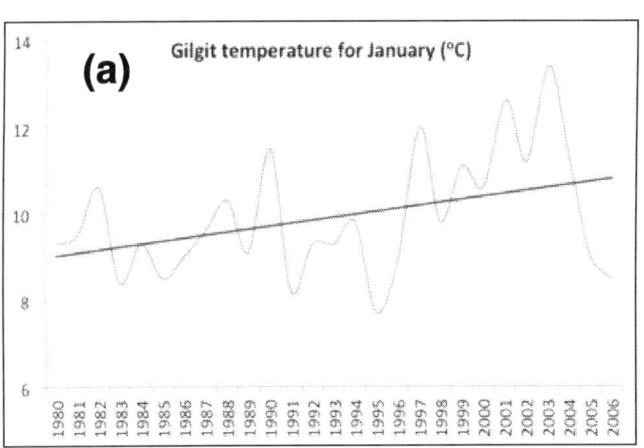

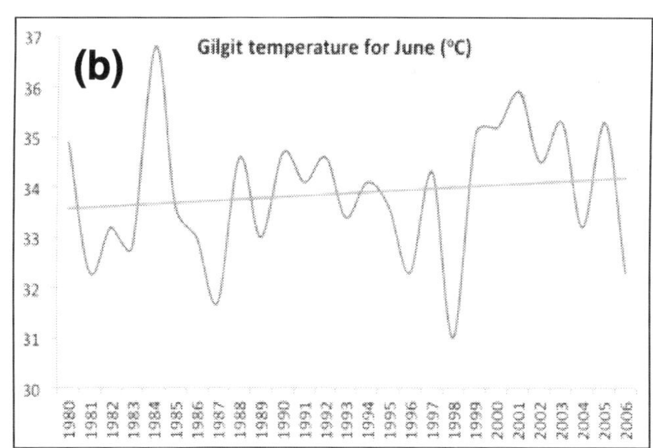

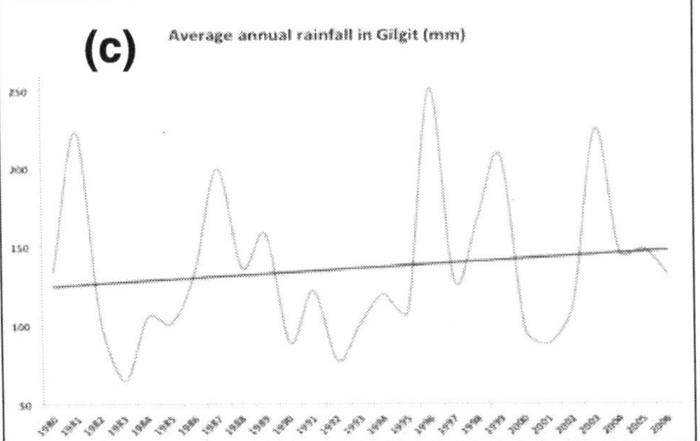

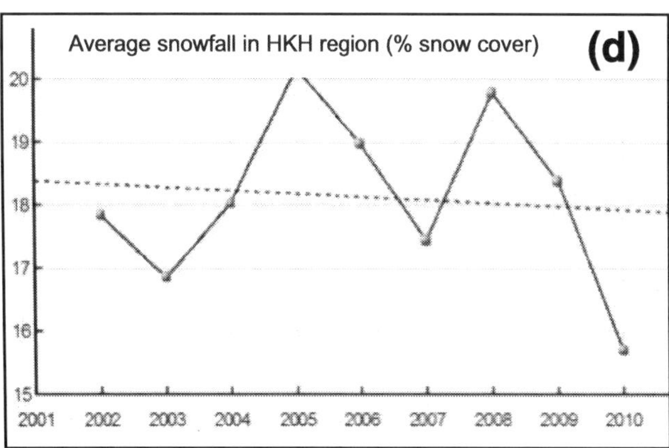

Figure 64.1a-d: Climate Trend in Gilgit-Baltistan.

Table 64.1: Nitrate-Nitrogen (N-NO_3) Level of the Sites

Age (Year)	Soil Depth			Mean	Variance	S.D
	12 cm	25 cm	50 cm			
0 (T0)	3.019	2.722	2.013	2.58	0.267	0.517
6 (T1)	2.739	1.947	2.31	2.33	0.157	0.396
8 (T2)	1.98	1.518	1.98	1.83	0.071	0.267
10 (T3)	24.09	3.465	3.415	10.32	142.141	11.922

Table 64.2: Organic Matter (OM per cent) Level of the Sites

Age (Year)	Soil Depth			Mean	Variance	S.D
	12 cm	25 cm	50 cm			
0 (T0)	1.1	0.9	0.71	0.009	0.038	0.195
6 (T1)	1.2	0.8	1.3	0.011	0.070	0.265
8 (T2)	0.6	0.95	0.85	0.008	0.033	0.180
10 (T3)	1.78	1.76	1.4	0.0165	0.046	0.214

Table 64.3: Electrical Conductivity (EC mg/L) Level of the Sites

Age (Year)	Soil Depth			Mean	Variance	S.D
	12 cm	25 cm	50 cm			
0 (T0)	0.432	0.209	0.488	0.38	0.022	0.148
6 (T1)	0.466	0.281	0.25	0.33	0.014	0.117
8 (T2)	0.508	0.254	0.134	0.30	0.036	0.191
10 (T3)	0.286	0.203	0.209	0.23	0.002	0.046

Table 64.4: Soil pH Level of the Sites

Age (Year)	Soil Depth			Mean	Variance	S.D
	12 cm	25 cm	50 cm			
0 (T0)	7.71	7.64	7.38	7.58	0.030	0.174
6 (T1)	7.42	7.61	7.51	7.51	0.009	0.095
8 (T2)	6.94	7.66	7.62	7.41	0.164	0.405
10 (T3)	7.44	7.54	7.67	7.55	0.013	0.115

The standard deviation result for the given data showed that T1 provides more consistent and reliable values in contrast to the other sample results.

Soil Texture

For all groups at all different depths, soil was sandy loam except for T3 at 12 cm of depth that has loamy soil texture (Table 64.5).

Table 64.5: Soil Texture Level of the Sites

Age (Year)	Soil Depth		
	12cm	25 cm	50 cm
0 (T0)	Sandy loam	Sandy loam	Sandy loam
6 (T1)	Sandy loam	Sandy loam	Sandy loam
8 (T2)	Sandy loam	Sandy loam	Sandy loam
10 (T3)	Loam	Sandy loam	Sandy loam

DISCUSSION

Results revealed that seabuckthorn has a profound effect on key soil properties including pH, EC, organic matter and nitrate nitrogen. Soils in the area being dry and barren are generally deficient in organic matter (Whiteman, 1985; Khan *at al.*, 2004). However, comparatively higher organic matter content in surface soils of older plantations is probably due to addition and accumulation of litter from seabuckthorn trees/ shrubs for longer periods than deeper soils, which in case of seabuckthorn plantations are rarely tilled, stirred or manipulated, allowing organic matter to accumulate in top soils.

Older plantations fix more atmospheric nitrogen than younger ones. Nitrate nitrogen content was high in surface soils and it decreased with increasing depth. Higher concentration of N-NO_3 in older plantations is attributed to their well developed root nodules capable of fixing up to 180 Kg atmospheric Nitrogen per ha per year compared to younger plants with under developed nodules (Lu, 1992). Similarly, less N-NO_3 in deeper soils could be due to high accumulation of organic matter through litter in surface soils (Di and Cameron, 2002), which seldom percolate to sub soils under the prevailing arid and semi arid climatic conditions (Robbins and Carter, 1980).

The pH value of sampled soils increased with the age of the forest. The pH attained highest value (7.55) in samples taken from a 10 year old plantation while its value was minimum (7.41) for soils of 8 years old plantation. Changes in pH of the sampled soil with increasing age and depth is a result of organic matter decomposition coupled with acidic parent materials, producing acidic compounds, in the surface and near to surface soils, whereas higher pH in deeper soils of older plantations is perhaps due to slow percolation of acidic materials under the prevailing semi arid climatic conditions of the area (Johnson and Raun, 1995).

EC decreased with increasing soil depth and age of the plantations due to varying soil temperature and parent material, as EC soil EC declines sharply towards the freezing point of water, and surface soils being better exposed to solar radiations may have higher soil temperature than deep soils. Gilgit soils being low in salts have slightly lower EC (Doerge *et al.*, 1999; Khan *et al.*, 2004).

Considering the fact that climate change is likely to increase the exposure of mountain areas to climatic risks *i.e.*, floods, GLOFs, storms, landslides, avalanches, and rockfalls, threatening both livelihoods and infrastructure, and millions of people living in the Himalayan catchments are predicted to face increased risk of floods followed by drought, desertification, food and water insecurity, framing local adaptation strategies to cope with adverse climatic impacts is a must for the mountain communities (Khan *et al.*, 2010; Karki *et al.*, 2012). Seabuckthorn plantations, a local adaptation strategy, has proved to be highly effective in restoring depleted soils, replenishing soil nutrients, combating soil erosion, stabilizing active slopes, and supplying cheap fodder and fertilizers to poor on marginal lands, besides offering increased income opportunities through seabuckthorn exploitation, utilization and animal husbandry.

CONCLUSION

Fast growth, wide adaptation, extensive root system, drought and frost tolerance, strong coppicing and suckering habits coupled with efficient nitrogen fixation make seabuckthorn particularly suitable for planting in marginal, depleted and eroded soils. Apart from controlling soil erosion effectively, seabuckthorn can increase land reclamation, slope stabilization, food and water security, and income generation opportunities for mountain people. Therefore, in the face of ever increasing climatic risks, mountain communities can be supported to raise seabuckthorn plantations on large scale as a part of an overall ecosystem based adaptation strategy to cope with adverse effects of climate change, while securing healthy food and freshwater for their families.

REFERENCES

1. Akkermans, A.D.L., Roelofsen, W., Blom, J., Hussdanel, K. and Harkink, R. 1983. Utilization of carbon and nitrogen compounds by *Frankia* in synthetic media and in root nodules of *Alnus glutinosa*, *Hippophae rhamnoides* and *Datisca cannabina*. *Can. J. Bot*. 61: 2793-2800.
2. Bon, G. 1983. Taxonomy and distribution of non-legume nitrogen-fixing systems. In: *Biological Nitrogen Fixation in Forest Ecosystems; Functions and Applications*. Edited by J.C. Gordon and C.I. Wheeler. Martinus Nijhoff, Dr. W. Junk Publishers, The Hague, pp. 55-87.
3. Di, H.J. and Cameron, K.C. 2002. Nitrate leaching in temperate agro ecosystems: Sources, factors and mitigating strategies. *Nutrient Cycling Agroecosyst* 64: 237-256.
4. Dobritsa, S.V. and Novik, S.N. 1992. Feedback regulation of nodule formation in *Hippophae rhamnoides*. *Plant Soil* 144: 45-50.
5. Doerge, T., Kitchen, N.R. and Lund, E.D. 1999. Soil electrical conductivity mapping site specific. Management Guidelines, Potash and Phosphate institute publication, SSMG, 30p.
6. Johnson, G.V. and Raun, W.R. 1995. Nitrate leaching in continuous winter wheat: Use of a soil plant buffering concept to account for fertilizer nitrogen. *J. Prod. Agric.* 8: 486-491.
7. Karki, M., Tolia, R.S., Mahat, T.J., Tuladhar, A. and Akash, A. 2012. Sustainable mountain development in Hindu Kush Himalaya from Rio 1990 to Rio 2012 and beyond. ICIMOD, Kathmandu.
8. Khan, B., Ablimit, A., Nawaz, M.A., Ali, R., Khan, M.Z., Din, J. and Karim, R. 2014. Pastoralist experience and tolerance of snow leopard, wolf and lynx predation in Karakoram Pamir Mountains. *J. Bio. and Env. Sci.* 5 (4): 214-229.
9. Khan, B. 2012. Pastoralism wildlife conservation conflict in climate change context-A study of climatic factors influencing fragile mountain ecosystems and pastoral livelihoods in the Karakoram Pamir Trans-border area between China and Pakistan. *Ph. D. Thesis,* submitted to Graduate University of the Chinese Academy of Sciences, China, 260p.
10. Khan, B., Khan, A. A. and Baig, R. 2011. Climate change and pastoralism in the Karakorum Mountains of Pakistan; Impacts and responses by

local herder communities. In Abstracts: *Convention on Impacts of Climate Change on Biodiversity in Gilgit-Baltistan*. Karakorum International University, Gilgit, 7p.

11. Khan, B., Ablimit, A., Mahmood, R. and Qasim, M. 2010. *Robinia pseudoacacia* leaves improve soil physical and chemical properties. *Journal of Arid Land* 2 (4): 266-271.

12. Khan, B., Khattak, R. A. and Hakeem, A. 2004. Studies on physico-chemical characteristics and fertility evaluation of Gilgit soils. *Journal of Agricultural Research* 42 (3-4): 305-312.

13. Kohler, T., Giger, M., Hurni, H., Ott, C., Wiesmann, U., Wymann, S. and Maselli, D. 2010. Mountains and climate change: A Global Concern. *Mountain Research and Development* 30 (1): 53-55.

14. Li, T.S.C. and Schroeder, W.R. 1996. Seabuckthorn (*Hippophae rhamnoides* L.): A multipurpose plant. *Hort. Technology* 6: 370-380.

15. Li, T.S.C. and Beveridge, T.H.J. (with contributions by B.D. Oomah, W.R. Schroeder and E. Small) 2003. Seabuckthorn (*Hippophae rhamnoides* L.): Production and utilization. NRC Research Press, Ottawa, Ontario, 133p.

16. Lu, R. 1992. Seabuckthorn: A multipurpose plant species for fragile mountains. Int. Centre for Integrated Mountain Development, Kathmandu, Nepal, 62p.

17. Mc-Cullough, B.D. and Heiser, D.A. 2008. On the accuracy of statistical procedures in Microsoft Excels 2007, Computational statistics and data analysis.

18. MoAFF. 2001. Seabuckthorn Fact Sheet. Ministry of Agriculture, Food and Fisheries. 162 Oriole Rd, Kamloops, British Colombia.

19. MoSTE. 2015. Indigenous and local knowledge and practices for climate resilience in Nepal: Mainstreaming climate change risk management in development. Ministry of Science, Technology and Environment (MoSTE), Kathmandu, Nepal.

20. Page, A.L., Miller, R.H. and Keeney, D.R. 1982. Method of soil analysis Part–II chemical and microbial properties. *Am. Soc. Agron.*, Madison, U.S.A.

21. Robbins, C.W. and Carter, D.L. 1980. Nitrate-nitrogen leached below the root zone during and following alfalfa. *J. Environ. Qual.* 9: 447-450.

22. Roots of Peace. 2008. Soil Testing: Perennial crop support series Jalalabad, Afghanistan. Publication No. 2008-001-AFG, p. 100.

23. Whiteman, P.T.S. 1985. Mountain Oases: A technical report of agricultural studies in Gilgit district, Northern Areas, Pakistan, pp. 28-40, FAO/UNDP.

Part VII

ANIMAL HUSBANDRY

Chapter 65

Effect of Seabuckthorn Cake on Rumen Metabolites and Microbial Population in Cattle

M. Hasanuzzaman*, V. K. Sharma, D. Rani and A. Parmar

Department of Animal Nutrition, COVAS, CSK HPKV, Palampur, Himachal Pradesh, India
(*E-mail: anikzaman@yahoo.com)

SUMMARY

The present experiment was aimed to see the effect of seabuckthorn (SBT) cake on rumen metabolites and microbial population in cattle. Three adult fistulated cattle (BW~250 kg) were maintained on three dietary combinations. The crude protein contents of the basal diet (control, T_1) were replaced with SBT cake CP at 50 (T_2) and 75 (T_3) per cent CP in the bull ration. In addition, green grasses and water was offered *ad-libitum* to all the treatment groups. The pH of the rumen liquor varied from 6.10±0.06 to 7.37±0.03. The highest and lowest values of pH were found at 12 h and 24 h in all the treatments. The TVFA concentration in rumen liquor varied 18.33±0.88 to 71.67±1.20 mEq/dl and reached to the peak levels at 24 h, 4 h, and 16 h post feeding in T_1, T_2 and T_3 diet treatment groups. The NH_3-N concentration in rumen liquor ranged from 14.00±0.58 to 30.67±0.67 mg/dl, with the highest concentration recorded at 4h post feeding in T_1 diet, 2 h post feeding in T_2 and T_3 diet, and lowest concentration was observed at 20 h post feeding in all the treatments. The total-N concentration in rumen liquor ranged from 95.20±5.83 to 247.93±7.06 mg/dl, with the highest concentration recorded at 12 h, 16 h, and 12 h post feeding in T_1, T_2 and T_3 diet, respectively and lowest concentration was observed at 24 h post feeding in T_1 and T_2, and 0 h post feeding in T_3 diet. The TCA-N concentration ranged from 69.58±1.67 to 179.20±5.60 mg/dl, with the highest concentration recorded at 12 h, 16 h, and 12 h post feeding in T_1, T_2, and T_3 diet, respectively and lowest concentration was observed at 24 h post feeding in all the treatments. The NPN concentration ranged from 12.58±4.47 to 68.73±2.06 mg/dl with the highest concentration recorded at 12 h post feeding in all the treatments and the lowest concentration was recorded at 24 h post feeding in T_1, and 0 h post feeding in T_2 and T_3 diets. The bacterial population (cellx10^{10}) ranged from 0.93±0.07 to 11.13±0.13 with peak level at 16 h, 20 h and 2 h post feeding and lowest values found at 24 h, 20 h and 24 h post feeding in T_1, T_2 and T_3 diets, respectively. The rumen protozoal population (cellx10^6) ranged from 1.33±0.03 to 4.27±0.03 per ml being highest at 12 h post feeding in T_1 and T_2 diet, and 20 h of post feeding in T_3 diet. It is concluded that T_2 diet ensured better rumen metabolites for growth and multiplication of rumen bacteria, protozoa.

***Keywords**: Seabuckthorn cake, Rumen metabolites, Microbial population and Cattle.*

INTRODUCTION

Ruminants solely depend on microbial population of reticulo-rumen for converting the plant biomass to high biological value animal products like milk and meat. The microbiology of the different segments of the digestive tract, especially the rumen is extremely complex due to the presence of large number of micro-organisms, the shifting populations and interactions among the various microbial groups. As the result of conductive environment in the rumen, anaerobic microorganism like bacteria (Hungate, 1966), fungi (Samanta and Wali, 1999) and protozoa (Agarwal *et al.*, 2000) grow and multiply in the system to secrete a huge array of enzymes for deriving the energy, protein and other nutrients present in complex plant bio-molecules. The populations of rumen bacteria, protozoa or protozoa are influenced by several factors like diet quality and quantity, frequency of feeding and feed additives

Seabuckthorn (*Hippophae* L.): *A Multipurpose Wonder Plant, Vol. 5: Advanced Technologies for Cultivation, Processing, Health Protection and Environmental Conservation* (Virendra Singh, *Editor-in-Chief*, 2018), p. 509–514, Astral International Pvt. Ltd., New Delhi, India.

(Hungate, 1966; Samanta and Wali, 1995). Similarly, the rumen metabolites concentration *viz.*, volatile fatty acids and different nitrogen fractions are governed by nature and quality of animal feed and post feeding interval (Chaudhary *et al.*, 1995; Santra and Karim, 2001; Samanta *et al.*, 2003).

MATERIALS AND METHODS

Animals, Feeds and Feeding

Three adult male crossbred cattle weighing around 250 kg, fitted with permanent rumen fistula were taken for the experimentation and de-wormed with broad spectrum anthelmintic. The animals were maintained on three combination of diet as in Table 65.1.

Table 65.1: Experimental Plan

Particulars	T_1 (Control)	T_2	T_3
Concentrate	100	65.40	48.10
Seabuckthorn cake	--	34.60	51.90

The concentrate ration was offered @1 per cent of the body weight in all dietary combinations of which 50 per cent was offered in the morning while rest offered in the evening. Besides, all the diets included *ad-libitum* grass and wheat bran access. Water was offered *ad-libitum* throughout the day. After an adaptation period of one month, the rumen liquor was collected at different post feeding intervals to monitor various rumen metabolites along with microbial population.

Table 65.2: Physical Composition of Bull Ration (per cent DM basis)

Particulars	T_1	T_2	T_3
Maize	35.00	24.00	23.60
Molasses	06.00	07.00	08.50
Mustard Oil Cake	15.00	05.00	02.00
Rice Polish	10.00	11.00	05.00
GNC	20.00	07.40	02.00
DORB	12.00	09.00	05.00
Mineral Mixture	02.00	02.00	02.00
SBT cake	–	34.60	51.90
Total	100	100	100

The natural grass was harvested at matured stage and chaffed in electrical chaff cutter having the chop length of 3.5 to 4.5 cm. The chemical composition of natural grass, wheat bran and the concentrate feed ingredients along with SBT cake were analyzed as per AOAC (2006). The physical and chemical composition of concentrate mixture and roughages are shown in Tables 65.2 and 65.3.

Analysis of Rumen Liquor

Rumen liquor was collected at 0, 2, 4, 8, 12, 16, 20 and 24 h of post feeding for consecutive three days. Immediately after collection, the rumen liquor was strained through four layer of muslin cloth and pH was recorded in a digital pH meter. Ammonia-N concentration (Conway, 1957), total volatile fatty acids (Barnett and Reid, 1957), TCA precipitable-N (Agarwal, 1999) and Total-N (AOAC, 1995) were determined in the rumen liquor samples. NPN (Soluble-N) was calculated after subtracted TCA-ppt-N from total-N of rumen liquor.

Microbial Population

Rumen liquor collected at different post feeding intervals was filtered through four layered muslin cloth and kept in a thermos where temperature was maintained at 39°C. The stored rumen liquor was further diluted in the anaerobic diluents; for bacterial inoculums the dilution was done up to 10^6. The protozoal motility was observed after taking 1 drop of liquor over a slide covered with cover slip under low power microscope. Hungate's roll tube technique (Hungate, 1966) as adopted by Samanta *et al.* (1999) was followed for enumeration of rumen bacteria. Rumen protozoa were identified according to Hungate (1966). The number of protozoa was calculated as described by Kamra *et al.* (1991).

Statistical Analysis

The data obtained from the experiment was analyzed as per the standard methods of statistical analysis (Snedecor and Cochran, 1967). The difference between the treatments means was tested by applying Duncan's multiple range test (DMRT, 1955).

RESULTS AND DISCUSSIONS

Rumen Metabolites

The diurnal variations different of rumen metabolites in cattle at different hours of feeding are presented in Table 65.4. The pH of the rumen liquor varied from 6.10±0.06 to 7.27±0.13, 6.10±0.06 to 7.37±0.03, and 6.80±0.06 to 7.30±0.17 in T_1, T_2 and T_3 group, respectively. The highest and lowest values of

Table 65.3: Chemical Composition of Bull Ration (per cent DM Basis)

Particulars	CP	EE	CF	T.ASH	AIA	NFE	Ca	P	ME
Control	18.01	05.38	05.06	06.69	04.79	61.00	01.42	00.87	2901
SBT 50	18.07	06.85	03.61	07.94	06.34	45.00	02.29	01.40	2924
SBT 75	18.42	06.96	02.73	08.03	07.00	38.00	02.90	01.65	2877

Table 65.4: Effect of Diet and Time on various Rumen Metabolites in cattle

Rumen Metabolites	*Treatment*	*Hours Post Feeding*						
		2 h	*4 h*	*8 h*	*12 h*	*16 h*	*20 h*	*24 h*
pH	T_1	6.83^{a}±0.09	6.63^{a}±0.09	6.17^{a}±0.09	6.10^{a}±0.06	6.30^{a}±0.06	6.90^{a}±0.06	7.27±0.13
	T_2	6.73^{a}±0.07	6.73^{a}±0.07	6.80^{b}±0.06	6.10^{a}±0.06	6.10^{a}±0.06	6.40^{b}±0.06	7.37±0.03
	T_3	7.23^{b}±0.12	7.17^{b}±0.09	7.00^{b}±0.00	6.80^{b}±0.06	6.93^{b}±0.03	6.83^{a}±0.03	7.30±0.17
	Level of sig.	*	*	***	***	***	**	NS
TVFA (mEq/L)	T_1	44.00^{a}±0.58	30.67^{a}±0.67	29.67^{a}±0.33	22.33^{a}±0.33	18.33^{a}±0.88	50.33^{a}±0.33	57.00^{a}±0.58
	T_2	51.00^{b}±0.58	71.67^{b}±1.20	63.67^{b}±0.33	65.00^{b}±0.00	59.67^{b}±0.88	70.67^{b}±0.67	53.67^{b}±0.33
	T_3	41.00^{c}±0.58	38.00^{c}±0.58	65.00^{b}±0.58	66.67^{b}±0.88	68.33^{c}±0.88	27.00^{c}±0.00	52.67^{b}±0.33
	Level of sig.	*	**	***	***	**	***	**
NH_3-N (mg/dl)	T_1	26.00^{a}±0.58	30.67^{a}±0.67	24.00^{a}±0.58	21.67^{a}±0.88	14.33±0.33	14.00±0.58	14.33^{a}±0.33
	T_2	30.00^{b}±1.16	20.67^{b}±0.67	29.67^{b}±0.33	16.33^{b}±0.33	16.00±0.58	14.33±0.33	20.67^{b}±0.67
	T_3	28.00^{ab}±0.58	23.67^{c}±0.33	26.00^{a}±0.58	22.33^{a}±0.33	14.33±0.33	14.33±0.88	21.00^{b}±0.58
	Level of sig.	*	*	**	**	NS	NS	***
Total-N (mg/dl)	T_1	139.41^{a}±4.86	186.67^{a}±9.33	205.76^{a}±11.95	247.93^{a}±7.06	224.00^{a}±3.23	198.21±6.19	117.26±6.92
	T_2	152.90^{a}±3.90	146.62^{b}±4.58	201.52^{a}±2.84	196.51^{b}±8.05	226.12^{a}±4.94	182.59±2.74	115.90±4.78
	T_3	119.38^{b}±2.55	173.86^{ab}±6.79	155.70^{b}±5.36	196.59^{b}±9.24	176.40^{b}±4.28	178.52±9.22	114.29±7.89
	Level of sig.	*	*	*	*	***	NS	NS
TCA-N (mg/dl)	T_1	119.89±11.76	130.33^{a}±4.07	157.57^{a}±6.29	179.20^{a}±5.60	159.18^{a}±3.61	154.59^{a}±2.74	108.35^{a}±2.49
	T_2	103.77±2.89	107.33^{b}±4.67	130.92^{b}±2.53	124.90^{b}±3.90	173.68^{a}±4.20	158.24^{a}±2.87	85.95^{b}±2.39
	T_3	89.85±6.80	119.30^{ab}±1.96	120.40^{b}±4.28	135.59^{b}±2.28	133.81^{b}±2.15	132.34^{b}±6.31	69.58^{c}±1.67
	Level of sig.	NS	*	*	***	**	*	**
Soluble-N or NPN (mg/dl)	T_1	19.51^{a}±8.50	56.34±11.91	48.19^{a}±6.97	68.73±2.06	64.82^{a}±3.61	43.61^{a}±4.30	12.58^{a}±4.47
	T_2	49.13^{b}±5.98	39.28±8.01	70.60^{b}±2.20	71.61±4.31	52.44^{b}±0.76	24.35^{b}±0.98	29.95^{ab}±3.31
	T_3	29.53^{ab}±4.62	54.56±6.14	35.30^{a}±1.96	61.01±7.97	42.59^{b}±2.21	46.18^{a}±3.72	44.71^{b}±8.16
	Level of sig.	*	NS	*	NS	*	*	*

a,b,c Values bearing different superscripts in a column within the same hour differ significantly; *P<0.05, **P<0.01, ***P<0.001.

pH were found at 12 h and 24 h in all the treatments. Mahouchi *et al.* (2003) reported lowest value of rumen pH in sheep at 1-2 h post feeding. In case of buffalo, the pH of the rumen represented significantly lower levels at 4 h post feeding either in presence or absence of protozoa (Chaudhary *et al.*, 2000; Kamra *et al.*, 2000). Grazing on either natural grassland or silvi-pasture system maintained a rumen pH of 7.30 to 7.96 (Samanta *et al.*, 1998). The other experiment by Samanta *et at.* (2005) showed that the cattle's grazing on natural pasture, the rumen pH was always highest above 7.0 irrespective of post feeding intervals. In the present investigation, T_3 group was highest irrespective of post feeding intervals as compared to other all groups. It might be due to higher secretion of alkalizing agents through saliva.

Total volatile fatty acids (TVFA) concentration of rumen liquor varied 18.33±0.88 to 57.00±0.58, 47.33±0.33 to 71.67±1.20, and 21.00±1.00 to 68.33±0.88 mEq/dl and reached to the peak levels at 24 h, 4 h, and 16 h of post feeding in T_1, T_2 and T_3 diet, respectively. This might be due to the greater availability of fermentable carbohydrates for their rapid and faster metabolism into volatile fatty acids at those particular feeding intervals. Irrespective of post feeding intervals, the rumen TVFA concentration was significantly (P<0.05) maximum in T_2 diet followed by T_3 and T_1while taking consideration of the diet only. Hence the animals on SBT cake based diet lead towards higher TVFA production. By virtue of huge microbial population, ruminants ferment the plant complex bio-molecules in to volatile fatty acids for deriving the energy. Though the concentration of VFA is highly dependent on the nature and type of diet, however, it generally varies from 60 to 120 mmol/litre (Houtert, 1993). In case of buffalo, Santra *et al.* (1996) reported the highest TVFA concentration (12.65 m mol/dl SRL) at 4 h post feeding. Devasena *et al.* (2004) recorded that the TVFA concentration ranges from 68.8 to 101.8 in sheep and goat whereas Pandya *et al.* (2009) reported that the TVFA concentration in SRL varies from 74 to 126 mEq/L in calves.

The rumen ammonia concentrate represents the balance between the degradation of nitrogenous compounds (including protein and non-protein nitrogenous compounds) and uptake by the microbes for the synthesis of microbial protein. In the present experiment, the rumen ammonia nitrogen (NH_3-N) concentration ranged from 14.00 ± 0.58 to 30.67 ± 0.67, 14.33 ± 0.33 to 30.00 ± 1.16, and 14.33 ± 0.88 to 28.00 ± 0.58

mg/dl in T_1, T_2 and T_3 treatment groups respectively. It was higher than the minimum values (5-7 mg/dl) for optimum microbial activity (Satter and Slyter, 1974) in SBT diet. The average range of NH_3-N is in corroborated with the findings of Samanta *et al.* (2006) and Devasena *et al.* (2004). The rumen NH_3-N concentration peaked ($P<0.05$) at 4h of post feeding in T_1 diet and 2 h of post feeding in T_2 and T_3 diet which is corroborated with the findings of Dahiya *et al.* (2011) in the calves.

The total-N concentration in rumen liquor ranged from 117.26±6.92 to 247.93±7.06, 115.90±4.78 to 226.12±4.94, and 95.20±5.83 to 196.59±9.24 mg/dl in treatment T_1, T_2 and T_3 groups respectively. Highest concentration recorded at 12 h, 16 h, and 12 h of post feeding in T_1, T_2 and T_3 diet and lowest N-concentration was observed at 24 h of post feeding in T_1 and T_2. This might be due to slow degradation of dietary protein and hydrolysis of NPN substances in the rumen. The subsequent changes of total-N concentration were observed as a zigzag manner might be due to changes in the rumen volume through inflow of saliva as well as outflow of digesta (Grubb and Dehority, 1975). The total-N concentration in the SRL is corroborated with the findings of Samanta *et al.* (2006), Rekha *et al.* (2006) and Lailer *et al.* (2011).

The TCA precipitable nitrogen (TCA-N) concentration ranged from 108.35±2.49 to 179.20±5.60, 85.95±2.39 to 173.68±4.20, and 69.58±1.67 to 135.59±2.28 mg/dl, with the highest concentration recorded at 12 h, 16 h, and 12 h of post feeding in T_1, T_2, and T_3 diet, respectively. The lowest concentration was observed at 24 h of post feeding in all the treatments. The concentration of TCA-N in SRL found higher than the other researcher might be due to the optimum proportion of N and energy provided by SBT cake required for microbial protein synthesis.

The Soluble-N (NPN) concentration ranged from 12.58±4.47 to 68.73±2.06, 26.22±4.06 to 71.61±4.31, and 19.13±5.14 to 61.01±7.97 mg/dl, in T_1, T_2, and T_3 diet, respectively with the highest concentration recorded at 12 h of post feeding in all the treatment groups. The lowest concentration was recorded at 24 h, 20 h and 2 h post feeding in T_1, T_2 and T_3 groups respectively. The NPN concentration in SBT diet group was found always higher inspite of few cases than the control group. It might be due to the synchronized availability of energy and microbial protein in the rumen. These results supported by the findings of Lailer *et al.* (2011) and Pandya *et al.* (2009).

Rumen Microbial Population

The microbial population (cellx10^{10}) in case of rumen bacteria ranged from 0.93±0.07 to 6.13±0.13, 6.03±0.03 to 11.13±0.13, and 2.80±0.06 to 6.20±0.10 per ml of SRL in T_1, T_2 and T_3 diets, respectively. The bacteria population attained peak level at 16 h, 20 h and 2 h of post feeding and lowest values found at 24 h, 20 h and 24 h of post feeding in T_1, T_2, and T_3 diets, respectively. The population of bacteria was higher both in T_2 and T_3 diet might be due to the positive balance of protein and energy in rumen where SBT cake was added. These results were similar as found by Rekha *et al.* (2006), Thakur (2006) and Chandanshive *et al.* (2007).

The rumen protozoal population (x10^6) ranged from 1.33±0.03 to 2.77±0.03, 1.53±0.03 to 2.70±0.00, and 1.87±0.03 to 4.27±0.03 per ml of SRL, respectively being highest at 12 h of post feeding in T_1 and T_2 diet, and 20 h of post feeding in T_3 diet. The concentration of protozoal population in the SRL was supported by the Murug (2007).

CONCLUSION

The highest and lowest values of pH were found at 12 h and 24 h in all the treatments. The TVFA concentration in rumen liquor reached to the peak levels at 24 h, 4 h, and 16 h post feeding in T1, T2 and T3 diet treatment groups. The NH3-N concentration in rumen liquor reached the highest concentration at 4h post

Table 65.5: Effect of Diet and Time on various Rumen Microbes in Cattle

Rumen Microbes	*Treatment*	*Hours Post Feeding*						
		2 h	*4 h*	*8 h*	*12 h*	*16 h*	*20 h*	*24 h*
Bacteria (Cell x 10^{10})	T_1	1.60^a±0.06	1.80^a±0.06	2.10^a±0.06	1.30^a±0.06	6.13^a±0.13	1.43^a±0.03	0.93^a±0.07
	T_2	10.27^b±0.27	7.13^b±0.09	9.67^b±0.03	9.40^b±0.21	9.17^b±0.17	6.03^b±0.03	11.13^b±0.13
	T_3	6.20^c±0.10	3.10^c±0.06	3.10^c±0.06	5.10^c±0.10	3.40^c±0.21	4.03^c±0.03	2.80^c±0.06
	Level of sig.	***	***	*	***	***	***	***
Protozoal Motility	T_1	++++	++++	++++	++++	+++	+++	++
	T_2	++++	++++	++++	+++	+++	+++	+++
	T_3	++++	++++	++++	++++	++++	++++	++++
Protozoa (Cell x 10^6)	T_1	1.33^a±0.03	2.20^a±0.10	1.37^a±0.07	2.77^a±0.03	1.53^a±0.03	2.13^a±0.03	1.87^a±0.03
	T_2	1.60^b±0.06	1.93^a±0.03	2.10^b±0.06	2.70^a±0.00	2.60^b±0.06	2.37^b±0.03	1.63^b±0.03
	T_3	1.87^c±0.03	4.13^b±0.07	2.80^c±0.06	2.03^b±0.03	4.30^c±0.15	4.27^c±0.03	2.87^c±0.03
	Level of sig.	*	***	***	***	***	**	**

a,b,c Values bearing different superscripts in a column within the same hour differ significantly; *P<0.05, **P<0.01, ***P<0.001.

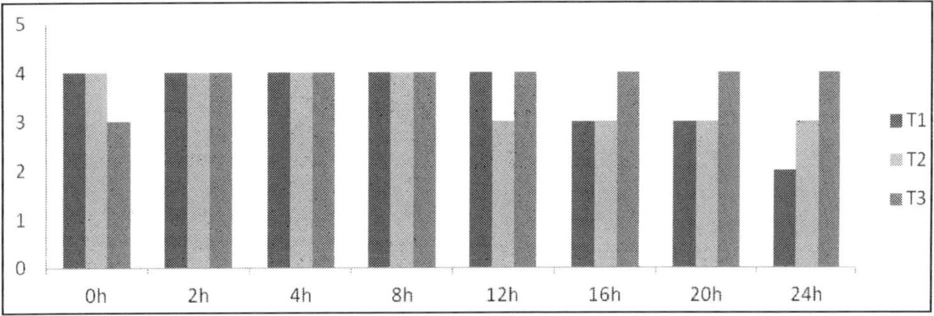

Figure 65.1: Effect of SBT Cake on Protozoa Motility in Cattle.

feeding in T1 diet, 2 h post feeding in T2 and T3 diet, and lowest of that was observed at 20 h post feeding in all the treatments. The total-N concentration in rumen reached highest concentration at 12 h, 16 h, and 12 h post feeding in T1, T2 and T3 diet, respectively and the lowest was observed at 24 h post feeding in T1 and T2, and 0 h post feeding in T3 diet. The TCA-N concentration reached to the highest concentration at 12 h, 16 h, and 12 h post feeding in T1, T2, and T3 diet, respectively and lowest was observed at 24 h post feeding in all the treatments. The highest NPN concentration recorded at 12 h post feeding in all the treatments and the lowest was at 24 h post feeding in all the treatment groups. The bacterial population (cellx1010) reached peak level at 16 h, 20 h and 2 h post feeding and lowest values found at 24 h, 20 h and 24 h post feeding in T1, T2 and T3 diets, respectively. The rumen protozoan population (cellx106) reached peak at 12 h post feeding in T1 and T2 diet, and 20 h of post feeding in T3 diet. It is concluded that T2 diet ensured better rumen metabolites for growth and multiplication of rumen bacteria, protozoa.

REFERENCES

1. Agarwal, N. 1999. Rumen liquor analysis. *Laboratory Manual of Animal Nutrition*. Division of Animal Nutrition. Indian Veterinary Research Institute, Izatnagar, UP, India, pp. 174-180.
2. Agarwal, N., Agarwal, I., Kamra, D. N. and Chaudhary, L. C. 2000. Diurnal variations of hydrolytic enzymes in different fractions of rumen contents of Murrah buffalo. *Journal of Applied Animal Research* 18: 73-80.
3. AOAC. 1995. Official Methods of Analysis, 16th edition, Association of Official Analytical Chemists, Washington, DC, USA.
4. AOAC. 2006. Official Methods of Analysis, 18th edition, Association of Official Analytical Chemists, Maryland.
5. Barnett, A. J. G. and Reid, R. L. 1957. Studies on the estimation of volatile fatty acids. *J. Agric. Sci.* 48: 315-324.
6. Chandanshive, S. V., Mani, V., Kaur, H. and Kewalramani, N. 2007. Zinc supplementation to ameliorate the adverse effect of Cadmium on rumen fermentation in cattle. *Indian J. Anim. Nutr.* 24 (2): 67-71.
7. Chaudhary, L. C., Kamra, D. N., Singh, R., Agarwal, N. and Pathak, N. N. 2000. Effect of feeding bentonite on rumen fermentation, enzyme activity and protozoal population in adult buffaloes (*Bubalus bubalis*). *Buffalo Journal* 1: 73-79.
8. Chaudhary, L. C., Srivastava, A. and Singh, K. K. 1995. Rumen fermentation pattern and digestion of structural carbohydrates in buffalo (*Bubalus bubalis*) calves as affected by ciliate protozoa. *Anim. Feed Sci. Tech.* 56: 111-117.
9. Conway, E. J. 1957. Micro-Diffusion Analysis and Volumetric Error, 4th edition, Cross by Lockwood and sons Ltd., London, England.
10. Devasena, B., Haritha, K., Ravi, A. and Prasad, R. 2004. Evaluation of fruit wastes in small ruminants. *Indian J. of Anim. Nutr.* 21 (2): 85-88.
11. Duncan, D.B. 1955. Multiple range and F tests. *Biometrics* 11: 1-42.
12. Grubb, J. A. and Dehority, B. A. 1975. Effect of an abrupt change in ration from all roughages to high concentrate diet upon rumen microbial number in sheep. *Applied Microbiol.* 30: 404-415.
13. Houtert, M. F. J. 1993. The production and metabolism of volatile fatty acids by ruminants fed roughages- A review. *Anim. Feed Sci. Tech.* 43: 189-225.
14. Hungate, R. E. 1957. Microorganisms in rumen of cattle fed on constant ration. *Can. J. Microbial.* 3: 289-311.
15. Hungate, R. E. 1966. *The Rumen and Its Microbes*. Academic press, London.
16. Kamra, D. N., Sawal, R. K., Pathak, N. N., Kewalramani, N. and Agarwal, N. 1991. Diurnal variations in ciliate protozoa in the rumen of blackbuck (*Anilope cervicapra*) fed green forages. *Letters Appl. Microbiol.* 13: 299-303.
17. Kamra, D. N., Singh, R., Chaudhary, L. C., Agarwal, N. and Pathak, N. N. 2000. Soap nut as natural

defaunating agent: Its effect on rumen fermentation and in sacco degradability of jowar hay in buffaloes. *Buffalo Journal* 1: 99-104.

18. Lailer, P.C., Dahya, S.S., Lal, D. and Chauhan, T.R. 2011. Rumen fermentation pattern in buffaloes fed complete feed blocks. *Indian J. Anim. Nutr.* 2: 177-180.
19. Mahouchi, M., Haddad, L., Kayouli, C., Thewis, A. and Beckers, Y. 2003. Effects of nature of nitrogen supplement on voluntary intake, rumen parameters and ruminal degradation of dry matter in sheep fed oat silage based diets. *Small Ruminant Research* 48: 181-187.
20. Pandya, P. R., Desai, M. C., Patel, G. R., Talpada, P. M., Pande, M. B. and Parnerkar, S. 2009. Economical rearing of crossbred calves on complete feeds, based on sugarcane bagasse and non conventional feeds. *Indian J. Anim. Nutr.* 26 (3): 211-215.
21. Rekha, N. A., Prasad, J. R., Ramana, J. V. and Ravi, A. 2006. Evaluation of groundnut haulms based complete rations with or without yeast culture in sheep. *Indian J. Anim. Nutri.* 23 (1): 1-4.
22. Samanta, A. K. and Wali, T. K. 1995. Rumen anaerobic fungi and their role in fiber degradation-A review. *Indian J. Dairy Sci.* 48: 179-184.
23. Samanta, A. K. and Wali, T. K. 1999. Enumeration and characterization of rumen anaerobic fungi in crossbred cattle. *Indian J. Anim. Nutr.* 16: 89-93.
24. Samanta, A. K., Singh, K. K., Das, M. M. and Kundu, S. S. 2006. Rumen fermentation pattern in crossbred cattle maintained on complete feed block. *Indian J. Anim. Nutri.* 23 (1): 50-52.
25. Samanta, A. K., Singh, K. K., Das, M. M., Maity, S. B. and Kundu, S. S. 2003. Effect of complete feed block on nutrient utilization and rumen fermentation in Barbari goats. *Small Ruminant Research* 48: 95-102.
26. Samanta, A. K., Singh, K. K., Das, M. M., Pailan, G. H., Verma, N. C. and Rai, S. 2005. Influence of roughage source on rumen metabolites and microbial population with special reference to anaerobic fungus in cattle. *Anim. Nutri. Feed Tech.* 5: 179-193.
27. Samanta, A. K., Upadhyay, V. S. and Pathak, P. S. 1998. Effect of silvipasture system on rumen fermentation and growth in crossbred heifers. *Indian J. Anim. Nutr.* 15: 18-20.
28. Samanta, A. K., Wali, T. K., Batish, V. K., Grover, S., Rajput, Y. S. and Mahanty, A.K. 1999. Characterization of anaerobic fungi in the rumen of riverine buffalo. *Indian J. Anim. Nutr.* 16: 275-278.
29. Santra, A. and Karim, S. A. 2001. Nutrient utilization, rumen fermentation characteristics and ciliate protozoa population in sheep and goat under stall feeding. *Indian J. Anim. Sci.* 71: 852-856.
30. Santra, A., Kamra, D. N. and Pathak, N. N. 1996. Influence of ciliate protozoa on biochemical changes and hydrolytic enzymes in the rumen of Murrah buffalo (*Bubalus bubalis*). *Buffalo Journal* 1: 95-100.
31. Satter, L. D. and Slyter, L. L. 1974. Effect of ammonia concentration on rumen microbial protein production *in vitro*. *Br. J. Nutr.* 32: 199-208.
32. Snedecor, G. W. and Cochran, W. G. 1967. Statistical Methods, 8th edition. Oxford and IBH publishing company, Calcutta.
33. Thakur, S. S., Tomar, S. K. and Malik, R. 2006. Effect of feeding total mixed rations on the performance of buffalo calves. *Indian J. Anim. Nutri.* 23 (1): 5-9.

Chapter 66

Evaluation of Efficacy of Seabuckthorn Leaves (*Hippophae* L.) as Protein Replacer in Broiler (Poultry) Diet and its Effect on Cost of Production

S.K. Ambatkar[1], V.K. Sharma[2*] and D. Wadhwa[2]

[1]Institute of Veterinary, Animal and Biomedical Sciences, Palmerston North, New Zealand
[2] Dr. G.C. Negi COVAS, CSKHPKV, Palampur, Himachal Pradesh, India
(*E-mail: vkcovas@gmail.com)

SUMMARY

There is acute shortage of feed stuffs to meet the demand of ever increasing population of livestock. As a nutritionist we have to exploit the more and more non-conventional feed resources (NCFR) to add in the list of feed stuffs to bridge the gap between the demand and supply. In this study an attempt was made to exploit the usefulness of the seabuckthorn leaves for poultry broiler production as protein replacer of the diet and the crude protein of basal diet of the poultry broilers was replaced at 3, 6, 9, 12 and 15 per cent levels of protein with the proteins of Seabuckthorn leaves. The leaves were accepted by the poultry birds at all the offered levels. The treatment groups T-3 and T-6 followed by T-9 perform significantly better than the T-C (Control) treatment group and other groups T-12 and T-15 as FCR in other groups was significantly higher ($P<0.05$) than T-3(1.40±0.03), T-6(1.44±0.03) and T-9(1.62±0.05).

***Keywords**: Seabuckthorn leaves, NCFR (Non-conventional feed resources), Protein replacer and Broiler production.*

ABBREVIATIONS

AIA: Acid insoluble ash, AOAC: Association of Official Analytical Chemist, BIS: Bureau of Indian Standards, Ca: Calcium, CF: Crude Fibre, CP: Crude protein, DM: Dry Matter, EE: Ether extract, wt.: Weight, g.: gram(s), FI: Feed Intake, Fig.: Figure, GIW: Gain in weight, I.U.: International units, Kcal.: Kilo Calories, Kg: Kilo Grams, ME: Metabolisable Energy, NFE: Nitrogen free extract, P: Phosphorus, P<0.05: Significant at 5 per cent level, p.: Page, pp.: Pages, FCR: Feed conversion ratio.

INTRODUCTION

The poultry industry has been one of the key contributors for animal production. India has become the second largest producer of eggs in the world (The Poultrysite, 2012) and fifth largest of broilers, producing some 2.3 million metric tons of meat and 47 billion eggs. Over recent years, poultry meat has been the fastest growing animal protein in India. Poultry is one of flourishing segments of the agriculture sector and feed is one of the largest items for the poultry production. Balanced and effective feeding is the most important requisite to a superior germplasm for economic production.

There is a deficit of dry, green fodder and concentrates for livestock feeding and more than 20 per cent livestock population do not get minimum dry matter requirement in different parts of India. To bridge this gap, alternative feeds for livestock and poultry must be discovered to meet the requirements. This can be possible by exploiting the NCFR (non- conventional feed resources) by use of some available plants and other agriculture by-products as feed for livestock and

Seabuckthorn (Hippophae L.): A Multipurpose Wonder Plant, Vol. 5: Advanced Technologies for Cultivation, Processing, Health Protection and Environmental Conservation (Virendra Singh, *Editor-in-Chief*, 2018), p. 515–525, Astral International Pvt. Ltd., New Delhi, India.

poultry. One of such plant which can be exploited as NCFR is "Seabuckthorn".

Seabuckthorn (*Hippophae* sp. of Family *Elaeagnaceae)* is a deciduous, thorny and nitrogen fixing plant (shrub) or a small tree, 2-4 m in height and grows abundantly in the dry temperate Himalayas-Lahoul-Spiti. Common seabuckthorn has branches that are dense, stiff and very thorny, (Rousi, 1971). The leaves are a distinct pale silvery-green, lanceolate, 3–8 cm long and less than 7 mm broad. Seabuckthorn leaves and fruit have been used as natural fresh feed for livestock and poultry for many years the significance has been well recognized in increasing weight, reproductive improvement and increased resistance to various diseases to livestock. This is supported by the experiences of the local herdsmen. In northern Ganehu province, a large area is covered by seabuckthorn foliage, which fattens their livestock (Rongsen, 1994). No systemic work, so far, has been carried out to study the effects of feeding seabuckthorn leaves, fruits, seeds and industrial by-products to the cattle and poultry to evaluate its nutrient utilization for various productive functions. Nutritionally the leaves and fruits are very rich source of the proteins, vitamins and minerals, keeping in view the usefulness of seabuckthorn leaves it will be of great importance to evaluate its nutrient utilization in commercial poultry birds. Feeding of the leaves may meet some amount of the nutrient availability to the commercial broiler birds benefiting the poultry farmer. So the attempt to replace protein of poultry feed with that of the protein of seabuckthorn leaves (*Hippophae rhamnoides* L.) was made in this experiment.

MATERIALS AND METHODS

The research work was carried out in the metabolic stall of the Department of Animal Nutrition, Dr. G.C. Negi, College of Veterinary and Animal Sciences CSK HPKV, Palampur, Himachal Pradesh, India. To achieve the objectives of research programme, the feeding experiments were conducted on 384 one day-old broiler chicks of Cobb strain.

Experimental Plan

To evaluate seabuckthorn leaves as protein replacer for efficient broiler production, the 384 one day old broiler chicks were divided into six groups of 16 chicks in each group with 4 replications. The control group (T-C) concentrate crude protein (CP) was replaced with 3, 6, 9, 12 and 15 per cent of CP of seabuckthorn leaves (*Hippophae* L.) in the T-3, T-6, T-9, T-12 and T-15 treatment groups respectively. The details of the treatments and feed supplementation are stated in Table 66.1.

Samples of feed ingredients like maize, groundnut cake, soybean cake flakes, sunflower cake, mustard cake and de-oiled rice bran were procured from the feed store of the Department of Animal Nutrition. The ingredients were oven dried and ground to pass through 1mm sieve and preserved in self sealed polythene bags. The seabuckthorn leaves were purchased from Shansha village in the Lahoul and Spiti region of Himachal Pradesh. The whole leaves were shed dried and ground to pass through 1mm sieve. All the ingredients were analyzed for dry matter (DM), ether extract (EE), crude fibre (CF), total ash (TA), acid insoluble ash (AIA/sand silica), nitrogen free extract (NFE), calcium (Ca) and phosphorus (P).

(a) Analytical Procedures

The proximate analysis of ingredients and feed (broiler starter and broiler finisher) and the seabuckthorn leaves was done as per the method given in AOAC (1985).

(b) Metabolisable Energy

The ME of different test diets was calculated as per the equation proposed by Lodhi *et al.* (1976) which is given as below:

ME (Kcal/kg) = 32.95 (per cent CP + per cent EE x 2.25 + per cent NFE) - 29.20

(c) Formulation of Test Feeds

The chemical composition of different feed ingredients used for formulation of feed has been given

Table 66.1: Treatments and Feed Supplementation Regime

Treatment Particular	*Labels*	*Supplementation Level*	
		Starter Phase	*Finisher Phase*
Control	T-C	Feed only	Feed only
3 per cent Crude protein of control feed replaced with crude protein of seabuckthorn leaves	T-3	3.46kg seabuckthorn leaves/quintal of feed	3.29kg seabuckthorn leaves/quintal of feed
6 per cent Crude protein of control feed replaced with crude protein of seabuckthorn leaves	T-6	6.92kg seabuckthorn leaves/quintal of feed	6.59kg seabuckthorn leaves/quintal of feed
9 per cent Crude protein of control feed replaced with crude protein of seabuckthorn leaves	T-9	10.39kg seabuckthorn leaves/quintal of feed	9.88kg seabuckthorn leaves/quintal of feed
12 per cent Crude protein of control feed replaced with crude protein of seabuckthorn leaves	T-12	13.84kg seabuckthorn leaves/quintal of feed	13.16kg seabuckthorn leaves/quintal of feed
15 per cent Crude protein of control feed replaced with crude protein of seabuckthorn leaves	T-15	17.31kg seabuckthorn leaves/quintal of feed	16.47kg seabuckthorn leaves/quintal of feed

in Table 66.2. The physical composition of different experimental diets on dry matter basis has been given in the Tables 66.3 and 66.4 for starter and finisher phase of broiler production respectively as per BIS (1992) standard.

Table 66.2: Chemical Composition of Feed Ingredients Used for Formulation of Feed

	DM (Per cent)	*CP (Per cent)*	*EE (Per cent)*	*CF (Per cent)*	*Ash (Per cent)*	*NFE (Per cent)*	*AIA (Per cent)*	*Ca (Per cent)*	*P (Per cent)*	*ME (Kcal/kg)*
Maize	93.01	9.87	3.90	2.20	1.55	82.48	0.39	0.35	0.75	3302.90
Groundnut cake	95.53	37.31	7.90	10.90	6.73	39.16	2.24	0.23	0.67	3076.20
Sunflower cake	93.79	25.63	10.90	10.60	8.29	43.58	1.62	0.43	0.96	3059.40
Mustard cake	91.30	35.10	1.43	9.20	9.02	53.25	1.20	0.89	1.11	2988.00
De-oiled rice bran	91.60	14.73	1.28	14.08	9.92	55.99	3.40	0.05	1.19	2395.90
Soya flakes cake	92.30	33.85	1.10	7.80	8.41	48.84	1.10	0.29	0.64	2777.90
Seabuckthorn leaves	90.06	17.34	10.90	5.17	7.15	59.45	0.02	3.67	0.19	3309.13

Table 66.3: Physical Composition of Experimental Diets for Starter Phase (per cent DM)

Ingredients Parts/100 kg	*T-C*	*T-3*	*T-6*	*T-9*	*T-12*	*T-15*
Maize	40.0	38.62	37.23	35.84	34.46	33.08
Groundnut cake	17.00	16.41	15.82	15.23	14.65	14.06
Sunflower cake	8.00	7.70	7.45	7.17	6.89	6.60
Mustard cake	5.00	4.83	4.65	4.48	4.31	4.13
De-oiled rice bran	10.75	10.38	10.01	9.63	9.26	8.89
Soya bean cake flakes	13.00	12.55	12.10	11.65	11.20	10.75
Molasses	3.00	2.90	2.80	2.69	2.59	2.48
Di-calcium phosphate	1.00	0.97	0.93	0.90	0.86	0.83
Lime powder	1.00	0.97	0.93	0.90	0.86	0.83
Premix*	0.25	0.24	0.23	0.22	0.22	0.21
Sago grain	1.00	0.97	0.93	0.90	0.86	0.83
Seabuckthorn leaves	-	3.46	6.92	10.39	13.84	17.31

Ingredients of Premix*(1). Merivite AB_2D_3K:25g (Vit A-82,500IU, Vit B_2-52 mg, Vit D_3-12000IU, Vit K- 10 mg, PO_4-395 mg/g (2). Nicomix BE: 25g (Vit B_1-4 mg, Vit B_{12}-40μg, Niacin-60 mg, Calcium Pentothenate-40 mg, Vit E-40 mg/g) (3). Trace Minerals: 100g (Ferric Oxide-2 g, Copper sulphate-2 g, Ferrous sulphate-10 g, Zn sulphate-0.6g, Di calcium phosphate-53.65g, Mangnese sulphate- 5.5g, Pottasium iodide-2.5g, Sodium thiosulphate-0.75g, Zinc oxide-1g) (4). Choline chloride: 100g.

Table 66.4: Physical Composition of Experimental Diets for Finsher Phase (per cent DM)

Ingredients Parts/100 kg	*T-C*	*T-3*	*T-6*	*T-9*	*T-12*	*T-15*
Maize	48.00	46.42	44.84	43.26	41.68	40.09
Groundnut cake	11.00	10.63	10.29	9.91	9.55	9.19
Sunflower cake	8.00	7.74	7.47	7.21	6.95	6.69
Mustard cake	5.00	4.84	4.67	4.51	4.34	4.19
De-oiled rice bran	4.75	4.59	4.44	4.28	4.13	3.98
Soya flakes	17.00	16.44	15.88	15.32	14.76	14.20
Molasses	3.00	2.90	2.80	2.70	2.62	2.50
Di-calcium phosphate	1.00	0.97	0.93	0.90	0.86	0.82
Lime powder	1.00	0.97	0.93	0.90	0.86	0.82
Premix*	0.25	0.24	0.23	0.23	0.22	0.21
Sago grain	1.00	0.97	0.93	0.90	0.87	0.84
Seabuckthorn leaves	-	3.29	6.59	9.88	13.16	16.47

Ingredients of Premix*(1) Merivite AB_2D_3K:25g (Vit A-82,500IU, Vit B_2-52 mg, Vit D_3-12000IU, Vit K- 10 mg, PO_4-395 mg/g (2) Nicomix BE: 25g (Vit B_1-4 mg, Vit B_{12}-40μg, Niacin-60 mg, Calcium Pentothenate-40 mg, Vit E-40 mg/g) (3) Trace Minerals: 100g (Ferric Oxide-2 g, Copper sulphate-2 g, Ferrous sulphate-10 g, Zn sulphate-0.6g, Di calcium phosphate-53.65g, Mangnese sulphate- 5.5g, Pottasium iodide-2.5g, Sodium thiosulphate-0.75g, Zinc oxide-1g) (4) Choline chloride: 100g.

(d) Management of Broiler Chicks

After procuring the Cobb strain broiler chicks from Uttam Hatcheries, Kangra, HP, the chicks were kept only on water for first 48 hours and then fed on crushed maize for next two days. They were wing banded, weighed and divided into 16 chicks in each of six treatments and 4 replications, having body weight (39.74±0.57 to 40.69±0.22 g.) with no significant difference (P<0.05). They were kept in 24 compartments which were thermostatically controlled. The standard management practices were followed for rearing the chicks in battery brooders up to 4 weeks of age (Starter phase) and thereafter shifted to Californian cages up to 6 weeks of age (Finisher phase). Clean drinking water and the respective experimental feeds were always available to the chicks during the entire course of study.

Observations Recorded

1. Initial body weight.
2. Body weight at bi-weekly interval.
3. Body weight at 6th week.
4. Feed consumption at bi-weekly interval.
5. Mortality was recorded every day.
6. Dressing per cent at the end of 6th week.
7. Feed conversion ratio and profit per kg gain in live weight were worked out.

Statistical Analysis

The data obtained from the experiment was analyzed using Software Graphpad InStat tm 1990-1994, by applying One Way Analysis of Variance.

RESULTS

Chemical Composition of Feed Ingredients and Test Diet

Data regarding the chemical composition of broiler starter and broiler finisher diets have been given in Tables 66.5 and 66.6 respectively. There was no appreciable difference in the chemical composition of the test diets and small difference was due to the adjustments of the different ingredients to keep the diets iso-proteinous and iso-caloric as far as possible.

Chemical Composition of Starter Phase Diets

The DM of the test diets varied from 89.35 to 89.97 per cent being minimum in T-12 and maximum in T-C

Table 66.5: Chemical Composition of Starter Phase Diets

Components	*T-C*	*T-3*	*T-6*	*T-9*	*T-12*	*T-15*
DM	89.97	89.52	89.53	89.91	89.35	89.89
CP	20.17	20.06	19.96	19.87	19.77	19.67
EE	5.95	4.37	4.59	4.82	5.06	5.30
CF	6.56	6.59	6.64	6.63	6.65	7.32
TA	5.06	5.08	5.08	5.08	5.09	5.09
NFE	62.26	63.90	63.75	63.60	63.43	62.62
AIA	1.24	1.19	1.15	1.11	1.05	1.02
Ca	2.22	2.80	2.78	2.81	2.85	2.81
P	1.00	0.98	0.94	0.90	0.87	0.84
ME (Kcal/kg)	3128	3061	3069	3079	3087	3075
ME:CP	1:155	1:153	1:154	1:155	1:156	1:156

Table 66.6: Chemical Composition of Finisher Phase Diets Chemical Composition of Finisher Phase Diets

Components	*T-C*	*T-3*	*T-6*	*T-9*	*T-12*	*T-15*
DM	89.99	91.73	91.91	91.89	92.11	92.10
CP	19.20	19.12	19.02	18.86	18.82	18.93
EE	3.93	4.16	4.39	4.62	4.84	5.09
CF	5.57	5.62	5.67	5.70	5.77	5.82
TA	4.52	4.57	4.62	4.60	4.62	4.60
NFE	66.78	66.53	66.30	66.26	65.84	65.56
AIA	0.98	0.79	0.90	0.87	0.83	0.71
Ca	2.86	3.15	2.86	2.54	2.92	2.48
P	0.86	0.84	0.81	0.79	0.77	0.74
ME (Kcal/kg)	3095	3101	3107	3117	3128	3123
ME:CP	1:161	1:162	1:163	1:165	1:166	1:165

treatment groups. As far as possible the diets formulated were kept iso-proteinous and iso-caloric. The protein concentration of the test diets ranged from 19.67 to 20.17 per cent, where it was highest in T-C and lowest in T-15 treatment group. Meatbolisable energy values varied from 3061 to 3128 kcal/kg. Energy was maximum in T-C treatment group and minimum in T-3 treatment group. The range of EE was found to be 4.37 to 5.95 per cent. Highest EE content was observed in T-C and lowest EE content was observed in T- 3 treatment group. The CF of the diets ranged from 6.56 to 7.32 per cent, which was maximum in T-15 and minimum in T-C. TA ranged from 5.06 to 5.09 per cent being minimum in T-C and maximum in T-15. NFE in diets ranged from 62.26 to 63.90 per cent, where it was maximum in T-3 and minimum in T-C. The AIA of the diets ranged from 1.02 to 1.24 per cent which was minimum in T-15 and maximum in T-C treatment group. The Ca in the diet ranged from 2.22 to 2.85 per cent being maximum for T-12 and minimum for T-C treatment group. P content in starter ranged from 0.84 to 1.00 per cent, where it was minimum for T-15 and maximum for T-C treatment group.

The DM of the finisher phase diets varied from 89.99 to 92.11 per cent being minimum in T-C and maximum in T-12 treatment groups. As far as possible the diets formulated were iso-proteinous and iso-caloric. The protein concentration of the test diets ranged from 18.82 to 19.20 per cent, where it was highest in T-C and lowest in T-12 treatment group. The metabolisable energy of the diet varied from 3095 to 3128 kcal/kg. Energy was maximum in T-12 treatment group and minimum in T-C treatment group. The range of EE was found to be 3.93 to 5.09 per cent. Highest EE content was observed in T-15 and lowest EE content was observed in T-C treatment group. The crude fibre of the diets ranged from 5.57 to 5.82 per cent, which was maximum in T-15 and minimum in T-C. TA ranged from 4.52 to 4.62 per cent being minimum in T-C and maximum in T-12. The NFE in the diets ranged from 65.56 to 66.78 per cent, where it was maximum in T-C and minimum in T-15. The AIA of the diets ranged from 0.71 to 0.98 per cent which was minimum in T-15 and maximum in T-C treatment group. The Ca in the diet ranged from 2.48 to 3.15 per cent being maximum for T-3 and minimum for T-15 treatment group. P content in finisher phase diets ranged from 0.74 to 0.86 per cent respectively, where it was minimum for T-15 and maximum for T-C treatment group.

Biological Performance of Broilers

Biological Performance of Broilers during Starter Phase

The weight at 4th week, gain in weight (GIW) and feed conversion ratio (FCR) has been presented in Table 66.7. Average weights at 3rd day varied from 39.74±0.57 to 40.69±0.22 g. No significant differences were observed in initial weight of chicks at 3rd day of age. It was found that the average weights at the 4th week varied from 278.35±10.49 (T-15) to 390.09±3.20 g (T-6) treatment group. These values were found to be highly significant ($P<0.05$) among the different groups *viz.*, T-C, T-3, T-6, T-12 and T-15, except for T-C and T-9 treatment group. The average GIW was Lowest in T-15 (247.90±12.37 g) group and highest in T-6 group (349.67±3.20 g). Significantly, ($P<0.05$) highest GIW were observed in T-3 and T-6 treatment group and significantly lowest ($P<0.05$) in T-12 and T-15 treatment groups. Average feed intake at the end of starter phase was highest for group T-C and it was significantly ($P<0.05$) higher than that of T-3, T-6, T-9, T-12, and T-15 treatment groups. No effect of diet was observed in T-3 and T-6 and T-12 and T-15 treatment group. Lowest feed intake was observed in T-15 which was significantly ($P<0.05$) lower than T-C, T-3, T-6 and T-9 treatment groups. FCR for the starter phase ranged from 1.40±0.05 (T-6) to 1.74±0.05 (T-15). Significantly higher FCR was observed in T-C and T-15 treatment groups followed by T-12 treatment group, as compared to T-3, T-6 and T-9 treatment groups. No appreciable effect on FCR in respect T-C and T-15 and T-3 and T-9 treatment was observed.

Biological Performance of Broilers during Finisher Phase

The weight at 6th week, GIW, FI and FCR, has been presented in Table 66.8. The final weight (g) of the starter phase was taken as initial weight at the finisher phase. Average final weights were found to differ significantly ($P<0.05$) among all the treatment groups being minimum 550.17±24.98 g (T-15) and maximum 1045.20±14.65 g

Table 66.7: Biological Performance of Broilers during Starter Phase

Particulars	*T-C*	*T-3*	*T-6*	*T-9*	*T-12*	*T-15*
Initial Wt (g)	40.62 ± 0.20	39.74 ± 0.57	40.42 ± 0.08	40.69 ± 0.22	40.62 ± 0.24	40.45 ± 0.24
Wt at 4th Week (g)	338.53^{a} ± 2.85	368.51^{b} ± 4.93	390.09^{c} ± 3.20	336.92^{ad} ± 10.29	303.85^{e} ± 5.13	278.35^{f} ± 10.49
GIW (g)	297.92^{a} ± 2.77	328.77^{b} ± 4.83	349.67^{c} ± 3.20	296.23^{a} ± 10.46	263.24^{d} ± 5.14	247.90^{e} ± 12.37
FI (g)	514.70^{a} ± 16.83	491.67^{b} ± 12.39	488.90^{b} ± 17.13	450.11^{c} ± 4.58	439.06^{d} ± 6.84	430.78^{d} ± 12.50
FCR	1.73^{a} ± 0.05	1.50^{b} ± 0.08	1.40^{c} ± 0.05	1.52^{b} ± 0.10	1.67^{d} ± 0.06	1.74^{a} ± 0.05

Values bearing different superscripts differ significantly. ($P<0.05$).

Table 66.8: Biological Performance of Broilers during Finisher Phase

Particulars	*T-C*	*T-3*	*T-6*	*T-9*	*T-12*	*T-15*
Wt at 4thWeek(g)	338.53^{a} ± 2.85	368.51^{b} ± 4.93	390.09^{c} ± 3.20	336.92^{ad} ± 10.29	303.85^{e} ± 5.13	278.35^{f} ± 10.49
Final Wt (g)	916.79^{a} ± 16.66	1020.60^{b} ±18.09	1045.20^{c} ± 14.65	874.32^{d} ± 18.11	732.09^{e} ± 20.51	550.17^{f} ± 24.98
GIW (g)	578.26^{a} ± 17.96	652.09^{b} ± 14.85	655.11^{b} ± 16.28	537.40^{c} ± 11.24	428.24^{d} ± 11.14	271.82^{e} ± 17.39
FI (g)	1131.80^{a} ± 12.13	882.77^{b} ± 5.87	956.80^{c} ± 10.65	900.75^{d} ± 16.75	990.07^{e} ± 23.89	732.46^{f} ± 9.50
FCR	1.96^{a} ± 0.12	1.35^{b} ± 0.02	1.46^{c} ± 0.07	1.68^{d} ± 0.06	2.31^{e} ± 0.15	2.70^{f} ± 0.24

Values bearing different superscripts differ significantly. (P<0.05).

(T-6). Average GIW at the end of the finisher phase found to be highly significant (P<0.05) among the different treatment groups except T-3 and T-6 and they varied from minimum 271.82±17.39 (T-15) to maximum 655.11±16.28 g (T-6) group. Feed intake on an average during finisher phase varied from 732.46±9.50 being minimum in T-15 to 1131.80±12.13 g being maximum in T-C treatment group. Feed intake was highly significant (P<0.05) among all the treatments. The FCR differed significantly (P< 0.05) among all the treatment groups. The minimum and maximum FCR values found in case of T-3 (1.96±0.12) and T-15 (2.70±0.24) respectively, which were highly significant (P<0.05).

Overall Biological Performance of Broilers

The overall final weight, GIW, FI, FCR, mortality and dressing percentage have been presented in Table 66.9. Graphs for GIW, FI and FCR for all phases are presented in Figures 66.1–66.3 respectively. Overall GIW ranged from minimum 509.72±14.62 (T-15) to maximum 1004.78±20.12 g (T-6) group. Overall GIW were highly significant (P<0.05) among different treatment groups. Average overall feed intake was highest in T-C and lowest in T-15 treatment group. These values were also found to be highly significant (P<0.05) among the different treatment groups. FCR for overall phase of broiler production from 3rd day to 6th week varied from 1.40±0.03 minimum in T-3 to 2.28±0.14 maximum

Table 66.9: Overall Biological Performances of Broilers

Particulars	*T-C*	*T-3*	*T-6*	*T-9*	*T-12*	*T-15*
Wt at 4thWeek(g)	338.53^{a} ± 2.85	368.51^{b} ± 4.93	390.09^{c} ± 3.20	336.92^{ad} ± 10.29	303.85^{e} ± 5.13	278.35^{f} ± 10.49
Final Wt (g)	916.79^{a} ± 16.66	1020.60^{b} ±18.09	1045.20^{c} ± 14.65	874.32^{d} ± 18.11	732.09^{e} ± 20.51	550.17^{f} ± 24.98
GIW (g)	578.26^{a} ± 17.96	652.09^{b} ± 14.85	655.11^{b} ± 16.28	537.40^{c} ± 11.24	428.24^{d} ± 11.14	271.82^{e} ± 17.39
FI (g)	1131.80^{a} ± 12.13	882.77^{b} ± 5.87	956.80^{c} ± 10.65	900.75^{d} ± 16.75	990.07^{e} ± 23.89	732.46^{f} ± 9.50
FCR	1.96^{a} ± 0.12	1.35^{b} ± 0.02	1.46^{c} ± 0.07	1.68^{d} ± 0.06	2.31^{e} ± 0.15	2.70^{f} ± 0.24

Values bearing different superscripts differ significantly. (P<0.05).

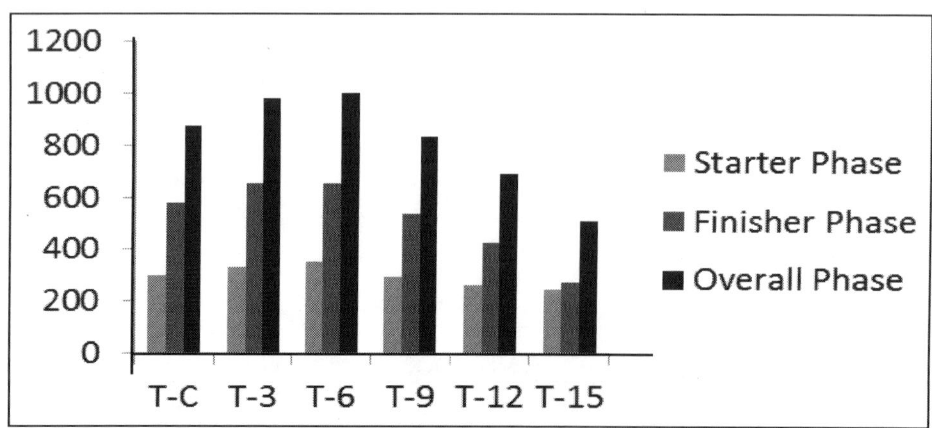

Figure 66.1: GIW (g.) for all Treatments during different Phases.

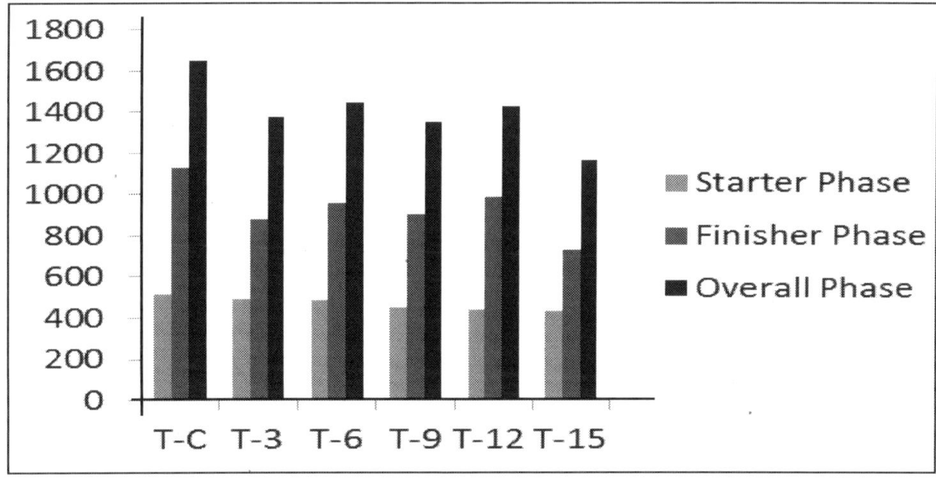

Figure 66.2: FI (g.) for all Treatments during all Phases.

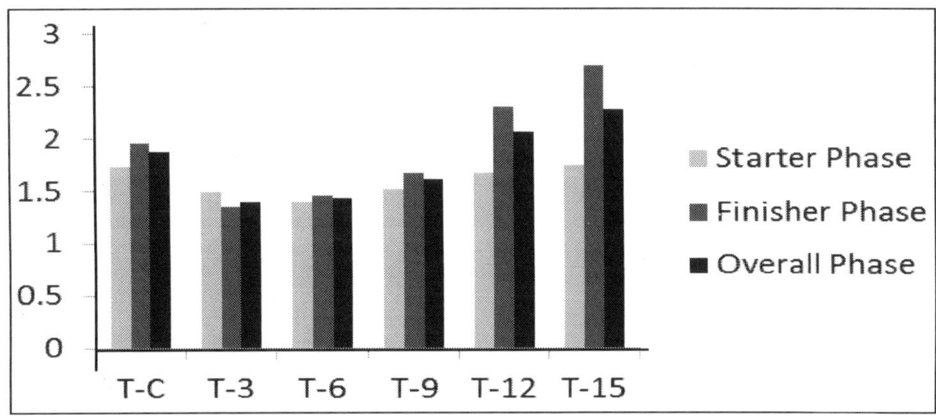

Figure 66.3: FCR for all Treatments during different Phases.

in T-15 treatment group. All these values found to be highly significant ($P<0.05$) among different treatment groups. Dressing percentage at the end of the 6th week varied from 51.14±0.63 (T-15) to 66.56±2.59 (T-6). Overall mortality during starter as well as finisher phase was recorded lowest in T-6 and T-12 (1.56 per cent) and highest mortality was recorded in T-3 group (10.94 per cent).

Economics of Broiler Production

The economics of broiler production by feeding seabuckthorn leaves as protein replacer in diets of broilers at different levels has been presented in Table 66.10 and graphic representation in Figure 66.4 and 66.5. The cost of feeding for one kg gain in weight for different groups was different. The T-15 group resulted in highest cost of feeding (Rs. 25.29) per kg GIW and T-6 group required lowest cost of feeding (Rs. 16.37) per kg GIW. Profit per kg gain in weight in Rs. was 52.06, 56.91, 57.03, 55.05, 50.05 and for T-C, T-3, T-6, T-9, T-12 and T-15 treatment groups respectively. Highest profit per kg gain in weight of broiler was in respect of T-6 treatment group (Rs. 57.03) and lowest profit per kg weight was gained in respect of T-15 treatment group (Rs. 48.11). The profit over control group (per cent) was also highest for T-6 group (9.55 per cent), whereas T-12 and T-15 groups showed a loss of 3.86 and 7.59 per cent with respect to control group.

Table 66.10: Economics of Broiler Production

Particular	*T-C*	*T-3*	*T-6*	*T-9*	*T-12*	*T-15*
Cost of feed/kg GIW (Rs.)	21.34	16.49	16.37	18.35	23.35	25.29
Expenditure/kg GIW (Rs.)	47.95	43.09	42.97	44.95	49.95	51.89
Cost of Sale (Rs.)	100	100	100	100	100	100
Profit/kg (Rs.)	52.06	56.91	57.03	55.05	50.05	48.11
Profit with respect to Control (per cent)	-	9.39	9.55	5.74	-3.86	-7.59

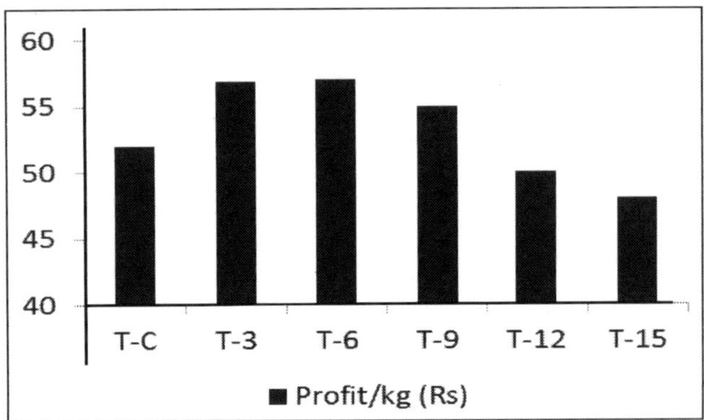

Figure 66.4: Overall Profit/kg (Rs) for different Treatments.

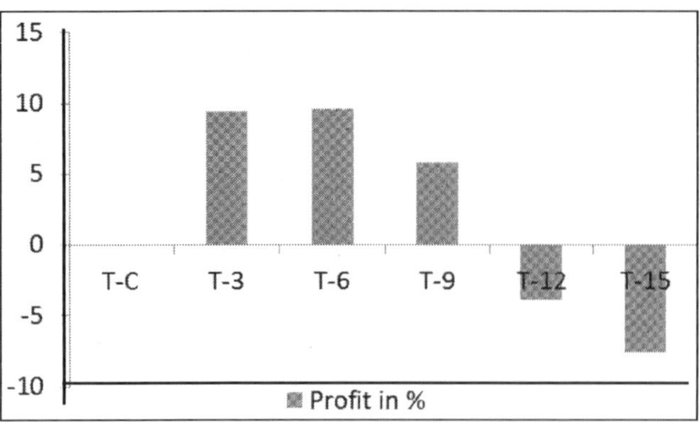

Figure 66.5: Profit (per cent) with Respect to Control Treatment.

DISCUSSION

Chemical Composition of Starter and Finisher Phase Test Tiets

There was no appreciable difference in the chemical composition of the test diets and small difference was due to the adjustments of the different ingredients to keep the diets iso-proteinous and iso-caloric as far as possible.

Chemical Composition of Starter Phase Diets

The DM of the starter phase diets varied from 89.35 to 89.97 per cent and this did not show any significant (P<0.05) effect. The protein content and metabolisable energy values (calculated) of the test diets ranged from 19.67 to 20.17 per cent and 3061 to 3128 kcal/kg respectively. These values found to be non-significant (P<0.05) among the different treatment groups and were nearer to BIS (1992) standards.

ME:CP ratio varied from 1:153 to 1:156 which was wider than the requirement of BIS (1992). The lower protein content and slightly higher energy content than the BIS (1992) were the reasons for these ratios.

The EE content varied from 4.37 to 5.95 per cent. The CF of the diets ranged from 6.56 to 7.32 per cent, which was slightly higher than required, due to higher CF content in the seabuckthorn leaves and these values of CF were increasing with the increased level of seabuckthorn leaves in diet, due to which the CF of the T-15 group was highest among all other treatment groups. The TA ranged from 5.06 to 5.09 per cent which were within the limits. NFE in diets ranged from 62.26 to 63.90 per cent. The AIA of the diets ranged from 1.02 to 1.24 per cent and was within the required limits. Ca ranged from 2.22 to 2.85 per cent and as per BIS (1992) requirement it was within the range. P ranged from 0.84 to 1.00 per cent respectively, where it was higher than the recommended levels, higher P levels were kept to maintain the Ca: P ratio which should be 2:1 to 2.5:1 as per recommendation (BIS, 1992).

Chemical Composition of Finisher Phase Diets

The DM of the test diets varied from 89.99 to 92.11 per cent. As the diets formulated were kept iso-proteinous and iso-caloric as far as possible, the protein concentration of the test diets ranged from 18.82 to 19.20 per cent and metabolisable energy varied from 3095 to 3128 kcal/kg. Energy was slightly higher than the requirement of BIS (1992) and these values within the different treatments varied due to adjustment made in different feed ingredients inclusion to keep the diets iso-nitrogenous and iso-caloric. All the values found to be non-significant (P<0.05).

The range of EE was found to be 3.93 to 5.09 per cent. The CF of the diets ranged from 5.57 to 5.82 per cent, which was within the recommended limits. TA values ranged from 4.52 to 4.62 per cent. NFE in diets ranged from 65.56 to 66.78 per cent. The AIA of the diets ranged from 0.71 to 0.98 per cent.Ca ranged from 2.48 to 3.15 per cent. P ranged from 0.74 to 0.86 per cent respectively. The higher levels of P were kept to maintain The Ca: P ratio within the range. All these values were nearer to the BIS (1992) standard and were found to be non-significant.

Chemical Composition of the Seabuckthorn Leaves

The chemical composition of the seabuckthorn leaves used as protein replacer in broiler diets was found to be 17.34 per cent. EE was found to be 10.90 per cent. Similar value has been reported by Sharma *et al.* (2009), Singh and Sharma (2008), and Sharma (2007). Energy of the seabuckthorn leaves was calculated as 3309.13 kcal/kg.

The value of CF was found to be 5.17 per cent, and it was lower than the value of CF reported by Dwivedi *et al.* (2003). The difference in CF was due to the CF content of different variety than the *Hippophae rhamnoides* which has been observed here.

The TA, Ca and P found to be 7.15, 3.67 and 0.19 per cent respectively. Where the values of TA, Ca and P were similar to those reported by Dwivedi *et al.* (2003).

NFE and AIA content of seabuckthorn leaves were, 59.45 and 0.02 per cent respectively.

Biological Performance of Broilers

There is no systematic studies of feeding seabuckthorn has been conducted so far but some work has been conducted by the different scientists with the use of feeding leaves for broiler production. The results of this study were in the same line to that of the experiments conducted by (Atteh, 2008), where he observed that by feeding ground stevia leaves or pure stevioside during 2nd week of starter and 2nd week of grower periods, higher (P<0.05) weight gain was observed.

Similarly when *Telfairia occidentalis* leaf meal (TOLM) was increased from 10 to 30 percent in the birds during the starter phase (Fasuyi and Nonyerem, 2007) observed that at 15 per cent TOLM in dietary inclusion levels the highest average weight gain was observed. This diet may be nutritionally beneficial in poultry feeding and subsequently reduce the use of expensive animal protein sources in practical feed formulation. Some work on Adampan leaf feeding to the poultry has also been undertaken by Rajdurai and Paramakkudy (1993) where in coconut meal was replaced up to 3 or 5 per cent with adampan (*Ipomoea* sp.) leaf and observed no effect on weight gain of the broilers. Similarly no effect on was observed by (Wisitiporn and Buakeeree, 2005) when fed lucerne based diet to the layers.

However our results were contrary to the observations given by Teguia *et al.*, 1997 in respect to sweet potato, perennial peanuts (groundnut) included diets. Similarly Atawodi (2003) observed the decrease in weight in poultry diet when fed cassava leaf meal and *Leucaena leucacephala* leaves.

In this study the overall feed intake significantly (P<0.05) decreased among different treatment groups. These findings were in agreement with (Lizama *et al.*, 1988) when up to 5 per cent of hydrilla and elodea was added to the chicken diet. However our results were contrary to the studies done by Rajdurai and Paramakkudy (1993), Lizama *et al.*, 1988 and Wisitiporn and Buakeeree, 2005 where no significant (P<0.05) effect of diet replacement with leaves of different plants as protein sources were used. Bhatt *et al.* (1991) also found significantly (P<0.05) low growth and intake due to feeding of almora leaf meal as a source of protein replaces of maize and sunflower oil meal.

FCR for the starter phase ranged from 1.40±0.05 (T-6) to 1.74±0.05 (T-12). Significantly higher FCR was observed in T-C and T-15 treatment groups, as compared to T-3, T-6, T-9 and T-12 treatment groups. No appreciable effect of FCR in respect T-C and T-15 and T-3 and T-9 treatment was observed.

FCR during finisher phase varied from 1.35±0.02 (T-3) to 2.70±0.24 (T-15). FCR for T-3 and T-6 did not vary significantly with that of T-C and T-9. Also T-15 group had significantly (P<0.05) higher FCR than all the treatments except T-12 and T-C group. Whereas FCR for T-12 varied significantly (P<0.05) with T-3 and T-9 and others were non-significant. Numerically lowest FCR was observed for T-3 and T-6 groups.

The FCR for overall phase from 3rd day to 6th week varied from 1.40 ± 0.03(T-3) to 2.28±0.14 (T-15). T-3, T-6 and T-9 were having significantly (P<0.05) lower FCR than T-15 group.

Hence, the overall feed conversion ratio is significantly (P<0.05) in T-3, T-6, T-9 and significantly higher (P<0.05) in T-12 and T-15 treatment group as compared to control treatment (T-C). These results are in agreement with the trail conducted by Atteh, 2008, where he observed that feed/gain ration of the birds was higher with feeding of broiler chicks with stevia leaves or stevioside where only up to 2 per cent diet was replaced with these ingredients. However no effect was observed by Rajdurai and Parmakkudddy, 1993. Natanman and Chandrasekaran, 1996 observed that on inclusion of subabul leaf meal up to 5 per cent has beneficial effect and if this increased to 10 per cent level, it reduced the FCR. In our study also when the levels of seabuckthorn leaves was increased, from 9 per cent to 12 or 15 per cent, the FCR decreased significantly (P<0.05). Teguia *et al.*, 1997, Odunsi *et al.*, 2002 and Lizama *et al.*, 1988 observed the decreased FCR with the feeding of sweet potato, sundried gliricidia leaf meal and hydrilla respectively. Radhakrishnaiah *et al.*, 1997 also concluded that up to 2 per cent of unprocessed and up to 4 per cent processed sesbania leaf meal does not affect the performance of the chicks. Similar trend is observed in our study that up to 9 per cent inclusion of seabuckthorn leaves in the diet of the broiler, the FCR remains better.

Dressing percentage at the end of the 6th week varied from 51.14±0.63 (T-15) to 66.56±2.59 (T-6) but they did not vary significantly with each other. There was significant increase (P<0.05) in the dressing per cent in all treatment group. Again up to 9 per cent of inclusion of seabuckthorn leaves the dressing percent increased 65.59±0.56 and 66.56±2.59 per cent in T-3 and T-6 treatment groups and decreased in all other groups as compared to control group which was 62.35±0.68 per cent.

Overall mortality during starter as well as finisher phase was recorded lowest *i.e.* T-6, T-12 and T-15 (1.56 per cent) and highest mortality was recorded in T-3 group (10.94 per cent). T-C and T-9 were having medium mortality *i.e.* 4.69 and 3.13 respectively.No effect of seabuckthorn inclusion at any level on mortality was observed as evident from the post mortem reports and similar results were observed by Wisitiporn and Buakeeree, 2005 when lucerne was fed to laying hens.

Economics of Broiler Production

The economics of broiler production upon feeding

seabuckthorn leaves as protein replacer in diets of broilers at different levels has been presented in table 4.6. The cost of feeding for different groups ranged from Rs. 16.37 to Rs. 25.29. The T-15 group resulted in highest cost of feeding per kg GIW due to higher FCR and T-6 group required lowest cost of feeding per kg GIW due to lowest FCR. Profit per kg in Rs. varied from 48.11 to 57.03. Highest profit per kg weight of broiler was gained for T-6 group while, lowest profit per kg weight was gained for T-15 group. The profit over control group (per cent) was also highest for T-6 group (9.55 per cent), whereas T-12 and T-15 groups were having loss of 3.86 and 7.59 per cent with respect to control group. Our results were similar to that of Trompiz *et al.*, 2007 where he conducted a trial to study the effect of cassava leaf meal and used upto 7.5 per cent cassava leaf meal without affecting the productive parameters, allowing a saving in the total feeding cost. The adjustments are needed to feed formulations at "safe" level, in order to minimize the risks of impairing performance as 70 per cent of the variable cost of poultry production is economically important (Walker and Gordon, 2003). Fasuyi and Nonyerem, 2007 fed 15 per cent teleferia occidentalis leaf meal and suggested that these diets are nutritionally beneficial in poultry feeding and subsequently reduce the cost of expensive animal protein as in our study.

CONCLUSION

It may be concluded that seabuckthorn leaves can be included in the diet of the broilers to replace CP of basal diet up to 9 per cent with overall scoring of the total biological parameters. So the results of this study are encouraging. Further research work on the same lines should be undertaken to ascertain the efficacy of use of Seabuckthorn leaves in the poultry production, as this single trial is not efficient to make a concrete conclusion to consider the Seabuckthorn leaves as feed protein replacer to an optimum levels.

REFERENCES

1. AOAC. 1985. *Official Methods of Analysis*. Association of Official Analytical Chemists, Arlington, VA, and USA, 24 (14): 868-878.
2. Atawodi, S. E., Mari, D., Atawodi, J. C. and Yahaya, Y. 2003. *African Journal of Biotechnology* 7 (3): 317-321.
3. Atteh, J. O., Onagbesan, O., Tona, M. K., Decuypere, E., Geuns, J. M. C. and Buyse, J. 2008. *Journal of Animal Physiology and Animal Nutrition*. The Authors, Journal compilation, Blackwell Publishing Ltd.
4. Bhatt, R. S., Sharma, M. and Katoch, B. S. 1991. *Indian journal of Animal Nutrition* 8 (2): 153-154.
5. BIS. 1992. *Bureau of Indian Standards Poultry feeds-specification*. Manas Bhavan Bahadur Shah Zafar Marg, New Delhi, 1100012.
6. Dwivedi, S. K., Singh, R. and Raut, B. 2003. Studies on distribution pattern of seabuckthorn through remote sensing and its genetic diversity in Ladakh. In: *International Seabuckthorn Association Congress*, Berlin (Germany); 14-18Sep 2003, p. 29.
7. Fasuyi, A. O. and Nonyerem, A. D. 2007. *African Journal of Biotechnology* 6 (8): 1055-1063.
8. Lizama, L. C., Marion, J. E. and McDowell, L. R. 1988. *Animal Feed Science and Technology* 20 (2): 155-161.
9. Lodhi, G. N., Singh, D. and Ichaponani, J. S. 1976. Variations in nutrient content of feeding stuff rich in protein and reassessment of the chemical methods of metabolisable energy estimation. *Journal of Agricultural Science* 86: 293-303.
10. Natanman, R. and Chandrasekaran, D. 1996. *Indian Veterinary Journal* 73 (10): 1042-1044.
11. Odunsi, A.A., Ogunleke, M.O., Alagbe, O.S. and Ajani, T.O. 2002. *International Journal of Poultry Science* 1 (1): 26-28.
12. Radhakrishnaiah, Y., Reddy, P. V. V. S., Reddy, P. S. and Reddy, D. V. 1997. *Indian Journal of Poultry Science* 32 (1): 14-22.
13. Rajdurai, H. and Paramakkuddy, K. 1993. *Poultry Advisor* 26 (7): 57-59.
14. Rongsen, L. 1994. New achievements in seabuckthorn development in China, ICIMOD. Kathmandu, Nepal, 62p.
15. Rousi, A. 1971. The genus *Hippophae* L. a taxonomic study. *Annals Botany Fennici* 8: 177-227.
16. Sharma, V. K. 2007. Nutritional evaluation of seabuckthorn for productive performance in cattle. Project Report submitted to ICAR 2006-2008, pp.1-58.
17. Sharma, V. K., Chauhan, J. S. and Rani, D. 2009. Fodder values of seabuckthorn. In: *Seabuckthorn Hippophae* Spp.: *The Golden Bush*. (Editors: Sanjai K. Dwivedi, T. Parimelazagan, Shashi Bala Singh, and Z. Ahmed), pp. 141-151, New Delhi.
18. Singh, V. and Sharma, V. K. 2008. Role of seabuckthorn (*Hippophae rhamnoides* L.) in development of animal husbandry. In: *Seabuckthorn (Hippophae L.) A Multipurpose Wonder Plant: Advances in Research and Development*. (Editor- In- Chief Virendra Singh), pp. 524-539, Daya Pub. House, New Delhi.
19. Teguia, A., Nijwe, R. M. and Foyette, C.N. 1997. *Animal Feed Science and Technology* 66 (1/4): 283-287.

20. *The Poultrysite* 2012. Digital, issue 14, p. 13.

21. Trompiz, J., Gómez, A., Rincón, H., Ventura, M., Bohórquez, N. and García, A. 2007. Effect of rations with cassava leaf flour on the productive performance of broilers Revista Cientifica de la Facultad de Ciencias Veterinarias de la Universidad del Zulia.

22. Walker, A. and Gordon, S. 2003. In: *Proceedings of the Nutrition Society* 62: 253-256.

23. Wisitiporn, S. and Kruan, B. 2005. Received: Aug 30, 2005; Revised: December 19, 2005; Accepted: Dec 27.

Chapter 67

To Increase the Poultry Broiler Production by Probiotics Addition in the Seabuckthorn Cake Based Ration

V.K. Sharma[1]*, Sachin Shama[1], Daisy Rani[1], Anu Parmar[1] and Virendra Singh[2]

[1]Department of Animal Nutrition, College of Veterinary and Animal Sciences
[2]Department of Biology and Environmental Sciences, CSK Himachal Pradesh Krishi Vishvavidyalaya, Palampur, Himachal Pradesh
(*E-mail: vkcovas@yahoo.co.in)

SUMMARY

The study was conducted on 300 (Nos), day-old broiler chicks of "Vancobb" strain dividing them into 5 groups and 4 replication of 15 poultry chicks each. The SBT cake has been used in poultry broiler production feed successfully upto 30 per cent (Sharma, 2010). To increase further the efficiency of poultry broiler on the feeding of 30 per cent level of seabuckthorn cake the probiotics like *Lactobacillus, Saccharomyces* and combination of both were added in the diet of the broilers. The CP (Crude protein)of the conventional control diet T_1 was replaced with 30 per cent CP of seabuckthorn cake in T_2 along with the addition of *Lactobacillus, Saccharomyces* and combination of both of these probiotics in T_3 T_4 and T_5 treatment groups respectively to study the effect on the growth and performance of the broiler.. The average initial body weight of the poultry broiler for different treatments ranged from 48.05 to 49.01 g on 5th day. The overall significant higher ($P<0.05$) GIW (gain in weight) of 1424.5 g was obtained in the T_3 treatment group where only *Lactobacillus* was added in the diets of the broilers with 30 per cent replacement of CP of control group with CP of SBT cake. The total FI (feed intake) up to 6th week of age was numerically lowest inT_3 group and highest in T_2 treatment group where only seabuckthorn cake was added to the diet of the broilers. The FCR (feed conversion ratio) of different treatments ranged from 2.12 (T_3) to 2.67 (T_2). These values for other groups were 2.52, 2.52 and 2.46 in respect to T_1, T_4 and T_5 treatment groups respectively. The FCR values was significant lowest ($P<0.05$) in T_3 where *Lactobacillus* was added along with the seabuckthorn cake followed by T_5 where *Lactobacillus+ Saccharomyces* was added to the diet along with 30 per cent SBT cake. This showed that supplementing probiotics can improve the FCR. The expenditure cost of rearing the poultry broiler ranged from Rs. 80.53 (T_1) to 87.99 (T_5) and the profit earned from the sale of the birds varies from Rs. 13.10 (T_4) to 19.47(T_1). The profit earned was highest in control group (T_1) and lowest in T5 group. It was concluded from this study that the probiotics improve the efficiency of poultry broiler production and growth. In other words from the mode of action of probiotics, it can be said that seabuckthorn cake has nothing adverse in its nutritional value and can be added in the diet of the poultry broiler production up to 30 per cent replacement of CP of conventional feed with SBT cake CP. It can also be used as the ulternate source of CP and EE (Ether extract) to the poultry production to bridge the gap between the demand and supply for the poultry production.

***Keywords**: Hippophae, Poultry, Probiotics, Seabuckthorn cake and Seabuckthorn feed.*

INTRODUCTION

Poultry is one of the fastest growing segments of the agriculture sector. Balanced and effective feeding is the most important requisite for economic livestock production. Feed accounts for approximately 65-70 per cent of total expenditure incurred in poultry rearing. So a small decrease in feed cost, or feed intake and improvement in FCR can be translated into large increase of profits. So we are looking for alternative feed sources and feed additives. Among feed additives Probiotics or Direct Fed Microbial has emerged to be

Seabuckthorn (Hippophae L.): A Multipurpose Wonder Plant, Vol. 5: Advanced Technologies for Cultivation, Processing, Health Protection and Environmental Conservation (Virendra Singh, *Editor-in-Chief*, 2018), p. 527–533, Astral International Pvt. Ltd., New Delhi, India.

a major alternative to antibiotics. Term "probiotic" was coined by Parker (1974) and he described it as live organism and substances that contribute to intestinal microbial balance. The term Probiotics is defined by united nation and WHO expert panel as live organism which when administered in adequate amount confers a health benefit on host. The microbial species used in the preparation of probiotics are either single strain or mixture of microbes (Raibaud and Raynaud, 1990). Leaves of seabuckthorn (*H. rhamnoides* subsp. *turkestanica*) is known to be a rich some of proteins, fat and minerals (Singh *et al.*, 1999). Among alternative feed sources seabuckthorn cake can be used which is a by-product, produced during the extraction of oil from seabuckthorn (SBT) seeds. Seabuckthorn oil is used in a number of cosmetic and neutraceutical industries and its cake is produced as a byproduct. As protein is the most expensive nutrient, by introducing new protein source in poultry/broiler feeds we can certainly decrease the cost of production. It has been found that the use of seabuckthorn (SBT) cake may improve the biological performance of broiler birds and decrease the cost of production when fully exploited to use. The use of seabuckthorn cake as source of protein by replacing the conventional protein sources has shown good results. Optimum growth has been recorded at 30 percent replacement of conventional protein sources with seabuckthorn cake. Among probiotics *lactobacillus* has shown good results at dose rate of 6.8×10^8 cfu/kg feed, *Saccharomyces* at dose rate of 5.8×10^7 cfu/kg feed and in combination shown optimum growth at dose rate of 3.4×10^8 cfu/kg and 2.9×10^7 cfu/kg of lactobacillus and *Saccharomyces* respectively. Therefore, the present study was planned to investigate the effects of probiotics supplementation on the utilization of seabuckthorn cake in broiler production.

MATERIALS AND METHODS

The main objectives of the investigation were as follows:

1. To test the biological performance upon feeding of the seabuckthorn (*Hippophae rhamnoides* subsp. *turkestanica*) cake along with probiotics to commercial broilers from day one to six weeks of age.
2. To study the nutrient availability upon the replacement of conventional concentrate crude protein with seabuckthorn cake along with probiotic.
3. To study the feed conversion ratio of broiler production.
4. To work out the cost benefit ratio for broiler productions.

The research work was carried out in the metabolic stalls of the Department of Animal Nutrition, Dr. G.C. Negi, College of Veterinary and Animal Sciences, CSK HPKV, Palampur. To achieve the objectives of research programme, the feeding experiment was conducted on 300 (Nos), day-old broiler chicks of "Vancobb" strain. The information regarding experimental plan, collection and analytical procedures adopted for different feed ingredients, seabuckthorn cake and formulated feeds; selection, distribution, management of broiler chicks, observations recorded and method adopted for statistical analysis are as follows:

Table 67.1: Experimental Plan

Treatment	*Experimental Group Diet*	*Dose of Probiotic (s) (CFU/Kg)*
T_1 (Control)	Control diet	Nil
T_2	30 percent crude protein of T_1 was replaced with the crude protein from seabuckthorn cake	Nil
T_3	T_2 diet + *Lactobacillus*	6.8×10^8
T_4	T_2 diet + *Saccharomyces*	5.8×10^7
T_5	T_2 diet + *Saccharomyces* + *Lactobacillus*	$2.9 \times 10^7 + 3.4 \times 10^8$

Total number of treatments: 5 (including control)

Total number of replications: 4

Total number of broilers per treatment: 60

Total number of broilers per replication: 15

Total number of broilers: 300

Experimental design: Randomized Block Design

The Proximate composition of the feed, faeces and redual feed was calculated by the Standard methods as reported in AOAC (1985) and the ME contents were calculated as per the equation proposed by Lodhi *et al.* (1976).

Formulation of Feed

The physical composition of experimental diets use for the broiler starter and finisher phase has been given in Tables 67.2 and 67.3 respectively. Based on the chemical composition of broiler starter and broiler finisher feeds were formulated to meet BIS (1992) standards and have been given in Tables 67.4 and 67.5 respectively. The difference in the chemical composition of different diets is due to the difference in the physical composition of the different diets in the treatment groups.

Management of Broiler Chicks

After arrival of the chicks from M/S Uttam Hatcheries, Kangra, HP, the chicks were kept on boiled tap water for first 48 hours and then fed on crushed maize for next two days. They were wing banded, weighed and allocated to 5 treatments with 4 replication of 15 chicks per replications, having same average body weight. They were kept in 20 compartments of thermostatically controlled battery brooders. The standard managemental practices were followed for

Table 67.2: Physical Composition of Experimental Diets for Broiler Starter Phase (per cent DM)

Particulars	T_1	T_2	T_3	T_4	T_5
Maize	34.00	25.38	25.38	25.38	25.38
Groundnut cake	17.00	12.69	12.69	12.69	12.69
Fish meal	5.00	3.74	3.74	3.74	3.74
De oiled rice bran	14.75	11.01	11.01	11.01	11.01
Soya flakes	24.00	17.91	17.91	17.91	17.91
Molasses	03.00	02.24	02.24	02.24	02.24
DCP	01.00	00.75	00.75	00.75	00.75
Lime powder	01.00	00.75	00.75	00.75	00.75
Premix	00.25	00.18	00.18	00.18	00.18
Seabuckthorn Cake	-	25.38	25.38	25.38	25.38
Lactobacillus	-	-	$6.8x10^8$	-	$3.4x10^8$
Sacchromyces	-	-	-	$5.8x10^7$	$2.9x10^7$
Total	100	100	100	100	100

*Premix: 1) Merivite AB2D3k:25g (Vit A-82,500IU, VitB_2-52mg, Vit D_3-12000IU, Vit K- 10 mg, PO_4-395mg/g); 2) Nicomix BE:25g (Vit B1-4mg, VitB_{12}-40µg, Niacin-60mg,Calcium Pentothenate-40mg, Vit E-40mg/g); 3) Trace Minerals: 100g(Ferric Oxide-2g,Copper sulphate-2g, Ferrous sulphate-10g, Zn sulphate-0.6g, Di calcium phosphate-53.65g, Mangnesesulphate- 5.5g, Pottasium iodide-2.5g, Sodium thiosulphate-0.75g, Zinc oxide-1g) and 4) Choline chloride:100g.

Table 67.3: Physical Composition of Experimental Diets for Broiler Finisher Phase (per cent DM)

Particulars	T_1	T_2	T_3	T_4	T_5
Maize	34.00	26.16	26.16	26.16	26.16
Groundnut cake	20.00	15.39	15.39	15.39	15.39
Fish meal	5.00	3.85	3.85	3.85	3.85
DORB	17.75	13.66	13.66	13.66	13.66
Soya flakes	18.00	13.85	13.85	13.85	13.85
Molasses	03.00	2.31	2.31	2.31	2.31
DCP	01.00	0.76	0.76	0.76	0.76
Lime powder	01.00	0.76	0.76	0.76	0.76
Premix	00.25	0.19	0.19	0.19	0.19
Seabuckthorn Cake	-	25.38	25.38	25.38	25.38
Lactobacillus	-	-	$6.8x10^8$	-	$3.4x10^8$
Sacchromyces	-	-	-	$5.8x10^7$	$2.9x10^7$
Total	100	100	100	100	100

*Premix: 1) Merivite AB2D3k:25g (Vit A-82,500IU, VitB_2-52mg, Vit D_3-12000IU, Vit K- 10 mg, PO_4-395mg/g); 2) Nicomix BE:25g (Vit B1-4mg, VitB_{12}-40µg, Niacin-60mg,Calcium Pentothenate-40mg, Vit E-40mg/g); 3) Trace Minerals: 100g(Ferric Oxide-2g,Copper sulphate-2g, Ferrous sulphate-10g, Zn sulphate-0.6g, Di calcium phosphate-53.65g, Mangnesesulphate- 5.5g, Pottasium iodide-2.5g, Sodium thiosulphate-0.75g, Zinc oxide-1g) and 4) Choline chloride:100g.

rearing the chicks in battery brooders up to 4 weeks of age (Starter phase) and thereafter they were shifted to Californian cages up to 6 weeks of age (Finisher phase). Clean drinking water and the respective experimental diets were always available to the chicks during the entire course of study.

Statistical Analysis

The data were analyzed by using one-way analysis of variance technique and Turkey's test (Statistical Package, Graph Pad Instat tm. of Russel, 1990).

RESULTS AND DISCUSSIONS

Data regarding the chemical composition of broiler starter and broiler finisher diets have been given in Tables 67.4 and 67.5 along with Seabuckthorn cake (SBT). The SBT cake used in the experiment had 90.06 per cent DM, 26.00 per cent CP, 4.50 per cent EE, 14.00 per cent CF, 2.50 per cent TA, 0.75 per cent AIA and 53.00 per cent NFE. The cake had 1.25 per cent calcium (Ca) and 0.52 per cent phosphorus (P). The ME of SBT cake was 2903.61 kcal/kg.

Table 67.4: Chemical Composition of Broiler Starter Diets (per cent DM)

Nutrient / Treatment	T_1	T_2	T_3	T_4	T_5	Seabuckthorn Cake
Dry matter	90.75	89.25	89.25	89.25	89.25	90.06
Crude protein	22.73	23.57	23.57	23.57	23.57	26.00
Ether extract	3.28	3.59	3.59	3.59	3.59	4.50
Crude fiber	6.22	8.2	8.2	8.2	8.2	14.00
Total ash	6.02	5.13	5.13	5.13	5.13	2.50
NFE	58.92	57.43	57.43	57.43	57.43	53.00
Acid insoluble ash	2.06	2.15	2.15	2.15	2.15	0.75
Calcium	2.20	1.98	1.98	1.98	1.98	1.25
Phosphorous	0.98	0.82	0.82	0.82	0.82	0.52
ME (kcal/kg)	2919.48	2916.32	2916.32	2916.32	2916.32	2903.61
ME: CP	128.44	123.73	123.73	123.73	123.73	111.68

Table 67.5: Chemical Composition of Broiler Finisher Diets (per cent DM)

Nutrient / Treatment	T_1	T_2	T_3	T_4	T_5
Dry matter	90.67	89.48	89.48	89.48	89.48
Crude protein	21.93	22.87	22.87	22.87	22.87
Ether extract	3.18	3.48	3.48	3.48	3.48
Crude fibre	6.61	8.32	8.32	8.32	8.32
Total ash	6.05	5.23	5.23	5.23	5.23
NFE	59.30	57.86	57.86	57.86	57.86
Acid insoluble ash	1.92	1.95	1.95	1.95	1.95
Calcium	1.96	1.89	1.89	1.89	1.89
Phosphorous	0.83	0.76	0.76	0.76	0.76
ME (kcal/kg)	2896.63	2898.86	2898.86	2898.86	2898.86
ME: CP	132.08	126.75	126.75	126.75	126.75

Table 67.6: Overall Biological Performances of Broilers Offered Diets from 5rd Day to 6 Weeks of Age

Nutrient / Treatment	T_1	T_2	T_3	T_4	T_5	CD
Initial wt (g)	48.49	49.01	48.74	48.05	48.65	NS
Final wt (g)	1328.8^{ad}	1281.3^{a}	1424.5^{bc}	1354.4^{abc}	1376.1^{cd}	94.778
GIW (g)	1280.8^{ae}	1233.2^{a}	1376.0^{bcde}	1306.8^{ac}	1327.4^{ad}	142.84
FI (g)	3237.4	3309.7	2926.7	3307.4	3283.1	NS
FCR	2.52^{a}	2.67^{a}	2.12^{b}	2.52^{a}	2.46^{a}	0.345
Dressing (per cent)	58.58^{a}	58.21^{a}	61.44^{bc}	59.38^{acd}	61.21^{bd}	2.62
Livability (per cent)	98.33	98.33	98.33	98.33	100	NS

Biological Performance of Broilers Offered Diets from 5rd day to 6th Weeks of Age

Average body weight of broilers at 6 weeks of age, GIW, FI, FCR, dressing percentage and livability has been presented in Table 67.6. The final body weight (g) of broilers at 6 weeks of age ranged from 1281.3 (T_2) to 1424.5 g (T_3). The highest body weight was observed in T_3 treatment group where seabuckthorn cake along with the *Lactobacillus* was fed to the broiler. These values were significantly higher ($P<0.05$) than T_2 and numerically higher than T_1, T_4 and T_5 treatment groups. The overall GIW (g) ranged from 1233.2 (T_2) to 1376.0 (T_3). The overall numerical GIW was obtained in the T_3 treatment group where *Lactobacillus added* in the diets of the broilers followed by T_5 treatment group where *Saccharomyces* + *Lactobacillus* was added along with the SBT cake in the diet. The minimum weight in gain was observed in T_2 treatment group where only SBT cake was added in the

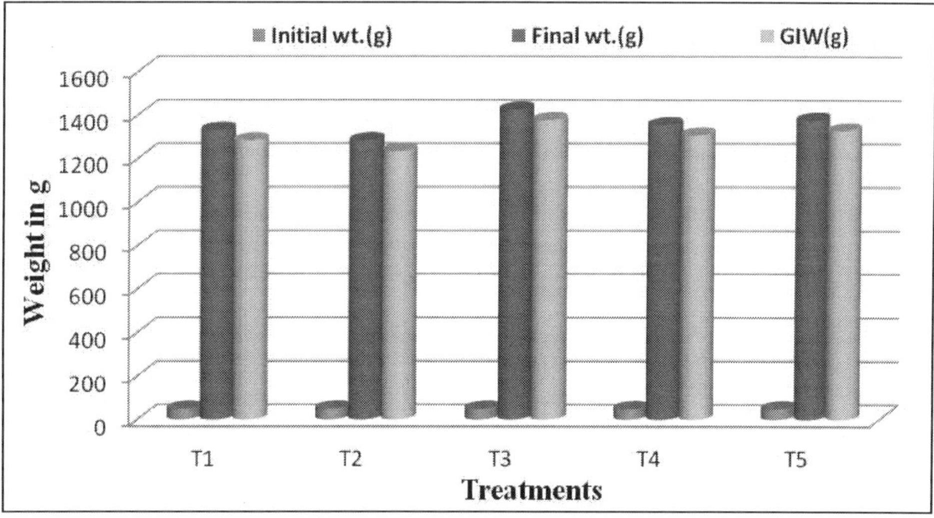

Figure 67.1: Overall Growth Performance of Broilers Offered Diets from 5th Day to 6th Week of Age.

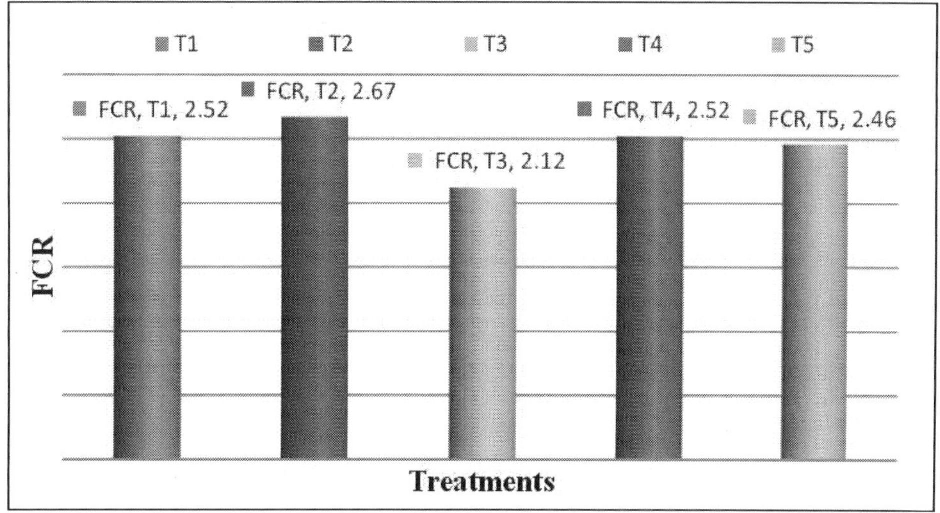

Figure 67.2: Overall FCR of Broilers Offered Diets from 5th Day to 6th Week of Age.

diet of the broilers. The graphic representation of growth performance of the broiler has been given in Figure 67.1.

The total FI up to 6th week of age ranged from 2926.7 ($T_{3)}$ to 3309.7 g (T_2). No significant ($P<0.05$) difference was observed among the different treatment groups. The feed intake of T_3 was numerically lowest where *Lactobacillus* was fed to the broiler and highest in T_2 treatment group where only seabuckthorn cake was added to the diet of the broilers.

The FCR of different treatments ranged from 2.12 (T_3) to 2.67 (T_2). These values were 2.52, 2.67, 2.12, 2.52 and 2.46 in respect to $T_{1,}$ $T_{2,}$ $T_{3,}$ T_4 and T_5 treatment groups respectively. The significant lowest ($P<0.05$) FCR was observed in treatment group T_3 where *Lactobacillus* was added along with the seabuckthorn cake followed by T_4 and T_5 where *Lactobacillus+ Saccharomyces* was added to the diet. This showed that supplementing probiotics can improve the FCR alone either alone (T_3) or in combination (T_5). The graphic representation of FCR has been given in Figure 67.2.

The dressing per cent values were found 58.58, 58.21, 61.44, 59.38 and 61.21 in respect of treatment $T_{1,}$ $T_{2,}$ T_3 T_4 and T_5 groups. These values were significantly higher ($P<0.05$) in T_3 and T_5 treatment groups as compared to T_4, T_1 and T_2 treatment groups respectively. No difference in the livability was observed in the different treatment groups during the experiment. The little per cent of death in different groups were natural and not due to any treatment effect as evident from the post-mortum reports. The graphic representation has been given in Figure 67.3.

Economics of Broiler Production Offered Diets Formulated by using Seabuckthorn Cake

The economics of broiler production upon replacing 30 per cent protein of conventional diet with protein from seabuckthorn cake with and without supplementation of probiotics have been presented in Table 67.7. Profit earned/kg wt. gain for different treatments of broiler production have been depicted in Figure 67.4.

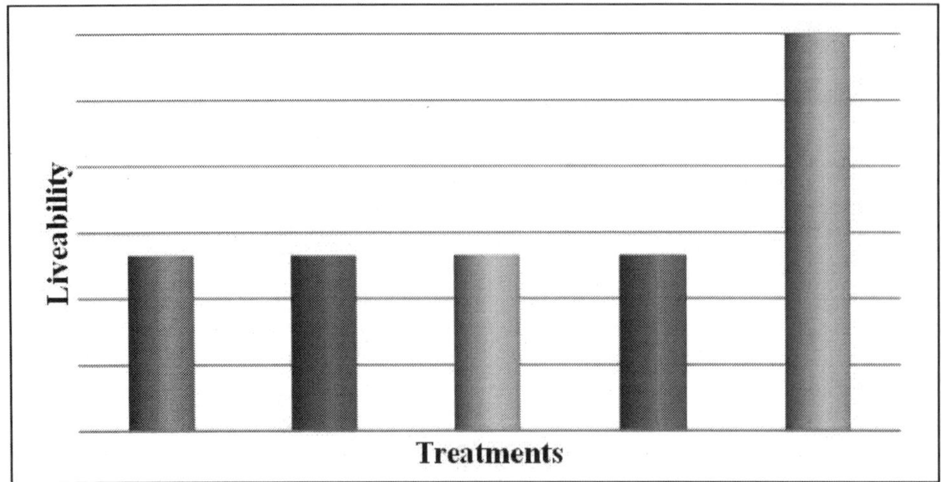

Figure 67.3: Overall Livability of Broilers Offered Diets from 5th Day to 6 Weeks of Age.

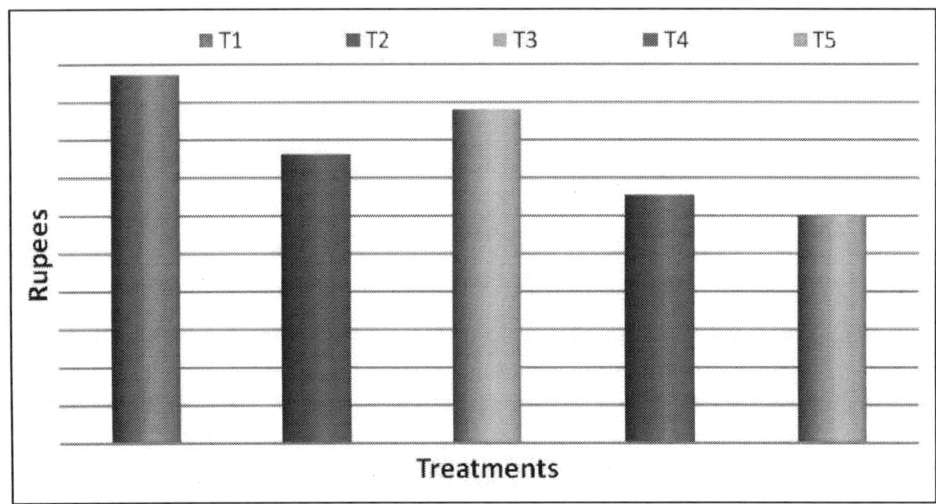

Figure 67.4: Profit Earned/kg wt. Gain for different Treatments of Broiler Production.

The cost of feed/Kg was Rs. 19.00, 19.50, 23.50, 21.54 and 22.52 in respect to T_1, T_2, T_3, T_4 and T_5 treatment groups respectively. The cost of feed in the different treatments was higher due to un-exploited use of SBT cake along with probiotics. T_1 resulted in lowest cost of production per Kg weight gain whereas, T_4 resulted in highest cost. Profit per Kg live weight varied from Rs. 13.10 (T_5) to 19.47 (T_1). On comparing the different treatments in which SBT cake was supplemented with *Lactobacillus* (T_3) showed maximum profit gain as compared to other probiotics added groups. Guleria (2010) revealed that poultry birds fed on diets containing 5 to 25 per cent SBT cake showed better biological performance and the diets having more than 25 per cent protein from SBT cake decreased the performance.

CONCLUSION

From this study it can be concluded that the Lactobacillus probiotics is better than other probiotics to increase the efficiency of the SBT cake in the poultry broiler production but there is no any appreciable profit in adding these probiotics till the cost of production of

Table 67.7: Economics of Broilers Offered Diets Formulated by using Seabuckthorn Cake

Particular / Treatment	T_1	T_2	T_3	T_4	T_5
Cost of feed (Rs./kg)	19.00	19.50	23.50	21.54	22.52
Cost of feed/Kg GIW (Rs.)	47.88	52.06	49.82	54.28	55.39
Expenditure/kg GIW (Rs.)	80.53	84.76	82.38	86.90	87.99
Market price (Rs.)/kg live wt.	100	100	100	100	100
Profit earned/kg live wt. (Rs.)	19.47	15.24	17.62	13.10	12.01

GIW: Gain in weight, FI: feed intake, FCR: feed conversion ratio, Figures bearing different superscripts with in a row differ significantly (P<0.05).

SBT cake is reduced when it is commercially exploited. However, the SBT cake can be used as alternate source of protein and oil in the broiler production.

REFERENCES

1. A.O.A.C. 1985. Official methods of Analysis, 14th edition. Association of official analytical chemists, Arlington, VA, USA, 24: 868-878.
2. BIS. 1992. Poultry feed specification. IS: 1374 (4_{th} revision). Bureau of Indian Standards, Manak Bhavan, 9, Bhadur Shah Zafar Marg, New Delhi-110002, India.
3. Lodhi, G.N., Singh, D. and Ichponani, J.S. 1976. Variation in nutrient content of feeding stuff rich in protein and reassessment of the chemical methods of metabolizable energy estimation. *Journal of Agricultural Sciences* 86 (8): 293-303.
4. Parker, R.B. 1974. Probiotics, the other half of antibiotics story. *Animal Nutrition Health* 29: 4-8.
5. Raiboud, P. and Raynaud, J.P. 1990. In: *European Association in Animal Production*, Wageningen, Netherland, 52: 296-275.
6. Sharma, V. K. 2010. Effect of Feeding SBT cake for broiler production. In: *13th Biennial Conference of Animal Nutrition Society of India* held at NIANP, Adugodi, Bangalore-560030 from December, 17-19, 2010, p. 46.

Annexure I

Some of Seabuckthorn Products/Technologies Developed by Various R&D Organizations of India

Sl. No.	*Name of Scientist & Address*	*E-mail & Mobile/Tel.*	*Name and Characteristics, Uses etc. of Seabuckthorn Products/ Technology/Machine*	*Photo of Seabuckthorn Product/ Technology/Machine*
1.	**Dr. Y.S. Dhaliwal,** **Dr. Ranjana Verma,** **Dr. Shailza Anand and** **Dr. V. Singh** Department of Food Science, Nutrition and Technology CSK Himachal Pradesh Agricultural University Palampur- 176062 (HP), India	ysdhaliwal44@ yahoo.co.in +919816082444 01894232397 Ext 644 (O) 01894232444 (R) 01894234274 (F)	Methodology for the preparation of different seabuckthorn based products *viz.* **RTS, Squash, Syrup, Fruit Toffee, Jam, Biscuits and Cake** have been standardized after conducting more than 40 trials. Studies on packaging, storage conditions and period on the nutritional and sensory characteristics of the prepared products have been conducted.	 Products Developed
2.	**Dr. Y.S. Dhaliwal,** **Dr. Ranjana Verma,** **Dr. Shailza Anand and** **Dr. V. Singh** Department of Food Science, Nutrition and Technology, CSK Himachal Pradesh Agricultural University Palampur- 176062 (HP), India	ysdhaliwal44@ yahoo.co.in +919816082444 01894232397 Ext 644 (O) 01894232444 (R) 01894234274 (F)	**Fruit leather** was prepared using pulp, sugar and pectin in different proportions in order to get non sticky and smooth leather. The prepared leather was packed in different packaging material like aluminium foil, cling film and butter paper for storage study and at room temperature as well as in refrigerator.	

Sl. No.	Name of Scientist & Address	E-mail & Mobile/Tel.	Name and Characteristics, Uses etc. of Seabuckthorn Products/ Technology/Machine	Photo of Seabuckthorn Product/ Technology/Machine
3.	**Dr. Y.S. Dhaliwal,** **Dr. Ranjana Verma,** **Dr. Shailza Anand** **Dr. V. Singh** Department of Food Science, Nutrition and Technology CSK Himachal Pradesh Agril. Univ. Palampur- 176062 (HP), India	ysdhaliwal44@yahoo.co.in +919816082444 01894232397 Ext 644 (O) 01894232444 (R) 01894234274 (F)	**Technology for preparation of tea from seabuckthorn leaves:** Tea from seabuckthorn leaves was standardized in the laboratory. Blending of green tea with seabuckthorn leaf tea has also been done.	
4.	**Dr. Y.S. Dhaliwal,** **Dr. Ranjana Verma,** **Dr. Shailza Anand** **Dr. V. Singh** Department of Food Science, Nutrition and Technology CSK Himachal Pradesh Agril. Univ. Palampur- 176062 (HP), India	ysdhaliwal44@yahoo.co.in +919816082444 01894232397 Ext 644 (O) 01894232444 (R) 01894234274 (F)	**Process for drying/dehydration of seabuckthorn berries:** Attempts have been made to standardize technology for dehydration of berries to prepare seabuckthorn powder. Utilization of seabuckthorn leaf powder and seabuckthorn berry powder in baked products has been done and level of berry powder in **baked products** has been standardized.	
5.	**Dr. Mandeep Sharma** **Dr. Rajesh Chahota** **Dr. Virendra Singh** Department of Veterinary Microbiology, DGCN COVAS, CSK Himachal Pradesh Agril. Univ. Palampur- 176062 (HP), India	Mandeepsharma289@hotmail.com MO: 09418045034	Standardized seabuckthorn oil and leaf extract based **anti-microbial agents** for domestic animals. Also developed wound healing agent.	 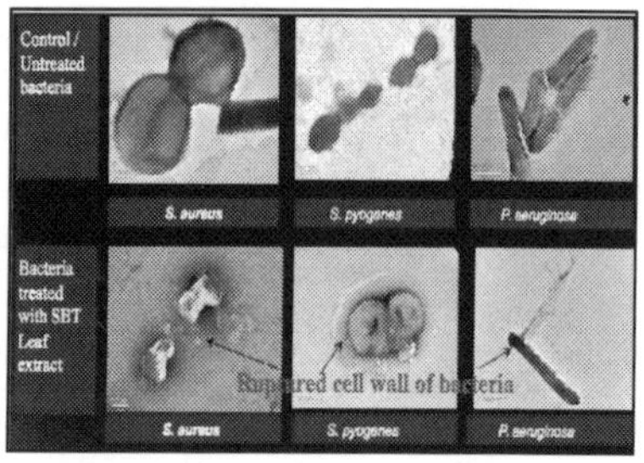

Sl. No.	Name of Scientist & Address	E-mail & Mobile/Tel.	Name and Characteristics, Uses etc. of Seabuckthorn Products/ Technology/Machine	Photo of Seabuckthorn Product/ Technology/Machine
6.	**Dr. V.K.Sharma** **Dr. Virendra Singh** Department of Animal Nutrition, DGCN COVAS, CSK Himachal Pradesh Agril. Univ. Palampur- 176062 (HP), India	ckcovas@yahoo.co.in MO: 09418425548	Variety of **seabuckthorn cakes** rich in protein, amino acids, fat, vitamin A, C, D, E, carotenoids, flavonoids and micro-minerals, which can be useful to milching animals and poultry.	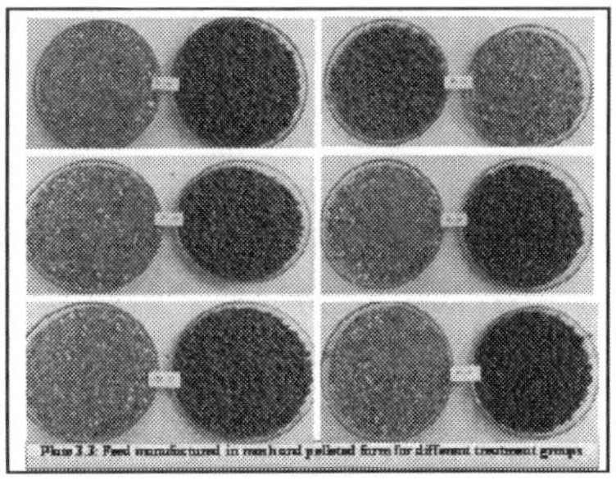
7.	**Dr. S.P. Taygi** **Dr.Virendra Singh** Department of Vet Surgery & Radiology, DGCN COVAS, CSK Himachal Pradesh Agricultural University Palampur- 176062 (HP), India	sptyagivet@gmail.com MO: 0941841225	Standardized **seabuckthorn oil dose and mixtures with strong gastric ulcer healing** in dogs. However, results are better for mixture with available drugs.	 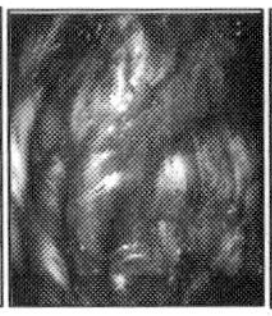 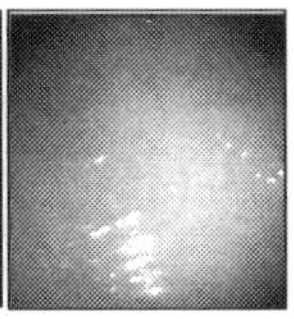
8.	**Dr.Virendra Singh** Department of Biology & Environmental Sciences, Engineering, CSK Himachal Pradesh Agricultural University, Palampur- 176062 (HP), India	virendrasingh1961@yahoo.com 09418045229	"**Drilbu**" a seabuckthorn cultivar of *Hippophae salicifolia* is mild thorny, high yielding (5.5 kg/ plant), rich in vitamin C (930 mg/100g) and omega 7 fatty acid, suitable for commercial cultivation on marginal and grasslands. Further 2 exotic seabuckthorn varieties have also been introduced.	
9.	**Dr. D.K.Vatsa** **Dr. Virendra Singh** Department of Agriculture Engineering, CSK Himachal Pradesh Agricultural University Palampur- 176062 (HP), India	erdkvatsa@rediffmail.com MO: 09418015941	**Five different fruit harvesters** for harvesting different types of seabuckthorn fruit developed. The fruit yield (30-45 kg/ person/day is 2-3 times higher than traditional methods of hitting with a stick (15-20 kg/ person/day). The Clipper with handle is most suitable for Indian seabuckthorn, whereas a wire loop is most suitable for Russian seabuckthorn.	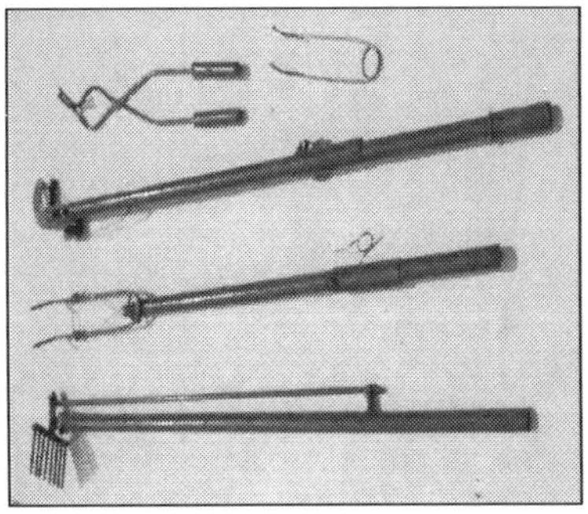

Sl. No.	*Name of Scientist & Address*	*E-mail & Mobile/Tel.*	*Name and Characteristics, Uses etc. of Seabuckthorn Products/ Technology/Machine*	*Photo of Seabuckthorn Product/ Technology/Machine*
10.	**Dr.Virendra Singh** Department of Biology & Environmental Sciences, Engineering CSK Himachal Pradesh Agricultural University Palampur- 176062 (HP), India	virendrasingh1961@yahoo.com 09418045229	**Propagation & Cultivation Technologies** Technologies for propagation from seeds and cuttings and plantation and cultivation methods (pits, spacing, pruning, fertilizers and pests and diseases control provided.	
11.	**Dr. D.S. Arya and Dr. Virendra Singh** Department of Pharmacology, AIIMS, New Delhi	dsarya16@hotmail.com MO: 09810210834	Seabuckthorn pulp and seed **oils doses standardized for cardioprotection**. The most optimum dose offering maximum cardioprotection on the basis of observed hemodynamic and biochemical variables with supporting evidence from histopathological and ultrastructural studies	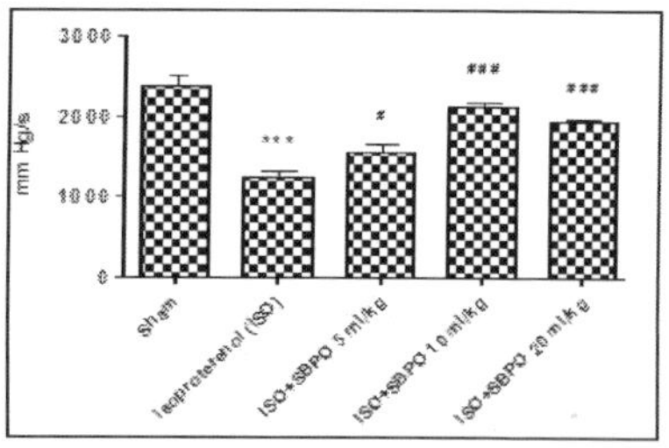
12.	**Dr. Geetha Surya-kumar** Scientist D DIPAS, Tiamrpur, Delhi 110054, India	geethasuryakumar@yahoo.com Tel: 011-23883311 MO: 9811687801	**DIP-HIP:** DIPAS has developed a potent new adjuvant, possessing the desired properties of safety, enhanced immunogenicity, and sustained dynamic response. DIP-HIP has been proven to be consistently superior to commercially available adjuvants, when evaluated comprehensively on the criteria of safety, immuno-potency, and bio-activity properties, as also on parameters of development, stability, with a simple "point-of-use" mixing procedure to produce a homogenous vaccine formulation with a long term antibody sustenance.	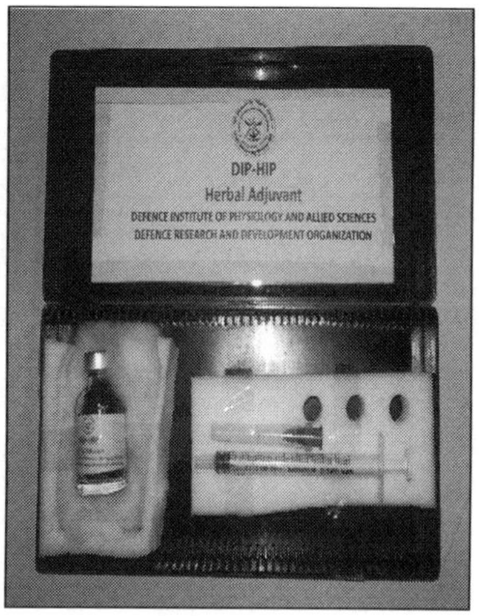

Sl. No.	*Name of Scientist & Address*	*E-mail & Mobile/Tel.*	*Name and Characteristics, Uses etc. of Seabuckthorn Products/ Technology/Machine*	*Photo of Seabuckthorn Product/ Technology/Machine*
13.	**Dr. Geetha Surya-kumar** Scientist D DIPAS, Tiamrpur, Delhi 110054, India	geethasuryakumar@yahoo.com Tel: 011-23883311 MO: 9811687801	**Herbo Healer:** A potent herbal wound healer in form of a topical ointment developed using Seabuckthorn extract, showed excellent healing properties to augment repair of normal as well as chronic wounds. Promotes rapid, aesthetic healing and has wide applications for skin tears, abrasions, incision, excision injuries, superficial/deep burns, scalds, bruises and diabetic wounds. Product exhibited enhanced healing via. wound contraction, re-epithelialization, cellular proliferation, collagen accumulation and neovascularization. Possesses anti-bacterial properties and found better than silver, iodine-based ointments. Also acts as a personal and preventative skin care, protects from dryness, heals skin's moisture barrier, provides soothing effects at harsh climatic conditions. It is pure herbal, rich in natural antioxidants, bioactive flavonoids, safe and economical.	
14.	**Dr. Geetha Surya-kumar** Scientist D DIPAS, Tiamrpur, Delhi 110054, India	geethasuryakumar@yahoo.com Tel: 011-23883311 MO: 9811687801	**Herbal Anti-Stress Adaptogen: 'DIP-91'** A herbal anti-stress adaptogen developed using Seabuckthorn extract. In passive multiple stress animal model (cold-hypoxia-restraint), 'DIP-91' reversed the hypoxia-induced changes and enhanced recovery from acute hypothermia. It strengthened oxygen delivery system, better utilization of free fatty acids, improved cell membrane permeability. Adaptogenic effect only under conditions of threat to the system and does not cause any metabolic alterations in body during resting state. Effective as health food supplements/ nutraceuticals, anti-stress agent to manage day-to-day stress and for overall health, well-being and immensely useful in adapting to adverse environmental conditions. Developed as a stand-	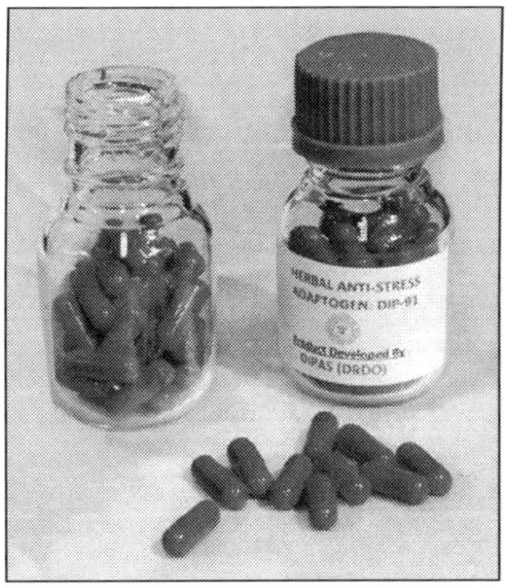

Sl. No.	*Name of Scientist & Address*	*E-mail & Mobile/Tel.*	*Name and Characteristics, Uses etc. of Seabuckthorn Products/ Technology/Machine*	*Photo of Seabuckthorn Product/ Technology/Machine*
			alone product, which possesses a non-cumulative potent adaptogenic activity, better than available products in the market *viz.* poly-herbal/multi-components preparations. Cost-effective, non-toxic, safe and free of heavy metal toxicity. Qualified amongst 240 products, selected for global commercialization under DRDO-FICCI- IC^2-ATAC cluster program.	
15.	**Dr. Geetha Surya-kumar** Scientist D DIPAS, Tiamrpur, Delhi 110054, India	geethasuryakumar@yahoo.com Tel: 011- 23883311 MO: 9811687801	**DIP-LIP:** DIP-LIP is a hypolipidemic and anti-atherogenic agent. This product is rich source of various bioactive substances such as phytosterols, Tocopherols, beta carotene and omega-3, omega-6 and omega-9 fatty acids. It reduces hypercholesterolemia and lowers atherogenic index. It enhances HDL cholesterol and is a good vasorelaxant. It can be used as a cardioprotective agent and also improves hypoxic tolerance.	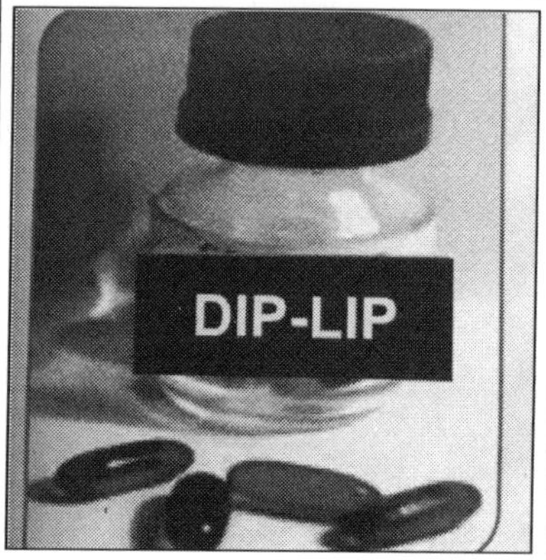
16.	**Dr. Tsering Stobdan** Defence Institute of High Altitude Research Defence Research & Development Organisation (DRDO) Leh Ladakh-194 101, J&K, India	ts_mbb@yahoo.com MO: 9419176057, 9622479992 Fax: 91-1982-252096	**Seabuckthorn Beverage** The beverage is rich in antioxidants and natural vitamins and contains no added colour and flavour. It is refreshing, rejuvenates the antioxidant system of the body, provides vital micronutrients and is fit for human consumption. No microbial growth detected even after 06 months of storage. Organoleptic studies conducted on human volunteers show that the product has high acceptability with the scores being above 8 out of 9.	

Sl. No.	Name of Scientist & Address	E-mail & Mobile/Tel.	Name and Characteristics, Uses etc. of Seabuckthorn Products/ Technology/Machine	Photo of Seabuckthorn Product/ Technology/Machine
17.	**Dr. Tsering Stobdan** Defence Institute of High Altitude Research Defence Research & Development Organisation (DRDO) Leh Ladakh-194 101, J&K, India	ts_mbb@yahoo.com MO: 9419176057, 9622479992 Fax: 91-1982-252096	**Seapricot Beverage** Seapricot beverage is a natural product based health drink prepared by mixing of optimized ratio of Seabuckthorn and apricot fruit pulp. The beverage is rich in antioxidants and natural vitamins and contains no added flavour. It is refreshing, rejuvenates the antioxidant system of the body, provides vital micronutrients and is fit for human consumption. Selflife is 06 months from date of manufacture. No microbial growth detected even after 06 months of storage. Organoleptic studies conducted on human volunteers show that the product has high acceptability with the scores being in the range of 8 out of 9.	
18.	**Dr. Tsering Stobdan** Defence Institute of High Altitude Research Defence Research & Development Organisation (DRDO) Leh Ladakh-194 101, J&K, India	ts_mbb@yahoo.com MO: 9419176057, 9622479992 Fax: 91-1982-252096	**Seabuckthorn Oil Soft Gel Capsules** Seabuckthorn oil soft-gel capsule is a formulation from supercritically extracted Seabuckthorn (*Hippophae rhamnoides*) seed oil, which is optimally enriched with bioactive compounds *viz.* Vit 'E', β-Carotene; along with essential (omega-3 & omega-6) and non-essential (omega-7 & omega-9) and others fatty acids; saturated and unsaturated [monounsaturated (MUFA) & polyunsaturated (PUFA)] fatty acids and minerals *viz.* calcium, phosphorous, iron, zinc, magnesium, selenium *etc.* that are a natural source of antioxidants. They carry all fat soluble vitamins *i.e.* vitamin A, D, E and K and also having another important function is to promote cognitive function, prevent heart disease and bone health.	

Sl. No.	Name of Scientist & Address	E-mail & Mobile/Tel.	Name and Characteristics, Uses etc. of Seabuckthorn Products/ Technology/Machine	Photo of Seabuckthorn Product/ Technology/Machine
19.	**Dr. Tsering Stobdan** Defence Institute of High Altitude Research Defence Research & Development Organisation (DRDO) Leh Ladakh-194 101, J&K, India	ts_mbb@yahoo.com MO: 9419176057, 9622479992 Fax: 91-1982-252096	**Herbal Antioxidant Supplement** The herbal antioxidant supplement is based on pulp of Seabuckthorn (*Hippophae rhamnoides* L.) and Aonla (*Emblica officinalis*), decoction of 15 indigenous herbs (Caper bush, Pushkarmool, Kuth, Revandchini, Yarrow *etc.*), fine powder of 11 indigenous aromatic & edible dye herbs (Rattanjot, Thyme, Mangistha, etc) and 04 other ingredients (Oregano, Wild mint, etc). The formulation is having bioactive compounds *viz.* Vit 'A', Vit 'B_1', Vit 'B_2', Vit 'B_3' Vit 'B_6', Vit 'B_9' Vit '$B1_2$', Vit 'C', Vit 'E', unsaturated fatty acids, carbohydrate, proteins, minerals *etc.* which acts as antioxidant with natural taste, colour and flavour and is suitable for direct consumption by human beings.	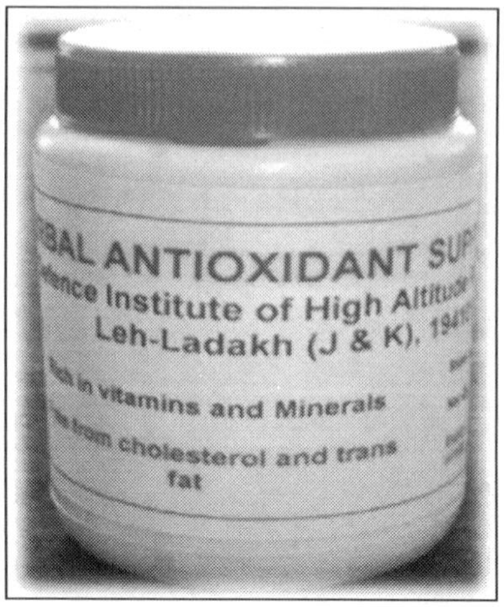
20.	**Dr. Tsering Stobdan** Defence Institute of High Altitude Research Defence Research & Development Organisation (DRDO) Leh Ladakh-194 101, J&K, India	ts_mbb@yahoo.com MO: 9419176057, 9622479992 Fax: 91-1982-252096	**Herbal Adaptogenic Appetizer** Herbal Adaptogenic Appetizer is a composite herbal nutraceutical comprising of Seabuckthorn (*Hippophae rhamnoides*) pulp, apricot (*Prunus armeniaca*) pulp and rhodiola (*Rhodiola imbricata*) root, mixed at standardized optimal ratio of 100:50:1, respectively, and innovating with other additives (Mint mix) to improve the aesthetic appeal, the additives being - Mint powder, Cumin powder, Coriander powder, dried Mango powder, Stevia extract and Himalyan black salt. The formulation is enriched with antioxidants *viz.* Vitamin C, Isorhamnitine, Vitamin E, β-Carotene; saturated and unsaturated fatty acids, minerals *viz.* calcium, phosphorous, iron, zinc, magnesium, selenium *etc.* and adaptogenic plant secondary metabolites like salidrosides, alkaloids, polyphenols, tannins and saponins which act as natural source of antioxidants,	

Sl. No.	*Name of Scientist & Address*	*E-mail & Mobile/Tel.*	*Name and Characteristics, Uses etc. of Seabuckthorn Products/ Technology/Machine*	*Photo of Seabuckthorn Product/ Technology/Machine*
			confer adaptogenic properties, act as a herbal appetizer and is suitable for direct consumption by human beings.	
21.	**Dr. Tsering Stobdan** Defence Institute of High Altitude Research Defence Research & Development Organisation (DRDO) Leh Ladakh-194 101, J&K, India	ts_mbb@yahoo.com MO: 9419176057, 9622479992 Fax: 91-1982-252096	**UV Protective Cream** UV protective cream with anti-photoageing and anti-blemishes properties is an herbal formulation that contains Seabuckthorn seed oil as a major ingredient. The formulation also contains uvinul t 150-ethyl hexyl triazone, uvinul mc 80-ethylhexyl methoxy cinnamate, tinisorb m-methylene bis-benzotriazolyl tetra methylbutyl phenol, uvinul a-diethyl amino hydroxyl benzoyl tetra methylbutylphenol, tinosorb-bis-ethylhexyloxyphenol methoxyphenyl triazine, petroleum jelly, isopropyl myristicate (IPM), light liquid paraffin (LLP), bees wax, emulsifying wax (E Wax), cetyl alcohol, glycerol monostrearate (GMS), stearic acid, Mucopolysaccharide polysulphate (MPS), Butylated hydroxytoluene (BHT), fragrance, and distilled water. The formulation does not contain zinc and has UV protective efficacy equivalent to SPF > 45 along with anti-blemishes and skin healing properties. The formulation can be used as a prophylactic for low humidity and UV mediated skin damage at high altitude.	 UV PROTECTIVE CREAM DRDO
22.	**Dr. Tsering Stobdan** Defence Institute of High Altitude Research Defence Research & Development Organisation (DRDO) Leh Ladakh-194 101, J&K, India	ts_mbb@yahoo.com MO: 9419176057, 9622479992 Fax: 91-1982-252096	**Seabuckthorn Herbal Tea** Seabuckthorn based herbal tea, having anti-oxidant and anti-stress properties, is formulated from indigenous high altitude medicinal plants of Ladakh Himalaya. These herbs are commonly used in *Amchi* system of medicine for various ailments such as stress, indigestion, gastritis, headache, cold, cough, fever, high blood pressure, body pain, high mountain sickness,	 MINISTRY OF DEFENCE डी. आर. डी. ओ. हर्बल चाय DRDO HERBAL TEA

Sl. No.	Name of Scientist & Address	E-mail & Mobile/Tel.	Name and Characteristics, Uses etc. of Seabuckthorn Products/ Technology/Machine	Photo of Seabuckthorn Product/ Technology/Machine
			memory lost, and weakness *etc.* since ages. The main composition of Herbal Tea is leaves of seabuckthorn which is rich in nutrients and bioactive substances such as carotenoids, free and esterified sterols, triterpenols, and isoprenols. The herbal tea has medicinal value *viz.* soothing, energizer, anti-stress, stimulant, and as health tonic. The herbal tea has its own natural colour, flavour and taste.	
23.	**Dr. Rakesh Sharma** University of Horticulture & Forestry, (Nauni) Solan (HP)-173230	drrakes@gmail.com MO: 9418061369	**Seabuckthorn-Apple Blended Wine** The high acid content of seabuckthorn which otherwise hinder the fermentation process has been reduced by blending with apple pulp. The developed wine contained higher amounts of ascorbic acid, total phenolics and has shown high antioxidant activity compared to apple wine and seabuckthorn wine.	 Seabuckthorn-Apple Blended Wine Seabuckthorn-Apple Blended Wine
24.	**Prof. Renu Deswal** Molecular Plant Physiology & Proteomics Laboratory, Department of Botany, University of Delhi, Delhi -110007	rdeswal@botany.du.ac.in, renudeswal@yahoo.co.in MO: 09711742276	**Antifreezer** A method was devised to purify the AFPs showing high antifreeze activity from seedlings and berry of *H.rhamnoides*. AFPs possess two major activities ice recrystallization inhibition and thermal hysteresis owing to the ability to control ice crystal growth and recrystallization leading to numerous applications like ✰ Maintaining texture of food during low temperature storage, ✰ Improving cold tolerance in freeze susceptible plants ✰ Aquaculture (cold tolerant fishes), ✰ Cryopreservation of cells, tissues, cell lines & organs ✰ Cryosurgery.	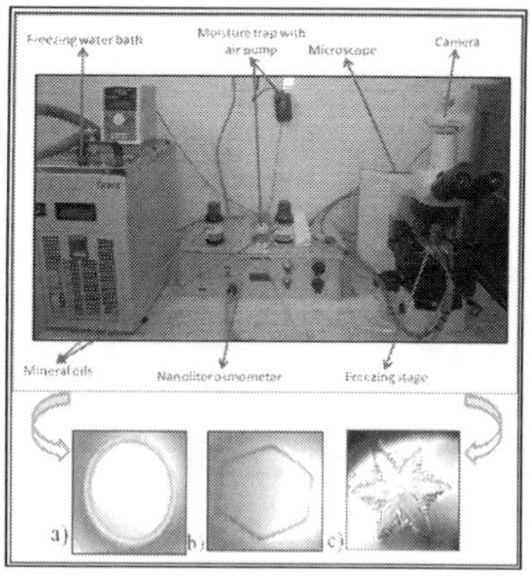 Freezing water bath; Moisture trap with air pump; Microscope; Camera; Mineral oils; Nanoliter osmometer; Freezing stage; a); b); c) Figure depicting the Nanoliter osmometer used to detect antifreeze activity. The ice crystals morphology in the presence of AFPs, showing b) hexagonal and c) flower shape while the ones in the absence of AFPs (a) control) showed disc shaped ice crystal.

Sl. No.	*Name of Scientist & Address*	*E-mail & Mobile/Tel.*	*Name and Characteristics, Uses etc. of Seabuckthorn Products/ Technology/Machine*	*Photo of Seabuckthorn Product/ Technology/Machine*
25.	**Prof. P. C. Sharma,** University School of Biotechnology, Guru Gobind Singh Indraprastha University, Dwarka Sector 16-C, New Delhi-10078.	prof.pcsharma@gmail.com Mobile: +91-9899088818 Telephone: +91-11-25302306	**Protocol for isolation of good quality RNA** from seabuckthorn tissues. This method is now being widely followed by different workers for seabuckthorn and other similar plants.	
26.	**Dr. Farhath Khanum** Defence Food Research Laboratory Mysore	farhathkhanum@gmail.com MO: 09449851403 0821 2470364	**Seabuckthorn Leaf Biscuits** The freeze dried seabuckthorn leaf-extract powder incorporated biscuits are highly nutritive, rich in polyphenols and flavonoids. They are tasty and crunchy and easy to carry. The effect of the biscuits was studied at high altitude in C-H-R animal models. The animals fed seabuckthorn biscuits mixed diet showed further improvement in adaptogenic activity as there was 37% increase in time to attain Trec 23°C and 42% faster recovery of Trec 37°C, in comparison to control rats fed normal feed. The recovery time of Trec 37°C of rats fed seabuckthorn biscuits mixed feed was significantly better even in comparison to animals fed on diet containing placebo biscuits.	
27.	**Dr. Farhath Khanum** Defence Food Research Laboratory Mysore	farhathkhanum@gmail.com MO: 09449851403 0821 2470364	**Seabuckthorn Leaf Tea Bags** Different parts of seabuckthorn especially the leaves are found to be rich source of vitamins, trace elements, polyphenols and flavonoids. The leaf extract of Seabuckthorn is known to possess a significant anti-inflammatory activity and has the potential for the treatment of arthritis. It has high quantities of polyphenols, which detoxify the free radical activities. The other ingredients were locally available cost-effective spices and herbs.	Herbal Tea

Sl. No.	Name of Scientist & Address	E-mail & Mobile/Tel.	Name and Characteristics, Uses etc. of Seabuckthorn Products/ Technology/Machine	Photo of Seabuckthorn Product/ Technology/Machine
			A cup of seabuckthorn leaf tea is found to be refreshing, stimulating and stress relieving apart from being a thirst quencher and a soothing and relaxing with very fine aroma. This herbal tea mix has been studied for its protective effect against the toxic effects of an organochlorine pesticide, hexachlorocyclohexane and the hyper cholestrolemia. The study revealed that the tea mix has the potency to reduce the oxidative stress induced by and the cholesterol-induced hypercholesterlia in experimental rats.	
28.	**Dr. Farhath Khanum** Defence Food Research Laboratory Mysore	farhathkhanum@ gmail.com MO: 09449851403 0821 2470364	**Seabuckthorn Fruit Squash** Seabuckthorn pulp is highly sour and highly perishable with an inherent problem of separation of liquid and solid contents. This problem of phase separation was solved successfully and Ready-to –use squash was prepared and analyzed for proximate composition, vitamins content, antioxidant activity, rheology and shelf stability. Seabuckthorn pulp-based squash have lovely colour, aroma and taste rich in antioxidants and spices with homogeneity.	
29.	**Dr. Farhath Khanum** Defence Food Research Laboratory Mysore	farhathkhanum@ gmail.com MO: 09449851403 0821 2470364	**Seabuckthorn Jam** Seabuckthorn pulp enriched jam has pleasant colour, flavour, homogeneous with packed antioxidants and possess good cutting edge.	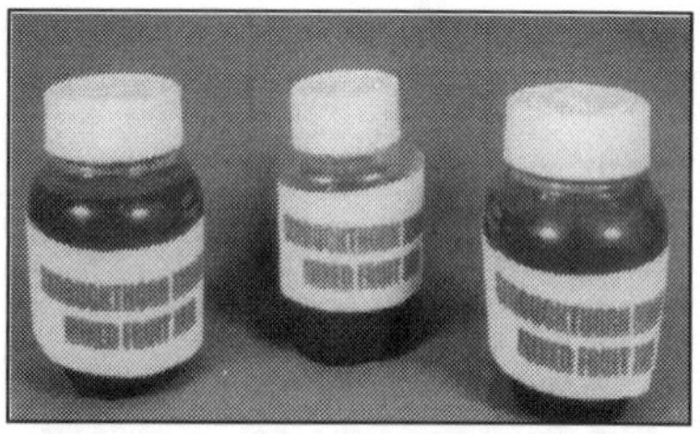
30.	**Dr. Farhath Khanum** Defence Food Research Laboratory Mysore	farhathkhanum@ gmail.com MO: 09449851403 0821 2470364	**Seabuckthorn Youghurt** A fermented milk product is prepared using strains of lactic acid bacteria blended with seabuckthorn fruit. It is easily digestible with enriched natural antioxidants and having the ability to protect and maintain the healthy gut micro flora	 SBT Yoghurt

Sl. No.	*Name of Scientist & Address*	*E-mail & Mobile/Tel.*	*Name and Characteristics, Uses etc. of Seabuckthorn Products/ Technology/Machine*	*Photo of Seabuckthorn Product/ Technology/Machine*
31.	**Dr. Farhath Khanum** Defence Food Research Laboratory Mysore	farhathkhanum@gmail.com MO: 09449851403 0821 2470364	**Seabuckthorn Juice Powder** A ready – to use juice powder that can be used instantly for preparation any food product like juice or yogurt *etc.*	
32.	**Dr. C. Arumughan** **RRL, Trivendrum, Kerala**	09447123437	**Concentrated Juice** Rich in vitamin C and other nutrients can be stored and transported easily. **Enzyme Technology of Oil Extraction** Enzyme method has been developed to extract seabuckthorn fruit pulp upto 65%. It is free from chemicals and safe.	
33.	**Prof. P. C. Sharma** University School of Biotechnology, Guru Gobind Singh Indraprastha University, Dwarka Sector 16-C, New Delhi-10078.	prof.pcsharma@gmail.com Mobile: +91-9899088818 Telephone: +91-11-25302306	Method for constructing **microsatellite enriched genomic library** and detection of microsatellite location **(Patent filed: 2523/DEL/2013)**. Large number of microsatellite markers developed using genomic library, ESTs and whole transcriptome of seabuckthorn	
34.	**Prof. P. C. Sharma** University School of Biotechnology, Guru Gobind Singh Indraprastha University, Dwarka Sector 16-C, New Delhi-10078.	prof.pcsharma@gmail.com Mobile: +91-9899088818 Telephone: +91-11-25302306	**Expressed Sequence Tags (ESTs): 3412 high quality ESTs** submitted to dbEST of NCBI with GenBank accession numbers: EY414394 - EY416137, FG35624 - FG357288, GT915047 - GT915671. These ESTS are valuable resource for gene discovery and other studies in seabuckthorn..	

Figure 1.1: Fusarium Wilt (*Fusarium* sp.). (p. 5)

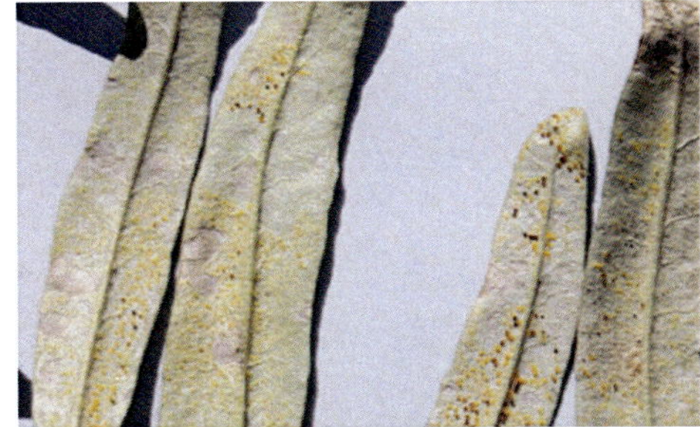

Figure 1.2: Powdery Mildew (*Phyllactinia hippophaes*). (p. 6)

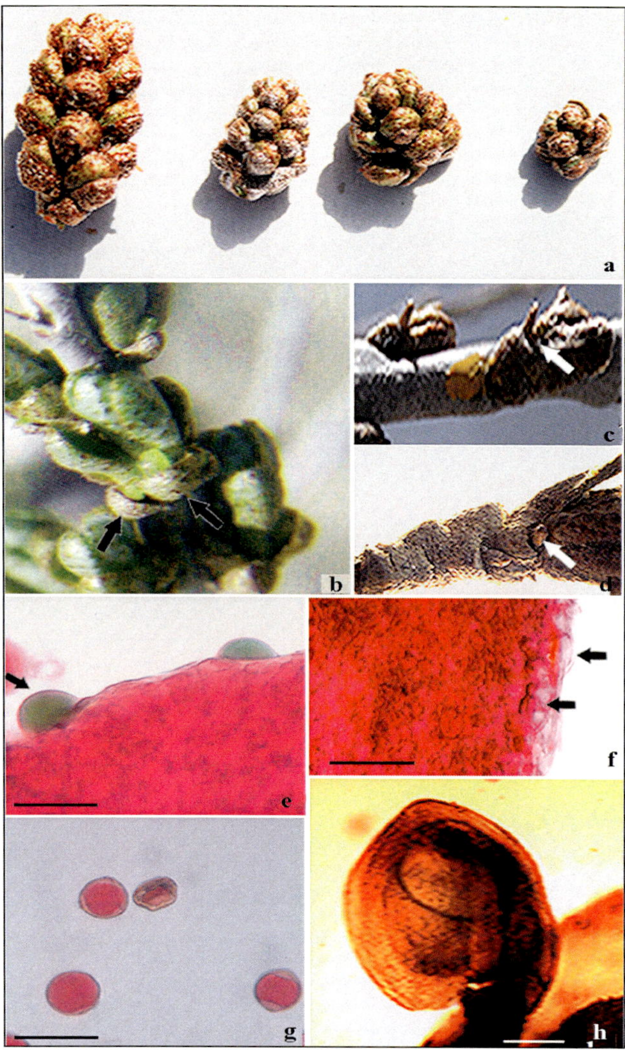

Figure 2.1 a-h: Male Inflorescences at different Developmental Stages. (p.14)

Figure 2.2 a-l: Visitors to Male and Female Plants of *H. rhamnoides* L. Mark the Variety of Biota. (p.14)

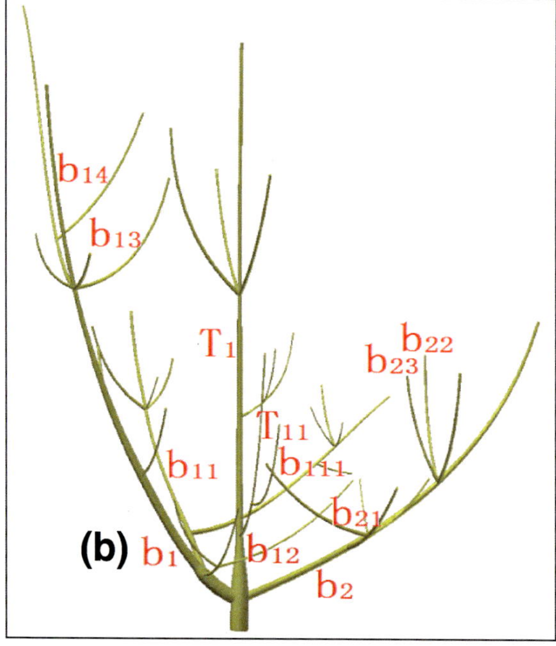

Figure 4.1: A SBT Tree and its 3D Model by Pro/E. (p. 26)

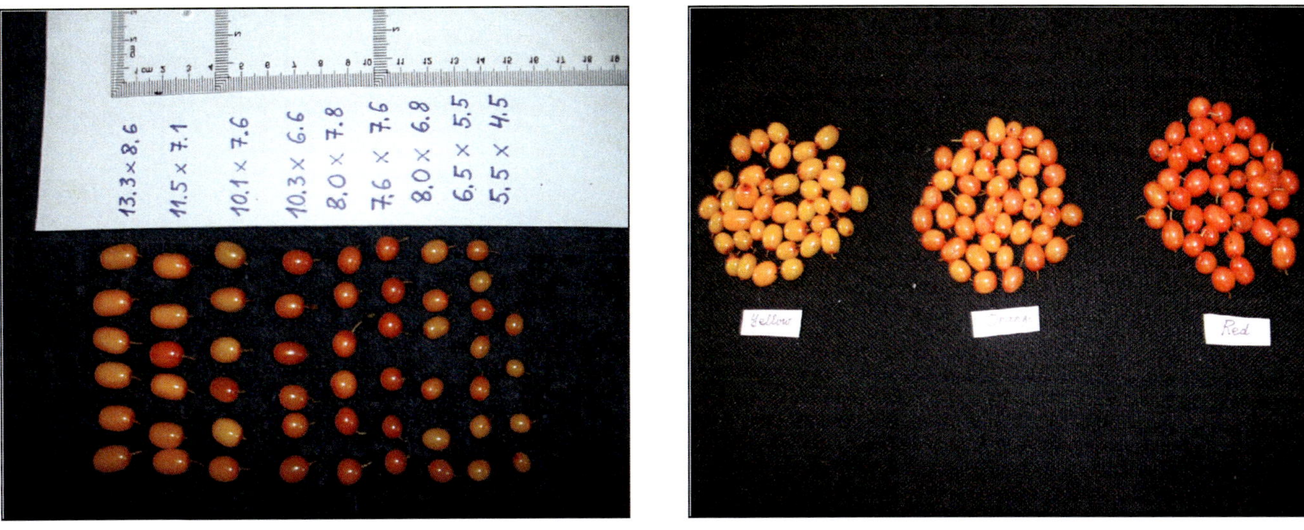

Figure 10.4: Variation in Morphological Characters of different Seabuckthorn Fruits. (p. 74)

A) RC-4

B) Harvest Moon

C) Prairie Sunset

D) Autumn Glow

Figure 11.2: Images of Elite Cultivars Developed and Released by the AAFC-ADC well Adapted to Canadian Prairies Growing Conditions. (p. 82)

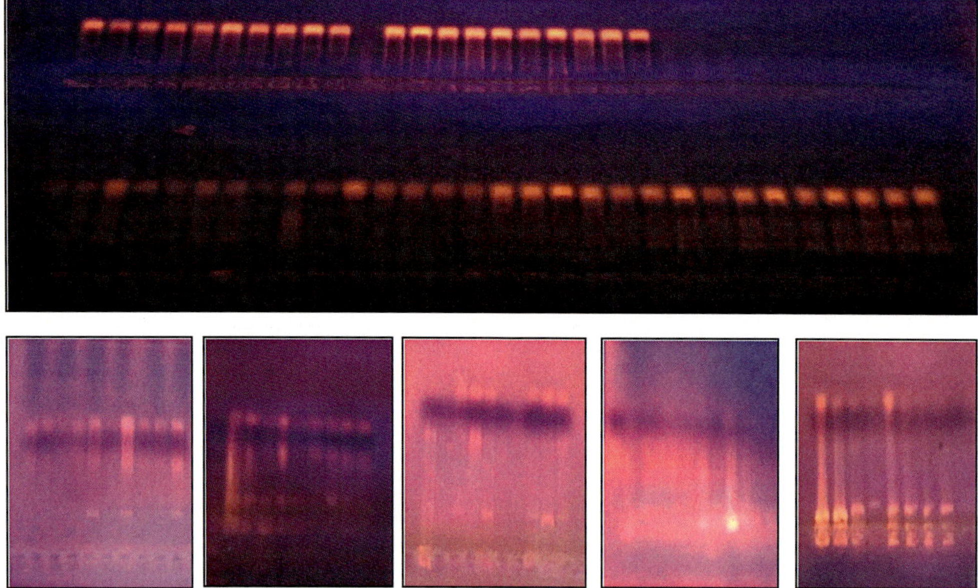

Figure 12.2. Genomic DNA of Seabuckthorn Ecotypes.
(Above photos belonged to Iran, below photo belonged to China individuals). (p. 89)

Figure 23.1: Ripe Berries of Eight Seabuckthorn Cultivars Used in this Study.
1: Siberia hongyun; 2: Wucifeng; 3: Zhuangyuanhuang; 4: Fangxiang; 5: Liusha-1; 6: Alexander-12; 7: Cross; 8: Jiuyuehuang. (p. 183)

Figure 25.1: Seabuckthorn 'Leikora' (Left) and 'Hergo' (Right) Seed, Seed Kernel and Seed Hull. (p. 198)

Day/ Groups	*I*	*II*	*III*	*IV*
Zero day	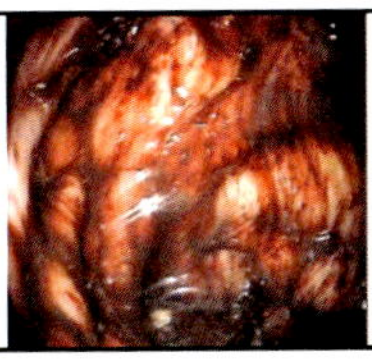	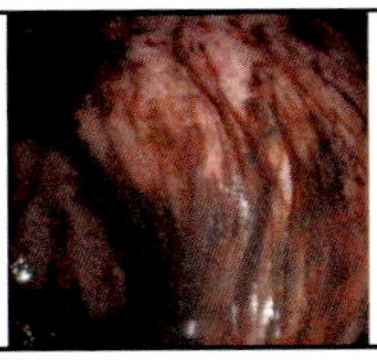	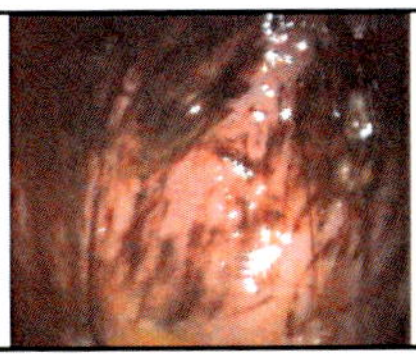	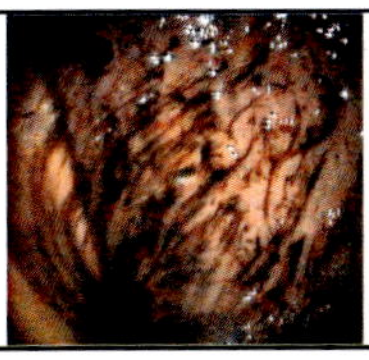
GUE Index	7.50±0.29	7.75±0.25	8.00±0.00	8.00±0.00
3rd day	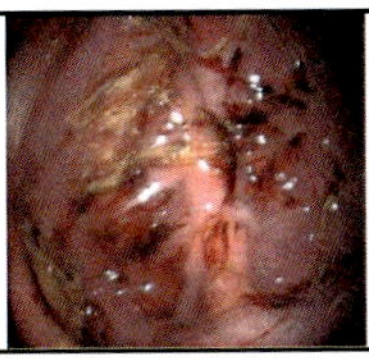	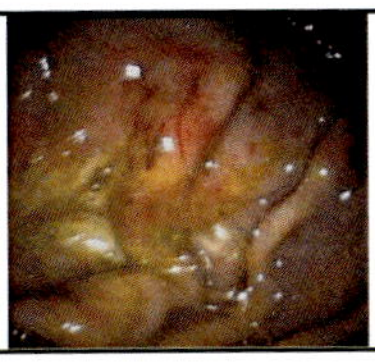	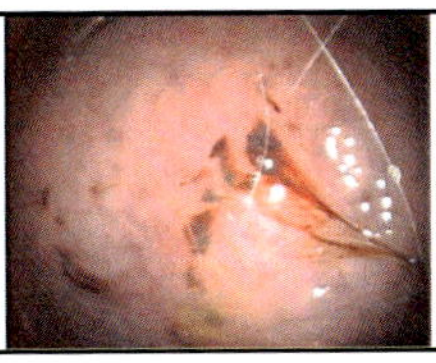	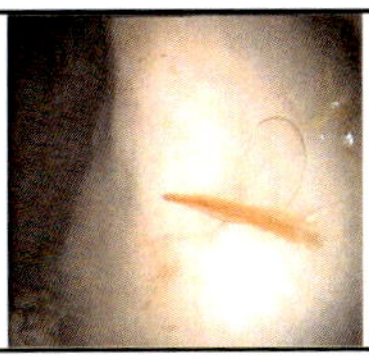
GUE Index	4.75±0.63	3.00±0.71	5.00±0.00	1.75±0.47
6th day	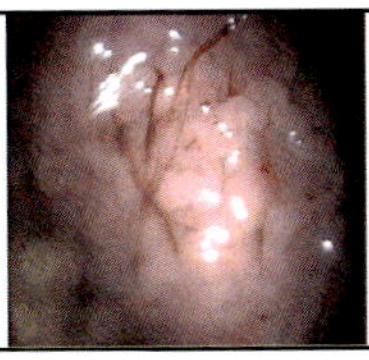	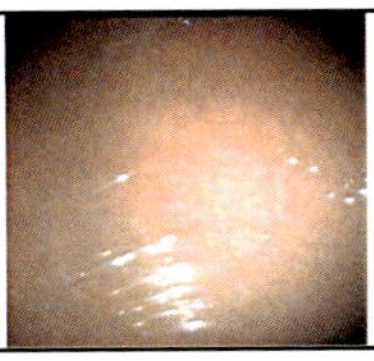	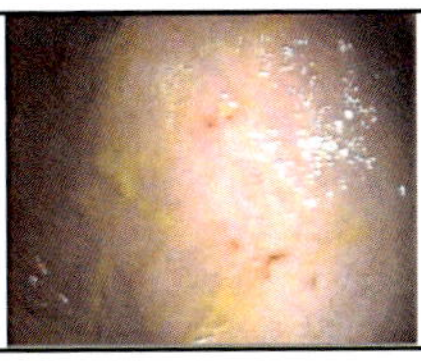	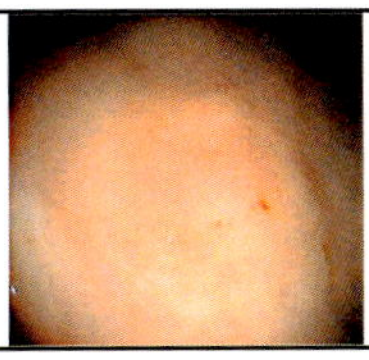
GUE Index	2.25±1.32	0.00±0.00	2.00±0.71	0.75±0.47
9th Day	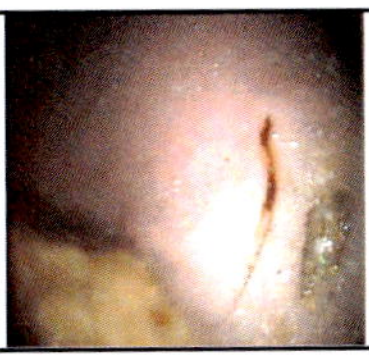	-	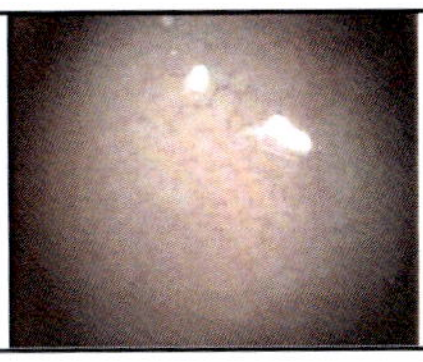	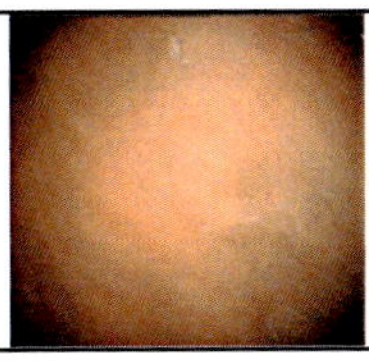
GUE Index	1.00±1.0	-	0.66±0.66	0.00±0.00
12th Day	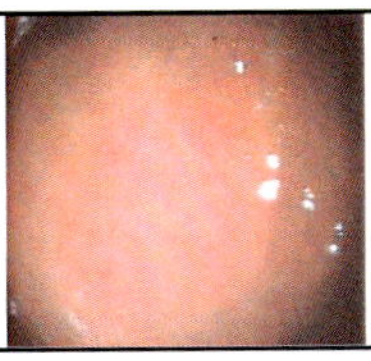	-	-	-
GUE Index	0.00±0.00	-	0.00±0.00	-

Figure 29.1: Endoscopic View of Gastric Mucosal Surface and GUE Indices of Dogs at Various Time Intervals. (p. 232)

Figure 32.1: Test Rat Restraint on a Wire Grid for 2h/day in Lab during Experiment. (p. 259)

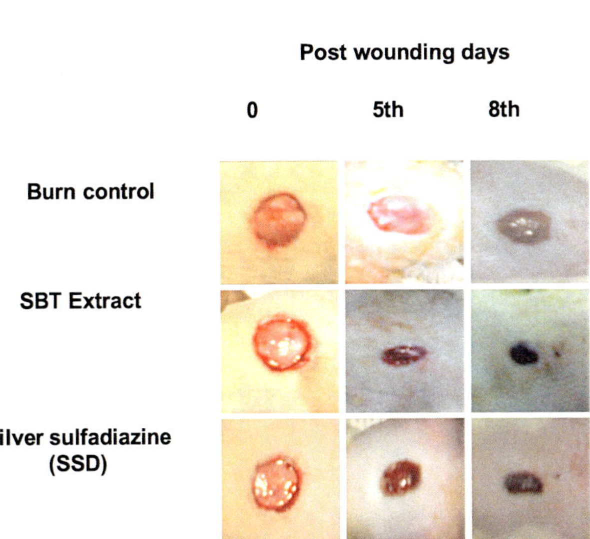

Figure 35.1: Visual Observation of Seabuckthorn (SBT) Extract Treated and Untreated Control and Reference Care (SSD) Treated Burn Wounds in rats. SSD: Silver sulfadiazine. (p. 283)

Figure 42.1: Leaves from Seabuckthorn. (p. 336)

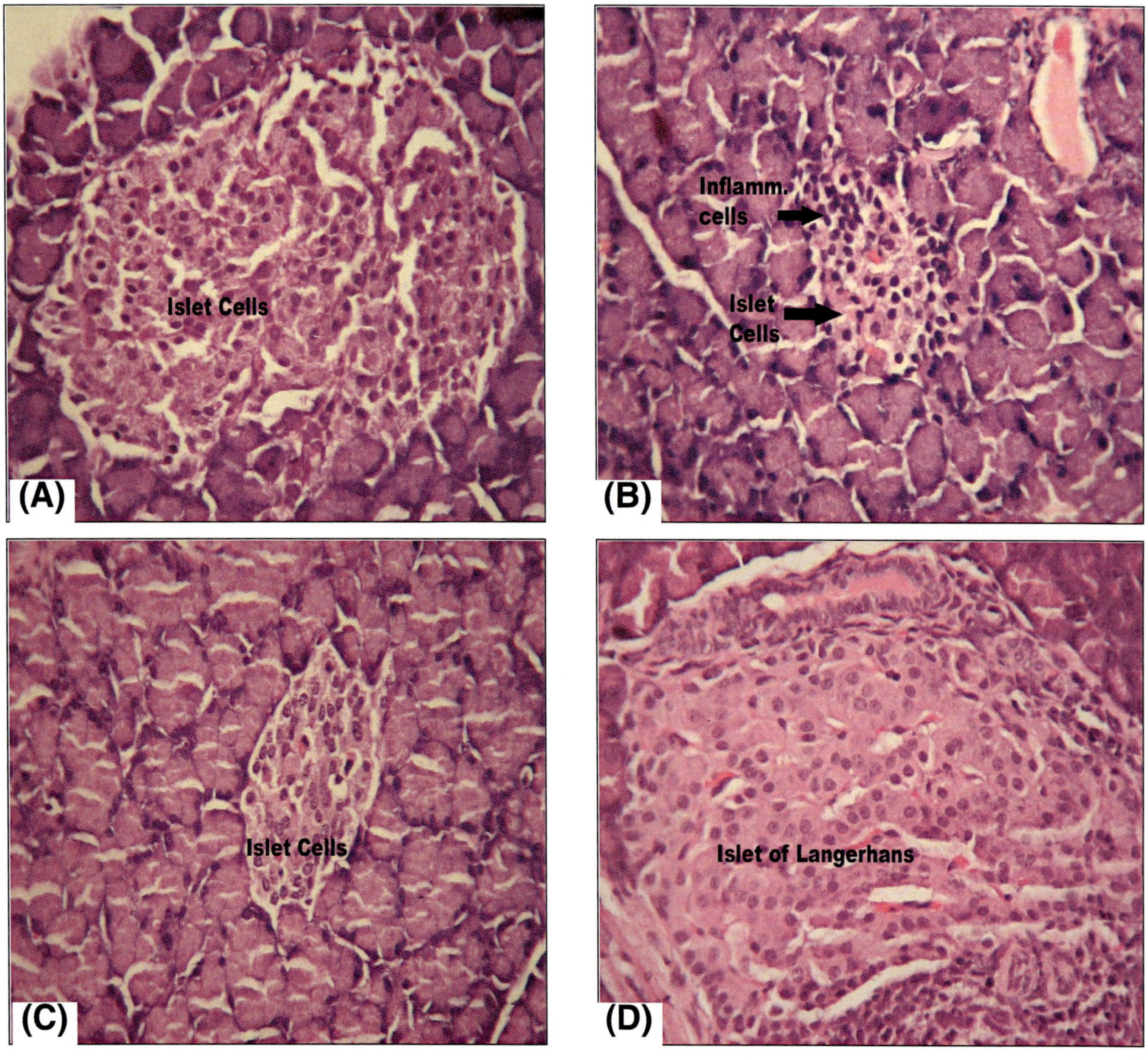

Figure 44.1: Hematoxylin and Eosin-Stained Sections of Rat Pancreas. (p. 351)

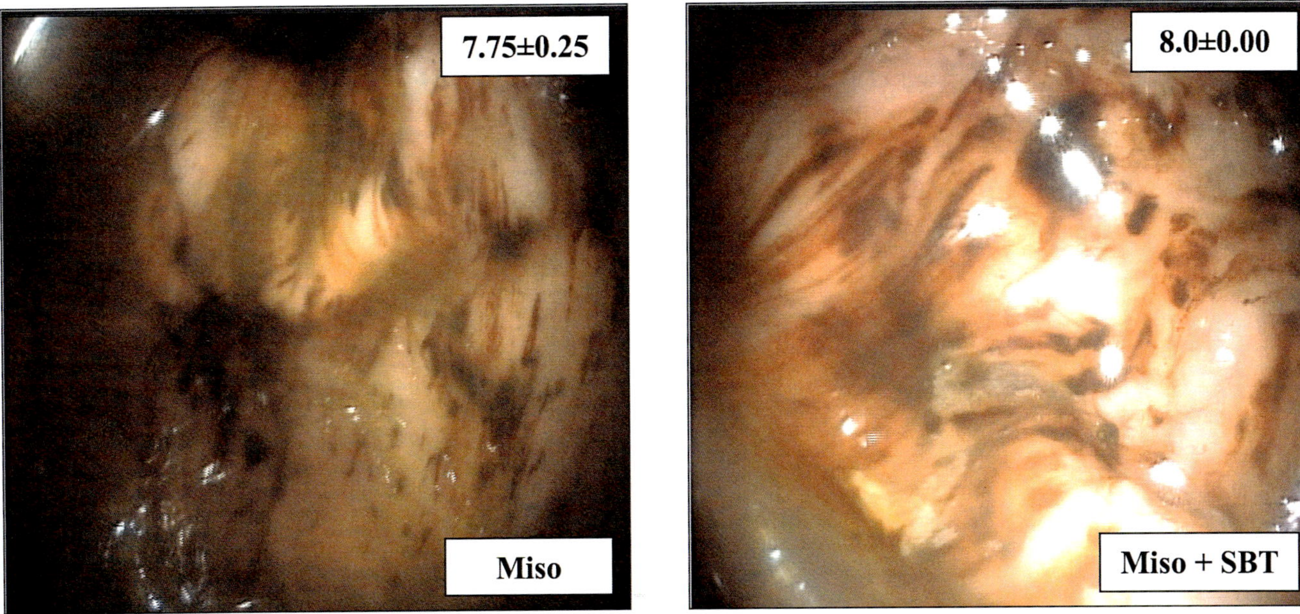

Multiple lesions of variable depths with active haemorrhagic base, thin mucus layer, adherent blood clots, hyperaemic mucosa with typical 'water fall' appearance.

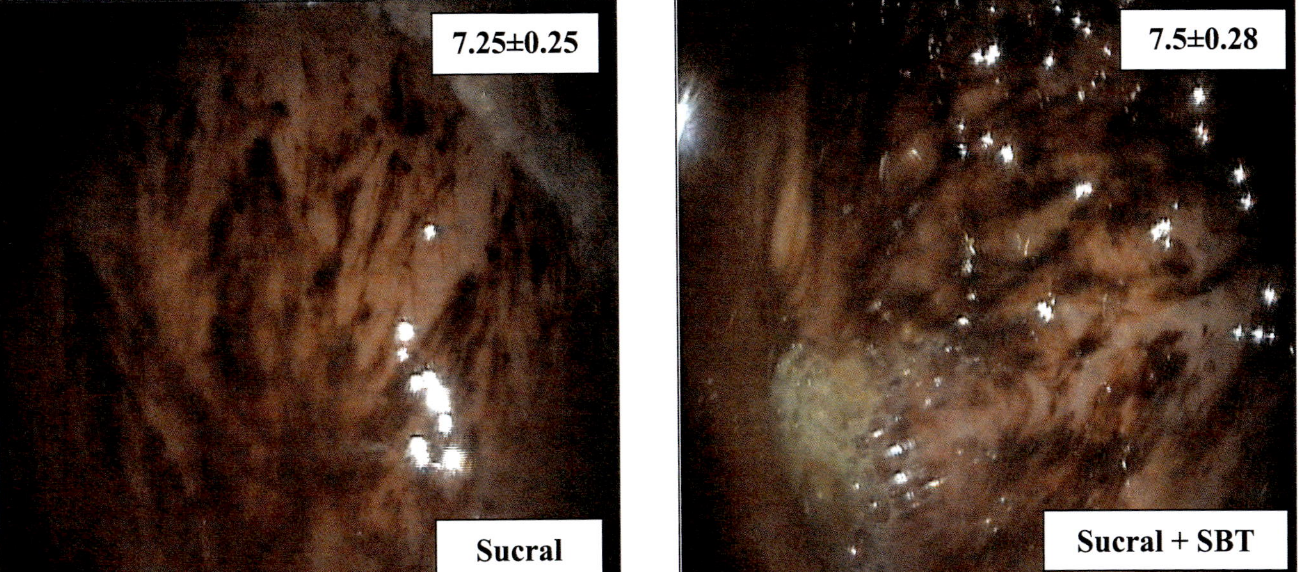

Figure 46.1: Endoscopic View of Gastric Mucosa in different Groups on Day 0. (p. 362)

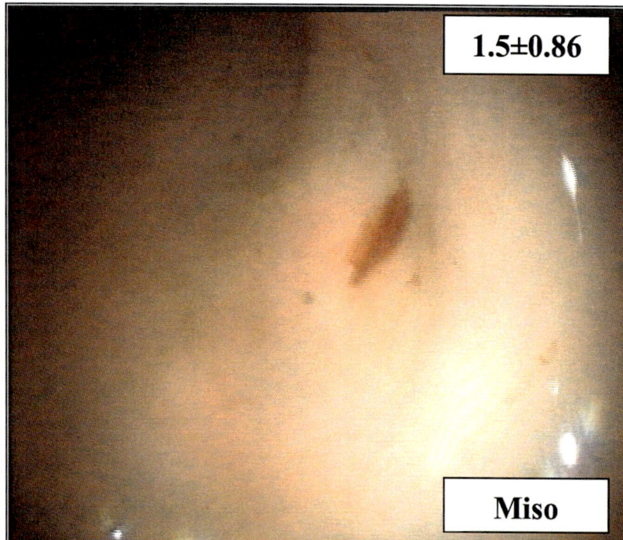

A few gastric erosions of variable degree

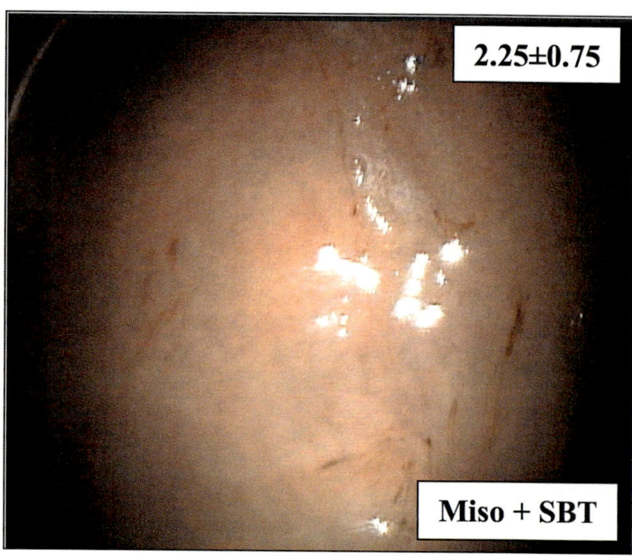

Multiple gastric lesions with erythema

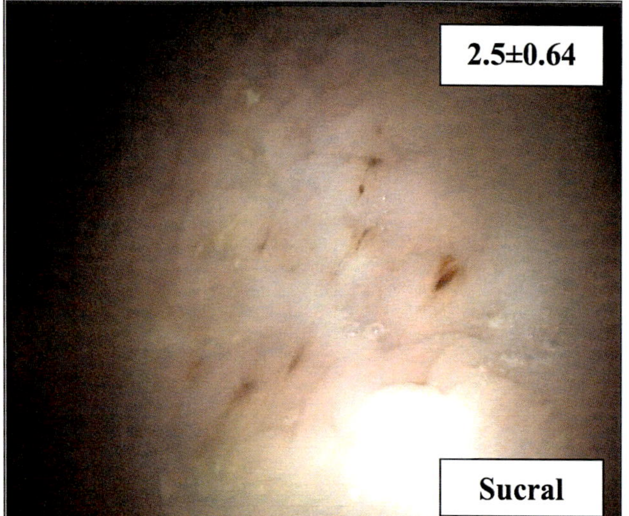

Multiple smaller gastric lesions with and without adherent blood clots

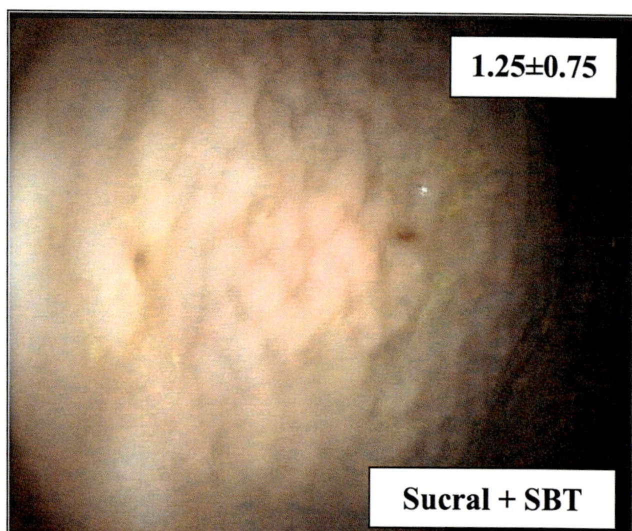

A few free floating blood clots

Figure 46.2: Endoscopic View of Gastric Mucosa in different Groups on Day 6. (p. 363)

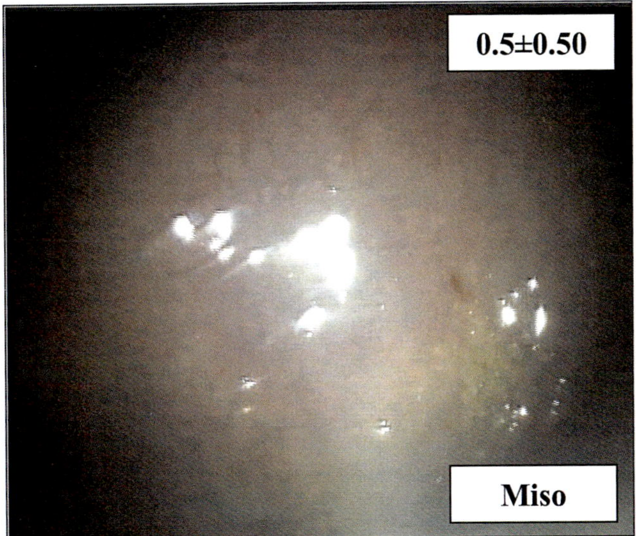

1-2 sub mucosal localised lesions

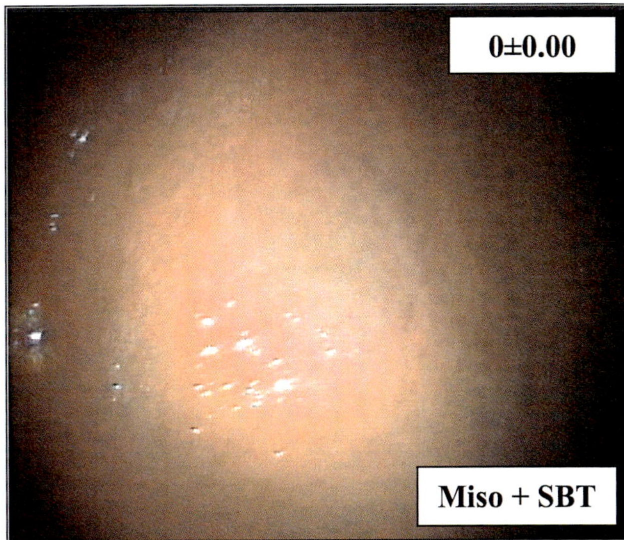

No lesion on gatric mucosa, no blood clots, hick and shiny mucus layer

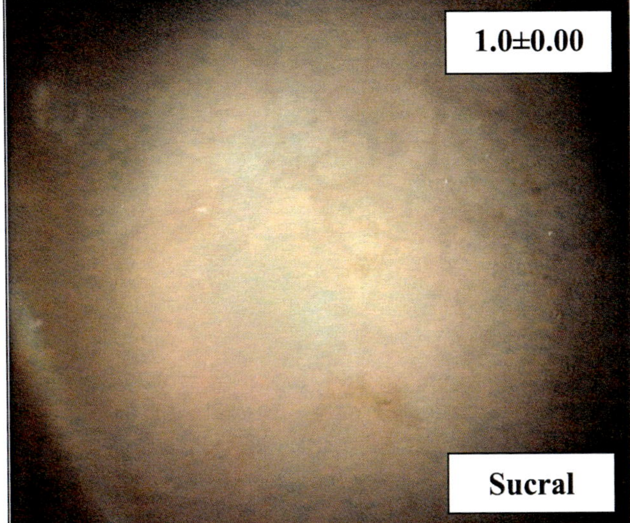

Solitary blood clot with well-formed mucosal layer

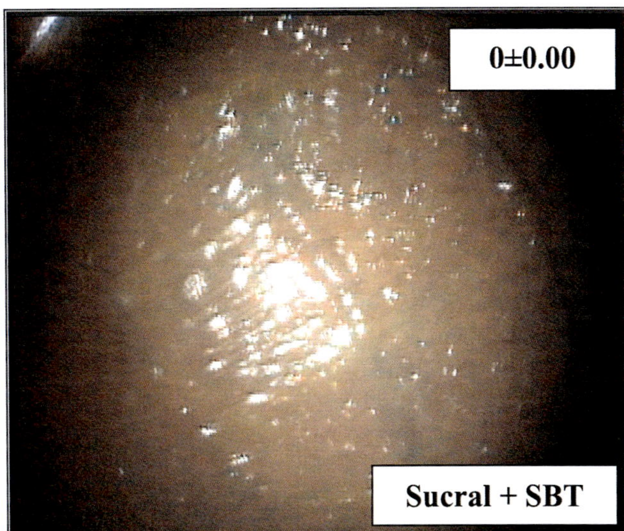

Thick and shiny mucus layer

Figure 46.3: Endoscopic View of Gastric Mucosa in different Groups on Day 12. (p. 364)

Figure 50.1: The Samples of Seabuckthorn Processing Products: Vinegar (Left) and Wine (Right). (p. 388)

Plate 57.1: Physical Appearance of Squash Prepared by using Various Proportions of Sand Pear and Seabuckthorn. (p. 440)

Plate 61.1: Sample of *H. salicifolia*. (p. 466)

Figure 62.1: Seabuckthorn (*Hippophae rhamnoides*) Forest, Spiti, HP. (p. 477)

Figure 63.2: Easily Eroded Bare Pisha Sandstone. (p. 488)

Figure 63.4: Growth State of Several SFDs in 2005 (Planted in 1996). (p. 489)

(p. 435-547)

Products Developed

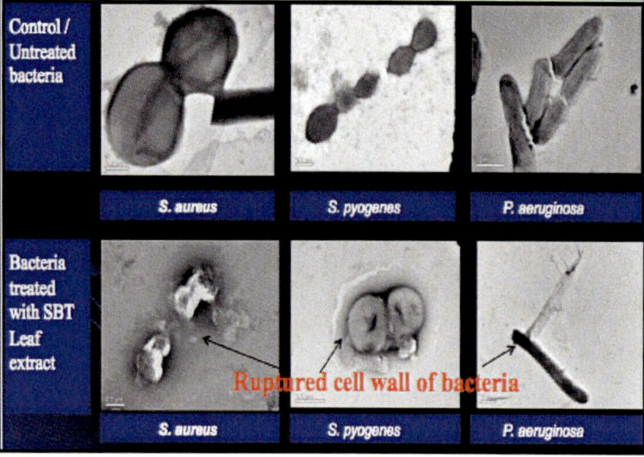
Control / Untreated bacteria
S. aureus
S. pyogenes
P. aeruginosa
Bacteria treated with SBT Leaf extract
Ruptured cell wall of bacteria
S. aureus
S. pyogenes
P. aeruginosa

Plate 3.3: Feed manufactured in mash and pelleted form for different treatment groups

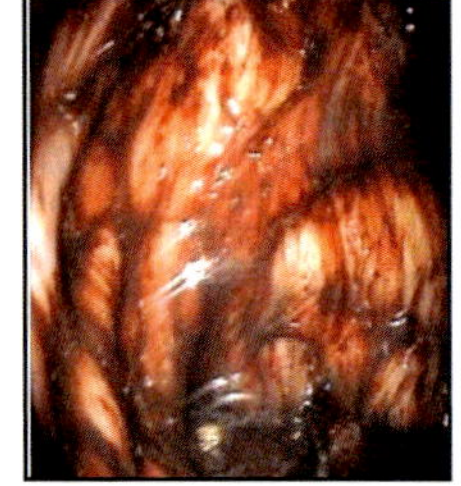

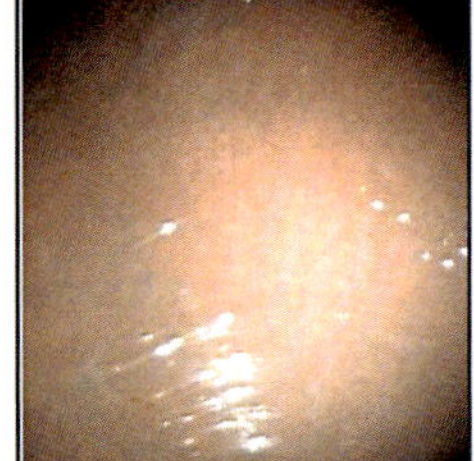

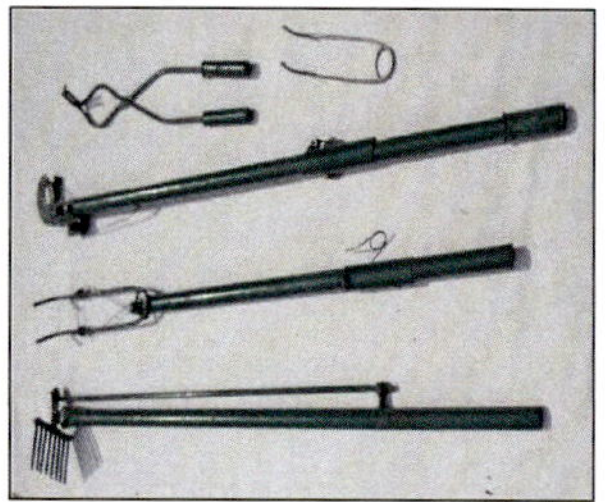

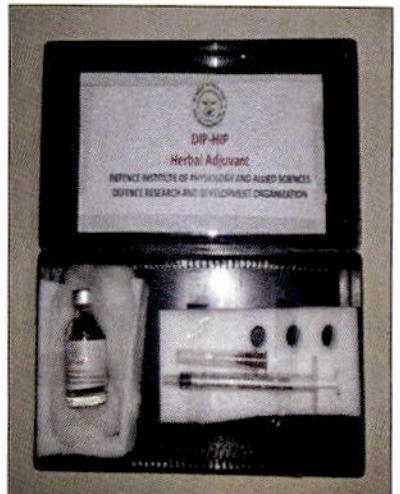

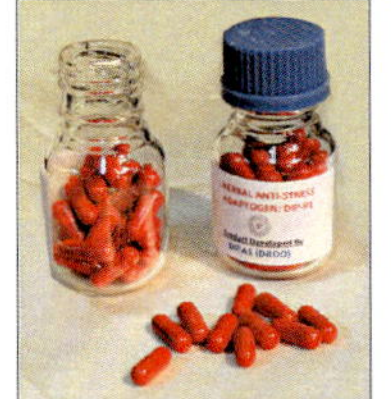

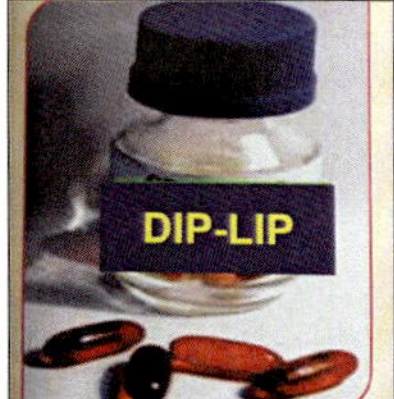
DIP-LIP

Multivit

Seapricot

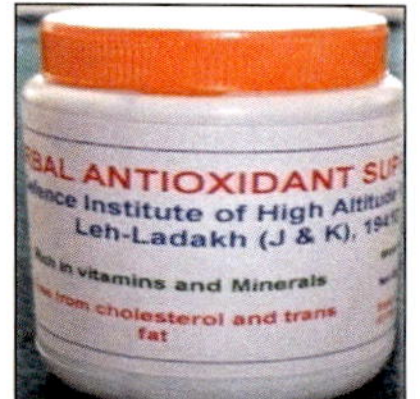
Leh-Ladakh (J & K)

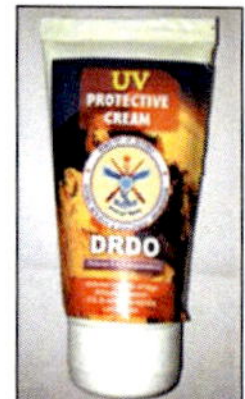
UV
DRDO

डी. आर. डी. ओ. हर्बल चाय
DRDO HERBAL TEA

Seabuckthorn-Apple Blended Wine
Seabuckthorn-Apple Blended Wine

Herbal Tea

SQUASH
SQUASH

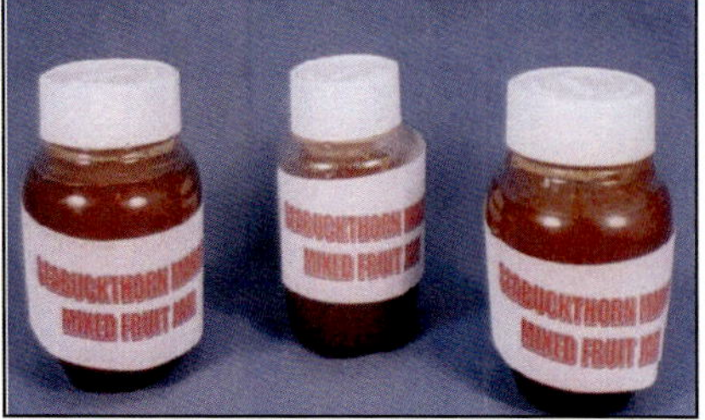

SBT Yoghurt

Seabuckthorn Juice Powder